The circle

Figure	Sketch	Circumference (c)	Area (A)
Circle	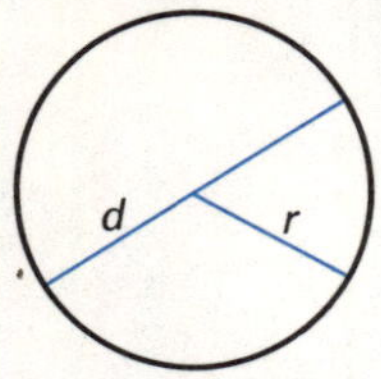	$c = 2\pi r$ $c = \pi d$	$A = \pi r^2$ $A = \frac{\pi d^2}{4}$

Perimeters and areas of plane figures

Figure	Sketch	Perimeter (P)	Area (A)
Triangle	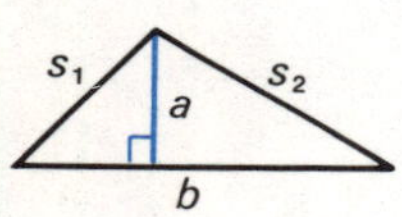	$P = s_1 + s_2 + b$	$A = \frac{1}{2} ba$
Trapezoid	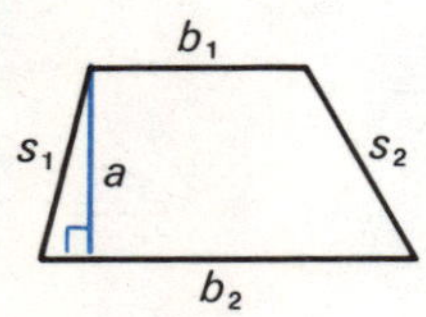	$P = s_1 + s_2 + b_1 + b_2$	$A = \frac{1}{2}(b_1 + b_2)a$
Parallelogram	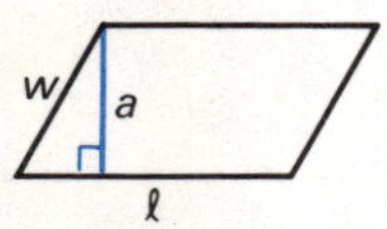	$P = 2\ell + 2w$	$A = \ell a$
Rhombus	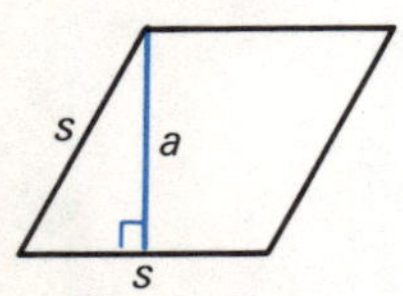	$P = 4s$	$A = sa$
Rectangle	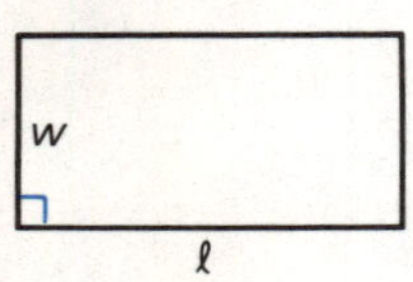	$P = 2\ell + 2w$	$A = \ell w$
Square	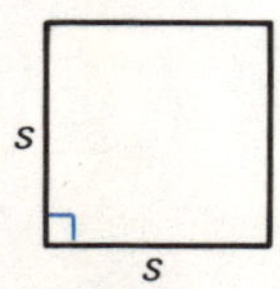	$P = 4s$	$A = s^2$

Basic Technical Mathematics

ADDISON-WESLEY PUBLISHING COMPANY

READING, MASSACHUSETTS • MENLO PARK, CALIFORNIA
DON MILLS, ONTARIO • WOKINGHAM, ENGLAND • AMSTERDAM
SYDNEY • SINGAPORE • TOKYO • MEXICO CITY
BOGOTÁ • SANTIAGO • SAN JUAN

Basic Technical Mathematics

STUART R. PORTER
JOHN F. ERNST
Monroe Community College

Jeffrey Pepper Sponsoring Editor

Kristin Belanger Art Editor
Ann E. Delacey Manufacturing Supervisor
Clark Dunbar Cover Photographer
Richard Hannus Cover Designer
Maureen Langer Text Designer
Herbert Nolan Production Manager
Laura Skinger Production Editor

Library of Congress Cataloging in Publication Data

Porter, Stuart R., 1932–
Basic technical mathematics.

Includes index.
1. Mathematics—1961– I. Ernst, John F.
II. Title.
QA39.2.P683 1985 510 84-8369
ISBN 0-201-05586-4

ABCDEFGHIJ-DO-898765

To our wives
Joyce and Gerry

Basic Technical Mathematics with Calculus is also available from Addison-Wesley. It includes an additional seven chapters of calculus to prepare students for more advanced mathematics courses.

Contents

Technical Calculus, also by Stuart R. Porter and John F. Ernst, will be available from Addison-Wesley in 1986.

Preface

This textbook is designed to provide students with the necessary mathematical skills to pursue a course of study in scientific or engineering technology or to prepare the student for calculus. The student should enter the course with two years of high-school mathematics (algebra and geometry). For the student with a weak background we have included Chapters 1 through 4, which provide a review of algebra and geometry. The text can be used in a one- or two-semester program.

Special Features

Graded Exercises/Review Exercises/Special Problem Sets We have attempted to write in a clear and precise style that is easy for the student to read and understand, but in our attempt to keep the book readable we have not sacrificed the preciseness of the language of mathematics. In each section there are clearly worked-out examples that illustrate the mathematical concepts. The examples are developed step by step with explanatory notes. At the end of each section there is an abundant set of graded exercises and at the end of each chapter there is a large set of review exercises. A chapter summary contains the important concepts, definitions, and theorems discussed in the chapter. As a challenge we have included a special set of problems entitled "Can you solve these?" at the end of each chapter. These problems may relate directly to the material in the chapter, may take the student to a related topic that could not be covered due to space limitations, or may ask the student to design a program for a calculator or computer that will solve a particular problem.

Real-World Applied Problems/Calculator Problems/Computer Problems There are answers for all the odd-numbered problems in the back of the book. There are many applied problems from different areas throughout the text. The applications are presented in such a manner that a student or instructor not familiar with the field or area can understand the problems. The inclusion of the scientific calculator in the text does not preclude the proper development of the mathematics. We have developed the mathematics and then shown how the calculator can be used to ease the computations. At key points we have included calculator key strokes to provide the student with examples of how the calculator can be used as a tool. In a few instances we have provided a program for programmable calculators to illustrate how the calculator can be used more efficiently. The algebraic calculator that was used for all key strokes and programs was Texas Instrument's TI-55-II and for RPN we used an HP 33E. Students should consult the instruction manual for their calculator for special situations. The text is written in such a manner that one could teach from or study from the text without using a calculator even though we have included the calculator material.

Units used in applications correspond to current SI or English practice.

A Brief Content Review

The material in Chapters 1 through 4 is review and can be omitted with the exception of Sections 3.2, 3.5, 4.1, 4.2, and 4.3. In Section 3.2 significant digits are discussed and rules are given for the manner in which approximate answers are expressed throughout the text. Section 3.5 provides the student with a technique for solving word problems. This technique is used in problem solving throughout the text. Section 4.3 is a quick review of the plane figures and places a slight emphasis on angles, providing helpful background material for trigonometry in Chapter 6. Chapter 5 is a brief introduction to the concept of function, providing the student with the basic terminology so that we can use and build upon the concept of function as we move through the text. Chapters 6, 7, and 8 provide an early introduction to trigonometry as is required by some technical areas. In the development of trigonometry both degree and radian measure have been stressed.

Chapter 9 is an introduction to imaginary and complex numbers. The applications in this chapter are directed primarily toward the field of electronics. Since this is a special topic and is not of interest to all areas of study, the chapter can be omitted without affecting the study of the remaining chapters.

Chapters 10, 11, and 12 deal with the study of solving equations. In Chapter 10 in addition to the standard algebraic methods of elimination and substitution for solving systems of linear equations, we have discussed Cramer's rule and matrix methods to obtain solutions.

In Chapter 13, we discuss the logarithmic and exponential functions, graphing of these functions, and logarithms to other bases, in addition to base 10 and base e. The techniques of solving linear, quadratic, and higher degree inequalities are discussed in Chapter 14. The sign method is used to solve inequalities of higher degree. The trigonometric identities such as the sum and difference of two angles, double angles, and half-angle formulas are developed in Chapter 15. We also discuss methods of solving trigonometric equations and inverse trigonometric functions.

Chapter 16 provides the student with an introduction to arithmetic and geometric sequences and series. An important topic in this chapter is the binomial theorem. It is perhaps the only time the student will see how we get all the terms or a specific term of a binomial expression.

Chapter 17 provides the student with the study of the straight line. The general development of the conic sections is presented starting with the centers (vertex) at the origin. The concept of translation has been developed for those who want to go beyond the basics.

Chapter 18 provides the student with a brief introduction to computer hardware and then introduces flowcharts and the BASIC language. We discuss the BASIC commands and introduce the student to loops. Chapter 18 is special in that it is designed so that it can be taught at any time in the course. The instructors who want their students to work with the computer throughout the course can start with Chapter 18. The instructor who wants nothing to do with computers can forget Chapter 18. All the programs in Chapter 18 have been run on Apple II and IIe computers.

There are two appendixes. Appendix A contains a brief discussion of the SI and British systems of measurements, and Appendix B contains tables of logarithmic and trigonometric functions.

The symbol, □, is used to mark the end of each example.

Supplementary Materials

The following supplements are available for *Basic Technical Mathematics:* a student supplement that contains the worked-out solutions to every third problem, special calculator solutions to problems, computer solutions to problems, and helpful study tips; an instructor's manual that contains the answers not included in the Answer section at the end of this text, answers to the Can you solve these? problems, chapter tests with their answers, and chapter commentaries. In addition there is a special set of computer programs that can be used in a laboratory setting or for classroom demonstration.

Acknowledgments

We would like to thank our friends and fellow faculty members, expecially Hubert Haefner and Susan Shrader, who have given us their criticisms and suggestions while we were developing this book. We would like to thank the following reviewers for their conscientiousness and their many thoughtful suggestions: Thomas G. Burgess, Pima Community College, Raymond Cook, San Bernardino Valley College, Hubert Haefner, Monroe Community College, McHenry B. Holt, Northern Virginia Community College, Larry R. Lance, Columbus Technical Institute, John R. Long, Midlands Technical College, Jeanne McKinnon, Central Piedmont Community College, David E. Stevens, Wentworth Institute of Technology, and John Tobey Jr. of North Shore Community College.

The solutions to the problems in the text and the teacher's guide are the results of the efforts of Paul Ford, Brown University; John Melder, Monroe Community College; Michele Scilla, Monroe Community College; John Von Hold, Genesee Community College; Michael Swick, Monroe Community College; and the authors.

Special thanks go to Barbara Erskine, an excellent secretary, who did an outstanding job of typing the manuscript.

We would also like to thank Cynthia Brankline, Patty Kaiser, Joanne Maxson, Lori Mulligen, and Nancy Schlinger for providing needed secretarial help at critical times.

We would also like to thank Ernest Passaro for his service to the project.

It is a pleasure to acknowledge the assistance given us by the staff of Addison-Wesley Publishing Company in the guidance provided from the inception to the completion of the project. A special thank you for Patricia Mallion who got us started and provided essential initial guidance. A special note of appreciation for Jeffrey Pepper, our editor, who provided encouragement and often-needed advice.

We would like to thank our wives Joyce and Gerry, and Brian, Lisa, Teri, and Tod. Without their support and their great sacrifices, this book could not have become a reality.

Rochester, New York S.R.P.
January 1985 J.F.E.

Contents

11 Quadratic Equations 396

12 Equations of Higher Degree 447

13 Exponential and Logarithmic Functions 481

14 Inequalities 535

15 Additional Topics in Trigonometry 579

16 Sequences and Series 610

17 Plane Analytic Geometry 652

18 Computers 714

1

Models of vessels are often "towed" in test tanks to determine their wave-making characteristics and therefore their propulsion requirements. The wave-making characteristics of both model and vessel should be similar if Froude's number, $N = \frac{V}{\sqrt{\ell g}}$ (V = velocity, ℓ = water-line length, and g = gravitational constant) is the same for both. What is Froude's number for an America's Cup yacht with a water-line length of 64 ft and an expected speed of 10 knots (g = 32.2 ft/s²)?

Number Systems

1.1 Numbers

The study of technical mathematics requires a knowledge of different kinds of numbers. These types range from the counting numbers to the complex numbers.

The set of **counting numbers** {1, 2, 3, 4, 5, 6, . . .} (also called **positive integers** or **natural numbers**) is the set of numbers to which we were introduced when we learned to count. Somewhere along the way a zero (0) was added to this set of counting numbers and the new set was called the set of **whole numbers** {0, 1, 2, 3, 4, 5, 6, . . .}. With the development of mathematical skills it becomes evident that not every problem has a solution that is a whole number.

EXAMPLE 1 If possible, find a whole number that is a solution to the following problems:

a. $3 + 2$ **b.** $4 - 7$

Solution:

a. For the problem $3 + 2$, we simply add, and the sum is the solution, the whole number 5.

b. For the problem $4 - 7$, there is no solution that is a whole number. It is not possible to subtract a larger number from a smaller one and find a solution that is a whole number. □

The problem "$4 - 7$" will have a solution if the set of numbers is extended to include the set of **negative integers** {. . . , −6, −5, −4, −3, −2, −1}. The number that is the solution to the problem "$4 - 7$" is the negative integer "-3." If the set of counting numbers, zero, and the set of negative integers, are combined into a single set, this new set is called the set of **integers** {. . . , −6, −5, −4, −3, −2, −1, 0, 1, 2, 3, 4, 5, 6, . . .}.

Prior to solving problems with the set of negative integers, most of us solved problems containing fractions or decimals. **Fractions** and **decimals** are members of the set of **rational numbers.** There are two common ways of defining rational numbers. Let us consider both of these definitions.

DEFINITION 1.1

> A **rational number** is a number of the form a/b, where a and b are integers and $b \neq 0$.

EXAMPLE 2 The following are examples of rational numbers:

$$\frac{1}{3}, \frac{2}{3}, \frac{-5}{4}, \frac{0}{4}, \frac{11}{-2}, \frac{25}{1}, \frac{1}{1}, \frac{6}{2}. \quad \square$$

DEFINITION 1.2

> A **rational number** is a number that can be expressed as a terminating or repeating decimal.

EXAMPLE 3

a. The following are examples of terminating decimals:

$0.50, \quad 0.75, \quad 1.34, \quad -3.856.$

b. The following are examples of repeating decimals:

$0.333\ldots, 0.666\ldots, -1.3434\ldots, -2.3478478\ldots$

Each of these can be written as

$0.\overline{3}, \quad 0.\overline{6}, \quad -1.\overline{34}, \quad -2.3\overline{478}.$

The bar indicates the group of digits that is repeating. □

With the aid of your calculator it is easy to change each of the rational numbers in Example 2 to rational numbers of the form in Example 3.

EXAMPLE 4 Convert the following fractions to their decimal equivalents:

a. $\frac{1}{3}$ **b.** $\frac{2}{3}$ **c.** $\frac{-5}{4}$

Solution: Using the calculator we get the following results:

a. $\frac{1}{3} = 1 \div 3 = 0.3333333.$

b. $\frac{2}{3} = 2 \div 3 = 0.6666666$ or $0.6666667.$

(Did the display on your calculator have a last digit of 6 or 7? If the last digit is a 6, the calculator truncates and does not automatically round off. If the last digit is a 7, the calculator automatically rounds off the answer.)

c. $\frac{-5}{4} = -5 \div 4 = -1.25.$ □

Numbers such as $\sqrt{2}$, $\sqrt{3}$, $\sqrt{5}$, or π cannot be expressed in the form a/b or as a terminating or repeating decimal. This set of numbers is called the set of **irrational numbers.**

DEFINITION 1.3

> A number that cannot be expressed in the form a/b, where a and b are integers and $b \neq 0$, is called an **irrational number.**

We can also say that an irrational number is a nonrepeating, nonterminating decimal expression.

EXAMPLE 5 Examples of irrational numbers are:

$$\frac{\sqrt{3}}{2},\ -\sqrt{7},\ \frac{3}{\sqrt{2}},\ \frac{\pi}{5},\ \sqrt{19},\ \sqrt{\frac{38}{2}}. \qquad \square$$

If the sets of all the numbers discussed — counting numbers, whole numbers, negative integers, integers, rational numbers, and irrational numbers — are placed in a single set, this collection is called the set of **real numbers.** It is impossible to list all the real numbers; therefore, the real numbers can be indicated by using a geometrical concept — the number line. A number line is a straight line that has an arbitrary point we call zero, or the origin, and by convention all the points to the left of zero are indicated with negative numbers and all the points to the right of zero are indicated with positive numbers (Fig. 1.1). The arrows indicate that the number line extends indefinitely to the right and left of zero. Each point on the number line can be named with a real number. Also, each real number can be represented as a point on the number line.

EXAMPLE 6 Draw a number line and indicate the position of the following real numbers on the number line:

$$-4\frac{1}{2},\quad \sqrt{2},\quad \frac{9}{2},\quad -1.75,\quad -\sqrt{7},\quad \pi,\quad \frac{\sqrt{214}}{2},\quad \frac{-7.5}{1}.$$

Solution: The solution is shown in Fig. 1.2. $\square$

FIGURE 1.1

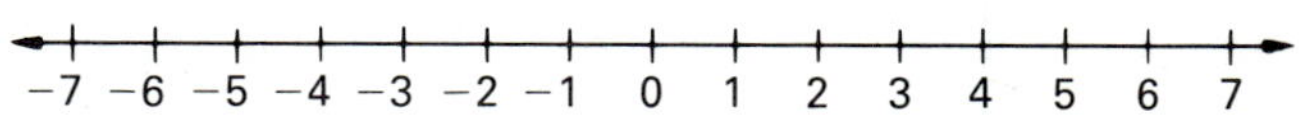

FIGURE 1.2

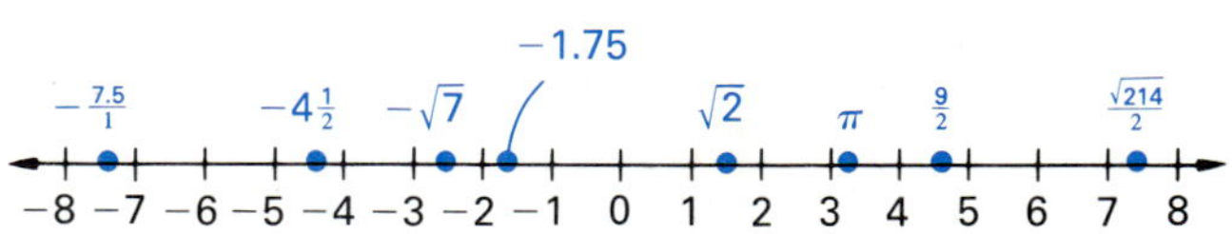

A solution to a problem may be of the form $\sqrt[n]{-a}$, where n is an even integer. A number of this type is called an **imaginary number.** However, $\sqrt[n]{-a}$, where n is an odd integer will be a real number.

EXAMPLE 7 Indicate whether the following numbers are real or imaginary:

a. $\sqrt{-9}$ **b.** $\sqrt[4]{-16}$ **c.** $\sqrt[3]{-27}$ **d.** $-\sqrt{4}$

Solution:

a. The symbols $\sqrt[2]{-9}$ and $\sqrt{-9}$ have the same meaning. They both mean the square root of -9. It is common practice to omit the 2 (index) indicating a square root. $\sqrt{-9}$ is an imaginary number since $n = 2$ is an even integer. Another way of writing this number is $\sqrt{-9} = \sqrt{9}j$ or $3j$. This will be discussed in greater detail in Chapter 9.

b. $\sqrt[4]{-16}$ is an imaginary number since 4 is an even integer.

c. $\sqrt[3]{-27}$ is a real number since 3 is an odd integer. In fact, $\sqrt[3]{-27} = -3$, since $(-3)(-3)(-3) = -27$.

d. An equivalent way of writing $-\sqrt{4}$ is $-(\sqrt{4})$ indicating that the operation of square root is performed and then the minus sign is assigned to the answer. Thus $-\sqrt{4} = -2$. □

Combining the set of real numbers and the set of imaginary numbers results in a set of numbers called the **complex numbers.**

DEFINITION 1.4

A **complex number** is a number that can be written in the form $a + bj$, where a and b are real numbers. If $b = 0$, then $a + bj$ is a real number. If $a = 0$, then we have a number of the form bj, which is called a **pure imaginary number.**

To gain a better understanding of the complete number system and the relationship between the numbers, see Fig. 1.3 on the next page.

1.1 Exercises

In Exercises 1–8, determine which, if any, represent a counting number.

1. $4 + 5$ **2.** $5 - 7$ **3.** $6 - 4$ **4.** $11 - 14$

5. $5 - 5$ **6.** $13 - 9$ **7.** $11 - 11$ **8.** $15 - 17$

In Exercises 9–20, determine if the numbers are rational or irrational.

9. $\frac{22}{7}$ **10.** $\sqrt{2}$

11. $3.14\overline{14}$ **12.** π

13. $3.141441444\ldots$ **14.** $\sqrt{16}$

15. $\frac{\sqrt{9}}{4}$ **16.** $\sqrt{3} + \sqrt{2}$

17. $3.14159\ldots$ **18.** $1.4142136\ldots$

19. $\sqrt{16} + \sqrt[5]{32}$ **20.** $\frac{0}{\sqrt{2}}$

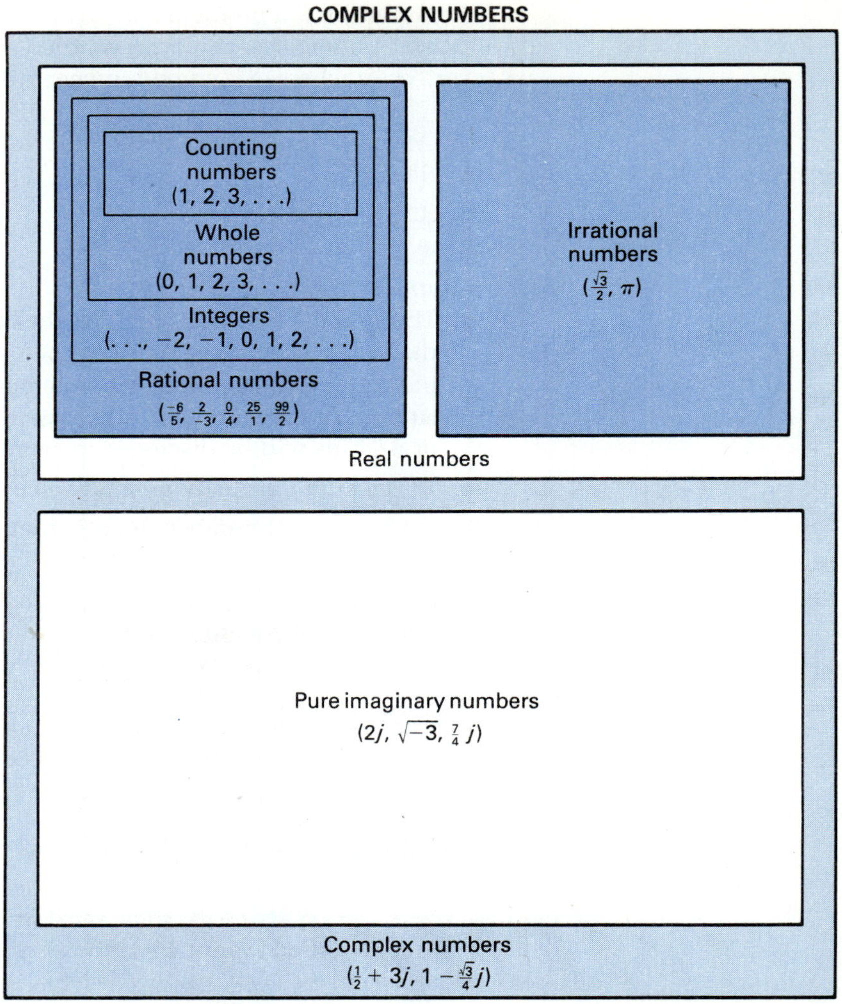

FIGURE 1.3

In Exercises 21–34, draw a number line and indicate the position of the real numbers on it.

21. 5 **22.** -8 **23.** $\frac{5}{4}$ **24.** -0.75

25. $\sqrt{9}$ **26.** $-\sqrt{2}$ **27.** $\frac{0}{14}$ **28.** $-5\frac{1}{2}$

29. $\sqrt{3}$ **30.** π **31.** $\frac{\sqrt{36}}{2}$ **32.** $\frac{18}{\sqrt{81}}$

33. $\frac{-\sqrt{121}}{\sqrt{121}}$ **34.** $\frac{\sqrt{49}}{\sqrt{7}}$

In Exercises 35–48, determine whether the numbers are real, pure imaginary, or complex.

35. $3 + 4j$ **36.** $7j$ **37.** $\frac{23}{3}$

38. $-17j$ **39.** $-11 + 9j$ **40.** $4\sqrt{3}$

41. $-\frac{3}{4}j$ **42.** $\sqrt[3]{-125}$ **43.** $\sqrt{7} + 5j$

44. $-\sqrt[3]{125}$ **45.** $-\sqrt{-36}$ **46.** $\sqrt[4]{-81}$

47. $\sqrt{\frac{0}{-4}}$ **48.** $-\sqrt{16} + \frac{\sqrt{-27}}{\sqrt{9}}$

49. Construct a chart, different from the one in Fig. 1.3, to illustrate the relationship between the sets of numbers in our number system. (*Hint:* Use a tree-type diagram.)

1.2 Relationships between numbers

Between any two real numbers, a and b, one of the following relationships must exist:

$a = b$ (a equals b)
$a < b$ (a is less than b)
$a > b$ (a is greater than b)

The symbols $<$ and $>$ are called signs of inequality.

EXAMPLE 1 For the following pairs of numbers insert the appropriate symbol, $<$, $>$, or $=$, between the numbers.

a. $3, 6$ **b.** $-3, -8$ **c.** $5, 0$

d. $-\pi, -3.14159 \ldots$ **e.** $-\sqrt{3}, -\sqrt{2}$ **f.** $4\frac{1}{3}, -2\frac{1}{7}$

Solution:

a. 3 is less than 6, therefore $3 < 6$.
b. -3 is greater than -8, therefore, $-3 > -8$.
c. 5 is greater than 0, therefore $5 > 0$.
d. $-\pi$ is equal to $-3.14159 \ldots$, therefore $-\pi = -3.14159 \ldots$
e. $-\sqrt{3}$ is less than $-\sqrt{2}$, therefore, $-\sqrt{3} < -\sqrt{2}$.
f. $4\frac{1}{3}$ is greater than $-2\frac{1}{7}$, therefore, $4\frac{1}{3} > -2\frac{1}{7}$. □

There are many real numbers that satisfy the expression of inequality $x < 2$. A few of these real numbers are -1, 0, 1/2, $-\sqrt{2}$, and $-\pi$. The set of all real numbers that satisfies such an expression is called the **solution set.** The number line is a method that can be used to show the set of numbers that makes this a true statement of inequality. In some problems it will be necessary to show that 2 is a member of the solution set. In this case we would write $x \leq 2$, read: "x is less than or equal to 2."

EXAMPLE 2 Illustrate the solution set for each of the following on a number line:

a. $x < 2$ **b.** $x > 4$ **c.** $x \geq -2$
d. $-3 \leq x$ **e.** $-2 < x < 4$ **f.** $-\pi \leq x \leq \pi$

Solution:

a. The expression $x < 2$ tells us that the solution set consists of all the real numbers less than 2, but not including 2. This is illustrated on a number line by drawing an open circle at 2 (meaning 2 is not included) and an arrow pointing to the left of 2, as shown in Fig. 1.4.

b. To illustrate the solution set for $x > 4$, draw an open circle at 4 and an arrow pointing to the right of 4 (Fig. 1.5).

c. The statement $x \geq -2$ tells us that the solution contains all the real numbers greater than and equal to -2. This solution set is illustrated on the graph by drawing a solid dot at -2 and an arrow pointing to the right of -2 (Fig. 1.6).

d. To illustrate the solution set of $-3 \leq x$, draw a solid dot at $x = -3$ and an arrow pointing to the right of -3, as shown in Fig. 1.7.

e. The statement $-2 < x < 4$ is different than the first four statements. This statement tells us to find all the real numbers greater than -2 and, at the same time, less than 4. The clue here is the word *and,* meaning that the members of the solution set must satisfy both statements of inequality. In simple language, it is saying that the real numbers that will satisfy the statement are the numbers between -2 and 4. We indicate this graphically by drawing open circles at -2 and 4 and connecting the circles with a line (Fig. 1.8).

f. The statement, $-\pi \leq x \leq \pi$ includes the end points $-\pi$ and π. Therefore, this solution set is illustrated by drawing solid dots at $-\pi$ and π and connecting the dots with a line, as shown in Fig. 1.9. □

FIGURE 1.4

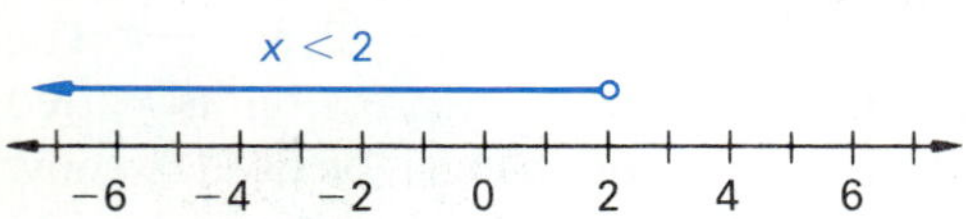

FIGURE 1.5

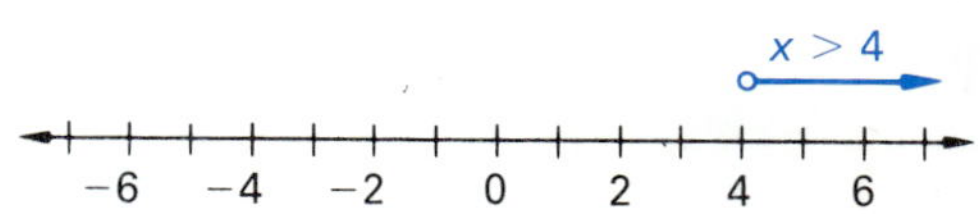

FIGURE 1.6

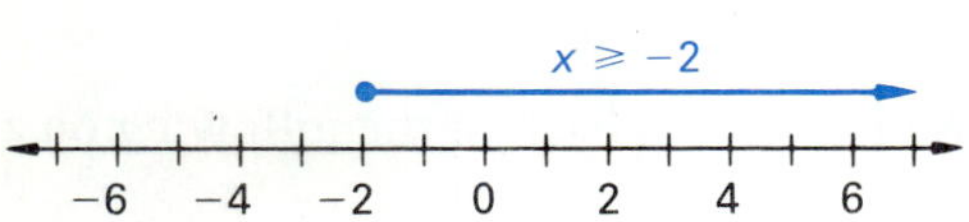

FIGURE 1.7

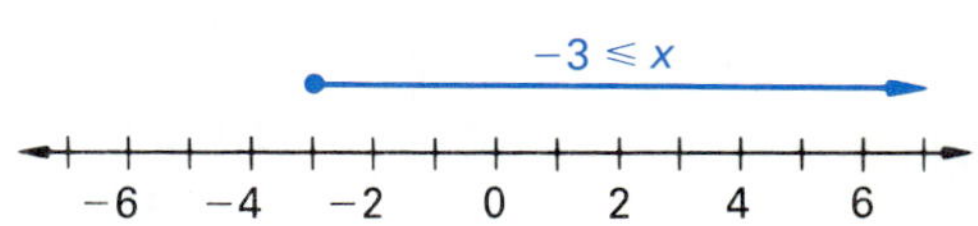

FIGURE 1.8

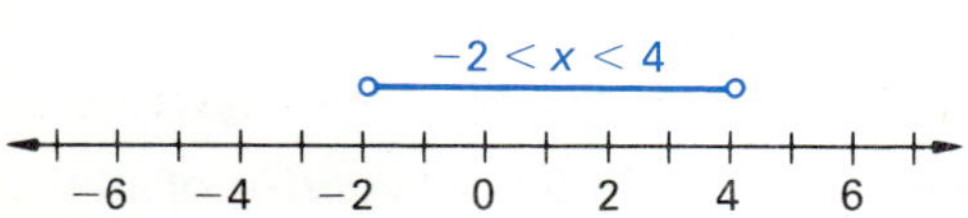

FIGURE 1.9

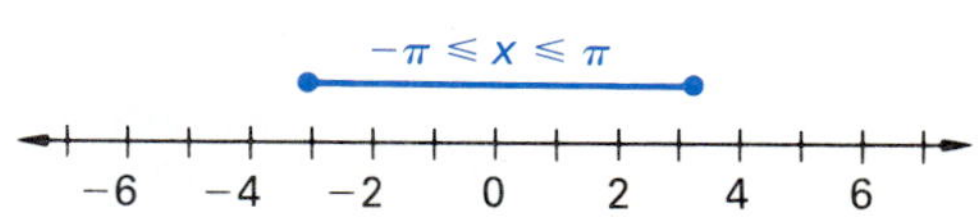

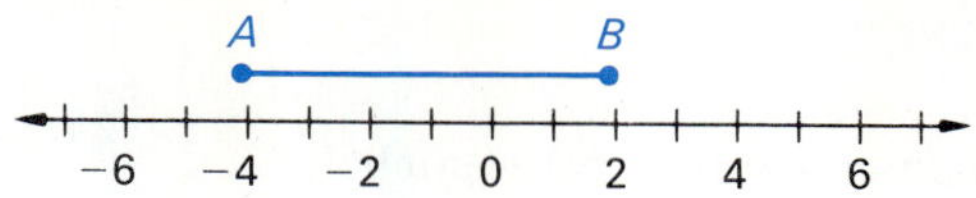

FIGURE 1.10

Absolute value

For the graph in Fig. 1.10, we ask: "What is the measure of the line segment connecting points A and B?" If by measure we mean to include the direction in which we are moving, then the measure of line segment AB, written $\overline{AB}$, is $+6$ units, whereas the measure of line segment BA, $\overline{BA}$, is -6 units. However, if the desired solution is the length of the line segment, then we disregard the sign and say 6 units. Another way of indicating the length would be to take the absolute value of the measure. The length AB would be equal to $|+6| = 6$, or the length of BA would be equal to $|-6| = 6$.

EXAMPLE 3 Find the absolute value of the following numbers:

a. $+3$ **b.** -3 **c.** $+\frac{3}{4}$ **d.** -3.427 **e.** $-\pi$ **f.** $-\sqrt{5}$

Solution:

a. $|+3| = 3$ **b.** $|-3| = 3$ **c.** $\left|\frac{+3}{4}\right| = \frac{3}{4}$

d. $|-3.247| = 3.247$ **e.** $|-\pi| = \pi$ **f.** $|-\sqrt{5}| = \sqrt{5}$ □

DEFINITION 1.5

The **absolute value** of the real number x, symbolized $|x|$, is

$$|x| = \begin{cases} x \text{ if } x > 0 \\ 0 \text{ if } x = 0 \\ -x \text{ if } x < 0 \end{cases}$$

From the definition we can see that an absolute value cannot be negative because when

$x > 0$, $|x| = x$, which is non-negative ($|3| = 3$)
$x = 0$, $|x| = 0$, which is non-negative, and
$x < 0$, $|x| = -x$, which is non-negative since x itself is negative ($|-3| = -(-3) = 3$).

1.2 Exercises

In Exercises 1–9, insert the appropriate symbol, $<, >$, or $=$, between the pairs of numbers.

1. $3, 14$
2. $-5, -21$
3. $-\sqrt{3}, -\sqrt{2}$
4. $-\sqrt{81}, -9$
5. $\sqrt{5}, 2.236068 \ldots$
6. $\frac{22}{7}, \pi$
7. $\frac{14}{4}, 3\frac{1}{4}$
8. $\frac{1}{4}-\frac{1}{3}, \frac{1}{3}-\frac{1}{4}$
9. $\frac{1}{2}+\frac{1}{3}, \frac{3}{2}-\frac{2}{3}$

In Exercises 10–21, graph the statements on a number line.

10. $x > 2$
11. $x < 1.4$
12. $x > -3$
13. $x < -2$
14. $x \geq -4$
15. $x \leq 4$
16. $-3 < x < 5$
17. $3 \leq x < 11$
18. $-5.4 < x < 4.3$
19. $-10.4 \leq x < 12.2$
20. $-\sqrt{5} \leq x \leq \sqrt{3}$
21. $-\sqrt{7} < x < \sqrt{9}$

In Exercises 22–29, write a statement using equality and inequality symbols which describe the set of points illustrated by each of the graphs.

22. 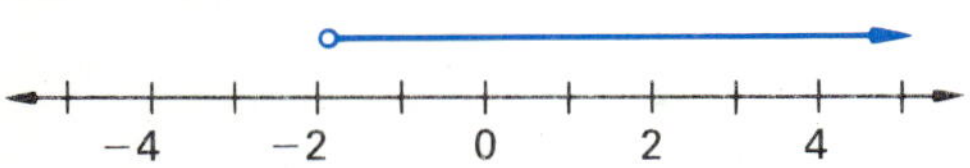

23. 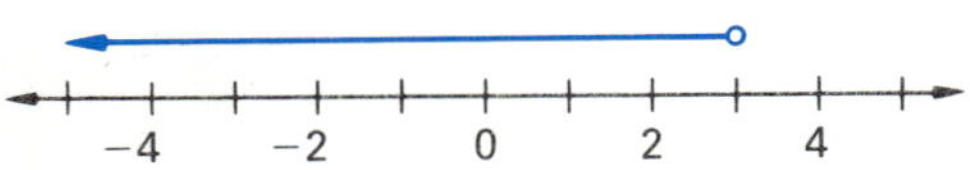

24. 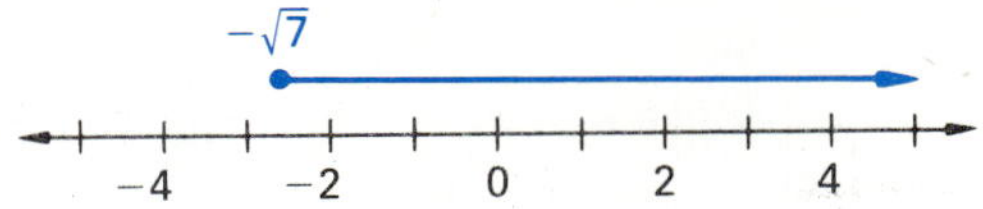

25. 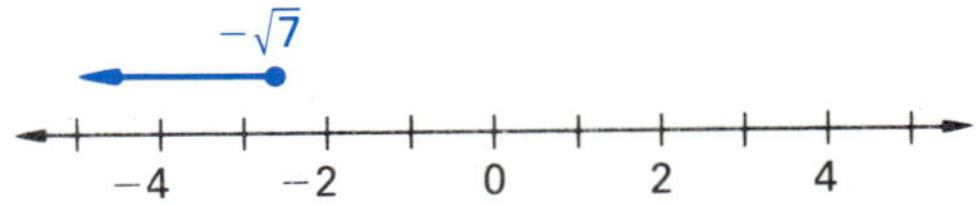

26. 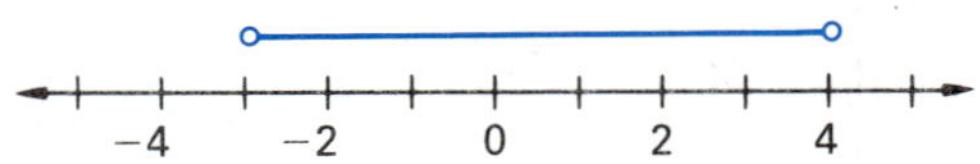

27. 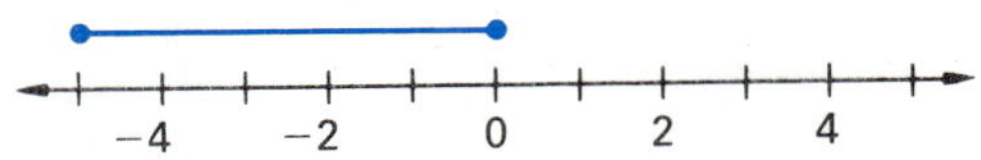

28. 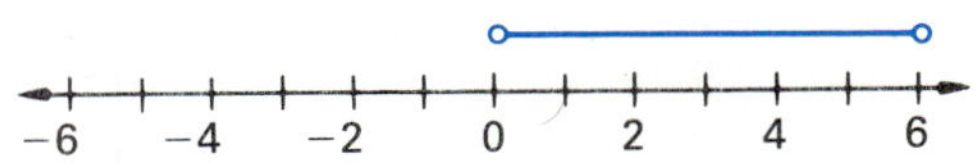

29. 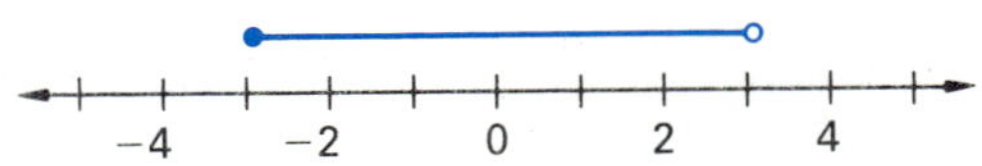

In Exercises 30–42, find the absolute value of the numbers.

30. $|-4|$
31. $|9|$
32. $|-\sqrt{11}|$
33. $|9-5|$
34. $\left|\frac{3}{7}-\frac{1}{5}\right|$
35. $|-\sqrt[3]{8}|$
36. $\left|3\frac{1}{7}-\pi\right|$
37. $|(-2)^2|$
38. $|(-2)^3|$
39. $|-2|^2$
40. $|(-2)^2-(-2)^3|$
41. $|-4.361|$
42. $|5.73-8.16|$

1.3 Properties and operations of real numbers

With the exception of one chapter, almost all the work in this text deals with the set of real numbers or expressions that represent real numbers. Thus, to ensure proper use of the real numbers we will discuss the set of axioms for addition and multiplication of real numbers. (An **axiom** is a statement that is accepted as being true without proof.) An axiom is like a rule in a game such as in baseball, where we all accept the fact that three strikes make an out.

Axioms for Addition and Multiplication of Real Numbers

Property	Addition	Multiplication
1. **Closure Property**	$a + b$ is a unique real number. $a + b$ is called the sum of a and b, and a and b are called the addends.	$a \cdot b$ is a unique real number. $a \cdot b$ is called the product of a and b, and a and b are called factors of the product.
2. **Commutative Property**	$a + b = b + a$. The sum of two numbers is unchanged when the order of the addends is reversed.	$a \cdot b = b \cdot a$. The product of two numbers is unaffected by interchanging the order of the factors.
3. **Associative Property**	$(a + b) + c = a + (b + c)$. The sum is the same regardless of the manner in which the addends are grouped.	$(a \cdot b) \cdot c = a \cdot (b \cdot c)$. The product is the same no matter how we group the factors.
4. **Identity Property**	There is a unique number, 0, such that $a + 0 = a$ and $0 + a = a$. 0 is called the **additive identity.**	There is a unique number, 1, such that $a \cdot 1 = a$ and $1 \cdot a = a$. 1 is called the **multiplicative identity.**
5. **Inverse Property**	For each number a there is a unique number, $-a$, such that $a + (-a) = 0$ and $(-a) + a = 0$. $-a$ is called the negative of a or the **additive inverse** of a.	For each nonzero number a, there is a unique number, $\frac{1}{a}$, such that $a \cdot \frac{1}{a} = 1$ and $\frac{1}{a} \cdot a = 1$. $\frac{1}{a}$ is called the **reciprocal** of a or the **multiplicative inverse of** a.
6. **Distributive Property**	$a \cdot (b + c) = a \cdot b + a \cdot c$ or $(b + c) \cdot a = b \cdot a + c \cdot a$. The product of a real number with a sum of real numbers is equal to the sum of the products of those numbers.	

EXAMPLE 1 Match each statement in column A with the property in column B that justifies the statement.

Column A	Column B
a. $6 \cdot (4+3) = 6 \cdot 4 + 6 \cdot 3$	**1.** Multiplicative identity
b. $3 + 4 = 4 + 3$	**2.** Distributive property
c. $3 \cdot (4 \cdot 7) = (3 \cdot 4) \cdot 7$	**3.** Commutative property of addition
d. $(4+3) \cdot 6 = 4 \cdot 6 + 3 \cdot 6$	**4.** Associative property of multiplication
e. $(6 \cdot 4) \cdot 8 = 8 \cdot (6 \cdot 4)$	**5.** Additive identity
f. $4 + 0 = 4$ and $0 + 4 = 4$	**6.** Associative property of addition
g. $3 + (4+7) = (3+4) + 7$	**7.** Commutative property of multiplication
h. $4 \cdot (1/4) = 1$ and $(1/4) \cdot 4 = 1$	**8.** Additive inverse
i. $2 \cdot 1 = 2$ and $1 \cdot 2 = 2$	**9.** Multiplicative inverse
j. $8 + (-8) = 0$ and $(-8) + 8 = 0$	

Solution:

a. 2 **b.** 3 **c.** 4 **d.** 2 **e.** 7 **f.** 5 **g.** 6
h. 9 **i.** 1 **j.** 8 □

All the properties discussed apply to addition and multiplication. The system would not be complete without defining subtraction and division.

DEFINITION 1.6

> The **subtraction** of b from a, written $a - b$, is defined as
>
> $$a - b = a + (-b).$$

That is, to subtract b from a, add the additive inverse of b to a.

DEFINITION 1.7

> The **division** of a by a nonzero number b, written $a \div b$ or $\frac{a}{b}$ is defined as
>
> $$\frac{a}{b} = c, \text{ if and only if } a = b \cdot c.$$

We also define $a \div b$ as $a \times \frac{1}{b}$.

We will now discuss the rules for signed numbers, so that we can add, subtract, multiply, and divide any real numbers. For the purpose of discussion it will be assumed that a and b are signed numbers.

Rules for Addition of Signed Numbers

Rule I If a and b have like signs (that is either both are positive or both are negative), compute $|a| + |b|$. Then place the common sign of a and b in front of the sum. The result is the sum of a and b.

Rule II If a and b have unlike signs (one positive and the other negative) and $|a| = |b|$, then $a + b = 0$.

Rule III If a and b have unlike signs and $|a| \neq |b|$, subtract the smaller absolute value from the larger absolute value. Then place the sign of the number having the larger absolute value in front of the difference. The resulting number is the sum of a and b.

EXAMPLE 2 Find the sum of the given signed numbers.

a. $(+4) + (+7)$ **b.** $(-4) + (-7)$ **c.** $(-4) + (+4)$
d. $(-7) + (+4)$ **e.** $(+3) + (+4) + (-5) + (-11)$

Solution:

a. Since the signs of the numbers are alike, we apply Rule I.

$$\begin{aligned}(+4) + (+7) &= +(|+4| + |+7|)\\ &= +(4 + 7)\\ &= +11\end{aligned}$$

b. Since the signs of the numbers are alike, we apply Rule I.

$$\begin{aligned}(-4) + (-7) &= -(|-4| + |-7|)\\ &= -(4 + 7)\\ &= -11\end{aligned}$$

c. Since $|-4| = |+4|$, we apply Rule II.

$$(-4) + (+4) = 0$$

d. Since the signs of the numbers are not alike, and $|-7| \neq |+4|$, we apply Rule III.

$$\begin{aligned}(-7) + (+4) &= -(|-7| - |+4|)\\ &= -(7 - 4)\\ &= -3\end{aligned}$$

$$\textbf{e. } \underbrace{(+3)+(+4)}+(-5)+(-11) =$$
$$\underbrace{(+7)+(-5)}+(-11) =$$
$$(+2)+(-11) = -9 \quad \square$$

The operation of subtraction of real numbers in terms of the operation of addition is defined in Definition 1.6.

EXAMPLE 3 Find the difference of the given signed numbers.

a. $(+6)-(+4)$ **b.** $(-6)-(-4)$ **c.** $(-6)-(+6)$
d. $(-6)-(-6)$ **e.** $(+4)-(-6)$

Solution: To find the difference in each of the problems, apply the definition of subtraction in the first step and then apply the rules of addition.

a. $(+6)-(+4) = (+6)+(-4)$ **Definition of subtraction**
$= +(|+6|-|-4|)$ **Rule III of addition**
$= +(6-4)$
$= 2$

b. $(-6)-(-4) = (-6)+(+4)$ **Definition of subtraction**
$= -(|-6|-|+4|)$ **Rule III of addition**
$= -(6-4)$
$= -2$

c. $(-6)-(+6) = (-6)+(-6)$ **Definition of subtraction**
$= -(|-6|+|-6|)$ **Rule I of addition**
$= -(6+6)$
$= -12$

d. $(-6)-(-6) = (-6)+(+6)$ **Definition of subtraction**
$= 0$ **Rule II of addition**

e. $(+4)-(-6) = (+4)+(+6)$ **Definition of subtraction**
$= +(|+4|+|+6|)$ **Rule I of addition**
$= +(4+6)$
$= +10$ $\square$

The rules for multiplication and division of signed numbers are very similar. Therefore we will discuss them together. To multiply (or divide) a and b, determine the absolute value of each number and compute $|a|\cdot|b|\left(\text{or } \frac{|a|}{|b|}\right)$.

If the signs of a and b are alike, the product (or quotient) is positive and if the signs are unlike, the product (or quotient) is negative.

Rules for Multiplication and Division of Signed Numbers

Rule IV If a and b have like signs, place a plus sign in front of the product (or quotient) of their absolute values. The resulting number is $+(|a| \cdot |b|)$ $\left(\text{or } +\left(\frac{|a|}{|b|}\right)\right)$.

Rule V If a and b have unlike signs, place a minus sign in front of the product (or quotient) of their absolute values. The resulting number is $-(|a| \cdot |b|)$ $\left(\text{or } -\left(\frac{|a|}{|b|}\right)\right)$.

EXAMPLE 4 Determine the product or the quotient of the given signed numbers.

a. $(+6) \cdot (+5)$ **b.** $(-6) \div (-2)$

c. $(+6) \cdot (-5)$ **d.** $(-6) \div (+2)$

Solution:

a. Applying Rule IV, we find

$$\begin{aligned}(+6) \cdot (+5) &= +(|+6| \cdot |+5|)\\ &= +(6 \cdot 5)\\ &= +30.\end{aligned}$$

b. Applying Rule IV, we find

$$\begin{aligned}(-6) \div (-2) &= +(|-6| \div |-2|)\\ &= +(6 \div 2)\\ &= +3.\end{aligned}$$

c. Applying Rule V, we find

$$\begin{aligned}(+6) \cdot (-5) &= -(|+6| \cdot |-5|)\\ &= -(6 \cdot 5)\\ &= -30.\end{aligned}$$

d. Applying Rule V, we find

$$\begin{aligned}(-6) \div (+2) &= -(|-6| \div |+2|)\\ &= -(6 \div 2)\\ &= -3. \quad \square\end{aligned}$$

The value of the expression $(35 - 15) \div 5$ is 4. The parentheses indicate that the operation of subtraction is performed first, $35 - 15 = 20$, then the difference, 20, is divided by 5. On the other hand, the value of the expression $35 - 15 \div 5$ is 32. Since there are no parentheses to tell us what to do first, we look at the operations. First perform all multiplications and divisions, as you move from left to right through the expression, then perform all additions and subtractions in order from left to right. Thus, in the expression $35 - 15 \div 5$, first divide 15 by 5 and then subtract the quotient, 3, from 35 giving the result of 32.

EXAMPLE 5 Find the value of the expression

$$-16 \div 4 - 8 \cdot (-2) + 7$$

Solution: Since no grouping is indicated with parentheses, perform all the operations of multiplication and division, then perform the operations of addition and subtraction.

$$\underbrace{-16 \div 4} - \underbrace{8 \cdot (-2)} + 7 =$$
$$\underbrace{-4 - (-16)} + 7 =$$
$$+12 + 7 = 19 \qquad \square$$

1.3 Exercises

In Exercises 1–8, determine the number indicated.

1. The additive inverse of 6
2. The reciprocal of $\frac{2}{3}$
3. The negative of -5
4. The multiplicative inverse of $\frac{1}{7}$
5. The additive inverse of -7
6. The reciprocal of -13
7. The additive inverse of $-\frac{1}{3}$
8. The negative of $\frac{3}{13}$

In Exercises 9–18 fill in each blank so that the resulting statement illustrates the indicated law. (a and b represent real numbers.)

9. The commutative property of multiplication: $8 \cdot 9 =$ _____.
10. The additive inverse of 4 is _____.
11. The associative property of multiplication: $4 \cdot (6 \cdot 7) =$ _____.
12. The multiplicative inverse of $\frac{-4}{7} =$ _____.
13. The distributive property: $3 \cdot (a + b) =$ _____.
14. The additive identity: $18 +$ _____ $= 18$.
15. The commutative property of addition: $3 + (a + b) =$ _____.
16. The multiplicative identity: $4 \cdot$ _____ $= 4$.
17. The associative property of addition: $3 + (14 + 19) =$ _____.
18. The distributive property: $(a + 4) \cdot b =$ _____.

In Exercises 19–28, identify the axiom that is illustrated in each case. (a and b represent real numbers.)

19. $0 + a = a$
20. $\frac{1}{a} \cdot a = 1$
21. $3 + a = a + 3$
22. $a + (b + 5) = (a + b) + 5$
23. $3 \cdot (2 + b) = 3 \cdot 2 + 3 \cdot b$
24. $(a + b) + 8 = 8 + (a + b)$
25. $5 + -5 = 0$
26. $[(2 \cdot a) \cdot b] \cdot 4 = 2 \cdot [a \cdot (b \cdot 4)]$
27. $(5 + 13) \cdot b = 5 \cdot b + 13 \cdot b$
28. $-15 \cdot \frac{1}{-15} = 1$

In Exercises 29–46, evaluate each of the given expressions by performing the indicated operations.

29. $7 + 6$
30. $8 + (-5)$
31. $(-8) + (+5)$
32. $(-6) + (-7)$
33. $15 - 9$
34. $(+18) - (+27)$
35. $(-18) - (+27)$
36. $(+18) - (-27)$
37. $(+18) - (+18)$
38. $(+18) - (-18)$
39. $(+4)(+3)$
40. $(-4)(-7)$
41. $(-15)(+3)$
42. $(+15)(-3)$
43. $18 \div 3$
44. $(-18) \div (+3)$
45. $(+21) \div (-3)$
46. $(-24) \div (-8)$

In Exercises 47–64, evaluate each of the given expressions by performing the indicated operations.

47. $3 - (-2) + (-7) - (+6)$
48. $5 + (-6) - (-11) + 13$
49. $-13 + (-12) - 14 + (-5)$
50. $(-4)(+3)(-5)$
51. $(-6)(-2)(-7)$
52. $(+7)(-6)(+8)$
53. $\dfrac{(-6)(8)}{-12}$
54. $\dfrac{-96}{(-3)(4)}$
55. $\dfrac{(-4)(-6)}{-8}$
56. $11 - 3 \cdot (7)$
57. $-2 \cdot (7 - 4) + 8$
58. $(-2)(7) - 4 + 8$
59. $\dfrac{8}{5 - 3} + 2 - 7$
60. $\dfrac{3 - (-4)}{9 - 2}$
61. $\dfrac{13 + (-2)}{3} - \dfrac{9 - 5}{-2}$
62. $(-6)(4) + 9 \div (-3) + 6$
63. $\dfrac{(+4)(-8) - 5 \cdot (-4)}{4 - 7}$
64. $7 + (-8)(-2) - (6) \div (-3) + 14 - (-7)$

1.4 Properties of zero

Zero appears to be an innocent number, but, it frequently causes problems for many students. To help you avoid problems, we will take a careful look at the various properties of zero.

Properties of Zero

1. For any real number a, $a + 0 = a$. This is the identity property of addition, which was discussed in the previous section.
2. For any real number a, $a - 0 = a$. That is, $4 - 0 = 4$.
3. For any real number a, $a \cdot 0 = 0$. That is $6 \cdot 0 = 0$. This is called the multiplicative property of zero.
4. If a and b are real numbers and $a \cdot b = 0$, then $a = 0$ or $b = 0$ or both equal zero. This will be an important rule in later chapters when solutions to equations are sought by the technique of factoring.
5. If $b \neq 0$, then $\frac{0}{b} = 0$. That is, $\frac{0}{5} = 0$. This property can be demonstrated by using the definition of division. $\frac{a}{b} = c$ if and only if $a = b \cdot c$. When $a = 0$ and $b \neq 0$, $a = b \cdot c$ becomes $0 = b \cdot c$. It can be seen that, for the statement to be true, c

must equal zero. There is no other real number that, when multiplied times b, will give the product of 0.

6. If $a \neq 0$, then $\frac{a}{0}$ is undefined. That is, $\frac{-6}{0}$ is undefined. This property can also be demonstrated using the definition of division. Let $a \neq 0$ and $b = 0$, then $a = b \cdot c$ becomes $a = 0 \cdot c$. There is no real number c that can be multiplied by zero to obtain a nonzero real number a. Thus, we say that $\frac{a}{0}$ is undefined.

7. If $a = 0$ and $b = 0$, then $\frac{a}{b} = \frac{0}{0}$ is indeterminate. Indeterminate expressions are studied in calculus. However, we will bend the rules slightly and use the definition of division to suggest why the quotient of $\frac{0}{0}$ is called indeterminate. With $a = 0$ and $b = 0$, $a = b \cdot c$ becomes $0 = 0 \cdot c$. The right side of the equation implies that c could be any real number. Thus, there is no unique solution to the expression $\frac{0}{0}$ and it is said to be indeterminate.

EXAMPLE 1 Evaluate

a. $\dfrac{2 \cdot 4 - 8}{9}$ **b.** $\dfrac{-5 + 11(-2)}{6 + -6}$ **c.** $\dfrac{4\sqrt{7} - 4\sqrt{7}}{\sqrt{7} - \sqrt{7}}$

Solution:

a.
$$\frac{2 \cdot 4 - 8}{9} = \frac{8 - 8}{9}$$
$$= \frac{0}{9}$$
$$= 0 \qquad \textbf{Property 5}$$

b.
$$\frac{-5 + 11(-2)}{6 + -6} = \frac{-5 + -22}{0}$$
$$= \frac{-27}{0}$$
$$= \text{Undefined} \qquad \textbf{Property 6}$$

c.
$$\frac{4\sqrt{7} - 4\sqrt{7}}{\sqrt{7} - \sqrt{7}} = \frac{0}{0}$$
$$= \text{Indeterminate} \qquad \textbf{Property 7}$$

□

1.4 Exercises

In Exercises 1–12, evaluate each of the given expressions by performing the indicated operations.

1. $3 + (6 - 6)$
2. $7(\sqrt{2} - \sqrt{2})$
3. $\sqrt{6} - \sqrt{0}$
4. $\frac{0}{4}$
5. $\frac{3-3}{3}$
6. $\frac{2 \cdot 3 - 6}{12}$
7. $\frac{-2 \cdot 3 + 7 \cdot 5}{5 - 5}$
8. $\frac{-4 + 7(-3)}{4 + -4}$
9. $\frac{(-5)(2) + 7}{-8 + 5}$
10. $\frac{\sqrt{3} - \sqrt{3}}{\sqrt{5}}$
11. $\frac{\sqrt{2}}{\pi - \pi}$
12. $\frac{2\sqrt{6} - 2\sqrt{6}}{\sqrt{6} - \sqrt{6}}$

In Exercises 13–16, let *a*, *b*, and *c* be any real numbers.

13. Is $\frac{a}{a}$ always equal to 1?
14. Is $\frac{a-4}{a-4}$ always equal to 1?
15. Is $\frac{a-b}{a-b}$ always equal to 1?
16. Is $\frac{a(b+c)}{a(b+c)}$ always equal to 1?
17. **a.** What is the result on your calculator when $4 \div 0$?
    b. Divide 4 by the smallest number you can find on the calculator. What is the quotient?

1.5 Introduction to the calculator

For most students the scientific calculator has become an essential tool for computation when solving problems. In this section we will develop techniques for performing basic operations on the scientific calculator. Throughout the remainder of the text, as new topics are developed, the use of the calculator in solving problems will be integrated.

There are two general types of scientific handheld calculators that are suitable for use in a technical mathematics course. The difference between the two types is the method of entry. A calculator with the algebraic entry will have an [=] key. The machine with Reverse Polish Notation (RPN) will have an [ENTER] key.

The two systems of entry have one common basic feature; they use real numbers in decimal form. The machines operate with decimal approximations of all real numbers correct to the capacity of the particular machine. Many scientific calculators will not handle imaginary numbers. For example, try finding the solution to $\sqrt{-4}$ on your calculator. The machine will either blink at you, say "error" or have some other way of saying "no solution on this machine, baby." Another way of discovering the error symbol is to press the keys [0] and [$1/x$].

Thus, whenever this error symbol appears, it means that you have performed an unacceptable operation.

At this time we will assume that you are somewhat familiar with the keyboard of your calculator. The first example will show how to add, subtract, multiply, and divide real numbers. The only way to develop an understanding is to turn on your calculator and hit the proper keys as you read.

EXAMPLE 1 Evaluate each of the given expressions using a calculator.

a. $7 + 6$ **b.** $17 - (-15)$ **c.** $(-18)(5)$ **d.** $18 \div (-3)$

Solution: An operation that is performed by the calculator will be indicated with a rectangle drawn around the symbol.

a. Algebraic: 7 $\boxed{+}$ 6 $\boxed{=}$ 13
RPN: 7 $\boxed{\text{ENTER}}$ 6 $\boxed{+}$ 13

b. Algebraic: 17 $\boxed{-}$ 15 $\boxed{+/-}$ $\boxed{=}$ 32
RPN: 17 $\boxed{\text{ENTER}}$ 15 $\boxed{\text{CHS}}$ $\boxed{-}$ 32

c. Algebraic: 18 $\boxed{+/-}$ $\boxed{\times}$ 5 $\boxed{=}$ -90
RPN: 18 $\boxed{\text{CHS}}$ $\boxed{\text{ENTER}}$ 5 $\boxed{\times}$ -90

d. Algebraic: 18 $\boxed{\div}$ 3 $\boxed{+/-}$ $\boxed{=}$ -6
RPN: 18 $\boxed{\text{ENTER}}$ 3 $\boxed{\text{CHS}}$ $\boxed{\div}$ -6 □

In Section 1.3 we stated a rule for the proper order of performing multiplication, division, addition, and subtraction of real numbers. We now extend this rule to include raising numbers to a power and finding roots of numbers.

The order in which the operations must be performed when moving from left to right is as follows: find roots or raise numbers to a power in the order in which they appear in the expression, then perform any multiplications or divisions in the order in which they appear in the expression, and then perform any additions or subtractions in the order in which they appear in the expression.

If parentheses are present in an expression, the expression in the innermost parentheses is computed first using the above rule, then the same rule is repeated to the next innermost parentheses, and so on, until all parentheses have been cleared and the expression is completely evaluated.

Rules for Evaluation of Expressions

1. Work from innermost parentheses (if any) out and perform operations in order from left to right.
2. Perform the operations of finding roots or raising a number to a power.
3. Perform multiplications and divisions.
4. Perform additions and subtractions.

The order of operations for most scientific calculators is such that they will automatically perform the operations in the order as stated in the above rule.

In the previous sections squares and square roots were used to illustrate various points. With a calculator you can easily perform these operations.

EXAMPLE 2 Perform the indicated operations using a calculator.

a. 12^2 **b.** $3^2 \cdot 5^2$ **c.** $\sqrt{7}$ **d.** $\sqrt{7} \cdot \sqrt{343}$

e. $\dfrac{\sqrt{2}}{\sqrt{3}}$ **f.** $\sqrt{\dfrac{2}{3}}$

Solution:

a. Algebraic: 12 $\boxed{x^2}$ 144

RPN*: 12 $\boxed{x^2}$ 144

b. Algebraic: 3 $\boxed{x^2}$ $\boxed{\times}$ 5 $\boxed{x^2}$ $\boxed{=}$ 225

RPN: 3 $\boxed{x^2}$ 5 $\boxed{x^2}$ $\boxed{\times}$ 225

c. Algebraic†: 7 $\boxed{\sqrt{x}}$ 2.6457513

RPN: 7 $\boxed{\sqrt{x}}$ 2.6457513

d. Algebraic: 7 $\boxed{\sqrt{x}}$ $\boxed{x}$ 343 $\boxed{\sqrt{x}}$ $\boxed{=}$ 49

RPN: 7 $\boxed{\sqrt{x}}$ 343 $\boxed{\sqrt{x}}$ $\boxed{\times}$ 49

e. Algebraic: 2 $\boxed{\sqrt{x}}$ $\boxed{\div}$ 3 $\boxed{\sqrt{x}}$ $\boxed{=}$ 0.8164966

RPN: 2 $\boxed{\text{ENTER}}$ $\boxed{\sqrt{x}}$ 3 $\boxed{\sqrt{x}}$ $\boxed{\div}$ 0.8164966

f. Algebraic: 2 $\boxed{\div}$ 3 $\boxed{=}$ $\boxed{\sqrt{x}}$ 0.8164966

RPN: 2 $\boxed{\text{ENTER}}$ 3 $\boxed{\div}$ $\boxed{\sqrt{x}}$ 0.8164966 □

---

* Some RPN calculators require shifts before performing certain operations.

† Some algebraic entry type calculators require stroking $\boxed{\text{2nd}}$ before $\boxed{\sqrt{x}}$.

With the algebraic type entry in (f) we may be tempted to do the following:

$2 \boxed{\div} 3 \boxed{\sqrt{x}} \boxed{=}$

The result would be $2 \div \sqrt{3}$. Due to the logic of the calculator, the calculator will find the $\sqrt{3}$ first, then divide 2 by $\sqrt{3}$. Thus, it is necessary to find the quotient first by stroking the $\boxed{=}$ key then the $\boxed{\sqrt{x}}$ key.

Understanding the logic, or the order of performance of operations, of your calculator is always important. However, it becomes more important when performing chain operations.

EXAMPLE 3 Perform the indicated operations using a calculator.

a. $(-11)(3)(-6)$ **b.** $\dfrac{(-6)(8)}{12}$

c. $(-6)(4) + 9 \div 3 + 6$ **d.** $\dfrac{(4)(-8) - 5(-4)}{4 - 7}$

Solution:

a. Algebraic: $11 \boxed{+/-} \boxed{\times} 3 \boxed{\times} 6 \boxed{+/-} \boxed{=}$ 198

RPN: $11 \boxed{\text{CHS}} \boxed{\text{ENTER}} 3 \boxed{\times} 6 \boxed{\text{CHS}} \boxed{\times}$ 198

b. Algebraic: $6 \boxed{+/-} \boxed{\times} 8 \boxed{\div} 12 \boxed{=}$ -4

RPN: $6 \boxed{\text{CHS}} \boxed{\text{ENTER}} 8 \boxed{\times} 12 \boxed{\div}$ -4

c. Algebraic: $6 \boxed{+/-} \boxed{\times} 4 \boxed{+} 9 \boxed{\div} 3 \boxed{+} 6 \boxed{=}$ -15

RPN: $6 \boxed{\text{CHS}} \boxed{\text{ENTER}} 4 \boxed{\times} 9 \boxed{\text{ENTER}} 3 \boxed{\div} \boxed{+} 6 \boxed{+}$ -15

The calculator multiplies $-6 \times 4 = -24$, sets this product aside, then divides $9 \div 3 = 3$, recalls the -24 and adds this to 3, which yields -21, and finally adds -21 and 6 for a result of -15.

d. The immediate response may be to work the problem in the following manner:

Algebraic: $4 \boxed{\times} 8 \boxed{+/-} \boxed{-} 5 \boxed{\times} 4 \boxed{+/-} \boxed{\div} 4 \boxed{-} 7 \boxed{=}$ -34

The answer, −34, is wrong. The calculator will multiply and divide, then it will add and subtract. To obtain the correct result, we must do the problem in two parts: Obtain the result in the numerator of the fraction and then the result in the denominator. Then find the quotient of the two results. This can be accomplished with the calculator using parentheses or the storage unit as the following examples illustrate:

Algebraic: $\boxed{(} 4 \boxed{\times} 8 \boxed{+/-} \boxed{-} 5 \boxed{\times} 4 \boxed{+/-} \boxed{)} \boxed{\div} \boxed{(} 4 \boxed{-} 7 \boxed{)} \boxed{=}$ 4

Algebraic: 4 [×] 8 [+/−] [−] 5 [×] 4 [+/−] [=]
[STO 1] 4 [−] 7 [=] −3 [1/x] [×] [RCL 1] = 4
RPN: 4 [ENTER] 8 [CHS] [×] 5 [ENTER] 4 [CHS]
[×] [−] 4 [ENTER] 7 [−] [÷] 4 □

In the remaining chapters of the text the key-stroke explanations will generally be limited to the algebraic scientific calculator.

1.5 Exercises

For Exercises 1–14, evaluate Problems 47–60 of Exercises 1.3 using your calculator.

In Exercises 15–34, evaluate each of the expressions using a scientific calculator.

15. $13^2 \cdot \pi^2$

16. $2 - (\sqrt{3})\,(3.1416)^2$

17. $\dfrac{\sqrt{11}}{\sqrt{47}}$

18. $\sqrt{\dfrac{11}{47}}$

19. $\dfrac{\sqrt{13} - \sqrt{61}}{\sqrt{83}}$

20. $\sqrt{\dfrac{81.7622 - 41.8765}{56.982}}$

21. $(-8.937)(13.7856)(-14.9762)$

22. $(\sqrt{-6.34})\,(-2.345)(-13.756)$

23. $(8.37)^2(-6.19)(13.79)$

24. $\dfrac{(-11.34)\,(18.7654)}{14.379}$

25. $\dfrac{-97.486}{(-35.75)(4.789)}$

26. $\dfrac{(-14.387)(14.372 - 14.372)}{-8.765}$

27. $\dfrac{\sqrt{14.79}\,(3.456 - 8.732)}{16.937 - 16.937}$

28. $(-34.86)(18.96 - 43.21) + 43.15$

29. $\dfrac{24.94}{106.42 - 376.41} + 67.39 - 79.46$

30. $\dfrac{\sqrt{34.89} - (-69.73)}{98.75 - 107.24}$

31. $\dfrac{136.785 + -207.456}{307.162} - \dfrac{346.751 - 961.88}{-43.617}$

32. $(-342.651)(876.477) + 909.706 \div (-3.04) + 8.569$

33. $\dfrac{(-4.76)(324.56) - (51.82)(61.487)}{6.852 - 13.796}$

34. $316.75 + (\sqrt{64}) \cdot (-314.86) - 6.817 \div (-5.147) + 143.782 \div (-714.86)$

Chapter summary

The reader should have an understanding of the numbers that are used in our daily lives as well as in a technical-mathematics course. We call these numbers collectively the **complex number system** (Fig. 1.11).

For any two real numbers a and b, one of the following is true: $a < b$, $a = b$, or $a > b$. The symbols $<$ and $>$ are called **signs of inequality.**

The **absolute value** of a number a is the distance between a and the origin on the number line. The absolute value of the number x is represented by the symbol $|x|$. Thus $|3| = 3$ and $|-3| = 3$.

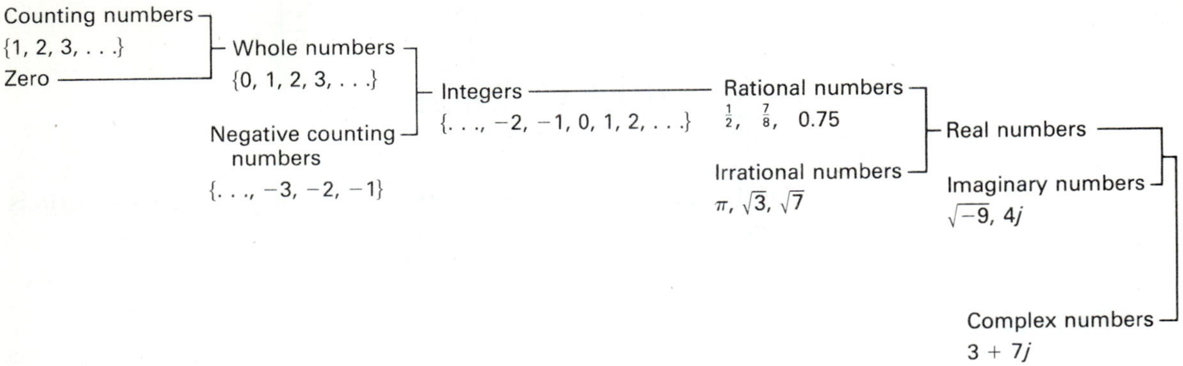

FIGURE 1.11

The axioms for addition and multiplication for any real numbers a, b, and c are

Closure	$a + b = c$	$a \cdot b = c$
Commutative	$a + b = b + c$	$a \cdot b = b \cdot a$
Associative	$(a + b) + c = a + (b + c)$	$a \cdot (b \cdot c) = (a \cdot b) \cdot c$
Identity	$a + 0 = 0 + a = a$	$a \cdot 1 = 1 \cdot a = a$
Inverse	$a + (-a) = 0$	$a \cdot \dfrac{1}{a} = 1, a \neq 0$
	$(-a) + a = 0$	$\dfrac{1}{a} \cdot a = 1, a \neq 0$

Distributive $\quad a(b + c) = ab + ac$ or $(b + c)a = ba + ca$

The definitions for subtraction and division are

Definition 1.6 The **subtraction** of b from a, written $a - b$, is defined as $a - b = a + (-b)$

Definition 1.7 The **division** of a by a nonzero number b, written $a \div b$ or a/b is defined as $a/b = c$, if and only if $a = b \cdot c$.

The rules for operations with signed numbers are

Addition

Rule I If a and b have like signs, compute $|a| + |b|$. Then place the sign of a and b in front of the sum. The result is the sum of a and b.

Rule II If a and b have unlike signs and $|a| = |b|$, then $a + b = 0$.

Rule III If a and b have unlike signs and $|a| \neq |b|$, subtract the smaller absolute value from the larger absolute value. Then place the sign of the number having the larger absolute value in front of the difference. The resulting number is the sum of a and b.

Subtraction

To find the difference apply the definition of subtraction and then apply the rules of addition.

Multiplication and division

Rule IV If a and b have like signs, place a plus sign in front of the product (or quotient). The resulting number is $+(a \cdot b)$ $\left(\text{or } +\left(\frac{a}{b}\right)\right)$.

Rule V If a and b have unlike signs, place a minus sign in front of the product (or quotient). The resulting number is $-(a \cdot b)$ $\left(\text{or } -\left(\frac{a}{b}\right)\right)$.

In working with zero the following properties are important: $a + 0 = a$, $a - 0 = a$, $a \cdot 0 = 0$, $\frac{0}{a} = 0$, $\frac{a}{0}$ is undefined, and $\frac{0}{0}$ is indeterminate.

Review Exercises

In Exercises 1–9, determine whether the following numbers are rational, irrational, real, pure imaginary, or complex. There may be more than one answer.

1. $\sqrt{144}$ **2.** $8.464674677\ldots$

3. $\sqrt[3]{-1728}$ **4.** $21j$

5. $\sqrt[4]{-64}$ **6.** $3.727272\ldots$

7. $\sqrt{\frac{0}{-98}}$ **8.** $4 + 7j$

9. $\sqrt{4} - \sqrt{16}j$

In Exercises 10–18, insert the appropriate symbol, $<$, $>$, or $=$ between the pairs of numbers.

10. $75, 34$ **11.** $-141, -32$

12. $-\sqrt{64}, -\sqrt{625}$ **13.** $\sqrt{13}, 3.6055512\ldots$

14. $3\frac{1}{4}, 3 + \frac{1}{4}$ **15.** $\frac{133}{4}, 33\frac{3}{4}$

16. $\frac{7}{8} - \frac{8}{7}, \frac{8}{7} - \frac{7}{8}$ **17.** $1\frac{5}{8} - \frac{13}{8}, 1\frac{7}{8} - \frac{15}{8}$

18. $\sqrt{3} - \sqrt{2}, \sqrt{2} - \sqrt{3}$

In Exercises 19–25, graph the following on a number line.

19. $x > 4$ **20.** $x < 5$

21. $x \geq -7$ **22.** $x \leq 9$

23. $-9 \leq x < 11$ **24.** $-\sqrt{7} \leq x \leq \sqrt{13}$

25. $-3\frac{1}{3} < x \leq 4\frac{1}{3}$

In Exercises 26–31, find the absolute value of the numbers.

26. $|-9|$ **27.** $|-\sqrt{9}|$ **28.** $\left|3\frac{1}{3} - 4\frac{1}{7}\right|$

29. $|-\sqrt[3]{27}|$ **30.** $|(-2)^3 + 2^3|$ **31.** $\left|2\frac{1}{3} - \frac{11}{4}\right|$

In Exercises 32–38, identify the axiom illustrated in each case. (x and y represent real numbers.)

32. $x + 3 = 3 + x$ **33.** $0 + x = x$

34. $x + -x = 0$ **35.** $\frac{1}{4} \cdot 4 = 1$

36. $(3 + x) + 7 = 3 + (x + 7)$

37. $4 \cdot (x + y) = 4 \cdot x + 4 \cdot y$

38. $(3x)(y) = 3(xy)$

In Exercises 39–58, evaluate each of the given expressions by performing the indicated operations.

39. $(-7) + (-14)$

40. $(-6)(-15)(+3)$

41. $(-75) \div (15)$

42. $-4(3 - 11) + 9$

43. $\dfrac{-13 + 7 + 6}{4 - 11}$

44. $\dfrac{11 - 18}{47 - 47}$

45. $\dfrac{-3(13 - 7) + 18}{11 - 3 \cdot 2 - 5}$

46. $5 + (-3)^2 + \dfrac{4 - 4}{6 - 3} + (-2)(-3)$

47. $\dfrac{(4 - 7)^2 + (-3)^2}{7 + (7 - 9)}$

48. $\dfrac{-13 - 6(-2) + 1}{11 + (9)(-2) + 7}$

49. $-43.765 + 32.1876$

50. $(23.14)(-316.71)(2176.31)$

51. $-3.47628 - (-1.7653)$

52. $\dfrac{(-3.782)(84.76)}{-3.472}$

53. $\dfrac{-34.76 + 18.95}{\pi - \pi}$

54. $\dfrac{2.345}{14.76 - 18.25} + 15.1 - 16.3$

55. $\dfrac{-6.415 + 6.415}{15.237} - 11.678$

56. $(42.35)(-13.78) \div 32.47$

57. $[\pi + (-6)(-7) \div 6] \div (3.25 + 8.76)$

58. $\dfrac{(3.12)^2 - (14.85)^2}{(2.31)^2 - (8.5)^2} - (15.95)^2$

Can you solve these?

1. Mt. Whitney has an elevation of 14,494 ft, while Death Valley's elevation is -282 ft. What is the difference in altitudes?

2. At noon the temperature was $-16°$F. The weather bureau recorded a change of $+13°$F between 12 noon and 3 P.M. and a change of $-14°$F between 3 P.M. and 6 P.M. What was the temperature at 6 P.M.?

3. An object moves along a horizontal line at the rate of 5 km/hr and passes point A on this line. Consider the rate positive or negative according to whether movement is to the right or left, the time positive or negative according to whether it is after or before the arrival of the object at A, and the distance from A positive or negative according to whether the object is to the right or left of A. Calculating with signed numbers, determine the position of the object:
   a. 3 hr before reaching A, if it was moving to the right.
   b. 2 hr after reaching A, if it was moving to the left.
   c. 4 hr before reaching A, if it was moving to the left.
   d. 7 hr before reaching A, if it was moving to the right.

4. Show that $\sqrt{2}$ cannot be written as a quotient of two integers.

5. **a.** Write a program for your calculator to evaluate the expression $3x^2 + 5\sqrt{x} - 2\sqrt[3]{x}$.
   b. Using your program find the value of the expression when x equals 2, 15, 83, and 179.

6. **a.** Write a program in BASIC language to evaluate the expression $5x^3 + 6\sqrt{x} - 7\sqrt[4]{x} + 2$.
   b. Using your program find the value of the expression when x equals 3, 11, 98, and 247.

2

The maximum range of a radar detection system is given by the formula:

$$R_{max} = K\sqrt[4]{P \cdot G^2 \cdot L^2 \cdot T \cdot F^{1/3} \cdot A_E}.$$

R_{max} = maximum range in nautical miles, $K = 5.44$, P = transmission peak power in watts, G = antenna gain as a power ratio, L = wave length in meters, T = pulse length in seconds, F = pulse repetition rate in pulses per second, N = noise factor (power ratio), and A_E = target area in meters. Find R_{max} when $P = 2000$ kW for an antenna 20 m high by 8 m wide, $G = 10^4$, $L = 10$ cm, $T = 1$ sec, $F = 1000$ pulses/sec, $N = 7.94$, and $A_E = 1\ m^2$.

Review of Algebra

2.1 Exponents

In Chapter 1 you were asked to evaluate 2^2 and 9^2. The assumption is that from past experience you would automatically say $2^2 = 2 \cdot 2 = 4$ and $9^2 = 9 \cdot 9 = 81$. We recognize that in the expression 9^2, 9 is the **base,** 2 is the **exponent** (or power), and 9^2 is generally read "9 to the second power" or "9 squared." The exponent, 2, tells us how many times the base, 9, appears as a factor. (Can you show that $-2^2 \neq (-2)^2$?)

This section will be devoted to developing the basic definitions of how to work with exponents.

DEFINITION 2.1

$$a^n = \underbrace{a \cdot a \cdot a \cdot \ \ldots \ \cdot a}_{n \text{ factors}} \qquad (n \text{ is a counting number})$$

EXAMPLE 1 Evaluate:

a. 3^4 **b.** 5^3 **c.** $\left(\frac{1}{7}\right)^3$ **d.** $(-4)^5$

Solution:

a. 3^4 is read "3 to the fourth power."

$$3^4 = 3 \cdot 3 \cdot 3 \cdot 3 = 81$$

b. 5^3 is read "5 cubed."

$$5^3 = 5 \cdot 5 \cdot 5 = 125$$

c. $\left(\frac{1}{7}\right)^3$ is read "one seventh cubed."

$$\left(\frac{1}{7}\right)^3 = \frac{1}{7} \cdot \frac{1}{7} \cdot \frac{1}{7} = \frac{1}{343}$$

d. $(-4)^5$ is read "-4 to the fifth power."

$$(-4)^5 = (-4)(-4)(-4)(-4)(-4) = -1024$$

(Recall the rule for multiplication with signed numbers.) □

The calculator can be used to raise a number to a power.

EXAMPLE 2 Evaluate:

a. 37.56^2 **b.** 1.5^6

Solution: For any number that is raised to the second power (squared) enter the number and stroke the $\boxed{x^2}$ key. If the number is to be raised to a power higher than 2, then use the $\boxed{y^x}$ key.

a. Algebraic: 37.56 $\boxed{x^2}$ 1410.7536

b. Algebraic: 1.5 $\boxed{y^x}$ 6 $\boxed{=}$ 11.390625 □

In addition to understanding the meaning of a^n, there are five definitions dealing with raising a quantity to a power and two definitions dealing with the quotient of quantities to a power. The first part of the discussion will be limited to positive exponents ($n > 0$ and $m > 0$).

To find the product of $a^4 \cdot a^5$ we can use Definition 2.1.

$$a^4 \cdot a^5 = \underbrace{\underbrace{a \cdot a \cdot a \cdot a}_{4 \text{ factors}} \cdot \underbrace{a \cdot a \cdot a \cdot a \cdot a}_{5 \text{ factors}}}_{9 \text{ factors}} = a^9$$

DEFINITION 2.2

$$a^n \cdot a^m = a^{n+m}$$

Definition 2.2 states when multiplying two quantities with the same base, each of which is raised to a power, all we need do is to add the exponents.

EXAMPLE 3 Find the products using Definition 2.2:

a. $a^3 \cdot a^2$ **b.** $3^8 \cdot 3^7$ **c.** $x^5 \cdot y^2$

Solution:

a. Since the bases are alike, Definition 2.2 can be applied:

$$a^3 \cdot a^2 = a^{3+2} = a^5$$

b. Since the bases are alike, Definition 2.2 can be applied:

$$3^8 \cdot 3^7 = 3^{8+7} = 3^{15}$$

c. For $x^5 \cdot y^2$ the bases are not alike, therefore, Definition 2.2 cannot be applied. □

(Note that we did not find a final numerical result for part (b) in Example 3. At this time our primary purpose is to develop an understanding of the definitions.)

The product of $(a^2)^3$ can be found by using Definitions 2.1 and 2.2.

$$\begin{aligned} (a^2)^3 &= a^2 \cdot a^2 \cdot a^2 && \textbf{Definition 2.1} \\ &= a^{2+2+2} && \textbf{Definition 2.2} \\ &= a^6 \end{aligned}$$

DEFINITION 2.3

$$(a^n)^m = a^{n \cdot m}$$

In this case a^n is raised to the mth power and a^n appears as m factors of a^n.

EXAMPLE 4 Evaluate using Definition 2.3:

a. $(a^2)^5$ **b.** $(8^3)^7$

Solution:

a. By Definition 2.3:

$$(a^2)^5 = a^{2 \cdot 5} = a^{10}$$

The same result can be obtained by applying Definitions 2.1 and 2.2.

$$\begin{aligned}(a^2)^5 &= a^2 \cdot a^2 \cdot a^2 \cdot a^2 \cdot a^2 && \textbf{Definition 2.1}\\ &= a^{2+2+2+2+2} && \textbf{Definition 2.2}\\ &= a^{10}\end{aligned}$$

b. By Definition 2.3:

$$(8^3)^7 = 8^{3 \cdot 7} = 8^{21}$$ □

We can use Definition 2.1 and the commutative property of multiplication to show that $(a \cdot b)^4 = a^4b^4$.

$$\begin{aligned}(a \cdot b)^4 &= a \cdot b \cdot a \cdot b \cdot a \cdot b \cdot a \cdot b\\ &= a \cdot a \cdot a \cdot a \cdot b \cdot b \cdot b \cdot b\\ &= a^4 \cdot b^4\end{aligned}$$

DEFINITION 2.4

$$(a \cdot b)^n = a^n \cdot b^n$$

The product of two numbers raised to a power is the same as the product of each of the numbers raised to that power.

EXAMPLE 5 Evaluate using Definition 2.4:

a. $(a \cdot b)^5$ **b.** $(4 \cdot 7)^3$

Solution:

a. By Definition 2.4:

$$(a \cdot b)^5 = a^5 \cdot b^5$$

b. By Definition 2.4:

$$\begin{aligned}(4 \cdot 7)^3 &= 4^3 \cdot 7^3 \\ &= 64 \cdot 343 \\ &= 21{,}952 \quad \square\end{aligned}$$

Using Definition 2.1, the quotient $\left(\frac{a}{b}\right)^3$ can be written as follows:

$$\begin{aligned}\left(\frac{a}{b}\right)^3 &= \frac{a}{b} \cdot \frac{a}{b} \cdot \frac{a}{b} \\ &= \frac{a^3}{b^3}.\end{aligned}$$

DEFINITION 2.5

$$\left(\frac{a}{b}\right)^n = \frac{a^n}{b^n}, \ b \neq 0$$

The quotient of two numbers raised to a power is the same as the quotient of each of the numbers raised to that power.

EXAMPLE 6 Evaluate using Definition 2.5:

a. $\left(\frac{a}{b}\right)^6$ **b.** $\left(\frac{7}{8}\right)^4$ **c.** $\left(\frac{1}{b}\right)^4$ **d.** $\left(\frac{1}{4}\right)^5$

Solution: By Definition 2.5:

a. $\left(\frac{a}{b}\right)^6 = \frac{a^6}{b^6}$

b. $\left(\frac{7}{8}\right)^4 = \frac{7^4}{8^4}$

c. $\left(\frac{1}{b}\right)^4 = \frac{1^4}{b^4}$ **(This is a special case of Definition 2.5, where $a = 1$.)**

$$= \frac{1}{b^4}$$

The same result may be obtained using Definition 2.1:

$$\left(\frac{1}{b}\right)^4 = \frac{1}{b} \cdot \frac{1}{b} \cdot \frac{1}{b} \cdot \frac{1}{b} = \frac{1}{b \cdot b \cdot b \cdot b} = \frac{1}{b^4}.$$

d. $\left(\frac{1}{4}\right)^5 = \frac{1}{4^5}$

The quotient $\frac{a^7}{a^3}$ can be written as:

$$\begin{aligned}\frac{a^7}{a^3} &= \frac{a \cdot a \cdot a \cdot a \cdot a \cdot a \cdot a}{a \cdot a \cdot a} \\ &= \frac{\overset{1}{\cancel{a}} \cdot \overset{1}{\cancel{a}} \cdot \overset{1}{\cancel{a}} \cdot a \cdot a \cdot a \cdot a}{\underset{1}{\cancel{a}} \cdot \underset{1}{\cancel{a}} \cdot \underset{1}{\cancel{a}}} \\ &= a \cdot a \cdot a \cdot a \\ &= a^4. \quad \square\end{aligned}$$

Definition 2.6 shows us a shorter method of arriving at the same result.

DEFINITION 2.6

$$\frac{a^m}{a^n} = \begin{cases} a^{m-n} & \text{if } m \geq n \text{ and } a \neq 0 \\ \dfrac{1}{a^{n-m}} & \text{if } m < n \text{ and } a \neq 0 \end{cases}$$

EXAMPLE 7 Evaluate using Definition 2.6 ($a \neq 0$):

a. $\frac{a^7}{a^3}$ **b.** $\frac{a^5}{a^{12}}$ **c.** $\frac{a^4}{a^4}$

Solution: By Definition 2.6:

a. $\frac{a^7}{a^3} = a^{7-3} = a^4$, since $7 > 3$.

b. $\frac{a^5}{a^{12}} = \frac{1}{a^{12-5}} = \frac{1}{a^7}$, since $5 < 12$.

c. $\frac{a^4}{a^4} = a^{4-4} = a^0$, since $4 = 4$. $\square$

This gives rise to an important definition:

DEFINITION 2.7

$$a^0 = 1, a \neq 0$$

Try this one on your calculator, find 3452^0. This definition tells us that any base, except 0, raised to the exponent zero is 1.

In part (b) of Example 7 the solution was $\frac{1}{a^7}$. Suppose that we had done the following:

$$\frac{a^5}{a^{12}} = a^{5-12} = a^{-7}.$$

This result implies that there is some relation between a^7 and a^{-7}. Thus we get the following definition:

DEFINITION 2.8

$$\left\{\begin{array}{c} a^{-n} = \dfrac{1}{a^n} \\ \text{or} \\ \dfrac{1}{a^{-n}} = a^n \end{array}\right\} \quad a \neq 0$$

EXAMPLE 8 Write the following with positive exponents:

a. a^{-3} **b.** $\frac{1}{a^{-5}}$ **c.** $\frac{3^{-2}}{5^{-4}}$ **d.** $\frac{x^{-3}}{x^4 y^{-2}}$

Solution:

a. $a^{-3} = \frac{1}{a^3},\ a \neq 0$ **Definition 2.8**

b. $\frac{1}{a^{-5}} = a^5,\ a \neq 0$ **Definition 2.8**

c. $\frac{3^{-2}}{5^{-4}} = \frac{3^{-2}}{1} \cdot \frac{1}{5^{-4}}$

$= \frac{1}{3^2} \cdot \frac{5^4}{1}$ **Definition 2.8**

$= \frac{5^4}{3^2}$

d. $\frac{x^{-3}}{x^4 y^{-2}} = \frac{x^{-3}}{x^4} \cdot \frac{1}{y^{-2}}$

$= \frac{1}{x^4 x^3} \cdot \frac{y^2}{1}$ **Definitions 2.8 and 2.6**

$= \frac{y^2}{x^7}$ □

From the results of Definitions 2.7 and 2.8 we can see that it is no longer necessary for m and n to be restricted to positive integers, as in Definitions 2.2, 2.3, 2.4, 2.5, and 2.6. Without formal proof or demonstration we will accept the fact that the properties of exponents will hold when the bases and exponents are not integers. This is illustrated in Example 9.

EXAMPLE 9 Evaluate using Definitions 2.1 through 2.8:

a. $\dfrac{\left(\dfrac{1}{3}\right)^4}{\left(\dfrac{1}{2}\right)^2}$ **b.** $\dfrac{(0.3)^2}{(0.5)^3}$ **c.** $\dfrac{(3^2)^{1/2}}{20^0}$

Solution:

a. $\dfrac{\left(\dfrac{1}{3}\right)^4}{\left(\dfrac{1}{2}\right)^2} = \dfrac{\dfrac{1}{3^4}}{\dfrac{1}{2^2}}$ **Definition 2.5**

$= \dfrac{1}{3^4} \cdot \dfrac{2^2}{1}$

$= \dfrac{2^2}{3^4}$

b. $\dfrac{(0.3)^2}{(0.5)^3} = \dfrac{(0.3)(0.3)}{(0.5)(0.5)(0.5)}$ **Definition 2.1**

$= \dfrac{0.09}{0.125}$

c. $\dfrac{(3^2)^{\frac{1}{2}}}{20^0} = \dfrac{3^{2\cdot\frac{1}{2}}}{1}$ **Definitions 2.3 and 2.7**

$= 3^1$ or 3 □

EXAMPLE 10 Evaluate using Definitions 2.1 through 2.8

a. $a^{1/3} \cdot a^{1/2}$ **b.** $\dfrac{a^{4/5}}{a^{2/3}}$ **c.** $(a^{1/3})^{3/5}$ **d.** $\left(\dfrac{a}{b}\right)^{-(1/2)}$

Solution:

a. $a^{\frac{1}{3}} \cdot a^{\frac{1}{2}} = a^{\frac{1}{3}+\frac{1}{2}}$ **Definition 2.2**

$= a^{\frac{5}{6}}$

b. $\dfrac{a^{\frac{4}{5}}}{a^{\frac{2}{3}}} = a^{\frac{4}{5}-\frac{2}{3}}$ **Definition 2.6 [$m > n$]**

$= a^{\frac{2}{15}}$

c. $(a^{\frac{1}{3}})^{\frac{3}{5}} = a^{\frac{1}{3}\cdot\frac{3}{5}}$ **Definition 2.3**

$= a^{\frac{1}{5}}$

d. $\left(\dfrac{a}{b}\right)^{-\frac{1}{2}} = \dfrac{1}{\left(\dfrac{a}{b}\right)^{\frac{1}{2}}}$ **Definition 2.8**

$= \dfrac{1}{a^{\frac{1}{2}}/b^{\frac{1}{2}}}$ **Definition 2.5**

$= \dfrac{b^{\frac{1}{2}}}{a^{\frac{1}{2}}}$ or $\left(\dfrac{b}{a}\right)^{\frac{1}{2}}$ □

Examples 11 and 12 illustrate how to simplify expressions by applying the definitions.

EXAMPLE 11 Simplify the following expressions by applying Definitions 2.1 to 2.8 and write the final result with positive exponents.

a. $(3^4x^2y^4z^{1/6})^3$ b. $\dfrac{6^4x^3y^7z^{5/3}}{6^3x^5y^7z^{2/3}}$

Solution:

a. $(3^4x^2y^4z^{1/6})^3 = (3^4)^3(x^2)^3(y^4)^3(z^{1/6})^3$ **Definition 2.4**

$= 3^{4\cdot 3}x^{2\cdot 3}y^{4\cdot 3}z^{(1/6)3}$ **Definition 2.3**

$= 3^{12}x^6y^{12}z^{1/2}$

$= 531{,}441x^6y^{12}z^{1/2}$

b. $\dfrac{6^4x^3y^7z^{5/3}}{6^3x^5y^7z^{2/3}} = \dfrac{6^{4-3}}{1}\cdot\dfrac{1}{x^{5-3}}\cdot\dfrac{y^{7-7}}{1}\cdot\dfrac{z^{\frac{5}{3}-\frac{2}{3}}}{1}$ **Definition 2.6**

$= \dfrac{6}{1}\cdot\dfrac{1}{x^2}\cdot\dfrac{y^0}{1}\cdot\dfrac{z^{\frac{3}{3}}}{1}$

$= \dfrac{6\cdot 1\cdot 1\cdot z}{1\cdot x^2\cdot 1\cdot 1}$ **Definition 2.7**

$= \dfrac{6z}{x^2}$ □

EXAMPLE 12 Simplify the following expressions by applying Definitions 2.1 to 2.8. Then use the calculator to find the numerical value.

a. $\dfrac{a^3b^{7/3}}{a^6b^{2/3}}$ $a = 3,\ b = 4$

b. $(13^3x^7y^{1/3})^{2/3}$ $x = 0.5,\ y = 1.3$

Solution:

a. $\dfrac{a^3b^{\frac{7}{3}}}{a^6b^{\frac{2}{3}}} = \dfrac{a^3}{a^6} \cdot \dfrac{b^{\frac{7}{3}}}{b^{\frac{2}{3}}}$

$= \dfrac{1}{a^{6-3}} \cdot \dfrac{b^{\frac{7}{3}-\frac{2}{3}}}{1}$ **Definition 2.6**

$= \dfrac{1}{a^3} \cdot \dfrac{b^{\frac{5}{3}}}{1}$

$= \dfrac{b^{\frac{5}{3}}}{a^3}$

Substituting $a = 3$ and $b = 4$

$= \dfrac{4^{5/3}}{3^3}$

Algebraic: 4 [y^x] [(] 5 [÷] 3 [)] [÷] 3 [y^x] 3 = 0.3733099
or 4 [y^x] [(] 5 [÷] 3 [)] [×] 3 [y^x] 3 [+/−] = 0.3733099
since $\dfrac{4^{5/3}}{3^3} = 4^{5/3} \cdot 3^{-3}$

b. $(13^3x^7y^{\frac{1}{3}})^{\frac{2}{3}} = (13^3)^{\frac{2}{3}}(x^7)^{\frac{2}{3}}(y^{\frac{1}{3}})^{\frac{2}{3}}$ **Definition 2.4**

$= 13^{\frac{3}{1}\cdot\frac{2}{3}}x^{\frac{7}{1}\cdot\frac{2}{3}}y^{\frac{1}{3}\cdot\frac{2}{3}}$ **Definition 2.3**

$= 13^2x^{\frac{14}{3}}y^{\frac{2}{9}}$

Substituting $x = 0.5$ and $y = 1.3$

$= 13^2(0.5)^{\frac{14}{3}}(1.3)^{\frac{2}{9}}$

Algebraic: 13 [x^2] [×] .5 [y^x] [(] 14 [÷] 3 [)] [×]
1.3 [y^x] [(] 2 [÷] 9 [)] [=] 7.0534372 □

2.1 Exercises

In Exercises 1–21, simplify each expression using Definitions 2.1 to 2.8. Write the answer with positive exponents.

1. $x^4 \cdot x^{13}$ **2.** $4^7 \cdot 4^{27}$ **3.** $(3^4x^{11})^5$

4. $\left(\dfrac{x^2}{y^3}\right)^4$ **5.** $\left(\dfrac{1}{x^3}\right)^5$ **6.** $b^y \cdot b^z$

7. $(a \cdot b)^0$ **8.** $(-2x)^3$ **9.** $\dfrac{a^7}{a^4}$

10. $\dfrac{b^8}{b^{11}}$ **11.** $\left(\dfrac{1}{z^{-3}}\right)^5$ **12.** $x^2y^3(x^4y^5)^2$

13. $\dfrac{15a^3b^2}{25a^2b^2}$ **14.** $\dfrac{27x^{13}y^4}{(3x^2y^5)^2}$

15. $(5^{3/4}a^{16}b^{5/4})^4$

16. $\dfrac{6x^{-4}y^7}{(2^6x^3y^2)^{1/3}}$ **17.** $\dfrac{(xy^2z^3)^n}{(x^ny^nz^n)^2}$

18. $(3x^2y^7z^{1/3})^0$ **19.** $(25^0a^2b^3c^{-1})^{-2}$

20. $\left(\dfrac{3a^2b^3}{xy}\right)^{-3}$ **21.** $\left(\dfrac{5x^3y^4}{3^{-2}x^5y^3z^2}\right)^{-4}$

In Exercises 22–38, simplify each expression using Definitions 2.1 through 2.8. Then use the calculator to find the numerical value of each expression.

22. $(11.23)^3$

23. $a^3 \cdot b^4$, $a = 1.2$, $b = 2$

24. $(3^3x^4)^5$, $x = 0.3$

25. $\left(\dfrac{x^{1.4}}{y^{3.2}}\right)^3$, $x = 2$, $y = 4$

26. $\left(\dfrac{1}{x^3}\right)^6$, $x = 2.5$

27. $\left(\dfrac{x^4}{x^4}\right)^{13}$, $x = 5.6$

28. $\left(\dfrac{x^4}{y^4}\right)^0$, $x = 5.6$, $y = 3.6$

29. $\dfrac{a^8}{a^5}$, $a = 3.4$

30. $\dfrac{x^5}{x^9}$, $x = 4.5$

31. $a^4b^5(a^2b^3)$, $a = 1.1$, $b = 0.01$

32. $a^2b^3(a^4b^5)^2$, $a = 1.1$, $b = 0.01$

33. $\dfrac{8x^3y^2}{12x^2y^5}$, $x = 3$, $y = 4$

34. $\dfrac{2^6a^3b^2}{6a^{-4}b^7}$, $a = 1.5$, $b = 2.05$

35. $\dfrac{a^nb^nc^n}{(ab^3c^4)^n}$, $a = 2$, $b = 3$, $c = 4$

36. $36^0x^3y^{-3}z^{-1}$, $x = 0.2$, $y = 1.1$, $z = 0.3$

37. $\dfrac{(27x^2y^3)^{1/3}}{(4x^5y^7)^{1/2}}$, $x = 2.1$, $y = 0.5$

38. $(15^5a^4b^{15})^{2/5}$, $a = 0.2$, $b = 0.3$

In Exercises 39–40, simplify the expressions.

39. In a problem dealing with electric potential we arrive at the following expression (units omitted):

$$\frac{(8.99)(10^9)(-3.00)(10^{-12})}{2.5}.$$

40. In comparing the relative strengths of the electric and magnetic interaction for particles one may arrive at the following expression (units omitted):

$$\frac{10^{-7}}{(9)(10^9)}.$$

2.2 Scientific notation

The scientist and the technician may work with extremely large and extremely small numbers. For example the spaceship Columbia flew at the rate of 17,800 mph in space. The heat on the bottom of the ship as it reentered the earth's atmosphere was estimated to be 2600°F. The size of a certain virus is 0.0000005 mm.

Each of these numbers can be expressed in a more convenient form, called **scientific notation**.

DEFINITION 2.9

> A number written in **scientific notation** is written as a product of a number from 1 to 10 (including 1 but not including 10) and a power of 10.

This can be represented symbolically as $p \times 10^k$, where $1 \leq p < 10$ and k can be any integer.

EXAMPLE 1 To illustrate Definition 2.9 we will write the three numbers given in the opening paragraph in scientific notation.

Solution:

	Number	Scientific notation
a.	17,800 mph	1.78×10^4 mph
b.	2600°F	2.6×10^3 °F
c.	0.0000005 mm	5.0×10^{-7} mm

□

Note that there is a single digit to the left of the decimal point in each case: 1, 2, and 5. Also note that the exponents of 10 are the integers 4, 3, and −7 and that the exponent is the same as the number of places the decimal point was moved in changing to scientific notation. In part (a) the decimal point was moved four places to the left (1.7,800.).

EXAMPLE 2 Express the following numbers in ordinary decimal notation:

a. 1.4×10^8 mi²

b. 6.0×10^{-6} mg

c. 9.46×10^{12} km

Solution: To change numbers that are written in scientific notation into ordinary decimal notation we reverse the procedure used in Example 1.

	Scientific notation	Ordinary notation
a.	1.4×10^8 mi²	140,000,000 mi²
b.	6.0×10^{-6} mg	0.000006 mg
c.	9.46×10^{12} km	9,460,000,000,000 km

Note that when the exponent is positive, as in parts (a) and (c), we move the decimal point to the right (1.40000000.) to change to an ordinary notation. When the exponent is negative, as in part (b), we move the decimal point to the left (.000006.). □

EXAMPLE 3 Write the numbers in scientific notation, perform the indicated operations, and express the result in scientific notation.

a. 34,000,000,000 + 25,000,000,000

b. 0.000000019 − 0.0000000027

Solution:

a. $34{,}000{,}000{,}000 + 25{,}000{,}000{,}000 = 3.4 \times 10^{10} + 2.5 \times 10^{10}$
Since the power of ten is the same for both numbers we can add the 3.4 and the 2.5 and attach the power of ten to the sum. Thus,

$$\begin{aligned} 3.4 \times 10^{10} + 2.5 \times 10^{10} &= (3.4 + 2.5) \times 10^{10} \\ &= 5.9 \times 10^{10} \end{aligned}$$

b. $0.000000019 - 0.0000000027 = 1.9 \times 10^{-8} - 2.7 \times 10^{-9}$
The powers of ten are not the same, therefore in order to be able to subtract we must write the numbers so that they have the same power of ten. We can select any convenient power so let us select 10^{-8}. Thus, we will write 2.7×10^{-9} as 0.27×10^{-8}.

$$\begin{aligned} 1.9 \times 10^{-8} - 0.27 \times 10^{-8} &= (1.9 - 0.27) \times 10^{-8} \\ &= 1.63 \times 10^{-8} \quad \square \end{aligned}$$

EXAMPLE 4 Write the numbers in scientific notation, perform the indicated operations, and express the result in scientific notation.

a. 0.0000075×0.00034

b. $3{,}800{,}000 \div 500$

Solution:

a.
$$\begin{aligned} 0.0000075 \times 0.00034 &= (7.5 \times 10^{-6}) \times (3.4 \times 10^{-4}) \\ &= (7.5 \times 3.4) \times (10^{-6} \times 10^{-4}) \\ &= 25.5 \times 10^{-10} \\ &= (2.55 \times 10^{1}) \times 10^{-10} \\ &= 2.55 \times 10^{-9} \end{aligned}$$

b.
$$\begin{aligned} \frac{3{,}800{,}000}{500} &= \frac{3.8 \times 10^{6}}{5 \times 10^{2}} \\ &= \frac{3.8}{5} \times \frac{10^{6}}{10^{2}} \\ &= 0.76 \times 10^{6-2} \\ &= 0.76 \times 10^{4} \\ &= (7.6 \times 10^{-1}) \times 10^{4} \\ &= 7.6 \times 10^{3} \quad \square \end{aligned}$$

Example 4 illustrates that scientific notation can be useful in performing calculations. Writing numbers in scientific notation can also be extremely helpful in estimating answers to problems. For example, in part (b) of Example 4 it can be seen after the problem is written in scientific notation that the answer is approximately 0.7×10^{4}. With all our machines it is important to be able to estimate answers to determine whether or not the answer is correct or at least in the right ball park.

The calculator can be used to solve problems in scientific notation as illustrated in the next example.

EXAMPLE 5 Perform the indicated operations on the calculator using scientific notation.

$$\frac{(8 \times 10^6)(7 \times 10^3)}{2 \times 10^{-4}}$$

Solution: Before we touch the calculator let us estimate the answer using scientific notation.

$$\frac{8 \times 7}{2} \times \frac{10^6 \times 10^3}{10^{-4}} = 28 \times 10^{13}$$
$$= 2.8 \times 10^{14}$$

The keys with the symbols [EE], [EXP], or [EEX] are used to enter a number in scientific notation. The number 8×10^6 is entered into the calculator in the following manner:

Keystroke	Display
8	8
[EE]	8 00
6	8 06

The solution using the calculator is:

Algebraic: 8 [EE] 6 [×] 7 [EE] 3 [÷] 2 [EE] 4 [+/−] [=] Display: 2.8 14

or 2.8×10^{14}.

The calculator result is the same as the estimate. □

EXAMPLE 6 Perform the indicated operations on the calculator using scientific notation.

$$\frac{(5.36 \times 10^{12})^3}{(1.82 \times 10^{-7})(4.63 \times 10^{-9})}$$

Solution: The solution using the calculator is:

5.36 [EE] 12 [y^x] 3 [÷] 1.82 [EE] 7 [+/−] [÷] 4.63

[EE] 9 [+/−] [=] Display: 1.8274352 53

Thus $$\frac{(5.36 \times 10^{12})^3}{(1.82 \times 10^{-7})(4.63 \times 10^{-9})} = 1.83 \times 10^{53}$$

Does this answer seem realistic? Let us estimate the answer using scientific notation. Express 5.36, 1.82 and 4.63 to the nearest integer. Thus,

$$\frac{(5.36 \times 10^{12})^3}{(1.82 \times 10^{-7})(4.63 \times 10^{-9})} \approx^* \frac{(5 \times 10^{12})^3}{(2 \times 10^{-7})(5 \times 10^{-9})}$$

$$\approx \frac{5^3 \times (10^{12})^3}{2 \times 5 \times 10^{-7} \times 10^{-9}}$$

$$\approx \frac{125 \times 10^{36}}{10 \times 10^{-16}}$$

$$\approx 1.25 \times 10^{53}$$

The estimated result, 1.25×10^{53}, shows that the calculator result is realistic. The estimated result is slightly smaller due to roundoff error. □

2.2 Exercises

In Exercises 1–12, write the number in scientific notation.

1. 5000	**2.** 367,000	**3.** 0.0004
4. 0.00345	**5.** 7	**6.** 27
7. 0.3	**8.** 0.33	**9.** 56,000,000
10. 632,400	**11.** 0.000000877	**12.** 7382.60

In Exercises 13–21, express the number in each statement in scientific notation.

13. One kilometer is approximately 39,370 in.

14. The mean distance from the earth to the sun is approximately 148,320,000 km.

15. One degree is approximately 0.01745 rad.

16. One atomic mass unit is approximately

0.00000000000000000000000166 g.

17. The mean distance from the earth to the moon is approximately 382,000 km.

18. One joule is approximately equal to 0.738 ft lb.

19. One radian is approximately equal to 206,265 seconds of arc.

20. A coulomb represents a grouping of

6,250,000,000,000,000,000

excess protons or electrons.

21. The average number of heartbeats per normal lifetime is 2,838,000,000.

In Exercises 22–33, write the numbers in ordinary decimal notation.

22. 4×10^3	**23.** 8.5×10^3
24. 5×10^{-4}	**25.** 5.67×10^{-4}
26. 9×10^0	**27.** 3×10^{-1}
28. 4.4×10^{-1}	**29.** 8.75×10^7
30. 3.4×10^{-9}	**31.** 6.53×10^{-7}
32. 8.76×10^0	**33.** 9.873×10^{10}

In Exercises 34–42, write the numbers in ordinary decimal notation.

34. Absolute zero is -4.60×10^2°F.

35. The half-life of uranium 235 is 7.5×10^8 years.

36. One milliliter is equal to 1×10^{-3} liters.

37. One yard is equal to 5.682×10^{-4} mi.

38. One centimeter is equal to 1×10^{-5} km.

39. The area of the U.S.S.R. is approximately 8.649×10^6 mi².

40. The area of the United States is approximately 3.615×10^6 mi².

* ≈ means "is approximately equal to."

41. The thickness of an oil film is approximately 5×10^{-7} cm.

42. One millimeter is equal to 1×10^{-5} hm.

In Exercises 43–50, use scientific notation to estimate the answer for each exercise and then perform the indicated operations.

43. $(5 \times 10^6) + (7 \times 10^6)$

44. $(6.1 \times 10^{-2}) + (5.3 \times 10^{-2}) - (4.8 \times 10^{-2})$

45. $\dfrac{(4 \times 10^5)(3.5 \times 10^3)}{2 \times 10^4}$

46. $\dfrac{(5 \times 10^7)(4.6 \times 10^9)}{6.3 \times 10^5}$

47. $\dfrac{(2.3 \times 10^3)(7.4 \times 10^5)}{5 \times 10^{-4}}$

48. $\dfrac{(3.7 \times 10^6)(8.6 \times 10^8)}{6.7 \times 10^{-5}}$

49. $\dfrac{(6.34 \times 10^{-5})(8.75 \times 10^7)^3}{(5.4 \times 10^7)(3.4 \times 10^3)^2}$

50. $\left[\dfrac{(3.4 \times 10^8)(6.8 \times 10^{-5})(2.7 \times 10^9)^2}{(86{,}000)^3(0.00048)^3}\right]^2$

In Exercises 51–54, solve the problems using scientific notation.

51. How long does it take a spaceship traveling 1.78×10^4 mph to travel the 2.4×10^5 mi from the earth to the moon?

52. From fluid kinematics we have the formula:

$$v = \frac{V}{A},$$

where v = velocity of the fluid, V = volume of the fluid flow per unit time, and A = cross-sectional area. If $V = 5.0 \times 10^{-2}$ ft^3/s and $A = 6.94 \times 10^{-3}$ ft^2, find v.

53. Using the formula from Exercise 52, find V if $A = 4.0 \times 10^{-2}$ m^2 and $v = 2.0$ m/s.

54. In chemistry the pH of a particular solution can be found by the expression

$$\text{pH} = \frac{E - (2.80 \times 10^{-1})}{5.9 \times 10^{-2}}$$

where E is the electric potential. Find the pH if $E = 4.75 \times 10^{-1}$.

2.3 Roots and radicals

In Section 2.1 the rules of exponents were discussed and illustrated. For example, in applying Definition 2.1 we find that $5^2 = 5 \cdot 5$ or 25. Using these rules find the value of $9^{1/2}$. Let us start with a statement that we know is true:

$$9 = 3^2 \qquad \textbf{Definition 2.1}$$
$$9^{\frac{1}{2}} = (3^2)^{\frac{1}{2}}$$
$$9^{\frac{1}{2}} = 3^{\frac{2}{1}\cdot\frac{1}{2}} \qquad \textbf{Definition 2.3}$$
$$9^{\frac{1}{2}} = 3$$

Thus, we conclude that $9^{1/2}$ could equal 3.

Is 3 the only solution? The answer is no, since $(-3) \times (-3) = 9$. Thus the solution is ± 3. Note that an equivalent method of writing $9^{1/2}$ is

$$9^{1/2} = \sqrt{9}$$
$$= \pm 3.$$

DEFINITION 2.10

> $x^{1/n} = \sqrt[n]{x}$, where $\sqrt{}$ is the **radical sign,** x is the **radicand,** and n is the **index** of the radical.

The index, n, tells what root we are seeking of the radicand. The index is written in every case, except $n = 2$, where we simply write $\sqrt{x}$.

EXAMPLE 1 Discuss and, if possible, find a real-number solution.

a. $\sqrt{16}$ **b.** $\sqrt[3]{-8}$ **c.** $\sqrt[4]{81}$
d. $\sqrt[5]{32}$ **e.** $\sqrt{36x^2}$ **f.** $\sqrt{-9}$

Solution:

a. $\sqrt{16}$ (square root of 16), $\sqrt{16} = \pm 4$
b. $\sqrt[3]{-8}$ (cube root of -8), $\sqrt[3]{-8} = -2$, since $-2 \cdot -2 \cdot -2 = -8$.
c. $\sqrt[4]{81}$ (fourth root of 81), $\sqrt[4]{81} = \pm 3$
d. $\sqrt[5]{32}$ (fifth root of 32), $\sqrt[5]{32} = 2$
e. $\sqrt{36x^2}$ (square root of $36x^2$), $\sqrt{36x^2} = \pm 6x$
f. $\sqrt{-9}$ (square root of -9). There is no real number that when multiplied by itself will give a product of -9. □

In parts (a), (c), and (e), where we are seeking an even root of a number, there are two possible answers, positive and negative. In parts (b) and (d), where an odd root of a number is sought, there is only one possible answer (in the set of real numbers), either positive or negative. To eliminate dual answers in the future, when finding an even root the answer will be the positive root. This positive root is called the **principal square root,** or the **principal root.**

DEFINITION 2.11

> If m and n are both integers with n a positive number, and if $\sqrt[n]{a}$ exists as a real number, then
>
> $$a^{m/n} = \sqrt[n]{a^m} = (\sqrt[n]{a})^m.$$

These two ways of writing $a^{m/n}$ in radical form are consistent with the two ways of writing $a^{m/n}$ with exponents.

$$a^{m/n} = (a^m)^{1/n} = (a^{1/n})^m$$

EXAMPLE 2 Write each of the following in radical form.

a. $3^{2/3}$ **b.** $7^{5/6}$ **c.** $4^{-3/2}$ **d.** $3^{2/2}$ **e.** $(-3)^{2/2}$

Solution:

a. $3^{2/3} = \sqrt[3]{3^2} = (\sqrt[3]{3})^2$

b. $7^{5/6} = \sqrt[6]{7^5} = (\sqrt[6]{7})^5$

c. $4^{-3/2} = \dfrac{1}{4^{3/2}} = \dfrac{1}{\sqrt{4^3}} = \dfrac{1}{\sqrt{64}} = \dfrac{1}{8}$

d. $(+3)^{2/2} = \sqrt{(+3)^2} = \sqrt{9} = +3$

e. $(-3)^{2/2} = \sqrt{(-3)^2} = \sqrt{9} = +3$

In parts (d) and (e) it becomes evident that $\sqrt{a^2}$ is a special situation. □

DEFINITION 2.12

$$\sqrt{a^2} = |a|$$

From the discussion thus far of the relation between the definitions of exponents and the definitions of radicals, we can obtain the following:

$(ab)^{1/n} = a^{1/n}b^{1/n}$ leads to

DEFINITION 2.13

$$\sqrt[n]{a \cdot b} = (\sqrt[n]{a})(\sqrt[n]{b})$$

$\left(\dfrac{a}{b}\right)^{1/n} = \dfrac{a^{1/n}}{b^{1/n}}$ leads to

DEFINITION 2.14

$$\sqrt[n]{\frac{a}{b}} = \frac{\sqrt[n]{a}}{\sqrt[n]{b}}$$

$(a^n)^m = a^{nm}$ leads to

DEFINITION 2.15

$$\sqrt[m]{\sqrt[n]{a}} = \sqrt[m \cdot n]{a}$$

EXAMPLE 3 Using the definitions for exponents and radicals, simplify the following expressions. (To simplify we mean to remove part or all of the expression from the radical, if possible.)

a. $\sqrt{54x^2}$ **b.** $3\sqrt[5]{64y^7}$ **c.** $\sqrt[4]{\dfrac{625}{243}}$ **d.** $\sqrt[3]{\sqrt{128}}$

Solution:

a. $\sqrt{54x^2} = \sqrt{6 \cdot 9x^2}$

$= \sqrt{6}\sqrt{3^2}\sqrt{x^2}$ **Definition 2.13**

$= \sqrt{6}\,(3)(x)$

$= 3x\sqrt{6}$

(The same result can be obtained using exponents.)

$\sqrt{54x^2} = (54x^2)^{1/2}$ **Definition 2.10**

$= (9 \cdot 6x^2)^{1/2}$

$= (3^2 \cdot 6 \cdot x^2)^{1/2}$

$= (3^2)^{1/2}(6)^{1/2}(x^2)^{1/2}$ **Definition 2.4**

$= 3 \cdot 6^{1/2} \cdot x$ **Definition 2.3**

$= 3x\sqrt{6}$

b. $3\sqrt[5]{64y^7} \neq 3\sqrt[5]{2^5y^5y^2}$

$= 3\sqrt[5]{2^5}\sqrt[5]{y^5}\sqrt[5]{y^2}$ **Definition 2.13**

$= 3 \cdot 2 \cdot y \cdot \sqrt[5]{y^2}$

$= 6y\sqrt[5]{y^2}$

c. $\sqrt[4]{\dfrac{625}{243}} = \dfrac{\sqrt[4]{5^4}}{\sqrt[4]{3^4 \cdot 3}}$ **Definition 2.14**

$= \dfrac{(5^4)^{1/4}}{(3^4 \cdot 3)^{1/4}}$ **Definition 2.10**

$= \dfrac{5}{3 \cdot 3^{1/4}}$ **Definition 2.3**

$= \dfrac{5}{3\sqrt[4]{3}}$ **Definition 2.10**

d. $\sqrt[3]{\sqrt{128}} = \sqrt[3 \cdot 2]{128}$ **Definition 2.15**

$= \sqrt[6]{2^6 \cdot 2}$

$= \sqrt[6]{2^6}\sqrt[6]{2}$ **Definition 2.13**

$= 2\sqrt[6]{2}$

(The same result can be obtained using exponents.)

$\sqrt[3]{\sqrt{128}} = (\sqrt{128})^{1/3}$ **Definition 2.10**

$= (128^{1/2})^{1/3}$ **Definition 2.10**

$= 128^{1/6}$ **Definition 2.3**

$= (2^6 \cdot 2)^{1/6}$

$= 2 \cdot 2^{1/6}$ **Definition 2.3**

$= 2\sqrt[6]{2}$ □

In part (c) the answer contains a radical in the denominator. Usually it is more convenient to work with or to evaluate the expression if there is no radical in the denominator. To eliminate the radical in the denominator we use a procedure called **rationalizing the denominator.**

EXAMPLE 4 Rationalize the denominator in each expression.

a. $\sqrt{\frac{2}{3}}$ **b.** $\frac{\sqrt{a}}{\sqrt{b^3}}$ **c.** $\sqrt[3]{\frac{3}{2x^2}}$

Solution: To rationalize a denominator with a radical having an index, n, multiply the numerator and denominator of the fraction by the smallest quantity that will raise the power of the denominator to a number that is "evenly" divided by n.

a. Since $n = 2$, we want the power of 3 to be divisible by 2. Multiply both the numerator and denominator by 3.

$$\sqrt{\frac{2}{3} \cdot \frac{3}{3}} = \sqrt{\frac{6}{3^2}}$$

$$= \frac{\sqrt{6}}{\sqrt{3^2}} \qquad \textbf{Definition 2.14}$$

$$= \frac{\sqrt{6}}{3}$$

b. $\frac{\sqrt{a}}{\sqrt{b^3}} = \sqrt{\frac{a}{b^3}}$ **Definition 2.14**

Since $n = 2$, multiply the numerator and denominator by b to make the power of b divisible by 2.

$$\sqrt{\frac{a}{b^3} \cdot \frac{b}{b}} = \sqrt{\frac{ab}{b^4}}$$

$$= \frac{\sqrt{ab}}{\sqrt{(b^2)^2}}$$

$$= \frac{\sqrt{ab}}{b^2}$$

c. Since $n = 3$, the power of 2 and x must be divisible by 3, thus, multiply both numerator and denominator by 2^2x.

$$\sqrt[3]{\frac{3}{2x^2} \cdot \frac{2^2x}{2^2x}} = \sqrt[3]{\frac{12x}{2^3 \cdot x^3}}$$

$$= \frac{\sqrt[3]{12x}}{2x} \qquad \square$$

To complete this discussion let us answer the question: Can we add, subtract, multiply, and divide radicals? The answer is yes, under certain conditions.

Expressions containing radicals can be added or subtracted if the radicals are the same, that is the radicals have the same radicand and the same index.

EXAMPLE 5 Simplify and combine where possible.

a. $\sqrt{3} + 5\sqrt{3}$ **b.** $8\sqrt[3]{5} - 14\sqrt[3]{625}$ **c.** $\frac{1}{2}\sqrt{24} + 3\sqrt{\frac{2}{3}} - 2\sqrt{\frac{3}{2}}$

Solution:

a. Since the radicands are the same in $\sqrt{3} + 5\sqrt{3}$ we can combine the like radicals, similar to combining like terms, or use the distributive law.

$$\sqrt{3} + 5\sqrt{3} = (1 + 5)\sqrt{3} = 6\sqrt{3}$$

b. In order to combine the radicals $8\sqrt[3]{5} - 14\sqrt[3]{625}$, the radicands must be the same. Can we express the 625 with a radicand of 5? Yes.

$$\sqrt[3]{625} = \sqrt[3]{5^3 \cdot 5} = 5\sqrt[3]{5}$$

Substituting we have:

$$\begin{aligned} 8\sqrt[3]{5} - 14 \cdot 5\sqrt[3]{5} &= 8\sqrt[3]{5} - 70\sqrt[3]{5} \\ &= (8 - 70)\sqrt[3]{5} \quad \textbf{Distributive property} \\ &= -62\sqrt[3]{5} \end{aligned}$$

c. In order to combine the radicals $\frac{1}{2}\sqrt{24} + 3\sqrt{\frac{2}{3}} - 2\sqrt{\frac{3}{2}}$, the radicands must be the same.

$$\sqrt{24} = \sqrt{4 \cdot 6} = 2\sqrt{6}$$

$$\sqrt{\frac{2}{3}} = \sqrt{\frac{2}{3} \cdot \frac{3}{3}} = \frac{\sqrt{6}}{\sqrt{9}} = \frac{\sqrt{6}}{3}$$

$$\sqrt{\frac{3}{2}} = \sqrt{\frac{3}{2} \cdot \frac{2}{2}} = \frac{\sqrt{6}}{\sqrt{4}} = \frac{\sqrt{6}}{2}$$

Substituting we have:

$$\begin{aligned} \frac{1}{\not{2}_1}(\not{2}^1\sqrt{6}) + \not{3}_1\left(\frac{\sqrt{6}}{\not{3}}\right) - \not{2}^1\left(\frac{\sqrt{6}}{\not{2}_1}\right) &= \sqrt{6} + \sqrt{6} - \sqrt{6} \\ &= \sqrt{6} \quad \square \end{aligned}$$

The radicands of two radicals may be multiplied or divided if the indexes of the radicals are the same.

EXAMPLE 6 If possible find the product or quotient of the following radicals and simplify.

a. $\sqrt{3x} \cdot \sqrt{5yz}$ **b.** $\frac{\sqrt{21xy}}{\sqrt{14yz}}$ **c.** $\frac{\sqrt[3]{4}}{\sqrt{2}}$

Solution:

a. Since the indexes of the radicals $\sqrt{3x} \cdot \sqrt{5yz}$ are the same, 2, we can multiply the radicands:

$$\sqrt{3x} \cdot \sqrt{5yz} = \sqrt{15xyz}.$$

b. The indexes of the radicals of $\dfrac{\sqrt{21xy}}{\sqrt{14yz}}$ are the same, 2. Therefore, we can divide the radicands:

$$\frac{\sqrt{21xy}}{\sqrt{14yz}} = \sqrt{\frac{21xy}{14yz}}$$
$$= \sqrt{\frac{3x}{2z}}$$

Note: The answer may be written $\dfrac{\sqrt{3x}}{\sqrt{2z}}$. By multiplying by $\dfrac{\sqrt{2z}}{\sqrt{2z}}$ we have $\dfrac{\sqrt{6xz}}{2z}$, which is a form preferred by some instructors. However, all three forms are correct results.

c. The indexes of the radicals of $\dfrac{\sqrt[3]{4}}{\sqrt{2}}$ are not the same, therefore, we cannot divide the radicands in this form. However, using our skill with exponents and radicals, the indexes can be changed. The smallest number that is divisible by both 2 and 3 is 6. Therefore the common index we seek is 6.

$$\frac{\sqrt[3]{4}}{\sqrt{2}} = \frac{\sqrt[3]{2^2}}{\sqrt{2}}$$

$$= \frac{\sqrt[6]{(2^2)^2}}{\sqrt[6]{2^3}} \text{ since } \left\{\frac{\sqrt[6]{2^4}}{\sqrt[6]{2^3}} = \frac{2^{4/6}}{2^{3/6}} = \frac{2^{2/3}}{2^{1/2}} = \frac{\sqrt[3]{2^2}}{\sqrt{2}}\right\} \qquad \textbf{Definition 2.11}$$

$$= \sqrt[6]{\frac{2^4}{2^3}} \qquad \textbf{Definition 2.14}$$

$$= \sqrt[6]{2}$$

Or, using exponents:

$$\frac{\sqrt[3]{4}}{\sqrt{2}} = \frac{4^{1/3}}{2^{1/2}}$$
$$= \frac{(2^2)^{1/3}}{2^{1/2}}$$
$$= \frac{2^{2/3}}{2^{1/2}}$$
$$= 2^{2/3-1/2}$$
$$= 2^{4/6-3/6}$$
$$= 2^{1/6} \qquad \square$$

2.3 Exercises

Write Exercises 1–8 in radical form.

1. The square root of 15.
2. The fifth root of 11.
3. The cube root of 13.
4. The square root of 17.
5. The seventh root of 2.
6. The cube root of $\frac{3}{32}$.
7. The square root of $\frac{5}{13}$.
8. The cube root of -31.

For Exercises 9–28, find the principal value of each of the following without using a calculator.

9. $\sqrt{36}$
10. $\sqrt{144}$
11. $\sqrt{121}$
12. $\sqrt[3]{125}$
13. $\sqrt[3]{64}$
14. $\sqrt[3]{-8}$
15. $\sqrt{0.01}$
16. $\sqrt{\frac{49}{81}}$
17. $\sqrt[5]{32}$
18. $\sqrt{9x^2y^4}$
19. $\sqrt{169x^4}$
20. $\sqrt[3]{8x^3y^6}$
21. $\sqrt[5]{a^{10}b^5z^{15}}$
22. $\sqrt[4]{81x^4y^8}$
23. $\sqrt[3]{-64x^9y^6}$
24. $\sqrt[5]{-32a^5b^{10}}$
25. $\sqrt[3]{8^2}$
26. $\sqrt[4]{9^2}$
27. $\sqrt[3]{\frac{-125x^3}{64y^3}}$
28. $\sqrt[4]{\frac{1296a^4}{625b^8}}$

In Exercises 29–48, use the definitions for exponents and radicals to simplify the expressions. Where possible, after expression has been simplified, use your calculator to compute the numerical results.

29. $\sqrt{12}$
30. $\sqrt{125}$
31. $\sqrt[3]{81}$
32. $\sqrt[3]{625}$
33. $\sqrt[4]{\frac{2}{7}}$
34. $\sqrt{\frac{24x^2}{18y^3}}$
35. $\sqrt{\frac{14x^3}{2x}}$
36. $\sqrt{\frac{9a^0 - b^0}{a^0 + (3b)^0}}$
37. $\sqrt{50a^3b^4}$
38. $\sqrt[3]{-24x^3y^6}$
39. $\sqrt[3]{\frac{9x^6y^2}{3y^2}}$
40. $\sqrt[n]{a^{n+1}}$
41. $\sqrt[n+1]{a^{2n+3}}$
42. $\sqrt[3n]{x^{6n}y^{7n}}$
43. $\sqrt[3k-1]{x^{6k}}$
44. $\sqrt{\sqrt{81x^4}}$
45. $\sqrt[3]{\sqrt{729}}$
46. $\sqrt[3]{\sqrt[3]{-x^9}}$
47. $\sqrt[3]{\sqrt[4]{\frac{(6x)^0 - y^0}{x^0 + (7y^2)^0}}}$
48. $\sqrt[3]{\sqrt[4]{x^{12n}}}$

In Exercises 49–70, perform the indicated operations and simplify the results.

49. $\sqrt{3} + 4\sqrt{3}$
50. $5\sqrt{8} + 7\sqrt{18}$
51. $7\sqrt{75} - 6\sqrt{27} + 2\sqrt{48}$
52. $11\sqrt{28} + 6\sqrt{1/7} - 5\sqrt{112}$
53. $2\sqrt{24} + 7\sqrt{54} - 5\sqrt{150}$
54. $5\sqrt[3]{24} - \sqrt[3]{81} + 2\sqrt[3]{375}$
55. $\sqrt{y} + \sqrt{y^3} + \sqrt{y^7}$
56. $\sqrt{3} \cdot \sqrt{5}$
57. $\sqrt{11} \cdot \sqrt{22}$
58. $3\sqrt{7} \cdot \sqrt{14}$
59. $\sqrt{xyz} \cdot \sqrt{wxy}$
60. $\sqrt[3]{4} \cdot \sqrt[3]{10}$
61. $\sqrt[4]{125} \cdot \sqrt[4]{5}$
62. $\sqrt[3]{9} \cdot \sqrt[3]{3}$
63. $\sqrt[3]{x^2yz^4} \cdot \sqrt[3]{xy^2z^5}$
64. $\frac{\sqrt{21}}{\sqrt{7}}$
65. $\frac{3\sqrt{12}}{\sqrt{3}}$
66. $\frac{\sqrt{55}}{\sqrt{35}}$
67. $\frac{\sqrt{12ab}}{\sqrt{14ac}}$
68. $\frac{\sqrt[3]{9}}{\sqrt{3}}$
69. $\frac{\sqrt[4]{a^5}}{\sqrt{a^3}}$
70. $\frac{\sqrt[3]{a^2b^4}}{\sqrt[4]{a^2b^3}}$

2.4 Addition and subtraction of algebraic expressions

A letter used to represent an unknown quantity is called a **variable.** In algebra traditionally the letters at the end of the alphabet were chosen as variables. In this text we will not limit ourselves to letters at the end of the alphabet, since many formulas in technical areas use many different letters.

DEFINITION 2.16

An **algebraic expression** is one of the following:

a. a constant

b. a variable

c. a combination formed by performing one or more algebraic operations (addition, subtraction, multiplication, division, raising to a power, and taking a root), on nonzero constants or variables.

To help understand the meaning of Definition 2.16, study Example 1.

EXAMPLE 1 Each of the following is an algebraic expression:

a. 3 **b.** x **c.** $x + y$

d. $3x - y$ **e.** $5y(y + 3y^2) - 3x$ **f.** $\dfrac{3x^2y - 5xz}{11y}$

g. $3x^2 - 7x^2$ **h.** $6\sqrt{3x} + 11\sqrt{3x}$ □

In part (d) of Example 1, the algebraic expression consists of two **terms** $3x$ and $-y$. A **term** is a product or quotient of one or more variables and constants. The **literal coefficient** x has a **numerical coefficient** of 3 and the numerical coefficient of y is -1 even though the 1 is not written $(-y = -1y)$. The algebraic expression in part (e) consists of two terms $5y(y + 3y^2)$ and $-3x$. The term $5y(y + 3y^2)$ has **factors** of $5y$ and $(y + 3y^2)$. The factor $y + 3y^2$ is composed of the two terms y and $3y^2$.

In part (f), $\dfrac{3x^2y - 5xz}{11y}$ is a term in which the numerator consists of two terms, $3x^2y$ and $-5xz$, and the denominator is a single term, $11y$.

An algebraic expression that contains only one term is called a **monomial,** one with two terms is called a **binomial,** one with three terms a **trinomial,** and one with two or more terms is called a **multinomial.**

Referring back to Example 1, parts (c), (d), (g), and (h) are binomial expressions. In addition, in (g) we say that the **literal coefficient*** of the constants 3 and 7 is x^2. Since the literal coefficients are the

* The word "literal" is derived from the same base as the word "letter."

same, we say that $3x^2$ and $7x^2$ are **like terms.** In part (h) the product under the radical is $3x$ in both terms; thus, $6\sqrt{3x}$ and $11\sqrt{3x}$ are classified as like terms.

DEFINITION 2.17

> Two or more terms that have the same literal coefficients are called **like terms.**

EXAMPLE 2 In each of the following, two of the three terms are like terms. Determine the pair of like terms and explain why the third term is not the same as the like terms.

a. $\frac{4}{3}\pi r^3,\ \pi r^2,\ 4\pi r^2$

b. $I^2R,\ \frac{1}{2}RI^2,\ IR$

c. $\frac{Ea-Eg}{2R},\ \frac{2(Ea-Eg)}{R},\ \frac{D-d}{L}$

Solution:

a. The term $\frac{4}{3}\pi r^3$ contains the factor r^3, which is not a factor of the terms πr^2 and $4\pi r^2$. Thus, πr^2 and $4\pi r^2$ are the like terms since the literal factor, r^2, is the same in both terms. (π is a numerical constant.)

b. The terms I^2R and $\frac{1}{2}RI^2$ contain identical literal factors. The term IR is different from the other terms since the factor I has an exponent of 1.

c. The terms $\frac{Ea-Eg}{2R}$ and $\frac{2(Ea-Eg)}{R}$ are like terms since $\frac{Ea-Eg}{2R}$ can be written $\frac{1}{2}\frac{Ea-Eg}{R}$. The only difference in the two terms is the numerical coefficient. The term $\frac{D-d}{L}$ is different since it does not contain the same literal coefficients. □

Now let us get back to the main objective of this section. In adding and subtracting algebraic expressions we combine like terms.

EXAMPLE 3 In each of the following algebraic expressions combine the like terms.

a. $3a + 4b + 9a$ **b.** $x^2y - 3x^2y + 7x^2y$ **c.** $6\sqrt{3x} + 11\sqrt{3x}$

Solution:

a. In the expression $3a + 4b + 9a$ the terms $3a$ and $9a$ are like terms. Thus, we can add them and obtain the expression $12a + 4b$, which is the simplified form of the expression $3a + 4b + 9a$.

b. In the expression $x^2y - 3x^2y + 7x^2y$, since all three terms are alike and the numerical coefficients are 1, -3, and $+7$, the result is $5x^2y$.

c. In the expression $6\sqrt{3x} + 11\sqrt{3x}$ the terms are alike since the same product is under both radicals. The simplified form of the expression $6\sqrt{3x} + 11\sqrt{3x}$ is $17\sqrt{3x}$. □

In algebra as in arithmetic the order in which operations are to be performed is essential. The tools that are used to show the order of operations are **parentheses** ( ), **brackets** [ ], and **braces** { }. The general practice is that parentheses are used first and innermost, then brackets, and then braces.

EXAMPLE 4 Simplify the following algebraic expressions.

a. $8 - (11 - 17)$

b. $5 + (3x - 4) - (2x + 5)$

c. $2x + 2 - [5x + (3 - 11x)]$

d. $3a - \{4c + [7a - 5c - 2(9a + 11c + 6)]\}$

Solution:

a.
$$\begin{aligned} 8 - (11 - 17) &= 8 - (-6) \\ &= 8 + 6 && \textbf{By definition of subtraction} \\ &= 14 \end{aligned}$$

b.
$$\begin{aligned} &5 + (3x - 4) - (2x + 5) \\ &= 5 + 1(3x - 4) + -1(2x + 5) && \textbf{Definition of subtraction} \\ &= 5 + 3x - 4 - 2x - 5 && \textbf{Distributive property} \\ &= 3x - 2x + 5 - 4 - 5 && \textbf{Commutative property} \\ &= x - 4 && \textbf{Combining like terms} \end{aligned}$$

c. In a problem such as this with more than one set of grouping symbols, remove the inner symbols first and work from the inside out until all grouping symbols are removed and the problem is simplified.

$$\begin{aligned} &2x + 2 - [5x + (3 - 11x)] \\ &= 2x + 2 - [5x + 3 - 11x] \\ &= 2x + 2 - [-6x + 3] && \textbf{Combining like terms} \\ &= 2x + 2 + -1[-6x + 3] && \textbf{Definition of subtraction} \\ &= 2x + 2 + 6x - 3 && \textbf{Distributive property} \\ &= 8x - 1 && \textbf{Combining like terms} \end{aligned}$$

d. Again, start by removing the parentheses and working out until all "fences" are removed.

$= 3a - \{4c + [7a - 5c + -2(9a + 11c + 6)]\}$	**Definition of subtraction**
$= 3a - \{4c + [7a - 5c - 18a - 22c - 12]\}$	**Distributive property**
$= 3a - \{4c + [-11a - 27c - 12]\}$	**Combining like terms**
$= 3a - \{4c - 11a - 27c - 12\}$	**Removing brackets**
$= 3a - \{-11a - 23c - 12\}$	**Combining like terms**
$= 3a + -1\{-11a - 23c - 12\}$	**Definition of subtraction**
$= 3a + 11a + 23c + 12$	**Distributive property**
$= 14a + 23c + 12$	**Combining like terms** □

The most common error committed by students in removing parentheses, braces, or brackets is not to change the sign of every term inside the parentheses, brackets, or braces that are preceded by a negative sign. To avoid making this mistake use the definition of subtraction as in the previous example and apply the distributive property.

An essential part of problem solving is being able to convert a verbal statement into a mathematical expression.

EXAMPLE 5 Write the following statements as mathematical expressions.

a. 6 less than 7 times a number

b. 11 more than twice a number

Solution:

a. The first thing we must do is select a letter to represent the number. Let us arbitrarily select the letter n. Seven times the number is written as $7n$. From this product we subtract 6. Thus, the solution is $7n - 6$.

(*Note:* If the problem had read 6 less 7 times a number, the answer would have been $6 - 7n$.)

b. Again we must select a letter to represent the number. In this example we will select x. Twice the number is represented as $2x$. To this product we add 11. The solution is $2x + 11$. □

2.4 Exercises

In Exercises 1–6, identify each of the terms of the algebraic expression.

1. $6x + y$
2. $3x - 5yz$
3. $\dfrac{a+b-c}{3}$
4. $3x^2 + \dfrac{4x^2y}{z}$
5. $-4xy + (6x + 4y)^2$
6. $\dfrac{ab}{c} + 7bc - 6b^2$

In Exercises 7–38, simplify the following algebraic expressions.

7. $3a + 5a + 7b$
8. $5a - 11b + 14b$
9. $2x - 3y + 9y$
10. $8y - 11x - 9y$
11. $15r - 7s + 8r - 9s - 17r$
12. $3s + 5r - 7r + 2s + 6r$
13. $3\pi r^2 - 4\pi r + 7\pi r - 5\pi r^2$
14. $6a^2b + 7ab^2 - 5a^2b + 4ab^2$
15. $\dfrac{1}{2}x + \dfrac{1}{3}x + \dfrac{1}{4}y$
16. $\dfrac{1}{5}v - \dfrac{1}{7}v + \dfrac{2}{3}t$
17. $\dfrac{5}{8}x^2 + \dfrac{3}{4}x^2 - \dfrac{1}{2}x^2$
18. $\dfrac{1}{6}\pi r^3 - \dfrac{4}{3}\pi r^3 + \dfrac{1}{9}\pi r^3$
19. $0.03r^2 - 1.45r^2 + 0.06r^2$
20. $0.125s^2 - 3.041s^2 + 0.047s^2$
21. $0.314r^2 + 0.0314r^2 - 3.14r^2$
22. $8.14s^2 + 11.031s^2 - 31.008s^2$
23. $4 + (7 + 11)$
24. $8 - (6 - 13)$
25. $3x + (5x - 7)$
26. $11x - (7x - 6)$
27. $4 + (5a - 7) + (8 - 2a)$
28. $7a - 2(4a + 6) - (5a - 9)$
29. $2x + [5y + 6x + (3y + 7x)]$
30. $8y - [3y + 2x + (x - 6y)]$
31. $3a - 5[2b + 7a - 3(b - a)]$
32. $2a - [3b - (5a + 2b) + (6a - 4b)]$
33. $5r - [2s - (8r - 3s) - (4s - 7r)]$
34. $6r - \{5s + 2[2r - 3s + (13s - 11r)]\}$
35. $6r - \{5s - 2[2r - 3s + (13s - 11r)]\}$
36. $6a - \{5b - 2[2a - 3b - (13b - 11a)]\}$
37. $-3a - 5\{2b + (5a - b) + 2[2a - (3b + 5a)]\}$
38. $4x - \{3y - 9(2x + y) - [4x - (5y - 7x)]\}$

In Exercises 39–48, write an algebraic expression for each of the statements.

39. Seven times a number.
40. Six more than 5 times a number.
41. The sum of negative eight and one-half a number.
42. A number whose square is decreased by 7.
43. The sum of the square root of five times the number and the square of the number.
44. The sum of the cubes of two numbers x and y.
45. The cube of the sum of two numbers x and y.
46. Two terms with one variable.
47. Two terms with three variables.
48. Three terms, two variables, and all numerical coefficients are equal.

In Exercises 49–51, simplify the expressions.

49. In electronics problems involving Kirchhoff's second law the following expressions were encountered:
    a. $4 - 0.1I_1 + 6 - 0.2I_2 - 2I_1$
    b. $5 - 0.5(I_2 - I_1) - 3(I_2 - I_1) + 6 - 0.2I_2$
    c. $2I_1 + 5 - 0.5(I_2 - I_1) - 3(I_2 - I_1) + 0.1I_1 - 4$
50. In a motion problem we arrive at the following expressions:
    a. $\dfrac{3\text{ m/s} - 2\text{ m/s}}{4s}$ b. $1/2(2\text{ m/s} + 3\text{ m/s})$
51. In strength of materials problems, the following expression is encountered:

$$\left(\frac{\pi d^2}{4}\right)\left(\frac{15\sigma^2}{8}\right) + \frac{\pi}{4}(D^2 - d^2)\sigma^2$$

2.5 Multiplication of algebraic expressions

In order to multiply algebraic expressions it is essential to be able to apply the definitions of exponents discussed in Section 2.1. This is illustrated in Example 1 in multiplying two monomials.

EXAMPLE 1 Find the following products:

a. $(3x^2y)(4xy^3)$

b. $(4x^2z^3)(-5x^3y^4z^5)$

c. $(-8x^3y^2)(-9x^{-5}y^4z^2)$

Solution:

a. $(3x^2y)(4xy^3) = 3 \cdot 4 \cdot x^2 \cdot x \cdot y \cdot y^3$ **The commutative property of multiplication allows us to write the factors in this order.**

$= 12 \cdot x^{2+1} \cdot y^{1+3}$ **Definition 2.2**

$= 12x^3y^4$

b. $(4x^2z^3)(-5x^3y^4z^5) = 4 \cdot -5 \cdot x^2 \cdot x^3 \cdot y^4 \cdot z^3 \cdot z^5$

$= -20x^5y^4z^8$

c. $(-8x^3y^2)(-9x^{-5}y^4z^2) = -8 \cdot -9 \cdot x^3 \cdot x^{-5} \cdot y^2 \cdot y^4 \cdot z^2$

$= 72x^{-2} \cdot y^6 \cdot z^2$

$= \dfrac{72y^6z^2}{x^2}$ □

To multiply a monomial and a binomial expression we use the distributive property: $a(b + c) = ab + ac$. This property is true when a, b, and c are algebraic expressions. As we will soon observe, in some cases it will be necessary to apply the distributive property more than once.

EXAMPLE 2 Find the product of $7x^2$ and $4x + 3y^2$.

Solution: $7x^2(4x + 3y^2)$ is of the same form as $a(b + c)$.

$$a(b+c) \quad = \quad ab + ac$$

$$\boxed{7x^2}\,(\boxed{4x} + \boxed{3y^2}) = \boxed{7x^2} \cdot \boxed{4x} + \boxed{7x^2} \cdot \boxed{3y^2}$$

$= 7 \cdot 4 \cdot x^2 \cdot x + 7 \cdot 3 \cdot x^2 \cdot y^2$

$= 28x^3 + 21x^2y^2$ □

The technique used to multiply two binomials is just an extension of the procedure used for multiplying a monomial and a binomial. Example 3 will illustrate this technique.

EXAMPLE 3 Find the product of $(x+3)(x-5)$.

Solution: $(x+3)\cdot(x-5)$ is of the same general form as $a\cdot(b+c)$.

$$a(b+c) \qquad = \qquad ab+ac$$

$$(x+3)\cdot(x\;-5) = (x+3)\cdot x + (x+3)\cdot -5$$

Distributive property

$$= x\cdot x + 3\cdot x + x\cdot -5 + 3\cdot -5$$

Distributive property

$$= x^2 + 3x + -5x + -15$$
$$= x^2 - 2x - 15 \quad \square$$

Now let us see if we can develop a general rule for finding the product of two binomial expressions.

In general:

$$\begin{aligned}(x+a)(x+b) &= (x+a)\cdot x + (x+a)\cdot b\\ &= x\cdot x + a\cdot x + x\cdot b + a\cdot b\\ &= x^2 + ax + bx + ab\\ &= x^2 + (a+b)x + ab\end{aligned}$$

Applying distributive property in reverse

Therefore

$$(x+a)(x+b) = x^2 + (a+b)x + ab$$

EXAMPLE 4 Find the product of:

a. $(x-3)(x+11)$

b. $(3x^2+7y^2)(5x^2-6y^2)$

Solution:

a. $(x-3)(x+11) = ?$ Using the general technique developed for finding the product of two binomial expressions we have the following result:

$$\begin{aligned}(x-3)(x+11) &= x^2 + (-3+11)x + -3\cdot 11\\ &= x^2 + 8x + -33\\ &= x^2 + 8x - 33.\end{aligned}$$

b. The binomials $(3x^2+7y^2)$ and $(5x^2-6y^2)$ are different from the expressions in part (a) in two ways: (1) there are literal factors in both binomials, and (2) each literal factor has a numerical coefficient. Using the second step in the development of our general product, we obtain the following result:

$$\begin{aligned}(3x^2 + 7y^2)(5x^2 - 6y^2) &= 3x^2 \cdot 5x^2 + 7y^2 \cdot 5x^2 + 3x^2 \cdot -6y^2 \\ &\quad + 7y^2 \cdot -6y^2 \\ &= 15x^4 + \underbrace{35x^2y^2 + -18x^2y^2}_{\text{like terms}} + -42y^4 \\ &= 15x^4 + 17x^2y^2 - 42y^4.\end{aligned}$$

In fact, when the literal factors are the same for both binomials as in part (b) the answer will be of this form. □

Now let us consider some special cases.

1. The product of $(x + a)(x + a)$ is a special case of $(x + a)(x + b)$. Recall that $(x + a)(x + b) = x^2 + ax + bx + ab$. From this result we can write

   $(x + a)(x + a) = x^2 + ax + ax + aa$ or $(x + a)^2 = x^2 + 2ax + a^2$.

   Note that $(x + a)^2 \neq x^2 + a^2$.

2. Another frequently used result is found by letting $b = -a$ in the product $(x + a)(x + b)$.

   $$(x + a)(x + b) = (x + a)(x + -a)$$

   or

   $$\begin{aligned}(x + a)(x - a) &= x^2 + (a + -a)x + a \cdot -a \\ &= x^2 + 0 \cdot x + -a^2 \\ &= x^2 - a^2\end{aligned}$$

   This result shows that $(x + a)(x - a) = x^2 - a^2$.

Before concluding this section let us consider two more examples.

EXAMPLE 5 Find the product of:

a. $(x + 3)^3$
b. $(x - 3)^2(x^2 + 5x + 3)$

Solution:

a. $(x + 3)^3$ can be written as:

$$\begin{aligned}&(x + 3)(x + 3)(x + 3) \\ &= (x + 3) \cdot (x^2 + 6x + 9) && \textbf{Applying special case 1} \\ &= (x + 3) \cdot x^2 + (x + 3) \cdot 6x + (x + 3) \cdot 9 && \textbf{Distributive property} \\ &= x^3 + 3x^2 + 6x^2 + 18x + 9x + 27 && \textbf{Distributive property} \\ &= x^3 + 9x^2 + 27x + 27 && \textbf{Combining like terms}\end{aligned}$$

As indicated in the solution, when a binomial is raised to a power it is better to write the expression as a product and then multiply. Note that $(x+3)^3 \neq x^3 + 3^3$.

b. To find the product of $(x-3)^2$ and (x^2+5x+3) first find the product of $(x-3)^2$ and then apply the distributive property.

$$(x-3)^2(x^2+5x+3) = (x-3)(x-3)(x^2+5x+3)$$
$$= (x^2-6x+9)(x^2+5x+3)$$
$$= (x^2-6x+9)\cdot x^2 + (x^2-6x+9)\cdot 5x + (x^2-6x+9)\cdot 3$$

Distributive property

$$= x^4 - 6x^3 + 9x^2 + 5x^3 - 30x^2 + 45x + 3x^2 - 18x + 27$$

Distributive property

$$= x^4 - 6x^3 + 5x^3 + 9x^2 - 30x^2 + 3x^2 + 45x - 18x + 27$$
$$= x^4 - x^3 - 18x^2 + 27x + 27$$

Combining like terms

□

2.5 Exercises

In Exercises 1–46, find the product of the expressions.

1. $(3x^2)(5x^4)$
2. $(8x^3)(9x^5)$
3. $(4xy)(7x^2y^3)$
4. $(-5x^2y)(-8x^3y^2)$
5. $(-3x^3y^4)(6x^8y^9)$
6. $(5x^2y^7)(-7x^3y^9)$
7. $(3ab)^2(-5ab)$
8. $(6a^5b^4)(-3ab^3)^2$
9. $(4p^2q)^2(6pq^3)$
10. $(-7p^2q^2)(4p^2q^3)^2$
11. $(4a)(ab)(-a^2b)$
12. $(-3ab)(-4a^2b)(ab^2)^2$
13. $8a(a+4)$
14. $5a(a^2-6)$
15. $5p^2(p^2+7p)$
16. $-6q^3(q^2-5q)$
17. $3ab(2a^2b-5a)$
18. $-7ab(3abc+8a^2)$
19. $(a+b)(a+b)$
20. $(x+7)(x+9)$
21. $(2x+5)(3x+7)$
22. $(4y+8)(6y+13)$
23. $(a+b)(a-b)$
24. $(x-5)(x+5)$
25. $(2x-3)(2x+3)$
26. $(4x+8a)(4x-8a)$
27. $(3x-4y)(2x+7y)$
28. $(3xy+7z)(5xy-8z)$
29. $(5a^2+2c)(7a^2b+3c^2)$
30. $(3ab-4c)(5a^2b^2-6c^2)$
31. $(3x+4)(x^2-2x+5)$
32. $(2x-1)(x^2-3x+7)$
33. $(2a+3)(2a^2+7a+8)$
34. $(5a-4)(3a^2+11a-9)$
35. $(2x^2-4x)(2x^2+5x-1)$
36. $(3x^4-7x^2)(2x^2-8x+3)$
37. $(2x+3y)^2$
38. $(5x+6y)^2$
39. $(3a-7b)^2$
40. $(p-qx^2)^2$
41. $(x+3)(x-4)(x+5)$
42. $(3+x)(4-x)(x+5)$
43. $(2a-3b)^3$
44. $3a(2a-3b)^3$
45. $5x(2x+1)^2(x-3)$
46. $[(2x-4)^2(x-3)]^2$

In Exercises 47–49, find the indicated products and simplify.

47. In working with equations of uniformly accelerated motion we arrive at the following expression:

$$\frac{\frac{1}{2}(v+v_0)(v-v_0)}{a}$$

48. In a problem dealing with internal energy and the specific heat of a gas we arrive at the following expression:

$$\frac{3}{4}R(T_2 - T_1) + R(T_2 - T_1)$$

(The subscripts indicate different temperatures.)

49. In determining the value of sales in a particular business the expression $(500 + 75n)(8 - 2n)$ is found.

50. Show that $x^2 + y^2 \neq (x + y)^2$.

2.6 Division of algebraic expressions

The expression $a^3 \div a^2$ can be equivalently written as the fraction $\frac{a^3}{a^2}$. The quotient (the answer to the division problem) can be found by applying Definition 2.6 of exponents. Thus:

dividend **quotient**

$$\frac{a^3}{a^2} = a^{3-2} = a$$

divisor

EXAMPLE 1 Find the quotient of each of the following:

a. $6x^4 \div 2x$
b. $15x^7y^5z^4 \div 3x^4y^2z^3$
c. $(4a^3 - 10a^2 + 6a) \div 2a$
d. $(8x^2y^3 - 24x^3y^2) \div 4xy^2$
e. $(x + 1) \div x^2$

Solution:

a. $$6x^4 \div 2x = \frac{6x^4}{2x} = \frac{6}{2} \cdot \frac{x^4}{x} = 3x^{4-1} = 3x^3$$

b. $$15x^7y^5z^4 \div 3x^4y^2z^3 = \frac{15x^7y^5z^4}{3x^4y^2z^3} = \frac{15}{3} \cdot \frac{x^7}{x^4} \cdot \frac{y^5}{y^2} \cdot \frac{z^4}{z^3} = 5 \cdot x^{7-4} \cdot y^{5-2} \cdot z^{4-3} = 5x^3y^3z$$

c. $(4a^3 - 10a^2 + 6a) \div 2a = \dfrac{4a^3 - 10a^2 + 6a}{2a}$

Recall from arithmetic that we can find the sum or difference of two fractions when the denominators are alike:

$$\frac{1}{3} + \frac{4}{3} = \frac{1+4}{3} = \frac{5}{3}.$$

Using this concept in reverse we can write

$$\begin{aligned}\frac{4a^3 - 10a^2 + 6a}{2a} &= \frac{4a^3}{2a} - \frac{10a^2}{2a} + \frac{6a}{2a}\\ &= \frac{4}{2}\cdot\frac{a^3}{a} - \frac{10}{2}\cdot\frac{a^2}{a} + \frac{6}{2}\cdot\frac{a}{a}\\ &= 2a^{3-1} - 5a^{2-1} + 3a^{1-1}\\ &= 2a^2 - 5a + 3 \qquad \text{(Recall that } a^0 = 1.)\end{aligned}$$

d. $$\begin{aligned}(8x^2y^3 - 24x^3y^2) \div 4xy^2 &= \frac{8x^2y^3 - 24x^3y^2}{4xy^2}\\ &= \frac{8x^2y^3}{4xy^2} - \frac{24x^3y^2}{4xy^2}\\ &= 2\,x^{2-1}\,y^{3-2} - 6x^{3-1}\,y^{2-2}\\ &= 2xy - 6x^2\end{aligned}$$

e. $$\begin{aligned}(x+1) \div x^2 &= \frac{x+1}{x^2}\\ &= \frac{x}{x^2} + \frac{1}{x^2}\\ &= \frac{1}{x^{2-1}} + \frac{1}{x^2}\\ &= \frac{1}{x} + \frac{1}{x^2}\end{aligned}$$

Note that since the divisor, x^2, is of higher degree than either term in the dividend, $x + 1$, the quotient has to be expressed as the sum of two fractions. □

The algebraic expressions $4a^3 - 10a^2 + 6a$, $8x^2y^3 - 24x^3y^2$ and $x + 1$ are called **polynomial expressions** or **polynomials.** These expressions are called polynomials since all the numerical coefficients are real numbers and the exponents of all the variables are positive integers. The expressions $3x^{1/2} + x^2$ and $\sqrt{-9}\,x^{1/4} + 6$ are not polynomial expressions.

The process of dividing two polynomials is very much the same as the procedure used to find the quotient of two real numbers. Let us review by considering the problem 147 divided by 4.

$$\begin{array}{lrl} & 36 & \\ & 4\overline{)147} & \\ 4\times 3 \longrightarrow & \underline{12} & \\ \textbf{difference} \longrightarrow & 27 & \longleftarrow \textbf{bring down 7} \\ 4\times 6 \longrightarrow & \underline{24} & \\ \textbf{difference} \longrightarrow & 3 & \longleftarrow \textbf{remainder} \end{array}$$

Dividend = Divisor × Quotient + Remainder
Where $0 \leq$ Remainder $<$ Divisor
$147 = 4 \times 36 + 3$

The result can also be expressed as $\dfrac{147}{4} = 36 + \dfrac{3}{4}$.

EXAMPLE 2 Divide $x^2 + 5x + 6$ by $x + 2$

Solution: In order to perform the division of two polynomials the degree of the divisor must be less than the degree of the dividend. In this example the degree of the divisor is one and the degree of the dividend is two.

$$x + 2\overline{)x^2 + 5x + 6}$$

Now we ask ourselves the question: By what term must we multiply the highest term of the divisor to obtain the highest term of the dividend? In this case we multiply x by x to obtain x^2. Thus, x is the first term of the quotient and is written above the corresponding term.

$$\begin{array}{r} x \\ x + 2\overline{)x^2 + 5x + 6} \end{array}$$

Multiplying the divisor by x, the result is $x^2 + 2x$, which is written under the appropriate like terms and subtracted. Since $3x$ is the same degree as the divisor we repeat the procedure

$$\begin{array}{lrl} & x + 3 & \\ & x + 2\overline{)x^2 + 5x + 6} & \\ \textbf{product } (x+2)\cdot x \longrightarrow & \underline{x^2 + 2x } & \\ \textbf{difference} \longrightarrow & 3x + 6 & \longleftarrow \textbf{bring down 6} \\ \textbf{product } (x+2)\cdot 3 \longrightarrow & \underline{3x + 6} & \\ \textbf{difference} \longrightarrow & 0 & \end{array}$$

to obtain 3 as the second term of the quotient. Now multiply the divisor by 3 and obtain the product $3x + 6$ and subtract. The difference is zero so there is no remainder. □

The quotient, $x + 3$, in Example 2 is a polynomial. When dividing two polynomials the quotient is a polynomial and if a remainder exists it is also a polynomial as illustrated in the next example.

EXAMPLE 3 Divide $2x^3 - 4x^2 + 5$ by $x^2 + x - 1$

Solution: Note that there is no first degree term in the dividend. Thus, in setting up the problem we either allow space for the first degree term or write it with a coefficient of zero.

$$\begin{array}{lr}
 & 2x - 6 \\
 & x^2 + x - 1\overline{)2x^3 - 4x^2 + 0x + 5} \\
\textbf{product } (x^2 + x - 1) \cdot 2x \longrightarrow & \underline{2x^3 + 2x^2 - 2x} \quad \downarrow \\
\textbf{difference} \longrightarrow & -6x^2 + 2x + 5 \\
\textbf{product } (x^2 + x - 1) \cdot (-6) \longrightarrow & \underline{+6x^2 + 6x + 6} \\
\textbf{difference} \longrightarrow & 8x - 1 \leftarrow \textbf{remainder}
\end{array}$$

Thus

$$\frac{2x^3 - 4x^2 + 5}{x^2 + x - 1} = 2x - 6 + \frac{8x - 1}{x^2 + x - 1}. \qquad \square$$

If the terms of both the divisor or the dividend are not in descending order (from highest degree to lowest degree) or in ascending order (from lowest degree to highest degree), they should be rearranged so that the terms of both are in the same order. Without proper ordering of the terms, it is difficult to perform the division.

EXAMPLE 4 Divide $7x + 2x^3 + 5$ by $x^4 + 5x$

Solution: First rearrange the terms of the dividend in descending order, that is $2x^3 + 7x + 5$:

$$x^4 + 5x\overline{)2x^3 + 7x + 5}.$$

At this point one could conclude that the division cannot be performed, since the degree of the divisor is greater than the degree of the dividend. What multiplied by x^4 will yield a product of $2x^3$? The number is $2x^{-1}$, which cannot be a term of a polynomial because x does not have a positive exponent. We could perform this division and the results would be similar to obtaining a decimal in division of integers. However, the result would be contrary to what was stated earlier, that the quotient is a polynomial. $\square$

2.6 Exercises

In Exercises 1–24, perform the division.

1. $2a^2 \div 3a^2$
2. $-32b^3 \div 8b^2$
3. $42a^3b^4 \div 6a^2b$
4. $-72c^2d^3 \div 9cd^2$
5. $36b^{14} \div 9b^3$
6. $-52p^2q^8 \div (-13pq^5)$
7. $(4xy - 8xy^2) \div 2xy$
8. $(51x^2y - 34x^3y) \div (-17x^2)$
9. $(6\pi rh + 18\pi r^2h) \div 3\pi rh$
10. $(-26m^2x^2 + 39mx^2) \div (-13mx)$
11. $(35x^2y^2 - 28x^2y + 14xy^2) \div (-7xy)$
12. $(121a^3b^2 - 55a^2b^3 - 22a^3b^3) \div 11a^2b^2$
13. $(m^2 + 7m + 12) \div (m + 3)$
14. $(6p^2 + 7p - 3) \div (3p - 1)$

15. $(9a^2 + 12ab + 4b^2) \div (3a + 2b)$
16. $(49a^2 - 25b^2) \div (7a - 5b)$
17. $(8x^2 - 2xy - 21y^2) \div (2x + 3y)$
18. $(27a^3 - 125) \div (3a - 5)$
19. $(64x^3 + 1) \div (4x + 1)$
20. $(x^3 - 18x - 35) \div (x - 5)$
21. $(y^3 - 25y + 66) \div (y + 6)$
22. $(z^4 - 5z^2 + 4) \div (z^2 - 1)$
23. $(q^4 - q^2 + 16) \div (q^2 - 3q + 4)$
24. $(b^5 - b^4 - b^3 + b^2 + 16b - 16) \div (b^2 + 3b + 4)$

In Exercises 25–40, perform the division and write all remainders in rational form.

25. $(a^2 - 7a + 2) \div (a - 3)$
26. $(b^2 + 1) \div (b - 1)$
27. $(6x^2 + 7x + 5) \div (3x + 5)$
28. $(16x^2 - 1) \div (2x - 1)$
29. $t^3 - 4t^2 + 7t + 4) \div (t + 2)$
30. $(8r^2 - r - 19r^3 + 15r^4 - 1) \div (5r^2 - 3r - 1)$
31. $(x^3 - 7x + 5x^2 + 2) \div (x^2 - x - 2)$
32. $(2x^3 - 9x^2y + 7xy^2 + 6y^3) \div (x - 3y)$
33. $(16x^4 - 96x^3y + 216x^2y^2 - 216xy^3 + 81y^4) \div (2x - 3y)$
34. $(r^5 - 13r^2 + 17r - 2r^4 + 5) \div (3r + 1 + r^3 - 4r^2)$
35. $1 \div (1 + x)$ to five terms ($|x| < 1$)
36. $1 \div (x + 1)$ to five terms ($|x| > 1$)
37. $(8.1\alpha^2\beta^2\gamma + 7.2\alpha^3\beta^2\gamma^2 - 3.6\alpha^4\beta^3\gamma^3) \div (0.09\alpha^2\beta^2\gamma)$
38. $[8(\theta + \phi)^2 - 16(\theta + \phi)^4 + 12(\theta + \phi)^6] \div [4(\theta + \phi)]$
39. $[8\pi(EI + P)^4 - 32\pi(EI + P)^2 + 96\pi(EI + P)] \div [16\pi(EI + P)^2]$
40. $\left(\frac{1}{8}\theta^3 - \frac{9}{4}\theta^2\phi + \frac{27}{2}\theta\phi^2 - 27\phi^3\right) \div \left(\frac{1}{2}\theta - \phi\right)$

2.7 Factoring

In this section we will do the reverse of multiplying algebraic expressions. This reverse operation of multiplication is called factoring.

The expression $3x^2 + 6y^2$ can be written as $3x^2 + 3 \cdot 2 \cdot y^2$. Now recall the distributive property, $a(b + c) = ab + ac$. By applying the distributive property in reverse, we see that the 3 corresponds to the a

$$ab + ac = a(b + c)$$

$$\textcircled{3}x^2 + \textcircled{3} \cdot 2 \cdot y^2 = 3(x^2 + 2y^2)$$

and we have similar expressions. This is known as **common-term factoring,** properly named since we are taking something that is common from each term, in this case a 3. The product $3(x^2 + 2y^2)$ is the prime factorization of $3x^2 + 6y^2$. By **prime factorization** or **product of the prime factors** we mean that there is no other number, letter or symbol, except 1, that is common to all the terms of the expression.

EXAMPLE 1 Express each of the following as a product of its prime factors.

a. $-4xy + 8y^2$
b. $7a^2b - 14ab^2 + 21a^2b^2$
c. $3x^2y + 6x + 5xy + 10$
d. $2x^2 - 6x - 3x + 9$

Solution:

a. $-4xy + 8y^2$ can be written as $-(2)(2)xy + (2)(2)(2)yy$. In this expression we have common to both terms $(2)(2)$ and y. Thus, we can write

$$-4xy + 8y^2 = 4y(-x + 2y)$$

or we could write the solution in the following way:

$$-4xy + 8y^2 = -4y(x - 2y).$$

Note the change in signs. Are both answers correct? Applying the distributive property:

$$\begin{aligned} 4y(-x + 2y) &= 4y \cdot -x + 4y \cdot 2y \\ &= -4xy + 8y^2 \quad \checkmark \text{ (check)} \end{aligned}$$

or

$$\begin{aligned} -4y(x - 2y) &= -4y \cdot x + -4y \cdot -2y \\ &= -4xy + 8y^2 \quad \checkmark \text{ (check)} \end{aligned}$$

Note that one must be extremely careful in working with negative signs. The problems you had with negative signs in multiplication will "haunt" you in factoring if you did not clear up the difficulties.

b. In examining the expression $7a^2b - 14ab^2 + 21a^2b^2$ we see that each term contains factors of 7, a, and b. Thus we can write:

$$\begin{aligned} &7a^2b - 14ab^2 + 21a^2b^2 \\ &= 7 \cdot a \cdot b \cdot a - 7 \cdot a \cdot b \cdot 2 \cdot b + 7 \cdot a \cdot b \cdot 3 \cdot a \cdot b \\ &= 7ab(a - 2b + 3ab). \end{aligned}$$

c. In examining the terms of the expression $3x^2y + 6x + 5xy + 10$ we see that there is no common factor in all four terms. However, the two terms $3x^2y$ and $6x$ have a common factor of $3x$ and the terms $5xy$ and 10 have a common factor of 5. Thus:

$$3x^2y + 6x + 5xy + 10 = 3x(xy + 2) + 5(xy + 2).$$

Note that by pairing up the terms and factoring a common term from each of the pairs of terms, the result is an expression with two terms, with a common factor of $xy + 2$. Factoring again:

$$3x(xy + 2) + 5(xy + 2) = (xy + 2)(3x + 5).$$

Thus

$$3x^2y + 6x + 5xy + 10 = (xy + 2)(3x + 5).$$

d. For the expression $2x^2 - 6x - 3x + 9$ you may be tempted to combine the like terms $-6x$ and $-3x$ and end up with the expression $2x^2 - 9x + 9$. We cannot factor this expression at this time. However, using the techniques developed in part (d) we can factor the expression

$$2x^2 - 6x - 3x + 9 = 2x(x - 3) + 3(-x + 3).$$

We have a problem at this point, since $(x-3)$ and $(-x+3)$ are not the same. The factors would be the same if -3 is used as the common term of $-3x+9$ rather than $+3$:

$$2x^2-6x-3x+9=2x(x-3)+-3(x-3).$$

Now we have like terms inside the parentheses and apply common-term factoring:

$$\begin{aligned}2x^2-6x-3x+9&=(x-3)(2x+-3)\\&=(x-3)(2x-3).\end{aligned}$$

This technique of common-term factoring will be helpful in factoring trinomials of the form ax^2+bx+c. In Example 3 of Section 2.5 we found the product of $(x+3)$ and $(x-5)$ to be $x^2-2x-15$. The second to last line of that example is the key to reversing the process of factoring $x^2-2x-15$. In the expression $x^2+3x-5x-15$ the sum of the two terms $3x$ and $-5x$ is $-2x$. The product of the numerical coefficients of $3x$ and $-5x$ is -15. Note that by applying common-term factoring as in (e) we have the following result:

$$\begin{aligned}x^2+3x-5x-15&=x(x+3)+-5(x+3)\\&=(x+3)(x-5).\end{aligned}$$ □

In general the trinomial ax^2+bx+c can be factored if we can find a pair of integers such that their product is ac and their sum is b. Part (d) of Example 1 is just such a case:

$$2x^2-6x-3x+9=2x^2-9x+9.$$

In this example $a=2$, $b=-9$, and $c=9$. Thus the product of $a\cdot c$, $(2)(9)=18$ is the same as $(-6)(-3)=18$, $b=-9$ is the same as the sum $-6+-3=-9$.

A summary of the technique for factoring a trinomial follows.

Method for Factoring a Trinomial

Assume that we want to factor ax^2+bx+c (where a, b, and c, are integers).

1. Multiply a and c (remember the signs).
2. Write down b (remember the sign).
3. Search for a pair of numbers whose product is ac and whose sum is b.
4. Write the bx term as the sum of two terms, the sum of whose coefficients is b.
5. Factor using common-term factoring.

EXAMPLE 2 Factor the quadratic expressions:

a. $x^2 + 8x + 15$ **b.** $x^2 - x - 20$

c. $3x^2 + 5x - 12$ **d.** $x^2 + 5x + 7$

Solution:

a. To factor the expression $x^2 + 8x + 15$ we need to find two numbers whose sum is 8 and whose product is 15. Also, since 8 and 15 are both positive, the numbers we are looking for are both positive.

Factors of 15	Sum of factors
$1 \cdot 15$	$1 + 15 = 16$
$3 \cdot 5$	$3 + 5 = 8$

Since $3 \cdot 5 = 15$ and $3 + 5 = 8$ we can write:

$$\begin{aligned} x^2 + 8x + 15 &= x^2 + 3x + 5x + 15 \\ &= x(x + 3) + 5(x + 3) && \textbf{Common-term factoring} \\ &= (x + \textcircled{3})(x + \textcircled{5}) && \textbf{Common-term factoring} \end{aligned}$$

b. In the expression $x^2 - x - 20$ the constant term is -20. Therefore, the numbers we are looking for must have opposite signs. In addition, since b is negative (-1), the larger number must be negative.

Factors of -20	Sum of factors
$-20 \cdot 1$	$-20 + 1 = -19$
$-10 \cdot 2$	$-10 + 2 = -8$
$-5 \cdot 4$	$-5 + 4 = -1$

Thus, the numbers we are looking for are -5 and 4.

$$\begin{aligned} x^2 - x - 20 &= x^2 - 5x + 4x - 20 \\ &= x(x - 5) + 4(x - 5) && \textbf{Common-term factoring} \\ &= (x + \textcircled{4})(x - \textcircled{5}) && \textbf{Common-term factoring} \end{aligned}$$

c. For the expression $3x^2 + 5x - 12$ we are looking for two numbers whose product is -36 and whose sum is 5. Since the product is negative, the numbers must have opposite signs, and since b is positive (5), the larger number must be positive.

Factors of -36	Sum of factors
$36 \cdot -1$	$36 + -1 = 35$
$18 \cdot -2$	$18 + -2 = 16$
$12 \cdot -3$	$12 + -3 = 9$
$9 \cdot -4$	$9 + -4 = 5$

The numbers are 9 and -4, so we write the bx term as the sum $9x - 4x$.

$$\begin{aligned} 3x^2 + 5x - 12 &= 3x^2 + 9x - 4x - 12 \\ &= 3x(x+3) - 4(x+3) \quad \textbf{Common-term factoring} \\ &= (x+3)(3x-4) \quad \textbf{Common-term factoring} \end{aligned}$$

d. For the expression $x^2 + 5x + 7$ we are looking for two numbers whose sum is 5 and whose product is 7. Since 5 and 7 are both positive the factors must be positive.

Factors of 7	Sum of factors
$1 \cdot 7$	$1 + 7 = 8$

There are no other factors of 7, and the sum of these factors is not equal to 5. Therefore, the expression $x^2 + 5x + 7$ is prime (it cannot be factored using the integers). In a later chapter you will be shown how this expression can be factored using real numbers. □

The quadratic expression $4x^2 + 20x + 25$ is a special form known as a **perfect square trinomial.** An expression of this type can be identified by observing that $\frac{1}{2}(20) = \sqrt{4 \cdot 25}$ or that $\frac{1}{2}b = \pm\sqrt{a \cdot c}$. In factoring this trinomial the two numbers we are looking for, whose sum is b, will both be 10. Thus,

$$\begin{aligned} 4x^2 + 20x + 25 &= 4x^2 + 10x + 10x + 25 \\ &= 2x(2x+5) + 5(2x+5) \quad \textbf{Common-term factoring} \\ &= (2x+5)(2x+5) \quad \textbf{Common-term factoring} \\ &= (2x+5)^2 \end{aligned}$$

Note that the coefficient of the x term is the square root of a (4), and the constant term (5) is the square root of c (25). Thus, when we recognize that a quadratic expression is a perfect square, in order to factor we find the square root of a and the square root of c and write the answer, $(\sqrt{a}x \pm \sqrt{c})^2 = (2x+5)^2$, where $\pm$ is determined by the sign of b.

EXAMPLE 3 Factor the following quadratic expressions:

a. $x^2 - 6x + 9$ **b.** $16a^2 + 56a + 49$

Solution:

a. To factor the expression $x^2 - 6x + 9$ we are looking for two numbers whose product is 9 and whose sum is -6. Since $\frac{1}{2}(-6) = -3$ and $-\sqrt{9} = -3$, the numbers we are looking for are -3 and -3. Thus, the expression is a perfect-square trinomial and the factors are $(x - 3)$, $(x - 3)$, and the solution is $(x - 3)^2$. $x^2 - 6x + 9 = (x - 3)^2$.

b. To factor $16a^2 + 56a + 49$ we need to look for two numbers whose product is 784 and whose sum is 56. Since $\frac{1}{2}(56) = 28$ and $\sqrt{784} = 28$, the two numbers we are looking for are 28 and 28. The expression is a perfect square trinomial and $16a^2 + 56a + 49 = (4a + 7)^2$. □

A quadratic expression of the form $x^2 - a^2$ is known as the **difference of two squares** and is factored as follows:

$$x^2 - a^2 = (x - a)(x + a).$$

EXAMPLE 4 Factor the following quadratic expressions:

a. $x^2 - 25$ **b.** $25x^2 - 64y^2$

Solution:

a. $x^2 - 25 = x^2 - 5^2$ **The difference of two squares**

Thus, $x^2 - 25 = (x - 5)(x + 5)$

b. $25x^2 - 64y^2 = (5x)^2 - (8y)^2$ **The difference of two squares**

Thus, $25x^2 - 64y^2 = (5x - 8y)(5x + 8y)$ □

2.7 Exercises

In Exercises 1–66, express each expression as a product of prime factors.

1. $3a + 12b$
2. $6a - 15b$
3. $7r + 14s$
4. $9r - 18s$
5. $2x - 12xy$
6. $5xy + 20x$
7. $5xy - 15x^2z + 20xz$
8. $4x^2y - 8xy^2 + 14x^2y^2$
9. $3ab + 12bc + 5ad + 20cd$
10. $2xy + 14yz + 7wx + 49wz$
11. $15rs - 25rt - 18ks + 30kt$
12. $38RI_1 - 19RI_2 - 34II_1 + 17II_2$
13. $15R_1I_1 - 30R_1I_2 - 35R_2I_1 + 70R_2I_2$
14. $28AB - 56AC + 26BD - 52CD$
15. $6A^2BD - 14ADC - 15A^2B^2 + 35ABC$
16. $P^2QTU - P^2QVW + PRSTU - PRSVW$
17. $\frac{1}{3}AB - \frac{2}{3}AC + \frac{2}{5}PB - \frac{4}{5}PC$
18. $\frac{3}{7}\pi r_1{}^2 - \frac{15}{49}\pi r_1{}^2 h - \frac{2}{9}\pi r_2{}^2 + \frac{10}{63}\pi r_2{}^2 h$
19. $0.06x^2y + 0.08xy^2 + 0.09xyz + 0.12y^2z$
20. $0.15x^2y^2 - 0.21wxyz + 0.35uvxy - 0.49uvwz$
21. $x^2 + 5x + 4$
22. $x^2 - 3x - 4$
23. $3a^2 + 6a + 3$
24. $5b^2 - 35b + 60$
25. $t^2 + 9t + 14$
26. $2r^2 + 7r - 4$
27. $r^2 - 35 + 2r$
28. $25 - a^2$
29. $-2 - 5b + 3b^2$
30. $6z^2 - 11z - 10$
31. $p^2 + 3p - 6$
32. $2p^2 + 2p - 4$
33. $11z + 5z^2 - 12$
34. $s^2 + s + 1$

35. $6s^2 + 16s - 6$
36. $5x^2 + 16xy + 3y^2$
37. $17p^2 + 6pq - 11q^2$
38. $3r^2 + 10rs + 3s^2$
39. $4w^2 - 20w + 25$
40. $7w^2 - 25w - 12$
41. $5x^2 + 15x - 28$
42. $8x^2 + 18x + 9$
43. $3z^2 + 10z + 3$
44. $2w^2 + 7w - 15$
45. $7R^2 - 16R + 4$
46. $x^2 + 6x + 9$
47. $w^2 + 14w + 49$
48. $4p^2 - 20p + 25$
49. $9p^2 + 42p + 49$
50. $16t^2 + 24t + 9$
51. $4s^2 - 12s + 9$
52. $12s^2 - 60s + 75$
53. $64m^2 - 80mn + 25n^2$
54. $49x^2 + 112xy + 64y^2$
55. $81w^2 - 72wz + 16z^2$
56. $121a^2 + 154ab + 49b^2$
57. $x^2 - 9$
58. $a^2 - 36$
59. $81r^2 - 16$
60. $5 - 5a^2$
61. $x^2y^2 - 25$
62. $100a^2b^2 - 49x^2y^2$
63. $(a + b)^2 - d^2$
64. $(z^2 + 16)^2 - 25y^2$
65. $p^2 (q + 3) - 16q - 48$
66. $3x^2 - 108$

2.8 Multiplication and division of fractions

To find the product of arithmetic fractions, find the product of the numerators and the product of the denominators and then reduce the resulting fraction to lowest terms. This same procedure can be used for multiplication of algebraic fractions.

EXAMPLE 1 Find the product of $\frac{5}{12}$ and $\frac{8}{15}$.

Solution:

$$\frac{5}{12} \cdot \frac{8}{15} = \frac{5 \cdot 8}{12 \cdot 15}$$
$$= \frac{40}{180}$$
$$= \frac{20 \cdot 2}{20 \cdot 9}$$
$$= \frac{2}{9} \qquad \text{(reducing to lowest terms)}$$

This method will always work. However, a great deal of time may be spent reducing the product fraction to lowest terms. A more practical method may be to factor the numerators and denominators of the fractions, reduce (cancel), and then multiply:

$$\frac{5}{12} \cdot \frac{8}{15} = \frac{\overset{1}{\cancel{5}}}{\underset{1}{\cancel{2}} \cdot \underset{1}{\cancel{2}} \cdot 3} \cdot \frac{\overset{1}{\cancel{2}} \cdot \overset{1}{\cancel{2}} \cdot 2}{3 \cdot \underset{1}{\cancel{5}}}$$
$$= \frac{2}{9} \quad \square$$

This process of reducing or canceling is accomplished by dividing both the numerator and the denominator by the common factors. In general, we make errors when we try to cancel when the operations of addition and subtraction are involved. For example in the problem $\frac{2+8}{2}$ we are tempted to make the error $\frac{2+8}{2} = 9$. It can be seen that the correct result is $\frac{2+8}{2} = \frac{10}{2} = 5$ and not 9. By using common-term factoring we could avoid the error:

$$\frac{2+8}{2} = \frac{\overset{1}{\cancel{2}}(1+4)}{\underset{1}{\cancel{2}}} = 5.$$

The second technique, reducing first and then multiplying, is by far the best method for multiplying algebraic fractions.

EXAMPLE 2 Find the product and simplify.

a. $\frac{2x+6}{2x+8} \cdot \frac{6x+12}{3x+9}$ **b.** $\frac{2x^2-7x-15}{x^2-25} \cdot \frac{x^2-2x-35}{7-x}$

Solution:

a.
$$\frac{2x+6}{2x+8} \cdot \frac{6x+12}{3x+9} = \frac{\overset{1}{\cancel{2}}\overset{1}{\cancel{(x+3)}}\,\overset{2}{\cancel{6}}(x+2)}{\underset{1}{\cancel{2}}(x+4)\,\underset{1}{\cancel{3}}\underset{1}{\cancel{(x+3)}}}$$

Common-term factoring and canceling

$$= \frac{2(x+2)}{x+4}$$

You can leave the answer in this form,

$$= \frac{2x+4}{x+4}$$

or you can multiply it out like this.

b.
$$\frac{2x^2-7x-15}{x^2-25} \cdot \frac{x^2-2x-35}{7-x}$$

$$= \frac{(2x+3)\overset{1}{\cancel{(x-5)}}}{\underset{1}{\cancel{(x-5)}}\underset{1}{\cancel{(x+5)}}} \cdot \frac{(x-7)\overset{1}{\cancel{(x+5)}}}{7-x}$$

Factoring and canceling

$$= \frac{(2x+3)(x-7)}{7-x}$$

Note that if we had $x-7$ in the denominator rather than $7-x$, we could cancel. Recall that $(a-b) = -(b-a)$ or $(a-b) = (-1)(b-a)$. Thus:

$$\frac{(2x+3)(x-7)}{7-x} = \frac{(2x+3)(-1)\overset{1}{\cancel{(7-x)}}}{\underset{1}{\cancel{7-x}}}$$

$$= -1(2x+3) \text{ or } -2x-3 \qquad \square$$

In our discussion of fractions we mentioned the special rule that states you can multiply a fraction by the same nonzero expression provided that you multiply both the numerator (top) and the denominator (bottom) by the same expression.* To divide one fraction by another, multiply the numerator and the denominator by the reciprocal of the denominator. The reason for this multiplication is to make the denominator equal to 1 and change the division to a multiplication problem.

EXAMPLE 3 Divide $\frac{2}{3}$ by $\frac{5}{7}$

Solution:

$$\frac{\frac{2}{3}}{\frac{5}{7}} = \frac{\frac{2}{3} \cdot \frac{7}{5}}{\frac{5}{7} \cdot \frac{7}{5}} = \frac{\frac{2}{3} \cdot \frac{7}{5}}{1} = \frac{14}{15} \qquad \left(\frac{7}{5} \text{ is the reciprocal of } \frac{5}{7}\right)$$ □

The reciprocal of the algebraic fraction $\frac{x+2}{x+3}$ is the fraction $\frac{x+3}{x+2}$. The product of an algebraic fraction and its reciprocal is 1, the same as that of an arithmetic fraction and its reciprocal.

EXAMPLE 4 Divide $\frac{(x+4)}{(x+2)}$ by $\frac{(x+5)}{3x}$.

Solution:

$$\frac{\frac{x+4}{x+2}}{\frac{x+5}{3x}} = \frac{\frac{x+4}{x+2}}{\frac{x+5}{3x}} \cdot \frac{\frac{3x}{x+5}}{\frac{3x}{x+5}}$$

$$= \frac{\frac{x+4}{x+2} \cdot \frac{3x}{x+5}}{1}$$

$$= \frac{(x+4)(3x)}{(x+2)(x+5)}$$ □

Note that in Examples 3 and 4 we could write the problems in the following manners:

$$\frac{2}{3} \div \frac{5}{7} \qquad \text{and} \qquad \frac{x+4}{x+2} \div \frac{x+5}{3x}.$$

* This is the same as multiplying the fraction by 1, which is the multiplicative identity. Therefore, we change the form of the fraction but not its value.

If we now invert the second fraction (the divisor) in both problems and multiply, we will obtain the same result.

$$\frac{2}{3} \div \frac{5}{7} = \frac{2}{3} \cdot \frac{7}{5} = \frac{14}{15}$$

$$\frac{x+4}{x+2} \div \frac{x+5}{3x} = \frac{x+4}{x+2} \cdot \frac{3x}{x+5} = \frac{(x+4)(3x)}{(x+2)(x+5)}$$

This is perhaps the more common way to divide fractions. Examples 3 and 4 showed why this technique works.

EXAMPLE 5 Perform the following division

$$\frac{x^2-25}{x^2+8x+15} \div \frac{x^2-2x-15}{3x+9}.$$

Solution:

$$\frac{x^2-25}{x^2+8x+15} \div \frac{x^2-2x-15}{3x+9}$$

$$= \frac{x^2-25}{x^2+8x+15} \cdot \frac{3x+9}{x^2-2x-15} \qquad \textbf{Invert the divisor}$$

$$= \frac{\overset{1}{\cancel{(x+5)}}\overset{1}{\cancel{(x-5)}}}{\underset{1}{\cancel{(x+5)}}\underset{1}{\cancel{(x+3)}}} \cdot \frac{3\overset{1}{\cancel{(x+3)}}}{\underset{1}{\cancel{(x-5)}}(x+3)} \qquad \textbf{Factor completely and cancel like factors}$$

$$= \frac{3}{x+3} \qquad \textbf{Multiply}$$

□

2.8 Exercises

In Exercises 1–15, find the product and express it in lowest terms.

1. $\frac{2}{3} \cdot \frac{6}{7}$

2. $\frac{8}{9} \cdot \frac{15}{20}$

3. $\frac{5p}{6m} \cdot \frac{12m^2}{10}$

4. $\frac{xy^2}{9xz} \cdot \frac{3x}{(xy)^2}$

5. $\frac{x}{6} \cdot \frac{x+3}{x-2}$

6. $\frac{x+3}{x+4} \cdot \frac{x+5}{x-6}$

7. $\frac{2x+6}{3x+12} \cdot \frac{x+4}{x+3}$

8. $\frac{x+5}{x+7} \cdot \frac{x^2+7x}{xy+5y}$

9. $\frac{a^2-b^2}{9} \cdot \frac{12}{a+b}$

10. $\frac{a+3}{p-4} \cdot \frac{p^2-16}{a^2-9}$

11. $\frac{x^2+6x+9}{3xyz} \cdot \frac{6xy}{x^2-9}$

12. $\frac{m^2-9n^2}{8pq} \cdot \frac{2p}{m^2-6mn+9n^2}$

13. $\frac{a^2-b^2}{a^2-2ab+b^2} \cdot \frac{a^2-b^2}{a^2+2ab+b^2}$

14. $\frac{3m^2-m-2}{9m^2+12m+4} \cdot \frac{3m^2+2m}{m^3-m^2}$

15. $\frac{p^2+p-12}{p^2+8p+16} \cdot \frac{3p^2+12p}{9-p^2}$

In Exercises 16–30, find the quotient and express it in lowest terms.

16. $\frac{3}{4} \div \frac{9}{16}$

17. $\frac{6}{7} \div \frac{15}{28}$

18. $\frac{3x}{5y} \div \frac{6x^2}{15y^3}$

19. $\frac{ab^2}{9ac} \div \frac{(ab)^2}{3a}$

20. $\frac{a}{6} \div \frac{a-2}{a+3}$

21. $\frac{a+3}{a+4} \div \frac{a+5}{a-6}$

22. $\frac{x+3}{x+4} \div \frac{xy+3y}{x^2+7x}$

23. $\frac{3x+9}{4x+16} \div \frac{x+3}{x+4}$

24. $\frac{p^2-q^2}{5} \div \frac{p-q}{15}$

25. $\frac{p^2-p-6}{p^2-9} \div \frac{p^2-4}{p+3}$

26. $\frac{m^2+7m-8}{m^2-64} \div \frac{m^2+2m+1}{8m-64}$

27. $\frac{x^2+2xy+y^2}{x^2-2xy-3y^2} \div \frac{x^2-y^2}{x-3y}$

28. $\frac{x^2+5x+6}{x^2+6x+9} \div \frac{x^2-9}{3-x}$

29. $\frac{6m^2-11m-10}{3m^2-m-2} \div \frac{2m^2+5m-25}{m^2+4m-5}$

30. $\frac{4x^2-1}{6x^2+5x-4} \div \frac{2x^2+x-1}{3x^2+7x+4}$

In Exercises 31–36, find the indicated result and express it in lowest terms.

31. $\frac{3x}{4y} \div \frac{2y}{3x} \cdot \frac{6x}{9y}$

32. $\frac{3x}{4y} \cdot \frac{2y}{3x} \div \frac{6x}{9y}$

33. $\frac{5x^2}{7y} \div \left(\frac{5x}{3z} \cdot \frac{2x^2}{21y}\right)$

34. $\frac{2a^2}{3b} \cdot \left(\frac{6a}{7b} \div \frac{12a^2}{35b^3}\right)$

35. $\frac{R^2-r^2}{r^2+Rr} \cdot \frac{R(R-r)}{(R-r)^2} \div \frac{R^2-3Rr+2r^2}{Rr-2r^2}$

36. $\frac{16I^4R^2-9}{4\left(I^2R+\frac{3}{4}\right)} \cdot \frac{I^4R^2-3I^2R-28}{2I^4R^2-32} \div \frac{8I^4R^2-62I^2R+42}{8I^2R-32}$

2.9 Addition and subtraction of fractions

The general procedure for adding or subtracting algebraic fractions is the same as the one used to add or subtract arithmetic fractions. Before the operations of addition or subtraction can be performed the numerators must be written over a common denominator. If the denominators are alike (common denominators) the operations can be performed as in Example 1.

EXAMPLE 1 Perform the indicated operations and simplify.

a. $\frac{2}{7} + \frac{3}{7}$ **b.** $\frac{x}{3yz} - \frac{5x}{3yz}$

Solution:

a.
$$\frac{2}{7} + \frac{3}{7} = \frac{2+3}{7}$$
$$= \frac{5}{7}$$

b. $\dfrac{x}{3yz} - \dfrac{5x}{3yz} = \dfrac{x - 5x}{3yz}$

$= \dfrac{-4x}{3yz}$ □

If the fractions do not have common denominators, then before the addition or subtraction can be performed it is necessary to find a common denominator. In fact, it is generally to our advantage to find the lowest common denominator. The lowest common denominator (lcd) is the smallest number or algebraic expression that is divisible by all the denominators. For example the lcd of the fractions $\frac{3}{2x^2}$, $\frac{5}{6xy^2}$, and $\frac{7}{4yz}$ is $12x^2y^2z$. What method do we use to determine that $12x^2y^2z$ is the lcd? The first thing we do is write each denominator as the product of its prime factors. Now examine each product $2x^2 = 2x^2$, $6xy^2 = 2 \cdot 3xy^2$, $4yz = 2^2 \cdot yz$ and select from the products the highest degree of each distinct factor that is represented. The product of these factors is the least common denominator. From the three terms above the distinct factors are 2, x, y, 3, and z. The highest degrees of these distinct factors are x^2, 3, y^2, 2^2, and z. The product of these factors is $12x^2y^2z$, which is the lcd.

EXAMPLE 2 Perform the indicated operations and simplify.

a. $\dfrac{3}{2x^2} + \dfrac{5}{6xy^2} + \dfrac{7}{4yz}$ **b.** $\dfrac{5}{x^2 - 1} - \dfrac{3}{x^2 + 2x + 1}$

Solution:

a. In the previous discussion we found the lcd of the fractions $\frac{3}{2x^2}$, $\frac{5}{6xy^2}$, and $\frac{7}{4yz}$ to be $12x^2y^2z$. Now we ask ourselves the following questions:

$2x^2 \cdot \underline{\ \ ?\ \ } = 12x^2y^2z \qquad 2x^2 \cdot \underline{6y^2z} = 12x^2y^2z$

$6xy^2 \cdot \underline{\ \ ?\ \ } = 12x^2y^2z \qquad 6xy^2 \cdot \underline{2xz} = 12x^2y^2z$

$4yz \cdot \underline{\ \ ?\ \ } = 12x^2y^2z \qquad 4yz \cdot \underline{3x^2y} = 12x^2y^2z$

We then multiply the numerator and denominator of each fraction by the expression that replaced the question mark. Recall that multiplying the numerator and denominator by the same number (other than zero) is the same as multiplying the fraction by 1 and will not change the value of the fraction.

$$\frac{3}{2x^2} \cdot \frac{6y^2z}{6y^2z} = \frac{18y^2z}{12x^2y^2z}$$

$$\frac{5}{6xy^2} \cdot \frac{2xz}{2xz} = \frac{10xz}{12x^2y^2z}$$

$$\frac{7}{4yz} \cdot \frac{3x^2y}{3x^2y} = \frac{21x^2y}{12x^2y^2z}$$

Thus:

$$\frac{3}{2x^2} + \frac{5}{6xy^2} + \frac{7}{4yz} = \frac{18y^2z}{12x^2y^2z} + \frac{10xz}{12x^2y^2z} + \frac{21x^2y}{12x^2y^2z}$$

$$= \frac{18y^2z + 10xz + 21x^2y}{12x^2y^2z}.$$

b. To find the difference of the two fractions $\frac{5}{x^2-1} - \frac{3}{x^2+2x+1}$, we must find the lcd. The first thing to do is to factor the denominators.

$x^2 - 1 = (x+1)(x-1)$ and $x^2 + 2x + 1 = (x+1)^2$

Thus, the lcd is $(x-1)(x+1)^2$, which also can be written as $(x-1)(x+1)(x+1) = (x^2-1)(x+1)$.

$$\frac{5}{x^2-1} - \frac{3}{x^2+2x+1} = \frac{5}{x^2-1} \cdot \frac{x+1}{x+1} - \frac{3}{(x+1)^2} \cdot \frac{x-1}{x-1} \quad (1)$$

$$= \frac{5x+5}{(x^2-1)(x+1)} - \frac{3(x-1)}{(x+1)^2(x-1)}$$

$$= \frac{5x+5-(3x-3)}{(x-1)(x+1)^2}$$

Note that this minus sign applies to all terms in the second fraction, not just the first term.

$$= \frac{5x+5-3x+3}{(x-1)(x+1)^2}$$

$$= \frac{2x+8}{(x-1)(x+1)^2}$$

$$= \frac{2(x+4)}{(x-1)(x+1)^2} \quad \square$$

In Step 1 why did we multiply the first fraction by $\frac{x+1}{x+1}$ and the second fraction by $\frac{x-1}{x-1}$? For the answer see the footnote on page 71.

Complex fractions

Now that we understand the operations of addition, subtraction, multiplication, and division we can use them to simplify complex fractions.

Working with complex fractions may cause concern to the student. The name is unfortunate because there is nothing complex or mysterious about working with complex fractions.

> *Simple rule:* To simplify a complex fraction, reduce both numerator and denominator to simple fractions, then perform the indicated division.

EXAMPLE 3 Simplify

$$\frac{\frac{1}{x}+3}{\frac{5}{y^2}}.$$

Solution: Following the rule, the first thing we want to do is to write the numerator as a simple fraction. To do this we must find the sum of $1/x$ and 3.

$$\begin{aligned}\frac{1}{x}+3 &= \frac{1}{x}+\frac{3x}{x}\\ &= \frac{1+3x}{x}\end{aligned}$$

Substituting we have:

$$\begin{aligned}\frac{\frac{1}{x}+3}{\frac{5}{y^2}} &= \frac{\frac{1+3x}{x}}{\frac{5}{y^2}} && \textbf{Dividing}\\ &= \frac{1+3x}{x}\cdot\frac{y^2}{5} && \textbf{Inverting and multiplying}\\ &= \frac{y^2+3xy^2}{5x}\end{aligned}$$

Therefore,

$$\frac{\frac{1}{x}+3}{\frac{5}{y^2}} = \frac{y^2+3xy^2}{5x}$$

□

EXAMPLE 4 Simplify

$$\frac{\frac{x}{y}+\frac{x+y}{x-y}}{\frac{x}{y}-\frac{x-y}{x+y}}.$$

Solution: The lcd for the numerator is $y(x-y)$ and the lcd for the denominator is $y(x+y)$. Multiply the fractions in the numerator and denominator by their respective lcd's.

$$\frac{\frac{x}{y}+\frac{x+y}{x-y}}{\frac{x}{y}-\frac{x-y}{x+y}}=\frac{\frac{x(x-y)}{y(x-y)}+\frac{y(x+y)}{y(x-y)}}{\frac{x(x+y)}{y(x+y)}-\frac{y(x-y)}{y(x+y)}}$$

$$=\frac{\frac{x^2-xy+yx+y^2}{y(x-y)}}{\frac{x^2+xy-yx+y^2}{y(x+y)}} \qquad \textbf{Expanding}$$

$$=\frac{\frac{x^2+y^2}{y(x-y)}}{\frac{x^2+y^2}{y(x+y)}} \qquad \textbf{Collecting like terms}$$

$$=\frac{x^2+y^2}{y(x-y)}\div\frac{x^2+y^2}{y(x+y)} \qquad \textbf{Dividing}$$

$$=\frac{\overset{1}{\cancel{x^2+y^2}}}{\underset{1}{\cancel{y}}(x-y)}\cdot\frac{\overset{1}{\cancel{y}}(x+y)}{\underset{1}{\cancel{x^2+y^2}}} \qquad \textbf{Inverting, multiplying, and canceling}$$

$$=\frac{x+y}{x-y} \qquad \square$$

2.9 Exercises

In Exercises 1–48 perform the indicated operations and simplify.

1. $\frac{2}{x}+\frac{3}{2x}$
2. $\frac{1}{a}+\frac{1}{b}$
3. $1+\frac{1}{p+q}$
4. $\frac{2}{m^2}+\frac{1}{m}$
5. $\frac{5}{ab^2}-\frac{1}{a^2b}$
6. $\frac{3a}{c^2}-\frac{2c}{a^2}$
7. $\frac{1}{x-1}-\frac{1}{x+1}$
8. $\frac{4}{x^2+x}+\frac{7}{4x+4}$
9. $\frac{1}{a+2}-\frac{2}{a+3}$
10. $\frac{3}{\theta-\pi}-\frac{4}{\theta+\pi}$
11. $\frac{x}{y-z}+\frac{x}{z-y}$
12. $\frac{1}{x}-\frac{1}{y}+\frac{1}{z}$
13. $\frac{a}{a-1}+\frac{2a}{1-a}$
14. $\frac{a}{R^2}-\frac{c}{R}+2R$

15. $\dfrac{m-1}{m+1}-\dfrac{m+1}{m-1}$

16. $\dfrac{p-q+r}{m-n}+\dfrac{p-q-r}{n-m}$

17. $\dfrac{3}{x^5y^2}-\dfrac{5}{xy^4z}$

18. $\dfrac{1}{a^2(a+1)}-\dfrac{1}{a^5(a-1)}$

19. $\dfrac{a}{a^2-4}+\dfrac{a}{2+a}$

20. $\dfrac{1}{x^2-1}-\dfrac{1}{(x-1)^2}$

21. $\dfrac{a-b}{c-d}+\dfrac{a-b}{d-c}$

22. $\dfrac{x}{y+y^2}-\dfrac{1}{y(y+y^2)}$

23. $\dfrac{1}{4x^2+4x+1}+\dfrac{1-x}{1-2x}$

24. $\dfrac{3(a+1)}{a^2+6a-16}-\dfrac{1}{a-2}$

25. $\dfrac{a^2}{p^2q^2}+\dfrac{b^2}{p^2r^2}-\dfrac{c^2}{q^2r^2}$

26. $\dfrac{3\alpha}{1-\alpha^2}+\dfrac{2}{\alpha+1}-\dfrac{3}{1-\alpha}$

27. $\dfrac{x-1}{x+3}+\dfrac{x-2}{x-3}-\dfrac{x^2}{x^2-9}$

28. $\dfrac{R}{R^2-6R-7}+\dfrac{2}{R-7}-\dfrac{3}{R+1}$

29. $\dfrac{a^2+b^2}{a^2-b^2}-\dfrac{a}{a+b}+\dfrac{b}{a-b}$

30. $\dfrac{1}{2x+2}+\dfrac{5}{3x-3}-\dfrac{3x-1}{1-x^2}$

31. $\dfrac{1}{x+y}-\left(\dfrac{1}{x-y}-\dfrac{1}{x^2-y^2}\right)$

32. $2-\left(\dfrac{3}{a}+\dfrac{4}{a-2}\right)$

33. $\dfrac{2}{t^2-12t+35}-\dfrac{1}{t^2-25}-\dfrac{1}{t^2-49}$

34. $\left(\dfrac{1}{a}+\dfrac{1}{a-1}\right)-\left(\dfrac{1}{x-1}-\dfrac{1}{x-2}\right)$

35. $5m-\dfrac{4m^2+7m-1}{2+m}-6$

36. $\left(p+\dfrac{1}{p}+2\right)\left(1-\dfrac{2}{p+1}\right)$

37. $\left(\dfrac{a}{b}+\dfrac{b}{c}+\dfrac{c}{a}\right)\left(\dfrac{abc}{8}\right)$

38. $\left[1-\left(\dfrac{x}{x-y}\right)^2\right]\left[1-\dfrac{x}{y}\right]$

39. $\left(\dfrac{a}{a-b}-\dfrac{2ab}{a^2-b^2}\right)\left(\dfrac{1}{a}-\dfrac{2}{a-b}\right)$

40. $\left(\dfrac{1}{x^2-4x+3}-\dfrac{4}{x^2-9}\right)\left(x+\dfrac{3(3x-1)}{x-7}\right)$

41. $\dfrac{3+\dfrac{1}{4}}{\dfrac{3}{4}-5}$

42. $\dfrac{5}{\dfrac{1}{3}+\dfrac{3}{7}}$

43. $\dfrac{m+\dfrac{1}{n}}{m-\dfrac{1}{n}}$

44. $\dfrac{\dfrac{i^2}{5}-5}{1+\dfrac{i}{5}}$

45. $\dfrac{I}{I-\dfrac{E}{R}}$

46. $\dfrac{\dfrac{E^2}{e^2}-1}{\dfrac{E^2+e^2}{2Ee}}$

47. $\dfrac{1-\dfrac{x-y}{x+y}}{1+\dfrac{x-y}{x+y}}$

48. $\dfrac{\dfrac{I-i}{I+i}+\dfrac{I+i}{I-i}}{\dfrac{I-i}{I+i}-\dfrac{I+i}{I-i}}$

Chapter summary

A thorough understanding of the definitions of exponents is essential so that the reader can solve algebraic problems and for future use in trigonometry, exponential functions, logarithms, and calculus.

Definition 2.1 $a^n = \underbrace{a \cdot a \cdot a \cdot \;\ldots\; \cdot a}_{n \text{ factors}}$ (n is a counting number)

Definition 2.2 $a^n \cdot a^m = a^{n+m}$

Definition 2.3 $(a^n)^m = a^{n \cdot m}$

Definition 2.4 $(a \cdot b)^n = a^n \cdot b^n$

Definition 2.5 $\left(\frac{a}{b}\right)^n = \frac{a^n}{b^n}, \quad b \neq 0$

Definition 2.6 $\frac{a^m}{a^n} = \begin{cases} a^{m-n} & \text{if } m \geq n,\ a \neq 0 \\ \frac{1}{a^{n-m}} & \text{if } m < n,\ a \neq 0 \end{cases}$

Definition 2.7 $a^0 = 1,\ a \neq 0$

Definition 2.8 $\left\{ \begin{matrix} a^{-n} = \frac{1}{a^n} \\ \text{or} \\ \frac{1}{a^{-n}} = a^n \end{matrix} \right\} \quad a \neq 0$

For the scientist and the technician who work with extremely small and large numbers it is convenient to work with these numbers in scientific notation. They should be able to add, subtract, multiply, divide, raise to a power, and take a root of a number in scientific notation. It is also important that they be able to express numbers that are in scientific notation in ordinary decimal notation.

Definition 2.9 A number written in **scientific notation** is written as a product of a number from 1 to 10 (including 1 but not including 10) and a power of 10.

The reader should understand the relationship between exponents and radicals. The definitions of radicals should be understood sufficiently well so that expressions can be written in such a form that operations can be performed and expressions simplified.

Definition 2.10 $x^{1/n} = \sqrt[n]{x}$, where $\sqrt{\ }$ is the radical sign, x is the radicand, and n is the index of the radical.

Definition 2.11 If m and n are both integers with n a positive number and if $\sqrt[n]{a}$ exists as a real number, then $a^{m/n} = \sqrt[n]{a^m} = (\sqrt[n]{a})^m$.

Definition 2.12 $\sqrt{a^2} = |a|$

Definition 2.13 $\sqrt[n]{a \cdot b} = (\sqrt[n]{a})(\sqrt[n]{b})$

Definition 2.14 $\sqrt[n]{\frac{a}{b}} = \frac{\sqrt[n]{a}}{\sqrt[n]{b}}$

Definition 2.15 $\sqrt[m]{\sqrt[n]{a}} = \sqrt[m \cdot n]{a}$

To be able to work with algebraic expressions you must have a working knowledge of the following terms. A **variable** is a letter used to represent an unknown quantity. A **term** is a product or quotient of one or more variables and constants.

Definition 2.16 An algebraic expression is one of the following: (a) a constant, (b) a variable, (c) a combination formed by performing one or more algebraic operations (addition, subtraction, multiplication, division, raising to a power, or taking a root), on nonzero constants or variables.

The algebraic expression $2x + 3xy/z$ consists of two terms $2x$ and $3xy/z$. The **literal coefficient** of the term $2x$ is x and the **numerical coefficient** of the term is 2. The term $z(x + 3)$ has **factors** z and $x + 3$.

Definition 2.17 Two or more terms that have the same literal coefficients are called **like terms.**

In working with algebraic expressions it is important to understand the order in which operations are performed. The tools that are used to show order of operations are **parentheses** ( ), **brackets** [ ], and **braces** { }.

In Section 2.5 the techniques of multiplying two monomials, a monomial and a binomial, two binomials, and a binomial and trinomial are developed.

In general:

$$a(b + c) = ab + ac$$
$$(x + a)(x + b) = x^2 + (a + b)x + ab$$
$$(x + a)(x + a) = x^2 + 2ax + a^2$$
$$(x + a)(x - a) = x^2 - a^2$$

The division of algebraic expressions can be performed in a manner similar to long division. There are several different forms of factoring:

$ax + ay = a(x + y)$	**Common-term**
$x^2 + 2xy + y^2 = (x + y)^2$	**Perfect-square trinomial**
$x^2 - y^2 = (x - y)(x + y)$	**Difference of two squares**
$x^2 + 5x + 6 = (x + 2)(x + 3)$	**Trinomial**

In finding the product of two or more fractions it is better to simplify (reduce) before multiplying the numerators and the denominators.

In finding the quotient of two fractions invert the divisor and then perform the operation of multiplication.

In finding the sum or difference of two fractions we must first find a common denominator.

To simplify a complex fraction, reduce both numerator and denominator to simple fractions and then perform the indicated division.

Review Exercises

In Exercises 1–9, simplify each expression by using the rules of exponents.

1. $a^3 \cdot a^{13}$ **2.** $b^5 \cdot b^{15}$ **3.** $(x^4y^3)^7$

4. $\left(\frac{m^7}{n^5}\right)^3$ **5.** $\left(\frac{a^3 \cdot b^4}{c^5}\right)^0$ **6.** $\left(\frac{1}{p^{-4}}\right)^7$

7. $\frac{35x^4y^{11}}{7x^3y^6}$ **8.** $\frac{32a^4b^9}{(4a^2b^5)^2}$ **9.** $\left(\frac{2m^3n^4}{pq}\right)^{-3}$

In Exercises 10–16, write the numbers in scientific notation.

10. 60,000 **11.** 3250 **12.** 13.3

13. 0.4 **14.** 0.0027 **15.** 0.00000579

16. The human eye is normally capable of responding to light waves whose lengths are in the range of 0.000038 to 0.000076 cm.

In Exercises 17–23, write the numbers in ordinary decimal notation.

17. 7.052×10^6 **18.** 9.931×10^{-4}

19. 4.32×10^0 **20.** 5.3×10^{-4}

21. 4.62×10^{-5} **22.** 3.45675×10^3

23. One gram is equal to 3.5×10^{-2} oz.

In Exercises 24–29, solve the problems using scientific notation. Write your answer in scientific notation.

24. $10^5 \times 10^9 \times 10^{-4}$ **25.** $\frac{10^{13} \times 10^{15}}{10^7 \times 10^9}$

26. $25{,}000 \times 0.002 \times 500 \times 0.00004$

27. The coefficient of linear expansion indicates the fractional change in the length of a rod per degree change in temperature. For tungsten the coefficient is 4.4×10^{-6} per degree Celsius. What is the fractional change in the length of a tungsten rod when the temperature rises from 10°C to 30°C?

28. A light year is the distance light can travel in a year's time. One light year is approximately 6×10^{12} mi. If a star is 375 light years away from earth, how far is it in miles?

29. One electron volt is the kinetic energy an electron acquires when it is accelerated in an electric field produced by a difference of potential of 1 volt. 1 electron volt $= 1.60 \times 10^{-19}$ J. How many joules equal five million electron volts?

In Exercises 30–35, use the definitions for exponents and radicals to simplify the expressions.

30. $\sqrt{27}$ **31.** $\sqrt[3]{32}$ **32.** $\sqrt[3]{a^{12}}$

33. $\sqrt[3]{\frac{129x^0 - y^0}{x^0 + (7b)^0}}$ **34.** $\sqrt[n]{a^{3n}}$ **35.** $\sqrt{\sqrt[3]{y^{6n}}}$

In Exercises 36–43, perform the indicated operations and simplify the results.

36. $\sqrt{7} + 2\sqrt{7}$ **37.** $4\sqrt{72} - 8\sqrt{50} + 3\sqrt{32}$

38. $\sqrt{m} + \sqrt{m^5} + \sqrt{m^7}$ **39.** $\sqrt{11} \cdot \sqrt{13}$

40. $\sqrt{13} \cdot \sqrt{26}$ **41.** $\sqrt{abc} \cdot \sqrt{abd}$

42. $\sqrt[3]{a^2bc^4} \cdot \sqrt[3]{ab^2c^5}$ **43.** $\frac{\sqrt{18x^2y^3}}{\sqrt{2xy}}$

In Exercises 44–53, simplify the following algebraic expressions.

44. $5x - 8x + 73x$

45. $3xy^2 + 7xy^2 - 18xy^2$

46. $6r - 7s + 14s - 8r$

47. $-4\phi\pi + 11RI - 2\phi\pi + 5RI$

48. $\frac{2}{3}a^2 - \frac{1}{7}a^2 + \frac{5}{6}a^2$

49. $\frac{2}{5}b^2 + \frac{7}{15}b^2 - \frac{11}{35}b^2$

50. $3x - (4x + 7)$

51. $-7x - (11x - 13)$

52. $4m - 6[3n + 8m - 5(n - m)]$

53. $8p - \{6q + 3[2p - 7q + (13q - 11p)]\}$

In Exercises 54–81, perform the indicated operation and simplify the expression.

54. $(4a^2)(7a^3)$ **55.** $(11r^2s^2)(-13r^3s^4)$

56. $72m^2n^3 \div 12mn^2$ **57.** $48p^4q^2 \div (-16p^3q)$

58. $\frac{8r}{9s} \cdot \frac{27s^3}{56r^2}$ **59.** $\frac{m+2}{m+3} \cdot \frac{m+4}{m-5}$

60. $\frac{3m}{5n} \div \frac{6m^2}{15n^3}$ **61.** $\frac{xy}{9xy} \div \frac{(xy)^2}{3x}$

62. $\frac{1}{x+2} + \frac{3}{x+3}$ **63.** $\frac{2x}{y^2} - \frac{2y}{x^2}$

64. $(3m - 4n)(2m + 7n)$

65. $(5R - 4)(3R^2 + 11R - 9)$

66. $(5x - 11y)^2$

67. $(x + 4)(x - 3)(x + 5)$

68. $(4pq - 8pq^2) \div 2pq$

69. $(102a^2y - 51a^3y) \div (-17a^2)$

70. $(8m^2 - 2mn - 21n^2) \div (2m + 3n)$

71. $(a^3 - 8a + 5a^2 - 12) \div (a^2 - a - 2)$

72. $\dfrac{m^2 - 36n^2}{6mn} \cdot \dfrac{3m}{m^2 - 12mn + 36n^2}$

73. $\dfrac{x^2 + 4x - 21}{x^2 + 2x - 15} \cdot \dfrac{x^2 + 6x + 5}{x^2 + 8x + 7}$

74. $\dfrac{x + 7}{x + 9} \div \dfrac{xy + 7y}{x^2 + 18x + 81}$

75. $\dfrac{8x^2 - 2}{6x^2 + x - 1} \div \dfrac{2x^2 + 7x - 4}{3x^2 + 11x - 4}$

76. $\dfrac{4x^2}{7y} \div \left(\dfrac{4x}{3z} \cdot \dfrac{2x^2}{21y}\right)$

77. $\dfrac{2R^2}{3I} \cdot \left(\dfrac{6R}{7I} \div \dfrac{12R^2}{35I^3}\right)$

78. $\dfrac{x - 2}{x - 4} + \dfrac{x - 3}{x + 4} - \dfrac{x^2}{x^2 - 16}$

79. $\dfrac{1}{2m + 2} + \dfrac{5}{3m - 3} - \dfrac{3m - 1}{1 - m^2}$

80. $\left(m + \dfrac{2}{m} + 3\right)\left(1 - \dfrac{2}{m + 1}\right)$

81. $\dfrac{\dfrac{x^2 + 1}{x} - 2}{\dfrac{x^2 + 3x + 5}{x} + 3}$

In Exercises 82–90, write each expression as a product of prime factors.

82. $3xy + 12yz + 5xw + 20zw$

83. $6x^2yw - 14xwz - 15x^2y^2 + 35xyz$

84. $6w^2 + 31w + 35$

85. $6r^2 - 6s^2$

86. $4x^2 + 3x - 10$

87. $64x^2 - 8x - 2$

88. $(x + y)^2 - z^2$

89. $10a^2 - 3ab - 18b^2$

90. $(x^2 + 25)^2 - 25y^2$

Can you solve these?

1. Is $(10^4)^3 = 10^{4^3}$? Explain your answer using rules of exponents and order of operations.
2. Describe a technique for finding $(1.25)^{13}$ with a calculator that has only an $\boxed{x^2}$ key (and does not have a $\boxed{y^x}$ key).
3. Use the $\boxed{y^x}$ key on a scientific calculator with algebraic logic to find the value of $(-5)^4$. Can you explain why the calculator indicates that you have committed an error? Show how you can obtain the answer with the aid of the calculator.
4. The half-life n and the rate of decay k of a radioactive substance are related by the equation $k^n = \dfrac{1}{2}$.
   a. Solve for k.
   b. The isotope tritium, which occurs naturally in water and has been used in dating wines, has a half-life of 12.4 years. Find the decay factor for tritium.
5. Prove: $\sqrt[m]{\sqrt[n]{a}} = \sqrt[mn]{a}$
6. Given: $\sqrt{5 + 2\sqrt{6}} = \sqrt{2} + \sqrt{3}$
   a. Verify on a calculator.
   b. Prove the relation.
7. Factor $x^2 + 2xy - 8y^2$ by writing $-8y^2$ in the form $y^2 - 9y^2$.
8. Consider two consecutive positive integers. Which is larger, the average of their squares or the square of their average?

3

The first stage thrust of the Saturn 5 rocket varied between 7,680,000 lb at ignition to 9,115,000 lb just before burnout. The weight of the rocket and the command module at launch was 6,391,120 lb. According to Newton's law the acceleration of an object is equal to the net force in the direction of the acceleration divided by its mass [$a = (F/m)$]. The mass of an object in the English system of units is expressed in terms of the slug, which is the weight of the object in pounds divided by the local acceleration of gravity in ft/s^2, or $m - \dfrac{w}{g}$. (The value of g near the surface of the earth is usually taken as 32.2 ft/s^2.)

The net upward force on the unit at ignition would be the thrust of its engines less its weight. What is the acceleration of the unit at ignition? What is this acceleration expressed in gs?

Linear Equations and Dimensional Analysis

3.1 Solving linear equations

In our review of algebraic concepts in Chapter 2 we worked with algebraic expressions. If we take two expressions and place an equal sign between the two expressions, we have an **equation.**

EXAMPLE 1

a. $x^2 + 5x + 6$ is an expression.

b. $2x + 3 = x + 4$ is an equation.

c. $4x - 12 = 4(x - 3)$ is an equation.

d. $\dfrac{x+4}{3} = \dfrac{x+5}{2}$ is an equation.

e. $3x + 2 = 3(x + 1)$ is an equation. □

An equation is a tool we use in solving problems. The question being asked by the equation is "What value (values) may be substituted in place of the variable (the letter) that will make the equation a true statement of equality?" In part (b) of Example 1 it can be shown that if we let $x = 1$ the expression on the right side and the expression on the left side will have the same value, 5. There are several accepted methods of indicating that 1 is a solution to the equation: (a) stating that the solution set contains the element 1, indicated $\{1\}$; (b) stating that 1 satisfies the equation; (c) stating that 1 is a **root** of the equation.

The equation $3x + 2 = 3(x + 1)$ is an example of an equation that has no roots. There are no numbers that can be substituted for the variable x and make the equation a true statement of equality.

The equation $4x - 12 = 4(x - 3)$ belongs to a special group of equations called **identities.** It is called an identity because the statement is true for all permissible values of the variable x.

Our challenge will be to determine the root(s) of a given equation. To solve an equation you write a series of equivalent equations until you arrive at the solution. By equivalent equation we mean that the form of the equation has been changed, but the solution set is the same. This series of equivalent equations is obtained by performing the same operations on both sides of the equation. There is one exception to the previous statement. We can never divide by zero.

EXAMPLE 2 Find the solution for the equations.

a. $x - 7 = 5$ **b.** $x + 3 = 2$

c. $\dfrac{x}{3} = 4$ **d.** $7x = 35$

Solution: To solve each of these equations it is necessary to isolate the variable on one side of the equal sign.

a. For the equation $x - 7 = 5$: to isolate x we need to add 7 to both sides of the equation.

$$x - 7 + 7 = 5 + 7 \qquad \textbf{Addition property}$$
$$x = 12$$

b. For the equation $x + 3 = 2$: to isolate x we need to subtract 3 from both sides of the equation.

$$x + 3 - 3 = 2 - 3 \qquad \textbf{Subtraction property}$$
$$x = -1$$

c. For the equation $\frac{x}{3} = 4$: to isolate x we need to multiply both sides by 3 $\left(\text{the reciprocal of } \frac{1}{3}\right)$.

$$\frac{x}{3} \cdot 3 = 4 \cdot 3 \qquad \textbf{Multiplication property}$$
$$x = 12$$

d. For the equation $7x = 35$: to isolate x we need to divide both sides by 7 (multiply by the reciprocal of 7).

$$\frac{7x}{7} = \frac{35}{7} \qquad \textbf{Division property}$$
$$x = 5 \quad \square$$

Rules for Solving Equations

Addition property:	If $a = b$, then $a + c = b + c$ for all real numbers a, b, and c.
Subtraction property:	If $a = b$, then $a - c = b - c$ for all real numbers a, b, and c.
Multiplication property:	If $a = b$, then $a \cdot c = b \cdot c$ for all real numbers a, b, and c.
Division property:	If $a = b$, then $a/c = b/c$ for all real numbers a, b, and c, $c \neq 0$.

EXAMPLE 3 Find the solution for the equations. Check your solution by replacing the variable in the original equation with the solution.

a. $2x + 5 = 17$ **b.** $2 = 5 + 7(p + 1)$

Solution:

a. $2x + 5 = 17$

$2x + 5 - 5 = 17 - 5$ **Subtract 5 from both sides of the equation (Subtraction property)**

$2x = 12$

$\frac{2x}{2} = \frac{12}{2}$ **Divide both sides of the equation by 2 (Division property)**

$x = 6$

Check

$2x + 5 = 17$

$2(6) + 5 = 17$

$12 + 5 = 17$

$17 = 17$ ✓

b. $2 = 5 + 7(p + 1)$

$2 = 5 + 7p + 7$ **Distributive property**

$2 = 5 + 7 + 7p$ **Commutative property**

$2 = 12 + 7p$

$2 - 12 = 12 - 12 + 7p$ **Subtract 12 from both sides of the equation**

$-10 = 7p$

$\frac{-10}{7} = \frac{7p}{7}$ **Divide both sides of the equation by 7**

$\frac{-10}{7} = p$

Check

$2 = 5 + 7(p + 1)$

$2 = 5 + 7\left(\frac{-10}{7} + 1\right)$

$2 = 5 - 10 + 7$

$2 = 2$ ✓ □

EXAMPLE 4 Find the solution for the equations. Check your solution by replacing the variable in the original equation with the solution.

$$\frac{2t}{3} + \frac{1}{3} = \frac{1}{4}$$

Solution: A technique used to simplify problems with fractions is to multiply each term by the least common denominator (lcd). In this example the lcd is 12.

$$\frac{2t}{3} + \frac{1}{3} = \frac{1}{4}$$

$$\overset{4}{\cancel{12}}\left(\frac{2t}{\underset{1}{\cancel{3}}}\right) + \overset{4}{\cancel{12}}\left(\frac{1}{\underset{1}{\cancel{3}}}\right) = +\overset{3}{\cancel{12}}\left(\frac{1}{\underset{1}{\cancel{4}}}\right)$$ **Multiply each term by lcd and cancel**

$$8t + 4 = 3$$ **Now we have an equation without fractions and can proceed as in the previous examples**

$$8t + 4 - 4 = 3 - 4$$ **Subtract 4 from both sides**

$$8t = -1$$

$$\frac{8t}{8} = \frac{-1}{8}$$ **Divide both sides by 8**

$$t = \frac{-1}{8}$$

Check

$$\frac{2t}{3} + \frac{1}{3} = \frac{1}{4}$$

$$\frac{\overset{1}{\cancel{2}}}{3}\left(\frac{-1}{\underset{4}{\cancel{8}}}\right) + \frac{1}{3} = \frac{1}{4}$$

$$\frac{-1}{12} + \frac{1}{3} = \frac{1}{4}$$

$$\frac{-1}{12} + \frac{1}{3} \cdot \frac{4}{4} = \frac{1}{4} \cdot \frac{3}{3}$$

$$\frac{-1}{12} + \frac{4}{12} = \frac{3}{12}$$

$$\frac{3}{12} = \frac{3}{12} \checkmark$$ □

EXAMPLE 5 Solve the equation and check your solution:

$$4x - 0.48 = 0.8x + 4.$$

Solution: This problem may be solved with the decimals, or you may multiply each term by 100 and eliminate the decimals. We will solve the problem with the decimals.

$$4x - 0.48 = 0.8x + 4$$

$$4x - 0.48 + 0.48 = 0.8x + 4 + 0.48$$ **Add 0.48 to both sides of the equation**

$$4x = 0.8x + 4.48$$

$$4x - 0.8x = 0.8x - 0.8x + 4.48$$ **Subtract 0.8x from both sides of the equation**

$$3.2x = 4.48$$

$$\frac{3.2x}{3.2} = \frac{4.48}{3.2}$$ **Divide both sides of the equation by 3.2**

$$x = 1.4$$

Check

$$
\begin{aligned}
4x - 0.48 &= 0.8x + 4 \\
4(1.4) - 0.48 &= 0.8(1.4) + 4 \\
5.6 - 0.48 &= 1.12 + 4 \\
5.12 &= 5.12 \checkmark \quad \square
\end{aligned}
$$

We sometimes encounter equations in which the variable is in the denominator of a fraction. Example 6 will illustrate how to handle that situation.

EXAMPLE 6 Solve the equation and check your solution.

$$\frac{3}{x} + 5 = \frac{2}{x}$$

Solution: In this problem x cannot be equal to zero. (Why?) Since $x \neq 0$, we can treat x as a real number and multiply both sides of the equation by x.

$$\frac{3}{x} + 5 = \frac{2}{x}$$

$$x\left(\frac{3}{x} + 5\right) = x\left(\frac{2}{x}\right)$$ **Multiply both sides of the equation by x**

$$x\left(\frac{3}{x}\right) + x(5) = x\left(\frac{2}{x}\right)$$

$$3 + 5x = 2$$

$$3 - 3 + 5x = 2 - 3$$ **Subtract 3 from both sides of the equation**

$$5x = -1$$

$$\frac{5x}{5} = -\frac{1}{5}$$ **Divide both sides of the equation by 5**

$$x = -\frac{1}{5}$$

Check

$$
\begin{aligned}
\frac{3}{x} + 5 &= \frac{2}{x} \\
\frac{3}{-1/5} + 5 &= \frac{2}{-1/5} \\
\frac{15}{-1} + 5 &= \frac{10}{-1} \\
-15 + 5 &= -10 \\
-10 &= -10 \checkmark \quad \square
\end{aligned}
$$

Often in mathematics and science courses we are given an equation expressed in terms of several variables and asked to express it in terms of one of those variables. To do this, treat each of the variables, except the one for which you are solving, as if they were constants. Then solve for the variable desired using the properties previously discussed. (These equations are called **literal equations.**)

EXAMPLE 7 The perimeter of a rectangle is given by the formula $P = 2\ell + 2w$. Solve for the length, ℓ.

Solution: To solve for ℓ we isolate ℓ on one side of the equal sign, placing everything else on the other side.

$$P = 2\ell + 2w$$

$$P - 2w = 2\ell + 2w - 2w$$ **Subtract $2w$ from both sides of the equation**

$$P - 2w = 2\ell$$

$$\frac{P - 2w}{2} = \frac{2\ell}{2}$$ **Divide both sides of the equation by 2**

$$\frac{P - 2w}{2} = \ell$$

or

$$\frac{P}{2} - w = \ell$$ □

3.1 Exercises

In Exercises 1–28, find the solution for each equation and check.

1. $r + 5 = 11$
2. $3x - 6 = 24$
3. $16 = 6n + 4$
4. $11 = 3 - 4R$
5. $12 = 14 + 2I$
6. $15m - 7 = -37$
7. $\frac{1}{2}x + \frac{1}{3} - \frac{2}{3}$
8. $\frac{1}{2}y + \frac{1}{3} = \frac{1}{4}$
9. $0.3a + 4.2 = 3.6$
10. $3r + 6 = 2r + 15$
11. $5x + 6 = 2x + 15$
12. $\frac{R}{4} + 2R = -\frac{2}{3}$
13. $27 - 6n = 4n - 7n$
14. $\frac{2}{3}(I + 4) = 5 + 3I$
15. $\frac{15}{y} + 4 = 9$
16. $12x - 1.2 = 3x + 1.5$
17. $\frac{4}{3}(R + 3) = 36$
18. $6s + 3(4 + s) = 8$
19. $\frac{1}{R} + \frac{2}{R} + \frac{3}{R} = 2$
20. $\frac{5}{y} + 1 = \frac{6}{y}$
21. $4.08 - 6m = 5.2 - 4m$
22. $\frac{2t - 1}{2t} = \frac{3}{t} + \frac{1}{2}$
23. $\frac{1}{a - 1} = \frac{2}{a - 2}$
24. $\frac{3}{x - 2} = \frac{2}{x - 3}$
25. $\frac{2x}{x - 5} = \frac{10}{x - 5}$
26. $\frac{6x}{x - 3} = \frac{18}{x - 3}$
27. $x(x - 3) + 4 = x^2 - 2(x + 4)$
28. $(x - 1)(x - 2) = (x + 3)(x - 4) - 3(x - 1)$

In Exercises 29–40, solve each of the formulas for the indicated letter.

29. $pv = c$ for p

30. $I = PRT$ for T

31. $A = \frac{1}{2}bh$ for b

32. $A + B + C = 180$ for C

33. $P = 2\ell + 2w$ for w

34. $s = \frac{1}{2}gt^2$ for g

35. $C = \frac{5}{9}(F - 32)$ for F

36. $A = \frac{1}{2}(b_1 + b_2)h$ for b_1

37. $V = \frac{1}{3}\pi r^2 h$ for h

38. $A = P(i + rt)$ for r

39. $F = \dfrac{G\,m_1 m_2}{r^2}$ for m_2

40. $S = \dfrac{a - r\ell}{1 - r}$ for r

41. It is necessary to reinforce concrete with steel mesh or rebars in order to have a floor or slab without joints. A formula that can be used to determine the type of steel mesh is

$$A = \frac{12tP}{100}.$$

A is the steel mesh cross-sectional area in square inches, t is the concrete's actual thickness in inches, and P is the percentage of steel required (15% for crack containment, 5% for reinforcement). Solve for t.

42. A formula used to determine the spacing for rebars in concrete is

$$S = \frac{100A}{Pt}.$$

S is the spacing between bars in inches, A is the rebar's cross-sectional area in square inches, t is the thickness of concrete in inches, and P is the percentage of steel required. Solve for P.

43. In working with fluid power a basic measurement for overall heat loss is to subtract useful energy from input energy:

$$E_L = E_1 - E_2.$$

The relationship may also be written:

$$E_L = 1.48QP(1 - \mu)$$

Where μ is system efficiency, Q is pump flow (gpm), and P is pump discharge pressure (psig). Solve for P.

44. The formula $I_1 + \frac{I_2}{2} + \frac{S_1}{\Delta t} - \frac{Q_1}{2} = \frac{S_2}{\Delta t} + \frac{Q_2}{2}$ was developed to indicate the water-storage levels in a flood-control area. I_1 and I_2 are inflow rates at times 1 and 2, Q_1 and Q_2 are outflow rates at times 1 and 2, S_1 and S_2 are storage values at times 1 and 2, and Δt is the time interval (time 2 − time 1).

a. Solve for S_1.
b. Solve for S_2 when I_1, Q_1, and S_1 are equal to zero.

3.2 Significant digits

The packing machine places 12 bottles of liquid detergent in the shipping carton.

Tom's measurement tells him that he needs a bolt 30. mm long to fasten a bracket to a chair.

In the first situation, 12 is an **exact** number, it tells us precisely how many bottles are placed in the carton. Generally, in counting, when defined, a number will be an exact number.

In the second situation, 30. mm is an **approximate** number. In general, all measurements are an approximation of an unknown number. We know that Tom's measurement was made to the nearest whole millimeter because directly after the 30. there is a decimal point. Thus we expect that the absolute value of the error would be less than 0.5 mm, or that the actual length lies somewhere between 29.5 mm and 30.5 mm.

There are three concepts that must be understood in order to work with approximate numbers:

1. How to round off approximate numbers expressed in decimal form.
2. How to determine the number of significant digits in an approximate number.
3. How to obtain the most accurate result when performing the four basic arithmetic operations on approximate numbers.

Rules for Rounding Off

Find the digit that is to be rounded, then:

Rule 1 If the digit to the right is less than 5, leave the digit in the place to be rounded unchanged and drop all the digits to the right.

Rule 2 If the digit to the right is 5 or greater, then add 1 to the digit in the place to be rounded and drop all the digits to the right.

EXAMPLE 1 Round off the following numbers to the nearest tenth.

a. 396.41 **b.** 396.452 **c.** 396.45
d. 396.48 **e.** 396.450 **f.** 396.35

Solution:

a. 396.41 to the nearest tenth is 396.4. **Rule 1**
b. 396.452 to the nearest tenth is 396.5. **Rule 2**
c. 396.45 to the nearest tenth is 396.5. **Rule 2**
d. 396.48 to the nearest tenth is 396.5. **Rule 2**
e. 396.450 to the nearest tenth is 396.5. **Rule 2**
f. 396.35 to the nearest tenth is 396.4. **Rule 2** □

Number of significant digits

The numbers 567, 56.7, 5670, and 0.000567 all have three significant digits, namely 5, 6, and 7. But 5607, 567.0, 56.70, and 0.0005670 have four significant digits, namely 5, 6, 7, and 0.

> In an approximate number a digit is significant if it is not a zero, if it is a zero between two nonzero digits, or if it is a terminal zero on the right side of the decimal point.

One problem we encounter with the above definition is determining the number of significant digits in a number such as 7500 mm. If the length is 7500 mm to the nearest hundred millimeters, then there are two significant digits. On the other hand if the length is 7500 mm to the nearest millimeter, then there are four significant digits. In order to determine whether or not the zeros are significant digits, we must know something about the number. In some textbooks a line is placed under or over the last significant zero. We will use the convention of underlining. For example, in the number $750\underline{0}$ the line under the zero indicates there are four significant digits: 7, 5, 0, and 0. Another way of indicating the number of significant digits is to write the number in scientific notation and retain the zeros that are significant digits. For example, 7500 would be written as 7.500×10^4 indicating four significant digits.

Calculating with approximate numbers

In calculating with approximate numbers the answer cannot be more accurate than the least accurate number in the data. However, it is advisable to carry one more place throughout the computation whenever possible and do the rounding at the end.*

EXAMPLE 2 Perform the indicated operations.

a. $3.43 + 1.124$ **b.** $5.42 - 2.56789$

Solution:

a. The sum of the approximate numbers 3.43 and 1.124 should be written as 4.55 rather than 4.554, since 3.43 is expressed to the nearest hundredth and 1.124 is expressed to the nearest thousandth. Thus, the sum is 4.554, which is then rounded to 4.55, the same as the least number of significant digits in either number.

$$\begin{array}{l} 3.43 \\ \underline{1.124} \\ 4.554 \text{ or } 4.55 \end{array}$$

* These are the rules most commonly followed by the technician. More precise rules are available.

b. The number 5.42 is expressed to the nearest hundredth while the number 2.56789 is expressed to the nearest hundred thousandth. We round the second number to the nearest thousandth or 2.568 and perform the subtraction,

$$\begin{array}{r} 5.420 \\ -2.568 \\ \hline 2.852 \end{array} \text{ or } 2.85$$

then round the answer to the nearest hundredth because 5.420 represents 5.42 for the purposes of subtraction. Therefore the answer is 2.85. □

EXAMPLE 3 Find the area of a desk top that is 0.75 m wide and 1.42 m long.

Solution: The area is found by using the formula $A = \ell w$.

$$\begin{aligned} A &= (0.75 \text{ m})(1.42 \text{ m}) \\ &= 1.065 \text{ m}^2 \quad \text{or} \quad 1.1 \text{ m}^2 \end{aligned}$$

The area of the desk top is 1.1 m^2 since the number of significant digits in 0.75 m is two and in 1.42 m is three. The area will be rounded to two significant digits. Thus $A = 1.1 \text{ m}^2$. □

EXAMPLE 4 The formula for the volume of a cylinder is $V = \pi r^2 h$. If the volume, V, is 12. in^3, and the radius, r, is 2. in., what is the height, h?

Solution: Solving $V = \pi r^2 h$ for h we have

$$h = \frac{V}{\pi r^2}.$$

Substituting the values $V = 12.\ \text{in}^3$, $r = 2.$ in., and $\pi = 3.1$ in the formula:

$$\begin{aligned} h &= \frac{12.\ \text{in}^3}{(3.1)(2.\ \text{in})^2} \\ &= 0.96 \text{ in.} \\ &= 1. \text{ in.} \end{aligned}$$

Why not 3.14159 for π? Since r has the least number of significant digits, 1, we should use at most one more for π, thus 3.1. To 1 significant digit $h = 1.$ in. Notice that in using the calculator the accuracy of the final answer in this problem would not have been affected if you had used π rather than 3.1. □

There are equations that contain numbers that are exact. For example, in the equation $V = 4/3\pi r^3$ the number 4/3 is exact and so is the 3.

We will indicate approximate numbers with a decimal point, i.e., 3.4 in., or 3. in., or 30. in. are all approximate numbers, whereas 3 in. or 30 in. are exact numbers. In a problem that contains all exact numbers and a constant such as π, we will use the value of π as determined by the calculator. In a problem that contains exact numbers, approximate numbers, and constants, accuracy will be obtained by calculating using one more digit than is in the least accurate number.

EXAMPLE 5 If two resistors R_1 and R_2 are hooked in parallel, the equivalent resistance R is given by the formula

$$R = \frac{R_1 \cdot R_2}{R_1 + R_2}.$$

If $R_1 = 2.3475$ and $R_2 = 1.3456$, find R.

Solution: With the given values of R_1 and R_2 it is preferable to use a calculator.

$$R = \frac{R_1 \cdot R_2}{R_1 + R_2}$$

$$= \frac{(2.3475)(1.3456)}{2.3475 + 1.3456} \quad \textbf{Substituting}$$

The key strokes on the calculator are

2.3475 [×] 1.3456 [÷] [(] 2.3475 [+] 1.3456 [)] [=] 0.8553237

Thus the solution to the correct number of significant digits is 0.85532. □

3.2 Exercises

In Exercises 1–10, round off the numbers to the decimal place indicated in parentheses.

1. 476.1 (tens)
2. 56.7 (units)
3. 345.2 (tens)
4. 2.34 (tenths)
5. 86.5 (units)
6. 46.75 (tenths)
7. 0.04655 (ten thousandths)
8. 475,000 (ten thousands)
9. 4615 (hundreds)
10. 0.055 (hundredths)

In Exercises 11–16, write each of the numbers in scientific notation (indicating significant digits).

11. $240{,}\underline{0}00$
12. $24{,}\underline{0}00$
13. 0.00450
14. $430{,}0\underline{0}0$
15. 230,000,000
16. 0.000203

In Exercises 17–26, perform the indicated operations. The numbers are approximate numbers.

17.
$$\begin{array}{r} 234.5 \\ 18.734 \\ +567.04 \\ \hline \end{array}$$

18.
$$\begin{array}{r} 56.941 \\ 0.81 \\ +176.4 \\ \hline \end{array}$$

19. $524.1 - 73.456$
20. $276.45 - 1.0476$
21. $34.75 - 6.187 + 87.942 - 605.703$
22. $106. - 90. + 764. - 375.$
23. $(76.4)(3.25)$
24. $(24\underline{0}0.)(104.)$
25. $264. \div 0.05$
26. $1001. \div 13.$

In Exercises 27–35, the measures are approximate numbers.

27. What is the total mass of four pieces of plate glass if the masses of the four pieces are 34.5 kg, 4.75 kg, 376.5 kg, and 10.04 kg?

28. What is the total length of three cells that are 0.034 mm, 0.0306 mm, and 0.0335 mm long?

29. If the circumference of a bicycle wheel is 81.681 in., what is the approximate diameter of the wheel ($c = \pi d$)?

30. A rocket is traveling at the rate of 7.8642 km/hr. How far will the rocket travel in 8. hrs ($d = rt$)?

31. Find R when $R_1 = 1.0034$ and $R_2 = 3.47621$ (see Example 5).

32. One solution to the quadratic equation $6.82x^2 + 8.72x - 7.13 = 0$ is

$$x = \frac{-(8.72) + \sqrt{(8.72)^2 - 4(6.82)(-7.13)}}{2(6.82)}.$$

Solve for x.

33. For $P = \dfrac{100\ R}{(0.500 + R)^2}$ and $R = 0.2514$, find P.

34. Stress in a beam is given by $S = \dfrac{My}{I}$. If $M = 19{,}200$, $y = 3.25$, and $I = 36.$, find S.

35. Given $I = \dfrac{400}{r^2}$. Find illuminance (a) if $r = 2.071$ m and (b) if $r = 2.831$ m.

3.3 Mathematical operations with dimensional units

There are two essential components in the solutions of technical problems: the numerical value and the dimension (units of measurement). For example, if the answer to a problem is 3. ft and the answer 3. is given, there is no way of determining whether it is 3. cm, 3. mi, 3. m, or 3. gal.

There are two systems of measurement used today. They are the **British System** and the **International System of Units (SI),** commonly referred to as the metric system. Today, the United States is the only industrial country in the world not using the SI system. Most of the major industries use the SI system for the purpose of international trade. Thus, even though it is not official, we are gradually moving toward the use of the SI system. **You will find tables and a discussion of both systems in Appendix A.**

In both systems of measurement there are basic dimensional units and all other units are derived from these basic units. Some of the basic units for both systems are given in Table 3.1. If the quantity symbol in Table 3.1 is treated as an algebraic quantity, then we can show how additional dimensional units can be derived (see Table 3.2).

In working with dimensional units it is frequently necessary to change from one unit to another in order to be able to perform the calculations. For example, it may be necessary to change 45. miles per

TABLE 3.1 **Quantities and their associated base units**

		British Units		SI (Metric) Units	
Quantity	Symbol	Name	Symbol	Name	Symbol
Length	L	foot	ft	meter	m
Mass (SI base unit)	M,m	slug		kilogram	kg
Force (British base unit)	F	pound	lb	newton	N
Time	t	second	s	second	s
Electric current	I,i	ampere	A	ampere	A
Temperature	T	fahrenheit	°F	celsius	°C
Amount of substance	n	mole	mol	mole	mol
Luminous intensity	I	candle power	cp	candela	cd

hour to kilometers per hour or cubic inches to cubic feet. To help do these conversions it is essential to identify unit fractions that we will call **conversion factors.** Examples of conversion factors are

$$\frac{5280 \text{ ft}}{1 \text{ mi}} \quad \frac{36 \text{ in.}}{1 \text{ yd}} \quad \frac{1 \text{ km}}{1000 \text{ m}} \quad \frac{1 \text{ m}^3}{1 \text{ k}\ell} \quad \frac{1000 \text{ g}}{1 \text{ kg}}.$$

TABLE 3.2

Quantity	Dimensional Unit	British Unit	SI Unit
Area = (length)^2	$L \times L = L^2$	ft^2	m^2
Volume = (length)^3	$L \times L \times L = L^3$	ft^3	m^3
Weight density = $\frac{\text{weight}}{\text{volume}}$	$\frac{F}{L^3}$	$\frac{\text{lb}}{\text{ft}^3}$	$\frac{\text{N}}{\text{m}^3}$
Mass density = $\frac{\text{mass}}{\text{volume}}$	$\frac{M}{L^3}$	$\frac{\text{slug}}{\text{ft}^3}$	$\frac{\text{kg}}{\text{m}^3}$
Velocity = $\frac{\text{distance}}{\text{time}}$	$\frac{L}{t}$	$\frac{\text{ft}}{\text{s}}$	$\frac{\text{m}}{\text{s}}$
Acceleration = $\frac{\text{change in velocity}}{\text{time}}$	$\frac{L}{t^2}$	$\frac{\text{ft}}{\text{s}^2}$	$\frac{\text{m}}{\text{s}^2}$
Pressure = $\frac{\text{force}}{\text{area}}$	$\frac{F}{L^2}$	$\frac{\text{lb}}{\text{ft}^2}$	Pa
Work = distance × force	$L \cdot F$	ft · lb	J
Power = $\frac{\text{work}}{\text{time}}$	$\frac{L \cdot F}{t}$	$\frac{\text{ft} \cdot \text{lb}}{\text{s}}$	W
Volt = $\frac{\text{work}}{\text{charge}}$	$\frac{L \cdot F}{t \cdot I}$	V	V

These conversion factors are identities such as 12 in. = 1 ft;

$$\text{thus } \frac{12\text{ in.}}{12\text{ in.}} = \frac{1\text{ ft}}{12\text{ in.}}$$

$$1 = \frac{1\text{ ft}}{12\text{ in.}}$$

Also note that

$$\frac{12\text{ in.}}{1\text{ ft}} = \frac{1\text{ ft}}{1\text{ ft}}$$

$$\frac{12\text{ in.}}{1\text{ ft}} = 1.$$

When multiplying by one of these conversion factors we are multiplying by the multiplicative identity, 1. Therefore even though we have changed the form of the expression, we have not changed its value.

EXAMPLE 1 Convert 45. ft^3 to cubic yards.

Solution: To convert ft^3 to yd^3 we need a unit fraction expressing yards in terms of feet. The fraction we need is $\frac{1\text{ yd}}{3\text{ ft}}$. It is necessary to multiply $\frac{1\text{ yd}}{3\text{ ft}} \times \frac{1\text{ yd}}{3\text{ ft}} \times \frac{1\text{ yd}}{3\text{ ft}}$ or $\frac{1\text{ yd}^3}{27\text{ ft}^3}$ so that we can cancel the ft^3.

$$45.\text{ ft}^3 = 45.\overset{1}{\cancel{\text{ft}^3}} \times \frac{1\text{ yd}^3}{27\underset{1}{\cancel{\text{ft}^3}}}$$

$$= \frac{45.}{27}\text{ yd}^3$$

$$= 1.7\text{ yd}^3. \qquad \square$$

We started with a measure of volume and ended with a measure of volume. Only the units were changed. The answer 1.7 yd^3 contains two significant digits, which is the same as the number of significant digits in 45. ft^3. The number of digits in 1 yd^3 and 27 ft^3 do not affect the answer since they are, by definition, exact numbers.

EXAMPLE 2 Convert 55. mi/hr (mph) to km/hr.

Solution: In order to do the above conversion we need the equality 1.0 mi = 1.6 km, which can be converted to a unit fraction by dividing both sides by 1.0 mi.

$$\frac{1.0\text{ mi}}{1.0\text{ mi}} = \frac{1.6\text{ km}}{1.0\text{ mi}}$$

$$1 = \frac{1.6\text{ km}}{1.0\text{ mi}}.$$

Why divide by 1.0 mi? We need miles in the denominator for our unit-conversion fraction to cancel the miles in the numerator of the quantity we are converting.

$$55.\ \frac{\text{mi}}{\text{hr}} = 55.\ \frac{\overset{1}{\cancel{\text{mi}}}}{\text{hr}} \times \frac{1.6\ \text{km}}{1.0\ \underset{1}{\cancel{\text{mi}}}}$$
$$= 88.\ \frac{\text{km}}{\text{hr}}.$$

We started with average speed and ended with average speed. □

EXAMPLE 3 Convert 55. mi/hr to feet per second.

Solution: To convert mi/hr to ft/s we need the following conversion factors:

$$\frac{5280\ \text{ft}}{1\ \text{mi}} \qquad \frac{1\ \text{hr}}{60\ \text{min}} \qquad \frac{1\ \text{min}}{60\ \text{s}}.$$

$$55.\ \frac{\text{mi}}{\text{hr}} = 55.\ \frac{\cancel{\text{mi}}}{\cancel{\text{hr}}} \times \frac{5280\ \text{ft}}{1\ \cancel{\text{mi}}} \times \frac{1\ \cancel{\text{hr}}}{60\ \cancel{\text{min}}} \times \frac{1\ \cancel{\text{min}}}{60\ \text{s}}$$
$$= \frac{(55.)(5280)\ \text{ft}}{(60)(60)\ \text{s}}$$
$$= 81.\ \text{ft/s}. \quad □$$

EXAMPLE 4 Convert $30\underline{0}$ British thermal units per minute (Btu/min) to joules per second (J/s).

Solution: $1.00\ \text{J} = 9.48 \times 10^{-4}\ \text{Btu}$;

$$\text{conversion factor} = \frac{1.00\ \text{J}}{9.48 \times 10^{-4}\ \text{Btu}}.$$

$$1\ \text{min} = 60\ \text{s; conversion factor} = \frac{1\ \text{min}}{60\ \text{s}}.$$

$$30\underline{0}\ \frac{\text{Btu}}{\text{min}} = 30\underline{0}\ \frac{\cancel{\text{Btu}}}{\cancel{\text{min}}} \times \frac{1.00\ \text{J}}{9.48 \times 10^{-4}\ \cancel{\text{Btu}}} \times \frac{1\ \cancel{\text{min}}}{60\ \text{s}}$$
$$= \frac{30\underline{0}\ \text{J}}{(9.48 \times 10^{-4})\ 60\ \text{s}}$$
$$= 5.27 \times 10^{3}\ \text{J/s}. \quad □$$

3.3 Exercises

In Exercises 1–27, perform the indicated conversions. The tables in Appendix A may be helpful.

1. 8.5 m to millimeters
2. 360. in^2 to square feet
3. 24. km to miles
4. 125. kg to pounds at standard acceleration
5. 440. yd to meters
6. 88. km/hr to miles per hour

7. $6\underline{0}$ mph to feet per second

8. 15. milliseconds to seconds

9. $13\underline{0}0$ A to milliamperes

10. 17. mm^3 to cubic centimeters

11. 32. ft^2 to square meters

12. 88. ft/s to kilometers per hour

13. 5.6 MΩ (megohms) to ohms

14. 6.3 kcal to calories

15. 145. Btu/min to joules per second

16. 4.7 pF (picofarads) to farads

17. $70{,}\underline{0}00$ Ω, to megohms

18. 6500. W to kilowatts

19. $4\underline{0}$ mpg to kilometers per liter

20. 55 Btu to joules

21. 175 hp (horse power) to watts

22. 15 cal to joules

23. A liter of wine occupies how many cubic inches? (A liter is approximately the volume of a cube 10.0 cm on a side.)

24. The density of lead is 11.3 g/cm^3. Express the density in terms of oz/in^3 at standard acceleration.

25. The density of cork is 0.13 oz/in^3. Express the density in terms of g/cm^3 at standard acceleration.

26. Convert 6 hr, 17 min, 13 sec to seconds.

27. A room is 4 yd, 2 ft, and 7 in. long. What is its length in meters? Express the answer to the nearest 0.001 m.

3.4 Formulas and dimensional analysis

In this section we will tie together the methods of working with equations developed in Section 3.1 and the methods of working with dimensional units developed in Section 3.3. The technician frequently works with special equations called formulas. A **formula** is a set of symbols expressing a mathematical rule. For example, $d = vt$, $V = IR$, $I = prt$, and $V = \ell wh$ are all formulas.

EXAMPLE 1 The formula $s = \frac{1}{2}gt^2$ expresses the relationship between the distance S that a freely falling object (starting at rest) will fall in t s. Find S when the acceleration of gravity $g = 9.80\,\frac{\text{m}}{\text{s}^2}$ and $t = 3.0$ s.

Solution:

$$S = \frac{1}{2}gt^2$$

$$= \frac{1}{2}\left(9.80\,\frac{\text{m}}{\text{s}^2}\right)(3.0\text{ s})^2$$

Substituting not just the numbers but also the dimensional units that are essential to determining the correct answer

$$= \frac{1}{2}\left(9.80\,\frac{\text{m}}{\cancel{\text{s}^2}_{\,1}}\right)(9.0\,\cancel{\text{s}^2}^{\,1})$$

$$= \frac{(9.80)(9.0)\text{ m}}{2}$$

$$= 44.10\text{ m.}$$

The distance is 44. m. The answer is expressed in meters, an appropriate unit of distance and to the proper number of significant digits. □

EXAMPLE 2 The width w of a rectangular solid can be calculated from the formula $w = \frac{V}{\ell h}$, where V = volume, ℓ = length, and h = height of the solid. Find the width of the rectangular solid whose volume is 5616 in³, whose length is 24 in., and whose height is 18 in.

Solution:

$$w = \frac{V}{\ell h}$$

$$= \frac{5616 \text{ in}^3}{(24 \text{ in})(18 \text{ in})} \qquad \textbf{Substituting}$$

$$\frac{5616 \text{ in}^3}{432 \text{ in}^2}$$

$$= 13 \text{ in}^{3-2}$$

$$= 13 \text{ in.}$$

Note that the rules of exponents apply equally well to dimensional units. □

EXAMPLE 3 If the weight of a moving truck is 6400 lb and it travels at 20 ft/s, find its kinetic energy.

Solution: The kinetic energy of a moving body is given by

$$KE = \frac{1}{2} mv^2,$$

where m is the mass and v is the magnitude of the velocity. If the weight, rather than the mass is given, you must find the mass by dividing the weight by g. (Be sure to use a value of g consistent with your units of weight.)

$$KE = \frac{1}{2} mv^2$$

$$= \frac{1}{2} \frac{6400 \text{ lb}}{\left(32.2 \frac{\text{ft}}{\text{s}^2}\right)} \left(\frac{20 \text{ ft}}{\text{s}}\right)^2 \qquad \textbf{Substituting (since weight is in lb, we use 32.2 ft/s}^2 \textbf{ for } g\textbf{)}$$

$$= \frac{1}{2} \left(\frac{198.8 \text{ lb}}{\text{ft s}^{-2}}\right)\left(\frac{400 \text{ ft}^2}{\text{s}^2}\right)$$

$$= \frac{39760. \text{ lb ft}^{2-1}}{\text{s}^{-2+2}}$$

$$= 39{,}800. \text{ lb ft}$$

or 3.98×10^4 ft lb, since ft lb is a common measurement for work or energy. □

EXAMPLE 4 Consider the formula known as Ohm's law, $I = E/R$. Compute I, where $E = 880$ V and $R = 22\ \Omega$.

Solution:

$$I = \frac{E}{R}$$

$$= \frac{880\text{ V}}{22\ \Omega}.$$

This is a case where we cannot cancel the units $\frac{\text{V}}{\Omega}$ because they do not have a common factor. However, $\frac{1\text{ V}}{1\ \Omega} = 1$ A, so

$$I = \frac{880\text{ V}}{22\ \Omega}$$

$$= 40\ \frac{\text{V}}{\Omega}$$

$$= 40\text{ A.}$$ □

3.4 Exercises

In Exercises 1–11, solve for the indicated quantity. In finding the solution pay strict attention to dimensional units and to significant digits.

1. Another form of Ohm's law is $E = IR$, where E is the voltage difference, I is the number of amperes of current, and R is the number of ohms resistance. Solve for E if $I = 1.60$ A and $R = 380.\ \Omega$.
2. For a uniformly accelerating object the velocity v is given by $v = at$. Find v when $a = 32$ ft/s^2 and $t = 4.50$ s.
3. The distance s that a freely falling object (starting from rest), will fall in a period of time t (in seconds) is given by the formula $s = \frac{1}{2}at^2$. Find s when $a = \frac{32\text{ ft}}{\text{s}^2}$ and $t = 3.50$ s.
4. The length ℓ of a rectangular solid can be calculated from the formula $\ell = \frac{V}{wh}$. (See Example 2.) Find the length of the rectangular solid whose volume is 5411.25 mm^3, whose width is 13.0 mm and whose height is 18.5 mm.
5. The formula for finding the volume of a regular cone is $V = \frac{1}{3}\pi r^2 h$. In this formula V = volume, r = radius, and h = height. Find the height of the cone whose volume is 78.28 in^3 and whose radius is 3 1/4 in.
6. If the weight of a moving car is 3200 lb and it travels at 23 ft/s, find its kinetic energy. (See Example 3.)
7. If the mass of a moving truck is 2880 kg and it travels at 6 m/s, find its kinetic energy. (See Example 3.)
8. The formula $v = v_0 + at$ enables us to find the velocity v at any instant t given the initial velocity v_0 and constant acceleration a. A bottled-nosed dolphin is observed to increase its speed uniformly from 1.2 m/s to 8.3 m/s in a time of 5 s. Find its average acceleration.

9. An equation that can be developed from the formula in Exercise 8 is

$$s = v_0 t + \frac{1}{2} at^2,$$

where $s =$ distance. Using the information from Exercise 8, determine how far the dolphin travels while accelerating.

10. A water beetle accelerates from rest at 0.8 m/s². How far will it have traveled in 1.5 s? (See Exercises 8 and 9.)

11. The time t (called the period) for one complete swing of a simple pendulum is given by the formula $t = 2\pi\sqrt{L/g}$, where $t =$ time in seconds, $L =$ length of the pendulum in feet, and $g =$ the acceleration constant 32.2 ft/s². Compute t for a pendulum whose length is 24. in.

3.5 Problem-solving techniques

In daily life, in the laboratory, in the workplace, or in the classroom, solving problems is more than just substituting a few numbers in a formula and grinding out an answer. It is extremely important to develop a proper technique for solving problems.

What we will be doing in this section is developing and illustrating an orderly procedure to follow that has proven to be quite effective. To perfect the procedure use it for all your problem solving.

Problem-Solving Method

1. Read the problem carefully. This is the obvious thing to do and the most important part of solving a problem. It is also often slighted. It is essential that you read the problem at least twice:
    - **a.** The first reading of the problem is to obtain a general overview. Do not stop and try to set up a formula or jot down facts at this time.
    - **b.** Read through a second time (and perhaps a third and a fourth) slowly and completely; and as you do, think ahead to the following steps.
2. Make a sketch. It is not always possible to do this. However, whenever possible, make a sketch and label it.
3. Write down all given information and identify it. This is necessary to get all essential facts in mind before looking for a solution. There are some phrases that have physical meanings. For example, the term "from rest" means that the initial velocity equals zero ($v_0 = 0$).
4. Write down the unknown or quantity asked for in the problem. Know what you are looking for; give it a name: $\ell =$ the length,

P = the perimeter, v = voltage, etc. Too frequently people cannot solve a problem because they do not know what they are looking for and consequently solve for the wrong quantity.

5. Write a basic equation or formula that relates the known and the unknown quantities. A particular problem may require more than one equation or formula to solve the problem.
6. Find a working equation by solving the basic equation or formula for the unknown quantity.
7. Substitute the data in the working equation, including the approximate units. It is important that you carry the units all the way through the problem as a check that you have solved the problem correctly.
8. Perform the indicated operations, arrive at a solution, and record the solution properly. After arriving at the solution, ask yourself, "Is my answer reasonable?" A quick estimate will many times reveal an error in your calculation.

The problem-solving method will be demonstrated by using relationships and formulas with which you are most likely familiar.

EXAMPLE 1 Find the area of a shop floor that is 15 m by 18 m.

Solution: Sketch:

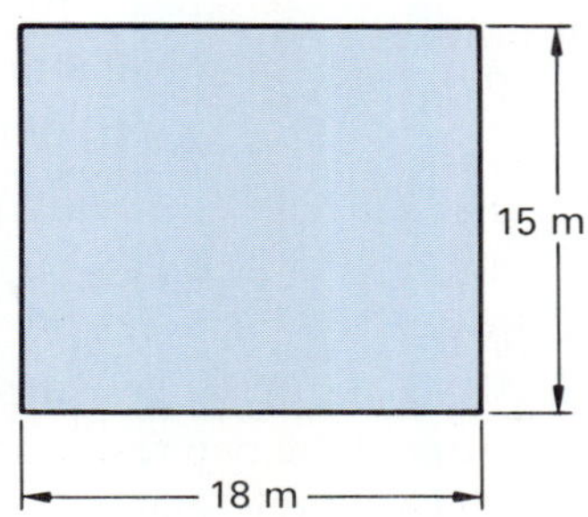

Data: $\left.\begin{array}{l}\ell = 18\text{ m}\\ w = 15\text{ m}\end{array}\right\}$ **Information we know**

$A = ?$ **Area is the unknown**

Basic equation: $A = \ell w$

Working equation: same

Substitution: $A = (18\text{ m})(15\text{ m})$
$= 270\text{ m}^2.$ □

EXAMPLE 2 The length and width of a tissue box are 250 mm and 120 mm, respectively. The volume of the box is 2,550,000 mm^3. Find the height of the box.

Solution: Sketch:

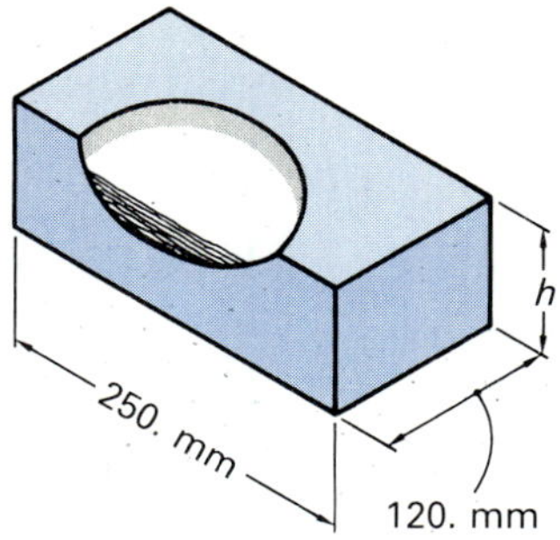

Data: $\ell = 250$ mm, $w = 120$ mm, $V = 2{,}550{,}000$ mm^3 — **Information we know**

$h = ?$ — **Identifies the unknown**

Basic equation: $V = \ell w h$

Working equation: $h = \dfrac{V}{\ell w}$

Found by solving the basic equation for h:

$$\frac{V}{\ell w} = \frac{\ell w h}{\ell w}$$

$$\frac{V}{\ell w} = h.$$

Substitution:
$$h = \frac{2{,}550{,}000 \text{ mm}^3}{(250 \text{ mm})(120 \text{ mm})}$$
$$= \frac{2{,}550{,}000 \text{ mm}^3}{30{,}000 \text{ mm}^2}$$
$$= 85 \text{ mm}. \quad \square$$

EXAMPLE 3 If two resistors are in a series circuit, the total resistance is the sum of the two resistances. If one resistance is three times that of the other, and the total resistance is 12 ohms, what is the resistance of each?

Solution: Sketch:

Data: $R_T = 12\ \Omega$

R = smaller resistance

$3R$ = larger resistance

Basic equation:
$$R_T = R + 3R$$
$$R_T = 4R$$
$$\frac{R_T}{4} = R$$

Working equation: $R = \dfrac{R_T}{4}$

Substitution: $R = \dfrac{12\Omega}{4}$

$R = 3\Omega$

Thus, the smaller resistance is 3Ω and the larger resistance is $3(3\Omega) = 9\Omega$. □

EXAMPLE 4 An airplane takes 4.0 hr to fly from city *A* to city *B*. On the return trip it has a tail wind so it flies $5\underline{0}.$ mph faster and makes the trip in 3.5 hr.

a. How fast did it fly on each part of the trip?

b. How far apart are the two cities?

Solution: Sketch:

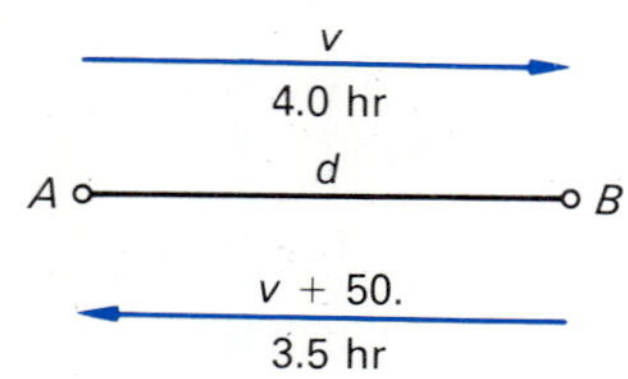

Data: Flying time from *A* to *B* is 4.0 hr; $t_1 = 4.0$ hr
Flying time from *B* to *A* is 3.5 hr; $t_2 = 3.5$ hr
Velocity of plane from *A* to *B* is v mph; $v_1 = v$ mph
Velocity of plane from *B* to *A* is $(v + 5\underline{0}.)$ mph;
$v_2 = (v + 5\underline{0}.)$ mph

Basic equation: $d = vt$

Working equation: Distance from *A* to *B*: $d_1 = v_1 \cdot t_1$
Distance from *B* to *A*: $d_2 = v_2 \cdot t_2$
Since $d_1 = d_2$ we can write:

$$v_1 t_1 = v_2 t_2$$

Substitution:

$$(v \text{ mph})(4.0 \text{ hr}) = [(v + 5\underline{0}.) \text{ mph}](3.5 \text{ hr})$$

$$4.0v \text{ miles} = 3.5v \text{ mi} + 175. \text{ mi} \qquad \textbf{(mph)(hr)} = \left(\frac{\textbf{miles}}{\textbf{hr}}\right)\textbf{(hr)} = \textbf{mi}$$

$$4.0v - 3.5v = 3.5v - 3.5v + 175.$$

$$0.5v = 175.$$

$$\frac{0.5v}{0.5} = \frac{175.}{0.5}$$

$$v = 350. \text{ mph}$$

a. $v_1 = 350.$ mph **Velocity from A to B: $v_1 = v$**

$v_2 = v + 5\underline{0}.$

$= 350. + 5\underline{0}.$

$= 4\underline{0}0.$ mph **Velocity from B to A: $v_2 = v + 5\underline{0}$**

b. Distance from A to B is (350. mph)(4.0 hr) = 1400. mi. Note that the distance from B to A is ($4\underline{0}0.$ mph)(3.5 hr) = 1400. mi, which is what we expected. □

3.5 Exercises

In Exercises 1–23, use the problem-solving method and the rules of calculations with measurements.

1. A rectangle has a length of 6.50 m and a width of 2.25 m. (a) Find its area. (b) Find its perimeter.

2. A rectangle has an area of 4200 cm². (a) If its length is 700 mm, what is its width? (b) Find its perimeter.

3. The dimensions of a rectangular box are 72.0 mm length, 34.0 mm width, and 16.0 mm height. (a) Find its volume. (b) Find its total surface area (including top and bottom).

4. A carpenter is asked to build a rectangular bin that holds 19,600 ft³. If its length and width are to be 56 ft and 25 ft, respectively, what must its height be?

5. Each side of a square piece of sheet metal measures 72.0 in. A 15.0 in. square is then cut from each corner. The metal is then folded to form a box without a top. Find the volume of the box.

6. The width of a room is 3/5 of its length. There are three doorways, each 3 ft wide. If 55 ft of moulding is to be put around the room (excluding doorways), what are the dimensions of the room?

7. Two resistors in a series circuit have a total resistance of 35. Ω. One resistance is 5. Ω less than twice the other. Find each resistance. (The total resistance equals the sum of individual resistances in a series circuit.)

8. Three resistors are in series. The second resistance is one fourth the first resistance, while the third is 3.0 more than the first. If the total resistance is 7.5 Ω, find the value of each resistance.

9. A truck driver takes 5.0 hr to cover a certain distance. On the return trip she drives 15. mph slower and takes 8.0 hr. How fast did she drive on each part of the trip? What distance did the driver cover in one direction?

10. A plane flies from airport A to airport B in 8.0 hr. On the return trip its velocity is 110. mph more than 0.94 times the original velocity and it makes the trip in only 1.5 hr. (a) How fast did the plane fly on each part of the trip? (b) How far did it travel?

11. The total surface area of a right circular cylinder is given by the formula: $S = 2\pi r(r + h)$. If the total surface area is 311. cm² and the radius is 32.5 mm, find the height.

12. A manufacturer is designing a cylindrical waste can whose height is 14. in. It must hold 4520 in³ of material. Determine the radius of the can.

13. The power P in any part of an electric circuit is given by $P = E^2/R$ watt, in which E is the electromotive force applied to that part of the circuit and R is the resistance of that part. What is the resistance of a circuit in which 1.21 W is expended at an electromotive force of 110. V?

14. The voltage drop E across any part of a circuit can be computed by the formula: $E = IZ$ volt, where I is the current in amperes through that part of the circuit and Z is the impedance in ohms of that part. What is the impedance of a circuit in which a voltage drop of 460. V is produced by a current flow of 0.115 A?

15. A satellite has a velocity of $40{,}00\underline{0}$ mi/hr. How many miles will it travel in one day?

16. A car, starting from rest, attains a velocity of $3\underline{0}$ m/s in 8.0 s. What is the acceleration of the car? ($v = at$.)

17. A horizontal force of 15.0 N moves a block 3.0 m along a flat surface. How much work is done by the force? ($W = Fs$)

18. What velocity does a 1.6-kg mass need to have a kinetic energy of 180. J? ($KE = 1/2\ mv^2$.)

19. To what height should a 150.-lb weight be lifted in order for its potential energy to be 1.5×10^5 ft lb? ($PE = wh$.)

20. A pole is located in a pond. One fifth of the length of the pole is in the sand, 4 m are in water, and two thirds of the length is in the air. How long is the pole?

21. An electronic circuit board is designed so that the width of the board is 180 mm less than the length. The perimeter of the board is 1700 mm. What are the dimensions of the circuit board?

22. The electric current in one transistor is double the current in a second transistor. The electric current in the third transistor is 0.003 A more than that in the second transistor. The sum of the three currents is 0.017 A. What is the current in each?

23. When a depth charge is exploded just below the surface of the water, the shock is felt by a ship 5 s before the explosion is heard by the crew. The velocity of the shock in the water is 1500 m/s and the velocity of the sound in air is 330 m/s. How far is the ship from the explosion?

Chapter summary

An **equation** consists of two expressions with an equal sign between them. Equations are tools used in problem solving. If g is a **solution** of an equation, the solution may be indicated by using set notation, $\{g\}$, by stating that g satisfies the equation, or by stating that g is a **root** of the equation. An **identity** is a statement that is true for all the permissible values of the variable.

To find the solution of an equation we use the following rules.

Rules for Solving Equations

Addition property:	If $a = b$, then $a + c = b + c$ for all real numbers a, b, and c.
Subtraction property:	If $a = b$, then $a - c = b - c$ for all real numbers a, b, and c.
Multiplication property:	If $a = b$, then $a \cdot c = b \cdot c$ for all real numbers a, b, and c.
Division property:	If $a = b$, then $\frac{a}{c} = \frac{b}{c}$ for all real numbers a, b, and c, $c \neq 0$.

It is essential that all solutions are checked in the original equation.

In numerical calculations we work with exact and **approximate** numbers. In performing calculations it is essential to determine the number of **significant digits** and how and when to round off approximate numbers in order to obtain the most accurate results.

Rules for Rounding Off

Find the digit in the number to be rounded, then:

Rule 1 If the digit to the right is less than 5, leave the digit in the place to be rounded unchanged and drop all the digits to the right.

Rule 2 If the digit to the right is 5 or greater, then add 1 to the digit in the place to be rounded and drop all the digits to the right.

In an approximate number a **digit is significant** if it is not a zero, if it is a zero between two nonzero digits, or if it is a terminal zero on the right side of the decimal point.

In calculating with approximate numbers the answer cannot be more accurate than the least accurate number in the data. However, it is advisable to carry one more place whenever possible throughout the computation and do the rounding at the end.

The **base units** in the British and the SI systems are found in Table 3.1. Table 3.2 shows how dimensional units can be created from the base units. In changing from one unit to another it is important to be able to identify the correct **conversion factor.** Remember that each conversion factor is equal to 1.

In order to be able to express answers in proper units we must be able to work with dimensional units when finding solutions to equations and formulas.

It is essential to be able to read a problem, analyze the information given, and solve the problem. In Section 3.5 a problem-solving procedure is outlined. The eight steps in the procedure should be followed each time a problem is solved to develop good problem-solving techniques.

Review Exercises

In Exercises 1–10, find the solution for each equation and check.

1. $x + 4 = 13$ **2.** $3R - 7 = 20$

3. $5a + 9 = 7a - 13$ **4.** $6y - 7 + 5y = 4y + 21$

5. $0.5z + 1.3 = 2.6 - 0.8z$

6. $3.4x + 3 = 1.2x - 4(2 - 2.3x)$

7. $\frac{1}{5}R + 4 = \frac{1}{3}R$ **8.** $\frac{4}{x} + 3 = \frac{7}{x}$

9. $\frac{2}{x+3} = \frac{5}{x-4}$

10. $y(y + 5) - 7 = (y + 5)(y - 3)$

In Exercises 11–20, perform the indicated operations. The measures are approximate numbers.

11. 137.5 lb + 13.73 lb **12.** 14.7 mm − 3.08 mm

13. 22.45 in. − 107.3 in. + 67.05 in. − 10.01 in.

14. (34.85 yd) × (1.35 yd)

15. (100 km)(0.32 km)

16. 3400 mm^2 ÷ 0.056 mm

17. 247.5 in^3 ÷ 1728 in^3

18. (432.7 in.)(4.03 in.) ÷ 144 in^2

19. (3.4 ft)(4.75 ft)(11.2 ft) ÷ 27 ft^3

20. (0.75 mm + 2.34 mm − 0.47 mm)(2.51 mm)

In Exercises 21–30, perform the indicated conversions.

21. 11.5 m to millimeters

22. 17.5 ft^2 to square inches

23. 75.4 mi to kilometers

24. 950 mA to amperes

25. 27.5 m to yards

26. 45 mph to feet per second

27. 65 mph to kilometers per hour

28. 75 km/liter to miles per gallon

29. 40 hp (horse power) to watts

30. 15 joules to calories

In Exercises 31–40, solve for the indicated quantity, paying strict attention to dimensional units and to significant digits.

31. For a uniformly accelerating object the velocity v is given by $v = at$. Find a when $v = 24.0$ ft/s and $t = 7.5$ s.

32. If the weight of a moving car is 3500. lb and its kinetic energy is 49,000. ft lb, find the rate at which the car is traveling. $\left(KE = \frac{1}{2}mv^2.\right)$

33. The area of a square is given by $A = s^2$. If the area is 36.9 cm^2, how many significant digits should be in the value for s? Why? Find s in mm.

34. The volume of a cone is given by the formula $V = \frac{1}{3}\pi r^2 h$. If $V = 36.44$ in^3 and $h = 2.03$ in., what value of π should be used in the computation of r? Find r.

35. If 312 bricks were used to edge a rectangular grassy patch whose length was twice its width, how many bricks were used on each side?

36. A car averaged 26. mpg in city driving and 4$\underline{0}$. mpg in highway driving. If the car used 15. gal of gas on a 460.-mi trip, how much of the trip was in the city?

37. A landlady earned \$2190 in one year from the rental of two rooms. Find the rent she charged per month per room if the rent on the first room was \$10 per month more than the rent for the second room and the first room was vacant for three months.

38. A student rode a bicycle for 3 hr then parked it and hiked up a mountain trail for 3 more hours. She rode 6 mph faster than she hiked and traveled 36 mi in all. How fast did she hike?

39. A man can exert a force of 100 lb on the end of a 20-ft lever. Where must he place the fulcrum under the lever to raise a 400-lb weight? $\left(d = \frac{f\ell}{f + w}.\right)$

40. A farmer is designing a bin to hold 1500. bushels of wheat. He knows that 1 bushel occupies approximately 1.244 ft^3. What should be the height of the bin if the base must be 12.5 ft by 15.4 ft?

Can you solve these?

1. How is relative error obtained? How is percent of relative error obtained? If the largest possible error of a measurement of 14.6 mi is 0.05 mi, what is its relative error? What is its percent of error? What is its degree of accuracy?

2. The equation $E = 8\ell wh^2$ is an algebraic model for energy E (in foot-pounds) that can be expected from sea waves, where ℓ is the length (in feet) of the wave (distance between successive crests), h is the height (in feet) of the wave (vertical distance from trough to crest), and w is the width (in feet) of the wave.

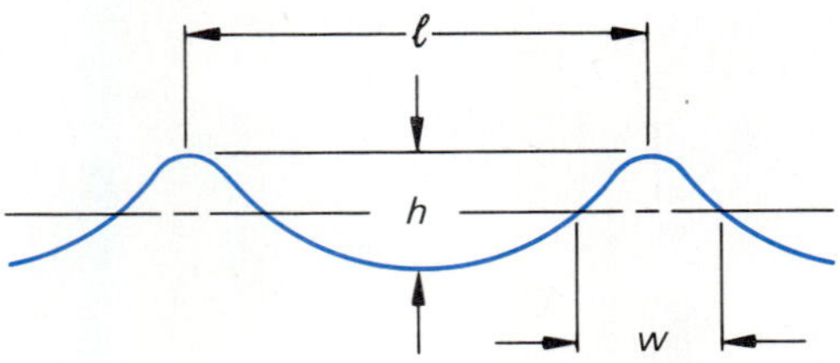

a. What are the units associated with the constant 8?
b. Solve this equation for w.
c. If a wave is 50 ft long and 8 ft high and has 400,000 ft lb of energy, how wide is the wave?

3. Solid-state digital watches use a vibrating quartz crystal to regulate the time. The formula $s = v/32{,}768$ is an algebraic model that gives the number of seconds elapsed, s, after v vibrations of the quartz crystal.
   a. What are the units associated with the constant 32,768?
   b. After the quartz crystal vibrates 1,048,567 times, how many seconds have elapsed?
   c. Solve the formula for v.
   d. How many vibrations does the quartz crystal make in 1 s?
   e. How many vibrations does the quartz crystal make in 5 s?

4. From atomic physics the Bohr theory of the hydrogen atom allows us to calculate the orbital radius r in centimeters by the mathematical equation

$$r = \frac{n^2 h^2}{4\lambda^2 m e^2}.$$

If $h = 6.6 \times 10^{-27}$, $e = 4.8 \times 10^{-10}$, $m = 9.1 \times 10^{-28}$, $\lambda = 9.12 \times 10^{-6}$ and n is the principal quantum number, which can assume any integral value, evaluate r when a) $n = 1$, b) $n = 2$, c) $n = 3$.

5. John stored his power boat on a trailer in the back yard. To take the pressure off the tires he wants to raise the trailer and place blocks under the axle to each wheel. The boat and trailer weigh 2800 lb. If he can place the fulcrum 1 ft from the axle and he has a lever that is 12 ft long, how much force must he exert on the end of the lever in order to lift one wheel (half the total weight) off the ground?

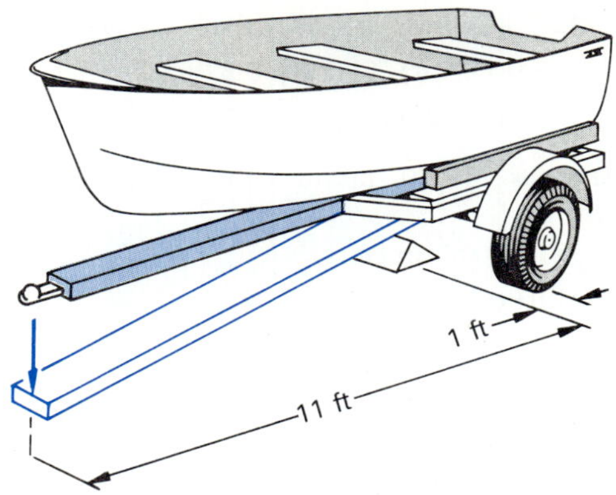

6. A rectangular swimming pool is 25 ft long. Every 40 gal of water increases the depth of water 1/8 in. What is the width of the pool? How many gallons are needed to fill the pool to a depth of 5 ft?

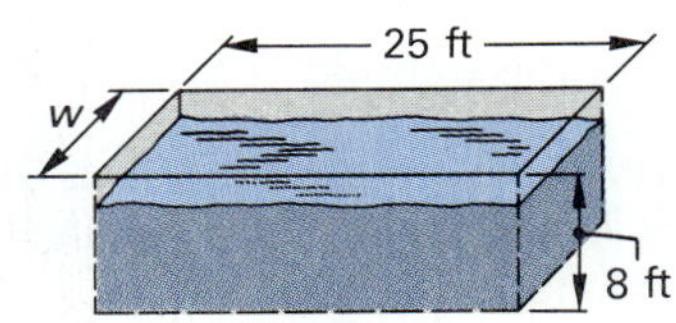

4

A concrete pier was originally constructed and then repaired at different times over a span of years. The newer sections are rectangular while the older section is lower, narrower, and has a trapezoidal top. How much concrete is needed to build the 1250-ft-long older section up and out so that it is flush with the newer sections? The inshore end of the older section of the pier is 9 ft above the bottom.

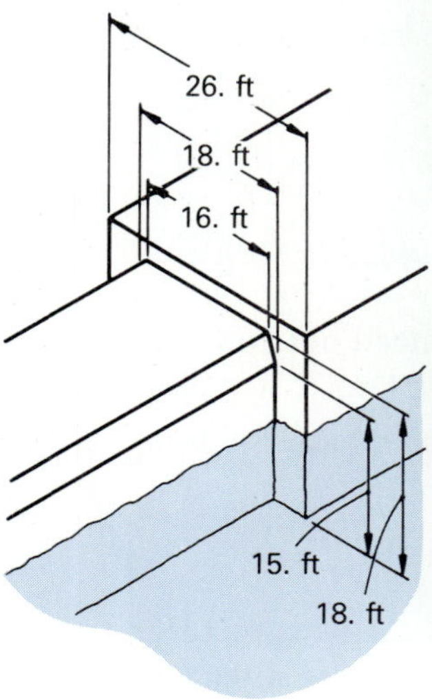

Ratio, Proportions, and Review of Geometry

4.1 Ratio and proportion

For design A of a metal storage shed the number of bolts required for assembly is 300. An alternate design, B, requires only 100 bolts. We can express a mathematical relationship between the two quantities by stating the difference of the two quantities as $300 - 100 = 200$ (design A requires 200 more bolts than design B), or $100 - 300 = -200$ (design B requires 200 fewer bolts than design A). Another way of expressing the relationship between the two quantities is by writing a special fraction called a **ratio:**

$$\frac{300 \text{ bolts}}{100 \text{ bolts}} = \frac{3}{1} \quad \text{and} \quad \frac{100 \text{ bolts}}{300 \text{ bolts}} = \frac{1}{3}.$$

The first ratio tells us that there are three times as many bolts in design A as design B, whereas the second ratio says that there are $1/3$ as many bolts in design B as design A.

DEFINITION 4.1

> A **ratio** is a number (without dimension) that expresses the numerical relationship between two quantities when the two quantities are compared by division. The ratio of a to b ($b \neq 0$) is written
>
> $$\frac{a}{b} \text{ or } a \text{ to } b \text{ or } a:b.$$

As the definition indicates, the results need not be expressed as a fraction; they could be expressed as 1 : 2 or 1 to 2. It is also common practice in technical work to express fractional ratios such as $\frac{1}{4}$ as 0.25 to 1, $\frac{3}{8}$ as 0.375 to 1, and $\frac{7}{4}$ as 1.75 to 1.

EXAMPLE 1 A wheel of a boat trailer is 14. in. in diameter and covers a distance of 44. in. as it rolls one turn along a flat surface (Fig. 4.1). The ratio of the distance around the wheel to the diameter is $\frac{44. \text{ in.}}{14. \text{ in.}}$ or $\frac{22.}{7.}$.

FIGURE 4.1

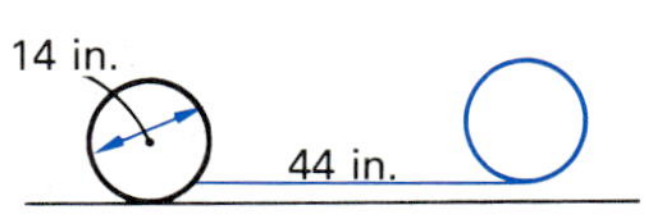

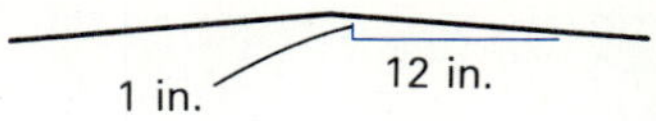

FIGURE 4.2

EXAMPLE 2 The roof of the Porter's family room rises 1 in. vertically for each 12 in. measured horizontally (Fig. 4.2). These measures are called **rise** and **run,** respectively. The ratio that expresses the comparison of the rise to the run is called the **slope** of the roof. The ratio or slope in this case is 1/12.

DEFINITION 4.2

> A **proportion** is a statement that two ratios are equal. $\frac{a}{b} = \frac{c}{d}$ is a proportion provided $b \neq 0$ and $d \neq 0$.

The four quantities of the proportion are called the **terms** of the proportion. The first and fourth terms (a and d) are called the **extremes** and the second and third terms (b and c) are called the **means** of the proportion.

$$\frac{a}{b} = \frac{c}{d}$$

$$(\cancel{b}d)\frac{a}{\cancel{b}} = (b\cancel{d})\frac{c}{\cancel{d}} \qquad \textbf{Multiply both sides by } \boldsymbol{bd}.$$

$$ad = bc. \qquad \textbf{Simplify}$$

From this result we can see that the product of the extremes is equal to the product of the means.

The result, $ad = bc$, can also be obtained by "cross"-multiplying:

$$\frac{a}{b} \times \frac{c}{d}$$

$$ad = bc.$$

This product lends itself nicely as a technique for solving for any one of the letters. Let us assume that b, c, and d are given and we want to find a; all we need to do is divide both sides by d:

$$ad = bc \qquad \textbf{Given}$$

$$\frac{a\cancel{d}}{\cancel{d}} = \frac{bc}{d} \qquad \textbf{Divide both sides by } \boldsymbol{d}$$

$$a = \frac{bc}{d}.$$

The same technique can be used to solve for any one of the variables.

EXAMPLE 3 If $\frac{x}{13} = \frac{4}{19}$, solve for x.

Solution:

$$\frac{x}{13} = \frac{4}{19} \qquad \textbf{Given}$$

$$(x)(19) = (13)(4) \qquad \textbf{Cross-multiply}$$

$$19x = 52 \qquad \textbf{Simplify}$$

$$\frac{\overset{1}{\cancel{19}}x}{\underset{1}{\cancel{19}}} = \frac{52}{19} \qquad \textbf{Divide both sides by 19}$$

$$x = \frac{52}{19}. \qquad \square$$

EXAMPLE 4 If $\frac{x}{11} = \frac{2/7}{3/4}$, solve for x.

Solution:

$$\frac{x}{11} = \frac{2/7}{3/4} \qquad \textbf{Given}$$

$$(x)\left(\frac{3}{4}\right) = (11)\left(\frac{2}{7}\right) \qquad \textbf{Cross-multiply}$$

$$\frac{3}{4}x = \frac{22}{7} \qquad \textbf{Simplify}$$

$$\left(\frac{4}{3}\right)\left(\frac{3}{4}\right)x = \left(\frac{4}{3}\right)\left(\frac{22}{7}\right) \qquad \textbf{Multiply both sides by the reciprocal of 3/4}$$

$$x = \frac{88}{21}. \qquad \square$$

EXAMPLE 5 Find the distance in kilometers from Buffalo, N.Y., to Chicago, Ill. via Detroit. The scale distance on a map of the United States was measured as 15.6 cm. The scale on the map is 1 cm = 56 km.

FIGURE 4.3

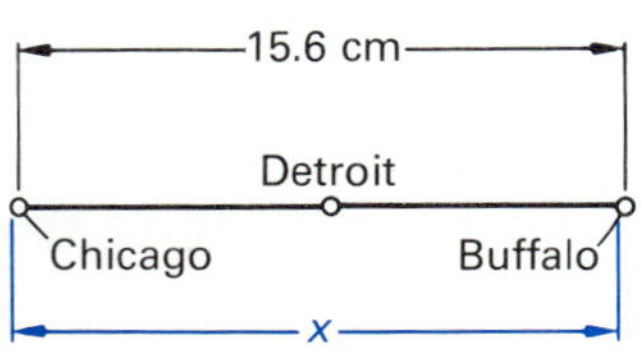

Solution: The scale distance on the map from Buffalo to Chicago is 15.6 cm. The scale on the map is 1 cm = 56 km. Let x = the actual distance in kilometers (Fig. 4.3). With this information we can set up two ratios.

1. The ratio of actual distance between cities, x, to actual distance of the scale, 56 km, is $x/56$ km.
2. The ratio of the distance on the map, 15.6 cm, to the scale, 1 cm, is 15.6 cm/1 cm. These two ratios are equal, so we can write the

proportion

$$\frac{x}{56 \text{ km}} = \frac{15.6 \overset{1}{\cancel{\text{cm}}}}{1 \underset{1}{\cancel{\text{cm}}}}$$

$$(x)(1) = (56 \text{ km})(15.6) \quad \textbf{Cross-multiply}$$
$$x = 873.6 \text{ km}.$$

Thus, the distance from Buffalo to Chicago is 874. km. □

EXAMPLE 6 If a spaceship travels 17,800 mph, use a proportion to determine how many hours are required to fly the 6,300,000 mi between a space station and Earth.

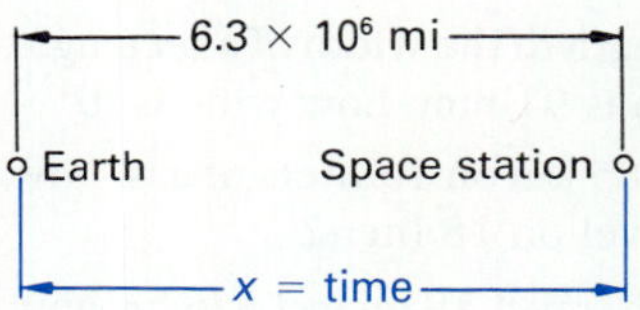

FIGURE 4.4

Solution: The spaceship travels at rate of 17,800 mph, thus, it travels 17,800 mi in one hour. Distance from Earth to the space station is 6,300,000 mi; x = number of hours it takes the spaceship to make the trip (Fig. 4.4). With this information we can write the following proportion:

$$\frac{\text{total number of hours}}{1 \text{ hr}} = \frac{\text{total number of miles}}{\text{miles in one hour}}$$

$$\frac{x}{1 \text{ hr}} = \frac{6.3 \times 10^6 \text{ mi}}{1.78 \times 10^4 \text{ mi}}$$

$$(1 \cancel{\text{hr}})\left(\frac{x}{1 \cancel{\text{hr}}}\right) = \frac{6.3 \times 10^6}{1.78 \times 10^4}(1 \text{ hr}) \quad \textbf{Multiply both sides by 1 hr}$$

$$x = 3.5393 \times 10^2 \text{ hr}.$$

The time required for the spaceship to make the trip is 3.5×10^2 hr. To find the value of x using a calculator we use the following key strokes.

Keystroke	Display
6.3	6.3
EE	6.3 00
6	6.3 06
÷	6.3 06
1.78	1.78
EE	1.78 00
4	1.78 04
=	3.53925802

Therefore $x = 3.5 \times 10^2$ hr. □

4.1 Exercises

In Exercises 1 – 15, express each as a ratio in simplest form.

1. 15 km to 48 km
2. 63 min to 1001 min
3. 42 days to 8 wk
4. 250ℓ to 2 kℓ
5. 24 m^2 to 96 m^2
6. 15 ft to 15 yd
7. 3 1/4 cm to 4 1/3 cm
8. 6 g to 14 1/3 g
9. The ratio of the height of a tree 75 ft tall and the length of its shadow, 105 ft long.
10. The slope of a roof that has a rise of 12 ft and a run of 4 ft.
11. The slope of a roof that has a rise of 3 1/2 m and a run of 2 1/2 m.
12. The ratio of the height of a woman 5′6″ tall and the length of her shadow 12′8″ long.
13. $6xy - 4y^2$ to $9x^2 - 6xy$.
14. $x^2 - 4y^2$ to $x^2 + 4xy + 4y^2$.
15. The wheel of a truck in an open copper mine is 3.00 m in diameter and has a circumference of 9.42 m. Find the ratio of the circumference to diameter.

In Exercises 16 – 23, solve each proportion for the unknown.

16. $\frac{x}{15} = \frac{2}{3}$
17. $\frac{t}{9} = \frac{5}{6}$
18. $\frac{11}{3} = \frac{y}{9}$
19. $\frac{7}{4} = \frac{x}{6}$
20. $\frac{x}{5} = \frac{3/10}{7/4}$
21. $\frac{y}{3} = \frac{2/3}{4/5}$
22. $\frac{x}{4} = \frac{0.25}{1}$
23. $\frac{x}{8} = \frac{0.375}{1}$

In Exercises 24 – 38, set up a proportion and solve.

24. The distance between two cities measures 2.5 cm on a map. The scale on the map is 1.5 cm = 24.7 km. How far apart are the cities?
25. The distance between two mountain tops on a map is 85 cm. The scale of the map is 1 : 4372. How far apart are the mountain tops?
26. Two cities that are exactly 15 mi apart appear as 3.456 in. apart on the map. What is the scale on the map?
27. A blueprint is made to a scale of 3 : 5. If a part appears to be 20.15 mm long on a blueprint, how long is the actual part?
28. An 8-ft-long board appears as 3 in. long on a blueprint. What is the scale of the blueprint?
29. If the ratio of the width of a rectangle to its length is 5/7 and its width is 62.5 m, how long is it?
30. If the ratio of the length to the width of a rectangle is 7/4 and its length is 91 mm, how wide is it?
31. If a truck can travel 85 km on 14 liters of gas, how far can the truck travel on 18 liters?
32. If a pleasure boat can travel 48 mi in 1 5/6 hr, how far can it travel in 2 3/4 hr?
33. If there are 12 g of nitric acid in 80 g of solution, how many grams of nitric acid would be in 110 g of solution?
34. How much alcohol is required to make 100 cc of a 3 : 5 alcohol/water solution?
35. A 35-mm color slide measures 23 mm by 34 mm. If you print a picture from a slide that has a length of 120 mm, what is its width?
36. If you enlarge a 35-mm slide so that the length of the enlargement is 5 in., what is its width? (See Exercise 35.)
37. The weight of a 3-ft piece of I-beam is 137 lb. A contractor needs to lift a 38-ft piece. How much weight will he be lifting?
38. If you were charged a commission of $1450 on the sale of a $25,000 house, what would be the commission (at the same rate) on a house selling for $73,850?

4.2 Variation

A student in fire-science technology is assigned the problem of determining the effect of water pressure on different sized fire hoses. He learns two important relationships: (1) the amount of water pressure in a hose has a direct relation to the intensity of the spray coming out of the nozzle, and (2) the life of the hose is decreased by increasing the pressure beyond a certain point. Let us analyze the two statements since they contain two mathematical ideas that are basic to solving the problem.

The first statement is an example of the mathematical concept called **direct variation** or, as it is sometimes called, **direct proportion.** By direct variation we mean that as the amount of pressure, p, changes, the intensity of the spray, I, changes. A simple formula that would express this relationship would be, $I = kp$, where k is a constant. In the formula it can be seen that as p increases or decreases so does I.

On the other hand, in the second example the life, L, of the hose decreases as p increases. This is an example of **inverse variation** or **inverse proportion.** A simple formula that would express this relation would be $L = k/p$, where k is a constant. In examining this equation we can see that since k is a constant, as p increases, L decreases, and as p decreases L increases. In both examples k is called the **constant of variation** or the **constant of proportionality.**

EXAMPLE 1 Write an algebraic statement that represents each of the statements.

a. S varies directly with t.

b. z varies inversely as w.

c. z is directly proportional to y.

d. A is inversely proportional to B.

e. x varies as y.

f. x varies inversely as y.

Solution:

a. $S = kt$ **b.** $z = k/w$ **c.** $z = ky$

d. $A = k/B$ **e.** $x = ky$ **f.** $x = k/y$ □

The first step in solving a problem on variation is determining the type of variation as illustrated in Example 1. A second important step may be solving for k, the constant of variation. This is illustrated in Example 2.

EXAMPLE 2 For each of the problems:

1. Write an algebraic statement.
2. Solve for k.
3. Having found k, write an algebraic statement of variation.

a. The mass of a steel rod is directly proportional to its length. A 2.40-cm-diameter rod with a length of 43.2 cm has a mass of 1.80 kg.

b. The strength of a horizontal wooden beam (rectangular cross-section), concentrated load applied in center, varies inversely with the distance between supports. A beam 3.6 m long and supported at each end has a safe load capacity of 378 N. What is the safe load capacity when the length is 12.5 m long?

Solution:

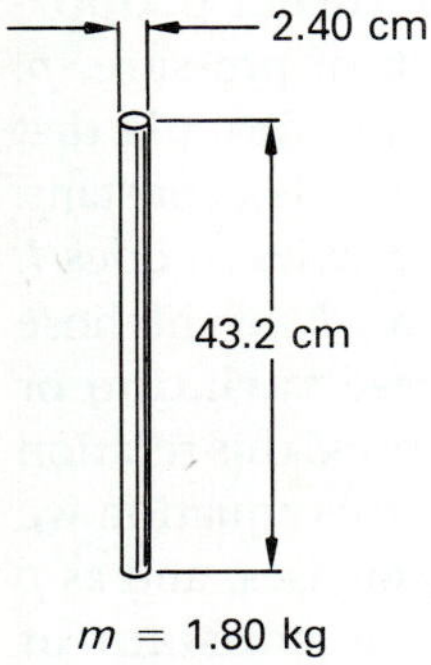

FIGURE 4.5

a. Let m = mass of steel rod: $m = 1.80$ kg,
ℓ = length of steel rod: $\ell = 43.2$ cm (Fig. 4.5), and
k = constant of proportionality.

Recognizing that it is a direct-variation problem, we can write the algebraic statement:

$$m = k\ell.$$

Since we want to solve for k, we would divide both sides of the equation by ℓ:

$$\frac{m}{\ell} = \frac{k\ell}{\ell}$$

$$\frac{m}{\ell} = k \qquad \textbf{Working equation}$$

$$k = \frac{1.80 \text{ kg}}{43.2 \text{ cm}} \qquad \textbf{Substituting for } \ell \textbf{ and } m.$$

$$k = 0.0417 \text{ kg/cm}.$$

Now substituting the value of k in our original equation we obtain a statement of direct variation for a particular value of k.

$$m = k\ell$$
$$m = (0.0417 \text{ kg/cm})\ell$$

We can now substitute any value for ℓ and obtain the corresponding value for m.

FIGURE 4.6

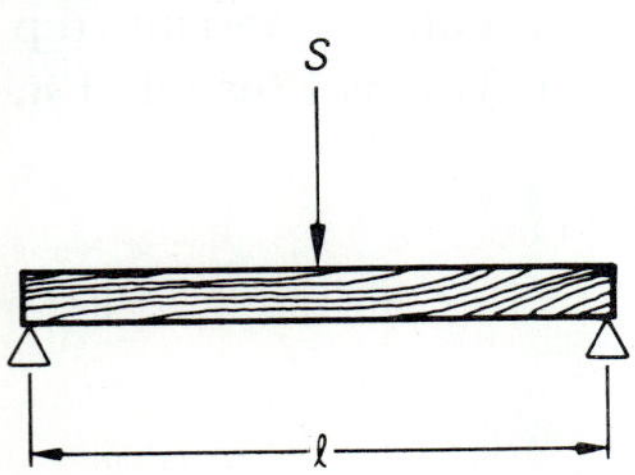

b. Let ℓ = length of the beam: $\ell = 3.6$ m,
S = Safe load capacity: $S = 378$ N (Fig. 4.6), and
k = constant of variation.

Recognizing that it is an inverse-variation problem, we can write the algebraic statement:

$$S = \frac{k}{\ell}$$

Since we want to solve for k we would multiply both sides by ℓ:

$$\ell(S) = \overset{1}{\cancel{\ell}}\left(\frac{k}{\underset{1}{\cancel{\ell}}}\right)$$

$$\ell S = k \quad \textbf{Working equation}$$
$$k = (3.6 \text{ m})(378 \text{ N}) \quad \textbf{Substitute for } \ell \textbf{ and } S$$
$$k = 1360.8 \text{ mN} \quad \text{or}$$
$$k = 1400. \text{ Nm}.$$

Now, substituting the value of k in our original equation we obtain a statement of inverse variation for a particular value of k:

$$S = \frac{1400. \text{ Nm}}{\ell}.$$

Thus, for a beam of length 12.5 m, the safe load capacity is found by substituting in the equation:

$$S = \frac{1400. \text{ Nm}}{\ell}$$

$$S = \frac{1400. \text{ Nm}}{12.5 \text{ m}} \quad \textbf{Substituting}$$

$$S = 112. \text{ N} \quad \text{or} \quad S = 110. \text{ N}. \quad \textbf{Significant digits} \quad \square$$

In technical problems we may encounter a variable that is dependent on two or more variables. For example, the volume of a cylinder is dependent upon the radius r of the base and the height h. The algebraic statement that expresses this is $V = \pi r^2 h$, where π is the constant of variation. We would read the statement as: The volume varies jointly with the square of the radius and the height. This is referred to as **joint variation.**

On the other hand, we could encounter the following problem in mechanical technology: Determine the speed of a gear on a shaft driven by a gear on a motor. The ratio of the number of teeth, T_1, of the gear on the shaft to the number of teeth, T_2, of the gear on the motor is the same as the inverted ratio of the speed, S_1, of the gear on the shaft to the speed, S_2, of the gear on the motor. This statement expressed in algebraic form is

$$\frac{T_1}{T_2} = \frac{S_2}{S_1}.$$

Solving for T_1 we have

$$\overset{1}{\cancel{T_2}}\left(\frac{T_1}{\underset{1}{\cancel{T_2}}}\right) = T_2\left(\frac{S_2}{S_1}\right) \quad \textbf{Multiply both sides by } T_2$$

$$T_1 = \frac{S_2}{S_1} T_2.$$

The problem could be stated as a combined joint- and inverse-variation problem with $k = 1$. The speed, S_1, of a gear on a shaft driven by a gear on a motor varies jointly as the number of teeth, T_2, on the gear on the motor and the speed, S_2, of the gear on the motor and inversely as the number of teeth, T_1, on the gear on the shaft. From this form of the problem we could arrive at the algebraic statement

$$S_1 = \frac{T_2 S_2}{T_1}.$$

Note that this equation is equivalent to the one developed above.

EXAMPLE 3 For each of the following statements write an equivalent algebraic statement.

a. T varies jointly as S and P.

b. R varies directly as T and inversely as P.

c. Z varies jointly as T and the square of L.

d. Q varies jointly as R and the square root of T and inversely as the square of L.

Solution:

a. $T = kSP$ **b.** $R = \frac{kT}{P}$ **c.** $Z = kTL^2$ **d.** $Q = k\frac{R\sqrt{T}}{L^2}$ □

Note that the word "jointly" indicates a product, whereas "inversely" indicates a quotient.

EXAMPLE 4 The time required to clear the snow in a driveway varies jointly as the length of the driveway, the width of the driveway, and the square of the average depth of the snow, and inversely with the number of snow blowers used. Using two snow blowers it takes 3/4 hr to clean a 600-ft driveway that is 14 ft wide when the snow has an average depth of 2 1/2 ft. How long will it take to clean 1.5 mi of a driveway that is 15. ft wide using five snow blowers, when the average depth of the snow is 3.5 ft?

Solution: This problem is a combination of joint and inverse variation. The first thing we want to do is write an algebraic statement. Secondly, we want to solve for k and thirdly, to find the specific value for the time.

Let t = time to clean the driveway,
w = width of the driveway,
ℓ = length of the driveway,
d = depth of the snow,
d^2 = square of the depth of the snow, and
n = number of snow blowers.

The algebraic statement would be

$$t = \frac{k\ell wd^2}{n}.$$

Solve the equation for k:

$$nt = \left(\frac{k\ell wd^2}{\cancel{n}_1}\right)\cancel{n}^{1} \qquad \textbf{Multiply both sides by } n$$

$$\frac{nt}{\ell wd^2} = \frac{k\ell wd^2}{\ell wd^2} \qquad \textbf{Divide both sides of the equation by } \ell wd^2$$

$$k = \frac{nt}{\ell wd^2}.$$

To find the value of k we substitute the values $t = 3/4$ hr, $\ell = 600$ ft, $w = 14$ ft, $d = 2\frac{1}{2}$ ft, $d^2 = 6.25$ ft^2, and $n = 2$ into the equation:

$$k = \frac{(2)\left(\frac{3}{4}\text{ hr}\right)}{(600\text{ ft})(14\text{ ft})(6.25\text{ ft}^2)}$$

$$= 0.0000286\,\frac{\text{hr}}{\text{ft}^4}$$

or

$$k = 2.86 \times 10^{-5}\,\frac{\text{hr}}{\text{ft}^4}.$$

Substituting the value for k in the equation:

$$t = \left(2.86 \times 10^{-5}\,\frac{\text{hr}}{\text{ft}^4}\right)\frac{\ell wd^2}{n}.$$

Using this equation we can determine the time required to clean the 1.5-mi driveway. Before we substitute we must change 1.5 mi to feet. 1 mi = 5280 ft, so 1.5 mi = 7920 ft.

$$t = \left(2.86 \times 10^{-5}\,\frac{\text{hr}}{\text{ft}^4}\right)\frac{(7920\text{ ft})(15\text{ ft})(3.5\text{ ft})^2}{5}$$

$$= \left(2.86 \times 10^{-5}\,\frac{\text{hr}}{\text{ft}^4}\right)\frac{(7920\text{ ft})(15\text{ ft})(12.25\text{ ft}^2)}{5}$$

$$= 8.32\text{ hr.}$$

Thus, the time required to clear the snow would be about 8.3 hr. □

The most difficult part of working with variation is recognizing the type: direct, inverse, joint, or combinations of two or more. Carefully examine the wording in the examples so that you recognize the key words.

4.2 Exercises

In Exercises 1–20, write an algebraic statement that represents each of the statements.

1. D varies directly as t.
2. A varies inversely as d.
3. C varies as d.
4. T varies inversely as w.
5. R varies jointly as V and W.
6. L varies directly as P and inversely as Q^2.
7. Z varies as the square of D and inversely as T.
8. G varies jointly as R and S^2 and inversely as the cube of P.
9. T varies jointly as R and the square root of x and inversely as the square of y.
10. The force, F, exerted by the wind against the wing of a plane increases with the area, A, of the wing.
11. The life, L, of a tire is less at higher speeds, V.
12. The area, A, of a circle varies as the square of the radius, r.
13. The area, A, of an equilateral triangle varies as the square of its altitude, a.
14. The length of time, t, required to empty a tank varies inversely as the square of the diameter, d, of the drain pipe.
15. Under certain conditions, the thrust, T, of a propeller varies jointly as the fourth power of its diameter, d, and the square of the number, n, of revolutions per second.
16. Newton's Law of Gravitation states that the force, F, exerted between two bodies varies jointly as their masses, m_1 and m_2, and inversely as the square of the distance, d, between them.
17. The resistance, R, of a wire to the flow of electricity varies directly as its length, ℓ, and inversely as the square of its diameter, d.
18. The strength, s, of a rectangular beam with a concentrated load in the center and supported at both ends varies jointly as its width, w, and the square of its depth, d, and inversely as its length, ℓ.
19. The time, t, in hours that it takes x men to build y houses varies directly as the number of houses and inversely as the number of men.
20. The number, n, of vibrations per second of a stretched cord varies directly as the square root of the stretching force, f, and inversely as the product of its length, ℓ, and diameter, d.

In Exercises 21–34, (a) Write an algebraic statement, (b) Solve for k, (c) Write the algebraic statement using the value of k, and, (d) Solve for the required quantity.

21. If y varies directly as the square of x, and $y = 3$ when $x = 4$, find the value of x that makes $y = 12$.
22. If the square of y varies directly as the cube of x, and $y = 4$ when $x = 2$, find the value of x when $y = 54$.
23. A photographer found that the exposure time needed for making an enlargement from a negative varies as the area of the enlargement. If he takes 10 s for a print 2 1/2 in. by 3 1/2 in., how long will it take for an enlargement 6 in. by 8 in.?
24. The volume, V, of an ideal gas varies directly as its temperature, T, and inversely as its pressure, p. If a certain amount of gas at constant temperature occupies 40 ft³ at a pressure of 15 psi, find its volume when the pressure is 20 psi.
25. The current, I, in amperes in an electric circuit varies inversely as the resistance, R, in ohms when the electromotive force is constant. If in a certain circuit I is 15 A when R is 2 Ω, find I when R is 0.2 Ω.
26. The force, P, of the wind on a plane surface at right angles to the direction of the wind varies jointly as the area, A, of the surface and the square of the speed, v, of the wind. If the force on 16 ft² is 8 lb when the speed of the wind is 10 mph, find the force on a surface 6 ft by 12 ft when the wind's speed is 50 mph.
27. If a 2-in. × 8-in. timber 16 ft in length will hold 500 lb when placed with the 2-in. side down, how much will a 1-in. × 8-in. timber hold when placed with the 1-in. side down? Use the formula developed in Exercise 18.
28. If a 2-in. × 6-in. timber will hold 1200 lb, how much will a 6-in. × 6-in. timber of the same length hold? (See Exercise 27.)
29. If a cord 4 m long and 2 mm in diameter vibrates n_1 times per second under a stretching force of 25 N, what stretching force would cause this cord to vibrate $2n_1$ times per second? Use the formula developed in Exercise 20.

30. The time required to dig a river channel varies jointly as its length, width, and square of its depth, and inversely with the number of machines used. If it takes 1600 hr to dig 1 km of a channel that is 30 m wide and 6 m deep using 40 earth-moving machines, how long will it take to dig a channel 4 1/2 km long that is 32 m wide and 10 m deep using 90 earth-moving machines?

31. The loss in pressure of a liquid flowing through a pipe varies as the length and inversely as the square of the diameter of the pipe. If a water pipe 89 m long and 0.29 m in diameter results in a drop in water pressure of 600 N/m^2, find the drop in pressure for a water pipe 264 m long whose diameter is 0.55 m.

32. Force, F, varies directly as mass, m, and inversely as distance, d, squared. If the force is 30 lb when the mass is 7 slugs and d is 3 ft, find the force when the mass is 3 slugs and distance is 5 ft.

33. $F = 37.4 \times 10^{-9}$ lb is the gravitational attraction between two bodies whose masses are 11 slugs and 8 slugs when the distance between the two bodies is 9 ft. Find F when $m_1 = 18$ slugs, $m_2 = 28$ slugs, and $d = 15$ ft. Use the formula developed in Exercise 16.

34. The kinetic energy, KE, of an irregularly shaped rotating body varies directly as its moment of inertia, I, and its velocity, v, squared and inversely as its distance from the center to the nearest point on the edge, d, squared. If $KE = 224$ ft lb when $I = 60$ slug ft^2, $v = 30$ ft/s, and $d = 12$ ft, what is KE when $I = 45$ slug ft^2, $v = 25$ ft/s, and $d = 7$ ft?

4.3 Plane figures

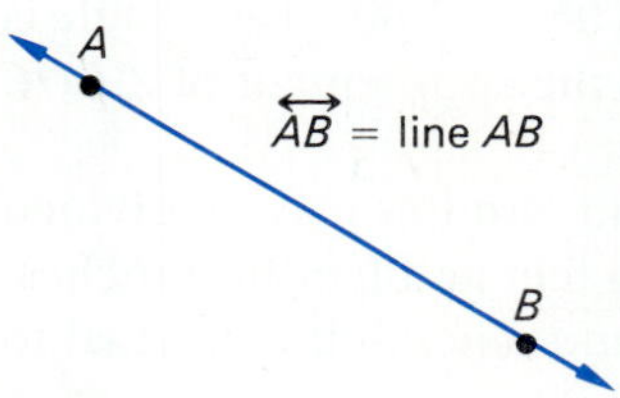

FIGURE 4.7

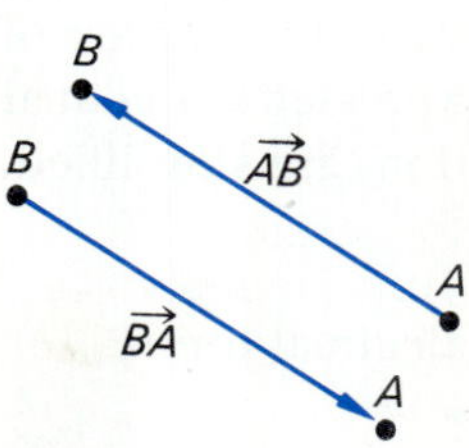

FIGURE 4.8

The design engineer for a new aircraft decides that the tail section should have the shape of a trapezoid. The surveyor finds the shape of a parcel of land to be a parallelogram and an isosceles triangle joined together. The scientist describes the base of a crystal in a residue to be a dodecagon. Do you recognize these geometric figures? Can you list their properties and determine their perimeter and area? The next few sections will provide a brief review of the essentials of geometry.

The line (straight line) and the point are the basis of many of the figures that are constructed on a plane. We think of **line** AB, symbolized $\overleftrightarrow{AB}$, as extending in either direction without bounds (Fig. 4.7), while **line segment** AB, symbolized $\overline{AB}$, is a measurable piece of line AB. The **directed line segment** AB, symbolized $\overrightarrow{AB}$ or **AB** means that we start at point A and move in the direction of point B, ending at point B. The directed line segment BA or **BA** means we start at point B and end at point A (Fig. 4.8). In physics such a directed line segment is called a **vector.**

FIGURE 4.9

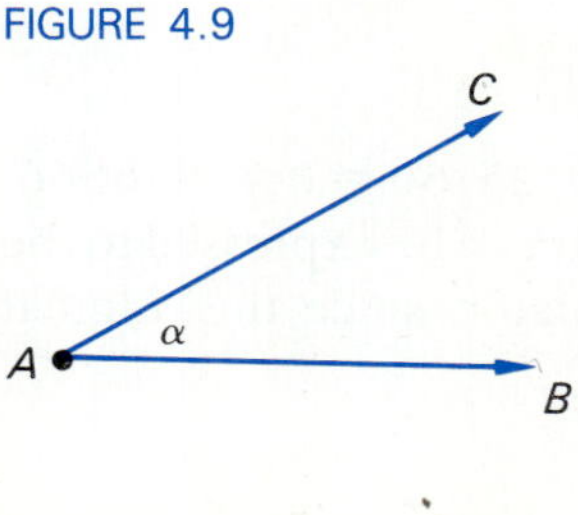

The union of two directed line segments, **AB** and **AC**, form angle BAC, symbolized $\angle BAC$ (Fig. 4.9). Sides **AB** and **AC** of $\angle BAC$ intersect at point A. Point A is called the **vertex** of $\angle BAC$. Other ways to designate $\angle BAC$ are $\angle A$ (using only the vertex letter), or α (using a character such as the Greek letter alpha inside the angle).

If we measure the opening between two directed line segments, with a common end point, by rotating in a counterclockwise direction,

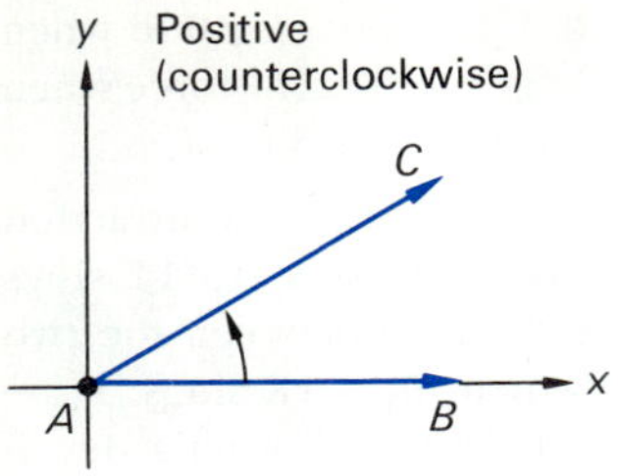

FIGURE 4.10

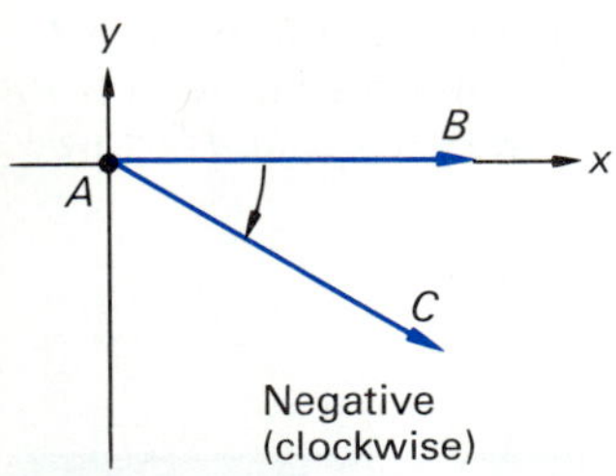

FIGURE 4.11

FIGURE 4.12

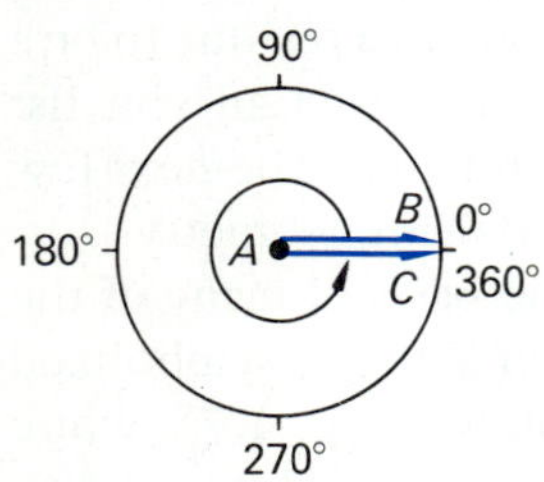

FIGURE 4.13

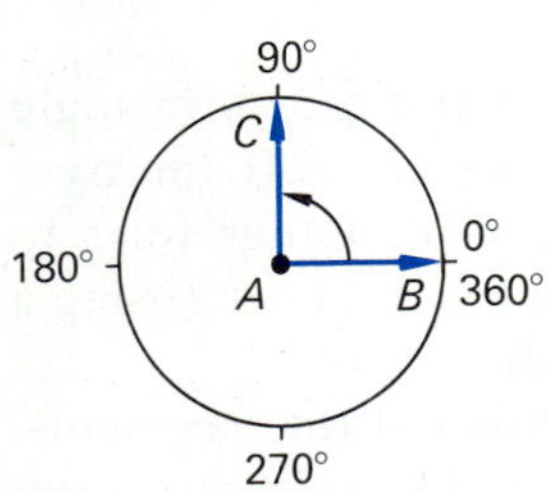

the measure of the angle is positive (Fig. 4.10). If we measure in a clockwise direction, the measure of the angle is negative (Fig. 4.11).

In Fig. 4.12, **AC** initially coincides with **AB**. Then **AC** is rotated counterclockwise one complete revolution, or 360°. The basic measure of an angle is the degree: $1^\circ = 1/360$ of a complete revolution. If **AC** is rotated one fourth of a revolution (Fig. 4.13) then the measure of $\angle BAC$ is 90°. Any angle whose measure is 90° is called a **right angle.** The measure of angle BAC is properly indicated by the symbolism $m(\angle BAC)$. To simplify the symbolism we will use the more common notation $\angle BAC$ to indicate either the angle or the measure of the angle.

In Fig. 4.14 angles AOB and BOC are adjacent angles. Two angles are **adjacent angles** if and only if they have a common vertex and a common side between them. Two angles are called **complementary angles** if the sum of their measures is 90°. Thus, angles AOB and BOC are complementary angles since $\angle AOB + \angle BOC = 22^\circ + 68^\circ = 90^\circ$. Each angle is a **complement** of the other. Thus, $\angle AOB$ is the complement of $\angle BOC$ and $\angle BOC$ is the complement of $\angle AOB$.

A rotation of **AC** halfway around the circle (Fig. 4.15) describes $\angle BAC$ with a measure of 180°. An angle with a measure of 180° is called a **straight angle.**

Two angles are called **supplementary angles** if the sum of their measures is 180°. Thus, $\angle AOB$ and $\angle BOC$ (Fig. 4.16) are supplementary angles, since $\angle AOB + \angle BOC = 75^\circ + 105^\circ = 180^\circ$. Each angle is a **supplement** of the other. Thus, $\angle AOB$ is the supplement of $\angle BOC$ and $\angle BOC$ is the supplement of $\angle AOB$.

Degrees may be divided into minutes and minutes may be divided into seconds, just as the yard is divided into feet and feet into inches. One degree is equal to 60 min ($1^\circ = 60'$) and one minute is equal to 60 s, ($1' = 60''$).

Let us assume that the measure of $\angle BOC$ is $31^\circ 29' 32''$. Calculations cannot be conveniently performed on a calculator with the measure of the angle expressed in this form. If the measure is expressed in decimal form, $\angle BOC = 31.49222222^\circ$, calculator calculations are simplified.

EXAMPLE 1 Change the following to degrees in decimal form.

a. $24^\circ 25'$ **b.** $31^\circ 29' 32''$

Solution:

a. $24^\circ 25' = \underline{\ \ ?\ \ }^\circ$.

$25'$ is equal to 25/60 of a degree. Dividing, $25/60 = 0.4166666\overline{6}$. Thus, $24^\circ 25' = 24.4166666\overline{6}^\circ$. The result can be expressed to the number of places provided by the calculator, since the original numbers were exact values.

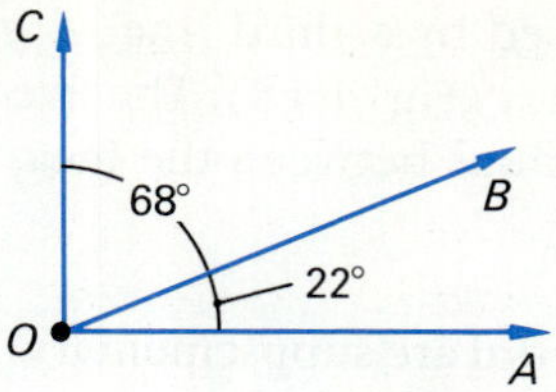

FIGURE 4.14

b. $31°29'32'' = \underline{\quad ? \quad}°$.
$32''$ is $32/60$ of a minute, and $32/60 = 0.5333333\overline{3}$. Adding this value to $29'$ we have $29.5333333\overline{3}'$. $29.5333333\overline{3}'$ is $29.5333333\overline{3}/60$ of a degree, and
$29.5333333\overline{3}°/60 = 0.4922222\overline{2}°$.
Adding this value to $31°$ we have
$31°29'32'' = 31.4922222\overline{2}°$. □

EXAMPLE 2 Write each of the following so that it does not contain a decimal. Write the answer in degrees, minutes, and seconds.

a. $47.9166666\overline{6}°$ **b.** $28.6402777\overline{7}°$

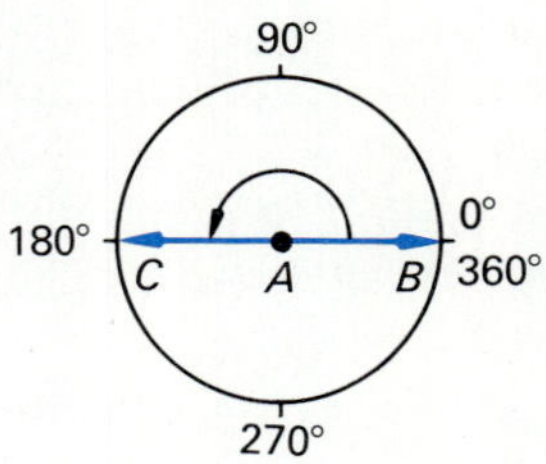

FIGURE 4.15

Solution: We defined $1' = 1°/60$ or $1' = 0.0166666\overline{6}°$. Thus, if we multiply both sides of the equation by 60 we have

$$60(1') = 60\left(\frac{1°}{60}\right).$$

$$60' = 1°$$

Therefore if we multiply the decimal part of the angle measure by 60, the resulting number will be the number of minutes. If there is a decimal after the first multiplication, multiply again by 60, since $60'' = 1'$ and this number will be the number of seconds.

a. To convert $47.9166666\overline{6}°$ multiply $0.9166666\overline{6}°$ by 60. The product is 55. Therefore $47.9166666\overline{6}° = 47°55'$.

b. To convert $28.6402777\overline{7}°$ multiply $0.6402777\overline{7}$ by 60. The result is $38.4166666\overline{6}'$. At this point we have $28°38.4166666\overline{6}'$. Now multiply $0.4166666\overline{6}$ by 60 and the result is $25''$. Therefore $28.6402777\overline{7}° = 28°38'25''$. □

FIGURE 4.16

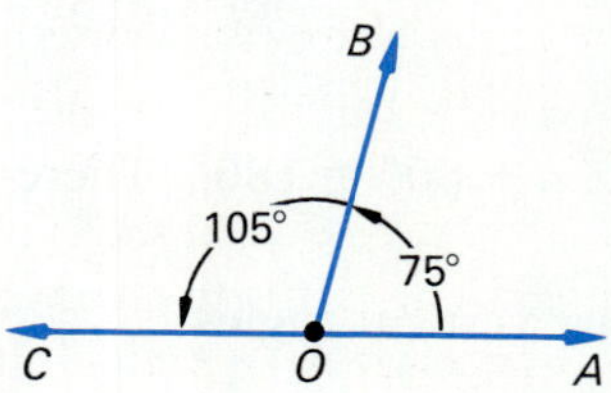

The measure of angles in degrees is practical in some areas but a more convenient unit of measure in theoretical areas as well as in applied areas is the **radian.** Radian measure will be discussed in detail in Chapter 6.

Two lines in a plane may: never intersect, in which case they are **parallel; coincide,** in which case they are the same line; or **intersect** in a single point. Lines $\overleftrightarrow{AC}$ and $\overleftrightarrow{BD}$ in Fig. 4.17 intersect at point O. The four angles formed by the intersection are α (alpha), β (beta), γ (gamma), and ϵ (epsilon). Angles β and ϵ are on opposite sides of the vertex and are called **vertical angles,** as are α and γ. Angles α and β are supplementary angles as are α and ϵ. Observe that $\angle\alpha + \angle\beta = 180°$ and $\angle\alpha + \angle\epsilon = 180°$, therefore, $\angle\beta = \angle\epsilon$ since they are supplements of the same angle α. Using similar reasoning can you conclude that $\angle\alpha = \angle\gamma$?

FIGURE 4.17

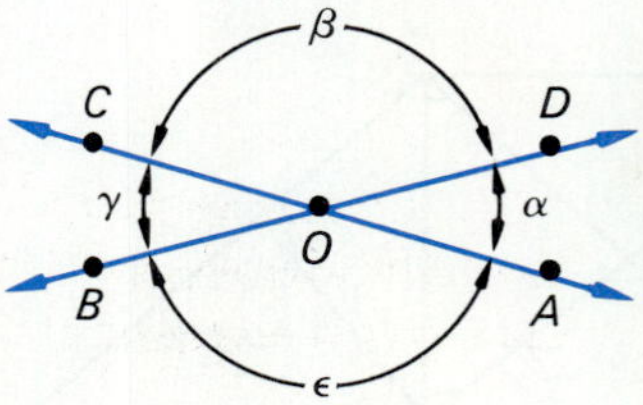

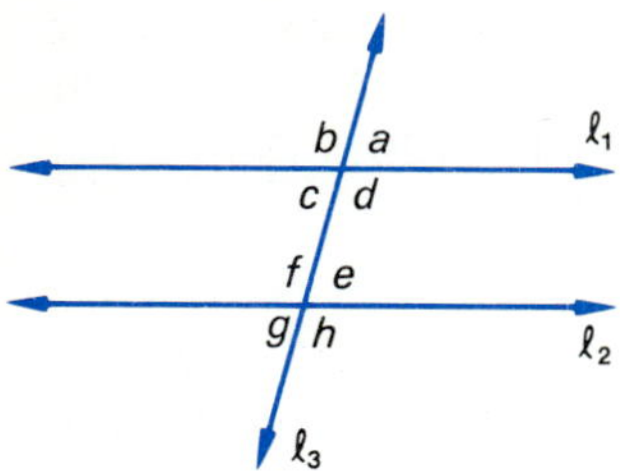

FIGURE 4.18

Two parallel lines ℓ_1 and ℓ_2 that are crossed by a third line, ℓ_3, called the **transversal,** form eight distinct angles (Fig. 4.18). The five properties that follow illustrate important relations between the measures of the angles.

1. Exterior angles on the same side of the traversal are supplementary:

   $\angle a + \angle h = 180°$ and $\angle b + \angle g = 180°$.

2. Interior angles on the same side of the transversal are supplementary:

   $\angle c + \angle f = 180°$ and $\angle d + \angle e = 180°$.

3. Corresponding angles are equal:

   $\angle a = \angle e$, $\angle b = \angle f$, $\angle c = \angle g$, and $\angle d = \angle h$.

4. Alternate exterior angles are equal:

   $\angle a = \angle g$ and $\angle b = \angle h$.

5. Alternate interior angles are equal:

   $\angle c = \angle e$ and $\angle d = \angle f$.

EXAMPLE 3 In Fig. 4.19, $\overline{AD}$ is parallel to $\overline{BC}$ and $\overline{BG}$ is parallel to $\overline{EF}$. Find the measure of

1. $\angle a$	**2.** $\angle b$	**3.** $\angle c$
4. $\angle d$	**5.** $\angle f$	**6.** $\angle g$

Solution:

1. $\angle a$ is the supplement of an 80° angle. $\angle a + 80° = 180°$. Therefore, $\angle a = 100°$
2. $\angle b = 80°$, since $\angle b$ and the 80° angle are vertical angles.
3. $\angle c = 100°$, since $\angle c$ and $\angle a$ are vertical angles.
4. $\angle d$ is the complement of a 50° angle, therefore $\angle d + 50° = 90°$ and $\angle d = 40°$.

FIGURE 4.19

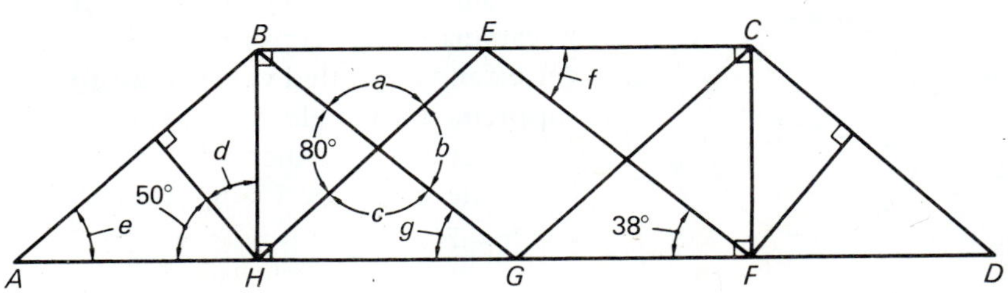

5. $\angle f = 38°$. If two parallel lines are cut by a transversal, then the alternate interior angles are equal.

6. $\angle g = 38°$. If two parallel lines are cut by a transversal, then the corresponding angles are equal. □

A **polygon** is a plane figure that has three or more sides. A polygon with three sides is called a **triangle.** Triangles may be classified according to the measure of their angles or the length of their sides. **The sum of the measures of the angles of a triangle is 180°.** A triangle that contains three **acute angles** (angles whose measures lie between 0° and 90°) is called an **acute triangle** (Fig. 4.20a). A triangle that contains an **obtuse angle** (an angle whose measure lies between 90° and 180°) is called an **obtuse triangle** (Fig. 4.20b). A triangle that contains a right angle is called a **right triangle** (Fig. 4.20c.). The side opposite the right angle is called the **hypotenuse** and the other two sides are called the **legs** (or the base and altitude).

A triangle in which no two sides are equal is called a **scalene triangle** (Fig. 4.21a). A triangle with two sides the same length is called an **isosceles triangle** (Fig. 4.21b). The angles opposite the equal sides have equal measure. A triangle with three equal sides is called an **equilateral triangle** (Fig. 4.21c). The three angles have equal measure (60° each) since they are opposite equal sides. The equilateral triangle is also called an **equiangular triangle.**

A polygon formed by the union of four line segments is called a **quadrilateral.** Quadrilaterals may be classified as to whether the figure

FIGURE 4.20

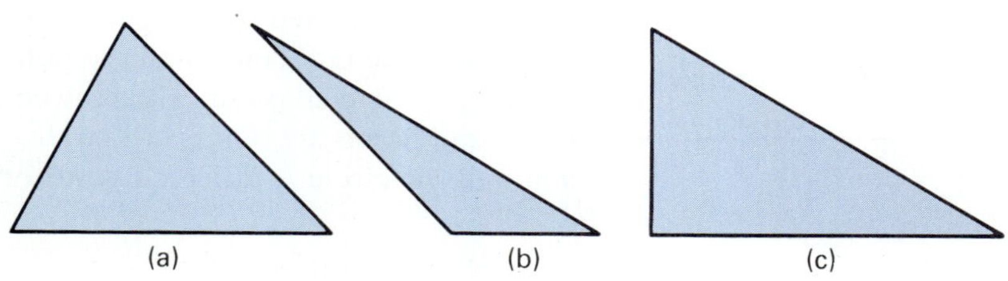

FIGURE 4.21

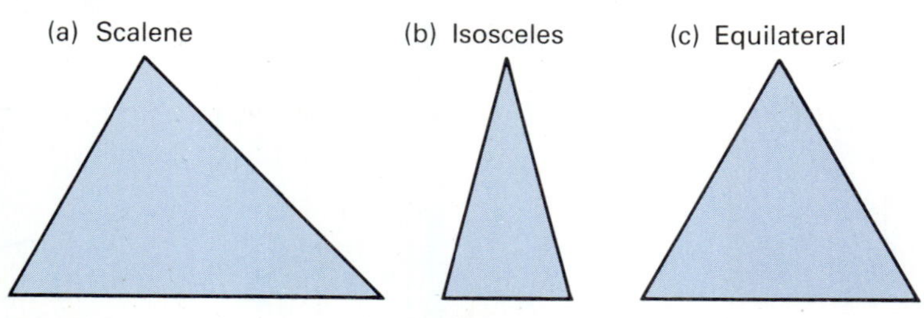

TRIANGLES

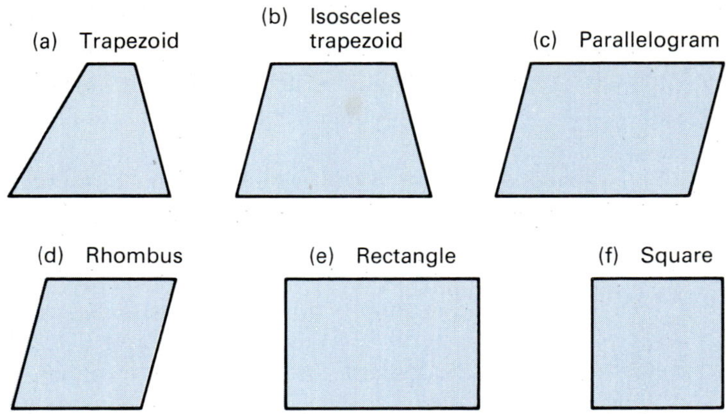

FIGURE 4.22

TABLE 4.1 **Common polygons**

Number of Sides	Name
3	Triangle
4	Quadrilateral
5	Pentagon
6	Hexagon
7	Heptagon
8	Octagon
9	Nonagon
10	Decagon
11	Dodecagon
12	Icosagon

has one or two pairs of opposite sides that are parallel lines and/or whether the figure has right angles. A quadrilateral with only one pair of parallel sides is called a **trapezoid** (Fig. 4.22a). The other two sides may or may not be of equal length. If the nonparallel sides are of equal length, it is called an **isosceles trapezoid** (Fig. 4.22b). A quadrilateral with both pairs of opposite sides parallel is called a **parallelogram** (Fig. 4.22c). A parallelogram that has four equal sides is called a **rhombus** (Fig. 4.22d). A parallelogram that contains a right angle is called a **rectangle** (Fig. 4.22e). A rectangle that contains four equal sides is called a **square** (Fig. 4.22f). A list of common polygons is found in Table 4.1. From this list it can be seen that a polygon is named by its number of sides.

A commonly used plane figure that is not a polygon is a circle (Fig. 4.23). A **circle** is a locus of all the points that are a given distance, the **radius,** from a fixed point called the **center.** A **chord** is a line segment whose end points are points of a circle. A chord that passes through the center of the circle is called a **diameter** of a circle. The portion of the

FIGURE 4.23

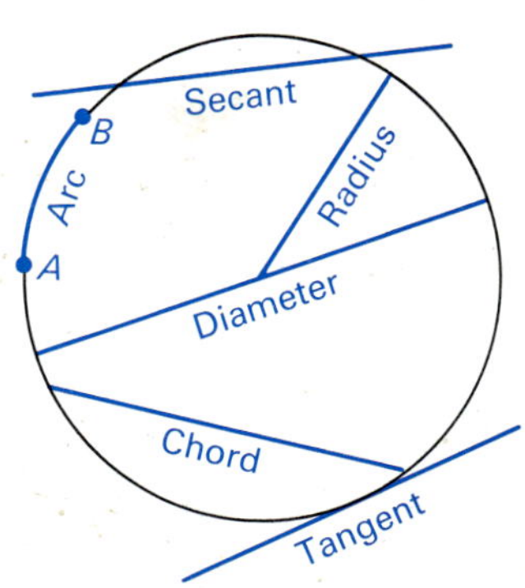

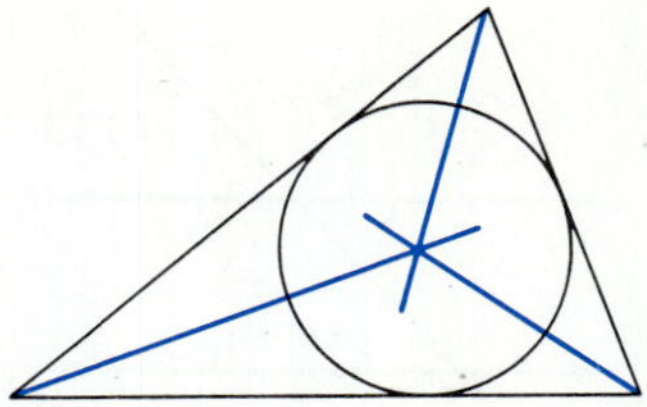

FIGURE 4.24

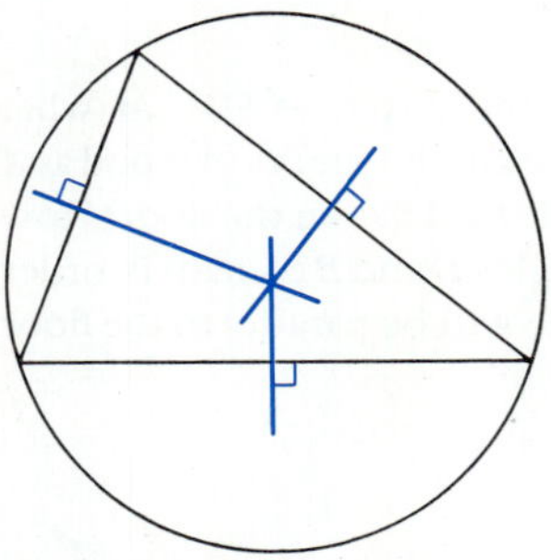

FIGURE 4.25

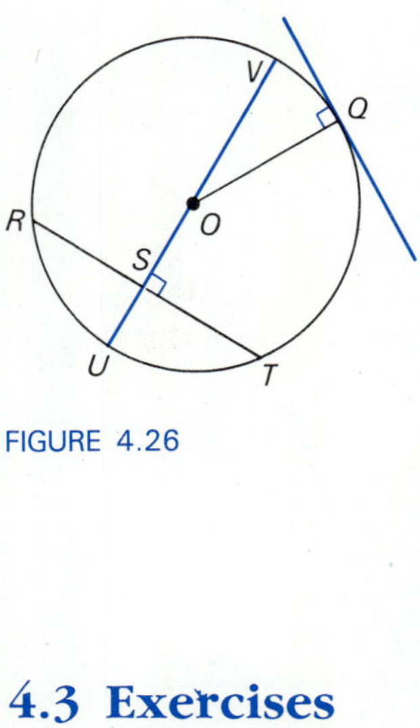

FIGURE 4.26

circle connecting points A and B is called **arc AB,** symbolized $\widehat{AB}$. A straight line that intersects a circle in two points is called a **secant.** A straight line that intersects a circle at one and only one point is called a **tangent.**

The distance around a circle is called the **circumference** and is a multiple of the diameter. The ratio of the circumference to the diameter is the irrational number π. Thus $\pi = c/d$ or equivalently $c = \pi d$.

To supplement the ideas presented in this section we include a list of geometrical facts.

1. The diagonals of a parallelogram bisect each other.
2. The diagonals of a rhombus are perpendicular.
3. A regular polygon is a polygon that is both equilateral and equiangular.
4. The sum of the measures of the interior angles of a polygon of n sides is equal to $(n-2)180°$.
5. Each interior angle of a regular polygon of n sides is equal to $[(n-2)/n]180°$.
6. The central angle of a regular polygon of n sides contains $360°/n$.
7. The intersection of the bisectors of the angles of a triangle determines the center of the inscribed circle (Fig. 4.24).
8. The intersection of the perpendicular bisectors of the sides of a triangle determines the center of the circumscribed circle (Fig. 4.25).
9. If a radius of a circle bisects a chord that is not a diameter, it is perpendicular to the chord (Fig. 4.26). (If $\overline{RS} = \overline{ST}$, then $\overline{OU} \perp \overline{RT}$.)
10. A tangent to a circle is perpendicular to the radius drawn to the point of contact (Fig. 4.26). Tangent at $Q \perp \overline{QO}$.
11. A diameter perpendicular to a chord bisects the chord and the arc subtended by the chord (Fig. 4.26). ($\overline{VU} \perp \overline{RT}$, so $\overline{RS} = \overline{ST}$ and $\widehat{RU} = \widehat{UT}$.)

4.3 Exercises

In Exercises 1–6, change each angle measure to degrees in decimal form.

1. 15°14′ **2.** 94°17′ **3.** 104°38′

4. 17°29′34″ **5.** 73°59′4″ **6.** 103°35′49″

In Exercises 7–12, write each angle measure so that it does not contain a decimal. Write the answer in degrees, minutes, and seconds.

7. 14.45° **8.** $28.61666\overline{6}°$

9. 102.05° **10.** $33.36888\overline{8}°$

11. $87.76333\overline{3}°$ **12.** $113.22027\overline{7}°$

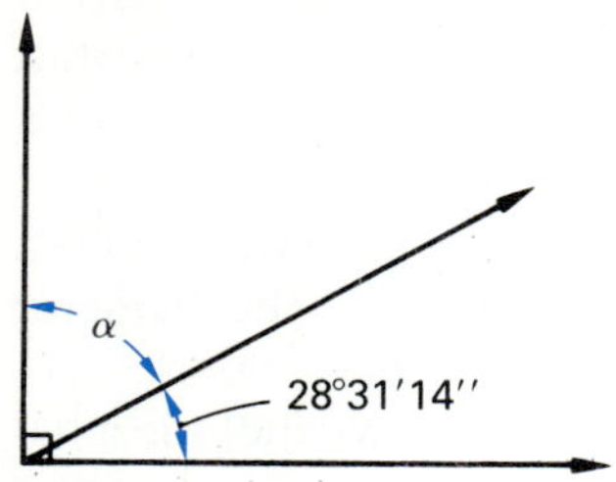

FIGURE 4.27

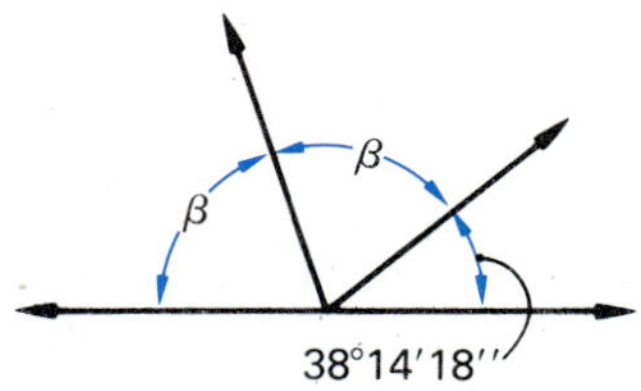

FIGURE 4.28

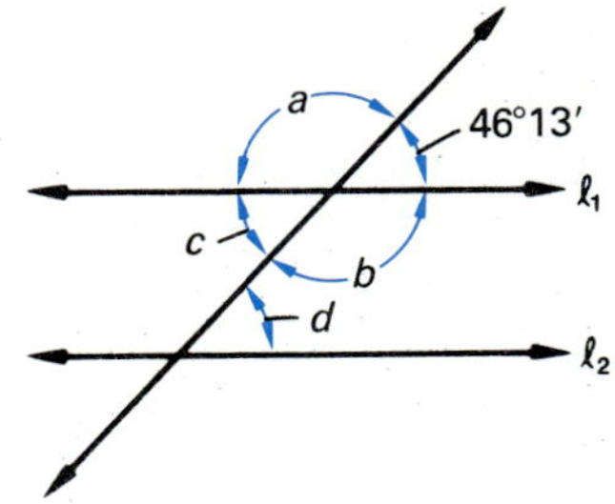

FIGURE 4.29

13. Find the measure of angle α in Fig. 4.27.

14. Find the measure of angle β in Fig. 4.28.

15. Find the measures of angles a, b, c, and d in Fig. 4.29 given that ℓ_1 and ℓ_2 are parallel.

16. A surveyor is asked to divide the plot of ground (Fig. 4.30) $ABCD$ into lots by constructing lines parallel to $\overline{AD}$. If $\angle ADC = 115°$ how many degrees must he make angles DEF and DGH? If $\overline{AD}$ and $\overline{BC}$ are perpendicular to $\overline{AB}$, how many degrees are in $\angle DCB$?

17. The timber $\overline{BF}$ (Fig. 4.31) supports a stairway and makes an angle of 150° with the floor $\overline{AB}$. What size must the carpenter make the angles CBD and CDB so that $\overline{CB} \perp \overline{AB}$ and $\overline{CD} \parallel \overline{AB}$?

18. In making the legs of a picnic table (Fig. 4.32) Jane nails two pieces of wood together so that $\overline{AO} = \overline{OC}$, $\overline{OD} = \overline{OB}$, and $\angle AOC = 50°$. At what exterior angle must she cut the pieces of wood at A and C so that they will stand flat on the floor? How many degrees must angles D and B contain in order that the top of the table will be parallel to the floor (line AC)?

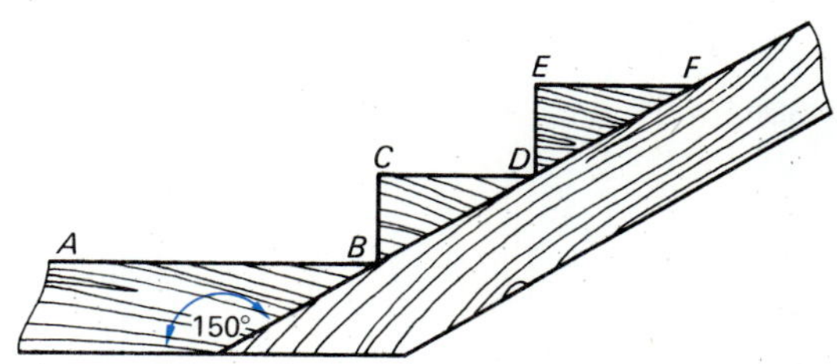

FIGURE 4.31

FIGURE 4.30

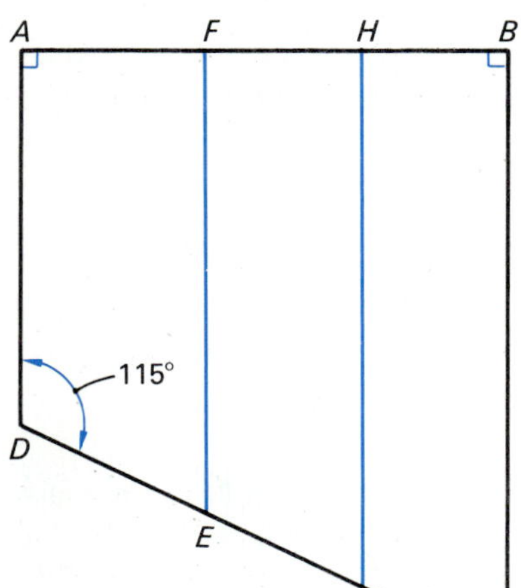

FIGURE 4.32

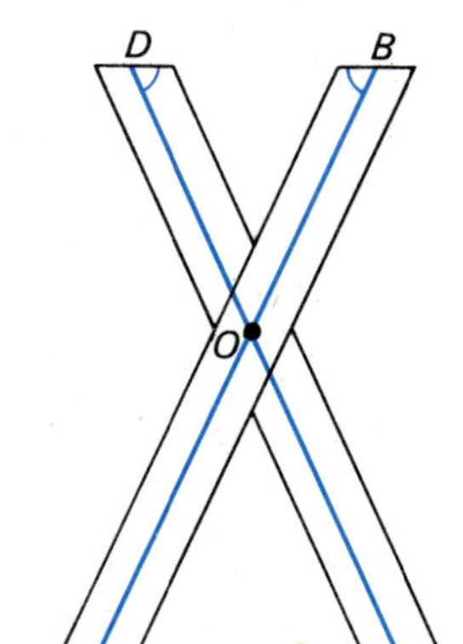

In Exercises 19–24, determine the measure of the missing angles and name the figure.

19.

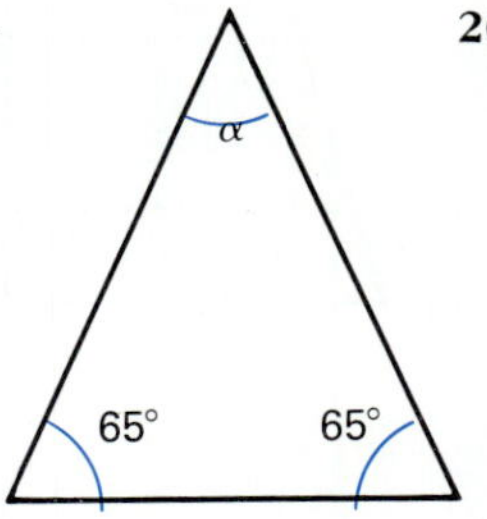

20.

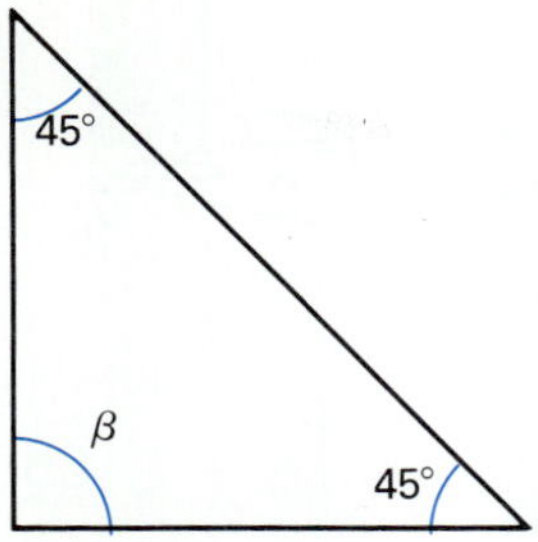

21.

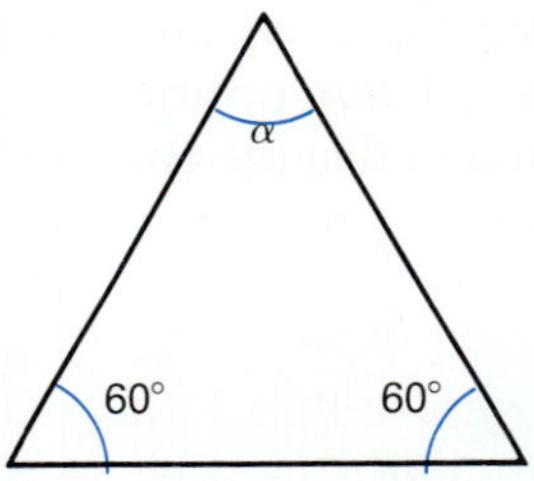

22.

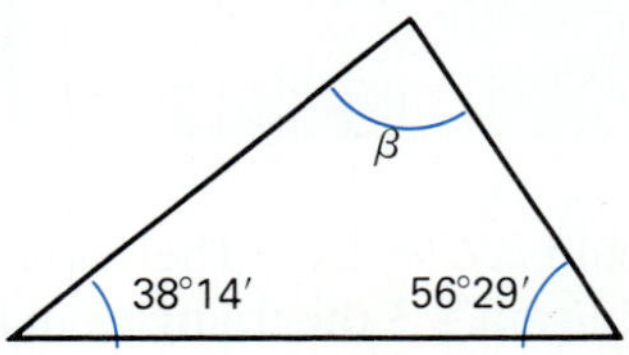

23.

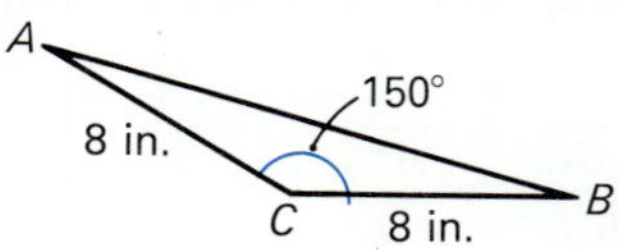

24.

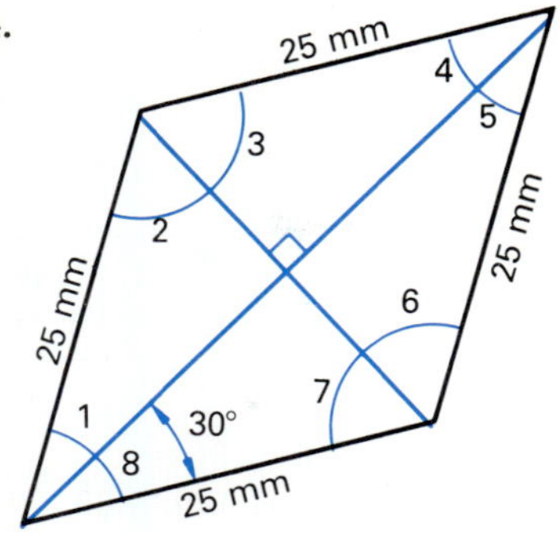

In Exercises 25 and 26, find the measures of the angles indicated on the figure.

25.

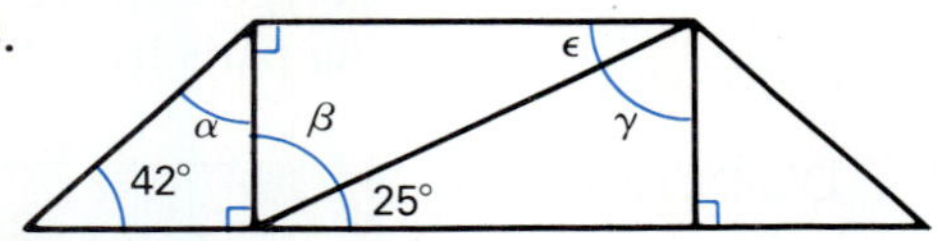

26.

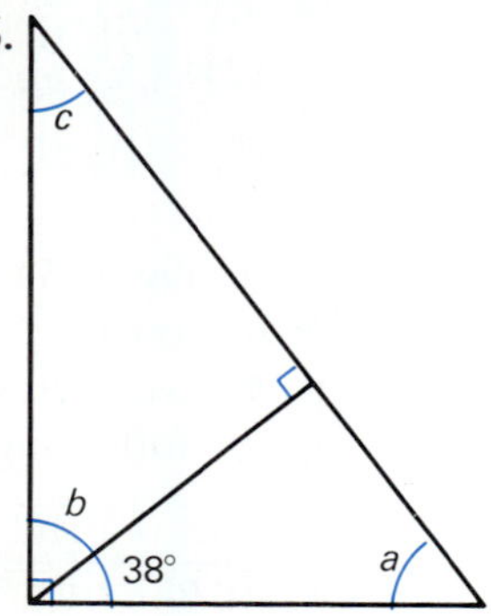

4.4 Similar polygons

A project engineer has the task of constructing the trapezoid-shaped tail section mentioned in Section 4.3. From the drawings of the design engineer how can he determine the actual size? One method of determining the actual size is by using the properties of similar polygons.

DEFINITION 4.3

Similar polygons are polygons that have corresponding angles with equal measure and corresponding sides that are proportional.

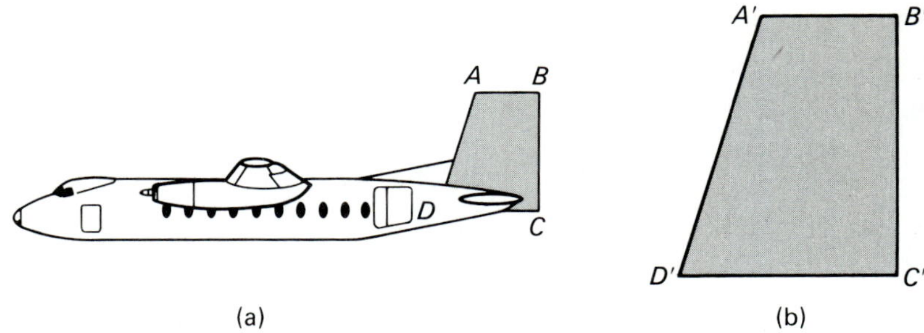

FIGURE 4.33

By corresponding sides and corresponding angles we are referring to those angles and sides that have the same relative position on the polygon. In Fig. 4.33 the trapezoids are labeled so that the corresponding parts have the same letters.

DEFINITION 4.4

> If the measure of the angles of $\triangle ABC$ are equal to the measures of the corresponding angles of $\triangle DEF$, then **$\triangle ABC$ is similar to $\triangle DEF$,** symbolized by $\triangle ABC \sim \triangle DEF$.

In Fig. 4.34 $\angle A = \angle A'$, $\angle B = \angle B'$, and $\angle C = \angle C'$. Therefore, by Definition 4.4 $\triangle ABC \sim \triangle A'B'C'$. By Definition 4.3 the triangles in Fig. 4.34 have corresponding sides that are proportional. Thus, we can set up the following proportions:

$$\frac{\overline{AB}}{\overline{A'B'}} = \frac{\overline{AC}}{\overline{A'C'}}, \qquad \frac{\overline{AB}}{\overline{A'B'}} = \frac{\overline{BC}}{\overline{B'C'}}, \qquad \frac{\overline{BC}}{\overline{B'C'}} = \frac{\overline{AC}}{\overline{A'C'}}.$$

DEFINITION 4.5

> If the corresponding sides of $\triangle ABC$ and $\triangle DEF$ are proportional, then **$\triangle ABC$ is similar to $\triangle DEF$** ($\triangle ABC \sim \triangle DEF$).

Note that in order for polygons with more than three sides to be similar the corresponding sides must be proportional *and* the measures

FIGURE 4.34

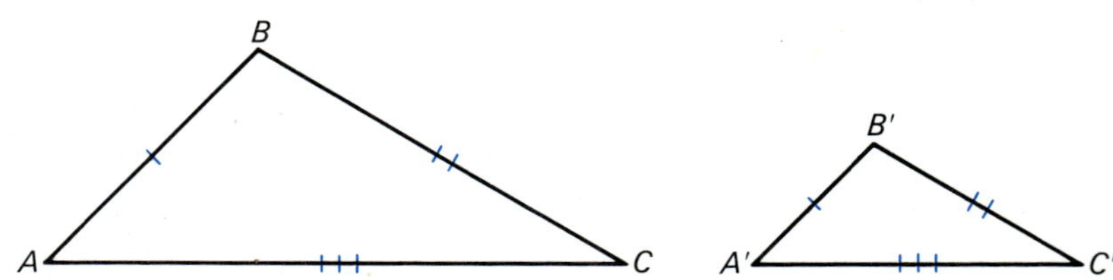

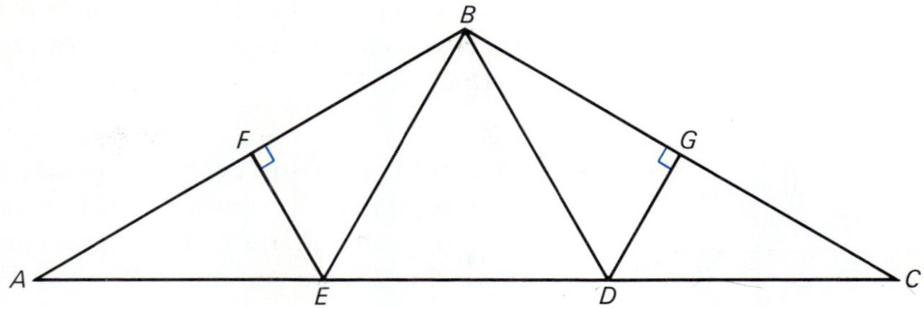

FIGURE 4.35

of the corresponding angles must be equal. For two triangles to be similar we need only to have the corresponding angles equal or the corresponding sides proportional.

EXAMPLE 1 In the sketch of the truss (Fig. 4.35) $\overline{BE} \parallel \overline{GD}$. Show that $\triangle BCE$ is similar to $\triangle CDG$.

Solution: Given $\overline{BE} \parallel \overline{GD}$, show that $\triangle BCE \sim \triangle CDG$. To show that the triangles are similar we must show that the corresponding angles are equal or the corresponding sides are proportional. Since $\angle BCE$ is common to both triangles we have one angle in each triangle with equal measure. Using the given fact that $\overline{BE} \parallel \overline{GD}$, we can state that

$$\angle CDG = \angle CEB \text{ and } \angle DGC = \angle EBC.$$

Why? If two parallel lines are cut by a transversal, then the corresponding angles are equal. Therefore, the measures of the angles of $\triangle BCE$ are equal to the measures of the corresponding angles of $\triangle CDG$. Thus, by Definition 4.4, $\triangle BCE \sim \triangle CDG$. □

EXAMPLE 2 In right $\triangle ABC$, $\angle C$ is a right angle and $\overline{CD} \perp \overline{AB}$ (Fig. 4.36). Show that $\triangle ABC \sim \triangle BCD$.

FIGURE 4.36

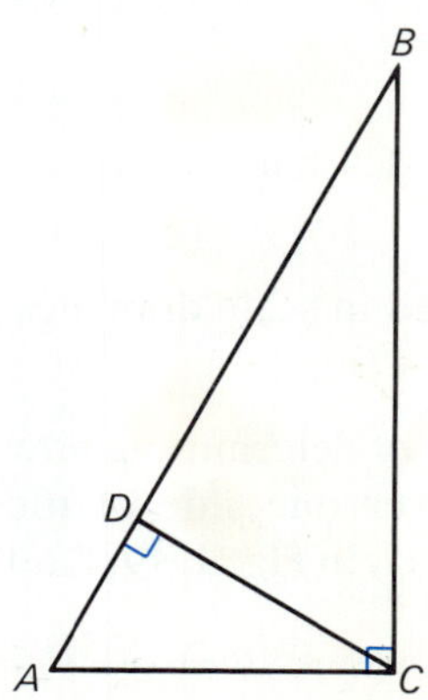

Solution: We can show that the triangles are similar by showing that the measure of the corresponding angles are equal. $\angle CDB$ is a right angle since $\overline{CD} \perp \overline{AB}$. The $\angle CDB = \angle BCA$, since all right angles are equal. Since $\angle DBC$ is common to both triangles, we can say that $\angle DBC = \angle ABC$. The sum of the measures of the angles of a triangle equals 180°. Thus $\angle CDB + \angle DBC + \angle BCD = 180°$ and

$$\angle BCA + \angle ABC + \angle CAB = 180°.$$

Since both statements are equal to 180°, they are equal to each other

$$\angle CDB + \angle DBC + \angle BCD = \angle BCA + \angle ABC + \angle CAB.$$

Substituting equal quantities from above we have

$$\angle BCA + \angle ABC + \angle BCD = \angle BCA + \angle ABC + \angle CAB.$$

Subtracting $\angle BCA$ and $\angle ABC$ from both sides we have $\angle BCD = \angle CAB$. Therefore the corresponding angles are equal and by Definition 4.4: $\triangle ABC \sim \triangle BCD$. □

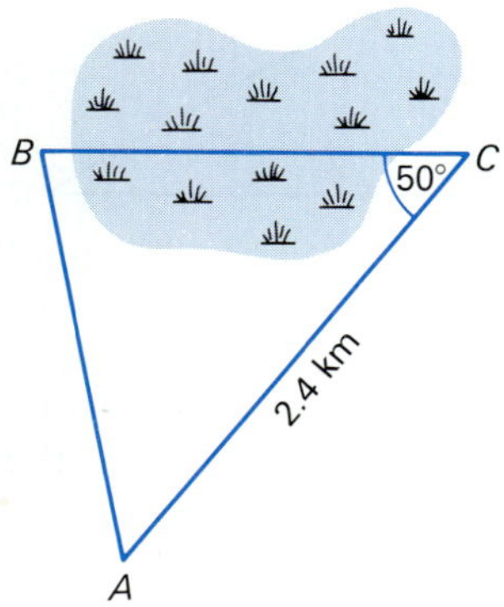

FIGURE 4.37

EXAMPLE 3 A surveyor has the task of determining the length of a swamp (Fig. 4.37). From point A he can clearly see both ends of the swamp, points B and C. He knows that the distance from point A to point C is 2.4 km and that $\angle BCA = 50°$.

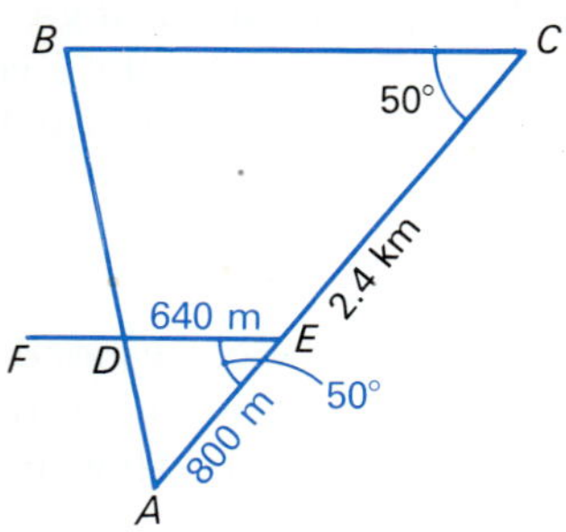

FIGURE 4.38

Solution: The surveyor can find the distance from point B to point C using similar triangles (Fig. 4.38). Standing at point A he sights to the right edge of the swamp, point C. At a distance of 800 m from the point along the line of sight he drives a stake in the ground, point E. He then constructs an angle of 50° with its vertex at point E, forming $\angle FEA$. He then sights to the left edge of the swamp, point B, and he names the point D where $\overleftrightarrow{AB}$ and $\overleftrightarrow{FE}$ intersect. He measures the length of $\overline{DE}$, which is 640 m. The lines $\overleftrightarrow{DE}$ and $\overleftrightarrow{BC}$ are parallel since they form equal corresponding angles with the transversal $\overleftrightarrow{AC}$. From the work in Example 2 we can conclude that when two angles of one triangle have the same measures as the corresponding angles of the other triangle, the third angles also have equal measures. From this we can conclude that $\angle ADE = \angle ABC$ and $\triangle ABC \sim \triangle ADE$. Since the triangles are similar the corresponding sides are proportional and we can set up the following proportion:

$$\frac{\overline{BC}}{\overline{DE}} = \frac{\overline{AC}}{\overline{AE}}.$$

Substituting:

$$\frac{\overline{BC}}{640 \text{ m}} = \frac{2.4 \text{ km}}{800 \text{ m}}$$

$$\overline{BC} = \frac{(2.4 \text{ km})(640 \overset{1}{\cancel{\text{m}}})}{800 \underset{1}{\cancel{\text{m}}}}$$

$$\overline{BC} = 1.92 \text{ km}.$$

The surveyor found the length of the swamp to be 1.9 km. □

A practical application of similar figures is used in scale drawings.

EXAMPLE 4 A sheet-metal mechanic has the task of determining how large a piece of material will be needed to cover one side of the trapezoidal tail piece. He has a scale drawing as given in Fig. 4.39. Find the dimensions of the actual tail piece.

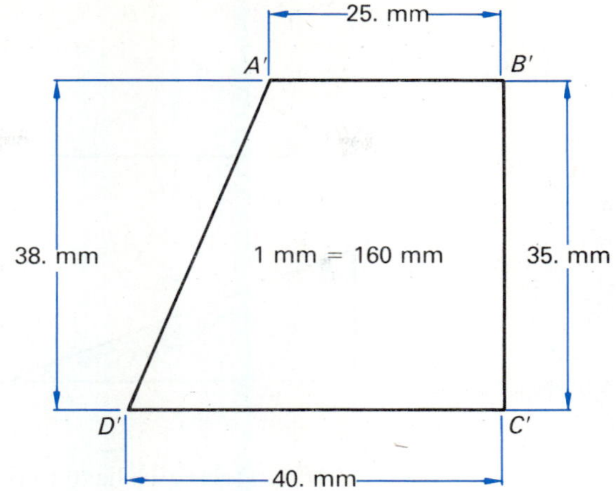

FIGURE 4.39

Solution: The scale drawing and the actual tail piece are similar polygons. Thus, to calculate the actual length of each side the mechanic can set up the following proportions using the lengths given in the diagram and the scale (1 mm = 160 mm).

$$\frac{\overline{AB}}{25.\text{ mm}} = \frac{160\text{ mm}}{1\text{ mm}} \qquad \overline{AB} = \frac{160\text{ mm}}{1\text{ mm}}(25.\text{mm})$$

$$\overline{AB} = 4000.\text{ mm}$$

$$\frac{\overline{BC}}{35.\text{ mm}} = \frac{160\text{ mm}}{1\text{ mm}} \qquad \overline{BC} = \frac{160\text{ mm}}{1\text{ mm}}(35.\text{ mm})$$

$$\overline{BC} = 5600.\text{ mm}$$

$$\frac{\overline{CD}}{40.\text{ mm}} = \frac{160\text{ mm}}{1\text{ mm}} \qquad \overline{CD} = \frac{160\text{ mm}}{1\text{ mm}}(40.\text{ mm})$$

$$\overline{CD} = 6400.\text{ mm}$$

$$\frac{\overline{DA}}{38.\text{ mm}} = \frac{160\text{ mm}}{1\text{ mm}} \qquad \overline{DA} = \frac{160\text{ mm}}{1\text{ mm}}(38.\text{ mm})$$

$$\overline{DA} = 6080.\text{ mm.}$$

Thus, the mechanic needs a piece of material at least 6.4 m wide and 5.6 m long. □

This technique of determining lengths from scale drawings can be used to determine actual distances between cities from a map.

Congruent triangles are a special case of similar triangles for which the corresponding sides are equal. (Congruent triangles have corresponding angles equal *and* corresponding sides equal.) The following list contains important statements about congruent triangles.

FIGURE 4.40

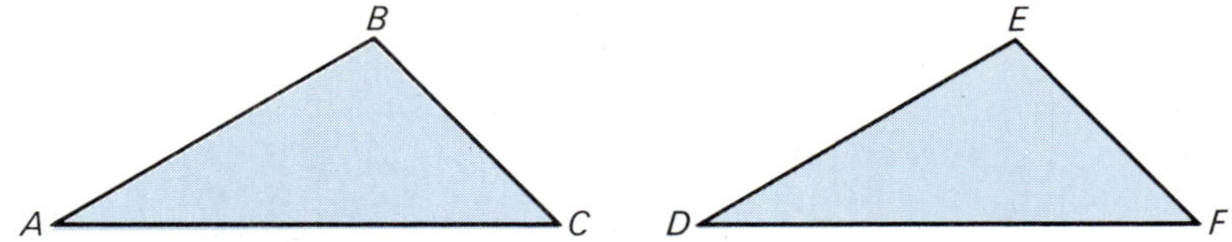

FIGURE 4.41

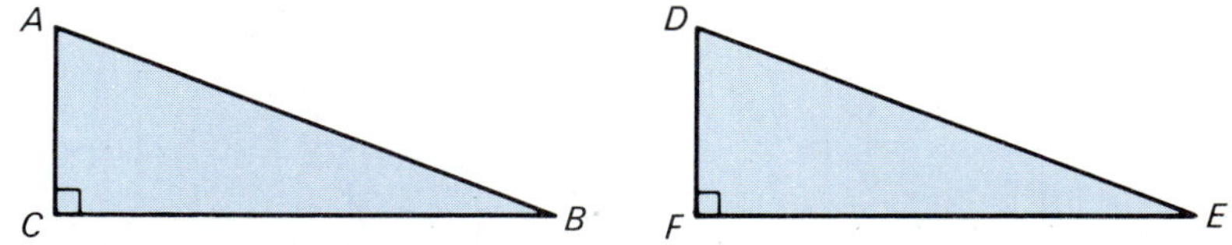

1. If two triangles have two sides and the included angle (SAS) of one equal respectively to two sides and the included angle of the other, then the triangles are congruent. In Fig. 4.40, $\angle A = \angle D, \overline{AB} = \overline{DE}$, and $\overline{AC} = \overline{DF}$.
2. If two triangles have the three sides of one equal, respectively, to the three sides (SSS) of the other, then the triangles are congruent. In Fig. 4.40, $\overline{AB} = \overline{DE}$, $\overline{AC} = \overline{DF}$, and $\overline{BC} = \overline{EF}$.
3. If two triangles have two angles and the included side (ASA) of one equal, respectively, to two angles and the included side of the other then the triangles are congruent. In Fig. 4.40, $\angle A = \angle D$, $\angle B = \angle E$, and $\overline{AB} = \overline{DE}$.
4. If two triangles have two angles and a side of one equal to two angles and the corresponding side of the second, then the triangles are congruent. In Fig. 4.40, $\angle A = \angle D$, $\angle B = \angle E$, and $\overline{BC} = \overline{EF}$.
5. If two right triangles have the hypotenuse and an acute angle of one equal, respectively, to the hypotenuse and acute angle of the other, the triangles are congruent. In Fig. 4.41, $\overline{AB} = \overline{DE}$ and $\angle A = \angle D$.
6. If two right triangles have the hypotenuse and a leg of one equal, respectively, to the hypotenuse and a leg of the other, then the triangles are congruent. In Fig. 4.41, $\overline{AB} = \overline{DE}$ and $\overline{AC} = \overline{DF}$.

EXAMPLE 5 Show that a diagonal of a parallelogram divides it into two congruent triangles.

FIGURE 4.42

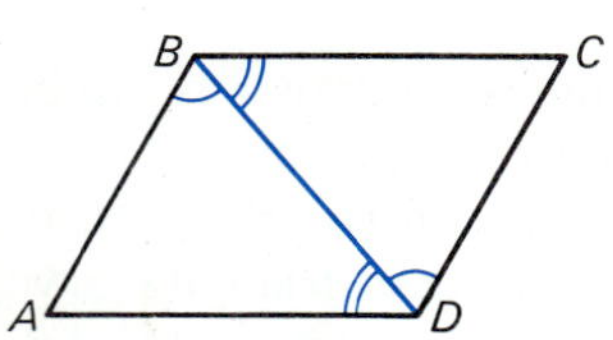

Solution: Draw any parallelogram $ABCD$ and a diagonal $\overline{BD}$ as in Fig. 4.42. $\angle DBC = \angle BDA$ and $\angle CDB = \angle ABD$, since they are alternate interior angles of the pairs of parallel lines $\overline{AB}$, $\overline{DC}$ and $\overline{BC}$, $\overline{AD}$. $\overline{BD} = \overline{BD}$, since any quantity is equal to itself. Therefore, $\triangle ABD \cong \triangle BCD$* since they have two angles and an included side, respectively, equal (Statement 3). □

* $\cong$ means "is congruent to."

4.4 Exercises

In Exercises 1 – 4, state the definition that tells why the triangles are similar, and then determine the values of the missing parts α, β, γ, ϕ, θ, a, b, t, x, y as indicated.

1.

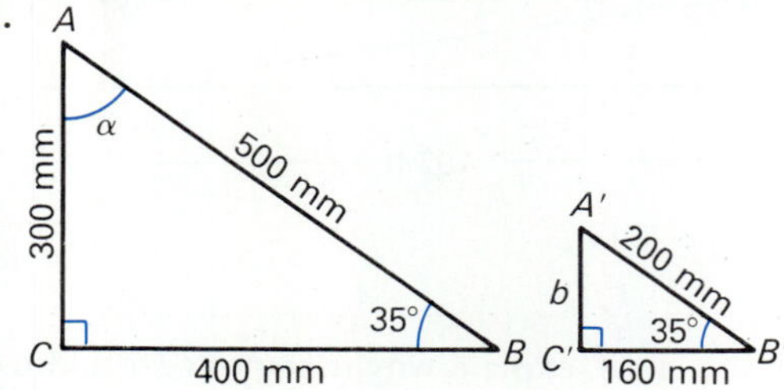

2.

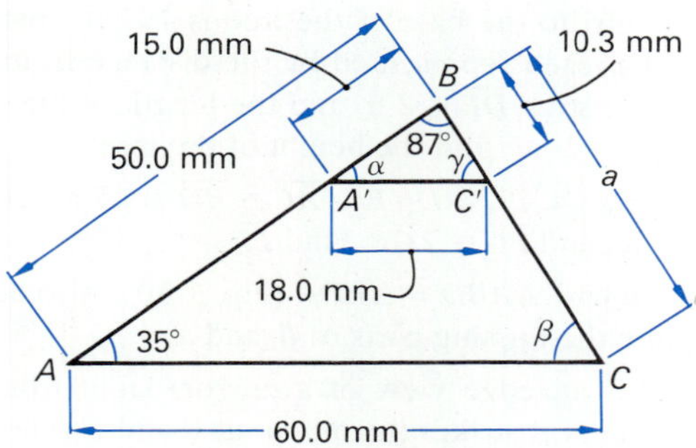

3.

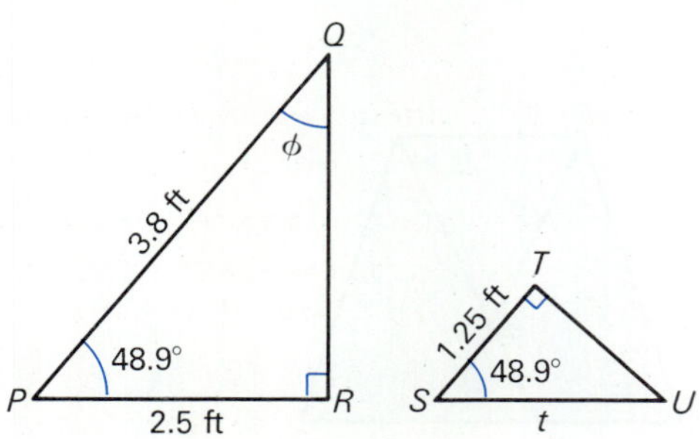

4.

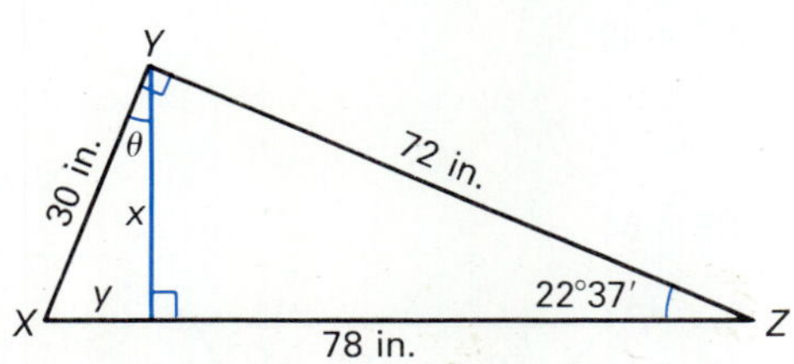

5. A tree 12 ft tall stands on the ground 542 ft from a vertical cliff. A boy whose eyes are 5′5″ above the ground level, standing 125 ft from the tree as shown in Fig. 4.43, sees the top of the cliff in line with the top of the tree. How high is the cliff?

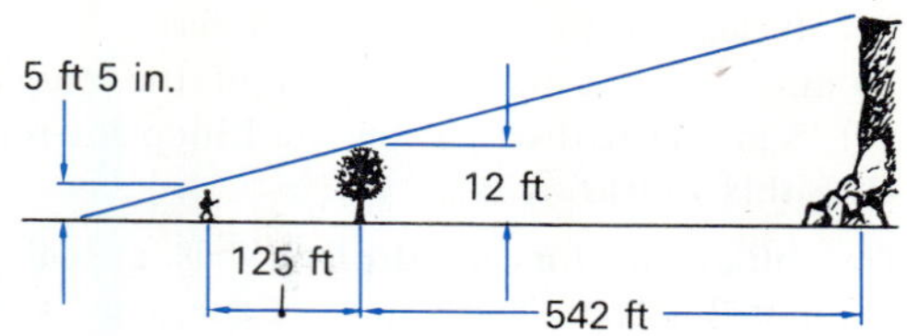

FIGURE 4.43

6. In making an aerial photograph of a city, the photographer made the picture to the scale 1 in. = 1600 ft. If the focal length (distance from lens to image) of the camera in Fig. 4.44 was 6 3/4 in., how high was the photographer when he made the picture?

7. A film 3 in. high is being projected on a screen. How high will the projection be if the light is 7 in. behind the film, and the light is 6 ft 5 in. from the screen?

8. Two similar polygons have a ratio of similitude (ratio between corresponding sides) of 5 : 12. If the sum of their perimeters is 360 in., find the perimeter of each in feet.

9. A flag pole casts a shadow of 65 m and a stick 1 1/2 m in length casts a shadow of 5 m. How tall is the flag pole?

10. In constructing the triangular truss, RST, in Fig. 4.45, the engineer found it necessary to add brace $\overline{XY}$, which is parallel to $\overline{RS}$. How long must the brace be if $\overline{ST} = 15$ m, $\overline{SY} = 8$ m, and $\overline{RS} = 11$ m?

FIGURE 4.44

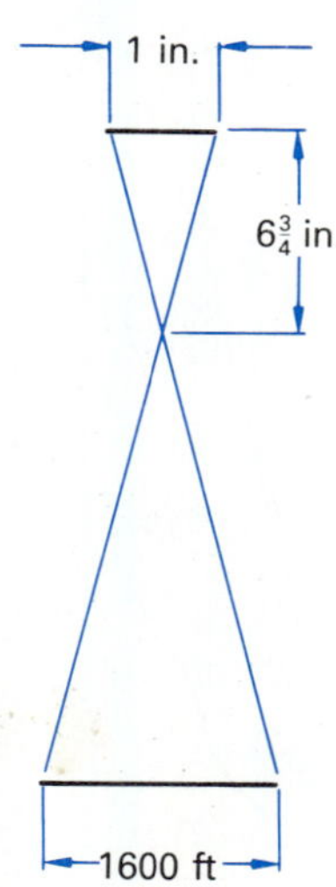

FIGURE 4.45

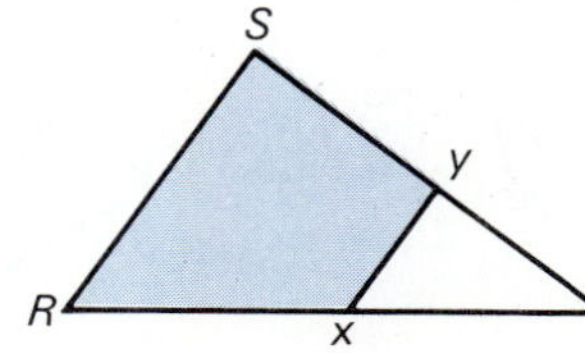

11. The blueprint for a solar panel has a scale of 20 mm = 1 1/2 m. The width of the panel is 4 1/5 m. What distance on the blueprint represents this width?

12. The blueprint for an airplane has a scale of 5 mm = 3 1/2 m. The length of the wing on the blueprint is 36 mm. What is the actual length of the wing?

13. On a map the scale is 1/2 in. = 75 mi. If the actual distance from Rochester, N.Y., to Philadelphia, Pa., is 330 mi, what is the distance between the two cities on the map?

14. On a map the scale is 1 mm = 8 km. If the distance on the map from Buffalo to San Francisco is 560 mm, what is the actual distance in kilometers?

15. In Fig. 4.46, $\angle ACB = \angle CAB = \angle ABC$ and $\overline{AB} =$ 275 km. What is the length of $\overline{AC}$? Explain your answer.

16. Use the relationships of similar triangles to find the distance $\overline{DE}$ across the lake as shown in Fig. 4.47 given $\angle CED = \angle ABC$. Explain how you determined the length of $\overline{DE}$.

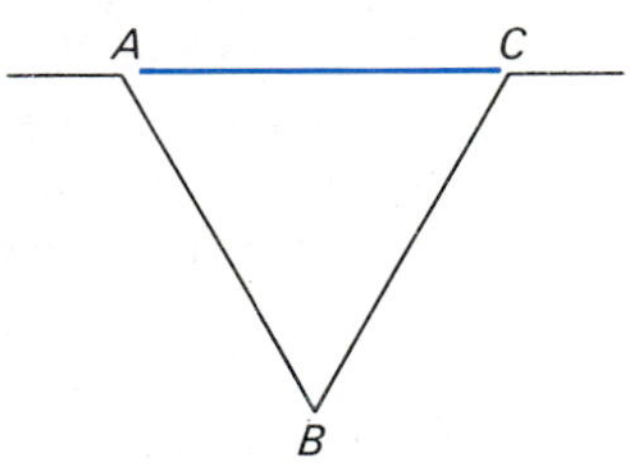

FIGURE 4.46

FIGURE 4.47

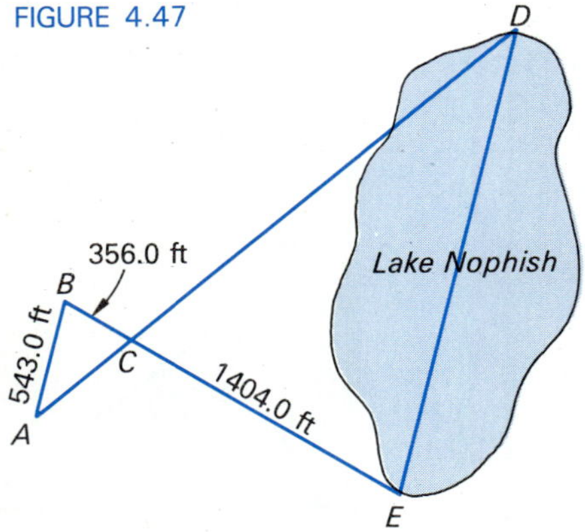

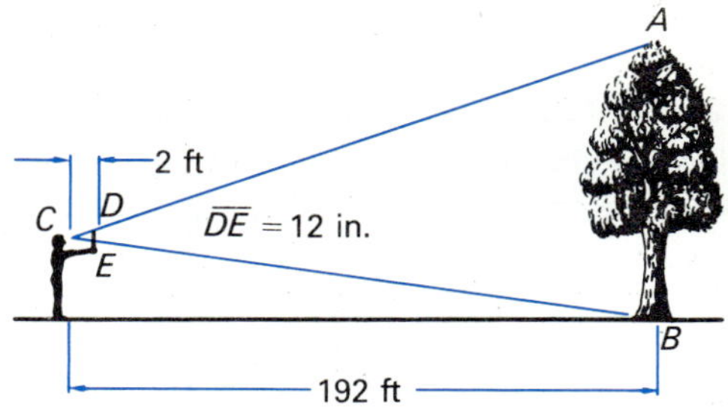

FIGURE 4.48

17. In Fig. 4.48, explain why triangles CDE and CAB are similar. If you assume that the distance from the girl to the base of the tree is 192 ft, that the girl's reach represented by the distance from her eye to stick $\overline{DE}$ is 2 ft, and the length of the stick $\overline{DE}$ is 12 in., find the height of the tree.

18. In Fig. 4.49, $\overline{AD} \parallel \overline{EF}$, $\overline{AC} = \overline{BD} = 15$ m, $\overline{AD} =$ 24 m, and $\overline{FG} = 2\overline{GB}$. Find $\overline{EF}$.

19. State why $\triangle ABD \cong \triangle BDC$ (Fig. 4.50). Also determine the missing parts α, β, and x.

20. MB is an edge view of a mirror. Light from an object at A strikes the mirror at C and is reflected

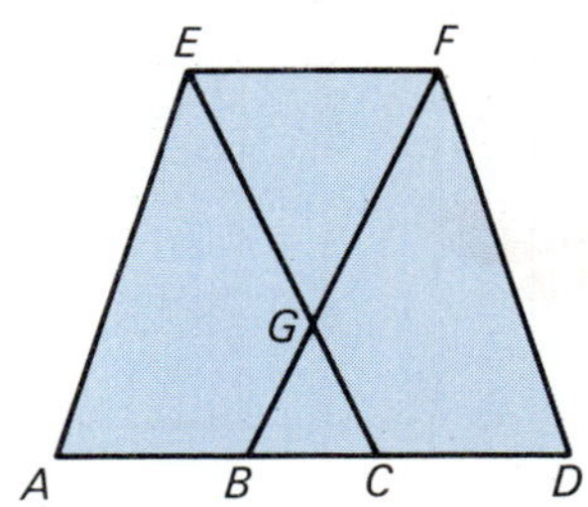

FIGURE 4.49

FIGURE 4.50

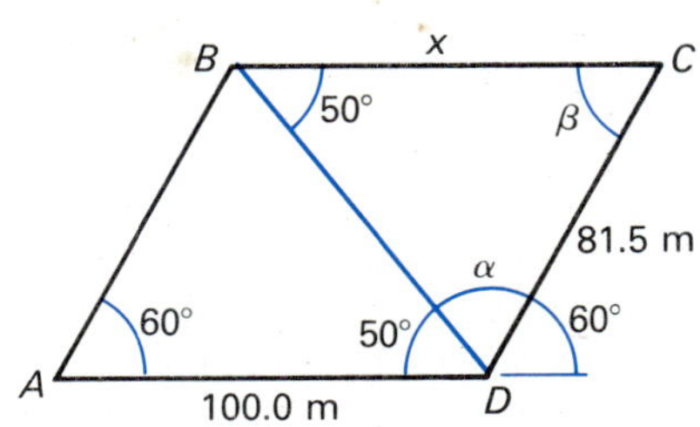

to the eye at E (Fig. 4.51). The mind projects the mirror to A', forming the image at A'. It is known that $\angle MCE = \angle ACB$ and that $\overline{CB} \perp \overline{AA'}$. Prove that $\overline{AB} = \overline{A'B}$.

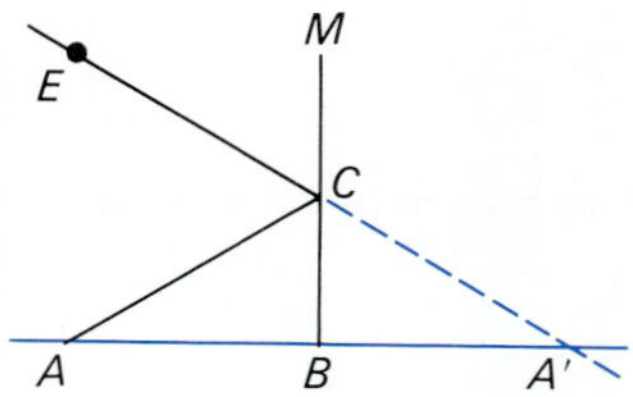

FIGURE 4.51

21. The stress on a member of a bridge assembly (Fig. 4.52) is found by determining the distance $\overline{OC}$. O is the intersection of $\overleftrightarrow{CA}$ and $\overleftrightarrow{DE}$. What is the distance $\overline{OC}$?

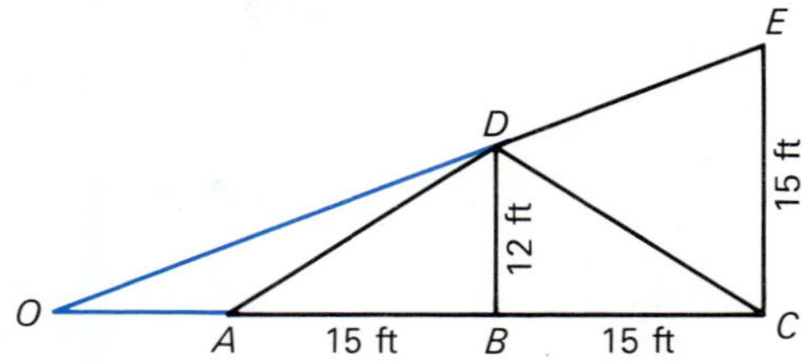

FIGURE 4.52

4.5 Perimeter and area of plane figures

The **perimeter** of a plane figure is the distance around the figure. For the tail section in Example 4 in Section 4.4, the perimeter is the sum of the lengths of the line segments $\overline{AB}$, $\overline{BC}$, $\overline{CD}$, and $\overline{DA}$. These lengths, respectively, are 4.0 m, 5.6 m, 6.4 m, and 6.1 m. The perimeter of the trapezoid is 22.1 m. The formulas used to find the perimeters and areas of plane figures are given in Table 4.2.

In finding the area of a plane figure such as a trapezoid, we say, "Find the area of the trapezoid." This is misleading in the sense that the trapezoid is the shape formed by four line segments, whereas the **area** is the region bounded by the line segments.

EXAMPLE 1 Find the area of the trapezoid-shaped tail section (Fig. 4.53).

FIGURE 4.53

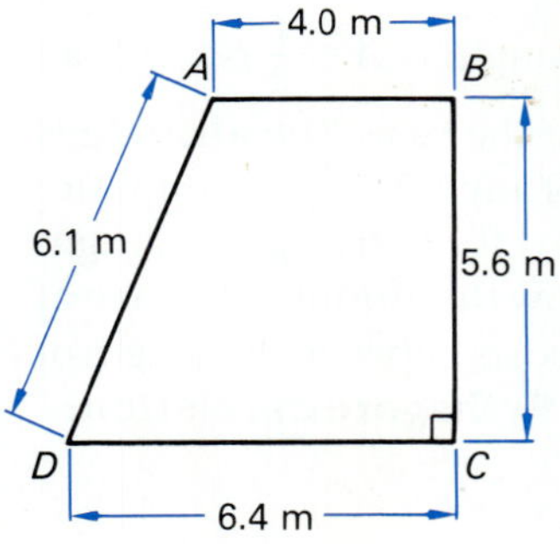

Solution: The formula for finding the area of a trapezoid is given in Table 4.2, $A = \frac{1}{2}(b_1 + b_2)a$. The lengths of the parallel sides of the trapezoid are b_1 and b_2. The perpendicular distance between these two sides is a, called the altitude. In our tail section $\overline{AB} \parallel \overline{CD}$, thus, let $b_1 = 4.0$ m and $b_2 = 6.4$ m. Since $\overline{BC} \perp \overline{DC}$ and $\overline{BC} \perp \overline{AB}$, we can let $a = \overline{BC}$, which is 5.6 m. Substituting in the equation:

$$\begin{aligned} A &= \frac{1}{2}(4.0 \text{ m} + 6.4 \text{ m})(5.6 \text{ m}) \\ &= \frac{1}{2}(10.4 \text{ m})(5.6 \text{ m}) \\ &= 29.12 \text{ m}^2 \end{aligned}$$

The area of the trapezoid is 29. m² to two significant digits. □

TABLE 4.2 **Perimeters and Areas of Plane Figures**

Figure	Sketch	Perimeter (P)	Area (A)
Triangle	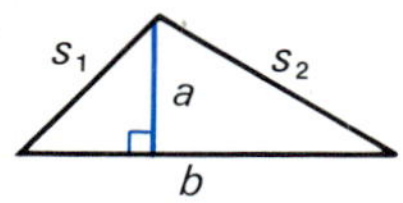	$P = s_1 + s_2 + b$	$A = \frac{1}{2} ba$
Trapezoid	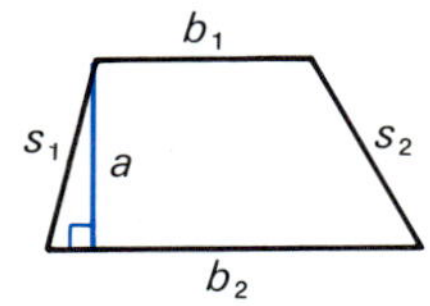	$P = s_1 + s_2 + b_1 + b_2$	$A = \frac{1}{2}(b_1 + b_2)a$
Parallelogram	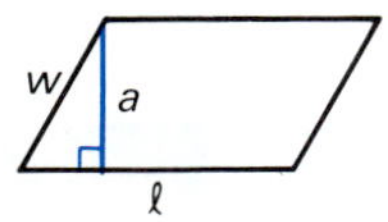	$P = 2\ell + 2w$	$A = \ell a$
Rhombus	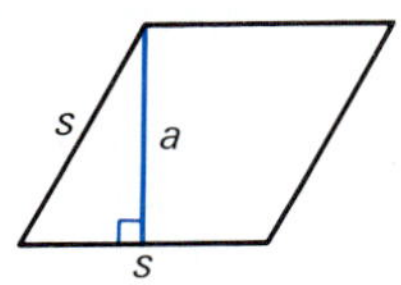	$P = 4s$	$A = sa$
Rectangle	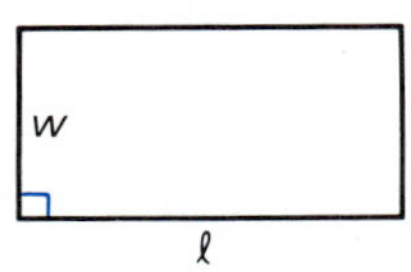	$P = 2\ell + 2w$	$A = \ell w$
Square	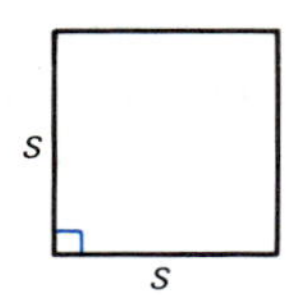	$P = 4s$	$A = s^2$

The formula for finding the area of a triangle is $A = \frac{1}{2} ba$. a is an altitude of the triangle, and b is the length of the base. The **altitude** is the perpendicular distance from the vertex of any angle to the opposite side (or opposite side extended) (Fig. 4.54). If the triangle is a right triangle, we have a special case in that we know the altitude, a, is one of the sides. In addition, a special relationship exists between the length of the sides of the right triangle, known as the **Pythagorean relation:**

FIGURE 4.54

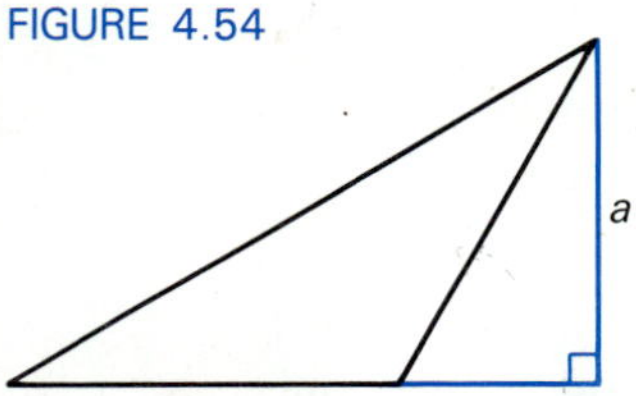

$$c^2 = a^2 + b^2.$$

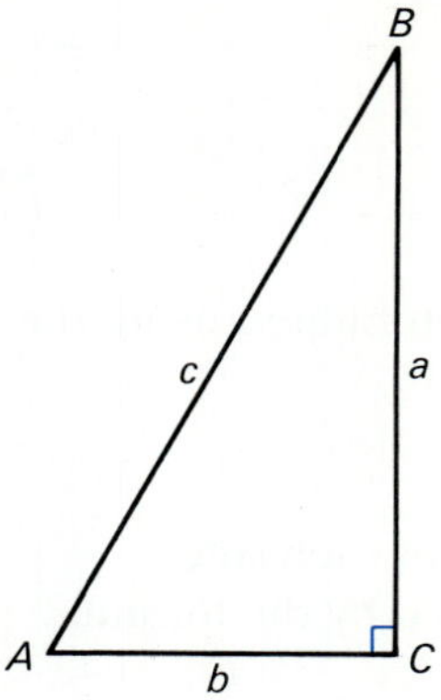

FIGURE 4.55

Figure 4.55 shows a conventional labeling of a right triangle. Note that the right angle is called $\angle C$ and the side opposite the right angle is called side c. A similar situation holds for the other two angles and sides: side a is opposite $\angle A$ and side b is opposite $\angle B$.

EXAMPLE 2 Find the perimeter and area of the right triangle in Fig. 4.56.

Solution: In order to find the perimeter of the triangle, we need to know the length of side b. The length of side b can be found by using the Pythagorean theorem.

$$c^2 = a^2 + b^2.$$

Solving for b

$$b^2 = c^2 - a^2$$
$$b = \sqrt{c^2 - a^2}.$$

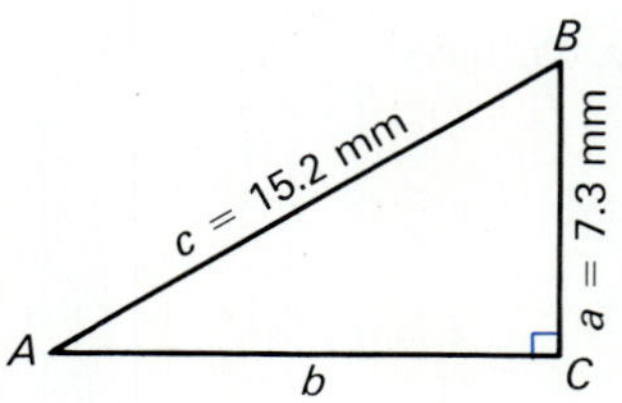

FIGURE 4.56

Substituting:

$$\begin{aligned} b &= \sqrt{(15.2\text{ mm})^2 - (7.3\text{ mm})^2} \\ &= \sqrt{231.04\text{ mm}^2 - 53.29\text{ mm}^2} \\ &= \sqrt{177.75\text{ mm}^2} \\ &= 13.3\text{ mm} \end{aligned}$$

Now we can find the perimeter using a variation of the formula given in Table 4.2, where $a = s_1$ and $c = s_2$.

$$\begin{aligned} P &= a + b + c \\ &= 7.3\text{ mm} + 13.3\text{ mm} + 15.2\text{ mm} \qquad \textbf{Substituting} \\ &= 36.\text{ mm}. \end{aligned}$$

We find the area by substituting in the formula

$$\begin{aligned} A &= \frac{1}{2}ba \\ &= \frac{1}{2}(13.3\text{ mm})(7.3\text{ mm}) \qquad \textbf{Substituting} \\ &= 49.\text{ mm}^2. \end{aligned}$$ □

EXAMPLE 3 Find the perimeter and area of the parallelogram in Fig. 4.57.

FIGURE 4.57

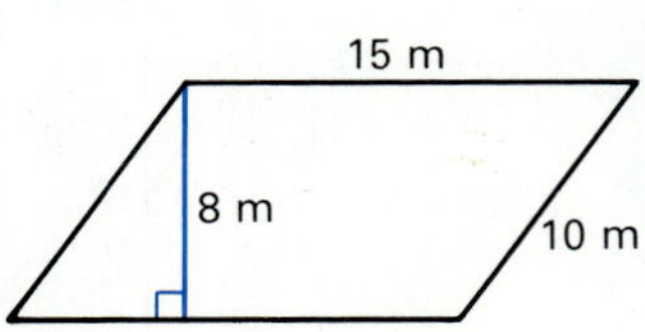

Solution: To find the perimeter we use the formula

$$\begin{aligned} P &= 2\ell + 2w \\ &= 2(15\text{ m}) + 2(10\text{ m}) \qquad \textbf{Substituting} \\ &= 30\text{ m} + 20\text{ m} \\ &= 50\text{ m}. \end{aligned}$$

The area is found by using the formula

$$\begin{aligned} A &= \ell a \\ &= (15 \text{ m})(8 \text{ m}) \quad \textbf{Substituting} \\ &= 120 \text{ m}^2. \quad \square \end{aligned}$$

The perimeter of a regular polygon can be determined using the formula

$$P = ns$$

where n is the number of sides and s is the length of each side.

The area of a regular polygon can be determined by the formula

$$A = n\left[\frac{1}{2}(as)\right].$$

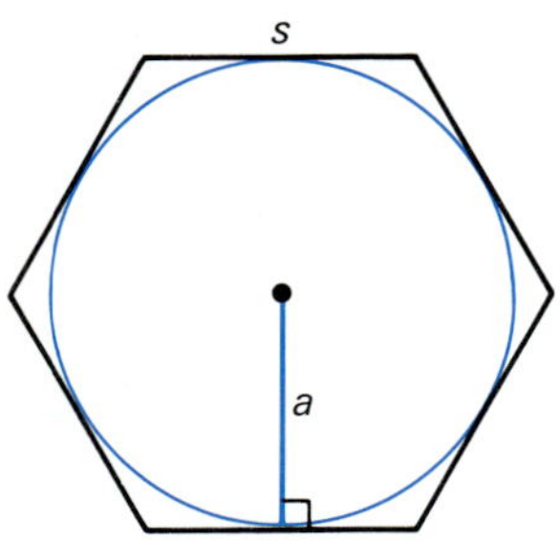

FIGURE 4.58

In the formula, a, called the **apothem,** is the radius of the inscribed circle, or the perpendicular distance from the center of the polygon to one side (Fig. 4.58).

If we rearrange the factors in the formula we have

$$A = \left(\frac{1}{2}a\right)(ns),$$

where

$$ns = P \quad \textbf{P is the perimeter}$$

$$A = \frac{1}{2}aP \quad \textbf{Substituting for ns and solving for A}$$

FIGURE 4.59

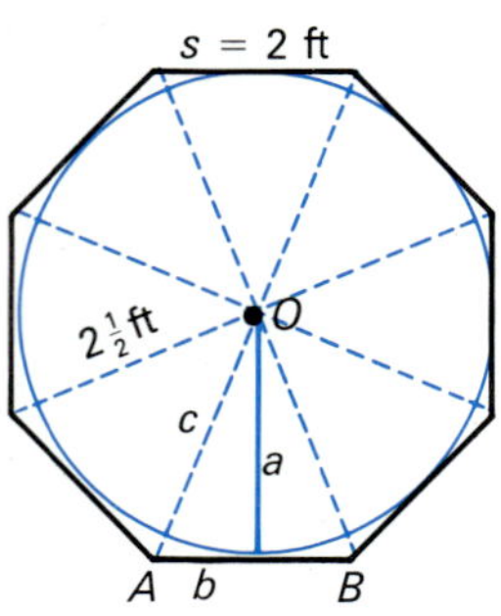

EXAMPLE 4 Find the perimeter and area of the octagonally shaped regular polygon in Fig. 4.59.

Solution: To find the perimeter we use the formula

$$\begin{aligned} P &= ns \\ &= 8(2 \text{ ft}) \quad \textbf{Substituting} \\ &= 16 \text{ ft}. \end{aligned}$$

The formula for finding the area is $A = 1/2\ aP$. Before we can find the area, we need to know the length of the apothem, a. The lines joining the vertices of the octagon and the center form eight congruent isosceles triangles. Thus, $\overline{AO} = \overline{BO} = 2\ 1/2$ ft. The altitude of an isosceles triangle bisects the base. Thus the apothem will bisect $\overline{AB}$ and can be found by using the Pythagorean theorem.

$$\begin{aligned} c^2 &= a^2 + b^2 \\ a &= \sqrt{c^2 - b^2} \quad \textbf{Solving for } \boldsymbol{a} \\ &= \sqrt{(2.5 \text{ ft})^2 - (1 \text{ ft})^2} \quad \textbf{Substituting} \\ &= \sqrt{6.25 \text{ ft}^2 - 1 \text{ ft}^2} \\ &= \sqrt{5.25} \text{ ft}. \end{aligned}$$

Substituting in the formula for the area:

$$A = \frac{1}{2} aP$$
$$= \frac{1}{2} (\sqrt{5.25} \text{ ft})(16 \text{ ft})$$
$$= 8\sqrt{5.25} \text{ ft}^2 \quad \text{or} \quad 18. \text{ ft}^2. \quad \square$$

This technique will work for finding the area of any **regular polygon.**

The formulas for finding the **circumference,** c (distance around), and the **area,** A, of a circle are found in Table 4.3.

EXAMPLE 5 In performing a pressure test an engineer needs to know the area of the shaded region in Fig. 4.60. Find the area of the shaded region.

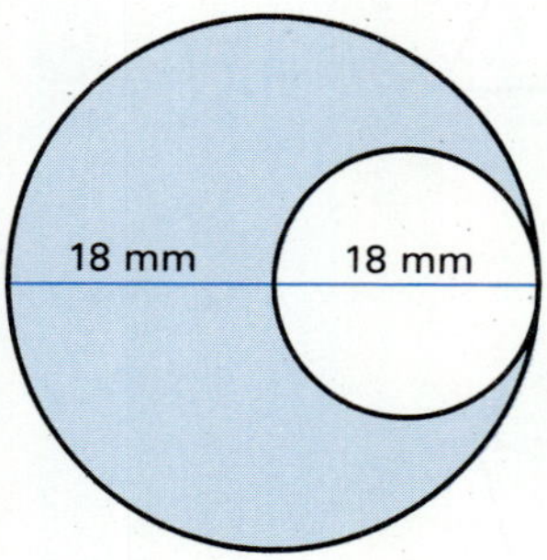

FIGURE 4.60

Solution: The area of the shaded region can be found by subtracting the area, A_1, of the smaller circle from the area, A_2, of the larger circle.

$A = A_2 - A_1$	
$= \pi {r_2}^2 - \pi {r_1}^2$	**r_2 and r_1 are the radii of the respective circles**
$= \pi({r_2}^2 - {r_1}^2)$	**Common-term factoring**
$= \pi[(18 \text{ mm})^2 - (9 \text{ mm})^2]$	**Substitute for r_2 and r_1.**
$= 243\pi \text{ m}^2$	**Square and then subtract.**
$= 763. \text{ mm}^2. \quad \square$	

It is left to the reader to show that we do not obtain the same result if we find the difference of r_1 and r_2 and then square.

TABLE 4.3

Figure	Sketch	Circumference (c)	Area (A)
Circle	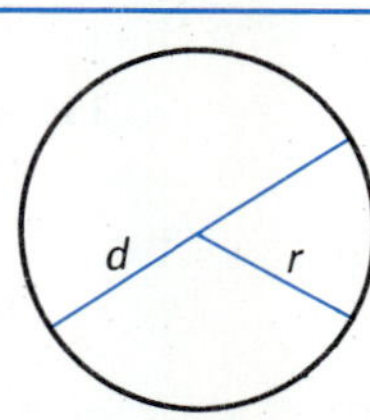	$c = 2\pi r$ $c = \pi d$	$A = \pi r^2$ $A = \frac{\pi d^2}{4}$

4.5 Exercises

In Exercises 1 – 10, find the perimeter and area of the figures.

1.

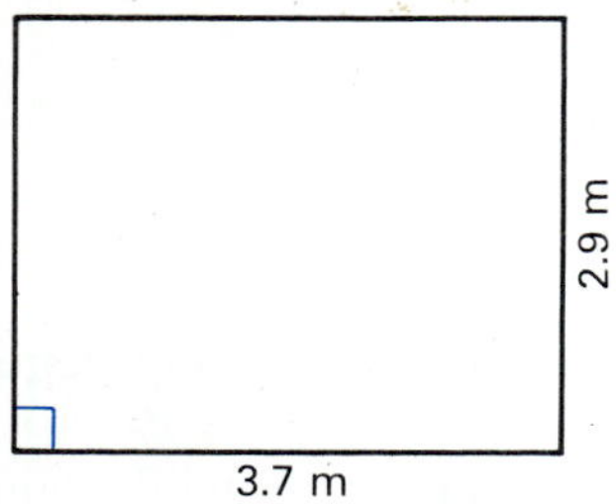

2.

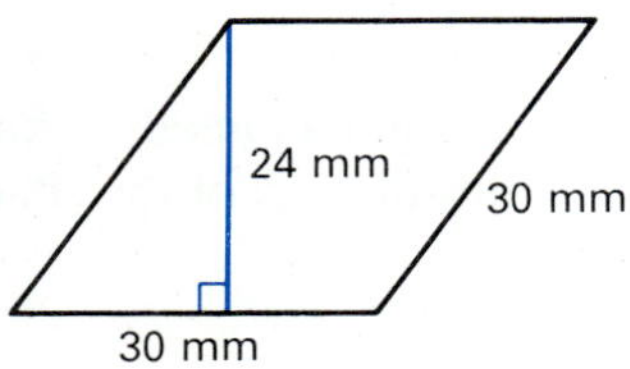

3.

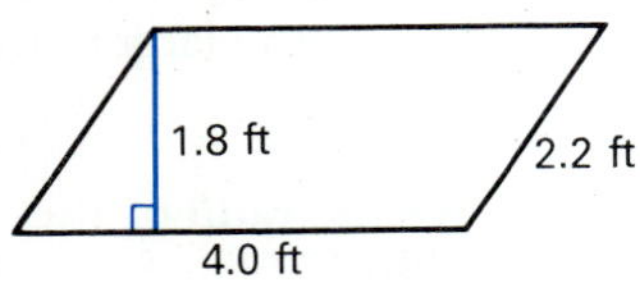

4.

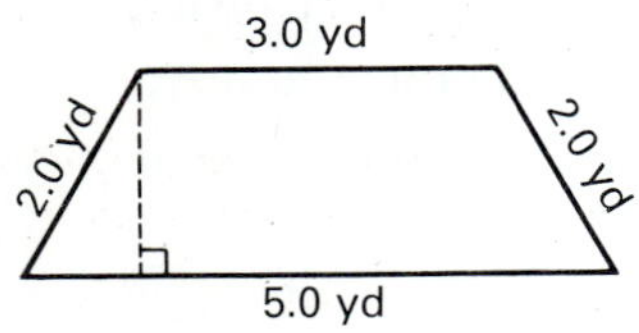

5.

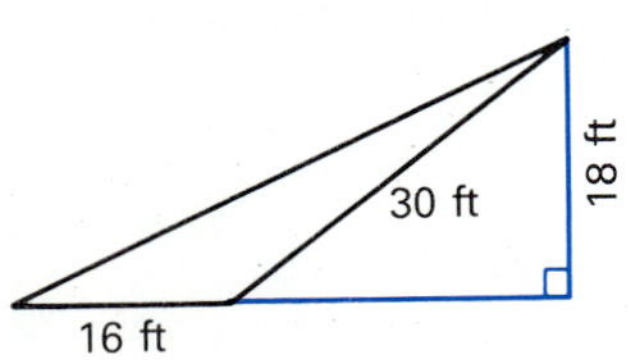

6.

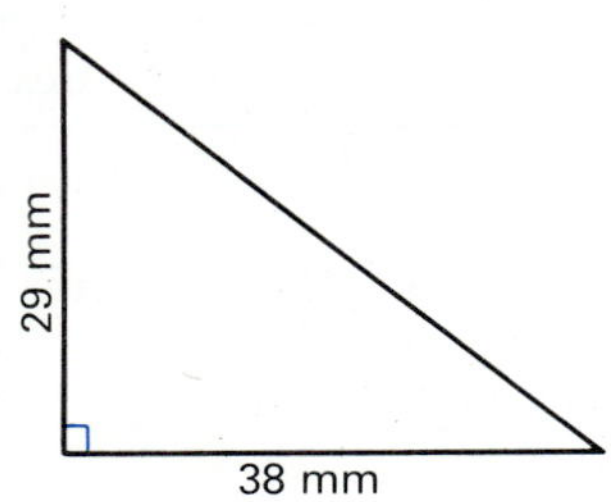

7.

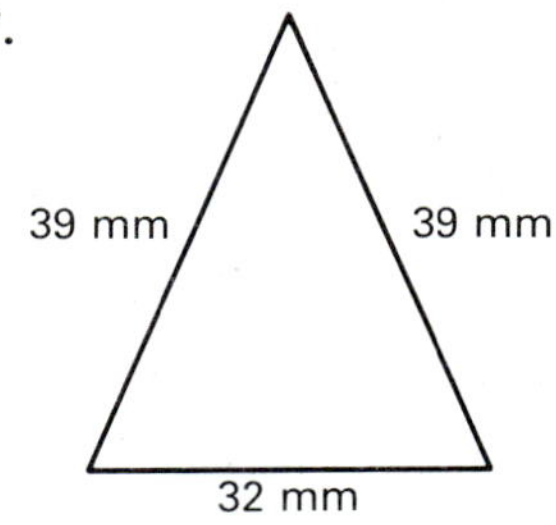

8.

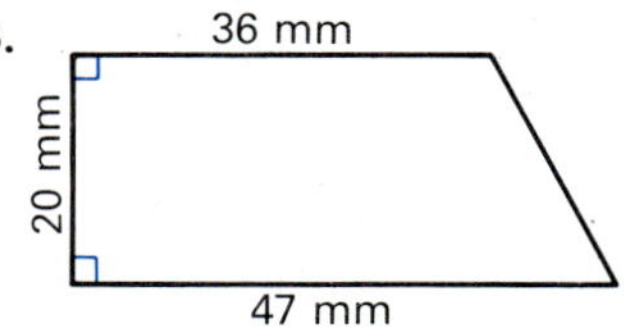

9.

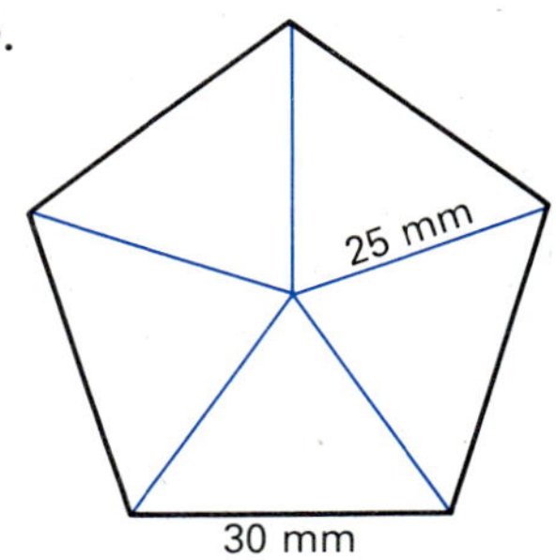

10.

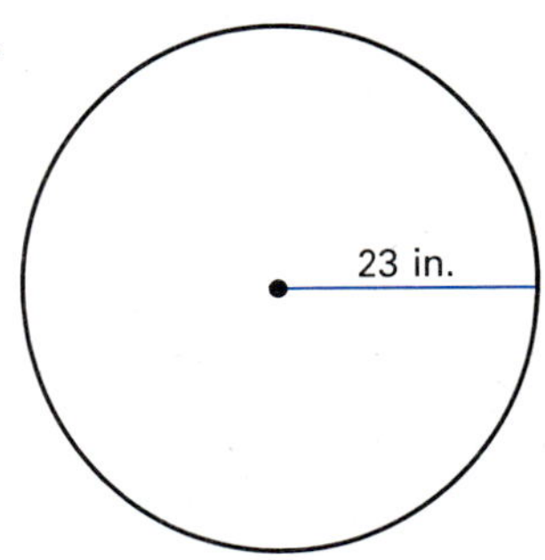

11. A bolt with a square head is cut from a round bronze rod as shown in Fig. 4.61. Four equal pieces are cut away.

a. What is the size of the central angle drawn in Fig. 4.61?

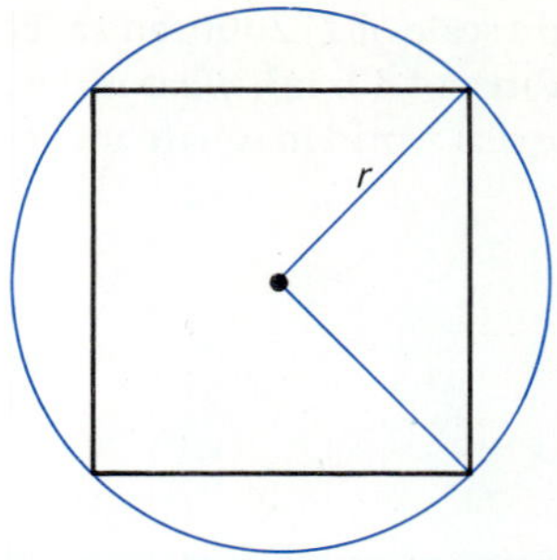

FIGURE 4.61

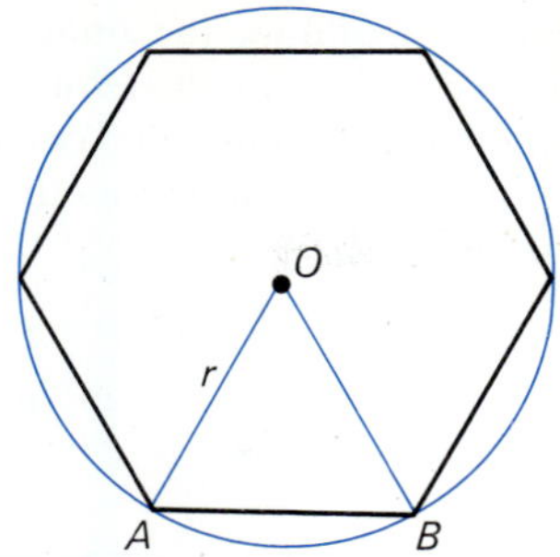

FIGURE 4.62

b. If $r = 1.000$ in., find the total area of the pieces that are cut away in making the head of the bolt.

12. A bolt with a hexagonal head is cut from a bronze rod as shown in Fig. 4.62.

a. What kind of triangle is *ABO*?

b. What is the size of each angle of the triangle?

c. If $r = 1.000$ in., find the length of the apothem from *O* to $\overline{AB}$.

d. Find the total area of the pieces that are cut away in making the head of the bolt.

13. A real estate developer bought a parcel of land that was subdivided as indicated in Fig. 4.63.

a. Find the area of Lot 1.

b. Find the area of Lot 2.

c. Find the area of Lot 3.

d. Find the area of Lot 4. (*Hint:* Lot 5 contains the missing data.)

e. In Lot 5 the measurement 114.86 ft was taken perpendicular to the lot line. What is the area of the parallelogram?

f. What kind of quadrilateral is Lot 6? Find its area.

g. Using the results of (a) – (f), find the total area of the parcel of land.

14. How many square feet of concrete walk will it take to go around a circular swimming pool 32. ft in diameter if the width of the walk is to be 4.0 ft?

15. If a circular disk with a diameter of 4.00 ft was cut from a piece of metal 4.00 ft square, what percent of the metal was wasted?

16. How many 3-in. diameter pipes will be needed to carry the flow of a 15-in. diameter pipe? The flow of water depends on the area of the cross-section of the pipe.

17. How do the areas of the bases of a cone compare if the radii are 3 in. and 9 in. respectively?

18. If you wish to double a square in area, by how much must you multiply each side?

FIGURE 4.63

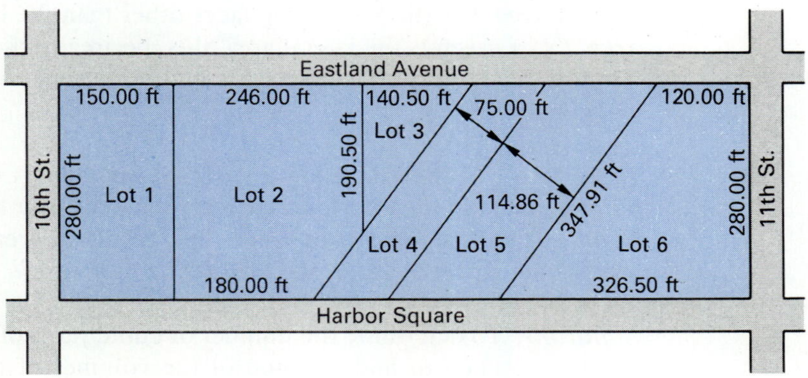

19. Without changing the rate of flow, the amount of flow of water through a water pipe was doubled. Does this mean that the radius of the pipe delivering the water was doubled? How much should its diameter be increased?

20. If a map is drawn to a scale of 1 : 2000, and a region on the map has an area of 3.6 in², what is the area of the region in the real world in square inches? In square feet?

4.6 Geometric solids, surface area, volume

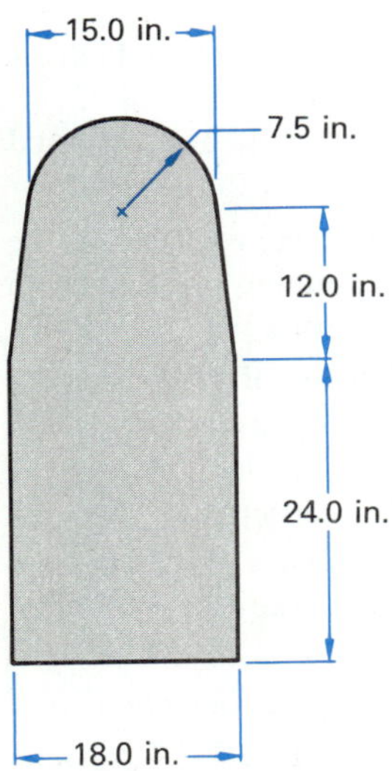

FIGURE 4.64

An engineer for a highway department has designed a divider to be placed between two lanes of a highway. A cross-section of the divider above ground is a trapezoid with a semicircle on its top as shown in Fig. 4.64. The base has a rectangular shape that extends 24.0 in. below the finished road surface. The lower base of the trapezoid is 18.0 in., the upper base, which is also the base of the semicircle is 15.0 in., and the distance between bases is 12.0 in.

The question facing the engineer is the cost. In order to determine the cost he needs to answer many questions. Two of these questions are: (1) What is the number of cubic yards of concrete required per linear foot of construction? (2) How many square feet of surface area are there for each linear foot of the exposed part of the divider?

Living in a three-dimensional world, questions such as the two suggested are common. Can you answer them? The first question deals with finding the volume of a solid figure and the second deals with the lateral area or surface of a solid figure.

The most common figure for which we find the volume is a right rectangular prism. The formula used is $V = \ell wh$, where ℓ times w is the area of the base. The volume of any prism can be found by multiplying the area of the base times the height or $V = Bh$. A **prism** is a polyhedron whose bases are parallel and equal, and whose faces are polygons. The **lateral area** L is the area of the faces other than the bases. The **total area** T of a prism is the lateral area plus the area of the bases.

A list of formulas that are used in finding volume and surface area is given in Table 4.4.

EXAMPLE 1 For the highway divider shown in Fig. 4.64: (1) Find the number of cubic yards of concrete required per linear foot of construction. (2) Find the number of square feet of surface area for each linear foot of the exposed part of the divider.

Solution: To determine the number of cubic yards of concrete in the divider we need to find the sum of the volumes of three geometric solids. The three solids are a rectanglar prism, a prism with a trapezoidal

TABLE 4.4

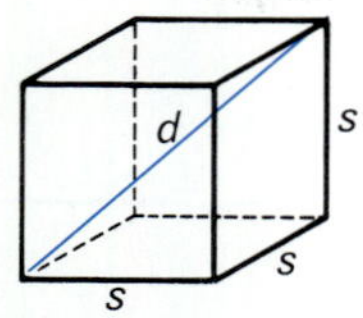

Cube

$d = s\sqrt{3}$

$L = 4s^2$

$T = 6s^2$

$V = s^3$

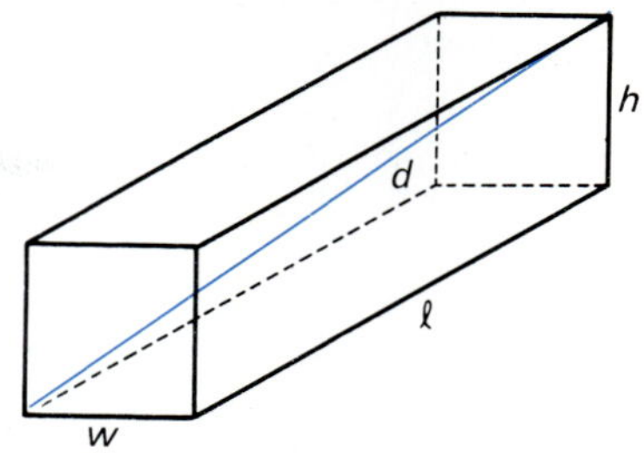

Right rectangular prism

$d = \sqrt{\ell^2 + w^2 + h^2}$

$L = 2(wh + \ell h)$

$T = 2(\ell w + wh + \ell h)$

$V = \ell wh$ or $V = Bh$

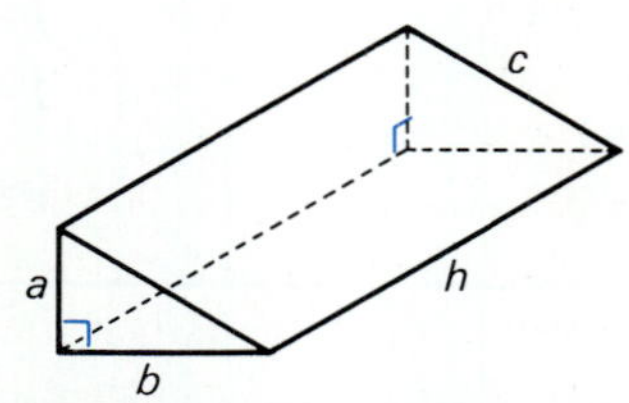

Right triangular prism

$L = ah + bh + ch$

$T = ah + bh + ch + ab$

$V = \frac{1}{2}abh$ or $V = Bh$

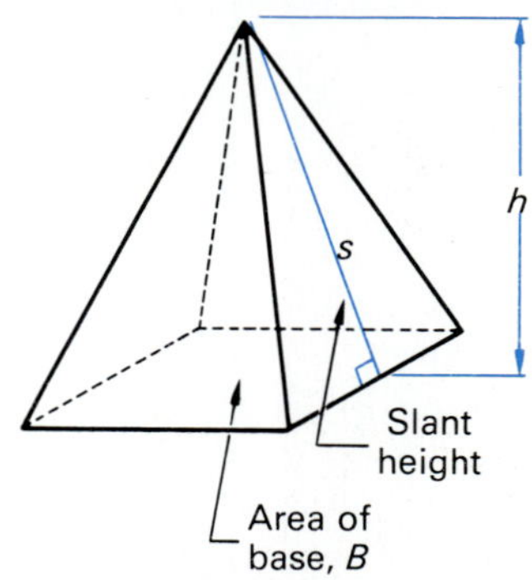

Pyramid

$L = \frac{1}{2}PS$ (P = Perimeter of base)

$T = \frac{1}{2}PS + B$

$V = \frac{1}{3}Bh$

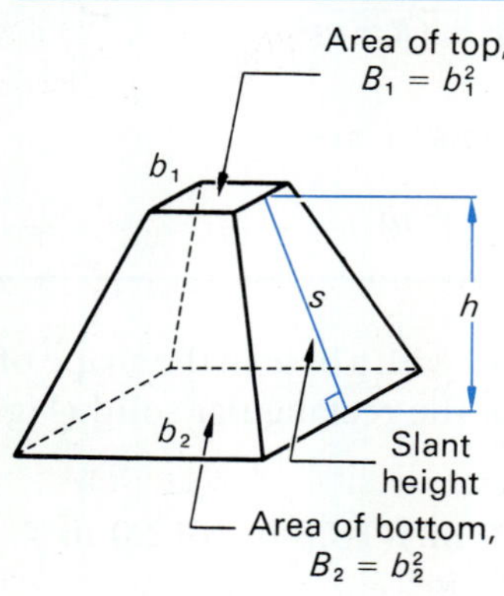

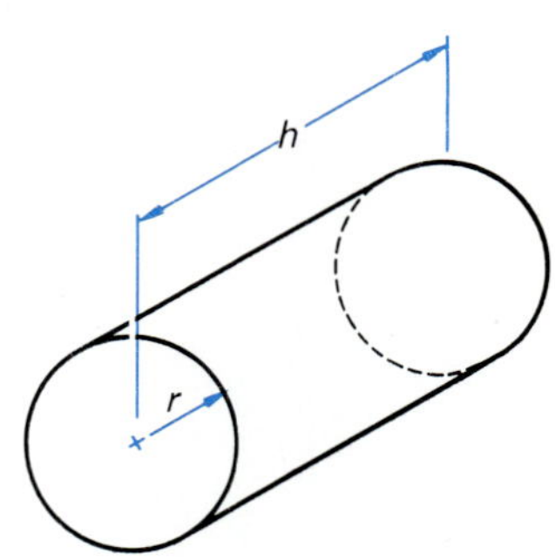

(Continued)

TABLE 4.4 (Continued)

Frustrum of square pyramid	Cylinder
$L = 2(b_1 + b_2)S$ $T = 2(b_1 + b_2)S + B_1 + B_2$ $V = \frac{h}{3}(B_1 + B_2 + \sqrt{B_1 B_2})$	$L = 2\pi rh$ $T = 2\pi rh + 2\pi r^2$ $V = \pi r^2 h$

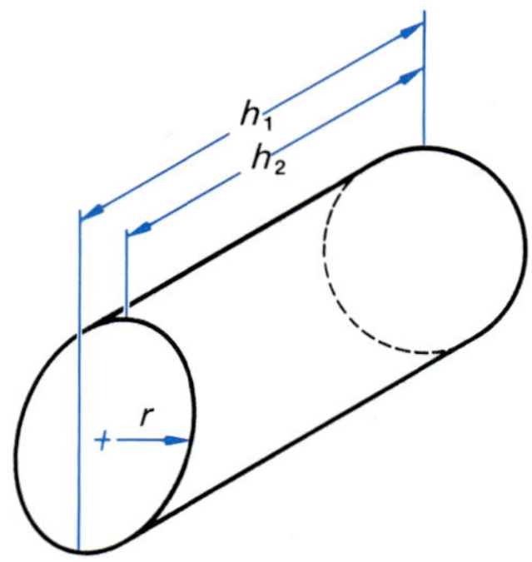

Portion of a cylinder

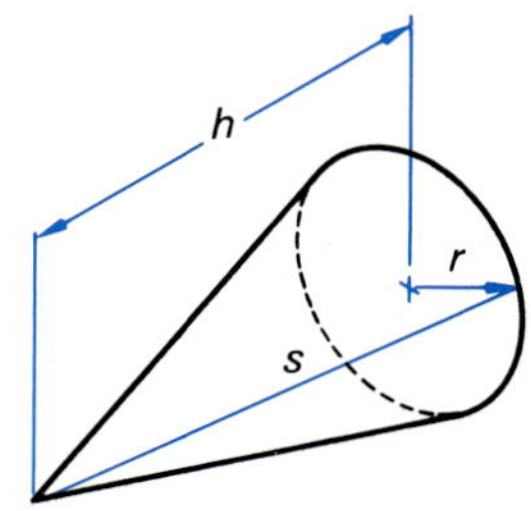

Cone

Portion of a cylinder	Cone
$L = \pi r(h_1 + h_2)$ $V = \frac{\pi r^2}{2}(h_1 + h_2)$	$L = \pi r\sqrt{r^2 + h^2}$ or $L = \pi rS$ $T = \pi r^2 + \pi rS$ $V = \frac{1}{3}\pi r^2 h$

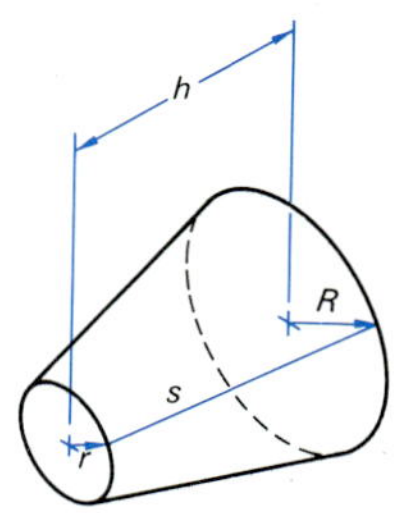

Frustrum of a cone

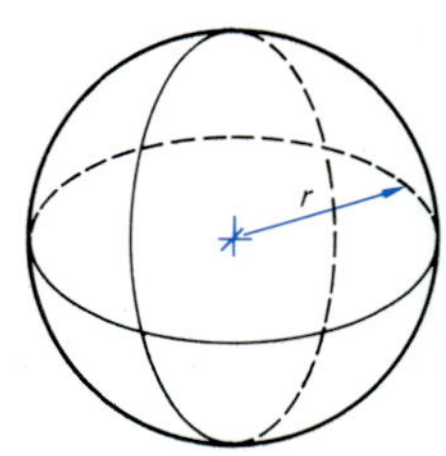

Sphere

Frustrum of a cone	Sphere
$L = \pi(R + r)\sqrt{h^2 + (R - r)^2}$ or $L = \pi(R + r)S$ $T = \pi(R + r)S + \pi R^2 + \pi r^2$ $V = \frac{\pi}{3}(R^2 + Rr + r^2)h$	$T = 4\pi r^2$ or $T = \pi d^2$ $V = \frac{4}{3}\pi r^3$

base, and a solid with a base in the shape of a semicircle (Fig. 4.65). The Volume, V_1, of the rectangular solid (Fig. 4.65c) is

$$\begin{aligned} V_1 &= \ell wh \\ &= (24.0 \text{ in.})(18.0 \text{ in.})(12.0 \text{ in.}) \\ &= 5184. \text{ in}^3. \end{aligned}$$

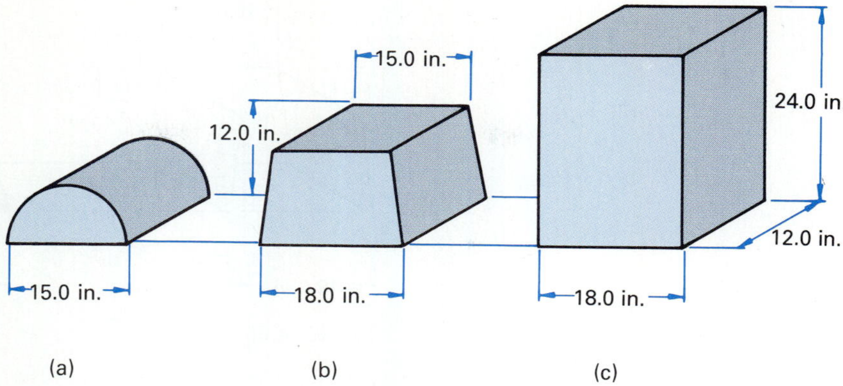

FIGURE 4.65

The volume, V_2, of the trapezoidal prism (Fig. 4.65b) is

$$V_2 = Bh, \text{ where } B = \frac{1}{2}(b_1 + b_2)a$$

$$= \frac{1}{2}(18.0 \text{ in.} + 15.0 \text{ in.})(12.0 \text{ in.})(12.0 \text{ in.})$$

$$= 2376. \text{ in}^3.$$

The volume, V_3, of the solid with a semicircular base (Fig. 4.65a) is

$$V_3 = Bh, \text{ where } B = \frac{\pi r^2}{2}$$

$$= \frac{\pi(7.50 \text{ in.})^2}{2}(12.0 \text{ in.})$$

$$= 1060. \text{ in}^3.$$

The total volume, V, is

$$V = V_1 + V_2 + V_3$$
$$= 5184. \text{ in}^3 + 2376. \text{ in}^3 + 1060. \text{ in}^3$$
$$= 8620. \text{ in}^3.$$

To determine the number of cubic yards, we must multiply the volume by the conversion factor

$$\left(\frac{1 \text{ yd}}{36 \text{ in.}}\right)^3 = \frac{1 \text{ yd}^3}{46656 \text{ in}^3}.$$

$$V = \frac{8620. \text{ in}^3}{1} \times \frac{1 \text{ yd}^3}{46656 \text{ in}^3}$$

$$= 0.185 \text{ yd}^3.$$

The amount of concrete required for one linear foot of divider is 0.185 yd^3.

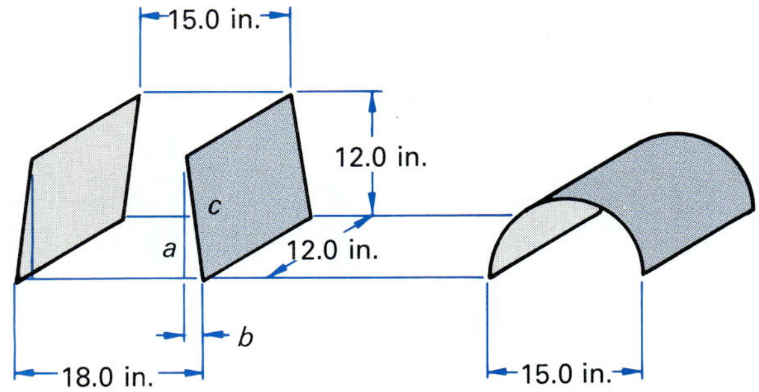

FIGURE 4.66

To determine the exposed surface area in square feet of the divider we need to find the area of the two congruent rectangles that form the sides of the trapezoid. Also, we must find the surface area of the cap, which is one half of a cylinder (Fig. 4.66).

In order to find the area of the rectangles we must find the length of side c of the right triangle in Fig. 4.66a. Using the Pythagorean theorem:

$$
\begin{aligned}
c^2 &= a^2 + b^2 \\
c &= \sqrt{a^2 + b^2} \\
&= \sqrt{(12.0 \text{ in.})^2 + (1.50 \text{ in.})^2} \\
&= 12.09 \text{ in.}
\end{aligned}
$$

Thus, the area of one rectangle is:

$$
\begin{aligned}
A_1 &= \ell w \\
&= (12.0 \text{ in.})(12.09 \text{ in.}) \\
&= 145.1 \text{ in}^2.
\end{aligned}
$$

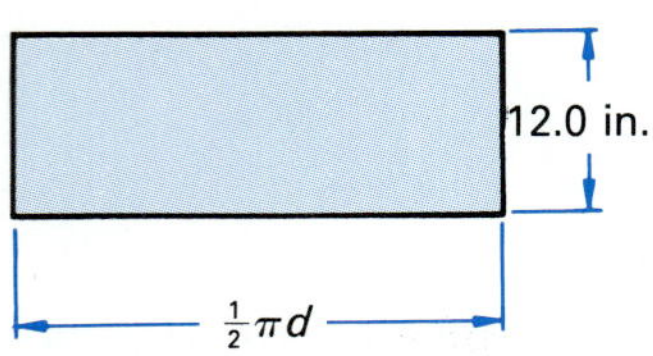

FIGURE 4.67

To find the area of the semicircular cap we can flatten out the cap and form a rectangle as in Fig. 4.67. The length of the rectangle is equal to one half the circumference of the circular arc. Thus

$$
\begin{aligned}
\ell &= \frac{1}{2}\pi d \\
&= \frac{1}{2}\pi(15.0 \text{ in.}) \\
&= 23.56 \text{ in.}
\end{aligned}
$$

The area of the semicircular cap is

$$
\begin{aligned}
A_2 &= (23.56 \text{ in.})(12.0 \text{ in.}) \\
&= 282.7 \text{ in}^2.
\end{aligned}
$$

The total surface area is

$$\begin{aligned} A &= 2A_1 + A_2 \\ &= 2(145.1 \text{ in}^2) + 282.7 \text{ in}^2 \\ &= 572.9 \text{ in}^2. \end{aligned}$$

To determine the number of square feet, we must multiply the area by the conversion factor

$$\left(\frac{1 \text{ ft}}{12 \text{ in.}}\right)^2 = \frac{1 \text{ ft}^2}{144 \text{ in}^2}.$$

$$\begin{aligned} A &= \frac{572.9 \text{ in}^2}{1} \times \frac{1 \text{ ft}^2}{144 \text{ in}^2} \\ &= 3.98 \text{ ft}^2. \end{aligned}$$

The exposed surface area for one linear foot of divider is 3.98 ft². □

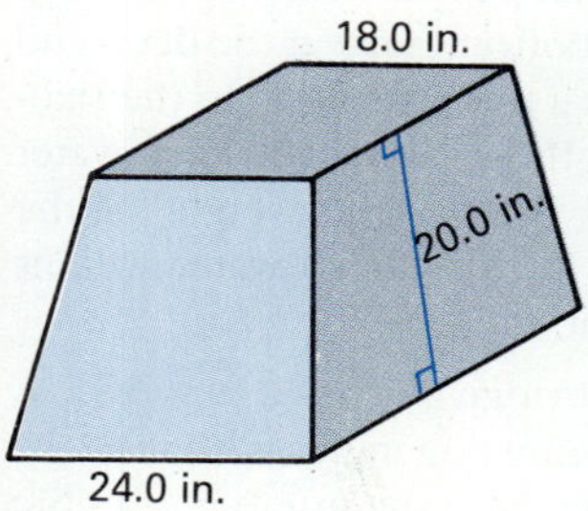

FIGURE 4.68

EXAMPLE 2 The frustrum of a right regular pyramid, Fig. 4.68, is used as the base of a large post. Each edge of the lower base is 24.0 in., each edge of the upper base is 18.0 in., and the slant height is 20.0 in. To the nearest inch: (a) What is the lateral area and (b) what is the volume of the base?

Solution: To find the lateral area we use the formula

$$\begin{aligned} L &= 2(b_1 + b_2)S \text{ from Table 4.4.} \\ &= 2(24.0 \text{ in.} + 18.0 \text{ in.})(20.0 \text{ in.}) \qquad \textbf{Substituting} \\ &= 1680. \text{ in}^2. \end{aligned}$$

To find the volume we use the formula

$$V = \frac{h}{3}(B_1 + B_2 + \sqrt{B_1 B_2}).$$

B_1 is the area of the upper base, thus

$$B_1 = 324.0 \text{ in}^2 = (18.0 \text{ in.})^2.$$

B_2 is the area of the lower base, thus,

$$B_2 = 576.0 \text{ in}^2 = (24.0 \text{ in.})^2.$$

FIGURE 4.69

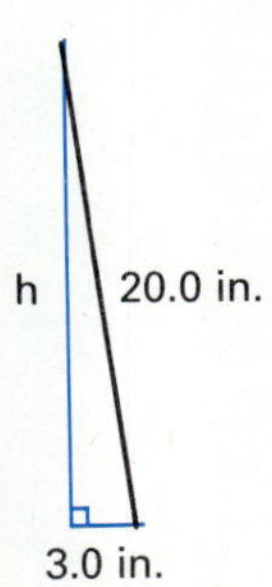

h is the perpendicular distance between the two bases. Taking a slice along the edge we see that h is the height of a right triangle (Fig. 4.69). Using the Pythagorean theorem we have

$$\begin{aligned} h &= \sqrt{c^2 - b^2} \\ &= \sqrt{(20.0 \text{ in.})^2 - (3.0 \text{ in.})^2} \\ &= 19.8 \text{ in.} \end{aligned}$$

Substituting in the formula we have

$$V = \frac{19.8 \text{ in}}{3}\left(324.0 \text{ in}^2 + 576.0 \text{ in}^2 + \sqrt{(324.0 \text{ in}^2)(576.0 \text{ in}^2)}\right)$$
$$= 8800. \text{ in}^3. \quad \square$$

What happens when we do not have a regular geometric solid? The best way to handle this situation is as we did in Example 1, divide the solid up into regular geometric solids.

4.6 Exercises

1. The radius of the earth is approximately 3958.6 mi. If three fourths of the earth's surface is water, how many square miles of land are there?
2. A silo is in the form of a right prism. The side of the octagonal base is 9.2 ft and the silo is 35.0 ft high. How many gallons of paint will be needed to paint the lateral surface of the silo if one gallon of paint will cover 250 ft^2 of surface?
3. How many cubic feet are there in a tank 8.00 ft long, 5.00 ft wide, and 3.00 ft deep?
4. If the total area of a cube is 78.60 in^2, what is its volume?
5. A farmer has decided to build a cylindrical silo 28 ft high. He wants the silo large enough to provide 1 ft^3 of ensilage for each cow each day. If 25 cows are to be fed for 150 days, what should the diameter of the silo be?
6. In a classroom 35 ft long, 25 ft wide, and 10 ft high, is there sufficient air space for 36 pupils viewing a filmstrip if approximately 300 ft^3 is the recommended allowance for each pupil?
7. On a grain combine the grain bin is in the shape of an inverted cone (Fig. 4.70). The diameter of the base of the cone is 58.0 in. and the altitude is 64.0 in. How many square inches of metal are in this open topped bin?
8. A rectangular swimming pool is 75 ft long and 25 ft wide. It is 12 ft deep at one end and 3 ft deep at the other end. The bottom is flat at the deep end for 10 ft and then has a uniform slope to the shallow end (Fig. 4.71). How many gallons of water does the pool hold? What is the weight of the water? (*Hint:* 1 $\text{ft}^3 = 7.5$ gal; fresh water weighs 62.4 lb per cubic foot.)
9. A concrete pier for a bridge is shaped as shown in Fig. 4.72, the bases being rectangles with semicircles added at the ends. Find the number of cubic yards of concrete needed to build one pier.

FIGURE 4.70

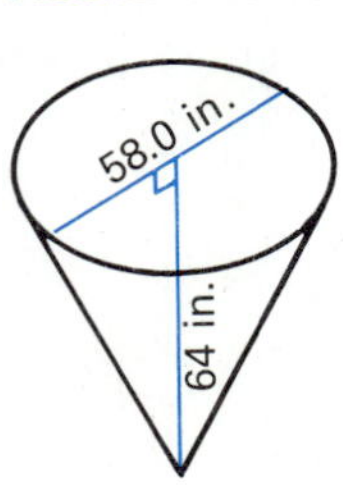

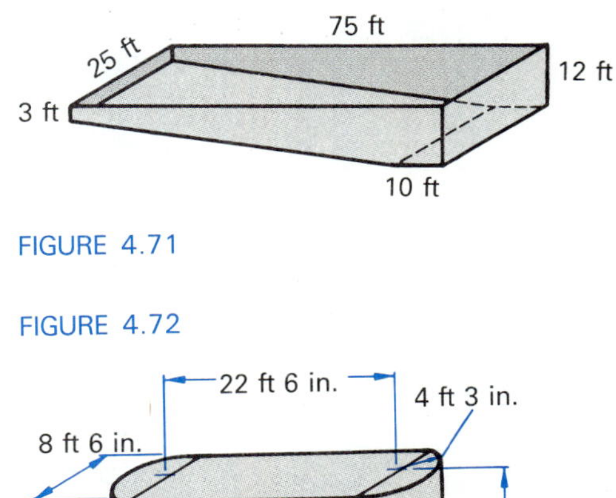

FIGURE 4.71

FIGURE 4.72

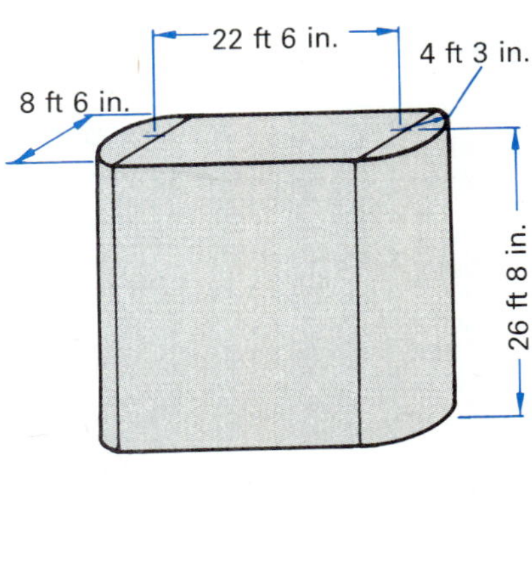

10. If oranges having a diameter of 2.0 in. sell for \$1.80 per dozen and those with a 3.2 in. diameter sell for \$3.60 per dozen, which are the most economical to buy?

11. A spherical shell of cast iron has an external diameter of 6.34 in., and the thickness of the shell is 0.523 in. Find its weight if 1 in^3 of cast iron weighs 0.2604 lb.

12. A concrete form used in making bridge supports has the shape of a frustrum of a cone with base radii of 12 ft and 18 ft. If the frustrum is 36 ft tall and has a 3-ft-diameter hollow cylindrical core, determine the volume of the concrete needed for pouring one support.

13. Determine the slant height and the lateral surface area of the cone frustrum in Exercise 12.

14. Compute the weight of a steel washer, shown in Fig. 4.73, if steel weighs 0.283 lb per cubic inch.

15. A water tank is shaped like a cylinder with a hemisphere below and a cone on the top. The diameter of the cylinder is 16.0 ft and its height 20.5 ft. The height of the cone is 3.4 ft (Fig. 4.74). Find the number of gallons of paint required to paint the entire tank (cone, cylinder, and hemisphere) two coats. A gallon of paint will cover 300 ft^2.

16. A motorist traveling in Europe ran out of gas. He walked to a station and was given a circular container 240 mm across the top, 300 mm across the bottom, and 380 mm high. He was told that it contained 30.0 liters of gasoline. Did it? How many liters did it contain?

FIGURE 4.73

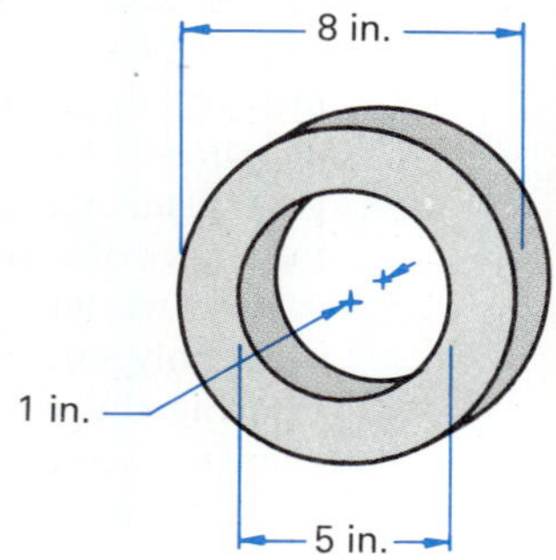

FIGURE 4.74

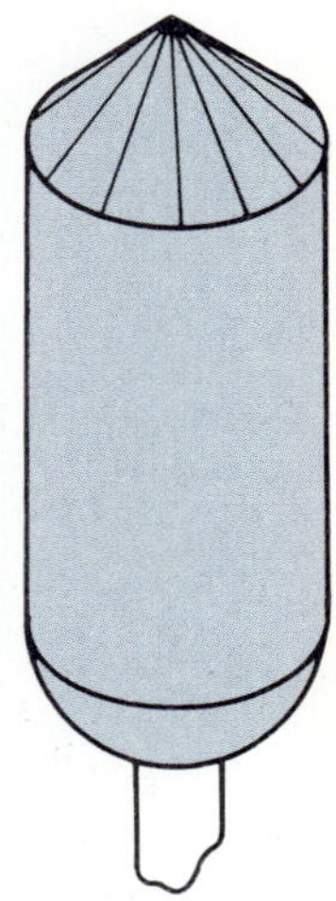

Chapter summary

A way of expressing a relationship between two quantities is by writing a special fraction called a ratio.

Definition 4.1 A **ratio** is a number (without dimension) that expresses the numerical relationship between two quantities, when the two quantities are compared by division. The ratio of a to b $(b \neq 0)$ is written a/b or a to b or $a:b$.

Definition 4.2 A **proportion** is a statement that two ratios are equal. $a/b = c/d$ is a proportion provided $b \neq 0$ and $d \neq 0$.

The four quantities of the proportion are called the **terms** of the proportion. The first and fourth terms (a and d) are called the **extremes** and the second and third terms (b and c) are called the **means** of the proportion. By "cross" multiplying we can see that the product of the extremes is equal to the product of the means; that is $ad = bc$.

In Section 4.2 we discussed three types of variation: (1) **direct variation** or direct proportion ($V = kt$, k is a constant), (2) **inverse variation** or inverse proportion ($L = k/p$, k is a constant) and (3) **joint variation** ($V = kr^2h$, k is a constant). The constant k in each variation statement is referred to as the **constant of variation** or constant of proportionality. It is also possible to have a problem with a combination of joint and inverse variation. In working problems dealing with variation it is important to read the problem carefully and first determine the type of variation problem: direct, inverse, joint, or combinations of two or more.

Sections 4.3 through 4.6 provide a brief review of the essentials of geometry. Section 4.3 discusses the figures that can be drawn on a plane (referred to as plane figures). These figures are lines, angles, triangles, quadrilaterals, polygons, and circles. To be able to work with plane figures you should understand the following terms: line ($\overleftrightarrow{AB}$), line segment ($\overline{AB}$), directed line segment ($\overrightarrow{AB}$), angle ($\angle BAC$), right angle (angle measure is 90°), adjacent angles, complementary angles (sum of measures is 90°), supplementary angles (sum of measures is 180°), intersecting lines, parallel lines, lines that coincide, polygon (plane figure with three or more sides), triangle (sum of the measures of the angles of a triangle is 180°), acute triangle, obtuse triangle, right triangle, scalene triangle, isosceles triangle, equilateral triangle, quadrilateral (four-sided polygon), trapezoid, parallelogram, rhombus, rectangle, and square. In studying the circle you must understand the words radius, center, chord, diameter, secant, tangent, and circumference.

In Section 4.4 we discussed similar polygons and the special case of similar triangles, congruent triangles.

Definition 4.3 **Similar polygons** are polygons that have corresponding angles with equal measure and corresponding sides that are proportional.

Definition 4.4 If the measures of the angles of $\triangle ABC$ are equal to the measures of the corresponding angles of $\triangle DEF$, then **$\triangle ABC$ is similar to $\triangle DEF$**, symbolized by $\triangle ABC \sim \triangle DEF$.

Definition 4.5 If the corresponding sides of $\triangle ABC$ and $\triangle DEF$ are proportional, then **$\triangle ABC$ is similar to $\triangle DEF$** ($\triangle ABC \sim \triangle DEF$).

Congruent triangles have corresponding angles equal *and* corresponding sides equal.

In Section 4.5 we introduced the formulas for finding the perimeters and the areas of the plane figures. These formulas are given in Tables 4.2 and 4.3.

In Section 4.6 we introduced geometric solids and the formulas for finding the surface areas and volumes of these solids. The formulas are found in Table 4.4.

Review Exercises

1. The teacher of a class made the statement that two thirds of her class were girls.
   a. Express the meaning of the ratio 2/3 as a comparison.
   b. What is the ratio of the number of boys to the entire class?
   c. If there are 27 pupils in the class, how many are girls?
2. An advertisement for raisins states, "In 10 out of 15 homes you will find Iron-Rich Raisins." If you visited 120 homes, how many homes would probably have Iron-Rich Raisins according to the information in the advertisement?
3. Express each as a ratio in simplest form:
   a. $r^2 - 9s^2$ to $r^2 + 6rs + 9s^2$
   b. $3x^3y - 12xy^3$ to $6x^3y + 12x^2y^2 + 6xy^3$
4. How much would 6 oz of spinach seed cost if the price of the seed is 78 cents per quarter pound?
5. If 1 in. on the road map represents a distance of 20 mi, what is the actual distance between Robesonia and Harrisburg, which are 1 13/16 in. apart?
6. A blueprint is made to a scale of 4:5. If a part appears as 30.17 mm long on a blueprint, how long is the actual part?
7. The Smith family traveled 315 mi in 7 1/2 hr. At the same average rate, how much farther could they have traveled in three more hours?
8. If a car traveling at an average rate of 45 mph reaches its destination in 6 hr, 20 min, how fast must a car travel to cover the same distance in 4 hr, 15 min?
9. Write an algebraic statement that represents each of the following:
   a. c varies directly as t.
   b. p varies jointly as t and r.
   c. z varies jointly as R and t^2 and inversely as the square of w.
   d. The area A of a sphere varies as the square of the radius r.
10. At a distance of 6 in. from the pivot, Jack grips a pair of pliers with a force of 56 lb (Fig. 4.75). The forces are inversely proportional to the distances from the pivot. What is the pressure on the wire if the wire is 1/2 in. from the pivot?

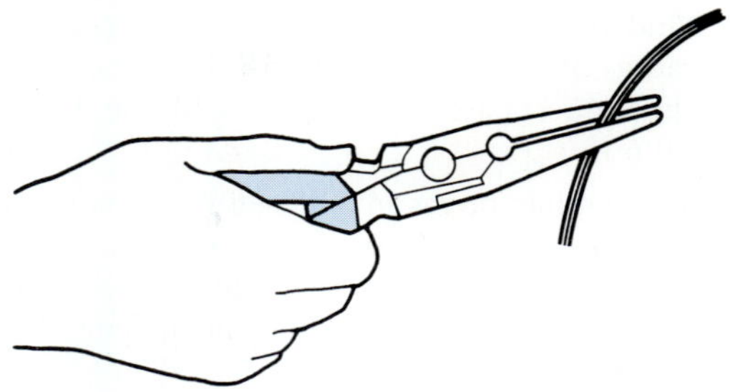

FIGURE 4.75

11. If y varies directly as the square of z, and $y = 3$ when $z = 5$, find the value of z that makes $y = 18$.
12. If the square of w varies directly as the cube of z, and $w = 8$ when $z = 4$, find the value of z when $w = 108$.
13. The compression ratio of an engine is the volume above the piston when it is at the bottom of the cylinder, compared with the volume above the piston when the piston is at the top of the cylinder. In the gasoline engine the compression ratio varies from 7:1 to 10:1 (Fig. 4.76) and the higher the compression ratio, the greater the efficiency. In a diesel engine the compression ratio is around 16:1. What is the compression ratio of an engine when the volumes mentioned above are 84 in^3 and 12 in^3, respectively?

FIGURE 4.76

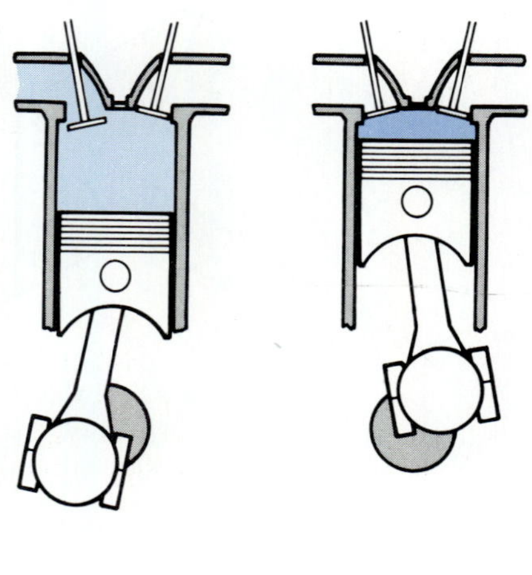

14. If the water pressure on a square inch of surface is 4.33 lb/in² when submerged 10 ft, what will be the pressure per square inch at a depth of 80 ft?

15. Find the volume of a frustrum of a cone if the radius of the lower base is 18.5 in., the radius of the upper base is 11.2 in., and the height is 10.6 in.

16. The volume of gas in a closed vessel is inversely proportional to the pressure on it. If the volume of a gas is 800 ft³ when it is under a pressure of 15 lb/in², what would its volume be under a pressure of 25 lb/in²?

17. If the edges of a cubical box are multiplied by 1 1/2, what effect does that have on the volume?

18. Two balls have diameters of 30 mm and 90 mm. How much greater is the volume of the second than that of the first?

19. What is the ratio of the surface areas of two similar solids if their volumes are represented by 160π and 540π?

20. A piece of tin 4.0 ft square was bent in half on a line parallel to one edge so that it formed a 90° angle (Fig. 4.77). Other pieces of tin were soldered across the ends to form a drinking trough, with the open top rectangular and the ends isosceles right triangles as shown.

- **a.** What is the depth of the trough?
- **b.** How much tin will it take to make it, including the ends?
- **c.** How many gallons of water are in it, when it is filled to a depth that is 2/3 of the total depth?

FIGURE 4.77

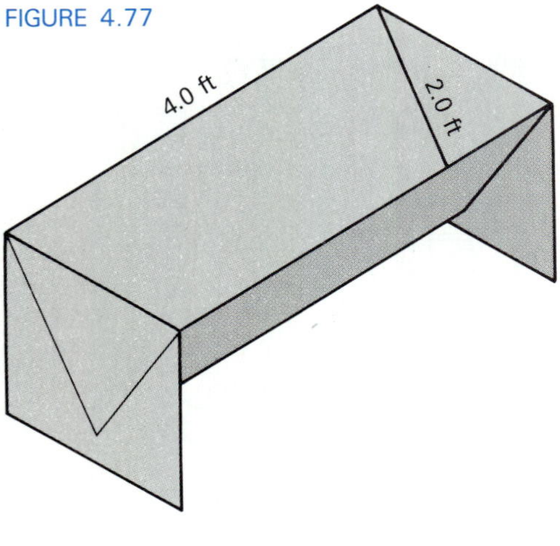

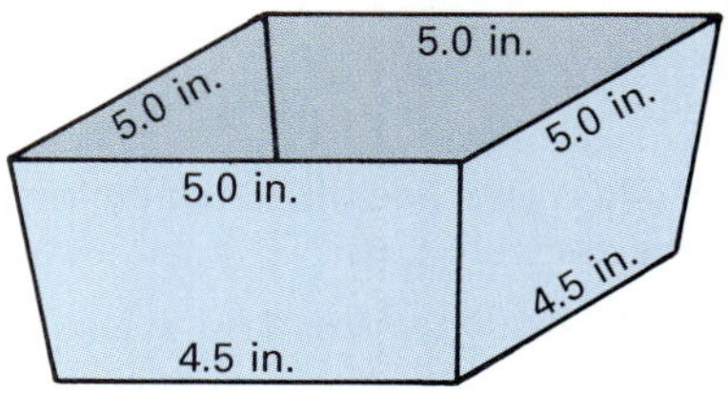

FIGURE 4.78

21. Find the area of an equilateral triangle whose side is 29.75 mm.

22. A cylindrical fruit can is 4.25 in. high and the diameter of its base is 3.40 in. Find the amount of metal needed to make the can. Neglect seams and overlapping area.

23. A manufacturer plans to make lamp shades in the shape of frustrums of regular square pyramids. For one shade, each side of the lower base is 6 1/8 in., each side of the upper base is 3 3/8 in., and the distance between the bases is 7 3/4 in. Find the lateral area.

24. The rain that falls on a horizontal flat roof of a home is carried to a cylindrical cistern 7.00 ft in diameter. If the roof is 22.0 ft by 28.0 ft, how great a fall of rain in inches would it take to fill the cistern to a depth of 8.00 ft?

25. A strawberry box shaped as in Fig. 4.78 is 5.0″ square at the top and 4.5″ square at the bottom. It is 3.0″ deep. Will it hold 1 qt of berries when it is level full, if 1 dry quart equals approximately 67 in³?

26. A spherical storage tank for gas is 50.0 ft in diameter. How many gallons will it hold? In this case assume that each cubic foot contains 7 1/2 gal.

27. A water tank is in the shape of a cylinder with a hemisphere attached below it. The diameter of the tank is 22.0 m and the height of the cylindrical part is 25.0 m. Compute the capacity of the tank in liters.

28. Find the cross-sectional area of the road cut shown in Fig. 4.79.

29. How many cubic feet of sand are there in a conical pile 9.5 ft high if the diameter of the base is 13.0 ft?

30. The picture of a child 3 ft, 9 in. tall is taken with a camera having a focal length of 6 in. If the child stands 9 ft, 6 in. from the camera, what should be the child's height in the picture?

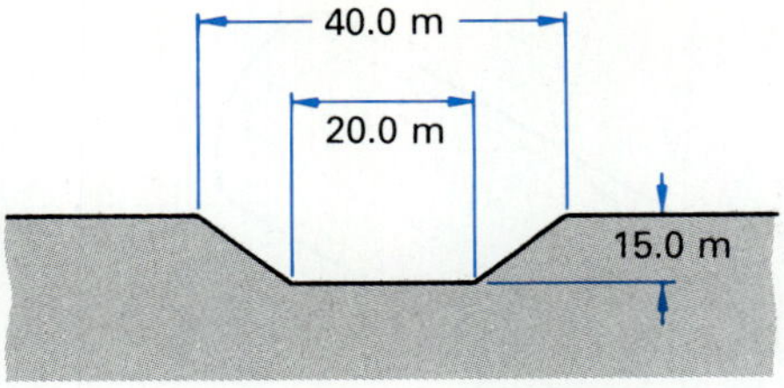

FIGURE 4.79

31. If a pine tree casts a shadow 41.6 m long at the same time of day that a 2.3 m post casts a shadow 5.2 m long, how tall is the tree?

32. Change the following to degrees in decimal form:

a. 24°13′ **b.** 143°19′32″ **c.** 122°17′49″

33. Change the following to degrees, minutes, and seconds:

a. 13.75° **b.** $23.4766\overline{6}°$ **c.** $92.220277\overline{7}°$

In Exercises 34 and 35, find the angles indicated on the figures:

34.

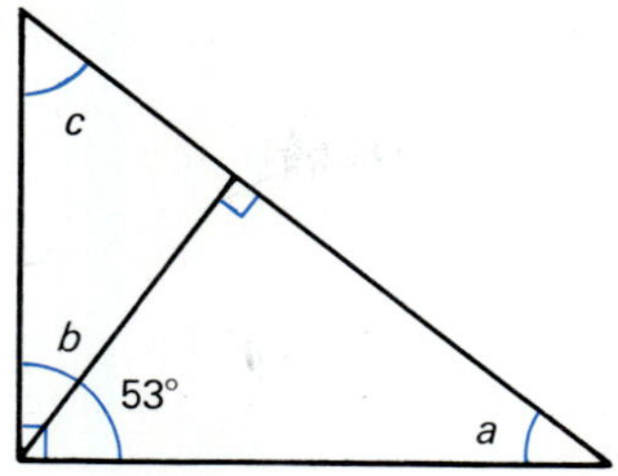

35.

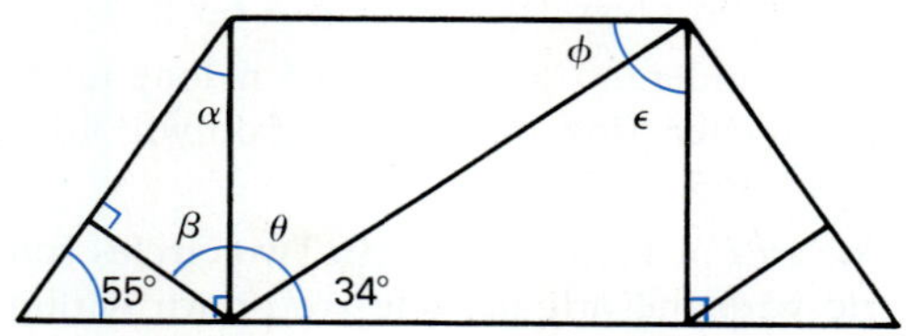

Can you solve these?

1. For the equilateral triangle in Fig. 4.80, assume that the side is s, the altitude is h, and the area is A. Show that $h = \dfrac{s\sqrt{3}}{2}$ and $A = \dfrac{s^2\sqrt{3}}{4}$.

2. A contractor is building a concrete walk with a width of 5 ft around the outside of an elliptical swimming pool (Fig. 4.81).

a. Find the circumference of the walk so that he can determine the number of 8-ft-long forms he needs.

b. Find the number of cubic yards of concrete needed if the walk will be 4 in. thick.

3. The diagonals of a rhombus intersect at right angles, and one half the product of these diagonals gives the area. Fig. 4.82 shows the diagonals to be 21.56 ft and 12.45 ft.

a. Use the diagonals to find the area of the rhombus.

b. Find the area of the rhombus using the formula given in Table 4.2 and compare your answers.

c. Show why the formula $A = \dfrac{1}{2} d_1 d_2$ is true.

4. If a camera is carried to a height of 3958.6 mi above the surface of the earth, what portion of the

FIGURE 4.80

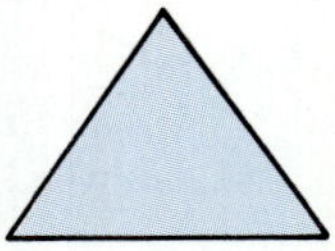

FIGURE 4.81

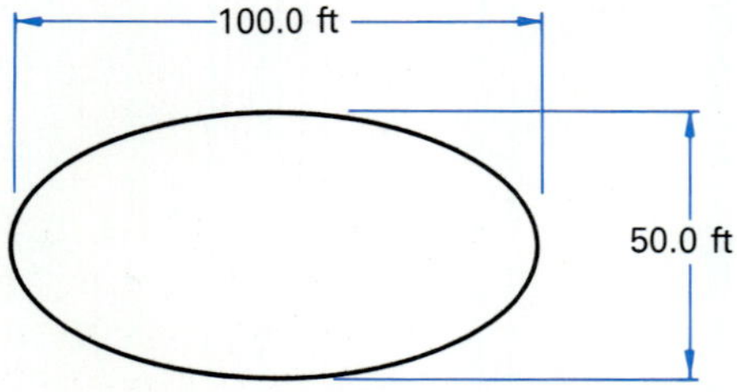

FIGURE 4.82

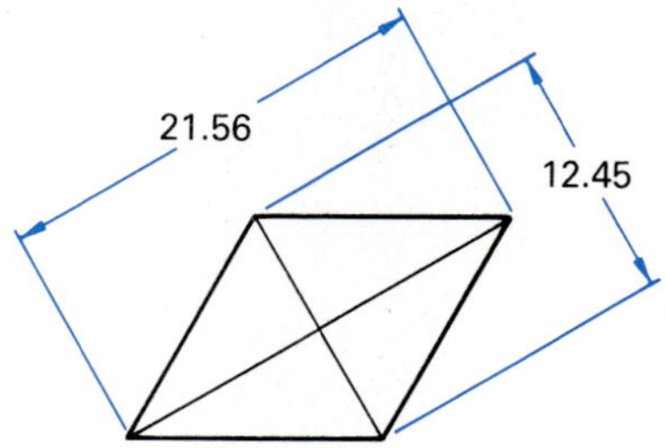

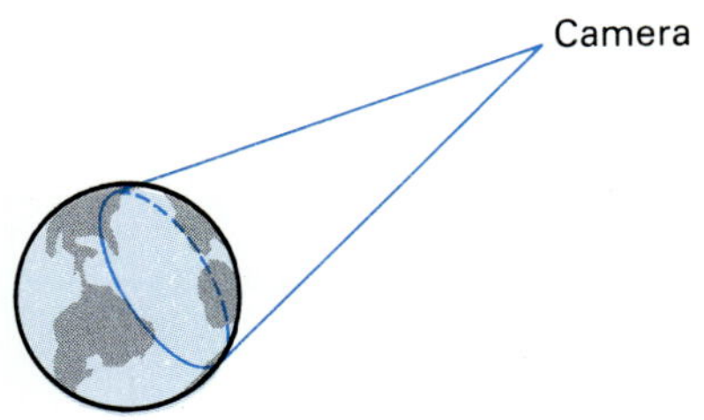

FIGURE 4.83

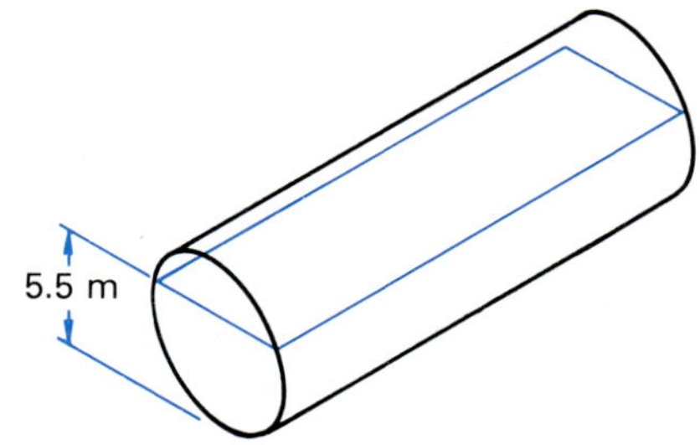

FIGURE 4.84

earth's surface would be available for a picture on a perfectly clear day? (See Fig. 4.83.)

5. The cylinder in Fig. 4.84 is 28.5 m long and 6.0 m in diameter. How many liters of oil will fill it to a depth of 5.5 m?

6. The area of a circle is 36 m^2. Two circles concentric with the original one are drawn within the circle so that they divide the original area into three equal parts. Find the radii of the two smaller circles.

7. If two chords intersect within a circle so that the ratio of the segments of one chord is 4 : 25 and the ratio of the segments of the other chord is 1 : 4, show that the ratio of the whole chords is 25 : 29.

5

In a study of human physiology the position of the pelvis as measured from its resting position was found to move from back to front during the course of one step according to the following function:

Time (s)	Displacement (mm)
0.2	0.0
0.3	24.5
0.4	20.5
0.5	9.5
0.6	−3.5
0.7	0.0

From the graph of this function determine at what times did the pelvis pass through the rest position during the cycle?

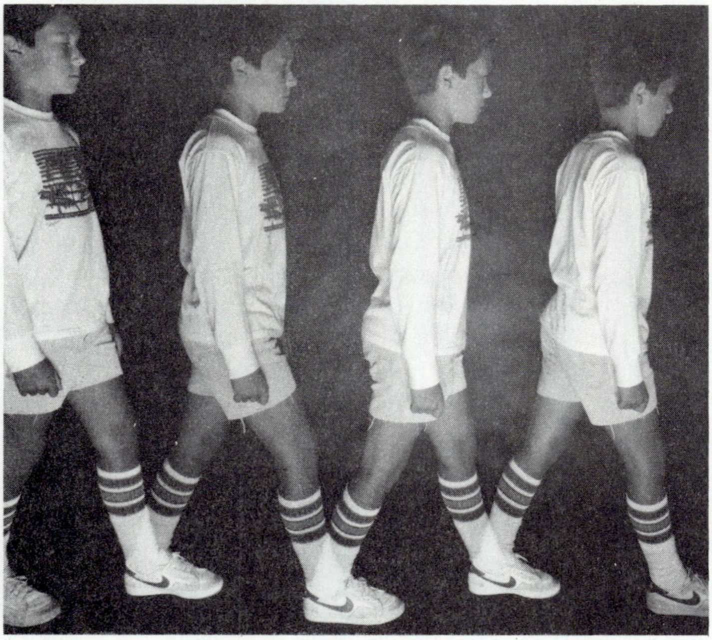

Functions and Graphs

5.1 Functions

The concept of a function is used in our everyday lives. The cost of running an automobile is a function of the cost of gasoline. The rate at which a rocket leaves the launch pad is a function of the thrust of the rocket. The depth of field in a photograph is a function of the lens opening.

In each of these situations we are saying that the first thing is dependent on the second. For example, if the price of gasoline goes up or down, it will cost more or less to run the car.

In mathematics the word "function" is used much the same way, but it is more restrictive. The cost of running a car is affected by other things, not just the price of fuel. While mathematicians may not agree that the above are functions, they would all agree that "The area of a square is a function of the length of its side." Thus, the mathematical function demands that one quantity is uniquely determined by one or more other quantities.

The area of the square is determined by the length of the side: $A = s^2$. We say that s is the **independent variable** and A is the **dependent variable.** The volume of a sphere is dependent on the length of its radius, $V = 4/3\pi r^3$. In this formula r is the independent variable and V is the dependent variable.

The relationship between quantities need not be exhibited as a formula, but may be exhibited in a table. Table 5.1 gives the relationship between the lap number and the velocity of a race car as it passes the judge's stand on each lap of a 12-lap race. Mathematically we say that the velocity of the race car is a function of the lap number. That is, there is one velocity that corresponds to each lap. The lap cannot be a function of the velocity since there is more than one lap that corresponds to a certain velocity, but only one velocity that corresponds to a given lap. For example the velocity of 99 mph was recorded for laps 3 and 7. Also the velocity of 101 mph was achieved in laps 4 and 11. In this illustration the lap is the independent variable and the velocity is the dependent variable.

DEFINITION 5.1

> A **function** is a rule that assigns for each value of the independent variable a unique value of the dependent variable.

A function can also be expressed as a set of ordered pairs. Using the data from Table 5.1, velocity V can be expressed as a function of lap number ℓ by writing ordered pairs: (1, 98), (2, 102), (3, 99), (4, 101), (5, 102), (6, 98), (7, 99), (8, 103), (9, 100), (10, 97), (11, 101),

TABLE 5.1

Lap	Velocity
1	98 mph
2	102 mph
3	99 mph
4	101 mph
5	102 mph
6	98 mph
7	99 mph
8	103 mph
9	100 mph
10	97 mph
11	101 mph
12	104 mph

(12, 104). The first number in each ordered pair represents the independent variable ℓ and the second number represents the dependent variable V as (ℓ, V).

EXAMPLE 1 Each of the following is an example of a function. In each case identify the independent variable(s) and the dependent variable:

a. $y = 4x + 7$ **b.** $S = 32t$ **c.** $y = \sqrt{x - 4}$

d.

A	-3	2	5	-5
B	4	9	14	4

e. $y = \dfrac{x}{x^2 + 3x + 1}$

f. $V = \dfrac{1}{3}\pi r^2 h$ **g.** $F = ma$ **h.** $2A + 3B - 4 = 0$

Solution:

a. For the function $y = 4x + 7$ the independent variable is x and the dependent variable is y. (By convention, x is taken as the independent variable and y the dependent variable.)

b. For the function $S = 32t$, the independent variable is t and the dependent variable is S.

c. For the function $y = \sqrt{x - 4}$, the independent variable is x and the dependent variable is y.

d. Unless other information is provided, in table form we will take the first element to be the independent variable. Thus, A is the independent variable and B the dependent variable.

e. For the function $y = \dfrac{x}{x^2 + 3x + 1}$ the independent variable is x and the dependent variable is y.

f. The function $V = \dfrac{1}{3}\pi r^2 h$ has two independent variables, r and h. The function could also be written $f(r, h) = \dfrac{1}{3}\pi r^2 h$. The dependent variable is V. There is no restriction on the number of independent variables. However, a function will have one and only one dependent variable. (Remember the symbol π represents a constant.)

g. The function $F = ma$ could also be written $F(m, a) = ma$. The independent variables are m and a, and the dependent variable is F.

h. With no added information, it is impossible to tell which is the independent variable in the function $2A + 3B - 4 = 0$. Thus, unless otherwise noted, we will select the independent variable by alphabetical order. In this case, A is the independent variable and B is the dependent variable. An exception is that in functions dealing with time, t is the independent variable. □

The dilemma created by part (h) in Example 1 can be eliminated if we use **functional notation.** For the function $y = \sqrt{x+3}$, we can write $f(x) = \sqrt{x+3}$ and read this as f of x or function of x equals $\sqrt{x+3}$. The notation $f(x)$ does not indicate multiplication or any other operation; it represents the dependent variable, in this case y. It is not necessary to restrict ourselves to the notation $f(x)$; $g(x)$, $h(x)$, $\phi(x)$, etc., can also be used for functional notation.

EXAMPLE 2 Let $y = f(x) = 3x^2 + 2x - 4$. Find the value of the function for

a. $x = 2$ **b.** $x = -3$

Solution:

a. $y = f(x) = 3x^2 + 2x - 4$. For $x = 2$,

$$f(2) = 3(2)^2 + 2(2) - 4$$
$$= 12.$$

In using functional notation it is easy to see that we replace the x with 2.

b. $y = f(x) = 3x^2 + 2x - 4$. For $x = -3$,

$$f(-3) = 3(-3)^2 + 2(-3) - 4$$
$$= 17. \quad \square$$

TABLE 5.2

Algebraic
LRN
CP
STO
0
5
×
RCL
0
x^2
×
RCL
0
−
4
×
RCL
0
x^2
+
2
×
RCL
0
−
3
=
RST
LRN

EXAMPLE 3 Let $y = f(x) = 3x^2 + 2x - 4$. Find the value of the function for

a. $x = t$ **b.** $x = 2t$ **c.** $x = t + 2$

Solution:

a. $y = f(x) = 3x^2 + 2x - 4$. For $x = t$,

$$f(t) = 3(t)^2 + 2(t) - 4$$
$$= 3t^2 + 2t - 4.$$

b. $y = f(x) = 3x^2 + 2x - 4$. For $x = 2t$,

$$f(2t) = 3(2t)^2 + 2(2t) - 4$$
$$= 3(4t^2) + 4t - 4$$
$$= 12t^2 + 4t - 4.$$

c. $y = f(x) = 3x^2 + 2x - 4$. For $x = t + 2$,

$$f(t+2) = 3(t+2)^2 + 2(t+2) - 4$$
$$= 3(t^2 + 4t + 4) + 2t + 4 - 4$$
$$= 3t^2 + 12t + 12 + 2t$$
$$= 3t^2 + 14t + 12. \quad \square$$

The evaluation of a function can be simplified with the use of a calculator.

EXAMPLE 4 For the function $f(x) = 5x^3 - 4x^2 + 2x - 3$, find $f(3.7)$ using a calculator.

Solution: If we have to evaluate the function for only a few values of x, we use key strokes. On the other hand, if we must evaluate the function for many values of x, we write a program. We will find the solution: a) using a series of key strokes to evaluate the function, and b) writing a program to evaluate the function.

a. 5 [×] 3.7 [y^x] 3 [−] 4 [×] 3.7 [x^2] [+]
2 [×] 3.7 [−] 3 [=] 202.905.

This method works well if we want to evaluate the function for a single value or if the value substituted for the variable is positive. With the above key strokes, $f(-3.7)$ would cause a calculator with algebraic logic to give an error signal. (The result of raising a negative number to a nonintegral power may be a complex number. The calculator does not distinguish between integral and nonintegral powers and therefore cannot handle the operation.) To avoid this problem the sequence of key strokes can be written for $f(-3.7)$ or for any other value for which x is negative.

5 [×] 3.7 [+/−] [x^2] [×] 3.7 [+/−] [−]
4 [×] 3.7 [+/−] [x^2] [+] 2 [×]
3.7 [+/−] [−] 3 [=] −318.425.

b. The key strokes for programming the function $f(x) = 5x^3 - 4x^2 + 2x - 3$ are given in Tables 5.2 and 5.3 for Algebraic Logic and RPN, respectively.

For the calculator with algebraic logic, when the [LRN] key is stroked the second time the calculator is out of the learn mode and you are ready to find values for $f(x)$. First stroke the [RST] key to instruct the calculator to return to the first step of the program.

[RST] 3.7 [R/S] 202.905 or $f(3.7) = 202.905$.
3.7 [+/−] [R/S] −318.425 or $f(-3.7) = -318.425$.
2.13 [R/S] 31.430385 or $f(2.13) = 31.430385$.

To find all the values we want for $f(x)$ all we need do is key in a value for x and stroke the [R/S] button.

For the calculator with RPN, when the [RUN] key is stroked the calculator is out of the learn mode and you are ready to find values for $f(x)$. First stroke [g] [RTN] to instruct the calculator to return to the beginning of the program. Then enter a value for x and stroke the [R/S] key to get a functional value.

4.25 [+/−] [R/S] = −467.578 or $f(-4.25) = -467.578$. □

TABLE 5.3

RPN
[PRGM]
[clear PRGM]
[STO]
0
5
[ENTER]
[RCL]
0
3
[y^x]
[×]
4
[ENTER]
[RCL]
0
[x^2]
[×]
[−]
2
[ENTER]
[RCL]
0
[×]
[+]
3
[−]
[GTO]
[RUN]

For information on writing a computer program to evaluate the function see the chapter on computers.

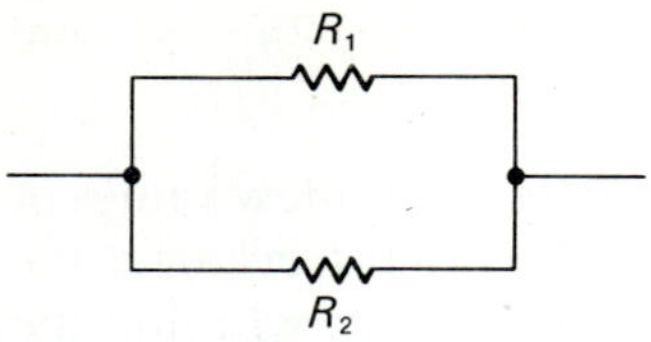

FIGURE 5.1

EXAMPLE 5 The total resistance of two resistors in parallel is a function of the individual resistance, R_1 and R_2 (Fig. 5.1).

$$R = g(R_1, R_2) = \frac{R_1 R_2}{R_1 + R_2}$$

Find:

a. $g(3, 4)$ **b.** $g(1/3, 2/7)$

Solution:

a.
$$R = g(R_1, R_2) = \frac{R_1 R_2}{R_1 + R_2}$$
$$g(3, 4) = \frac{3 \cdot 4}{3 + 4}$$
$$= \frac{12}{7}$$

b.
$$g\left(\frac{1}{3}, \frac{2}{7}\right) = \frac{\frac{1}{3} \cdot \frac{2}{7}}{\frac{1}{3} + \frac{2}{7}}$$
$$= \frac{\frac{2}{21}}{\frac{7+6}{21}}$$
$$= \frac{2}{21} \times \frac{21}{13}$$
$$= \frac{2}{13}. \quad \square$$

The set of real numbers that can be substituted in a function for the independent variable and give us real numbers for the dependent variable is called the **domain** of the function. The set of real numbers that we obtain for the dependent variable is called the **range** of the function.

EXAMPLE 6 The formula $P = 64d$ is used to determine the pressure P on objects that are immersed in salt water d feet below a free surface. Find the domain D and the range R.

Solution: For the function $P = 64d$ the depth d would always be measured as a positive number of feet. Thus, $D = \{d | d \geq 0\}$. Since d is non-negative, P must be non-negative and $R = \{P | P \geq 0\}$. The symbol $\{d | d \geq 0\}$ is read "the set of all elements d such that d is greater than or equal to zero."

The part after the bar (which corresponds to the words "such that") is the condition d must meet in order to be a member of the set D. Also, set R represents all the real numbers that meet the condition of being greater than or equal to zero. $\square$

EXAMPLE 7 Find the domain and range for the function $f(x) = x^2 + 5$.

Solution: We can replace the independent variable x with any real number and obtain real values for $f(x)$. Thus, the domain is $D = \{x | x \in \mathbb{R}\}$. ($x \in \mathbb{R}$ means x is an element of the real numbers.)

If we let $x = -3$, then

$$\begin{aligned} f(-3) &= (-3)^2 + 5 \\ &= 9 + 5 \\ &= 14. \end{aligned}$$

We know, as can be seen from the substitution, that squaring a number will yield a positive product. Thus, the functional value will always be five greater than the number squared. The smallest value the function can attain is 5, which occurs when $x = 0$. Thus, $R = \{f(x) | f(x) \geq 5\}$. □

EXAMPLE 8 Find the domain and range for the function: $y = \dfrac{1}{x-2}$.

Solution: In order to determine the domain of the function $y = \dfrac{1}{x-2}$, ask the question, "What real numbers may be substituted for x and *not* give us real values for y?" An obvious answer is $x = 2$. For $x = 2$ the denominator would be zero, and division by zero is undefined. Since 2 is the only restriction, we state the domain in a negative sense, $D = \{x | x \neq 2\}$. This statement means that all real numbers, except 2, may be substituted for x and obtain real values for y.

Are there any restrictions on the range? Can y be equal to all real numbers? By looking at the function it is difficult to tell, but if we solve for x, the solution becomes apparent.

$$y = \frac{1}{x-2}$$

$$(x-2)y = (x-2)\frac{1}{x-2} \qquad \textbf{Multiply both sides by } x - 2$$

$$\begin{aligned} xy - 2y &= 1 \\ xy &= 2y + 1 \end{aligned}$$

$$x = \frac{2y+1}{y}.$$

From this result we can see that we could substitute any real number, except zero, for y and obtain a real number for x. Thus $R = \{y | y \neq 0\}$. □

Remember, the secret to finding the domain and range is to look for the exceptional value. Look for those real numbers for which the function will not be defined. (Two of these cases are numbers that will cause division by zero or the square root of a negative number.)

5.1 Exercises

In Exercises 1 – 8, identify the independent variable(s) and the dependent variable.

1. $P = 64d$

2. $F = \frac{9}{5}C + 32$

3. $C = \frac{5}{9}(F - 32)$

4. $y = f(x) = \sqrt{4x - 1}$

5. $s = f(r, \theta) = r \cdot \theta$

6. $V = f(r, h) = \frac{1}{3}\pi r^2 h$

7. $F = \theta(m, a) = ma$

8. $SA = \theta(\ell, w, h) = 2\ell w + 2\ell h + 2wh$

In Exercises 9 – 25, find the value of the functions.

9. $f(x) = 3x^2 + 7x - 5$:

a. $f(3)$ **b.** $f(-4)$ **c.** $f(a + h)$

10. $f(x) = 2x^2 - 7x + 11$:

a. $f(-2)$ **b.** $f(5)$ **c.** $f(a + h)$

11. $f(t) = \frac{t + 5}{t - 3}$:

a. $f(2)$ **b.** $f(7.4)$ **c.** $f(-3.7)$

12. $g(R) = \frac{R^2 - R + 6}{R - 3}$:

a. $g(2)$ **b.** $g(3)$ **c.** $g(3/8)$

13. $f(a, b) = a^2 + 3ab + b^2$:

a. $f(2, 3)$ **b.** $f(-1, 3)$ **c.** $f(1/3, 1/7)$

14. $f(x, y) = x^2 + 2xy - y^2$:

a. $f(1, 1)$ **b.** $f(0, -4)$ **c.** $f(0.24, 0.13)$

15. $f(t) = 3t^2 + 2t - \sqrt{t}$:

a. $f(3.217)$ **b.** $f(5.613)$ **c.** $f(\pi)$

16. $f(z) = z^2 - \sqrt{z}$:

a. $f(5)$ **b.** $f(0.375)$ **c.** $f(2/3)$

17. The circumference of a circle is given by $C(r) = 2\pi r$, where r is the length of the radius (see Fig. 5.2). Find:

a. $C(2.34 \text{ in.})$
b. $C(6.41 \text{ m})$
c. $C(5/11 \text{ in.})$

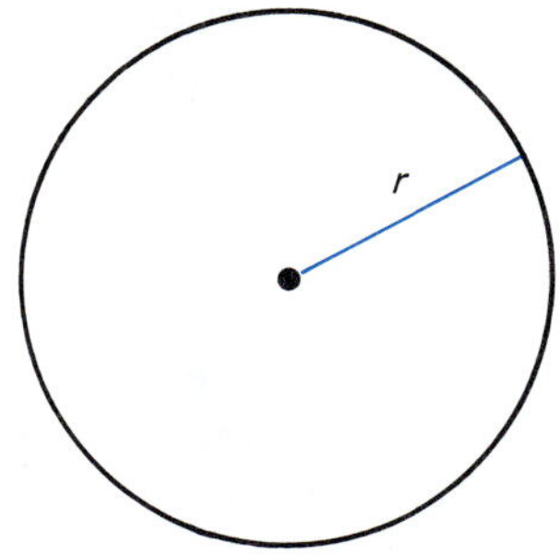

FIGURE 5.2

18. The area of a circle is given by $A(r) = \pi r^2$, where r is the length of the radius. Find:

a. $A(2.34 \text{ in.})$ **b.** $A(6.41 \text{ m})$
c. $A(5/11 \text{ in.})$

19. The total surface area of a cube is given by the function $f(s)$, where s is the length of the side of the cube (see Fig. 5.3) and $f(s) = 6s^2$. Find:

a. $f(3.75 \text{ m})$ **b.** $f(6.05 \text{ in.})$
c. $f(13.42 \text{ mm})$

20. If distance $s = f(t) = -16t^2 + 64t - 48$, where t is in seconds and s is in feet, find:

a. $f(0)$ **b.** $f(1)$
c. $f(2.05)$ **d.** $f(3)$

21. Resistance R is a function of voltage V given by $R = \phi(V) = \frac{V}{0.21}$. Find:

a. $\phi(1.57)$ **b.** $\phi(3.75)$

22. The measure of the angle θ in radians is given by $\theta = f(s, r) = s/r$, where s is the length of the arc

FIGURE 5.3

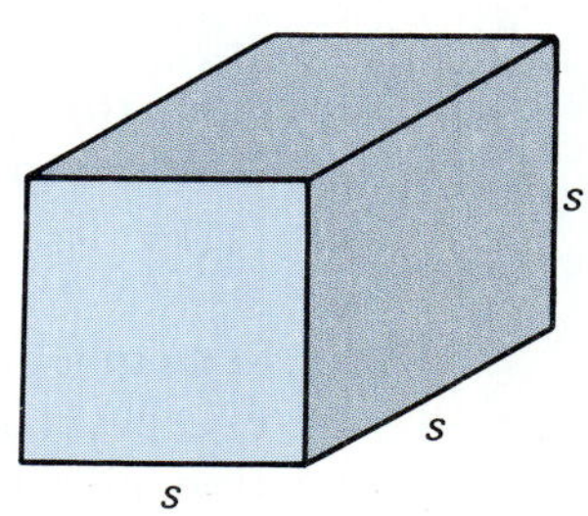

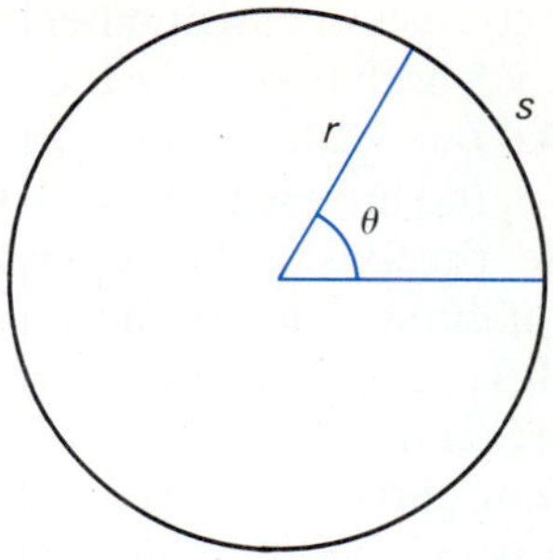

FIGURE 5.4

determined by $\angle\theta$ and r is the length of the radius of the circle (see Fig. 5.4). Find:

a. $f(4.71, 3)$ **b.** $f(15.71, 5)$

23. The tax rate R is given by $R(\ell, v) = \frac{\ell}{v}$, where ℓ is the amount of tax levied and v is the assessed value of taxable property. Find:

a. $R(375000, 8132000)$
b. $R(17000000, 9300000000)$

24. Horsepower HP is given by $\phi(F, d, t) = \frac{Fd}{550t}$, where F is the force in pounds, d is the distance in feet, and t is the time in seconds. Find:

a. $\phi(340, 8.7, 4.8)$ **b.** $\phi(650, 7.6, 3.7)$

25. Resistance R, in ohms, is given by $\theta(P, L, A) = \frac{PL}{A}$, where P is the resistivity in ohm meters, L is the length in meters, and A is the cross-section of the wire in square meters. Find:

a. $\theta(17, 7.5, 1 \times 10^{-5})$
b. $\theta(32, 12.5, 1.5 \times 10^{-5})$

In Exercises 26–39, find the domain and range for each function.

26. $y = 3x + 4$ **27.** $R = 6S - 7$

28. $f(t) = t^2 + 5$ **29.** $\phi(s) = s^2 - 3$

30. $SA = f(r) = 4\pi r^2$ **31.** $V = f(r) = \frac{4}{3}\pi r^3$

32. $y = f(x) = \frac{3}{x}$ **33.** $y = f(x) = \frac{4}{x - 1}$

34. $y = f(x) = \frac{x + 1}{x - 1}$ **35.** $y = f(x) = \frac{x - 1}{1 - x}$

36. $s = f(t) = \sqrt{t + 2}$ **37.** $s = f(t) = \sqrt{t - 2}$

38. $p = f(\ell) = \sqrt{3 + \ell}$ **39.** $p = f(\ell) = \sqrt{3 - \ell}$

In Exercises 40–42, find a function that describes:

40. The number of seconds in an hour.

41. The Celsius reading corresponding to $x°$ Fahrenheit.

42. A printer agrees to print a brochure for a sum of \$300 plus 15 cents for each copy. Express the cost as a function of the number of copies.

5.2 Rectangular coordinates

We have all heard the statement, "A picture is worth a thousand words." This is especially true for the technician, who, without a graph may not be able to interpret the data that were collected. In this section we will develop the basic tools for constructing graphs of functions.

In our discussion of functions, such as $y = f(x) = x + 6$, we found that there are many ordered pairs that satisfy the function. Some of the ordered pairs that satisfy $y = f(x) = x + 6$ are $(0, 6)$, $(-6, 0)$, $(3, 9)$, $(-3, 3)$, and $(-5, 1)$. One way to gain an idea of the many ordered pairs that satisfy the equation is to draw a graph of the equation on the coordinate plane. To do this we must understand the rectangular coordinate system.

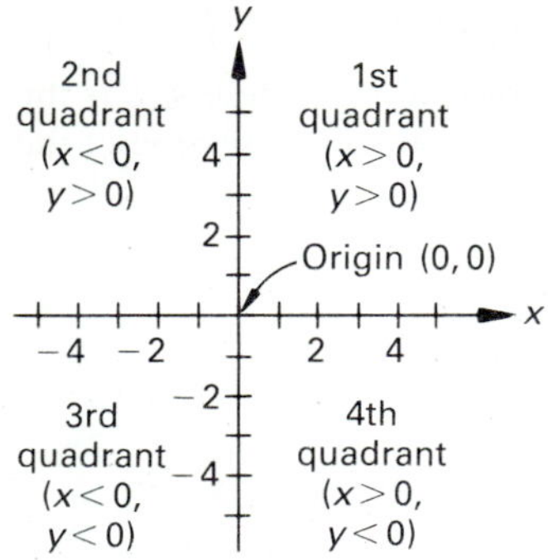

FIGURE 5.5

The **rectangular coordinate system** consists of two number lines drawn perpendicular to each other. In Fig. 5.5 both lines have the same scale. In some cases we may wish to make one scale larger than the other. We call the horizontal line the axis of the independent variable, in this case the x-axis, and the vertical line, the axis of the dependent variable, in this case the y-axis. The point of intersection of the x and y axes is called the **origin.** Note that the scales to the right and above the origin are positive and the scales to the left and below the origin are negative. The axes divide the plane into four parts, which we call the first, second, third, and fourth **quadrants** as indicated in Fig. 5.5.

Every point on the plane can be represented by an ordered pair (x, y). We call the ordered pair (x, y) a set of coordinates. The first coordinate, x, is called the **abscissa,** and the second coordinate, y, is called the **ordinate.** The origin is represented by the ordered pair $(0, 0)$.

EXAMPLE 1 Determine the coordinates of points A, B, C, D, E, F, G, and H in Fig. 5.6.

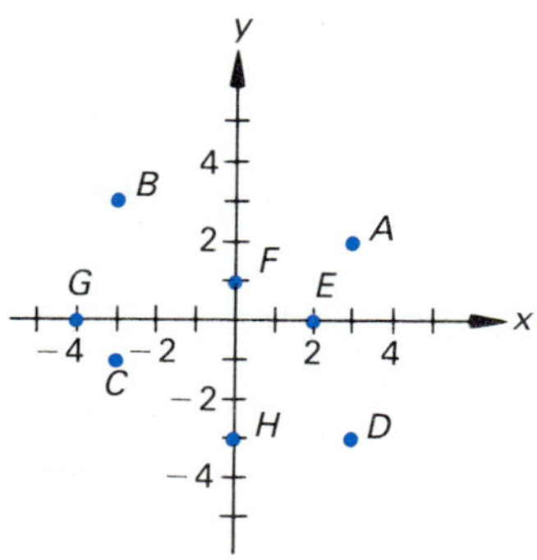

FIGURE 5.6

Solution: Point A has an x value of 3 and a y value of 2; therefore, its ordered pair is $(3, 2)$. Similarly, the other ordered pairs are: $B(-3, 3)$, $C(-3, -1)$, $D(3, -3)$, $E(2, 0)$, $F(0, 1)$, $G(-4, 0)$, and $H(0, -3)$. The two points, E and G, that are on the x-axis have zero for the ordinate and the two points, F and H, that are on the y-axis have a zero for the abscissa. □

EXAMPLE 2 Graph the points $A(0, 6)$, $B(-6, 0)$, $C(3, 9)$, $D(-3, 3)$, and $E(-5, 1)$.

Solution: See Fig. 5.7. □

FIGURE 5.7

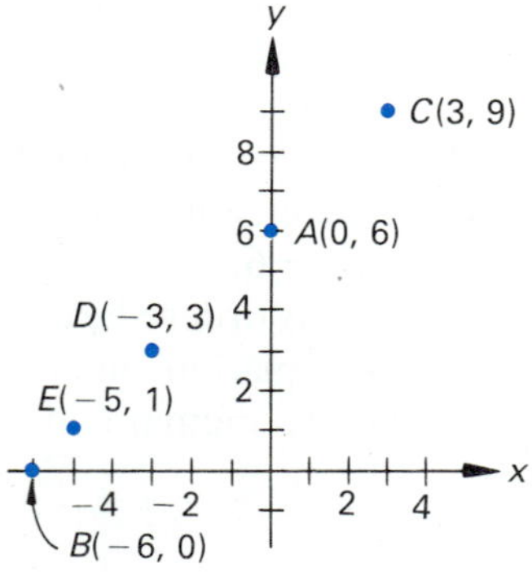

In the study of analytic geometry and calculus it is essential that you are able to determine the distance between two points. The following discussion contains the development of the **distance formula.**

We shall denote the distance between two points $P_1\ (x_1, y_1)$ and $P_2\ (x_2, y_2)$ by $\overline{P_1P_2}$ and make two preliminary comments: $\overline{P_1P_2} \geq 0$, and $\overline{P_1P_2} = \overline{P_2P_1}$.

Let us take a look at a specific example before getting to the general formula. Consider the three points $A(2, 5)$, $B(6, 5)$, and $C(6, 2)$ (Fig. 5.8). The points A and B lie on the same horizontal line; thus we find $\overline{AB} = 4$ by counting (on the same axis). Also, $\overline{BA} = 4$ by counting (on the same axis). $\overline{AB}$ can also be found by taking the absolute value of the difference of the abscissas, thus, $\overline{AB} = |6 - 2|$ or $\overline{BA} = |2 - 6|$. From previous experience we know that $|6 - 2| = |2 - 6| = 4$. Since B and C lie on a line parallel to the y-axis, we can use similar arguments to show that $\overline{CB} = 3$.

$\overline{AC}$ cannot be found by the techniques used to find $\overline{AB}$ and $\overline{CB}$ since

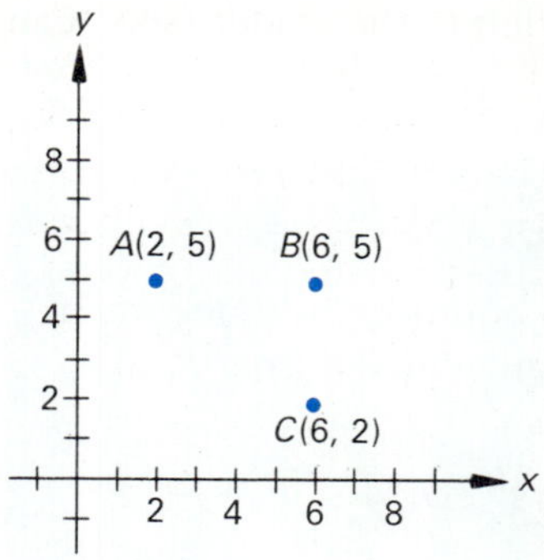

FIGURE 5.8

the line segment AC is not parallel to either one of the axes. Since $\triangle ABC$ is a right triangle, we can use the Pythagorean theorem to find $\overline{AC}$:

$$\begin{aligned}(\overline{AC})^2 &= (\overline{AB})^2 + (\overline{BC})^2\\ &= 4^2 + 3^2\\ &= 16 + 9\\ &= 25.\end{aligned}$$

Taking the square root of both sides we have $\overline{AC} = 5$.

Let P_1, P_2, and P be points with coordinates (x_1, y_1), (x_2, y_2), and (x_1, y_2), respectively (see Fig. 5.9). From the figure we see that $\overline{PP_2} = |x_2 - x_1|$ and $\overline{P_1P} = |y_2 - y_1|$. Using the Pythagorean theorem we have:

$$\begin{aligned}\overline{P_1P_2}^2 &= \overline{PP_2}^2 + \overline{P_1P}^2\\ &= |x_2 - x_1|^2 + |y_2 - y_1|^2\\ &= (x_2 - x_1)^2 + (y_2 - y_1)^2 && \textbf{For any real number } |z|^2 = z^2\\ \overline{P_1P_2} &= \sqrt{(x_2 - x_1)^2 + (y_2 - y_1)^2} && \textbf{Take the principal square root of both sides}\end{aligned}$$

This proves the following theorem:

THEOREM 5.1

DISTANCE FORMULA

If $P_1(x_1, y_1)$ and $P_2(x_2, y_2)$ are any two points in the plane, then the distance, $\overline{P_1P_2}$, between them is given by

$$\overline{P_1P_2} = \sqrt{(x_2 - x_1)^2 + (y_2 - y_1)^2}.$$

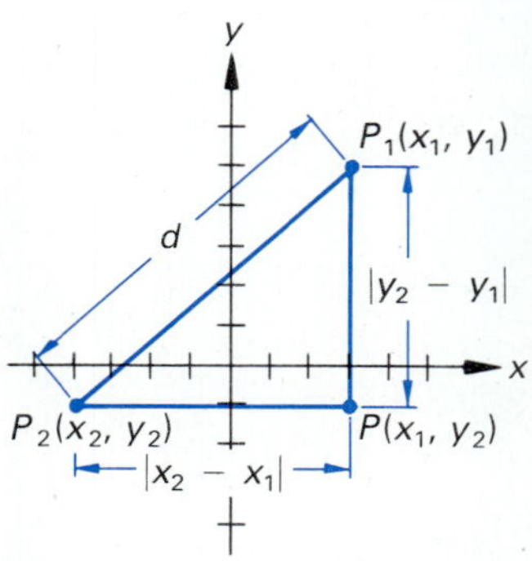

FIGURE 5.9

EXAMPLE 3 Show that the triangle with vertices $A(-2, 6)$, $B(4, 8)$, and $C(2, 4)$ is isosceles.

Solution: To show that triangle ABC is isosceles, we must show that two of its sides are equal in length. Using the distance formula, we find that:

$$\begin{aligned}\overline{AB} &= \sqrt{[4-(-2)]^2 + (8-6)^2}\\ &= \sqrt{6^2 + 2^2}\\ &= \sqrt{40}.\\ \overline{AC} &= \sqrt{[2-(-2)]^2 + (4-6)^2}\\ &= \sqrt{4^2 + (-2)^2}\\ &= \sqrt{20}.\\ \overline{BC} &= \sqrt{(2-4)^2 + (4-8)^2}\\ &= \sqrt{(-2)^2 + (-4)^2}\\ &= \sqrt{20}.\end{aligned}$$

FIGURE 5.10

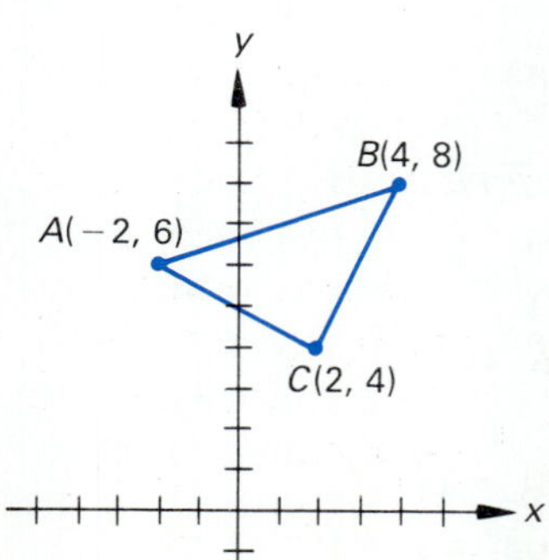

Therefore, $\overline{AC} = \overline{BC}$ and $\triangle ABC$ is isosceles. The triangle is shown in Fig. 5.10. □

The proof of the following theorem is left to the reader (see "Can you solve these?").

THEOREM 5.2

MIDPOINT FORMULA

If $P_1(x_1, y_1)$ and $P_2(x_2, y_2)$ are any two points in the plane, then the coordinates of the midpoint of the line segment from P_1 to P_2 are given by

$$\left(\frac{x_1 + x_2}{2}, \frac{y_1 + y_2}{2}\right).$$

EXAMPLE 4 Find the coordinates of the midpoint m of the line segment joining $A(-2, 6)$ and $B(4, 8)$.

Solution: Using Theorem 5.2,

$$m = \left(\frac{-2 + 4}{2}, \frac{6 + 8}{2}\right)$$
$$= (1, 7). \quad \square$$

5.2 Exercises

In Exercises 1 and 2, plot the given points on a cartesian coordinate system.

1. $A(3, 5)$, $B(-3, -7)$, $C(-3, 4)$, $D(5, -4)$, $E(-2, 0)$, $F(3, 0)$, $G(0, -2)$, $H(0, 5)$, $I(0, \sqrt{2})$
2. $A(2, 5)$, $B(-3, 2)$, $C(7/2, 3/2)$, $D(-4.5, -3.5)$, $E(-3, 0)$, $F(4, 0)$, $G(0, -5)$, $H(3, 6)$

In Exercises 3 and 4, find the coordinates of the given points.

3.

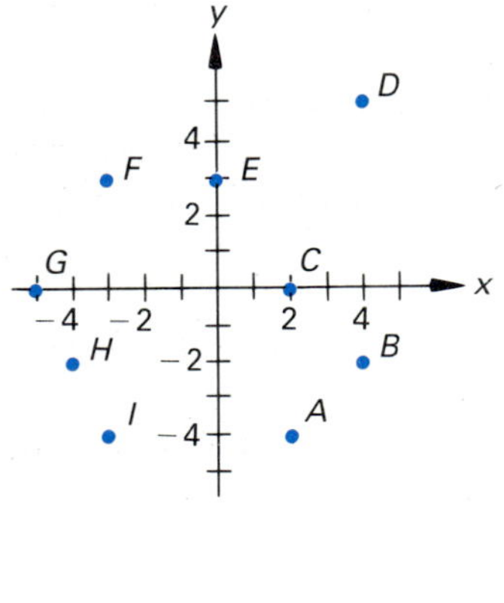

4.

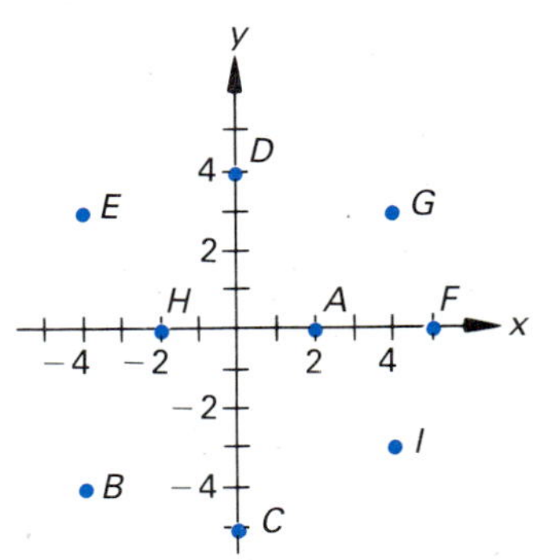

In Exercises 5–12, find $\overline{P_1P_2}$.

5. $P_1(3, 7)$; $P_2(-2, 5)$
6. $P_1(-5, 6)$; $P_2(4, 3)$
7. $P_1(-7, -11)$; $P_2(6, 3)$
8. $P_1(2, 3)$; $P_2(-4, -7)$
9. $P_1(3.4, 7.5)$; $P_2(6.3, 7.4)$
10. $P_1(-6.5, -7.4)$; $P_2(-5.4, 6.4)$

11. $P_1\left(3\frac{1}{2}, -7\frac{1}{2}\right)$; $P_2\left(-3\frac{1}{2}, 7\frac{1}{2}\right)$

12. $P_1(-0.54, 0.44)$; $P_2(7.5, -8.4)$

In Exercises 13–18, a) graph the points, and b) find the coordinates of the midpoint of the line segment joining them.

13. $P_1(2, 7)$; $P_2(5, 2)$ **14.** $P_1(9, 6)$; $P_2(2, 2)$

15. $P_1(2, 5)$; $P_2(-3, 4)$ **16.** $P_1(-3, -4)$; $P_2(2, 3)$

17. $P_1(-\pi, -6)$; $P_2(\sqrt{3}, -3)$

18. $P_1(\sqrt{2}, 0)$; $P_2(-\sqrt{2}, \pi)$

In Exercises 19–30, use reasoning similar to that used in Example 3.

19. a. Show that the triangle with vertices $A(0, 1)$, $B(2, 3)$, and $C(3, 0)$ is isosceles.

b. Find the perimeter of triangle ABC.

20. a. Show that the triangle with vertices $A(-4, 5)$, $B(4, 4)$, and $C(0, -2)$ is isosceles.

b. Find the perimeter of triangle ABC.

21. Show that the points $A(-5, -1)$, $B(-1, 3)$, and $C(-1, -5)$ are the vertices of a right triangle.

22. Find the area of the right triangle in Exercise 21.

23. Show that the points $A(-1, 1)$, $B(6, 0)$, and $C(3, -3)$ are the vertices of a right triangle.

24. Find the area of the right triangle in Exercise 23.

25. Show that the points $A(-5, -1)$, $B(-2, 2)$, $C(3, -3)$, and $D(0, -6)$ are vertices of a rectangle.

26. Find all the points on the y-axis that are a distance of 5 from the point $(3, 5)$.

27. The ordinate of a point (or points) is 2 and its distance from $(7, 3)$ is $\sqrt{122}$. Find the abscissa of the point (or points).

28. What is the length of the longest straight line that can be drawn on a sheet of $8\frac{1}{2}$-by-11-in. typing paper?

29. Use a convenient coordinate system to show that the diagonals of a parallelogram bisect each other.

30. If guy wires are to run from the top of a 100-m tall antenna to the ground 20 m from the base of the antenna, how long must the wires be?

31. Using the information given in Examples 3 and 4, find the area of the isosceles triangle ABC.

5.3 Graph of a function

FIGURE 5.11

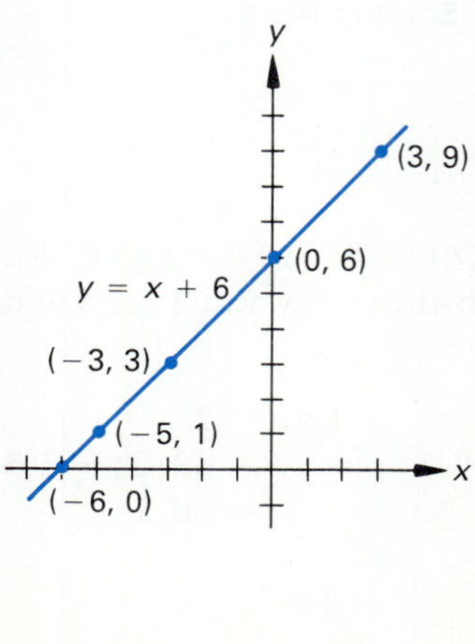

In the previous section we found that the ordered pairs $(0, 6)$, $(-6, 0)$, $(3, 9)$, $(-3, 3)$, and $(-5, 1)$ all satisfy the function $y = f(x) = x + 6$. Then, in Fig. 5.7 we plotted the points. With a straight edge we can draw one line that passes through all these points (Fig. 5.11). We label the points and write the function alongside the line so that when more than one line appears on a graph the lines and points can be identified. When extended indefinitely in either direction, this straight line passes through all the points on the plane that satisfy the equation $y = x + 6$. Since there are an infinite number of ordered pairs of points that satisfy the equation, the graph in Fig. 5.11 is the only way the solution can be illustrated.

All equations of the form $ax + by = c$, where $a \neq 0$ and $b \neq 0$, are called **linear equations in two variables.**

The graphs of linear equations are straight lines. Since only two points are needed to draw a straight line, only two points are needed to graph a linear equation. However, it is a good idea to have a third point as a checkpoint.

To obtain these ordered pairs it is often helpful to solve the equation for the dependent variable y. After solving the equation for y, we substitute values for the independent variable x in the equation and find the corresponding values for the dependent variable.

> The following suggestions can simplify the graphing of functions:
>
> 1. Determine which variable is the dependent variable and which is the independent variable.
> 2. Solve the equation for the dependent variable.
> 3. Find the domain and the range of the function. This will indicate the restrictions on the function, thus showing where the curve does not exist.
> 4. Construct a table of values. (For a linear function only three pairs of values are required. For nonlinear functions find enough pairs of values to be able to sketch a smooth curve.
> 5. From the table of values determine an appropriate scale for the axes of the graph.
> 6. Plot points, label points, especially points on horizontal and vertical axes, and label the graph.

EXAMPLE 1 Graph $2S - 64t + 7 = 0$.

Solution: From our discussion in Section 5.1 we know that the dependent variable of $2S - 64t + 7 = 0$ is S and the independent variable is t. Thus we name the vertical axis the S-axis and the horizontal axis the t-axis.

$$2S - 64t + 7 = 0$$

$$2S - 64t + 64t + 7 - 7 = 64t - 7 \qquad \textbf{Solving for } S$$

$$\frac{2S}{2} = \frac{64t}{2} - \frac{7}{2}$$

$$S = 32t - 3.5.$$

There are no restrictions on t or S. $D = \{t|t \in \mathbb{R}\}$ and $R = \{S|S \in \mathbb{R}\}$. To determine a table of values we can substitute any value for t and obtain a corresponding value for S.

Let $t = -2$
$$S = 32(-2) - 3.5 = -67.5.$$

Let $t = 0$
$$S = 32(0) - 3.5 = -3.5.$$

Let $t = 2$
$$S = 32(2) - 3.5 = 60.5.$$

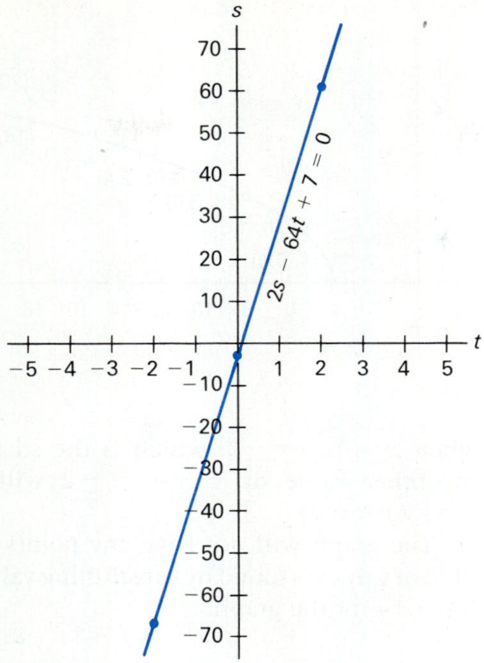

t	S
−2	−67.5
0	−3.5
2	60.5

FIGURE 5.12

Since the values for S are extremely large, it will be convenient to use a unit scale on the horizontal axis and a scale of 10 on the vertical axis (Fig. 5.12). □

If two different scales are used, we must be extremely careful in analyzing a graph. The different scales will produce a graph that is visually distorted compared to a graph where the scales are the same.

EXAMPLE 2 Graph $y = \sqrt{x - 4}$.

Solution: The dependent variable is y and the independent variable is x. In order to obtain real values for y, $x - 4$ must be greater than or equal to 0, or $x \geq 4$. The domain $D = \{x | x \geq 4\}$. Therefore y must always be greater than 0, thus, $R = \{y | y \geq 0\}$. From these restrictions we know that there will be no points to the left of 4 and below the y-axis.

The table of values is found by selecting values for x and solving for y. (See Fig. 5.13 for the graph.) □

EXAMPLE 3 Graph $y = x^2 - 2$.

Solution: The independent and dependent variables are x and y, respectively. We can substitute any value for x; thus $D = \{x | x \in \mathbb{R}\}$.

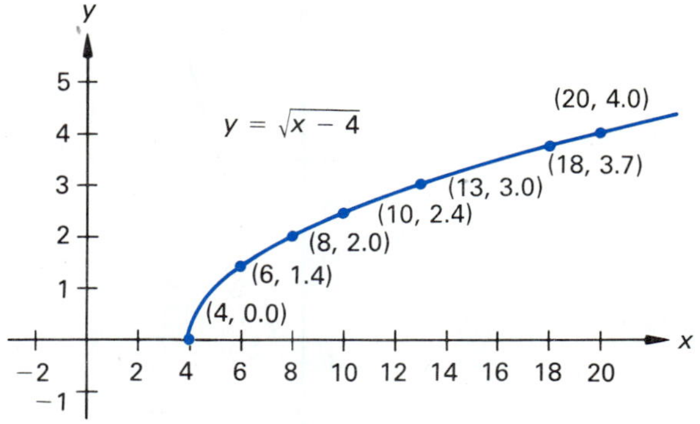

x	y
4	0.0
6	1.4
8	2.0
10	2.4
13	3.0
18	3.7
20	4.0

FIGURE 5.13

When $x = 0$, $y = -2$, which is the smallest value y can attain. For any other value of x, $y = x^2 - 2$ will be larger than -2. Thus, $R = \{y | y \geq -2\}$.

The graph will not have any points below the line $y = -2$. The table of values is found by substituting values for x and solving for y. See Fig. 5.14 for the graph. □

EXAMPLE 4 Graph $y = f(x) = \dfrac{1}{x - 2}$.

Solution: In Section 5.1 we found the domain and range for the function $y = \dfrac{1}{x - 2}$ to be $D = \{x | x \neq 2\}$ and $R = \{y | y \neq 0\}$. From this we

FIGURE 5.14

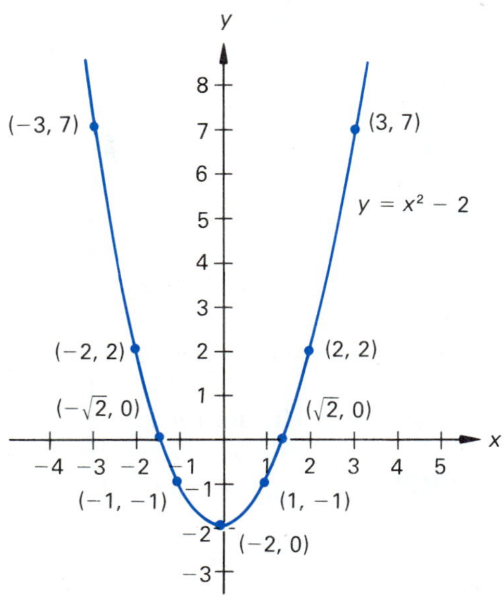

x	y
-3	7
-2	2
$-\sqrt{2}$	0
-1	-1
0	-2
1	-1
$\sqrt{2}$	0
2	2
3	7

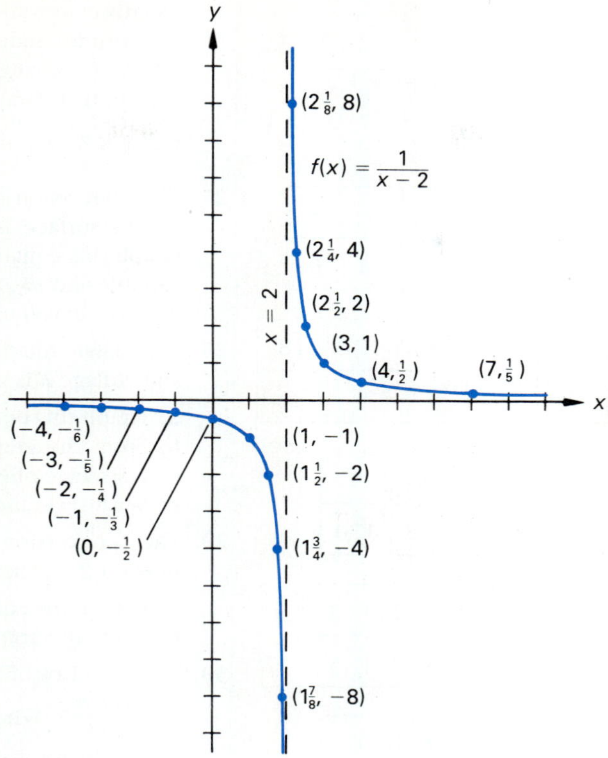

FIGURE 5.15

x	y
−4	−1/6
−3	−1/5
−2	−1/4
−1	−1/3
0	−1/2
1	−1
1 1/2	−2
1 3/4	−4
1 7/8	−8
2 1/8	8
2 1/4	4
2 1/2	2
3	1
4	1/2
7	1/5

know that the curve will not exist when $x = 2$ and $y = 0$. The lines $x = 2$ and $y = 0$ are special lines for this function called **asymptotes.** In sketching the curve as we plot larger and larger positive and negative values the curve gets closer and closer to the asymptotes $x = 2$ and $y = 0$, but never touches the lines. We say the curve is asymptotic to the line. Substituting values for x and solving for y we arrive at the table of values. Additional points are needed to show both branches of the curve and the effect of the asymptotes. See Fig. 5.15 for the graph. □

5.3 Exercises

In Exercises 1–29, graph the function. Determine whether or not asymptotes exist. If they do exist, use the asymptotes to help graph the function. Where appropriate, write a program for the calculator to help create the table of values.

1. $f(x) = x + 3$

2. $f(x) = 2x - 1$

3. $R = 5 - 2S$

4. $Z = 4y - 2$

5. $2A + 3B = 6$

6. $S = 2t - 5$

7. $y = \frac{1}{4}x^2 - 3$

8. $y = \frac{1}{2}x^2 + 4$

9. $B = A^2 - A - 6$

10. $S = 2t^2 + 5t + 2$

11. $R = \frac{1}{S}$

12. $f(S) = \frac{1}{S + 2}$

13. $f(x) = \frac{1}{x - 5}$

14. $f(t) = \frac{t + 5}{t - 3}$

15. $f(R) = \dfrac{R-3}{R+4}$

16. $y = \sqrt{2-x}$

17. $S = \sqrt{t-1}$

18. $y = \sqrt{x+2}$

19. $y = x^3 - 1$

20. $y = x^3 - 3x + 1$

21.

A	0	1	−1	2	−2	3	4	5
B	−5	−4	−6	−3	−7	−2	−1	0

22.

t	0	$\frac{1}{2}$	2	$4\frac{1}{2}$	8	$12\frac{1}{2}$	18
S	0	1	2	3	4	5	6

23.

t	0	1	2	3	4	5
V	10	42	74	106	138	170

24.

x	−3	−2	−1	0	1	2	3
y	3	2	1	0	1	2	3

25.

L	1	2	3	4	5	6
V	98	102	99	101	102	98
L	7	8	9	10	11	12
V	99	103	100	97	101	104

26. In biology there is an approximate rule for temperate climates, called the bioclimatic rule, that states that in spring and early summer periodic phenomena, such as blossoming for a given species, appearance of certain insects, and ripening of fruit usually comes about four days later for each 500 ft of altitude. Stated as a formula: $d = 4(a/500)$, where d = change in days and a = change in altitude in feet. Graph the equation for

$0 \leq a \leq 3500$.

27. The conversion of kilograms into pounds on the earth's surface is approximately $W_{\text{lb}} = 2.2\ m_{\text{kg}}$. Graph this equation letting W_{lb} be the dependent variable and m_{kg} the independent variable in kilograms. Can you use this graph to estimate weights?

28. The classic relationship of current I, resistance R, and voltage E is given by Ohm's law as $E = IR$.

a. Graph this equation when $I = 1.2$ A.

b. From the graph what voltage is needed at a resistance of 6.0 Ω.

c. What resistance produces 6.3 V?

29. The conversion of kilometers to miles is $m = 0.62d$ (where d is distance in kilometers).

a. Graph this equation.

b. Using the graph convert 40 mi to kilometers.

30. Newton's Law of Mutual Attraction (Gravitation) is $F = \dfrac{km_1m_2}{r^2}$. When $km_1m_2 = 10^{-9}$:

a. Graph the equation.

b. What value of r produces a force of 4×10^{-11}?

31. Coulomb's Law (electrostatic force between charged particles) is $F = \dfrac{q_1q_2}{4\pi\epsilon r^2}$. When $q_1 = 4 \times 10^{-6}$ and $q_2 = -\pi\epsilon \times 10^{-6}$, graph the equation.

In Exercises 32–37, use your calculator to determine a table of values and then graph the function.

32. $y = 3^x,\ 0 \leq x \leq 10$

33. $y = 6^x,\ -2 \leq x \leq 4$

34. $y = \ln x,\ x > 0$

35. $y = \log x,\ x > 0$

36. $y = e^x,\ -3 \leq x \leq 3$

37. $y = 10^x,\ -3 \leq x \leq 3$

5.4 Inverse functions

In Section 5.1 we said that a statement of equality such as $y = x + 2$ is a function. In fact, we can call this statement a **one-to-one function.** That is, for each value substituted in place of x we obtain exactly one value for y. The discussion in this section will be limited to one-to-one functions.

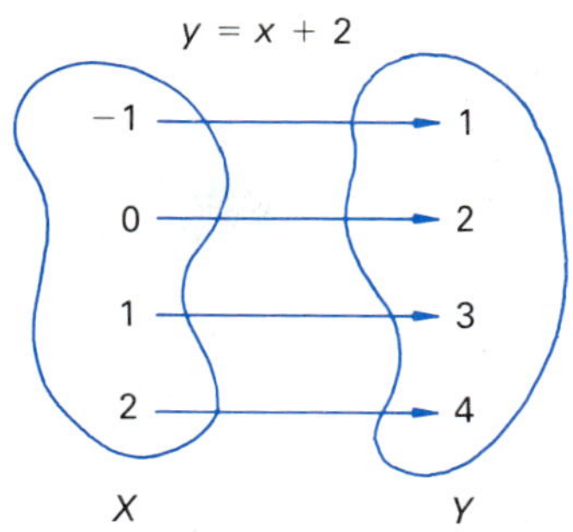

FIGURE 5.16

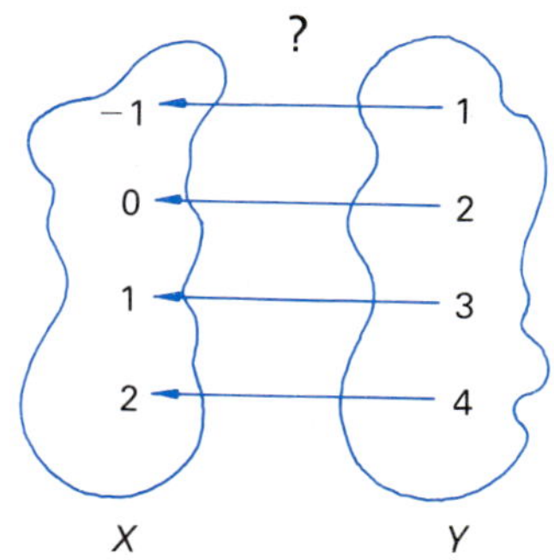

FIGURE 5.17

Figure 5.16 is an illustration of using the rule $x + 2$ on Set X to obtain Set Y. The rule, formally stated for the function f, is $y = f(x) = x + 2$. From this figure we can see that the domain is $\{-1, 0, 1, 2\}$ and the range is $\{1, 2, 3, 4\}$. Using the domain and range we can write the ordered pairs $(-1, 1)$, $(0, 2)$, $(1, 3)$, and $(2, 4)$, which represent the function.

A question now arises: Is it possible to obtain the number in Set X knowing Set Y and the rule $x + 2$? Other ways of asking the same question are: Can we go backwards, or undo what we have done? Can a rule be found that would take the elements in Set Y and give us the elements in Set X? or Can we find the ordered pairs $(1, -1)$, $(2, 0)$, $(3, 1)$, and $(4, 2)$? Figure 5.17 illustrates this idea.

The question mark represents the rule called an **inverse function** of f and indicated by the symbol f^{-1}. (The -1 is not a negative exponent but is part of the symbol meaning inverse function.)

RULE 5.1

> **The inverse of a function** f can be found by interchanging the independent variable x and the dependent variable y and then solving for y. The resulting statement of equality will be the inverse function of f, indicated by f^{-1}.

EXAMPLE 1 Find the inverse function of $y = f(x) = x + 2$.

Solution: Applying the rule we interchange the role of x and y; thus $x = y + 2$. Now solving for y:

$$x - 2 = y + 2 - 2$$
$$x - 2 = y \qquad \text{or} \qquad y = x - 2.$$

In other words $f^{-1}(x) = x - 2$. Thus $f^{-1}(x) = x - 2$ is the inverse function of $f(x) = x + 2$ or, stated more formally, if $f(x) = x + 2$, then $f^{-1}(x) = x - 2$. □

(*Note:* Finding the inverse is not accomplished simply by subtracting 2 instead of adding 2.)

EXAMPLE 2 Find the inverse function of $y = f(x) = 3x - 5$. Sketch the function and the inverse function on the same set of axes.

Solution: Applying the rule to $y = 3x - 5$, we interchange x and y. Thus $x = 3y - 5$. Now solving for y:

$$x + 5 = 3y - 5 + 5$$
$$x + 5 = 3y$$
$$\frac{x+5}{3} = \frac{\cancel{3}y}{\cancel{3}}.$$

Therefore

$$y = \frac{x+5}{3}.$$

Thus $y = \frac{x+5}{3}$ is the inverse function of $y = 3x - 5$ or, more formally, if $f(x) = 3x - 5$, then $f^{-1}(x) = \frac{x+5}{3}$.

The graphs of $f(x)$ and $f^{-1}(x)$ are mirror images of each other with respect to the line $y = x$. This is illustrated in Fig. 5.18. □

FIGURE 5.18

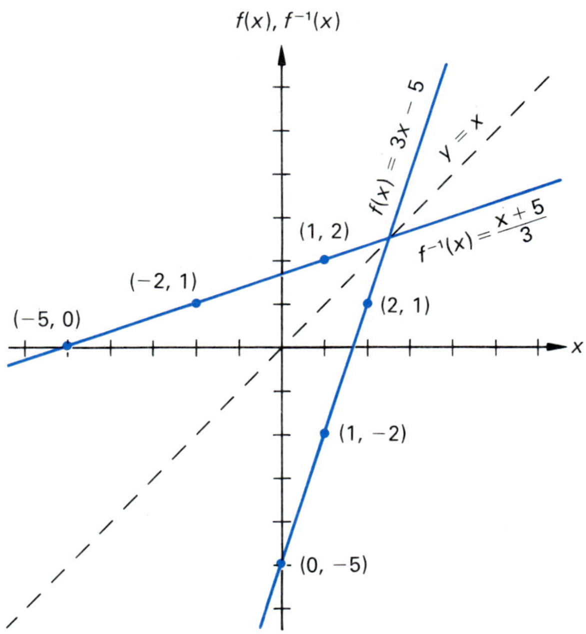

$f(x) = 3x - 5$

x	$f(x)$
0	−5
1	−2
2	1

$f^{-1}(x) = \frac{x+5}{3}$

x	$f^{-1}(x)$
−5	0
−2	1
1	2

EXAMPLE 3

a. Find the inverse function of $y = f(x) = \dfrac{2}{x-3}$.

b. Determine the domain and range of the function and of its inverse.

Solution:

a. Applying the rule we interchange the x and y:

$$x = \frac{2}{y-3}.$$

Solving for y

$$(y-3)x = \overset{1}{\cancel{(y-3)}}\,\frac{2}{\underset{1}{\cancel{y-3}}}, \qquad y \neq 3$$

$$xy - 3x = 2$$
$$xy - 3x + 3x = 2 + 3x$$
$$\frac{xy}{x} = \frac{2+3x}{x}$$
$$y = \frac{2+3x}{x} \qquad \text{or} \qquad f^{-1}(x) = \frac{2+3x}{x}.$$

b. For $y = \dfrac{2}{x-3}$, it can be seen that $x \neq 3$. Thus, $D = \{x|x \neq 3\}$. If we solve the equation for x, we would have $x = \dfrac{2+3y}{y}$. From this equation it can be seen that $y \neq 0$; thus $R = \{y|y \neq 0\}$. For the inverse function $y = \dfrac{2+3x}{x}$, $x \neq 0$; therefore $D = \{x|x \neq 0\}$. Solving the inverse function for x we would have $x = \dfrac{2}{y-3}$ and $y \neq 3$; therefore $R = \{y|y \neq 3\}$. □

Note the range of the inverse of the function is the same as the domain of the original function. Also, the domain of the inverse of the function is the same as the range of the original function.

5.4 Exercises

In Exercises 1 – 18, find the inverse function of each of the functions.

1. $y = f(x) = x + 5$

2. $y = f(x) = x - 4$

3. $y = f(x) = 2x + 7$

4. $y = f(x) = 3x - 5$

5. $y = f(x) = 2(x - 4)$

6. $y = f(x) = 3(2 - x)$

7. $y = f(x) = \dfrac{x+4}{2}$

8. $y = f(x) = \dfrac{6+x}{5}$

9. $y = f(x) = \dfrac{9-2x}{3}$

10. $y = f(x) = \dfrac{2x-5}{4}$

11. $y = f(x) = \frac{1}{2}(x + 7)$

12. $y = f(x) = \frac{1}{5}(2x + 3)$

13. $y = f(x) = \frac{1}{x + 2}$

14. $y = f(x) = \frac{1}{x - 5}$

15. $y = f(x) = \frac{2}{x + 1}$

16. $y = f(x) = \frac{4}{2x + 5}$

17. $y = f(x) = \frac{5}{3x + 2}$

18. $y = f(x) = \frac{4}{4x - 1}$

In Exercises 19–30, sketch the graph of the functions in Exercises 1–12. On the same set of axes sketch the inverse of each of the functions.

In Exercises 31–36, find the domain and range of each of the functions in Exercises 13–18 and then find the domain and range of its inverse function.

37. Given the function $y = f(x) = \frac{2}{11x - 9}$, determine the domain and range. Now find a function that has the same range as the original function's domain and same domain as the original function's range.

38. Given the function $y = f(x) = \frac{2x}{x - 3}$, determine the domain and range. Now find a function that has the same range as the original function's domain and same domain as the original function's range.

39. By examining the sketches in Exercises 19–30, can you determine the geometric relation between the function and its inverse?

Chapter summary

It is essential that the reader develop an understanding of the concept of function. We work with functions in algebra, trigonometry, analytic geometry, and calculus.

Definition 5.1 A **function** is a rule that assigns for each value of the independent variable a unique value of the dependent variable.

A function can also be expressed as a set of ordered pairs or by using functional notation. For example the function $y = x + 4$ can be written as $f(x) = x + 4$.

The set of real numbers that can be substituted in a function for the independent variable and give us real numbers for the dependent variable is called the **domain** of the function. The set of real numbers that we obtain for the dependent variable is called the **range** of the function.

To be able to graph a function we must understand the rectangular coordinate system. The important terms of the system are *axis of the independent variable* (x-axis), *axis of the dependent variable* (y-axis), *origin, quadrants* (first, second, third, and fourth). Every point on the plane can be represented by an ordered pair (x, y). We call the ordered pair (x, y) *a set of coordinates*. The first coordinate, x, is called the *abscissa* and the second coordinate, y, is called the *ordinate*. The origin is represented by the ordered pair $(0, 0)$.

Theorem 5.1 **Distance Formula** If $P_1(x_1, y_1)$ and $P_2(x_2, y_2)$ are any two points in the plane, then the distance, $\overline{P_1P_2}$ between them is given by

$$\overline{P_1P_2} = \sqrt{(x_2 - x_1)^2 + (y_2 - y_1)^2}.$$

Theorem 5.2 **Midpoint Formula** If $P_1(x_1, y_1)$ and $P_2(x_2, y_2)$ are any two points in the plane, then the coordinates of the midpoint of the line segment from P_1 to P_2 are given by

$$\left(\frac{x_1 + x_2}{2}, \frac{y_1 + y_2}{2}\right).$$

In order to simplify the graphing of functions the following suggestions are made:

1. Determine which variable is the dependent variable and which is the independent variable.
2. Solve the equation for the dependent variable.
3. Find the domain and the range of the function. This will indicate the restrictions on the function, thus showing where the curve does not exist.
4. Construct a table of values. (For a linear function, only two pairs of values are required. For nonlinear functions find enough pairs of values so a smooth curve can be sketched.)
5. From the table of values determine an appropriate scale for the axes of the graph.
6. Plot points, label points, especially points on horizontal and vertical axes, and label the graph.

In plotting points of a curve if the points get closer and closer to a line but never touch the line for larger and larger values of the independent variable, then we say the line is an **asymptote** and the curve is asymptotic to the line.

If we have a **one-to-one function,** we can find the inverse of the function.

Rule 5.1 The inverse of a function f can be found by interchanging the independent variable x and the dependent variable y and then solving for y. The resulting statement of equality will be the inverse function of f, indicated by f^{-1}.

The graphs of $f(x)$ and $f^{-1}(x)$ are mirror images of each other with respect to the line $y = x$. Note that the range of the inverse of the function is the same as the domain of the original function. Also the domain of the inverse of the function is the same as the range of the original function.

Review Exercises

In Exercises 1 – 7, find the value of the functions.

1. $f(x) = 3x + 5$ a. $f(-3)$ b. $f(2)$ c. $f(h+3)$
2. $g(t) = 5t - 4$ a. $g(-4)$ b. $g(3)$ c. $g(k-5)$
3. $h(t) = t^2 + 4$ a. $h(-3)$ b. $h(3)$ c. $h(z+2)$
4. $k(a) = \dfrac{5}{a}$ a. $k(-2)$ b. $k(25)$ c. $k(m-3)$
5. The volume of a cube is given by $V(s) = s^3$, where s is the length of a side. Find:
   a. $V(2.35 \text{ mm})$
   b. $V(7.34 \text{ m})$
   c. $V(6/7 \text{ in})$
6. The total surface area of a rectangular solid is given by $T(\ell, w, h) = 2(\ell w + wh + \ell h)$ Find:
   a. $T(3 \text{ cm}, 4 \text{ cm}, 5 \text{ cm})$
   b. $T(4.3 \text{ m}, 7.5 \text{ m}, 8.3 \text{ m})$
   c. $T(0.060 \text{ mm}, 0.040 \text{ mm}, 0.030 \text{ mm})$
7. Force F is a function of mass m and acceleration a given by $F(m, a) = ma$. Find:
   a. $F(0.300 \text{ kg}, 10 \text{ m/s}^2)$
   b. $F(21.000 \text{ kg}, 75{,}000 \text{ m/s}^2)$

In Exercises 8 – 17, find the domain and range for each function.

8. $y = x + 13$
9. $2B = 6A - 5$
10. $f(t) = t^2 - 5$
11. $f(t) = t^2 + 5$
12. $f(x) = \dfrac{4}{x-1}$
13. $f(R) = \dfrac{2}{R+3}$
14. $f(L) = \dfrac{L+1}{L-1}$
15. $f(L) = \dfrac{L-3}{L+3}$
16. $g(s) = \sqrt{s+5}$
17. $g(s) = \sqrt{5-s}$

In Exercises 18 – 21, find a function that describes:

18. The number of liters in a gallon.
19. The area of a square as a function of its side s.
20. The cost of water used in a home as a function of the number of gallons used.
21. The distance a marble rolls down an inclined plane varies directly as the square of the time during which it rolls.

In Exercises 22 – 27, find $\overline{P_1P_2}$ and the midpoint of P_1P_2.

22. $P_1(4, 2)$; $P_2(7, 5)$
23. $P_1(-3, 5)$; $P_2(6, 7)$
24. $P_1(-4, -3)$; $P_2(4, 3)$
25. $P_1(4, -7)$; $P_2(-6, 3)$
26. $P_1(-2.4, 3.5)$; $P_2(-7.1, -6.4)$
27. $P_1(0, -3.4)$; $P_2(3.5, -4.1)$

In Exercises 28 – 37, graph the function.

28. $y = f(x) = x + 7$
29. $y = g(x) = x - 3$
30. $B = 3A^2 - 4A + 5$
31. $B = 4A^2 - 5A + 7$
32. $f(t) = \dfrac{t+3}{t-2}$
33. $f(t) = \dfrac{t-5}{t+7}$
34. $y = 12x - x^3$
35. $y = x^4 - 4x^2 + 2$
36. $S = \sqrt{5-t}$
37. $S = \sqrt{t-5}$

In Exercises 38 – 46, a) Find the inverse function of each function, b) Find the domain and range of each of the functions and then find the domain and range of its inverse function, and c) Sketch the graph of each of the functions and its inverse function on the same set of axes.

38. $y = f(x) = x - 7$
39. $y = f(x) = x + 6$
40. $y = f(x) = 5x - 4$
41. $y = f(x) = 6x + 5$
42. $y = f(x) = \dfrac{2x+4}{3}$
43. $y = f(x) = \dfrac{1}{3}(2x + 6)$
44. $y = f(x) = \dfrac{3}{x}$
45. $y = f(x) = \dfrac{1}{x+2}$
46. $y = f(x) = \dfrac{-1}{x+2}$
47. An automobile and a truck leave the same point at the same time traveling due south at 48 mph and due east at 30 mph, respectively. Find an expression for their distance apart d, measured in miles, at the end of t hours.
48. A sea plane flying 300 statute miles per hour at an altitude of 600 ft passes over a buoy. Find an expression for its distance d measured in feet, from the buoy t seconds later.
49. Find the lengths of the sides and thereby prove that the points are vertices of an isosceles triangle: $A(2, 3)$, $B(6, 7)$, $C(6, -1)$.
50. a. Prove that the points $(1, 5)$, $(2, 4)$, and $(-3, 5)$ are on a circle with center $(-1, 2)$.
    b. What is the radius of the circle?

51. An average driver traveling r miles per hour requires f feet to stop his car, where r and f are related as in the following table:

r	20	30	40	50	60	70
f	40	80	125	180	250	340

a. Represent the data graphically.

b. From the graph estimate the number of feet needed to stop when traveling at the rate of 55 mph.

52. A taxi charges $1.15 as soon as you enter the car, then 25 cents after each 1/6 mile.

a. Write a function $f(x)$ expressing the cost of riding x miles in this cab.

b. Construct a table of values.

c. Graph the function.

Can you solve these?

1. Show that the midpoints of the sides of a rectangle are the vertices of a rhombus. [*Hint:* Place the rectangle so that its vertices are at the points (0, 0), (a, 0), (0, b), and (a, b)].
2. Use a convenient coordinate system to show that the midpoint of the hypotenuse of a right triangle is equidistant from its vertices.
3. Write a BASIC program that will find the distance between any two points in a plane.
4. Write a program for your programmable calculator that will find the distance between any two points in a plane.
5. Prove Theorem 5.2. The proof requires two parts: a) Show that $\overline{P_1M} = \overline{MP_2}$ and b) Show that $\overline{P_1M} + \overline{MP_2} = \overline{P_1P_2}$, that is, that the points are collinear.
6. The perpendicular bisector of the line segment $\overline{PQ}$ is the line that is perpendicular to $\overline{PQ}$ and that divides $\overline{PQ}$ into two line segments of the same length. If P and Q are $(-2, 5)$ and $(4, 9)$, respectively, find the equation of the perpendicular bisector of $\overline{PQ}$.
7. A rectangular box with a square base and open top is to be made from 48 ft² of lumber. Express the volume V of the box as a function of a side x, measured in feet, of the square base. Find, by means of a graph, the approximate dimensions and volume of the largest box that can be made.

6

The 260-ft Penn Central Railroad Bridge crossing the Genesee River at Rochester Harbor is a swing bridge built on a pier in the middle of the river. The bridge is supported by six $2\frac{1}{2}$-ft-diameter vertical wheels that, as the bridge turns, roll around the circumference of a horizontal 40-ft-diameter plate mounted on the support pier. How many revolutions do the support wheels make as the bridge rotates ninety degrees from full open to full close?

Introduction to Trigonometry

6.1 Introduction to radian measure

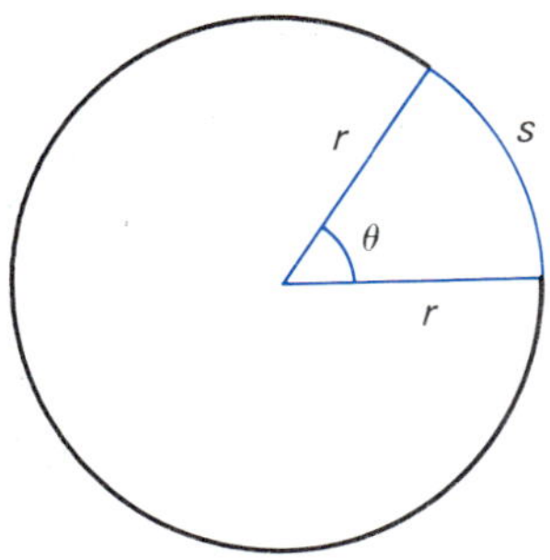

FIGURE 6.1

In Chapter 4 we discussed angles measured in degrees and mentioned that there was another measure of angles. This other form is called **radian measure.** In Fig. 6.1 the measure of the central angle θ is determined by the ratio of the length of the arc s subtended by the angle θ and the radius r of the circle or $\theta = s/r$. In this case $s = r$, thus the measure of θ is exactly one radian.

$$\theta = \frac{s}{r} \text{ radian}$$

$$= \frac{r}{r} \text{ radian}$$

$$= 1 \text{ radian.}$$

DEFINITION 6.1

> One **radian** is the measure of a central angle of a circle that subtends an arc equal in length to the radius of the circle. (See Fig. 6.1.)

Figure 6.2 illustrates several angles measured in radians.

The length of the arc s may be determined by the formula

$$s = r\theta,$$

where θ = measure of central angle in **radians** (abbreviated rad). The measure of the central angle can be determined by the formula

$$\theta = \frac{s}{r} \text{ rad.}$$

FIGURE 6.2

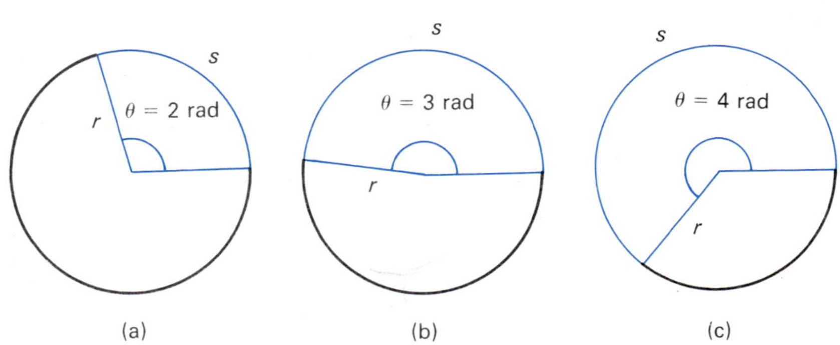

For a complete circle $s = 2\pi r$, which is the circumference of the circle

$$\theta_{\text{circle}} = \frac{s}{r} = \frac{2\pi \cancel{r}^{1}}{\cancel{r}_{1}} = 2\pi \text{ rad.}$$

Thus, we can express the measure of the central angle of a complete circle as 2π radians. We also know that $\theta_{\text{circle}} = 360$ degrees, which gives us the following relationship:

$$\theta_{\text{circle}} = 2\pi \text{ rad} = 360°.$$

$$2\pi \text{ rad} = 360°.$$

Dividing both sides by 2π we have

$$\frac{\cancel{2\pi}^{1} \text{ rad}}{\cancel{2\pi}_{1}} = \frac{360°}{2\pi}$$

$$1 \text{ rad} = \frac{180°}{\pi};$$

also,

$$\frac{1 \text{ rad}}{\text{rad}} = \frac{180°}{\pi \text{ rad}}.$$

$$1 = \frac{180°}{\pi \text{ rad}}.$$

If we divide both sides by 360° we have

$$\frac{2\pi \text{ rad}}{360°} = \frac{360°}{360°}$$

$$\frac{\pi \text{ rad}}{180°} = 1.$$

These two relationships provide us with conversion formulas for changing degrees to radians and from radians to degrees. The reader should note that a radian is a dimensionless quantity, since a radian is the ratio of one distance to another. When radians are used, it is often customary not to indicate units for the angle.

$$\frac{180^\circ}{\pi} = 1 \qquad \frac{\pi}{180^\circ} = 1$$

EXAMPLE 1 Convert from degrees to radians.

a. 45° **b.** 150°

Solution:

a. Using the conversion factor we have

$$45^\circ = \overset{1}{\cancel{45^\circ}}\left(\frac{\pi}{\underset{4}{\cancel{180^\circ}}}\text{ rad}\right)$$

$$= \frac{\pi}{4}\text{ rad.}$$

When expressing an angle in radians it is customary to write the angle with no words or symbols, since it is a dimensionless quantity. Thus

$$45^\circ = \frac{\pi}{4}.$$

The answer in this form, $\pi/4$, is called **exact,** whereas if we had changed to a decimal form and divided by 4 the answer would be **approximate.**

b. Using the conversion factor we have

$$150^\circ = \overset{5}{\cancel{150^\circ}}\left(\frac{\pi}{\underset{6}{\cancel{180^\circ}}}\right)$$

$$= \frac{5}{6}\pi. \qquad \square$$

Note that there is no symbol after $5/6\pi$, which means that the angle is in radian measure. As you encounter numbers such as 2, 1.4, etc., as measures of angles it will mean 2 rad, 1.4 rad, etc. When the discussion relates to a measure in degrees, it will be indicated by 2° or 2 degrees.

EXAMPLE 2 Convert from radians to degrees:

a. $\frac{\pi}{6}$ **b.** $\frac{3\pi}{2}$

Solution:

a. $\frac{\pi}{6} = \frac{\pi}{6}\left(\frac{180°}{\pi}\right)$ **b.** $\frac{3\pi}{2} = \frac{3\pi}{2}\left(\frac{180°}{\pi}\right)$

$= 30°.$ $= 270°.$ □

Without a great deal of work you can see some relationships develop, i.e., $0° = 0$, $90° = \pi/2$, $180° = \pi$, and $360° = 2\pi$. It is left to the reader to check these using the conversion factors.

The conversion factors do not provide us with exact answers when we convert radians to degrees, or degrees to radians using an approximate value for π. In these cases we use our calculators and obtain approximate answers.

EXAMPLE 3 Convert from radians to degrees expressing the answer in approximate decimal form.

a. 1 **b.** 2.3

Solution:

a. $1 = 1\left(\frac{180°}{\pi}\right)$

$= 57.29578°$ **Using the calculator value for π. For calculations without the calculator 1 rad is generally expressed as 57.3°.**

$= 57.30°$ **(to 4 significant digits)**

b. $2.3 = 2.3\left(\frac{180°}{\pi}\right)$

$= 131.78029$

$= 130°$ **(to 2 significant digits).** □

FIGURE 6.3

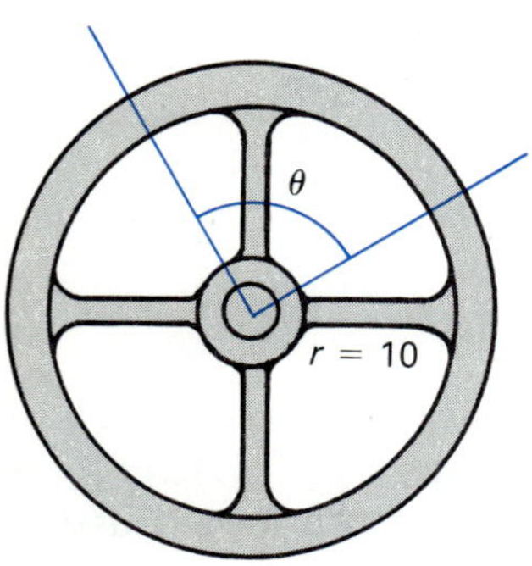

EXAMPLE 4 On a flywheel with a 10-ft radius (Fig. 6.3), how long is an arc associated with a central angle of 1.6 rad?

Solution: To determine the length of the arc we can use the formula $s = r\theta$, with $r = 10$ ft and $\theta = 1.6$ rad. (θ must be in radian measure.)

Substituting, we have

$$s = (10 \text{ ft})(1.6)$$
$$= 16 \text{ ft}.$$

Thus, the central angle of 1.6 rad subtends an arc of 16 ft on the flywheel. □

6.1 Exercises

In Exercises 1–14, change the degree measures to radian measures. In each case give the answer in exact form.

1. 15° **2.** 30° **3.** 45° **4.** 60°

5. 75° **6.** 120° **7.** 135° **8.** 150°

9. 210° **10.** 240° **11.** −225° **12.** −60°

13. 300° **14.** 315°

In Exercises 15–24, change the degree measures to radian measures. In each case give the answer in approximate form to the nearest thousandth.

15. 330° **16.** 22.5° **17.** 212.5° **18.** 4°

19. −120° **20.** −60° **21.** 39° **22.** 190°

23. 290° **24.** 100°

In Exercises 25–44, change the radian measures to degree measures. If the answer is not exact, express it to the nearest thousandth.

25. $\frac{\pi}{8}$ **26.** $\frac{\pi}{6}$ **27.** $\frac{\pi}{4}$ **28.** $\frac{\pi}{3}$

29. $\frac{5\pi}{12}$ **30.** $\frac{2\pi}{3}$ **31.** $\frac{3\pi}{4}$ **32.** $\frac{5\pi}{6}$

33. $\frac{7\pi}{6}$ **34.** $\frac{4\pi}{3}$ **35.** $\frac{5\pi}{4}$ **36.** $-\frac{\pi}{3}$

37. $\frac{5\pi}{3}$ **38.** $\frac{7\pi}{4}$ **39.** $\frac{11\pi}{6}$ **40.** 13

41. 1.74 **42.** 6.282 **43.** 1 **44.** 143

45. Through how many radians does the minute hand of a clock rotate in 15 minutes? Convert to degrees.

46. Through how many radians does the minute hand of a clock rotate in 45 minutes? Convert to degrees.

47. On a flywheel with a 70-mm radius, how long is an arc subtended by a central angle of 2.10 rad?

48. A telescope is designed so that it can distinguish two points 1.0 mm apart at a distance of 1.6 km. The angular separation of the two points at that distance is 6.25×10^{-6} rad. What is the angular separation in degrees?

49. The sensitivity of a measuring instrument is given as 10μ A per degree deflection. What is the sensitivity in amperes per radian?

50. Suppose that a pendulum of length 250 mm swings through an arc of length 150 mm. (See Fig. 6.4.) Through how many degrees does the pendulum swing?

51. An engineer is to design a cloverleaf for an exit from an interstate highway. A part of the cloverleaf can be thought of as an arc of a circle with radius of 420 m. The central angle of this arc is 135°. Find the length of this part of the cloverleaf.

52. For a simple beam acted on by a concentrated load, the angle of deflection in radians is given by $\theta = P\ell^2/16EI$. Find θ in (a) radians, and (b) degrees when $P = 46{,}000$ N, $\ell = 0.800$ m, and $EI = 10^{14}$ Nm^2.

53. While rounding a curve one front wheel of a truck rotates 286°40′ and a second front wheel rotates 5.00 rad. Which wheel is closer to the inside of the curve? (*Hint:* Outside wheels rotate more.)

54. How many radians is 2π?

FIGURE 6.4

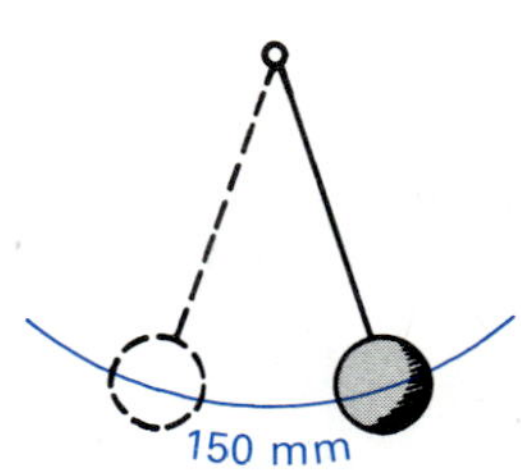

6.2 The trigonometric functions

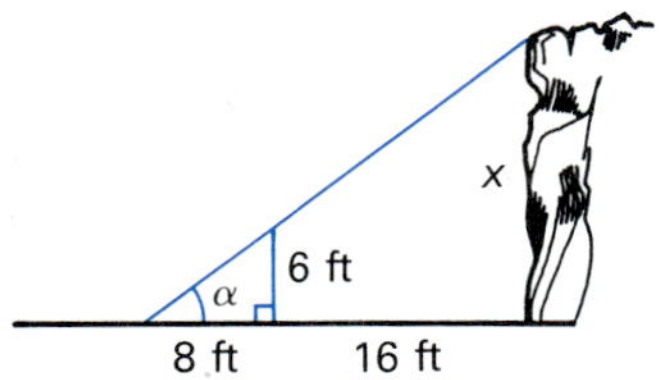

FIGURE 6.5

From our study of similar triangles in Section 4.4, we could find the height of the cliff in Fig. 6.5 to be 18′. A surveyor with a transit would measure the angle α and use right-triangle trigonometry to determine the height of the cliff. In order to understand how the surveyor can determine the height of the cliff using right-triangle trigonometry it is necessary to understand the six basic trigonometric functions.

By using similar triangles in Fig. 6.5 we can determine the additional lengths in Fig. 6.6. The measure of angle θ in Fig. 6.6 determines the position of the terminal side of the angle. Thus, the terminal side is dependent (dependent variable) on the measure of the angle (independent variable). If we pick two points (8, 6) and (24, 18) on the terminal side of the angle θ and drop perpendiculars to the initial ray as in Fig. 6.6, we illustrate the similar triangles that were used to construct the figure. Recall from earlier discussions that corresponding parts of similar triangles are proportional. Thus

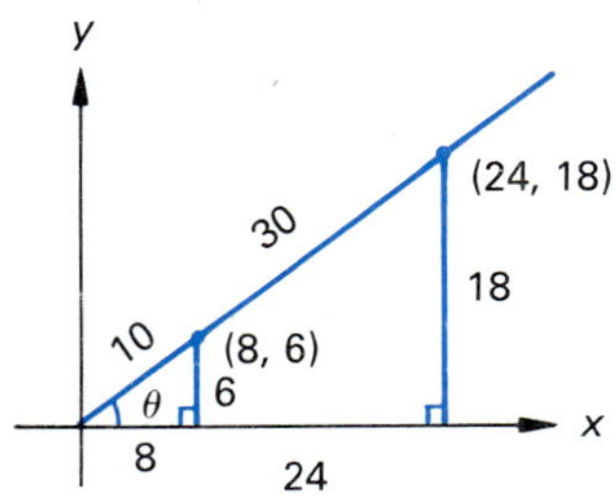

FIGURE 6.6

$$\frac{6}{8} = \frac{18}{24} = \frac{3}{4} \qquad \frac{8}{10} = \frac{24}{30} = \frac{4}{5} \qquad \frac{6}{10} = \frac{18}{30} = \frac{3}{5}.$$

From these results we can see that the measure of angle θ determines the ratio and not the position of the point on the terminal ray or the size of the triangle. If we designate the sides of the triangle as in Fig. 6.7 we can create six distinct ratios of the sides. Each of these ratios is determined by one of the six trigonometric functions, which are given in Table 6.1.

EXAMPLE 1 Find the six trigonometric functions of θ if the terminal side passes through (3, 5). Give the answers in exact form.

Solution: Figure 6.8 shows a sketch of θ. Since (3, 5) is on the terminal side of the angle, it is a convenient point to draw a perpendicular to the x-axis. Thus, $x = 3$ and $y = 5$. We can find r using the Pythagorean theorem:

FIGURE 6.7

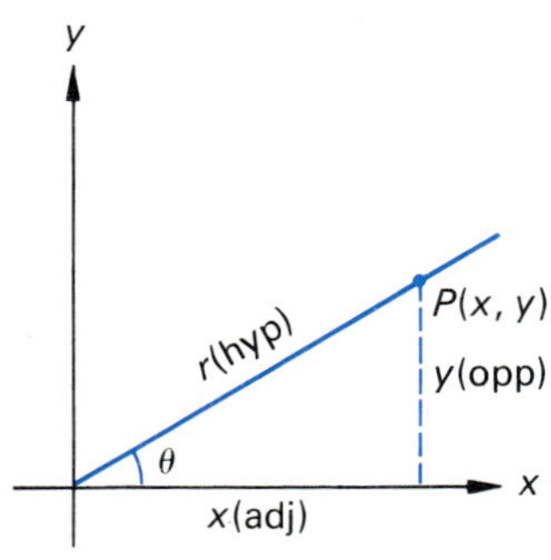

$$\begin{aligned} r &= \sqrt{(3)^2 + (5)^2} \\ &= \sqrt{34}. \end{aligned}$$

The functional values are

$$\sin\theta = \frac{y}{r} = \frac{5}{\sqrt{34}}, \qquad \csc\theta = \frac{r}{y} = \frac{\sqrt{34}}{5},$$

$$\cos\theta = \frac{x}{r} = \frac{3}{\sqrt{34}}, \qquad \sec\theta = \frac{r}{x} = \frac{\sqrt{34}}{3},$$

$$\tan\theta = \frac{y}{x} = \frac{5}{3}, \qquad \cot\theta = \frac{x}{y} = \frac{3}{5}. \qquad \square$$

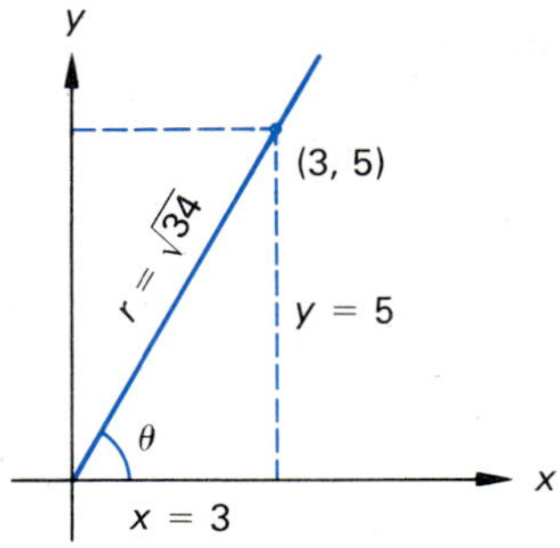

FIGURE 6.8

TABLE 6.1 **Definitions of trigonometric functions**

Function	Abbreviation for function value	Defining ratio for function value	Defining ratio for right triangle*
Sine	$\sin\theta$	$\sin\theta = \dfrac{y}{r}$	$\sin\theta = \dfrac{\text{opp}}{\text{hyp}}$
Cosine	$\cos\theta$	$\cos\theta = \dfrac{x}{r}$	$\cos\theta = \dfrac{\text{adj}}{\text{hyp}}$
Tangent	$\tan\theta$	$\tan\theta = \dfrac{y}{x}$	$\tan\theta = \dfrac{\text{opp}}{\text{adj}}$
Cotangent	$\cot\theta$	$\cot\theta = \dfrac{x}{y}$	$\cot\theta = \dfrac{\text{adj}}{\text{opp}}$
Secant	$\sec\theta$	$\sec\theta = \dfrac{r}{x}$	$\sec\theta = \dfrac{\text{hyp}}{\text{adj}}$
Cosecant	$\csc\theta$	$\csc\theta = \dfrac{r}{y}$	$\csc\theta = \dfrac{\text{hyp}}{\text{opp}}$

* Abbreviations are side opposite to $\theta =$ opp, side adjacent to $\theta =$ adj, and hypotenuse = hyp.

From Table 6.1 and from Example 1 note that $\csc\theta$ is the reciprocal of $\sin\theta$, $\sec\theta$ is the reciprocal of $\cos\theta$, and $\cot\theta$ is the reciprocal of $\tan\theta$. Therefore

$$\csc\theta = \frac{1}{\sin\theta}, \quad \sin\theta \neq 0, \quad \text{or} \quad \sin\theta = \frac{1}{\csc\theta}, \quad \csc\theta \neq 0,$$

$$\sec\theta = \frac{1}{\cos\theta}, \quad \cos\theta \neq 0, \quad \text{or} \quad \cos\theta = \frac{1}{\sec\theta}, \quad \sec\theta \neq 0,$$

$$\cot\theta = \frac{1}{\tan\theta}, \quad \tan\theta \neq 0, \quad \text{or} \quad \tan\theta = \frac{1}{\cot\theta}, \quad \cot\theta \neq 0.$$

EXAMPLE 2 Find the five remaining trigonometric functions, given that $\sin\theta = 7/9$ and θ is a first quadrant angle. Give the answers in exact form.

Solution: Given that $\sin\theta = 7/9$, we know that two possible paired values for y and r are $y = 7$ and $r = 9$. We can find x using the Pythagorean theorem:

$$x = \sqrt{r^2 - y^2}.$$

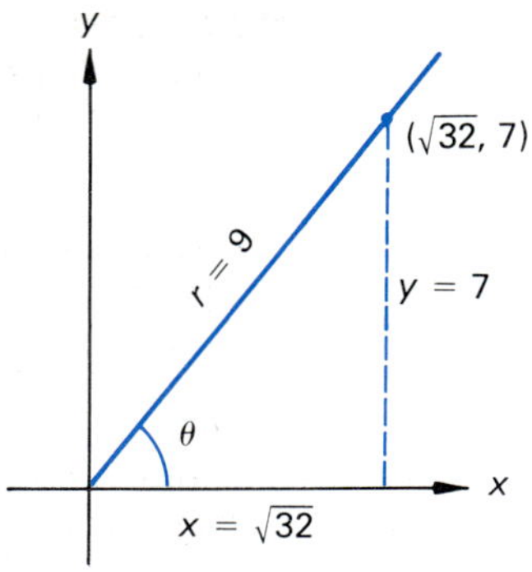

FIGURE 6.9

Since the angle is in the first quadrant, we want the positive root; thus $x = \sqrt{9^2 - 7^2}$ or $\sqrt{32}$. We can locate the point $(\sqrt{32}, 7)$ on the terminal side of the angle. From this point drop a perpendicular to the x-axis and form the reference triangle and label as in Fig. 6.9.

Using $x = \sqrt{32}$, $y = 7$, and $r = 9$ we can indicate the other trigonometric functions of θ:

$$\cos\theta = \frac{\sqrt{32}}{9}, \quad \csc\theta = \frac{9}{7}, \quad \cot\theta = \frac{\sqrt{32}}{7},$$

$$\tan\theta = \frac{7}{\sqrt{32}}, \quad \sec\theta = \frac{9}{\sqrt{32}}. \quad \square$$

EXAMPLE 3 Use the given trigonometric function to find the indicated trigonometric functions.

$\tan\theta = 0.3$, find $\cos\theta$ and $\csc\theta$.

Solution: For the case where the trigonometric function is equal to a fraction or a whole number we follow the procedure in Example 2. For $\tan\theta = 0.3$, we can easily change 0.3 to a fraction by multiplying $\frac{0.3}{1}$ by $\frac{10}{10}$. The result is the ratio $\frac{3}{10}$, and $\tan\theta = \frac{3}{10}$. With this information we know that the terminal side of the angle passes through the point (10, 3), (Fig. 6.10). Dropping a perpendicular from (10, 3) to the x-axis we have a right triangle with $x = 10$, $y = 3$, and $r = \sqrt{109}$. Therefore $\cos\theta = \frac{10}{\sqrt{109}}$, and $\csc\theta = \frac{\sqrt{109}}{3}$. $\square$

When two angles are placed in standard position and their terminal sides coincide, we say that the two angles are **coterminal.** For example, in Fig. 6.11, $\alpha = 30°$ and $\beta = 390°$ are coterminal since $390° = 360° + 30°$. Note that there is more than one coterminal angle for each angle. In this case, if $\theta = -330°$, then α and θ are coterminal angles.

FIGURE 6.10

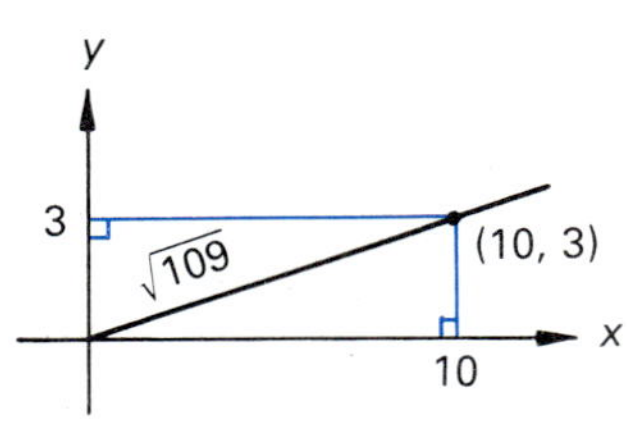

FIGURE 6.11

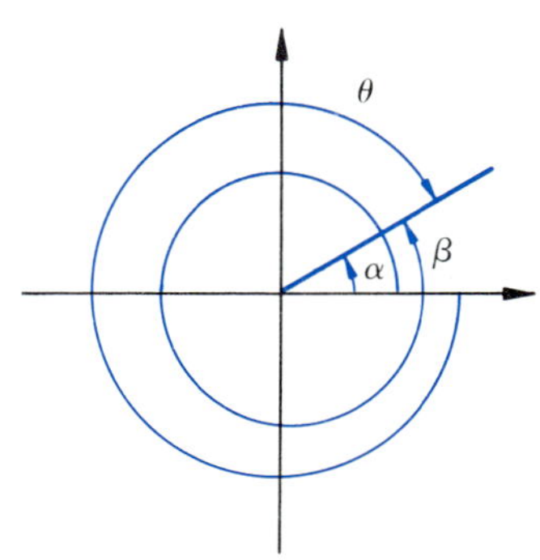

6.2 Exercises

In Exercises 1–9, find the six trigonometric functions of θ if the terminal side passes through the given point. Give the answers in exact form.

1. (3, 4) **2.** (8, 6) **3.** (4, 4)
4. (2, 1) **5.** (12, 15) **6.** $(\sqrt{3}, 1)$
7. $(\sqrt{2}, \sqrt{2})$ **8.** (7, 5) **9.** (1, 2)

In Exercises 10–18, use the given trigonometric function to find the indicated trigonometric functions.

10. $\sin\theta = 3/5$; find $\tan\theta$ and $\sec\theta$

11. $\tan\theta = 2/3$; find $\cos\theta$ and $\csc\theta$

12. $\cos\theta = 1/2$; find $\sin\theta$ and $\cot\theta$

13. $\cot\theta = 0.4$; find $\sin\theta$ and $\sec\theta$

14. $\csc\theta = 5/2$; find $\sin\theta$ and $\tan\theta$

15. $\sec\theta = 1.5$; find $\csc\theta$ and $\cot\theta$

16. $\sin\theta = \sqrt{3}/2$; find $\cot\theta$ and $\cos\theta$

17. $\cos\theta = \sqrt{2}/2$; find $\csc\theta$ and $\tan\theta$

18. $\tan\theta = \sqrt{3}$; find $\sec\theta$ and $\sin\theta$

In Exercises 19–30, find two angles coterminal to the given angle and sketch the angles.

19. 25° **20.** 120° **21.** $\frac{\pi}{2}$ **22.** 2

23. −45° **24.** 210° **25.** $-\pi$ **26.** 3

27. −135° **28.** 150° **29.** $\frac{\pi}{6}$ **30.** $\frac{-\pi}{3}$

31. From the definitions in Table 6.1 show that

a. $\tan\theta = \dfrac{\sin\theta}{\cos\theta}$ **b.** $\cot\theta = \dfrac{\cos\theta}{\sin\theta}$

32. From the definitions in Table 6.1 show that

a. $\sec\theta = \dfrac{1}{\cos\theta}$ **b.** $\csc\theta = \dfrac{1}{\sin\theta}$

6.3 Evaluating the trigonometric functions

The discussion in Section 6.2 provided us with a value for each of the six trigonometric functions for a given ratio in the first quadrant. What we want to do in this section is to develop techniques for finding the trigonometric functional value for any acute angle. Then we will use these to determine the trigonometric functional values for any angle.

EXAMPLE 1 Find the exact values for the six trigonometric functions of a 60° angle.

FIGURE 6.12

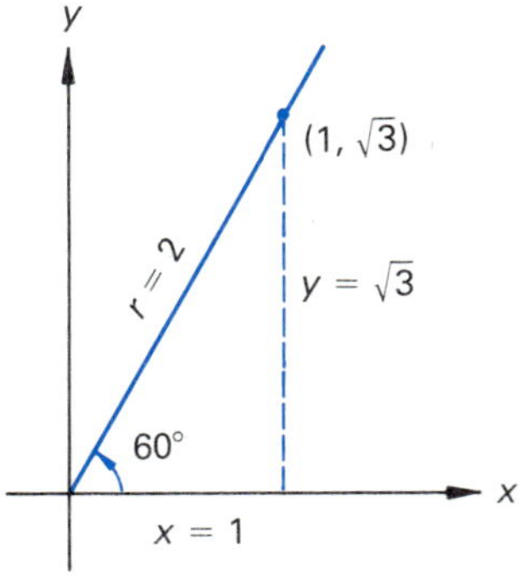

Solution: From geometry we recall that a 30°–60° right triangle has sides in the ratio $1:2:\sqrt{3}$. The most convenient point to pick is $(1, \sqrt{3})$ as shown in Fig. 6.12. The values of the six functions can be written by inspecting the figure:

$$\sin 60° = \frac{\sqrt{3}}{2}, \qquad \csc 60° = \frac{2}{\sqrt{3}},$$
$$\cos 60° = \frac{1}{2}, \qquad \sec 60° = \frac{2}{1},$$
$$\tan 60° = \frac{\sqrt{3}}{1}, \qquad \cot 60° = \frac{1}{\sqrt{3}}. \qquad \square$$

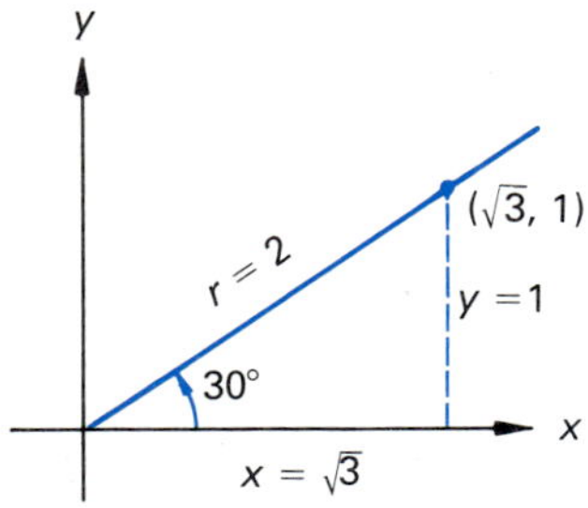

FIGURE 6.13

EXAMPLE 2 Find the exact values for the six trigonometric functions for a 30° angle.

Solution: In Fig. 6.12 the other acute angle of the triangle is 30° (since the sum of the angles of a triangle is 180°). If we interchange the x and y values, we have the reference triangle given in Fig. 6.13. By inspection we can write the six functions:

$$\sin 30^\circ = \frac{1}{2}, \qquad \csc 30^\circ = \frac{2}{1},$$

$$\cos 30^\circ = \frac{\sqrt{3}}{2}, \qquad \sec 30^\circ = \frac{2}{\sqrt{3}},$$

$$\tan 30^\circ = \frac{1}{\sqrt{3}}, \qquad \cot 30^\circ = \frac{\sqrt{3}}{1}. \qquad \square$$

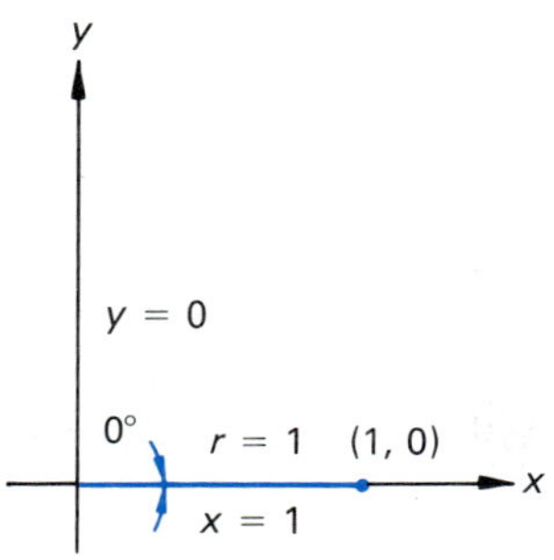

FIGURE 6.14

From Fig. 6.14 we see that there is no rotation, $x = r$ and $y = 0$. Thus

$$\sin \theta = \frac{y}{r}.$$

$$\sin 0^\circ = \frac{0}{r}$$

$$= 0.$$

From Fig. 6.15 we see that the rotation is 90° and $x = 0$, $y = r$. Thus

$$\tan \theta = \frac{y}{x}.$$

$$\tan 0^\circ = \frac{y}{0}$$

$$= \text{undefined}.$$

Using methods similar to Examples 1 and 2 we can also determine the values for the six functions for 0°, 45°, and 90°. These values are summarized in Table 6.2.

Note from Examples 1 and 2 and Table 6.2 that $\sin 30^\circ = \cos 60^\circ$ or that $\sin 30^\circ = \cos (90^\circ - 30^\circ)$. That is, the sine of the angle is equal to the cosine of the complement of the angle. When this is true we say the functions are **cofunctions.** Thus, sine and cosine are cofunctions, tangent and cotangent are cofunctions, and so are secant and cosecant.

It is not always possible to express the value of all trigonometric functions exactly, as was done for 30°, 45°, and 60°. For most other numbers we have to express the functional values as decimal approximations. These values have been calculated for us to as many decimal places as we choose to use by using the technique of infinite series. This technique is used internally in calculators and computers to give decimal approximations for trigonometric functions. In the text we will use calculators to give us the values for the trigonometric functions.

FIGURE 6.15

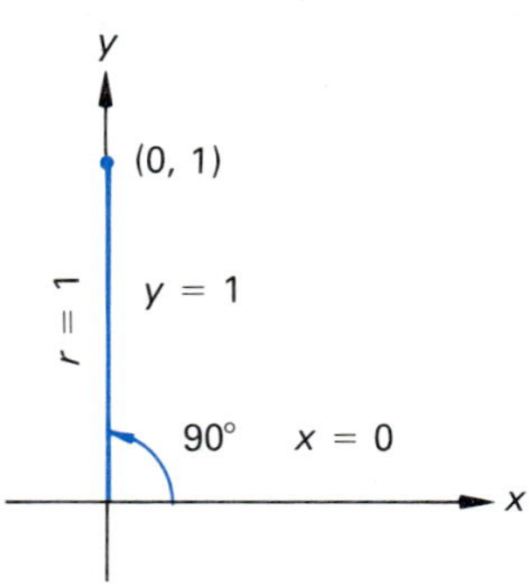

TABLE 6.2

θ	0°	30°	45°	60°	90°
sin	0	$\frac{1}{2}$	$\frac{1}{\sqrt{2}}$	$\frac{\sqrt{3}}{2}$	1
cos	1	$\frac{\sqrt{3}}{2}$	$\frac{1}{\sqrt{2}}$	$\frac{1}{2}$	0
tan	0	$\frac{1}{\sqrt{3}}$	1	$\sqrt{3}$	undefined
cot	undefined	$\sqrt{3}$	1	$\frac{1}{\sqrt{3}}$	0
sec	1	$\frac{2}{\sqrt{3}}$	$\frac{2}{\sqrt{2}}$	2	undefined
csc	undefined	2	$\frac{2}{\sqrt{2}}$	$\frac{2}{\sqrt{3}}$	1

On a calculator trigonometric keys labeled [SIN], [COS], and [TAN] are used to compute function values for these functions. To find sin 0, the 0 value is entered on the keyboard and then the [SIN] key is stroked. The calculator must know whether the angle measure is in radians or degrees. Many calculators when first turned on will be in **degree mode.** Most calculators have a key labeled [DRG] (check your instruction book), which is used to change the operation of the calculator to the **radian mode** and back to the degree mode.

We have discussed two units of measurement for angles, degrees and radians. A third unit found on the calculator is the grad. A grad is equal to 1/400 of a circle, thus $90° = 100$ grads and $180° = 200$ grads. The grad is used in some countries with the SI system. We are mentioning this for information only; grads will not be used in this text.

In solving problems dealing with degrees the number of significant digits is dependent on the degree of accuracy of the angle measure as shown in Table 6.3. For convenience in calculations we will follow the rules established in the table unless otherwise noted.

For radians we will use the same number of significant digits for the trigonometric functional value as used in the radian value.

TABLE 6.3

Angles to the nearest	Significant digits for trigonometric functions
degree	2
ten minutes or tenth of a degree	3
one minute or hundredth of a degree	4

EXAMPLE 3 Compute to three decimal places using a scientific calculator.

a. sin 14.3° **b.** sin 14.3

Solution:

a. The degree symbol next to the 3 in the expression sin 14.3° tells us that we want the sine of 14.3 degrees. Make sure the calculator is in the degree mode, then enter 14.3 and stroke the [SIN] key. The calculator will display 0.246999. Round off to 0.247.

b. Since there is no degree symbol next to the 3, we are asked to find the sine of 14.3 rad. Make sure the calculator is in the radian mode and enter 14.3, then stroke the [SIN] key. The calculator will display 0.986772. Round off to 0.987. □

Note that the trigonometric keys will operate only on the value in the display. Therefore, to find the trigonometric value of a result of a calculation, the calculation must be completed and be in the display before stroking the trigonometric key.

EXAMPLE 4 Compute to five decimal places:

a. $\cos[30° + 2(360°)]$ **b.** $\tan \dfrac{3\pi}{7}$

Solution:

a. Place the calculator in the degree mode. Compute [30 + 2(360)] and observe the result 750 in the display, then stroke the [COS] key. The result to five decimal places is 0.86603.

b. Place the calculator in the radian mode. Compute $\dfrac{3\pi}{7}$ and observe the result 1.3463969 in the display, then stroke the [TAN] key. The result to five decimal places is 4.38129. □

Now that we can find the trigonometric functional values of sine, cosine, and tangent, we can perform the inverse operation and determine the angle when given the functional value. In Section 5.4 we discussed this concept for simple algebraic functions. The procedure for trigonometric functions is easier since we need only learn a few steps on the calculator or learn how to read a table. (A rigorous definition of inverse trigonometric functions will be left until Chapter 15.)

The sequence of key strokes for finding the angle θ is different on different calculators. The keys on the calculators may be: [ARC], [INV], [SIN^{-1}], [COS^{-1}], and [TAN^{-1}]. In addition, it may be necessary to use a color-coded key or the [2nd] key. Consult your handbook for specific details. For example, if the trigonometric function of θ is $\sin\theta = 0.5$, then θ may be found on most calculators by using one of the following sequences:

0.5 [INV] [SIN] or 0.5 [ARC] [SIN] or 0.5 [SIN^{-1}].

If the calculator is in the degree mode, the number 30 will appear on the display. If it is in the radian mode, the number 0.5235988 will appear.

EXAMPLE 5 Use your calculator to verify the following. The measure in degrees will be given to the nearest tenth and the measure in radians will be given to the nearest hundredth.

a. If $\sin \theta = 0.45879$, then in degrees, $\theta = 27.309056$, or $\theta = 27.3°$; in radians, $\theta = 0.4766329$, or $\theta = 0.48$.

b. If $\tan \theta = 11.32001$, then in degrees, $\theta = 84.951645$, or $\theta = 85.0°$; in radians, $\theta = 1.4826859$, or $\theta = 1.48$.

Each of these results can be checked by taking the value obtained for θ (not rounded off) and stroking the function key. For example,

27.309056 [SIN] 0.45879. □

The discussion up to this point has been limited to the three functions: sine, cosine, and tangent. The discussion of the reciprocal functions cosecant, secant, and cotangent has been delayed because they do not appear on the calculator. In Section 6.2 we discovered that

$$\sin \theta = \frac{1}{\csc \theta}, \qquad \sec \theta = \frac{1}{\cos \theta}, \qquad \cot \theta = \frac{1}{\tan \theta}.$$

With these relationships we can use the calculator to calculate the cosecant θ, secant θ, and cotangent θ even though there are no keys with these names.

EXAMPLE 6 Compute to five decimal places:

a. csc 32.7° **b.** sec 0.27

Solution:

a. The statement of equality $\csc \theta = \frac{1}{\sin \theta}$ provides us with a method for solving the problem. Find the value of $\sin \theta$ and then find the reciprocal of this value by stroking the [1/x] key. (In other words $\csc 32.7° = \frac{1}{\sin 32.7°}$.) To find csc 32.7°, make sure that the calculator is in the degree mode, then stroke the following keys:

32.7 [SIN] [1/x].

The calculator will display 1.851028. Round off to 1.85103. *Warning:* Be sure that you see a display after stroking the [SIN] key and before you stroke the [1/x] key. If you stroke them too rapidly, you may obtain a wrong value.

b. The statement of equality $\sec \theta = \frac{1}{\cos \theta}$ provides us with a method for solving the problem. Find the value of $\cos \theta$ and then find the reciprocal of this value by stroking the [1/x] key. (In other words $\sec 0.27 = \frac{1}{\cos 0.27}$.) To find sec 0.27, make sure that the calculator is in the radian mode and stroke the following keys:

0.27 [COS] [1/x].

The calculator will display 1.037591, which rounded to five decimal places is 1.03759. □

The procedure is the same regardless of whether the function is cosecant, secant, or cotangent. Note that in this case **you are finding the reciprocal value of the function and not the inverse of the function.**

If $\csc \theta = 1.83608$, what is the measure of θ to the nearest tenth of a degree? The method used to answer this question is to find the inverse of the cosecant function. To provide a better understanding of the technique for finding the inverse of the cosecant, secant, and cotangent functions, let us review how we found the functional values. *The steps were as follows:*

1. *Enter the angle measure in the proper mode.*
2. *Stroke the key of the reciprocal, [SIN], [COS], or [TAN].*
3. *Stroke the [1/x] key.*

To find the measure of the angle, we do the procedure in reverse with a few modifications.

EXAMPLE 7 Find θ to the nearest tenth of a degree given

$\csc \theta = 1.83608.$

Solution: Remember that

$$\sin \theta = \frac{1}{\csc \theta} \quad \text{or} \quad \frac{1}{1.83608}.$$

To find θ we use the following steps:

	Display
1. Enter the functional value	1.83608
2. Stroke [1/x] key	0.544639
3. Stroke [INV] [SIN]	32.999969

Round to the nearest tenth of a degree: $\theta = 33.0°$ □

The thing to remember about this procedure is to find the reciprocal first and then find the inverse of the function. By finding the reciprocal first we are changing the functional value of cosecant, secant, or cotangent to sine, cosine, or tangent, respectively, before finding the inverse. This is essential since we have only sine, cosine, and tangent on the calculator.

EXAMPLE 8 Use your calculator to verify the following. The measure in degrees will be given to the nearest tenth and the measure in radians will be given to the nearest hundredth.

a. If $\cot\theta = 2.41421$, then in degrees, $\theta = 22.50003$, or $\theta = 22.5$; in radians, $\theta = 0.3926996$, or $\theta = 0.39$.

b. If $\sec\theta = 1.12738$, then in degrees, $\theta = 27.49981$, or $\theta = 27.5°$; in radians, $\theta = 0.4799622$, or $\theta = 0.48$.

Each of the results may be checked by finding the reciprocal functional value for θ (not rounded off) and then stroking the [1/x] key.

For example, 22.50003 [TAN] [1/x] 2.41421, which checks part (a). □

6.3 Exercises

In Exercises 1–24, compute the functional value to five decimal places.

1. sin 12°	**2.** sin 78.4°	**3.** cos 78°
4. cos 11.6°	**5.** tan 19°	**6.** tan 34.3°
7. tan 0.05	**8.** tan 1.01	**9.** cos 0.7
10. cos 0.4	**11.** sin 1.01	**12.** sin 0.29
13. sin 26.9°	**14.** cos 73.1°	**15.** tan 56°
16. tan 1.4	**17.** cos 1.3	**18.** sin 1.2
19. sin 21.7°	**20.** sin 0.38	**21.** cos 68.3°
22. cos 1.19	**23.** tan 37°	**24.** tan 0.65

25. Use your knowledge of geometry to find the exact functional values for the six trigonometric functions for a 45° angle.

26. Use your knowledge of geometry to find the exact functional values for the six trigonometric functions for a 90° angle.

27. Use your knowledge of geometry to find the exact functional values for the six trigonometric functions for a 0° angle.

In Exercises 28–39, compute each of the following to five decimal places.

28. $\sin[23° + 2(360°)]$	**29.** $\cos[17° + 2(360°)]$
30. $\tan[34° + 2(180°)]$	**31.** $\sin[53.2° + 2(360°)]$
32. $\cos[67° + 2(360°)]$	**33.** $\tan[59° + 2(180°)]$
34. $\sin\dfrac{4\pi}{9}$	**35.** $\cos\dfrac{2\pi}{2}$
36. $\tan\dfrac{4}{5}\pi$	**37.** $\sin\left(\dfrac{\pi}{3} + 2\pi\right)$
38. $\cos\left(\dfrac{\pi}{6} + 2\pi\right)$	**39.** $\tan\left(\dfrac{\pi}{4} + \pi\right)$

In Exercises 40–47, find the angle measure to the nearest tenth of a degree.

40. $\sin\theta = 0.22835$	**41.** $\cos\theta = 0.22835$
42. $\tan\theta = 0.24377$	**43.** $\sin\theta = 0.50000$
44. $\cos\theta = 0.86603$	**45.** $\tan\theta = 0.57735$
46. $\sin\theta = 0.86603$	**47.** $\cos\theta = 0.50000$

In Exercises 48–51, find the angle measure to the nearest hundredth of a radian.

48. $\tan\theta = 1.73205$ **49.** $\sin\theta = 0.70711$

50. $\cos\theta = 0.70711$ **51.** $\tan\theta = 1.00000$

In Exercises 52–63, compute the functional value to five decimal places.

52. csc 12° **53.** csc 78.4° **54.** sec 78°

55. sec 11.6° **56.** cot 19° **57.** cot 34.3°

58. cot 0.05 **59.** cot 1.01 **60.** sec 0.7

61. sec 1.47 **62.** csc 1.2 **63.** csc 0.01

In Exercises 64–71, find the angle measure to the nearest tenth of a degree.

64. $\csc\theta = 19.01732$ **65.** $\csc\theta = 1.00061$

66. $\sec\theta = 1.00061$ **67.** $\sec\theta = 52.09027$

68. $\cot\theta = 28.63625$ **69.** $\cot\theta = 0.01746$

70. $\cot\theta = 1.73205$ **71.** $\cot\theta = 0.57735$

In Exercises 72–75, find the angle measure to the nearest hundredth of a radian.

72. $\sec\theta = 1.15470$ **73.** $\sec\theta = 2.00000$

74. $\csc\theta = 2.00000$ **75.** $\csc\theta = 1.15470$

A roadway with radius of curvature R (in meters) is banked at an angle θ. The maximum speed v (in kilometers per hour) at which a vehicle can travel without skidding or leaving the road is given by

$$v = \sqrt{\frac{1800\,R\tan\theta}{121}}.$$

In Exercises 76–79, find to the nearest whole number the maximum safe speed for each θ and R.

76. $\theta = 2.1°$, $R = 930$ m

77. $\theta = 0.106$, $R = 540$ m

78. $\theta = 7.2°$, $R = 510$ m

79. $\theta = 0.056$, $R = 520$ m

80. When a light ray passes from air to another medium, it is refracted (bent) according to Snell's law. The ratio of the sine of the angle of incidence to the sine of the angle of refraction is a constant for any two given media.

$$n = \frac{\sin\theta_i}{\sin\theta_r},$$

where θ_i and θ_r are the respective angles with respect to the vertical (also called normal) (see Fig. 6.16) and n is a constant called the index of refraction of the medium the light ray is entering. To the nearest hundredth, find the index of refraction n for each θ_i and θ_r pairs.

a. water $\theta_i = 43.2°$, $\theta_r = 32.1°$

b. oil $\theta_i = 70.0°$, $\theta_r = 38.2°$

FIGURE 6.16

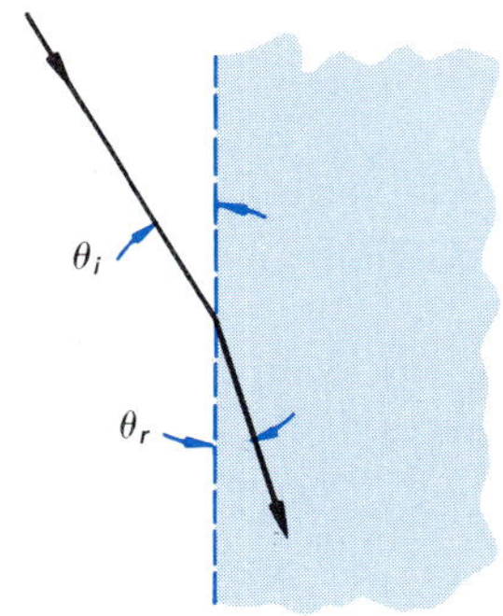

6.4 The right triangle and applications

The word trigonometry means triangle measurement, and the subject was invented for the purpose of measuring triangles.

Now that the basic trigonometric ratios have been discussed, the simplest application of these is the right triangle. For uniformity in our discussions, unless noted otherwise, we will label the right triangle as in Fig. 6.17. Capital letters A, B, and C will indicate the angles, with C naming the right angle, and the sides opposite the angles will be named with the corresponding lower case letters a, b, and c.

FIGURE 6.17

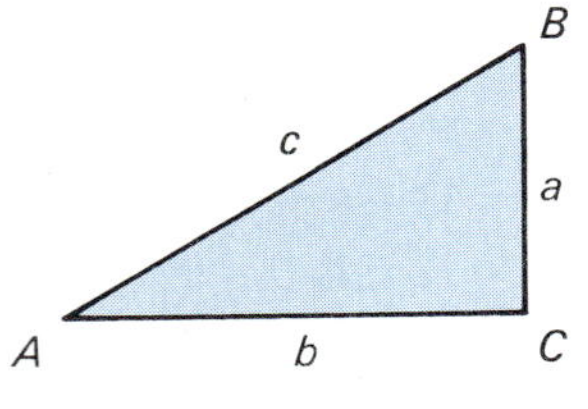

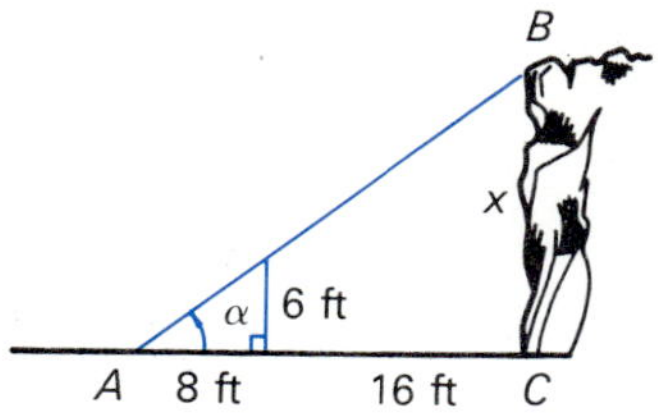

FIGURE 6.18

In the introduction to Section 6.2 it was suggested that a surveyor could find the height of a cliff by measuring angle α (Fig. 6.18) and use right-triangle trigonometry rather than similar triangles.

EXAMPLE 1 Find the height of the cliff to the nearest foot in Fig. 6.18 knowing that $\angle A = 36.9°$ and $\overline{AC} = 24$ ft.

Solution: If we take the given information and construct a right triangle as in Fig. 6.19, the problem boils down to finding the length of side a. We know that $\angle A = 36.9°$ and the side adjacent to $\angle A$, $b = 24$ ft.

Now let us recall the trigonometric ratios that were defined in Section 6.2 and listed in Table 6.1. Two ratios in the table that include all three quantities are $\tan \theta = \text{opp/adj}$ and $\cot \theta = \text{adj/opp}$. Since we have the tangent key on the calculator, we will work with the tangent ratio. (The cotangent ratio would provide us with the same result.) Translating the ratio into the symbols of our problem:

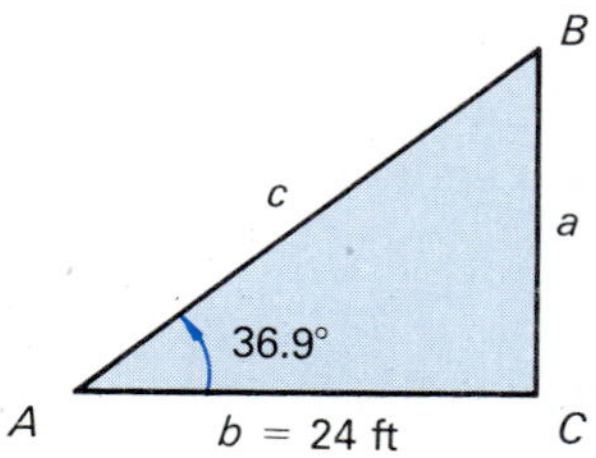

FIGURE 6.19

$$\tan A = \frac{a}{b}.$$

$$\begin{aligned} a &= b \tan A && \textbf{Solving for } \boldsymbol{a} \\ &= 24\ (\tan 36.9°) && \textbf{Substituting} \\ &= 18.02 \text{ ft.} \end{aligned}$$

Thus, the height of the cliff is 18.0 ft, which is the same as the result obtained using similar triangles. □

For the triangle in Fig. 6.19 we can find the length of side C and the measure of $\angle B$. Since $\angle A$ and $\angle B$ are complementary angles, $\angle B = 90° - \angle A$ or $\angle B = 53.1°$.

The length of side C in Fig. 6.20 can be found two ways: (1) using right-triangle trigonometry or (2) using the Pythagorean theorem.

Method 1

Since we want to find the length of the hypotenuse, we must use a trigonometric ratio that contains the hypotenuse. Limiting ourselves to the functions on the calculator, we could select sine or cosine. Selecting sine we have:

FIGURE 6.20

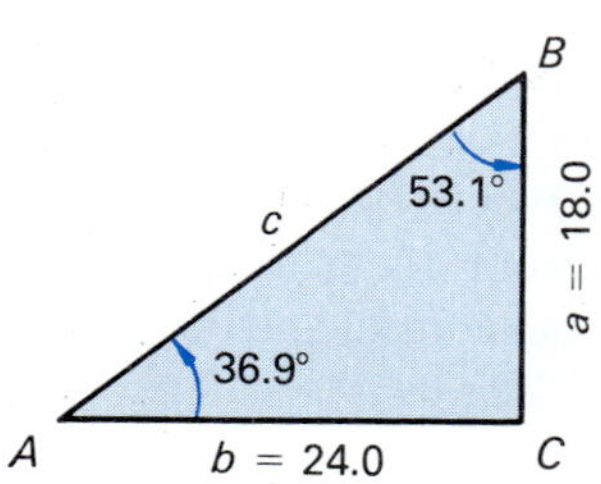

$$\sin \theta = \frac{\text{opp}}{\text{hyp}}.$$

Translating to the labeling of Fig. 6.20:

$$\sin A = \frac{a}{c}.*$$

* Since we know the measure of $\angle B$, we could select the ratio $\sin B = b/c$ and arrive at the same answer.

Solving for c we have

$$c = \frac{a}{\sin A}.$$

Substituting:

$$\begin{aligned} c &= \frac{18.02}{\sin 36.9^\circ} \\ &= 30.01. \end{aligned}$$

The length of c to the nearest tenth is 30.0 ft.

Method 2
Using the Pythagorean theorem, $c^2 = a^2 + b^2$,

$$\begin{aligned} c &= \sqrt{a^2 + b^2} \\ &= \sqrt{(18.02)^2 + (24)^2} && \textbf{Substituting} \\ &= 30.0 \text{ ft.} \end{aligned}$$

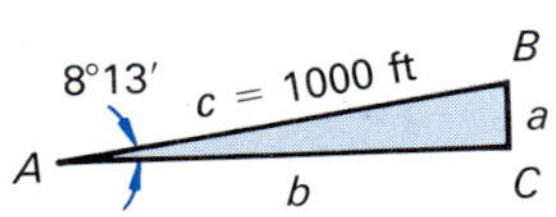

FIGURE 6.21

This illustrates that the Pythagorean theorem could be used as a check on the calculation by trigonometric ratios, or the calculation by the trigonometric ratio can be used as a check on the value calculated by the Pythagorean theorem when calculating the length of the third side of a right triangle.

EXAMPLE 2 The angle of inclination of a road is $8^\circ 13'$. Find to the nearest foot the rise for 1000 ft measured along the surface of the road.

Solution: First draw and label a sketch. From the information given it is known that $\angle A = 8^\circ 13'$ and the length of side c is $1000'$ (Fig. 6.21). The rise for 1000 ft of roadway is the length of side a. To find the length of side a we can use the sine ratio:

$$\sin A = \frac{a}{c}$$

$$\begin{aligned} a &= c\,(\sin A) && \textbf{Solving for } \boldsymbol{a} \\ &= 1000\,(\sin 8^\circ 13') && \textbf{Substituting} \\ &= 1000\,(\sin 8.217^\circ) && \textbf{Converting to decimal degrees} \\ &= 142.92. \end{aligned}$$

The rise to the nearest foot is 143. ft. □

FIGURE 6.22

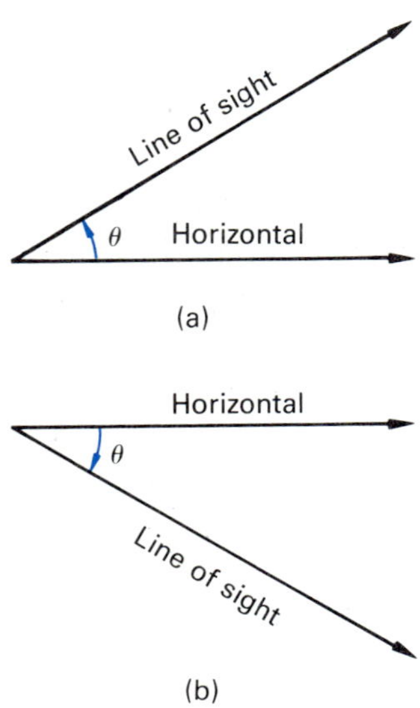

In navigation and surveying, names have been developed for certain angles. The **angle of elevation** is the angle between the horizontal and the line of sight to an object above the horizontal (Fig. 6.22a). The **angle of depression** is the angle between the horizontal and the line of sight to an object below the horizontal (Fig. 6.22b).

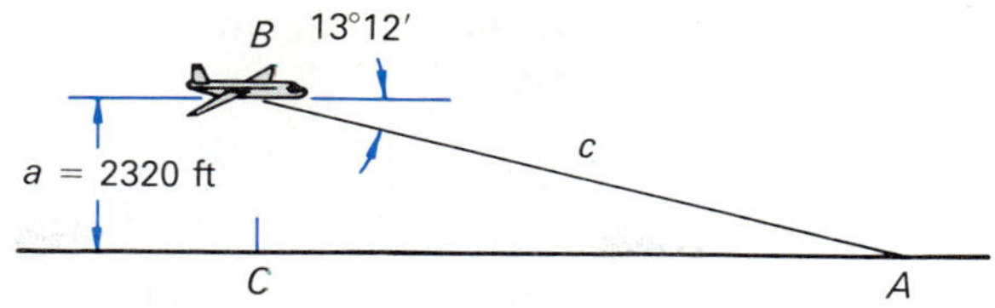

FIGURE 6.23

EXAMPLE 3 An observer in an airplane sees a ground light at the airport shown in Fig. 6.23 at an angle of depression of 13°12′. How far is the light from the observer if he is at a height of 2320 ft?

Solution: Label the triangle as in Fig. 6.23. From geometry we know that $\angle A = 13°12'$ (the same as the angle of depression). To find how far the observer is from the light, we want to find the length of side c. Using the sine ratio we have

$$\sin A = \frac{a}{c}$$

$$c = \frac{a}{\sin A} \qquad \textbf{Solving for } c$$

$$= \frac{2320}{\sin 13°12'} \qquad \textbf{Substituting}$$

$$= 10{,}160.$$

Therefore, the light is 10,160. ft away (to four significant digits). □

EXAMPLE 4 Find the area of a right triangle if one of the perpendicular sides is 16.18 ft and the angle opposite it is 58°24′.

FIGURE 6.24

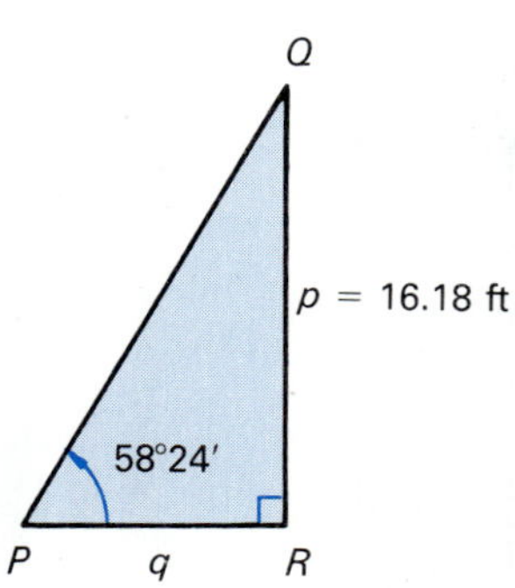

Solution: Sketch the triangle and label as in Fig. 6.24. The formula for finding the area of a triangle is $A = 1/2\ ba$. We know the altitude $p = 16.18'$. In order to find the area we must find the length of side q, which is the base of the triangle. We can do this using the tangent ratio.

$$\tan P = \frac{p}{q}$$

$$q = \frac{p}{\tan P} \qquad \textbf{Solving for } q$$

$$= \frac{16.18}{\tan 58°24'} \qquad \textbf{Substituting}$$

Now substituting in the formula for finding the area of the triangle, we have:

$$A = \frac{1}{2}\left(\frac{(16.18)}{\tan 58°24'}\right)(16.18) \qquad \text{or} \qquad 80.528.$$

To four significant digits the area is 80.53 ft². □

EXAMPLE 5 In Fig. 6.25 find the length of side x ($\overline{BD}$).

Solution: Since triangle ABD is not a right triangle, we cannot apply any of the trigonometric ratios directly to this triangle. Triangle ABC is a right triangle, and using the tangent ratio we have

$$\tan A = \frac{a}{b}.$$

Also from the drawing we know that $a = x + 70$ and $A = 39°$. Substituting, we have

$$\tan 39° = \frac{x + 70}{b}.$$

Solving for x, we have

$$x = b(\tan 39°) - 70.$$

If we can determine the length of side b, we can find x. Using the tangent ratio and triangle ADC we have:

$$\tan 18° = \frac{70}{b} \qquad \text{or} \qquad b = \frac{70}{\tan 18°}.$$

Substituting for b, we have

$$x = \frac{70}{\tan 18°}(\tan 39°) - 70 \qquad \text{or} \qquad 104.\text{ mm}.$$

Therefore to two significant digits $x = 100.$ mm. □

FIGURE 6.25

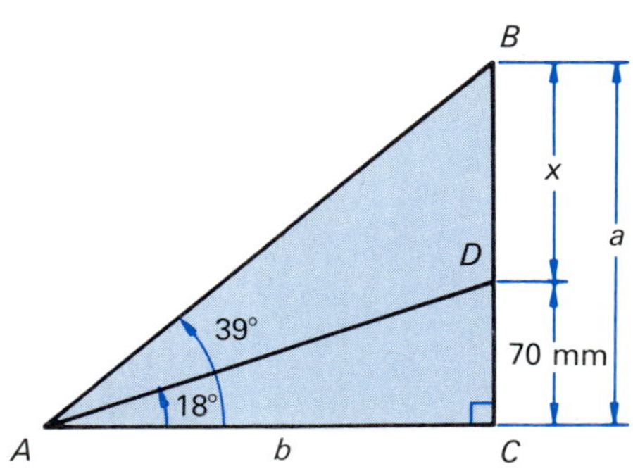

6.4 Exercises

In Exercises 1 – 12, solve each right triangle ABC (Fig. 6.26). Find the measures of all angles to the nearest tenth of a degree and length of sides to three significant digits.

1. $A = 15.3°$, $c = 16.4$ in.
2. $B = 31.5°$, $a = 9.24$ m
3. $B = 64.2°$, $c = 31.4$ mm
4. $A = 47.3°$, $b = 82.3$ mm
5. $A = 45°$, $b = 1.82$ mm
6. $A = 73.4°$, $a = 2.06$ in.
7. $B = 18.2°$, $a = 3.04$ ft
8. $B = 30°$, $b = 17.4$ mm
9. $B = 45°$, $b = 22.3$ in.
10. $B = 52.4°$, $c = .384$ m
11. $A = 60°$, $a = 14.7$ m
12. $A = 63.7°$, $c = 15.9$ in.

In Exercises 13 – 16, find the area of each right triangle PQR (Fig. 6.27), to three significant digits.

13. $P = 29.4°$, $p = 3.45$ m
14. $Q = 0.790$, $p = 32.1$ in.
15. $Q = 1.02$, $p = 11.8$ mm
16. $P = 83.4°$, $p = 74.3$ m
17. The string attached to a kite is 175 ft long. The angle formed by the line of the string and the horizontal is 63.3°. How high is the kite when the string runs taut?
18. The Washington Monument is approximately 556 ft high. When the shadow of the monument is 159 ft in length, what is the angle of elevation of the sun?
19. Find the size of the acute angles of a right triangle whose sides are in the ratio 3 : 4 : 5.
20. Find the area of the parallelogram whose sides are 15.00 in. and 25.00 in. if the angle between these sides is 72°45′.
21. At a distance of 14,265 ft from the foot of the Empire State Building the angle of elevation to its top is 5.42°. Find the height of the building.
22. A steep gabled roof on a building has ends in the shape of an isosceles triangle (Fig. 6.28). These two ends are to be painted. The contractor estimates the rafters to be 8 m long and that they come together at an angle of 90°. Find the total exterior area to be painted.
23. A person at point B observes an inaccessible object due east at point C. He walks in a direction 47.0° north of east for a distance of 750 ft and observes the object is due south.
    a. What is his distance from the object?
    b. What was his original distance from the object?
24. An airplane pilot finds the angle of depression of a beacon light on the ground to be 14°20′. His altimeter shows that he is 400 m high. How far does he estimate his ground distance to be from the beacon?
25. From a window 26 m above the ground, the angle of elevation to the top of a nearby building is 24°20′ and the angle of depression to the bottom of the building is 14°40′. Find the height of the building.
26. What is the angle of the roof with the attic floor if the ridge pole is 14.3 ft above the attic floor and the attic is 74.5 ft wide?

FIGURE 6.26

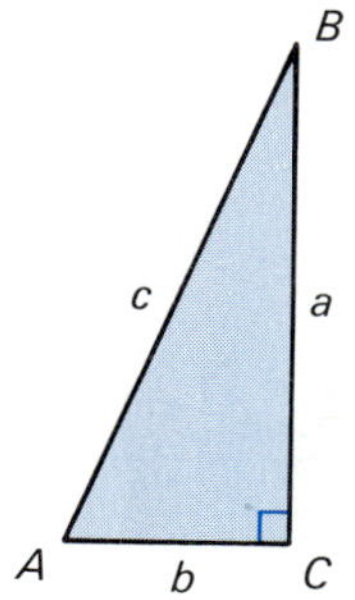

FIGURE 6.27

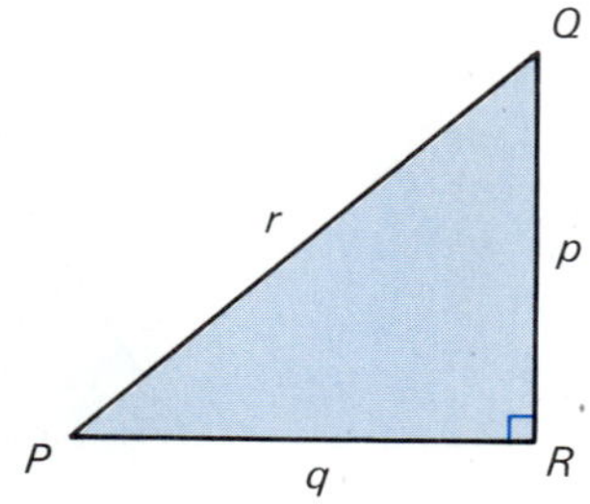

FIGURE 6.28

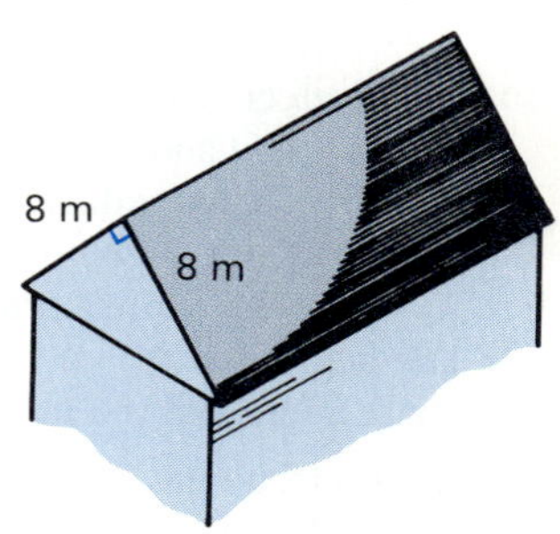

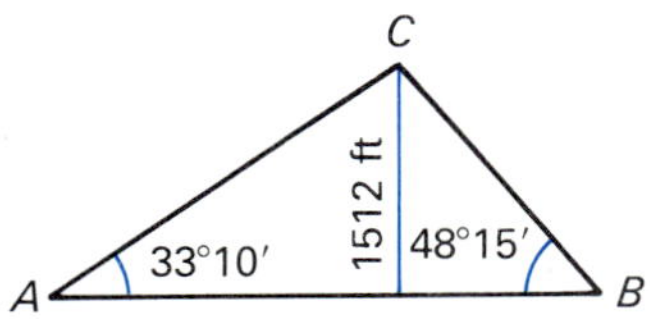

FIGURE 6.29

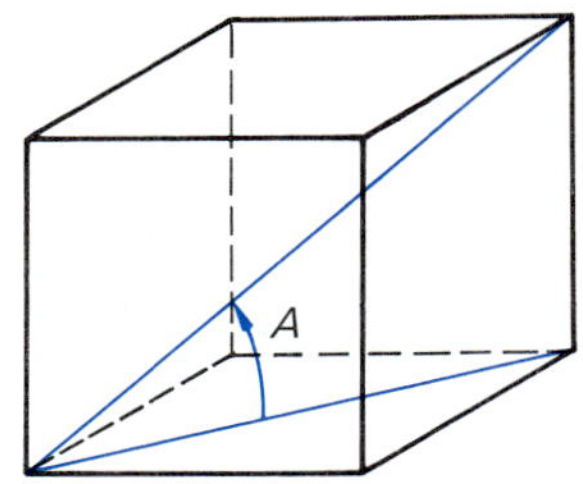

FIGURE 6.30

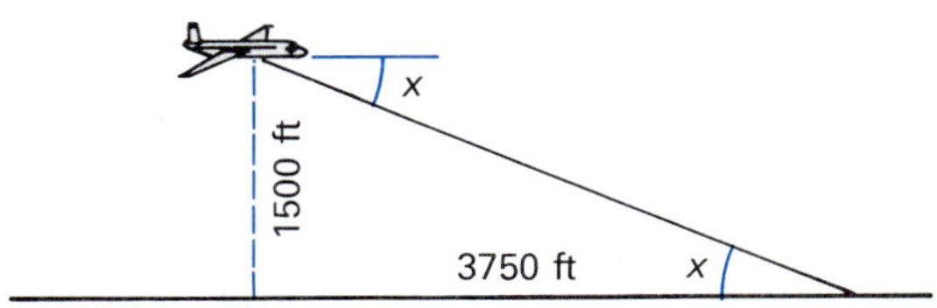

FIGURE 6.31

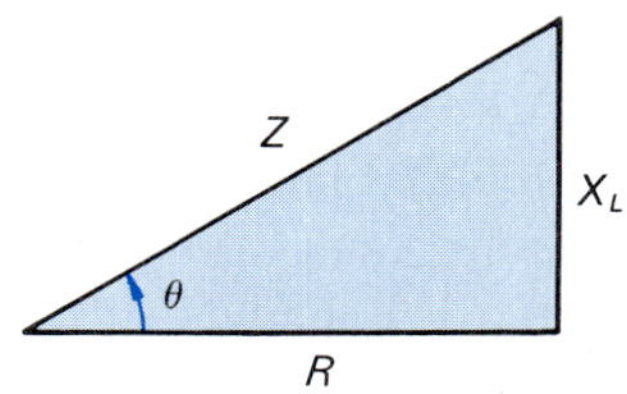

FIGURE 6.32

27. The angles of elevation of the top of a mountain from A and B on opposite sides of its base are 33°10′ and 48°15′. (See Fig. 6.29.) The top of the mountain is known to be 1512 ft higher than the two points A and B. Find the length of a tunnel through the mountain from A to B.

28. What angle A does the diagonal of the cube in Fig. 6.30 make with the diagonal of one of the faces, the two diagonals being drawn from the same vertex?

29. An airplane pilot wishes to land at a point on the field that he estimates to be 3750 ft from where he is. Will he be able to descend to the field in a straight line, or must he circle it, if a safe glide angle is 10° and his altitude is 1500 ft? (See Fig. 6.31.)

30. An AC electric circuit has an inductive reactance $X_L = 10.0\ \Omega$ and a resistance $R = 22.3\ \Omega$. In such a circuit the **impedance** Z, in ohms, can be represented by the hypotenuse of a right triangle with X_L and R as its other sides, as shown in Fig. 6.32. The angle θ is called the **phase angle.** Find the phase angle and the impedance.

31. An AC circuit has an impedance $Z = 22.0\ \Omega$. If the circuit has a resistance $R = 18.9\ \Omega$, find the inductive reactance X_L and the phase angle θ. (See Exercise 30.)

32. The right triangle shown in Fig. 6.33a is called a **power triangle,** and is used in solving AC-circuit problems. The hypotenuse represents the **apparent power P_A,** measured in volt-amperes (VA). The other sides represent **real power P,** measured in watts (W), and the **reactive power P_R,** measured in vars. The angle θ is called the **phase angle.** In a particular circuit the real power is

FIGURE 6.33

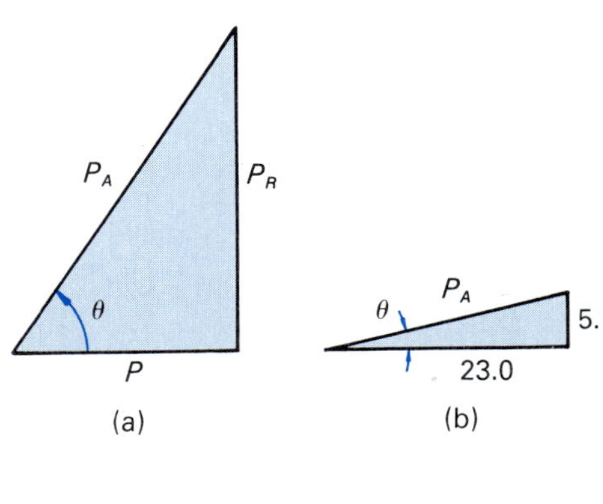

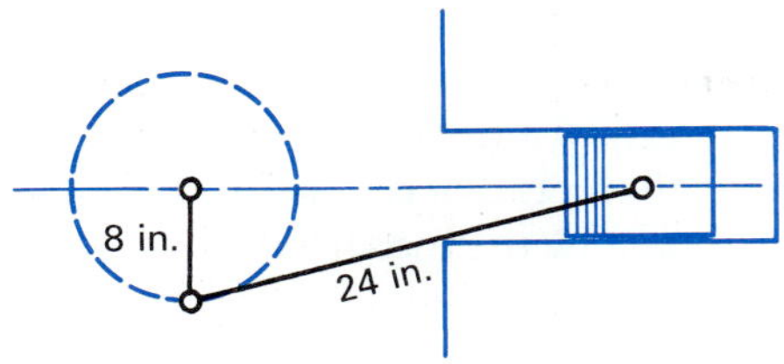

FIGURE 6.34

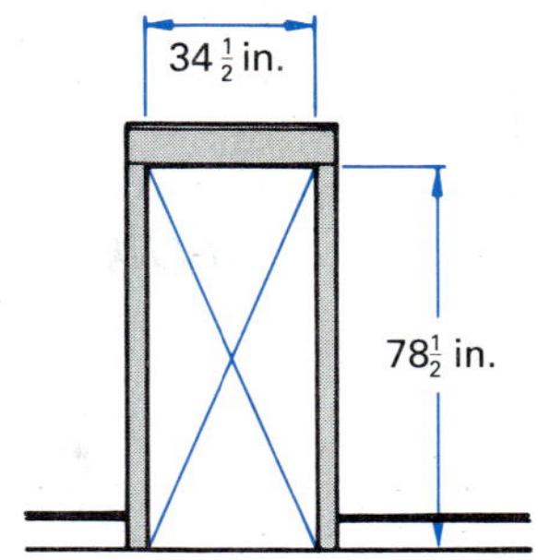

FIGURE 6.35

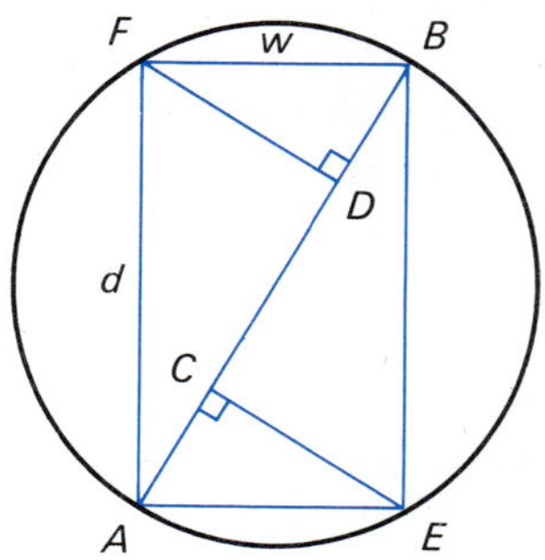

FIGURE 6.36

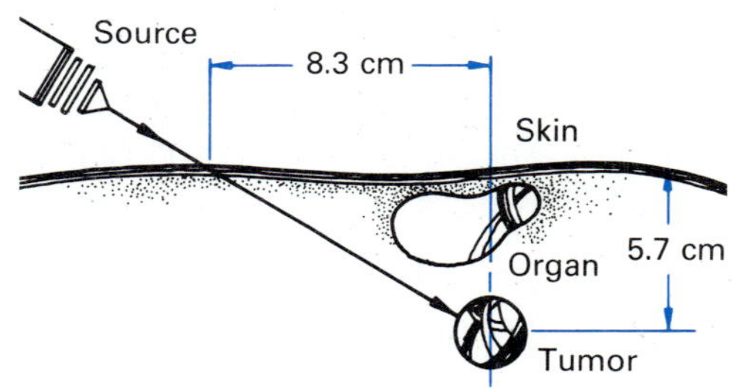

FIGURE 6.37

$P = 23.0$ W and the reactive power is $P_R = 5.1$ vars. Find the apparent power P_A and the phase angle (Fig. 6.33b).

33. If the real power in a circuit is 1.73 W and the phase angle θ is 66.5°, find the apparent power and the reactive power. (See Exercise 32.)

34. An escalator inclined at an angle of 42.6° moves at a constant speed of 0.6 m/s. A person steps on the escalator and reaches the top in 42. s. Through what vertical height has the person been lifted?

35. A piston and rod assembly has the dimensions shown in Fig. 6.34. Find the linear advance of the piston in the cylinder as the throw arm moves from the position shown to a position in the line with the cylinder.

36. A block mason is setting a door frame. To ensure square corners, he measures the diagonals of the rectangular opening. What is the length of each diagonal if the doorjamb is square? (See Fig. 6.35.)

37. A weight hangs on a 15 m chain from a high ceiling. If the weight is pulled sideways so that the chain forms an angle of 24.8° with its vertical position, how much is the weight raised vertically?

38. It has been found that the strongest beam that can be cut from a log (see Fig. 6.36) may be obtained by dividing the diameter into three equal parts, $\overline{AC}$, $\overline{CD}$, and $\overline{DB}$. Then, perpendiculars $\overline{CE}$ and $\overline{DF}$ are drawn. The rectangle *AEBF* represents the cross section of such a beam. If the diameter $\overline{AB}$ of the log is 24″, find the width w and the depth d of the beam.

39. A right triangle is inscribed in a circle of radius 56 mm. One angle of the triangle is 64°. Find the lengths of the two sides.

40. A beam of gamma rays is to be used to treat a tumor known to be 5.7 cm beneath the patient's skin. To avoid damaging a vital organ, the radiologist moves the source over 8.3 cm (Fig. 6.37). (a) At what angle to the patient's skin must the radiologist aim the gamma-ray source to hit the tumor? (b) How far will the beam have to travel through the patient's body before reaching the tumor?

6.5 Signs of trigonometric functions

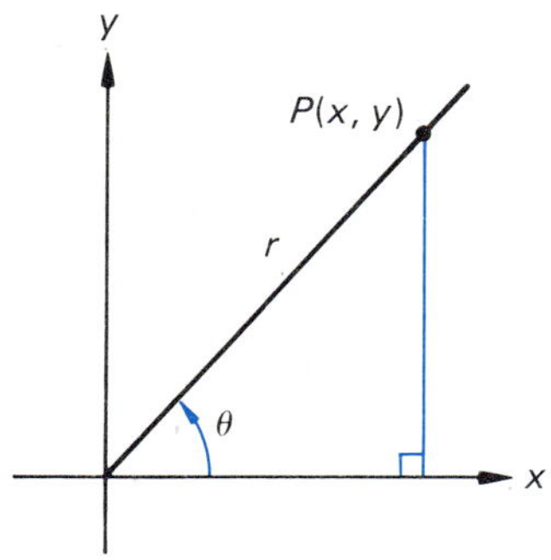

FIGURE 6.38

The discussion of the trigonometric functions has been limited to angles in the first quadrant. Recall that the values of the six trigonometric functions were determined by a ratio of two sides of a right triangle. The right triangle was determined by the terminal side of the angle and the x-axis. This right triangle is referred to as the **reference triangle.** The acute angle of the right triangle formed by the x-axis and the terminal side of the angle is called the **reference angle** (Fig. 6.38). Recall the ratios $\sin\theta = y/r$, $\cos\theta = x/r$, and $\tan\theta = y/x$. The radius vector r is always positive, as are x and y in the first quadrant. Therefore, all six ratios are positive in the first quadrant. The question is what happens in the other quadrants?

An angle in standard position is called a **quadrantal angle** if its terminal side corresponds with an axis (positive or negative). Thus, the quadrantal angles measured in a positive rotation are $0°$ (0), $90°$ $(\pi/2)$, $180°$ (π), $270°$ $(3\pi/2)$, and $360°$ (2π).

An angle in standard position is said to be in the quadrant in which the terminal side is located. For example, a $127°$ angle in standard position is called a second-quadrant angle, since its terminal side is in the second quadrant $(90° < 127° < 180°)$.

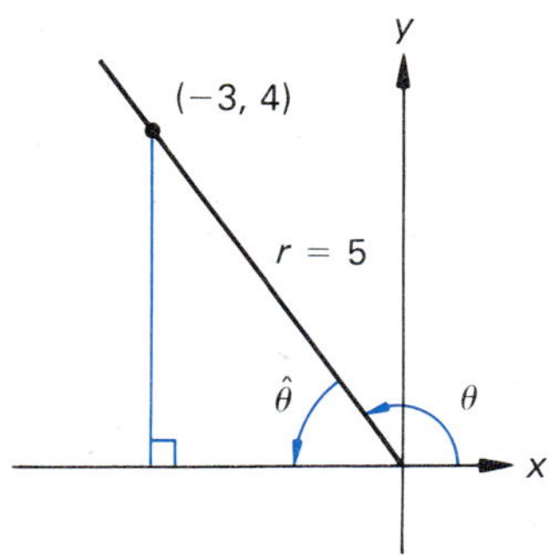

FIGURE 6.39

EXAMPLE 1 Determine the trigonometric functions of θ if the terminal side of θ passes through $(-3, 4)$.

Solution: We know that $x = -3$ and $y = 4$ and from the Pythagorean theorem we find that $r = 5$ (Fig. 6.39). Therefore, the six trigonometric functions of θ are

$$\sin\theta = \frac{4}{5}, \qquad \cos\theta = \frac{-3}{5}, \qquad \tan\theta = \frac{4}{-3},$$

$$\csc\theta = \frac{5}{4}, \qquad \sec\theta = \frac{5}{-3}, \qquad \cot\theta = \frac{-3}{4}. \qquad \square$$

FIGURE 6.40

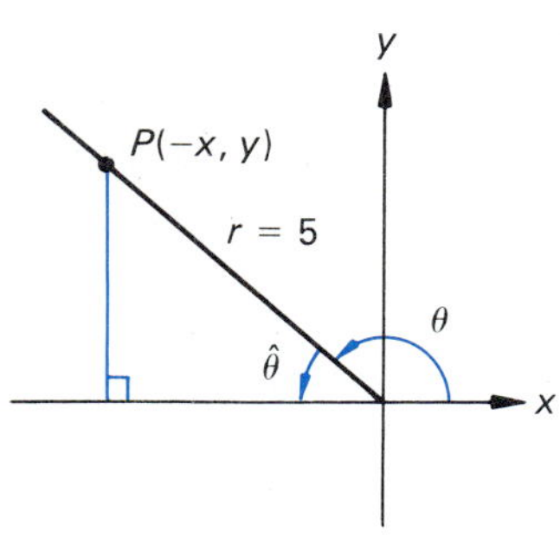

For any second-quadrant angle (Fig. 6.40) the reference triangle would be determined by picking a point on the terminal side of the angle and constructing a perpendicular to the x-axis. The x-value (abscissa) of the ordered pair is negative and the y-value (ordinate) is positive. By referring to the figure we see that sin and its reciprocal, csc, are positive in the second quadrant, whereas cos, tan, and their reciprocals, sec and cot, are negative in the second quadrant.

For any third-quadrant angle (Fig. 6.41) the reference triangle would be determined by picking a point on the terminal side of the angle and constructing a perpendicular to the x-axis. The x-value

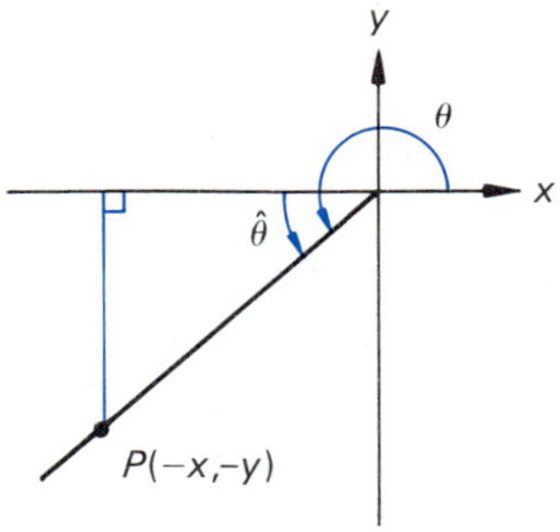

FIGURE 6.41

TABLE 6.4 **Signs of trigonometric functions**

	Function					
Quadrant	sin θ	cos θ	tan θ	cot θ	sec θ	csc θ
1st Quadrant $0 < \theta < 90°$ or $0 < \theta < \pi/2$	+	+	+	+	+	+
2nd Quadrant $90° < \theta < 180°$ or $\pi/2 < \theta < \pi$	+	−	−	−	−	+
3rd Quadrant $180° < \theta < 270°$ or $\pi < \theta < 3\pi/2$	−	−	+	+	−	−
4th Quadrant $270° < \theta < 360°$ or $3\pi/2 < 0 < 2\pi$	−	+	−	−	+	−

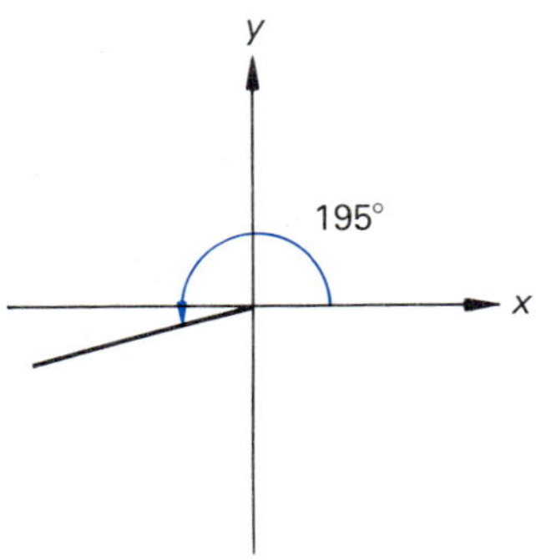

FIGURE 6.42

(abscissa) of the ordered pair is negative and the y-value (ordinate) of the ordered pair is also negative. By examining the figure we can see that sin, cos, and their reciprocals, csc, and sec, are negative, whereas tan and its reciprocal, cot, are positive.

Using the same method of analyzing a fourth-quadrant angle it can be shown that cos and its reciprocal are positive, whereas sin, tan, and their reciprocals are negative. These results are summarized in Table 6.4.

EXAMPLE 2 Determine the algebraic sign of the following trigonometric functions.

a. sin 195° **b.** cos 135° **c.** tan 315°

Solution:

a. The terminal side of the 195° angle in standard position is in the third quadrant (Fig. 6.42). The algebraic sign of the sine of any third-quadrant angle is negative. Thus, sin 195° is negative.

b. The terminal side of the 135° angle is in the second quadrant (Fig. 6.43). The algebraic sign of the cosine of any second-quadrant angle is negative. Thus, cos 135° is negative.

c. The terminal side of the 315° angle is in the fourth quadrant (Fig. 6.44). The algebraic sign of the tangent of any fourth-quadrant angle is negative. Thus, tan 315° is negative. □

FIGURE 6.43

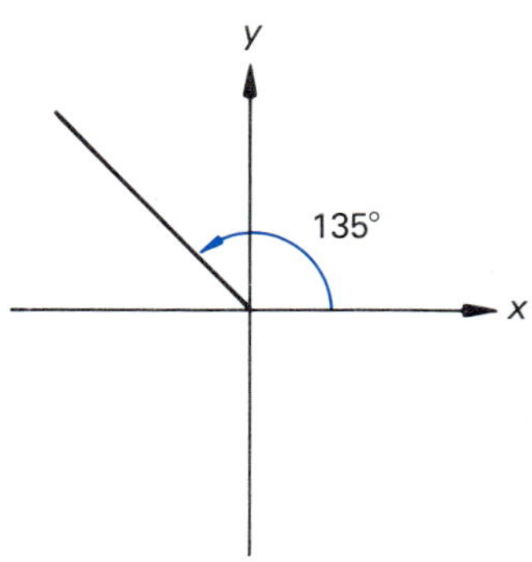

EXAMPLE 3 Determine the algebraic sign of the following trigonometric functions:

a. sin 1.63 **b.** cos 3.76 **c.** tan 0.93

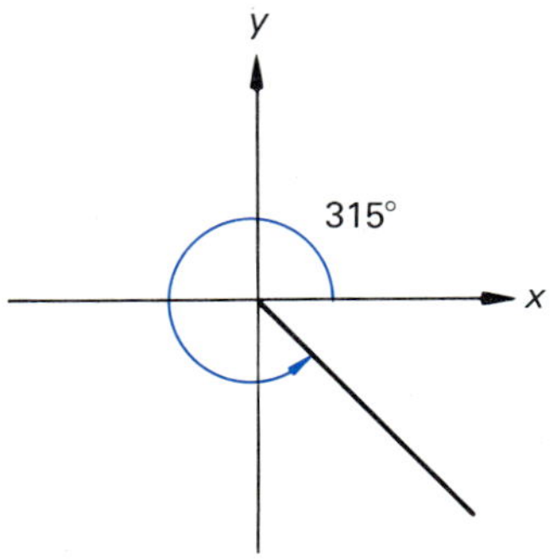

FIGURE 6.44

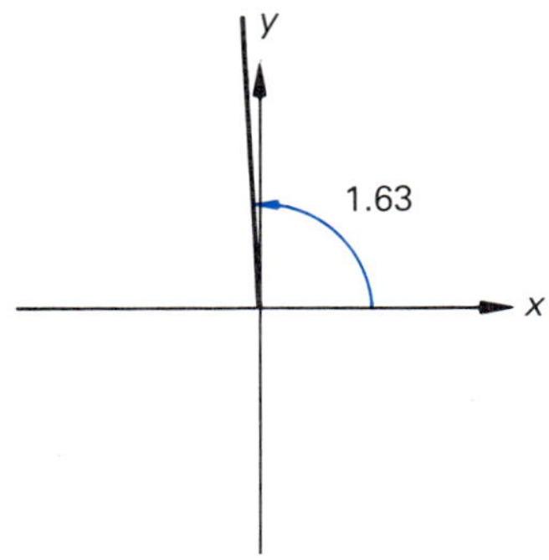

FIGURE 6.45

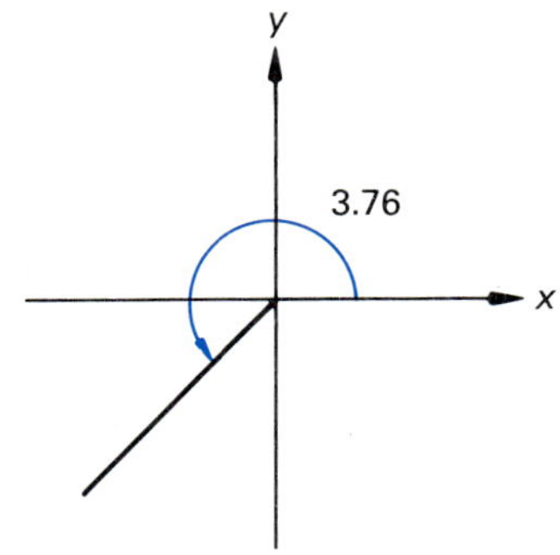

FIGURE 6.46

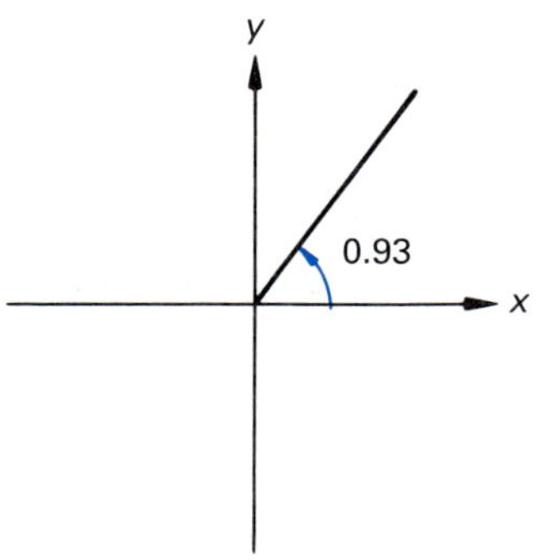

FIGURE 6.47

Solution:

a. The terminal side of an angle of 1.63 rad is in the second quadrant (Fig. 6.45). From Table 6.3 the algebraic sign of the sine of a second-quadrant angle is positive. Therefore, sin 1.63 is positive.

b. The terminal side of an angle of 3.76 rad is in the third quadrant (Fig. 6.46). From Table 6.3 the algebraic sign of the cosine of a third-quadrant angle is negative. Therefore, cos 3.76 is negative.

c. The terminal side of an angle of 0.93 rad is in the first quadrant (Fig. 6.47). Since all trigonometric functions are positive in the first quadrant, tan 0.93 is positive. □

6.5 Exercises

In Exercises 1 – 36, determine the algebraic sign of the trigonometric functions.

1. sin 75° **2.** cos 124° **3.** tan 195°
4. sin 295° **5.** cos 220° **6.** tan 250°
7. sin 140° **8.** cos 82° **9.** tan 350°
10. sin 225° **11.** cos 358° **12.** tan 22.5°
13. sin 1.67 **14.** cos 4.50 **15.** tan 6.10
16. sin 0.84 **17.** cos 2.25 **18.** tan 3.45
19. sin 5.25 **20.** cos 0.79 **21.** tan 0.52
22. sin 3.93 **23.** cos 5.76 **24.** tan 2.36
25. csc 85° **26.** sec 135° **27.** cot 240°
28. csc 1.68 **29.** sec 3.25 **30.** cot 4.85
31. csc 210° **32.** sec 300° **33.** cot 45°
34. csc 4.95 **35.** sec 0.25 **36.** cot 150°

In Exercises 37 – 44, find the six trigonometric functions of θ for which the terminal side of θ passes through the given point. Find exact values.

37. $(-6, 8)$ **38.** $(6, -8)$ **39.** $(-3, -4)$
40. $(-1, -2)$ **41.** $(4, -5)$ **42.** $(-3, 5)$
43. $(2, 5)$ **44.** $(-2, 3)$

In Exercises 45 – 56, determine the quadrant(s) in which the terminal side of θ lies, subject to the following conditions:

45. $\sin\theta > 0$, $\cos\theta > 0$ **46.** $\tan\theta < 0$, $\sin\theta > 0$
47. $\cos\theta < 0$, $\tan\theta < 0$ **48.** $\cos\theta < 0$, $\sin\theta > 0$
49. $\sin\theta < 0$, $\tan\theta > 0$ **50.** $\cos\theta > 0$, $\csc\theta > 0$
51. $\cot\theta > 0$, $\sin\theta > 0$ **52.** $\tan\theta > 0$, $\sec\theta < 0$
53. $\sec\theta > 0$, $\csc\theta < 0$ **54.** $\cot\theta > 0$, $\sin\theta < 0$
55. $\cot\theta < 0$, $\sec\theta < 0$ **56.** $\cot\theta < 0$, $\cos\theta < 0$

6.6 Trigonometric functions of any angle

In Section 6.2 we showed that the value of a trigonometric function is determined by a reference triangle. In Section 6.5 it was shown that if the reference triangle was located in different quadrants, the ratio may have a positive or negative value. Now we will put these two ideas together to determine the value of a trigonometric function of any angle.

Let us consider four angles θ_1, θ_2, θ_3, and θ_4 in standard position determined by the points $(\sqrt{3}, 1)$, $(-\sqrt{3}, 1)$, $(-\sqrt{3}, -1)$, and $(\sqrt{3}, -1)$. (See Fig. 6.48.) In each case $r = 2$. The reference triangles are formed by constructing perpendiculars to the x-axis. In each case the reference

FIGURE 6.48

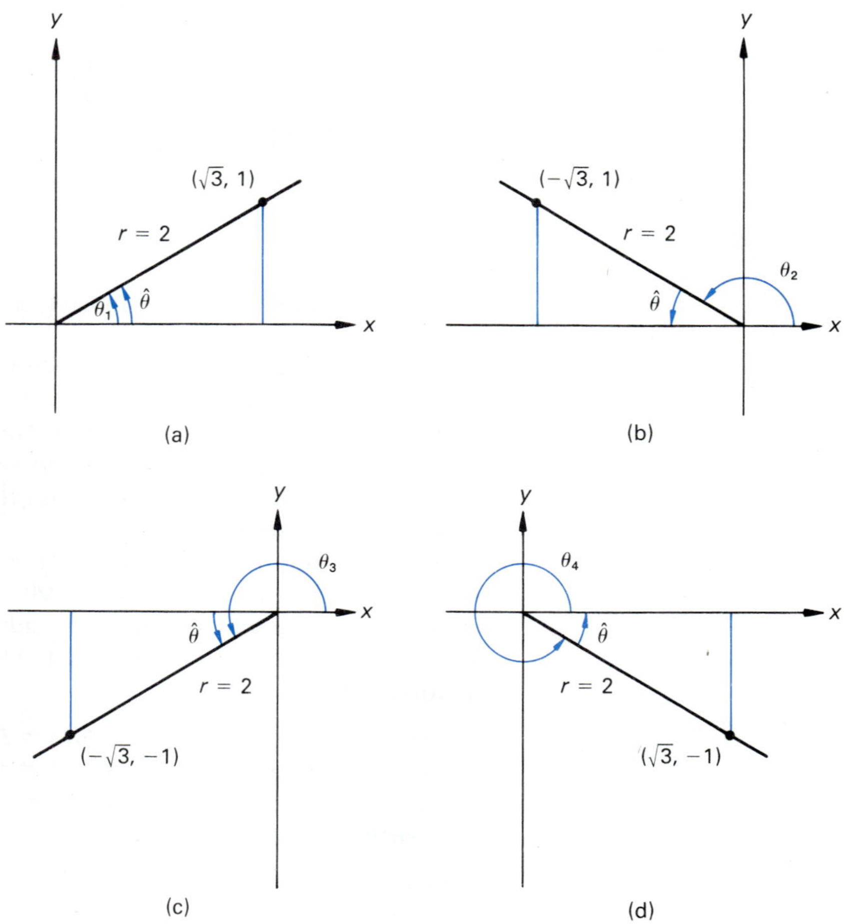

angle is indicated with the symbol $\hat{\theta}$. *The reference angle* $\hat{\theta}$ *is formed with the x-axis and the terminal side of the angle.* The reference angle $\hat{\theta}$ in the first quadrant and θ_1 are identical.

Therefore $F(\theta_1) = F(\hat{\theta})$, where F represents any one of the six trigonometric functions. Also in this illustration:

$$|F(\theta_1)| = |F(\theta_2)| = |F(\theta_3)| = |F(\theta_4)| = |F(\hat{\theta})|,$$

or the numerical measure of the functional values of each of these angles has the same numerical value. Notice that $\hat{\theta}$ is always a positive acute angle.

EXAMPLE 1 Use the data from Fig. 6.48 to show that the numerical values for $|\sin \theta_1|$, $|\sin \theta_2|$, $|\sin \theta_3|$, $|\sin \theta_4|$, and $|\sin \hat{\theta}|$ are the same.

Solution:

$$|\sin \theta_1| = \left|\frac{1}{2}\right| = \frac{1}{2} \qquad |\sin \theta_2| = \left|\frac{1}{2}\right| = \frac{1}{2} \qquad |\sin \theta_3| = \left|\frac{-1}{2}\right| = \frac{1}{2}$$

$$|\sin \theta_4| = \left|\frac{-1}{2}\right| = \frac{1}{2} \qquad |\sin \hat{\theta}| = \left|\pm\frac{1}{2}\right| = \frac{1}{2}$$

Thus, the numerical value in each case is 1/2. □

Why all the discussion and concern about the reference angle? There are two reasons: *(1) In order to find the value of a trigonometric function using trigonometric tables, the angle θ must be expressed in terms of the acute reference angle $\hat{\theta}$. (2) The reference angle is an essential part of the technique for finding the angle given the functional value.*

In examining the tables in the Appendix it can be seen that the tables are restricted so that $0° \le \theta \le 90°$. Thus, it is essential for us to be able to express an angle in any quadrant in terms of its reference angle. Generalizing what is illustrated in Fig. 6.48, we have the following formulas:

First-quadrant angle:	$F(\theta_1) = F(\hat{\theta})$,
Second-quadrant angle:	$F(\theta_2) = \pm F(180° - \theta) = \pm F(\hat{\theta})$,
Third-quadrant angle:	$F(\theta_3) = \pm F(\theta - 180°) = \pm F(\hat{\theta})$,
Fourth-quadrant angle:	$F(\theta_4) = \pm F(360° - \theta) = \pm F(\hat{\theta})$.

The sign depends on whether or not the function is positive or negative in that quadrant.

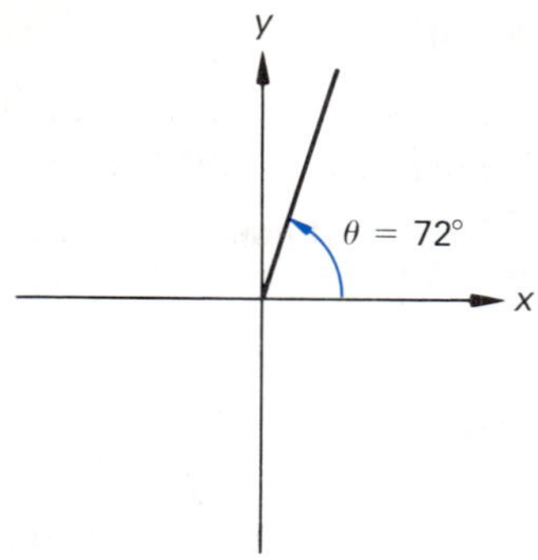

FIGURE 6.49

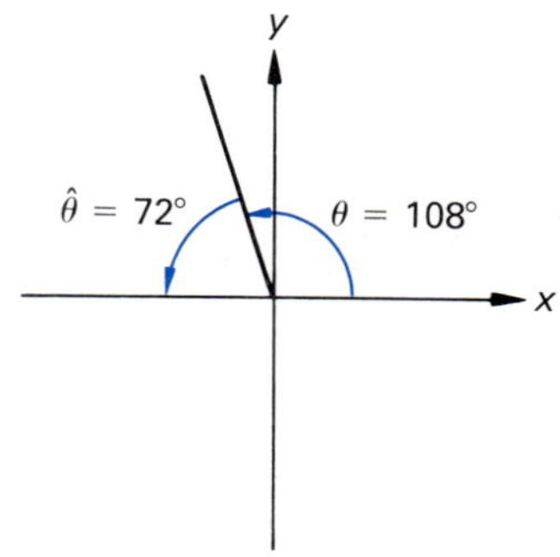

FIGURE 6.50

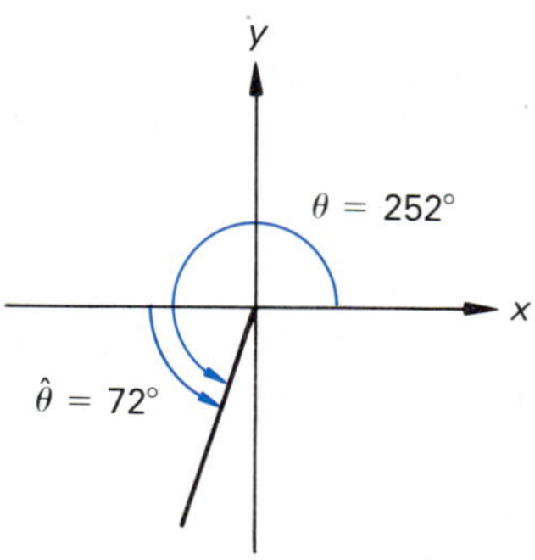

FIGURE 6.51

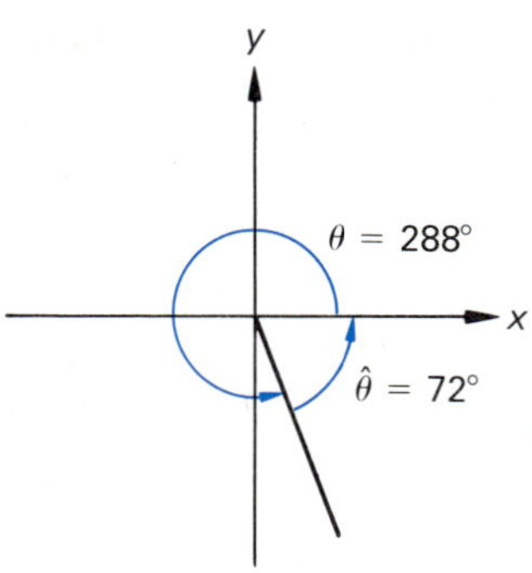

FIGURE 6.52

EXAMPLE 2 Find cos θ to five decimal places when θ is:

a. 72° (Fig. 6.49) **b.** 108° (Fig. 6.50)
c. 252° (Fig. 6.51) **d.** 288° (Fig. 6.52)

Solution:

a. Cos 72° = ? Using a calculator the answer to five decimal places is cos 72° = 0.30902.

b. Cos 108° = ? An angle of 108° in standard position has its terminal side in the second quadrant and cos θ is negative in the second quadrant. Substituting in the formula:

$$F(\theta_2) = \pm F(180° - \theta_2) = F(\hat{\theta}),$$
$$\cos 108° = -\cos(180° - 108°)$$
$$= -\cos 72°$$
$$= -0.30902.$$

Using the calculator directly:

$$\cos 108° = -0.30902$$

c. Cos 252° = ? An angle of 252° in standard position has its terminal side in the third quadrant, and cos θ is negative in the third quadrant. Substituting in the formula:

$$F(\theta_3) = \pm F(\theta_3 - 180°) = \pm F(\hat{\theta}),$$
$$\cos 252° = -\cos(252° - 180°)$$
$$= -\cos 72°$$
$$= -0.30902.$$

Using the calculator directly:

$$\cos 252° = -0.30902$$

d. Cos 288° = ? An angle of 288° in standard position has its terminal side in the fourth quadrant, and cos θ is positive in the fourth quadrant. Substituting in the formula:

$$\begin{aligned}F(\theta_4) &= \pm F(360^\circ - \theta_4) = \pm F(\hat{\theta}),\\ \cos 288^\circ &= \cos(360^\circ - 288^\circ)\\ &= \cos 72^\circ\\ &= 0.30902.\end{aligned}$$

Using the calculator directly:

$\cos 288^\circ = 0.30902.$ □

From the example it can be seen that the cosine functions of the first- and fourth-quadrant angles that have the same reference angle have the same algebraic values. Also, the cosine functions of second- and third-quadrant angles that have the same reference angle have the same algebraic values.

At this stage you may be saying, "I always use a calculator and therefore do not need to worry about all of this." The previous ideas and examples are helpful in understanding a second reason to study reference angles: finding the angle given the functional value.

EXAMPLE 3 Given $\tan \theta = 0.75355$, find θ to the nearest tenth of a degree for $0 < \theta < 360^\circ$ (all angles θ between 0° and 360°).

Solution: Given $\tan \theta = 0.75355$, there are two possible answers for θ. Since $\tan \theta$ is positive in the first and third quadrants, θ could be a first- or a third-quadrant angle.

Using the calculator we find that 0.75355 [INV] [TAN] 36.999852° or, to the nearest tenth of a degree, 37.0°. The reference angle is 37.0°, which is the same as the first-quadrant angle. How do we find the third-quadrant angle? It can be found by constructing a reference angle, Fig. 6.53. From the diagram we can see that

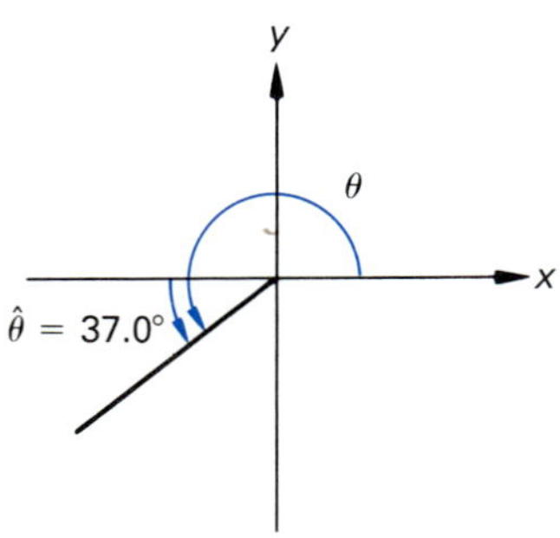

FIGURE 6.53

$$\begin{aligned}\theta &= 180^\circ + \hat{\theta}\\ \theta &= 180^\circ + 37.0^\circ\\ \theta &= 217.0^\circ.\end{aligned}$$

The two values of θ that make this a true statement are:

$\theta = 37.0^\circ$ and $\theta = 217.0^\circ$. □

In general to determine the angle θ knowing the quadrant and the reference angle $\hat{\theta}$, we can use the following rules:

First-quadrant angle:	$\theta = \hat{\theta}$,
Second-quadrant angle:	$\theta = 180^\circ - \hat{\theta}$,
Third-quadrant angle:	$\theta = 180^\circ + \hat{\theta}$,
Fourth-quadrant angle:	$\theta = 360^\circ - \hat{\theta}$.

Why was the restriction $0^\circ < \theta < 360^\circ$ placed on θ in Example 3? The angles 397.0° and 577.0° are, respectively, $360^\circ + 37.0^\circ$ and

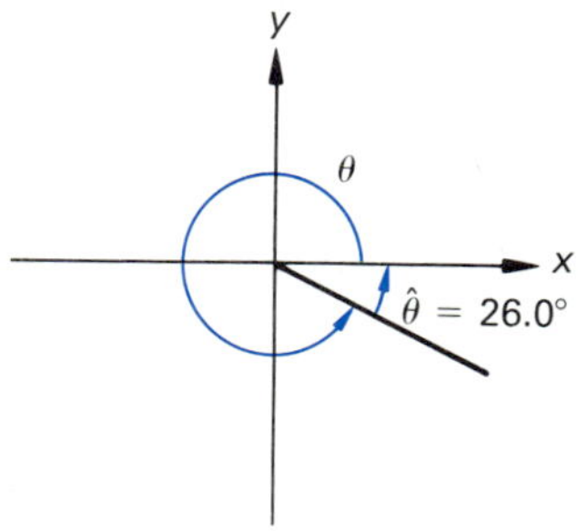

FIGURE 6.54

$360° + 217.0°$. The angles $397.0°$ and $37.0°$ are coterminal angles, as are $577.0°$ and $217.0°$. There are an unlimited number of angles that could be created by adding or subtracting multiples of $360°$.

EXAMPLE 4 Given $\sin\theta = -0.43775$, find θ to the nearest tenth of a degree, for $0° < \theta < 360°$.

Solution: For $\sin\theta = -0.43775$, $\theta = ?$ There are two possible values for θ since $\sin\theta$ is negative in the third and fourth quadrants. Using a calculator we obtain:

0.43775 [+/−] [INV] [SIN] − 25.96041;

to the nearest tenth of a degree, $-26.0°$. The calculator will give us only one value, in this case, $-26.0°$, which in standard position is a negative rotation of $26.0°$, and is in the fourth quadrant. This tells us we have a reference angle $\hat{\theta}$ of $26.0°$ (see Fig. 6.54).

Third-quadrant angle: $\theta_3 = 180° + 26.0°$,
$\theta_3 = 206.0°$,

Fourth-quadrant angle: $\theta_4 = 360° - 26.0°$,
$\theta_4 = 334.0°$. □

The same technique is used to determine θ when the angle is measured in radians, rather than degrees.

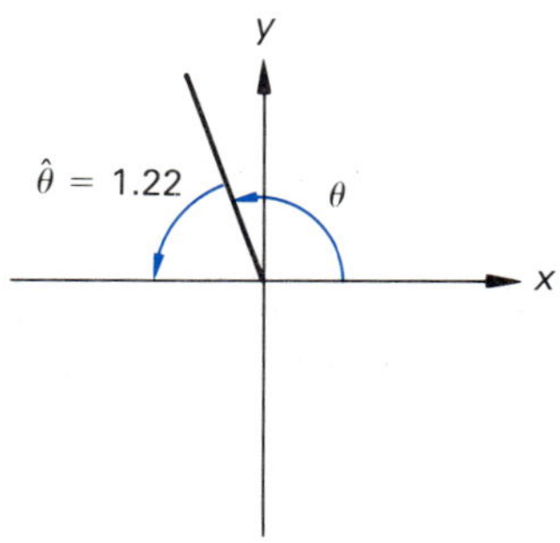

FIGURE 6.55

EXAMPLE 5 Given $\cos\theta = -0.34215$, find θ to the nearest hundredth of a radian, for $0 < \theta < 2\pi$.

Solution: For $\cos\theta = -0.34215$, $\theta = ?$ There are two possible values for θ since the cosine function is negative in the second and third quadrants. Using a calculator in radian mode, we obtain

0.34215 [+/−] [INV] [COS] 1.9200004,

which to the nearest hundredth is 1.92 radians. This result is a second quadrant angle, thus, $\theta_2 = 1.92$ rad.

To determine the third-quadrant angle we must determine the reference angle $\hat{\theta}$ (Fig. 6.55).

$$\begin{aligned}\hat{\theta} &= \pi - \theta_2\\ &= \pi - 1.92\\ &= 1.22 \text{ (nearest hundredth)}\end{aligned}$$

Now using our formula (see Fig. 6.56), we have

$$\begin{aligned}\theta_3 &= \pi + \hat{\theta}\\ &= \pi + 1.22\\ &= 4.36 \text{ rad.}\end{aligned}$$ □

FIGURE 6.56

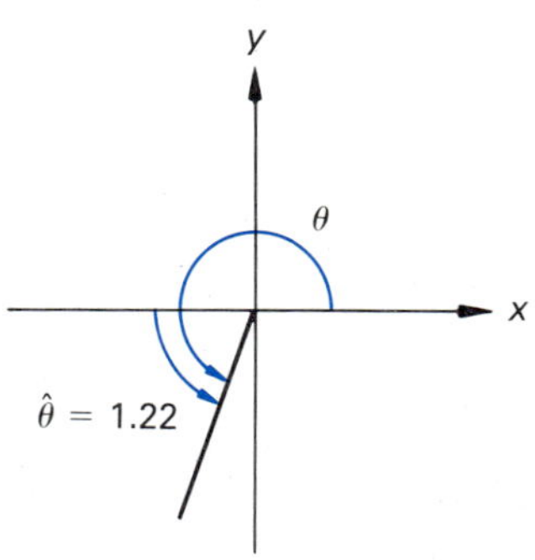

EXAMPLE 6 Find $\sin\theta$ when $\cos\theta = -0.44265$ and $\tan\theta < 0$.

Solution: To find $\sin\theta$, we need to know what angle θ will yield $\cos\theta = -0.44265$ and $\tan\theta < 0$. Since the cosine and tangent functions are both negative, we are looking for a second-quadrant angle. Using the calculator we find:

0.44265 [+/−] [INV] [COS] 116.27308°

Since this is a second-quadrant angle, the final result can be found by finding the sine of this angle.

$$\sin 116.27308° = 0.8966945$$

To five decimal places we have

$$\sin\theta = 0.89669$$

The same result would be found if the calculator was in the radian mode rather than the degree mode. □

6.6 Exercises

In Exercises 1–30, determine the values of the trigonometric functions to five decimal places (if using tables, to four decimal places).

1. $\sin 137.3°$ **2.** $\cos 193.4°$

3. $\tan 276.8°$ **4.** $\sin 211.5°$

5. $\cos 138.4°$ **6.** $\tan 124.2°$

7. $\csc 297.2°$ **8.** $\sec 215.5°$

9. $\cot 137.4°$ **10.** $\sin 285°$

11. $\cos 311.4°$ **12.** $\tan 268.4°$

13. $\csc 100°$ **14.** $\sec 175°$

15. $\cot 294.7°$ **16.** $\sin 482°$

17. $\cos 552.4°$ **18.** $\tan 664.7°$

19. $\sin(-48.2°)$ **20.** $\cos(-94.7°)$

21. $\tan(-194.5°)$ **22.** $\sin(-93.7°)$

23. $\cos(-13.8°)$ **24.** $\tan(-11.7°)$

25. $\sin 2.25$ **26.** $\cos 3.84$

27. $\tan 1.93$ **28.** $\sin 3.24$

29. $\cos 4.98$ **30.** $\tan 4.25$

In Exercises 31–42, find θ to the nearest tenth of a degree for $0° < \theta < 360°$.

31. $\sin\theta = 0.74664$ **32.** $\cos\theta = 0.60876$

33. $\tan\theta = 0.75082$ **34.** $\sin\theta = 0.68835$

35. $\cos\theta = -0.74548$ **36.** $\tan\theta = -1.21310$

37. $\sin\theta = -0.54464$ **38.** $\cos\theta = -0.76604$

39. $\tan\theta = 1.42814$ **40.** $\sin\theta = -0.93969$

41. $\cos\theta = 0.50754$ **42.** $\tan\theta = -0.62487$

In Exercises 43–48, find θ to the nearest hundredth of a radian for $0 < \theta < 2\pi$.

43. $\sin\theta = 0.81915$ **44.** $\cos\theta = -0.93042$

45. $\tan\theta = -1.11061$ **46.** $\sin\theta = -0.34202$

47. $\cos\theta = -0.17365$ **48.** $\tan\theta = 1.42815$

In Exercises 49–54, determine the functional value that satisfies the given conditions:

49. Find $\sin\theta$ when $\cos\theta = -0.58779$ and $\tan\theta > 0$

50. Find $\sin\theta$ when $\tan\theta = -0.70021$ and $\cos\theta < 0$

51. Find $\cos\theta$ when $\sin\theta = 0.17365$ and $\tan\theta > 0$

52. Find $\tan\theta$ when $\cos\theta = -0.78801$ and $\sin\theta > 0$

53. Find $\tan\theta$ when $\sin\theta = -0.57358$ and $\cos\theta > 0$

54. Find $\cos\theta$ when $\tan\theta = 1.23490$ and $\sin\theta < 0$

55. Use your calculator or reference triangles to show that the values in Table 6.5 are correct.

56. The range x of a projectile fired with velocity v_0 at an angle α with the horizontal is given by $x = (v_0^2 \sin 2\alpha)/g$, where g = gravitational accelera-

TABLE 6.5

Function	Angle 0°	90°	180°	270°	360°
$\sin\theta$	0	1	0	−1	0
$\cos\theta$	1	0	−1	0	1
$\tan\theta$	0	undefined	0	undefined	0
$\cot\theta$	undefined	0	undefined	0	undefined
$\sec\theta$	1	undefined	−1	undefined	1
$\csc\theta$	undefined	1	undefined	−1	undefined

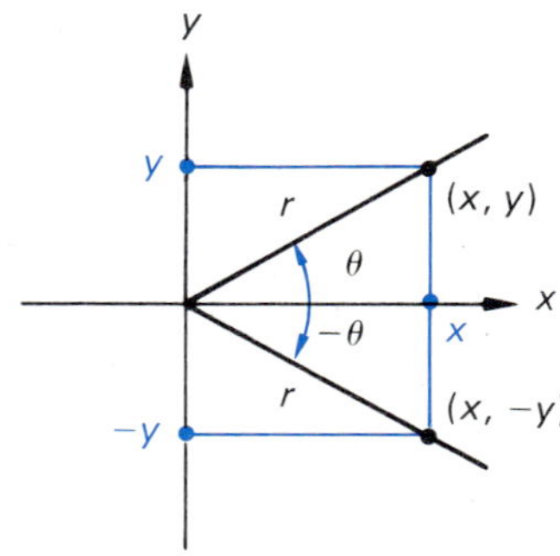

FIGURE 6.57

tion. Find x in terms of v_0 and g when

a. $\alpha = 15°$ **b.** $\alpha = 45°$ **c.** $\alpha = 75°$

d. Determine why x is the greatest when $\alpha = 45°$, which is the angle of maximum range.

57. A formula for finding the area of a triangle, knowing sides a and b and angle C is $A = 1/2\, ab \sin C$. Find the area of a triangle for which $a = 34.5'$, $b = 60.4'$, and $C = 148.3°$.

58. In calculating the area of a triangular tract of land a surveyor used the formula in Exercise 57. He used the values $a = 324$ m, $b = 196$ m, and $c = 118.7°$. Find the area.

59. In a special circuit containing a capacitance, the voltage $V = 45 \sin\left(\omega t - \frac{\pi}{2}\right)$. Find V when

a. $\omega t = \frac{3\pi}{4}$ **b.** $\omega t = 0$

60. In an AC circuit with resistance R, the current $I = I_{max} \sin \omega t$, where I_{max} = maximum current and ωt = angle of rotation. Find I when I_{max} = 10A and

a. $\omega t = \frac{\pi}{2}$ **b.** $\omega t = 0$

61. From Fig. 6.57 we see that $\sin\theta = y/r$ and $\sin(-\theta) = -y/r$. From this we conclude that $\sin(-\theta) = -\sin\theta$ and we say sin is an **odd function.** Likewise, $\cos\theta = x/r$ and $\cos(-\theta) = x/r$. From this we conclude that $\cos(-\theta) = \cos\theta$ and we say that cosine is an **even function.** Verify the following equations in the same way:

$\tan(-\theta) = -\tan\theta$ $\cot(-\theta) = -\cot\theta$
$\sec(-\theta) = \sec\theta$ $\csc(-\theta) = -\csc\theta$

In Exercises 62–67, find the value of the functions using the rules developed in Exercise 61.

62. $\sin(-75°)$ **63.** $\cos(-135°)$
64. $\tan(-110°)$ **65.** $\cot(-185°)$
66. $\sec(-295°)$ **67.** $\csc(-132°)$

6.7 Applications

An understanding of trigonometry is essential for the navigator and the surveyor. In navigation or surveying, direction is specified by stating either the bearing, the relative bearing, or the heading. The **bearing** expresses the angle between a north–south line and an object. The **relative bearing** expresses the angle between the line through an object and the line through another object. (See Fig. 6.58.) In either case the direction of the angular measurement must be specified. (The mariner's compass is commonly used as a measuring reference. It

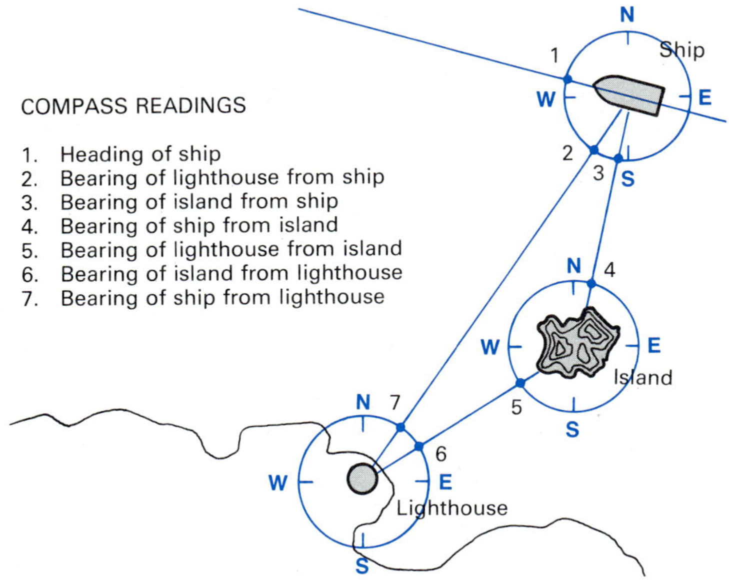

FIGURE 6.58

assumes a circle of 360° with 0° (also 360°) at north and the angular measure increasing in a clockwise direction making east lie at 90°, south lie at 180° and west lie at 270°. Compare this with the method of angular measure based on the *xy*-axis system that is conventionally used for most other purposes. The **heading** is the direction of travel according to the mariner's compass of a ship or aircraft.

EXAMPLE 1 A sailboat is sailing on the course shown in Fig. 6.59.

a. Determine the heading $\angle NAB$.
b. Determine the relative bearing, $\angle EBC$ of point C.

Solution:

a. We can determine the measure of $\angle NAB$ by finding the measure of $\angle BAD$ and subtracting $\angle BAD$ from 90°, since $\angle NAC = 90°$. To find $\angle BAD$, draw a perpendicular line from point B to $\overline{AC}$ (Fig. 6.60). Since $\overline{BD}$ is the shorter leg of the triangle BDC, $\overline{BD}$ must be one half the hypotenuse (theorem in geometry). Thus, $\overline{BD} =$ 150 m.

$$\sin \angle BAD = \frac{150 \text{ m}}{175 \text{ m}} \quad \text{or} \quad 0.85714$$

$$\angle BAD = 59.0° \quad \textbf{(nearest tenth)}$$

$$\angle NAB = 90° - 59.0° \quad \text{or} \quad 31.0°.$$

Thus, the heading is 31.0°.

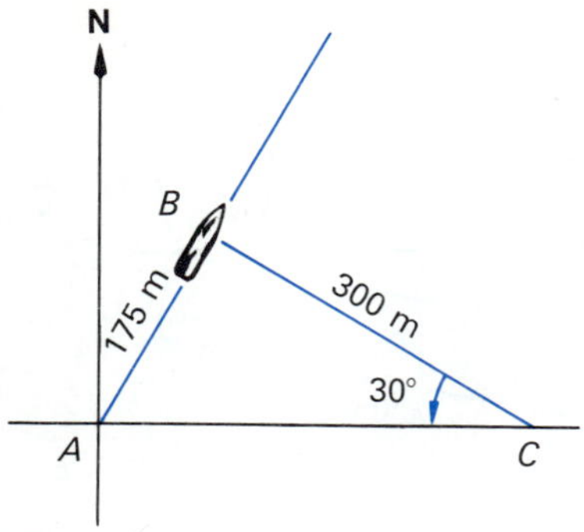

FIGURE 6.59

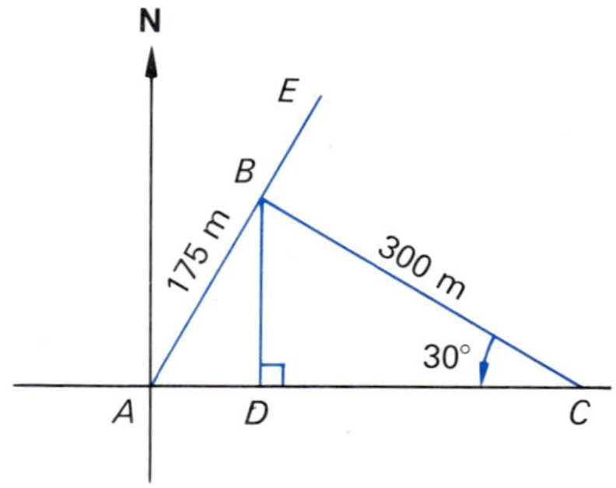

FIGURE 6.60

b. There are two ways of finding the relative bearing.

First Method

$\angle ABE$ is a straight angle; therefore $\angle ABD + \angle DBC + \angle CBE = 180°$. We know that for triangle ABD, $\angle BAD = 59.0°$ and $\angle ADB = 90°$, therefore $\angle ABD = 180° - (59.0° + 90.0°)$ or $\angle ABD = 31.0°$. We also know that for the triangle BDC, $\angle DCB = 30°$ and $\angle BDC = 90°$; therefore $\angle DBC = 180° - (30° + 90°)$ or $\angle DBC = 60.0°$. Substituting, we have $31.0° + 60.0° + \angle CBE = 180°$ or $\angle CBE = 89.0°$. Thus, the relative bearing is 89.0°.

Second Method

There is a theorem in geometry that states: "The measure of the exterior angle of a triangle is equal to the sum of the measures of the two nonadjacent interior angles. Since we know that the measure of $\angle BAD$ is 59.0° and the measure of $\angle ACB$ is 30.0°, by the theorem $\angle CBE = 59.0° + 30.0°$ or $\angle CBE = 89.0°$. □

Industrial and engineering problems

The engineer uses trigonometry to solve many industrial problems. They may be problems such as finding the length of a steel rod needed to fit a certain curve, finding the area of a sector of a circle, or finding the angular velocity of a point on the rim of a pulley.

EXAMPLE 2 Determine the total length of steel rod needed to meet the conditions established in Fig. 6.61.

Solution: First make a sketch indicating the essential parts of the figure (Fig. 6.62). The length of rod needed is equal to the length of $\overline{AB}$ plus the length of $\overline{ED}$ plus the length of the circular arc BGD. We know that $\overline{AB} = 120$ mm and $\overline{ED} = 140$ mm, therefore we need to find the

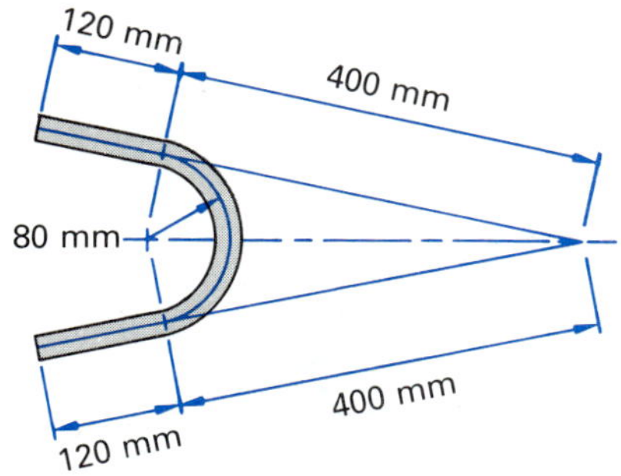

FIGURE 6.61

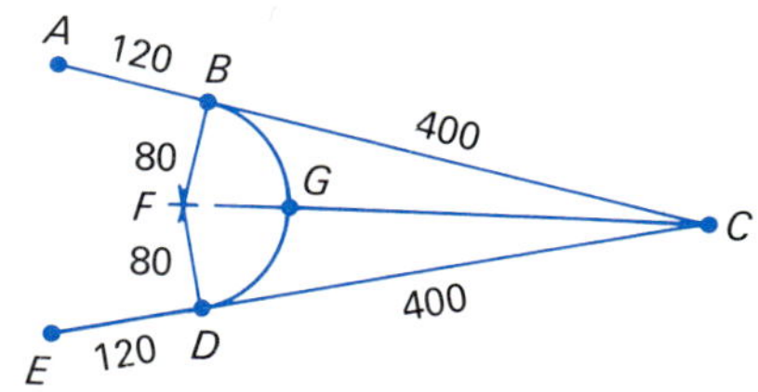

FIGURE 6.62

length of $\widehat{BGD}$ (arc BGD). We can find the length of $\widehat{BGD}$ by using the arc-length formula $s = \theta r$. (This formula was discussed in Section 6.1.)

To use the arc-length formula we need to know the measure of $\angle BFD$ in radians. The line segments $\overline{AC}$ and $\overline{EC}$ are tangent to the circular arc BGD, therefore the radii $\overline{FB}$ and $\overline{FD}$ are perpendicular to the lines $\overline{AC}$ and $\overline{EC}$ forming the triangles CDF and CBF, which are congruent. The $\angle BFD = 2\angle CFD$. By right-triangle trigonometry we have (change calculator to radian mode):

$$\tan \angle CFD = \frac{400 \text{ mm}}{80 \text{ mm}} \quad \text{or} \quad 1.373$$

$$\angle BFD = 2(1.373) \quad \text{or} \quad 2.75.$$

Substituting in the formula $s = r\theta$, we have:

$$\begin{aligned} \text{length of } \widehat{BGD} &= (80 \text{ mm})(2.747) \\ &= 219.8 \text{ mm} \\ \text{length of rod} &= 219.8 \text{ mm} + 120 \text{ mm} + 140 \text{ mm} \\ &= 480. \text{ mm}. \end{aligned}$$ □

It is sometimes necessary to calculate the area of circular sectors as in Fig. 6.63. The area of a circle is $A = \pi r^2$. Noting that the central angle of a circle measures 2π rad we note that the area of the circle can also be expressed as $A = \frac{1}{2}(2\pi)r^2$. The central angle of a semicircle measures π rad or half the central angle of a circle. The area of a semicircle is half the area of a circle or

FIGURE 6.63

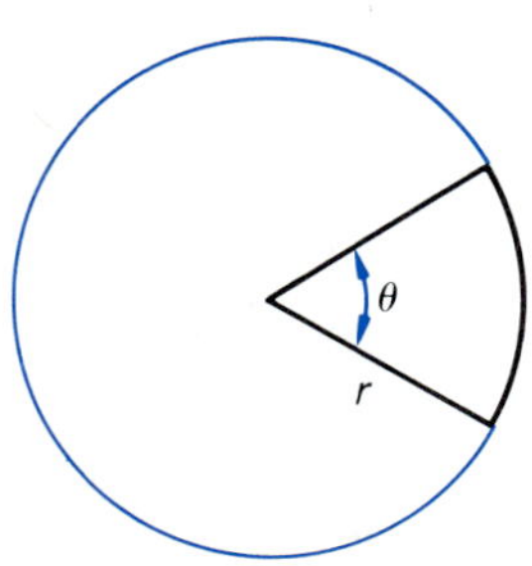

$$\frac{1/2(2\pi)r^2}{2} = \frac{1}{2}(\pi)r^2.$$

The central angle of a quarter circle measures $\pi/2$ or one fourth the central angle of a circle. The area of a quarter circle is one fourth the area of a circle or

$$\frac{1/2(2\pi)r^2}{4} = \frac{1}{2}\left(\frac{\pi}{2}\right)r^2.$$

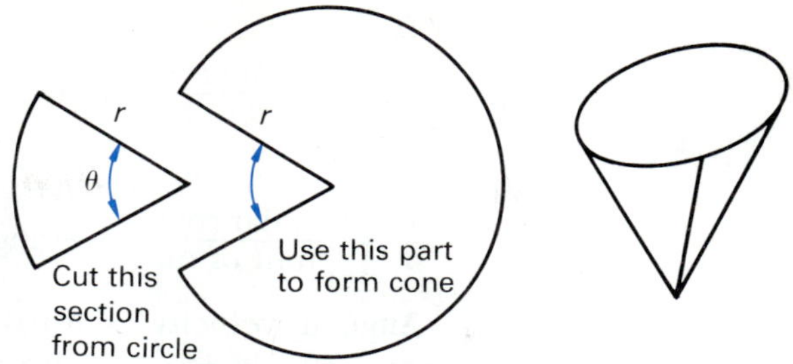

FIGURE 6.64

In each case the area of the portion of the circle is one half the central angle (expressed in radians) times r^2.

Area of Circular Sector

$A = \frac{1}{2}\theta r^2$, where θ is measured in radians.

EXAMPLE 3 Find the outside surface area of a cone formed from a circle with a diameter of 6.00 m by cutting a circular sector with a central angle of 1.5 rad from the circle and joining the cut edges (of the remaining sector), as shown in Fig. 6.64.

Solution: The area can be found using the formula $A = \frac{1}{2}\theta r^2$. We know $r = 3.00$ m since the diameter is 6.00 m. The central angle of the sector cut away is 1.50 rad, therefore, the central angle of the circular sector forming the cone is $2\pi - 1.50$ rad. Substituting in the formula we have:

$$A = \frac{1}{2}(2\pi - 1.50)(3.00 \text{ m})^2$$
$$= 21.5 \text{ m}^2.$$

The outside surface area of the cone is 21.5 m². □

Angular and linear velocities

If we start with the statement $s = r\theta$ and divide each member by time t we have

$$\frac{s}{t} = \frac{r\theta}{t} \text{ or } r\frac{\theta}{t},$$

where

$$\frac{s}{t} \text{ is } \frac{\text{distance}}{\text{unit of time}} = \text{velocity } v,$$

and

$$\frac{\theta}{t} \text{ is } \frac{\text{angular rotation}}{\text{unit of time}} = \text{angular velocity } \omega.$$

Angular velocity is usually represented by the Greek letter ω (omega). Replacing s/t with v and θ/t with ω, we have $v = r\omega$. This equation states that the velocity of a point along a circular path is equal to the product of the radius of the path and the angular velocity ω (in radians per unit of time).

Velocity along Circular Path

$v = r\omega$, where ω is measured in radians per unit of time

EXAMPLE 4 A wheel is rotating at 7 rad/s. The wheel has a 14-in. radius. What is the velocity of a point on its rim in inches per second?

Solution: We know that $\omega = 7$ rad/s and $r = 14$ in. Substituting in the formula: $v = r\omega$

$$\begin{aligned} v &= (14 \text{ in.})(7 \text{ rad/s}) \\ &= 98 \text{ in./s}. \end{aligned}$$

A point on the rim of a wheel is moving at the rate of 98 in./s. (Remember, a radian is a dimensionless quantity.) □

EXAMPLE 5 A pulley 120 mm in diameter turns at 500 rpm. Find its angular velocity in radians per second and the velocity of the belt (rpm = revolutions per minute).

Solution: We know that the angular velocity is 500 rpm. To change from rpm to rad/s we use the conversion factors $\left(\frac{1 \text{ min}}{60 \text{ s}}\right)$ and $\left(\frac{2\pi \text{ rad}}{1 \text{ rev}}\right)$.

Thus

$$\begin{aligned} \omega &= 500 \frac{\text{rev}}{\text{min}} \left(\frac{1 \text{ min}}{60 \text{ s}}\right)\left(\frac{2\pi \text{ rad}}{1 \text{ rev}}\right) \\ &= 52.36 \frac{\text{rad}}{\text{s}}. \end{aligned}$$

The angular velocity is 52.4 rad/s. The velocity of the belt is the same as the velocity of a point on the rim of the pulley

$$\begin{aligned} v &= \omega r \\ &= \left(52.36 \frac{\text{rad}}{\text{s}}\right)(60 \text{ mm}) \\ &= 3142. \frac{\text{rad mm}}{\text{s}}. \end{aligned}$$

Since radians are dimensionless, the answer is commonly written:

$$v = 3140. \text{ mm/s}. \qquad \square$$

6.7 Exercises

1. A pendulum 1 m long oscillates through an angle that subtends an arc length of 0.12 m. Determine the angle and express the results in degrees and radians.

2. Find the length of an arc through which a pendulum 1 m long swings if it swings through a total angle of 10°.

3. Determine the linear velocity in miles per hour of a point on the earth's equator if the diameter of the earth is approximately 7920 mi.

4. A satellite in a circular orbit 800 mi above the earth makes one complete orbit every 90 min. What is its linear speed? Use 4000 mi for the length of the radius of the earth.

5. A bullet is discovered in the wall of a room with an entry point 2.5 m above the floor. The path of the bullet in the wall is inclined upward with an 18° angle of elevation. Determine how far from the wall the gun was fired if **a.** the gun was fired from the floor, **b.** the gun was fired from 1.2 m above the floor.

6. A ship is sailing on the course shown in Fig. 6.65. **a.** Determine the heading ($\angle NAB$), **b.** Determine $\overline{AC}$, **c.** Determine measure of $\angle ABC$, and **d.** Determine the bearing of line BC.

7. A ship is sailing on the course shown in Fig. 6.66. **a.** Determine the heading ($\angle NAB$), **b.** Determine the relative bearing ($\angle DBC$), and **c.** Determine the bearing of line BC.

8. A surveyor measures the angle of elevation to the top of a building to be 49°. She then moves 75 ft closer to the building and measures the angle of elevation to be 65°. How high is the building? (See Fig. 6.67 on page 224.)

FIGURE 6.65

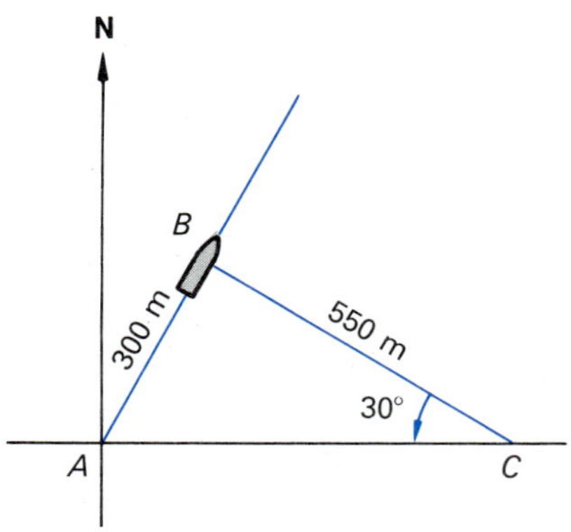

FIGURE 6.66

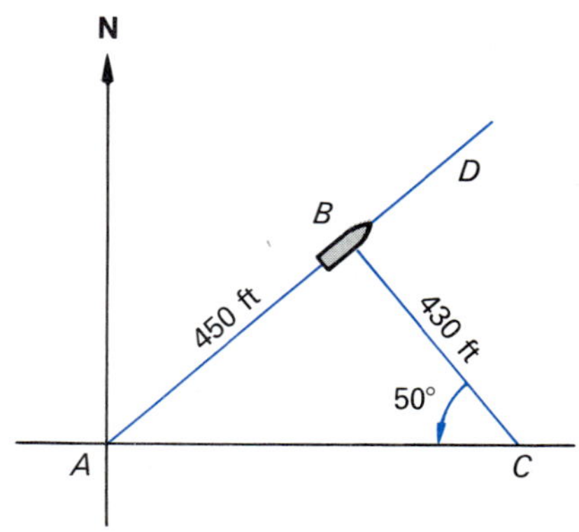

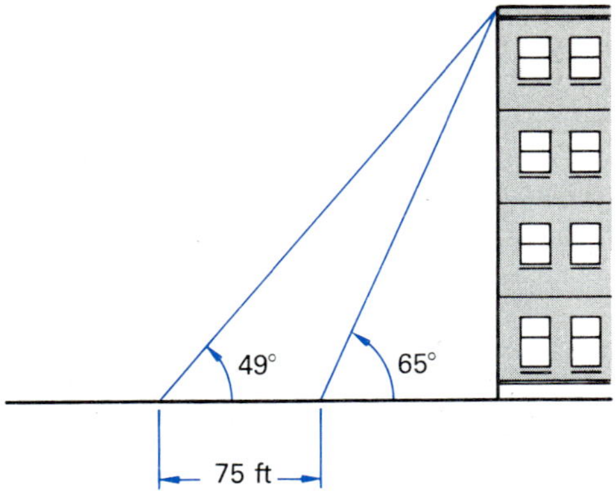

FIGURE 6.67

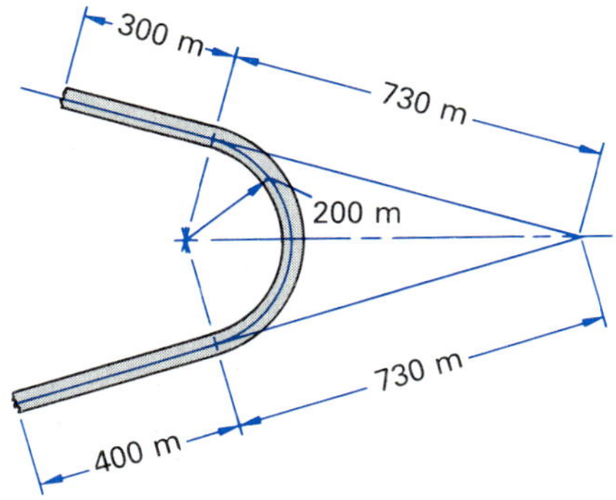

FIGURE 6.68

9. Determine the length of the road along the center line in Fig. 6.68.

10. Determine the total length of steel rod needed to meet the conditions established in Fig. 6.69.

11. A circular sector whose central angle is 240° is cut from a circular piece of sheet metal of diameter 180 mm. A cone is then formed by bringing the two straight edges of the sector together. What is the lateral surface area of the cone?

12. For the road in Fig. 6.68 determine the number of cubic meters of concrete needed for the curved section. The thickness of the concrete is 350 mm and the width of the road is 9.60 m.

13. A conical tent is made from a circular piece of canvas 18.0 ft in diameter, with a central angle of 150° removed. What is the surface area of the tent?

14. A wind-driven electrical generator has a 24.0-ft-diameter propeller. **a.** Find the angular velocity in radians per second at which the tip of the propeller will reach the speed of sound (approximately 1087 ft/s). **b.** Express the angular velocity in revolutions per second and rpm.

15. A satellite is orbiting the earth with a linear velocity of 15,742 mph. The orbit is approximately a circle with radius of 5265 mi. Find the angular velocity of the satellite in revolutions per hour.

16. A spacecraft is circling the moon once every 1.45 hr. If the altitude of the craft is constant at 120 mi and the radius of the moon is 1080 mi, compute the linear velocity of the spacecraft.

17. The wheels of a scooter have a 200-mm diameter. If the wheels are turning at the rate of 28 rpm, compute the distance in meters the scooter will travel in 3 min.

18. A seesaw rotates through a 40° angle. If the end of the seesaw travels 1 m, how long is it?

19. A 10-in. phonograph record plays at a constant speed of 33 1/3 rpm. What is the linear velocity of a point on its rim? What is the angular velocity of the same point?

20. A car is traveling at 40 mph. Its wheels have a 14-in. radius. Find the angle in radians through which a wheel rotates in 10 s.

FIGURE 6.69

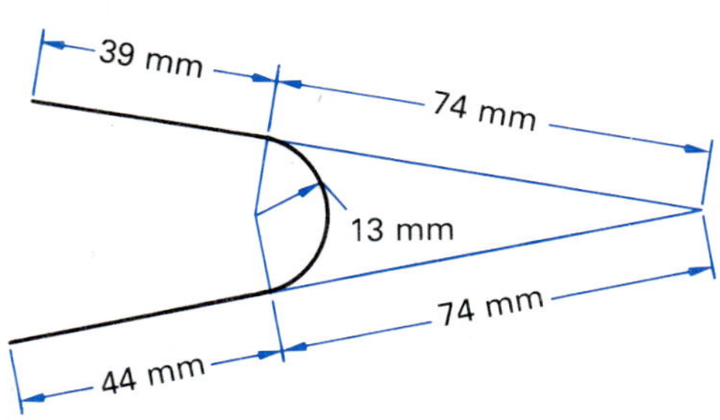

FIGURE 6.70

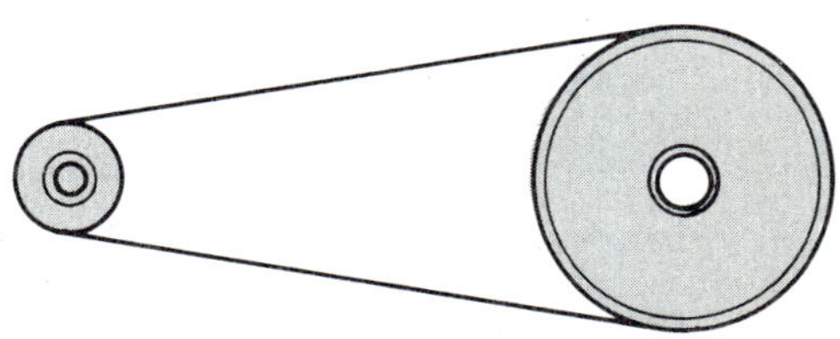

21. A small pulley 60 mm in diameter is connected by a belt to a larger pulley 150 mm in diameter as shown in Fig. 6.70. The small pulley is turning at 120 rpm.

a. Find the angular velocity of the small pulley in radians per second.

b. Find the linear velocity of the rim of the large pulley.

c. What is the linear velocity of the rim of the small pulley? Explain.

d. Find the angular velocity of the large pulley in radians per second.

e. How many rpm is the large pulley turning?

22. The wheels of a bicycle have a 26-in. diameter. When the bicycle is being ridden so that the wheels rotate at 12 rpm, how far will the bike travel in 1 min?

Chapter summary

Two common measures of angles are radians and degrees.

Definition 6.1 One **radian** is the measure of the central angle of a circle that subtends an arc equal in length to the radius of the circle.

The length of the arc s may be determined by the formula $s = r\theta$ (where θ is in radians).

Important formulas relating degrees and radians are:

$$2\pi \text{ rad} = 360°, \qquad 1 = \frac{180°}{\pi}, \qquad \frac{\pi}{180°} = 1.$$

To change from radians to degrees we multiply by the conversion factor $180°/\pi$. To change from degrees to radians we multiply by the conversion factor $\pi/180°$.

A measure given in radians will have no symbol after the measure. For example, $2\pi/3$ is a measure in radians. A measure in degrees will have the degree symbol, °, or the word degree after the measure. For example we write 3° or 3 degrees.

When converting from degrees to radians in exact form, the answer will contain the symbol π.

The six trigonometric functions of an angle can be determined by knowing one point on the terminal side of the angle. The names, abbreviations, and defining ratios are given in Table 6.1. From the defined ratios we can determine other relations between the functions. Some of these are

$$\csc\theta = \frac{1}{\sin\theta}, \qquad \sin\theta \neq 0$$

$$\sec\theta = \frac{1}{\cos\theta}, \qquad \cos\theta \neq 0$$

$$\cot\theta = \frac{1}{\tan\theta}, \qquad \tan\theta \neq 0.$$

In Section 6.3 we learned how to find exact values for the trigonometric functions for certain angles. These are given in Table 6.2. We also learned how

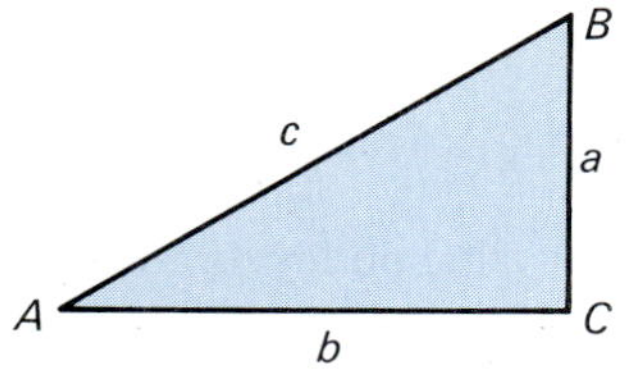

FIGURE 6.71

to find the value of any trigonometric function using the calculator. The degree of accuracy of trigonometric functions is dependent on the accuracy of the angle measured (Table 6.3). The technique of finding the angle given the functional value, using the calculator, was discussed. In general we find the inverse of the functional value on the calculator. For example if $\cos \theta = 0.8526401$, to find θ we do the following: 0.8526401 [INV] [COS] and the display shows $\theta = 31.5°$.

In Section 6.4 we learned how trigonometric ratios can be used to solve right-triangle problems. To solve the right-triangle problems, sketch the triangle and label the figure properly (Fig. 6.71). Two terms that are helpful in solving right-triangle problems are **angle of elevation** and **angle of depression.** The angle of elevation is the angle between the horizontal and the line of sight to an object above the horizontal. The angle of depression is the angle between the horizontal and the line of sight to an object below the horizontal.

The **reference triangle** and **reference angle** $\hat{\theta}$ are helpful tools in discussing angles that are not in the first quadrant. Using the reference triangle we can determine the sign of all the trigonometric functions for each quadrant. The signs (+ or −) of all the functions are given in Table 6.4.

By using the reference triangle and the reference angle we can (1) find the value of any trigonometric function using trigonometric tables, and (2) find the angle or angles given any functional value.

In general to determine the angle θ given a functional value and knowing the quadrant and the reference angle $\hat{\theta}$, we can use the following rules:

First-quadrant angle $\theta = \hat{\theta}$
Second-quadrant angle $\theta = 180° - \hat{\theta}$
Third-quadrant angle $\theta = 180° + \hat{\theta}$
Fourth-quadrant angle $\theta = 360° - \hat{\theta}$.

Section 6.7 deals with applications of trigonometric functions.

Review Exercises

In Exercises 1–5, change the degree measures to radian measures. In each case give the answer in exact form and in approximate form to the nearest thousandth.

1. 70° **2.** 160° **3.** 220° **4.** 310° **5.** −80°

In Exercises 6–10, change the radian measures to degree measures. If the answer is not exact, express it to the nearest thousandth.

6. $\frac{\pi}{12}$ **7.** $\frac{11\pi}{12}$ **8.** 1.45 **9.** 2 **10.** $-\frac{\pi}{4}$

In Exercises 11–15, find the six trigonometric functions of θ if the terminal side passes through the given point. Give the answer in exact form.

11. $(5, 7)$ **12.** $(-3, 2)$ **13.** $(-\sqrt{2}, -\sqrt{2})$
14. $(4, -3)$ **15.** (π, π)

In Exercises 16–20, use the given trigonometric function to find the indicated trigonometric functions. Give each answer in exact form.

16. $\sin \theta = 4/5$; find $\cot \theta$ and $\cos \theta$.

17. $\tan \theta = -\frac{\sqrt{3}}{2}$; find $\cos \theta$ and $\csc \theta$.

18. $\cos \theta = 0.5$; find $\sin \theta$ and $\tan \theta$.

19. $\cot \theta = 1/7$; find $\csc \theta$ and $\cos \theta$.

20. $\csc \theta = \frac{\sqrt{5}}{-2}$; find $\cos \theta$ and $\cot \theta$.

In Exercises 21–30, compute the functional value to five decimal places.

21. sin 18.6° **22.** cos 96.4° **23.** tan 1.72
24. csc 113° **25.** sec 195.4° **26.** cot 286°
27. cos 3.45 **28.** tan 142° **29.** sin 2.57
30. cos 194.2°

In Exercises 31–40, find the angle measure in degrees to the nearest tenth and in radian measure to the nearest hundredth. ($0 \le \theta < 360°$, $0 \le \theta < 2\pi$).

31. $\sin\theta = 0.23456$ **32.** $\cos\theta = -0.39875$

33. $\tan\theta = 0.24008$ **34.** $\csc\theta = -4.93707$

35. $\sec\theta = -3.86370$ **36.** $\cot\theta = -1.45501$

37. $\cos\theta = 0.76604$ **38.** $\sin\theta = -0.90631$

39. $\sin\theta = -1.00000$ **40.** $\tan\theta = 1.00000$

In Exercises 41–45, solve each right triangle *ABC*. Find the angles to the nearest tenth of a degree and the length of sides to three significant digits.

41. $A = 38.5°$, $c = 19.4$ mm

42. $A = 48.7°$, $b = 3.92$ in.

43. $B = 24.7°$, $c = 24.2$ in.

44. $B = 68.9°$, $a = 3.42$ mm

45. $A = 0.75$, $a = 14.5$ m

46. The most efficient operating angle for a conveyor used to elevate crushed ore in a smelter is 30°. If the ore is to be elevated 215 ft, what length conveyor is needed?

47. A surveyor at point P is trying to determine his position between two buildings of known height (Fig. 6.72). **a.** Determine his distance from building A. **b.** What is the distance between the two buildings?

FIGURE 6.72

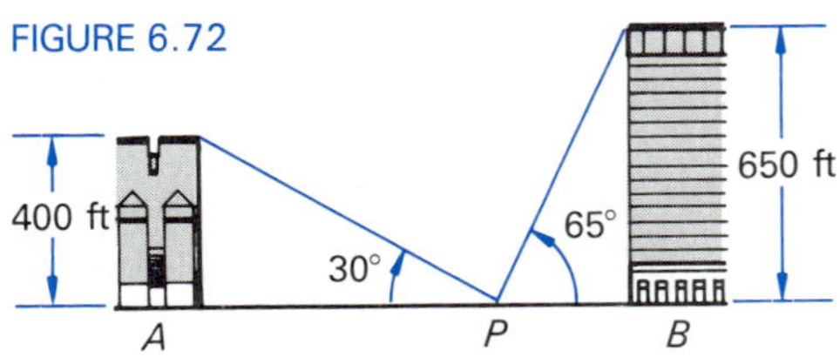

48. Find the length of water pipe needed to run from point P to point Q in Fig. 6.73.

FIGURE 6.73

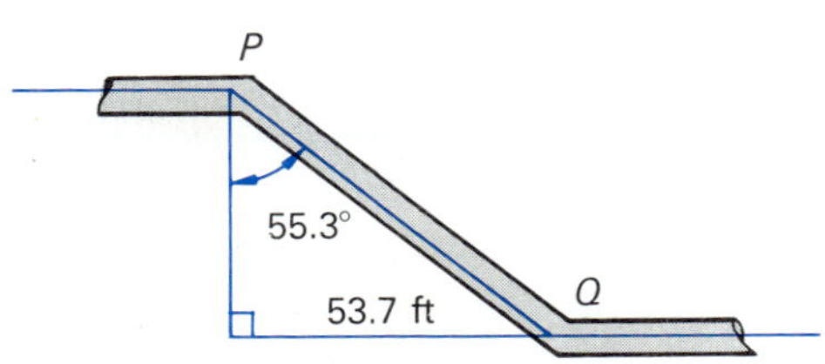

49. With respect to the harbor master's observation post, the resultant velocity v of a ship (in knots) is given by $v = \sqrt{67 - 28\cos\theta}$. For $v = 9$, determine θ.

50. The large hand on a wall clock is 19 in. long.

a. Through what angle (in both radians and degrees) does the hand move in 20 min?

b. How many inches does its tip move in 20 min?

51. Suppose a bike wheel has a radius of 13 in. If the bike is rolling at 10 mph, through what angle does a spoke turn in 1 min?

52. If the bike wheel in Exercise 51 made 200 revolutions in a minute, how fast was the bike traveling?

53. One leg of a right triangle is 92 mm long. The tangent of the angle between the other leg and the hypotenuse is 1.56. How long is the other leg?

54. The angle of depression of a car from a balloon is 67.5°. If the balloon is 3000 ft from the car, how high is the balloon?

55. The angle of elevation to the top of a tower from the tip of its shadow (along level ground) is $\pi/6$. If the tower is 60 ft tall, how long is the shadow?

56. The angle of elevation to a mountain peak from point A is 65°. The angle of elevation to the same mountain peak from point B, 100 ft farther away from the peak, is 61°. How high is the mountain, to the nearest foot?

57. The distance a spring is displaced during simple harmonic motion is given by $d = 8\cos(100\,t)°$. Find the displacement, in centimeters, if:

a. $t = 0$ s **b.** $t = 1.3$ s **c.** $t = 3.5$ s

58. In a certain circuit containing a capacitance the voltage $V = 45\sin(\omega t - \pi/2)$. Find V when

a. $\omega t = 3\dfrac{\pi}{4}$ **b.** $\omega t = 0$.

59. The cross-section of the rivet in Fig. 6.74 is in the shape of a segment of a circle. The area of a circular segment is given by $A = (r^2/2)(\theta - \sin\theta)$, where θ is in radians. Find A for the bolt head if $r = 2$ cm and $\theta = 115°$.

FIGURE 6.74

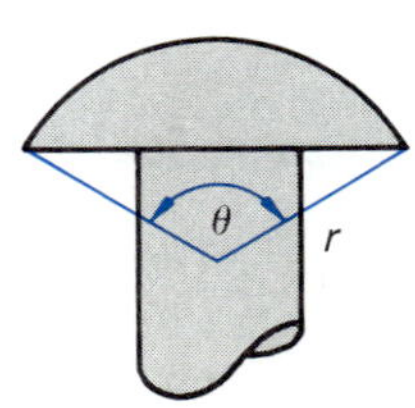

60. Three concentric circles have radii of 2.0, 3.5, and 4.0 m. What is the area of the region between the inner and the middle circles bounded by radii forming a central angle of 78.0°?

Can you solve these?

1. A formula for the area A of a parallelogram with sides a and b and included angle θ is $A = ab \sin \theta$. Find the area of a piece of plastic, where $a = 1.5$ m, $b = 0.90$ m, and $\theta = 118°$.

2. Demonstrate why the formula $A = ab \sin \theta$ will give the area of a parallelogram.

3. One of the formulas used to calculate the value of $\cos x$ comes from calculus:

$$\cos x = 1 - \frac{x^2}{2} + \frac{x^4}{2 \cdot 3 \cdot 4} - \frac{x^6}{2 \cdot 3 \cdot 4 \cdot 5 \cdot 6} + \cdots .$$

The formula continues indefinitely, increasing in accuracy. The angle x must be in radians. Using your calculator, check the formula results with the four terms given against the calculator value for $\cos x$ for:

a. $x = \pi/3$ **b.** $x = \pi/4$.

4. The area of a segment of a circle can be found using the formula $A = \frac{1}{2} r^2(\theta - \sin \theta)$. A segment of a circle is the region bounded by a chord and its related arc (Fig. 6.75). Use this formula to find the number of gallons of oil in a tank of radius 1.5 m and length 3 m. The depth of oil in the tank is 1.2 m (Fig. 6.76).

5. If a wheel with a 100-mm radius must have five holes equally spaced at a distance of 40 mm from the center, find the x and y coordinates of each hole.

6. A triangular piece of property has perimeter p of 900 ft (Fig. 6.77). (That is $a + b + c = 900$ ft.)

a. Find a, b, and c.

b. Show that

$$a = \frac{900}{1 + \dfrac{1}{\tan 35°} + \dfrac{1}{\sin 35°}}.$$

c. Show that the area

$$A = \frac{1}{2}\left(\frac{900}{1 + \dfrac{1}{\tan 35°} + \dfrac{1}{\sin 35°}}\right)^2 \left(\frac{1}{\tan 35°}\right).$$

FIGURE 6.75

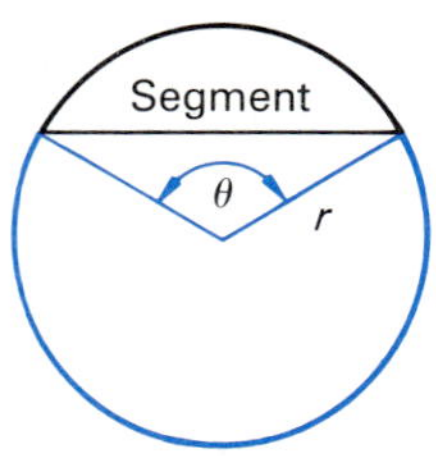

FIGURE 6.76

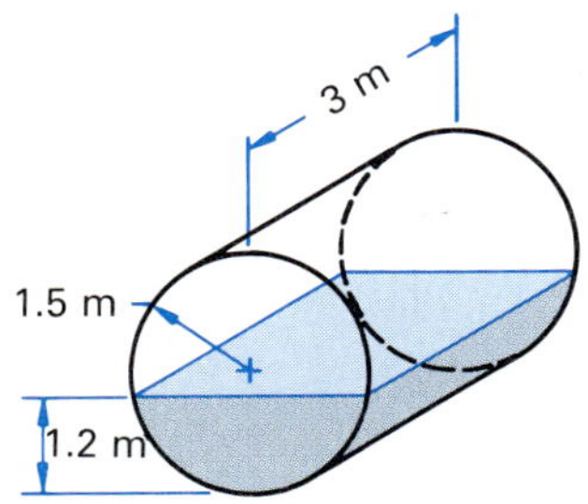

FIGURE 6.77

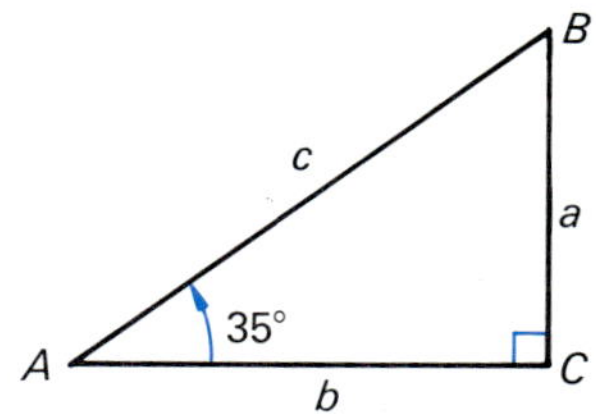

7

A car is stuck in the mud up to its axles. A cable attached to the car and to a tree 100 ft away is stretched tight. A person between the car and the tree pulls the cable 2 degrees out of the line with a force of 50 lb. How much tension is placed on the cable?

Vectors and Oblique Triangles

7.1 Introduction to vectors

In studying quantities in physics and technical areas you will soon discover that one kind of quantity can be described by means of a number. The number of screws in a bracket, the length of a bridge, the height of a building, and the value of a trigonometric ratio are examples of this kind of quantity. Most quantities that you have encountered are of this type. Quantities that are undirected and can be fully described by a number, are called **scalars** or **scalar quantities.**

There is another type of quantity that possesses not only magnitude, but also direction. In other words, these quantities are directed magnitudes. A force acting in a specified direction is an example of this second kind.

DEFINITION 7.1

Quantities having both direction and magnitude are called **vector quantities.**

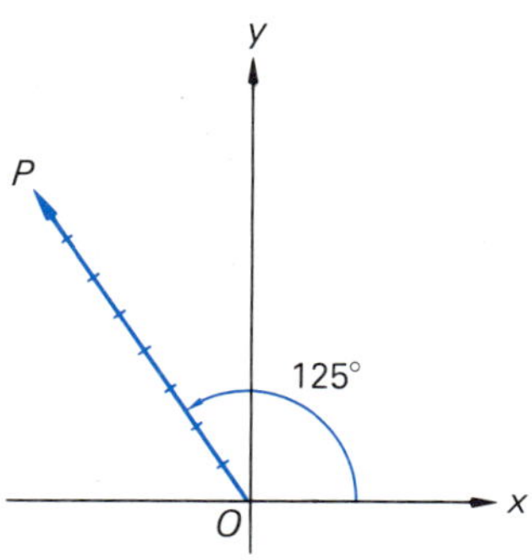

FIGURE 7.1

A vector quantity will be represented in the text in bold face type **V**. In the case of the symbol **OP** the letter *O* indicates the initial point of the vector and the letter *P* indicates the terminal point.

A vector quantity is represented in a diagram by means of an arrow whose length (to a suitable scale) indicates the **magnitude** or size of the vector quantity, and whose head indicates the **direction** of the vector. The vector **OP** has a magnitude of 80 lb, which will be indicated by the symbol $|\mathbf{OP}| = 80$ lb or $OP = 80$ lb. The direction of the vector is $125°$ with respect to the positive x-axis (Fig. 7.1).

Vectors may be drawn from any point in the plane. However for calculation purposes it is most common to draw them in standard position, that is, with the initial point at the origin of the coordinate

FIGURE 7.2

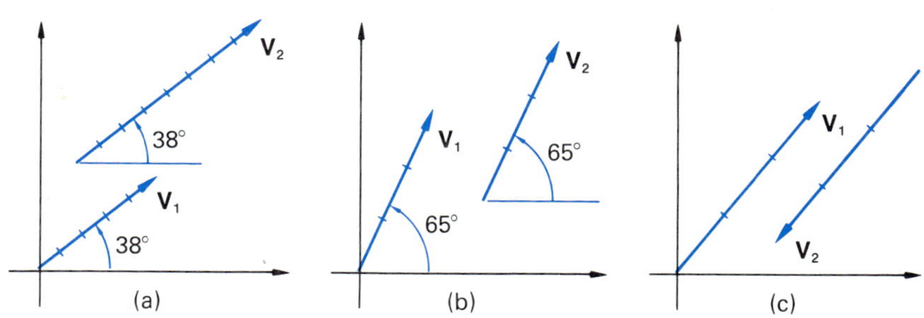

system. When a vector is in standard position it can be identified by its length and the angle it makes with the positive x-axis. In this form it is easy to use the definitions of the trigonometric functions developed in Chapter 6. We will use the following notation to indicate a vector.

$\mathbf{V} = |\mathbf{V}| \underline{/\theta}$, where $|\mathbf{V}|$ represents the magnitude and θ the direction of the vector. (In Fig. 7.1 **OP** = 80 lb $\underline{/125^\circ}$.)

DEFINITION 7.2

a. Two vectors $\mathbf{V}_1$ and $\mathbf{V}_2$ are **parallel** if $\mathbf{V}_1$ and $\mathbf{V}_2$ have the same direction (Fig. 7.2a).

b. Two vectors $\mathbf{V}_1$ and $\mathbf{V}_2$ are **equal** if the vectors have the same direction and magnitude (Fig. 7.2b).

c. $\mathbf{V}_1 = -\mathbf{V}_2$ if and only if $|\mathbf{V}_1| = |\mathbf{V}_2|$ and $\mathbf{V}_1$ and $\mathbf{V}_2$ are in opposite directions (Fig. 7.2c).

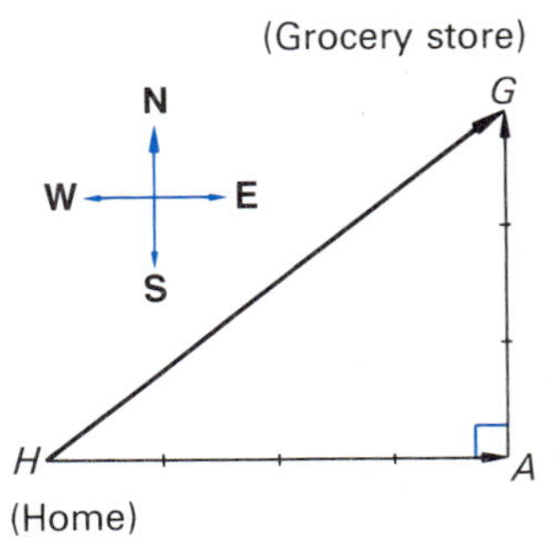

FIGURE 7.3

For example, you ride your bicycle to the grocery store. To accomplish this you travel four blocks east and then three blocks north. This is illustrated with two vectors **HA** and **AG** (Fig. 7.3). A third vector can be drawn from the starting point to the end point (from H to G). This vector represents the resultant displacement. We call vector **HG** the resultant displacement vector or simply the **resultant vector** or **resultant.** How far did the two displacements, four blocks east and three blocks north, displace you from the starting point? From Fig. 7.3 it can be seen that the displacement is the hypotenuse of a right triangle. Using the Pythagorean theorem we have

$$\begin{aligned} |\mathbf{HG}| &= \sqrt{|\mathbf{HA}|^2 + |\mathbf{AG}|^2} \\ &= \sqrt{(4)^2 + (3)^2} \\ &= 5 \text{ blocks.} \end{aligned}$$

The magnitude of vector **HG** is 5. In symbols $|\mathbf{HG}| = 5$ or $HG = 5$. From this result we can see that the magnitude of the resultant HG is **not** the sum of the magnitudes of the vectors **HA** and **AG**. We will discuss the sum of the vectors in the next section.

FIGURE 7.4

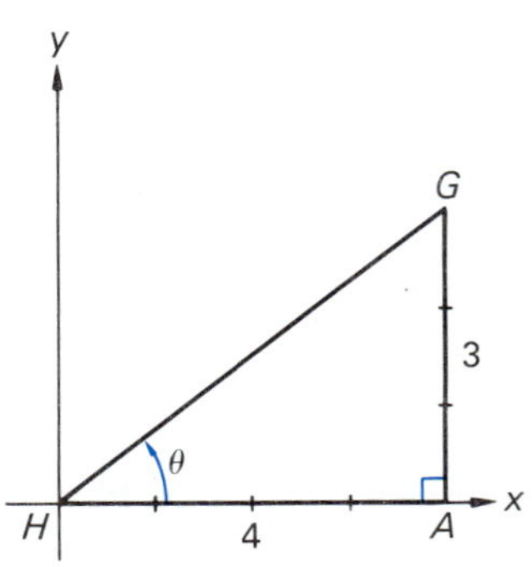

The direction of the resultant can be determined by using right-triangle trigonometry. In many cases it may be helpful to place the figure in standard position with respect to the rectangular coordinate axes as in Fig. 7.4. Therefore $\tan \theta = 3/4$; $\theta = 36.9^\circ$ (nearest tenth). Thus the resultant vector **HG** = 5 blocks $\underline{/36.9^\circ}$.

In the previous example the vector **HG** is called the resultant and the two vectors **HA** and **AG** are called **component vectors** of **HG**. In orienting the figure to the coordinate axes we refer to **HA** as the ***x*-component** and **AG** as the ***y*-component,** since they are in the direction of the x-axis and y-axis, respectively.

EXAMPLE 1 A wagon is being pulled with a force of 35 lb by a person holding the handle at an angle of 40° with the ground (Fig. 7.5). Find the components of the force in both the horizontal and the vertical directions.

Solution: We are given the resultant vector **V**, which has a magnitude of 35. lb and a direction of 40°. Figure 7.6 illustrates this vector **V** in standard position with respect to the x- and y-axes. We are asked to find the component of the force in the x-direction (or x-component) and the component of the force in the y-direction (or y-component). The components can be determined geometrically by constructing perpendiculars to the x- and y-axes from point P (Fig. 7.7). This process determines the component vectors $\mathbf{V}_x$ and $\mathbf{V}_y$. These component vectors are referred to as the **projection** of the vector **V** onto the x- and y-axes.

We can now use right-triangle trigonometry to determine $\mathbf{V}_x$ and $\mathbf{V}_y$.

$$\cos\theta = \frac{|\mathbf{V}_x|}{|\mathbf{V}|} \quad \text{or} \quad |\mathbf{V}_x| = 35.\cos 40°$$

$$|\mathbf{V}_x| = 27.\ \text{lb.}$$

$$\cos(90° - \theta) = \frac{|\mathbf{V}_y|}{|\mathbf{V}|} \quad \text{or} \quad |\mathbf{V}_x| = 35.\cos(90° - 40°)$$

$$|\mathbf{V}_y| = 35.\cos 50°$$
$$|\mathbf{V}_y| = 22.\ \text{lb.}$$

Thus $\mathbf{V}_x = 27.\ \text{lb}\ \underline{/0°}$ and $\mathbf{V}_y = 22.\ \text{lb}\ \underline{/90°}$. Since $\mathbf{V}_x$ and $\mathbf{V}_y$ always lie parallel to the x- and y-axis, respectively, we will indicate these vectors as $\mathbf{V}_x = 27.$ lb and $\mathbf{V}_y = 22.$ lb. The magnitude of $\mathbf{V}_y$ can also be determined by using the sine function, since $\cos(90° - \theta) = \sin\theta$, thus $|\mathbf{V}_y| = |\mathbf{V}|\sin\theta$. □

FIGURE 7.5

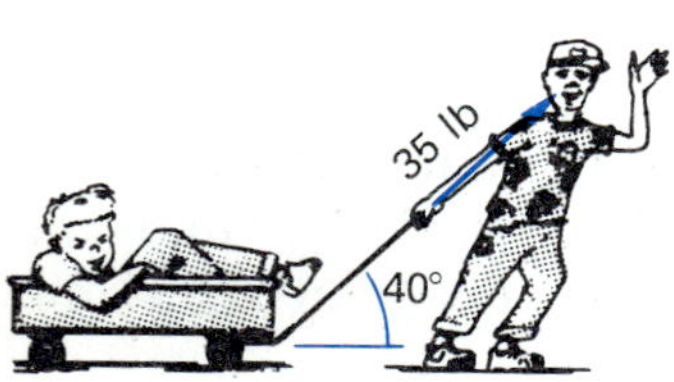

FIGURE 7.6

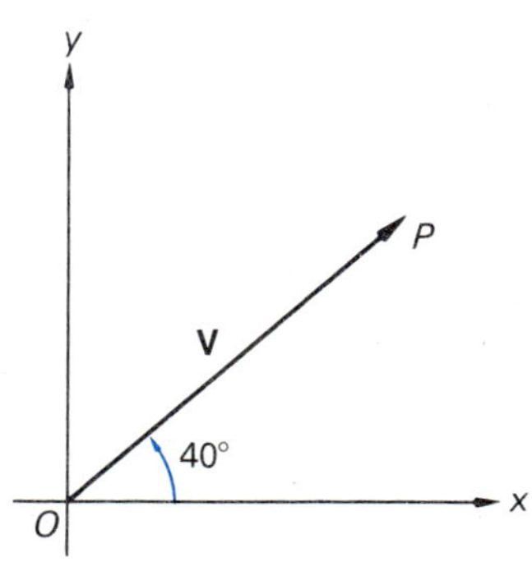

FIGURE 7.7

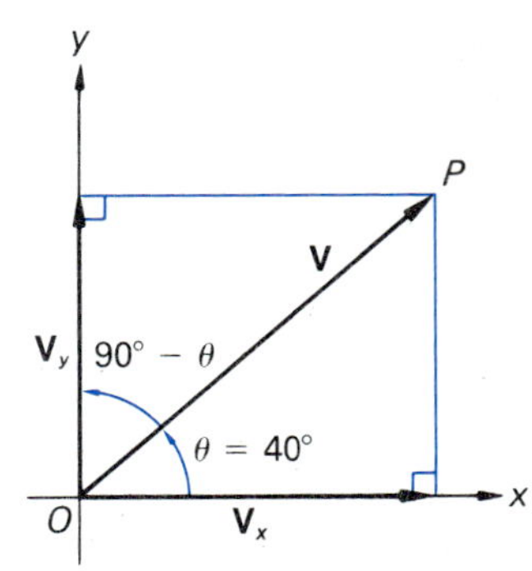

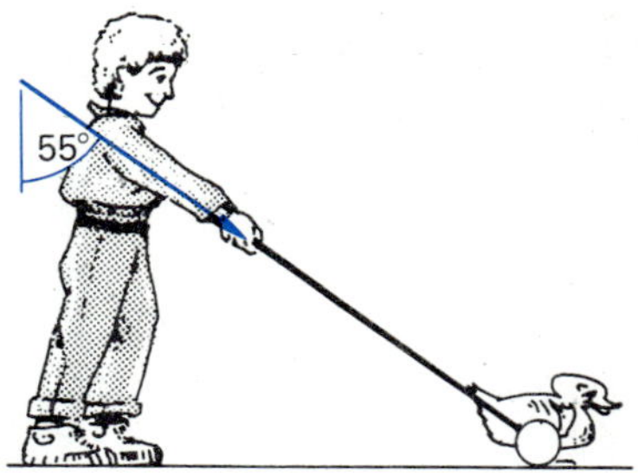

FIGURE 7.8

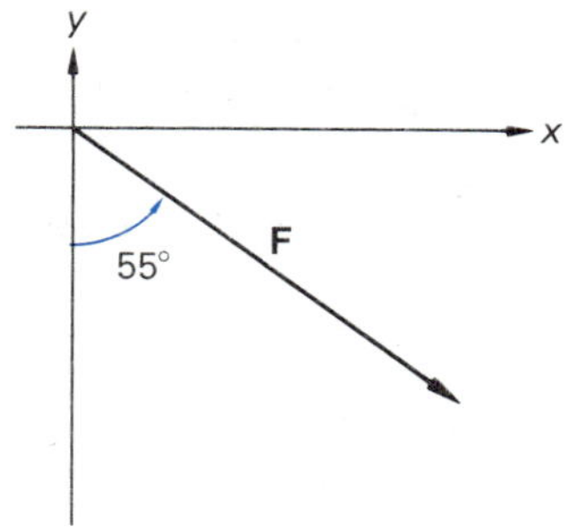

FIGURE 7.9

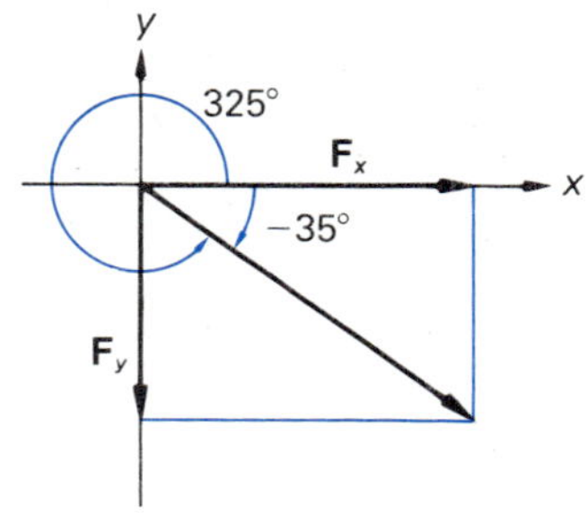

FIGURE 7.10

DEFINITION 7.3

> The magnitude of the horizontal component of a vector **V** of magnitude $|\mathbf{V}|$ acting at angle θ with the horizontal is $|\mathbf{V}_x| = |\mathbf{V}| \cos \theta$ and the magnitude of the vertical component is $|\mathbf{V}_y| = |\mathbf{V}| \sin \theta$.

EXAMPLE 2 A child pushes on the handle of a toy with a force of 8.0 lb at an angle of 55° with his body (Fig. 7.8). How much of his force acts horizontally (to push the toy) and how much acts vertically?

Solution: The resultant vector **F** is illustrated in standard position in Fig. 7.9. (One might incorrectly jump to the conclusion that $|\mathbf{F}_x| = 8.0 \cos 55°$. This is incorrect since when we place the vector in standard position, we must consider the angle in relation to the positive x-axis.) The correct identification of the vector is a force of 8.0 lb at a direction of 325° or −35° (Fig. 7.10). The magnitude of the horizontal component is

$$|\mathbf{F}_x| = 8.0 \cos 325° = 6.6 \text{ lb}.$$

The magnitude of the vertical component is

$$|\mathbf{F}_y| = 8.0 \sin 325° = -4.6 \text{ lb}$$

(The negative sign indicates that this is a downward force.)

Thus $|\mathbf{F}_x| = 6.6$ lb and $|\mathbf{F}_y| = -4.6$ lb. □

FIGURE 7.11

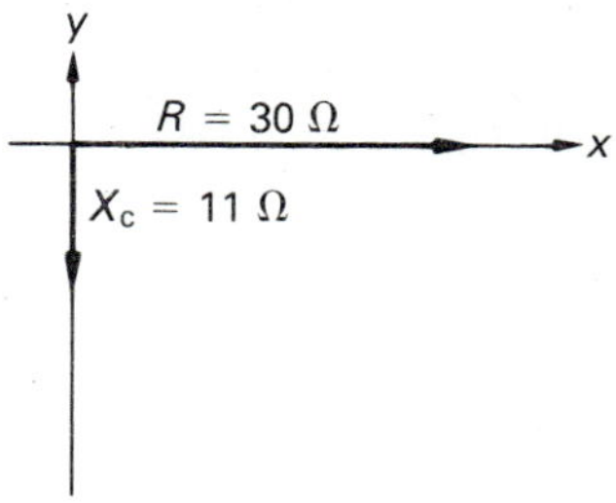

EXAMPLE 3 An AC circuit has a resistance **R** of 30 Ω, acting at 0° and a capacitive reactance $\mathbf{X}_c$ of 11 Ω acting at 270° (Fig. 7.11). These two vectors form the components of a resultant vector **Z**, which is called the impedance of the circuit. What is the impedance **Z**?

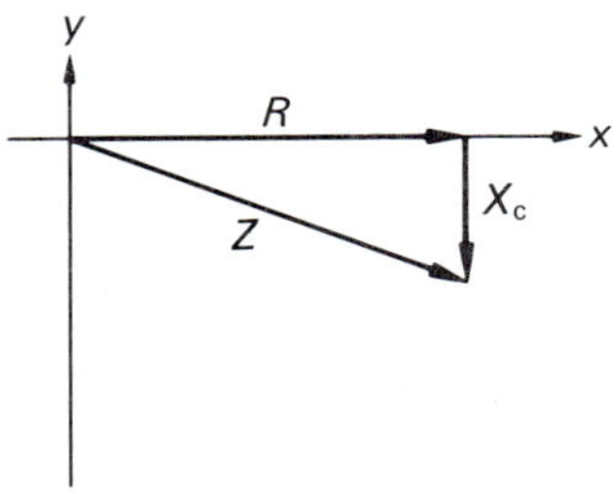

FIGURE 7.12

Solution: The resultant $\mathbf{Z}$ is the hypotenuse of a right triangle. To construct the right triangle we move the vector $\mathbf{X}_c$ so that its tail is against the head of $\mathbf{R}$, and then draw the resultant $\mathbf{Z}$ (Fig. 7.12). Find the magnitude using the Pythagorean theorem:

$$\begin{aligned} |\mathbf{Z}| &= \sqrt{|\mathbf{R}|^2 + |\mathbf{X}_c|^2} \\ &= \sqrt{30^2 + 11^2} \\ &= 32.0\ \Omega. \end{aligned}$$

Find the direction of $\mathbf{Z}$ using the tangent function:

$$\begin{aligned} \tan\theta &= \frac{-11}{30} \\ &= -0.367 \\ \theta &= 20.1^\circ. \end{aligned}$$

From the diagram we know that the direction is a fourth-quadrant angle, thus $\theta = 360^\circ - 20.1^\circ$ or $\theta = 339.9^\circ$. Thus

$$\mathbf{Z} = 32.0\ \Omega \ \underline{/339.9^\circ}. \qquad \square$$

DEFINITION 7.4

Given vector components $|\mathbf{V}_x|$ and $|\mathbf{V}_y|$, the magnitude of the vector is given by

$$|\mathbf{V}| = \sqrt{|\mathbf{V}_x|^2 + |\mathbf{V}_y|^2},$$

and the angle that the vector makes with the positive x-axis is given by

$$\tan\theta = \frac{|\mathbf{V}_y|}{|\mathbf{V}_x|}.$$

7.1 Exercises

In Exercises 1–10, indicate which are scalar quantities and which are vector quantities. Briefly state why you selected scalar or vector quantity.

1. The flow of electricity in a wire from switch to bulb.
2. The weight of a spaceship.
3. The flight of a spaceship in a lunar orbit.
4. The area of a computer chip.
5. The height of the CN tower in Toronto.
6. The force exerted by a spring.
7. The amount of frozen water in a pipe.
8. The number of seconds it takes to read this statement.
9. The force exerted by stretching a spring from point A to point B.
10. The sound from the television to your ear.

In Exercises 11–14, draw the vectors on the same coordinate system, each having the origin as the initial point.

11. $\mathbf{V}_1 = 140 \ \underline{/33^\circ}$

12. $\mathbf{V}_2 = 220 \ \underline{/115^\circ}$

13. $\mathbf{V}_3 = 160 \ \underline{/270^\circ}$

14. $\mathbf{V}_4 = 130 \ \underline{/220^\circ}$

In Exercises 15–18, draw the vectors for Exercises 11–14 using the indicated point as the initial point instead of the origin.

15. $\mathbf{V}_1$, $P(5, 5)$ **16.** $\mathbf{V}_2$, $P(0, 10)$

17. $\mathbf{V}_3$, $P(-5, 0)$ **18.** $\mathbf{V}_4$, $P(5, -5)$

In Exercises 19–30, determine the horizontal and vertical components of the vector quantities. Sketch the vector and its components.

19. $\mathbf{V} = 75$ cm/s $\angle 85^\circ$ **20.** $\mathbf{V} = 120$ cm/s $\angle 130^\circ$

21. $\mathbf{F} = 25$ lb $\angle 1.72$ **22.** $\mathbf{F} = 38$ lb $\angle 72^\circ$

23. $\mathbf{D} = 48$ km $\angle 220^\circ$ **24.** $\mathbf{D} = 148$ mi $\angle 35^\circ$

25. $\mathbf{F} = 145$ dynes $\angle -38^\circ$

26. $\mathbf{F} = 228$ dynes $\angle 138^\circ$

27. $\mathbf{A} = 32.2$ cm/s^2 $\angle 55^\circ$

28. $\mathbf{A} = 32.2$ cm/s^2 $\angle -42.5^\circ$

29. $\mathbf{F} = 0.25$ kg $\angle 185^\circ$

30. $\mathbf{F} = 24.5$ kg $\angle 0.750$

In Exercises 31–40, find the magnitude of the resultant vector and the angle it makes with the positive x-axis. Sketch the resultant and its component vectors.

31. $\mathbf{F}_x = 35$ lb, $\mathbf{F}_y = 47$ lb

32. $\mathbf{F}_x = -42$ lb, $\mathbf{F}_y = 55$ lb

33. $\mathbf{D}_x = 340$ km, $\mathbf{D}_y = -135$ km

34. $\mathbf{D}_x = 85$ mi, $\mathbf{D}_y = 55$ mi

35. $\mathbf{F}_x = -17$ dynes, $\mathbf{F}_y = 34$ dynes

36. $\mathbf{F}_x = -118$ dynes, $\mathbf{F}_y = -75$ dynes

37. $\mathbf{V}_x = -42$ cm/s, $\mathbf{V}_y = -78$ cm/s

38. $\mathbf{V}_x = 110$ ft/s, $\mathbf{V}_y = 115$ ft/s

39. $\mathbf{A}_x = -18$ ft/s^2, $\mathbf{A}_y = 24$ yd/s^2

40. $\mathbf{A}_x = 24$ cm/s^2, $\mathbf{A}_y = -12$ cm/s^2

41. A motorboat heads due east at a rate of 12 mph, and the current is flowing north at 4 mph. Sketch and find the resultant velocity of the boat.

42. Two boys are hauling a wagonload of wood. One boy is pushing with an effective force of 32 lb in a horizontal direction to the right, and the other boy is pulling with an effective force of 50 lb in the same direction. What single force would give the same result?

43. The lift of an airplane wing (Fig. 7.13) is 750 lb. The drag is 300 lb. What is the magnitude and the direction of the resulting force?

44. The force of 800 Newtons is acting on the wing of an airplane at an angle of 25° with the vertical (Fig. 7.14). Find the vertical and horizontal components.

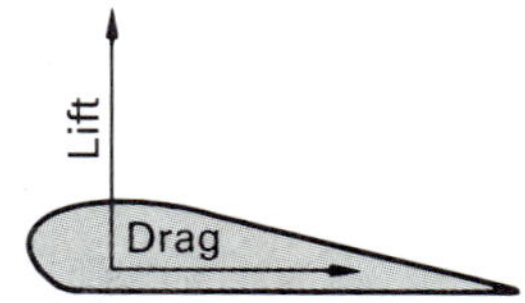

FIGURE 7.13

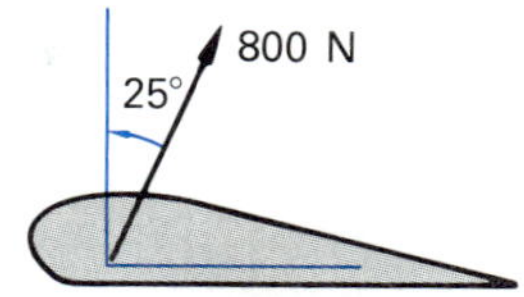

FIGURE 7.14

45. In moving a load of cement, Lisa is exerting a force of 85 lb on the handles of her wheelbarrow. As indicated by **AD** in Fig. 7.15, part of this force is used to lift the legs of the wheelbarrow off the ground; the rest of the force, indicated by **AB**, is used in pushing the wheelbarrow forward. The 85-lb force is applied at an angle of 50° with the horizontal.

  a. Find the force used to lift the legs of the wheelbarrow off the ground.

  b. Find the force used to push the wheelbarrow forward.

46. An AC circuit has two sources in series. Find the total voltage if $V_1 = 15$ V at 0° and $V_2 = 26$ V at 90°.

47. An AC circuit has a resistance $\mathbf{R}$ of 25 Ω acting at 0°, and a capacitive reactance $\mathbf{X}_c$ of 38 Ω acting at 270°. Find the impedance $\mathbf{Z}$ that is the result of these components.

FIGURE 7.15

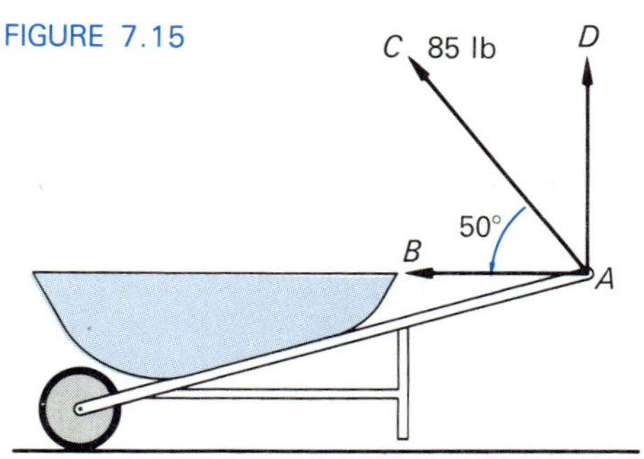

7.2 Vector addition

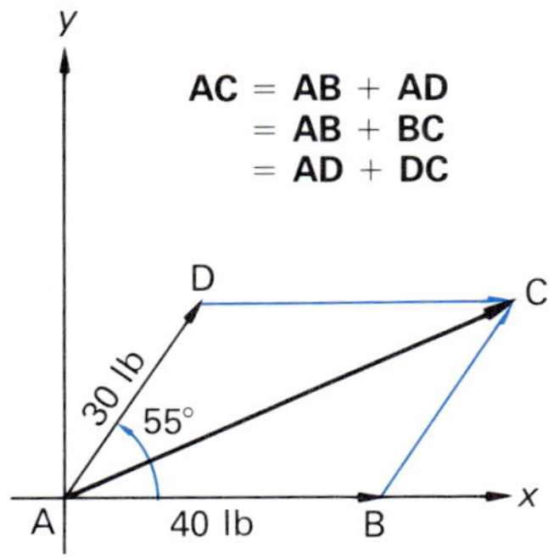

FIGURE 7.16

Sometimes it is desirable to combine two or more vectors into one resultant vector, that is, to find one vector that represents (in magnitude and direction) the one force having the same effect as the combined original forces.

Suppose that two forces, one of 40 lb and the other of 30 lb, act upon point *A*, the first in the direction of 0° and the other 55° (Fig. 7.16). A parallelogram of forces *ABCD* can be constructed. The diagonal of the parallelogram **AC** is the resultant vector. The resultant **AC** can also be obtained by constructing vector **BC** at the end of the vector **AB**, **BC** being equal to **AD**. This completes a triangle of forces *ABC,* and as before **AC** is the resultant. The resultant **AC** is the sum of **AB** and **BC** or **AC** = **AB** + **BC**.

Any number of vectors **A**, **B**, **C**, **D**, **E**, etc., can be added by placing them end to end in any order, the tail of the second at the head of the first, the tail of the third at the head of the second and so on, until all the vectors are used. The vector drawn from the tail of the first to the head of the last will be the resultant vector, indicating the sum of the vectors. Figures 7.17a and 7.17b show how this is done. Vectors $\mathbf{A}_1$, $\mathbf{B}_1$, $\mathbf{C}_1$, $\mathbf{D}_1$, and $\mathbf{E}_1$ are parallel to **A**, **B**, **C**, **D**, and **E**, respectively.

Another way to obtain the resultant of several vectors is to resolve each into its *x*- and *y*-components and then add the *x*-components to get the *x*-component of the resultant, and add the *y*-components to get the *y*-component of the resultant.

Figure 7.18 shows how this can be accomplished. The three vectors **OA** = 2.5 cm $\underline{/15^\circ}$, **OB** = 5.0 cm $\underline{/55^\circ}$, and **OC** = 3.0 cm $\underline{/70^\circ}$ acting at the point *O* are first resolved into their *x*- and *y*-components. The *x*-components of **OA**, **OB**, and **OC** are, respectively, **OD**, **OE**, and **OF**. Their *y*-components are **OG**, **OH**, and **OJ**. Along the *x*-axis lay off **FK** = **OE**, **KL** = **OD**. Therefore **OL** = **OD** + **OE** + **OF**, the sum of the three *x*-components. On a vertical line through *L*, lay off **LM**, **MN**, and **NP** equal, respectively, to **OG**, **OH**, and **OJ**. Thus **LP** equals the sum of the *y*-components. Vector **OP** is the resultant or sum of the three given

FIGURE 7.17

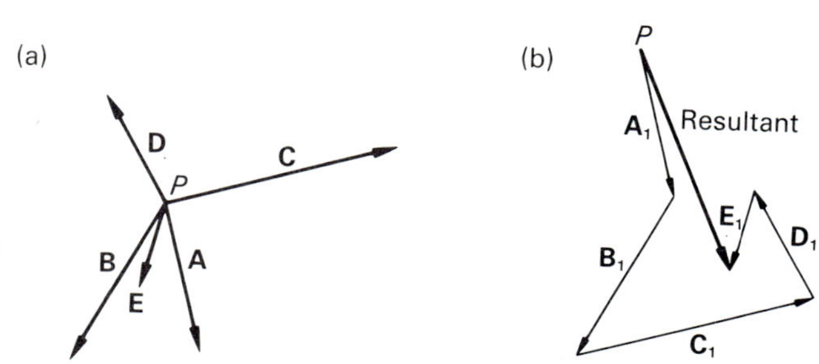

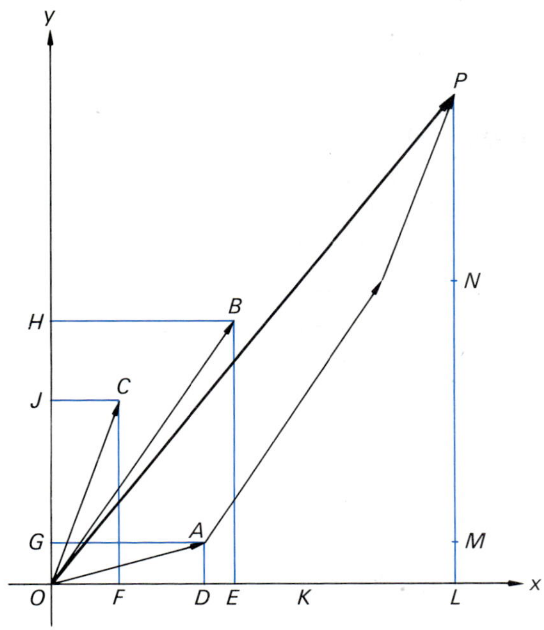

FIGURE 7.18

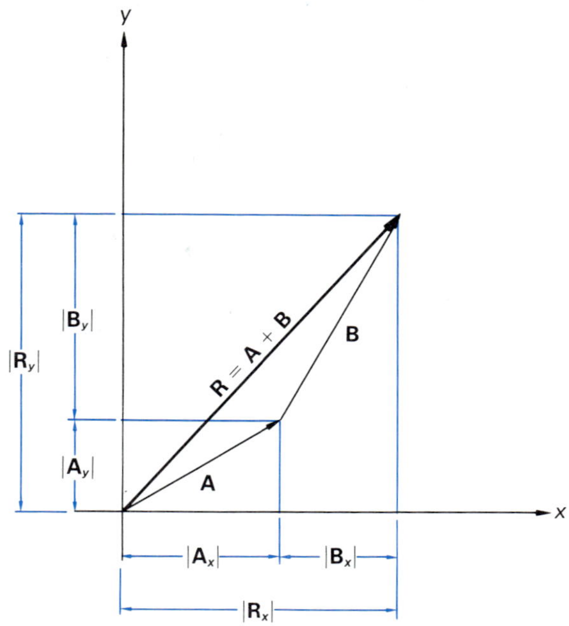

FIGURE 7.19

vectors whose magnitude measures 10 cm and whose angle with the positive x-axis is 50°. This shows that the horizontal and vertical components of the resultant are the sums of the horizontal and vertical components of the individual vectors. That is, $\mathbf{OL} = \mathbf{OD} + \mathbf{OE} + \mathbf{OF}$ and $\mathbf{LP} = \mathbf{OG} + \mathbf{OJ} + \mathbf{OH}$.

In Section 6.1 we learned how to find the horizontal and vertical components of the individual vectors using trigonometric functions. In Example 1 we will illustrate how these ideas can be combined to find the sum of vectors or the resultant vector.

EXAMPLE 1 Given the vectors $\mathbf{A} = 30$ mm $\underline{/30°}$ and $\mathbf{B} = 40$ mm $\underline{/60°}$, find the resultant vector, indicating magnitude and direction.

Solution: Vectors **A** and **B** are sketched in Fig. 7.19. To find the resultant vector **R**, $\mathbf{R} = \mathbf{A} + \mathbf{B}$, we first find the horizontal and the vertical components of the vectors **A** and **B**.

	Horizontal component	Vertical component
$\mathbf{A} = 30$ mm $\underline{/30°}$	$\lvert\mathbf{A}_x\rvert = \lvert\mathbf{A}\rvert \cos\theta$ $= 30 \cos 30°$ $= 26.0.$	$\lvert\mathbf{A}_y\rvert = \lvert\mathbf{A}\rvert \sin\theta$ $= 30 \sin 30°$ $= 15.0.$
$\mathbf{B} = 40$ mm $\underline{/60°}$	$\lvert\mathbf{B}_x\rvert = \lvert\mathbf{B}\rvert \cos\theta$ $= 40 \cos 60°$ $= 20.0.$	$\lvert\mathbf{B}_y\rvert = \lvert\mathbf{B}\rvert \sin\theta$ $= 40 \sin 60°$ $= 34.6.$

The horizontal and vertical components of the resultant vector are the respective sums of the horizontal and vertical components,

$$|\mathbf{A}_x| + |\mathbf{B}_x| = |\mathbf{R}_x| \quad \text{and} \quad |\mathbf{A}_y| + |\mathbf{B}_y| = |\mathbf{R}_y|,$$
$$|\mathbf{A}_x| + |\mathbf{B}_x| = 26.0 + 20.0 \quad \text{and} \quad |\mathbf{A}_y| + |\mathbf{B}_y| = 15.0 + 34.6$$
$$|\mathbf{R}_x| = 46.0 \qquad |\mathbf{R}_y| = 49.6$$

To find the magnitude of the resultant $\mathbf{R} = \mathbf{A} + \mathbf{B}$ we use the Pythagorean theorem, since $|\mathbf{R}_x|$ and $|\mathbf{R}_y|$ are the legs of the right triangle.

$$\begin{aligned} |\mathbf{R}| &= \sqrt{|\mathbf{R}_x|^2 + |\mathbf{R}_y|^2} \\ &= \sqrt{(46.0)^2 + (49.6)^2} \qquad \textbf{Substituting} \\ &= 67.6 \text{ mm.} \end{aligned}$$

To find the direction of the resultant vector $\mathbf{R} = \mathbf{A} + \mathbf{B}$, we use the tangent function.

$$\begin{aligned} \tan\theta &= \frac{|\mathbf{R}_y|}{|\mathbf{R}_x|} \\ &= \frac{49.6}{46.0} \\ &= 1.07826\ \ldots \qquad \textbf{(Calculator result)} \end{aligned}$$

Now 1.07826 . . . [INV] [TAN] 47.2°. Since this value is positive, θ could be a first- or third-quadrant angle. From our drawing (Fig. 7.19) we know that θ is a first-quadrant angle. Thus $\theta = 47.2°$. Therefore, the resultant vector $\mathbf{R} = 68.\text{ mm } \underline{/47°}$. □

EXAMPLE 2 Given the vectors **OA** = 2.5 cm $\underline{/15°}$, **OB** = 5.0 cm $\underline{/55°}$, and **OC** = 3.0 cm $\underline{/70°}$, as in Fig. 7.20, find the resultant vector, indicating the magnitude and direction.

Solution: In order to find the resultant vector, we first find the horizontal and vertical components of the vectors **OA**, **OB**, and **OC**.

	Horizontal component	Vertical component
OA = 2.5 cm $\underline{/15°}$	$\lvert\mathbf{OD}\rvert = \lvert\mathbf{OA}\rvert \cos\theta$ $= 2.5 \cos 15°$ $= 2.41$ cm	$\lvert\mathbf{OG}\rvert = \lvert\mathbf{OA}\rvert \sin\theta$ $= 2.5 \sin 15°$ $= 0.647$
OB = 5.0 cm $\underline{/55°}$	$\lvert\mathbf{OE}\rvert = \lvert\mathbf{OB}\rvert \cos\theta$ $= 5.0 \cos 55°$ $= 2.87$ cm	$\lvert\mathbf{OH}\rvert = \lvert\mathbf{OB}\rvert \sin\theta$ $= 5.0 \sin 55°$ $= 4.10$ cm
OC = 3.0 cm $\underline{/70°}$	$\lvert\mathbf{OF}\rvert = \lvert\mathbf{OC}\rvert \cos\theta$ $= 3.0 \cos 70°$ $= 1.03$ cm	$\lvert\mathbf{OJ}\rvert = \lvert\mathbf{OC}\rvert \sin\theta$ $= 3.0 \sin 70°$ $= 2.82$ cm

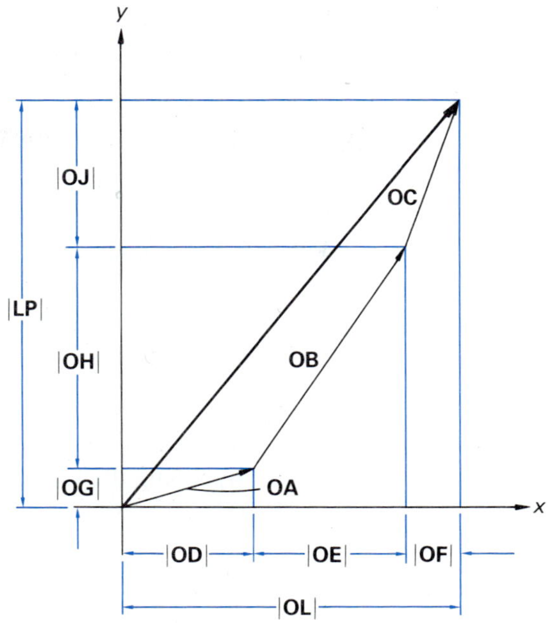

FIGURE 7.20

From our prior discussion we know that the horizontal and vertical components of the resultant vector will be the respective sums of the horizontal and vertical components. Thus

$$\begin{aligned}|\mathbf{OL}| &= |\mathbf{OD}| + |\mathbf{OE}| + |\mathbf{OF}| \\ &= 2.41 \text{ cm} + 2.87 \text{ cm} + 1.03 \text{ cm} \\ &= 6.31 \text{ cm},\end{aligned}$$

and

$$\begin{aligned}|\mathbf{LP}| &= |\mathbf{OG}| + |\mathbf{OH}| + |\mathbf{OJ}| \\ &= 0.647 \text{ cm} + 4.10 \text{ cm} + 2.82 \text{ cm} \\ &= 7.57 \text{ cm}.\end{aligned}$$

To find the magnitude of the resultant, **OP**, we use the Pythagorean theorem, since **OL** and **LP** are legs of a right triangle.

$$\begin{aligned}|\mathbf{OP}| &= \sqrt{|\mathbf{OL}|^2 + |\mathbf{LP}|^2} \\ &= \sqrt{(6.31)^2 + (7.57)^2} \\ &= 9.9 \text{ cm} \qquad \textbf{(two significant digits)}\end{aligned}$$

To find the direction of the resultant vector **OP**, we use the tangent function.

$$\begin{aligned}\tan\theta &= \frac{|\mathbf{LP}|}{|\mathbf{OL}|} \\ &= \frac{7.57}{6.31} \\ &= 1.199683 \qquad \textbf{(calculator reading)}\end{aligned}$$

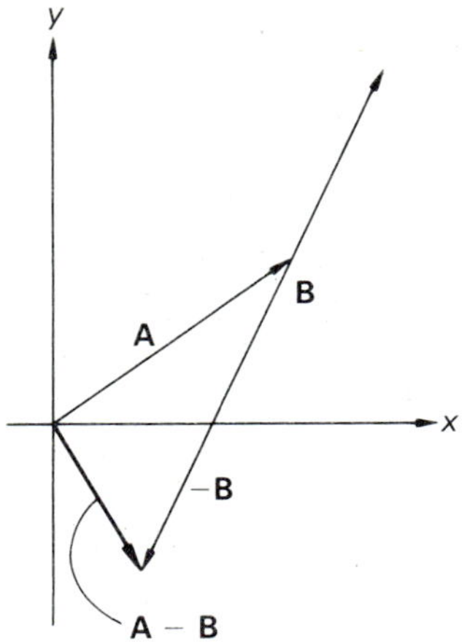

FIGURE 7.21

INV TAN 50.2. Since this value is positive, θ could be a first- or third-quadrant angle. From Fig. 7.20 we know that θ is a first-quadrant angle. Thus, $\theta = 50^\circ$. Therefore the resultant vector

$$\mathbf{OP} = 9.9 \text{ cm } \angle 50^\circ. \quad \square$$

Comparing the outcomes of finding the resultant by construction and using trigonometry we see the results in this case are very close. We will use trigonometry, since graphing is generally not this accurate.

EXAMPLE 3 Given the vectors $\mathbf{A} = 15 \text{ cm } \angle 35^\circ$ and $\mathbf{B} = 18 \text{ cm } \angle 64^\circ$, find $\mathbf{A} - \mathbf{B}$. (See Fig. 7.20.)

Solution: To find the difference of two vectors $\mathbf{A} - \mathbf{B}$ all we need to do is apply the rule for subtraction of signed numbers. That is $\mathbf{A} - \mathbf{B} = \mathbf{A} + (-\mathbf{B})$. Using the same technique as in Example 1, we can find the horizontal and vertical components.

	Horizontal component	Vertical component	
$\mathbf{A} = 15 \text{ cm } \angle 35^\circ$	$\lvert\mathbf{A}_x\rvert = \lvert\mathbf{A}\rvert \cos\theta = 15 \cos 35^\circ$ $= 12.3$	$\lvert\mathbf{A}_y\rvert = \lvert\mathbf{A}\rvert \sin\theta = 15 \sin 35^\circ$ $= 8.60$	Components of **A**
$\mathbf{B} = 18 \text{ cm } \angle 64^\circ$	$\lvert\mathbf{B}_x\rvert = \lvert\mathbf{B}\rvert \cos\theta = 18 \cos 64^\circ$ $= 7.89$	$\lvert\mathbf{B}_y\rvert = \lvert\mathbf{B}\rvert \sin\theta = 18 \sin 64^\circ$ $= 16.2$	Components of **B**
	$\lvert\mathbf{A}_x\rvert + (-\lvert\mathbf{B}_x\rvert) = 12.3 + (-7.89)$ $= 4.4$	$\lvert\mathbf{A}_y\rvert + (-\lvert\mathbf{B}_y\rvert) = 8.6 + (-16.2)$ $= -7.6$	Add **A** and −**B**

$$\tan\theta = \frac{-7.6}{4.4} = -1.73$$
$$\theta = 60.$$

Since the vertical component is negative, the resultant is in the fourth quadrant, thus $\theta = 300^\circ$ (see Fig. 7.21).

$$|\mathbf{A} - \mathbf{B}| = \sqrt{(4.4)^2 + (-7.6)^2} = 8.8$$

Therefore $\mathbf{A} - \mathbf{B} = 8.8 \text{ cm } \angle 300^\circ$. $\square$

7.2 Exercises

In Exercises 1–15, estimate the vector sums and differences by drawing the given vectors.

$\mathbf{A} = 4 \text{ in. } \angle 0^\circ$, $\mathbf{B} = 3 \text{ in. } \angle 90^\circ$, $\mathbf{C} = 5 \text{ in. } \angle 45^\circ$
$\mathbf{D} = 6 \text{ in. } \angle 135^\circ$, $\mathbf{E} = 4 \text{ in. } \angle 180^\circ$

1. **A + B**
2. **A + C**
3. **C + D**
4. **D + E**
5. **A + C + E**
6. **A + D + E**
7. **B + C + E**
8. **C + D + E**
9. **A − B**
10. **B − C**
11. **E − A**
12. **A − D**

13. $\mathbf{A}+\mathbf{B}+\mathbf{C}+\mathbf{E}$

14. $\mathbf{A}+\mathbf{B}+\mathbf{C}+\mathbf{D}$

15. $\mathbf{B}+\mathbf{C}+\mathbf{D}+\mathbf{E}$

16. Show that $\mathbf{A}+\mathbf{A}+\mathbf{A}+\mathbf{A}=4\mathbf{A}$

In Exercises 17–20, use the result of Exercise 16 and the definitions for **A**, **B**, **C**, **D**, and **E** to find the sum or difference by drawing the vectors.

17. $2\mathbf{A}+3\mathbf{B}$

18. $3\mathbf{C}-2\mathbf{D}$

19. $4\mathbf{D}-5\mathbf{E}$

20. $2\mathbf{A}+3\mathbf{E}$

In Exercises 21–38, find the resultant vector using trigonometric functions and the Pythagorean theorem.

21. $\mathbf{R}_x = 3.45$ cm
$\mathbf{R}_y = 2.75$ cm

22. $\mathbf{R}_x = 4.75$ in.
$\mathbf{R}_y = 6.75$ in.

23. $\mathbf{A} = 3.4 \text{ ft } \underline{/180^\circ}$
$\mathbf{B} = 5.4 \text{ ft } \underline{/74^\circ}$

24. $\mathbf{A} = 27.1 \text{ yd } \underline{/125^\circ}$
$\mathbf{B} = 11.4 \text{ yd } \underline{/97^\circ}$

25. $\mathbf{V}_1 = 11 \text{ m } \underline{/185^\circ}$
$\mathbf{V}_2 = 27 \text{ m } \underline{/118^\circ}$

26. $\mathbf{V}_1 = 22 \text{ in. } \underline{/295^\circ}$
$\mathbf{V}_2 = 17 \text{ in. } \underline{/215^\circ}$

27. $\mathbf{A} = 0.71 \text{ in. } \underline{/93.5^\circ}$
$\mathbf{B} = 0.44 \text{ in. } \underline{/192.4^\circ}$

28. $\mathbf{A} = 3.45 \text{ ft } \underline{/138.4^\circ}$
$\mathbf{B} = 8.75 \text{ ft } \underline{/298.8^\circ}$

29. $\mathbf{A} = 17 \text{ cm } \underline{/118^\circ}$
$\mathbf{B} = 24 \text{ cm } \underline{/159^\circ}$

30. $\mathbf{A} = 29 \text{ m } \underline{/185^\circ}$
$\mathbf{B} = 16 \text{ m } \underline{/265^\circ}$

31. $\mathbf{A} = 0.42 \text{ in. } \underline{/17.9^\circ}$
$\mathbf{B} = 0.71 \text{ in. } \underline{/93.5^\circ}$
$\mathbf{C} = 0.44 \text{ in. } \underline{/192.4^\circ}$

32. $\mathbf{A} = 2.75 \text{ ft } \underline{/2.2^\circ}$
$\mathbf{B} = 3.45 \text{ ft } \underline{/138.4^\circ}$
$\mathbf{C} = 8.75 \text{ ft } \underline{/298.8^\circ}$

33. $\mathbf{V}_1 = 4.85 \text{ m } \underline{/0.82}$
$\mathbf{V}_2 = 0.67 \text{ m } \underline{/1.82}$
$\mathbf{V}_3 = 1.73 \text{ m } \underline{/3.01}$

34. $\mathbf{V}_1 = 2.14 \text{ mm } \underline{/6.14}$
$\mathbf{V}_2 = 11.1 \text{ mm } \underline{/3.75}$
$\mathbf{V}_3 = 7.4 \text{ mm } \underline{/0.96}$

35. $\mathbf{V}_1 = 5 \text{ ft } \underline{/24^\circ}$
$\mathbf{V}_2 = 24 \text{ ft } \underline{/137^\circ}$
$\mathbf{V}_3 = 4 \text{ ft } \underline{/203^\circ}$
$\mathbf{V}_4 = 9 \text{ ft } \underline{/313^\circ}$

36. $\mathbf{V}_1 = 13.0 \text{ cm } \underline{/39.4^\circ}$
$\mathbf{V}_2 = 3.0 \text{ cm } \underline{/76.8^\circ}$
$\mathbf{V}_3 = 9.4 \text{ cm } \underline{/214.5^\circ}$
$\mathbf{V}_4 = 18.6 \text{ cm } \underline{/156.4^\circ}$

37. $\mathbf{V}_1 = 11.7 \text{ mm } \underline{/3.2^\circ}$
$\mathbf{V}_2 = 2.4 \text{ mm } \underline{/84.6^\circ}$
$\mathbf{V}_3 = 21.3 \text{ mm } \underline{/173.4^\circ}$
$\mathbf{V}_4 = 9.4 \text{ mm } \underline{/234.5^\circ}$
$\mathbf{V}_5 = 3.5 \text{ mm } \underline{/314.5^\circ}$

38. $\mathbf{V}_1 = 16.75 \text{ in. } \underline{/93.47^\circ}$
$\mathbf{V}_2 = 27.65 \text{ in. } \underline{/187.45^\circ}$
$\mathbf{V}_3 = 3.47 \text{ in. } \underline{/137.08^\circ}$
$\mathbf{V}_4 = 11.59 \text{ in. } \underline{/268.47^\circ}$
$\mathbf{V}_5 = 6.95 \text{ in. } \underline{/347.65^\circ}$

7.3 Applications of vectors

EXAMPLE 1 Find the resultant effect of the three men pulling the car in Fig. 7.22.

Solution: We wish to find the resultant effect of the three forces on the car. To help understand the problem it is best to set up x- and y-axes

FIGURE 7.22

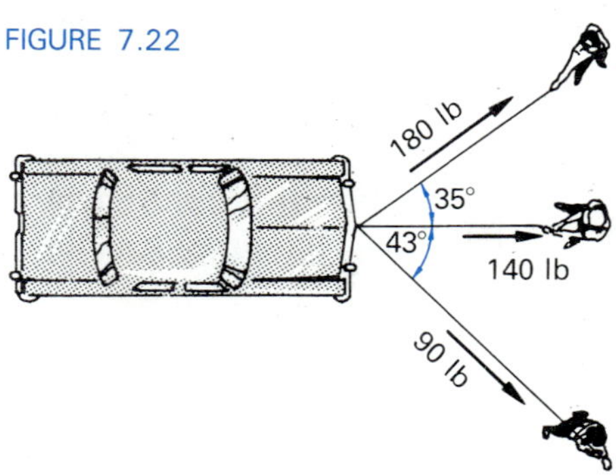

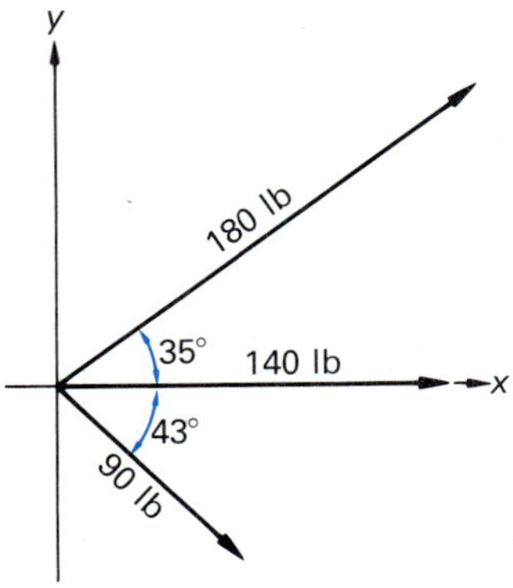

FIGURE 7.23

as shown in Fig. 7.23. We shall find the resultant by adding the x-components together to find the x-component of the resultant and the y-components together to find the y-component of the resultant.

Vector	x-component	y-component
180 lb $\underline{/35°}$	$180 \cos 35° = 147.4$ lb	$180 \sin 35° = 103.2$ lb
140 lb $\underline{/0°}$	$140 \cos 0° = 140.0$ lb	$140 \sin 90° = 0.0$ lb
90 lb $\underline{/-43°}$	$90 \cos(-43°) = 65.8$ lb	$90 \sin(-43°) = -61.4$ lb
Resultant	x-component = 353.2 lb	y-component = 41.8 lb

The results show that the resultant force **R** has x- and y-components of 353.2 and 41.8 lb, respectively. The resultant is shown in Fig. 7.24. To find the magnitude we have

$$|\mathbf{R}| = \sqrt{(353.2)^2 + (41.8)^2}$$
$$= 356. \text{ lb.}$$

The direction of the resultant **R** is found by determining the angle θ.

$$\tan \theta = \frac{41.8}{353.2}$$
$$\theta = 6.75°.$$

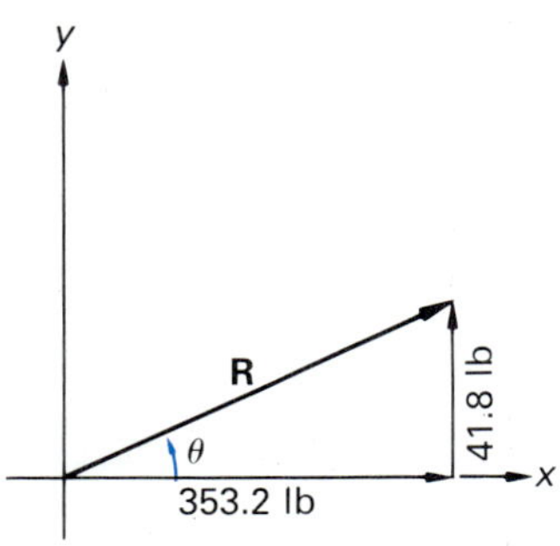

FIGURE 7.24

The result tells us that as a result of the pulling, there is a force of 356. lb in a direction of 6.75° with respect to our positive x-axis. Example 1 is another illustration of how we can use the component method for finding the sum of vectors. □

A physical principle called Newton's first law is frequently used to solve force problems involving bodies in equilibrium. This principle states that the resultant force on a body in equilibrium must be zero. By equilibrium we mean that the body is either initially at rest and remains at rest, or the body is moving with constant velocity, which implies both constant speed and unchanging (straight-line) direction.

Newton's third law states that for any two particles that exert forces on each other, the forces must be equal in magnitude and opposite in direction.

If we combine the statement of the first and third Newtonian laws for the special case of a body that has only two forces acting on it, we can show that for this type of body to be in equilibrium the forces must be equal, opposite, and directed along the line through the two points of application.

FIGURE 7.25

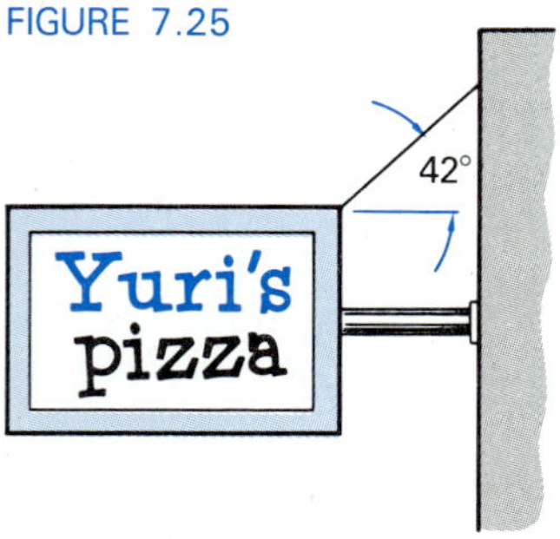

EXAMPLE 2 A 325-lb sign is supported from a wall by a cable and a brace as shown in Fig. 7.25. What are the magnitudes of the forces in the cable and in the brace?

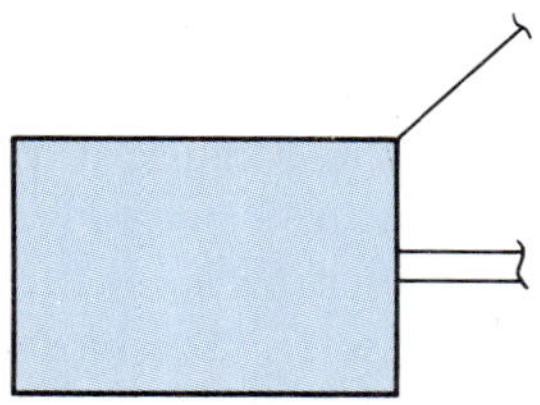

FIGURE 7.26

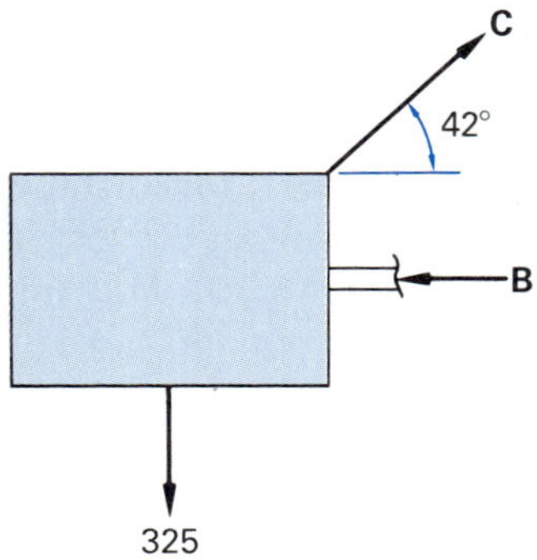

FIGURE 7.27

Solution: To help understand the problem we draw what is called a free-body diagram. It is used in all branches of science and engineering. To form a **free-body diagram** we first isolate the body we are interested in from its surroundings. We essentially cut it away. Then we replace the effects of the surroundings by force vectors. After this we can use vector algebra to determine unknown forces.

In this case we can form a free-body diagram of the sign by cutting the supporting cable and brace as shown in Fig. 7.26. Although we cannot see it, we also assume that we have removed the sign from the effects of the earth's gravitational field, wind forces, magnetic effects, etc.

The remaining step in forming the free-body diagram is to replace the effects of the surroundings by force vectors as shown in Fig. 7.27. Here we have had to make several assumptions. We have assumed that the weight of the cable is negligible compared to the forces generated in it by supporting the sign. This means that the cable has forces applied to it only at the end where it attaches to the wall and at the end where it attaches to the sign. This makes a cable of two-force member and means that according to the combined first and third laws that the directions of the forces on the cable must be along a line through the attachment points and thus along the line of the cable itself. Then by application of the third law we can say that if the force of the sign on the cable is directed along the line of the cable, the force of the cable on the sign must be along the same line in the opposite direction. The physical characteristics of the cable tell us that the cable must be in tension or in other words that the sign must pull on the cable so that the cable must pull on the sign. (Any flexible member can only resist tensile loads. It cannot withstand a compressive force. Have you ever tried to push something with a rope?)

All of the above discussion justifies doing something that may have seemed obvious from the beginning. We can now replace the cable on the sign by a force vector **C** on the free-body diagram. The only part of the vector **C** that is unknown is its magnitude $|\mathbf{C}|$. We know its direction has a reference angle of $42°$.

When we neglect the weight of the brace as being too small to consider when compared to the other forces in it, we see that the brace is also a two-force member. We know then that the force that it puts on the sign must be along the line through the end points of the brace. Since the brace is rigid it can withstand a force of either tension or compression. We cannot make a determination then about whether the direction of the force vector should be away from or toward the sign. From previous experience we will assume that the sign pushes on the brace so that the brace pushes back on the sign and we will replace the effect of the brace on the sign with the force vector **B** on the free-body diagram pointing toward the sign. (Don't worry about deciding whether a force due to an attached member is a push or a pull. Make a guess as to whether the force is a push or a pull and solve the problem

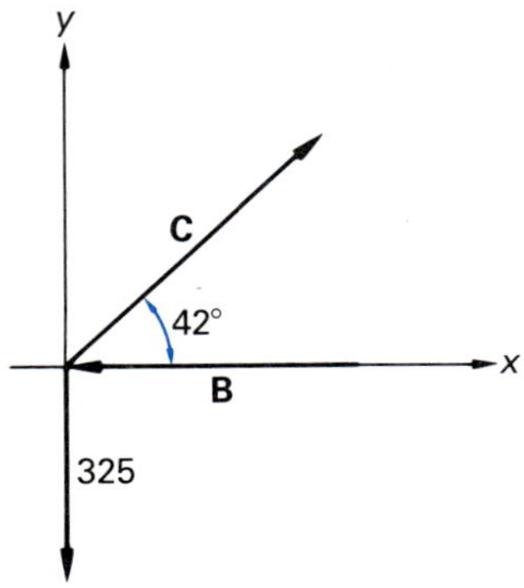

FIGURE 7.28

on that basis. If you guessed correctly the solution for that vector will be found to be positive. If you guessed incorrectly the solution for that vector will have a negative magnitude, which, when you consider the definition of the negative of a vector, is equally valid.)

The remaining 325-lb force on the free-body diagram represents the weight of the sign. When we isolate the sign to draw its free-body diagram we also isolate it from the effects of the earth's gravitational field so the effects of the field (the weight of the sign) must be represented on the free-body diagram of the sign.

Taking all the forces acting on the free-body diagram of the sign and superimposing them on a xy-axis system we have the diagram shown in Fig. 7.28. We have a choice of finding the unknown magnitudes $|\mathbf{C}|$ and $|\mathbf{B}|$ by graphical methods, or by right-triangle trigonometry using components. The latter approach is the one most commonly used.

Since the sign is in equilibrium (it remains at rest) we know that the resultant force acting on it must be zero. It can be shown that if a vector is zero, every one of its components must be zero. We can find the x-component of the resultant force by adding the x-components of all of the individual forces; this sum must be equal to zero for the sign to be in equilibrium. The sum of the y-components must also be zero for the sign to be in equilibrium. Table 7.1 shows the components of all the forces acting on the free-body diagram of the sign.

Adding the x-components and setting the sum equal to zero gives

$$-|\mathbf{B}| + 0.743\,|\mathbf{C}| + 0 = 0.$$

Adding the y-components and setting the sum equal to zero gives

$$0 + 0.669\,|\mathbf{C}| - 325 = 0.$$

From this equation

$$|\mathbf{C}| = \frac{325}{0.669} = 485.8.$$

Substituting this result into the first equation gives

$$|\mathbf{B}| = 0.743\,(485.8) = 360.9.$$

The force of the cable on the sign is 486. lb and the force of the brace on the sign is 361. lb. Also the force of the sign on the cable is 468. lb and the force of the sign on the brace is 361. lb. □

FIGURE 7.29

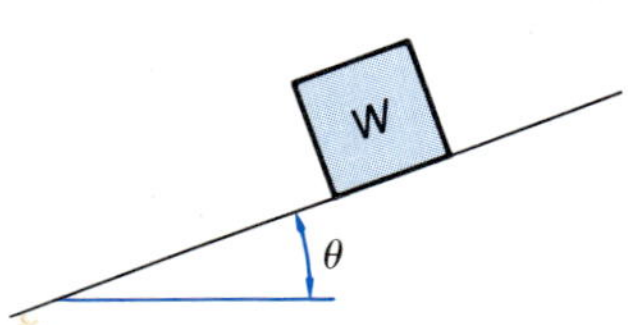

The force that holds an object at rest on an inclined plane or the acceleration with which the object slides on the plane can be found using vector calculations. Figure 7.29 illustrates a typical situation. An object of weight W rests on an incline. The incline makes an angle θ with the horizontal. θ is sometimes called the angle of the incline or the angle of inclination.

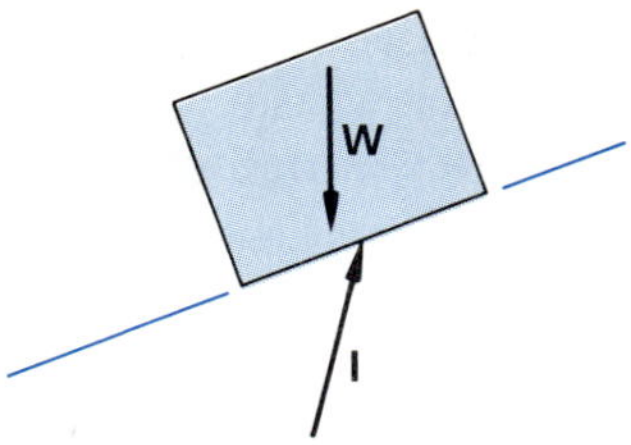

FIGURE 7.30

TABLE 7.1

Force	x-component	y-component
B	$\lvert\mathbf{B}\rvert \cos 180° = -\lvert\mathbf{B}\rvert$	$\lvert\mathbf{B}\rvert \sin 180° = 0$
C	$\lvert\mathbf{C}\rvert \cos 42° = 0.743\,\lvert\mathbf{C}\rvert$	$\lvert\mathbf{C}\rvert \sin 42° = 0.669\,\lvert\mathbf{C}\rvert$
Weight of sign	$325 \cos(-90°) = 0$	$325 \sin(-90°) = -325$

We can analyze the effects on the object by forming a free-body diagram of it as in Fig. 7.30. The weight of the object is shown as the earth's gravitational pull on the object by the vector **W**. The force that the inclined plane puts on the object due to its contact is shown as the vector **I**, which is unknown both in magnitude and direction. To simplify dealing with this type of problem and to more closely match the physical constraints of the situation, the force **I** is commonly resolved into its components **N** perpendicular to the plane and **f** parallel to the plane, as in Fig. 7.31. Since the directions of these components are known we have merely replaced the unknown magnitude and direction of **I** with the two unknown magnitudes of **N** and **f** and have neither gained nor lost regarding the ultimate solution. The force **N** is referred to as the normal force (because it is perpendicular or normal to the plane) and is the force with which the plane supports the object. The force **f** is the friction force of the plane on the object and must always be pointed opposite to the impending motion of the object. If there were other forces on the object which were tending to push it up the plane, the friction force would have to be pointed in the opposite direction.

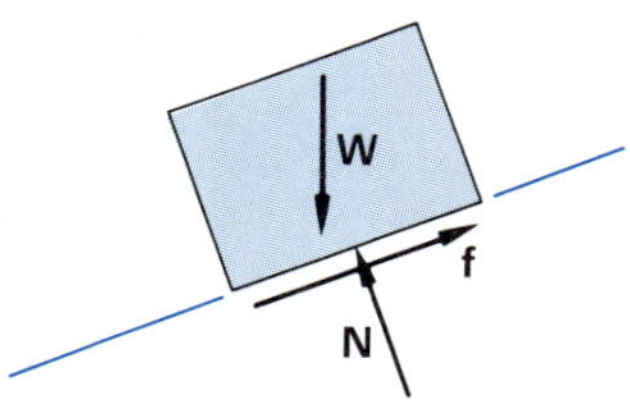

FIGURE 7.31

Since motion of the object is assumed always to be along the surface of the plane, it is convenient to choose an xy-axis system such that the x-axis is parallel to the plane (Fig. 7.32). (Can you show that the two angles labeled θ have the same measure?) Resolving the three forces present into components produces Table 7.2.

If the body has a tendency to start from rest and begin to slide along the plane, or if it changes speed while it is sliding, we cannot say that it is in equilibrium. We are assuming that the body neither lifts off the surface of the incline nor sinks into the surface. This allows us to say that, no matter what motion the body has along the plane (the x

FIGURE 7.32

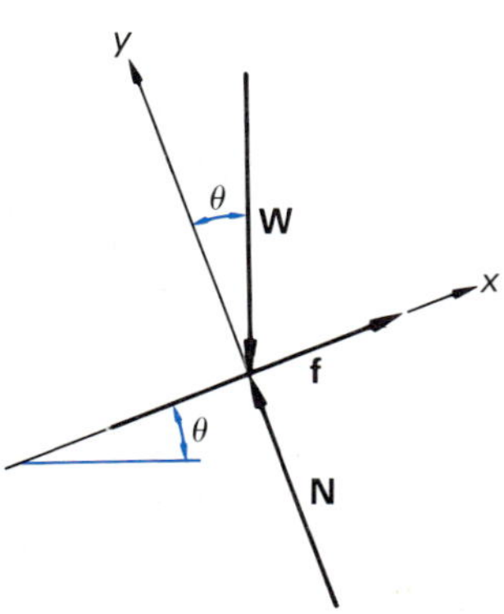

TABLE 7.2

Force	x-component	y-component
W	$\lvert\mathbf{W}\rvert \cos(90° - \theta) = -\lvert\mathbf{W}\rvert \sin\theta$	$\lvert\mathbf{W}\rvert \sin(90° - \theta) = -\lvert\mathbf{W}\rvert \cos\theta$*
f	$\lvert\mathbf{f}\rvert \cos 0° = \lvert\mathbf{f}\rvert$	$\lvert\mathbf{f}\rvert \sin 0° = 0$
N	$\lvert\mathbf{N}\rvert \cos 90° = 0$	$\lvert\mathbf{N}\rvert \sin 90° = \lvert\mathbf{N}\rvert$

* These results can be found by right-triangle trigonometry or by using the sum-of-the-angles formula, which is discussed in Chapter 15.

direction), it is always at rest with respect to motion perpendicular to the plane (the y direction). Therefore for this situation we can make the statement that the sum of the y-components of the forces acting on the object must be zero.

$$-|\mathbf{W}| \cos \theta + 0 + |\mathbf{N}| = 0 \qquad \text{or} \qquad |\mathbf{N}| = |\mathbf{W}| \cos \theta.$$

It has been experimentally determined that the friction force existing between solid objects that are sliding on each other or that are about to slide is proportional to the normal force between the objects. The proportionality constant is the **coefficient of friction** μ (Greek letter mu). Thus $|\mathbf{f}| = \mu |\mathbf{N}|$. (This equation is not valid unless the objects are actually sliding or are just about to slide. It represents a limiting value on the friction force so that $0 \le |\mathbf{f}| \le \mu |\mathbf{N}|$.) If we have the special case where the angle of inclination is such that the object is just on the verge of sliding, but does not actually move, we can use the friction equation and in addition claim equilibrium in the x-direction. Setting the sum of the x-components equal to zero gives

$$-|\mathbf{W}| \sin \theta + |\mathbf{f}| + 0 = 0 \qquad \text{or} \qquad |\mathbf{f}| = |\mathbf{W}| \sin \theta.$$

(Remember that this equation is only valid when the object remains at rest or slides with constant velocity.)

Dividing the x-component equation by the y-component equation gives

$$\frac{|\mathbf{f}|}{|\mathbf{N}|} = \frac{|\mathbf{W}| \sin \theta}{|\mathbf{W}| \cos \theta}$$

$$\frac{|\mathbf{f}|}{|\mathbf{N}|} = \tan \theta,$$

and by rearranging the friction equation

$$\frac{|\mathbf{f}|}{|\mathbf{N}|} = \mu$$

so that for this special case

$$\mu = \tan \theta.$$

This equation is used with an adjustable inclined plane to experimentally determine the coefficient of friction between objects composed of various materials. The object is placed on the plane and the angle of the plane is increased until the object barely begins to slide. The angle of inclination is measured at that time and the value of the coefficient of friction is calculated from the above equation.

FIGURE 7.33

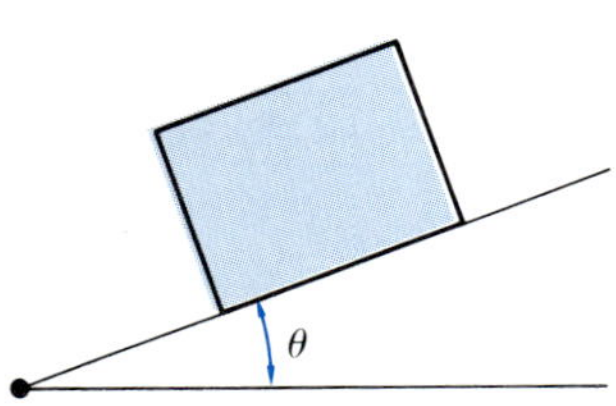

EXAMPLE 3 A block rests on an adjustable plane (Fig. 7.33). The block has a mass of 1 kg and it begins to slip on the plane when the angle of inclination reaches 22°. What is the coefficient of friction between

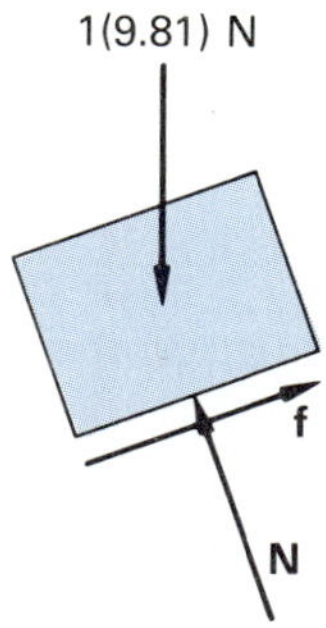

FIGURE 7.34

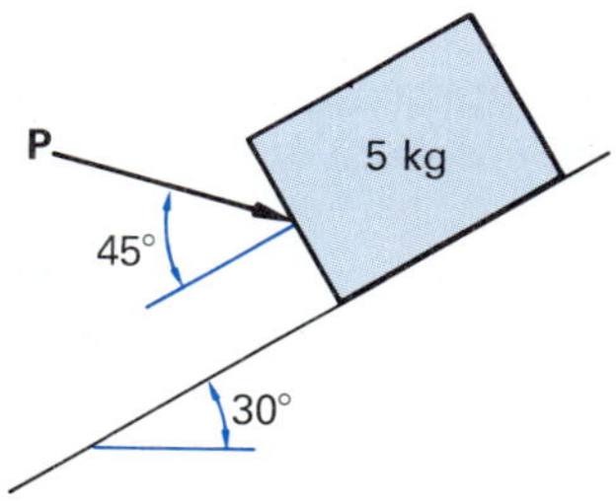

FIGURE 7.35

FIGURE 7.36

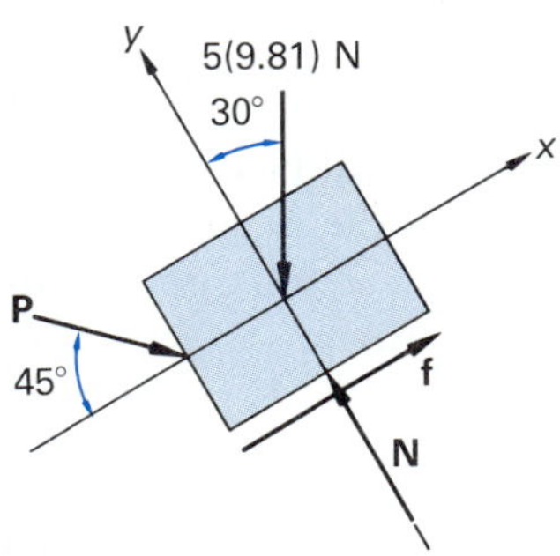

the block and the plane? What is the normal force and the friction force on the block when $\theta = 19°$?

Solution: Since the block begins to slip when $\theta = 22°$ we can use

$$\begin{aligned}\mu &= \tan\theta\\ &= \tan 22°\\ &= 0.404.\end{aligned}$$

Allowing for experimental error we would take $\mu = 0.4$ for this case.

Figure 7.34 shows the free-body diagram for the block. When θ is 19° the angle of the plane is less than the angle where slippage is impending so the block remains at rest and we have equilibrium in both the y-direction and in the x-direction. (We cannot use $|\mathbf{f}| = \mu|\mathbf{N}|$ because we do not have slipping or impending slipping.)

Remember that the kilogram is a mass unit even though the "weights" of bodies are often given in kilograms in the SI system. To convert from mass to weight we need $W = mg$, where g is the local gravitational acceleration; g is 9.81 in the SI system at the earth's surface.

From the equilibrium equations

$$\begin{aligned}|\mathbf{N}| &= |\mathbf{W}|\cos\theta & \qquad |\mathbf{f}| &= |\mathbf{W}|\sin\theta\\ &= 9.81\cos 19° & &= 9.81\sin 19°\\ &= 9.28\text{ N}. & &= 3.19\text{ N}.\end{aligned}$$

Thus the normal force and friction force are 9.28 N and 3.19 N, respectively. □

EXAMPLE 4 A 5-kg block rests on an inclined plane as shown in Fig. 7.35. The coefficient of friction between the block and the plane is 0.3 and a force **P** is applied to the block as shown. If **P** has different magnitudes of 25 N, 5 N, or 100 N will the block move up the plane, down the plane, or remain at rest?

Solution: The free-body diagram for the block is shown in Fig. 7.36. The components of all the forces on the free-body diagram are shown in Table 7.3.

TABLE 7.3

Force	x-component	y-component
WEIGHT	$5(9.81)\cos(-120°) = -24.5$	$5(9.81)\sin(-120°) = -42.5$
f	$\lvert\mathbf{f}\rvert\cos 0° = \lvert\mathbf{f}\rvert$	$\lvert\mathbf{f}\rvert\sin 0° = 0$
N	$\lvert\mathbf{N}\rvert\cos 90° = 0$	$\lvert\mathbf{N}\rvert\sin 90° = \lvert\mathbf{N}\rvert$
P	$\lvert\mathbf{P}\rvert\cos(-45°) = 0.707\lvert\mathbf{P}\rvert$	$\lvert\mathbf{P}\rvert\sin(-45°) = -0.707\lvert\mathbf{P}\rvert$

We know that whether the block is sliding on the plane or not we have equilibrium in the y-direction so the sum of the y-components must be zero.

$$-42.5 + 0 + |\mathbf{N}| - 0.707\,|\mathbf{P}| = 0.$$

For the case where $|\mathbf{P}| = 25$ N, solving the above equation for $|\mathbf{N}|$ gives $|\mathbf{N}| = 60.2$.

If the block is not sliding on the plane, we can use the equation of equilibrium for the x-direction to determine the friction force.

$$-24.5 + |\mathbf{f}| + 0 + 0.707\,|\mathbf{P}| = 0.$$

For the case where $|\mathbf{P}| = 25$ N, solving the above equation for $|\mathbf{f}|$ gives $|\mathbf{f}| = 6.83$.

From the friction equation we know that $0 \leq |\mathbf{f}| \leq \mu\,|\mathbf{N}|$ or that $|\mathbf{f}|$ must be between 0 and $(0.3)(60.2) = 18.1$. Since our value of $|\mathbf{f}| = 6.83$ falls within the acceptable range our assumption that the block is at rest is valid. Therefore, if $|\mathbf{P}| = 25$ N, the block remains at rest on the plane.

If we look at the case where $|\mathbf{P}| = 5$ N, using the same approach gives $|\mathbf{N}| = 46.0$ from y-direction equilibrium, and $|\mathbf{f}| = 21.0$ from x-direction equilibrium.

Checking the limiting values for $|\mathbf{f}|$ we find that $|\mathbf{f}|$ must be between 0 and $(0.3)(46.0) = 13.8$. Since the value of $|\mathbf{f}|$ for the block to remain at rest, 21.0, is greater than the maximum allowable value for $|\mathbf{f}|$, 13.8, our assumption that the block remains at rest must be invalid and the block must slide on the plane. The positive value of $|\mathbf{f}|$ from x-direction equilibrium indicates that the block slides down the plane.

For the case where $|\mathbf{P}| = 100$ N, we know from y-direction equilibrium that $|\mathbf{N}| = 113.$ and from assumed x-direction equilibrium that $|\mathbf{f}| = -46.2$. The friction equation requires that $|\mathbf{f}|$ be between 0 and $(0.3)(113.) = 33.9$. The fact that $|\mathbf{f}|$ for assumed equilibrium is negative shows that the assumption is again invalid and the negative value for $|\mathbf{f}|$ indicates that the block must be sliding in a direction opposite to the friction force on it, or, in other words, up the plane. □

7.3 Exercises

1. Calculate the force (neglecting friction) needed to pull an object up an inclined plane if $W = 875$ lb and

   a. $\theta = 30.0°$ **b.** $\theta = 13.5°$ **c.** $\theta = 43.5°$

2. A canoe leaves shore and travels as follows on a large, still lake: it goes 2.00 mi at 180°, then 0.50 mi at 150°, and finally 1.00 mi at 30° with all angles measured counterclockwise from east. How far must it go if it is to follow a straight-line path back to the starting point?

3. An airplane is taking off from an airport with a velocity of 200 mph at an angle of 30° above horizontal. a) How fast is it rising? b) How fast is it moving parallel to the earth?

4. An airplane is flying due east at a velocity of 375 mph relative to the surrounding air. However, the air itself is part of a strong wind that is blowing straight west with a velocity of 75 mph relative to the ground. Find the velocity of the airplane relative to the ground.

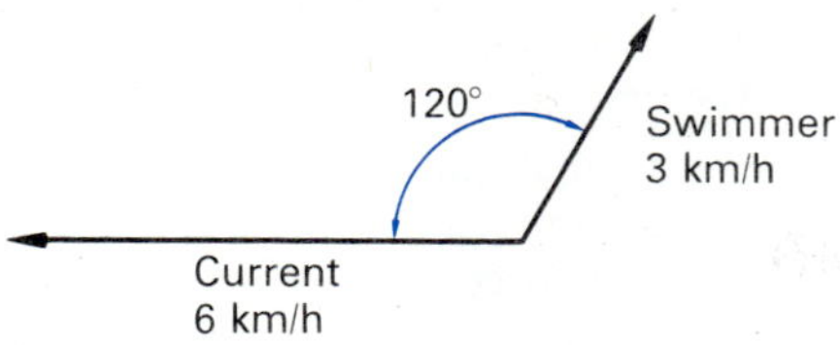

FIGURE 7.37

5. Tod is swimming across the Rhine River, which he knows has a 6 km/h current. To compensate for the current he heads in a direction upstream at an angle of 120° with the current (Fig. 7.37). If he swims through the water at a velocity $v = 3$ km/h, what is the resultant velocity R (the magnitude of the total velocity and its angle with the current)?

6. Teri is swimming across the Genesee River, which she knows has a current of 1.8 km/h. If she swims at the rate of 3.8 km/h, at what angle with respect to the shore must she swim to end up directly across the river from the starting point?

7. A rocket traveling at 1000 km/h ascends at an angle of 80° with the horizontal. What are the horizontal and vertical components of the velocity?

8. Find the resultant effect of three tugboats pulling a larger tanker as in Fig. 7.38.

FIGURE 7.38

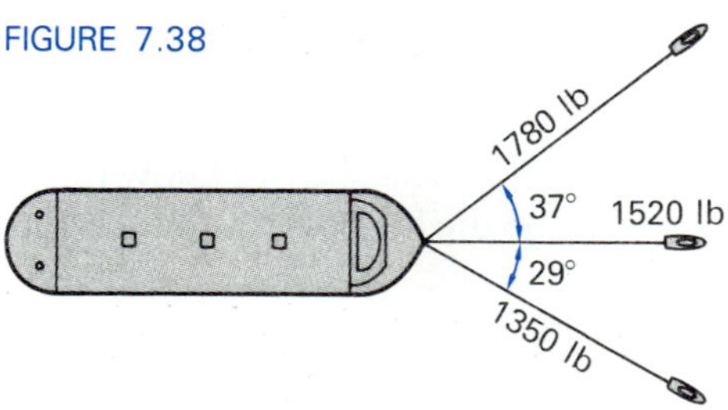

FIGURE 7.39

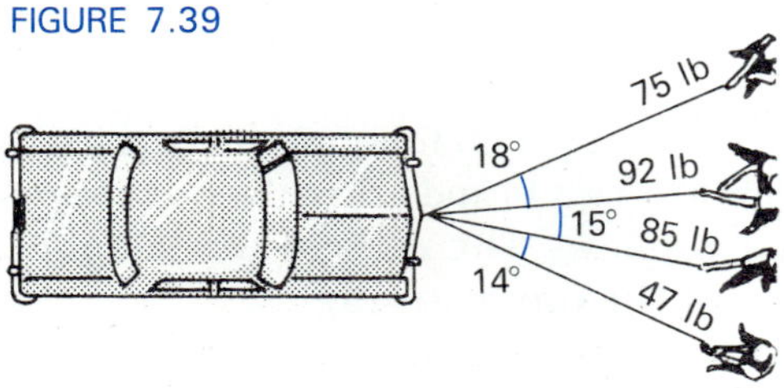

9. Find the resultant effect of four people pulling on the car in Fig. 7.39.

10. A Boston taxi driver (who should be fined for littering) is traveling at 25 km/h and throws a candy wrapper out the window perpendicular to his direction at a speed of 5 km/h. What is the velocity of the candy wrapper as it leaves his hand (the speed and direction with respect to the taxi's direction)?

11. The heading of a boat traveling 15 mph up river makes an angle of 115° with the current, which is 5 mph. What is the boat's resultant velocity (the speed and direction with respect to its heading)?

12. The heading of a boat traveling 25.0 km/h up river makes an angle of 115° with the current, which is 8.3 km/h. What is the boat's resultant velocity (the speed and direction with respect to the river)?

13. A truck weighing 6500 kg must be pulled up a ramp that is inclined 10° with the horizontal. If friction is neglected, what force in newtons must be exerted barely to move the truck?

14. A 60 kg box sits on a 37° incline. Find the normal force that acts on the box. How large a force must friction supply if the box is not to slide?

15. A 50-kg box sits on a 37° incline. Find the normal force that acts on it. How large a force must friction supply if the box is not to slide?

16. A refrigerator rests on an adjustable loading ramp. The refrigerator has a mass of 455 kg and it begins to slip on the plane when the angle of inclination reaches 12°. (a) What is the coefficient of friction between the refrigerator and the plane? (b) What are the normal force and the friction force on the refrigerator when $\theta = 8°$?

17. The movers have a washer that has a mass of 140 kg on the inclined plane as shown in Fig. 7.40 (on p. 250). The coefficient of friction between the washer and the plane is 0.6 and a mover is applying a force **P** at an angle of 35° with the inclined plane. If at different times $|\mathbf{P}|$ is 65 N, 23 N, or 150 N, will the washer move up the plane, down the plane, or remain at rest?

18. Sam's Meat Market is placing a new sign in the front of the shop. The sign has a mass of 205 kg and is supported from the wall by a cable and a brace. The cable makes an angle of 32° with the wall and the brace is perpendicular to the wall.

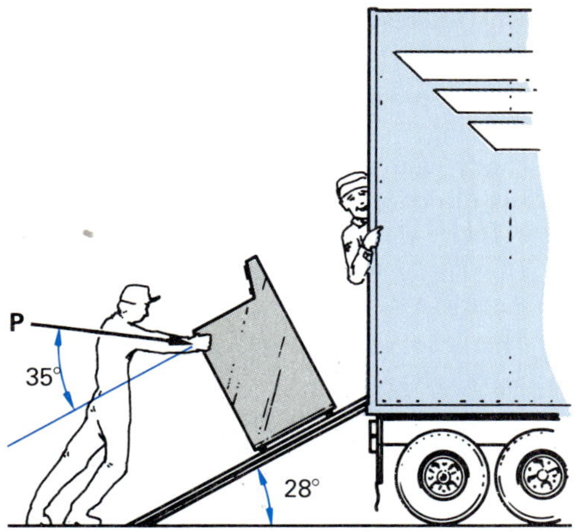

FIGURE 7.40

What are the magnitudes of the forces in the cable and the brace?

19. When a football player pushes a 200-lb blocking dummy on level ground he must exert a force of 80 lb. What force parallel to the slope must he exert to push it up a 5° grade?

20. A 170-lb tightrope walker causes a 6-in. sag in a 50-ft cable, which was initially stretched tight. Assuming the walker is in the center of the cable and that we can neglect the weight of the cable, what is the tension produced in the cable?

21. A 1730-N block attached to a cable is being loaded onto a truck by picking it up with a backhoe and driving it over the truck bed to be lowered. The operator misjudges his height and the block hits the back of the truck and the lifting cable is 15° out of plumb before the backhoe stops. What is the tension in the cable?

7.4 The law of sines

In our previous work with trigonometric ratios, we were able to solve for the unknown parts of a right triangle when only two parts were known and we knew that one angle was a right angle. In this and the next section we will develop formulas to determine the length of the sides and measures of the angles of an **oblique triangle** (no right angles).

In solving an oblique triangle, it is necessary to know three pieces of information, one of which must be the length of a side. The combinations can be:

1. One side and any two angles;
2. Two sides and an angle opposite one of them;
3. Two sides and an included angle;
4. Three sides.

FIGURE 7.41

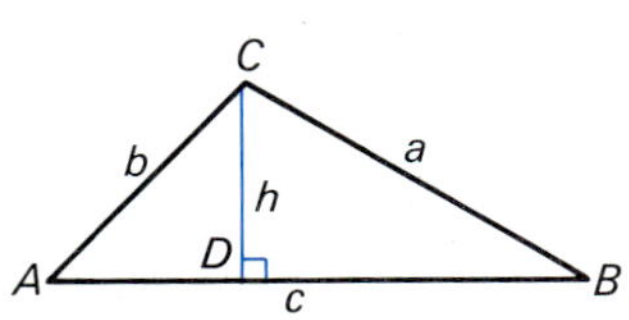

The law of sines, the proof of which follows, can be used to solve oblique triangles for combinations (1) and (2) of the previous list. The Law of Sines may be stated as follows: **The sides of any triangle are in proportion to the sines of the opposite angles.**

For the sake of uniformity we will label side a opposite angle A, side b opposite angle B, and side c opposite angle C, as in Fig. 7.41. By dropping a perpendicular to side c from vertex C we form two right triangles. Label the perpendicular h. For the right triangle ACD:

$$\sin A = \frac{h}{b}$$

and for the right triangle CBD:

$$\sin B = \frac{h}{a}.$$

Solving both of these equations for h, we have

$$h = b \sin A \qquad \text{and} \qquad h = a \sin B.$$

Since both are equal to h we can set them equal to each other:

$$b \sin A = a \sin B.$$

Multiplying both sides by $\frac{1}{\sin A \sin B}$,

$$\frac{a}{\sin A} = \frac{b}{\sin B}.$$

By drawing a perpendicular from B to $\overline{AC}$ and using a similar sequence of steps produces

$$\frac{a}{\sin A} = \frac{c}{\sin C}.$$

Also, by drawing a perpendicular from a to $\overline{BC}$ and using a similar sequence of steps produces

$$\frac{b}{\sin B} = \frac{c}{\sin C}.$$

These ratios may be developed from any triangle. However, in the case of an obtuse triangle it may be necessary to extend a side to form the right triangle. By combining the results of the three proportions we have Theorem 7.1.

THEOREM 7.1

LAW OF SINES

In any triangle ABC

$$\frac{a}{\sin A} = \frac{b}{\sin B} = \frac{c}{\sin C}.$$

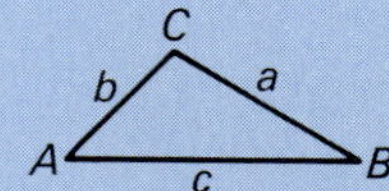

EXAMPLE 1 Given $A = 24°$, $B = 82°$, and $a = 17.$ mm find the measures of the remaining three parts of the triangle.

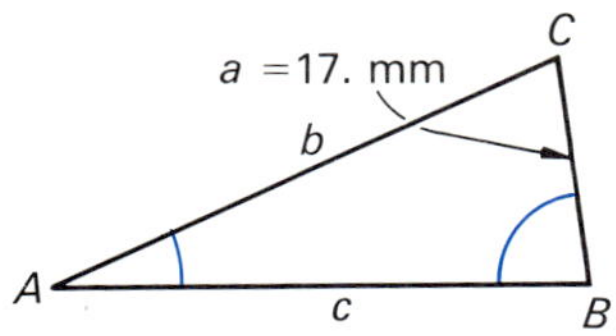

FIGURE 7.42

Solution: We first sketch and label the triangle as in Fig. 7.42. To find angle C we use the fact that

$$C = 180° - (A + B).$$

Substituting, $C = 180° - (24° + 82°)$ or $C = 74°$. Since we know A, a, and B, this suggests using $\frac{a}{\sin A} = \frac{b}{\sin B}$. Solving for b we have

$$b = \frac{a \sin B}{\sin A}.$$

Substituting,

$$b = \frac{17 \sin 82°}{\sin 24°} \quad \text{or} \quad b = 41.\text{ mm}.$$

To find side c we can use $\frac{a}{\sin A} = \frac{c}{\sin C}$ since we know A, c, and a. Solving for c we have

$$c = \frac{a \sin C}{\sin A}.$$

Substituting,

$$c = \frac{17 \sin 74°}{\sin 24°} \quad \text{or} \quad c = 40.\text{ mm}. \qquad \square$$

Example 1 illustrated the case where we know the measures of one side and two angles of a triangle. Example 2 will illustrate the case where two sides and an angle opposite one of these sides are known.

EXAMPLE 2 Given $a = 10$, $b = 3$, and $B = 50°$, find the measures of the remaining three parts of the triangle.

Solution: Knowing a, b, and B suggests using $\frac{a}{\sin A} = \frac{b}{\sin B}$; solving for $\sin A$ we have

$$\sin A = \frac{a \sin B}{b}.$$

FIGURE 7.43

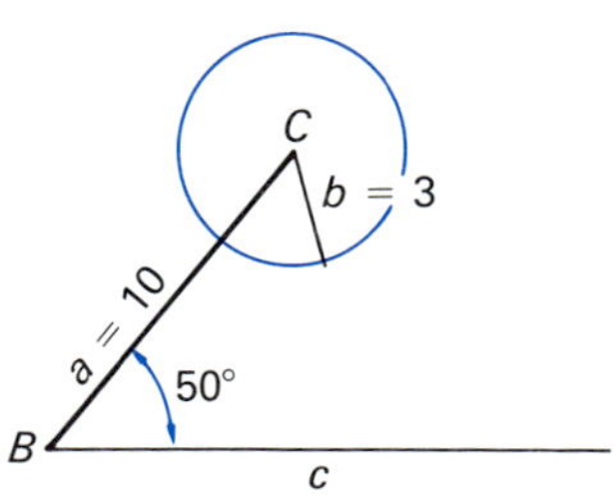

Substituting

$$\sin A = \frac{10 \sin 50°}{3} = 2.55.$$

Since $\sin A > 1$ (impossible) there is no solution. In Fig. 7.43 we can see that side b would never intersect side c. Therefore there can be no triangle. $\square$

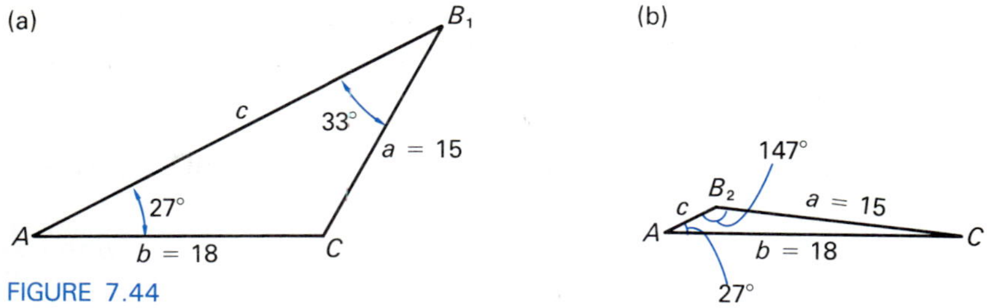

FIGURE 7.44

EXAMPLE 3 Given $a = 15$, $b = 18$, and $A = 27°$, determine angle B.

Solution: With the given information we would use $\dfrac{a}{\sin A} = \dfrac{b}{\sin B}$ and solve for $\sin B$. Thus, $\sin B = \dfrac{b \sin A}{a}$. Substituting,

$$\sin B = \frac{18 \sin 27°}{15}$$
$$= 0.545.$$

Since the sine function is positive in the first and second quadrants, there are two possible answers for angle B:

$$B_1 = 33°; \qquad B_2 = 180° - 33°, \quad \text{or} \quad 147°.$$

Figures 7.44a and 7.44b illustrate the two triangles. □

The situation of having two possible triangles arises when we are given two sides and an angle opposite one of these sides. If the side opposite the angle, in this case side a, is less than the other side, side b, ($a < b$), then there are two triangles. If $a > b$ then there is only one solution.

Examples 1, 2, and 3 illustrate the possible results using the law of sines: (1) a unique solution, (2) no solution, or (3) two solutions. In case (3) one of the two solutions may not be possible, depending on the information given.

EXAMPLE 4 A surveyor needed to know the three sides and the three angles of a triangular plot of ground as shown in Fig. 7.45. Find the missing parts.

FIGURE 7.45

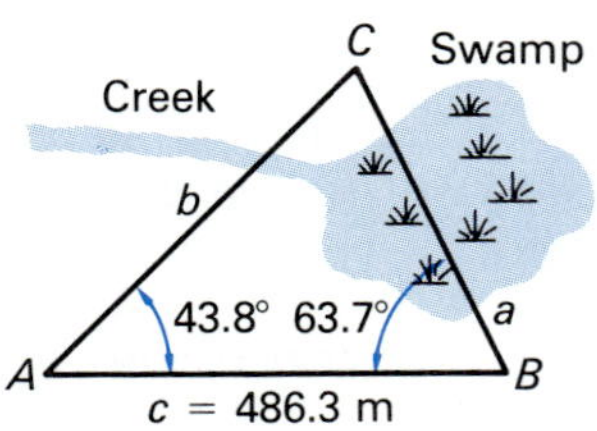

Solution: From the figure we know that $A = 43.8°$, $B = 63.7°$, and $c = 486.3$ m. Knowing A and B we can find C using the fact that

$$C = 180° - (A + B).$$

Substituting,

$$C = 180° - (43.8° + 63.7°)$$
$$= 72.5°.$$

Now we can use

$$\frac{a}{\sin A} = \frac{c}{\sin C} \qquad \text{and} \qquad \frac{b}{\sin B} = \frac{c}{\sin C}$$

to solve for a and b, respectively. Solving for a and b

$$a = \frac{c \sin A}{\sin C} \qquad b = \frac{c \sin B}{\sin C}.$$

Substituting,

$$a = \frac{(486.3)(\sin 43.8°)}{\sin 72.5°} = 352.9 \text{ m.} \qquad b = \frac{(486.3)(\sin 63.7°)}{\sin 72.5°} = 457.1 \text{ m.}$$

Thus, $C = 72.5°$, $a = 353.$ m, and $b = 457.$ m. □

7.4 Exercises

In Exercises 1–20, solve each triangle for the missing parts if possible.

1. $A = 68°$, $B = 79°$, $b = 8.0$ cm

2. $A = 48°$, $C = 100°$, $c = 18$ cm

3. $A = 60°$, $a = 9.0$ in., $b = 7.0$ in.

4. $A = 40°$, $C = 35°$, $a = 10$ in.

5. $a = 5.14$ mm, $b = 4.31$ mm, $A = 64.3°$

6. $b = 672$ km, $c = 630$ km, $B = 82.7°$

7. $C = 30.5°$, $a = 4.0$ in., $c = 10.5$ in.

8. $B = 40°$, $a = 18$ in., $c = 10$ in.

9. $b = 11.1$ cm, $c = 15.8$ cm, $B = 150.7°$

10. $a = 0.432$ cm, $c = 0.765$ cm, $C = 105.2°$

11. $B = 112°30'$, $C = 27°40'$, $b = 426$ ft

12. $C = 97°30'$, $c = 5.0 \times 10^{-4}$ m, $b = 6.0 \times 10^{-5}$ m

13. $C = 45°$, $b = 2.5$ in., $c = 2.5$ in.

14. $B = 30°$, $b = 5$ ft, $c = 10$ ft

15. $A = 1.7$, $B = 1.1$, $a = 53$ cm

16. $B = .48$, $b = 0.72$, $c = 0.75$ cm

17. $c = 1.7 \times 10^4$ m, $b = 2.3 \times 10^5$ m, $B = 82.3°$

18. $C = 30.9°$, $b = 9.3$ in., $c = 4.0$ in.

19. $B = 119°$, $C = 47°$, $b = 8$ ft

20. $B = 50°$, $b = 8$ ft, $c = 8$ ft

21. Two points, B and C, are on opposite sides of a swamp. To measure the distance across the swamp, point A, from which B and C were visible, was chosen. The distances $\overline{AB}$ and $\overline{AC}$ were measured to be 945.6 ft and 887.8 ft, respectively, and the angle ABC was found to be $56°44'$. Find the distance across the swamp.

22. A distant tower on shore is sighted from points at opposite ends of a ship 650 ft long (Fig. 7.46). If the sight lines make angles of 87.4° and 87.8° with the line joining the two points, how far is the tower from the nearest of the two points?

23. Two fire lookout towers P and Q are 28.5 km apart. Tower P spots a fire at R and measures angle QPR to be 49.3°, whereas tower Q measures angle PQR to be 43.7°. Which tower is closer to the fire and by how much?

FIGURE 7.46

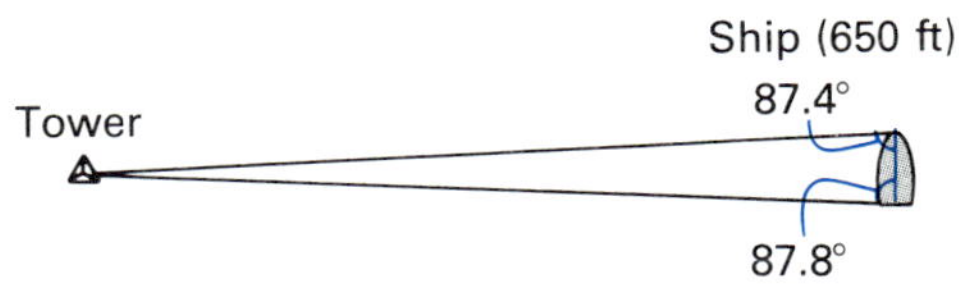

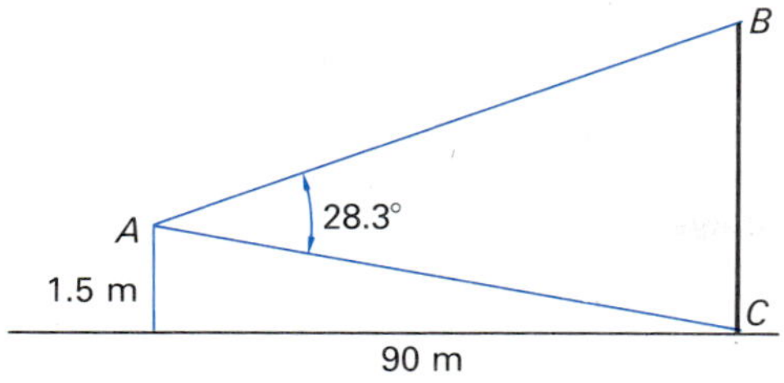

FIGURE 7.47

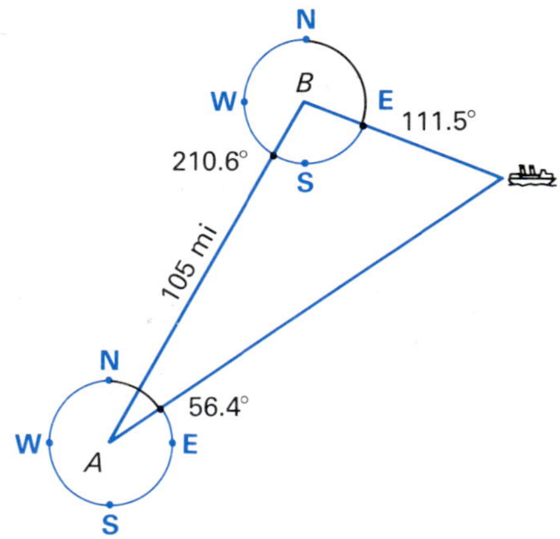

FIGURE 7.48

24. An antenna 25 m high is on the top edge of a building. From a point on the ground the angles of elevation to the top and bottom of the antenna are 58.7° and 46.3°. What is the distance from the point on the ground to the top of the antenna?

25. From point A on top of a building the angle of depression to point C on the ground is observed to be 56°, while from a window at point B (18 m directly below A) the angle of depression to point C is 43°. Find the distance from point B to point C.

26. A surveyor sights to the top and bottom of an antenna from point A. Determine the height of the antenna $\overline{BC}$. (See Fig. 7.47.)

27. A radio operator on a Coast Guard patrol boat A received a distress call from a ship, which he located at 56.4°. At the same time an operator on a second Coast Guard boat B picked up the call from a position which he located at 111.5°. He also located the patrol boat at A at 210.6°. If the two patrol boats were 105 mi apart, how far was the ship in distress from the nearer of the two Coast Guard boats? (See Fig. 7.48.)

7.5 The law of cosines

From the previous section we observed that oblique triangles can be solved by the law of sines only when either (1) a side and any two angles or (2) two sides and an angle opposite one of them is known. We will need another formula in order to solve oblique triangles when either two sides and an included angle are known or when three sides are known. This formula is called the **Law of Cosines.** The Law of Cosines states that **the square of any side is equal to the sum of the squares of the other two sides decreased by twice their product times the cosine of the angle between them.**

FIGURE 7.49

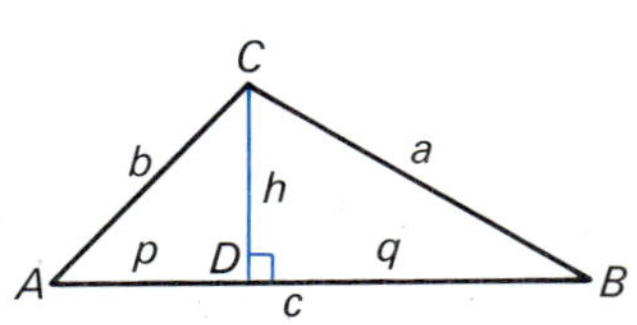

To develop $a^2 = b^2 + c^2 - 2bc\cos A$, which is one form of the Law of Cosines, we have triangle ABC and h, the altitude from C to $\overline{AB}$. What is the value of a^2 in right triangle BCD?

Using the Pythagorean theorem and Fig. 7.49

$$a^2 = h^2 + q^2 \qquad [1]$$

and

$$q = c - p. \qquad [2]$$

Squaring both sides,

$$q^2 = c^2 - 2cp + p^2. \qquad [3]$$

Substituting [3] into [1],

$$a^2 = h^2 + c^2 - 2cp + p^2. \qquad [4]$$

In right triangle ADC

$$h^2 + p^2 = b^2. \qquad [5]$$

Substituting [5] into [4]

$$a^2 = b^2 + c^2 - 2cp. \qquad [6]$$

In right triangle ADC

$$\cos A = \frac{p}{b} \quad \text{or} \quad p = b \cos A. \qquad [7]$$

Substituting [7] into [6]

$$a^2 = b^2 + c^2 - 2bc \cos A. \qquad [8]$$

If we go back and draw the altitude from point A or C and follow the same procedure, we derive the following equations:

$$b^2 = a^2 + c^2 - 2ac \cos B \quad \text{and} \quad c^2 = a^2 + b^2 - 2ab \cos C.$$

These three equations give us Theorem 7.2.

THEOREM 7.2

LAW OF COSINES

In any triangle ABC

$$a^2 = b^2 + c^2 - 2bc \cos A,$$
$$b^2 = a^2 + c^2 - 2ac \cos B,$$
$$c^2 = a^2 + b^2 - 2ab \cos C.$$

EXAMPLE 1 Determine the missing parts of the triangle in Fig. 7.50.

Solution: Since we are given the two sides and the included angle, we can use the Law of Cosines.

FIGURE 7.50

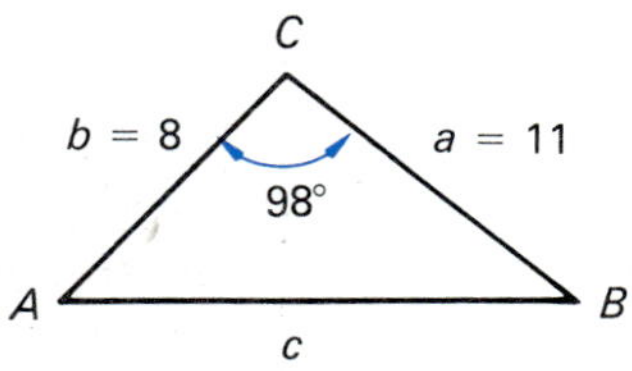

$$c^2 = a^2 + b^2 - 2ab \cos C.$$

Substituting,

$$c^2 = (11)^2 + (8)^2 - 2(11)(8)(\cos 98^\circ)$$
$$= 121 + 64 - 2(11)(8)(-0.139).$$
$$c = 14.47 = 14.5 \quad \textbf{(to the nearest tenth).}$$

Now that we have found side c, angle A is found most easily by using the Law of Sines.

$$\begin{aligned}\sin A &= \frac{a \sin C}{c}\\ &= \frac{11 \sin 98^\circ}{14.47}\\ &= 0.7528,\\ A &= 48.8^\circ.\end{aligned}$$

The sine function is positive in the first and second quadrants, but we use only the first-quadrant solution. (Why?)

$$B = 180^\circ - (98^\circ + 48.8^\circ) \quad \text{or} \quad 33.2^\circ. \quad \square$$

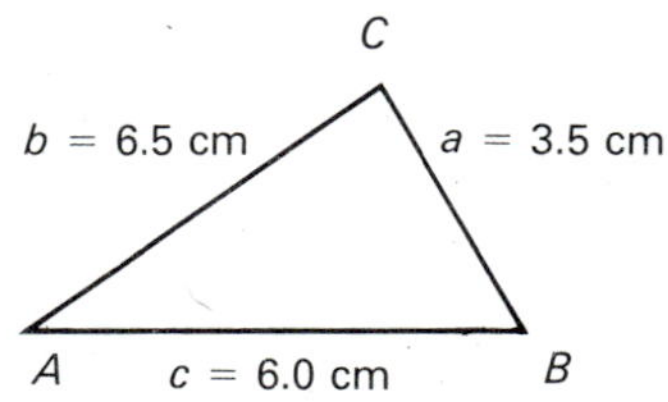

FIGURE 7.51

EXAMPLE 2 Determine the missing parts of the triangle in Fig. 7.51.

Solution: Given the lengths of the three sides we can find an angle by using the Law of Cosines.

$$a^2 = b^2 + c^2 - 2bc \cos A.$$

Solving for $\cos A$ we have

$$\cos A = \frac{b^2 + c^2 - a^2}{2bc}.$$

Substituting,

$$\cos A = \frac{(6.5)^2 + (6.0)^2 - (3.5)^2}{2(6.5)(6.0)} \quad \text{or} \quad 0.8462,$$

$$A = 32.2^\circ.$$

At this point we can use either the Law of Cosines or the Law of Sines to determine the other two angles.

To find angle B we use $b^2 = a^2 + c^2 - 2ac \cos B$. Solving for $\cos B$,

$$\cos B = \frac{a^2 + c^2 - b^2}{2ac}.$$

Substituting,

$$\cos B = \frac{(3.5)^2 + (6.0)^2 - (6.5)^2}{2(3.5)(6.0)} \quad \text{or} \quad 0.1429,$$

$$\begin{aligned}B &= 81.8^\circ,\\ C &= 180^\circ - (32.2^\circ + 81.8^\circ) \quad \text{or} \quad 66.0^\circ.\end{aligned}$$

Thus, to two significant digits $A = 32.^\circ$, $B = 82.^\circ$, and $C = 66.^\circ$. $\square$

FIGURE 7.52

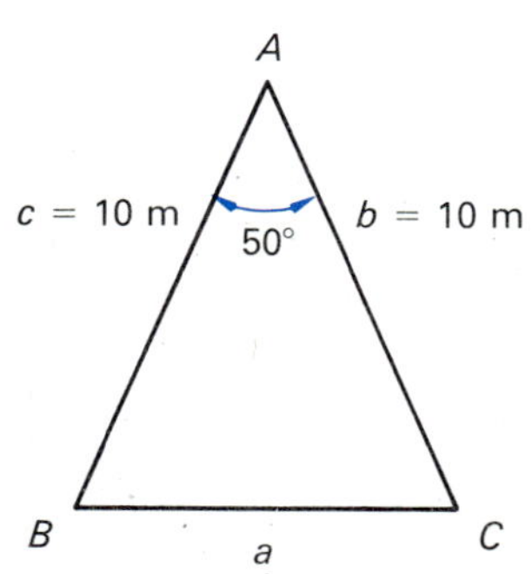

EXAMPLE 3 The attic walls of an A-framed house are in the form of an isosceles triangle. The equal legs are each 10 m long and the included angle is 50°. Find the width of the floor. (See Fig. 7.52.)

Solution: Since we know two sides and the included angle and want to find the third side, we use the Law of Cosines.

$$a^2 = b^2 + c^2 - 2bc \cos A$$

Substituting,

$$\begin{aligned} a^2 &= 10^2 + 10^2 - 2(10)(10) \cos 50^\circ \\ &= 71.442 \\ a &= 8.45 \text{ m.} \end{aligned}$$ □

7.5 Exercises

In Exercises 1–12, solve each triangle for the missing parts.

1. $a = 12$, $b = 10$, $C = 78^\circ$

2. $a = 9$, $b = 10$, $C = 47^\circ$

3. $b = 45$, $c = 40$, $A = 59.9^\circ$

4. $b = 38$, $c = 35$, $A = 42.1^\circ$

5. $a = 52$, $c = 32$, $B = 104.3^\circ$

6. $a = 17$, $c = 30$, $B = 23.5^\circ$

7. $a = 20$, $b = 25$, $c = 28$

8. $a = 21$, $b = 22$, $c = 25$

9. $a = 17.8$, $b = 10.3$, $c = 14.2$

10. $a = 0.12$, $b = 0.08$, $c = 0.05$

11. $a = 6.743$, $b = 4.765$, $c = 8.512$

12. $a = 41.37$, $b = 59.73$, $c = 61.43$

13. If $a = 31.6$, $b = 53.4$, and $c = 34.9$, find the measure of the smallest angle of the triangle correct to the nearest minute.

14. A plot of land in the form of a parallelogram has sides of 43 m and 68 m. If one diagonal measures 72 m, find the angles of the parallelogram.

15. A reconnaissance plane finds that the distances from the plane to two radar stations, A and B, are 3592.7 m and 4591.2 m, respectively. If the angle between the two lines of sight to A and B is 123.4°, find the distance between the two stations (to the nearest tenth of a meter).

16. The two ships shown in Fig. 7.53 sail from the same dock P at the same time. One travels on a course of 035° at 22 knots and the other on a course of 158° at 20 knots. How far apart will they be in 2 1/2 hr?
(*Note:* one knot = one nautical mile per hour.)

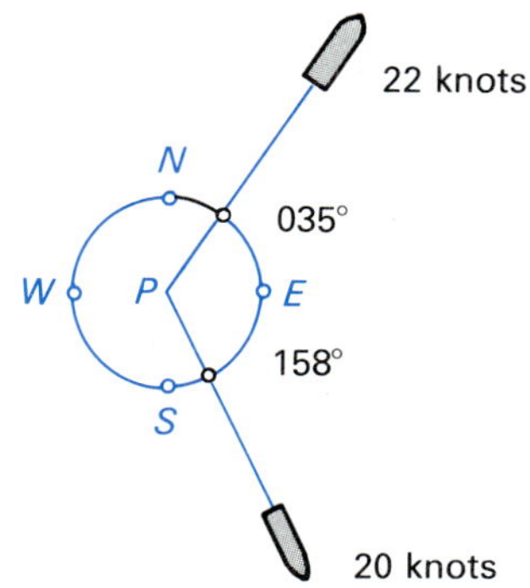

FIGURE 7.53

17. To pull out a car stuck in the sand, forces of 415 lb and 325 lb were applied, acting at an angle of 49°36′ with one another. What was the magnitude of the resultant force applied to the car?

18. Two magnetic forces of 60 N and 85 N act on an object to produce a resultant force of 95 N. Find the angle that the resultant makes with each force.

19. Pat hikes into the woods in a northeast direction for 8 mi then turns due east and hikes 7 mi. How far is she from her starting point?

20. Two forces $F_1 = 675$ lb and $F_2 = 1250$ lb act on a point of a bridge girder at an angle of 85° to each other. Find the resultant force **R** (the magnitude and the direction measured from F_1).

21. A guy wire runs 17.8 m from the top of a pole to the level ground. The original wire is replaced with a 25.7-m wire that runs from the top of the pole to a spot on the ground 10.8 m farther from the pole than the old wire. What angle does the new wire make with the ground?

Chapter summary

Quantities that are undirected and can be fully described by a number, are called **scalars** or **scalar quantities.**

Definition 7.1 Quantities having both direction and magnitude are called **vector quantities.**

In a diagram vector **A** is represented by an arrow whose length indicates the **magnitude** or size of the vector quantity and whose head indicates the **direction** of the vector. If vector **OP** has a magnitude of 20 lb, this can be symbolized as $|\mathbf{OP}| = 20$ lb or $\mathbf{OP} = 20$ lb. A vector is indicated using the following notation, $\mathbf{V} = |\mathbf{V}| \underline{/\theta}$, where $|\mathbf{V}|$ represents the magnitude and θ the direction of the vector.

Definition 7.3 The magnitude of the horizontal component of a vector **V** of magnitude $|\mathbf{V}|$ acting at angle θ with the horizontal is

$$|\mathbf{V}_x| = |\mathbf{V}| \cos \theta$$

and the magnitude of the vertical component is

$$|\mathbf{V}_y| = |\mathbf{V}| \sin \theta.$$

Definition 7.4 Given vector components $|\mathbf{V}_x|$ and $|\mathbf{V}_y|$, the magnitude of the vector is given by

$$|\mathbf{V}| = \sqrt{|\mathbf{V}_x|^2 + |\mathbf{V}_y|^2},$$

and the angle that the vector makes with the positive x-axis is given by

$$\tan \theta = \frac{|\mathbf{V}_y|}{|\mathbf{V}_x|}.$$

The sum of two or more vectors can be found by geometric construction, but with a maximum accuracy of about 90%. With the use of trigonometric functions the sum can be found to an accuracy consistent with the accuracy of measurement. To find the sum we determine the horizontal and vertical components of each of the vectors and then find the sum of all the horizontal components and the sum of all the vertical components. Then, using the Pythagorean theorem we can find the magnitude of the resultant vector. The direction of the resultant vector is found by solving the equation

$$\tan\theta = \frac{\text{sum of the vertical components}}{\text{sum of the horizontal components}}$$

for θ.

Section 7.3 deals with application of vectors.

The **Law of Sines** and the **Law of Cosines** are rules for determining the other parts of an oblique triangle when three pieces of information are given, one of which must be a side. The combinations can be

1. One side and any two angles (Law of Sines);
2. Two sides and an angle opposite one of them (Law of Sines);
3. Two sides and an included angle (Law of Cosines);
4. Three sides (Law of Cosines).

Theorem 7.1 Law of Sines

In any triangle ABC

$$\frac{a}{\sin A} = \frac{b}{\sin B} = \frac{c}{\sin C}.$$

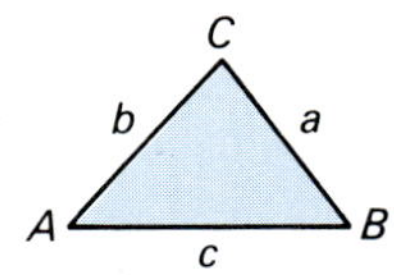

When using the Law of Sines one may obtain the following results (1) a unique solution, (2) two solutions (this happens when given $\angle A$, sides a and c, and $a < c$), or (3) no solution.

Theorem 7.2 Law of Cosines

In any triangle ABC

$$a^2 = b^2 + c^2 - 2bc\cos A,$$
$$b^2 = a^2 + c^2 - 2ac\cos B,$$
$$c^2 = a^2 + b^2 - 2ab\cos C.$$

Review Exercises

In Exercises 1–8, determine the horizontal and vertical components of the vector quantities. Sketch the vector and its components.

1. $\mathbf{V} = 47$ cm/s $\underline{/68^\circ}$
2. $\mathbf{V} = 125$ cm/s $\underline{/135^\circ}$
3. $\mathbf{F} = 35$ lb $\underline{/210^\circ}$
4. $\mathbf{F} = 48$ lb $\underline{/300^\circ}$
5. $\mathbf{D} = 120$ km $\underline{/-30^\circ}$
6. $\mathbf{D} = 85$ km $\underline{/-120^\circ}$
7. $\mathbf{a} = 32.2$ cm/s^2 $\underline{/1.72}$
8. $\mathbf{a} = 32.2$ cm/s^2 $\underline{/0.52}$

In Exercises 9–16, find the magnitude of the resultant vector and the angle it makes with the positive x-axis. Sketch, showing the resultant and the component vectors.

9. $\mathbf{F}_x = 241$ lb, $\mathbf{F}_y = 18$ lb
10. $\mathbf{F}_x = 85$ lb, $\mathbf{F}_y = 45$ lb
11. $\mathbf{D}_x = -14$ km, $\mathbf{D}_y = 15$ km

12. $\mathbf{D}_x = -21$ km, $\mathbf{D}_y = 24$ km
13. $\mathbf{F}_x = -0.472$ N, $\mathbf{F}_y = -0.965$ N
14. $\mathbf{F}_x = -98.76$ N, $\mathbf{F}_y = -45.72$ N
15. $\mathbf{V}_x = 0.007$ m/s, $\mathbf{V}_y = -0.028$ m/s
16. $\mathbf{V}_x = 4.2 \times 10^5$ m/s, $\mathbf{V}_y = 9.8 \times 10^3$ m/s
17. A swimmer heads due west at a rate of 3 mph, and the current of the river is flowing north at 2 mph. Find and sketch the resultant with respect to a point on the opposite bank of the river.
18. An AC circuit has two sources in series. Find the total voltage if $\mathbf{V}_1 = 22$ V at 0° and $\mathbf{V}_2 = 33$ V at 90°.
19. An AC circuit has a resistance $\mathbf{R}$ of 33 Ω acting at 0°, and a capacitive reactance $\mathbf{X}_c$, of 38 Ω acting at 270°. Find the impedance $\mathbf{Z}$ that is the resultant of these vectors.
20. A stone is thrown into the air with a speed of 22 m/s directed at an angle of 42° with the horizontal. Find the magnitudes of the horizontal and vertical vector components of the stone's initial velocity.
21. A truck weighing 20,000 lb is placed on a 7° inclined plane. Find the magnitudes of the vector components of the weight that are (a) parallel and (b) perpendicular to the inclined surface.

In Exercises 22–25, determine the resultant vector.

22. $\mathbf{A} = 8.7$ ft $\underline{/125°}$
$\mathbf{B} = 9.8$ ft $\underline{/65°}$
23. $\mathbf{A} = 34.2$ cm $\underline{/47°}$
$\mathbf{B} = 67.4$ cm $\underline{/88°}$
24. $\mathbf{V}_1 = 27.85$ $\underline{/10°}$
$\mathbf{V}_2 = 37.21$ $\underline{/170°}$
$\mathbf{V}_3 = 64.25$ $\underline{/260°}$
25. $\mathbf{V}_1 = 0.472$ $\underline{/20°}$
$\mathbf{V}_2 = 0.365$ $\underline{/40°}$
$\mathbf{V}_3 = 0.596$ $\underline{/50°}$
26. A 42-kg weight is on a ramp inclined at 31°. If there is no friction, what force, parallel to the ramp, is necessary to keep the weight at rest?
27. A 2400-lb car is on a hill inclined at 12°. How much force, pushing up the hill, is required to keep the car from rolling? (Assume no friction.)
28. A force of 45 lb is required to keep a weight of 80 lb from sliding down an inclined plane. What is the angle of inclination? (Assume no friction.)
29. A manual lawnmower requires a force of 50 N parallel to the ground in order to move it through the grass. If the handle is inclined at 38° to the ground, how much force is necessary along the handle (Fig. 7.54)?

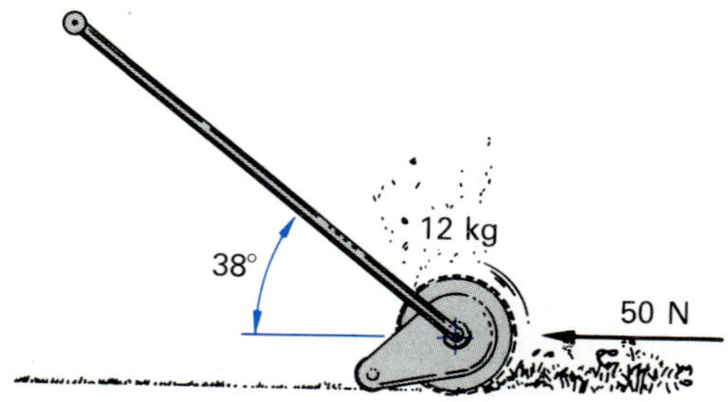

FIGURE 7.54

30. If the lawnmower in Exercise 29 weighs 12 kg and is pushed up a hill of 8°, how much force exerted along the handle is necessary to keep it moving through the grass? The handle is still inclined 38° to the slope, and you must take into account both the 50 N from Exercise 29 and the weight of the lawnmower (Fig. 7.55).

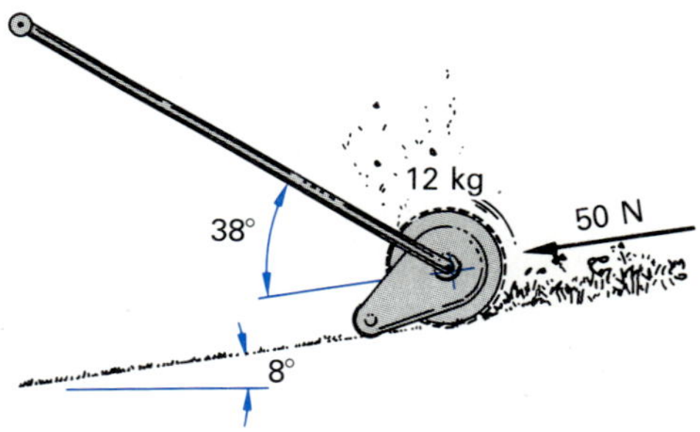

FIGURE 7.55

In Exercises 31–42, solve each triangle for the missing parts.

31. $A = 38.2°$, $B = 67.1°$, $b = 42$
32. $A = 72.3°$, $C = 84.5°$, $c = 72$
33. $a = 11.24$, $b = 14.72$, $A = 23°$
34. $a = 24.7$, $b = 28.65$, $A = 28°$
35. $a = 19$, $c = 15$, $C = 40°$
36. $b = 29.4$, $c = 26.7$, $C = 36°$
37. $A = 50.9°$, $b = 45$, $c = 40$
38. $B = 42.1°$, $a = 38.7$, $c = 35.6$
39. $a = 21.4$, $b = 22.8$, $c = 25.6$
40. $a = 33.7$, $b = 28.9$, $c = 51.9$
41. $C = 146.3°$, $a = 210$, $b = 120$
42. $a = 149$, $b = 233$, $c = 330$
43. A hiker leaves her camp and walks directly west 5.5 mi and then turns and walks 3 mi in a direc-

tion 40° south of east. She then decides to walk directly to her camp. How far and in what direction must she walk?

44. From town A a straight road leads to town B, a distance of 8 mi. From B another road leads directly to C, a distance of 7 miles. The angle between the two roads is 70°. A road is to be built directly from A to C. Find the distance from A to C and the direction with respect to $\overline{AB}$.

45. The maximum angle between the 18-in. handles of a pair of shrubbery clippers is 68°. When it is completely open, what is the distance between the ends of the handles?

46. A pass receiver on a football team gets the signal directing him to run to the right. He has to run 20 ft parallel to the line of scrimmage, turn at an angle of 120°, then run 40 ft to point A in order to be ready for the pass. If he starts 6 ft behind the position of the ball, and everything is in his favor for a successful pass, how many yards did he gain assuming that he was tackled at point A?

47. Two boys whose homes are 1 mi apart on a north–south line looked at a plane directly above their homes. The boy nearest to the plane read an elevation of 47°10′; the boy farthest away read 38°40′. How far away was the plane from the nearer boy?

48. A pilot wishes to fly northwest. If her airspeed is 180 km/h and there is a 60 km/h wind from due south, how many degrees west of north should she fly so that the resultant direction will be northwest?

49. From a land-based radar station it is determined that the bearing of ship A is 70° and it is a distance of 3.5 km from the station, while ship B has a bearing of 110° and is a distance of 5.2 km from the station. Find the distance between the two ships.

Can you solve these?

1. One side of a triangle is 3.472 times as long as another side x. The angle between them is 32°01′.

a. Find the other two angles.

b. Find the third side in terms of the side x.

2. At the local carnival there is a ferris wheel. There are 18 seats equally spaced on the wheel. If the radius of the wheel is 18 ft, find the distance between any two adjacent seats on the wheel.

3. **a.** Show that the area of $\triangle ABC$ is $\frac{1}{2}\,ab \sin C$.

b. Find the area of a triangle with sides 9 cm, 8 cm, and 15 cm.

4. Prove that if ABC is a right triangle, the Law of Cosines reduces to the Pythagorean theorem.

5. In a triangle, sides a and b and angle A are given. Prove the following:

a. If $b \sin A > a$, there is no triangle with the given parts.

b. If $b \sin A = a$, the parts determine a right triangle.

c. If $b \sin A < a < b$, there are two triangles with the given parts.

6. An ant is crawling in the $+x$-direction on a sheet of paper with a velocity of 40 mm/s relative to the paper. At the same time the paper is being moved in the $+x$-direction with a velocity 60 mm/s relative to the desk top. What is the velocity of the ant relative to the desk top? Repeat if the ant is crawling in the $-x$-direction.

7. Write a program for the calculator or the computer to solve problems using the Law of Sines.

8. Write a program for the calculator or the computer to solve problems using the Law of Cosines.

9. What is the maximum acceleration a dragster can attain assuming purely frictional contact between the tires and the road? Assume total weight transfer to the rear wheels.

8

In order to hold wear rates in internal combustion engines to acceptable limits, a rule of thumb in design has been to limit cruising piston speeds to 2500 ft/min. For an engine with stroke s in inches, connecting rod length ℓ in inches, and turning at n RPM, an approximate expression for piston velocity is

$$v = \frac{\pi s n}{12}\left(\sin\theta + \frac{s}{48\ell}\sin 2\theta\right).$$

For an engine with a 2.50-in. stroke and 6.25-in. connecting rods, graph the function in parentheses so that the RPM for a piston velocity of 2500 ft/min can be found.

Graphs of Trigonometric Functions

8.1 Periodic functions

The calendar year is cyclic or periodic. The year in most climates is broken up into four distinct seasons (fall, winter, spring, and summer), which make up one period.

In mathematics there are many periodic functions. From a graph it can easily be seen that the values of the function are repeated at regular intervals. An example is the graph in Fig. 8.1.

After a careful examination we can see that:

$$f(-4) = f(-2) = f(0) = f(2) = f(4) = 0$$

and

$$f(-3) = f(-1) = f(1) = f(3) = 1.$$

This function goes through one complete **cycle** from $x = -4$ to $x = -2$, from $x = -2$ to $x = 0$, from $x = 0$ to $x = 2$, and from $x = 2$ to $x = 4$. With this information we can say that the **period** p is 2 units.

DEFINITION 8.1

A function f is **periodic** if there is a positive number p such that for every real number x, whenever x is in the domain of f, so is $x + p$ and $f(x + p) = f(x)$. The smallest such positive number p is then called the **period** of the function.

In the previous example $p = 2$, since

$$f(-4) = f(-4 + 2) = f(-2) = 0$$
$$f(0) = f(0 + 2) = f(2) = 0$$
$$f(1) = f(1 + 2) = f(3) = 1.$$

FIGURE 8.1

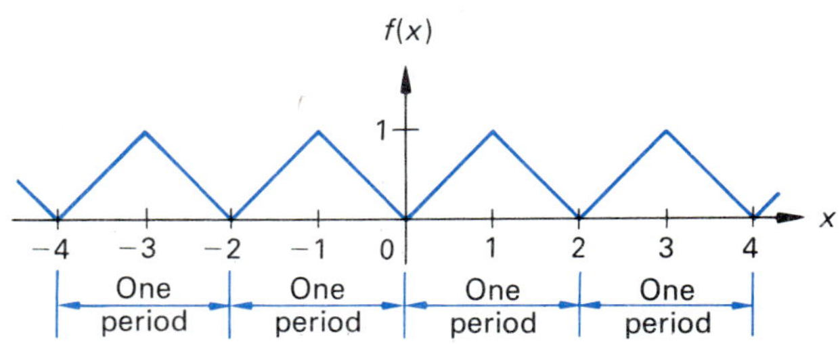

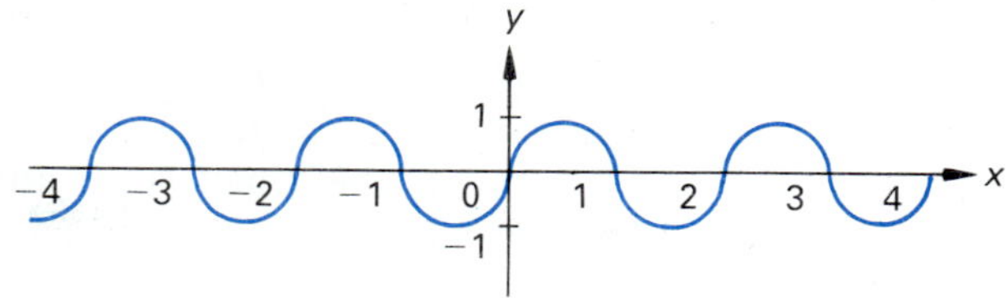

FIGURE 8.2

EXAMPLE 1 For the function in Fig. 8.2 determine the period p.

Solution: The function goes through one complete cycle during each of the intervals from -4 to -2, -2 to 0, 0 to 2, and 2 to 4. Thus, the period is 2 units. Using the definition we see that

$$f(-4) = f(-4 + 2) = f(-2) = 0 \qquad \text{or}$$
$$f(1.5) = f(1.5 + 2) = f(3.5) = -1. \qquad \square$$

We may be tempted to let $p = 1$ in Example 1. To show that this is not true let $x = 0.5$ and by definition

$$f(x + p) = f(x)$$

But

$$f(0.5 + 1) \neq f(0.5)$$

since

$$f(0.5 + 1) = -1 \qquad \text{and} \qquad f(0.5) = 1.$$

All six trigonometric functions are periodic. To illustrate that the sine function and the cosine function both have a period of 2π we will use the unit circle. Every central angle determines a point of the unit circle. Recall that the central angle of the entire circle is 2π. Central angles θ and $\theta + 2\pi$ determine the same point P on the unit circle (Fig. 8.3). Recall from right-triangle trigonometry that $x = r \cos \theta$ and $y = r \sin \theta$. Since we are dealing with a unit circle with $r = 1$, $x = \cos \theta$, and $y = \sin \theta$. Therefore, any point P on the circle with coordinates (x, y)

FIGURE 8.3

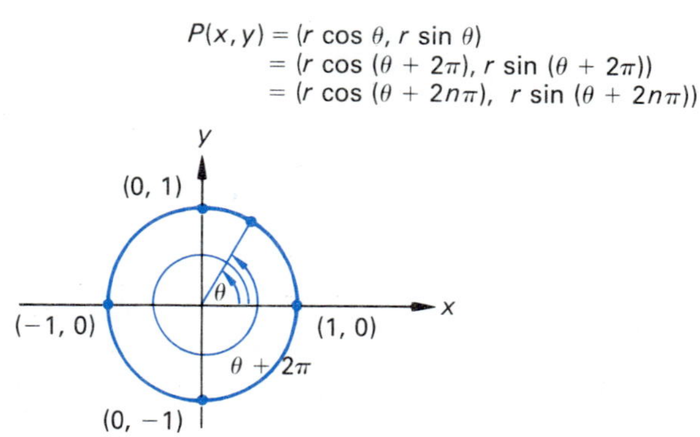

will also have coordinates $(\cos\,\theta,\ \sin\,\theta)$. It follows that $\cos\,\theta = \cos\,(\theta + 2n\pi)$ and $\sin\,\theta = \sin\,(\theta + 2n\pi)$. This fact can also be illustrated with the calculator.

Is 2π the smallest positive number p? The functional value, $\sin\,\theta = 1$, occurs only when $y = 1$ and also when $P = (0,\ 1)$. This happens only when $\theta = \frac{\pi}{2}$ or when $\theta = \frac{\pi}{2} + 2k\pi$ (where k is an integer). We conclude that the period p of the sine function is 2π.

By similar reasoning, $\cos\,\theta = -1$ only when $x = -1$ or when $P = (-1,\ 0)$. This is true only when $\theta = \pi$ or $\theta = \pi + 2k\pi$. Thus, $\cos\,\pi = \cos\,(\pi + 2k\pi)$, where k is an integer, and cosine has a period of 2π.

When graphing the sine and cosine functions using the above information, once you have drawn the graph for an interval of 2π, it is easy to extend the graph for any values of θ since the rest of the graph repeats.

8.1 Exercises

In Exercises 1–12, determine (a) whether the function is periodic and (b) if it is, determine the period p.

1.

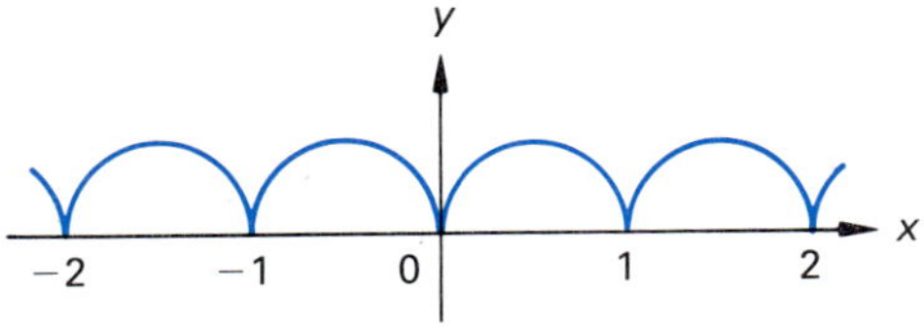

2.

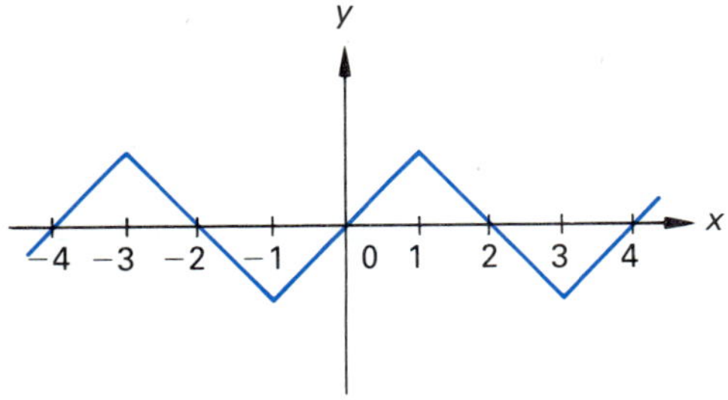

3.

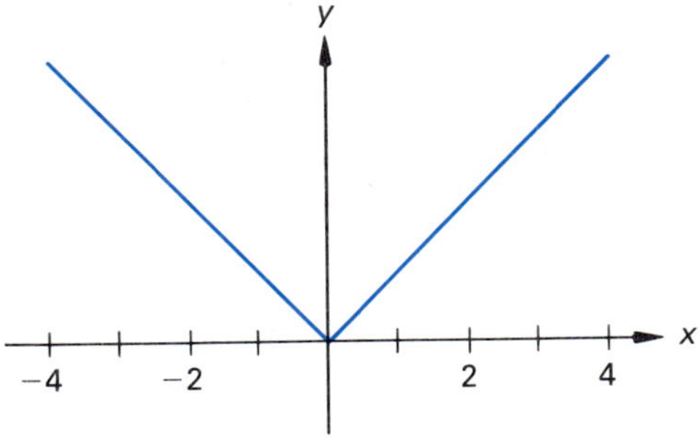

4.

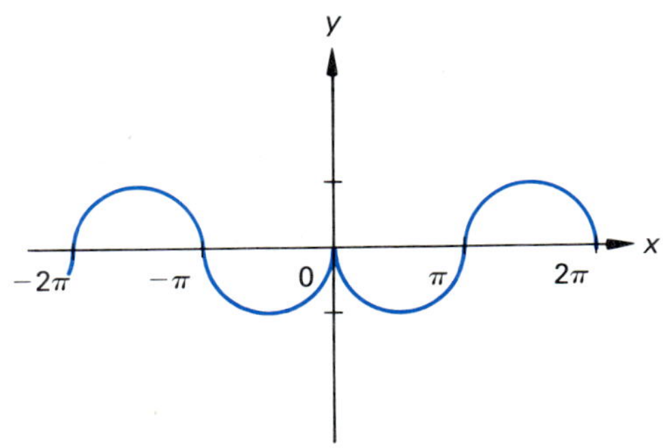

5.

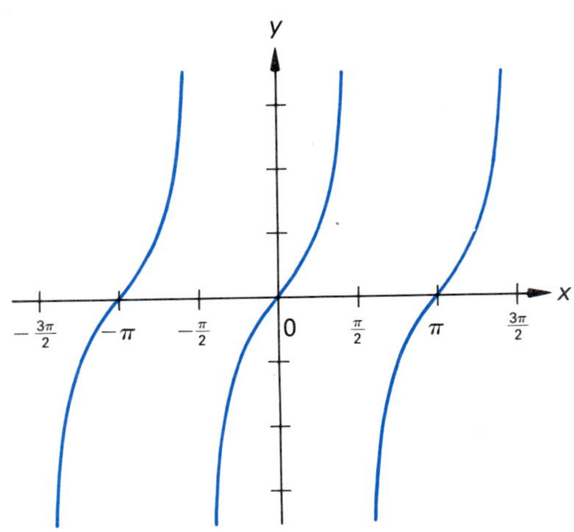

6.

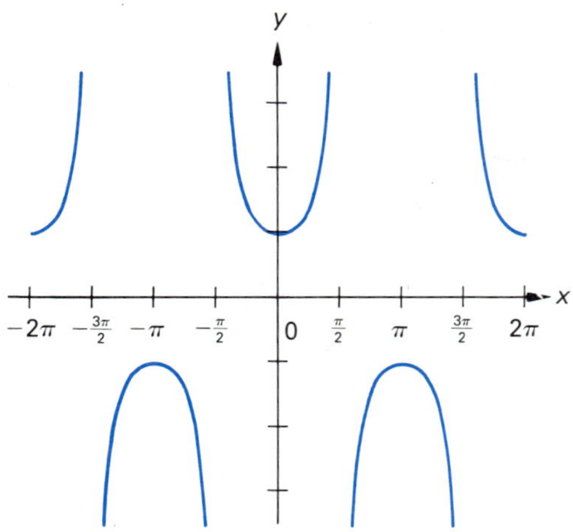

7.

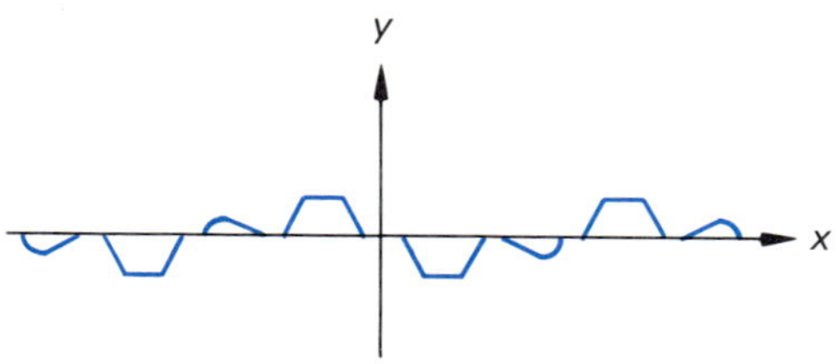

8.

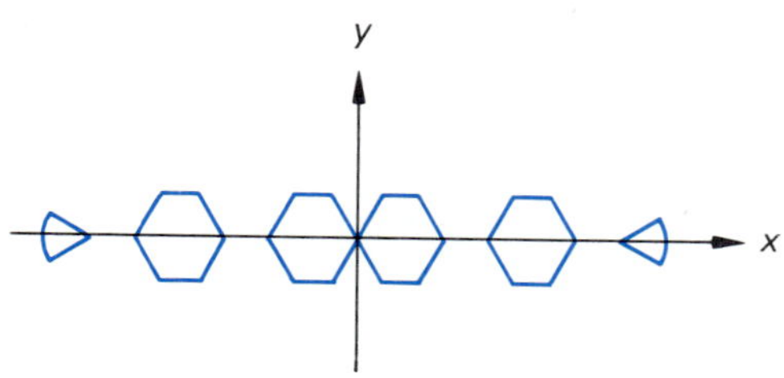

9.

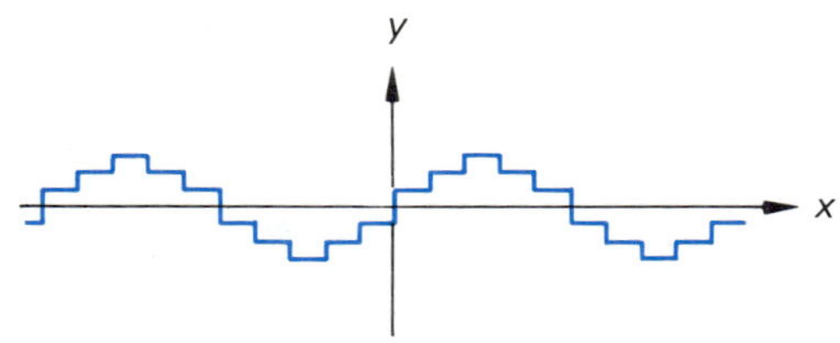

10.

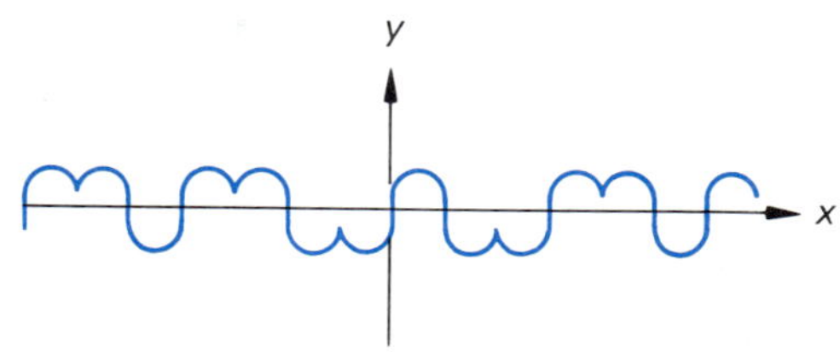

11.

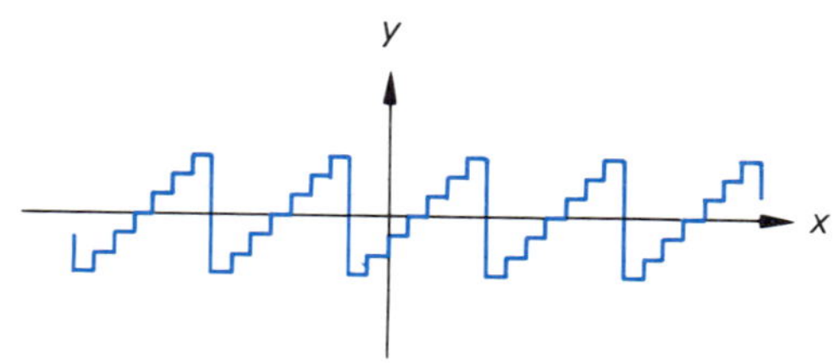

12.

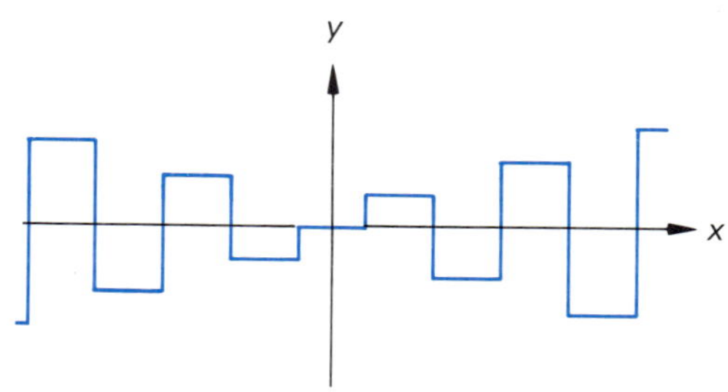

8.2 Graphs of the trigonometric functions

In this section we will sketch the basic graphs of the six trigonometric functions. These graphs will help you understand the properties of the functions. All six functions are periodic. Thus, the concepts developed in Section 8.1 will be used to facilitate the construction of the graphs.

In our graphing, the independent variable will be the angle (whether it is expressed in degrees or radians) and the dependent variable will be the calculated value of the trigonometric function.

EXAMPLE 1 Construct the graph of $y = \sin \theta$ for the interval $0 \le \theta \le 2\pi$ ($0° \le \theta \le 360°$). Determine coordinates for intervals of $\pi/12$ rad (or 15°).

TABLE 8.1 $y = \sin \theta$

θ			θ		
Radians	Degrees	$\sin \theta$	Radians	Degrees	$\sin \theta$
0	0°	0	π	180°	0
$\frac{\pi}{12}$	15°	0.2588	$\frac{13\pi}{12}$	195°	−0.2588
$\frac{\pi}{6}$	30°	0.5000	$\frac{7\pi}{6}$	210°	−0.5000
$\frac{\pi}{4}$	45°	0.7071	$\frac{5\pi}{4}$	225°	−0.7071
$\frac{\pi}{3}$	60°	0.8660	$\frac{4\pi}{3}$	240°	−0.8660
$\frac{5\pi}{12}$	75°	0.9659	$\frac{17\pi}{12}$	255°	−0.9659
$\frac{\pi}{2}$	90°	1.0000	$\frac{3\pi}{2}$	270°	−1.0000
$\frac{7\pi}{12}$	105°	0.9659	$\frac{19\pi}{12}$	285°	−0.9659
$\frac{2\pi}{3}$	120°	0.8660	$\frac{5\pi}{3}$	300°	−0.8660
$\frac{3\pi}{4}$	135°	0.7071	$\frac{7\pi}{4}$	315°	−0.7071
$\frac{5\pi}{6}$	150°	0.5000	$\frac{11\pi}{6}$	330°	−0.5000
$\frac{11\pi}{12}$	165°	0.2588	$\frac{23\pi}{12}$	345°	−0.2588
π	180°	0	2π	360°	0

Solution: The calculator was used to determine the values for Table 8.1. The graph of $y = \sin \theta$ is sketched in Fig. 8.4. □

In examining the graph of the curve $y = \sin \theta$ we note that for the interval $0 < \theta < \pi$ the curve is above the horizontal axis, which is what we would expect since the functional values are positive in the first and second quadrants, as we learned in Chapter 6. For the interval $\pi < \theta < 2\pi$ the curve is below the horizontal axis corresponding to the fact that $\sin \theta$ is negative in the third and fourth quadrants.

Note from our definition of periodic functions in Section 8.1 that as the curve reaches the value of 2π, it has gone through one complete cycle.

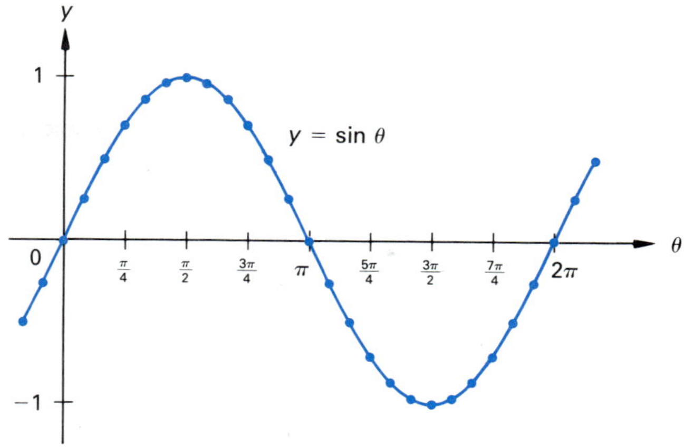

FIGURE 8.4

DEFINITION 8.2

> A **cycle** of a periodic function is a portion of the graph of the function from any point A on the graph to the first point B at which the graph starts repeating itself. (See Fig. 8.5.)

DEFINITION 8.3

> The **period** of a trigonometric function is the number of degrees or radians needed to complete one cycle. (See Fig. 8.6.)

From Table 8.1 and Fig. 8.4 we note the following **properties of the sine function:**

1. **The domain is the set of all real numbers.** The curve does not stop at 0 and 2π, it continues to the right and left indefinitely. (See Fig. 8.7.)

FIGURE 8.5

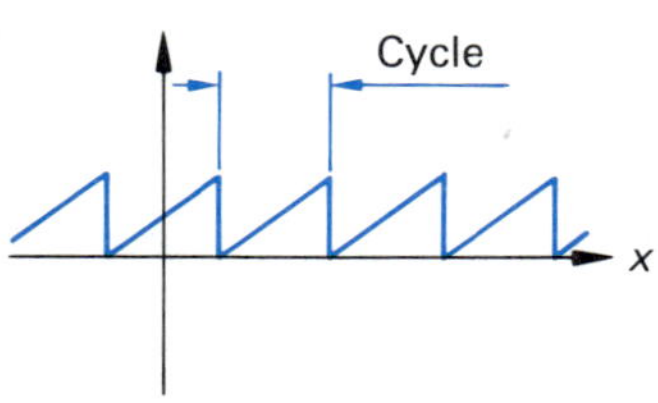

FIGURE 8.6

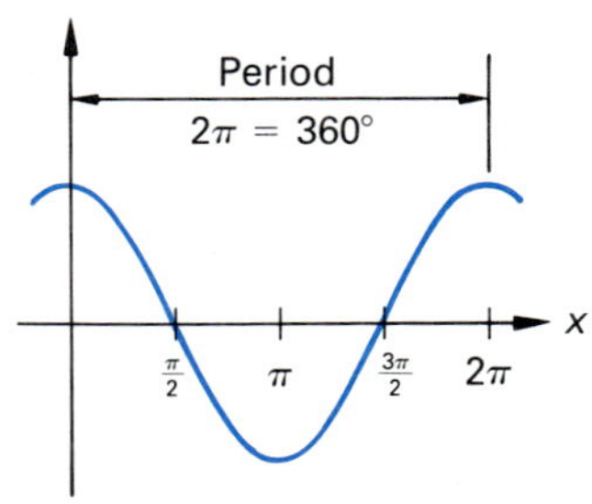

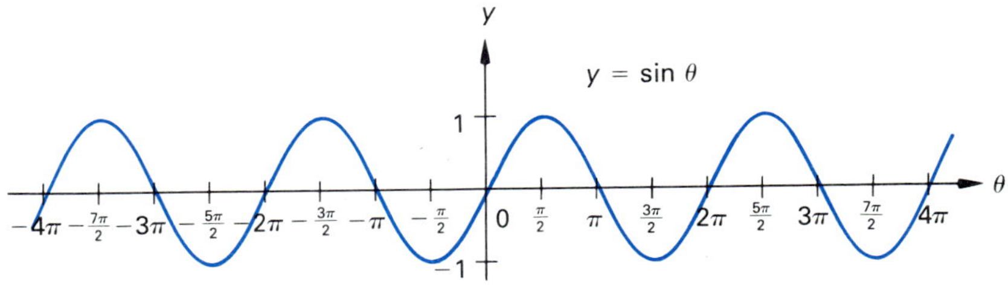

FIGURE 8.7

2. **The range is the closed interval $-1 \leq y \leq 1$.** The function achieves values between 1 and -1 inclusive, but never larger than 1 or smaller than -1.
3. **The *y*-intercept is 0** (where the curve crosses the vertical axis).
4. **The θ intercepts are 0, $\pm\pi$, $\pm 2\pi$, $\pm 3\pi$, . . .** (where the curve crosses the horizontal axis, Fig. 8.7).
5. **$\sin(\theta + \pi) = -\sin\theta$, for all θ.**

$$\sin\left(\frac{5\pi}{12} + \pi\right) = \sin\frac{17\pi}{12} = -0.9659 = -\sin\frac{5\pi}{12}$$

$$\sin\left(\frac{\pi}{2} + \pi\right) = \sin\frac{3\pi}{2} = -1 \qquad = -\sin\frac{\pi}{2}$$

6. **$\sin(-\theta) = -\sin\theta$, for all θ.** (The sine is an odd function.)

$$\sin\left(-\frac{23\pi}{12}\right) = -0.2588 = -\sin\frac{23\pi}{12}$$

$$\sin\left(-\frac{\pi}{12}\right) = -0.2588 = -\sin\frac{\pi}{12}$$

7. **$\sin(\pi - \theta) = \sin\theta$, for all θ.**

$$\sin\left(\pi - \frac{\pi}{4}\right) = \sin\left(\frac{3\pi}{4}\right) = 0.7071 \qquad = \sin\frac{\pi}{4}$$

$$\sin\left(\pi - \frac{4\pi}{3}\right) = \sin\left(-\frac{\pi}{3}\right) = -0.8660 = \sin\frac{4\pi}{3}$$

8. **The sine function is periodic with period 2π.**

Figure 8.7 illustrates again the fact that the curve goes through one complete cycle in 2π rad.

EXAMPLE 2 Construct the graph of $y = \cos\theta$ for the interval $0 \leq \theta \leq 2\pi$ ($0° \leq \theta \leq 360°$). Determine coordinates for intervals of $\pi/12$ rad (or 15°).

Solution: The calculator was used to determine the values for Table 8.2. The graph of $y = \cos\theta$ is sketched in Fig. 8.8. □

The graph of $\cos\theta$ illustrates the fact that the function is positive in the first and fourth quadrants. In these quadrants the curve is above the horizontal axis. It also shows that the function is negative in the second and third quadrants since the curve lies below the horizontal axis.

Also, note from our definition of periodic functions in Section 8.1 that as the curve reaches the value of 2π, it has gone through one complete cycle.

From Table 8.2 and Fig. 8.8 we note the following **properties of the cosine function.**

TABLE 8.2 $y = \cos\theta$

θ			θ		
Radians	Degrees	$\cos\theta$	Radians	Degrees	$\cos\theta$
0	0°	1.0000	π	180°	−1.0000
$\frac{\pi}{12}$	15°	0.9659	$\frac{13\pi}{12}$	195°	−0.9659
$\frac{\pi}{6}$	30°	0.8660	$\frac{7\pi}{6}$	210°	−0.8660
$\frac{\pi}{4}$	45°	0.7071	$\frac{5\pi}{4}$	225°	−0.7071
$\frac{\pi}{3}$	60°	0.5000	$\frac{4\pi}{3}$	240°	−0.5000
$\frac{5\pi}{12}$	75°	0.2588	$\frac{17\pi}{12}$	255°	−0.2588
$\frac{\pi}{2}$	90°	0	$\frac{3\pi}{2}$	270°	0
$\frac{7\pi}{12}$	105°	−0.2588	$\frac{19\pi}{12}$	285°	0.2588
$\frac{2\pi}{3}$	120°	−0.5000	$\frac{5\pi}{3}$	300°	0.5000
$\frac{3\pi}{4}$	135°	−0.7071	$\frac{7\pi}{4}$	315°	0.7071
$\frac{5\pi}{6}$	150°	−0.8660	$\frac{11\pi}{6}$	330°	0.8660
$\frac{11\pi}{12}$	165°	−0.9659	$\frac{23\pi}{12}$	345°	0.9659
π	180°	−1.0000	2π	360°	1.0000

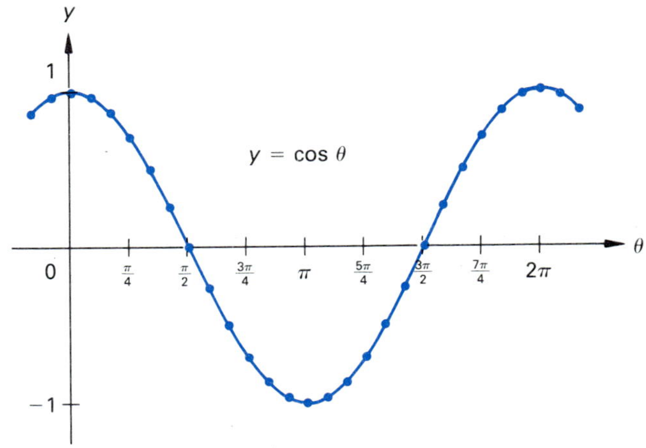

FIGURE 8.8

1. **The domain is the set of all real numbers.** The curve does not stop at 0 and 2π, it continues to the right and left indefinitely (see Fig. 8.9).
2. **The range is the closed interval $-1 \leq y \leq 1$.** The function achieves values between -1 and 1 inclusive, but never larger than 1 or smaller than -1.
3. **The y-intercept is 1.**
4. **The θ intercepts are $\pm\frac{\pi}{2}, \pm\frac{3\pi}{2}, \pm\frac{5\pi}{2}$, ...** (See Fig. 8.8.)
5. **$\cos(\pi + \theta) = \cos(\pi - \theta) = -\cos\theta$, for all θ.**

$$\cos\left(\pi + \frac{\pi}{4}\right) = \cos\frac{5\pi}{4} = -0.7071 = -\cos\frac{\pi}{4}$$

$$\cos\left(\pi - \frac{\pi}{4}\right) = \cos\frac{3\pi}{4} = -0.7071 = -\cos\frac{\pi}{4}$$

$$\cos\left(\pi + \frac{5\pi}{6}\right) = \cos\frac{11\pi}{6} = 0.8660 = -\cos\frac{5\pi}{6}$$

$$\cos\left(\pi - \frac{5\pi}{6}\right) = \cos\frac{\pi}{6} = 0.8660 = -\cos\frac{5\pi}{6}$$

6. **$\cos(-\theta) = \cos\theta$, for all θ.** Cosine is an even function.

$$\cos\left(\frac{-\pi}{3}\right) = 0.5000 = \cos\left(\frac{\pi}{3}\right)$$

$$\cos\left(\frac{-7\pi}{12}\right) = -0.2588 = \cos\left(\frac{7\pi}{12}\right)$$

7. **The cosine function is periodic with period 2π.**

If we shifted cosine curve in Fig. 8.9 $\pi/2$ rad to the right and placed it on the sine curve in Fig. 8.7, it would fit exactly. From studying Fig.

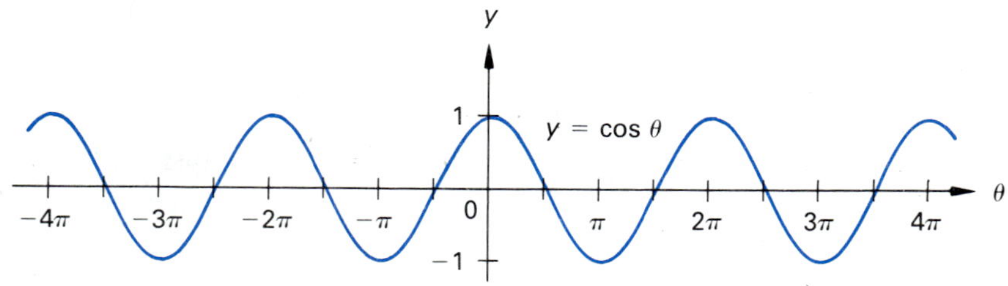

FIGURE 8.9

8.7 we can see that $\sin\left(\frac{\pi}{2}-\theta\right)=\cos\theta$. This result will be proved in Chapter 15.

To help construct the graph of the tangent function, recall that $\tan\theta=\frac{\sin\theta}{\cos\theta}$, $\cos\theta\neq 0$. However, $\cos\theta=0$ when $\theta=\pm\frac{\pi}{2}, \pm\frac{3\pi}{2}, \pm\frac{5\pi}{2}, \ldots$ and the tangent function is not defined for these values of θ. The $\tan\theta$ is 0 when $\sin\theta$ is 0, that is, when $\theta=0, \pm\pi \pm 2\pi, \ldots$.

The value π is the smallest positive number p for which $\tan(\theta+p)=\tan\theta$ for all values of θ. Since $\tan 0=\tan\pi=0$, $\tan\left(\frac{\pi}{3}\right)=\tan\left(\frac{\pi}{3}+\pi\right)=1.7321$, and $\tan\left(-\frac{\pi}{2}\right)=\tan\frac{\pi}{2}$, which is undefined. From the previous statement we know that for $\frac{-\pi}{2}<\theta<\frac{\pi}{2}$, $\tan\theta$ is defined and takes on unique values for each value of θ in the interval. Therefore, the tangent function has a period of π, and the graph repeats every π units.

EXAMPLE 3 Construct the graph of $y=\tan\theta$ for the interval $-\frac{\pi}{2}<\theta<\frac{\pi}{2}$. Determine coordinates for intervals of $\frac{\pi}{12}$ rad (or 15°).

Solution: From Table 8.3 it can be seen as we get closer to $\theta=-\pi/2$ the functional value, $\tan\theta$, becomes larger in the negative direction. In fact, if we calculate $\tan\theta$ for values of θ that are closer to $-\pi/2$, we would obtain negative values for the tangent function whose absolute values increase without bound. For example, if $\theta=-74\pi/150$, then $\tan\theta=-47.7$ and if $\theta=-149\pi/300$, then $\tan\theta=-95.5$. The same type of results can be shown for values of θ close to $\pi/2$. For values of θ that are closer and closer to $-\pi/2$ and $\pi/2$ the absolute values increase without bound or, in other words, become undefined at $-\pi/2$ and $\pi/2$. As a result we call the dotted lines on the graph, $\theta=-\pi/2$ and $\theta=\pi/2$, asymptotes. The asymptotes will occur wherever the function is undefined, that is $\theta=\pm\frac{\pi}{2}, \pm\frac{3\pi}{2}, \pm\frac{5\pi}{2}, \ldots$. □

TABLE 8.3 $y = \tan\theta$

θ			θ		
Radian	Degrees	$\tan\theta$	Radian	Degrees	$\tan\theta$
$-\frac{\pi}{2}$	$-90°$	Undefined	$\frac{\pi}{12}$	$15°$	0.2679
$\frac{-5\pi}{12}$	$-75°$	-3.7321	$\frac{\pi}{6}$	$30°$	0.5774
$-\frac{\pi}{3}$	$-60°$	-1.7321	$\frac{\pi}{4}$	$45°$	1.0000
$-\frac{\pi}{4}$	$-45°$	-1.0000	$\frac{\pi}{3}$	$60°$	1.7321
$-\frac{\pi}{6}$	$-30°$	-0.5774	$\frac{5\pi}{12}$	$75°$	3.7321
$-\frac{\pi}{12}$	$-15°$	-0.2679	$\frac{\pi}{2}$	$90°$	Undefined
0	$0°$	0			

From Table 8.3 and Fig. 8.10 we note the following **properties of the tangent function.**

1. **The domain is the set of real numbers except** $\pm\frac{\pi}{2}, \pm\frac{3\pi}{2}, \pm\frac{5\pi}{2}, \pm\frac{7\pi}{2}, \ldots$.
2. **The range is the set of all real numbers.**
3. **The y-intercept is 0.**
4. **The θ-intercepts are** $0, \pm\pi, \pm 2\pi, \pm 3\pi, \ldots$.
5. **Tan $(\theta + \pi) = \tan\theta$ for all θ in the domain of the tangent. That is, the tangent is a periodic function with period π.** (See Fig. 8.11.)
6. **Tan** $(-\theta) = \dfrac{\sin(-\theta)}{\cos(-\theta)}$ **or** $\tan(-\theta) = \dfrac{-\sin\theta}{\cos\theta}$; **therefore** $\tan(-\theta) = -\tan\theta$ **for all θ in the domain of the tangent. (Thus, the tangent is an odd function.)**
7. **Tan θ is undefined (approaches infinity) as θ approaches** $\pm\frac{\pi}{2}, \pm\frac{3\pi}{2}, \pm\frac{5\pi}{2}, \ldots$ **from the left or the right.** (See Fig. 8.11.)

FIGURE 8.10

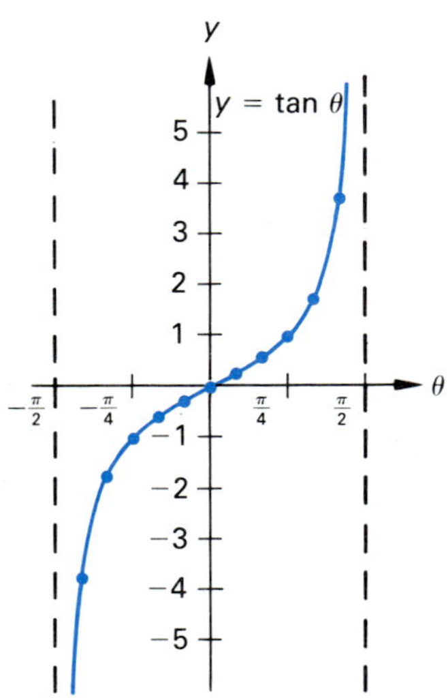

The graphs of the secant, cosecant, and cotangent functions are given in Figures 8.12, 8.13, and 8.14, respectively. It will be left to the reader to make a list of the properties of each of the functions.

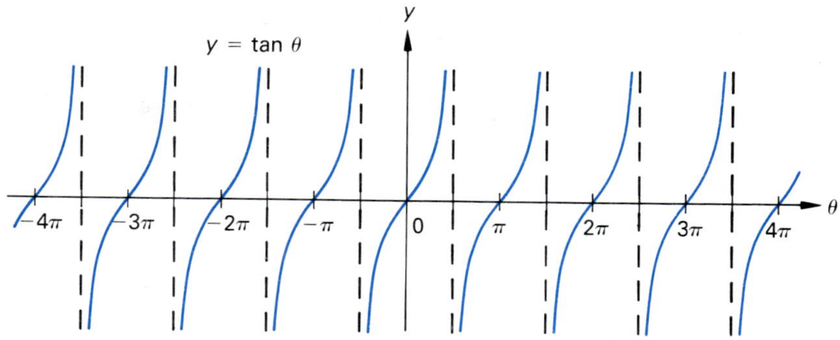

FIGURE 8.11

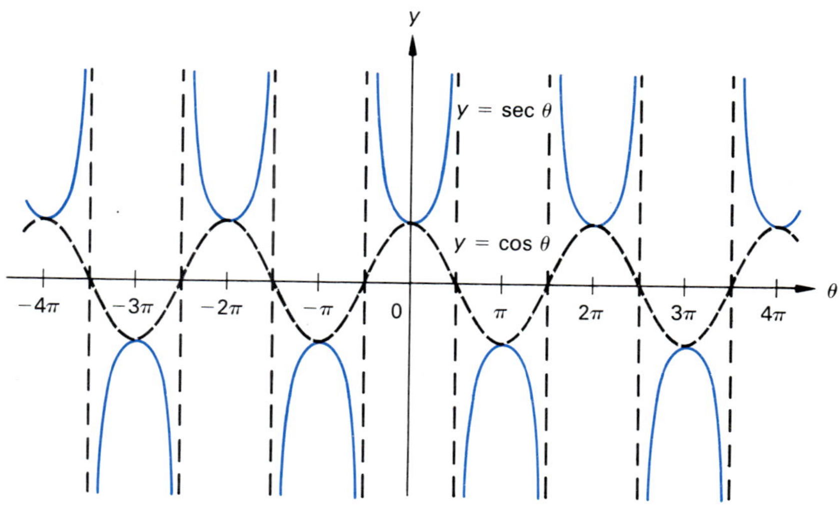

FIGURE 8.12

FIGURE 8.13

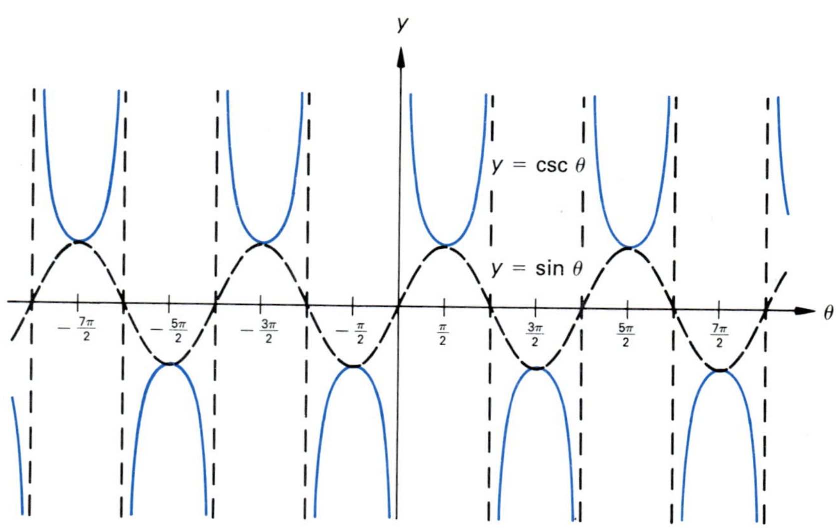

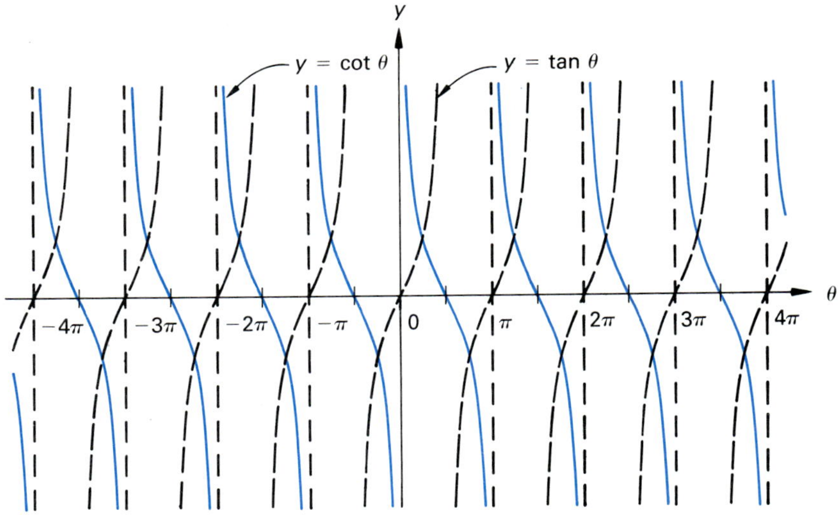

FIGURE 8.14

It is helpful in graphing the functions to recall that:

$\sec \theta = \dfrac{1}{\cos \theta}, \cos \theta \neq 0$ **The secant function is undefined where $\cos \theta = 0$.**

$\csc \theta = \dfrac{1}{\sin \theta}, \sin \theta \neq 0$ **The cosecant function is undefined where $\sin \theta = 0$.**

$\cot \theta = \dfrac{1}{\tan \theta}, \tan \theta \neq 0$ **The cotangent function is undefined where $\tan \theta = 0$.**

The cosecant and cotangent functions are undefined when $\theta = 0, \pm\pi, \pm 2\pi, \pm 3\pi, \ldots$, that is, when the sine and tangent functions are 0.

8.2 Exercises

In Exercises 1–6, make a list of the properties of the

1. sine function
2. cosine function
3. tangent function
4. cosecant function
5. secant function
6. cotangent function

In Exercises 7–18, graph each of the functions for the indicated interval.

7. $y = \sin \theta$, for values of θ at intervals of $\frac{\pi}{15}$ rad, $-\pi \le \theta \le \pi$.
8. $y = \cos \theta$, for values of θ at intervals of 15°, $-180° \le \theta \le 180°$.
9. $y = \tan \theta$, for values of θ at intervals of 15°, $-270° < \theta < 270°$.
10. $y = \sin \theta$, for values of θ at intervals of 30°, $0 \le \theta \le 360°$.
11. $y = \cos \theta$, for values of θ at intervals of $\frac{\pi}{6}$ rad, $-2\pi \le \theta \le 0$.
12. $y = \tan \theta$, for values of θ at intervals of $\frac{\pi}{3}$ rad, $-\frac{\pi}{2} < \theta < \frac{7\pi}{2}$.

13. $y = \sec \theta$, for values of θ at intervals of $\frac{\pi}{6}$ rad, $-\frac{\pi}{2} \le \theta \le \frac{3\pi}{2}$.

14. $y = \cot \theta$, for values of θ at intervals of $\frac{\pi}{4}$ rad, $\frac{\pi}{2} \le \theta \le \frac{5\pi}{2}$.

15. $y = \csc \theta$, for values of θ at intervals of $45°$, $-180° \le \theta \le 360°$.

16. $y = \sin \theta$, for values of θ at intervals of $\frac{\pi}{4}$ rad, $\frac{-5\pi}{2} \le \theta \le \frac{5\pi}{2}$.

17. $y = \cos \theta$, for values of θ at intervals of $60°$, $0° \le \theta \le 480°$.

18. $y = \tan \theta$, for values of θ at intervals of $\frac{\pi}{4}$ rad, $-\frac{5\pi}{2} < \theta < \frac{5\pi}{2}$.

In Exercises 19–22, graph the functions on the same set of axes for the indicated interval. Determine coordinates for intervals of $\frac{\pi}{12}$ rad (or $15°$).

19. $y = \sin \theta$ and $y = \csc \theta$ for $-360° \le \theta \le 360°$.

20. $y = \sin \theta$ and $y = \csc \theta$ for $-270° \le \theta \le 270°$.

21. $y = \cos \theta$ and $y = \sec \theta$ for $\frac{-3\pi}{2} \le \theta \le \frac{3\pi}{2}$.

22. $y = \tan \theta$ and $y = \cot \theta$ for $-\pi \le \theta \le \pi$.

In Exercises 23–40, which trigonometric functions (sine, cosine, tangent, secant, cosecant, cotangent) have the following properties?

23. The domain is the set of real numbers.

24. The domain is the set of real numbers excluding 0, $\pm\pi, \pm 2\pi, \pm 3\pi, \ldots$.

25. The function has period of π.

26. The y-intercept is 1.

27. The range is the set of all real numbers.

28. There is no y-intercept.

29. The domain is the set of real numbers excluding $\pm\frac{\pi}{2}, \pm\frac{3\pi}{2}, \pm\frac{5\pi}{2}, \ldots$.

30. The range is $-1 \le y \le 1$.

31. The range is $|y| \ge 1$.

32. The function has a period of 2π.

33. The θ-intercepts are $0, \pm\pi, \pm 2\pi, \pm 3\pi, \ldots$.

34. The θ-intercepts are $\pm\frac{\pi}{2}, \pm\frac{3\pi}{2}, \pm\frac{5\pi}{2}, \ldots$.

35. The y-intercept is 0.

36. $f(-\theta) = f(\theta)$.

37. $f(-\theta) = -f(\theta)$.

38. $f(\theta + \pi) = f(\theta)$.

39. $f(\theta + \pi) = -f(\theta)$.

40. $f(\pi - \theta) = f(\theta)$.

8.3 Sinusoids ($Y = A \sin \theta$ and $y = A \cos \theta$)

In the discussion of the cosine function in Section 8.2 we discovered that if the cosine curve is shifted $\pi/2$ rad to the right, it will fit exactly on the sine curve. Thus, we call the family of curves that look like the sine curve **sinusoids.** The word sinusoids can be split into two parts, "sinus-" coming from the same origin as "sine," and "-oid," a suffix meaning "like." We encounter many "sine-like" curves in electronic testing devices, alternating currents, wave actions, and forces in rotating mechanisms.

Before getting into specific applications we will discuss the general forms of the sine and cosine curves. The most general form of the curves is expressed by the functions

$$y = A \sin (B\theta + C) \text{ and } y = A \cos (B\theta + C).$$

TABLE 8.4
$y = 2 \sin \theta$

θ	$\sin \theta$	$2 \sin \theta$
0	0	0
$\frac{\pi}{4}$	0.707	1.41
$\frac{\pi}{2}$	1.000	2.00
$\frac{3\pi}{4}$	0.707	1.41
π	0	0
$\frac{5\pi}{4}$	-0.707	-1.41
$\frac{3\pi}{2}$	-1.000	-2.00
$\frac{7\pi}{4}$	-0.707	-1.41
2π	0	0

Each of the constants A, B, and C has an individual effect on the curve. Therefore, we will examine individual functions with each of the constants. [For example: $y = A \sin \theta$, $y = \sin B\theta$, $y = \sin(\theta + C)$.] Then we will consider the general form, $y = A \sin(B\theta + C)$ and $y = A \cos(B\theta + C)$.

EXAMPLE 1 For $0 \le \theta \le 2\pi$ and intervals of $\pi/4$ sketch the graph of $y = 2 \sin \theta$.

Solution: In Table 8.4, use the values in the columns headed θ and $2 \sin \theta$ to sketch the graph of $y = 2 \sin \theta$ (Fig. 8.15).

To help illustrate the difference between $y = 2 \sin \theta$ and $y = \sin \theta$, we sketched $y = \sin \theta$ on the same axes. From the sketch we observe that the only difference between the two curves is the range (distance above and below the horizontal axis). Note that the range is multiplied by a factor of 2, the coefficient of $\sin \theta$ for the function $y = 2 \sin \theta$. □

If the function being sketched were $y = 4 \sin x$, the range would be multiplied by 4. When graphing the curve of $y = A \sin x$, its distance above and below the horizontal axis will be A units (Fig. 8.16).

EXAMPLE 2 For $0 \le \theta \le 2\pi$ and intervals of $\pi/4$ sketch the graph of $y = -0.80 \cos \theta$.

Solution: In Table 8.5, use the values in the columns headed θ and $-0.80 \cos \theta$ to sketch the graph of $y = -0.80 \cos \theta$ (Fig. 8.17).

FIGURE 8.15

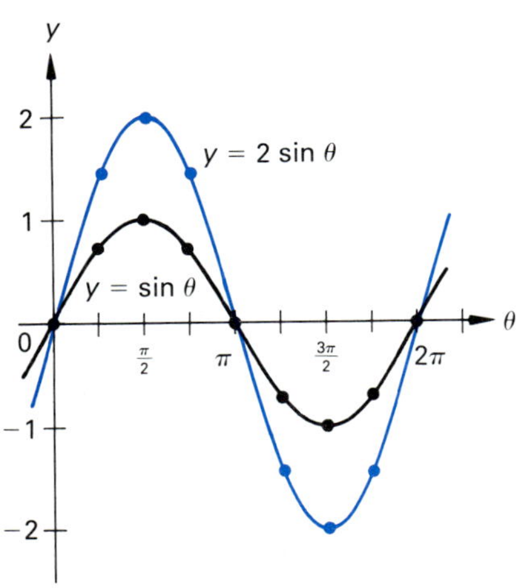

FIGURE 8.16

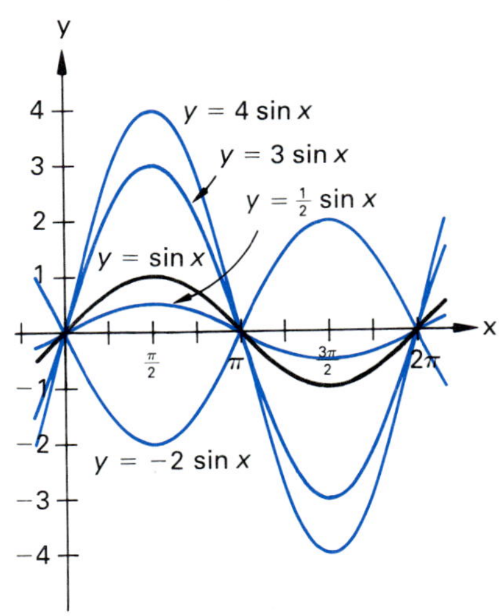

TABLE 8.5
$y = -0.80 \cos \theta$

θ	$\cos \theta$	$-0.80 \cos \theta$
0	1.000	−0.80
$\frac{\pi}{4}$	0.707	−0.57
$\frac{\pi}{2}$	0	0
$\frac{3}{4}\pi$	−0.707	0.57
π	−1.000	0.80
$\frac{5\pi}{4}$	−0.707	0.57
$\frac{3\pi}{2}$	0	0
$\frac{7\pi}{4}$	0.707	−0.57
2π	1.000	−0.80

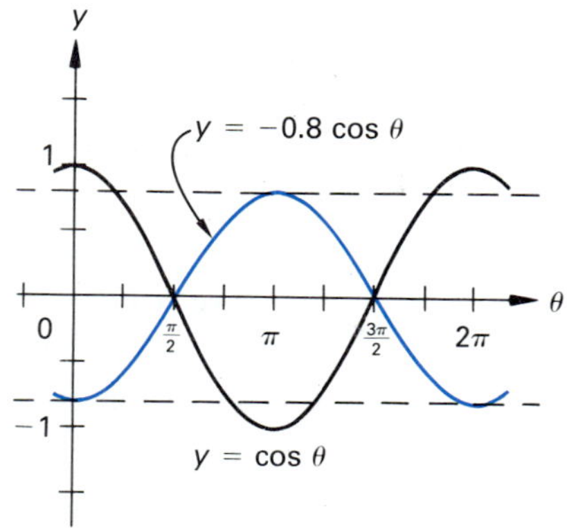

FIGURE 8.17

Again to help analyze the curve, $y = \cos \theta$ is sketched on the same axes (Fig. 8.17). Note that the function $y = -\cos \theta$ is a mirror image of $y = \cos \theta$. Therefore, where $y = \cos \theta$ is positive, $y = -0.80 \cos \theta$ is negative; also, when $y = \cos \theta$ is negative, $y = -0.80 \cos \theta$ is positive. The curves have the same shape and period. The only other difference is that $y = -0.80 \cos \theta$ has a range of $-0.80 \leq y \leq 0.80$, while $y = \cos \theta$ has a range of $-1 \leq y \leq 1$. □

When graphing the curve of $y = A \cos x$, the displacement above and below the horizontal axis will be A units. The function $y = A \cos x$ is sketched in Fig. 8.18 for $A = 1/2, 1, -2, 3$, and 4.

The distance a sinusoid curve rises and falls above and below its horizontal axis is called the **amplitude.** The amplitudes of the functions $y = \sin \theta$, $y = 3 \cos \theta$, and $y = -0.80 \cos \theta$ are 1, 3, and 0.80, respectively.

FIGURE 8.18

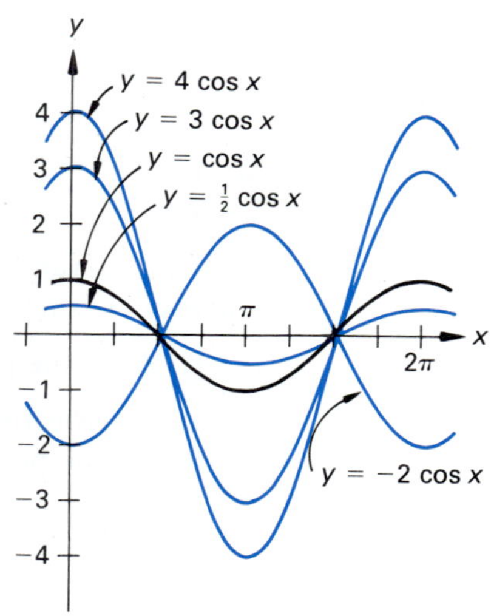

TABLE 8.6 $y = \sin 3\theta$

θ	3θ	$\sin 3\theta$
0	0	0
$\frac{\pi}{6}$	$\frac{\pi}{2}$	1
$\frac{\pi}{3}$	π	0
$\frac{\pi}{2}$	$\frac{3\pi}{2}$	-1
$\frac{2\pi}{3}$	2π	0
$\frac{5\pi}{6}$	$\frac{5\pi}{2}$	1
π	3π	0
$\frac{7\pi}{6}$	$\frac{7\pi}{2}$	-1
$\frac{4\pi}{3}$	4π	0
$\frac{3\pi}{2}$	$\frac{9\pi}{2}$	1
$\frac{5\pi}{3}$	5π	0
$\frac{11\pi}{6}$	$\frac{11\pi}{2}$	-1
2π	6π	0

Properties of $y = A \sin \theta$ and $y = A \cos \theta$ are

1. The amplitude of $y = A \sin \theta$ and $y = A \cos \theta$ is given by the absolute value of A.
2. If the coefficient of the function is negative, the graph of the curve is a mirror image (with respect to the horizontal axis) of the same function with a positive coefficient.
3. The period of both of these functions is 2π rad or $360°$.

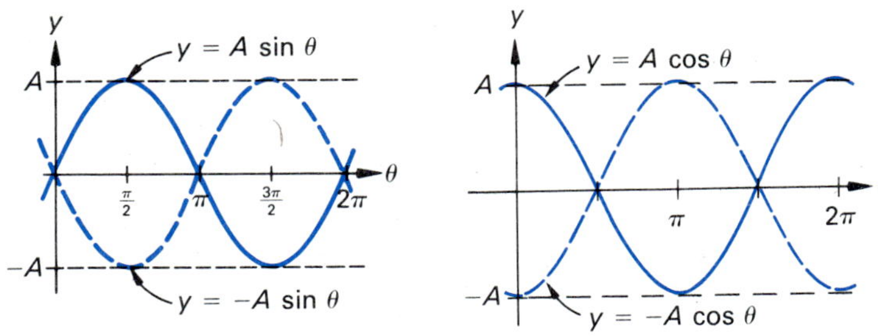

$y = \sin Bx$ and $y = \cos Bx$

To determine the effect that B has on the curve we will examine functions of the form $y = A \sin (Bx + C)$ and $y = A \cos (Bx + C)$, where $A = 1$ and $C = 0$.

EXAMPLE 3 Sketch the graph of

a. $y = \sin 3\theta$ for $0 \le \theta \le 2\pi$

b. $y = \sin \frac{1}{3}\theta$ for $0 \le \theta \le 6\pi$

Solution:

a. $y = \sin 3\theta$

The function $y = \sin 3\theta$ makes three complete cycles in 2π rad, whereas $y = \sin \theta$ makes one complete cycle (see Table 8.6 and Fig. 8.19). Thus, the **period** of $y = \sin 3\theta$ is $\frac{2\pi}{3}$. The amplitude of the curve is 1. From this we see that the 3 in the function $y = \sin 3\theta$ will affect only the period of the curve.

FIGURE 8.19

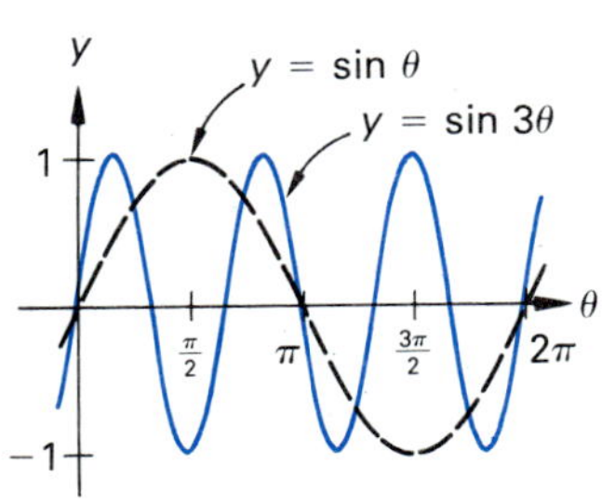

b. $y = \sin \frac{1}{3}\theta$

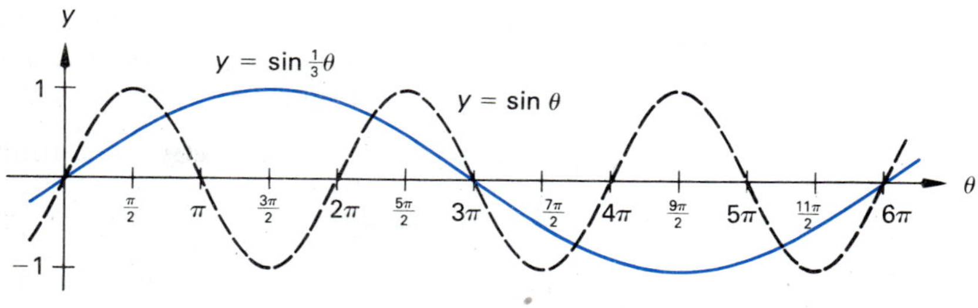

FIGURE 8.20

From Fig. 8.20 it can be seen that $y = \sin\frac{1}{3}\theta$ goes through one cycle in 6π rad, whereas $y = \sin\theta$ goes through three cycles. Thus, the period of $y = \sin\frac{1}{3}\theta$ is 6π. Note that even though the period is 6π, the interval from $0 \le \theta \le 6\pi$ can be divided into four regions corresponding to the four quadrants ($0 \le \theta \le \frac{3\pi}{2}$, $\frac{3\pi}{2} \le \theta \le 3\pi$, $3\pi \le \theta \le \frac{9\pi}{2}$, and $\frac{9\pi}{2} \le \theta \le 6\pi$). (See also, Table 8.7.) Also note that the values 0, $\frac{3\pi}{2}$, 3π, $\frac{9\pi}{2}$, and 6π determine key points in sketching the curve. □

TABLE 8.7 $y = \sin\frac{1}{3}\theta$

θ	$\frac{1}{3}\theta$	$\sin\frac{1}{3}\theta$
0	0	0
$\frac{\pi}{2}$	$\frac{\pi}{6}$	0.50
π	$\frac{\pi}{3}$	0.87
$\frac{3\pi}{2}$	$\frac{\pi}{2}$	1
2π	$\frac{2\pi}{3}$	0.87
$\frac{5\pi}{2}$	$\frac{5\pi}{6}$	0.50
3π	π	0
$\frac{7\pi}{2}$	$\frac{7\pi}{6}$	−0.50
4π	$\frac{4\pi}{3}$	−0.87
$\frac{9\pi}{2}$	$\frac{3\pi}{2}$	−1
5π	$\frac{5\pi}{3}$	−0.87
$\frac{11\pi}{2}$	$\frac{11\pi}{6}$	−0.50
6π	2π	0

From the graphs in Example 2, we can see that for a function of the type $y = \sin B\theta$, if B is a counting number, the period will be less than 2π and if B is a fraction of the form $1/n$, where n is a counting number, the period will be greater than 2π.

Property of $y = \sin B\theta$ and $y = \cos B\theta$

The **period** of $y = \sin B\theta$ and $y = \cos B\theta$ can be determined by dividing 2π by $|B|$:

$$\text{Period} = \frac{2\pi}{|B|}.$$

The period of $y = \cos\frac{2\theta}{3}$ is

$$\text{Period} = \frac{2\pi}{2/3} = 3\pi.$$

Generally it is important to be able to sketch one or two periods (cycles) of a curve. The following are a few key ideas that will help in sketching $y = A \sin B\theta$ and $y = A \cos B\theta$.

1. Determine the amplitude. (Amplitude $= |A|$.)
2. Determine the period. $\left(\text{Period} = \dfrac{2\pi}{|B|}.\right)$
3. Each period is divided into four equal intervals corresponding to the four quadrants. The end points of these intervals are key points in sketching the function. To determine these five points, do the following: Divide the period by 4, $\dfrac{\text{period}}{4} = \dfrac{2\pi}{4|B|}$. Multiply the result, $\dfrac{2\pi}{4|B|}$, by 0, 1, 2, 3, and 4. Substitute these values $\left[0, \dfrac{2\pi}{4|B|}, 2\left(\dfrac{2\pi}{4|B|}\right), 3\left(\dfrac{2\pi}{4|B|}\right), 4\left(\dfrac{2\pi}{4|B|}\right)\right]$ into the function to obtain the five key points, $(0, f(0))$, $\left(\dfrac{2\pi}{4|B|}, f\left(\dfrac{2\pi}{4|B|}\right)\right)$, $\left(2\left(\dfrac{2\pi}{4|B|}\right), f\left(2\dfrac{2\pi}{4|B|}\right)\right)$ $\left(3\left(\dfrac{2\pi}{4|B|}\right), f\left(3\dfrac{2\pi}{4|B|}\right)\right)$, and $\left(4\left(\dfrac{2\pi}{4|B|}\right) f\left(4\dfrac{2\pi}{4|B|}\right)\right)$. (See Fig. 8.21.)

EXAMPLE 4 Sketch $y = 4 \cos \frac{3}{5}\theta$.

Solution: To help sketch the function $y = 4 \cos \frac{3}{5}\theta$ determine the key ideas as suggested in the previous discussion.

1. Amplitude $= 4$.
2. Period $= \dfrac{2\pi}{3/5}$ or $\dfrac{10\pi}{3}$.
3. The five key points are found by:
   a. Dividing period by 4, $\frac{1}{4}\left(\dfrac{10\pi}{3}\right) = \dfrac{5\pi}{6}$.
   b. Multiplying $\dfrac{5\pi}{6}$ by 0, 1, 2, 3, and 4 to get 0, $\dfrac{5\pi}{6}, \dfrac{5\pi}{3}, \dfrac{5\pi}{2}$, and $\dfrac{10\pi}{3}$.

FIGURE 8.21

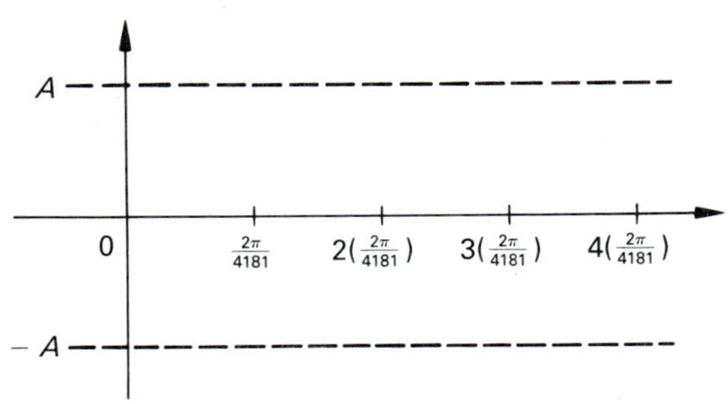

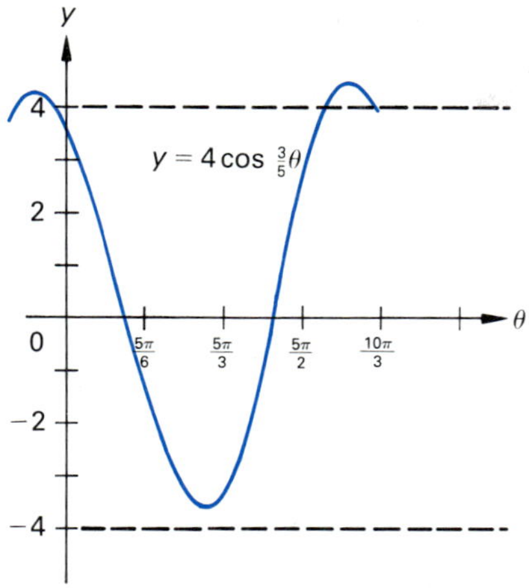

FIGURE 8.22

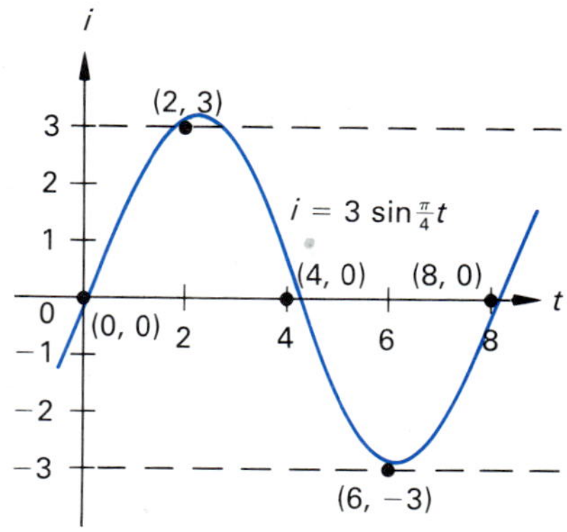

FIGURE 8.23

c. Now, substituting the values $0, \frac{5\pi}{6}, \frac{5\pi}{3}, \frac{5\pi}{2}$, and $\frac{10\pi}{3}$ for θ in the function $y = 4\cos\frac{3}{5}\theta$, we can write the points $(0, 4)$, $\left(\frac{5\pi}{6}, 0\right)$, $\left(\frac{5\pi}{3}, -4\right)$, $\left(\frac{5\pi}{2}, 0\right)$, and $\left(\frac{10\pi}{3}, 4\right)$.

4. Using these results we sketch $y = 4\cos\frac{3}{5}\theta$. (See Fig. 8.22.) □

EXAMPLE 5 A control system current is given by $i = 3\sin\frac{\pi}{4}t$. Sketch one cycle of the function.

Solution: For the function $i = 3\sin\frac{\pi}{4}t$ we determine that

1. $A = 3$ and the amplitude is 3.

2. $B = \frac{\pi}{4}$ and the period $= \dfrac{2\pi}{\frac{\pi}{4}}$ or 8.

3. The five key values are $0\left(\frac{8}{4}\right) = 0$, $1\left(\frac{8}{4}\right) = 2$, $2\left(\frac{8}{4}\right) = 4$, $3\left(\frac{8}{4}\right) = 6$, and $4\left(\frac{8}{4}\right) = 8$. Thus the key points on the graph would be $(0, 0)$, $(2, 3)$, $(4, 0)$, $(6, -3)$, and $(8, 0)$.

4. Using these results we sketch $i = 3\sin\frac{\pi}{4}t$. (See Fig. 8.23.) □

8.3 Exercises

In Exercises 1–12, determine (a) the amplitude and (b) the period of each function.

1. $y = 3 \sin \theta$

2. $y = \frac{1}{2} \cos \theta$

3. $y = 0.75 \cos \theta$

4. $y = 1.3 \sin \theta$

5. $y = \sin 3\theta$

6. $y = \cos \frac{\theta}{2}$

7. $y = -6 \cos \frac{\theta}{4}$

8. $y = 2 \sin \pi\theta$

9. $y = \frac{1}{3} \sin \frac{\theta}{3}$

10. $y = -\pi \cos \frac{\pi}{2} \theta$

11. $y = -\frac{1}{3} \cos \frac{1}{3} \theta$

12. $y = \frac{5}{3} \sin \frac{5}{3} \theta$

In Exercises 13–24, sketch the graph of each function for two complete cycles starting with the origin.

13. $y = 2 \sin \theta$

14. $y = \sin 2\theta$

15. $y = \cos \frac{\theta}{2}$

16. $y = \frac{1}{2} \cos \theta$

17. $y = 3 \cos 2\theta$

18. $y = 2 \cos 2\theta$

19. $y = 0.5 \sin \frac{\pi}{4} \theta$

20. $y = 4 \sin 0.5\theta$

21. $y = -\frac{1}{2} \sin 2\theta$

22. $y = -2 \sin \frac{\theta}{2}$

23. $y = -\frac{2}{3} \cos \frac{2}{3} \theta$

24. $y = -\pi \cos \frac{\pi}{2} \theta$

In Exercises 25–28, plot the graphs of the functions and label the major points $(0, \frac{\pi}{2}, \pi, \frac{3\pi}{2}, 2\pi$, etc.).

25. $y = 3 \sin \theta$ for $-\pi \leq \theta \leq \pi$ using values of θ at intervals of $\frac{\pi}{12}$.

26. $y = 4 \cos \theta$ for $-\pi \leq \theta \leq \pi$ using values of θ at intervals of $\frac{\pi}{12}$.

27. $y = \frac{1}{2} \cos \theta$ for two cycles starting with $-\frac{3\pi}{2}$.

28. $y = 0.75 \sin \theta$ for two and one-half cycles starting with $-270°$.

29. Find an interval on which $\sin 2\theta$ increases from -1 to 1.

30. Find an interval on which $\cos 4\theta$ decreases from 1 to -1.

31. A control system current is given by $i = 2 \sin \frac{\pi}{2} t$. Sketch one cycle of the function.

32. A servomotor voltage is given by $v = 12 \cos \frac{\pi}{3} t$. Sketch one cycle of the function.

8.4 General sinusoids ($y = A \sin (Bx + C)$ and $y = A \cos (Bx + C)$)

$y = \sin (x + C)$ and $y = \cos (x + C)$

For the general curve $y = A \sin (Bx + C)$ when $C = 0$, we have seen in the previous discussion that the constants A and B affect the amplitude and the period of the curve, respectively. It was also observed that neither has an effect on the other. Now we want to determine the effect that the constant C has on the curve.

The value of C has the effect of shifting the curve to the right or left. For example, in graphing $y = \sin x$ the origin is considered the starting point for sketching the curve (Fig. 8.24). Now let us sketch $y =$

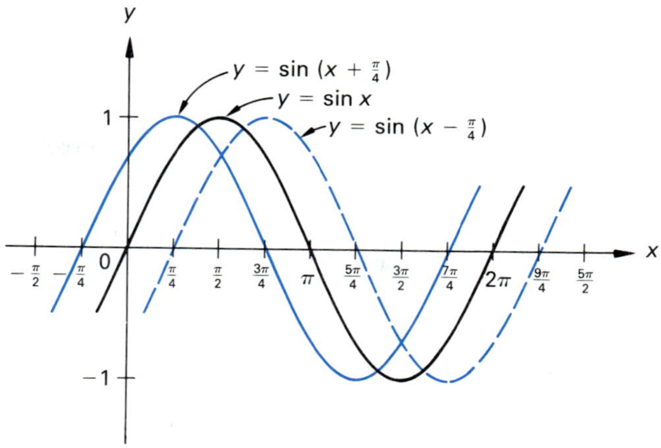

FIGURE 8.24

$\sin\left(x + \frac{\pi}{4}\right)$ and $y = \sin\left(x - \frac{\pi}{4}\right)$ on the same set of axes with $y = \sin x$. Tables 8.8 and 8.9 provide key points for graphing the functions.

From the graph and the tables it can be seen that the starting point is shifted or horizontally displaced $\pi/4$ units to the left or right of the origin, accordingly, as $\pi/4$ is positive or negative. (Note that the curves are "parallel.") This horizontal displacement is called the **phase shift** of the function.

EXAMPLE 1 Given $y = 2\cos\left(x - \frac{\pi}{3}\right)$, sketch one cycle of the curve on each side of the origin.

TABLE 8.8 $y = \sin\left(x + \frac{\pi}{4}\right)$

x	$x + \frac{\pi}{4}$	$\sin\left(x + \frac{\pi}{4}\right)$
$-\frac{\pi}{4}$	0	0
$\frac{\pi}{4}$	$\frac{\pi}{2}$	1
$\frac{3\pi}{4}$	π	0
$\frac{5\pi}{4}$	$\frac{3\pi}{2}$	−1
$\frac{7\pi}{4}$	2π	0

TABLE 8.9 $y = \sin\left(x - \frac{\pi}{4}\right)$

x	$x - \frac{\pi}{4}$	$\sin\left(x - \frac{\pi}{4}\right)$
$\frac{\pi}{4}$	0	0
$\frac{3\pi}{4}$	$\frac{\pi}{2}$	1
$\frac{5\pi}{4}$	π	0
$\frac{7\pi}{4}$	$\frac{3\pi}{2}$	−1
$\frac{9\pi}{4}$	2π	0

Solution: If we substitute θ for $\left(x-\frac{\pi}{3}\right)$ in $y=3\cos\left(x-\frac{\pi}{3}\right)$, the result is $y=3\cos\theta$, which has an amplitude of 3 and a period of 2π. Sketch the function $y=3\cos\theta$ (Fig. 8.25).

The function to be sketched, $y=3\cos\left(x-\frac{\pi}{3}\right)$, has the same amplitude and period as $y=3\cos\theta$. However, it has a phase shift of $\pi/3$ units. Because the sign of $\pi/3$ is negative we move the key points $\pi/3$ units to the right and sketch the required curve. □

EXAMPLE 2 Given $y=\cos\left(3x+\frac{3\pi}{2}\right)$, sketch one cycle on each side of the origin.

Solution: If we factor the number 3 from the expression $3x+\frac{3\pi}{2}$, the expression becomes $3\left(x+\frac{\pi}{2}\right)$. If we let $\theta=x+\frac{\pi}{2}$, then substitute, we have $y=\cos 3\theta$, which has a period of $\frac{2\pi}{3}$ and amplitude of 1. The sketch of $y=\cos 3\theta$ is shown in Fig. 8.26.

The original function has a phase shift of $\pi/2$ units to the left, since the sign of $\pi/2$ is positive. With this information plot the key points and sketch the graph. □

Note that in Example 2 when performing the common-term factoring we are actually dividing $3\pi/2$ by 3. In the general form we divide $-C$ by B, $-C/B$. (We are solving the expression $BX+C=0$ for x.) Therefore, the phase shift can be determined by the formula

$$\text{phase shift} = \frac{-C}{B}.$$

FIGURE 8.25

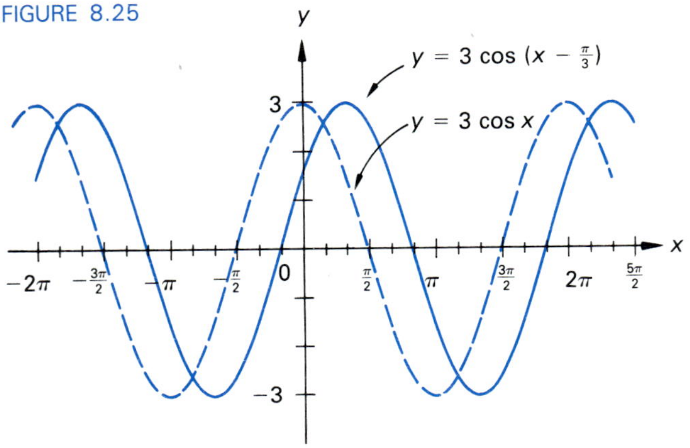

FIGURE 8.26

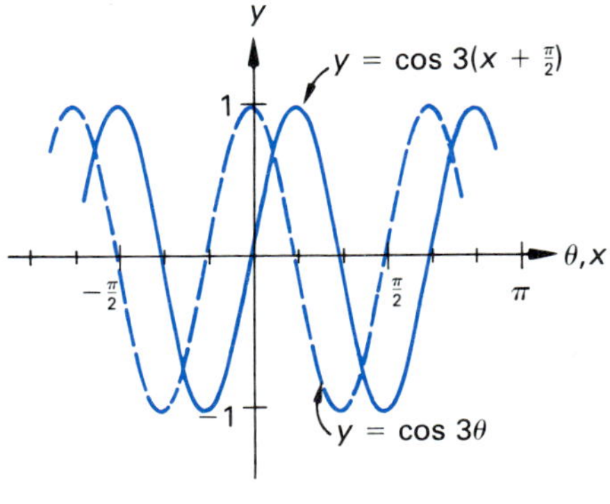

$y = A \sin (Bx + C)$ and $y = A \cos (Bx + C)$

To sketch the graphs of curves of the form $y = A \sin (Bx + C)$ and $y = A \cos (Bx + C)$, it is helpful to determine the following facts:

1. Amplitude $= |A|$.
2. Period $= \dfrac{360°}{|B|}$ or Period $= \dfrac{2\pi}{|B|}$.
3. Phase shift $= \dfrac{-C}{B}$.
4. A new cycle of the curve will begin at $\dfrac{-C}{B} + \dfrac{360°}{|B|}$ or $\dfrac{-C}{B} + \dfrac{2\pi}{|B|}$. This suggests that an initial cycle could begin at $\dfrac{-C}{B}$.
5. Key points are $-\dfrac{C}{B}$, $-\dfrac{C}{B} + \dfrac{2\pi}{4|B|}$, $-\dfrac{C}{B} + 2\left(\dfrac{2\pi}{4|B|}\right)$, $\dfrac{-C}{B} + 3\left(\dfrac{2\pi}{4|B|}\right)$, $-\dfrac{C}{B} + 4\left(\dfrac{2\pi}{4|B|}\right)$. (See Fig. 8.27.)

EXAMPLE 3 Sketch $y = 3 \sin (2\theta + 60°)$. List the amplitude, the period, and the phase shift. Show one cycle and label the graph.

Solution: For the function $y = 3 \sin (2\theta + 60°)$:

$$\text{Amplitude} = |3| = 3 \qquad \text{Period} = \frac{360°}{2} = 180° \qquad \text{Phase shift} = \frac{-60°}{2} = -30°.$$

Curve will recycle at $\dfrac{-60°}{2} + \dfrac{360°}{2} = 150°$.

FIGURE 8.27

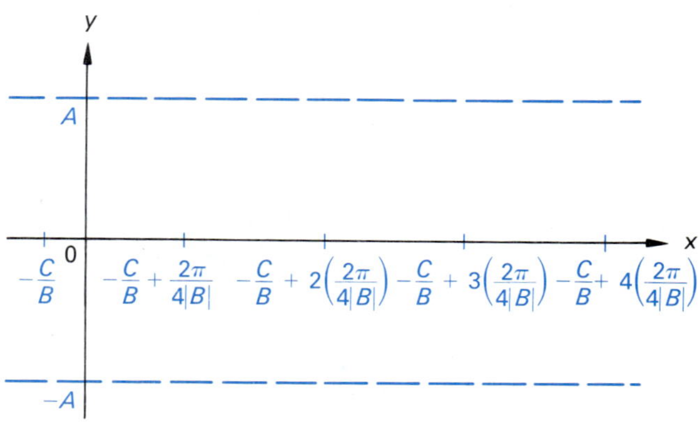

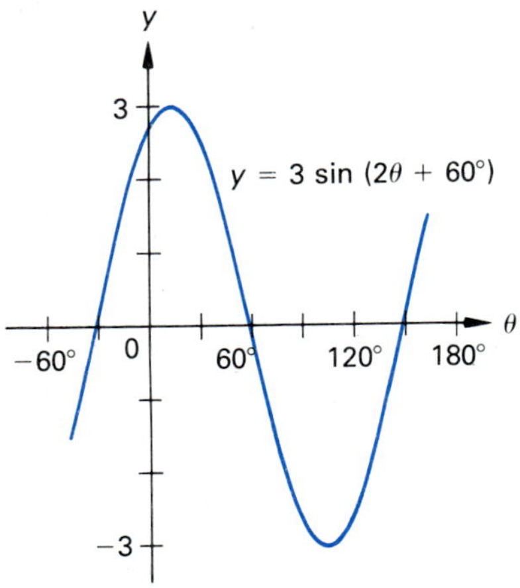

FIGURE 8.28

TABLE 8.10 $y = 3 \sin (2\theta + 60°)$

θ	$2\theta + 60°$	$3 \sin (2\theta + 60°)$
−30°	0°	0
15°	90°	3
60°	180°	0
105°	270°	−3
150°	360°	0

Divide the period 180° by 4 and the result, 45°, will help to locate the key points of the curve. A natural starting point in graphing one cycle would be $-C/B = -30°$. Thus we determine the key values to be $-30°$, $-30° + 45° = 15°$, $-30° + 2(45°) = 60°$, $-30° + 3(45°) = 105°$, and $-30° + 4(45°) = 150°$. The function values are given in Table 8.10. See Fig. 8.28 for the sketch. □

EXAMPLE 4 Sketch $y = 20 \sin (20\pi\theta - \pi)$. List the amplitude, the period, and the phase shift. Show one cycle and label the graph.

Solution: For the function $y = 20 \sin (20\pi\theta - \pi)$:

$$\text{Amplitude} = |20| = 20 \qquad \text{Period} = \frac{2\pi}{20\pi} = \frac{1}{10} \qquad \text{Phase shift} = \frac{-(-\pi)}{20\pi} = \frac{1}{20}.$$

Curve will recycle at $\frac{\pi}{20\pi} + \frac{2\pi}{20\pi} = \frac{3}{20}$.

Divide the period 1/10 by 4 and the result, 1/40, will help to locate the key points of the curve. A natural starting point in graphing one cycle would be $-\frac{C}{B} = \frac{1}{20}$. Thus we determine the key values to be $\frac{1}{20}$, $\frac{1}{20} + \frac{1}{40} = \frac{3}{40}$, $\frac{1}{20} + 2\left(\frac{1}{40}\right) = \frac{1}{10}$, $\frac{1}{20} + 3\left(\frac{1}{40}\right) = \frac{5}{40}$, and $\frac{1}{20} + 4\left(\frac{1}{40}\right) = \frac{3}{20}$. The function values are given in Table 8.11 and sketched in Fig. 8.29. □

EXAMPLE 5 A capacitor voltage is given by the function $v = 9 \cos (5t + 3)$. Sketch one cycle of the function.

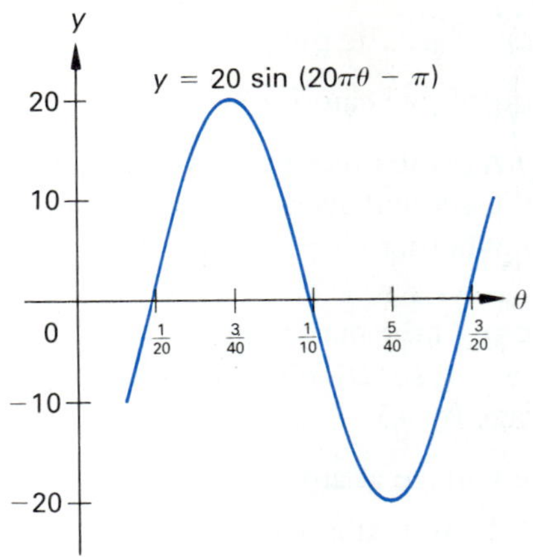

FIGURE 8.29

TABLE 8.11 $y = 20 \sin (20\pi\theta - \pi)$

θ	$20\pi\theta - \pi$	$20 \sin (20\pi\theta - \pi)$
$\frac{1}{20}$	0	0
$\frac{3}{40}$	$\frac{\pi}{2}$	20
$\frac{1}{10}$	π	0
$\frac{5}{40}$	$\frac{3\pi}{2}$	−20
$\frac{3}{20}$	2π	0

Solution: For the function $v = 9 \cos (5t + 3)$ the amplitude $= |9|$, or 9, the period $= \frac{2\pi}{5}$, and the phase shift $= \frac{-3}{5}$.

The key values can be found using the calculator. We know that each value is

$$\frac{2\pi}{4B} = \frac{2\pi}{4(5)} = 0.3142$$

greater than the preceding value. The five values are $-\frac{3}{5}$, $-\frac{3}{5} + 0.3142$, $-\frac{3}{5} + 2(0.3142)$, $-\frac{3}{5} + 3(0.3142)$, and $-\frac{3}{5} + 4(0.3142)$. On the calculator evaluate $\frac{2\pi}{4|B|} = \frac{2\pi}{4(5)}$ and store this value. Then calculate $\frac{-C}{B} = \frac{-3}{5}$, which is the value of the first key point. Then add the stored value to the display; this is the value of the second key point. Repeat the process of adding the stored value to the display to obtain values of the other key points. The key values are −0.6, −0.286, 0.028, 0.343, and 0.657 (Fig. 8.30). □

FIGURE 8.30

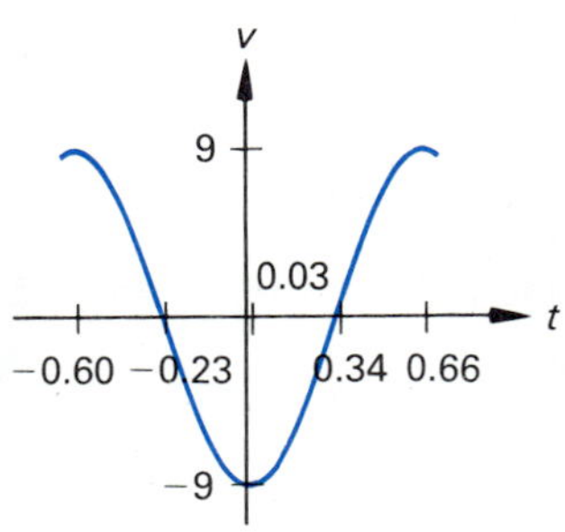

In technical applications we often encounter sinusoidal functions in which the independent variable is time. As a convenience we may write functions of this form as $y = A \sin (\omega t + C)$ or $y = A \cos (\omega t + C)$ instead of $y = A \sin (Bx + C)$ or $y = A \cos (Bx + C)$. The quantity ω is called the angular frequency of the function. The period T of the function is the time required for the function to complete one cycle. Therefore $T = \frac{2\pi}{\omega}$. Since the frequency (or linear frequency) f of the function is related to the period by $f = \frac{1}{T}$, the angular frequency can be written as $\omega = 2\pi f$ or $\omega = \frac{2\pi}{T}$. The phase shift of the function is given by $\frac{-C}{\omega}$. (If the units require, we can use 360° instead of 2π.)

If two sine functions or two cosine functions have the same angular frequency we may say that the second leads or lags the first by the amount $C_2 - C_1$ depending on whether $C_2 - C_1$ is positive or negative.

EXAMPLE 6 In a particular circuit the maximum values of the voltage and current are 175 V and 113 A, respectively, the frequency is 50 Hz and the current leads the voltage by 45°.

a. Write the equation for the voltage at any time t.

b. Write the equation for the current at any time t.

c. What is the instantaneous value of the current when the voltage has reached 15° of its cycle?

d. Plot one cycle of both curves starting with $t = 0$, using the equations from parts (a) and (b).

Solution: We are given the following information:

$$\begin{aligned} \text{maximum voltage} &= E_{max} = 175 \text{ V}, \\ \text{maximum current} &= I_{max} = 113 \text{ A}, \\ \text{frequency} &= f = 50 \text{ Hz}, \\ \text{phase angle} &= \theta = 45° \text{ or } \frac{\pi}{4}, \\ \text{voltage angle} &= \phi = 15°, \\ \text{angular frequency} &= \omega = 2\pi f. \end{aligned}$$

a. Substituting the values in the equation

$$\begin{aligned} e &= E_{max} \sin \omega t \\ &= 175 \sin (2\pi)(50)t \\ &= 175 \sin 100\pi t. \end{aligned}$$

b. Substituting the values in the equation

$$\begin{aligned} i &= I_{max} (\omega t + \theta) \\ &= 113 \sin \left[(2\pi)(50)t + \frac{\pi}{4} \right] \\ &= 113 \sin \left(100\pi t + \frac{\pi}{4} \right). \end{aligned}$$

c. The instantaneous values of the voltage and current for any angle θ of the voltage are:

$$e = E_{max} \sin \phi$$

and

$$i = I_{max} \sin (\phi + \theta).$$

Substituting given values in the equation

$$
\begin{aligned}
i &= I_{\max} \sin(\phi + \theta) \\
&= 113 \sin(15^\circ + 45^\circ) \\
&= 113 \sin 60^\circ \\
&= 97.9 \text{ A} \quad \textbf{(the instantaneous value of the current).}
\end{aligned}
$$

d.

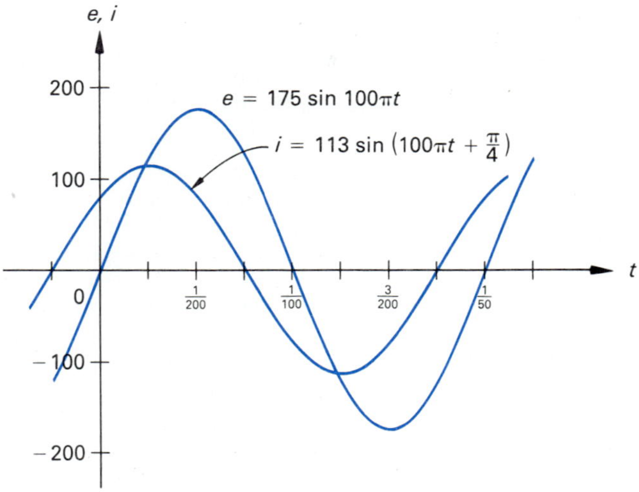

□

8.4 Exercises

In Exercises 1–20, determine (a) the period, (b) the phase shift, (c) the amplitude, (d) where the curve will recycle, and (e) sketch the graph.

1. $y = \sin\left(\theta - \frac{\pi}{2}\right)$

2. $y = \cos(\theta - \pi)$

3. $y = \sin(\theta - 30^\circ)$

4. $y = \cos(\theta - 90^\circ)$

5. $y = \cos\left(3\theta - \frac{\pi}{3}\right)$

6. $y = \sin\left(2\theta - \frac{\pi}{4}\right)$

7. $y = \cos(5\theta + 15^\circ)$

8. $y = \sin(4\theta + 12^\circ)$

9. $y = 3 \sin\left(\theta - \frac{\pi}{2}\right)$

10. $y = \frac{1}{2} \cos\left(3\theta - \frac{\pi}{2}\right)$

11. $y = 10 \sin(2\pi\theta - 2\pi)$

12. $y = 0.01 \cos(0.1\theta - 20^\circ)$

13. $y = -30 \cos(15\pi\theta + \pi)$

14. $y = -60 \sin\left(30\pi\theta + \frac{\pi}{2}\right)$

15. $y = 25 \sin(3\theta - 120^\circ)$

16. $y = -25 \sin(3\theta - 120^\circ)$

17. $y = 15 \cos\left(\frac{\theta}{2} + 360^\circ\right)$

18. $y = -\cos\left(\frac{\theta}{3} - 180^\circ\right)$

19. $y = 35 \cos(\theta - 720^\circ)$

20. $y = 100 \sin(\theta + 90^\circ)$

In Exercises 21–24, write an equation for each specified trigonometric function with given amplitude, period, and phase shift, respectively.

21. sine, 5, 3π, $\frac{\pi}{4}$ to the right.

22. sine, 3, $\frac{2\pi}{3}$, $\frac{\pi}{3}$ to the left.

23. cosine, 2, $\frac{\pi}{2}$, $\frac{\pi}{6}$ to the left.

24. cosine, 4, π, $\frac{\pi}{2}$ to the right.

25. A capacitor voltage is given by $v = 1.5 \cos(3t + 2)$. Sketch one cycle of the function.

26. A wire vibrates according to the function $y = 0.2 \sin(4t - 1)$. Sketch one cycle of the function.

27. A 30-Hz alternator generates 7.2 kV at 750 A. The current leads the voltage by an angle of 25°.

a. Write the equation for the current at any time t.
b. How much of the voltage cycle will have been completed the first time that the instantaneous current rises to 465 A?
c. Sketch the graph of the equation in part (a) for one cycle starting with $t = 0$.

28. A 45-Hz alternator generates 3.2 kV with a current of 190 A. The voltage is lagging the current by 20°.

a. Write the equation for the current at any time t.
b. What is the instantaneous value of the current when the voltage has completed 182° of its cycle?

29. The maximum values of a voltage and a current are 150 V and 12 A, respectively. The frequency is 700 Hz and the current lags the voltage by 30° at any time t.

a. Write the equation for the voltage at any time t.
b. Write the equation for the current at any time t.
c. What is the instantaneous value of the current when the voltage has reached 15° of its cycle?
d. Plot one cycle of both curves starting with $t = 0$, using the equations from parts (a) and (b).

30. The hum you hear on a radio when it is not tuned to a station is a sound wave of 60 cycles per second.

a. Is the 60 cycles per second the period or the frequency?
b. If it is the period, find the frequency. If it is the frequency, find the period.
c. The wavelength of a sound wave is defined to be the distance the wave travels in one period. If sound travels at 1100 ft/s, find the wavelength of a 60-cycle-per-second sound wave.

31. The lowest musical note the human ear can hear is about 16 cycles per second. In order for the pipe on a church organ to generate this note, the pipe must be exactly half as long as the wavelength. How long an organ pipe would be required to generate a 16-cycle-per-second note?

8.5 Graphs of tangent, cosecant, secant, and cotangent

In Section 8.1 we graphed the functions $y = \tan x$, $y = \csc x$, $y = \sec x$, and $y = \cot x$. In this section we will learn how to graph the transformed functions

$$y = A \tan (Bx + C), \qquad y = A \csc (Bx + C),$$
$$y = A \sec (Bx + C) \qquad y = A \cot (Bx + C).$$

The graphs of these functions will show characteristics similar to those of sine and cosine (amplitude, period, and phase shift). But there are some basic differences.

One of the big differences is that not one of the graphs in this section is a smooth continuous curve as are the sine and cosine curves. Each is discontinuous; it has points where the function is undefined. At the point where the curves are discontinuous there will be vertical asymptotes. Also, due to the fact that the curves are discontinuous, the cycles do not have end points.

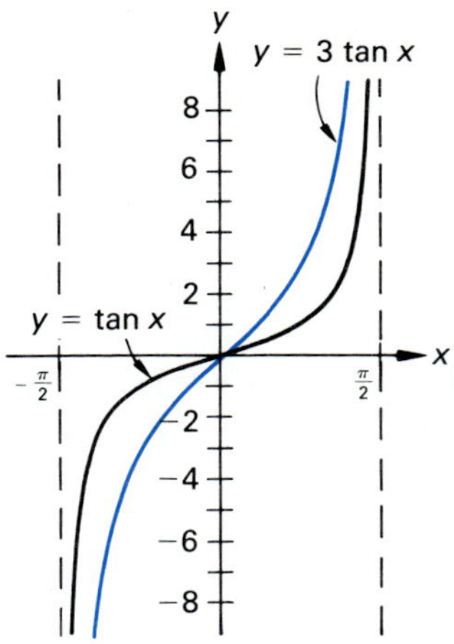

FIGURE 8.31

TABLE 8.12 $y = 3 \tan x$

x	$-\frac{3\pi}{8}$	$-\frac{\pi}{4}$	$-\frac{\pi}{8}$	0	$\frac{\pi}{8}$	$\frac{\pi}{4}$	$\frac{3\pi}{4}$
$\tan x$	−2.4	−1	−0.4	0	0.4	1	2.4
$3 \tan x$	−7.2	−3	−1.2	0	1.2	3	7.2

The concept of amplitude does not apply to these functions. The coefficient A will cause points on the graph to be stretched or shrunk as $A > 1$ and $A < 1$. There are no maximum or minimum points on the curve since the end points of the cycle are not defined.

EXAMPLE 1 Sketch the graph of $y = 3 \tan x$ for $-\frac{\pi}{2} < x < \frac{\pi}{2}$. Label key points on the horizontal axis.

Solution: To help sketch $y = 3 \tan x$, review the properties of $y = \tan x$ in Section 8.2. Recall that the period of $\tan x$ is π and it is undefined at $-\frac{\pi}{2}$ and $\frac{\pi}{2}$. Sketch the graph of $y = \tan x$ (Fig. 8.31). Now stretch the guide points by a factor of 3 as indicated in Table 8.12 and draw a smooth curve through the resulting points. □

EXAMPLE 2 Sketch the graph of $y = -\frac{1}{3} \cot x$ for $0 < x < \pi$. Label key points on the horizontal axis.

Solution: As in Example 1 review the properties of $y = \cot x$ in Section 8.2. Recall that the period of $\cot x$ is π and the function is undefined at 0 and π.

Sketch the graph of $y = \cot x$ (Fig. 8.32). Now shrink the guide points by a factor of 1/3. The negative sign in front of the 1/3 will reflect each of these points across the x-axis as illustrated in Table 8.13. □

FIGURE 8.32

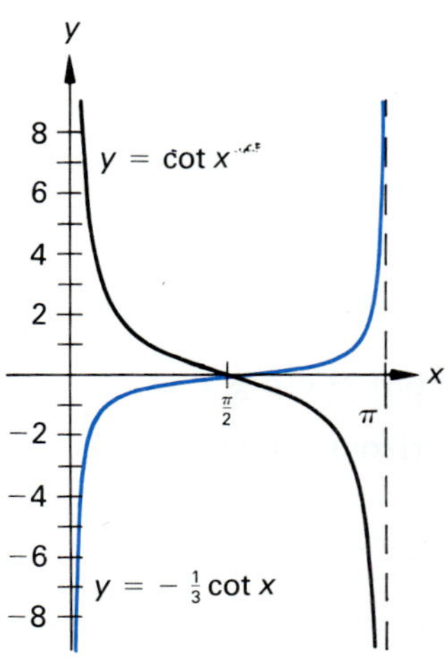

TABLE 8.13 $y = -\frac{1}{3} \cot x$

x	$\frac{\pi}{8}$	$\frac{\pi}{4}$	$\frac{3\pi}{8}$	$\frac{\pi}{2}$	$\frac{5\pi}{8}$	$\frac{3\pi}{4}$	$\frac{7\pi}{8}$
$\cot x$	2.4	1	0.4	0	−0.4	−1	2.4
$-\frac{1}{3} \cot x$	−0.8	−0.3	−0.1	0	0.1	0.3	0.8

Period

In Section 8.2 we learned that the period of the functions $y = \tan x$ and $y = \cot x$ is π. For functions of the form $y = \tan Bx$ and $y = \cot Bx$ the period is determined by the formula

$$\text{period} = \frac{\pi}{|B|}.$$

It was also learned that the period of the functions $y = \csc x$ and $y = \sec x$ is 2π. For functions of the type $y = \csc Bx$ and $y = \sec Bx$ the period is determined by the formula

$$\text{period} = \frac{2\pi}{|B|}.$$

EXAMPLE 3 Determine the period of

a. $y = \tan \frac{3}{2} x$, **b.** $y = \cot 7x$,

c. $y = \csc \frac{3}{5} x$, **d.** $y = \sec \frac{\pi}{2} x$.

Solution: Using the formulas in the previous paragraphs

a. $y = \tan \frac{3}{2} x$

$$\text{period} = \frac{\pi}{3/2} = \frac{2\pi}{3}.$$

b. $y = \cot 7x$

$$\text{period} = \frac{\pi}{7}.$$

c. $y = \csc \frac{3}{5} x$

$$\text{period} = \frac{2\pi}{3/5} = \frac{10\pi}{3}.$$

d. $y = \sec \frac{\pi}{2} x$

$$\text{period} = \frac{2\pi}{\pi/2} = 4.$$

□

EXAMPLE 4 Sketch one cycle of the graph of $y = \csc \frac{3}{5} x$. Determine the period and indicate key points on the horizontal axis.

Solution: The period of $y = \csc \frac{3}{5} x$ is $\frac{10\pi}{3}$. To help sketch $y = \csc \frac{3}{5} x$, first sketch $y = \sin \frac{3}{5} x$. To sketch the sine curve we need the following key facts: amplitude is 1, period is $\frac{10\pi}{3}$, and phase shift is 0. Also the key points on the curve are (0, 0), $\left(\frac{5\pi}{6}, 1\right)$, $\left(\frac{5\pi}{3}, 0\right)$, $\left(\frac{5\pi}{2}, -1\right)$, and $\left(\frac{10\pi}{3}, 0\right)$.

Remember that $\csc\theta = \frac{1}{\sin\theta}$. The function $y = \csc\frac{3}{5}x$ will be undefined when $\sin x = 0$. Therefore, the lines $x = 0$, $x = \frac{5\pi}{3}$, and $x = \frac{10\pi}{3}$ are asymptotes. The sketch of $y = \csc\frac{3}{5}x$ is given in Fig. 8.33. □

EXAMPLE 5 Sketch one cycle of the graph of $y = \sec\pi x$. Determine the period and indicate key points on the horizontal axis.

Solution: The period of $y = \sec\pi x$ is 2. To help sketch $y = \sec\pi x$, first sketch $y = \cos\pi x$. To sketch the cosine curve we need the following key facts: amplitude is 1, period is 2, and phase shift is 0. Also the key points on the curve are (0, 1), (1/2, 0), (1, −1), (3/2, 0), and (2, 1). Remember that $\sec\theta = \frac{1}{\cos\theta}$. The function $y = \sec\pi x$ will be undefined when $\cos\theta = 0$. Therefore, the lines $x = \frac{1}{2}$ and $x = \frac{3}{2}$ are asymptotes. The sketch of $y = \sec\pi x$ is given in Fig. 8.34. □

FIGURE 8.33

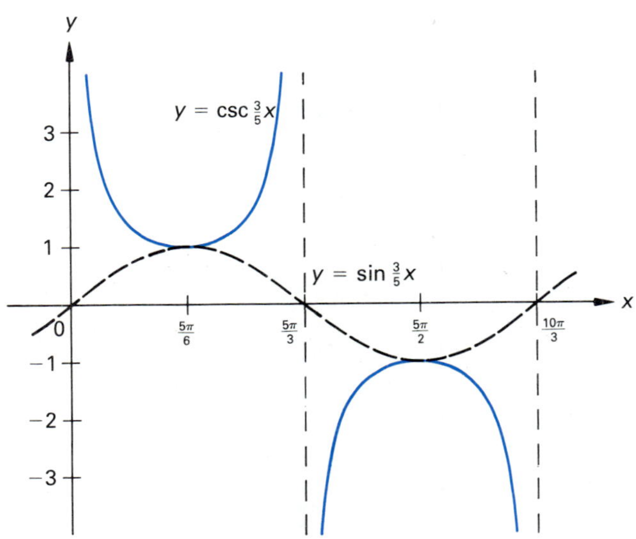

FIGURE 8.34

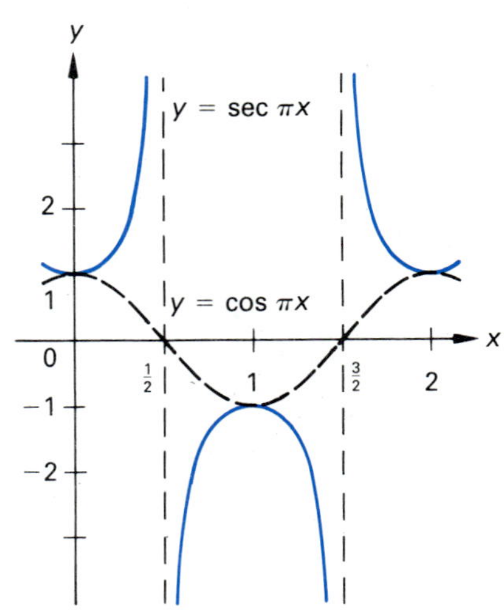

Phase shift

In Section 8.4 we learned that the phase shift is a horizontal displacement of the curve to the right or to left. For any curve with asymptotes, the asymptotes are also translated the same distance.

$$\text{phase shift} = \frac{-C}{B}.$$

EXAMPLE 6 Sketch one cycle of the graph of $y = \tan\left(2x - \frac{\pi}{3}\right)$. Specify the period, phase shift, and label key points on the horizontal axis.

Solution: For the function $y = \tan\left(2x - \frac{\pi}{3}\right)$ the period $= \frac{\pi}{2}$ and the phase shift $= \frac{-\left(-\frac{\pi}{3}\right)}{2} = \frac{\pi}{6}$. The normal interval for sketching one cycle of the tangent curve is $-\frac{\pi}{2} < x < \frac{\pi}{2}$. To determine a convenient starting point solve the equation $2x - \frac{\pi}{3} = -\frac{\pi}{2}$ for x: $x = -\frac{\pi}{12}$. To sketch one cycle of the curve with a phase shift of $\frac{\pi}{6}$ and a period of $\frac{\pi}{2}$ it is convenient to begin the cycle at $-\frac{\pi}{12}$ and end the cycle at $-\frac{\pi}{12} + \frac{\pi}{2} = \frac{5\pi}{12}$. See Fig. 8.35 for the sketch. □

FIGURE 8.35

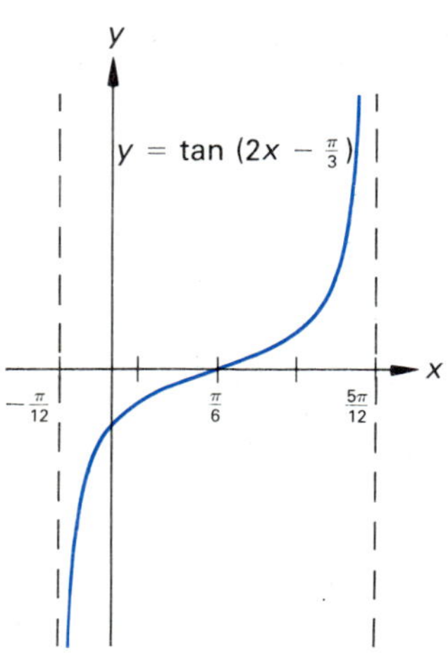

FIGURE 8.36

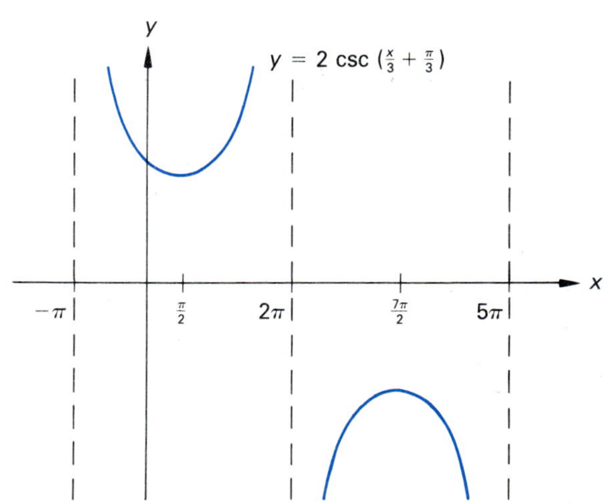

EXAMPLE 7 Sketch one cycle of the graph of $y = 2\csc\left(\frac{x}{3} + \frac{\pi}{3}\right)$. Specify the period, phase shift, and label key points on the horizontal axis.

Solution: For the function $y = 2\csc\left(\frac{x}{3} + \frac{\pi}{3}\right)$ the period $= \frac{2\pi}{1/3} = 6$ and the phase shift $= \frac{-\pi/3}{1/3} = -\pi$. The coefficient, 2, has the effect of raising the lowest point of the curve that lies above the x-axis to 2 units above the axis. Also it has the effect of lowering the highest point of the curve below the x-axis to 2 units below the axis.

To sketch one cycle of the curve with a phase shift of $-\pi$ and period of 6π it is convenient to begin the cycle at $-\pi$ and end the cycle at 5π. See Fig. 8.36 for the sketch. □

8.5 Exercises

In Exercises 1–24, sketch one cycle of the function.

1. $y = 2\tan\theta$

2. $y = -\frac{1}{4}\tan\theta$

3. $y = -\frac{1}{3}\cot\theta$

4. $y = 3\cot\theta$

5. $y = 3\csc\theta$

6. $y = \frac{1}{2}\csc\theta$

7. $y = 4\sec\theta$

8. $y = -\frac{1}{3}\sec\theta$

9. $y = \frac{3}{2}\tan\theta$

10. $y = \frac{3}{2}\cot\theta$

11. $y = \frac{4}{3}\csc\theta$

12. $y = \frac{2}{3}\sec\theta$

13. $y = \tan\left(\theta + \frac{\pi}{4}\right)$

14. $y = \frac{1}{2}\tan\left(\frac{\theta}{2} + \frac{\pi}{12}\right)$

15. $y = \frac{3}{2}\cot\left(2\theta - \frac{2\pi}{3}\right)$

16. $y = \cot\left(2x + \frac{2\pi}{3}\right)$

17. $y = -\csc\left(2\theta + \frac{\pi}{2}\right)$

18. $y = \frac{1}{2}\csc\left(3x + \frac{3\pi}{4}\right)$

19. $y = \sec\left(\theta - \frac{\pi}{6}\right)$

20. $y = -3\sec(2\theta + 2\pi)$

21. $y = 2\tan\left(\theta - \frac{\pi}{4}\right)$

22. $y = 3\cot\left(\frac{\theta}{2} - \frac{\pi}{3}\right)$

23. $y = 3\csc\left(2\theta - \frac{\pi}{2}\right)$

24. $y = 4\sec\left(\theta + \frac{\pi}{6}\right)$

In Exercises 25–32, sketch the graph of each function over the given interval.

25. $y = 2\tan\left(\frac{\theta}{2} + \frac{\pi}{6}\right), -4\pi \le \theta \le 4\pi$

26. $y = \frac{1}{2}\tan\left(\frac{\theta}{3} - \frac{\pi}{6}\right), -2\pi \le \theta \le 2\pi$

27. $y = 3\cot(2\theta - 135^\circ), -180^\circ \le \theta \le 180^\circ$

28. $y = 2\cot\left(\frac{\theta}{3} + 30^\circ\right), -540^\circ \le \theta \le 540^\circ$

29. $y = -3\csc\left(3\theta - \frac{3\pi}{2}\right), -2\pi \le \theta \le 2\pi$

30. $y = 2\csc\left(2\theta + \frac{\pi}{3}\right), -\pi \le \theta \le \pi$

31. $y = \frac{1}{3}\sec(2\theta - 180^\circ), -180^\circ \le \theta \le 180^\circ$

32. $y = 2\sec\left(\frac{1}{3}x - 60^\circ\right), 0^\circ \le \theta \le 540^\circ$

In Exercises 33–40, write an equation for each of the trigonometric functions with given coefficient A, period, and phase shift, respectively.

33. tangent, $\frac{1}{3}, \frac{\pi}{2}, \frac{\pi}{12}$ to the left

34. cotangent, 2, 2π, $\frac{\pi}{3}$ to the right

35. secant, -3, 720°, 30° to the left

36. cosecant, $\frac{2}{3}$, $\frac{\pi}{4}$, $\frac{\pi}{5}$ to the left

37. cosecant, 3, π, $\frac{\pi}{2}$ to the right

38. secant, $\frac{1}{2}$, 90°, 36° to the right

39. cotangent, $-\frac{3}{4}$, $\frac{\pi}{3}$, $\frac{\pi}{4}$ to the left

40. tangent, $\frac{3}{2}$, 3π, π to the right

8.6 Composite trigonometric functions

There are many applications that involve combinations of trigonometric functions or combinations of other functions with trigonometric functions. As an example, musical notes are composed of a sum of many sinusoids. The designers of the amplification systems that carry the musical notes deal with functions that are sums and differences of sine or cosine functions.

The graph of such combinations of trigonometric functions may be sketched by a technique known as **addition of ordinates.** This technique will be illustrated in Examples 1 and 2.

EXAMPLE 1 Sketch the graph of $y = 4 + 2 \cos x$ for the interval $0 \leq x \leq 2\pi$.

FIGURE 8.37

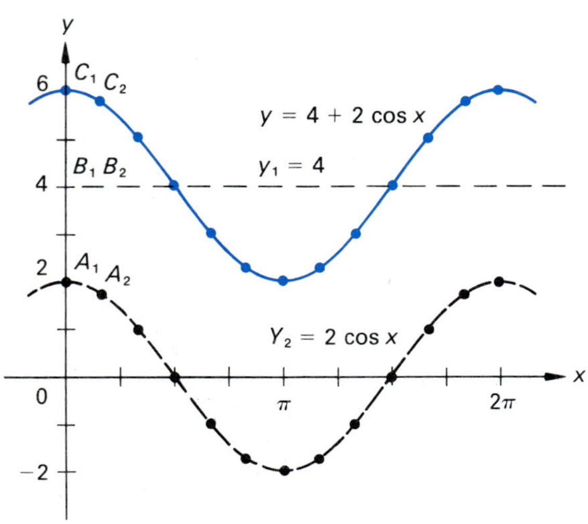

TABLE 8.14 $y = 4 + 2 \cos x$

x	y_1	y_2	$y = y_1 + y_2$
0	4	2.0	6.0
$\frac{\pi}{4}$	4	1.4	5.4
$\frac{\pi}{2}$	4	0	4.0
$\frac{3\pi}{4}$	4	-1.4	2.6
π	4	-2.0	2.0
$\frac{5\pi}{4}$	4	-1.4	2.6
$\frac{3\pi}{2}$	4	0	4.0
$\frac{7\pi}{4}$	4	1.4	5.4
2π	4	2.0	6.0

Solution: The equation $y = 4 + 2 \cos x$ can be thought of as the sum of two functions: $y = y_1 + y_2$, where $y_1 = 4$ and $y_2 = 2 \cos x$.

To illustrate the concept of addition of ordinates we will first sketch the graphs of $y_1 = 4$ and $y_2 = 2 \cos x$ on the same set of axes (Fig. 8.37). To help understand the concept, Table 8.14 contains columns for the values of y_1, y_2, and $y_1 + y_2$. On the graph the points for y_1, y_2, and y when $x = 0$ are indicated by A_1, B_1, and C_1, respectively. If we measured the distance from the x-axis to A_1 and added this distance to B_1, we would be at point C_1. □

In Example 1 the student should recognize from the figure that the graph of the composite function is the graph of the cosine function raised 4 units. Thus, $y_1 = 4$, the constant function, has this effect on the graph. If it had been a negative constant, it would have lowered the graph that number of units. Note also that since $y_2 = 2 \cos x$ is a periodic function, the composite function is periodic.

EXAMPLE 2 Sketch the graph of $y = \sin x + \cos x$ for the interval $0° \leq x \leq 360°$.

Solution: The function $y = \sin x + \cos x$ can be separated into two functions, $y_1 = \sin x$ and $y_2 = \cos x$. To help sketch the function we developed Table 8.15 providing values for y_1, y_2, and $y_1 + y_2$. The graph of the composite function is given in Fig. 8.38. □

TABLE 8.15 $y = \sin x + \cos x$

x	0°	45°	90°	135°	180°	225°	270°	315°	360°
y_1	0	0.7	1	0.7	0	−0.7	−1	−0.7	0
y_2	1	0.7	0	−0.7	−1	−0.7	0	0.7	1
$y = y_1 + y_2$	1	1.4	1	0	−1	−1.4	−1	0	1

FIGURE 8.38

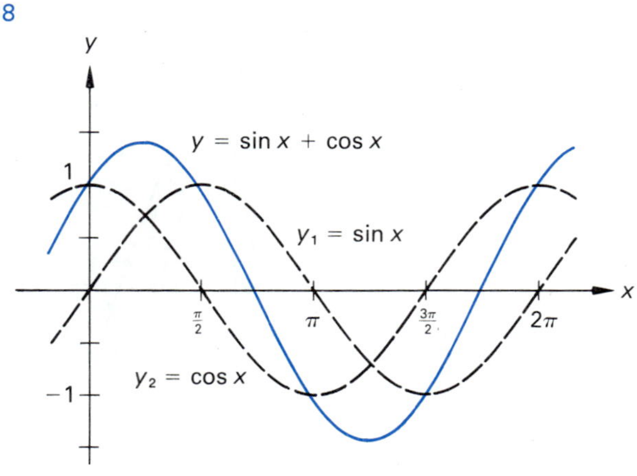

TABLE 8.16 $y = \sin 2x + 2 \sin (x + 45°)$

x	0°	45°	90°	135°	180°	225°	270°	315°	360°
y_1	0	1	0	−1	0	1	0	−1	0
y_2	1.4	−2	−1.4	0	−1.4	−2	−1.4	0	1.4
$y = y_1 + y_2$	1.4	3.0	1.4	−1.0	−1.4	−1.0	−1.4	−1.0	1.4

x	405°	450°	495°	540°	585°	630°	675°	720°
y_1	1	0	−1	0	1	0	−1	0
y_2	2	1.4	0	−1.4	−2	−1.4	0	1.4
$y = y_1 + y_2$	3.0	1.4	−1	−1.4	−1.0	−1.4	−1.0	1.4

EXAMPLE 3 Sketch the graph of $y = \sin 2x + 2 \sin (x + 45°)$ for the interval $0° \leq x \leq 720°$.

Solution: The function $y = \sin 2x + 2 \sin (x + 45°)$ can be separated into two functions: $y_1 = \sin 2x$ and $y_2 = 2 \sin (x + 45°)$. To help sketch the curve we have developed Table 8.16 providing values for y_1, y_2, and $y_1 + y_2$. The graph of the composite function is given in Fig. 8.39. □

In Fig. 8.39 if the graph were extended in either direction, it would be easy to see that the composite function is periodic. Are all composite trigonometric functions periodic? No. If any of the functions of the composite function has a variable amplitude, [for example, if $y = y_1 + y_2$ and $y_1 = x \sin (Bx + C)$ or $y_2 = x^2 \cos (Bx + C)$] then the composite function is not periodic. This will be illustrated in Example 5. Also, composite functions of the type $y = \sin \pi t + \cos 2t$ are not periodic.

FIGURE 8.39

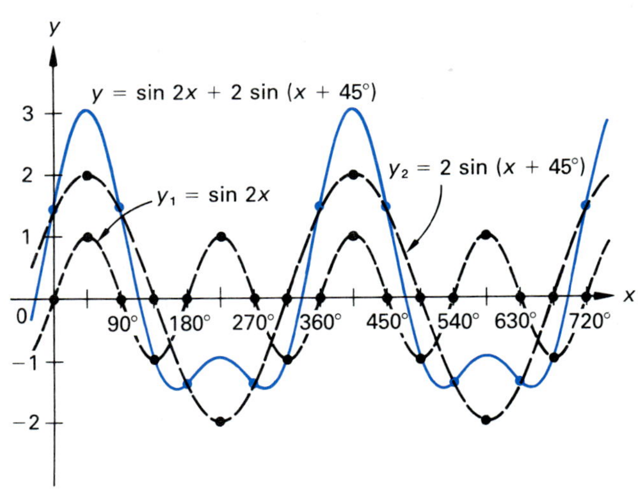

In Example 4 we illustrate how to find the period of composite functions that are periodic.

To find the period of the composite function, we first find the periods of the component functions. We then determine the lcm (least common multiple) of the periods of the component functions. From this we can also determine the number of cycles of each component function in the period of the composite function.*

EXAMPLE 4 Find the period of each of the composite functions.

a. $y = 2 \sin 3x + \cos x$

b. $y = \cos \frac{\pi}{2} x + \sin \frac{\pi}{3} x$

Solution:

a. For the function $y = 2 \sin 3x + \cos x$ the periods of the component functions are $2\pi/3$ and 2π, respectively. The lcm of $2\pi/3$ and 2π is 2π (since $2\pi = 6\pi/3$ and $6\pi/3$ is evenly divisible by $2\pi/3$ and 2π). For the period 2π of the composite function, $y_1 = 2 \sin 3x$ will go through three cycles and $y_2 = \cos x$ will go through one cycle.

b. For the function $y = \cos \frac{\pi}{2} x + \sin \frac{\pi}{3} x$ the periods of the component functions are 4 and 6, respectively. The lcm of 4 and 6 is 12. For the period 12 of the composite function, $y_1 = \cos \frac{\pi}{2} x$ will go through three cycles and $y_2 = \sin \frac{\pi}{3} x$ will go through two cycles. □

The process of determining the ordinate values for composite functions can be simplified using a programmable calculator or computer. The use of the programmable calculator will be illustrated in Example 5.

EXAMPLE 5 When a weight hanging on a spring is given an initial displacement, y_0 and an initial velocity v_0, its position is given by

$$y = y_0 \cos \omega t + \frac{v_0}{\omega} \sin \omega t,$$

where the angular frequency ω depends on the weight and the elasticity of the spring. In a graph illustrate the spring action for one cycle of the composite function when $y_0 = 0.75$ m, $v_0 = 2.4$ m/s, and $\omega =$ 3.0 rad/s.

Solution: Substituting the given values into the function

$$y = y_0 \cos \omega t + \frac{v_0}{\omega} \sin \omega t,$$

* The lcm of x and y is the smallest number that is divisible by x and y.

TABLE 8.17

Algebraic key strokes
LRN
CP
STO
0
0.75
×
(
3
×
RCL
0
)
COS
+
0.8
×
(
3
×
RCL
0
)
SIN
=
RST
LRN

TABLE 8.18

RPN key strokes
clear program
180
×
π
÷
3
×
STO 1
COS
0.75
×
RCL 1
SIN
0.8
×
+
GTO $\phi\phi$

TABLE 8.19 $y = 0.75 \cos 3.0t + 0.8 \sin 3.0t$

t	0	$\frac{\pi}{12}$	$\frac{\pi}{6}$	$\frac{\pi}{4}$	$\frac{\pi}{3}$	$\frac{5\pi}{12}$	$\frac{\pi}{2}$	$\frac{7\pi}{12}$	$\frac{2\pi}{3}$
y	0.75	1.10	0.80	0.04	−0.75	−1.10	−0.80	−0.04	0.75

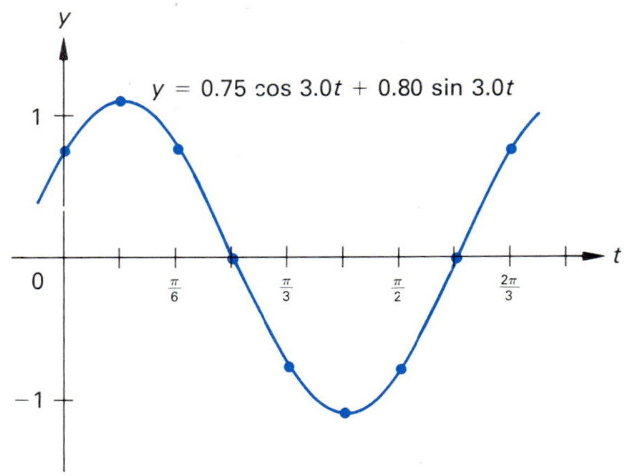

FIGURE 8.40

we have

$$y = 0.75 \text{ m} \cos (3.0 \text{ rad/s})t + \frac{2.4 \text{ m/s}}{3.0 \text{ rad/s}} \sin (3.0 \text{ rad/s})t$$

or

$$y = 0.75 \cos 3.0t + 0.8 \sin 3.0t.$$

From this equation we see that the periods of the component functions are the same $(2\pi/3)$. Thus the period of the composite function is $2\pi/3$. To obtain a smooth curve we will graph the function at intervals of $\pi/12$. To help determine these values we will use the program for the calculator given in Table 8.17 or 8.18. The results are given in Table 8.19 and the curve is sketched in Fig. 8.40. □

The effect of a variable coefficient on a sinusoidal function may cause the amplitude to decrease or increase as one calculates larger and larger values of the independent values. In the physical world an equation of this type could describe the decrease and/or increase in the amplitude of the swing of a pendulum, a spring action, or an electrical energy flow.

EXAMPLE 6 Graph the function $y = \frac{2}{x+2} \sin \pi x$ for the interval $0 \leq x \leq 6$.

TABLE 8.20 $y=\dfrac{2}{x+2}\sin \pi x$

x	0	0.5	1	1.5	2	2.5	3	3.5	4	4.5	5	5.5	6
y	0	0.8	0	−0.57	0	0.44	0	−0.36	0	0.31	0	−0.27	0

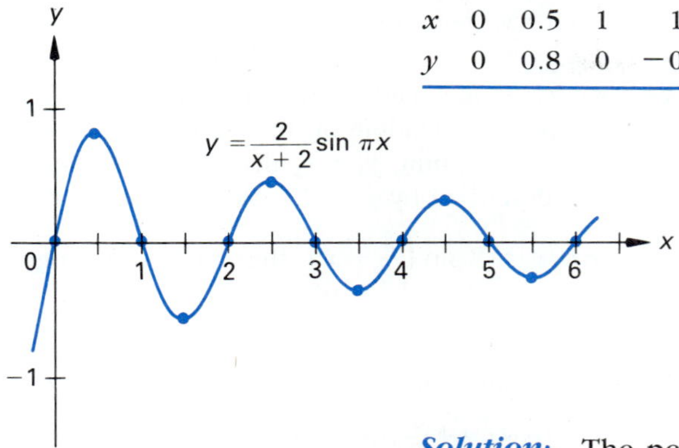

FIGURE 8.41

Solution: The period of the function $y=\dfrac{2}{x+2}\sin \pi x$ is 2 and the curve will go through three cycles in the indicated interval. Using the calculator we calculate values for x for each half unit, shown in Table 8.20. In this case the variable coefficient has a damping effect on the curve and causes the amplitude to decrease as x increases (Fig. 8.41). □

8.6 Exercises

In Exercises 1–20, for the composite function

a. Determine the period of each component function.

b. Determine the period of the composite function.

c. Find the number of cycles each component function will go through for one period of the composite function.

d. Sketch the curve for one period of the composite function.

1. $y=3+2\sin x$

2. $y=5-\cos 2x$

3. $y=-0.5+0.4\cos\dfrac{\theta}{2}$

4. $y=-0.8+0.9\sin\dfrac{\theta}{3}$

5. $y=x+2\sin x$

6. $y=x-4\cos x$

7. $y=2x+3\cos x$

8. $y=5x-3\sin x$

9. $y=\sin x-\cos x$

10. $y=\sin x+2\cos x$

11. $y=2\cos x+2\sin x$

12. $y=3\cos x-\sin x$

13. $y=4\cos x+3\sin x$

14. $y=5\sin x-6\cos x$

15. $y=\cos 2x+\sin 2x$

16. $y=\sin\dfrac{x}{2}+\cos\dfrac{x}{2}$

17. $y=\cos 3x+2\cos(x+45°)$

18. $y=\sin 2x+3\sin(x+30°)$

19. $y=\sin 3x+\sin\left(x+\dfrac{\pi}{4}\right)$

20. $y=3\cos x+\cos\left(x-\dfrac{\pi}{3}\right)$

In Exercises 21–24, use point-by-point plotting to sketch the graph of each function over $0\le x\le 4.5$ for values of x for each quarter unit (0, 0.25, 0.50, . . .).

21. $y=\sin \pi t-2\cos t$

22. $y=0.5\cos x+2\sin \pi x$

23. $y=2\sin\dfrac{\pi}{2}x+\cos 3x$

24. $y=4\sin 2x-2\cos\dfrac{\pi}{4}x$

In Exercises 25–28, graph each function over the specified interval.

25. $y=\dfrac{3}{x+3}\cos \pi x,\ 0\le x\le 4$

26. $y = \dfrac{4}{x+3} \sin 2x,\ 0 \le x \le 3\pi$

27. $y = \dfrac{6+x}{5} \sin \dfrac{\pi}{2} x,\ 0 \le x \le 8$

28. $y = \dfrac{3.4-x}{2.1} \cos \dfrac{\pi}{3} x,\ 0 \le x \le 8$

29. In a graph, illustrate the spring action for one cycle of the composite function when $y_0 = 2.3$ m, $v_0 = 3.4$ m/s, and $\omega = 1.7$ rad/s. Use the function given in Example 5.

30. The instantaneous power in an AC circuit containing an inductance is given by $P = EI$, where $E = 150 \cos \omega t$ and $I = 15 \sin \omega t$. (a) Sketch P vs. the angle ωt for one cycle by finding values for E and I and computing the product EI. (b) Show that the curve in (a) is equivalent to the one described by $P = 1125 \sin 2\omega t$ by sketching this curve and comparing it with the curve in (a).

31. The motion of a vibrating spring is given by $d = 20 \cos (t + \pi) + 10 \sin \left(t - \dfrac{\pi}{2}\right)$. Sketch one cycle. When does it reach its maximum displacement?

32. The horizontal position of a point on a rolling wheel is given by $x = -r\phi + r \cos \phi$, where r is the radius and ϕ is the angle of rotation in radians. Sketch x vs. ϕ for one cycle when $r = 0.25$ m.

33. Two waves in a medium will reinforce or weaken each other depending on their phase difference. Show by graphing $y_1 + y_2$ that when one water wave described by $y_1 = 2.5 \cos x$ meets another wave, $y_2 = 3 \sin \left(\dfrac{\pi}{2} + x\right)$, the result is a stronger wave.

34. The acceleration of a piston is expressed as $a = 10\omega^2 \left(\cos \omega t + \dfrac{1}{2} \cos 2\omega t\right)$, where $\omega = 10^2$ rad/s. Sketch one cycle of the function.

35. The power dissipated by an electric resistor is $P = ei$, where P is the power in watts, e is the instantaneous voltage in volts, and i is the instantaneous current in amps. If $e = 120 \sin 120\pi t$ and $i = 5 \sin 120\pi t$, graph the power loss for one voltage cycle.

Chapter summary

The graphing of trigonometric functions is helped by understanding what is meant by a periodic function.

Definition 8.1 A function f is **periodic** if there is a positive number p such that for every real number x whenever x is in the domain of f, so is $x + p$ and $f(x + p) = f(x)$. The smallest such positive number p is then called the **period** of the function.

FIGURE 8.42

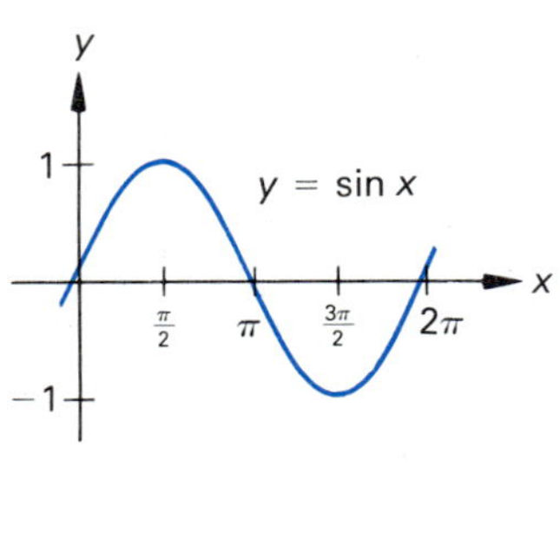

Definition 8.2 A **cycle** of a periodic function is a portion of the graph of the function from any point A on the graph to the first point B at which the graph starts repeating itself. (See Fig. 8.42.)

Definition 8.3 The **period** of a trigonometric function is the number of degrees or radians needed to complete one cycle. (See Fig. 8.43.)

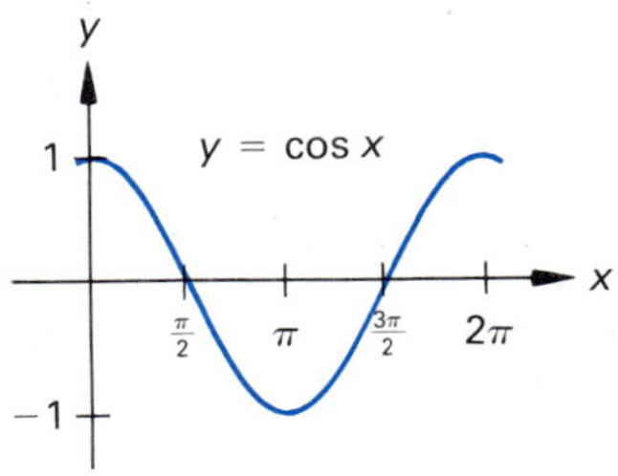

FIGURE 8.43

The domain of $y = \sin x$ is the set of all real numbers, the range is the closed interval $-1 \leq x \leq 1$, the y-intercept is 0, the x-intercepts are $0, \pm\pi, \pm 2\pi, \pm 3\pi, \ldots$, and the function is periodic with period of 2π.

The domain of $y = \cos x$ is the set of all real numbers, the range is the closed interval $-1 \leq x \leq 1$, the y-intercept is 1, the x-intercepts are $\pm\pi/2, \pm 3\pi/2, \pm 5\pi/2, \pm 7\pi/2, \ldots$, and the function is periodic with period 2π.

The domain of $y = \tan x$ is the set of all real numbers except $\pm\pi/2, \pm 3\pi/2, \pm 5\pi/2, \pm 7\pi/2, \ldots$, the range of the function is all real numbers, the y-intercept is 0, the x-intercepts are $0, \pm\pi, \pm 2\pi, \pm 3\pi, \ldots$, the tangent is undefined as x approaches $\pm\pi/2, \pm 3\pi/2, \pm 5\pi/2, \ldots$ from the left or the right, and the function is periodic with period π. (See Fig. 8.44.)

Sinusoids are curves that look like sine curves. The most general form of the curves is expressed by the functions $y = A \sin (Bx + C)$ and $y = A \cos (Bx + C)$. Each of the constants A, B, and C have an individual effect on the curve.

The constant A helps us determine the amplitude of the function:

$$\text{amplitude} = |A|.$$

The constant B helps us determine the period of the function:

$$\text{period} = \frac{2\pi}{|B|}.$$

The constant C helps us determine the phase shift of the function:

$$\text{phase shift} = \frac{-C}{B}.$$

To graph a sinusoid determine the amplitude, period, and phase shift. A natural starting point in graphing one cycle would be $-C/B$. Divide the period $2\pi/|B|$ by 4 and add this value to $-C/B$ to obtain a second key point on the graph, add $2\pi/4|B|$ to the previous result to get a third key point, again to obtain a fourth key point, and so on.

FIGURE 8.44

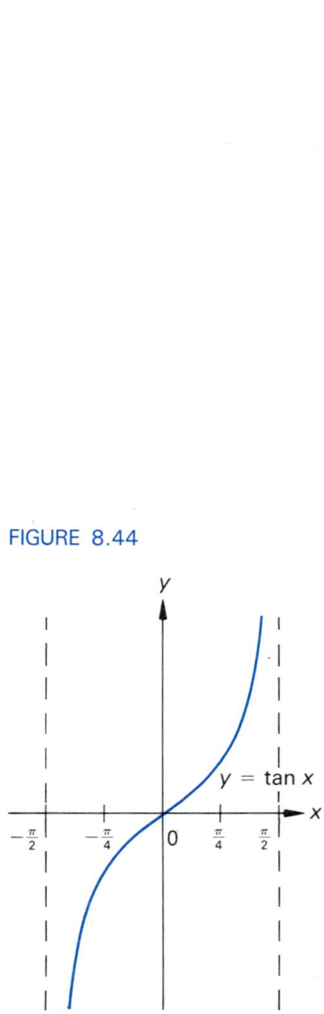

The graphs of $y = A \tan(Bx + C)$, $y = A \cot(Bx + C)$, $y = A \sec(Bx + C)$, and $y = A \csc (Bx + C)$ are all discontinuous functions; they do not have smooth curves like the sine and cosine functions. The concept of amplitude does not apply because there are no maximum or minimum points on these curves. For the tangent and cotangent functions the period is determined using the formula period $= \pi/|B|$. For the secant and cosecant functions the period is determined using the formula period $= 2\pi/|B|$. The phase shift for all the functions is found using phase shift $= -C/B$. For any curve with asymptotes, the asymptotes are translated the same distance as the phase shift.

Section 8.6 discusses the techniques used in graphing combinations of trigonometric functions and the combination of trigonometric functions and constant functions. One common technique of sketching this type function is by the technique known as **addition of ordinates.** We simply break the complex function into simple components, find ordinate values for each component, find the sum of the ordinate values, and thus determine the point on the graph.

Review Exercises

In Exercises 1–10, graph each of the functions for the indicated interval.

1. $y = 3 \sin \theta$ for values of θ at intervals of $15°$, $-270° \le \theta \le 270°$
2. $y = \frac{1}{2} \cos \theta$ for values of θ at intervals of $\frac{\pi}{6}$, $-2\pi \le \theta \le 2\pi$
3. $y = 2 \tan \theta$ for values of θ at intervals of $\frac{\pi}{4}$, $-\pi \le \theta \le 3\pi$
4. $y = 0.5 \sin \frac{\pi}{4}\theta$, $-\frac{\pi}{4} \le \theta \le 3\pi$
5. $y = -3 \cos 3\theta$, $-\frac{\pi}{2} \le \theta \le \frac{5}{2}\pi$
6. $y = 5 \tan \frac{\theta}{2}$, $0 < \theta \le 4\pi$
7. $y = 2 \csc 0.5\theta$, $-270° \le \theta \le 270°$
8. $y = 3 \sec 0.75\theta$, $-45° \le \theta \le 360°$
9. $y = \frac{1}{2} \cot 3\theta$, $-\frac{3\pi}{2} \le \theta \le \frac{3\pi}{2}$
10. $y = -2 \tan \frac{\theta}{2}$, $0 \le \theta \le 180°$

In Exercises 11–16, determine (a) the period, (b) the phase shift, (c) the amplitude, (d) where the curve will recycle, and (e) sketch the graph.

11. $y = 2 \sin\left(3\theta + \frac{\pi}{4}\right)$
12. $y = \frac{1}{2} \sin\left(\pi\theta - \frac{\pi}{2}\right)$
13. $y = 4 \cos\left(\frac{1}{3}\theta + 30°\right)$
14. $y = 2.4 \cos\left(\frac{\pi}{2}\theta - \frac{\pi}{4}\right)$
15. $y = 16 \sin (0.01\theta + 15°)$
16. $y = 13 \cos (0.03\theta - 18°)$

In Exercises 17–21, sketch the graph of each function over the given interval.

17. $y = \frac{1}{2} \tan\left(3\theta + \frac{\pi}{4}\right)$, $-\pi \le \theta \le \pi$
18. $y = 2 \tan\left(\frac{1}{2}\theta - \frac{\pi}{3}\right)$, $0 \le \theta \le 2\pi$
19. $y = \frac{1}{3} \cot (2\theta - 60°)$, $-90° \le \theta \le 270°$
20. $y = -4 \csc (2\theta - 60°)$, $-90° \le \theta \le 270°$
21. $y = 0.4 \sec\left(0.4\theta - \frac{\pi}{12}\right)$, $-4\pi \le \theta \le 4\pi$

In Exercises 22–27, for the composite function:

a. Determine the period of each component function.
b. Determine the period of the composite function.
c. Determine the number of cycles each component function will go through for one period of the composite function.
d. Sketch the curve for two periods of the composite function.

22. $y = -2 + 0.5 \sin 2x$
23. $y = 3 - 2 \cos \frac{x}{2}$
24. $y = \frac{1}{2} \sin (2\theta + 22°) + 2 \cos (\theta - 15°)$
25. $y = 2 \sin\left(\frac{\pi}{3}\theta + \frac{\pi}{2}\right) - \frac{1}{2} \cos\left(\frac{\pi}{4}\theta\right)$
26. $y = 2 \cos 2\theta - 3 \cos\left(\frac{\theta}{2} + 30°\right)$
27. $y = 3 \sin\left(\frac{\pi}{6}\theta - \frac{\pi}{3}\right) + 2 \sin\left(\frac{\pi}{3}\theta - \frac{\pi}{6}\right)$
28. The motion of a vibrating spring is given by $d = 15 \sin (\theta + \pi) + 5 \cos\left(\theta - \frac{\pi}{2}\right)$. Sketch one cycle. When does the spring reach its maximum displacement?
29. The motion of a piston is given by $s = \sin 5t + \cos (5t - \pi)$. Sketch one period. When is the maximum displacement reached?
30. A current in an AC circuit is given by $i = -30 \sin (30\pi t + \pi)$, where i is in amperes and 30π is in rad/s. What are the amplitude, period, phase shift, and frequency of this waveform?
31. A voltage in an AC circuit is given by $v = 5 \sin 60\pi t$. Find the amplitude and period, and sketch one cycle.
32. The range r of a cannon on a tank with muzzle velocity v, fired at an angle of elevation ϕ is given by $r = (v^2 \sin 2\phi)/g$ where g is a constant. Sketch

a graph of r vs. ϕ. What angle determines the maximum value for r?

33. The surface of a lake when viewed horizontally will trace out approximately sinusoidal curves. Assuming that the x-axis coincides with the surface of the lake during a perfect calm, write the equation when the waves are 5 ft from crest to trough and 20 ft in wavelength. Draw a graph of the sinusoidal curve.

34.

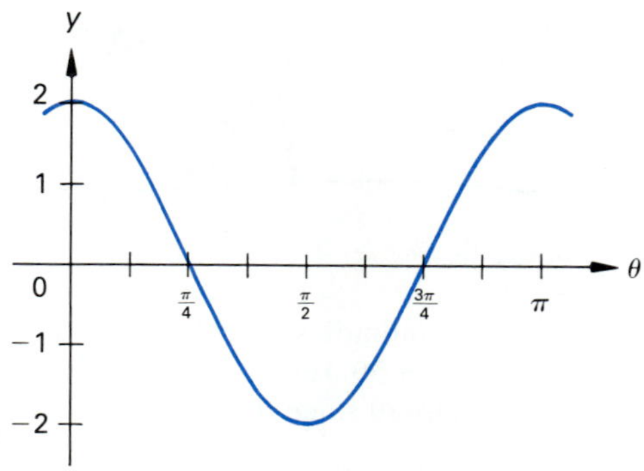

FIGURE 8.45

a. What is the amplitude of the graph in Fig. 8.45?
b. What is the period?
c. Write an equation for the graph in Fig. 8.45.

35.

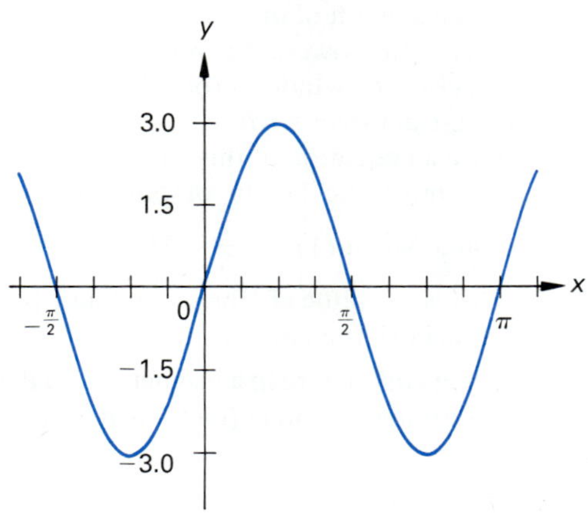

FIGURE 8.46

a. What is the amplitude of the graph in Fig. 8.46?
b. What is the period?
c. Write an equation for the graph in Fig. 8.46.

36. Write an equation that describes the function graphed in Fig. 8.47.

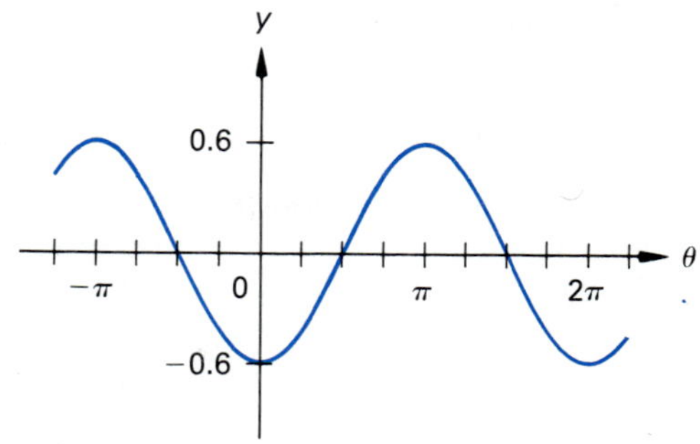

FIGURE 8.47

37. Write an equation that describes the function graphed in Fig. 8.48.

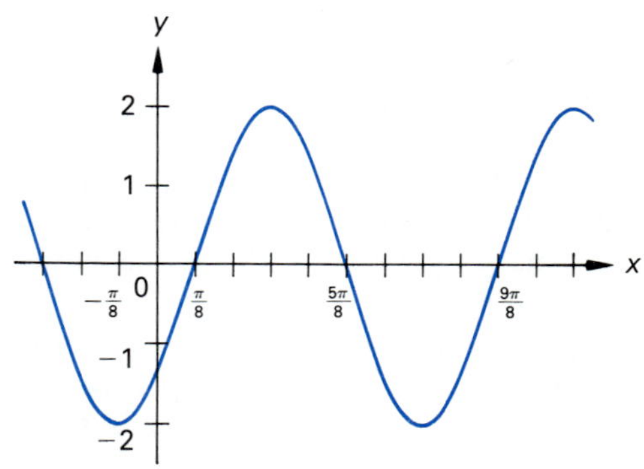

FIGURE 8.48

38. Write an equation that describes the function graphed in Fig. 8.49.

FIGURE 8.49

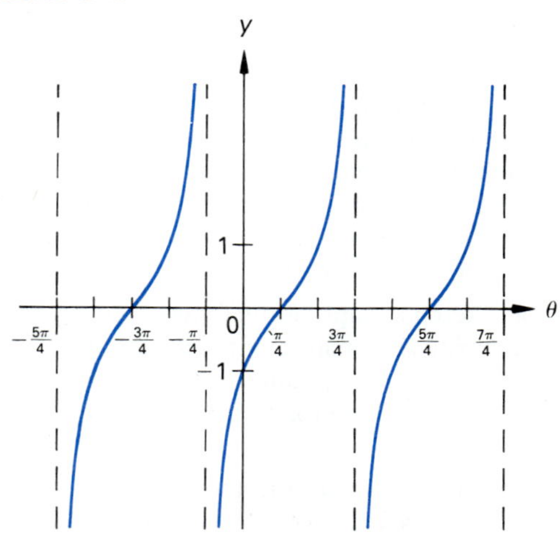

39. Two AC generators produce currents that are given in terms of time t by the equations:

$$I_1 = \sqrt{3} \sin 120\pi t \qquad \text{and} \qquad I_2 = -\cos 120\pi t.$$

If the output of the second is added to the output of the first, find (a) the maximum current output, (b) when it occurs, and (c) determine the phase shift produced for the composite function.

40. During the respiratory cycle (inhale and then exhale) of a person at rest, the rate of flow y (in liters per second) of air in and out of a person's lungs at time t (in seconds) is

$$y = \frac{1}{2} \sin \frac{2\pi}{5} t.$$

- **a.** How long does it take to complete one respiratory cycle?
- **b.** How many respiratory cycles are completed in one-half minute?
- **c.** Graph two complete cycles starting with $t = 0$.
- **d.** Explain the positive and negative values of y.
- **e.** Find y when $t = 2.5$ s, to the nearest hundredth.

41. The motion of each piston in a reciprocating engine closely approximates simple harmonic motion. The speed of each piston is $v = \omega y_0 \sin\left(\omega t + \frac{\pi}{2}\right)$, where ω is the angular speed of the crank shaft in radians per unit time and y_0 is the maximum displacement of the piston from the midpoint of its path. Graph the speed vs. time in seconds of the engine at 8500 rpm. The stroke (distance between extreme positions) of each piston is 2.3 s in.

42. A pebble becomes wedged between the treads of your tire. As you drive away the distance of the pebble from the road surface varies sinusoidally with the distance you have traveled. The period is the time needed for the wheel to make one complete revolution. Assume that the diameter of the wheel is 0.6 m and it makes one revolution in 0.035. (Use cosine function.)

- **a.** Sketch a graph of the function.
- **b.** Write an equation of the function.
- **c.** Predict the distance of the pebble from the surface of the road when you have traveled for 3 s.

43. The distance from the ground to the tips of the blades of a windmill vary sinusoidally with time.

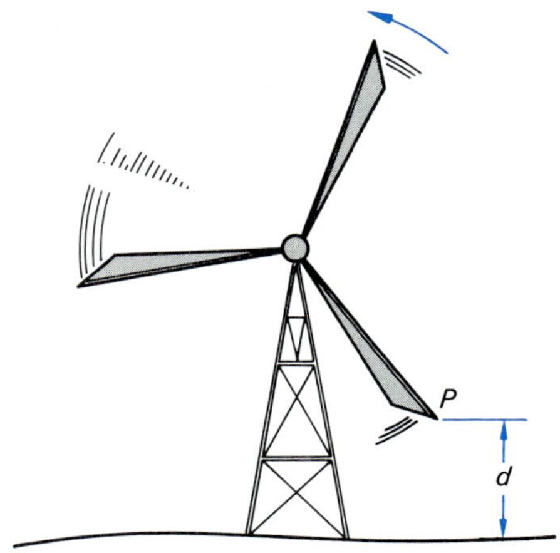

FIGURE 8.50

When the windmill starts, P is in the position shown in Fig. 8.50. Let $t =$ the number of seconds that have elapsed since the windmill started. You find that it takes point P $\frac{1}{25}$ second to reach the top, 60 ft above the ground and that the tip of the blade makes a revolution once every one tenth of a second. The diameter of the circle described by the blades of the windmill is 50 ft. (Use sine function.)

- **a.** Sketch a graph of this sinusoid.
- **b.** What is the lowest that point P will go as the blades of the windmill rotate? Why is the number greater than zero?
- **c.** Write an equation of this sinusoid.
- **d.** Determine the height of point P above the ground when (1) $t = \frac{1}{5}$ s, (2) $t = \frac{1}{4}$ s.
- **e.** What is the value of t the second time point P is 30 ft above the ground?

44. A rotating coil of wire in a magnetic field develops an electromotive force (emf) that is expressed

FIGURE 8.51

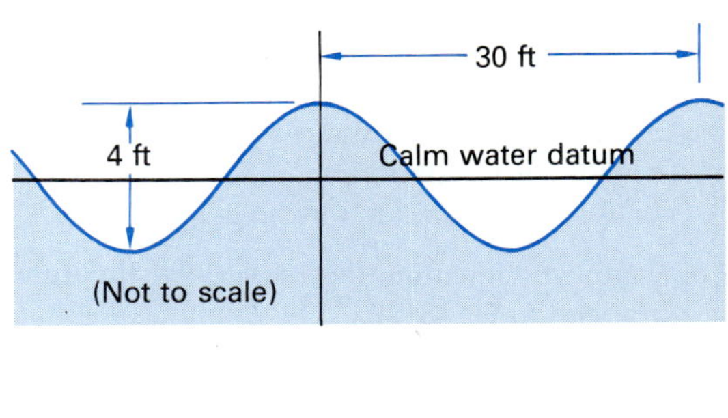

by emf $= nBA \cos\left(\omega t - \frac{\pi}{2}\right)$, where $n =$ number of turns of wire in the coil, $B =$ magnetic density, and $A =$ area of the coil. If $n = 100$, $B = 3 \times 10^{-3}\, t$, $A = 2\text{ m}^2$ and $\omega = 60$ rad/s: (a) sketch the graph of emf vs. t for one cycle, (b) express emf in terms of the sine function.

45. The angular acceleration of a simple pendulum is given by $\alpha = 2 \sin\left(2t + \frac{\pi}{2}\right)$, where $t =$ time in seconds. (a) Find the period of the pendulum's acceleration curve and sketch α vs. t for one cycle. (b) Express α in terms of the cosine function.

46. On a windy day the waves on the surface of Lake Ontario trace out approximately a sine or cosine curve if viewed horizontally. Find the equation describing the surface waves that are 4 ft from crest to trough and 30 ft from one crest to the next (Fig. 8.51).

Can you solve these?

1. Draw a graph with amplitude and period half that of the graph of $y = \sin\theta$. (Draw the graph from $-90°$ to $360°$.) What is the equation of the graph you have drawn?

2. Show that $\cos(90° - \theta) = \sin\theta$ by sketching graphs of both functions.

3. In Fig. 8.52 suppose that $r = 0.1$ m, $\theta = 0$ when $t = 0$, and point P completes 1 revolution in 3 s with uniform speed.

a. What is the angular speed of the radius to point P in rad/s?

b. What is the equation for the position of Q (the point directly under P on the x-axis) as a function of time?

FIGURE 8.52

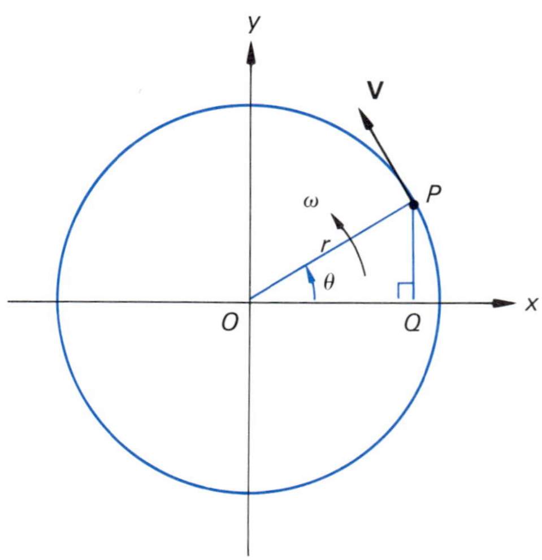

c. What is the frequency of the motion of Q?

d. What is the position of Q when $t = 0.60$ s? When $t = 1.50$ s? When $t = 1.75$ s?

e. Graph two cycles of the position of Q.

4. A saw-tooth waveform can be closely approximated by the following sum of sine functions:

$$\sin x - \frac{\sin 3x}{3^2} + \frac{\sin 5x}{5^2} - \ . \ . \ . \ .$$

To see the saw tooth forming, graph the first term, then the first two terms, and finally the first three terms on one set of axes. Write a program for your calculator or computer to facilitate the graphing.

5. Show that $f(x) = A \sin(Bx + C)$ has period $2\pi/B$ and phase shift $-C/B$ (assume $B > 0$). (*Hint:* Let the argument vary between 0 and 2π.)

6. The equation $x = 3^{-t} \cos \pi t$ is an algebraic model for a certain damped harmonic oscillator, where x is displacement and t is time. Graph the equation and comment on the nature of the motion ($t \geq 0$). Write a program for the calculator or computer to determine coordinates and to make the graphing easier.

7. Jean Fourier (1768-1830) a French mathematician investigated infinite series of sines and cosines. In this problem you will learn one of the things that he discovered. Let

$$f(\theta) = \cos\theta - \frac{1}{3}\cos 3\theta + \frac{1}{5}\cos 5\theta - \frac{1}{7}\cos 7\theta + \ . \ . \ .$$

a. Let $g(\theta)$ be the first three terms of $f(\theta)$; that is,

$g(\theta) = \cos\theta - \frac{1}{3}\cos 3\theta + \frac{1}{5}\cos 5\theta$. Sketch a graph of g for values of θ for the interval $0° \leq \theta \leq 360°$.

b. Let $h(\theta)$ be the first 10 terms of $f(\theta)$. Write a computer program to calculate values of $h(\theta)$ for each 10° for the interval $0° \leq \theta \leq 360°$.

c. Use the output from the program in part (b) to plot a graph of h. If there is a plotter available for the computer, use it.

d. Modify the program in part (b) so that it prints values for $h(\theta)$ for each 0.1° from 80° through 90°. Use the results to help refine the graph in part (c).

e. By comparing the graphs of g and h you should be able to determine the effect of adding more terms. Use your conclusion to sketch a graph of f for $0° \leq \theta \leq 360°$.

f. Compute $f(\pi/4)$. How does this number seem to relate to the graph of f?

9

Electrical circuits and mechanical systems can be tuned to respond more strongly to one frequency than to others by varying the values of the circuit components (usually capacitors in electrical circuits and shock absorbers in mechanical systems).

The current in a series AC circuit containing an inductive reactance ($X_L = 2\pi fL$), a capacitative reactance $\left(X_C = \dfrac{1}{2\pi fC}\right)$ and a resistance R is given by $i = v/Z$, where v is the source voltage and Z is a complex number called the impedance of the circuit. The impedance $Z = R + j(X_L - X_C)$.

Choose fixed values for v, R, and X_L. Choose a series of values for X_C and use phasor arithmetic to calculate the corresponding values for i. Graph i vs. X_C to find the value of X_C that causes the circuit to resonate (i becomes a maximum).

Imaginary and Complex Numbers

9.1 Introduction

In Chapter 1, as part of the discussion of number systems, the reader was introduced to the number $\sqrt{-9}$. The $\sqrt{-9}$ was defined as an **imaginary number.** It was also shown that an imaginary number such as $\sqrt{-9}$ could be expressed as

$$\begin{aligned}\sqrt{-9} &= \sqrt{(-1)9}\\ &= \sqrt{-1}\sqrt{9}\\ &= (\sqrt{-1})(3)\\ &= 3j.\end{aligned}$$

DEFINITION 9.1

$$j = \sqrt{-1}.$$

In most mathematics textbooks $\sqrt{-1} = i$. We use the letter j to avoid confusing the i for imaginary numbers with the i used for current in electrical-circuit theory, which is standard practice in electrical textbooks. The reason for this will be made evident in the last section of this chapter when applications are discussed dealing with alternating currents.

DEFINITION 9.2

If n is a positive real number, then

$$\sqrt{-n} = j\sqrt{n},$$

and $j\sqrt{n}$ is called an **imaginary** number.

EXAMPLE 1 Write the following as a product of a real number and the imaginary unit j.

a. $\sqrt{-25}$ **b.** $\sqrt{-7}$ **c.** $\sqrt{-a},\ a > 0$

Solution:

a. $\sqrt{-25} = \sqrt{25}\sqrt{-1}$ **Definition 9.2**

$= 5j.$

b. $\sqrt{-7} = \sqrt{7}\sqrt{-1}$

$= \sqrt{7}\,j,$ **which is generally written as $j\sqrt{7}$.**

c. Since $a > 0$

$$\begin{aligned}\sqrt{-a} &= \sqrt{a}\sqrt{-1}\\ &= j\sqrt{a} \quad \square\end{aligned}$$

DEFINITION 9.3

$$j^2 = -1.$$

EXAMPLE 2 Find the following products:

a. $(3j)^2$ **b.** $\sqrt{-9}\sqrt{-25}$

Solution:

a.
$$\begin{aligned}(3j)^2 &= (3^2)(j^2) && \textbf{Rules of exponents}\\ &= (9)(-1) && \textbf{Definition of } j^2\\ &= -9\end{aligned}$$

b. Before performing any operations on imaginary numbers they must be written in j-notation.

$$\begin{aligned}\sqrt{-9}\sqrt{-25} &= \sqrt{-1}\sqrt{9}\sqrt{-1}\sqrt{25}\\ &= (3j)(5j) && \textbf{Definition 9.2}\\ &= 15j^2 && \textbf{Commutative and associative laws}\\ &= 15(-1) && \textbf{Definition 9.3}\\ &= -15. \quad \square\end{aligned}$$

Recall that in dealing with radicals in Chapter 2 the following multiplication could be performed:

$$\begin{aligned}\sqrt{9}\sqrt{25} &= \sqrt{9(25)}\\ &= \sqrt{225}\\ &= 15.\end{aligned}$$

This procedure will not work when the numbers under the radicals are negative. For example:

$$\begin{aligned}\sqrt{-9}\sqrt{-25} &\neq \sqrt{(-9)(-25)}\\ &\neq \sqrt{225}\\ &\neq 15.\end{aligned}$$

This result is a contradiction of the result obtained in Example 2 part (b). It is for this reason that we must convert to j-notation before performing operations on imaginary numbers. It should also be noted that

$$\begin{aligned}\sqrt{(-4)(-9)} &= \sqrt{36}\\ &= 6.\end{aligned}$$

TABLE 9.1

$j = \sqrt{-1}$	$= j$	$j^7 = j^6 j = (j^2)^3 j = (-1)^3 j$	$= -j$
j^2	$= -1$	$j^8 = (j^2)^4 = (-1)^4$	$= 1$
$j^3 = j^2 j = -1j$	$= -j$	$j^9 = j^8 j = (j^2)^4 j = (-1)^4 j$	$= j$
$j^4 = (j^2)^2 = (-1)^2$	$= 1$	$j^{10} = (j^2)^5 = (-1)^5$	$= -1$
$j^5 = j^4 j = (j^2)^2 j = (-1)^2 j$	$= j$	$j^{11} = j^{10} j = (j^2)^5 j = (-1)^5 j$	$= -j$
$j^6 = (j^2)^3 = (-1)^3$	$= -1$	$j^{12} = (j^2)^6 = (-1)^6$	$= 1$

The product of two negative integers is a positive integer and the root of a positive integer is a real number. Therefore $\sqrt{(-4)(-9)}$ **cannot be written as**

$$\begin{aligned}\sqrt{-4}\sqrt{-9} &\neq (2j)(3j)\\ &\neq -6.\end{aligned}$$

By writing $\sqrt{(-4)(-9)}$ as $\sqrt{-4}\sqrt{-9}$ we are changing a real number to an imaginary number, which we cannot do.

In the previous discussion we defined $j^2 = -1$. Let us now examine what happens when we raise j to higher powers. All the powers of j are based on Definitions 9.1, 9.2, and 9.3, and on the laws of exponents $a^m \cdot a^n = a^{m+n}$. Table 9.1 contains the first 12 powers of j.

Table 9.1 shows that the powers of j go through repetitive cycles of $j, -1, -j$, and 1. The table also shows that regardless of the power of j, we can write it in one of the four forms using the definition $j^2 = -1$. *In general, for any positive integer* n, j^n *can be expressed as* $(j^2)^{n/2}$ *if* n *is an even positive integer or as* $(j^2)^{\frac{n-1}{2}}(j)$ *if* n *is an odd positive integer*. The object is to never leave j to a higher power than one.

EXAMPLE 3 Simplify the following:

a. j^{17}

b. j^{23}

c. $(-2j^{18})(5j^6)$

Solution:

a. For j^{17}, we can write

$$\begin{aligned}j^{17} &= (j^2)^{\frac{17-1}{2}}(j) && \textbf{From general statement}\\ &= (-1)^8 j && \textbf{Definition of } j^2\\ &= j.\end{aligned}$$

b. For j^{23}, we can write

$$\begin{aligned}j^{23} &= (j^2)^{\frac{23-1}{2}}(j) && \textbf{From general statement}\\ &= (-1)^{11} j && \textbf{Definition of } j^2\\ &= (-1)j\\ &= -j.\end{aligned}$$

c. The product $(-2j^{18})(5j^6)$ can be written as

$$\begin{aligned}(-2j^{18})(5j^6) &= (-2)(5)(j^{18})(j^6) && \textbf{Commutative and associative laws}\\ &= (-10)j^{24} && \textbf{Property of exponents}\\ &= (-10)(j^2)^{24/2} \\ &= (-10)(-1)^{12} && \textbf{Definition of } j^2\\ &= -10. \quad \square\end{aligned}$$

In Example 4 it will be shown how to divide a real number by an imaginary number.

EXAMPLE 4 Divide:

a. 28 by $\sqrt{-7}$

b. 17 by $\sqrt{-5}$

Solution: In dealing with imaginary numbers it is more convenient to work with numbers that do not contain imaginary numbers in the divisor (denominator). Thus, our objective will be to remove the imaginary number from the denominator. To accomplish this we use the definition of j^2.

a. $28 \div \sqrt{-7} = \dfrac{28}{\sqrt{-7}}$.

Since the denominator contains an imaginary number, can we change it to a real number? Yes, since $\sqrt{-7} = j\sqrt{7}$ and if we multiply j by j, the result is -1. Thus,

$$\begin{aligned}\frac{28}{\sqrt{-7}} &= \frac{28}{j\sqrt{7}} && \textbf{Definition 9.2}\\ &= \frac{28}{j\sqrt{7}} \cdot \frac{j}{j} && \textbf{Multiplying by 1}\\ &= \frac{j28}{-\sqrt{7}}. && j^2 = -1\end{aligned}$$

At this stage we have accomplished our goal. However, we can also eliminate the radical in the denominator by multiplying numerator and denominator by $\sqrt{7}$.

$$\begin{aligned}\frac{28}{\sqrt{-7}} &= \frac{j28}{-\sqrt{7}} \cdot \frac{\sqrt{7}}{\sqrt{7}} && \textbf{Rationalizing the denominator}\\ &= \frac{j28\sqrt{7}}{-7} && (\sqrt{a})(\sqrt{a}) = a\\ &= -4j\sqrt{7} && \textbf{Reducing the fraction}\end{aligned}$$

Therefore,

$28 \div \sqrt{-7} = -4j\sqrt{7}$

b. Following the procedure developed in part (a) we have

$$\frac{17}{\sqrt{-5}} = \frac{17}{j\sqrt{5}} \qquad \textbf{Definition 9.2}$$

$$= \frac{17}{j\sqrt{5}} \cdot \frac{j\sqrt{5}}{j\sqrt{5}} \qquad \textbf{Rationalizing the denominator}$$

$$= \frac{17j\sqrt{5}}{-5} \qquad \textbf{Multiplication of fractions and } j^2 = -1$$

$$= \frac{-17j\sqrt{5}}{5} \qquad \textbf{Multiplying numerator and denominator by } -1$$

□

EXAMPLE 5 Find the sum of the following numbers:

$\sqrt{-27}$, $\sqrt{-8}$, $\sqrt{-12}$.

Solution:

$$\sqrt{-27} + \sqrt{-8} + \sqrt{-12} = j\sqrt{27} + j\sqrt{8} + j\sqrt{12}$$

Definition 9.2

$$= j \cdot 3\sqrt{3} + j \cdot 2\sqrt{2} + j \cdot 2\sqrt{3}$$

Simplifying radicals

$$= j(3\sqrt{3} + 2\sqrt{2} + 2\sqrt{3})$$

j **is a common-term factor**

$$= j(5\sqrt{3} + 2\sqrt{2}).$$

Adding "like" terms □

9.1 Exercises

In Exercises 1–8, write each of the following as a product of a real number and the imaginary unit j. Leave the answers in simplest radical form.

1. $\sqrt{-64}$ **2.** $\sqrt{-81}$ **3.** $\sqrt{-32}$ **4.** $\sqrt{-125}$

5. $\sqrt{-a^4}$ **6.** $\sqrt{-x^8}$ **7.** $\sqrt{-p^6}$ **8.** $\sqrt{-q^{12}}$

In Exercises 9–46, perform the indicated operations and simplify. Leave the answers in radical form.

9. j^{13} **10.** j^{14} **11.** j^{31} **12.** j^{56}

13. j^{63} **14.** j^{175} **15.** j^{302} **16.** j^{1001}

17. j^0 **18.** j^{-2} **19.** j^{-29} **20.** j^{-58}

21. $(\sqrt{-121})(\sqrt{-144})$ **22.** $(\sqrt{-72})(\sqrt{-50})$

23. $(\sqrt{-96})(\sqrt{-24})$ **24.** $(\sqrt{-15})(\sqrt{-12})$

25. $(\sqrt{-8})(\sqrt{24})$ **26.** $(\sqrt{-51})(\sqrt{17})$

27. $\sqrt{(-4)(-9)}$ **28.** $\sqrt{(-12)(-48)}$

29. $(\sqrt{-8})(\sqrt{-15})(\sqrt{-24})$

30. $(\sqrt{-14})(\sqrt{+21})(\sqrt{-18})$

31. $(\sqrt{-11})(\sqrt{-44})(\sqrt{-20})$

32. $(\sqrt{-45})(\sqrt{-35})(\sqrt{-7})$

33. $\dfrac{\sqrt{5}}{\sqrt{-14}}$ **34.** $\dfrac{\sqrt{-7}}{\sqrt{28}}$ **35.** $\dfrac{\sqrt{35}}{\sqrt{5}}$

36. $\dfrac{\sqrt{1001}}{\sqrt{-11}}$ **37.** $\dfrac{\sqrt{-8}}{\sqrt{-32}}$ **38.** $\dfrac{\sqrt{-28}}{\sqrt{-14}}$

39. $\sqrt{-18} + \sqrt{-50} + \sqrt{-32}$

40. $\sqrt{-24} + \sqrt{-150} + \sqrt{-96}$

41. $3\sqrt{-80} - \sqrt{-45}$

42. $4\sqrt{-4} + 4\sqrt{-9} - 5\sqrt{16}$

43. $\sqrt{-25} + \sqrt{-144} - \sqrt{169}$

44. $\sqrt{-289} + \sqrt{-361} - \sqrt{-400}$

45. $j + j^2 + j^3 + j^4$ **46.** $\dfrac{1}{j} + \dfrac{1}{j^2} + \dfrac{1}{j^3} + \dfrac{1}{j^4}$

9.2 Basic operations with complex numbers

If we combine an imaginary number and a real number, the result is a number called a complex number.

DEFINITION 9.4

A **complex number** is a number of the form $a + jb$, where a and b are real numbers and $j = \sqrt{-1}$.

Let $Z = a + jb$; a is called the **real part** of Z and b is called the **imaginary part** of Z. When $a = 0$ and $b \neq 0$, the complex number $(a + jb)$ may be expressed in the form $0 + jb = jb$ and is called a **pure imaginary number.** When $b = 0$ and $a \neq 0$, the complex number $(a + jb)$ represents the real number $a + 0j = a$.

Complex numbers that differ only in the signs of the imaginary parts are called **conjugate complex numbers** (or complex conjugates) and either is called the conjugate of the other. Thus, if $Z_1 = a + jb$ and $Z_2 = a - jb$, then Z_1 is the complex conjugate of Z_2 and Z_2 is the complex conjugate of Z_1. (Also, if $Z_1 = jb$ and $Z_2 = -jb$, then Z_1 and Z_2 are conjugates of each other.) A real number is its own conjugate.

DEFINITION 9.5

Two complex numbers are equal if and only if their real parts are equal and their imaginary parts are equal. That is $a + jb = c + jd$ if and only if $a = c$ and $b = d$.

From Definition 9.5 we can conclude that if the complex number $a + jb = 0$, then $a = 0$ and $b = 0$.

EXAMPLE 1 For what real values of x and y will the complex numbers Z_1 and Z_2 be equal?

a. $Z_1 = x + 3j$, $Z_2 = 2 - 3jy$.

b. $Z_1 = 2x + 7jy$, $Z_2 = 6 - 49j + jx$.

Solution:

a. For the complex numbers $Z_1 = x + 3j$ and $Z_2 = 2 - 3jy$ to be equal, Definition 9.5 states that the corresponding parts must be equal. Therefore

$$x = 2 \qquad \text{and} \qquad 3 = -3y$$
$$y = -1.$$

Thus

$$Z_1 = 2 + 3j \qquad \text{and} \qquad Z_2 = 2 - 3(-1)j$$
$$= 2 + 3j$$

b. For $Z_1 = 2x + 7jy$ and $Z_2 = 6 - 49j + jx$ to be equal,

$$2x = 6 \quad \text{and} \quad 7y = -49 + x \qquad \textbf{By Definition 9.5}$$
$$x = 3 \quad \text{and} \quad 7y = -49 + 3$$
$$7y = -46$$
$$y = \frac{-46}{7}.$$

Thus

$$Z_1 = 2(3) + 7\left(\frac{-46}{7}\right)j \qquad \text{and} \qquad Z_2 = 6 - 49j + 3j$$
$$Z_1 = 6 - 46j \qquad \text{and} \qquad Z_2 = 6 - 46j. \qquad \square$$

It can be demonstrated that the definition of equality of complex numbers is the only possible definition that is consistent with the definition of real numbers. Therefore, if $a + jb = c + jd$, then $a - c = j(d - b)$ and squaring both sides $(a - c)^2 = -(d - b)^2$, which is only possible when $(a - c) = 0$ and $(d - b) = 0$. Thus, $a = c$ and $b = d$ when two complex numbers are equal.

Since the complex numbers are a new set of numbers, we must define how to perform the fundamental operations of addition, subtraction, multiplication, and division with complex numbers.

Addition and subtraction

In order to add (or subtract) two complex numbers such as $a + jb$ and $c + jd$, add (or subtract) the real parts and the imaginary parts separately.

DEFINITION 9.6

$$(a + jb) + (c + jd) = (a + c) + j(b + d).$$
$$(a + jb) - (c + jd) = (a - c) + j(b - d).$$

EXAMPLE 2 Perform the indicated operations:

a. $(6 + 3j) + (8 - 5j)$ **b.** $(5 + 4j) - (6 - 7j)$

Solution:

a. $(6 + 3j) + (8 - 5j) = (6 + 8) + (3 + (-5))j$ **Definition of addition**

$= 14 - 2j$ **Combining like terms**

b. $(5+4j)-(6-7j)=(5-6)+(4-(-7))j$ **Definition of subtraction**

$=-1+11j$ **Combining like terms** □

Multiplication

To multiply two complex numbers, such as $a+jb$ and $c+jd$, find the product as you would for any product of two binomials. Then evaluate the powers of j according to the rules in Section 9.1 and combine the real and imaginary parts.

$$\begin{aligned}(a+jb)(c+jd) &= ac+jad+jbc+j^2bd\\ &= ac+j(ad+bc)+(-1)bd\\ &= (ac-bd)+j(ad+bc).\end{aligned}$$

DEFINITION 9.7

$$(a+jb)(c+jd)=(ac-bd)+j(ad+bc).$$

EXAMPLE 3 Find the product of $(3+7j)$ and $(4-5j)$

Solution:

$(3+7j)(4-5j)=12-15j+28j-35j^2$	**Definition of multiplication**
$=12+13j-35(-1)$	**Definition of j^2**
$=12+35+13j$	**Commutative law**
$=47+13j$	**Combining like terms** □

Division

To express the quotient of $a+jb$ and $c+jd$, where $c+jd\neq 0$, as a single complex number multiply both numerator and denominator of the indicated quotient $\dfrac{a+jb}{c+jd}$ by $c-jd$, the conjugate of the denominator.

$$\begin{aligned}\frac{a+jb}{c+jd} &= \frac{a+jb}{c+jd}\cdot\frac{c-jd}{c-jd}\\ &= \frac{ac-jad+jbc-j^2bd}{c^2-jcd+jcd-j^2d^2}\\ &= \frac{(ac+bd)+j(bc-ad)}{c^2+d^2}\\ &= \frac{ac+bd}{c^2+d^2}+j\frac{bc-ad}{c^2+d^2}.\end{aligned}$$

DEFINITION 9.8

$$\frac{a+jb}{c+jd}=\frac{ac+bd}{c^2+d^2}+j\frac{bc-ad}{c^2+d^2} \quad \text{if } c+jd \neq 0.$$

EXAMPLE 4 Find the quotient of $6j$ and $5-7j$.

Solution:

$$\frac{6j}{5-7j}=\frac{6j}{5-7j}\cdot\frac{5+7j}{5+7j}$$ **Multiply numerator and denominator by the conjugate of $5-7j$**

$$=\frac{30j+42j^2}{25+49}$$ **Multiply**

$$=\frac{-42+30j}{74}$$

$$=\frac{-21}{37}+\frac{15}{37}j. \quad \square$$

EXAMPLE 5 Find the quotient of $4-5j$ by $2+3j$

Solution:

$$\frac{4-5j}{2+3j}=\frac{4-5j}{2+3j}\cdot\frac{2-3j}{2-3j}$$ **Multiplying numerator and denominator by the conjugate of $2+3j$**

$$=\frac{8-12j-10j+15j^2}{4+9}$$ **Multiplying and remembering that $(c+jd)(c-jd)=c^2+d^2$**

$$=\frac{(8-15)+(-12j-10j)}{4+9}$$ **Collecting like terms, $j^2=-1$**

$$=\frac{-7-22j}{13}$$

$$=\frac{-7}{13}-\frac{22}{13}j. \quad \square$$

9.2 Exercises

In Exercises 1–4, find the real numbers x and y that will make the complex numbers Z_1 and Z_2 equal.

1. $Z_1=2x+jy$, $Z_2=3+5j$

2. $Z_1=4x+6jy$, $Z_2=12-42j+jx$

3. $Z_1=(x-4)+j(3y-5)$, $Z_2=-3y+7j$

4. $Z_1=(2y-5)+j(7x+6)$, $Z_2=9-8j$

In Exercises 5–26, perform the indicated operations and write the answer in the form $a+jb$.

5. $(5-6j)+(6+5j)$

6. $(4-j)+(-9-3j)$

7. $(4 + 11j) + (8 + 6j)$
8. $(8 + j) - (4 + j)$
9. $(8 - \sqrt{-16}) - (3 + \sqrt{-36})$
10. $(9 + \sqrt{-18}) - (\sqrt{-72} + 1)$
11. $(2 + 5j)(4 + 2j)$
12. $(8 - 7j)(7 + 8j)$
13. $(3 + 6j)^2$
14. $(-5 + 4j)^2$
15. $(2 - \sqrt{-3})(3 + \sqrt{-12})$
16. $(8 - 6\sqrt{-5})(-2 + \sqrt{-20})$
17. $(4j - 2j^2 + 3j^3)(2j^4 - j^5)$
18. $(2j^5 - 5j^4 + 2j^3 - 5j^2)^2$
19. $(3 + 2j) \div (1 - 2j)$
20. $(4 - 5j) \div (4 + 6j)$
21. $(3 - \sqrt{-5}) \div (5 + 2\sqrt{-5})$
22. $(9 + \sqrt{-32}) \div (\sqrt{-2} + 3)$
23. $(8 - j^3) \div (3 - j)$
24. $(64 + j^3) \div (4 + j)$
25. $1 \div (4 + 3j)$
26. $1 \div (6 + 3\sqrt{-50})$

27. Two AC voltages are given by the expressions $\frac{3}{2} + j(y + 1)$ and $(7 + x) - 3j$. If the voltages are equal, what are the values of x and y?

28. What can be said about a complex number that is equal to its conjugate?

29. The impedance Z in an AC circuit is given by $Z = V/I$, where V = voltage and I = current. Find Z when $V = 2.3 - 0.4j$ and $I = -1.2j$.

30. Find V based on the formula in Exercise 29 if $Z = +0.4 + 1.3j$ and $I = 2.2j$.

31. What type of a number is (a) the sum of a complex number and its conjugate? (b) The difference of a complex number and its conjugate?

32. The total impedance Z_T of an AC circuit containing two impedances Z_1 and Z_2 in parallel is given by:

$$Z = \frac{Z_1 Z_2}{Z_1 + Z_2}$$

Find Z_T when $Z_1 = 2 - j$ and $Z_2 = 3 + j$.

33. Find Z_T using the formula in Exercise 32 if $Z_1 = 1.73 - j$ and $Z_2 = 0.5 + 0.9j$.

34. If three resistors in an AC circuit are connected in parallel, the total impedance Z may be found by using the formula:

$$\frac{1}{Z} = \frac{1}{Z_1} + \frac{1}{Z_2} + \frac{1}{Z_3}$$

or the formula

$$Z = \frac{Z_1 \cdot Z_2 \cdot Z_3}{Z_1 Z_2 + Z_1 Z_3 + Z_2 Z_3}.$$

If $Z_1 = 2 - j$, $Z_2 = 2 + j$, and $Z_3 = 4j$, find Z using both formulas.

9.3 Polar form of a complex number

Before discussing the polar form of complex numbers, we will explain the technique of graphing complex numbers.

If a geometric representation of a point P is graphed as an ordered pair (a, b) as if on a rectangular plane, but labeled as $a + jb$, then the plane is considered as the **complex plane.**

A complex number is graphed on a complex plane. A complex plane may be divided into four quadrants with a horizontal and a vertical axis similar to a real plane. By convention the horizontal axis is called the **real axis** and the vertical axis is called the **imaginary axis** (Fig. 9.1).

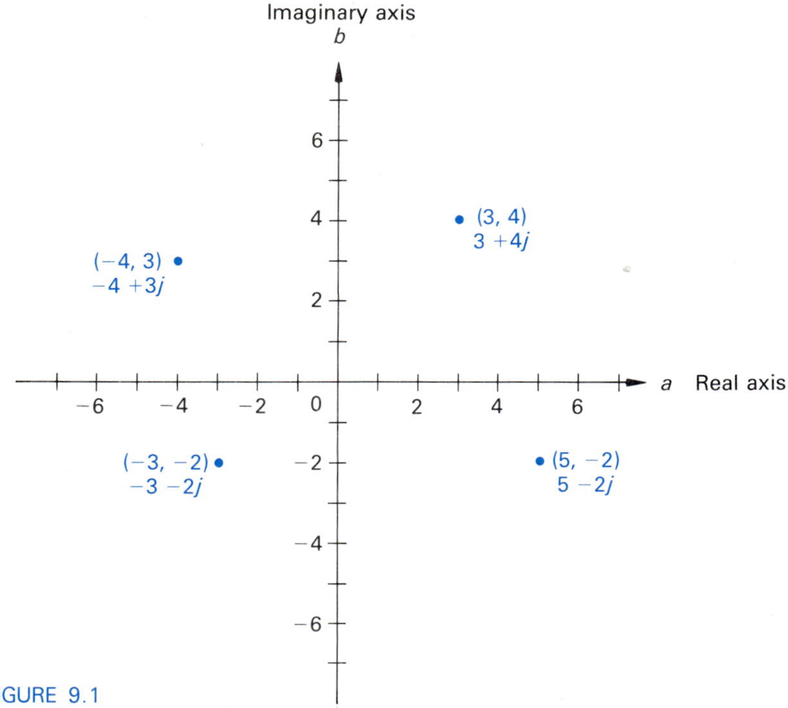

FIGURE 9.1

To illustrate the technique of graphing we have plotted the complex numbers $3 + 4j$, $5 - 2j$, $-4 + 3j$, and $-3 - 2j$ in Fig. 9.1.

A complex number is not a vector, and a vector is not a complex number. However, this representation is sometimes used since a complex number appears to be of a vector form and can be easily illustrated as a vector on a graph.

For any complex number $(a + jb)$, let P be the point on the complex plane that represents the number and draw the line segment $\overline{OP}$ (Fig. 9.2). The length r of $\overline{OP}$ is $r = \sqrt{a^2 + b^2}$. The length r is called the **magnitude, absolute value,** or **modulus** of the complex number $(a + jb)$. The angle θ with the a-axis as the initial side is called the **argument** or the **amplitude** of the complex number.

FIGURE 9.2

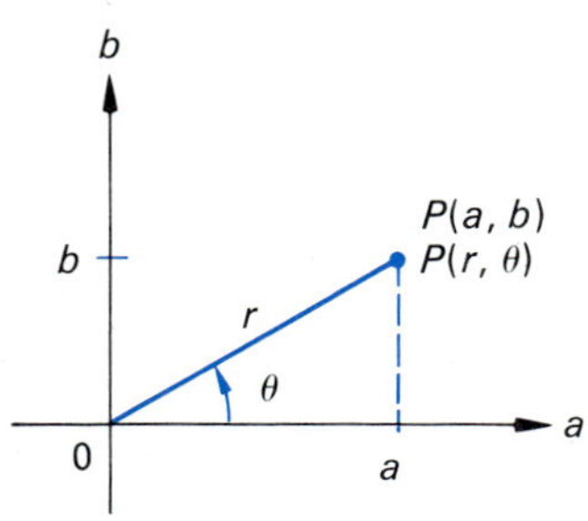

The **absolute value** of any complex number $(a + jb)$ is denoted by $|a + jb|$. Therefore $|a + jb| = \sqrt{a^2 + b^2}$. Thus, the absolute value of any complex number is a non-negative real number.

The argument. θ may be replaced by any angle that is coterminal with it, $\theta + k \cdot 360°$, where k is an integer. For convenience we will select the smallest non-negative value for the argument of the complex number.

Any point P on the complex plane can be identified with its rectangular coordinates, (a, b), or its **polar coordinates, (r, θ)** (Fig. 9.2).

From Fig. 9.2 and our knowledge of right-triangle trigonometry, we have

$$\cos\theta = \frac{a}{r} \qquad \text{and} \qquad \sin\theta = \frac{b}{r}.$$

Thus

$$a = r\cos\theta \qquad \text{and} \qquad b = r\sin\theta. \qquad [1]$$

Now substituting equations [1] into $a + jb$, we have

$$a + jb = r\cos\theta + jr\sin\theta$$

or

$$a + jb = r(\cos\theta + j\sin\theta) \qquad [2]$$

The right side of Eq. [2] is known as the **trigonometric form** or the **polar form** of the complex number and the left side is the **rectangular form.** The polar form may be abbreviated as $r \text{ cis } \theta$ or $r\,\underline{/\theta}$.

If the complex number is given in polar form, it may be converted to rectangular form by using Eq. [1].

EXAMPLE 1 Graph $3(\cos 210° + j\sin 210°)$ and write the complex number in rectangular form.

Solution: To graph $3\,\underline{/210°}$, first construct the angle $\theta = 210°$, Fig. 9.3. On the terminal side of the angle mark off 3 units and locate point P. The point P represents the complex number. From the information on the figure, using Eq. [1] we have

$$\begin{aligned} a &= 3\cos 210° \\ &= -2.60 \end{aligned}$$

and

$$\begin{aligned} b &= 3\sin 210° \\ &= -1.50 \end{aligned}$$

Therefore, the rectangular form of $3\,\underline{/210°}$ is $-2.60 - 1.50j$. □

FIGURE 9.3

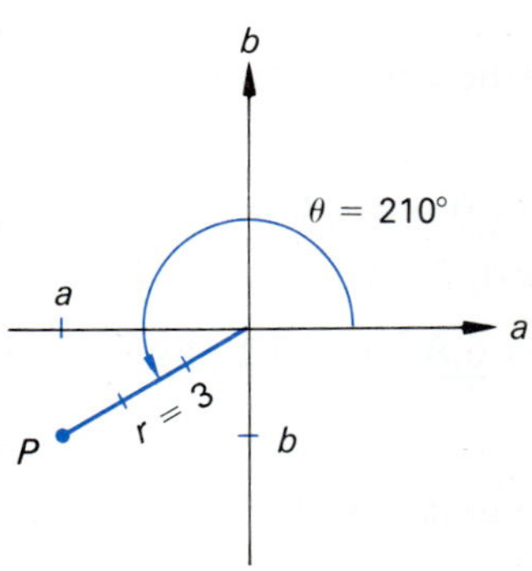

When the complex number is in rectangular form it can be converted to polar form using the equations:

$$r = \sqrt{a^2 + b^2} \qquad \text{and} \qquad \tan\theta = b/a. \qquad [3]$$

Remember that the signs of a and b determine the quadrant in which the terminal side of θ lies. Be careful.

EXAMPLE 2 Graph the complex numbers and write the polar form of each, using the smallest non-negative argument.

a. $1 + j\sqrt{3}$ **b.** $2 - 2j\sqrt{3}$

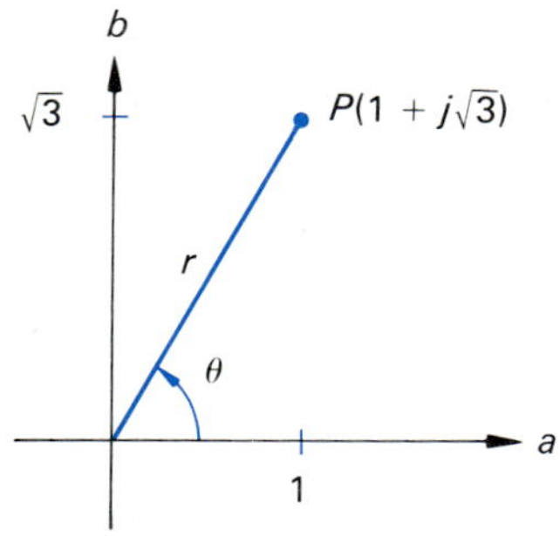

FIGURE 9.4

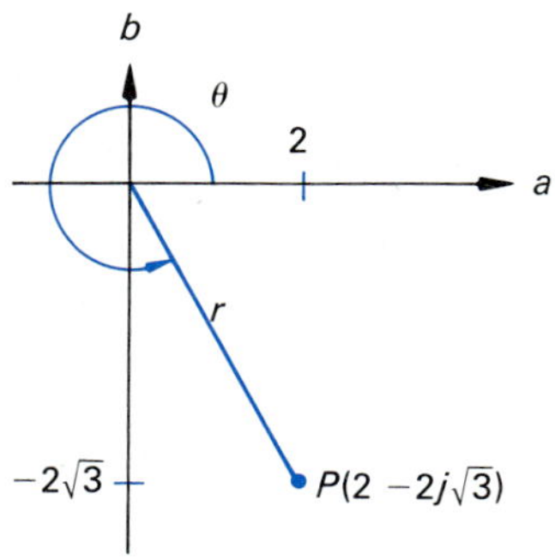

FIGURE 9.5

Solution:

a. In Fig. 9.4 point P represents $(1 + j\sqrt{3})$. The point is in the first quadrant. Knowing that $a = 1$ and $b = \sqrt{3}$ and using Eq. [3], we have

$$r = \sqrt{1^2 + (\sqrt{3})^2} \quad \text{and} \quad \tan \hat{\theta} = \frac{\sqrt{3}}{1}$$
$$= 2 \qquad\qquad \hat{\theta} = 60°$$

Since the complex number is in the first quadrant $\theta = \hat{\theta}$. Therefore, $1 + j\sqrt{3}$ in polar form is $2(\cos 60° + j \sin 60°)$ or $2 \underline{/60°}$.

b. In Fig. 9.5 the point P represents $(2 - 2j\sqrt{3})$. The point is in the fourth quadrant. Knowing that $a = 2$ and $b = -2\sqrt{3}$ and using Eq. [3] we have

$$r = \sqrt{2^2 + (-2\sqrt{3})^2} \quad \text{and} \quad \tan \theta = \frac{-2\sqrt{3}}{2}$$
$$= \sqrt{16} \qquad\qquad = -\sqrt{3}$$
$$= 4. \qquad\qquad \hat{\theta} = -60°.$$

Since the complex number is in the fourth quadrant $\theta = 360° - 60°$ or $300°$. Therefore, the equation of $2 - 2j\sqrt{3}$ in polar form is $4(\cos 300° + j \sin 300°)$ or $4 \underline{/300°}$. □

To ensure that you have the correct angle, it is necessary to sketch a graph of the complex number.

Changing complex numbers from polar to rectangular coordinates and from rectangular to polar coordinates can be done using a calculator. This will be illustrated in Examples 3 and 4 using the Texas Instruments model 55 II. The procedure may vary slightly on other calculators, so check your instruction book.

EXAMPLE 3 Change $3 - 7j$ to polar form using a calculator.

Solution: To change $3 - 7j$ to polar form follow the following steps:

—Enter 3 (the real part of the complex number)
—Press [X ⇄ Y].
—Enter 7 [+/−] (the imaginary part of the complex number)
—Place calculator into degree mode.
—Press [INV] [2nd] [P ↔ R]. The value of 0 is displayed: $-66.8°$.
—Press [X ⇄ Y], the r value is displayed: 7.62.

Thus, $3 - 7j$ can be expressed as $7.62 \underline{/-66.8°}$ or $7.62 \underline{/293.2°}$ in polar form. □

EXAMPLE 4 Change $7.3 \underline{/115.°}$ to rectangular form using a calculator.

Solution: To change $7.3\ \underline{/115.^\circ}$ to rectangular form follow the following steps:

—Enter 7.3 (the r value, the modulus)

—Press $\boxed{X \rightleftarrows Y}$.

—Enter 115° (θ, the argument, making sure calculator is in the degree mode.)

—Press $\boxed{\text{2nd}}$ $\boxed{P \leftrightarrow R}$. The imaginary part of the complex number is displayed: 6.6.

—Press $\boxed{X \rightleftarrows Y}$, the real part of the complex number is displayed: -3.1.

Thus, $7.3\ \underline{/115.^\circ}$ can be expressed as $-3.1 + 6.6j$ in rectangular form. □

9.3 Exercises

In Exercises 1–12, graph each complex number in the complex plane.

1. $2 + 3j$ **2.** $-3 + 5j$ **3.** $-6 - 4j$

4. $3 - 5j$ **5.** $-4.3 + 3.4j$ **6.** $-3j$

7. 4.7 **8.** -6.3 **9.** $2.3j$

10. $-3 - 3j\sqrt{3}$

11. $2\sqrt{3} - 4j\sqrt{3}$

12. $3 + 2j - 4j^2$

In Exercises 13–21, graph each complex number and change to the polar form. (Give exact answers.)

13. $2 + 2j$ **14.** $3 - 3j$ **15.** $\sqrt{3} - j$

16. $-2\sqrt{3} + 2j$ **17.** $-4 - 4j$ **18.** 5

19. -5 **20.** $-4j$ **21.** $4j$

In Exercises 22–30, write each complex number in polar form. (Express answers to three significant digits.)

22. $-4 + 3j$ **23.** $-4 - 3j$ **24.** $4 - 12j$

25. $-11 - 6j$ **26.** $5 + 7j$ **27.** $3 + 11j$

28. $-8 - 7j$ **29.** $5 - 9j$ **30.** $-3 - 5j$

In Exercises 31–45, change each complex number from polar form to rectangular form. Find exact answers if the reference angle is 30°, 45°, 60°, or quadrantal; otherwise give answers to three significant digits.

31. $3\ \underline{/45^\circ}$ **32.** $2.3\ \underline{/90^\circ}$ **33.** $6\ \underline{/210^\circ}$

34. $2\ \underline{/193.4^\circ}$ **35.** $2.5\ \underline{/73^\circ}$ **36.** $3\ \underline{/135^\circ}$

37. $1\ \underline{/150^\circ}$ **38.** $2\ \underline{/300^\circ}$ **39.** $2\ \underline{/330^\circ}$

40. $1.3\ \underline{/118.7^\circ}$ **41.** $3.0\sqrt{2.0}\ \underline{/267.5^\circ}$

42. $\sqrt{5}\ \underline{/180^\circ}$ **43.** $\sqrt{3.0}\ \underline{/275.6^\circ}$

44. $5.0\ \underline{/2.5^\circ}$ **45.** $8.7\ \underline{/94.7^\circ}$

The amount of alternating current in a circuit is restricted by the impedance Z (in ohms). If a resistor, inductor, and capacitor are connected in series the impedance is $Z = R + j(X_L - X_C)$, with magnitude $|Z|$ and direction θ given, respectively, by

$$|Z| = \sqrt{R^2 + (X_L - X_C)^2} \quad \text{and} \quad \theta = \tan^{-1}\left(\frac{X_L - X_C}{R}\right).$$

In Exercises 46–50 compute the magnitude and direction of Z (to the nearest whole number) for each set of data.

46. $R = 2500$, $X_L = 90$, $X_C = 0$

47. $R = 300$, $X_L = 0$, $X_C = 175$

48. $R = 60$, $X_L = 75$, $X_C = 40$

49. $R = 40$, $X_L = 5$, $X_C = 80$

50. $R = 200$, $X_L = 80$, $X_C = 75$

9.4 Products, quotients, powers, and roots of complex numbers

In the field of electronics, when complex numbers are used it may be necessary to multiply, divide, or raise these numbers to a power. For example, it may be necessary to combine several operations. When two impedances Z_1 and Z_2 are connected in parallel, the equivalent impedance is

$$Z_p = \frac{Z_1 Z_2}{Z_1 + Z_2}.$$

Before working a problem that involves this formula, we will develop each operation separately.

EXAMPLE 1 In an AC circuit the current $I = 3$ amps $\underline{/25^\circ}$ and the impedance $Z = 4\ \Omega\ \underline{/42^\circ}$. Find the voltage E, where $E = IZ$.

Solution: In Section 9.2 we learned how to multiply two complex numbers in rectangular form. We will use this knowledge to find the product of two complex numbers in polar form.

First write the numbers in rectangular form

$$3\text{ A}\ \underline{/25^\circ} = 3\cos 25^\circ + 3j\sin 25^\circ.$$
$$4\ \Omega\ \underline{/42^\circ} = 4\cos 42^\circ + 4j\sin 42^\circ.$$

We determine E by finding the product of I and Z. Thus,

$$E = (3\cos 25^\circ + 3j\sin 25^\circ)(4\cos 42^\circ + 4j\sin 42^\circ).$$

Finding the product in the same manner as finding the product of two binomial expressions, we have

$$E = 3\cdot 4\cos 25^\circ \cos 42^\circ + 3\cdot 4\cdot j\cos 25^\circ \sin 42^\circ + 3\cdot 4\cdot j\sin 25^\circ \cos 42^\circ + 3\cdot 4\cdot j^2\sin 25^\circ \sin 42^\circ.$$

By common-term factoring and definition of j^2 we have

$$E = 3\cdot 4\cos 25^\circ \cos 42^\circ + 3\cdot 4\,j(\cos 25^\circ \sin 42^\circ + \sin 25^\circ \cos 42^\circ) - 3\cdot 4\sin 25^\circ \sin 42^\circ.$$

By common-term factoring and associative and commutative properties we have

$$E = 3\cdot 4\,[\cos 25^\circ \cos 42^\circ - \sin 25^\circ \sin 42^\circ] + 3\cdot 4\,j[\sin 25^\circ \cos 42^\circ + \cos 25^\circ \sin 42^\circ]. \quad [1]$$

In Chapter 15 the following statements will be shown to be true:

$$\cos(A + B) = \cos A \cos B - \sin A \sin B \quad [2]$$

and

$$\sin (A + B) = \sin A \cos B + \cos A \sin B. \quad \textbf{[3]}$$

From statement [2] it can be seen that the expression inside the first set of brackets in [1] can be expressed as cos $(25° + 42°)$. Also, the expression inside the second set of brackets in [1] can be expressed as sin $(25° + 42°)$ using statement [3].

Substituting we have

$$\begin{aligned} E &= 3 \cdot 4 \cos (25° + 42°) + 3 \cdot 4j \sin (25° + 42°) \\ &= 12 \cos 67° + 12j \sin 67° \\ &= 12 \underline{/67°}. \quad \square \end{aligned}$$

The result of Example 1 implies that the product of two complex numbers in polar form can be found by multiplying the moduli and adding the amplitudes of the two numbers. The product and sum are the modulus and the amplitude respectively of the product.

DEFINITION 9.9

If $Z_1 = r_1 (\cos \theta_1 + j \sin \theta_1)$ and $Z_2 = r_2 (\cos \theta_2 + j \sin \theta_2)$, then $Z_1 Z_2 = r_1 r_2 [\cos (\theta_1 + \theta_2) + j \sin (\theta_1 + \theta_2)]$.

Or, if $Z_1 = r_1 \underline{/\theta_1}$ and $Z_2 = r_2 \underline{/\theta_2}$, then $Z_1 Z_2 = r_1 r_2 \underline{/\theta_1 + \theta_2}$

To show that this statement holds for any two complex numbers.

Let

$$\begin{aligned} Z_1 &= r_1 (\cos \alpha + j \sin \alpha) \\ Z_2 &= r_2 (\cos \beta + j \sin \beta) \end{aligned}$$

be any two numbers in polar form. Thus

$$\begin{aligned} Z_1 \cdot Z_2 &= r_1 (\cos \alpha + j \sin \alpha) \cdot r_2 (\cos \beta + j \sin \beta) \\ &= r_1 r_2 [\cos \alpha \cos \beta + j \cos \alpha \sin \beta \\ &\quad + j \sin \alpha \cos \beta + j^2 \sin \alpha \sin \beta] \\ &= r_1 r_2 [(\cos \alpha \cos \beta - \sin \alpha \sin \beta) \\ &\quad + j(\sin \alpha \cos \beta + \cos \alpha \sin \beta)]. \end{aligned}$$

Since

$$\cos \alpha \cos \beta - \sin \alpha \sin \beta = \cos (\alpha + \beta)$$

and

$$\sin \alpha \cos \beta + \cos \alpha \sin \beta = \sin (\alpha + \beta),$$

we have

$$Z_1 \cdot Z_2 = r_1 r_2 [\cos (\alpha + \beta) + j \sin (\alpha + \beta)],$$

or

$$Z_1 \cdot Z_2 = r_1 r_2 \underline{/\alpha + \beta}.$$

This technique can be extended to find the product of more than two complex numbers by first multiplying two of the numbers and then multiplying their product by the third and so on.

EXAMPLE 2 Find the product of the complex numbers in polar form.

a. $Z_1 = 3\,\underline{/40^\circ}$, $Z_2 = 8\,\underline{/32^\circ}$

b. $Z_1 = 4\,\underline{/150^\circ}$, $Z_2 = 5\,\underline{/225^\circ}$

Solution:

a. $Z_1 \cdot Z_2 = 3 \cdot 8\,\underline{/40^\circ + 32^\circ}$ **By Definition 9.9**

$= 24\,\underline{/72^\circ}$.

b. $Z_1 \cdot Z_2 = 4 \cdot 5\,\underline{/150^\circ + 225^\circ}$ **By Definition 9.9**

$= 20\,\underline{/375^\circ}$

$= 20\,\underline{/15^\circ}$ **(since 375° and 15° are coterminal angles).** □

For the simple AC circuit in Example 1, if we know the current I and the voltage E, can we determine the impedance Z?

EXAMPLE 3 In an AC circuit the current $I = 3\text{ A}\,\underline{/25^\circ}$ and the voltage $E = 12\text{ V}\,\underline{/67^\circ}$. Find the impedance Z, where $Z = E/I$.

Solution: Since these are the same numbers that were used in Example 1, we know that $Z = 4\;\Omega\,\underline{/42^\circ}$. Substituting in the formula $Z = E/I$, we have

$$Z = \frac{12\,\underline{/67^\circ}}{3\,\underline{/25^\circ}}.$$

Note that if we divide the moduli and subtract the arguments, we obtain the correct value for Z. That is,

$$Z = \frac{12}{3}\,\underline{/67^\circ - 25^\circ}$$
$$= 4\,\underline{/42^\circ}. \quad \square$$

This example suggests the following definition:

DEFINITION 9.10

If $Z_1 = r_1 (\cos\theta_1 + j\sin\theta_1)$ and $Z_2 = r_2(\cos\theta_2 + j\sin\theta_2)$, then $\frac{Z_1}{Z_2} = \frac{r_1}{r_2}[\cos(\theta_1 - \theta_2) + j\sin(\theta_1 - \theta_2)]$. Or, if $Z_1 = r_1\,\underline{/\theta_1}$ and $Z_2 = r_2\,\underline{/\theta_2}$, then $\frac{Z_1}{Z_2} = \frac{r_1}{r_2}\,\underline{/\theta_1 - \theta_2}$, $Z_2 \neq 0$.

It can be shown that for any two complex numbers

$$Z_1 = r_1 (\cos \alpha + j \sin \alpha) \text{ and } Z_2 = r_2 (\cos \beta + j \sin \beta)$$

that

$$\frac{Z_1}{Z_2} = \frac{r_1}{r_2} [\cos (\alpha - \beta) + j \sin (\alpha - \beta)].$$

This demonstration will be left as an exercise.

EXAMPLE 4 Find the quotient of Z_1 divided by Z_2.

a. $Z_1 = 18 \underline{/75°}, \quad Z_2 = 5 \underline{/41°}$

b. $Z_1 = 84.1 \underline{/18.°}, \quad Z_2 = 15.0 \underline{/-34.°}$

Solution:

a.
$$\begin{aligned} \frac{Z_1}{Z_2} &= \frac{18 \underline{/75°}}{5 \underline{/41°}} \\ &= 3.6 \underline{/75° - 41°} \qquad \textbf{By Definition 9.10} \\ &= 3.6 \underline{/34°}. \end{aligned}$$

b.
$$\begin{aligned} \frac{Z_1}{Z_2} &= \frac{84.1 \underline{/18°}}{15.0 \underline{/-34°}} \\ &= 5.6 \underline{/18° - (-34°)} \qquad \textbf{By Definition 9.10} \\ &= 5.6 \underline{/52°}. \qquad \square \end{aligned}$$

Using the rule developed for multiplication of complex numbers in polar form we can find Z^2, Z^3, Z^4, and Z^5. For example, if $Z = r \underline{/\theta}$, then

$$\begin{aligned} Z^2 &= (r \underline{/\theta})(r \underline{/\theta}) & Z^3 &= (r^2 \underline{/2\theta})(r \underline{/\theta}) \\ &= r^2 \underline{/\theta + \theta} & &= r^3 \underline{/2\theta + \theta} \\ &= r^2 \underline{/2\theta} & &= r^3 \underline{/3\theta} \\ Z^4 &= (r^3 \underline{/3\theta})(r \underline{/\theta}) & Z^5 &= (r^4 \underline{/4\theta})(r \underline{/\theta}) \\ &= r^4 \underline{/3\theta + \theta} & &= r^5 \underline{/4\theta + \theta} \\ &= r^4 \underline{/4\theta} & &= r^5 \underline{/5\theta} \end{aligned}$$

From these few examples one may conclude that $Z^n = r^n \underline{/n\theta}$. This result is known as De Moivre's Theorem.*

THEOREM 9.1

DE MOIVRE'S THEOREM

$$Z^n = r^n (\cos n\theta + j \sin n\theta) \qquad \text{or} \qquad Z^n = r^n \underline{/n\theta}.$$

* The theorem is named for Abraham De Moivre (1667-1754), who was born in France and spent most of his life in London. He was a friend of Newton and a fellow of the Royal Society.

De Moivre's theorem can be extended to find the roots of a number by accepting the fact that it will work for fractional exponents. That is

$$\sqrt[n]{Z} = Z^{1/n} = r^{1/n} \angle \frac{\theta}{n}.$$

To determine whether or not this is correct, we raise both sides to the nth power.

$$(\sqrt[n]{Z})^n = (r^{1/n})^n \angle n \cdot \frac{\theta}{n}.$$

The result is $Z = r \angle \theta$, which is what we would expect.

DEFINITION 9.11

$$\sqrt[n]{Z} = r^{1/n}\left(\cos\frac{\theta}{n} + j\sin\frac{\theta}{n}\right) \quad \text{or} \quad \sqrt[n]{Z} = r^{1/n} \angle \frac{\theta}{n}.$$

The only problem is that $r^{1/n} \angle \frac{\theta}{n}$ is not the only nth root of Z. Since there are many angles that are coterminal with θ, ($\theta + k \cdot 360°$, where k is an integer) other roots can be found.

EXAMPLE 5 Find all the cube roots of $Z = 27 \angle 30°$.

Solution: To find the first cube root of $Z = 27 \angle 30°$ we use Definition 9.11.

$$\begin{aligned}\sqrt[3]{Z} &= \sqrt[3]{27} \angle \frac{30°}{3}\\ &= 3 \angle 10°.\end{aligned}$$

Since $30° + 1(360°) = 390°$ is coterminal with $30°$, $Z = 27 \angle 390°$. Thus, another possible cube root of Z is

$$\begin{aligned}\sqrt[3]{Z} &= \sqrt[3]{27} \angle \frac{390°}{3}\\ &= 3 \angle 130°.\end{aligned}$$

This value is different from the first value obtained. Another coterminal angle is $30° + 2(360°) = 750°$ and $Z = 27 \angle 750°$. Thus

$$\begin{aligned}\sqrt[3]{Z} &= \sqrt[3]{27} \angle \frac{750°}{3}\\ &= 3 \angle 250°.\end{aligned}$$

The next coterminal angle would be $30° + 3(360°) = 1110°$, which may produce a fourth cube root of Z.

$$\sqrt[3]{Z} = \sqrt[3]{27} \angle \frac{1110°}{3}$$
$$= 3 \angle 370°.$$

However, $370°$ and $10°$ are coterminal angles, therefore $3 \angle 370°$ is not a unique root. The three unique cube roots of $27 \angle 30°$ are $3 \angle 10°$, $3 \angle 130°$ and $3 \angle 250°$. □

In general there are n distinct nth roots of a complex number. If $Z = r \angle \theta$, then

$$\sqrt[n]{Z} = \sqrt[n]{r} \angle \frac{\theta + k \cdot 360°}{n},$$

where k is an integer ranging from 0 through $n - 1$.

EXAMPLE 6 Find the three cube roots of $Z = 8 \angle -30°$.

Solution: For $Z = 8 \angle -30°$ the first cube root is:

$$\sqrt[3]{Z} = \sqrt[3]{8} \angle \frac{-30°}{3}$$
$$= 2 \angle -10°.$$

The second cube root is

$$\sqrt[3]{Z} = \sqrt[3]{8} \angle \frac{-30° + 360°}{3}$$
$$= 2 \angle 110°.$$

The third cube root is

$$\sqrt[3]{Z} = \sqrt[3]{8} \angle \frac{-30° + 2(360°)}{3}$$
$$= 2 \angle 230°.$$

Note: The modulus is the same for all three roots, however, the arguments differ by $360°/3 = 120°$. We could find the first root and then add $120°$ to the argument to obtain the second root and then add $120°$ to the argument of the second root to obtain the third root. □

The roots of a complex number in rectangular form can be found by changing the number to polar form and then applying the technique used in Example 6.

9.4 Exercises

In Exercises 1–12, find $Z_1 \cdot Z_2$.

1. $Z_1 = 4 \angle 35°$, $Z_2 = 5 \angle 42°$
2. $Z_1 = 8 \angle 127°$, $Z_2 = 5 \angle 75°$
3. $Z_1 = 2 (\cos 145° + j \sin 145°)$, $Z_2 = \cos 115° + j \sin 115°$
4. $Z_1 = 6 (\cos 150° + j \sin 150°)$, $Z_2 = \cos 210° + j \sin 210°$
5. $Z_1 = 4.7 \angle 216°$, $Z_2 = 5.6 \angle 285°$
6. $Z_1 = 6.4 \angle 290.°$, $Z_2 = 7.3 \angle 320.°$
7. $Z_1 = 1 + 2j$, $Z_2 = 3 - j$
8. $Z_1 = 5 + 7j$, $Z_2 = -2 - 3j$
9. $Z_1 = 8.5 \angle 85.6°$, $Z_2 = 9.4 \angle 105.4°$
10. $Z_1 = 24.5 \angle 111.4°$, $Z_2 = 27.6 \angle 151.4°$
11. $Z_1 = 7.4 (\cos (-48.3°) + j \sin (-48.3°))$, $Z_2 = 9.5 (\cos (-107.3°) + j \sin (-107.3°))$
12. $Z_1 = 2.017 (\cos (-114.3°) + j \sin (-114.3°))$, $Z_2 = 3.105 (\cos 107.3° + j \sin 107.3°)$

In Exercises 13–24, find $Z_1 \div Z_2$.

13. $Z_1 = 8 \angle 47°$, $Z_2 = 2 \angle 38°$
14. $Z_1 = 15 \angle 85°$, $Z_2 = 3 \angle 72°$
15. $Z_1 = 3.8 (\cos 138° + j \sin 138°)$, $Z_2 = 1.9 (\cos 87° + j \sin 87°)$
16. $Z_1 = 9.6 (\cos 182° + j \sin 182°)$, $Z_2 = 3.2 (\cos 108° + j \sin 108°)$
17. $Z_1 = 14.71 \angle 116.9°$, $Z_2 = 17.85 \angle 127.8°$
18. $Z_1 = 9.08 \angle 192.5°$, $Z_2 = 8.04 \angle 211.3°$
19. $Z_1 = 1 + 2j$, $Z_2 = -2 - 3j$
20. $Z_1 = 5 + 7j$, $Z_2 = 3 - j$
21. $Z_1 = 0.014 \angle 108.7°$, $Z_2 = 0.041 \angle -164.2°$
22. $Z_1 = 0.094 \angle 137.2°$, $Z_2 = 0.347 \angle 113.2°$
23. $Z_1 = 0.214 (\cos 782° + j \sin 782°)$, $Z_2 = 0.317 (\cos 325° + j \sin 325°)$
24. $Z_1 = 98.72 (\cos (-142.7°) + j \sin (-142.7°))$, $Z_2 = 46.89 (\cos (-138.6°) + j \sin (-138.6°))$

In Exercises 25–36, find Z_1^2 for Exercises 1–12.

In Exercises 37–48, find Z_2^3 for Exercises 1–12.

For Exercises 49–56, find the indicated root of the complex number.

49. cube roots of $64 \angle 48°$
50. cube roots of $125 \angle 63°$
51. fourth roots of $16 \angle 96°$
52. fourth roots of $81 \angle 48°$
53. square roots of i
54. square roots of $-i$
55. cube roots of 8
56. cube roots of -27.
57. In an AC circuit the current $I = 8 \text{ A} \angle 32°$ and the impedance $Z = 4 \ \Omega \angle 72°$. Find the voltage E, where $E = IZ$.
58. In an AC circuit the current $I = 3.0 \text{ A} \angle 17.2°$ and the voltage $E = 13 \text{ V} \angle 18.4°$. Find the impedance Z, where $Z = E/I$.
59. Impedances $Z_1 = 4 + 5j$ and $Z_2 = 1 + 3j$ are connected in parallel. Find the equivalent impedance $Z = \dfrac{Z_1 Z_2}{Z_1 + Z_2}$.
60. Impedances $Z_1 = 7 + 5j$ and $Z_2 = 11 - 3j$ are connected in parallel. Find the equivalent impedance $Z = \dfrac{Z_1 Z_2}{Z_1 + Z_2}$.

9.5 An application to alternating current (AC) circuits

Currents and voltages in most AC circuits are sinusoidal quantities. To analyze the behavior of these circuits it is necessary to add, subtract, multiply, and divide quantities of the form $y = A \cos (Bt + C)$. It has been found that as long as the frequencies of the sinusoids to be manipulated (the *B*-terms) are the same, the sinusoids can be represented as complex numbers called **phasors.** The necessary mathematical manipulations are made simpler by permitting them to be carried out using complex arithmetic.

EXAMPLE 1 Two 60-Hz voltage sources are connected in parallel. The resulting output is the sum of the two voltages: $v_1 = 117.\ \sin 337.t$ volts and $v_2 = 117.\ \sin (337.t + 60°)$. What is the output?

Solution: The angular frequency, ω, which is equal to $2\pi f = 2\pi$ (60 Hz) = 377. rad/s is the same for the two voltages so they may be added by treating them as phasors. The phasor is a complex number with its magnitude the same as the amplitude of the sinusoid and its argument the same as the phase angle. Therefore the phasor representation of $v_1 = 117.\ \sin 377.t$ is $V_1 = 117.\ \underline{/0°}$ and the phasor representation of $v_2 = 117.\ \sin (377.t + 60°)$ is $V_2 = 117.\ \underline{/60°}$. To add two complex numbers in polar form we convert them to rectangular form:

$$
\begin{aligned}
V_1 &= 117.\ \underline{/0°} = 117. \quad + \quad 0.j \\
V_2 &= 117.\ \underline{/60°} = \underline{\ 58.50 + 101.3j\ } \\
& \qquad V_T = 175.5 \quad + 101.3j.
\end{aligned}
$$

Converting V_T to polar form gives $V_T = 203.\ \underline{/30.0°}$. The sum of $v_1 = 117.\ \sin\ 377.t$ and $v_2 = 117\ \sin\ (377.t + 60°)$ is $v_T = 203.\ \sin (377.t + 30.0°)$. □

In addition to treating currents and voltages as phasors, it has been found that it is possible to represent the common-circuit components by complex numbers and then treat them the same as pure resistances are treated in Ohm's law. In this case Ohm's law is stated as $i = v/Z$, where Z is called the **impedance** of the circuit and has units of ohms. The common circuit elements and their impedances are shown in Table 9.2.

The impedances of combined circuit elements are treated the same as combined resistors in DC circuits. If the elements are hooked in series (one after the other) the combined impedance is the sum of the individual impedances: $Z_s = Z_1 + Z_2 + Z_3 + \ldots$. If the elements are hooked in parallel (one next to the other) the combined impedance is given by $\frac{1}{Z_p} = \frac{1}{Z_1} + \frac{1}{Z_2} + \frac{1}{Z_3} + \ldots$. In the special case of just two impedances hooked in parallel the preceding equation can be rearranged to give

$$Z_p = \frac{Z_1 Z_2}{Z_1 + Z_2}.$$

EXAMPLE 2 Given a resistor $R = 100\ \Omega$, an inductor $L = 7$ millihenry (7×10^{-3} H) and a capacitor $C = 47$ microfarad (47×10^{-6} F)

TABLE 9.2

Circuit element	Symbol and units	Resistance or reactance	Impedance in rectangular form	Impedance in polar form
Resistor	R (ohm)	Resistance $R = R$	$Z_R = R = R + 0j$	$Z_R = R \underline{/0^\circ}$
Inductor	L (henry)	Inductive Reactance $X_L = \omega L$	$Z_L = jX_L = 0 + jX_L$	$Z_L = X_L \underline{/90^\circ}$
Capacitor	C (farad)	Capacitive Reactance $X_C = \dfrac{1}{\omega C}$	$Z_C = -jX_C = 0 - jX_C$	$Z_C = X_C \underline{/-90^\circ}$

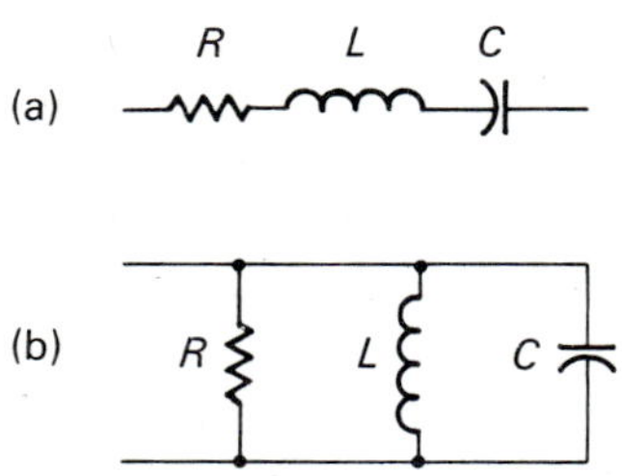

FIGURE 9.6

in a 60-Hz circuit. What is the combined impedance if they are a) connected in series (Fig. 9.6a), and b) if they are connected in parallel (Fig. 9.6b)?

Solution: The impedances of the individual elements are

$$Z_R = R = 100. + 0j\ \Omega$$
$$Z_L = j\omega L = j[2\pi(60)][7 \times 10^{-3}] = 0 + 2.639j\ \Omega$$
$$Z_C = -j\frac{1}{\omega C} = -j\frac{1}{[2\pi(60)][47 \times 10^{-6}]} = 0 - 56.44j\ \Omega.$$

a. The combined impedance for the series circuit is

$$\begin{aligned} Z_S &= Z_R + Z_L + Z_C \\ &= (100. + 0j) + (0 + 2.639j) + (0 - 56.44j) \\ &= 100. - 53.8j\ \Omega \\ &= 114. \underline{/-28.3^\circ}\ \Omega. \end{aligned}$$

b. The combined impedance for the parallel circuit is

$$\begin{aligned} \frac{1}{Z_p} &= \frac{1}{Z_R} + \frac{1}{Z_L} + \frac{1}{Z_C} \\ &= \frac{1}{100.} + \frac{1}{2.639j} + \frac{1}{-56.44j} \end{aligned}$$

Substituting

$$= \frac{1 \underline{/0^\circ}}{100. \underline{/0^\circ}} + \frac{1 \underline{/0^\circ}}{2.639 \underline{/90^\circ}} + \frac{1 \underline{/0^\circ}}{56.44 \underline{/-90^\circ}}$$

Changing numerator and denominator to polar form

$$= 0.010 \angle 0° + \frac{1}{2.639} \angle -90° + \frac{1}{56.44} \angle 90°$$ **Dividing**

$$= 0.010 - 0.3789j + 0.01772j$$ **Changing back to rectangular form**

$$= 0.010 - 0.3612j$$ **Combining like terms**

$$Z_p = \frac{1}{0.0101 - 0.3612j}$$ **Taking the reciprocal of both sides**

$$= \frac{1 \angle 0°}{0.3613 \angle -88.4°}$$ **Changing to polar form**

$$= 2.77 \angle 88.4° \ \Omega$$ **Dividing**

$$= 0.0766 + 2.77j \ \Omega.$$ □

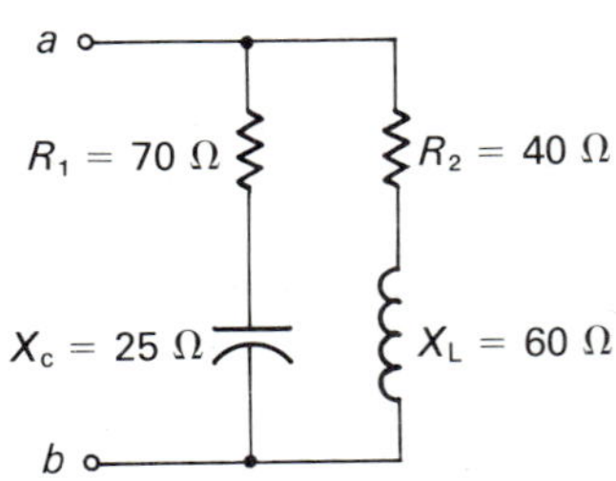

FIGURE 9.7

EXAMPLE 3 Find the combined impedance of the circuit shown in Fig. 9.7.

Solution: R_1 and X_c are hooked in series so they can be replaced by $Z_1 = R_1 - jX_c = 70 - 25j$. R_2 and X_L are also in series so they can be replaced by $Z_2 = R_2 + jX_L = 40 + 60j$. The circuit can then be represented as shown in Fig. 9.8. A single impedance Z that can represent parallel impedances Z_1 and Z_2 is given by the formula

$$Z = \frac{Z_1 Z_2}{Z_1 + Z_2}$$

$$= \frac{(70 - 25j)(40 + 60j)}{(70 - 25j) + (40 + 60j)}$$ **Substituting**

$$= \frac{(74.3 \angle -19.7°)(72.1 \angle 56.3°)}{110. + 35.0j}$$ **Changing numerator to polar form and finding the sum in the denominator**

$$= \frac{5360. \angle 36.6°}{115. \angle 17.7°}$$ **Finding product in the numerator and changing denominator to polar form**

$$= 46.6 \angle 18.9°.$$ **Dividing**

The combined impedance $Z = 47.\ \Omega \angle 19.°$. □

FIGURE 9.8

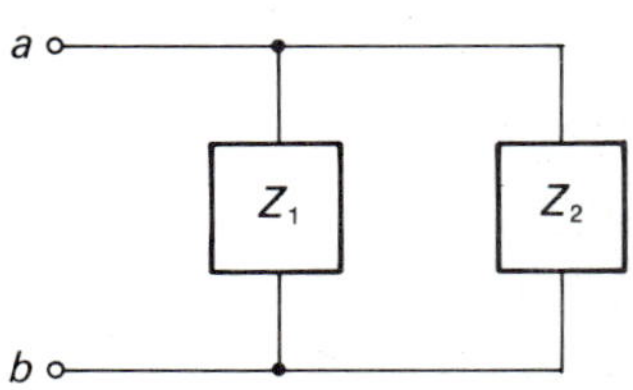

EXAMPLE 4 If an AC voltage source $v = 110$ volt is connected to the circuit of Example 3 at point ab, what is the current supplied to the circuit?

Solution: From Ohm's law we have $i = \frac{v}{Z}$. The voltage $v = 110$ volt can be written in phasor form as $V = 110\ \angle 0^\circ$, and $Z = 46.6\ \angle 18.9^\circ$ from Example 3.

$$I = \frac{110\ \angle 0^\circ}{46.6\ \angle 18.9^\circ}$$
$$= 2.3\ \angle -19.^\circ$$

The current supplied will be 2.3 amp at the same frequency as the voltage and lagging the voltage by 19°. □

9.5 Exercises

In Exercises 1–8, determine (a) the equivalent or combined impedance of the circuit, (b) the current supplied, and (c) the phase angle between the voltage and the current.

1.

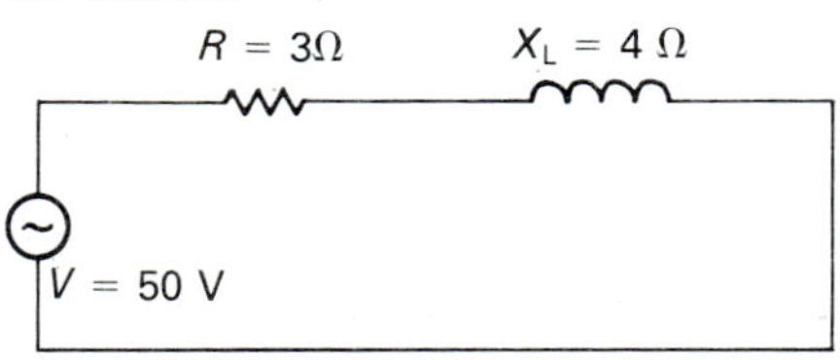

2.

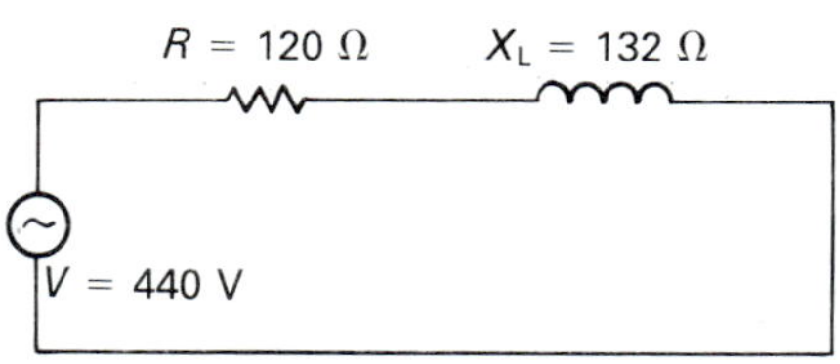

3.

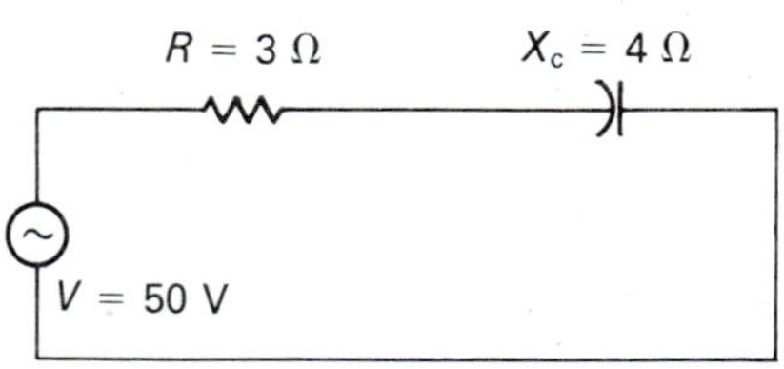

4.

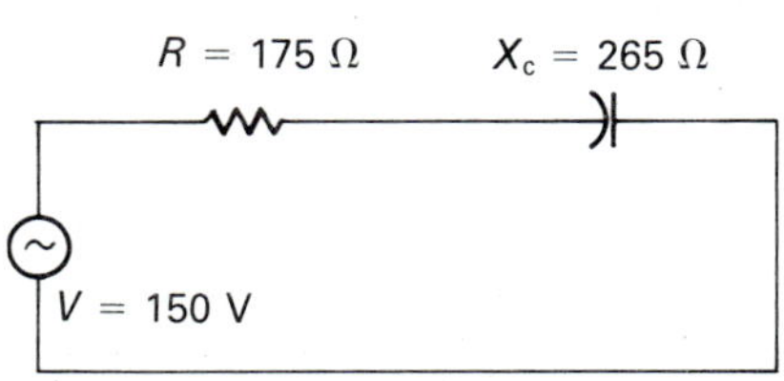

5.

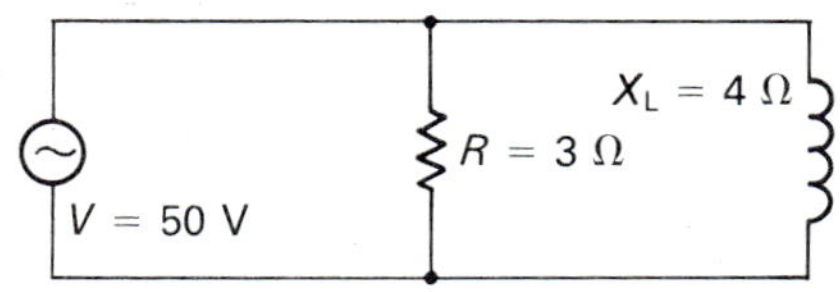

6.

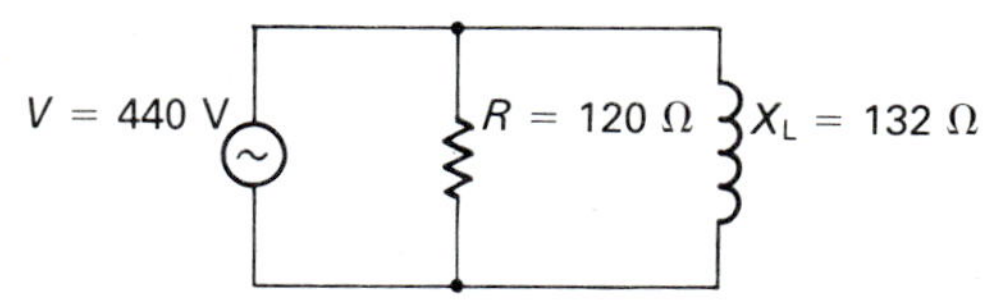

7.

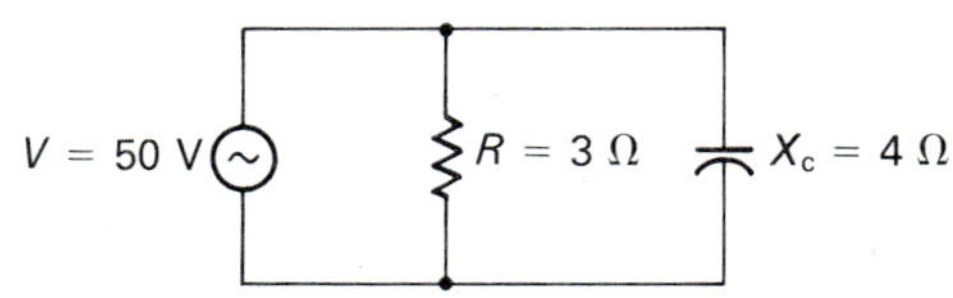

8.

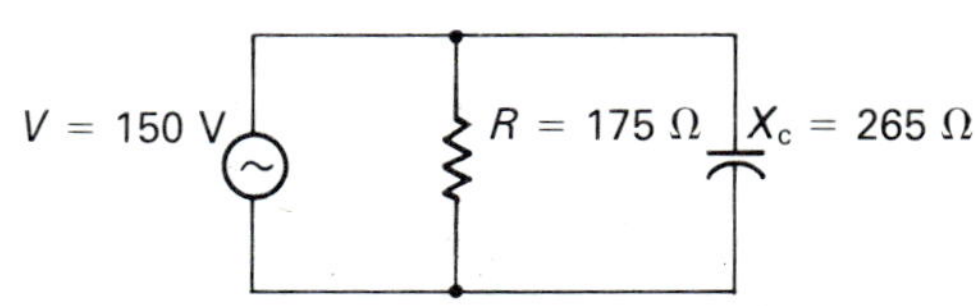

In Exercises 9–16, determine the impedance Z_P for the impedances Z_1 and Z_2 connected in parallel.

9. $Z_1 = 161\ \Omega\ \angle 9.3^\circ$, $Z_2 = 60.\ \Omega\ \angle 58.3^\circ$

10. $Z_1 = 158.6\ \Omega\ \angle 47.2^\circ$, $Z_2 = 155.0\ \Omega\ \angle -17.7^\circ$

11. $Z_1 = 83.9 - 44.4j\ \Omega$, $Z_2 = 40 + 50j\ \Omega$

12. $Z_1 = 286 - 190j\ \Omega$, $Z_2 = 107 - 28.6j\ \Omega$

13. $Z_1 = 50.5\ \Omega\ \angle 15^\circ$, $Z_2 = 110 + 0j\ \Omega$

14. $Z_1 = 345\ \Omega\ \angle 18^\circ$, $Z_2 = 0 - 90j\ \Omega$

15. $Z_1 = 241\ \Omega\ \underline{/-8^\circ}$, $Z_2 = 0 + 80j\ \Omega$

16. $Z_1 = 40.3\ \Omega\ \underline{/-10^\circ}$, $Z_2 = 28\ \Omega\ \underline{/0^\circ}$

In Exercises 17–20, find the equivalent impedance of the circuit.

17.

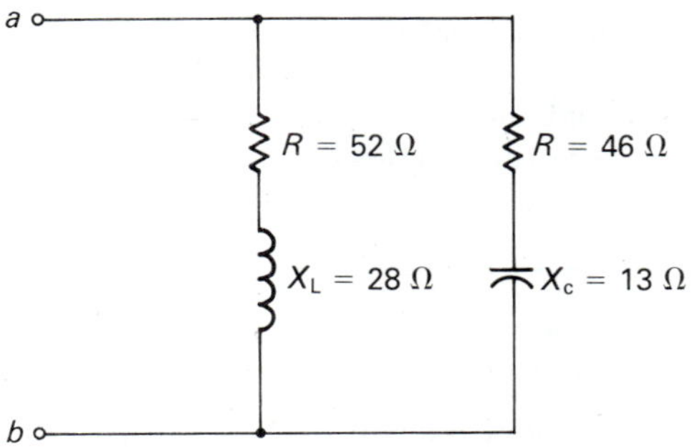

18.

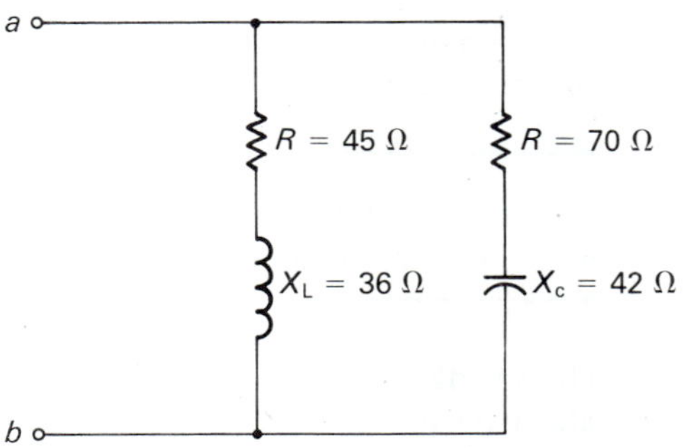

19.

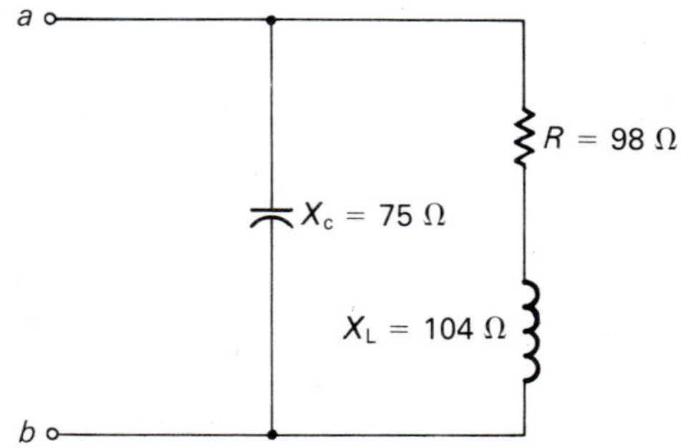

20.

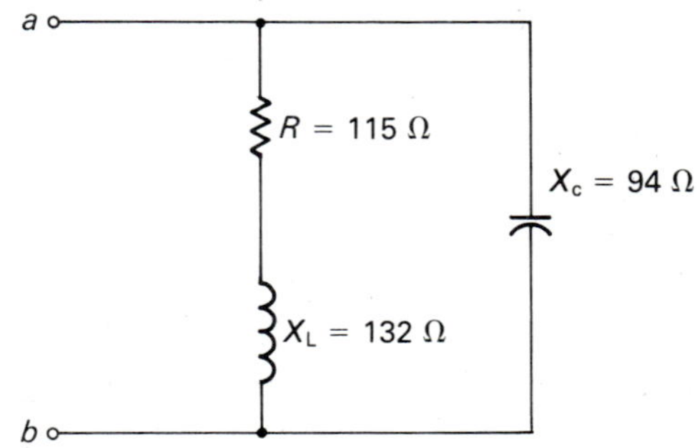

21. The joint impedance of two parallel impedances is $58.3\ \Omega\ \underline{/-38.2^\circ}$. One of the impedances is $138.\ \Omega\ \underline{/32^\circ}$. What is the other impedance?

22. What impedance must be connected in parallel with $74.3 + j54.5\ \Omega$ to produce $33.7 + j145.4\ \Omega$?

Chapter summary

The $\sqrt{-3}$ is an **imaginary number.**

Definition 9.1 $j = \sqrt{-1}$.

Definition 9.2 If n is a positive real number, then

$$\sqrt{-n} = j\sqrt{n},$$

and $j\sqrt{n}$ is called an **imaginary** number.

Definition 9.3 $j^2 = -1$.

Before performing any operations on imaginary numbers they must be written in j-notation.

For any positive integer n, j^n can be expressed as $(j^2)^{n/2}$ if n is an even positive integer or as $(j^2)^{\frac{n-1}{2}}(j)$ if n is an odd positive integer.

Definition 9.4 A **complex number** is a number of the form $a + jb$, where a and b are real numbers and $j = \sqrt{-1}$.

Let $Z = a + jb$, then a is called the **real part** of Z and jb is called the **imaginary part** of Z.

Complex numbers that differ only in the signs of the imaginary parts are called **conjugate complex numbers** (or complex conjugates). If $Z_1 = a + jb$ and $Z_2 = a - jb$, then Z_1 is the conjugate of Z_2 and Z_2 is the conjugate of Z_1.

Definition 9.5 Two complex numbers are equal if and only if their real parts are equal and their imaginary parts are equal. That is:

$$a + jb = c + jd \text{ if and only if } a = c \text{ and } b = d.$$

In order to add (or subtract) two complex numbers such as $a + jb$ and $c + jd$, add (or subtract) the real parts and the imaginary parts.

Definition 9.6 $(a + jb) + (c + jd) = (a + c) + j(b + d)$.
$(a + jb) - (c + jd) = (a - c) + j(b - d)$.

To multiply two complex numbers find the product as you would for any product of two binomials. Then evaluate the powers of j according to the rules in Section 9.1 and combine the real and imaginary parts.

Definition 9.7 $(a + jb)(c + jd) = (ac - bd) + j(ad + bc)$.

To express the quotient of $a + jb$ and $c + jd$, where $c + jd \neq 0$, as a single complex number multiply both numerator and denominator by the conjugate of the denominator.

Definition 9.8 $\dfrac{a + jb}{c + jd} = \dfrac{ac + bd}{c^2 + d^2} + j\dfrac{bc - ad}{c^2 + d^2}$ if $c + jd \neq 0$.

A complex number can be graphed on a complex plane. The horizontal axis is the real axis and the vertical axis is the imaginary axis. For any point on the plane the distance from the origin to the point is called the **magnitude, absolute value,** or **modulus** of the complex number $a + jb$. The angle of the line joining the origin and the point it makes with the positive real axis is called the **argument** or the **amplitude** of the complex number. The absolute value of any complex number $(a + jb)$ is denoted by $|a + jb|$. Therefore $|a + jb| = \sqrt{a^2 + b^2}$. The argument θ is given by $\tan \theta = \dfrac{b}{a}$.

Any point P on the complex plane can be identified with its rectangular coordinates (a, b) or its polar coordinates (r, θ). Thus from right-triangle trigonometry we have $\cos\theta = a/r$ and $\sin\theta = b/r$; thus $a = r\cos\theta$ and $b = r\sin\theta$, and $a + jb = r(\cos\theta + j\sin\theta)$.

The right side of $a + jb = r(\cos\theta + j\sin\theta)$ is the **trigonometric form** or the **polar form** of the complex number and the left side is the **rectangular form.** The polar form may be abbreviated as $r \text{ cis } \theta$ or $r \underline{/\theta}$. One can change from the rectangular to polar form using right-triangle trigonometry and the Pythagorean theorem or using a calculator.

If it is necessary to multiply, divide, raise to a power, or take a root of complex numbers, the process may be simpler if the complex number is changed to polar form.

Definition 9.9 If $Z_1 = r_1(\cos\theta_1 + j\sin\theta_1)$ and $Z_2 = r_2(\cos\theta_2 + j\sin\theta_2)$, then $Z_1Z_2 = r_1r_2[\cos(\theta_1 + \theta_2) + j\sin(\theta_1 + \theta_2)]$.
Or, if $Z_1 = r_1 \underline{/\theta_1}$ and $Z_2 = r_2 \underline{/\theta_2}$, then $Z_1Z_2 = r_1r_2 \underline{/\theta_1 + \theta_2}$.

Definition 9.10 If $Z_1 = r_1(\cos\theta_1 + j\sin\theta_1)$ and $Z_2 = r_2(\cos\theta_2 + j\sin\theta_2)$, then $\dfrac{Z_1}{Z_2} = \dfrac{r_1}{r_2}[\cos(\theta_1 - \theta_2) + j\sin(\theta_1 - \theta_2)]$.

Or, if $Z_1 = r_1 \underline{/\theta_1}$ and $Z_2 = r_2 \underline{/\theta_2}$, then $\dfrac{Z_1}{Z_2} = \dfrac{r_1}{r_2} \underline{/\theta_1 - \theta_2}$, $Z_2 \neq 0$.

Theorem 9.1 De Moivre's Theorem

$$Z^n = r^n(\cos n\theta + j\sin n\theta) \quad \text{or} \quad Z^n = r^n \underline{/n\theta}.$$

Definition 9.11 $\sqrt[n]{Z} = r^{1/n}\left(\cos\dfrac{\theta}{n} + j\sin\dfrac{\theta}{n}\right)$ or $\sqrt[n]{Z} = r^{1/n} \underline{/\dfrac{\theta}{n}}$.

Remember that for $\sqrt[n]{Z}$ there are n distinct roots.

Review Exercises

In Exercises 1–12, carry out the indicated operations and simplify. Leave the answers in radical form.

1. j^{15}
2. j^{-14}
3. $(\sqrt{-324})(\sqrt{-361})$
4. $(\sqrt{-1058})(\sqrt{-192})$
5. $\sqrt{-484}\sqrt{507}$
6. $\sqrt{(-6)(-24)}$
7. $\sqrt{(-8)(-98)}$
8. $(\sqrt{-48})(\sqrt{-75})(\sqrt{-90})$
9. $\dfrac{\sqrt{343}}{\sqrt{-49}}$
10. $\dfrac{\sqrt{-135}}{\sqrt{-14}}$
11. $\sqrt{-288} + \sqrt{-363} + \sqrt{-162}$
12. $\sqrt{-578} - \sqrt{-676} + \sqrt{-729}$

In Exercises 13–18, perform the indicated operations and write the answer in the form $a + jb$.

13. $(3 - 5j) + (8 + 7j)$
14. $(-19 + 31j)^2$
15. $(8 + 5\sqrt{-6})(17 - 7\sqrt{-216})$
16. $(3j^7 - 8j^{13} + 7j^{17} - 5j^{15})^2$
17. $(5 - j^{14}) \div (18 - j^{25})$
18. $1 \div (13 + 4\sqrt{-72})$

In Exercises 19–24, graph each complex number and change to polar form. (Answer to the nearest tenth.)

19. $-8 + 11j$
20. $4 - 8j$
21. $5 + 14j$
22. $5\sqrt{3} - 4j\sqrt{3}$
23. $11j$
24. -3

In Exercises 25–30, change each complex number in polar form to rectangular form. Find exact answers if the reference angle is 30°, 45°, 60°, or quadrantal; otherwise give answers to three significant digits.

25. $3.4 \angle 240^\circ$ **26.** $3 \angle 138^\circ$

27. $1.40 \angle 120.^\circ$ **28.** $5.0 \angle 258.4^\circ$

29. $5.0 \angle 3.4^\circ$ **30.** $9.3 \angle 113.8^\circ$

In Exercises 31–36, perform the indicated operation on the complex numbers.

31. $(5 \angle 28^\circ) \cdot (15 \angle 87^\circ)$

32. $(133 \angle 173^\circ) \div (15 \angle 29^\circ)$

33. $(7.5 \angle 83.4^\circ)^2$

34. $(5.6 \angle 148.3^\circ)^3$

35. $(14.7 \angle -32.1^\circ)(19.8 \angle 51.7^\circ)$

36. $(0.040 \angle 122.3^\circ)(0.016 \angle -37.4^\circ)$

37. Find the cube roots of $216 \angle 114^\circ$

38. Find the cube roots of -512.

39. From Fig. 9.9 determine (a) the phase angle between voltage and current, (b) the impedance of the circuit, and (c) the current through the circuit.

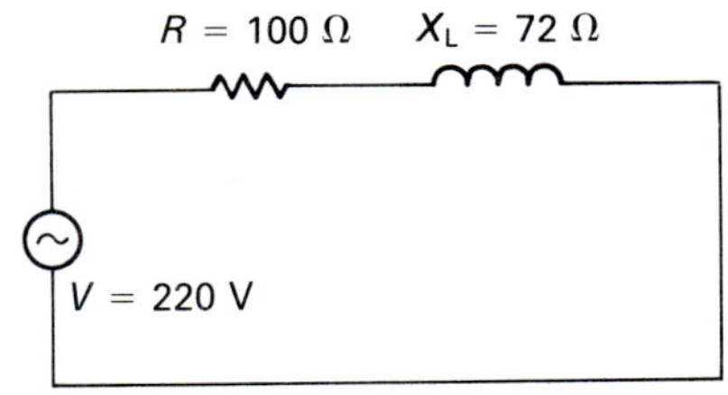

FIGURE 9.9

40. The equivalent impedance of a circuit may be found using the formula

$$Z_T = Z_S + \frac{Z_1 Z_2}{Z_1 + Z_2}$$

Find Z_T when $Z_S = 15.4 + 20.6j\ \Omega$, $Z_1 = 50 + 11.8j\ \Omega$, and $Z_2 = 40.0 - 80.0j\ \Omega$. Sketch the circuit.

Can you solve these?

1. The characteristic impedance of a *T*-section filter is

$$Z_{OT} = \sqrt{Z_1 Z_2 + \frac{Z_1^{\,2}}{4}},$$

where Z_1 is the full series arm impedance and Z_2 is the shunt impedance of the filter section. If $Z_1 = 30.0 \angle 86.0^\circ$ and $Z_2 = 10.0 \angle -90^\circ\ \Omega$, find Z_{OT}.

2. Complex numbers can be written in exponential form as $re^{j\theta}$, where $r(\cos\theta + j\sin\theta) = re^{j\theta}$. Find A^7 using polar form and exponential form and show that the results are the same. [$A = 6(\cos 120^\circ + j\sin 120^\circ)$.]

3. Under what circumstances does the conjugate of a complex number equal its reciprocal?

4. Although $2/6 = 1/3$, $(-27)^{2/6} \neq (-27)^{1/3}$. This awkward situation arises because $(-27)^{1/3} = -3$ but $(-27)^{2/6} = ((-27)^2)^{1/6} = (729)^{1/6} = 3$. Use De Moivre's theorem to unravel the mystery of this problem. (*Hint:* find the three distinct roots of -27 and the six distinct roots of 729.)

5. Show that for numbers (not necessarily complex)

a. If $\dfrac{1}{Z} = \dfrac{1}{Z_1} + \dfrac{1}{Z_2}$, then $Z = \dfrac{Z_1 Z_2}{Z_1 + Z_2}$

b. If $\dfrac{1}{Z} = \dfrac{1}{Z_1} + \dfrac{1}{Z_2} + \cdots + \dfrac{1}{Z_n}$ then

$$Z = \frac{Z_1 Z_2 \cdots Z_n}{Z_1 Z_2 \cdots Z_{n-1} + Z_2 \cdots Z_{n-1} Z_n + \cdots + \cdots Z_{n-2} Z_{n-1} Z_n}.$$

6. Write a computer program to change a complex number in rectangular form to polar form.

7. Write a computer program to change a number in polar form to rectangular form.

10

Strain gauges are thin foil devices that are cemented to objects such as aircraft structural members, automobile chassis, and skyscraper girders to measure their deflections under load. When they deflect with the member to which they are attached, their change in length causes a measurable change in resistance. The change in resistance of the strain gauges is directly related to the deflection of the member, which is in turn related to the operating stress in the member.

Two-dimensional stress s and strain e are related in metals by the equations

$$e_x = \frac{1}{E}(s_x - v s_y) \qquad \text{and} \qquad e_y = \frac{1}{E}(s_y - v s_x),$$

where E is the modulus of elasticity (30×10^6 psi for steel) and v is Poisson's ratio (0.30 for steel). If two crossed strain gauges on a steel member indicate strains of

$$e_x = 2.27 \times 10^{-3} \text{ in./in. and } e_y = 0.533 \times 10^{-6} \text{ in./in.},$$

what are the corresponding stresses s_x and s_y in the member?

Systems of Linear Equations and Determinants

10.1 Introduction to linear equations

The river was rising at a constant rate of 3 in. per hour when the weather bureau issued a flood watch. At midnight the river level gauge indicated 8 in. above the normal level. Table 10.1 shows the height of the river above the normal level for the next 12 hours after midnight. (Midnight is taken as 0 hours.) A graph of Table 10.1 is shown in Fig. 10.1. When the points of Table 10.1 are plotted in Fig. 10.1, it becomes apparent that the curve through the points is a straight line. Analyzing the graph we see that the height varies directly with time. The height at any time is the height at midnight plus the change in height for that period of time. The height at any time can therefore be represented by the equation $h = 8 + 3t$. By convention $h = 8 + 3t$ is usually written as $h = 3t + 8$. Substituting points from Table 10.1 will verify that the equation describes the variation in river height with time.

A form of the equation of a line is $y = mx + b$, where m and b are constants and x and y are, respectively, the independent and dependent variables. This form of the equation of a line is called the **slope–intercept form.**

DEFINITION 10.1

> The slope–intercept form of an equation of a line is
>
> $$y = mx + b.$$

TABLE 10.1

Time (hr)	Level (in.)
0	8
1	11
2	14
3	17
4	20
5	23
6	26
7	29
8	32
9	35
10	38
11	41
12	44

FIGURE 10.1

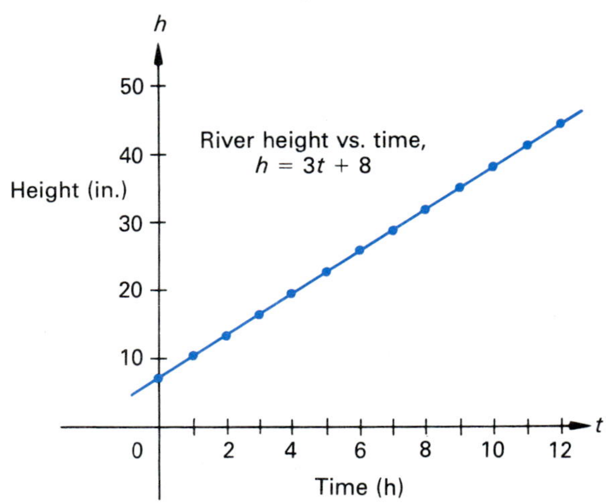

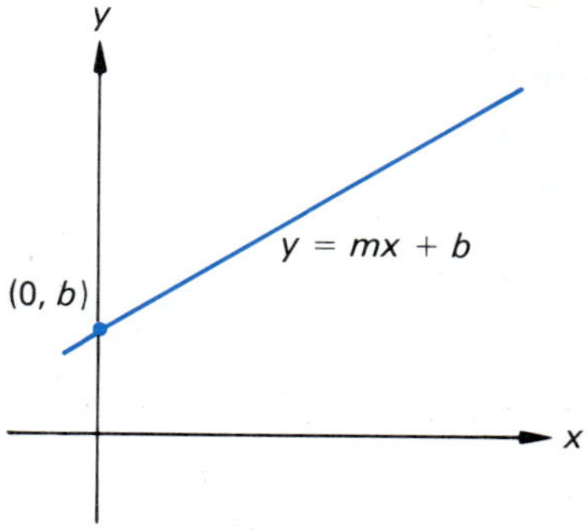

FIGURE 10.2

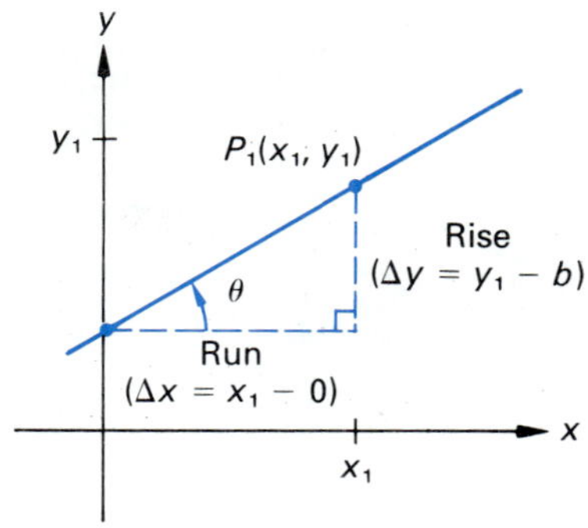

FIGURE 10.3

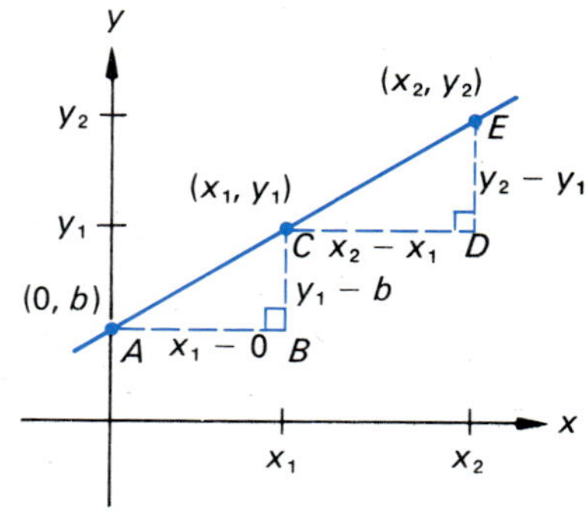

FIGURE 10.4

In examining $y = mx + b$ we see that if $x = 0$ then $y = b$. The line intersects the y-axis at the point $(0, b)$ and b is called the ***y*-intercept** (Fig. 10.2).

If we choose any point $P_1(x_1, y_1)$ on the line, the coordinates of the point must satisfy the equation of the line. Substituting into the equation we have $y_1 = mx_1 + b$. Solving for m we obtain $m = (y_1 - b)/x_1$ (Fig. 10.3). The quantity $y_1 - b$ is the vertical distance, or the change in y (Δy or delta y) as we go from $(0, b)$ to (x_1, y_1) along the line. Δy is sometimes called the **rise.**

The quantity x_1, which is the same as $x_1 - 0$, is the horizontal distance, or the change in x (Δx or delta x), as we go from $(0, b)$ to (x_1, y_1) along the line. Δx is sometimes called the **run.** The ratio m is called the **slope** of the line. Taking θ as the angle the line makes with the horizontal, the right triangle in Fig. 10.3 shows that the ratio $\dfrac{\Delta y}{\Delta x} = \dfrac{\text{rise}}{\text{run}}$ is the same as the ratio $\dfrac{\text{opposite side}}{\text{adjacent side}}$. As long as the scales are the same for both axes, we can also write the slope m as $m = \tan \theta$.

In Fig. 10.4, triangles ABC and CDE are similar triangles. From geometry the ratios of corresponding sides of similar triangles are equal. Therefore

$$\frac{y_1 - b}{x_1} = \frac{y_2 - y_1}{x_2 - x_1}.$$

Since

$$m = \frac{y_1 - b}{x_1},$$

we can also say that

$$m = \frac{y_2 - y_1}{x_2 - x_1}.$$

DEFINITION 10.2

The **slope** of a straight line containing points $P_1(x_1, y_1)$ and $P_2(x_2, y_2)$ is defined by the equation

$$m = \frac{y_2 - y_1}{x_2 - x_1},$$

where $x_2 - x_1 \neq 0$.

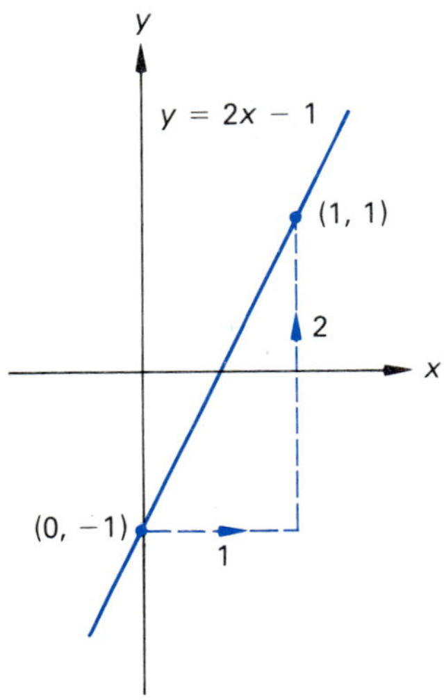

FIGURE 10.5

Thus we can calculate the slope of a line given any two points on the line.

EXAMPLE 1 Graph $y = 2x - 1$ using the slope and the y-intercept.

Solution: $y = 2x - 1$ is the same form as $y = mx + b$, the slope–intercept form of a straight line. The y-intercept is $b = -1$. The line $y = 2x - 1$ crosses the y-axis at $(0, -1)$. The slope of the line is 2 $(m = 2)$. The slope can be equivalently written as $m = \frac{2}{1}$. Recall that the slope is the $\frac{\text{rise}}{\text{run}}$. This tells us that the line rises, or y increases by 2 units, for every 1 unit of run, or 1 unit of increase in x. With this information we can now graph $y = 2x - 1$.

As in Fig. 10.5, first plot the y-intercept $(0, -1)$. Since the slope is positive, starting with the point $(0, -1)$, move one unit to the right and then two units up. This provides us with a second point $(1, 1)$, which is also a point on the line $y = 2x - 1$. Now with two points we can draw the line. [Note that moving two units up and then one unit to the right would also have provided us with the point $(1, 1)$]. □

FIGURE 10.6

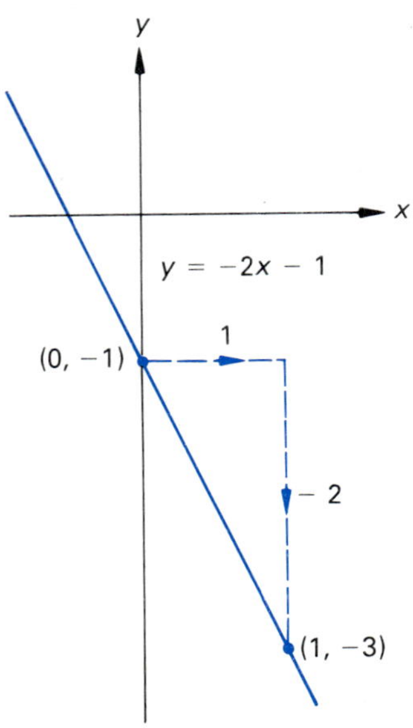

EXAMPLE 2 Graph $y = -2x - 1$ using the slope and the y-intercept.

Solution: $y = -2x - 1$ is a straight line with y-intercept $b = -1$. The graph of the equation passes through the point $(0, -1)$. This equation differs from the equation of Example 1 only in that the slope is a negative number. Equivalent ways of writing this slope are $m = -2$, $m = \frac{-2}{1}$, or $m = \frac{2}{-1}$.

The slope $m = \frac{-2}{1}$ has a negative rise and a positive run, or y decreases by 2 units $(\triangle y = -2)$ while x increases by 1 unit $(\triangle x = 1)$. Using this information we can graph the equation as shown in Fig. 10.6. Writing the slope in the form $m = \frac{2}{-1}$, the rise is a positive 2 and the run is a negative 1, or y increases 2 units while x decreases 1 unit. With

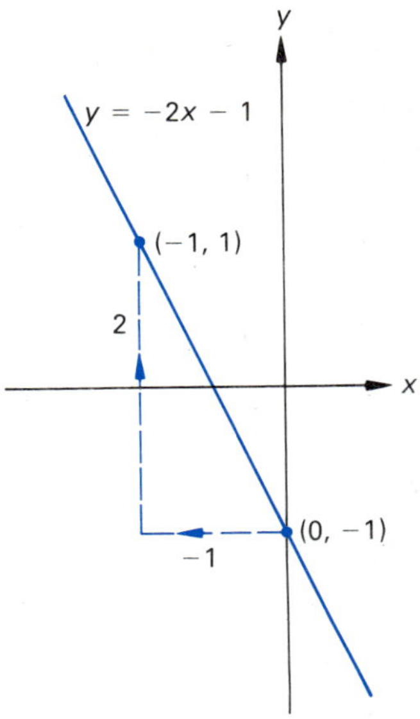

FIGURE 10.7

this information we can draw the graph of the equation as in Fig. 10.7. Note that the graphs in Fig. 10.6 and Fig. 10.7 are identical. □

If we compare the results of Example 1 (Fig. 10.5) and Example 2 (Fig. 10.6), we see that the line with **positive slope** goes up (rises) as it moves from left to right. A line with a **negative slope** goes down (drops) as it moves from left to right.

The graphs of the equations $y = 2x - 1$ and $y = -2x - 1$ are straight lines. All equations that can be represented by a straight line are called **linear equations.** We also know that $y = 2x - 1$ and $y = -2x - 1$ are linear equations since the exponents of the variables x and y in both equations are 1.

EXAMPLE 3 Graph $x + 3y - 2 = 0$ using the slope and the y-intercept.

Solution: The equation $x + 3y - 2 = 0$ is linear and can be arranged in the form $y = mx + b$.

$$x + 3y - 2 = 0$$

$$3y = -x + 2 \qquad \textbf{Rearranging}$$

$$y = \frac{-1}{3}x + \frac{2}{3} \qquad \textbf{Dividing by 3}$$

The y-intercept is $b = \frac{2}{3}$, and the slope $m = -\frac{1}{3}$. The equation is graphed in Fig. 10.8. □

EXAMPLE 4 Graph $2x + y + 1 = 0$ and $-6x - 3y + 2 = 0$ using slope and y-intercept.

Solution: Both equations are linear and can be put into the form $y = mx + b$.

$$2x + y + 1 = 0$$

$$y = -2x - 1 \qquad \textbf{Rearranging}$$

$$-6x - 3y + 2 = 0$$

$$-3y = 6x - 2 \qquad \textbf{Rearranging}$$

$$y = -2x + \frac{2}{3} \qquad \textbf{Dividing by } -3$$

FIGURE 10.8

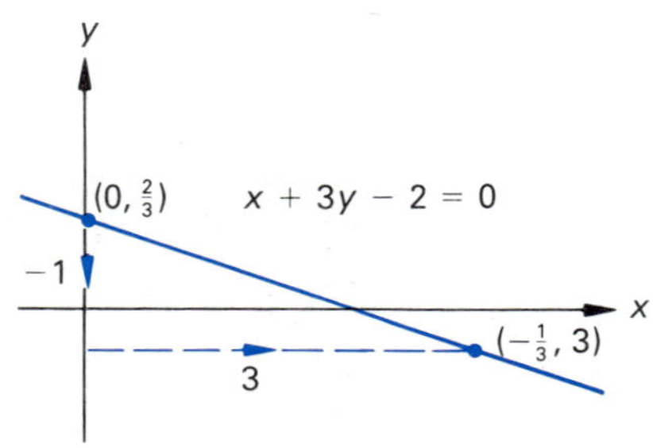

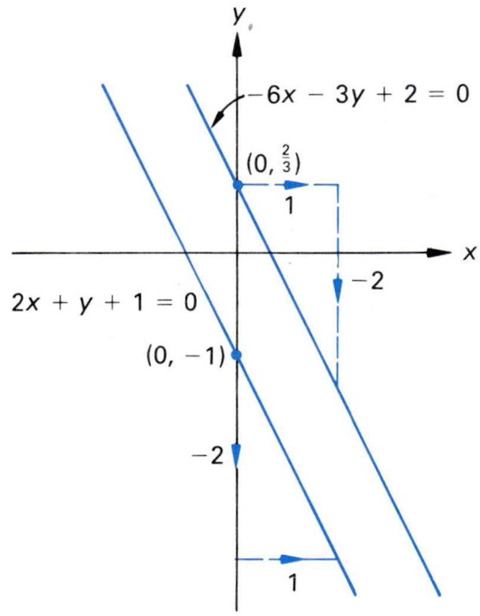

FIGURE 10.9

Both lines have the same slope, $m = -2$, but their y-intercepts are $b_1 = -1$ and $b_2 = 2/3$, respectively. The lines are graphed in Fig. 10.9 and from the figure we can see that they are parallel. □

DEFINITION 10.3

> Two lines are parallel if and only if their slopes are equal. That is, $L_1 \parallel L_2$ if and only if $m_1 = m_2$.

TABLE 10.2

x	y
−3	−2
0	−2
3	−2

EXAMPLE 5 Sketch the graph of $y = -2$.

Solution: The equation $y = -2$ is linear since the exponent of y is 1. Therefore, we can write the equation as $y = 0x - 2$. With the equation in this form we recognize that $m = 0$ and $b = -2$. To sketch the graph we determine sets of ordered pairs as in Table 10.2. Plotting these points as in Fig. 10.10 we see that the line $y = -2$ is parallel to the x-axis. *The graph of any equation of the form* y = a *(where* a *is a real number) will be parallel to the* x*-axis.* □

FIGURE 10.10

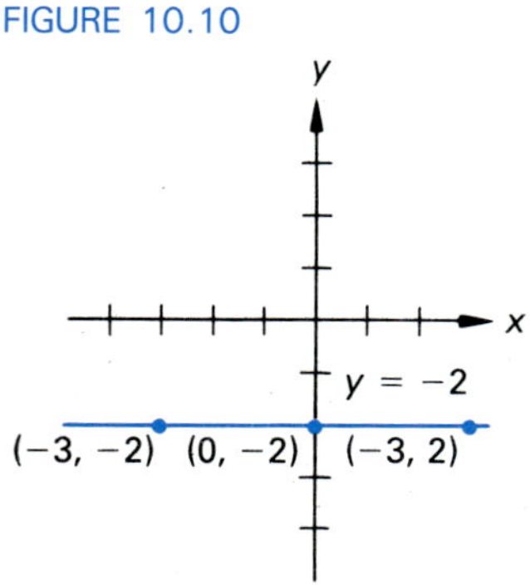

EXAMPLE 6 Sketch the graph of $x = 3$.

Solution: The equation $x = 3$ is linear since the exponent of x is 1. Therefore, we can write the equation as $0y = -x + 3$. With the equation in this form we recognize that $m = \frac{-1}{0}$ and $b = \frac{3}{0}$ are both undefined. To sketch the graph we determine sets of ordered pairs as in Table

TABLE 10.3

x	y
3	−2
3	0
3	2

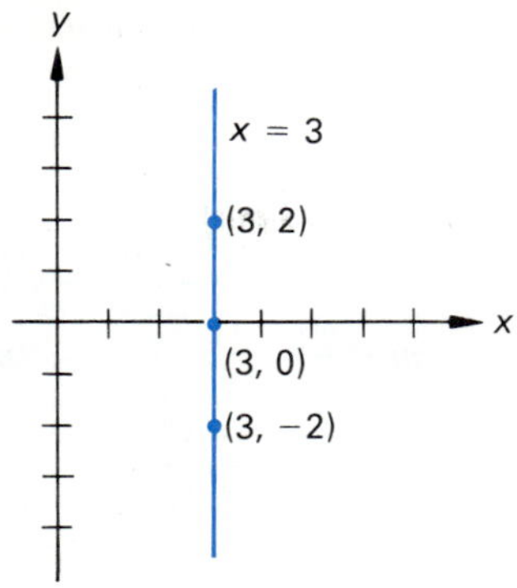

FIGURE 10.11

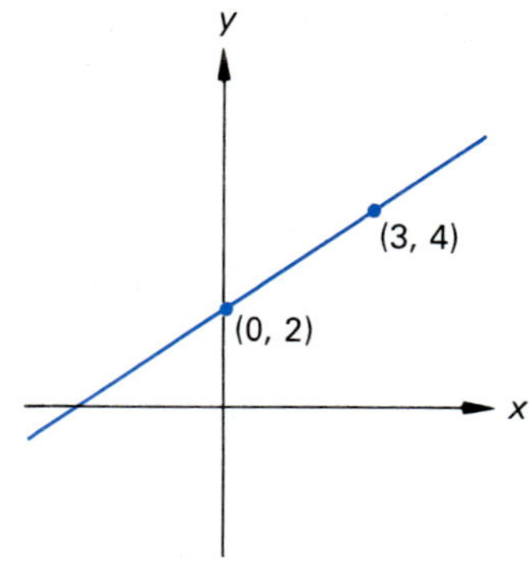

FIGURE 10.12

10.3. Plotting these points as in Fig. 10.11 we can see that the line $x = 3$ is parallel to the y-axis. *The graph of any equation of the form* x = a *(where* a *is a real number) will be parallel to the* y*-axis.* □

EXAMPLE 7 Write the equation for the line in Fig. 10.12 using the slope–intercept form.

Solution: On inspection we see that the line crosses the y-axis at (0, 2) so the y-intercept is $b = 2$. The line passes through point $(x_1, y_1) = (3, 4)$. From the definition, m can be written as

$$m = \frac{y_1 - b}{x_1}$$

$$= \frac{4 - 2}{3}$$

$$= \frac{2}{3}.$$

Substituting the values 2/3 and 2 for m and b, respectively, in the equation $y = mx + b$, the result is the equation of the line,

$$y = \frac{2}{3}x + 2.$$ □

FIGURE 10.13

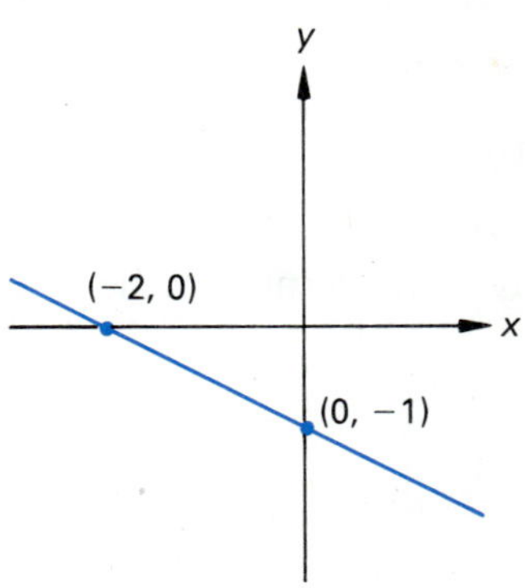

EXAMPLE 8 Write the equation for the line in Fig. 10.13 using the slope–intercept form.

Solution: By inspecting the figure we see that the line crosses the y-axis at $(0, -1)$ so the y-intercept $b = -1$. Since we know two points on the line, $(x_1, y_1) = (-2, 0)$ and $(x_2, y_2) = (0, -1)$, we can either calculate the slope from the equation in Definition 10.2 or by the method used in Example 7.

From Definition 10.2	From Example 7
$m = \dfrac{y_2 - y_1}{x_2 - x_1}$	$m = \dfrac{y_1 - b}{x_1}$
$= \dfrac{-1 - 0}{0 - (-2)}$	$= \dfrac{0 - (-1)}{-2}$
$= \dfrac{-1}{2}$	$= \dfrac{1}{-2}$
$= -\dfrac{1}{2}.$	$= -\dfrac{1}{2}.$

By either method the slope $m = -\frac{1}{2}$. Substituting the values $b = -1$ and $m = -\frac{1}{2}$ we can write the equation of the line, $y = -\frac{1}{2}x - 1$. □

EXAMPLE 9

a. Write the equation for the line in Fig. 10.14.

b. What is the angle the line makes with the horizontal axis?

Solution:

a. The y-intercept of the line in Fig. 10.14 is not apparent. We do have two points $(x_1, y_1) = (3, 2)$ and $(x_2, y_2) = (4, 6)$ so we can calculate the slope from Definition 10.2:

$$m = \frac{y_2 - y_1}{x_2 - x_1} = \frac{6 - 2}{4 - 3} = 4$$

FIGURE 10.14

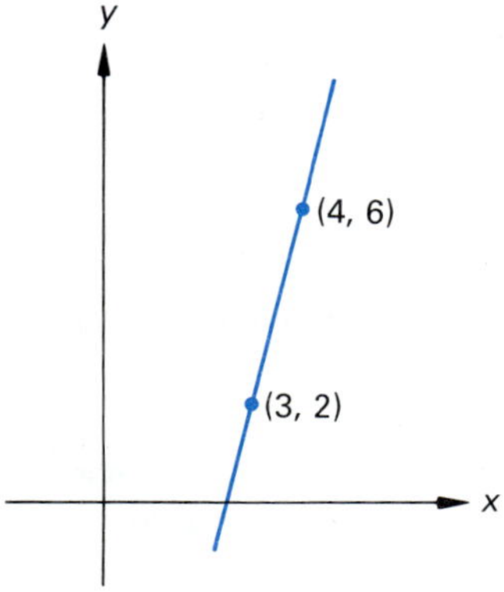

We also know that the slope can be written as $m = \frac{y_1 - b}{x_1}$. Since we already know m we have

$$4 = \frac{2 - b}{3} \qquad 12 = 2 - b \qquad b = -10.$$

Therefore the equation of the line $y = mx + b$ is $y = 4x - 10$. We can check our solution by making sure both points on the line satisfy the equation. *Check:*

For (3, 2):	For (4, 6):
$y = 4x - 10$	$y = 4x - 10$
$2 = 4(3) - 10$	$6 = 4(4) - 10$
$2 = 2$ ✓	$6 = 6$ ✓

b. The angle, θ, that the line makes with the horizontal axis is given by the equation $m = \tan \theta$.

$$\tan \theta = 4$$
$$\theta = 76.0°.$$

The line $y = 4x - 10$ makes an angle of $76.0°$ with the horizontal, or x-, axis. □

10.1 Exercises

In Exercises 1–24, identify the slope and the y-intercept of the lines and sketch the graph of the lines.

1. $y = 3x + 2$ **2.** $y = x - 1$ **3.** $y = \frac{1}{2}x - 1$

4. $y = \frac{3}{4}x + 2$ **5.** $y = 3x - 2$ **6.** $y = x + 1$

7. $y = \frac{1}{2}x + 1$ **8.** $y = \frac{3}{4}x - 2$

9. $2x + 4y + 1 = 0$ **10.** $6x + 3y + 2 = 0$

11. $-2x + 4y + 1 = 0$ **12.** $-6x - 3y + 2 = 0$

13. $y = \frac{1}{3}x$ **14.** $y = 3x$

15. $x + y = 0$ **16.** $2x - y = 0$

17. $y = 4$ **18.** $y = 1$

19. $x = -1$ **20.** $x = \frac{1}{2}$

21. $\frac{1}{2}y = \frac{-3}{4}x + 2$ **22.** $\frac{2}{3}y = \frac{-1}{5}x - \frac{1}{3}$

23. $2y + \frac{5}{2}x + 3 = 0$ **24.** $\frac{3}{2}y + \frac{3}{4}x + \frac{2}{5} = 0$

In Exercises 25–30, determine whether or not the lines are parallel without graphing them.

25. $x + y + 1 = 0$
$2x + 2y - 2 = 0$

26. $y = \frac{1}{2}x - 3$
$2y = x - 6$

27. $2x + 3y = -4$
$3x + 2y = -1$

28. $2x + 3y - 6 = 0$
$3x + 2y - 4 = 0$

29. $\frac{1}{2}y = 2x + 3$
$2y = 8x + 12$

30. $y - 2x = 3$
$2y - 4x = 5$

In Exercises 31–40, write the equation of the line.

31. $m = 2$, $b = 2$ **32.** $m = 3$, $b = -1$

33. $m = -6$, $b = 3$ **34.** $m = -2$, $b = 8$

35. $m = 1$, $b = 0$ **36.** $m = -1$, $b = 0$

37. $m = 0$, $b = 4$ **38.** $m = 0$, $b = -4$

39. m is undefined, $x = 5$

40. m is undefined, $x = -4$

In Exercises 41–46, write the equation of the line with the given y-intercept and parallel to the given line.

41. $b = 2$; $y = \frac{1}{2}x - 3$

42. $b = 2$; $y = 2x + 3$

43. $b = \frac{3}{2}$; $3x + 2y + 1 = 0$

44. $b = 0$; $2x + 3y + 4 = 0$

45. $b = 4$; $y = -1$

46. $b = \frac{2}{3}$; $y = -\frac{2}{3}$

In Exercises 47–58, write the equation of the given line.

47.

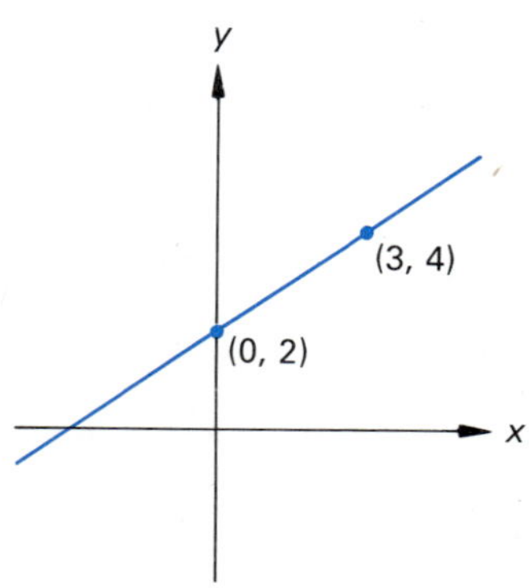

48.

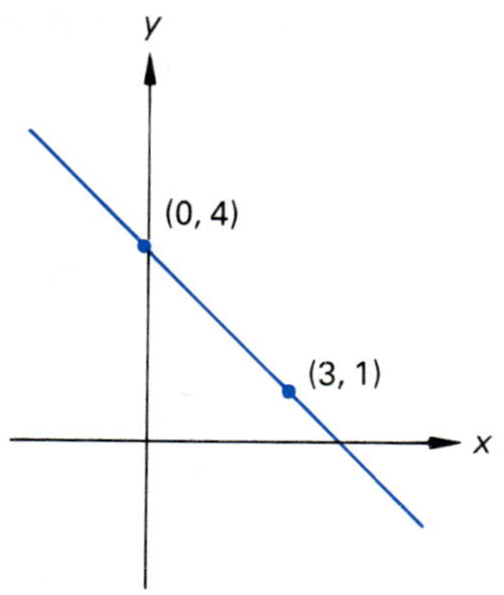

49.

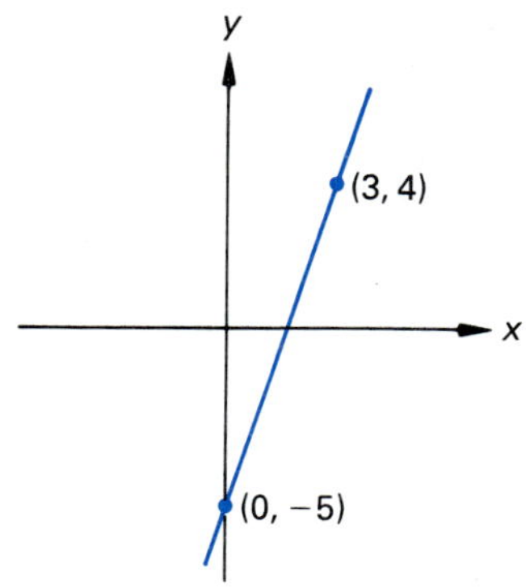

50.

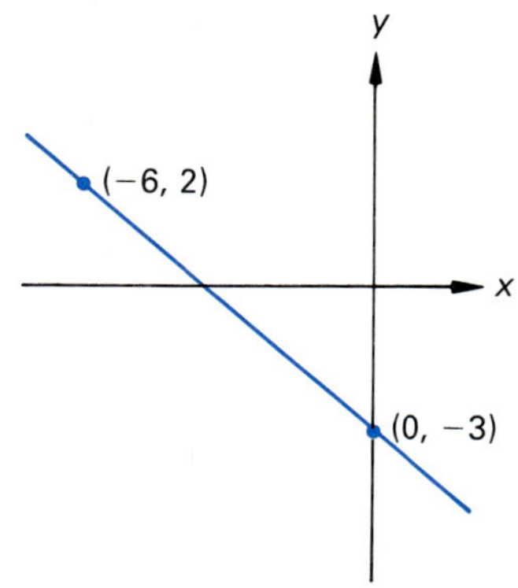

51.

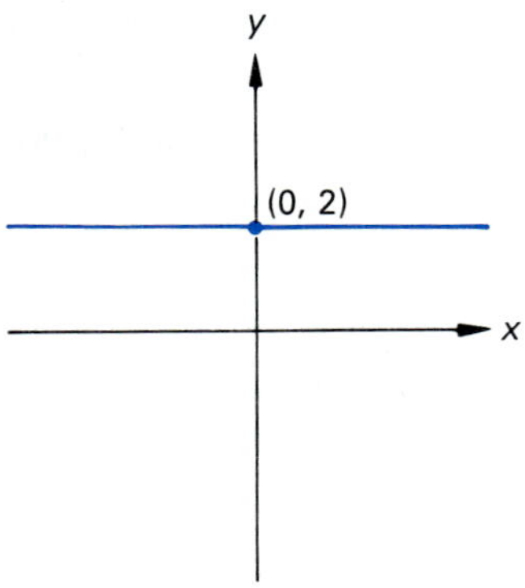

52.

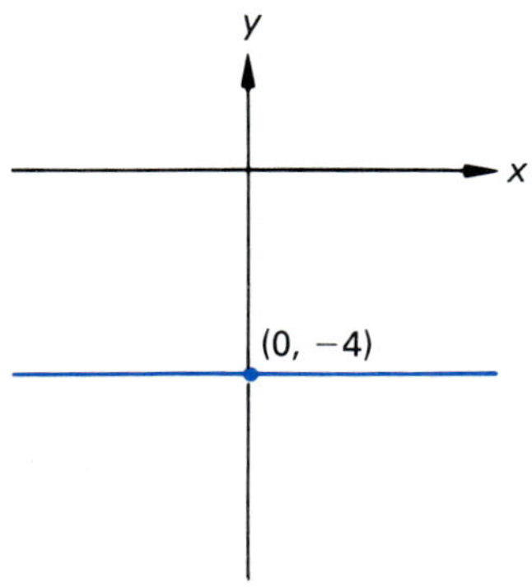

53.

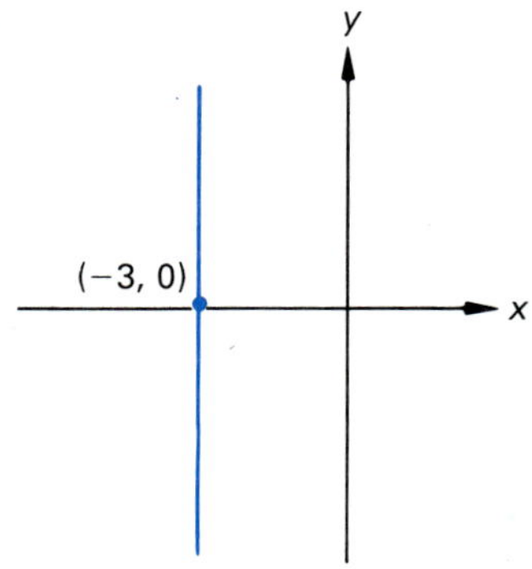

54.

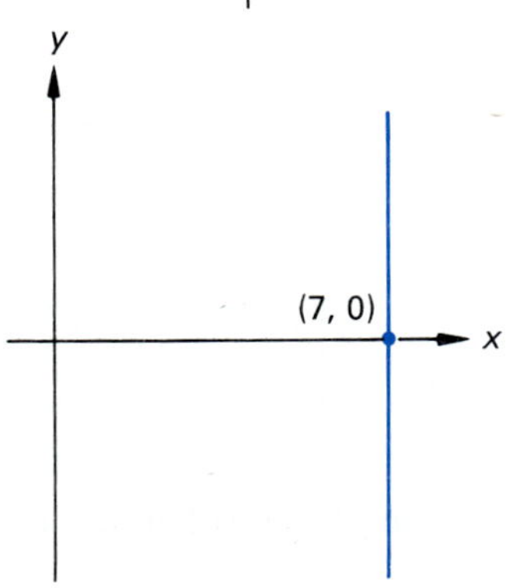

55.

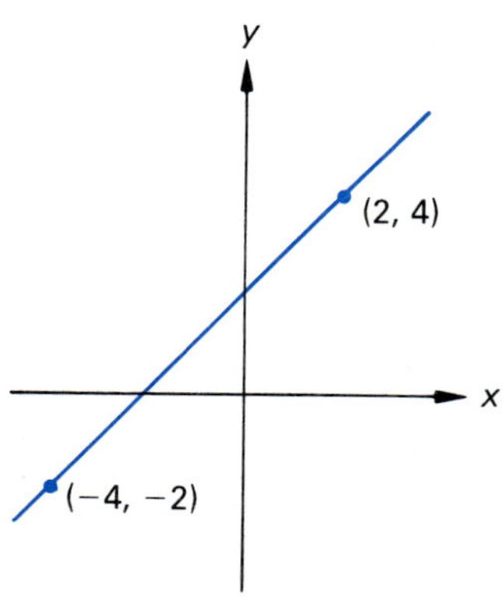

56.

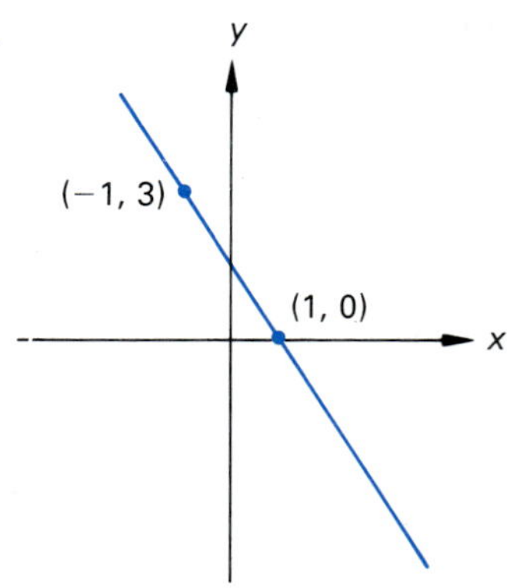

57.

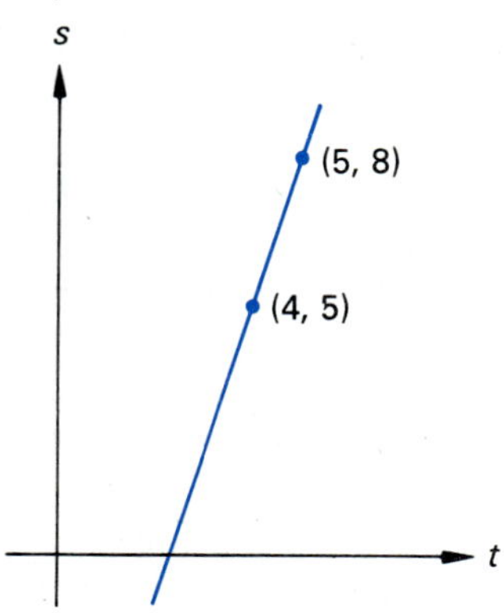

58.

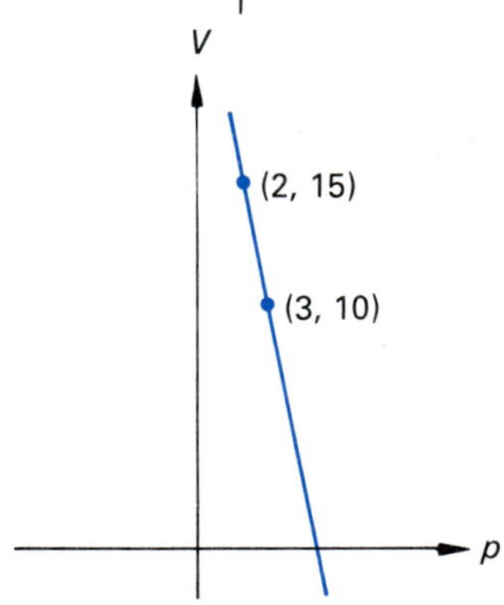

In Exercises 59–66, find the angle the line makes with the horizontal axes in Exercises 49–56.

67. A ramp rises 10 m in a 100-m run. If the ramp goes up and to the right from an origin at the foot of the ramp, what equation describes points on the surface of the ramp?

68. A parking lot has a drainage slope of 1 ft in 50 ft. What is the equation of the surface of the lot to the right of the drain if the origin of the axis system is on the drain?

69. A parking lot has a drainage slope of 1 ft in 50 ft. What is the equation of the surface of the lot to the left of the drain if the origin of the axis system is on the drain?

70. An A-frame house has a roof composed of lines *AB* and *AC* as shown in Fig. 10.15.

a. What are the equations of line *AB* for axes with origins at *A*, *B*, and *O*, respectively?

b. What are the equations of line *AC* for axes with origins at *A*, *C*, and *O*, respectively?

71. A raised A-frame house has a roof composed of lines *AB* and *AC* as shown in Fig. 10.16.

a. What are the equations of line *AB* for axes with origins at point *D* and *E*, respectively?

b. What are the equations of line *AC* for axes with origins at points *F* and *E*, respectively?

72. If the resistance *R* is a constant, is Ohm's law, $i = v/R$, a straight line? Explain.

73. A loop equation for a circuit is $-R_1 i_1 + (R_1 + R_2) i_2 = 0$. If i_1 is the independent variable and i_2 is the dependent variable, what is the slope of the line?

74. The equation for calculating degrees Fahrenheit when given degrees Celsius is $F = \frac{9}{5}C + 32$. Sketch the linear equation indicating slope and *F*-intercept.

75. The equation for calculating degrees Celsius when given degrees Fahrenheit is $C = \frac{5}{9}(F - 32)$. Sketch the linear equation indicating slope and *C*-intercept.

FIGURE 10.15

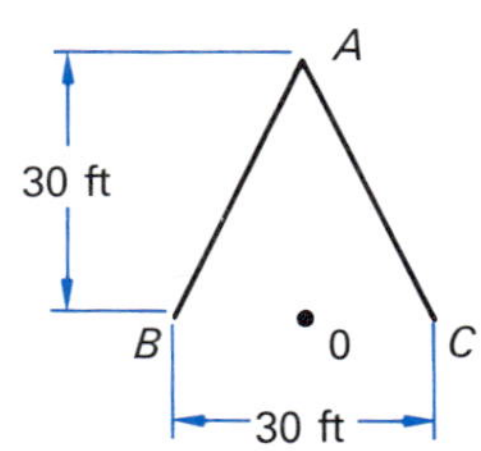

FIGURE 10.16

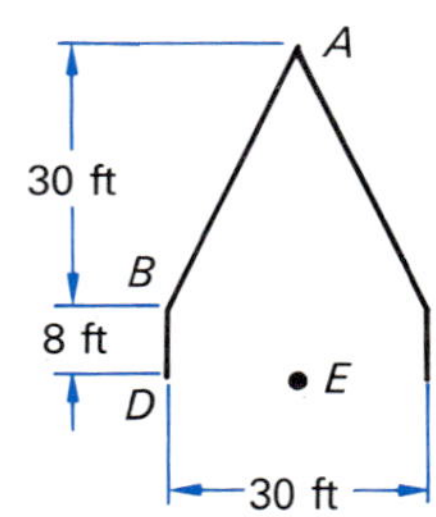

10.2 Solving two linear equations in two unknowns algebraically

A **system of equations** is a group of equations whose variables all relate to a common situation. For example, the forces in members *A* and *B* of a roof truss could be related by the equations

$$0.8A - 0.6B = 4,$$
$$0.6A + 0.8B = 0.$$

This pair of equations forms a system of two linear equations in two unknowns. The **solution** to a system of two linear equations in two unknowns is the ordered pair (or pairs) that makes the equations true simultaneously. The solution of a system of equations in two unknowns may be found by graphing or algebraic methods. Graphical solutions to systems of linear equations are useful because they provide a visual insight into the character of the system of equations. The graphical solutions are limited by the difficulty of producing graphs that can be read with the desired accuracy. Algebraic solutions are a means of obtaining more accurate solutions. The two algebraic methods of solving equations that we will discuss in this section are elimination and substitution.

EXAMPLE 1 Find the solution of the system of equations using the substitution method:

$$\begin{aligned} 2x + y &= 4 \\ x - 2y &= -3. \end{aligned}$$

Solution: To solve by the substitution method, first solve either one of the equations for either one of the variables. For instance, solving $2x + y = 4$ for x gives

$$x = \frac{4 - y}{2}.$$

Then substitute this expression for the variable into the remaining equation and solve for the other variable. Substituting $x = \frac{4 - y}{2}$ into $x - 2y = -3$ for x gives

$$\begin{aligned} \frac{4 - y}{2} - 2y &= -3 \\ 4 - y - 4y &= -6 && \textbf{Multiplying both sides by 2} \\ -5y &= -10 && \textbf{Combining like terms} \\ y &= 2. \end{aligned}$$

Having obtained a value for one of the variables, substitute this value into the expression obtained earlier to obtain a value for the other variable. Substituting 2 for y in the equation $x = \frac{4 - y}{2}$, we have

$$x = \frac{4 - 2}{2}$$

$$x = 1.$$

The solution for the system $2x + y = 4$ and $x - 2y = -3$ is the ordered pair (1, 2). To check the solution we substitute (1, 2) into both equations.

Check:

$$\begin{aligned} 2x + y &= 4 \\ 2(1) + 2 &= 4 \\ 4 &= 4. \checkmark \\ x - 2y &= -3 \\ 1 - 2(2) &= -3 \\ -3 &= -3. \checkmark \end{aligned}$$

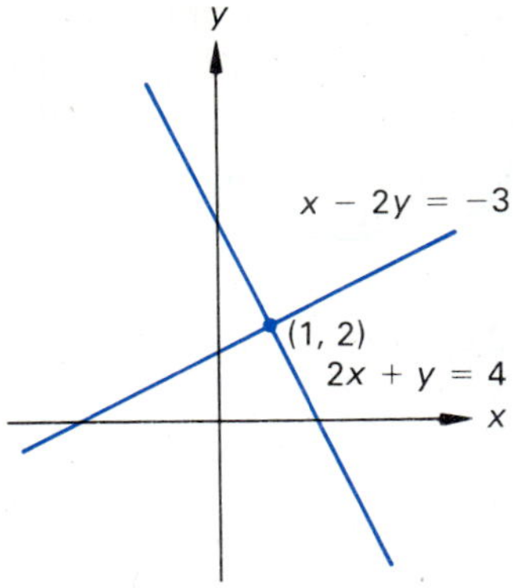

FIGURE 10.17

The solution checks. It is left as an exercise for the reader to show that solving for y first instead of x or starting with the second equation instead of the first equation would have produced the same solution. □

If we graph the equations $2x + y = 4$ and $x - 2y = -3$ as shown in Fig. 10.17, we see that the lines intersect at a unique point, represented by a single ordered pair.

EXAMPLE 2 Solve the system of equations using the substitution method:

$$\begin{aligned} 2x + y &= 2 \\ 4x + 2y &= -1. \end{aligned}$$

Solution: Since the y term in the first equation has a coefficient of 1, it is easier to solve for y in the first equation.

$$\begin{aligned} 2x + y &= 2 \\ y &= 2 - 2x. \end{aligned}$$

Substituting $2 - 2x$ for y in the equation $4x + 2y = -1$ we have

$$\begin{aligned} 4x + 2(2 - 2x) &= -1 \\ 4x + 4 - 4x &= -1 \\ 4 &= -1. \end{aligned}$$

This result is a contradiction since 4 cannot equal -1. Therefore, there is no solution to the system of equations, which we indicate by writing the words "no solution," or using the symbol ∅. The symbol ∅ in set notation means the empty set, which is another way of saying "no solution." □

If we graph the equations $2x + y = 2$ and $4x + 2y = -1$ as shown in Fig. 10.18, we see that the lines are parallel (they do not intersect).

A system of two linear equations in two unknowns in which the lines are parallel and have no points in common is called an **inconsistent system.** An inconsistent system has no solution (no ordered pairs of numbers that will satisfy both equations).

FIGURE 10.18

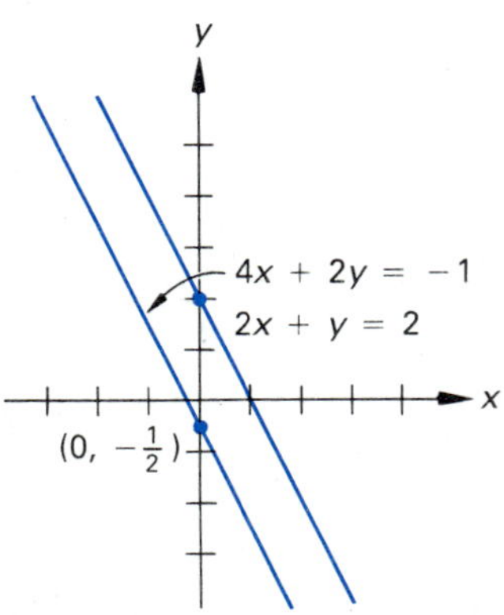

EXAMPLE 3 Solve the system of equations using the elimination method:

$$\begin{aligned} 2x + y &= 3 \\ x - 2y &= 9. \end{aligned}$$

Solution: To solve the equations by the elimination method we will attempt to add or subtract the equations to eliminate one of the variables. We observe that if the x-term in the second equation had a coefficient of 2, we could subtract the second equation from the first and eliminate the variable x. The result would be a linear equation with a single variable y. This is not just wishful thinking; we can derive a $2x$ term in the second equation by multiplying the entire equation by 2. (Remember that the equality is preserved when both sides of an equation are multiplied by the same nonzero factor.)

$$\left.\begin{aligned} 2x + y &= 3 \\ 2(x - 2y) &= 2(9) \end{aligned}\right\} \quad \text{is the same as} \quad \begin{cases} 2x + y = 3 \\ x - 2y = 9 \end{cases}$$

$$\begin{aligned} 2x + y &= 3 \\ 2x - 4y &= 18 \qquad \textbf{Subtracting} \\ \hline 5y &= -15 \\ y &= -3. \end{aligned}$$

Now the value obtained for y can be substituted into either of the original equations to solve for x. Substituting $y = -3$ into $2x + y = 3$ gives

$$\begin{aligned} 2x + (-3) &= 3 \\ 2x &= 6 \end{aligned}$$

or

$$x = 3.$$

The solution $(3, -3)$ can be checked by substituting into the remaining equation. Substituting $(3, -3)$ into $x - 2y = 9$:

$$\begin{aligned} x - 2y &= 9 \\ 3 - 2(-3) &= 9 \\ 9 &= 9. \checkmark \quad \square \end{aligned}$$

It is left to the reader to graph the system and show that the lines intersect at a unique point.

EXAMPLE 4 Solve the following system of equations using the elimination method:

$$\begin{aligned} x - 2y &= -1 \\ -2x + 4y &= 2. \end{aligned}$$

Solution:

$$\left.\begin{aligned} 2(x - 2y) &= 2(-1) \\ -2x + 4y &= 2 \end{aligned}\right\} \quad \text{is the same as} \quad \begin{cases} x - 2y = -1 \\ -2x + 4y = 2 \end{cases}$$

so

$$\begin{aligned} -2x + 4y &= 2 \\ 2x - 4y &= -2 \qquad \textbf{Adding} \\ \hline 0 &= 0. \end{aligned}$$

This result is true but does not provide us with a unique solution. Instead it tells us that any ordered pair (x, y) that satisfies one equation

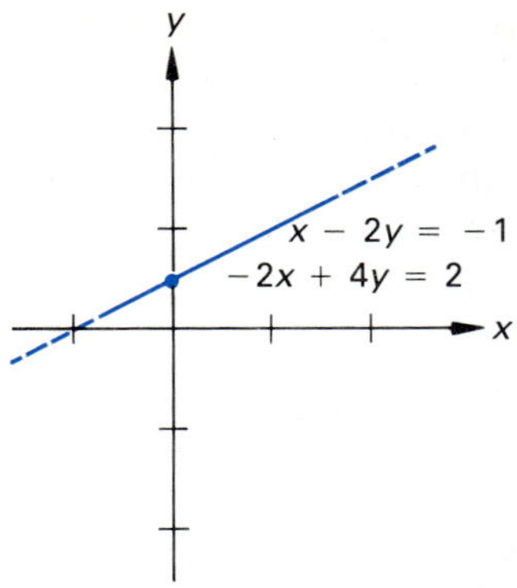

FIGURE 10.19

also satisfies the other equation. The solution is indicated by stating there is no unique solution. This says that there are infinitely many ordered pairs that satisfy both equations. □

If we graph the equations $x - 2y = -1$ and $-2x + 4y = 2$ as shown in Fig. 10.19 we see that the lines coincide, forming a single line.

A system of two linear equations in two unknowns in which the graphical solution is a single line is a **dependent system.** A dependent system of equations has an infinite number of ordered pairs that satisfy both equations.

EXAMPLE 5 A system of equations relating the forces in members A and B of a roof truss is

$$
\begin{aligned}
0.800A - 0.600B &= 4000\\
0.600A + 0.800B &= 0.
\end{aligned}
$$

Solve the equations using the substitution method.

Solution: (The equations demand accuracy to three significant digits in the solution. We will keep this in mind even though for convenience we will not carry all the trailing zeros or decimal points.) Observing the system of equations,

$$0.8A - 0.6B = 4000, \qquad [1]$$
$$0.6A + 0.8B = 0, \qquad [2]$$

Eq. [2] appears to be easier to solve for one of the variables in terms of the other. Solving for A

$$0.6A + 0.8B = 0$$
$$A = \frac{-0.8B}{0.6}.$$

Substituting into Eq. [1]

$$
\begin{aligned}
0.8\left(\frac{-0.8B}{0.6}\right) - 0.6B &= 4000\\
-1.067B - 0.6B &= 4000\\
-1.667B &= 4000\\
B &= -2400\\
A &= \frac{-0.8}{0.6}(-2400)\\
A &= 3200.
\end{aligned}
$$

The solution to the system is $A = 3200$ and $B = -2400$. The check is left to the reader. □

EXAMPLE 6 Twice the current in electric circuit A is 3 amperes greater than three times the current in the adjoining circuit B. When current B is subtracted from current A and the difference of the two currents is multiplied by a resistance of 5 Ω, the result is a voltage drop of 7 V. What are the currents?

Solution: Let

$$\begin{aligned} i_A &= \text{current in circuit } A, \\ i_B &= \text{current in circuit } B, \\ 2i_A &= \text{twice the current in circuit } A, \text{ and} \\ 3i_B &= \text{three times the current in circuit } B. \end{aligned}$$

Knowing that twice the current in circuit A is 3 amperes greater than three times the current in circuit B we can write:

$$2i_A = 3i_B + 3.$$

Knowing $5(i_A - i_B) =$ voltage drop of 7 V, we can write

$$5i_A - 5i_B = 7.$$

We can solve the system of two equations in two unknowns using the elimination method:

$$2i_A - 3i_B = 3 \qquad [1]$$
$$5i_A - 5i_B = 7. \qquad [2]$$

To eliminate i_A we multiply Eq. [1] by 5 and Eq. [2] by -2 and add:

$$10i_A - 15i_B = 15 \qquad 5 \times [1] \qquad [3]$$
$$-10i_A + 10i_B = -14 \qquad -2 \times [2] \qquad [4]$$
$$-5i_B = 1 \qquad [3] + [4]$$
$$i_B = \frac{-1}{5}.$$

Substituting for i_B in Eq. [1] we have

$$2i_A - 3\left(-\frac{1}{5}\right) = 3$$

$$2i_A + \frac{3}{5} = 3$$

$$2i_A = \frac{12}{5}$$

$$i_A = \frac{6}{5}.$$

The currents are $i_A = \frac{6}{5}$ A and $i_B = -\frac{1}{5}$ A.

The check is left to the reader. □

10.2 Exercises

In Exercises 1–12, use the substitution method to solve the systems of equations. Check solutions.

1. $2x + y = -10$
$x - 4y = -14$

2. $x + y = 21$
$3x - 2y = 8$

3. $-5x + y = 1$
$7x - y = 5$

4. $x - y = 4$
$3x - 2y = 1$

5. $7x - y = 1$
$-2x + y = 19$

6. $x + 2y = -3$
$-x + y = -30$

7. $-5x + 2y = -11$
$x - 3y = -3$

8. $3x - y = -4$
$-5x + 4y = 7$

9. $x + \frac{1}{3}y = 2$
$3x - 7y = -18$

10. $\frac{4}{5}x - y = 1$
$5x + 11y = 11$

11. $0.4x + 0.5y = 3.7$
$x - 0.7y = -0.5$

12. $0.07x + 0.09y = 0.011$
$-0.03x + y = 0.203$

In Exercises 13–26, use the elimination method to solve the systems of equations. Check solutions.

13. $-5x + 2y = -11$
$2x - 3y = 22$

14. $3x - 2y = -4$
$-5x + 4y = 0$

15. $2x - 3y = -3$
$4x + 3y = 39$

16. $5x + 2y = 6$
$9x + 3y = 3$

17. $2x + 5y = 12$
$3x - 2y = 18$

18. $4x + 7y = -16$
$-5x + 3y = 20$

19. $8x - 5y = 15$
$-3x + 7y = -21$

20. $-3x + 8y = 56$
$5x + 2y = 14$

21. $2x + 5y = 13$
$4x - 7y = -8$

22. $4x - 7y = 7$
$2x - 14y = -7$

23. $3x + 2y = -5.5$
$2x + 5y = 0$

24. $6x + 5y = -8.5$
$7x + 2y = 3.5$

25. $8x + 5y = 2.75$
$3x + 2y = 0.75$

26. $4x + 3y = 3.25$
$6x + 7y = -3.25$

In Exercises 27–36, use algebraic methods to solve the systems of linear equations. From the results determine whether any of the systems are inconsistent or dependent.

27. $2x + 3y = 6$
$-4x - 6y = -8$

28. $x - y = 3$
$-x + y = 0$

29. $2x + 5y = 2$
$4x + 10y = 4$

30. $3x - 2y = -1$
$-3x + 2y = 1$

31. $3x + 2y = -4$
$-3x + 2y = -4$

32. $6x + 5y = 18$
$6x - 5y = 18$

33. $3.6x - 1.4y = 0.5$
$-1.8x + 0.7y = 0.2$

34. $1.8x + 2.0y = 2.3$
$3.6x + 4.0y = -4.6$

35. $-1.8x + 1.1y = 2.1$
$5.4x - 3.3y = -6.2$

36. $5.1x + 7.2y = -0.6$
$1.7x + 2.4y = -0.2$

Exercises 37–40 are the equilibrium equations for the joints in the truss shown in Fig. 10.20. Solve for the unknown forces in each system of equations.

37. $0.866A + B = 0$
$7500 + 0.500A = 0$

38. $-0.866C + 0.866H = 0$
$-0.500C - 5000 - 0.500H - 5000 = 0$

39. $-0.866(-15{,}000) + 0.866C + 0.866D = 0$
$-0.500(-15{,}000) + 0.500C - 0.500D - 5000 = 0$

40. $-0.866(13{,}000) - 0.866D + 0.866F + 0.866(13{,}000) = 0$
$0.500D + 5000 + 0.500F = 0$

In Exercises 41–44, recall that the conversion from degrees Celsius to degrees Fahrenheit is given by $F = \frac{9}{5}C + 32$. Find both the Fahrenheit and Celsius temperatures under the circumstances given.

41. Fahrenheit temperature is twice Celsius temperature.

FIGURE 10.20

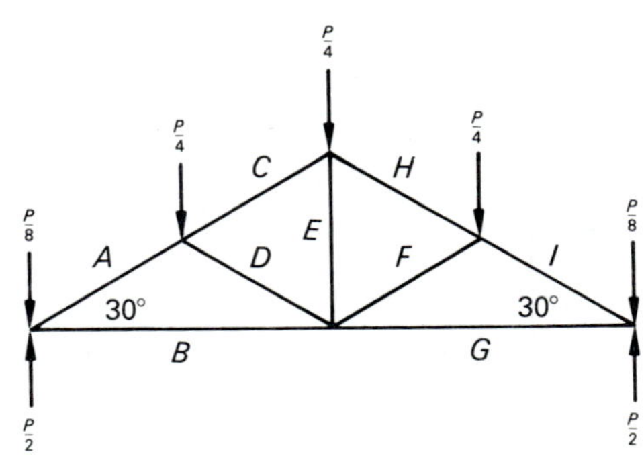

42. Celsius temperature is one-third Fahrenheit temperature.

43. The temperature in degrees Celsius is five less than half the temperature in degrees Fahrenheit.

44. Three times the temperature in degrees Fahrenheit is four more than the temperature in degrees Celsius.

In Exercises 45 and 46, solve for the conditions in Exercises 41 and 42 if the temperature conversion formula is given as $C = \frac{5}{9}(F - 32)$.

47. A restaurant serves a salad that contains lettuce and bean sprouts in a ratio of 4 : 1 by weight. If the chef wants to make a 100-lb batch of salad, how much produce should be ordered?

48. Twice the water flow in the hot-water pipe is the same as three times the flow in the cold-water pipe. The combined flow is 1600 liters/hr. What is the flow in each pipe?

49. Three times the force in the left member of a structure is 1000 N less than twice the force in the right member. The sum of the two forces multiplied by 7 is 5000 N. What are the forces?

50. If voltage v_1 is subtracted from voltage v_2 and the difference is divided by a 50-Ω resistance, the result is a current of 3 A. If v_2 is twice v_1, what are the voltages?

10.3 Solving two linear equations in two unknowns by determinants

Another method for solving systems of linear equations is the use of determinants. Before we can solve systems of equations with determinants it is essential that we know what a determinant is and how to evaluate it.

A **determinant** is a unique value that is associated with a square array of numbers. A 2 × 2 (read 2-by-2) array of four numbers arranged in two rows and two columns and bordered by two vertical lines as shown in Fig. 10.21 is called a 2 × 2 determinant. Each number in the determinant is called an **element** of the determinant and is designated by its position at the intersection of a row and a column of the determinant. If we use i to designate a row in the determinant ($i = 1, 2, \ldots,$ indicates 1st, 2nd, . . . , row) and use j to designate a column in the determinant ($j = 1, 2, \ldots,$ indicates 1st, 2nd, . . . , column) we can specify any element in the determinant as a_{ij}.

FIGURE 10.21

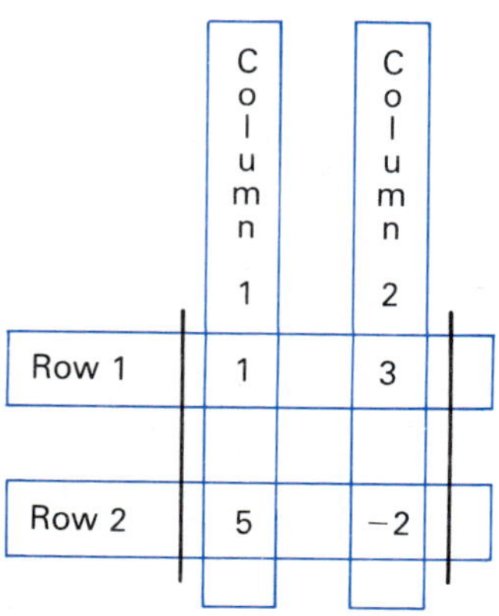

EXAMPLE 1 Use the a_{ij} notation to designate the elements of the determinant

$$\begin{vmatrix} 1 & 3 \\ 5 & -2 \end{vmatrix}$$

Solution: a_{11} indicates the element at the intersection of the first row and first column of the determinant. Therefore, $a_{11} = 1$. a_{12} indicates the element at the intersection of the first row and second column of the determinant, so $a_{12} = 3$. In similar fashion, $a_{21} = 5$ and $a_{22} = -2$. □

Using the subscript notation, a 2×2 determinant can be written as

$$\begin{vmatrix} a_{11} & a_{12} \\ a_{21} & a_{22} \end{vmatrix}$$

DEFINITION 10.4

A 2×2 determinant, A, has a unique value given by

Principal Diagonal

$$A = \begin{vmatrix} a_{11} & a_{12} \\ a_{21} & a_{22} \end{vmatrix} = a_{11}a_{22} - a_{21}a_{12}.$$

Secondary Diagonal

($a_{11}a_{22}$ is the product of the elements of the principal diagonal and $a_{21}a_{12}$ is the product of the elements of the secondary diagonal.)

EXAMPLE 2 What is the value of the 2×2 determinants?

a. $\begin{vmatrix} 1 & 3 \\ 5 & -2 \end{vmatrix}$ **b.** $\begin{vmatrix} 1 & -2 \\ 3 & 5 \end{vmatrix}$

Solution: According to Definition 10.4:

a. $$\begin{vmatrix} 1 & 3 \\ 5 & -2 \end{vmatrix} = (1)(-2) - (5)(3) = -17.$$

b. $$\begin{vmatrix} 1 & -2 \\ 3 & 5 \end{vmatrix} = (1)(5) - (3)(-2) = 11.$$

We see that even though the numbers in the determinants of parts (a) and (b) are the same, the values of the determinants are different because the numbers are located in different positions in the determinants. □

We can use determinants to solve systems of equations using a technique called Cramer's rule. To show this, let us consider a system of two linear equations in two unknowns.

$$ax + by = c, \qquad [1]$$
$$dx + ey = f, \qquad [2]$$

where a, b, c, d, e, and f are constants. To solve such a system of equations by the elimination method we could eliminate the y-term by multiplying Eq. [1] by e, multiplying Eq. [2] by $-b$ and adding the results:

$$
\begin{array}{rll}
aex + bey = ce & e \times [1] & [3] \\
-bdx - bey = -bf & -b \times [2] & [4] \\
\hline
aex - bdx = ce - bf & [3] + [4] & \\
(ae - bd)x = ce - bf & \textbf{Factoring} & \\
x = \dfrac{ce - bf}{ae - bd}. & &
\end{array}
$$

The reader can find the value for y by using the same techniques we used to solve for x:

$$y = \frac{af - cd}{ae - bd}.$$

Examining the solutions for x and y,

$$x = \frac{ce - bf}{ae - bd}$$

and

$$y = \frac{af - cd}{ae - bd},$$

we see the similarity between the expressions in the numerators and denominators of the fractions and the value of a 2×2 determinant

$$\begin{vmatrix} a_{11} & a_{12} \\ a_{21} & a_{22} \end{vmatrix} = a_{11}a_{22} - a_{21}a_{12}.$$

The denominators for x and y are the same, so if we call them D and call the numerators N_x and N_y, respectively, we have

$$x = \frac{N_x}{D}$$

and

$$y = \frac{N_y}{D},$$

where

$D = ae - db$, which is the value of the determinant $\begin{vmatrix} a & b \\ d & e \end{vmatrix}$.

$N_x = ce - fb$, which is the value of the determinant $\begin{vmatrix} c & b \\ f & e \end{vmatrix}$.

$N_y = af - dc$, which is the value of the determinant $\begin{vmatrix} a & c \\ d & f \end{vmatrix}$.

This result is called **Cramer's rule.**

DEFINITION 10.5

CRAMER'S RULE

The solution to a system of two linear equations in two unknowns in standard form

$$\begin{aligned} ax + by &= c \\ dx + ey &= f \end{aligned}$$

is given by

$$x = \frac{N_x}{D} \qquad y = \frac{N_y}{D}$$

where $D = \begin{vmatrix} a & b \\ d & e \end{vmatrix}$, $N_x = \begin{vmatrix} c & b \\ f & e \end{vmatrix}$, $N_y = \begin{vmatrix} a & c \\ d & f \end{vmatrix}$,

and $D \neq 0$.

We see that with the equations written as shown, the elements of D are the coefficients of the unknowns x and y. The elements of N_x are the elements of D with the x-coefficients replaced by the constants and the elements of N_y are the elements of D with the y-coefficients replaced by the constants.

$$\boxed{2}x + \boxed{1}y = \boxed{4}$$
$$\boxed{1}x - \boxed{2}y = \boxed{-3}$$

FIGURE 10.22

EXAMPLE 3 Use Cramer's rule to solve the system of equations:

$$\begin{aligned} 2x + y &= 4 \\ x - 2y &= -3. \end{aligned}$$

Solution: Arranging the equations as in Fig. 10.22

$$D = \begin{vmatrix} 2 & 1 \\ 1 & -2 \end{vmatrix} \quad \text{Determinant of coefficients of variables}$$

$$= (2)(-2) - (1)(1)$$
$$= -5.$$

$$N_x = \begin{vmatrix} 4 & 1 \\ -3 & -2 \end{vmatrix} \quad x\text{-column of } D \text{ replaced by constant column}$$

$$= (4)(-2) - (-3)(1)$$
$$= -5.$$

$$N_y = \begin{vmatrix} 2 & 4 \\ 1 & -3 \end{vmatrix} \quad y\text{-column of } D \text{ replaced by constant column}$$

$$= (2)(-3) - (1)(4)$$
$$= -10.$$

$$x = \frac{N_x}{D} \qquad y = \frac{N_y}{D}$$
$$= \frac{-5}{-5} \qquad = \frac{-10}{-5}$$
$$x = 1. \qquad y = 2.$$

The solution to the system of equations is the ordered pair (1, 2). To check the solution, substitute (1, 2) into both of the original equations. □

$$\boxed{2}x + \boxed{1}y = \boxed{2}$$
$$\boxed{4}x + \boxed{2}y = \boxed{-1}$$

FIGURE 10.23

EXAMPLE 4 Use Cramer's rule to solve the system of equations:

$$2x + y - 2 = 0$$
$$4x + 2y + 1 = 0.$$

Solution: To use Cramer's rule, the equations of the system must be written with the constant terms to the right of the equal sign. Rearranging we have Fig. 10.23.

$$D = \begin{vmatrix} 2 & 1 \\ 4 & 2 \end{vmatrix}$$
$$= (2)(2) - (4)(1)$$
$$= 0.$$

Therefore, we cannot solve this system of equations using Cramer's rule. If we attempt to continue the solution procedure, we find

$$N_x = \begin{vmatrix} 2 & 1 \\ -1 & 2 \end{vmatrix}$$
$$= (2)(2) - (-1)(1)$$
$$= 5.$$
$$N_y = \begin{vmatrix} 2 & 2 \\ 4 & -1 \end{vmatrix}$$
$$= (2)(-1) - (4)(2)$$
$$= -10.$$

$$x = \frac{N_x}{D} \qquad y = \frac{N_y}{D}$$
$$= \frac{5}{0} \qquad = \frac{-10}{0}$$
$$x = \text{undefined}. \qquad y = \text{undefined}.$$

Values for x *and* y *with denominators of zero and at least one nonzero numerator when using Cramer's rule are indicative of an inconsistent system of equations. (By use of the graphical representation of the linear equations, the lines are parallel and have different* y*-intercepts.)*

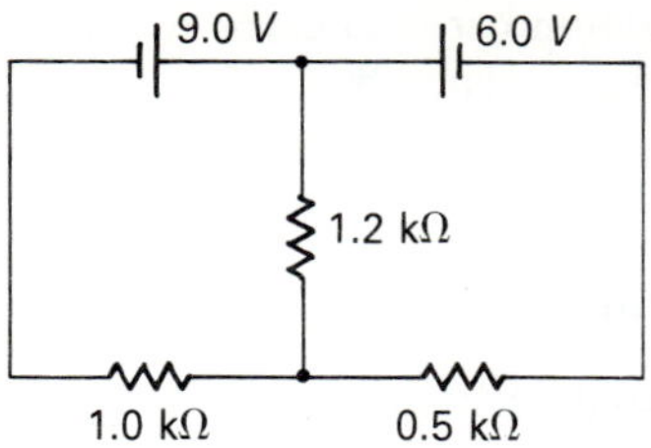

FIGURE 10.24

Values for x and y with both numerator and denominator zero when using Cramer's rule are indicative of a dependent system of equations (equations with the same slope and same y-intercept). □

EXAMPLE 5 Use Cramer's rule to find the loop currents in the system of equations representing the circuit in Fig. 10.24.

$$2.2\text{ k } i_1 - 1.2\text{ k } i_2 = 9.0$$
$$-1.2\text{ k } i_1 + 1.7\text{ k } i_2 = -6.0$$
$$(1\text{ k }\Omega = 10^3\ \Omega)$$

Solution:

$$D = \begin{vmatrix} 2.2\times10^3 & -1.2\times10^3 \\ -1.2\times10^3 & 1.7\times10^3 \end{vmatrix}$$
$$= (2.2\times10^3)(1.7\times10^3) - (-1.2\times10^3)(-1.2\times10^3)$$
$$= 3.74\times10^6 - 1.44\times10^6$$
$$= 2.30\times10^6$$

$$N_{i_1} = \begin{vmatrix} 9.0 & -1.2\times10^3 \\ -6.0 & 1.7\times10^3 \end{vmatrix}$$
$$= (9.0)(1.7\times10^3) - (-6.0)(-1.2\times10^3)$$
$$= 15.3\times10^3 - 7.20\times10^3$$
$$= 8.10\times10^3.$$

$$N_{i_2} = \begin{vmatrix} 2.2\times10^3 & 9.0 \\ -1.2\times10^3 & -6.0 \end{vmatrix}$$
$$= (2.2\times10^3)(-6.0) - (-1.2\times10^3)(9.0)$$
$$= -13.2\times10^3 + 10.8\times10^3$$
$$= -2.40\times10^3.$$

$$i_1 = \frac{N_{i_1}}{D} = \frac{8.10\times10^3}{2.30\times10^6} = 3.52\times10^{-3} \qquad i_2 = \frac{N_{i_2}}{D} = \frac{-2.40\times10^3}{2.30\times10^6} = -1.04\times10^{-3}$$

Therefore $i_1 = 3.5\times10^{-3}$ A and $i_2 = -1.0\times10^{-3}$ A. The check is left to the reader. □

10.3 Exercises

In Exercises 1–20, use Cramer's rule to solve the systems of linear equations.

1. $3x + y = 6$
$-2x + y = 1$

2. $-x + 3y = 3$
$x - 2y = 0$

3. $5x - 2y = -9$
$3x + 7y = 11$

4. $3x + 5y = 16$
$-3x + 2y = 19$

5. $5x + y = -17$
$2x - 7y = 8$

6. $x - 5y = 0$
$x + 5y = -10$

7. $2x + y = 40$
$-2x + y = 0$

8. $2x + 7y = 1$
$2x + 9y = -5$

9. $y = -4x - 2$
$y = -7x + 1$

10. $y = -3x - 3$
$y = -2x - 11$

11. $y = 2x + 1$
$y = \frac{1}{2}x - 4$

12. $y = -\frac{1}{2}x - \frac{1}{2}$
$y = -\frac{1}{4}x - \frac{9}{4}$

13. $3x - y = 2.0$
$2x + y = 5.5$

14. $2x + y = 0.0$
$4x - y = 1.0$

15. $x + 4y = 23.5$
$2x - 12y = 17.0$

16. $2x - 10y = -15.0$
$-x + 4y = 9.0$

17. $1.6x + 2.4y = 3.0$
$8.2x - 1.3y = 5.5$

18. $3.1x + 5.7y = 0.0$
$-6.5x + 4.7y = 3.7$

19. $4.4x - 6.1y = 1.6$
$3.2x + 4.0y = -2.7$

20. $6.2x + 4.8y = -7.7$
$8.3x + 2.6y = -5.1$

In Exercises 21–24, use Cramer's rule to solve the systems of linear equations. From the results state whether the systems are consistent, inconsistent, or dependent.

21. $1.6x + 2.1y = 7.3$
$4.8x + 6.3y = 3.7$

22. $2.0x - 5.0y = 4.0$
$3.0x - 7.5y = 6.0$

23. $2.7x + 3.9y = -1.5$
$0.9x + 1.3y = -0.5$

24. $2.7x - 3.1y = 1.6$
$8.1x - 9.3y = 4.8$

25. Stresses in an aluminum bar connected to a prestressed steel sleeve are related by

$$5\ S_S + 2\frac{1}{2}\ S_A = 0$$
$$S_S - 2\ S_A = 80{,}000 \text{ MPa}.$$

What are the stresses in MPa?

26. Force transfer due to tightening a bolt in a sleeve produces stresses S_B in the bolt and S_S in the sleeve related by

$$\frac{\pi}{4}[d_2^2 - d_1^2]S_S + \frac{\pi}{4}d_1^2S_B = 0$$
$$S_B = 3S_S + 6600.$$

If $d_1 = \frac{1}{2}$ in. and $d_2 = 1$ in., what are S_S and S_B (in pounds per square inch)?

27. Forces on a beam produce the equilibrium equations.

$$R_1 - 500 - (10)(500) + R_2 = 0$$
$$-10R_1 + 4(500) - (5)(10)(500) - 100 + 10R_2 = 0.$$

What are the reaction forces R_1 and R_2 (in newtons)?

28. Forces on a beam produce the equilibrium equations

$$R_1 - 1000 + R_2 = 0$$
$$-25R_1 + 15(1000) - 50 + 15R_2 = 0.$$

What are the reaction forces R_1 and R_2?

29. The loop currents in a circuit are given by

$$-4 + 6(i_1 - i_2) + 5i_1 = 0$$
$$6(i_2 - i_1) + 3 + 2\frac{1}{2}\ i_2 = 0$$

What are the currents i_1 and i_2?

30. The loop equations for a circuit are

$$3i_1 + 6 + 7(i_1 - i_2) - 12 + 5i_1 = 0$$
$$12 + 7(i_2 - i_1) + 4 + 8i_2 + 2i_2 = 0.$$

What are the currents i_1 and i_2?

31. Four bolts with one nut on each weigh a total of 3 oz. Six bolts with two nuts on each weigh a total of 5 oz. What are the individual weights of the nuts and bolts?

32. A dozen bolts with single nuts weigh a total of

12 oz. A gross of the same bolts with double nuts weighs 160 oz. What are the individual weights of the nuts and bolts?

33. An asymmetrical parking lot is 100 m wide and has both sides at the same height. The left side slopes down with a drop of 1 m in 50 m. The right side slopes down with a drop of 1 m in 75 m. With respect to the left side, where is the low point for the location of the drain?

34. If the right side of the parking lot in Exercise 33 is 1/2 m higher than the left side, where should the drain be located?

10.4 Solving three linear equations in three unknowns algebraically

A group of equations such as

$$\begin{aligned} x + y + z &= -2 \\ 2x - y - z &= 5 \\ -x + 3y + 4z &= -11 \end{aligned}$$

is a system of three linear equations in three unknowns. Graphical methods are not practical for the solution of equations in three unknowns because of the complexity of drawing figures in three-dimensional space. However, the algebraic techniques used for solving systems of two equations can be used to solve systems of three equations in three unknowns.

EXAMPLE 1 Use the substitution method to solve the system of three linear equations in three unknowns:

$$x + y + z = -2 \qquad [1]$$
$$2x - y - z = 5 \qquad [2]$$
$$-x + 3y + 4z = -11. \qquad [3]$$

Solution: First solve any one of the equations for any variable that is convenient. Generally the most convenient variable to solve for is a variable with a coefficient of 1. Let us choose to solve Eq. [3] for x:

$$x = 3y + 4z + 11. \qquad [4]$$

Substituting this value of x into Eqs. [1] and [2]:

$3y + 4z + 11 + y + z = -2$	**Substituting for x in [1]**	
$4y + 5z = -13$	**Rearranging**	[5]
$2(3y + 4z + 11) - y - z = 5$	**Substituting for x in [2]**	
$5y + 7z = -17.$	**Rearranging**	[6]

Equations [5] and [6] form a system of two linear equations in two unknowns, which can be solved as in Section 10.2. We will solve this system using the substitution method. However, due to the nature of the system the reader may prefer to use the elimination method.

$$y = \frac{-5z - 13}{4} \qquad \textbf{Solving [5] for } y \qquad [7]$$

$$5\left(\frac{-5z - 13}{4}\right) + 7z = -17 \qquad \textbf{Substituting for } y \textbf{ in [6]}$$

$$\frac{-25z - 65}{4} + 7z = -17 \qquad \textbf{Removing parentheses}$$

$$-25z - 65 + 28z = -68 \qquad \textbf{Multiplying each term by 4}$$

$$3z = -3$$

$$z = -1$$

$$y = \frac{-5(-1) - 13}{4} \qquad \textbf{Substituting for } z \textbf{ in [6]}$$

$$y = -2$$

$$x = 3(-2) + 4(-1) + 11 \qquad \textbf{Substituting for } y \textbf{ and } z \textbf{ in [4]}$$

$$x = 1$$

The solution to the system of three linear equations is the ordered triple $(1, -2, -1)$. The reader can check the solution by substituting the values for x, y, and z into the three original equations to show that the ordered triple satisfies all three equations. □

EXAMPLE 2 Use the elimination method to solve the system of three linear equations in three unknowns:

$$x + y + z = -2 \qquad [1]$$

$$2x - y - z = 5 \qquad [2]$$

$$-x + 3y + 4z = -11. \qquad [3]$$

Solution: If we can eliminate one of the variables, then we will have a system of two equations in two unknowns.

Let us eliminate the x-terms first (we could just as well begin with either the y- or z-terms). Since the coefficient of the x-term in Eq. [1] is 1 and the coefficient of the x-term in Eq. [3] is -1 we can add the two equations to eliminate the x from the resulting equation:

$$\begin{array}{rl} x + y + z = -2 & [1] \\ -x + 3y + 4z = -11 & [3] \\ \hline 4y + 5z = -13. & \textbf{Adding} \qquad [4] \end{array}$$

Now produce another equation with the x-term eliminated by working with either of the unused combinations, Eqs. [1] and [2] or Eqs. [2] and [3]. Let us multiply Eq. [3] by 2 and add the result to Eq. [2].

$$\begin{array}{rl} 2x - y - z = 5 & [2] \\ -2x + 6y + 8z = -22 & 2 \times [3]. \\ \hline 5y + 7z = -17. & \textbf{Adding} \qquad [5] \end{array}$$

Equations [4] and [5] now form a system of two linear equations in two unknowns.

$$\begin{aligned}
4y + 5z &= -13 && [4]\\
5y + 7z &= -17 && [5]\\
20y + 25z &= -65 && 5 \times [4]\\
20y + 28z &= -68 && 4 \times [5]\\
\hline
-3z &= 3 && \textbf{Subtracting}\\
z &= -1 &&\\
5y + 7(-1) &= -17 && \textbf{Substituting for } z \textbf{ in [5]}\\
5y &= -10 &&\\
y &= -2 &&\\
-x + 3(-2) + 4(-1) &= -11 && \textbf{Substituting for } y \textbf{ and } z \textbf{ in [3]}\\
-x &= -1 &&\\
x &= 1. &&
\end{aligned}$$

The solution to the system of three linear equations in three unknowns is $x = 1, y = -2, z = -1$, or the ordered triple $(1, -2, -1)$. The check is left to the reader. □

EXAMPLE 3 Use the elimination method to solve the system of three linear equations in three unknowns that describes the loop currents in a circuit:

$$\begin{aligned}
2000\, i_1 - 1000\, i_2 &= 9\\
-1000\, i_1 + 3000\, i_2 - 1000\, i_3 &= 9\\
-1000\, i_2 + 3000\, i_3 &= 0.
\end{aligned}$$

Solution: As the first step of a solution it is good practice to write the equations of the system in an orderly fashion, placing like terms in similar positions and inserting zero-coefficients for missing terms:

$$\begin{aligned}
2000\, i_1 - 1000\, i_2 + \quad 0\, i_3 &= 9 && [1]\\
-1000\, i_1 + 3000\, i_2 - 1000\, i_3 &= 9 && [2]\\
0\, i_1 - 1000\, i_2 + 3000\, i_3 &= 0. && [3]
\end{aligned}$$

We observe that the term i_1 in Eq. [3] is missing so that if we eliminate the i_1 term from the other pair of equations we will have two equations in two unknowns.

$$\begin{aligned}
-2000\, i_1 + 6000\, i_2 - 2000\, i_3 &= 18 && 2 \times [2] && [4]\\
2000\, i_1 - 1000\, i_2 + \quad 0\, i_3 &= 9 && \textbf{Repeating [1]} &&\\
\hline
5000\, i_2 - 2000\, i_3 &= 27. && [1] + [4] && [5]
\end{aligned}$$

With this result we have a system of two equations in two unknowns (Eqs. [3] and [5]).

$$\begin{aligned}
-1000\, i_2 + 3000\, i_3 &= 0 && [3]\\
5000\, i_2 - 2000\, i_3 &= 27. && [5]
\end{aligned}$$

To eliminate i_2 from this set of equations we can multiply Eq. [3] by 5 and add the result to Eq. [5]:

$$\begin{array}{rll} -5000\, i_2 + 15{,}000\, i_3 = 0 & 5 \times [3] & [6] \\ 5000\, i_2 - 2{,}000\, i_3 = 27 & [5] & \\ \hline 13{,}000\, i_3 = 27 & [5] + [6] & \end{array}$$

$i_3 = 2.077 \times 10^{-3}$ **Solving for i_3**

$-5000\, i_2 + 15{,}000(2.077 \times 10^{-3}) = 0$

Substituting for i_3 in [6]

$i_2 = 6.231 \times 10^{-3}$

$-2000\, i_1 + 6000(6.231 \times 10^{-3}) - 2000(2.077 \times 10^{-3}) = 18$

Substituting for i_2 in [4]

$i_1 = 7.616 \times 10^{-3}$.

The solution to the system of equations is $i_1 = 7.62 \times 10^{-3}$ A, $i_2 = 6.23 \times 10^{-3}$ A, and $i_3 = 2.08 \times 10^{-3}$ A ($i_1 = 7.62$ mA, $i_2 = 6.23$ mA, and $i_3 = 2.08$ mA). The check is left to the reader. □

10.4 Exercises

In Exercises 1–20, use algebraic methods to solve the systems of three linear equations in three unknowns.

1. $6x - 7y = 6$
   $8x + 3y + 5z = 3$
   $-5x - 3y + 2z = -7$

2. $7x - z = -5$
   $4x + 2y - 5z = 2$
   $-2x + y = 0$

3. $7x + 4y = 8$
   $x + 3z = -5$
   $2x + y = 0$

4. $-2y - 3z = 7$
   $8x - 3y - 9z = 7$
   $-5x + 2y - 8z = 9$

5. $-2x + 5y + 8z = -2$
   $x + 4z = 4$
   $-x - y - z = 1$

6. $6x - 6y + 7z = 7$
   $-5y - z = 4$
   $-7x + 7y + 8z = 8$

7. $-8x - y - 8z = -1$
   $-4x - 8y + 3z = 3$
   $-2x - 3y + z = 8$

8. $-3x - 2y + 6z = 1$
   $3x + y + 4z = 8$
   $6x + y - 5z = 2$

9. $-8x + 4y + 5z = -6$
   $-3x + 9y + z = -8$
   $-3x - 4y + 7z = -7$

10. $-5x + 4y + 3z = -4$
    $5x + 3y + 3z = 1$
    $-3x - 2y - 2z = 4$

11. $2x + y - z = -1$
    $x - 3y - 3z = 2$
    $7x - 9y - 4z = -3$

12. $5x + 4y + 3z = 9$
    $-2x + y + 5z = -2$
    $2x - y + 6z = -9$

13. $3x + 7y + 8z = -6$
    $-2x - y - 9z = -7$
    $-3x + 5y + 6z = -4$

14. $-2x + y + 4x = 2$
    $5x - y + 8z = -2$
    $4x - y + 3z = -2$

15. $-2x - 3y - 4z = 3$
    $5x - 3y + 3z = 3$
    $5x - 4y - 9z = 4$

16. $2x - 3y - 4z = 3$
    $4x + y + 2z = -5$
    $3x - 2y - z = 2$

17. $2x - 2y + z = -2$
    $-x - 6y - 8z = 1$
    $6x - 6y + 7z = -6$

18. $-4x - 8y + 9z = -1$
    $7x + 4y - 8z = 4$
    $-4x - 3y + 4z = -1$

19. $-8x - 4y + 5z = -6$
    $-3x + 9y + z = -8$
    $-3x - 4y + 7z = -7$

20. $7x - 7y - z = -9$
    $-4x - 9y + z = -7$
    $-x + 4y - 4z = -4$

21. The loop equations for a circuit are

$$-100 + 40\,i_1 + 33\,1/3(i_1 - i_2) = 0$$
$$33\,1/3(i_2 - i_1) + 50(i_2 - i_3) = 0$$
$$50(i_3 - i_2) + 20\,i_3 = 0.$$

What are the loop currents i_1, i_2, and i_3?

22. The loop equations for a circuit are

$$-200 + 25\,i_1 + 300\,(i_1 - i_2) = 0$$
$$300(i_2 - i_1) + 80\,i_2 + 30(i_2 - i_3) = 0$$
$$30(i_3 - i_2) + 60\,i_3 = 0.$$

What are the loop currents i_1, i_2, and i_3?

23. The node equations for a circuit are

$$\frac{1}{4}(v_1 - 100) + \frac{1}{20}(v_1 - v_2) + \frac{1}{40}(v_1 - v_2) = 0$$
$$\frac{1}{20}(v_2 - v_1) + \frac{1}{40}(v_2 - v_1) + \frac{1}{6}(v_2 - v_3) = 0$$
$$\frac{1}{6}(v_3 - v_2) + \frac{1}{10}v_3 = 0.$$

What are the node voltages v_1, v_2, and v_3?

24. The node equations for a circuit are

$$\frac{1}{15}(v_1 - 100) + \frac{1}{3}(v_1 - v_2) + \frac{1}{4}(v_1 - v_3) = 0$$
$$\frac{1}{3}(v_2 - v_1) + \frac{1}{7}v_2 = 0$$
$$\frac{1}{4}(v_3 - v_1) + \frac{1}{9}v_3 + \frac{1}{18}v_3 = 0.$$

What are the node voltages v_1, v_2, and v_3?

25. The section forces in a truss are related by

$$500 + F_2 \cos 30° + F_3 \cos 30° + F_1 = 0$$
$$1000 + F_3 \sin 30° - F_2 \sin 30° = 0$$
$$-10(1000) + (10 \tan 30°)(F_2 \cos 30°) + (10 \tan 30°)(F_3 \cos 30°) = 0.$$

What are forces F_1, F_2, and F_3?

26. The section forces in a truss are related by

$$500 + F_1 \cos 30° + F_3 = 0$$
$$1000 + F_1 \sin 30° + F_2 - 500 = 0$$
$$-10(1500) + (10 \tan 30°)(F_1 \cos 30°) + 10F_2 = 0$$

What are the forces F_1, F_2, and F_3?

27. An alloy is composed of three metals, A, B, and C. The percentage of metal A present is twice the percentage of metal C and the percentage of metal B present is triple the percentage of metal A. What are the percentages of the individual metals?

28. A compound is made up of parts x, y, and z. The proportion of part x to part y is 5 to 1 and the proportion of part y to part z is 3 to 1. What percentage of the total is furnished by each part?

29. A nut, bolt, and washer assembly has a mass of 0.044 kg. Five nuts and bolts have a mass of 0.200 and ten assemblies of a bolt with double washers and double nuts has a mass of 0.560 kg. What are the masses of the individual nuts, bolts, and washers?

30. Design the height dimensions for three supports such that S_3 is twice as tall as S_1, the height of S_2 is 3 m more than S_1, and the height of S_3 is 5 m more than S_2.

10.5 Solving three linear equations in three unknowns by determinants

Cramer's rule can be used to solve systems of three linear equations in three unknowns. After arranging the unknowns in order on the left of the equal sign and the constants on the right, the determinants for the denominator and numerators of the solution are formed in similar patterns as for systems of two linear equations (Section 10.3). Before we can solve a system of three equations in three unknowns we need to know how to evaluate a 3×3 determinant.

The technique for evaluating a 2×2 determinant cannot be used to evaluate a 3×3 determinant. The procedure shown, called **expansion by cofactors,** is a method of reducing the 3×3 determinant to a series of 2×2 determinants, which are then evaluated by Definition 10.4.

Consider the 3×3 determinant

$$A = \begin{vmatrix} a_{11} & a_{12} & a_{13} \\ a_{21} & a_{22} & a_{23} \\ a_{31} & a_{32} & a_{33} \end{vmatrix}$$

The **minor,** M_{ij}, of element a_{ij} is the determinant formed by the remaining elements when the ith row and the jth column are eliminated. The **cofactor,** A_{ij}, of element a_{ij} is the product $(-1)^{i+j} \times M_{ij}$.

EXAMPLE 1 Given the 3×3 determinant

$$\begin{vmatrix} 2 & -6 & -2 \\ -5 & -3 & 1 \\ -1 & 4 & 3 \end{vmatrix}$$

determine

a. minor M_{13} **b.** cofactor A_{13}

c. minor M_{32} **d.** cofactor A_{32}

Solution:

$$\begin{matrix} 2 & -6 & -2 \\ -5 & -3 & 1 \\ -1 & 4 & 1 \end{matrix}$$

FIGURE 10.25

a. To determine minor M_{13} we eliminate the first row and the third column as in Fig. 10.25. The minor is the determinant of the remaining elements.

$$\begin{aligned} M_{13} &= \begin{vmatrix} -5 & -3 \\ -1 & 4 \end{vmatrix} \\ &= (-5)(4) - (-1)(-3) \\ &= -23. \end{aligned}$$

b. The cofactor

$$\begin{aligned} A_{ij} &= (-1)^{i+j} \times M_{ij} \\ A_{13} &= (-1)^{1+3} \times M_{13} \\ &= (-1)^4(-23) \\ &= -23. \end{aligned}$$

c. The minor M_{32} is the determinant of the remaining elements after we eliminate the third row and the second column as in Fig. 10.26.

FIGURE 10.26

$$\begin{matrix} 2 & -6 & -2 \\ -5 & -3 & 1 \\ -1 & 4 & 3 \end{matrix}$$

$$\begin{aligned} M_{32} &= \begin{vmatrix} 2 & -2 \\ -5 & 1 \end{vmatrix} \\ &= (2)(1) - (-5)(-2) \\ &= -8. \end{aligned}$$

d. The cofactor

$$\begin{aligned} A_{32} &= (-1)^{3+2} \times M_{32} \\ &= (-1)^5(-8) \\ &= 8. \quad \square \end{aligned}$$

DEFINITION 10.6

MINOR OF a_{ij}

If A is any $n \times n$ determinant, and M_{ij} is the $(n-1) \times (n-1)$ determinant formed from A by deleting the ith row and the jth column of A, then M_{ij} is called the minor of the element a_{ij} of A.

DEFINITION 10.7

COFACTOR OF a_{ij}

If A is any $n \times n$ determinant and M_{ij} is the minor of the element a_{ij} of A, then

$$A_{ij} = (-1)^{i+j} \times M_{ij}$$

is called the cofactor of the element a_{ij} of A.

Any determinant can be evaluated using expansion by cofactors. To evaluate a determinant by cofactors, any row or any column may be selected. Each element in that row or column is multiplied by its own cofactor and the sum of the products so obtained is the value of the determinant.

DEFINITION 10.8

To evaluate an $n \times n$ determinant A by expansion by cofactors using the

$$i\text{th row: } A = a_{i1}A_{i1} + a_{i2}A_{i2} + \cdots + a_{in}A_{in}.$$

$$j\text{th column: } A = a_{1j}A_{1j} + a_{2j}A_{2j} + \cdots + a_{nj}A_{nj}.$$

EXAMPLE 2 Evaluate

$$\begin{vmatrix} 2 & -6 & -2 \\ -5 & -3 & 1 \\ -1 & 4 & 3 \end{vmatrix}$$

using expansion by cofactors.

a. Expanding about the first row.

b. Expanding about the second column.

Solution:

a. Evaluating the 3×3 determinant $\begin{vmatrix} 2 & -6 & -2 \\ -5 & -3 & 1 \\ -1 & 4 & 3 \end{vmatrix}$ by expanding about the first row we have, from Definition 10.8

$$\begin{aligned}
A &= a_{i1}A_{i1} + a_{i2}A_{i2} + \cdots + a_{in}A_{in}, \ (n = 3,\ i = 1) \\
&= a_{11} \cdot A_{11} + a_{12} \cdot A_{12} + a_{13} \cdot A_{13} \\
&= (2)\left[(-1)^{1+1}\begin{vmatrix} -3 & 1 \\ 4 & 3 \end{vmatrix}\right] + (-6)\left[(-1)^{1+2}\begin{vmatrix} -5 & 1 \\ -1 & 3 \end{vmatrix}\right] \\
&\quad + (-2)\left[(-1)^{1+3}\begin{vmatrix} -5 & -3 \\ -1 & 4 \end{vmatrix}\right] \\
&= (2)[(-1)^2(-9-4)] + (-6)[(-1)^3(-15-(-1))] \\
&\quad +(-2)[(-1)^4(-20-3)] \\
&= (2)[(1)(-13)] + (-6)[(-1)(-14)] + (-2)[(1)(-23)] \\
&= (2)[-13] + (-6)[14] + (-2)[-23] \\
&= -26 - 84 + 46 \\
&= -64.
\end{aligned}$$

Therefore $\begin{vmatrix} 2 & -6 & -2 \\ -5 & -3 & 1 \\ -1 & 4 & 3 \end{vmatrix} = -64.$

b. Expanding about the second column

$$\begin{aligned}
A &= a_{1j}A_{1j} + a_{2j}A_{2j} + \cdots + a_{nj}A_{nj}, \ (n = 3,\ j = 2) \\
&= a_{12} \cdot A_{12} + a_{22} \cdot A_{22} + a_{32} \cdot A_{32} \\
&= (-6)\left[(-1)^{1+2}\begin{vmatrix} -5 & 1 \\ -1 & 3 \end{vmatrix}\right] + (-3)\left[(-1)^{2+2}\begin{vmatrix} 2 & -2 \\ -1 & 3 \end{vmatrix}\right] \\
&\quad + (4)\left[(-1)^{3+2}\begin{vmatrix} 2 & -2 \\ -5 & 1 \end{vmatrix}\right] \\
&= (-6)[(-1)(-14)] + (-3)[(1)(4)] + (4)[(-1)(-8)] \\
&= (-6)[14] + (-3)[4] + (4)[8] \\
&= -84 - 12 + 32 \\
&= -64. \quad \square
\end{aligned}$$

The value of a given determinant is the same no matter what row or what column we choose to use for the expansion.

As mentioned at the beginning of this section, Cramer's rule can also be used to solve systems of three linear equations in three unknowns.

DEFINITION 10.9

CRAMER'S RULE

The solution to a system of three linear equations in three unknowns with the equations in standard form

$$\begin{aligned} a_{11}x + a_{12}y + a_{13}z &= c_1 \\ a_{21}x + a_{22}y + a_{23}z &= c_2 \\ a_{31}x + a_{32}y + a_{33}z &= c_3 \end{aligned}$$

is given by

$$x = \frac{N_x}{D}, \qquad y = \frac{N_y}{D}, \qquad z = \frac{N_z}{D},$$

where

$$D = \begin{vmatrix} a_{11} & a_{12} & a_{13} \\ a_{21} & a_{22} & a_{23} \\ a_{31} & a_{32} & a_{33} \end{vmatrix}, \qquad N_x = \begin{vmatrix} c_1 & a_{12} & a_{13} \\ c_2 & a_{22} & a_{23} \\ c_3 & a_{32} & a_{33} \end{vmatrix},$$

$$N_y = \begin{vmatrix} a_{11} & c_1 & a_{13} \\ a_{21} & c_2 & a_{23} \\ a_{31} & c_3 & a_{33} \end{vmatrix}, \qquad N_z = \begin{vmatrix} a_{11} & a_{12} & c_1 \\ a_{21} & a_{22} & c_2 \\ a_{31} & a_{32} & c_3 \end{vmatrix},$$

and $D \neq 0$.

EXAMPLE 3 Solve the system of three linear equations in three unknowns using determinants (Cramer's rule):

$$\begin{aligned} x + y + z &= -2 \\ 2x - y - z &= 5 \\ -x + 3y + 4z &= -11. \end{aligned}$$

Solution: Applying Cramer's rule to the system of three equations in three unknowns we have $x = \frac{N_x}{D}$, $y = \frac{N_y}{D}$, and $z = \frac{N_z}{D}$;

$$D = \begin{vmatrix} 1 & 1 & 1 \\ 2 & -1 & -1 \\ -1 & 3 & 4 \end{vmatrix},$$ the determinant of the coefficients.

Expanding about the first row:

$$\begin{aligned} D &= (1)\left[(-1)^{1+1}\begin{vmatrix} -1 & -1 \\ 3 & 4 \end{vmatrix}\right] + (1)\left[(-1)^{1+2}\begin{vmatrix} 2 & -1 \\ -1 & 4 \end{vmatrix}\right] + (1)\left[(-1)^{1+3}\begin{vmatrix} 2 & -1 \\ -1 & 3 \end{vmatrix}\right] \\ &= (1)[(1)(-1)] + (1)[(-1)(7)] + (1)[(1)(5)] \\ &= -1 - 7 + 5 \\ &= -3. \end{aligned}$$

$$N_x = \begin{vmatrix} -2 & 1 & 1 \\ 5 & -1 & -1 \\ -11 & 3 & 4 \end{vmatrix}$$ **D, with the x-column replaced by the column of constants**

Expanding about the first row:

$$\begin{aligned} N_x &= (-2)\left[(-1)^{1+1}\begin{vmatrix} -1 & -1 \\ 3 & 4 \end{vmatrix}\right] + (1)\left[(-1)^{1+2}\begin{vmatrix} 5 & -1 \\ -11 & 4 \end{vmatrix}\right] + (1)\left[(-1)^{1+3}\begin{vmatrix} 5 & -1 \\ -11 & 3 \end{vmatrix}\right] \\ &= (-2)[(1)(-1)] + (1)[(-1)(9)] + (1)[(1)(4)] \\ &= 2 - 9 + 4 \\ &= -3. \end{aligned}$$

$$N_y = \begin{vmatrix} 1 & -2 & 1 \\ 2 & 5 & -1 \\ -1 & -11 & 4 \end{vmatrix}$$ **D, with the y-column replaced by the column of constants**

Expanding about the first row:

$$\begin{aligned} N_y &= (1)\left[(-1)^{1+1}\begin{vmatrix} 5 & -1 \\ -11 & 4 \end{vmatrix}\right] + (-2)\left[(-1)^{1+2}\begin{vmatrix} 2 & -1 \\ -1 & 4 \end{vmatrix}\right] + (1)\left[(-1)^{1+3}\begin{vmatrix} 2 & 5 \\ -1 & -11 \end{vmatrix}\right] \\ &= (1)[(1)(9)] + (-2)[(-1)(7)] + (1)[(1)(-17)] \\ &= 9 + 14 - 17 \\ &= 6. \end{aligned}$$

$$N_z = \begin{vmatrix} 1 & 1 & -2 \\ 2 & -1 & 5 \\ -1 & 3 & -11 \end{vmatrix}$$ **D, with the z-column replaced by the column of constants**

Expanding about the first row:

$$\begin{aligned} N_z &= (-1)\left[(-1)^{1+1}\begin{vmatrix} -1 & 5 \\ 3 & -11 \end{vmatrix}\right] + (1)\left[(-1)^{1+2}\begin{vmatrix} 2 & 5 \\ -1 & -11 \end{vmatrix}\right] + (-2)\left[(-1)^{1+3}\begin{vmatrix} 2 & -1 \\ -1 & 3 \end{vmatrix}\right] \\ &= (1)[(1)(-4)] + (1)[(-1)(-17)] + (-2)[(1)(5)] \\ &= -4 + 17 - 10 \\ &= 3. \end{aligned}$$

$$x = \frac{N_x}{D} = \frac{-3}{-3} = 1. \qquad y = \frac{N_y}{D} = \frac{6}{-3} = -2. \qquad z = \frac{N_z}{D} = \frac{3}{-3} = -1.$$

The solution to the system of three linear equations in three unknowns is the ordered triple $(1, -2, -1)$. The check is left to the reader. □

10.5 Exercises

In Exercises 1–12, evaluate the 3-X-3 determinant using expansion by cofactors.

1. $\begin{vmatrix} -8 & -5 & 5 \\ -6 & -3 & 5 \\ -9 & 5 & 4 \end{vmatrix}$

2. $\begin{vmatrix} 4 & 4 & -3 \\ 0 & -5 & -2 \\ 4 & -2 & -3 \end{vmatrix}$

3. $\begin{vmatrix} -1 & -7 & -3 \\ -1 & 4 & -8 \\ 2 & 1 & -2 \end{vmatrix}$

4. $\begin{vmatrix} -2 & 5 & 6 \\ 1 & 2 & 1 \\ 0 & -2 & -1 \end{vmatrix}$

5. $\begin{vmatrix} 0 & 0 & -1 \\ 2 & 4 & 1 \\ -1 & 6 & 2 \end{vmatrix}$

6. $\begin{vmatrix} 2 & -3 & 0 \\ -1 & -2 & 0 \\ -3 & 4 & 5 \end{vmatrix}$

7. $\begin{vmatrix} 5 & 3 & -2 \\ 9 & -6 & -7 \\ 7 & -9 & 8 \end{vmatrix}$

8. $\begin{vmatrix} 9 & 7 & 0 \\ 5 & -2 & -7 \\ 2 & 6 & -3 \end{vmatrix}$

9. $\begin{vmatrix} -8 & -7 & -2 \\ -5 & -2 & 0 \\ -8 & -3 & 4 \end{vmatrix}$

10. $\begin{vmatrix} -4 & -3 & -1 \\ -7 & 3 & 4 \\ 0 & 8 & -3 \end{vmatrix}$

11. $\begin{vmatrix} -9 & -1 & 1 \\ 2 & 1 & 2 \\ -3 & 8 & 0 \end{vmatrix}$

12. $\begin{vmatrix} 2 & -3 & -2 \\ -8 & 0 & -5 \\ -1 & -7 & 6 \end{vmatrix}$

In Exercises 13–33, use Cramer's rule to solve the systems of equations.

13. $7x + 4y = 8$
$x + 3z = -5$
$2x + y = 0$

14. $-2y - 3x = 7$
$8x - 3y - 9z = 7$
$-5x + 2y - 8z = 9$

15. $6x - 7y = 6$
$8x + 3y + 5z = 3$
$-5x - 3y + 2z = -7$

16. $7x - z = -5$
$4x + 2y - 5z = 2$
$-2x + y = 0$

17. $-2x + 5y + 8z = -2$
$x + 4z = 4$
$-x - y - z = 1$

18. $6x - 6y + 7z = 7$
$-5y - z = 4$
$-7x + 7y + 8z = 8$

19. $-8x - y - 8z = -1$
$-4x - 8y + 3z = 3$
$-2x - 3y + z = 8$

20. $-3x - 2y + 6z = 1$
$3x + y + 4z = 8$
$6x + y - 5z = 2$

21. $-8x + 4y + 5z = -6$
$-3x + 9y + z = -8$
$-3x - 4y + 7z = -7$

22. $-5x + 4y + 3z = -4$
$5x + 3y + 3z = 1$
$-3x - 2y - 2z = 4$

23. $2x + y - z = -1$
$x - 3y - 3z = 2$
$7x - 9y - 4z = -3$

24. $5x + 4y + 3z = 9$
$-2x + y + 5z = -2$
$2x - y + 6z = -9$

25. $3x + 7y + 8z = -6$
$-2x - y - 9z = -7$
$-3x + 5y + 6z = -4$

26. $-2x + y + 4z = 2$
$5x - y + 8z = -2$
$4x - y + 3z = -2$

27. A network of guy wires (Fig. 10.27) is used to support the 50 kg mass until it can finally be anchored in position. The forces in the wires are described by the system of equations.

$$0.392\ T_1 + 0.314\ T_2 + 0.500\ T_3 - 491. = 0$$
$$0.235\ T_1 - 0.628\ T_2 = 0$$
$$-0.889\ T_1 - 0.712\ T_2 + 0.866\ T_3 = 0$$

Find the forces in the wires.

28. A chemical compound is composed of three elements, *A*, *B*, and *C*. The percentage of element *A* present is triple the percentage of element *C* and the percentage of element *B* present is twice the percentage of element *A*. What are the percentages of the individual elements?

FIGURE 10.27

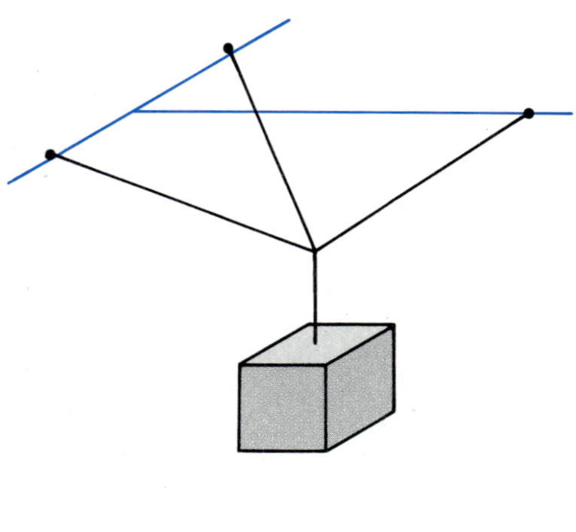

29. A mixture is made up of parts x, y, and z. The proportion of part x to part y is 4 to 1 and the proportion of part y to part z is 5 to 1. What percentage of the total is furnished by each part?

30. An assembly using three different sizes of bolts, B_1, B_2, and B_3 is capable of supporting 120,000 N. The assembly may be made up of combinations of $2B_1$, $3B_2$, and $3B_3$ or $3B_1$, $2B_2$, and $2B_3$ or of $2B_1$, $2B_2$, and $5B_3$. What are the load ratings of the individual bolts?

31. A medical treatment requires a total dosage of 16 ml of medicine. It can be given as one massive dose followed by ten small doses or as one massive dose followed by three intermediate doses and four small doses or as five intermediate doses followed by six small doses. What size are the doses?

32. A 15-m length of 0.1-m-×-0.1-m steel angle is to be cut into three lengths. The first length is 1 m longer than the second length and is 2 m shorter than twice the third length. What are the individual lengths?

33. A car with an adult passenger and a full fuel tank weighs 2792 lb. The same car with the same adult passenger and a half tank of fuel weighs 2522 lb. The car with the adult, one child and a fifth of a tank of fuel weighs 2410 lb. The child weighs one quarter as much as the adult. What is the weight of the car, the full gas tank, the adult, and the child?

10.6 Matrix operations

In working with fluid-flow problems, electrical-network problems, or structural-truss problems, it is often necessary to solve systems of simultaneous equations that are larger than 3×3. It can be seen from the preceding sections that when we increase the number of simultaneous equations to be solved from two to three, the number of steps required for the solution increases dramatically. This increase is even more dramatic as the number of equations increases. Using Cramer's rule a set of two equations requires 8 multiplications or divisions for a solution, a set of three equations requires 27, a set of four requires 124, a set of five requires 725, and a set of ten equations requires 39,916,810 multiplications and divisions! To obtain a faster method for solving larger sets of simultaneous equations we turn to matrix methods. (It requires more operations to solve 2×2 or 3×3 systems of equations using matrix methods than with Cramer's rule; 4×4 systems require about the same number of operations, and a set of 10 simultaneous equations requires only about 2100 multiplications and divisions using matrix methods.) In this section we will introduce some elementary matrix operations and in the next section we will develop a method for finding the inverse of a matrix and solving simultaneous equations.

A **matrix** (plural: matrices) is a rectangular array of numbers (called elements) arranged in rows and columns. It may consist of a single element, a single row, a single column, equal numbers of rows and columns, or unequal numbers of rows and columns. An $m \times n$ matrix (or a matrix of order $m \times n$) has m rows and n columns. Table 10.4 contains examples of matrices.

TABLE 10.4

Matrix	Order	Special name
(27)	1×1	Single element
$(x \quad y \quad z)$	1×3	Row matrix (or row vector)
$\begin{pmatrix} 5 \\ -3 \\ 0 \end{pmatrix}$	3×1	Column matrix (or column vector)
$\begin{pmatrix} 1 & -6 & 3 \\ -2 & 1 & 4 \\ 7 & 5 & 8 \end{pmatrix}$	3×3	Square matrix
$\begin{pmatrix} 1 & a & -7 \\ b & d & 5 \end{pmatrix}$	2×3	

As in determinants, an element of a matrix may be denoted using subscript notation a_{ij}, where i represents the row in which the element is located and j represents the column. The element a_{ij} lies at the intersection of the ith row and the jth column. (A square matrix and the determinant consisting of the same array of elements are not the same. The determinant assigns a single value to the array, whereas the matrix makes no such attempt.)

For two matrices to be equal they must be of the same order and each element of one must equal the corresponding element of the other.

EXAMPLE 1 Are the following matrices equal?

a. $A = (7 \quad 9)$, $B = \begin{pmatrix} 7 \\ 9 \end{pmatrix}$

b. $A = \begin{pmatrix} 4 & -6 & 3 \\ 2 & 1 & -5 \end{pmatrix}$, $B = \begin{pmatrix} 4 & -6 & 3 \\ 2 & -1 & -5 \end{pmatrix}$

c. $A = \begin{pmatrix} 1 & 2 \\ a & 3 \\ 4 & b \end{pmatrix}$, $B = \begin{pmatrix} c & 2 \\ 5 & 3 \\ d & 6 \end{pmatrix}$

Solution:

a. $(7 \quad 9) \neq \begin{pmatrix} 7 \\ 9 \end{pmatrix}$

The matrices are not of the same order. A is a 1×2 matrix while B is a 2×1 matrix.

b. $\begin{pmatrix} 4 & -6 & 3 \\ 2 & 1 & -5 \end{pmatrix} \neq \begin{pmatrix} 4 & -6 & 3 \\ 2 & -1 & -5 \end{pmatrix}$

The matrices are both of order 2×3, but the elements of A are not equal to the corresponding elements of B ($a_{22} \neq b_{22}$).

c. $\begin{pmatrix} 1 & 2 \\ a & 3 \\ 4 & b \end{pmatrix} \stackrel{?}{=} \begin{pmatrix} c & 2 \\ 5 & 3 \\ d & 6 \end{pmatrix}$

Matrix A will be equal to matrix B if $a = 5$ and $b = 6$ and $c = 1$ and $d = 4$. If all of these conditions are not met, the matrices are not equal. □

To add two matrices A and B, they must be of the same order. If they are, then the sum matrix C is also of that order and each element of C is the sum of the corresponding elements of A and B. (In other words, $c_{ij} = a_{ij} + b_{ij}$ for all possible combinations of $1 \leq i \leq m$ and $1 \leq j \leq n$.)

EXAMPLE 2 If possible, add the following matrices.

a. $\begin{pmatrix} 1 & 4 \\ 5 & -3 \end{pmatrix} + \begin{pmatrix} 2 \\ -1 \end{pmatrix}$ **b.** $\begin{pmatrix} 1 & -5 \\ 2 & 3 \\ -1 & 6 \end{pmatrix} + \begin{pmatrix} 2 & 1 \\ -5 & 0 \\ 7 & 3 \end{pmatrix}$

Solution:

a. $\begin{pmatrix} 1 & 4 \\ 5 & -3 \end{pmatrix} + \begin{pmatrix} 2 \\ -1 \end{pmatrix}$

Cannot be added. The matrices are not of the same order.

b.
$$\begin{pmatrix} 1 & -5 \\ 2 & 3 \\ -1 & 6 \end{pmatrix} + \begin{pmatrix} 2 & 1 \\ -5 & 0 \\ 7 & 3 \end{pmatrix} = \begin{pmatrix} 1+2 & -5+1 \\ 2+(-5) & 3+0 \\ -1+7 & 6+3 \end{pmatrix}$$
$$= \begin{pmatrix} 3 & -4 \\ -3 & 3 \\ 6 & 9 \end{pmatrix}. \quad \square$$

The negative of a matrix is formed by multiplying each element of the matrix by -1. Matrices are subtracted by adding the negative matrix. A matrix may be multiplied by a scalar by multiplying every element of the matrix by the scalar.

EXAMPLE 3 Let $A = \begin{pmatrix} 1 & 0 \\ 4 & 2 \end{pmatrix}$ and $B = \begin{pmatrix} -1 & 3 \\ 5 & 6 \end{pmatrix}$

Find

a. $2A$, **b.** $3A - 4B$

Solution:

a. To find $2A$ we multiply every element of A by 2.

$$2A = \begin{pmatrix} 2(1) & 2(0) \\ 2(4) & 2(2) \end{pmatrix} \quad \text{or} \quad \begin{pmatrix} 2 & 0 \\ 8 & 4 \end{pmatrix}$$

b. To find $3A - 4B$ we use the definition of subtraction and write $3A - 4B$ as $3A + (-4B)$.

$$\begin{aligned} 3A + (-4B) &= \begin{pmatrix} 3(1) & 3(0) \\ 3(4) & 3(2) \end{pmatrix} + \begin{pmatrix} -4(-1) & -4(3) \\ -4(5) & -4(6) \end{pmatrix} \\ &= \begin{pmatrix} 3 & 0 \\ 12 & 6 \end{pmatrix} + \begin{pmatrix} 4 & -12 \\ -20 & -24 \end{pmatrix} \\ &= \begin{pmatrix} 3+4 & 0-12 \\ 12-20 & 6-24 \end{pmatrix} \\ &= \begin{pmatrix} 7 & -12 \\ -8 & -18 \end{pmatrix}. \end{aligned}$$ □

In order to multiply two matrices the number of columns of the first must be the same as the number of rows of the second. The result of multiplying two matrices is a matrix that has the number of rows of the first and the number of columns of the second. For example if A is a 3×4 matrix and B is a 4×2 matrix we can symbolically write

$$3 \times 4 | 4 \times 2.$$

The upper bracket indicates that A has four columns and B has four rows so it is possible to multiply A times B. The lower bracket indicates that the product AB will be a 3×2 matrix.

In general we can multiply two matrices of order $m \times n$ and $n \times p$, respectively. (The number of columns of the first, n, is equal to the number of rows of the second.) The product matrix will be of order $m \times p$.

EXAMPLE 4 Given the row vector (matrix) $A = (1 \quad 2 \quad 3)$ and the column vector (matrix) $B = \begin{pmatrix} 4 \\ 5 \\ 6 \end{pmatrix}$ find

a. AB and **b.** BA.

Solution:

a. $AB = (1 \quad 2 \quad 3)\begin{pmatrix} 4 \\ 5 \\ 6 \end{pmatrix}$

$1 \times 3 | 3 \times 1$ **Order of matrices**

The dimensions of the matrices tell us we can multiply them and the product will be a 1×1 matrix.

$$(1 \quad 2 \quad 3)\begin{pmatrix}4\\5\\6\end{pmatrix} = (1(4) + 2(5) + 3(6))$$
$$= (4 + 10 + 18)$$
$$= (32).$$

Therefore $AB = (32)$.

b. $BA = \begin{pmatrix}4\\5\\6\end{pmatrix}(1 \quad 2 \quad 3)$

$3 \times 1|1 \times 3$ **Order of matrices**

The dimensions of the matrices tell us that we can multiply them and the product will be a 3×3 matrix.

$$\begin{pmatrix}4\\5\\6\end{pmatrix}(1 \quad 2 \quad 3) = \begin{pmatrix}4(1) & 4(2) & 4(3)\\ 5(1) & 5(2) & 5(3)\\ 6(1) & 6(2) & 6(3)\end{pmatrix}$$
$$= \begin{pmatrix}4 & 8 & 12\\ 5 & 10 & 15\\ 6 & 12 & 18\end{pmatrix}.$$

Therefore $BA = \begin{pmatrix}4 & 8 & 12\\ 5 & 10 & 15\\ 6 & 12 & 18\end{pmatrix}$. □

We can see from Example 4 that $AB \neq BA$. In general, matrix multiplication is not commutative.

EXAMPLE 5 Find the product: $\begin{pmatrix}1 & 3\\ 2 & -1\\ 0 & 9\end{pmatrix}\begin{pmatrix}4 & 7 & -3 & -2\\ 5 & -1 & 0 & 8\end{pmatrix}$

Solution: Since the matrices are of order 3×2 and 2×4 we can multiply them and their product will be a 3×4 matrix.

$$\begin{pmatrix}1 & 3\\ 2 & -1\\ 0 & 9\end{pmatrix}\begin{pmatrix}4 & 7 & -3 & -2\\ 5 & -1 & 0 & 8\end{pmatrix} = \begin{pmatrix}1(4)+3(5) & 1(7)+3(-1) & 1(-3)+3(0) & 1(-2)+3(8)\\ 2(4)+(-1)(5) & 2(7)+-1(-1) & 2(-3)+(-1)(0) & 2(-2)+-1(8)\\ 0(4)+9(5) & 0(7)+9(-1) & 0(-3)+9(0) & 0(-2)+9(8)\end{pmatrix}$$
$$= \begin{pmatrix}19 & 4 & -3 & 22\\ 3 & 15 & -6 & -12\\ 45 & -9 & 0 & 72\end{pmatrix}.$$

Note that the element in the first-row, first-column position of the result is formed by multiplying the first row of the first matrix term by term times the first column of the second matrix and adding the results. The second-row, third-column element is formed by multiplying the second row of the first matrix times the third column of the second. Each element of the product is formed in this way. Therefore

$$\begin{pmatrix} 1 & 3 \\ 2 & -1 \\ 0 & 9 \end{pmatrix}\begin{pmatrix} 4 & 7 & -3 & -2 \\ 5 & -1 & 0 & 8 \end{pmatrix} = \begin{pmatrix} 19 & 4 & -3 & 22 \\ 3 & 15 & -6 & -12 \\ 45 & -9 & 0 & 72 \end{pmatrix}.$$

Note also that it is impossible to multiply the matrices in the reverse order. □

10.6 Exercises

In Exercises 1–4, what must be done to make the matrices equal?

1. $\begin{pmatrix} 1 & 2 \\ 0 & -1 \end{pmatrix} = \begin{pmatrix} a & b \\ d & c \end{pmatrix}$

2. $\begin{pmatrix} 1 & -6 & 3 \\ 2 & 0 & -2 \end{pmatrix} = \begin{pmatrix} a & d & f \\ c & b & e \end{pmatrix}$

3. $\begin{pmatrix} 1 \\ a \\ 4 \end{pmatrix} = \begin{pmatrix} a \\ x \\ 4 \end{pmatrix}$

4. $(5 \quad 1 \quad b) = (b \quad c \quad b)$

In Exercises 5–12, state the order of the indicated matrix.

$A = (1 \quad 0 \quad -3)$, $B = (4 \quad -1 \quad 2)$, $C = \begin{pmatrix} 1 & 2 \\ 4 & 3 \end{pmatrix}$,

$D = \begin{pmatrix} -3 & 1 \\ 6 & 5 \end{pmatrix}$, $E = \begin{pmatrix} -2 & 6 & 3 \\ 1 & 0 & -1 \\ 5 & 8 & 4 \end{pmatrix}$,

$F = \begin{pmatrix} 9 & 0 & -7 \\ 2 & -2 & 0 \\ 1 & 6 & 5 \end{pmatrix}$, $G = \begin{pmatrix} 2 & 6 \\ 1 & 3 \\ -1 & 5 \end{pmatrix}$,

$H = \begin{pmatrix} 0 & -6 \\ -1 & 0 \\ 3 & 8 \end{pmatrix}$

5. A 6. B 7. D 8. C

9. E 10. F 11. H 12. G

In Exercises 13–20, perform the operations using the matrices used in exercises 5–12.

13. $E + F$ 14. $G + H$ 15. $H - G$

16. $F - E$ 17. $5A + 2B$ 18. $3A - 2B$

19. $2D - 3C$ 20. $4C + 2D$

In Exercises 21–30, a) state the order of each matrix, b) multiply the matrices in the order given (if possible), and c) multiply the matrices in the opposite order (if possible).

21. $\begin{pmatrix} 2 & 4 \\ 1 & 3 \end{pmatrix}\begin{pmatrix} 5 & 6 \\ 0 & 9 \end{pmatrix}$

22. $\begin{pmatrix} -4 & 6 \\ 2 & -1 \end{pmatrix}\begin{pmatrix} 0 & 3 \\ 5 & 7 \end{pmatrix}$

23. $(1 \quad 0 \quad 3)\begin{pmatrix} 3 & 2 \\ 1 & 4 \\ 5 & 7 \end{pmatrix}$

24. $\begin{pmatrix} 1 & 0 & 3 \\ 3 & 1 & 5 \end{pmatrix}\begin{pmatrix} 2 \\ 4 \\ 7 \end{pmatrix}$

25. $\begin{pmatrix} 2 & 8 & 7 \\ 6 & 2 & 1 \end{pmatrix}\begin{pmatrix} -1 & 3 \\ 5 & 4 \\ 9 & 0 \end{pmatrix}$

26. $\begin{pmatrix} 2 & -3 \\ 6 & 1 \\ -4 & 5 \end{pmatrix}\begin{pmatrix} 8 & 0 & 3 \\ -1 & 5 & 9 \end{pmatrix}$

27. $(3 \quad 0 \quad 8)\begin{pmatrix} 1 & 2 & 3 \\ -5 & 0 & 4 \\ 2 & 6 & -2 \end{pmatrix}$

28. $\begin{pmatrix} 2 & -4 & 5 \\ 3 & 6 & -2 \\ -5 & 1 & 1 \end{pmatrix}\begin{pmatrix} 9 \\ 2 \\ -1 \end{pmatrix}$

29. $\begin{pmatrix} 1 & 2 & 3 \\ -2 & 1 & 0 \\ 5 & -4 & 6 \end{pmatrix}\begin{pmatrix} x \\ y \\ z \end{pmatrix}$

30. $\begin{pmatrix} 3 & 4 & -2 \\ 1 & -3 & 5 \\ 2 & -1 & 6 \end{pmatrix}\begin{pmatrix} a \\ b \\ c \end{pmatrix}$

10.7 The inverse matrix: simultaneous equations

In the previous section it was stated that matrix methods could be used to solve simultaneous equations. To see that this is true we must first write the system of equations in matrix form. Consider the system of simultaneous equations

$$\begin{aligned} 2x - y - 2z &= -6 \\ 4x + 2y + z &= 1 \\ 2x + y - z &= -1 \end{aligned}$$

The system of equations can be written as a matrix equation $AX = C$, where

$$A = \begin{pmatrix} 2 & -1 & -2 \\ 4 & 2 & 1 \\ 2 & 1 & -1 \end{pmatrix}, \quad X = \begin{pmatrix} x \\ y \\ z \end{pmatrix}, \quad \text{and} \quad C = \begin{pmatrix} -6 \\ 1 \\ -1 \end{pmatrix}.$$

Substituting, we have

$$\begin{pmatrix} 2 & -1 & -2 \\ 4 & 2 & 1 \\ 2 & 1 & -1 \end{pmatrix} \begin{pmatrix} x \\ y \\ z \end{pmatrix} = \begin{pmatrix} -6 \\ 1 \\ -1 \end{pmatrix}.$$

We see that matrix A consists of the coefficients of the unknowns of the equations. The column vectors X and C are matrices composed of the unknowns and of the constants, respectively. (The reader is urged to multiply A and X to verify that $AX = C$.)

If it were possible to divide both sides of the equation by A, we would obtain the matrix of unknowns X (equal to a matrix of constants). Unfortunately matrix division is not defined. Instead we have to use an alternate approach. *First we define the* **multiplicative identity matrix *I*** *as a square matrix having ones as the elements on the principal diagonal (from upper left to lower right) and zeroes as all the remaining elements.* Thus, for the set of 3×3 matrices the identity matrix I is

$$I = \begin{pmatrix} 1 & 0 & 0 \\ 0 & 1 & 0 \\ 0 & 0 & 1 \end{pmatrix}.$$

The reader can demonstrate that for any 3×3 matrix A, $AI = A$ and $IA = A$. If a system of simultaneous equations has a solution, it is possible to find an inverse matrix A^{-1}, which has the property $A^{-1}A = I$ and $AA^{-1} = I$. For this to happen A and A^{-1} must be square matrices of the same order. By using the inverse matrix A^{-1} it is possible to solve the matrix equation that represents the system of simultaneous equations.

$$AX = C$$

$$A^{-1}AX = A^{-1}C \qquad \textbf{Premultiplying by the inverse of } A$$

$$IX = A^{-1}C \qquad \textbf{Recalling } A^{-1}A = I$$

$$X = A^{-1}C \qquad \textbf{Recalling } IX = X$$

For a system of n simultaneous equations A is of order $n \times n$, X is of order $n \times 1$ and C is of order $n \times 1$. The inverse of A, A^{-1}, must also be of order $n \times n$. Therefore the product $A^{-1}C$ must exist and is an $n \times 1$ matrix of constants. This then must correspond element by element with the $n \times 1$ matrix of unknowns X and the system of equations is solved.

There are several methods for finding the inverse matrix. We will adopt the Gauss elimination method, which is favored because it requires the fewest number of operations and because it is readily adaptable to repetitive operations.

To perform the Gauss elimination method of matrix inversion we write the matrix to be inverted followed by an identity matrix of the same order. For the matrix

$$A = \begin{pmatrix} 2 & -1 & -2 \\ 4 & 2 & 1 \\ 2 & 1 & 1 \end{pmatrix}$$

we write

$$\left(\begin{array}{ccc|ccc} 2 & -1 & -2 & 1 & 0 & 0 \\ 4 & 2 & 1 & 0 & 1 & 0 \\ 2 & 1 & -1 & 0 & 0 & 1 \end{array}\right).$$

Our objective is to obtain the inverse of matrix A. If A has an inverse, this objective can be achieved by performing the same operations on the rows of both matrices A and I. As a result of the operations matrix A will be changed to the identity matrix and the identity matrix will be changed to A^{-1}.

Operations that can be performed on the matrices are

1. Interchange any two rows.
2. Multiply any row by a nonzero constant.
3. Add a multiple of one row to any other row.

The general procedure that we will follow to change A to the identity matrix is to obtain a 1 in the a_{11} position and then get zeroes in all the other rows of the first column. Next we want to obtain a 1 in the a_{22} position and get zeroes in all the other rows of the second column. We continue this process until there are ones on the principal diagonal and zeroes in every other position.

To find the inverse of our example we first multiply the first row by 1/2, which is the multiplicative inverse of the a_{11} element, 2:

$$\left(\begin{array}{rrr|rrr} 2 & -1 & -2 & 1 & 0 & 0 \\ 4 & 2 & 1 & 0 & 1 & 0 \\ 2 & 1 & -1 & 0 & 0 & 1 \end{array}\right). \qquad \textbf{Given}$$

$$\left(\begin{array}{ccc|ccc} \frac{1}{2}(2) & \frac{1}{2}(-1) & \frac{1}{2}(-2) & \frac{1}{2}(1) & \frac{1}{2}(0) & \frac{1}{2}(0) \\ 4 & 2 & 1 & 0 & 1 & 0 \\ 2 & 1 & -1 & 0 & 0 & 1 \end{array}\right)$$

$$\left(\begin{array}{rrr|rrr} 1 & -\frac{1}{2} & -1 & \frac{1}{2} & 0 & 0 \\ 4 & 2 & 1 & 0 & 1 & 0 \\ 2 & 1 & -1 & 0 & 0 & 1 \end{array}\right).$$

We now replace row 2 with the sum of itself and -4 times row 1. (-4 is the additive inverse of the element in the second row of column 1; do not change row 1 while doing this.)

$$\left(\begin{array}{ccc|ccc} 1 & -\frac{1}{2} & -1 & \frac{1}{2} & 0 & 0 \\ 4+(-4)(1) & 2+(-4)\left(-\frac{1}{2}\right) & 1+(-4)(-1) & 0+(-4)\left(\frac{1}{2}\right) & 1+(-4)(0) & 0+(-4)(0) \\ 2 & 1 & -1 & 0 & 0 & 1 \end{array}\right)$$

$$\left(\begin{array}{rrr|rrr} 1 & -\frac{1}{2} & -1 & \frac{1}{2} & 0 & 0 \\ 0 & 4 & 5 & -2 & 1 & 0 \\ 2 & 1 & -1 & 0 & 0 & 1 \end{array}\right).$$

Next replace row 3 with the sum of itself and -2 times row 1. (-2 is the additive inverse of the element in the third row of column 1; do not change row 1 while doing this.)

$$\left(\begin{array}{ccc|ccc} 1 & -\frac{1}{2} & -1 & \frac{1}{2} & 0 & 0 \\ 0 & 4 & 5 & -2 & 1 & 0 \\ 2+(-2)(1) & 1+(-2)\left(-\frac{1}{2}\right) & -1+(-2)(-1) & 0+(-2)\left(\frac{1}{2}\right) & 0+(-2)(0) & 1+(-2)(0) \end{array}\right)$$

$$\left(\begin{array}{rrr|rrr} 1 & -\frac{1}{2} & -1 & \frac{1}{2} & 0 & 0 \\ 0 & 4 & 5 & -2 & 1 & 0 \\ 0 & 2 & 1 & -1 & 0 & 1 \end{array}\right)$$

We see that the result of these operations has been to obtain a 1 on the principal diagonal and zeroes in all the other positions in the first column. We now repeat the same sequence of operations on the second column. First we multiply the second row by 1/4, which is the multiplicative inverse of the a_{22} element, 4. This will obtain a 1 on the principal diagonal in the second column:

$$\left(\begin{array}{ccc|ccc} 1 & -\frac{1}{2} & -1 & \frac{1}{2} & 0 & 0 \\ 0 & 1 & \frac{5}{4} & -\frac{1}{2} & \frac{1}{4} & 0 \\ 0 & 2 & 1 & -1 & 0 & 1 \end{array}\right).$$

Now replace row 3 with the sum of itself and -2 times row 2. (Row 2 is the principal diagonal row for the second column; -2 is the additive inverse of the element in the third row of column 2.)

$$\left(\begin{array}{ccc|ccc} 1 & -\frac{1}{2} & -1 & \frac{1}{2} & 0 & 0 \\ 0 & 1 & \frac{5}{4} & -\frac{1}{2} & \frac{1}{4} & 0 \\ 0 & 0 & -\frac{3}{2} & 0 & -\frac{1}{2} & 1 \end{array}\right).$$

Now replace row 1 with the sum of itself and 1/2 times row 2. (1/2 is the additive inverse of the element in the first row of column 2.)

$$\left(\begin{array}{ccc|ccc} 1 & 0 & -\frac{3}{8} & \frac{1}{4} & \frac{1}{8} & 0 \\ 0 & 1 & \frac{5}{4} & -\frac{1}{2} & \frac{1}{4} & 0 \\ 0 & 0 & -\frac{3}{2} & 0 & -\frac{1}{2} & 1 \end{array}\right).$$

We have now obtained a 1 on the principal diagonal and zeroes in all the other positions of column 2. We now perform a similar series of operations on the third column. First multiply row 3 by $-2/3$:

$$\left(\begin{array}{ccc|ccc} 1 & 0 & -\frac{3}{8} & \frac{1}{4} & \frac{1}{8} & 0 \\ 0 & 1 & \frac{5}{4} & -\frac{1}{2} & \frac{1}{4} & 0 \\ 0 & 0 & 1 & 0 & \frac{1}{3} & -\frac{2}{3} \end{array}\right).$$

Next replace row 1 with the sum of itself and 3/8 times row 3:

$$\left(\begin{array}{ccc|ccc} 1 & 0 & 0 & \frac{1}{4} & \frac{1}{4} & -\frac{1}{4} \\ 0 & 1 & \frac{5}{4} & -\frac{1}{2} & \frac{1}{4} & 0 \\ 0 & 0 & 1 & 0 & \frac{1}{3} & -\frac{2}{3} \end{array}\right).$$

Finally replace row 2 with the sum of itself and $-5/4$ times row 3:

$$\left(\begin{array}{ccc|ccc} 1 & 0 & 0 & \frac{1}{4} & \frac{1}{4} & -\frac{1}{4} \\ 0 & 1 & 0 & -\frac{1}{2} & -\frac{1}{6} & \frac{5}{6} \\ 0 & 0 & 1 & 0 & \frac{1}{3} & -\frac{2}{3} \end{array}\right).$$

The operations have transformed the matrix A into an identity matrix and the identity matrix into the inverse matrix A^{-1}:

$$A^{-1} = \begin{pmatrix} \frac{1}{4} & \frac{1}{4} & -\frac{1}{4} \\ -\frac{1}{2} & -\frac{1}{6} & \frac{5}{6} \\ 0 & \frac{1}{3} & -\frac{2}{3} \end{pmatrix}.$$

The reader is urged to perform the multiplications to verify that $A^{-1}A = I$ and $AA^{-1} = I$.

To solve the set of simultaneous equations $AX = C$ we now need only to solve $X = A^{-1}C$. Recalling X and C we have

$$\begin{pmatrix} x \\ y \\ z \end{pmatrix} = \begin{pmatrix} \frac{1}{4} & \frac{1}{4} & -\frac{1}{4} \\ -\frac{1}{2} & -\frac{1}{6} & \frac{5}{6} \\ 0 & \frac{1}{3} & -\frac{2}{3} \end{pmatrix} \begin{pmatrix} -6 \\ 1 \\ -1 \end{pmatrix}$$

$$= \begin{pmatrix} \left(\frac{1}{4}\right)(-6) + \left(\frac{1}{4}\right)(1) + \left(-\frac{1}{4}\right)(-1) \\ \left(-\frac{1}{2}\right)(-6) + \left(-\frac{1}{6}\right)(1) + \left(\frac{5}{6}\right)(-1) \\ (0)(-6) + \left(\frac{1}{3}\right)(1) + \left(-\frac{2}{3}\right)(-1) \end{pmatrix} = \begin{pmatrix} -1 \\ 2 \\ 1 \end{pmatrix}.$$

Therefore $x = -1$, $y = 2$, and $z = 1$ is the solution set of the system of simultaneous equations:

$$2x - y - 2z = -6$$
$$4x + 2y + z = 1$$
$$2x + y - z = -1.$$

Steps for Finding the Inverse of a Square Matrix

1. Write down the matrix to be inverted and write an identity matrix of the same order next to it.
2. Start with the leftmost column of the matrix A to be inverted and complete Steps 3 to 5 before proceeding to the next column.
3. Identify the column to be operated on.
4. Multiply the row containing the principal diagonal element, for the column being operated on, by the multiplicative inverse of the principal diagonal element. Replace the row by the results of the multiplication.
5. For each remaining row in the column being operated on, replace each element of the row by the sum of itself and the additive inverse of the element in the column and row being worked on times the principal diagonal row.
6. After Steps 3 to 5 have been performed on each column of the original matrix A, it has been transformed into an identity matrix and the original identity matrix has been transformed into the inverse matrix A^{-1}.

A square matrix will not have an inverse when a zero is found on the main diagonal and it cannot be removed. If the matrix does not have an inverse, the matrix is called **singular matrix.** One method to determine whether or not a zero on the main diagonal is not removable is to start the inversion process over after rearranging the order of the rows of the matrix. If after all possible combinations of rearrangement of the rows a division by zero is necessary, the matrix is singular. The matrix of coefficients of a solvable system of simultaneous equations will always have an inverse. Remember that if the rows of the matrix of coefficients must be rearranged, the matrix of constants must be rearranged accordingly. (The matrix of unknowns remains the same.) It is also possible to determine whether or not a matrix is singular by evaluating the determinant of the matrix. If the determinant of the matrix is not equal to zero, the matrix is **nonsingular.**

EXAMPLE 1 Solve the system of simultaneous linear equations using matrix inversion.

$$\begin{aligned} x + y + z &= -2 \\ 2x - y - z &= 5 \\ -x + 3y + 4z &= -11 \end{aligned}$$

Solution: The system of equations can be written in matrix form as

$$\begin{pmatrix} 1 & 1 & 1 \\ 2 & -1 & -1 \\ -1 & 3 & 4 \end{pmatrix} \begin{pmatrix} x \\ y \\ z \end{pmatrix} = \begin{pmatrix} -2 \\ 5 \\ -11 \end{pmatrix}$$

or $AX = C$.

We apply the steps for matrix inversion to A:

$$\left(\begin{array}{ccc|ccc} 1 & 1 & 1 & 1 & 0 & 0 \\ 2 & -1 & -1 & 0 & 1 & 0 \\ -1 & 3 & 4 & 0 & 0 & 1 \end{array}\right). \qquad \textbf{Step 1}$$

For Step 3, start with column 1. Since there is a 1 in position a_{11} we can skip the multiplication called for in Step 4. For the second row of the first column the additive inverse of a_{21} is -2 and for the third row of the first column the additive inverse of a_{31} is 1.

$$\left(\begin{array}{ccc|ccc} 1 & 1 & 1 & 1 & 0 & 0 \\ 2+(-2)(1) & -1+(-2)(1) & -1+(-2)(1) & 0+(-2)(1) & 1+(-2)(0) & 0+(-2)(0) \\ -1+(1)(1) & 3+(1)(1) & 4+(1)(1) & 0+(1)(1) & 0+(1)(0) & 1+(1)(0) \end{array}\right) \qquad \textbf{Step 5}$$

$$\left(\begin{array}{ccc|ccc} 1 & 1 & 1 & 1 & 0 & 0 \\ 0 & -3 & -3 & -2 & 1 & 0 \\ 0 & 4 & 5 & 1 & 0 & 1 \end{array}\right) \qquad \textbf{Combining terms}$$

Going to the second column the multiplicative inverse of the a_{22} element, -3, is $1/-3$. Replace row 2 with the product of $-1/3$ and row 2:

$$\left(\begin{array}{ccc|ccc} 1 & 1 & 1 & 1 & 0 & 0 \\ 0 & 1 & 1 & -\frac{2}{3} & -\frac{1}{3} & 0 \\ 0 & 4 & 5 & 1 & 0 & 1 \end{array}\right). \qquad \textbf{Step 4}$$

For the first row of the second column the additive inverse of a_{12} is -1 and for the third row of the second column the additive inverse of a_{32} is -4:

$$\left(\begin{array}{ccc|ccc} 1+(-1)(0) & 1+(-1)(1) & 1+(-1)(1) & 1+(-1)\left(\frac{2}{3}\right) & 0+(-1)\left(-\frac{1}{3}\right) & 0+(-1)(0) \\ 0 & 1 & 1 & \frac{2}{3} & -\frac{1}{3} & 0 \\ 0+(-4)(0) & 4+(-4)(1) & 5+(-4)(1) & 1+(-4)\left(-\frac{2}{3}\right) & 0+(-4)\left(-\frac{1}{3}\right) & 1+(-4)(0) \end{array}\right). \qquad \textbf{Step 5}$$

$$\left(\begin{array}{ccc|ccc} 1 & 0 & 0 & \frac{1}{3} & \frac{1}{3} & 0 \\ 0 & 1 & 1 & \frac{2}{3} & -\frac{1}{3} & 0 \\ 0 & 0 & 1 & -\frac{5}{3} & \frac{4}{3} & 1 \end{array}\right). \quad \textbf{Combining terms}$$

Since the element in the a_{33} position is a 1, all that needs to be done is to get a zero in the a_{23} position. The additive inverse of a_{23} is -1.

$$\left(\begin{array}{ccc|ccc} 1 & 0 & 0 & \frac{1}{3} & \frac{1}{3} & 0 \\ 0+(-1)(0) & 1+(-1)(0) & 1+(-1)(1) & \frac{2}{3}+(-1)-\frac{5}{3} & -\frac{1}{3}+(-1)\frac{4}{3} & 0+(-1)(1) \\ 0 & 0 & 1 & -\frac{5}{3} & \frac{4}{3} & 1 \end{array}\right)$$

$$\left(\begin{array}{ccc|ccc} 1 & 0 & 0 & \frac{1}{3} & \frac{1}{3} & 0 \\ 0 & 1 & 0 & \frac{7}{3} & -\frac{5}{3} & -1 \\ 0 & 0 & 1 & -\frac{5}{3} & \frac{4}{3} & 1 \end{array}\right).$$

Matrix A has been completely transformed to an identity matrix so

$$A^{-1} = \begin{pmatrix} \frac{1}{3} & \frac{1}{3} & 0 \\ \frac{7}{3} & -\frac{5}{3} & -1 \\ -\frac{5}{3} & \frac{4}{3} & 1 \end{pmatrix}$$

Recalling

$$C = \begin{pmatrix} -2 \\ 5 \\ -11 \end{pmatrix}.$$

To solve the equations we use $X = A^{-1}C$

$$\begin{pmatrix} x \\ y \\ z \end{pmatrix} = \begin{pmatrix} \frac{1}{3} & \frac{1}{3} & 0 \\ \frac{7}{3} & -\frac{5}{3} & -1 \\ -\frac{5}{3} & \frac{4}{3} & 1 \end{pmatrix} \begin{pmatrix} -2 \\ 5 \\ -11 \end{pmatrix}$$

$$\begin{pmatrix} x \\ y \\ z \end{pmatrix} = \begin{pmatrix} \left(\frac{1}{3}\right)(-2) + \left(\frac{1}{3}\right)(5) + (0)(-11) \\ \left(\frac{7}{3}\right)(-2) + \left(-\frac{5}{3}\right)(5) + (-1)(-11) \\ \left(-\frac{5}{3}\right)(-2) + \left(\frac{4}{3}\right)(5) + (1)(-11) \end{pmatrix}$$

$$= \begin{pmatrix} 1 \\ -2 \\ -1 \end{pmatrix}.$$

Therefore $x = 1$, $y = -2$, and $z = -1$ is the solution set for the given system of simultaneous equations. □

EXAMPLE 2 Use the inverse matrix to solve the system of simultaneous equations:

$$\begin{aligned} 2x + y - z &= -1 \\ 4x + 2y + z &= 1 \\ 2x - y - 2z &= -6. \end{aligned}$$

Solution:

$$\left(\begin{array}{ccc|ccc} 2 & 1 & -1 & 1 & 0 & 0 \\ 4 & 2 & 1 & 0 & 1 & 0 \\ 2 & -1 & -2 & 0 & 0 & 1 \end{array}\right). \qquad \textbf{Step 1}$$

We need a 1 in the a_{11} position, so we multiply row 1 by 1/2:

$$\left(\begin{array}{ccc|ccc} 1 & \frac{1}{2} & -\frac{1}{2} & \frac{1}{2} & 0 & 0 \\ 4 & 2 & 1 & 0 & 1 & 0 \\ 2 & -1 & -2 & 0 & 0 & 1 \end{array}\right)$$

For the first column $a_{21} = 4$ and $a_{31} = 2$, so the appropriate additive inverses are -4 and -2:

$$\left(\begin{array}{ccc|ccc} 1 & \frac{1}{2} & -\frac{1}{2} & \frac{1}{2} & 0 & 0 \\ 4 + (-4)(1) & 2 + (-4)\left(\frac{1}{2}\right) & 1 + (-4)\left(-\frac{1}{2}\right) & 0 + (-4)\left(\frac{1}{2}\right) & 1 + (-4)(0) & 0 + (-4)(0) \\ 2 + (-2)(1) & -1 + (-2)\left(\frac{1}{2}\right) & -2 + (-2)\left(-\frac{1}{2}\right) & 0 + (-2)\left(\frac{1}{2}\right) & 0 + (-2)(0) & 1 + (-2)(0) \end{array}\right)$$

$$\begin{pmatrix} 1 & \frac{1}{2} & -\frac{1}{2} \\ 0 & 0 & 3 \\ 0 & -2 & -1 \end{pmatrix} \begin{pmatrix} \frac{1}{2} & 0 & 0 \\ -2 & 1 & 0 \\ -1 & 0 & 1 \end{pmatrix}.$$

For the second column $a_{22} = 0$ so we cannot proceed because division by zero is undefined. However if we change the order of the rows by interchanging the first and third rows of the original matrix and change the matrix of constants correspondingly, we obtain the system of equations used to illustrate the first part of this section. As was shown, this system can be solved by matrix inversion giving the results $x = -1$, $y = 2$, and $z = 1$. □

10.7 Exercises

In Exercises 1–4, write the systems of simultaneous equations in matrix form.

1. $x + y + z = -2$
$2x + y + 2z = 0$
$-x - 2y - 3z = 5$

2. $2x + 3y + z = -1$
$x - 2y - z = -8$
$3x - 3y + 2z = 1$

3. $x + y + z = 7$
$y - z = 7$
$2x - 2y - z = 0$

4. $2x + y - z = 2$
$4x + y - 2z = -3$
$x + z = 2$

In Exercises 5–8, find the inverse of the matrix of the coefficients in Exercises 1–4.

In Exercises 9–12, use the results of Exercises 5–8 to find the solution set for the system of equations in Exercises 1–4.

In Exercises 13–24, find the solution sets for the system of equations using matrix methods.

13. $-x + 4y - z = 7$
$x - y + 4z = -1$
$3x + 9y - 2z = -5$

14. $x + 2y + 3z = 5$
$3x + 3y - z = 0$
$x + 3y - 2z = -3$

15. $4x - 5y + 5z = 2$
$x + y + 3z = 1$
$9x - 3y + 2z = -1$

16. $x + 5z = -6$
$2x + 3y + 4z = -9$
$3x + 4y = -7$

17. $2x - y + 2z = 2$
$5x + 4y + 3z = -7$
$x + y + z = -1$

18. $4x - y + z = 9$
$5x + y + 9z = 9$
$6x + 5y - z = 7$

19. $x + 5z = -4$
$8x + 7y - z = 9$
$9x - 4y + z = 8$

20. $4x + 4y + 5z = -7$
$4x + 3y + 6z = 8$
$3x + y + 9z = 4$

21. The loop equations for a circuit are

$$300(i_2 - i_1) + 80\, i_2 + 30(i_2 - i_3) = 0$$
$$30(i_3 - i_2) + 60\, i_3 = 0$$
$$-200(i_2 - i_1) + 80\, i_2 + 30(i_2 - i_3) = 0.$$

What are the currents i_1, i_2, and i_3?

22. The node equations for a circuit are

$$\frac{1}{3}(v_2 - v_1) + \frac{1}{7}v_2 = 0$$
$$\frac{1}{15}(v_1 - 100) + \frac{1}{3}(v_1 - v_2) + \frac{1}{4}(v_1 - v_3) = 0$$
$$\frac{1}{4}(v_3 - v_1) + \frac{1}{9}v_3 + \frac{1}{18}v_3 = 0.$$

What are the node voltages v_1, v_2, and v_3?

23. The section forces in a truss are related by

$$1000 + F_1 \sin 30° + F_2 - 500 = 0$$
$$-10(1500) + (10 \tan 30°)(F_1 \cos 30°) + 10F_2 = 0$$
$$1000 + F_1 \sin 30° + F_2 - 500 = 0.$$

What are the forces F_1, F_2, and F_3?

24. Design the height dimensions for three supports such that S_3 is three times as tall as S_1, the height of S_2 is 4 more than S_1 and the height of S_3 is 3 more than S_2.

Chapter summary

In Chapter 10 we have dealt with linear equations and the methods for solving them. The **slope–intercept** form of an **equation for a line,** $y = mx + b$, where m is the **slope** and b is the **y-intercept** was presented. The slope was defined as the **rise/run** of the line. Using the slope and the y-intercept we can graph a line or write the equation of a line. We found that two lines are parallel if they have the same slope and different y-intercepts and that two linear equations represent the same line if their slopes and y-intercepts are the same. If two lines cross at a unique point, then the system of equations is a **consistent system** and that point is the **solution** of the system. A system of linear equations that graphs as parallel lines is an **inconsistent system.** A system of linear equations that graphs as the same line is a **dependent system.**

Systems of n linear equations in n unknowns can be solved by the **substitution method,** by the **elimination method,** or by using **determinants.**

A determinant is a single numerical value that can be expressed as an $n \times n$ array of numbers. The value of a 2×2 determinant is

$$A = \begin{vmatrix} a_{11} & a_{12} \\ a_{21} & a_{22} \end{vmatrix} = a_{11}a_{22} - a_{21}a_{12}$$

The value of a 3-×-3 determinant is found using expansion by cofactors: $A = a_{i1}A_{i1} + a_{i2}A_{i2} + a_{i3}A_{i3}$ or $A = a_{1j}A_{1j} + a_{2j}A_{2j} + a_{3j}A_{3j}$ where a_{ij} is an **element** of the determinant and A_{ij} is the **cofactor** of the element. $A_{ij} = (-1)^{i+j} \times M_{ij}$, where M_{ij} is the **minor** of the element a_{ij}.

Systems of linear equations are solved using determinants and **Cramer's rule.** For a system of two linear equations in two unknowns

$$\begin{aligned} a_{11}x + a_{12}y &= c_1 \\ a_{21}x + a_{22}y &= c_2, \end{aligned}$$

$$x = \frac{N_x}{D}, \qquad y = \frac{N_y}{D}, \qquad D \neq 0,$$

$$D = \begin{vmatrix} a_{11} & a_{12} \\ a_{21} & a_{22} \end{vmatrix}, \qquad N_x = \begin{vmatrix} c_1 & a_{21} \\ c_2 & a_{22} \end{vmatrix}, \qquad N_y = \begin{vmatrix} a_{11} & c_1 \\ a_{21} & c_2 \end{vmatrix}.$$

For a system of three linear equations in three unknowns

$$\begin{aligned} a_{11}x + a_{12}y + a_{13}z &= c_1 \\ a_{21}x + a_{22}y + a_{23}z &= c_2 \\ a_{31}x + a_{32}y + a_{33}z &= c_3, \end{aligned}$$

$$x = \frac{N_x}{D}, \qquad y = \frac{N_y}{D}, \qquad z = \frac{N_z}{D}, \qquad D \neq 0,$$

$$D = \begin{vmatrix} a_{11} & a_{12} & a_{13} \\ a_{21} & a_{22} & a_{23} \\ a_{31} & a_{32} & a_{33} \end{vmatrix}, \qquad N_x = \begin{vmatrix} c_1 & a_{12} & a_{13} \\ c_2 & a_{22} & a_{23} \\ c_3 & a_{32} & a_{33} \end{vmatrix},$$

$$N_y = \begin{vmatrix} a_{11} & c_1 & a_{13} \\ a_{21} & c_2 & a_{23} \\ a_{31} & c_3 & a_{33} \end{vmatrix}, \qquad N_z = \begin{vmatrix} a_{11} & a_{12} & c_1 \\ a_{21} & a_{22} & c_2 \\ a_{31} & a_{32} & c_3 \end{vmatrix}.$$

Review Exercises

In Exercises 1 – 8, identify the slope and y-intercept of the lines and sketch the graph of the lines.

1. $y = -x + 4$
2. $y = x - 3$
3. $y = \frac{1}{2}x + 2$
4. $y = -\frac{1}{4}x - 2$
5. $y = 3$
6. $x = -4$
7. $3x + 6y - 2 = 0$
8. $2x + 4y + 1 = 0$

In Exercises 9 – 16, write the equation of the line.

9. $m = -4,\ b = 0$
10. $m = 5,\ b = 0$
11. $m = \frac{1}{2},\ b = 3$
12. $m = \frac{1}{6},\ b = 1$
13. $m = 0,\ b = -2$
14. $m = 0,\ b = 4$
15. $m = 1.32,\ b = 4.17$
16. $m = -2.61,\ b = 4.88$

In Exercises 17 – 24, solve the system of two linear equations in two unknowns using algebraic methods.

17. $\begin{aligned} x - y &= 0 \\ 2x - y &= 6 \end{aligned}$
18. $\begin{aligned} -x - 3y &= 3 \\ 2x - 3y &= 3 \end{aligned}$
19. $\begin{aligned} 5x - y &= -1 \\ x + y &= 1 \end{aligned}$
20. $\begin{aligned} 4x + y &= 1 \\ 9x + 5y &= -6 \end{aligned}$
21. $\begin{aligned} -x - y + 2 &= 0 \\ x + 2y &= 0 \end{aligned}$
22. $\begin{aligned} 3x - y + 4 &= 0 \\ x + y &= 0 \end{aligned}$
23. $\begin{aligned} 2x - y - 8 &= 0 \\ -3x - 2y + 5 &= 0 \end{aligned}$
24. $\begin{aligned} x - y - 4 &= 0 \\ 3x - y - 2 &= 0 \end{aligned}$

In Exercises 25 – 32, use algebraic methods to solve the system of linear equations. From the results determine whether any of the systems are inconsistent or dependent. Graph the equations.

25. $\begin{aligned} 8x - 5y &= 6 \\ 3x - 2y &= 8 \end{aligned}$
26. $\begin{aligned} 7x + 8y &= -3 \\ 6x + 7y &= 8 \end{aligned}$
27. $\begin{aligned} x + 9y + 1 &= 0 \\ x + 9y &= 1 \end{aligned}$
28. $\begin{aligned} -7x - y + 4 &= 0 \\ 14x + 2y - 16 &= 0 \end{aligned}$
29. $\begin{aligned} x + 2y &= -1 \\ 4x + 3y &= 4 \end{aligned}$
30. $\begin{aligned} x + 3y &= 0 \\ x + 2y &= -6 \end{aligned}$
31. $\begin{aligned} x - y + 3 &= 0 \\ -3x + 3y - 9 &= 0 \end{aligned}$
32. $\begin{aligned} x + 3y - 1 &= 0 \\ -x - 3y &= -1 \end{aligned}$

In Exercises 33 – 38, evaluate the 2×2 determinants.

33. $\begin{vmatrix} 3 & -5 \\ 9 & 2 \end{vmatrix}$
34. $\begin{vmatrix} -9 & 1 \\ -5 & 0 \end{vmatrix}$
35. $\begin{vmatrix} -1 & -3 \\ 0 & 8 \end{vmatrix}$
36. $\begin{vmatrix} -1 & 1 \\ -7 & 5 \end{vmatrix}$
37. $\begin{vmatrix} 1 & 0 \\ -4 & -1 \end{vmatrix}$
38. $\begin{vmatrix} 9 & 0 \\ -5 & -1 \end{vmatrix}$

In Exercises 39 – 50, solve the systems of equations using Cramer's rule.

39. $\begin{aligned} 8x - 3y &= -3 \\ 7x - 2y &= 3 \end{aligned}$
40. $\begin{aligned} x + 3y &= 1 \\ x + 2y &= 4 \end{aligned}$
41. $\begin{aligned} 2x - 3y + 2 &= 0 \\ 3x - 4y - 8 &= 0 \end{aligned}$
42. $\begin{aligned} -3x + 5y - 2 &= 0 \\ 3x - 4y + 4 &= 0 \end{aligned}$
43. $\begin{aligned} 7x - y - 8 &= 0 \\ -6x + y + 4 &= 0 \end{aligned}$
44. $\begin{aligned} x - y + 8 &= 0 \\ -7x + 8y - 4 &= 0 \end{aligned}$
45. $\begin{aligned} 6x - 7y &= 4 \\ -2x + y &= -4 \end{aligned}$
46. $\begin{aligned} -2x - 5y &= 8 \\ -7x + y &= -9 \end{aligned}$
47. $\begin{aligned} -3x - 4y &= -4 \\ 2x + 3y &= 2 \end{aligned}$
48. $\begin{aligned} -2x - 9y &= -7 \\ 5x - y &= -6 \end{aligned}$
49. $\begin{aligned} x + 2y &= -2 \\ -x + 5y &= 9 \end{aligned}$
50. $\begin{aligned} -4x - 5y &= 7 \\ -9x - 7y &= 3 \end{aligned}$

In Exercises 51 – 58, solve the systems of three equations in three unknowns using algebraic methods.

51. $\begin{aligned} -7x - 6y + 3z &= 3 \\ -x - 9y + 6z &= 3 \\ -4x - 9y &= 9 \end{aligned}$
52. $\begin{aligned} -x - 5z &= 4 \\ 8x + 7y - z &= 9 \\ 9x - 4y + z &= 8 \end{aligned}$
53. $\begin{aligned} 5x + y + 9z &= 9 \\ 6x + 5y - z &= 7 \\ 4x - y + z &= 9 \end{aligned}$
54. $\begin{aligned} -8x + 4y + 5z &= -6 \\ -3x + 9y + z &= -8 \\ -3x - 4y + 7z &= -7 \end{aligned}$
55. $\begin{aligned} 8x + y - 8z &= 1 \\ -4x - 8y + 3z &= 3 \\ -2x - 3y + z &= 8 \end{aligned}$
56. $\begin{aligned} x - y + 4z &= -1 \\ 3x + 9y - 2z &= -5 \\ -x + 4y - z &= 7 \end{aligned}$
57. $\begin{aligned} -5x + 4y + 3z &= 6 \\ 6x + 3y + 7z &= -5 \\ 3x + 2y + 2z &= 2 \end{aligned}$
58. $\begin{aligned} 5x + y + 6z &= 9 \\ x + 6y - 8z &= -5 \\ -7x + 9y - 8z &= 9 \end{aligned}$

In Exercises 59 – 66, evaluate the 3×3 determinant using expansion by cofactors.

59. $\begin{vmatrix} -1 & -9 & 0 \\ -4 & -3 & 7 \\ 3 & 0 & -5 \end{vmatrix}$
60. $\begin{vmatrix} -1 & -1 & 3 \\ 0 & -1 & -1 \\ 7 & -6 & 0 \end{vmatrix}$
61. $\begin{vmatrix} -2 & -9 & 5 \\ 3 & 5 & 3 \\ 1 & 3 & 3 \end{vmatrix}$
62. $\begin{vmatrix} 7 & -8 & 1 \\ -3 & -2 & -4 \\ 4 & 3 & -3 \end{vmatrix}$
63. $\begin{vmatrix} 4 & -5 & 0 \\ -5 & 6 & 0 \\ 5 & -6 & 5 \end{vmatrix}$
64. $\begin{vmatrix} 2 & 7 & 2 \\ 9 & 8 & 2 \\ 0 & 0 & -3 \end{vmatrix}$
65. $\begin{vmatrix} 3 & 4 & -8 \\ 3 & -9 & 4 \\ 4 & -1 & -3 \end{vmatrix}$
66. $\begin{vmatrix} 2 & -3 & -1 \\ -7 & 5 & 3 \\ -1 & 3 & 4 \end{vmatrix}$

In Exercises 67–74, solve the systems of three linear equations in three unknowns using Cramer's rule.

67. $2r - t = 0$
$r + 2s + t = 3$
$-2s - 3t = 0$

68. $3r + 4s = -7$
$r + 5t = -6$
$2r + 3s + 4t = -9$

69. $2x - 3y + 4z = 1$
$4x - 5y + 2z = -1$
$x - y + 5z = 2$

70. $6x - y - 3z = -8$
$-7x + 9y - 7z = -9$
$-x + 4y + 2z = -1$

71. $9x + 5y - z = 5$
$-2x + y - 7z = 4$
$x - 3y + 3z = 1$

72. $-9x + 3y - 2z = 1$
$x + y + 3z = 1$
$4x - 5y + 5z = 2$

73. $4x + 4y + 5z = -7$
$4x - 3y + 6z = 8$
$3x + y + 9z = 4$

74. $5x + 3y - 2z = 1$
$2x + 2y + z = 4$
$x - 2y + 4z = 9$

Exercises 75–80 are systems of joint equations used for finding the forces in members of trusses. Solve for the unknown forces.

75. $-F_1 \cos 56° + F_2 \cos 56° + 4000 = 0$
$F_1 \sin 56° + F_2 \sin 56° - 5000 = 0$

76. $-F_5 \cos 45° - F_6 \cos 45° + 10{,}000 = 0$
$F_5 \sin 45° - F_6 \sin 45° - 10{,}000 = 0$

77. $-F_A \cos 58° + 5000 + 4720 \cos 58° - F_B \cos 58° = 0$
$F_A \sin 58° - 4720 \sin 58° - F_B \sin 58° = 0$

78. $-5000 - F_2 \cos 58° + F_3 \cos 58° + 5000 = 0$
$-F_1 - F_2 \sin 58° - F_3 \sin 58° = 0$
$8(5000) - 5 F_2 \sin 58° - 5 F_3 \sin 58° = 0$

79. $F_A + F_B \cos 40° + F_C \cos 9° = 0$
$10{,}000 + F_B \sin 40° - F_C \sin 9° = 0$
$-10 F_A + 7 F_B \cos 40 + 7 F_C \cos 9 - 20(10{,}000) = 0$

80. $-F_{BC} \cos 45° - F_{CF} \cos 45° - F_{EF} = 0$
$F_{BC} \sin 45° - F_{CF} \sin 45° - 15{,}000 - 10{,}000 = 0$
$5 F_{BC} \cos 45° + 5 F_{CF} \cos 45° - 10(10{,}000) - 5 F_{EF} = 0$

Exercises 81–87 are systems of loop or node equations used to find the currents or voltages in electric circuits. Solve for the unknown currents or voltages.

81. $-100 + 4i_1 + 40(i_1 - i_2) + 6i_1 + 10i_1 = 0$
$5i_2 + 15i_2 + 40(i_2 - i_1) = 0$

82. $-100i_1 + 4i_1 + 40(i_1 - i_2) + 6i_1 + 40 + 10i_1 = 0$
$5i_2 + 15i_2 + 40(i_2 - i_1) = 0$

83. $\frac{1}{10}(v_1 - 100) + \frac{1}{20}v_1 + \frac{1}{5}(v_1 - v_2) + \frac{1}{20}(v_1 - v_2) = 0$

$\frac{1}{5}(v_2 - v_1) + \frac{1}{20}(v_2 - v_1) + \frac{1}{16}v_2 = 0$

84. $-60 + 60i_1 + 30(i_1 - i_2) = 0$
$30(i_2 - i_1) - 50 + 30i_2 + 60(i_2 - i_3) = 0$
$60(i_3 - i_2) + 40i_3 + 100 = 0$

85. $-120 + 30i_1 + 60(i_1 - i_2) = 0$
$60(i_2 - i_1) - 40 + 10i_2 + 30(i_2 - i_3) = 0$
$30(i_3 - i_2) + 60i_3 - 10 = 0$

86. $\frac{1}{30}(v_1 - 90) + \frac{1}{30}v_1 + \frac{1}{10}(v_1 - v_2) = 0$

$\frac{1}{10}(v_2 - v_1) + \frac{1}{30}v_2 + \frac{1}{30}(v_2 - v_3) = 0$

$\frac{1}{30}(v_3 - v_2) + \frac{1}{30}v_3 = 0$

87. $\frac{1}{30}(v_1 - 120) + \frac{1}{10}((v_1 + 40) - v_2) + \frac{1}{60}v_1 = 0$

$\frac{1}{10}(v_2 - (v_1 + 40)) + \frac{1}{30}v_2 + \frac{1}{60}(v_2 - v_3) = 0$

$\frac{1}{60}(v_3 - v_2) + \frac{1}{10}(v_3 - (-10)) = 0$

88. An architect is laying out a parking lot. She finds she can put five parking spaces for compact cars and six spaces for standard cars into a 100-ft row. She can alternatively put 10 compact and 2 standard spaces in the same row. What widths is she using as the basis for her layout?

89. "Grab bag" assortments of fishing lures all contain \$6 worth of lures. Selection *A* contains three plugs and one jig. Selection *B* has a plug, two spoons, and a jig. Selection *C* has five jigs and two spoons. What are the types of lures worth?

90. Grandpa Joe's walking stick is 2 cm longer than Uncle Paul's, whose stick in turn is 5 cm shorter than Brother Will's. The sticks average 1 m in length. How long are the individual sticks?

In Exercises 91–96, perform the indicated operations on the matrices if possible.

$$A = \begin{pmatrix} 4 & 2 \\ -1 & 0 \\ 3 & 6 \end{pmatrix} \qquad B = \begin{pmatrix} -1 & 0 \\ 2 & 4 \\ 5 & 3 \end{pmatrix}$$

$$C = \begin{pmatrix} 2 & 61 \\ 4 & -13 \end{pmatrix} \qquad D = \begin{pmatrix} 2 & 6 & -1 \\ 4 & -1 & 5 \end{pmatrix}$$

$$E = \begin{pmatrix} 1 & 4 & 2 \\ 6 & 3 & 1 \\ 2 & -5 & 3 \end{pmatrix} \qquad F = \begin{pmatrix} -3 & 0 & 2 \\ 1 & 1 & 0 \\ 5 & 4 & 3 \end{pmatrix}$$

91. $A + B$ **92.** $2C - 3D$ **93.** $A \times D$

94. $D \times E$ **95.** $2E - F$ **96.** $B \times F$

In Exercises 97–100, find the solution set for the systems of equations using matrix methods.

97.
$$\begin{aligned} x + 6y - 8z &= -5 \\ 7x - 9y + 8z &= -9 \\ 5x + y + 6z &= 9 \end{aligned}$$

98.
$$\begin{aligned} 4x - 5y + 2z &= -1 \\ x - y + 5z &= 2 \\ 2x - 3y + 4z &= 1 \end{aligned}$$

99.
$$\begin{aligned} x - 4y + z &= -7 \\ -x + y - 4z &= 1 \\ 3x + 9y - 2z &= -5 \end{aligned}$$

100.
$$\begin{aligned} 6x - y - 9z &= 7 \\ x + 2y &= -1 \\ x - 4y - z &= 5 \end{aligned}$$

Can you solve these?

1. A line that goes "up and to the right" has a positive slope. Use the definition of slope to show that a line going "down and to the left" also has a positive slope.

2. Write the equation of the line that has a slope $m = 3/5$ and passes through point $(2, -1)$. Write the general equation of a line with slope m and passing through the point (x_1, y_1).

3. Write the equation of the line that passes through points $(3, 2)$ and $(5, -1)$. Write the general equation of a line that passes through the points (x_1, y_1) and (x_2, y_2).

4. Show that the order in which a system of equations is written does not affect its solution by Cramer's rule. In other words, show that the system

$$\begin{aligned} ax + by &= c \\ dx + ey &= f \end{aligned}$$

has the same solution for x and y as the system

$$\begin{aligned} dx + ey &= f \\ ax + by &= c. \end{aligned}$$

5. A determinant containing just one element is equal to that element. Using this information show that a 2×2 determinant can be evaluated using expansion by cofactors.

6. Interchanging any two rows or any two columns of a determinant changes the sign of its value but not its magnitude. Using this information, show that the order of a system of three equations in three unknowns makes no difference in its solution by Cramer's rule.

11

In designing a doctor's examining table it is planned to make the table 7 ft long. For design purposes the load on the table is assumed to be evenly distributed across the table top at 70 lb/ft of length (a factor of safety is included). The legs are initially planned to be located 1 ft from either end of the table.

The table top must be designed to withstand the bending moment generated in it by the load. The bending moment is given by

$$M = -\frac{w}{2\ell}x^2 \ (0 \le x \le d)$$

and

$$M = -\frac{w}{2\ell}x^2 + \frac{w}{2}x - \frac{w}{2}d \ (d \le x \le (\ell - d)),$$

where the origin is at the left end of the table, w is the design load, ℓ is the table length, and d is the distance of the legs from the ends of the table.

What is the algebraically maximum bending moment and the algebraically minimum bending moment and where do they occur? When is the bending moment zero? Where should the legs be located to make the absolute values of the maximum and minimum bending moments the same?

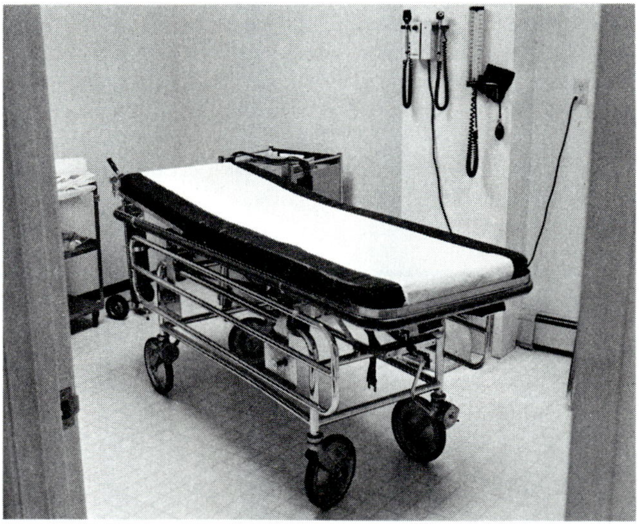

Quadratic Equations

11.1 The quadratic equation

DEFINITION 11.1

An equation of the form

$$ax^2 + bx + c = 0,$$

where a, b, and c are real numbers, $a \neq 0$, is called a **general quadratic equation** in the variable x.

Any quadratic equation of the form of $ax^2 + bx + c = 0$ is said to be of **standard form.** For example,

$$6x^2 + 7x + 13 = 0$$

$$\frac{3}{2}x^2 + 7 = 0$$

$$4x^2 - 3x = 0$$

are quadratic equations in standard form. The equation

$$(x + 3)(x - 5) = 3x$$

is a quadratic equation and can be expressed in standard form:

$$(x + 3)(x - 5) = 3x \qquad \textbf{Multiplying terms and}$$
$$x^2 - 5x - 15 = 0. \qquad \textbf{combining like terms}$$

The quadratic equation $\frac{1}{3}x^2 = \frac{1}{4}x + 13$ can be expressed in standard form.

$$\frac{1}{3}x^2 - \frac{1}{4}x - 13 = 0. \qquad \textbf{Rearranging like terms}$$

The equations are also called second-degree equations in the variable x, since the polynomial $ax^2 + bx + c$ is of degree 2. The name quadratic comes from the Latin word for square, as in x^2.

On the other hand

$$3x^2 + 5x = \frac{3}{x}$$

is not a quadratic equation because of the x in the denominator. Also

$$(x^2 + 5)(x - 3) = 3x$$

is not a quadratic equation. When the left side is multiplied we have

$$x^3 - 3x^2 + 5x - 15 = 3x,$$

which is an equation of the third degree.

In this chapter we will discuss the quadratic equation and the various ways of finding the solutions (roots) of the equation. Also, quadratic equations will be used to solve applied problems.

The ability to work the problems presented in the next section depends on being able to factor quadratic expressions. The techniques of factoring were presented in Section 2.7.

11.2 Solving quadratic equations by factoring

The solution of a quadratic equation by factoring depends upon the following theorem of algebra:

THEOREM 11.1

If the product of two numbers is zero, then one (or both) of the numbers is zero; that is, if $a \cdot b = 0$ then $a = 0$ or $b = 0$ (or both $a = 0$ and $b = 0$).

The important consequence of this statement is that when the product of two numbers is zero, at least one of the two must be zero. For example,

$$\text{if} \quad (x-5)(x+7) = 0,$$
$$\text{then} \quad x-5=0 \text{ or } x+7=0.$$

We now have two linear expressions that can be solved. Thus, $x = 5$ and $x = -7$ are both solutions of the original equation.

Beware! A common mistake made by students is to take an equation such as

$$(x+5)(x-7) = 3$$

and set

$$x+5=3 \text{ or } x-7=3.$$

These two statements are not true because the expression is not equal to zero.

It will be left to the student to solve for x and show that the values -2 and 10 cannot be solutions.

The following set of rules is an excellent guide to solving a quadratic equation by factoring.

To Solve Quadratic Equations by Factoring:

1. Write the quadratic equation in standard form.

$$ax^2 + bx + c = 0 \text{ (must be equal to zero)}$$

2. Factor $ax^2 + bx + c$.
3. Set each factor equal to 0 and solve the resulting linear equations.
4. Check the results by substituting back into the original equation.

EXAMPLE 1 Solve $x^2 - 3x - 10 = 0$.

Solution: The equation is in standard form so we must factor $x^2 - 3x - 10 = 0$.

$$x^2 - 3x - 10 = 0$$
$$(x - 5)(x + 2) = 0 \qquad \textbf{Factoring}$$
$$x - 5 = 0 \quad \text{or} \quad x + 2 = 0 \qquad \textbf{Setting each factor equal to 0}$$
$$x = 5 \quad \text{or} \quad x = -2 \qquad \textbf{Solving for } \boldsymbol{x}$$

The solutions of the quadratic equation $x^2 - 3x - 10 = 0$ are $x = 5$ and $x = -2$. To check the solutions we substitute each solution back into the original equation.

$$x = 5: \quad (5)^2 - 3(5) - 10 = 0$$
$$25 - 15 - 10 = 0$$
$$0 = 0. \checkmark$$
$$x = -2: \quad (-2)^2 - 3(-2) - 10 = 0$$
$$4 + 6 - 10 = 0$$
$$0 = 0. \checkmark$$

Both values check, therefore $x = 5$ and $x = -2$ are solutions to the quadratic equation. □

The solutions of a quadratic equation are called roots of the equation. Thus, 5 and −2 are roots of $x^2 - 3x - 10 = 0$.

EXAMPLE 2 Solve $I^2 + 5I = 0$

Solution: The equation $I^2 + 5I = 0$ is a special case since there is no constant term, $c = 0$. The equation is in standard form. Since there is no constant term, we can use common-term factoring.

$$I(I + 5) = 0 \qquad \textbf{Common-term factoring}$$
$$I = 0 \quad \text{or} \quad I + 5 = 0 \qquad \textbf{Setting factors equal to zero}$$
$$I = 0 \quad \text{or} \quad I = -5 \qquad \textbf{Solving for } \boldsymbol{I}$$

The roots of the equation are 0 and -5. To check solutions we substitute each back into the original equation.

$$I = 0: \quad (0)^2 + 5(0) = 0 \qquad\qquad I = -5: \quad (-5)^2 + 5(-5) = 0$$
$$0 = 0. \checkmark \qquad\qquad 25 - 25 = 0$$
$$0 = 0. \checkmark$$

Both values check, therefore $I = 0$ and $I = -5$ are solutions to the quadratic equation. □

EXAMPLE 3 Solve $R^2 - 9 = 0$.

Solution: The equation $R^2 - 9 = 0$ is another special case of the quadratic equation, the coefficient of the linear term is zero, $b = 0$. The solutions of the equation can be determined by two different methods.

Method 1

$R^2 - 9 = 0$

$(R - 3)(R + 3) = 0$ **Factoring difference of two squares**

$R - 3 = 0$ or $R + 3 = 0$ **Setting the factors equal to zero**

$R = 3$ or $R = -3$ **Solving for *R***

Method 2

$R^2 - 9 = 0$

$R^2 - 9 + 9 = 0 + 9$ **Solving the equation for R^2 by adding 9 to both sides of the equation**

$R^2 = 9$

$\sqrt{R^2} = \sqrt{9}$ **Taking the square root of both sides (Definition of square root.)**

$R = \pm 3$

Thus $R = 3$ or $R = -3$.

The solutions of the equation are $R = 3$ and $R = -3$. The check is left to the reader. □

EXAMPLE 4 Solve $X^2 + 81 = 0$.

Solution: The equation $X^2 + 81 = 0$ is in standard form. Remember that this expression cannot be factored as we factored the difference of two squares.

$X^2 + 81 = 0$

$X^2 + 81 - 81 = 0 - 81$ **Solving the equation for X^2 by adding -81 to both sides**

$X^2 = -81$

$X = \pm\sqrt{-81}.$ **Taking the square root of both sides**

The result in this case is an imaginary number. Using the results of Chapter 9 we can write

$$X = \pm 9j.$$

The solutions of the equation are $X = 9j$ and $X = -9j$. The check is left to the reader. □

EXAMPLE 5 Solve $9x^2 + 12x = -4$.

Solution: The equation is not in standard form.

$9x^2 + 12x = -4$	
$9x^2 + 12x + 4 = 0$	**Adding 4 to both sides of the equation**
$(3x + 2)(3x + 2) = 0$	**Factoring (perfect square trinomial)**
$3x + 2 = 0 \quad \text{or} \quad 3x + 2 = 0$	**Setting each factor equal to zero**
$x = -\dfrac{2}{3} \quad \text{or} \quad x = -\dfrac{2}{3}.$	**Solving for *x***

For this problem the solution to the equation is one unique value, $x = -2/3$. This will always be the case when the factors are identical. We say that the root, $-2/3$, is of multiplicity two (i.e., the root appears twice). The check is left to the reader. □

EXAMPLE 6 Solve $3x^2 - 12x - 36 = 0$.

Solution: For the equation $3x^2 - 12x - 36 = 0$, 3 is a common factor of the numerical coefficients.

$3(x^2 - 4x - 12) = 0$	**Common-term factoring**
$\dfrac{3(x^2 - 4x - 12)}{3} = \dfrac{0}{3}$	**Dividing both sides by 3 to simplify the expression**
$x^2 - 4x - 12 = 0$	
$(x - 6)(x + 2) = 0$	**Factoring**
$x - 6 = 0 \quad \text{or} \quad x + 2 = 0$	**Setting each factor equal to zero**
$x = 6 \quad \text{or} \quad x = -2.$	**Solving for *x***

Thus the solutions of the equation are $x = 6$ and $x = -2$. The check is left to the reader. □

You can divide both sides of an equation by the same quantity only if you are sure that the quantity is not zero. This means in practice that you can only divide both sides by a numerical factor and not by an expression containing a variable. Otherwise, you may not obtain the complete solution.

EXAMPLE 7 Solve $k(3k+2) = 5(3k+2)$.

Solution: Seeing the factor $3k + 2$ on both sides of the equation you may be tempted to divide both sides of the equation by $3k + 2$. This is not permitted since $3k + 2 = 0$ when $k = -2/3$, and, as we know, dividing by zero is not permitted. To solve the equation we need to have both quantities on the same side of the equal sign.

$k(3k+2) - 5(3k+2) = 0$	**Subtracting 5(3k + 2) from both sides**
$(3k+2)(k-5) = 0$	**Common-term factoring**
$3k + 2 = 0 \quad \text{or} \quad k - 5 = 0$	**Setting both factors equal to zero**
$k = -\dfrac{2}{3} \quad \text{or} \quad k = 5.$	**Solving for k**

Thus, the solutions of the equation are $k = -2/3$ and $k = 5$. The same result could have been obtained by multiplying, expressing the result in the form $ax^2 + bx + c = 0$, and then factoring. □

From the previous examples we have observed that a quadratic equation may have 0, 1, or 2 distinct real solutions (real roots). In Section 11.4 we will discuss how to determine whether there will be 0, 1, or 2 real solutions.

EXAMPLE 8 Determine a quadratic equation that has roots 3 and $-2/3$.

Solution: To determine a quadratic equation whose roots are 3 and $-2/3$ think in reverse of the procedure used to determine the roots. Let $x = 3$ or $x = -2/3$.

$x = 3 \quad \text{or} \quad 3x = -2$	**Multiplying both sides of x = −2/3 by 3 to eliminate the fraction**
$x - 3 = 0 \quad \text{or} \quad 3x + 2 = 0$	**Subtracting 3 and adding 2 to both sides of the respective expressions**
$(x-3)(3x+2) = 0.$	**Since both expressions equal zero their product equals zero.**

Thus $3x^2 - 7x - 6 = 0$. A quadratic equation whose roots are 3 and $-2/3$ is $3x^2 - 7x - 6 = 0$. □

EXAMPLE 9 A contractor is preparing a corner for the footing of a new building. (The footing of a building is the base on which the walls are built.) He has a board 12 ft long. He plans to use all of the board to construct a right triangle as a guide to ensure a square corner. His plan calls for making the hypotenuse 5 ft in length. How should he cut the remaining piece to ensure that he ends up with a right triangle?

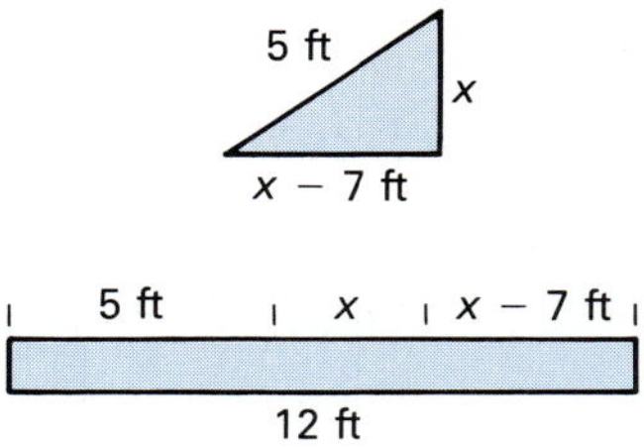

FIGURE 11.1

Solution: The problem provides us with the following facts (see Fig. 11.1):

12 ft = length of the board,
5 ft = length of the hypotenuse of right triangle, and
7 ft = sum of the lengths of the other two sides of the triangle.
Let x = the length of one of the sides of the triangle.
Then $7 - x$ = length of the third side.

Since we are dealing with a right triangle, we use the Pythagorean formula: $c^2 = a^2 + b^2$.

$5^2 = x^2 + (x-7)^2$	**Substituting**
$25 = x^2 + x^2 - 14x + 49$	**Squaring**
$2x^2 - 14x + 24 = 0$	**Collecting like terms**
$x^2 - 7x + 12 = 0$	**Dividing each term by 2**
$(x-4)(x-3) = 0$	**Factoring**
$x - 4 = 0$ or $x - 3 = 0$	**Setting each factor equal to zero**
$x = 4$ or $x = 3$.	**Solving for x**

The result tells us that the carpenter should cut the remaining piece of wood so that one leg of the triangle will be 3 ft and the other leg 4 ft. □

*A **quadratic function** is a function of the form*

$$f(x) = ax^2 + bx + c \text{ where } a \neq 0.$$

Thus, $f(x) = 3x^2 - 7x - 6$ is a quadratic function that (see Example 8) takes on the value 0 when $x = 3$ and when $x = -2/3$.

EXAMPLE 10 Find the value(s) of x for which $f(x) = 0$:

$$f(x) = (2x-3)(x-7).$$

Solution: To find the values of x that will make $f(x) = 0$, we set the factors $2x - 3$ and $x - 7$ equal to zero:

$$2x - 3 = 0 \quad \text{or} \quad x - 7 = 0$$

$$x = \frac{3}{2} \quad \text{or} \quad x = 7.$$

Now substitute these values in the function and we have

$$f\left(\frac{3}{2}\right) = \left(2\left(\frac{3}{2}\right) - 3\right)\left(\frac{3}{2} - 7\right)$$
$$= (3-3)\left(\frac{3}{2} - 7\right)$$
$$= 0$$

and

$$f(7) = (2(7) - 3)(7 - 7)$$
$$= 0.$$

Therefore, for $x = 3/2$ and $x = 7$, $f(x) = 0$ or 3/2 and 7 are zeroes of $f(x) = (2x - 3)(x - 7)$. □

11.2 Exercises

In Exercises 1–30, solve the quadratic equations and check the solutions.

1. $x^2 - 5x + 6 = 0$

2. $x^2 - 11x + 18 = 0$

3. $x^2 - 2x = 15$

4. $x^2 + 2x = 15$

5. $2x^2 - 7x - 15 = 0$

6. $3x^2 - 4x - 15 = 0$

7. $5x^2 - 18x = 8$

8. $6x^2 - 19x = 20$

9. $9R^2 + 2 = -11R$

10. $7R^2 - 2 = -5R$

11. $R^2 - 64 = 0$

12. $R^2 - 121 = 0$

13. $I^2 - 125 = 0$

14. $I^2 - 13 = 0$

15. $Y^2 + 18Y + 81 = 0$

16. $Y^2 + 12Y + 36 = 0$

17. $16Y^2 + 24Y + 9 = 0$

18. $9Y^2 + 30Y + 25 = 0$

19. $Z^2 + 169 = 0$

20. $Z^2 + 289 = 0$

21. $n^2 + 13n = 0$

22. $n^2 + 11 = 0$

23. $5t^2 + 15t = 0$

24. $11t^2 + 1001t = 0$

25. $3c^2 = 7c + 6$

26. $c(c - 2) = 3c - 2(c + 1)$

27. $c(c + 4) - 4 = c - 3(c - 1)$

28. $3(x - 4) = x(x - 4)$

29. $\dfrac{x^2}{8} + \dfrac{3x}{4} = -1$

30. $\dfrac{x^2}{4} + x = 3$

In Exercises 31–40, determine the quadratic equation that has the indicated roots.

31. 3 and -5

32. 5 and -3

33. $\dfrac{2}{5}$ and 3

34. $\dfrac{2}{3}$ and 4

35. only 5

36. only $\dfrac{3}{4}$

37. $\dfrac{1}{3}$ and $\dfrac{1}{2}$

38. -0.2 and -0.4

39. $-1 - 2j$ and $-1 + 2j$

40. only $-\dfrac{3}{2}$

In Exercises 41–50, find each value(s) of x for which $f(x) = 0$.

41. $f(x) = (x - 5)(x + 3)$

42. $f(x) = (x + 11)(x - 7)$

43. $f(x) = (2x + 11)(3x - 5)$

44. $f(x) = (3x + 8)(2x - 9)$

45. $f(x) = x^2 - 7x + 12$

46. $f(x) = x^2 - x - 12$

47. $f(x) = x^2 + 22x + 121$

48. $f(x) = x^2 - \dfrac{x}{4} - \dfrac{1}{8}$

49. $f(x) = \dfrac{x^2}{9} - \dfrac{2x}{3} + 1$

50. $f(x) = 13x^2 + 169x$

51. A rectangular heating duct is to be formed by bending an 80-cm-wide piece of sheet metal into a rectangular-shaped opening. The cross-sectional area of the duct opening must be 300 cm². What are the height and width of the duct? (See Fig. 11.2.)

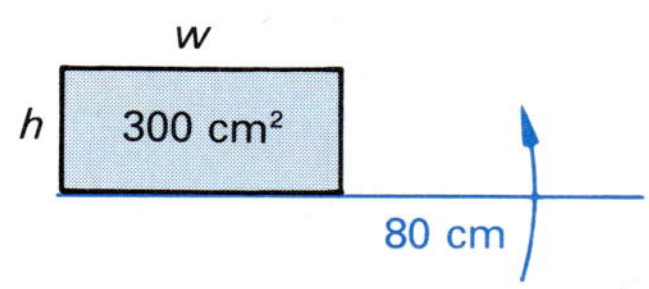

FIGURE 11.2

52. The Pepper family uses about 10 tons of coal each winter to heat their home. They want to build a bin 10 ft high and with the length 3 ft more than the width. Allowing 40 ft^3 to the ton, what dimensions should they use for the bin so that it will hold 10 tons of coal?

53. Find the length of a side of a right triangle, which is 4.00 m shorter than the other side and 8.00 m shorter than the hypotenuse.

54. The net power output of an electric generator is the gross power EI produced, less the power loss I^2R due to internal resistance. What is the current I in a generator that gives a net output of 3600 W at $E = 120$ V, if the internal resistance R is 1 Ω?

11.3 Completing the square

We have seen that we can solve $x^2 = a$ by taking the square root of both sides:

$$\sqrt{x^2} = \sqrt{a}$$
$$x = \pm\sqrt{a},$$

where $x = \pm\sqrt{a}$ means $x = +\sqrt{a}$ or $x = -\sqrt{a}$. What this means is that we must take both the positive and negative square root to get the complete solution to the equation.

The solution of the equation $(x + 5)^2 = 64$ can be found by taking the square root of both sides of the equation:

$x + 5 = \pm 8$	**Taking square root of both sides**
$x = -5 \pm 8$	**Solving for x**
$x = -5 + 8$ or $x = -5 - 8$	
$x = 3$ or $x = -13$.	

The roots of the equation $(x + 5)^2 = 64$ are 3 and -13.

The right side of the equation $(x + 5)^2 = 64$ is a perfect square. Thus, if we can take any quadratic equation and write it in the form

$$(x + h)^2 = k^2,$$

where h and k are constants, we can solve for x by taking the square root of both sides. Quadratic equations that are not in this form can be put into this form by a technique called **completing the square.** That is, we make the left side of the equation a perfect square. This is illustrated in Example 1.

EXAMPLE 1 Solve the quadratic equation $x^2 + 8x - 9 = 0$ by the method of completing the square.

Solution: To solve the equation $x^2 + 8x - 9 = 0$ by completing the square we do the following:

$$x^2 + 8x - 9 + 9 = 0 + 9 \quad \textbf{Adding 9 to both sides}$$
$$x^2 + 8x = 9 \quad \textbf{Simplifying}$$

The left side can become a perfect square by adding the correct constant. The constant can be found by dividing the coefficient of the linear term by 2, 8/2, and squaring the quotient, $(8/2)^2$. We add this number to both sides of the equation, so that we do not change the equality.

$$x^2 + 8x + \left(\frac{8}{2}\right)^2 = 9 + \left(\frac{8}{2}\right)^2$$
$$x^2 + 8x + (4)^2 = 9 + 16$$
$$(x + 4)^2 = 25 \quad \textbf{Writing the left side as a perfect square}$$
$$x + 4 = \pm 5 \quad \textbf{Taking square root of both sides}$$
$$x + 4 = 5 \quad \text{or} \quad x + 4 = -5$$
$$x = -4 + 5 \quad \text{or} \quad x = -4 - 5 \quad \textbf{Solving for } \boldsymbol{x}$$
$$x = 1 \quad \text{or} \quad x = -9.$$

The roots of the equation $x^2 + 8x - 9 = 0$ are 1 and -9. We could have arrived at the same result by factoring. The check is left to the reader. □

EXAMPLE 2 Solve the quadratic equation $3x^2 + 12x - 15 = 0$ by the method of completing the square.

Solution: In Example 1 the coefficient of the second-degree term was 1. For the equation $3x^2 + 12x - 15 = 0$ the coefficient is 3. To use the method of completing the square the coefficient of the second-degree term must be 1.

$$\frac{3x^2}{3} + \frac{12x}{3} - \frac{15}{3} = \frac{0}{3} \quad \textbf{Dividing each term by 3}$$
$$x^2 + 4x = 5 \quad \textbf{Adding 5 to both sides}$$
$$x^2 + 4x + \left(\frac{4}{2}\right)^2 = 5 + \left(\frac{4}{2}\right)^2 \quad \textbf{Dividing 4 by 2, squaring, and adding to both sides}$$
$$(x + 2)^2 = 5 + 4 \quad \textbf{Factoring}$$
$$x + 2 = \pm 3 \quad \textbf{Taking square root of both sides}$$
$$x = -2 \pm 3. \quad \textbf{Solving for } \boldsymbol{x}$$

The roots of the quadratic equation $3x^2 + 12x - 15 = 0$ are 1 and -5. The check is left to the reader. □

Steps in Solving Quadratic Equations by Completing the Square

We will illustrate the steps with the equation

$$4x^2 + 3x - 5 = 0$$

1. Divide both sides of the equation by the coefficient of x^2 (if it is not already 1).

$$\frac{4x^2}{4} + \frac{3x}{4} - \frac{5}{4} = \frac{0}{4}$$

$$x^2 + \frac{3x}{4} - \frac{5}{4} = 0$$

2. Move the constant term to the right side of the equal sign and all terms containing xs to the left:

$$x^2 + \frac{3}{4}x = \frac{5}{4}.$$

3. Complete the square by adding the square of one half the coefficient of the first-degree term to both sides:

$$x^2 + \frac{3}{4}x + \left[\frac{1}{2}\left(\frac{3}{4}\right)\right]^2 = \frac{5}{4} + \left[\frac{1}{2}\left(\frac{3}{4}\right)\right]^2$$

$$x^2 + \frac{3}{4}x + \left(\frac{3}{8}\right)^2 = \frac{5}{4} + \frac{9}{64}.$$

4. Factor the left side and combine the terms on the right side:

$$\left(x + \frac{3}{8}\right)^2 = \frac{89}{64}.$$

5. Take the square root of both sides and solve for x:

$$x + \frac{3}{8} = \pm\frac{\sqrt{89}}{8}.$$

The roots of the equation $4x^2 + 3x - 5 = 0$ are

$$\frac{-3 + \sqrt{89}}{8} \quad \text{and} \quad \frac{-3 - \sqrt{89}}{8}.$$

EXAMPLE 3 Solve the quadratic equation $3x^2 - 5x + 7 = 0$ by completing the square.

Solution: We will solve the equation $3x^2 - 5x + 7 = 0$ by following the suggested steps.

$$x^2 - \frac{5}{3}x + \frac{7}{3} = \frac{0}{3}$$ **Step 1, dividing each term by 3**

$$x^2 - \frac{5}{3}x = -\frac{7}{3}$$ **Step 2, moving constant term to the right side**

$$x^2 - \frac{5}{3}x + \left[\frac{1}{2}\left(\frac{-5}{3}\right)\right]^2 = -\frac{7}{3} + \left[\frac{1}{2}\left(\frac{-5}{3}\right)\right]^2$$ **Step 3, completing the square**

$$x^2 - \frac{5}{3}x + \left(\frac{-5}{6}\right)^2 = -\frac{7}{3} + \frac{25}{36}$$

$$\left(x - \frac{5}{6}\right)^2 = -\frac{59}{36}$$ **Step 4, factoring left side and combining like terms on the right**

$$x - \frac{5}{6} = \pm\frac{\sqrt{-59}}{6}$$ **Step 5, taking the square root of both sides and solving for *x***

$$x = \frac{5}{6} \pm j\frac{\sqrt{59}}{6}$$

The roots of the equation $3x^2 - 5x + 7 = 0$ are

$$\frac{5 + j\sqrt{59}}{6} \quad \text{and} \quad \frac{5 - j\sqrt{59}}{6}. \qquad \square$$

11.3 Exercises

In Exercises 1–10, find the roots of the quadratic equation by taking the square root of both sides.

1. $(x - 8)^2 = 36$
2. $(x - 4)^2 = 64$
3. $2x^2 = 18$
4. $2x^2 = 50$
5. $(x + 5)^2 = 15$
6. $(x + 7)^2 = 13$
7. $(3x + 5)^2 = 49$
8. $(4x + 1)^2 = 121$
9. $(y - 3)^2 = -16$
10. $(y + 5)^2 = -81$

In Exercises 11–30, solve each quadratic equation by completing the square.

11. $x^2 + 3x - 4 = 0$
12. $x^2 + 2x - 8 = 0$
13. $x^2 + 7x - 3 = 0$
14. $x^2 + 10x - 4 = 0$
15. $2R^2 + 3R = 9$
16. $2R^2 - 4R - 6 = 0$
17. $5A^2 = 1 - A$
18. $1/3 - 2A - 3A^2 = 0$
19. $3x^2 - 9x = 4$
20. $3x^2 + 15x = 2$
21. $x(x - 5) = 3(1 - x)$
22. $(x - 3)(x + 4) = x$
23. $y^2 - 6y + 9 = 0$
24. $3y^2 - 30y + 75 = 0$
25. $Z^2 + 5Z + 7 = 0$
26. $3Z^2 + 2Z + 1 = 0$
27. $(x - 4)(x + 5) = x$
28. $x(x + 7) = 5(1 - x)$
29. $ax^2 + bx + c = 0$
30. $mx^2 + nx + c = 0$

11.4 The quadratic formula

The method of solving a quadratic equation by completing the square provides us with a technique for solving any quadratic equation. In this section we will solve the general quadratic equation

$$ax^2 + bx + c = 0 \quad (a,\ b, \text{ and } c \text{ are any numbers and } a \neq 0)$$

by completing the square. The final result will be a formula called the **quadratic formula.** The quadratic formula can be used to solve any quadratic equation. Start with

$$ax^2 + bx + c = 0$$

$$x^2 + \frac{b}{a}x + \frac{c}{a} = \frac{0}{a}$$

Step 1, dividing each term by *a*

$$x^2 + \frac{b}{a}x = -\frac{c}{a}$$

Step 2, subtracting constant term from both sides of equation

$$x^2 + \frac{b}{a}x + \left[\frac{1}{2}\left(\frac{b}{a}\right)\right]^2 = -\frac{c}{a} + \left[\frac{1}{2}\left(\frac{b}{a}\right)\right]^2$$

Step 3, completing the square

$$x^2 + \frac{b}{a}x + \left(\frac{b}{2a}\right)^2 = -\frac{c}{a} + \frac{b^2}{4a^2}$$

$$\left(x + \frac{b}{2a}\right)^2 = \frac{-4ac + b^2}{4a^2}$$

Step 4, factoring left side and combining like terms on the right

$$x + \frac{b}{2a} = \pm\frac{\sqrt{b^2 - 4ac}}{2a}$$

Step 5, taking square root of both sides

Therefore

$$x = -\frac{b}{2a} + \frac{\sqrt{b^2 - 4ac}}{2a} \quad \text{or} \quad x = \frac{-b}{2a} - \frac{\sqrt{b^2 - 4ac}}{2a}$$

$$x = \frac{-b + \sqrt{b^2 - 4ac}}{2a} \quad \text{or} \quad x = \frac{-b - \sqrt{b^2 - 4ac}}{2a}.$$

The solutions to $ax^2 + bx + c = 0$ are

$$x = \frac{-b + \sqrt{b^2 - 4ac}}{2a} \quad \text{or} \quad x = \frac{-b - \sqrt{b^2 - 4ac}}{2a}.$$

DEFINITION 11.2

The **quadratic formula** for solving $ax^2 + bx + c = 0$, where $a \neq 0$, is

$$x = \frac{-b \pm \sqrt{b^2 - 4ac}}{2a}.$$

The formula can be used to solve any quadratic equation in standard form without going through the steps of completing the square. All we need to do is substitute the appropriate values for a, b, and c in the formula.

EXAMPLE 1 Solve $2x^2 + 3x - 1 = 0$ using the quadratic formula.

Solution: For the equation $2x^2 + 3x - 1 = 0$, $a = 2$, $b = 3$, and $c = -1$.

Substituting in the quadratic formula

$$x = \frac{-b \pm \sqrt{b^2 - 4ac}}{2a}$$

we have

$$x = \frac{-3 \pm \sqrt{(3)^2 - 4(2)(-1)}}{2(2)}$$

$$= \frac{-3 \pm \sqrt{9 + 8}}{4} \quad \text{or} \quad \frac{-3 \pm \sqrt{17}}{4}.$$

Therefore

$$x = \frac{-3 + \sqrt{17}}{4} \quad \text{and} \quad x = \frac{-3 - \sqrt{17}}{4}$$

are the solutions of the quadratic equation. The solutions can be checked by substituting the results back into the original equation.

Check: $2x^2 + 3x - 1 = 0$ if $x = \dfrac{-3 + \sqrt{17}}{4}$

$$2\left(\frac{-3 + \sqrt{17}}{4}\right)^2 + 3\,\frac{(-3 + \sqrt{17})}{4} - 1 = 0$$

$$\not{2}\left(\frac{9 - 6\sqrt{17} + 17}{\underset{8}{\not{16}}}\right) + \frac{-9 + 3\sqrt{17}}{4} - 1 = 0$$

$$\frac{9 - 6\sqrt{17} + 17 - 18 + 6\sqrt{17} - 8}{8} = 0 \quad 0 = 0. \checkmark$$

We leave the check of the other root to the reader. □

EXAMPLE 2 Solve $5x^2 - 6x - 2 = 0$ using the quadratic formula.

Solution: For the equation $5x^2 - 6x - 2 = 0$, $a = 5$, $b = -6$, and $c = -2$. Substituting in the equation

$$x = \frac{-b \pm \sqrt{b^2 - 4ac}}{2a},$$

we have

$$\begin{aligned}
x &= \frac{-(-6) \pm \sqrt{(-6)^2 - 4(5)(-2)}}{2(5)} \\
&= \frac{6 \pm \sqrt{36 + 40}}{10} \\
&= \frac{6 \pm \sqrt{76}}{10} \\
&= \frac{6 \pm \sqrt{(4)(19)}}{10} && \textbf{Factoring} \\
&= \frac{6 \pm 2\sqrt{19}}{10} && \textbf{Taking square root of 4} \\
&= \frac{2(3 \pm \sqrt{19})}{10} && \textbf{Common-term factoring} \\
&= \frac{3 \pm \sqrt{19}}{5} && \textbf{Dividing numerator and denominator by 2}
\end{aligned}$$

Therefore

$$x = \frac{3 + \sqrt{19}}{5} \quad \text{and} \quad x = \frac{3 - \sqrt{19}}{5}$$

are the solutions of the quadratic equation. The check is left to the reader. □

EXAMPLE 3 Solve $x^2 + x + 1 = 0$ using the quadratic formula.

Solution: In the equation $x^2 + x + 1 = 0$, $a = 1$, $b = 1$, and $c = 1$. Substituting in the equation we have

$$\begin{aligned}
x &= \frac{-1 \pm \sqrt{(1)^2 - 4(1)(1)}}{2(1)} \\
&= \frac{-1 \pm \sqrt{1 - 4}}{2} \\
&= \frac{-1 \pm \sqrt{-3}}{2}.
\end{aligned}$$

Recall that $\sqrt{-1} = j$, therefore

$$x = \frac{-1 + j\sqrt{3}}{2} \quad \text{and} \quad \frac{-1 - j\sqrt{3}}{2}$$

are the roots of the equation $x^2 + x + 1 = 0$. □

EXAMPLE 4 Solve the quadratic equation $32y^2 - 64y - 256 = 0$ using the quadratic formula.

Solution: Before we attempt to solve the equation $32y^2 - 64y - 256 = 0$ we will simplify the equation by factoring and then dividing both sides by 32. The resulting equation is $y^2 - 2y - 8 = 0$.

Thus, $a = 1$, $b = -2$, and $c = -8$. Substituting into the formula we have

$$y = \frac{-(-2) \pm \sqrt{(-2)^2 - 4(1)(-8)}}{2(1)}$$

We changed the variable to y since we are solving the equation for y and not for x.

$$y = \frac{2 \pm \sqrt{4 + 32}}{2}$$

$$= \frac{2 \pm 6}{2}$$

Therefore

$$y = \frac{2 + 6}{2} \quad \text{or} \quad y = \frac{2 - 6}{2}$$

$$= 4 \qquad\qquad = -2.$$

The solutions of the equation $32y^2 - 64y - 256 = 0$ are $y = 4$ and $y = -2$. The check is left to the reader. □

11.4 Exercises

In Exercises 1–34, solve the quadratic equations using the quadratic formula.

1. $x^2 - 16x - 36 = 0$

2. $x^2 + 16x - 36 = 0$

3. $x^2 + 14x + 24 = 0$

4. $x^2 - 14x + 24 = 0$

5. $6x^2 - 12x - 5 = 0$

6. $6x^2 - 7x - 5 = 0$

7. $2y^2 + 4y = -3$

8. $2y^2 + 7y = -3$

9. $6y^2 = 19y - 3$

10. $6y^2 = 9y - 3$

11. $8x^2 + 15 = 22x$

12. $8x^2 + 15 = 26x$

13. $-4t + 1 = -t^2$

14. $-7t + 2 = -t^2$

15. $2t^2 - 10t + 11 = 0$

16. $3t^2 - 6t + 1 = 0$

17. $R^2 + 2R - 4 = 0$

18. $2R^2 + 3R - 7 = 0$

19. $Z^2 - 2Z + 5 = 0$

20. $Z^2 - 6Z + 34 = 0$

21. $Z^2 + 4Z + 13 = 0$

22. $Z^2 + Z + 2 = 0$

23. $A^2 - 5A + 1 = 0$

24. $2A^2 - 5A - 3 = 0$

25. $A(3A - 11) = 20$

26. $A(6A + 7) = 3$

27. $(B + 1)(B + 5) = 13$

28. $(3B + 1)(B - 3) = 7$

29. $2B^2 - \sqrt{3}\,B + 1 = 0$

30. $\sqrt{2}\,B^2 - 3B + \sqrt{18} = 0$

31. $9.8x^2 + 14.0x - 8.0 = 0$

32. $-4x^2 + 13x - 10 = 0$

33. $980x^2 + 100x - 50 = 0$

34. $1.2x^2 + 2.5x - 1.0 = 0$

35. The rotational motion of a body is given by $\theta = \frac{1}{2}\alpha t^2 + \omega_0 t$. Find the values of t (in seconds) for $\alpha = 3$, $\omega_0 = 5$, and $\theta = 4$.

36. The reactance X of an AC circuit is given by $X = \omega L - \frac{1}{\omega C}$. For $X = 5$, $L = 3$, and $C = 0.02$, find the values of ω in rad/s.

37. The cost (in dollars) of building a shelf is given by $C = \frac{\ell^2}{10} - 3\ell$, where ℓ is the length of the shelf (in inches). How long can the shelf be made if the cost is to be kept under \$125?

38. An object is thrown downward with an initial velocity of 5 ft/s. The relation between the distance s it travels and time t is given by $s = 5t + 16t^2$. How long does it take the object to fall 74 ft?

39. A filter has a transfer function given as

$$\frac{E_0}{E_1} = S^2 - S - 1.293.$$

What positive value of S will cause the value of $\frac{E_0}{E_1}$ to be 0.707?

40. The ionization constant of acids is given by

$$K = \frac{\alpha^2}{1 - \alpha}$$

Solve for α.

11.5 The roots of a quadratic equation

In problem solving it may be helpful to have information about the roots of a quadratic equation. That is, are the roots real or imaginary, equal or unequal, rational or irrational?

Let r_1 and r_2 denote the roots of the general quadratic equation $ax^2 + bx + c = 0$. Using the quadratic formula we have

$$r_1 = \frac{-b + \sqrt{b^2 - 4ac}}{2a} \qquad r_2 = \frac{-b - \sqrt{b^2 - 4ac}}{2a}.$$

We can see by examining these two roots that the expression under the radical, $b^2 - 4ac$, will affect the roots. The expression $b^2 - 4ac$ is called the **discriminant** of the quadratic formula.

EXAMPLE 1 Examine the roots of the quadratic equation

$$x^2 + 5x + 6 = 0.$$

Solution: Solving the equation using the quadratic formula we have

$$r_1 = \frac{-5 + \sqrt{1}}{2} = -2. \qquad r_2 = \frac{-5 - \sqrt{1}}{2} = -3.$$

The roots of the equation are real, unequal, rational numbers. We obtain this type of result since the coefficients a, b, and c of the equation are rational numbers and the discriminant, $b^2 - 4ac$ equals 1, which is a perfect square. □

EXAMPLE 2 Examine the roots of the quadratic equation

$$x^2 + 6x + 9 = 0.$$

Solution: Solving the equation using the quadratic formula we have

$$r_1 = \frac{-6 + \sqrt{36 - 36}}{2} \qquad r_2 = \frac{-6 - \sqrt{36 - 36}}{2}$$
$$= -3. \qquad\qquad = -3.$$

The roots of the equation are real, equal, rational numbers. We obtain this type of result since the coefficients a, b, and c of the equation are rational numbers and the discriminant, $b^2 - 4ac$, equals 0. □

EXAMPLE 3 Examine the roots of the quadratic equation

$$x^2 + 5x + 3 = 0.$$

Solution: Solving the equation using the quadratic formula we have

$$r_1 = \frac{-5 + \sqrt{13}}{2} \qquad r_2 = \frac{-5 - \sqrt{13}}{2}$$

The roots of the equation are real, unequal, irrational numbers. We obtain this type of result since the coefficients a, b, and c of the equation are rational numbers and the discriminant, $b^2 - 4ac$, equals 13, which is not a perfect square. □

EXAMPLE 4 Examine the roots of the quadratic equation

$$x^2 + 3x + 3 = 0.$$

Solution: Solving the equation using the quadratic formula we have

$$r_1 = \frac{-3 + \sqrt{-3}}{2} \qquad r_2 = \frac{-3 - \sqrt{-3}}{2}.$$

The roots of the equation are imaginary and unequal. We obtain this type of result since the coefficients a, b, and c, of the equation are rational numbers and the discriminant, $b^2 - 4ac$, equals -3, which is negative. □

We have summarized these results in Tables 11.1 and 11.2.

EXAMPLE 5 Determine the values of the constant m for which the equation $mx^2 - 3x = m - 3$ will have equal roots.

TABLE 11.1 **Quadratic equations with real coefficients**

If $b^2 - 4ac$ is	Positive	Zero	Negative
The roots of $ax^2 + bx + c = 0$ are	real and unequal	real and equal	imaginary and unequal

TABLE 11.2 **Quadratic equations with rational coefficients**

If $b^2 - 4ac$ is	Positive and a perfect square	Positive but not a perfect square	Zero	Negative
The roots of $ax^2 + bx + c = 0$ are	real, rational, and unequal	real, irrational, and unequal	real, rational, and equal	imaginary and unequal

Solution: Write the given equation in standard form:

$$mx^2 - 3x = m - 3$$
$$mx^2 - 3x - m + 3 = 0.$$

If the roots are to be equal, the discriminant, $b^2 - 4ac$, must be equal to zero. Since $a = m$, $b = -3$, and $c = -m + 3$

$$(-3)^2 - 4(m)(-m + 3) = 0 \quad \textbf{Substituting in } b^2 - 4ac = 0$$
$$9 + 4m^2 - 12m = 0 \quad \textbf{Multiplying}$$
$$4m^2 - 12m + 9 = 0$$
$$(2m - 3)(2m - 3) = 0 \quad \textbf{Factoring}$$
$$m = \frac{3}{2}.$$

Therefore, if $m = 3/2$, the equation $mx^2 - 3x = m - 3$ will have equal roots. □

11.5 Exercises

In Exercises 1–20, state whether the discriminant is positive, negative, or zero and describe the roots of each equation.

1. $3x^2 + 5x - 2 = 0$
2. $9x^2 + 12x + 4 = 0$
3. $0.3y^2 - 0.1y + 1 = 0$
4. $2y^2 - 0.2y + 0.6 = 0$
5. $\frac{1}{3}t^2 + \frac{1}{4}t - \frac{1}{7} = 0$
6. $\frac{3}{5}t^2 + \frac{2}{7}t - \frac{1}{3} = 0$
7. $4R^2 - 5 = R + 3$
8. $6R^2 + 13 = 13R$
9. $9x^2 + 12x + 4 = 0$
10. $x^2 - 8x + 16 = 0$
11. $y^2 - 7y = 0$
12. $3y^2 + 11y = 0$
13. $2x^2 - \sqrt{3}\,x + 1 = 0$
14. $\sqrt{2}\,x^2 + \sqrt{3}\,x - \sqrt{8} = 0$
15. $x^2 + \sqrt{5}\,x - 5 = 0$
16. $\sqrt{2}\,x^2 - x + \sqrt{2} = 0$
17. $x^2 + 2jx - 5 = 0$
18. $2jx^2 + 5x + j = 0$
19. $x^2 + 4jx - 4 = 0$
20. $9x^2 - 6jx - 1 = 0$

In Exercises 21–26, find the value(s) of k for which the equations will have equal roots.

21. $kx^2 - 12x + 4 = 0$
22. $2kx^2 - 20x + 25 = 0$
23. $9x^2 - 30x + k = 0$
24. $kx^2 - 8x + 16k = 0$

25. $k(y^2 + y + 1) = y + 1$

26. $y^2 + 2(k + 4)y + 16k = 0$

27. In solving for the area of a rectangular region, John arrived at the equation $x^2 + 5x + 7 = 0$. His instructor informed him that this could not be the correct equation. Why not?

In Exercises 28 and 29, determine B so that the given equation has just one root.

28. $x^2 + Bx + 16 = 0$

29. $3x^2 + Bx + 7 = 0$

30. Show that the equation $x^2 + Bx - 2 = 0$ always has real roots (regardless of the coefficient B).

11.6 The graph of a quadratic function

Graphical solution of quadratic equations

Polynomials of degree two, such as $ax^2 + bx + c$, define a class of functions called quadratic functions. The graph of a quadratic function is a parabola.

DEFINITION 11.3

A function f defined by $y = f(x) = ax^2 + bx + c$, where a, b, and c are real numbers and $a \neq 0$, is called a quadratic function.

EXAMPLE 1 The function f defined by $f(x) = x^2 + x - 20$ is a quadratic function with $a = 1$, $b = 1$, and $c = -20$. □

EXAMPLE 2 The function g defined by $g(x) = x^2$ is a quadratic function with $a = 1$, $b = 0$, and $c = 0$. □

EXAMPLE 3 The function h defined by $h(x) = \dfrac{1}{3x^2 + 7x}$ is not a quadratic function, since $\dfrac{1}{3x^2 + 7x}$ cannot be expressed as a second-degree polynomial. □

EXAMPLE 4 The function k, defined by $k(x) = -2x^2 + 3x + 20$ is a quadratic function with $a = -2$, $b = 3$, and $c = 20$. □

To graph the function of Example 1, we must determine sets of ordered pairs that satisfy the function. Table 11.3 demonstrates how we can obtain ordered pairs that satisfy $f(x) = x^2 + x - 20$. (Table 11.4 contains a BASIC program to determine the integer ordered pairs.) The ordered pairs are graphed in Fig. 11.3. With the aid of these ordered pairs we can sketch the curve as in Fig. 11.4.

TABLE 11.3

x	$y = f(x) = x^2 + x - 20$	$y = f(x)$	Ordered pairs
-6	$f(-6) = (-6)^2 + (-6) - 20$	10	$(-6, 10)$
-5	$f(-5) = (-5)^2 + (-5) - 20$	0	$(-5, 0)$
-4	$f(-4) = (-4)^2 + (-4) - 20$	-8	$(-4, -8)$
-3	$f(-3) = (-3)^2 + (-3) - 20$	-14	$(-3, -14)$
-2	$f(-2) = (-2)^2 + (-2) - 20$	-18	$(-2, -18)$
-1	$f(-1) = (-1)^2 + (-1) - 20$	-20	$(-1, -20)$
$-\frac{1}{2}$	$f\left(-\frac{1}{2}\right) = \left(-\frac{1}{2}\right)^2 + \left(-\frac{1}{2}\right) - 20$	$-20\frac{1}{4}$	$\left(-\frac{1}{2}, -20\frac{1}{4}\right)$
0	$f(0) = (0)^2 + (0) - 20$	-20	$(0, -20)$
1	$f(1) = (1)^2 + (1) - 20$	-18	$(1, -18)$
2	$f(2) = (2)^2 + (2) - 20$	-14	$(2, -14)$
3	$f(3) = (3)^2 + (3) - 20$	-8	$(3, -8)$
4	$f(4) = (4)^2 + (4) - 20$	0	$(4, 0)$
5	$f(5) = (5)^2 + (5) - 20$	10	$(5, 10)$

TABLE 11.4

```
1Ø   FOR X = -6 TO 5
2Ø   LET Y = X * X + X - 2Ø
3Ø   PRINT "X = ";X,"Y = ";Y
4Ø   NEXT X
5Ø   END
```

FIGURE 11.3

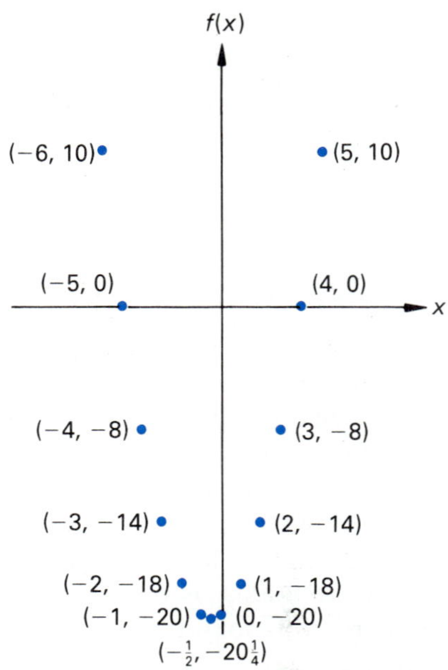

FIGURE 11.4

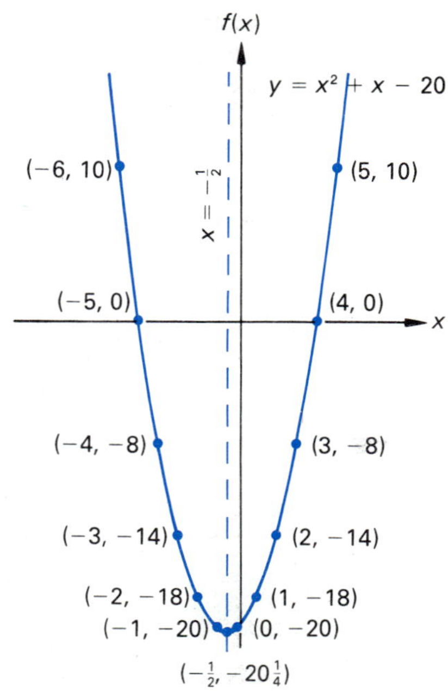

Examining the graph of $y = x^2 + x - 20$ we can note special properties that will also be true for all quadratic functions. Note the sets of points $(-6, 10)$, $(5, 10)$; $(-5, 0)$, $(4, 0)$; $(-4, -8)$, $(3, -8)$; $(2, -18)$, $(1, -18)$; and $(-1, -20)$, $(0, -20)$. The first point of each ordered pair is on the left branch of the curve and the second point is directly opposite it on the right branch of the curve. One can almost imagine a line dividing the curve into two identical halves. In fact, $x = -1/2$ is the **line of symmetry** that cuts through the center of the parabola at the point $\left(-\frac{1}{2}, -20\frac{1}{4}\right)$. This low point is called the **vertex** of the parabola. The vertex is the maximum point (highest point) or the minimum point (lowest point) on the curve.

RULE 11.1

> The abscissa (x-value) of the vertex of the parabola
>
> $$f(x) = ax^2 + bx + c$$
>
> can be determined using the formula
>
> $$x = -\frac{b}{2a}$$
>
> The ordinate (y-value) of the vertex is found by substituting x into the function.

For the function $f(x) = x^2 + x - 20$

$$x = \frac{-(1)}{2(1)} = -\frac{1}{2}. \qquad f\left(-\frac{1}{2}\right) = \left(-\frac{1}{2}\right)^2 + \left(-\frac{1}{2}\right) - 20 = -20\frac{1}{4}.$$

The ordered pair $\left(-\frac{1}{2}, -20\frac{1}{4}\right)$ is the vertex of the parabola.

The roots of the quadratic equation $x^2 + x - 20 = 0$ are the abscissas of the points on the graph where the parabola crosses the x-axis (or horizontal axis). The abscissas of the x-intercepts are those values of x for which $f(x) = 0$. For this function 4 and -5 are the roots since the points $(4, 0)$ and $(-5, 0)$ are on the x-axis, $f(4) = 0$, and $f(-5) = 0$.

A key point on the graph is $(0, -20)$, which is the ***y*-intercept,** the point where the curve crosses the y-axis. For the general function $f(x) = ax^2 + bx + c$, the y-intercept is found by letting $x = 0$. Then $f(0) = c$, giving us the ordered pair $(0, c)$.

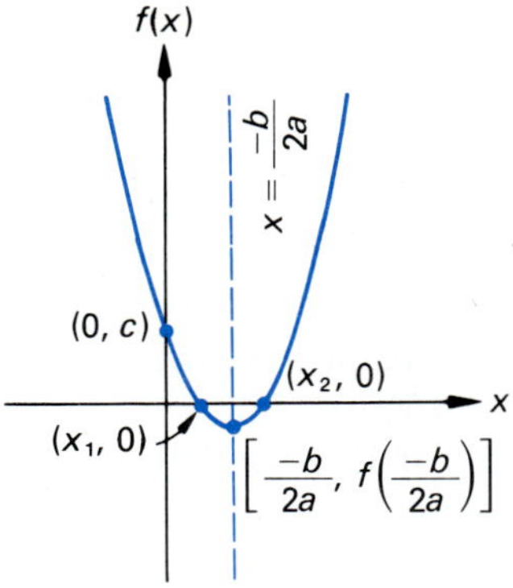

FIGURE 11.5

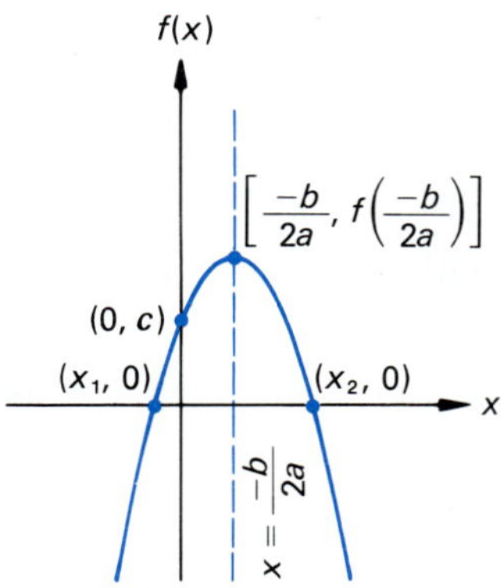

FIGURE 11.6

Summarizing the major points of the previous discussion we suggest that the reader determine the following points whenever possible to simplify the graphing of a quadratic function.

1. The vertex of the parabola is $\left(\frac{-b}{2a}, f\left(\frac{-b}{2a}\right)\right)$.
2. If a (the coefficient of the second-degree term) is positive, the vertex is a minimum and the parabola opens up (Fig. 11.5). If a is negative, then the vertex is a maximum and the parabola opens down (Fig. 11.6).
3. Find the x-intercepts by factoring $ax^2 + bx + c = 0$ or using the quadratic formula. (As illustrated in the following examples there may be 2, 1, or 0 intercepts.)
4. The y-intercept is $(0, c)$.
5. The line of symmetry is $x = \frac{-b}{2a}$.
6. Use the concept of symmetry to graph additional points.

EXAMPLE 5 Graph the quadratic function $f(x) = -2x^2 + 3x + 20$ and determine the roots of the quadratic equation.

Solution: Determining the key points for $f(x) = -2x^2 + 3x + 20$ as suggested we have

1. Vertex (0.75, 21.125).

$$\text{Since } \frac{-b}{2a} = \frac{-(3)}{2(-2)} = 0.75,$$

$$f(0.75) = -2(0.75)^2 + 3(0.75) + 20 = 21.125.$$

2. The parabola opens down since $a = -2$. Also the vertex (0.75, 21.125) is the maximum point on the curve.
3. x-intercepts are $(-2.5, 0)$ and $(4, 0)$.

$$-2x^2 + 3x + 20 = 0$$

$$(-x + 4)(2x + 5) = 0 \quad \textbf{Factoring}$$

$$x = 4,\ x = -\frac{5}{2} \quad \textbf{Solving for } x$$

4. The y-intercept is $(0, 20)$.

5. The line of symmetry is $x = 0.75$, since $x = \dfrac{-b}{2a}$.

With this information we can sketch the curve. To develop a smoother curve we determine the points on the curve for $x = 2$, 3, and 5. The corresponding ordered pairs are (2, 18), (3, 11), and (5, −15). Now that we have these points and know that the line of symmetry is $x = 0.75$, we can determine the respective symmetric points (−0.5, 18), (−1.5, 11), and (−3.5, −15).

Note that the ordinates of (2, 18) and (−0.5, 18) are both the same, 18, and the abscissas 2 and −0.5 are both the same distance, 1.25 units, from the line of symmetry.

The curve is sketched in Fig. 11.7. The roots of the quadratic equation $-2x^2 + 3x + 20 = 0$ are 4 and −5/2, the points where the curve crosses the x-axis. □

EXAMPLE 6 Graph the quadratic function

$$f(x) = 2x^2 + 4x + 3$$

and determine the roots of the quadratic equation.

Solution: Determining the key points for $f(x) = 2x^2 + 4x + 3$ as suggested we have

1. Vertex (−1, 1).
2. The parabola opens up since $a = 2$. Thus, the vertex (−1, 1) is a minimum point on the curve.
3. Since $2x^2 + 4x + 3$ is not factorable, using the quadratic formula we have

   $$x = \frac{-4 \pm \sqrt{16 - 24}}{4} \qquad \text{or} \qquad x = \frac{-4 \pm \sqrt{-8}}{4}$$

   since $b^2 - 4ac < 0$, the roots are not real numbers and there are no x-intercepts.
4. The y-intercept is (0, 3).
5. The line of symmetry is $x = -1$.
6. Calculate points for $x = 0$, 1, and their respective symmetric points. The result is (0, 3), (−2, 3), (1, 9), and (−3, 9).

The curve is sketched in Fig. 11.8. The quadratic equation

$$2x^2 + 4x + 3 = 0$$

does not have any real roots since the curve does not cross the x-axis. □

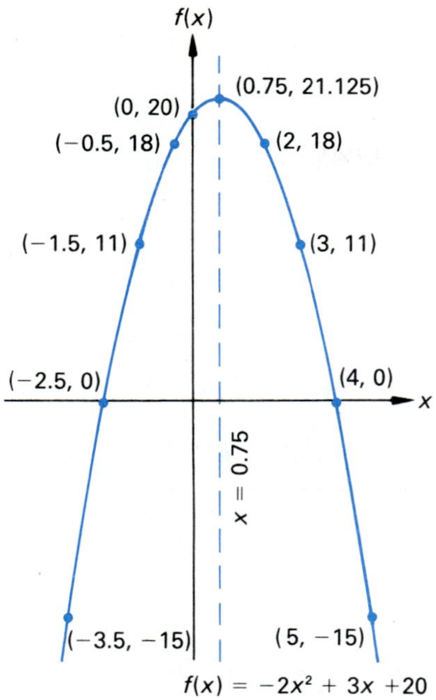

FIGURE 11.7

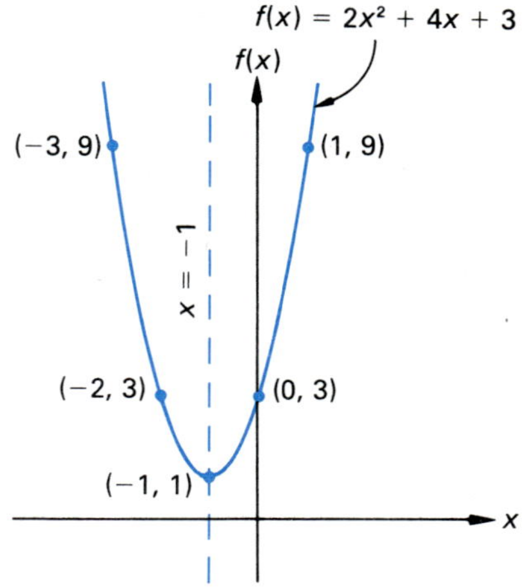

FIGURE 11.8

EXAMPLE 7 Graph the quadratic function

$$f(x) = x^2$$

Solution: We can determine the following facts about the function $f(x) = x^2$:

1. Vertex $(0, 0)$.
2. Parabola opens up since $a = 1$, and the vertex is the minimum point on the curve.
3. The x-intercept is $(0, 0)$.
4. The y-intercept is $(0, 0)$.
5. The line of symmetry is $x = 0$.
6. Some points that satisfy the function are $(1, 1)$, $(-1, 1)$, $(2, 4)$, $(-2, 4)$, $(3, 9)$, and $(-3, 9)$.

The curve is sketched in Fig. 11.9. The curve touches the x-axis at only one point, therefore the quadratic equation $x^2 = 0$ has only one real distinct root (it has two roots but both values are the same). □

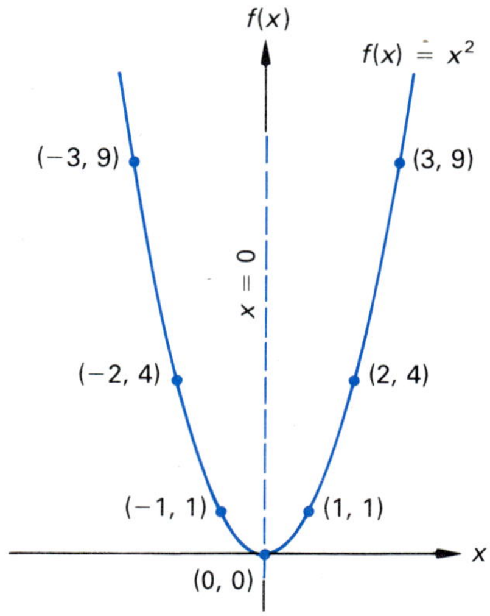

FIGURE 11.9

For the quadratic equations in Examples 5, 6, and 7, we found that there were two, zero, and one real distinct roots. The curve crossed the horizontal axis at two points, did not cross, and touched at only one point, respectively. If we had calculated the discriminant, the results would have been $b^2 - 4ac > 0$, $b^2 - 4ac < 0$, and $b^2 - 4ac = 0$, respectively.

11.6 Exercises

In Exercises 1–20, graph the quadratic function and find the roots of the quadratic equation to the nearest tenth.

1. $f(x) = x^2 + 6x + 5$

2. $f(x) = x^2 - 6x + 5$

3. $f(x) = x^2 - 5x + 4$

4. $f(x) = 4x^2 + 12x + 9$

5. $f(x) = -3x^2 + 2x + 5$

6. $f(x) = x^2 + x + 2$

7. $f(x) = -2x^2 + 3x + 5$

8. $f(x) = -4x^2 - 4x + 3$

9. $f(x) = 2x^2 + 3x + 2$

10. $f(x) = -2x^2 + 5x - 6$

11. $f(x) = x^2 - 4x$

12. $f(x) = 3x - x^2$

13. $f(x) = 4x^2 - 20x + 25$

14. $f(x) = -9x^2 + 42x - 49$

15. $f(x) = 3(x + x^2) + x - 5$

16. $f(x) = 4(1 - x^2) + 2x - 3$

17. $f(x) = -9x^2 + 12x - 4$

18. $f(x) = 4x^2 - 9$

19. $f(x) = 2x^2 + 5x + 2$

20. $f(x) = \frac{1}{3}x^2 + 3$

In Exercises 21–30, without plotting the graph, decide by means of the discriminant whether the graph will cross the x-axis at two distinct points, will touch the x-axis at one distinct point, or will not touch the x-axis at all.

21. $f(x) = 4x^2 + 12x + 9$

22. $f(x) = -x^2 + 7x - 12$

23. $f(x) = x^2 - 2x - 15$

24. $f(x) = 3x^2 + 3x + 4$

25. $f(x) = 4x^2 + 3x + 2$

26. $f(x) = -x^2 + 2x + 5$

27. $f(x) = 2x^2 + 3x - 4$

28. $f(x) = 16x^2 + 25$

29. $f(x) = 13x^2 + 11x - 5$

30. $f(x) = 8x^2 + 11x - 8$

In Exercises 31–34, find the value of m for which the graph of each function will (a) touch the x-axis at one point, (b) cross the x-axis in two distinct points, or (c) will not touch the x-axis at all.

31. $f(x) = mx^2 - 2x + 1$

32. $f(x) = 2x^2 - 4x + m$

33. $f(x) = 4x^2 - 3mx + 9$

34. $f(x) = 4x^2 + 4mx + 1$

11.7 Quadratic equations involving trigonometric functions

Equations involving trigonometric functions such as

$$\cos A = \frac{1}{2} \qquad \text{or} \qquad \tan^2 A - \tan A = 2$$

are known as trigonometric equations. We want to know all the values of A in the domain of the function that will satisfy the given equation. Before discussing quadratic equations, let us examine some simple examples to illustrate how we write the solutions to trigonometric equations.

EXAMPLE 1 Suppose that $0 \le A < 2\pi$. Determine the solution to each equation.

a. $\sin A = -1$ **b.** $\cos A = \frac{1}{2}$ **c.** $\cos A = 3$

Solution:

a. $\sin A = -1$, $\sin \frac{3\pi}{2} = -1$. $\frac{3\pi}{2}$ is the only number in the interval $0 \le A < 2\pi$ whose sine is -1. Therefore, the root of the equation is $\frac{3\pi}{2}$.

b. The cosine is positive in the first and fourth quadrants. There are two possible values that satisfy $\cos A = \frac{1}{2}$:

$$\cos\frac{\pi}{3} = \frac{1}{2} \qquad \text{and} \qquad \cos\frac{5\pi}{3} = \frac{1}{2}.$$

Therefore, the roots of the equation are $\frac{\pi}{3}$ and $\frac{5\pi}{3}$.

c. For all A, $-1 \leq \cos A \leq 1$. Thus, $\cos A$ can never be 3. Therefore, the equation $\cos A = 3$ has no solution. □

Recall from Chapter 8 that the trigonometric functions are periodic functions. The sine and cosine functions have a period of 2π and the tangent function has a period of π. Therefore, if we want the roots of all the equations in Example 1, not just for A in the interval $0 \leq A < 2\pi$, we must add multiples of 2π to the solutions.

EXAMPLE 2 Let A be any real number, then:

a. The roots of $\sin A = -1$ are all the numbers of the form $\frac{3\pi}{2} + 2k\pi$, where k is any integer. A solution in this form is called the **general solution.** Some of the numbers that are in this set are $\frac{3\pi}{2}$ when $k = 0$, $\frac{7\pi}{2}$ when $k = 1$, and $\frac{-\pi}{2}$ when $k = -1$.

b. The roots of $\cos A = \frac{1}{2}$, written in general form are $\frac{\pi}{3} + 2k\pi$ or $\frac{5\pi}{3} + 2k\pi$, where k is any integer. □

We are now ready for the discussion of trigonometric equations of the quadratic form.

EXAMPLE 3 Solve the equation $2\sin^2\theta + \sin\theta - 1 = 0$.

a. Find the solution of the equation for $0 \leq \theta < 2\pi$.

b. Find the general solution of the equation.

Solution:

a. If we let $x = \sin\theta$, then the equation $2\sin^2\theta + \sin\theta - 1 = 0$ takes the form

$$2x^2 + x - 1 = 0$$
$$(2x - 1)(x + 1) = 0. \quad \textbf{Factoring}$$

Now replace x with $\sin\theta$ to obtain

$$(2\sin\theta - 1)(\sin\theta + 1) = 0$$
$$2\sin\theta - 1 = 0 \quad \text{or} \quad \sin\theta + 1 = 0 \qquad \textbf{Setting each factor equal to zero}$$

$$\sin\theta = \frac{1}{2} \quad \text{or} \quad \sin\theta = -1 \qquad \textbf{Solving for } \sin\theta$$

$$\theta = \frac{\pi}{6}, \frac{5\pi}{6} \qquad \theta = \frac{3\pi}{2}. \qquad \textbf{Solving for } \theta$$

Check: $2\sin^2\theta + \sin\theta - 1 = 0$

$$\theta = \frac{\pi}{6}$$

$$2\left(\sin\frac{\pi}{6}\right)^2 + \sin\frac{\pi}{6} - 1 = 0$$

$$2(0.5)^2 + (0.5) - 1 = 0$$

$$0.5 + 0.5 - 1 = 0. \checkmark$$

$$\theta = \frac{5\pi}{6}$$

$$2\left(\sin\frac{5\pi}{6}\right)^2 + \sin\frac{5\pi}{6} - 1 = 0$$

$$2(0.5)^2 + (0.5) - 1 = 0$$

$$0.5 + 0.5 - 1 = 0. \checkmark$$

$$\theta = \frac{3\pi}{2}$$

$$2\left(\sin\frac{3\pi}{2}\right)^2 + \sin\frac{3\pi}{2} - 1 = 0$$

$$2(-1)^2 + (-1) - 1 = 0$$

$$2 - 1 - 1 = 0. \checkmark$$

The roots for $2\sin^2\theta + \sin\theta - 1 = 0$ in the interval $0 \le \theta < 2\pi$ are $\frac{\pi}{6}, \frac{5\pi}{6}$, and $\frac{3\pi}{2}$.

b. Since the sine function has a period of 2π, the general solution of $2\sin^2\theta + \sin\theta - 1 = 0$ contains the roots $\frac{\pi}{6} + 2k\pi$, $\frac{5\pi}{6} + 2k\pi$, and $\frac{3\pi}{2} + 2k\pi$, where k is an integer. □

EXAMPLE 4 Solve the equation $\tan^2\theta + 3\tan\theta + 1 = 0$. Find the general solution of the equation and express the roots to the nearest tenth of a degree.

Solution: Substituting x for $\tan\theta$ the equation becomes $x^2 + 3x + 1 = 0$. The quadratic equation is not factorable therefore we must use the quadratic formula to solve for $\tan\theta$. From the equation we see that $a = 1$, $b = 3$, and $c = 1$. Substituting we have

$$\tan\theta = \frac{-3 \pm \sqrt{3^2 - 4(1)(1)}}{2(1)}$$

$$= \frac{-3 \pm \sqrt{5}}{2}$$

$$\tan\theta = \frac{-3 + \sqrt{5}}{2} \quad \text{or} \quad \tan\theta = \frac{-3 - \sqrt{5}}{2}$$

$$= -0.3820 \quad \text{or} \quad = -2.6180$$

$$\hat{\theta} = 20.9^\circ \quad \text{or} \quad \hat{\theta} = 69.1^\circ.$$

Recall that the tangent function is negative in the second quadrant. Therefore, the second-quadrant angles to the nearest tenth of a degree are

$$\theta = 159.1° \qquad \text{and} \qquad \theta = 110.9°.$$

To the nearest tenth of a degree the general solution is $159.1° + k(180°)$ and $110.9° + k(180°)$, where k is an integer. Note that we need only consider angles in the first and second quadrants, since the tangent function has a period of $180°$. □

EXAMPLE 5 Solve the equation: $\tan^2 x = 3$. Find the general solution and express the roots in exact form.

Solution:

$$\tan^2 x = 3$$

$$\tan x = \pm\sqrt{3} \qquad \textbf{Taking square root of both sides}$$

$$\tan x = \sqrt{3} \quad \text{or} \quad \tan x = -\sqrt{3}$$

$$x = \frac{\pi}{3} + k\pi \quad \text{or} \quad x = \frac{2\pi}{3} + k\pi,$$

where k is an integer.

The roots of the equation are $\frac{\pi}{3} + k\pi$ and $\frac{2\pi}{3} + k\pi$. Since radians were not specified, the answer could be expressed as $60° + k \cdot 180°$ and $120° + k \cdot 180°$. □

11.7 Exercises

In Exercises 1–12, solve the equation for all values of θ such that $0 \le \theta \le 2\pi$. Do not use calculators or tables. Check answers.

1. $\cos\theta = 0$

2. $\sin\theta = 1$

3. $\tan\theta + \sqrt{3} = 0$

4. $2\sin\theta + \sqrt{3} = 0$

5. $\sqrt{3}\tan\theta = 1$

6. $2\cos\theta - 1 = 0$

7. $\sqrt{2}\cos\theta - 1 = 0$

8. $\tan\theta - 1 = 0$

9. $\sin\theta = 0$

10. $2\cos\theta + \sqrt{3} = 0$

11. $2\sin\theta + 1 = 0$

12. $\tan x \sec x = \tan x$

In Exercises 13–24, find the general solution to each of the equations. Do not use calculators or tables. Check answers.

13. $\sin^2\theta + \sin\theta - 2 = 0$

14. $\cos^2\theta + 2\sqrt{2}\cos\theta + 2 = 0$

15. $2\cos^2\theta + \sqrt{3}\cos\theta = 0$

16. $\sqrt{3}\sin^2\theta - 2\sin\theta = 0$

17. $4\sin^2\theta - 3 = 0$

18. $4\cos^2\theta - 1 = 0$

19. $2\csc^2\theta - 5\csc\theta + 2 = 0$

20. $1 - \tan^2\theta = 0$

21. $3\cot^2\theta - 1 = 0$

22. $4 - \sec^2 x = 0$

23. $2\sin^2\theta + 3\sin\theta = -1$

24. $2\cos^2 x - 3\cos x - 2 = 0$

In Exercises 25–36, find the general solution to the equation and express the roots to the nearest tenth of a degree.

25. $6\sin^2\theta + \sin\theta - 2 = 0$

26. $12\cos^2\theta - \cos\theta - 6 = 0$

27. $4\tan^2\theta - 12\tan\theta + 9 = 0$

28. $\tan^2\theta + \tan\theta = 1$

29. $\cos^2\theta - \cos\theta - 1 = 0$

30. $2\cos^2\theta - 5\cos\theta = -2$

31. $2\sin^2 x\cos x = \cos x$

32. $3\tan^2 x\cot x = \cot x$

33. $\cos^2 A - 2\cos A = 2$

34. $\sin^2 A + \sin A = 1$

35. $-3\sin^2 A - 3\sin A + 5 = 0$

36. $7\cos^2 A + 5\cos A = 2$

37. The volume V of a conical tank of height h and vertex angle θ is given by

$$V = \frac{\pi h^3(\tan\theta)^2}{3}. \qquad 0^\circ < \theta < 90^\circ$$

To the nearest tenth of a degree, find the vertex angle of a conical tank if $V = 393$ cm³ and $h =$ 15 cm. (See Fig. 11.10.)

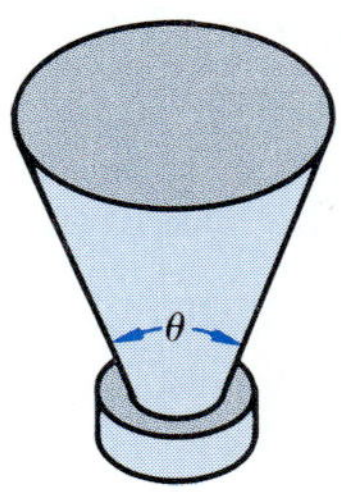

FIGURE 11.10

38. The area A of the trapezoid shown in Fig. 11.11 is given by $A = s^2(1 + \cos\theta)\sin\theta$. It can be shown that the greatest possible area of the trapezoid occurs when $2\cos^2\theta + \cos\theta - 1 = 0$ is solved for the interval $0^\circ < \theta < 90^\circ$. Find the angle θ that gives the greatest area and compute the greatest area.

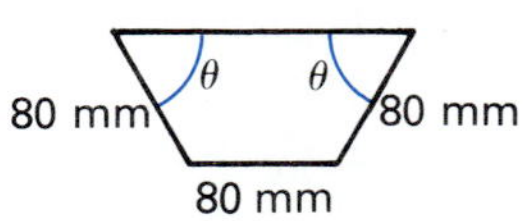

FIGURE 11.11

39. The volume V of a cone inscribed in a sphere with a 1-m radius is given by

$$V = \frac{8\pi}{3}\sin^2\theta\cos^4\theta.$$

It can be shown that the cone with maximum volume occurs when θ is the solution of

$$1 - 3\sin^2\theta = 0, \qquad 0^\circ < \theta < 90^\circ.$$

To the nearest tenth, find θ for the maximum volume cone and compute the maximum volume. (See Fig. 11.12.)

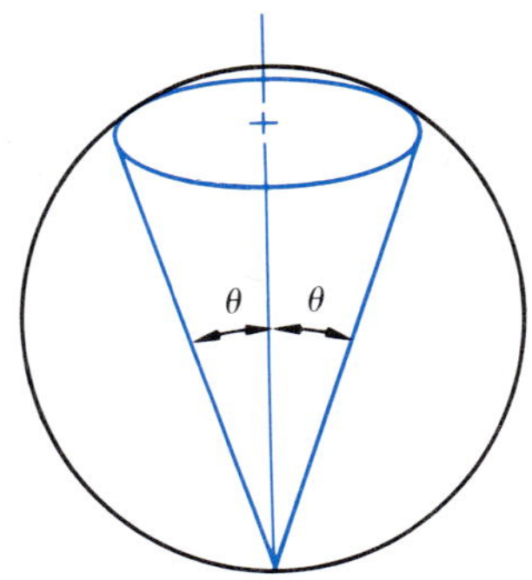

FIGURE 11.12

40. The displacement of a particle is given by $s = 2\sin^2 t - \sin t + 1$, where s is the distance in meters and t is the time in seconds. Determine the time when $s = 2$ m.

11.8 Equations that can lead to quadratic equations

There are many equations that are not quadratic. However, with algebraic techniques the equations can be changed to quadratic form and solved. In this section we will examine equations in quadratic form, fractional equations, and equations involving radicals.

Equations that are in quadratic form

In Section 11.1 we said that an algebraic equation of the form

$$ax^2 + bx + c = 0, \qquad a \neq 0$$

is a quadratic equation in standard form. An equation of the form $x^4 - 5x^2 + 6 = 0$ can be put into quadratic form by the technique of changing the variable. This is illustrated in Example 1.

EXAMPLE 1 Solve the equation $x^4 - 5x^2 + 6 = 0$.

Solution: Since the equation

$$x^4 - 5x^2 + 6 = 0$$

can be rewritten as

$$(x^2)^2 - 5x^2 + 6 = 0$$

we can change the variable by letting $q = x^2$. Substituting, we have

$$q^2 - 5q + 6 = 0.$$

The new equation in terms of the variable q is in quadratic form. We can solve for q by factoring the equation:

$$(q-3)(q-2) = 0$$
$$q = 3 \qquad \text{or} \qquad q = 2.$$

Since $q = x^2$, we can substitute to determine the value of x.

$$x^2 = 3 \qquad \text{or} \qquad x^2 = 2.$$

Thus $x = \pm\sqrt{3}$, or $x = \pm\sqrt{2}$. The roots of the equation are $\sqrt{3}$, $-\sqrt{3}$, $\sqrt{2}$, and $-\sqrt{2}$. The check is left to the reader. □

EXAMPLE 2 Solve the equation $x^{1/3} - 3x^{1/6} + 2 = 0$.

Solution: The equation can be rewritten as

$$(x^{1/6})^2 - 3x^{1/6} + 2 = 0.$$

We can write the equation in quadratic form by letting $q = x^{1/6}$. Substituting, we have

$$q^2 - 3q + 2 = 0.$$

Solving the equation by factoring, we have

$$(q-2)(q-1) = 0$$
$$q = 2 \qquad \text{or} \qquad q = 1.$$

Since $q = x^{1/6}$, then $x^{1/6} = 2$ or $x^{1/6} = 1$. Now x can be found by raising both sides of the equation to the sixth power.

$$(x^{1/6})^6 = (2)^6 \quad \text{or} \quad (x^{1/6})^6 = (1)^6$$
$$x = 64 \quad \text{or} \quad x = 1.$$

Check:

$x = 64$	$x = 1$
$(64)^{1/3} - 3(64)^{1/6} + 2 = 0$	$(1)^{1/3} - 3(1)^{1/6} + 2 = 0$
$4 - 3(2) + 2 = 0$	$1 - 3(1) + 2 = 0$
$0 = 0.$ ✓	$0 = 0.$ ✓

Thus, the roots of the equation $x^{1/3} - 3x^{1/6} + 2 = 0$ are 1 and 64. □

EXAMPLE 3 Solve the equation $x + 3x^{1/2} = 4$.

Solution: Rewriting the equation we have

$$(x^{1/2})^2 + 3x^{1/2} - 4 = 0.$$

Let $q = x^{1/2}$, then $q^2 = x$; substituting, we have

$$q^2 + 3q - 4 = 0.$$

The equation is quadratic in q and can be solved by factoring

$$(q + 4)(q - 1) = 0$$
$$q = -4 \quad \text{or} \quad q = 1.$$

Now, substituting for q, we have

$$x^{1/2} = -4 \quad \text{or} \quad x^{1/2} = 1$$
$$x = 16 \quad \text{or} \quad x = 1. \qquad \textbf{Squaring both sides}$$

Check:

$x = 16$	$x = 1$
$16 + 3(16)^{1/2} = 4$	$1 + 3(1)^{1/2} = 4$
$16 + 12 \neq 4.$ **x**	$1 + 3 = 4.$ ✓

Thus, the only root of the equation $x + 3x^{1/2} = 4$ is 1. □

Whenever the new equation has roots that are not roots of the original equation, these roots are called **extraneous roots** of the original equation. In Example 2, 16 is an extraneous root.

EXAMPLE 4 Solve the equation $\dfrac{1}{x^4} - \dfrac{2}{x^2} - 8 = 0.$

Solution: Rewriting the equation, we have

$$\left(\frac{1}{x^2}\right)^2 - 2\left(\frac{1}{x^2}\right) - 8 = 0.$$

The equation is in quadratic form. Let $q = \frac{1}{x^2}$; substituting, we have

$$q^2 - 2q - 8 = 0.$$

The equation can be solved by factoring

$$(q-4)(q+2) = 0$$
$$q = 4 \quad \text{or} \quad q = -2.$$

Since $q = \frac{1}{x^2}$, we have

$$\frac{1}{x^2} = 4 \quad \text{or} \quad \frac{1}{x^2} = -2$$

$$x^2 = \frac{1}{4} \quad \text{or} \quad x^2 = -\frac{1}{2}$$ **Equivalently**

$$x = \pm\frac{1}{2} \quad \text{or} \quad x = \pm j\sqrt{\frac{1}{2}}.$$ **Taking square root of both sides**

Check:

$$x = \pm\frac{1}{2}$$

$$\frac{1}{\left(\pm\frac{1}{2}\right)^4} - \frac{2}{\left(\pm\frac{1}{2}\right)^2} - 8 = 0$$
$$16 - 2(4) - 8 = 0$$
$$0 = 0. \checkmark$$

$$x = \pm j\sqrt{\frac{1}{2}}$$

$$\frac{1}{\left(\pm j\sqrt{\frac{1}{2}}\right)^4} - \frac{2}{\left(\pm j\sqrt{\frac{1}{2}}\right)^2} - 8 = 0$$
$$4 - 2(-2) - 8 = 0$$
$$0 = 0. \checkmark$$

The roots of the equation $\frac{1}{x^4} - \frac{2}{x^2} - 8 = 0$ are $\pm\frac{1}{2}$ and $\pm j\sqrt{\frac{1}{2}}$. □

EXAMPLE 5 Solve the equation $3(y-2)^2 - 5(y-2) - 2 = 0$.

Solution: Let $u = y - 2$. Substituting into the equation, we have

$$3u^2 - 5u - 2 = 0.$$

The new equation is quadratic in u and can be solved by factoring:

$$(3u+1)(u-2)=0$$

$$u=-\frac{1}{3} \quad \text{or} \quad u=2.$$

Now, substituting for u, we can solve for y:

$$-\frac{1}{3}=y-2 \quad \text{or} \quad 2=y-2$$

$$y=1\frac{2}{3} \quad \text{or} \quad y=4.$$

Check:

$$y=1\frac{2}{3}$$

$$3\left(1\frac{2}{3}-2\right)^2-5\left(1\frac{2}{3}-2\right)-2=0$$

$$\frac{1}{3}+\frac{5}{3}-2=0. \checkmark$$

$$y=4$$

$$3(4-2)^2-5(4-2)-2=0$$

$$12-10-2=0. \checkmark$$

The roots of the equation $3(y-2)^2-5(y-2)-2=0$ are $1\frac{2}{3}$ and 4. □

Equations involving radical expressions

Solving equations with radicals will be easier if we remember the definitions of exponents from Section 2.1 and the definitions of radicals from Section 2.3. To solve equations with radicals we must raise both sides of the equation to the same power to eliminate the radicals.

EXAMPLE 6 Solve the equation $\sqrt{x+3}+2x=0$.

Solution: If we square both sides of the equation $\sqrt{x+3}+2x=0$ in present form, the result would contain a radical. That is,

$$(\sqrt{x+3}+2x)^2=x+3+4x\sqrt{x+3}+4x^2.$$

It is possible to prevent this from happening if we can isolate the radical on one side of the equal sign before squaring. Thus

$$\sqrt{x+3}=-2x.$$

Now square both sides and we have

$$x + 3 = 4x^2$$
$$4x^2 - x - 3 = 0. \qquad \textbf{Collecting like terms}$$

Factoring we have

$$(4x + 3)(x - 1) = 0$$
$$x = -\frac{3}{4} \quad \text{or} \quad x = 1.$$

Check:

$$x = -\frac{3}{4} \qquad\qquad x = 1$$
$$\sqrt{-\frac{3}{4} + 3} + 2\left(-\frac{3}{4}\right) = 0 \qquad\qquad \sqrt{1 + 3} + 2(1) = 0$$
$$\frac{3}{2} - \frac{3}{2} = 0. \checkmark \qquad\qquad 2 + 2 \neq 0. \text{ x}$$

Thus, the only root of the equation $\sqrt{x + 3} + 2x = 0$ is $-\frac{3}{4}$. The value $x = 1$ is an extraneous root; it will satisfy the intermediate form of the equation $4x^2 - x - 3 = 0$. □

EXAMPLE 7 Solve the equation $\sqrt{u - 2} = 2 + \sqrt{2u + 3}$

Solution: Since there are two radicals in the equation we cannot isolate the radical. The best approach in this situation is to have one radical on each side of the equal sign. By doing this we have a simpler expression with which to work. Squaring both sides,

$$\sqrt{u - 2} = 2 + \sqrt{2u + 3}$$

we have

$$u - 2 = 4 + 4\sqrt{2u + 3} + 2u + 3$$
$$-u - 9 = 4\sqrt{2u + 3} \qquad \textbf{Collecting like terms and isolating the radical}$$
$$u^2 + 18u + 81 = 16(2u + 3) \qquad \textbf{Squaring both sides}$$
$$u^2 - 14u + 33 = 0 \qquad \textbf{Collecting like terms}$$
$$(u - 11)(u - 3) = 0 \qquad \textbf{Factoring}$$
$$u = 11 \quad \text{or} \quad u = 3.$$

Check:

$$u = 11 \qquad\qquad u = 3$$
$$\sqrt{11 - 2} = 2 + \sqrt{2(11) + 3} \qquad\qquad \sqrt{3 - 2} = 2 + \sqrt{2(3) + 3}$$
$$3 \neq 2 + 5. \text{ x} \qquad\qquad 1 \neq 2 + 3. \text{ x}$$

Since neither value checks, the equation $\sqrt{u - 2} = 2 + \sqrt{2u + 3}$ has no solution. □

11.8 Exercises

In Exercises 1–20, solve the equations and check.

1. $x^4 - 11x^2 + 10 = 0$ **2.** $x^4 - 7x^2 + 10 = 0$

3. $x^4 - 9x^2 + 14 = 0$ **4.** $x^4 - 14x^2 + 33 = 0$

5. $x^5 - 8x^3 + 15x = 0$ **6.** $x^6 - 12x^4 + 11x^2 = 0$

7. $4x^4 - 13x^2 + 3 = 0$ **8.** $5x^4 - 16x^2 + 11 = 0$

9. $2x + 8\sqrt{x} - 10 = 0$ **10.** $x + 7\sqrt{x} - 8 = 0$

11. $\frac{1}{x^2} + \frac{9}{x} + 8 = 0$ **12.** $\frac{1}{x^2} - \frac{8}{x} + 7 = 0$

13. $\frac{1}{x^4} - \frac{13}{x^2} + 12 = 0$ **14.** $\frac{1}{x^4} - \frac{2}{x^2} - 15 = 0$

15. $x^{1/3} - 7x^{1/6} + 6 = 0$

16. $x^{2/3} - 6x^{1/3} - 7 = 0$

17. $2x - 5x^{1/2} + 2 = 0$

18. $x^{5/3} - 16x^{4/3} + 15x = 0$

19. $(x-5)^2 + 10(x-5) + 16 = 0$

20. $(x+3)^2 - 5(x+3) + 6 = 0$

In Exercises 21–34, solve the fractional equations and check.

21. $\frac{x}{4} + \frac{2}{x} = 1$ **22.** $x = \frac{4}{x-3}$

23. $\frac{x}{5-x} = \frac{2}{x-5}$ **24.** $\frac{x}{3-x} = \frac{3}{x-3}$

25. $\frac{3}{x-2} + \frac{x-4}{x} = 5$

26. $\frac{x-2}{x+1} = \frac{4x-3}{2x}$

27. $\frac{2}{3x+4} - \frac{1}{3x-4} = \frac{2x}{9x^2-16}$

28. $\frac{1}{2x+5} - \frac{1}{2x-5} = \frac{2}{4x^2-25}$

29. $\frac{4}{x+5} + \frac{9}{x+1} = 1$

30. $\frac{3}{3x+7} + \frac{2}{5x+5} = \frac{1}{3x-1}$

31. $\frac{1}{2x+3} + \frac{2}{x-4} = \frac{3}{2x^2-5x-12}$

32. $\frac{1}{3x+4} + \frac{2}{x-5} = \frac{3}{3x^2-11x-20}$

33. $\frac{x+2}{x} + \frac{1}{x+4} = \frac{-2x-6}{x(x+4)}$

34. $\frac{2}{x^2-9} - \frac{1}{x(x-3)} = \frac{2}{x^2}$

In Exercises 35–46, solve the radical equation and check.

35. $\sqrt{x+3} = 4$

36. $4 - \sqrt{3x} = 1$

37. $\sqrt{x} = 2 - x$

38. $\sqrt{x} = x - 2$

39. $x + 2 = 2\sqrt{2x-7}$

40. $\sqrt{x+4} - \sqrt{2x} = 1$

41. $\sqrt{2x+4} = 1 + \sqrt{2x}$

42. $\sqrt{3x-1} + \sqrt{3x+6} = 7$

43. $\sqrt{2x+1} + \sqrt{2x-1} = 4$

44. $\sqrt{3-2x} - 3 = \sqrt{2+2x}$

45. $3\sqrt{2x+1} = 3$

46. $3 + 4\sqrt{2x-3} = 3 + 4\sqrt{17}$

47. Two pipes can fill a tank in seven hours when used together. Alone, one can fill the tank three hours faster than the other. How long will it take each pipe to fill the tank alone?

48. A new printing press can do a job in one hour less than an older press. Together they can do the same job in 1.2 hours. How long will it take each to do the job alone?

49. An object is 60 cm from a screen. In order to focus the image of the object on the screen, a converging lens with a focal length of 12 cm is to be used. The lens is placed between the object and the screen at a distance of p cm from the object, where

$$\frac{1}{p} + \frac{1}{60-p} = \frac{1}{12}$$

Find p.

50. If a resistance R and reactance X are connected in parallel, the impedance Z of the circuit is given by

$$Z = \frac{RX}{\sqrt{R^2 + X^2}}$$

Solve the equation for R.

11.9 Solutions of systems of equations involving quadratics

In Chapter 10 we discussed systems of linear equations in two and three unknowns. In this section we shall study systems of equations where the equations in the system may be linear and quadratic or quadratic and quadratic. The different types of systems will be illustrated in the following examples.

If one of the equations is linear and the other quadratic, the best approach is to solve the linear equation for one of the variables and then substitute the result into the quadratic equation. At this point in the process we will have a quadratic equation in terms of one variable.

EXAMPLE 1 Solve the system of equations algebraically.

$$x + y = 7$$
$$x^2 + y^2 = 25.$$

Solution: Solve the equation $x + y = 7$ for the variable x. Thus, $x = 7 - y$. Now substitute $7 - y$ for x in the quadratic equation.

$x^2 + y^2 = 25$	
$(7 - y)^2 + y^2 = 25$	**Substituting for x**
$2y^2 - 14y + 24 = 0$	**Squaring and collecting like terms**
$y^2 - 7y + 12 = 0$	**Dividing both sides of equation by 2**
$(y - 4)(y - 3) = 0$	**Factoring**
$y = 4$ or $y = 3$	

To determine the values of x, knowing y, we substitute the values of y into the equation $x = 7 - y$.

When

$y = 4$	$y = 3$
$x = 7 - 4$	$x = 7 - 3$
$x = 3$	$x = 4$

Check:

	$(3, 4)$	$(4, 3)$
	$(3)^2 + (4)^2 = 25$ ✓	$(4)^2 + (3)^2 = 25$ ✓
and	$3 + 4 = 7$ ✓	$4 + 3 = 7.$ ✓

The solutions of the system of equations are the ordered pairs $(3, 4)$ and $(4, 3)$. □

TABLE 11.5 $y = \pm\sqrt{25 - x^2}$

x	<-5	-5	-4	-3	-2	-1	0	1	2	3	4	5	>5
y	imaginary	0	± 3	± 4	± 4.6	± 4.9	± 5	± 4.9	± 4.6	± 4	± 3	0	imaginary

A graphical solution of a system of two equations in two unknowns may be found by plotting the graphs of the two equations on the same set of axes and to the same scale. The points of intersection of the two curves are the solutions to the system. The graphical method will give us only real solutions and in many cases only general solutions. The technique is illustrated in Example 2.

EXAMPLE 2 Solve the following system of equations graphically.

$$x^2 + y^2 = 25$$
$$xy = 7.$$

Solution: To determine ordered pairs to graph $x^2 + y^2 = 25$ we solve the equation for y.

$$y = \pm\sqrt{25 - x^2}$$

In order to obtain real values for y by inspection we can see that x must be greater than or equal to -5 and less than or equal to 5. This can be written as $-5 \leq x \leq 5$, which is the domain of the relation. Using this as a guide we construct the table of values (Table 11.5).

For the equation $xy = 7$ we solve for y to determine the ordered pairs for graphing:

$$y = \frac{7}{x}$$

With the equation in this form we can see that $x \neq 0$. The line $x = 0$ is an asymptote (see Section 5.3). If we solve for x, $y \neq 0$, and the line $y = 0$ is also an asymptote. For large positive and negative values of x and y the curve is asymptotic to the x- and y-axes, respectively. With this information as a guide we calculate ordered pairs as in Table 11.6.

The graph of the equation $x^2 + y^2 = 25$ is a circle and the graph of the equation $xy = 7$ is a hyperbola. These figures will be discussed in detail in Chapter 17. The two figures are graphed on the same set of axes in Fig. 11.13 (on p. 436). Points $P_1(1.5, 4.8)$, $P_2(4.8, 1.5)$, $P_3(-4.8, -1.5)$, and $P_4(-1.5, -4.8)$ are the points of intersection of the two curves. □

TABLE 11.6 $y = \dfrac{7}{x}$

x	-7	-6	-5	-4	-3	-2	-1	0	1	2	3	4	5	6	7
y	-1	-1.2	-1.4	-1.8	-2.3	-3.5	-7	undefined	7	3.5	2.3	1.8	1.4	1.2	1

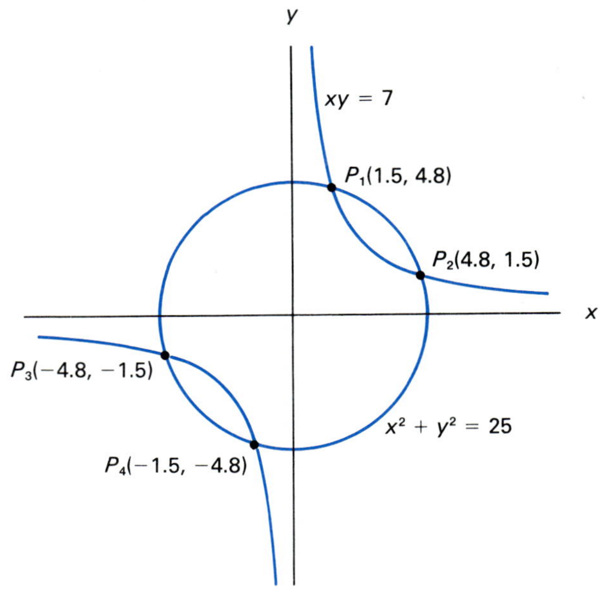

FIGURE 11.13

Note: The graphical method will provide approximate solutions. Also, if two curves are tangent to each other the system has two identical solutions for this point.

EXAMPLE 3 Solve the following system of equations algebraically.

$$x^2 - xy + 2y^2 = 8$$
$$xy = 4.$$

Solution: Solve the equation $xy = 4$ for x, $x = \dfrac{4}{y}$. Now substitute $\dfrac{4}{y}$ for x in the equation $x^2 - xy + 2y^2 = 8$.

$$\left(\frac{4}{y}\right)^2 - \left(\frac{4}{y}\right)(y) + 2y^2 = 8$$

$$\frac{16}{y^2} - 4 + 2y^2 = 8$$

$$2y^4 - 12y^2 + 16 = 0 \qquad \textbf{Multiplying both sides by } y^2 \textbf{ and collecting like terms}$$

$$y^4 - 6y^2 + 8 = 0 \qquad \textbf{Dividing by 2}$$

$$(y^2 - 4)(y^2 - 2) = 0 \qquad \textbf{Factoring}$$

$$y^2 = 4 \quad \text{or} \quad y^2 = 2$$

$$y = \pm 2 \quad \text{or} \quad y = \pm\sqrt{2}$$

Substituting into the equation $x = \frac{4}{y}$ when

$$\begin{array}{ll} y = \pm 2 & y = \pm\sqrt{2} \\ x = \dfrac{4}{\pm 2} & x = \dfrac{4}{\pm\sqrt{2}} \\ x = \pm 2 & x = \pm 2\sqrt{2}. \end{array}$$ **Rationalizing the denominator**

The solutions to the system of equations are the ordered pairs (2, 2), (−2, −2), $(2\sqrt{2}, \sqrt{2})$, and $(-2\sqrt{2}, -\sqrt{2})$. The check is left to the reader. □

EXAMPLE 4 Solve the following system of equations algebraically.

$$3x^2 - 2y^2 = 10 \quad [1]$$
$$x^2 + 3y^2 = 7. \quad [2]$$

Solution: This system can be solved by multiplying Eq. [2] by −3 and adding to Eq. [1].

$$\begin{aligned} 3x^2 - 2y^2 &= 10 && \textbf{[1]} \\ -3x^2 - 9y^2 &= -21 && \mathbf{-3 \times [2]} \\ -11y^2 &= -11 && \textbf{Adding} \\ y^2 &= 1 \\ y &= \pm 1. \end{aligned}$$

Substituting the y-values in Eq. [1] to obtain the values for x.

$$\begin{array}{cc} y = 1 & y = -1 \\ 3x^2 - 2(1)^2 = 10 & 3x^2 - 2(-1)^2 = 10 \\ x^2 = 4 & x^2 = 4 \\ x = \pm 2. & x = \pm 2. \end{array}$$

The solutions of the system of equations are the ordered pairs (2, 1), (−2, 1), (2, −1), and (−2, −1). The check is left to the reader. □

EXAMPLE 5 The area of a right-triangle support for a solar panel is 1.5 m^2 and its hypotenuse is 2.5 m. Find the length of the other two sides of the triangle.

FIGURE 11.14

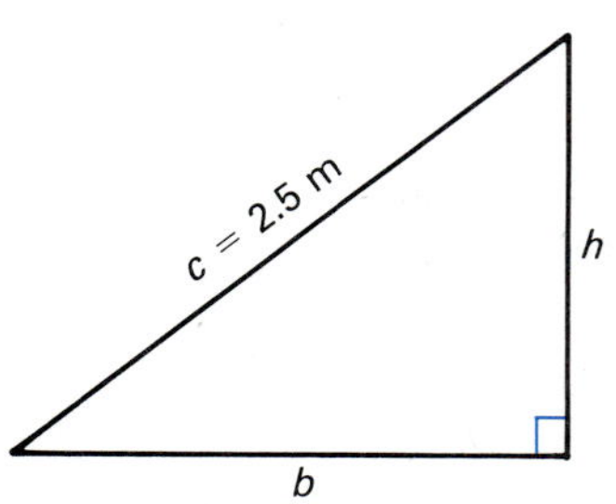

Solution: Sketch the right triangle and label as in Fig. 11.14. The problem provides us with the following information: the hypotenuse of the right triangle is 2.5 m and the area of the right triangle is 1.5 m^2. Let

h = height of the triangle, and
b = length of the base.

Substituting in the formula for the area of a triangle, $A = \frac{1}{2}bh$, we get the equation

$$1.5 = \frac{1}{2}bh \qquad \text{or} \qquad 3 = bh \qquad [1]$$

Substituting $c = 2.5$ m into the Pythagorean theorem $(c^2 = b^2 + h^2)$ we get the equation:

$$(2.5)^2 = b^2 + h^2 \qquad \text{or} \qquad 6.25 = b^2 + h^2. \qquad [2]$$

We can solve for h in Eq. [1] and substitute the value into Eq. [2] and obtain values for b.

$$h = \frac{3}{b}$$ **Solving for h**

$$6.25 = b^2 + \left(\frac{3}{b}\right)^2$$ **Substituting**

$$b^4 - 6.25b^2 + 9 = 0$$ **Squaring and collecting like terms**

$$b^2 = \frac{-(-6.25) \pm \sqrt{(-6.25)^2 - 4(1)(9)}}{2(1)}$$ **Solving for b using the quadratic formula**

$$b^2 = 4 \qquad \text{or} \qquad b^2 = 2.25$$
$$b = \pm 2 \qquad \text{or} \qquad b = \pm 1.5$$

Since b is the length of a side of the triangle, b cannot be negative. When

$$b = 2 \qquad\qquad b = 1.5$$
$$h = \frac{3}{2} \qquad\qquad h = \frac{3}{1.5}$$
$$h = 1.5. \qquad\qquad h = 2.$$

The lengths of the two sides of the triangle are 1.5 m and 2 m. □

11.9 Exercises

In Exercises 1–12, solve the systems of equations graphically.

1. $x^2 + y^2 = 25$, $4y - 3x = 0$
2. $x^2 + y^2 = 25$, $2x + y = 10$
3. $xy = 12$, $x + y = 7$
4. $y^2 = 4x$, $y + 2x = 4$
5. $y^2 = 2x$, $x^2 = 2y$
6. $x^2 + y^2 = 25$, $4y^2 - 9x = 0$
7. $x^2 + y^2 = 10$, $x^2 + 9y^2 = 18$
8. $x^2 - y^2 = 5$, $2x^2 + y^2 = 22$
9. $x^2 + 2y^2 = 33$, $y^2 - x^2 = 15$
10. $y^2 = 9x$, $y^2 - x^2 = 8$
11. $x^2 - y^2 = 7$, $x^2 + y^2 = 25$
12. $xy = 2$, $y^2 - 4x^2 = 16$

In Exercises 13–24, solve the systems of equations algebraically.

13. $x + y = 7$, $x^2 + y^2 = 25$
14. $x^2 - y^2 = 14$, $x + y = 7$
15. $x^2 + 2y^2 = 8$, $x + 2y = 1$
16. $x^2 + y^2 = 36$, $x + y = 5$
17. $y^2 - 4x^2 = 16$, $2x - y = 2$
18. $4x^2 + y^2 = 17$, $x - 3y + 1 = 0$
19. $2x^2 - y^2 = -1$, $x^2 + y^2 = 13$
20. $3x^2 - 2y^2 = 10$, $x^2 + 3y^2 = 7$

21. $y^2 = 6x + 12$
$y^2 = -4x + 4$

22. $x^2 - y^2 = 9$
$y^2 = 2x + 6$

23. $x^2 + y^2 = 5$
$xy = 2$

24. $xy = 2$
$y^2 - 4x^2 = 16$

25. A rectangular piece of tin has an area of 216 cm². An open box containing 224 cm³ can be made from it by cutting a 2-cm square from each corner and turning up the ends and sides. What are the dimensions of the piece of tin?

26. Find the dimensions of a rectangular door having a diagonal of 85 in. and a perimeter of 226 in.

27. The area of a triangular wedge is 210 mm². The hypotenuse of the right triangle is 29 mm. Find the length of the legs.

28. The perimeter of a rectangle is 206 mm and its area is 2640 mm². Find its length and width.

29. The hypotenuse of a right triangle is 85 m long. If each leg were increased by 3 m, the hypotenuse would be increased by 4 m. Find the length of the sides of the original triangle.

30. The sum of the circumferences of two circles is 22π mm, and their combined area is 65π mm². What are the radii of the two circles?

11.10 Applications

There are many problems where the solution depends on solving a quadratic equation.

EXAMPLE 1 An architect is designing a solar house with a glass-enclosed area on the south side. The client decides that she wants a swimming pool in this enclosed area. It will serve two purposes, recreation and a heat-storage mass.

The client wants the length of the pool area to be $2\frac{1}{2}$ times the width. She also wants a walk around the pool that is 1 m wide on one side and one end and $1\frac{1}{2}$ m wide on the other side and end. With these restrictions the architect has an area of $43\frac{3}{4}$ m² for the pool. What are the dimensions of the pool and the pool area (pool and walk)?

FIGURE 11.15

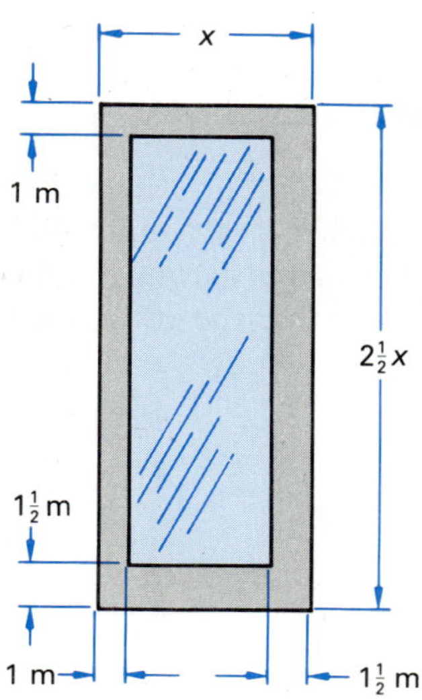

Solution: We first sketch the pool and the surrounding walk as in Fig. 11.15. Letting x = width of the pool area,

$$2\frac{1}{2}x = \text{length of the pool area,}$$

$$x - 2\frac{1}{2} = \text{width of pool,}$$

$$2\frac{1}{2}x - 2\frac{1}{2} = \text{length of pool, and}$$

$$\left(x - 2\frac{1}{2}\right)\left(2\frac{1}{2}x - 2\frac{1}{2}\right) = \text{area of the pool (width} \times \text{length).}$$

Since the area to be covered by the pool is $43\frac{3}{4}$ m^2, we have

$$\left(x-\frac{5}{2}\right)\left(\frac{5}{2}x-\frac{5}{2}\right)=43\frac{3}{4}$$

$$\frac{5}{2}x^2-\frac{35}{4}x+\frac{25}{4}=\frac{175}{4} \qquad \textbf{Multiplying}$$

$$2x^2-7x+5=35 \qquad \textbf{Multiplying each term by 4/5}$$

$$2x^2-7x-30=0.$$

Solving for x using the quadratic formula:

$$a=2,\ b=-7, \text{ and } c=-30$$

$$x=\frac{-(-7)\pm\sqrt{(-7)^2-4(2)(-30)}}{2(2)} \qquad \textbf{Substituting}$$

$$=\frac{7\pm\sqrt{289}}{4}$$

$$=\frac{7+17}{4} \quad \text{or} \quad x=\frac{7-17}{4}$$

$$x=6 \quad \text{or} \quad =-\frac{10}{4}.$$

We reject $-10/4$ since x is a positive measurement. Therefore, the width of the pool area is 6 m and the length is 15 m. The width of the pool is $3\frac{1}{2}$ m and the length is $12\frac{1}{2}$ m.

Check: Width of the pool × length of pool $=43\frac{3}{4}$

$$3\frac{1}{2}\times 12\frac{1}{2}=43\frac{3}{4}.$$

Also the length of the pool area is $2\frac{1}{2}$ times the width. □

EXAMPLE 2 Two motor boats travel at right angles to each other after leaving the same dock at the same time. One hour after starting out the boats are 15 km apart. If one travels 9 km/hr faster than the other, what is the speed of each boat?

Solution: Let

$$x\,\frac{\text{km}}{\text{hr}}=\text{speed of one boat and}$$

$$(x+9)\,\frac{\text{km}}{\text{hr}}=\text{speed of other boat.}$$

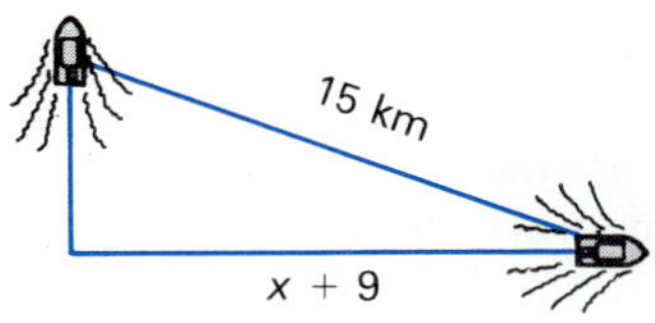

FIGURE 11.16

Since the boats are traveling at right angles to each other, their relative positions can be illustrated with a right triangle as in Fig. 11.16. The hypotenuse of the right triangle is the distance the two boats are apart at the end of one hour. Having assigned values to the two legs and the hypotenuse of the right triangle, we can solve for x using the Pythagorean theorem.

$$a^2 + b^2 = c^2$$

$$(x)^2 + (x+9)^2 = (15)^2 \qquad \textbf{Substituting}$$

$$x^2 + x^2 + 18x + 81 = 225 \qquad \textbf{Squaring}$$

$$2x^2 + 18x - 144 = 0 \qquad \textbf{Collecting like terms}$$

$$x^2 + 9x - 72 = 0. \qquad \textbf{Dividing each term by 2}$$

To solve we use the quadratic formula

$$x = \frac{-9 \pm \sqrt{(9)^2 - 4(1)(-72)}}{2(1)} \quad \text{or} \quad \frac{-9 \pm \sqrt{369}}{2}$$

$$x = 5.10 \quad \text{or} \quad x = -14.1.$$

Since rate of speed is positive we reject -14.1. Therefore, one boat traveled at the rate of 5.10 km/hr and the other boat traveled at the rate of 14.1 km/hr. The check is left to the reader. (It may not check exactly due to rounding off.) □

EXAMPLE 3 The rotational motion of a body is given by $\phi = \frac{1}{2}\alpha t^2 + \beta t$.

Find the value of t (in seconds) for $\alpha = 4$, $\beta = 6$, and $\phi = 10$.

Solution: The first thing we want to do is substitute the values for α, β, and ϕ into the equation:

$$\phi = \frac{1}{2}\alpha t^2 + \beta t$$

$$10 = \frac{1}{2}(4)t^2 + 6t$$

$$10 = 2t^2 + 6t.$$

The result is a quadratic equation in terms of t.

$$t^2 + 3t - 5 = 0.$$

Solving by means of the quadratic formula we have

$$t = \frac{-3 \pm \sqrt{(3)^2 - 4(1)(-5)}}{2(1)} = \frac{-3 \pm \sqrt{29}}{2}$$

$$t = 1.19 \quad \text{or} \quad t = -4.19.$$

We reject the answer $t = -4.19$. Thus the answer is $t = 1.19$ s. The check is left to the reader. □

11.10 Exercises

1. A wire stretches from the top of an 80 ft high pole to a stake in the ground 120 ft from the foot of the pole. Find the length of the wire.

2. A 10-m ladder is placed against a wall with the foot of the ladder 2 m from the base of the wall. How high above the ground will the ladder reach? (See Fig. 11.17.)

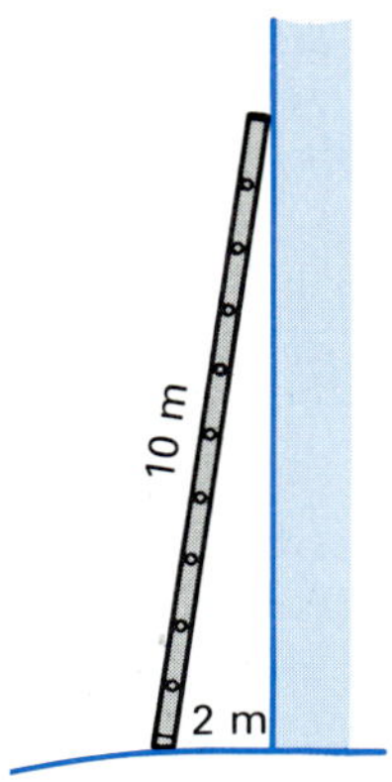

FIGURE 11.17

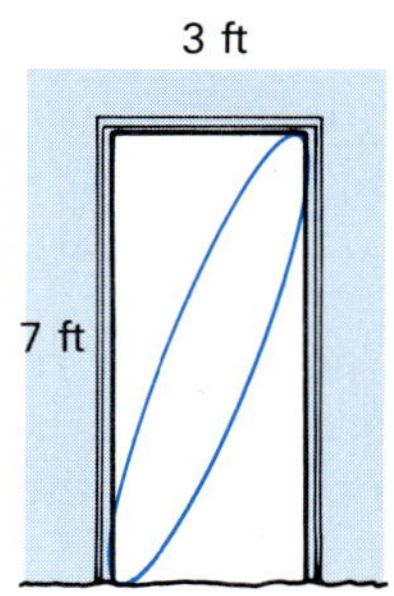

FIGURE 11.18

3. A doorway measures 3 ft by 7 ft. What is the diameter of the largest thin metal circular disk that will pass through the doorway? (See Fig. 11.18.)

4. An electrician's tool box measures 16 in. in length, 12 in. in width and 10 in. high. What is the length of the longest extension bit holder that may be placed in the tool box?

5. A rectangular swimming pool is surrounded by a 2-m-wide walk on two ends and one side. The walk is 1 m wide on the other side. The length of the pool area (pool and walk) is 3 m more than twice the width and the area of the pool is 52 m^2. (a) Find the length and the width of the pool. (b) Find the area of the pool area.

6. A page being designed for a textbook is to have 20-mm margins on the top and bottom and right side. The left side of the page is to have a margin of 40 mm. The height of the page is to be 80 mm more than the width. The area of the printed part of the page is to be 8800 mm^2. Find the width and height of the entire page.

7. Find the area of a right triangle with sides x m, $(x + 7)$ m, and $(x + 8)$ m long.

8. Find the dimensions of a rug that will cover 60% of the floor and be equidistant from the walls of a room 12 by 15 ft.

9. The radius r of a circular arch of height h and span b is given by the formula: $r = \dfrac{(b^2 + 4h^2)}{8h}$. Find h if $b = 20$ and $r = 26$.

10. The height h of a ball t seconds after it is thrown vertically upward is given in feet by $h = 64t - 16t^2$. a) Find the length of time the ball will be in the air. b) Find the maximum height of the ball.

11. The formula $s = v_0 t + \dfrac{1}{2}gt^2$ gives the distance s in feet a body has fallen in t seconds in a vacuum with an initial velocity downward of v_0 feet per second. If $g = 32$ ft/s^2 and $v_0 = 80$ ft/s, how long will it take a body to fall (a) 384 ft? (b) 1200 ft?

12. A rectangular piece of metal is 60 mm longer than it is wide. An open box containing 405,000 mm^3 is made by cutting a 30-mm square from each corner and turning up the sides and ends (Fig. 11.19). Find the dimensions of the box.

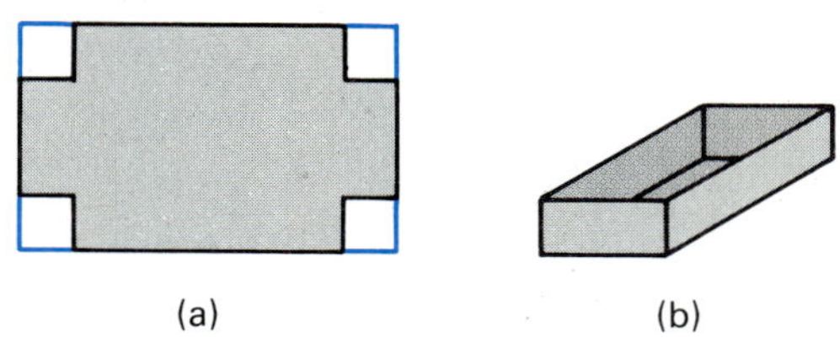

FIGURE 11.19

13. An open box is made from a square piece of tin by cutting a 4-in. square from each corner and turning up the sides. Find the area of the original square if the box is to contain (a) 16 in^3, (b) 400 in^3.

14. The illumination from any light at a stated distance may be expressed as a constant multiple k of the intensity of the light divided by the square of the distance. Two lamps of intensity 4 and 9 candles are placed 20 in. apart. Where should a screen be placed between them in order to be equally illuminated on both sides?

15. If the weekly profit p (in thousands of dollars) obtained from manufacturing computers is given by $p = -x^2 + 16x - 24$, where x is the number of

computers produced, how many computers must be produced to derive a weekly profit of \$40,000?

16. The cost c (in dollars) of building a counter is given by

$$c = \frac{\ell^2}{10} - 3\ell,$$

where ℓ is the length of the counter (in meters). How long a counter can be made if the cost is to be kept to \$280?

17. For a car traveling at a speed of v mph the least number of feet d necessary to stop the car is given by the formula $d = 0.044v^2 + 1.1v$. Estimate the speed of a car requiring 165 ft to stop after danger is realized.

18. In calculating the current in a simple-series electric circuit with inductance L henrys, resistance R ohms, and capacitance C farads, it becomes necessary to solve the equation $Lx^2 + Rx + \frac{1}{C} = 0$. Find x in terms of L, R, and C.

19. From a proportion in geometry a relationship can be developed between the span of the arch s, the height of the span h, and the radius of the arch r shown in Fig. 11.20. This relation is

$$h(2r - h) = \frac{s^2}{4}.$$

Find h if $r = 16.00$ and $s = 25.00$.

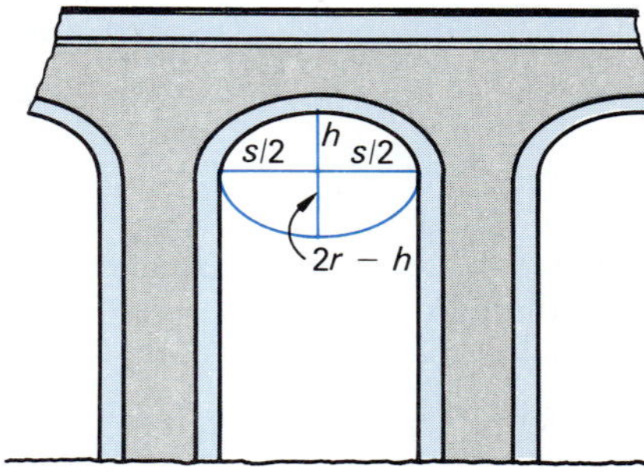

FIGURE 11.20

20. To get to the cottage Barbara rides her moped for 5 km and then jogs the remaining 3 km through the woods. She rides 14 km/hr faster than she jogs (both at constant rates). If it takes 45 min to make the entire trip, how fast does she jog?

Chapter summary

The **quadratic equation in standard form is $ax^2 + bx + c = 0$.** The quadratic equation can be solved by **factoring** (Section 11.2), **completing the square** (Section 11.3), and with the **quadratic formula** (Section 11.4).

$$x = \frac{-b \pm \sqrt{b^2 - 4ac}}{2a} \text{ is the quadratic formula.}$$

The roots of a quadratic equation will be real and unequal if $b^2 - 4ac > 0$, real and equal if $b^2 - 4ac = 0$, or complex if $b^2 - 4ac < 0$. The expression $b^2 - 4ac$ is called the **discriminant** of the quadratic formula.

The graph of a quadratic function $f(x) = ax^2 + bx + c$ is a parabola. If the graph of the function intersects the horizontal axis at: two points, then there are two real distinct roots; one point, then there are two real equal roots; no points, then there are two distinct complex roots. There can never be only one complex root. Section 11.6 describes in detail the graphing of a quadratic function.

Section 11.7 contains a discussion of the techniques for solving trigonometric functions that are in quadratic form. We can use factoring, completing the square, or the quadratic formula to solve these equations.

There are equations that can be changed to quadratic form by the method of substitution. These equations may be in the form of expressions raised to fractional powers or powers greater than 2.

In Section 11.9 techniques are demonstrated on how to solve systems of equations involving quadratic equations. The techniques are substitution, graphing, and addition or subtraction.

There are applications of quadratic equations relating to area, volume, speed, motion, profit, cost, and electricity.

Review Exercises

In Exercises 1–10, solve the quadratic equations and check the solutions.

1. $x^2 - 7x - 44 = 0$
2. $x^2 + x - 72 = 0$
3. $6R^2 - R - 35 = 0$
4. $5R^2 - 13R - 6 = 0$
5. $3t^2 + 7t - 5 = 0$
6. $2t^2 + 5t - 4 = 0$
7. $\frac{1}{9}R^2 - \frac{5}{3}R + 6 = 0$
8. $\frac{1}{4}R^2 + 4R + 15 = 0$
9. $3x^2 + 7x + 5 = 0$
10. $2x^2 + 4x + 2 = 0$

In Exercises 11–16, determine the quadratic equation that has the indicated roots.

11. 4 and 3
12. $\frac{2}{3}$ and -3
13. 4 and 4
14. $\frac{1}{5}$ and $\frac{1}{7}$
15. $-3 + 4j$ and $-3 - 4j$
16. $3j$ and $-3j$

In Exercises 17–20, solve each quadratic equation by completing the square.

17. $x^2 + 6x - 55 = 0$
18. $2x^2 - x - 21 = 0$
19. $10x^2 + 7x - 12 = 0$
20. $\alpha x^2 + \beta x + \gamma = 0$

In Exercises 21–26, state whether the discriminant is positive, negative, or zero and describe the roots of each equation.

21. $2x^2 + 5x - 7 = 0$
22. $\frac{1}{9}x^2 + \frac{1}{6}x + \frac{1}{16} = 0$
23. $0.1R^2 + 0.01R + 0.02 = 0$
24. $x^2 + 2x - 35 = 0$
25. $6x^2 + 13x - 28 = 0$
26. $3x^2 + 4x + 4 = 0$

In Exercises 27–36, graph the quadratic function and find the roots of the quadratic equation to the nearest tenth.

27. $f(x) = x^2 - 11x + 18$
28. $f(x) = x^2 + 4x - 21$
29. $f(t) = t^2 + 7t + 5$
30. $f(t) = t^2 + 5t + 7$
31. $f(R) = 8R^2 + 2R - 15$
32. $f(R) = 15R^2 + R - 2$
33. $f(t) = 2t^2 + 2t + 3$
34. $f(t) = 2t^2 + 3t - 2$
35. $f(x) = 9x^2 - 16$
36. $f(x) = \frac{1}{4}x^2 - 4$

In Exercises 37–46, find the solutions of the equations for $0 \le \theta \le 2\pi$.

37. $\sqrt{2}\sin\theta - 1 = 0$
38. $\sqrt{3}\tan\theta + \sqrt{3} = 0$
39. $2\sin^2\theta + \sqrt{3}\sin\theta = 0$
40. $2\sin^2\theta - 3\sin\theta - 2 = 0$
41. $6\cos^2\theta + \cos\theta - 2 = 0$
42. $9\tan^2\theta - 12\tan\theta + 4 = 0$
43. $2\cos^2\theta\sin\theta = \sin\theta$
44. $\cos^2\theta + \cos\theta = 1$
45. $\sqrt{3}\tan^2\theta\cot\theta = \cot\theta$
46. $7\sin^2\theta + 5\sin\theta = 2$

In Exercises 47–60, solve the equations and check.

47. $x^4 - 13x^2 + 36 = 0$
48. $9x^4 - 37x^2 + 4 = 0$
49. $x^{1/3} - 10x^{1/6} + 9 = 0$
50. $\frac{1}{x^2} + \frac{8}{x} + 7 = 0$
51. $\frac{x}{3} - \frac{1}{x} = 1$

52. $\frac{2}{x-1}+\frac{x-3}{x}=6$

53. $\frac{2}{x^2-4}-\frac{1}{x(x-2)}=\frac{1}{x^2}$

54. $\sqrt{x-5}=4x$

55. $\sqrt{2x+7}=5x$

56. $\sqrt{3x-4}=11x$

57. $\sqrt{2x+1}=1+3x$

58. $\sqrt{2x+3}+\sqrt{x-5}=4$

59. $\sqrt{3x-4}-\sqrt{x-5}=1$

60. $\sqrt{x-1}+\sqrt{2x+1}=2$

In Exercises 61–66, solve the following systems of equations algebraically.

61. $y=x^2+2$
$x^2+y^2=4$

62. $x+y=3$
$x^2+4y^2=9$

63. $xy=3$
$2x+y=5$

64. $x^2+y^2=5$
$x^2+3y^2=9$

65. $y=x^2-5$
$x^2+4y^2=16$

66. $x^2+3y^2=16$
$4x^2+y^2=20$

67. If two electrical resistors are connected in series, the equivalent resistance is $R_s=R_1+R_2$. If they are connected in parallel, the equivalent resistance is $R_p=\frac{R_1R_2}{R_1+R_2}$. What two resistors should be used so that the series resistance is 2 Ω and the parallel resistance $=\frac{1}{2}$ Ω?

68. If two springs are connected in parallel as shown in Fig. 11.21, the equivalent spring coefficient $K_p=K_1+K_2$. If they are connected in series, the equivalent spring coefficient is $K_g=\frac{K_1K_2}{K_1+K_2}$.

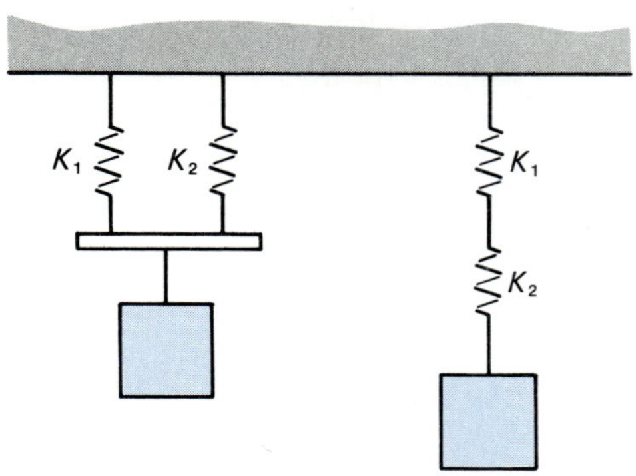

FIGURE 11.21

What two springs should be used if the parallel spring factor is 3 lb/in and the series spring factor is 1/2 lb/in?

69. A rectangular heating duct (Fig. 11.22) is to be formed by bending a 68-cm-wide piece of sheet metal. The area inside the rectangle is to be 228 cm². Find the width and height of the duct.

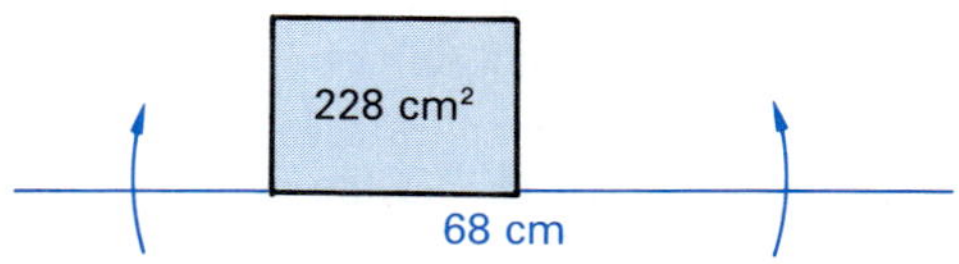

FIGURE 11.22

70. A page being designed for a textbook is to have 30-mm margins on the top, bottom, and right side. The left side of the page is to have a margin of 50 mm. The height of the page is to be 30 mm more than the width. The area of the printed part of the page is to be 23,400 mm². Find the width and height of the entire page.

71. A computer programmer and her assistant working together, complete a special program in four days. The programmer working alone can finish the job in six days less than the assistant working alone. How long does it take each person to do the job alone?

72. A free-lance writer works a certain number of days for a newspaper to earn \$480. If she had been paid \$8 less per day, she would have earned the same amount in two more days. What was her daily rate of pay?

73. The equation $\frac{1}{d}=\frac{pq}{p+q}$ is used in optometry. Solve the equation for q.

74. The time T it takes a pendulum of length ℓ to swing through one complete oscillation is given in the algebraic model

$$T=2\pi\sqrt{\frac{\ell}{9.8}}$$

a. Solve the equation for ℓ.

b. If T is measured in seconds and ℓ in meters, how long would a pendulum have to be in order to have an oscillation time of 1 s?

Can you solve these?

1. The formula

$$S = \frac{n(n+1)}{2}$$

gives the sum of the first n natural numbers 1, 2, 3, ... How many consecutive numbers must be added to obtain a sum of 325?

2. The sum of the roots of a quadratic equation is $r_1 + r_2 = -b/a$. The product of the roots of a quadratic equation is $r_1 r_2 = c/a$. Use the quadratic formula to develop these two formulas.

3. Show that if $y = ax^2 + bx + c$ $(a \neq 0)$ has real coefficients, then the maximum or minimum point of its graph has the coordinates $x = -\frac{b}{2a}$, $y = \frac{4ac - b^2}{4a}$, and that the point is a maximum when a is negative or is a minimum when a is positive.

4. Write a computer program to determine the solution to the following system of quadratic equations:

$$x^2 + y^2 = r^2$$
$$ax^2 + by^2 = 1.$$

5. If an isosceles triangle has an altitude h and a perimeter p, find an expression in terms of h and p for the base b.

6. A monument consists of two cubical blocks of granite, the smaller resting on the larger. The total height of the monument is 10 m. The total volume of the monument is 270 m^3. Find the exposed surface area of the monument.

12

Analysis of the response of a stereo-system speaker requires evaluation of a transfer function that contains the polynomial function

$$P(s) = s^3 + 761.s^2 + 147. \times 10^3 s + 995. \times 10^3.$$

What are the roots of the function expressed to three significant digits?

Equations of Higher Degree

12.1 Introduction

In Chapters 3 and 10 we discussed methods for solving linear equations. In Chapter 11 we demonstrated the various techniques for determining the solutions of quadratic equations. Linear and quadratic equations (first-degree and second-degree equations, respectively) belong to a class of equations called **polynomial equations.**

DEFINITION 12.1

> Any function of the form
>
> $$P(x) = a_nx^n + a_{n-1}x^{n-1} + \cdots + a_1x + a_0, \qquad [1]$$
>
> where n is a positive integer and the coefficients $a_n, a_{n-1}, \ldots, a_2, a_1, a_0$ are real or complex numbers, $a_n \neq 0$, is called a **polynomial function.**

If we set the polynomial function [1] equal to zero, then we have a polynomial equation of **degree** n.

$$a_nx^n + a_{n-1}x^{n-1} + a_{n-2}x^{n-2} + \cdots + a_1x + a_0 = 0. \qquad [2]$$

EXAMPLE 1 Determine the degree of the following polynomials:

a. $3x + 1$ **b.** $4x^2 - 7x + 5$
c. $6x^3 + 7x^2 + 5x + 6$ **d.** $x^4 + 7x^2 + 5x + 1$

Solution:

a. $3x^1 + 1$ is of degree 1.

b. $4x^2 - 7x + 5$ is of degree 2.

c. $6x^3 + 7x^2 + 5x + 6$ is of degree 3.

d. $x^4 + 7x^2 + 5x + 1$ is of degree 4.

The blue exponents 1, 2, 3, and 4 indicate the highest power of the variable and thus the degree of the polynomial. □

Our major objective in this chapter will be to determine the roots of a polynomial equation. The **solutions** or **roots** of an equation are also called the **zeroes** of the polynomial function.

EXAMPLE 2 Show that the roots of the polynomial equation

$$x^2 + 5x + 6 = 0$$

are also the zeroes of the polynomial function

$$P(x) = x^2 + 5x + 6.$$

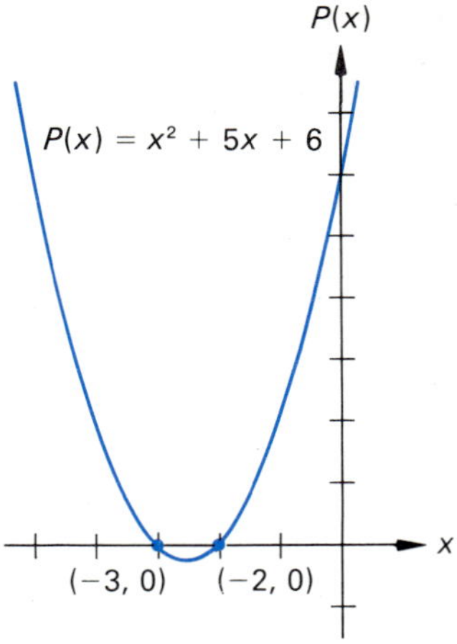

FIGURE 12.1

Solution: The polynomial equation $x^2 + 5x + 6 = 0$ in factored form is $(x + 2)(x + 3) = 0$. Solving, the roots are $x = -2$ and $x = -3$. Now, substituting into the polynomial function, we have

$$P(-2) = (-2)^2 + 5(-2) + 6 = 0$$

and

$$P(-3) = (-3)^2 + 5(-3) + 6 = 0.$$

Thus, the roots -2 and -3 of the polynomial equation

$$x^2 + 5x + 6 = 0$$

are zeroes of the polynomial function, $P(x) = x^2 + 5x + 6$ (Fig. 12.1). □

To help make the process of finding roots easier we will attempt to develop techniques to answer the following questions about *n*th-degree polynomials:

1. How many roots does a polynomial equation have within the set of complex numbers?
2. How many roots of the polynomial equation are real numbers?
3. What relation exists if any between the roots and the factors of the polynomial equation?
4. Are the roots of the equation rational or irrational numbers?
5. How many of the roots are rational numbers if the coefficients of the polynomials are integers?

12.2 Synthetic division

If it is known that $x - r$ is a factor of a polynomial equation of degree n, then by the process of division the polynomial equation can be expressed as a product of $x - r$ and a polynomial of degree $n - 1$. That is, if $x - 1$ is a factor of the polynomial equation $x^3 + 4x^2 + x - 6$, then this equation can be expressed as $(x - 1)(x^2 + 5x + 6)$. As a result of the division it is easier to express the polynomial as a product of its prime factors.

The process of division in algebra is similar to long division in arithmetic. The procedure for dividing one polynomial by another is discussed in detail in Section 2.6.

To simplify the following discussion we will use the following notation: $P(x) =$ the dividend, $D(x) =$ the divisor, $Q(x) =$ the quotient, and $R(x) =$ the remainder. The result of dividing two polynomials can be stated as

$$\frac{P(x)}{D(x)} = Q(x) + \frac{R(x)}{D(x)} \quad \text{if } D(x) \neq 0$$

or

$$P(x) = Q(x)D(x) + R(x). \qquad [1]$$

Equation [1] is the definition of division, which states that the dividend is equal to the quotient times the divisor plus the remainder. This equation can be used to check the results of the division. If we divide the polynomial $P(x) = x^3 + 4x^2 + x - 6$ by $D(x) = x - 1$, the result is $Q(x) = x^2 + 5x + 6$ and $R(x) = 0$. Checking the results by substitution into Eq. [1] we have

$$\begin{aligned} P(x) &= Q(x)D(x) + R(x) \\ x^3 + 4x^2 + x - 6 &= (x^2 + 5x + 6)(x - 1) + 0 \\ &= x^3 + 4x^2 + x - 6. \checkmark \end{aligned}$$

Multiplying and collecting like terms

The process of division of polynomials may be simplified by a method called **synthetic division.** The method will be illustrated by dividing $P(x) = 2x^3 + 9x^2 + 7x - 6$ by $D(x) = x + 3$. To begin, arrange the coefficients of $P(x)$ in order on a line (in order of descending power), inserting a zero for the coefficient of any missing power of x:

$$2 \quad 9 \quad 7 \quad -6.$$

Now set the divisor equal to zero and solve for x.

$$x + 3 = 0 \qquad \text{or} \qquad x = -3.$$

The number -3 will be the divisor for this method. To the right of the -6 place the divisor

$$2 \quad 9 \quad 7 \quad -6 \qquad \underline{|-3} \text{ (divisor).}$$

Now skip a line and draw a line under the dividend and bring 2 down to line 3.

Line 1	2	9	7	−6	$\underline{|-3}$
Line 2					
Line 3	2				

Multiply the 2 on line 3 times the divisor, -3, and place the product, -6, under the 9 on line 2.

Line 1	2	9	7	−6	$\underline{|-3}$
Line 2		−6			
Line 3	2				

Find the sum of 9 and -6, and place the sum on line 3, under the -6.

Line 1	2	9	7	-6	$\underline{\lfloor -3}$
Line 2		-6			
Line 3	2	3			

Now multiply 3 by the divisor, -3, and place the product under the 7 on line 2.

Line 1	2	9	7	-6	$\underline{\lfloor -3}$
Line 2		-6	-9		
Line 3	2	3			

Find the sum of 7 and -9, and place the sum on line 3 under the -9.

Line 1	2	9	7	-6	$\underline{\lfloor -3}$
Line 2		-6	-9		
Line 3	2	3	-2		

Now multiply the -2 on line 3 by the divisor, -3, and place the product on line 2 under the -6.

Line 1	2	9	7	-6	$\underline{\lfloor -3}$
Line 2		-6	-9	6	
Line 3	2	3	-2		

Find the sum of -6 and 6 and place under the 6 on line 3.

Line 1	2	9	7	-6	$\underline{\lfloor -3}$
Line 2		-6	-9	6	
Line 3	2	3	-2	0	

With the 0, the result of the addition in the last column, placed in line 3 and under the 6, we have completed the division. How is the result read? The dividend, $P(x) = 2x^3 + 9x^2 + 7x - 6$, is a polynomial of degree three and the divisor, $D(x) = x + 3$, is of degree one; therefore the quotient will be of degree two. The first three numbers in line 3 will be the coefficients of $Q(x)$. Therefore the quotient $Q(x) = 2x^2 + 3x - 2$. The fourth or last number in line 3 is the remainder; therefore $R(x) = 0$.

If the dividend is a fourth-degree polynomial, then line 1 will be five numbers across and so will line 3. However, the first four numbers of line 3 will be the coefficients of the third-degree polynomial, which is the quotient. The last, or fifth, number will be the remainder.

Writing the results of the problem using the definition of division we have

$$\begin{aligned} P(x) &= Q(x)D(x) + R(x) \\ 2x^3 + 9x^2 + 7x - 6 &= (2x^2 + 3x - 2)(x + 3) + 0 \\ &= 2x^3 + 9x^2 + 7x - 6. \checkmark \end{aligned}$$

In this particular case the divisor of the synthetic division, -3, is also a zero of the polynomial. This is true when the remainder is zero.

EXAMPLE 1 Divide $x^3 + 4x - 7$ by $x - 3$.

Solution: Arrange the coefficients of $P(x)$, $P(x) = x^3 + 4x - 7$, in order of descending power of the variable, inserting a zero for the coefficient of any missing power of x. The divisor is 3, found by letting $x - 3 = 0$ and solving for x. Write line 1, the divisor, leave space for line 2, draw a line and place the first coefficient of $P(x)$ on line 3.

Line 1	1	0	4	−7	|3
Line 2					
Line 3	1				

Now multiply 1 by 3 and write the product under the 0 on line 2.

Line 1	1	0	4	−7	|3
Line 2		3			
Line 3	1				

Now add 0 and 3, place the sum on line 3 and multiply the sum, 3, by 3 and place the product under 4 on line 2.

Line 1	1	0	4	−7	|3
Line 2		3	9		
Line 3	1	3			

Now add 4 and 9 and place the sum, 13, on line 3. Multiply 13 by 3 and place the product under −7 on line 2. Do the final addition and the division is complete.

Line 1	1	0	4	−7	|3
Line 2		3	9	39	
Line 3	1	3	13	32	

The result of the synthetic division is a quotient of $x^2 + 3x + 13$ and a remainder of 32.

Check:

$$\begin{aligned} x^3 + 4x - 7 &= (x^2 + 3x + 13)(x - 3) + 32 \\ &= x^3 + 4x - 7. \checkmark \quad \square \end{aligned}$$

The opening sentence of this section and the examples up to this point show divisors of the form $x - r$, where the coefficient of x is 1. For synthetic division we have no problems performing the division when the divisor is in the form $x - r$. Example 2 will address the problem when the divisor is of the form $ax - r$.

EXAMPLE 2 Divide $2x^4 + 11x^3 + 2x^2 - 55x - 60$ by $2x + 3$.

Solution: The divisor, $2x + 3$, is not of the form $x - r$ since the coefficient of the x-term is 2. Whenever this happens we can set the divisor equal to zero and solve for x. Thus, $2x + 3 = 0$, or $x = -\frac{3}{2}$.

Now we are ready to divide using synthetic division.

$$\begin{array}{cccccc}
2 & 11 & 2 & -55 & -60 & \underline{\big|-\frac{3}{2}} \\
 & -3 & -12 & 15 & 60 & \\
\hline
2 & 8 & -10 & -40 & 0 & \\
\left(-\frac{3}{2}\right)(2) & \left(-\frac{3}{2}\right)(8) & \left(-\frac{3}{2}\right)(-10) & \left(-\frac{3}{2}\right)(-40) & &
\end{array}$$

The result of the synthetic division is the quotient

$$2x^3 + 8x^2 - 10x - 40$$

and the remainder is 0.

Check:

$$2x^4+11x^3+2x^2-55x-60=(2x^3+8x^2-10x-40)\left(x+\frac{3}{2}\right)+0$$

$$= 2x^4 + 11x^3 + 2x^2 - 55x - 60. \checkmark \quad \square$$

Problems arise that have divisors that are not of the form $x - r$. They may be of the form $(x - r)^2$ or $(x - r)(x + s)$ or $(x + r)(x - s)$. Example 3 will illustrate how we can do the division with synthetic division.

EXAMPLE 3 Divide $2x^4 + 3x^3 - 18x^2 - 44x - 24$ by $(x + 2)^2$.

Solution: The divisor $(x + 2)^2$ can be expressed as $(x + 2)(x + 2)$. From arithmetic we know that the quotient for $12 \div [(2)(3)]$ can be found by dividing 12 by 2 and the result by 3 (i.e., $12 \div 2 = 6$ and $6 \div 3 = 2$, thus $12 \div [(2)(3)] = 2$). Using this idea we set up the problem in the usual manner for synthetic division.

$$\begin{array}{rrrrrl}
2 & 3 & -18 & -44 & -24 & \underline{\big|-2} \\
 & -4 & +2 & +32 & +24 & \\
\hline
2 & -1 & -16 & -12 & 0 &
\end{array}$$

The quotient from the first division becomes the dividend for the second division.

$$\begin{array}{rrrrl}
2 & -1 & -16 & -12 & \underline{\big|-2} \\
 & -4 & 10 & +12 & \\
\hline
2 & -5 & -6 & 0 &
\end{array}$$

We could have continued with the original problem and it would look like this:

$$\begin{array}{rrrrrl}
2 & 3 & -18 & -44 & -24 & \underline{\big|-2} \\
 & -4 & +2 & +32 & +24 & \\
\hline
2 & -1 & -16 & -12 & 0 & \underline{\big|-2} \\
 & -4 & +10 & +12 & & \\
\hline
2 & -5 & -6 & 0 & &
\end{array}$$

The result of the division is

$$2x^4+3x^3-18x^2-44x-24=(2x^2-5x-6)(x+2)(x+2)+0$$
$$=2x^4+3x^3-18x^2-44x-24. \checkmark \quad \square$$

This technique can be used when the divisors for synthetic division are not zeroes of the polynomial; however, one must be careful when reading the answer.

12.2 Exercises

In Exercises 1–25, use synthetic division to find the quotient and remainder. Check each of the results using the equation $P(x) = Q(x)D(x) + R(x)$.

1. $(6x^2 - 2x - 20) \div (x - 4)$
2. $(2x^2 - 7x - 1) \div (x - 4)$
3. $(x^3 + 2x^2 - 3x) \div (x + 1)$
4. $(z^3 - 6z^2 + 13z + 2) \div (z - 3)$
5. $(-x^3 + 2x + 2) \div (x + 2)$
6. $(-y^3 + 3y + 1) \div (y + 2)$
7. $(2x^3 - 6x^2 + 3x + 2) \div (x - 2)$
8. $(3x^3 - x^2 + 4) \div (x - 1/2)$
9. $(x^4 + 3x^3 + x^2 + 4) \div (x + 2)$
10. $(x^4 + x^3 + 4) \div (x + 2)$
11. $(2z^4 + 4z^3 + 3z^2 + 11z - 7) \div (z + 4)$
12. $(3x^5 - 5x^3 + 1) \div (x - 5)$
13. $(3x^5 - 9x^2 + 6x) \div (x - 1)$
14. $(-y^5 + 20y^2 + 8) \div (y - 3)$
15. $(9R^4 - 15R^3 + 13R^2 + 3R + 1) \div \left(R - \frac{1}{3}\right)$
16. $(2R^3 + 0.3R^2 - 0.45R - 0.167) \div (R - 0.2)$
17. $(2x^4 + 6x^3 - 8x + 10) \div \left(x + \frac{1}{2}\right)$
18. $(6x^4 - 10x^3 - 4x^2 + 3x + 3) \div \left(x + \frac{1}{3}\right)$
19. $(y^5 - 32) \div (y - 2)$
20. $(y^6 + a^6) \div (y + a)$
21. $(-Z^7 + 4.3Z^4 - 7.2Z^2) \div (Z + 2)$
22. $(4Z - 3Z^3 - 0.384) \div (Z + 1.2)$
23. $(-3c^3z + z^4 - c^2z^2 - 81z^4) \div (z + 3c)$
24. $(2x^4 - 12x^3 + 27x^2 - 54x + 81) \div (x - 3)^2$
25. $(2x^5 + 3x^4 - 11x^3 - 5x^2 + 2x + 6) \div [(x - 2)(x + 3)]$

12.3 The remainder theorem and the factor theorem

The process of synthetic division is a mechanical method of finding the roots of polynomial equations. In order to take some of the guesswork out of this process we will discuss several theorems. These theorems will not only help solve polynomial equations, but they will also help in graphing polynomial functions.

As shown below, the result of dividing $(x^3 - 2x^2 + 2x - 3)$ by $(x - 2)$ shows a quotient of $(x^2 + 2)$ and a remainder of $+1$.

$$\begin{array}{rrrr|l} 1 & -2 & 2 & -3 & \underline{2} \\ & 2 & 0 & 4 & \\ \hline 1 & 0 & 2 & 1 & \end{array}$$

From the definition of division we can write the following statement of equality:

$$(x^3 - 2x^2 + 2x - 3) = (x^2 + 2)(x - 2) + 1.$$

This statement is an identity that is true for all values of x. If in the same way any division of the polynomial, $P(x)$, by $(x - r)$, where r is a constant, results in a quotient $Q(x)$ and a constant remainder R, then we can write

$$P(x) = Q(x)(x - r) + R.$$

This statement is true for all values of x and in particular it will be true when $x = r$.

Substituting r in the equation, we obtain

$$P(r) = (r - r)Q(r) + R$$
$$P(r) = 0 \cdot Q(r) + R$$

or

$$P(r) = R.$$

The general discussion of division of a polynomial by $(x - r)$ provides us with a proof of an important theorem called the remainder theorem.

THEOREM 12.1

THE REMAINDER THEOREM

If a polynomial is divided by $x - r$, where r is any constant, until a constant remainder R independent of x is obtained, then $P(r) = R$.

EXAMPLE 1 Let $P(x) = x^3 + 4x - 7$ and $r = 3$. Determine the remainder R using the remainder theorem.

Solution: In Example 1 in Section 12.2 we found that $R = 32$.

$$\begin{array}{rrrr|l} 1 & 0 & 4 & -7 & \underline{3} \\ & 3 & 9 & 39 & \\ \hline 1 & 3 & 13 & 32 & \end{array}$$

Since $r = 3$, a constant, the hypothesis of the remainder theorem is met.

Now substituting $x = 3$ in the polynomial we have

$$\begin{aligned} P(x) &= x^3 + 4x - 7 \\ P(3) &= (3)^3 + 4(3) - 7 \\ &= 32. \end{aligned}$$

The result, 32, as a result of the remainder theorem, checks with the remainder obtained by synthetic division. □

EXAMPLE 2 Let $P(x) = 2x^3 + 11x^2 - 7x - 7$ and $D(x) = 2x + 1$.

a. Find the remainder by synthetic division.

b. Check the result of part (a) using the remainder theorem.

Solution:

a. To find the remainder by synthetic division arrange the coefficients of the dividend in proper order. Let $2x + 1 = 0$ and solve for x, $x = -1/2$. Thus, $-1/2$ is our divisor.

$$\begin{array}{rrrr|l} 2 & 11 & -7 & -7 & -\frac{1}{2} \\ & -1 & -5 & 6 & \\ \hline 2 & 10 & -12 & -1 & \end{array}$$

From the synthetic division we see that the remainder is -1.

b. Using the remainder theorem we may be tempted to substitute -1 for x. If we do this, the result is

$$\begin{aligned} P(-1) &= 2(-1)^3 + 11(-1)^2 - 7(-1) - 7 \\ &= 9. \end{aligned}$$

The result is not the same as the remainder obtained by synthetic division. Why? The hypothesis of the remainder theorem says that the divisor must be of the form $x - r$. In this example our divisor is not of this form, the x-term has a coefficient of 2. The best way to avoid the problem, whether the divisor is $x - 4$ or $3x - 5$, is to solve for x and substitute this value for x into $P(x)$. Since $D(x) = 2x + 1$, solving for x we have $x = -1/2$. Substituting, we have

$$\begin{aligned} P\left(-\frac{1}{2}\right) &= 2\left(-\frac{1}{2}\right)^3 + 11\left(-\frac{1}{2}\right)^2 - 7\left(-\frac{1}{2}\right) - 7 \\ &= -1. \end{aligned}$$

The result checks with the result obtained by synthetic division and the remainder is -1. □

The special case of the remainder theorem when $R = 0$ is called the **factor theorem.** For example, for the polynomial function $P(x) = x^2 + 5x + 6$, we know from the study of quadratic equations that $x+2$ is a factor of x^2+5x+6. Using the remainder theorem we see that

$$\begin{aligned} P(-2) &= (-2)^2 + 5(-2) + 6 \\ &= 0. \end{aligned}$$

This leads us to a formal statement of the factor theorem.

THEOREM 12.2

THE FACTOR THEOREM

If $P(x)$ is a polynomial and $P(r) = 0$ then $x - r$ is a factor of $P(x)$. r is a zero of $P(x)$.

The example above demonstrates the factor theorem for one case. We will write a proof that states it is true in all cases.

Proof:
By hypothesis r is a zero of $P(x)$; that is, $P(r) = 0$ and $R = 0$. Thus we have

$$P(x) = (x - r)Q(x) + 0.$$

Therefore $x - r$ is a factor of $P(x)$. □

EXAMPLE 3 Determine whether $x - 5$ is a factor of $x^2 - 6x + 5$.

Solution: To use the factor theorem we solve $x - 5 = 0$ for x. Substituting $x = 5$ in $P(x)$:

$$\begin{aligned} P(x) &= x^2 - 6x + 5 \\ P(5) &= (5)^2 - 6(5) + 5 \\ &= 0. \end{aligned}$$

Since $R = 0$, there is no remainder when we divide $x^2 - 6x + 5$ by $x - 5$. Thus, the factor theorem tells us that $x - 5$ is a factor of $x^2 - 6x + 5$. □

EXAMPLE 4 Determine whether $x + 1$ is a factor of $x^5 - 1$.

Solution: Solving $x + 1 = 0$ for x we obtain $x = -1$ and substituting for x in $P(x)$ we have

$$\begin{aligned} P(x) &= x^5 - 1 \\ P(-1) &= (-1)^5 - 1 \\ &= -2. \end{aligned}$$

Therefore, $x + 1$ is not a factor of $x^5 - 1$ because the remainder is not equal to zero. □

EXAMPLE 5 Determine under what conditions $x + a$ is a factor of $x^n + a^n$.

Solution: By the factor theorem in order for $x + a$ to be a factor of $x^n + a^n$, we find by substitution of $x = -a$ that $(-a)^n + a^n = 0$. For this sum to equal zero, $(-a)^n$ must equal $-(a)^n$, which will be true only when n is odd. For example, $(-2)^3 = -(2)^3$. Therefore, the condition for $x + a$ to be a factor of $x^n + a^n$ is that n be odd. □

12.3 Exercises

In Exercises 1–10, (a) find the remainder by synthetic division and (b) check the result of part (a) using the remainder theorem.

1. $P(x) = 3x^2 - 2x - 4$, $D(x) = x - 2$

2. $P(x) = x^3 - 2x^2 + 4x - 7$, $D(x) = x - 3$

3. $P(x) = x^3 - 3x^2 + 9$, $D(x) = x + 2$

4. $P(x) = 3x^3 - 5x^2 + x + 2$, $D(x) = x - 2$

5. $P(x) = 2x^4 - x^3 + 2x^2 + x - 8$, $D(x) = x + 1$

6. $P(x) = x^4 - 3x^3 - 3x^2 + x - 4$, $D(x) = x - 3$

7. $P(x) = x^4 - 3x^3 - 3x^2 + x + 4$, $D(x) = x - 1$

8. $P(x) = 4x^5 - 5x^4 - x + 8$, $D(x) = x - 1/2$

9. $P(x) = 7x^3 - x^2 + 7x - 1$, $D(x) = x - j$

10. $P(x) = x^3 + 4x^2 - 2x - 5$, $D(x) = x + 3j$

In Exercises 11–20, use the factor theorem to determine whether the first quantity is a factor of the second quantity.

11. $x - 2$, $x^3 + 2x^2 - 5x - 6$

12. $x + 3$, $x^3 - 4x^2 + 9$

13. $x + 1$, $x^3 - 4x^2 - 2x - 2$

14. $x - 2$, $x^3 - x^2 - 5x + 6$

15. $x + 2$, $x^4 - 3x - 5$

16. $x + 3$, $x^3 + 27$

17. $x + 2$, $x^4 + 16$

18. $x - 3$, $x^3 + 27$

19. $x + 2j$, $x^3 - x^2 + 4x - 4$

20. $x - j$, $x^4 + 4x^3 + 2x^2 + 4x + 1$

21. a. If $P(x) = 2x^3 + 10x^2 - 13x - 30$ and $P(x) = (x - 2)Q(x)$, find $Q(x)$.
b. Find all the roots of $P(x)$.

22. a. If $P(x) = 3x^3 + 7x^2 - 3x - 7$ and $P(x) = (x + 1)Q(x)$, find $Q(x)$.
b. Find all the roots of $P(x)$.

23. a. If $P(x) = 2x^3 - 21x^2 + 64x - 105$ and $P(x) = (x - 7)Q(x)$, find $Q(x)$.
b. Find all the roots of $P(x)$.

24. a. If $P(x) = x^3 + 12x^2 + 39x + 18$ and $P(x) = (x + 6)Q(x)$, find $Q(x)$.
b. Find all the roots of $P(x)$.

25. Determine the values of r for which division of $x^2 - 3x - 1$ by $x - r$ has a remainder of 3.

26. Determine the values of r for which division of $3x^2 + 5x - 12$ by $x - r$ has a remainder of -4.

In Exercises 27–32, use the factor theorem to determine whether or not each of the following statements is true.

27. $x - y$ is a factor of $x^n - y^n$ if n is any natural number.

28. $x - y$ is not a factor of $x^n + y^n$ if n is any natural number.

29. $x + y$ is a factor of $x^n - y^n$ if n is even.

30. $x + y$ is a factor of $x^n + y^n$ if n is odd.

31. $x + y$ is not a factor of $x^n - y^n$ if n is odd.

32. $x + y$ is not a factor of $x^n + y^n$ if n is even.

33. Verify that $3x - 1$ is a factor of $P(x) = 3x^3 + 14x^2 + 7x - 4$. May we conclude that $P(1/3) = 0$?

34. Verify that $x^2 + 1$ is a factor of $P(x) = 3x^3 - 5x^2 + 3x - 5$. May we conclude that $P(-1) = 0$?

35. Find the value of k for which $x - 3$ is a factor of $kx^3 - 6x^2 + 2kx - 12$.

36. Find the value of k for which $x - 9$ is a factor of $3x^4 - 40x^3 + 130x^2 + kx + 27$.

12.4 Factors and roots of polynomial equations

The discussion in this section will deal with theorems that will help determine the factors and roots of polynomial functions.

THEOREM 12.3

> In the synthetic division of
>
> $$P(x) = a_nx^n + a_{n-1}x^{n-1} + a_{n-2}x^{n-2} + \cdots + a_1x + a_0$$
>
> by $x - r$, where $a_n > 0$,
>
> **1.** if $r > 0$ and all the numbers in the third row are positive, then r is an upper bound for the positive zeros (roots) of $P(x)$;
>
> **2.** if $r < 0$ and the signs of the numbers in the third row alternate in sign, then r is a lower bound for the negative zeros (roots) of $P(x)$.

If the hypotheses of the theorem are met, then we can conclude the following when performing synthetic division: (1) If all the numbers in the third row are positive, then there is no positive real number larger than r that is a zero (root) of the equation. (2) If the numbers in the third row alternate sign from positive to negative (+, −, +, −, . . .), then there is no negative real number smaller than r that can be a zero (root) of the equation. This is illustrated in Example 1.

EXAMPLE 1 Find an upper and lower bound for the zeros (roots) of $P(x) = 2x^3 - x^2 - 22x - 24$.

Solution: Let us guess that 5 might be an upper bound for the roots. Using synthetic division we have

$$\begin{array}{rrrr|l} 2 & -1 & -22 & -24 & \underline{5} \\ & 10 & 45 & 115 & \\ \hline 2 & 9 & 23 & 91. & \end{array}$$

Since all the numbers in the third row are positive, 5 is an upper bound and there is no zero (root) larger than 5.

Now let us guess a lower bound. In this case pick -4.

$$\begin{array}{rrrr|l} 2 & -1 & -22 & -24 & \underline{-4} \\ & -8 & +36 & -56 & \\ \hline 2 & -9 & +14 & -80 & \end{array}$$

The signs in the third row alternate; therefore there is no zero (root) smaller than -4. We say that -4 is a lower bound. We can also conclude that all real zeroes (roots) of the polynomial will be between -4 and 5. Note that there can be other upper and lower bounds for the function; that is 4.5, 6.0, and 6.5 are all upper bounds. □

Does every polynomial have a root? This question was answered by Karl Friedrich Gauss in 1799. At the age of 22 Gauss proved the following theorem:

THEOREM 12.4

THE FUNDAMENTAL THEOREM OF ALGEBRA

Every polynomial function of the form

$$P(x) = a_nx^n + a_{n-1}x^{n-1} + a_{n-2}x^{n-2} + \cdots + a_1x + a_0,$$

$n \geq 1$, $a_n \neq 0$, has at least one complex root.

The proof of the theorem is beyond this text; however, the theorem is essential in understanding the following theorem:

THEOREM 12.5

Every polynomial function of the form

$$P(x) = a_nx^n + a_{n-1}x^{n-1} + a_{n-2}x^{n-2} + \cdots + a_1x + a_0,$$

$n \geq 1$, $a_n \neq 0$, has exactly n roots.

EXAMPLE 2 $P(x) = (x-5)^4(x-3)(x+7)^3$ is a polynomial function of eighth degree. The eight roots are 5, 5, 5, 5, 3, -7, -7, and -7. Note that any root that occurs m times is counted as m roots. We say that 5 is a root of multiplicity 4 and -7 is a root of multiplicity 3. □

EXAMPLE 3 Form a polynomial equation (in lowest degree) with the roots 1/2, $-2/3$, and 4, and having integral coefficients.

Solution: An equation having the desired roots is

$$\left(x - \frac{1}{2}\right)\left(x + \frac{2}{3}\right)(x - 4) = 0.$$

The equation obtained by multiplying these factors together will have fractional coefficients. The fractional coefficients may be cleared away by multiplying the equation by the lcd. There is less work involved if the lcd is introduced immediately. The lcd of 2 and 3 is 6. The lcd turns out to be the same value as a_n. We now can write

$$6\left(x-\frac{1}{2}\right)\left(x+\frac{2}{3}\right)(x-4)=0$$

or

$$(2)(3)\left(x-\frac{1}{2}\right)\left(x+\frac{2}{3}\right)(x-4)=0$$

or

$$2\left(x-\frac{1}{2}\right)3\left(x+\frac{2}{3}\right)(x-4)=0$$

or

$$(2x-1)(3x+2)(x-4)=0.$$

Multiplying these factors, the product is the desired equation.

A polynomial equation with integral coefficients with roots of 1/2, −2/3, and 4 is $6x^3-23x^2-6x+8=0$. □

EXAMPLE 4 Find the roots of the polynomial equation

$$x^2-4x+5=0.$$

Solution: To solve the equation we use the quadratic formula.

$$\begin{aligned} x &= \frac{-(-4)\pm\sqrt{(-4)^2-4(1)(5)}}{2(1)} \\ &= \frac{4\pm\sqrt{-4}}{2} \\ &= \frac{4\pm 2j}{2} \end{aligned}$$

$$x = 2+j \qquad \text{or} \qquad x = 2-j.$$

The roots of the polynomial equation are $2+j$ and $2-j$. □

We see from this solution that for this polynomial equation with real coefficients the solution is made up of a complex number and its conjugate. This brings us to the next theorem.

THEOREM 12.6

CONJUGATE-ROOTS THEOREM

If $P(x)=a_nx^n+a_{n-1}x^{n-1}+a_{n-2}x^{n-2}+\cdots+a_1x+a_0$ is a polynomial of degree $n\geq 1$ with real coefficients and if $a+bj$, $b\neq 0$ is a root of $P(x)$, then the complex conjugate $a-bj$ is also a root of $P(x)$.

12.4 Exercises

In Exercises 1–12, (a) by inspection find the zeroes of each of the following functions and (b) give the multiplicity of each root.

1. $P(x) = (x+3)(x-2)^3(x+5)^2$
2. $P(x) = (x+1)^4(x-4)^5$
3. $P(x) = (x-6)(2x+5)^3$
4. $P(x) = (2x+7)^2(3x-5)^3(5x-3)^4$
5. $P(x) = (3x+2)(x^2+6x+9)^2$
6. $P(x) = (x^2+7x+10)(x^2+11x+24)$
7. $P(x) = (x^2+10x+25)^3(x^2-2x-35)$
8. $P(x) = (x^2-16)^2(x^2-9)(6x^2+x-12)$
9. $P(x) = (x \pm j)(2x^2-x-15)$
10. $P(x) = (x \pm j)(3x^2-19x-14)$
11. $P(x) = [x-(2-3j)][x-(2+3j)](x+3)^2$
12. $P(x) = [x-(3+2j)][x-(3-2j)](x-5)^3$

In Exercises 13–22, find (a) an upper and (b) a lower bound for the zeroes of the following functions.

13. $P(x) = x^3+6x^2+5x-12$
14. $P(x) = x^3+2x^2-x-2$
15. $P(x) = x^3-x^2-5x-3$
16. $P(x) = 2x^3+x^2-2x-1$
17. $P(x) = x^4-5x^2+4$
18. $P(x) = x^4-6x^2-8x-3$
19. $P(x) = x^5+5x^4+4x^3-x^2-5x-4$
20. $P(x) = x^5+2x^4+x^3+8x^2+16x+8$
21. $P(x) = x^5+5x^2-7$
22. $P(x) = x^5-3x^3+24$

In Exercises 23–40, form polynomial equations with integral coefficients with the given roots. Write the answer in the form $a_nx^n + a_{n-1}x^{n-1} + \cdots + a_0 = 0$.

23. 3, 13
24. $-2, 1, 3$
25. 1, 2, 3
26. $-1, 0, 4, 5$
27. $-3, 0, 1, 1$
28. $-1, -1, 2, 2$
29. $\frac{1}{2}, \frac{1}{3}, -1$
30. $\frac{1}{3}, \frac{1}{3}, -3$
31. $-\frac{2}{3}, \frac{1}{4}, \frac{1}{2}$
32. $-\frac{3}{2}, -\frac{3}{2}, \frac{1}{3}$
33. $2, 1 \pm \sqrt{3}$
34. $1 \pm \sqrt{2}, \pm j\sqrt{3}$
35. $\pm 3, \pm\sqrt{2}, \pm\frac{1}{3}j$
36. $\frac{1}{4}, 1 \pm j\sqrt{3}, \pm\sqrt{5}$
37. $\frac{1}{2}, 1 \pm j\sqrt{5}, \pm 2$
38. -2 is a triple root.
39. 2 is a double root and -1 is a triple root.
40. 1 is a triple root and $\pm j$ are roots.
41. The equation $y^4 - 4y^3 - 4y^2 + 24y - 9 = 0$ has 3 as a double root. Find the other roots.
42. The equation $R^4 + 6R^3 + 14R^2 + 14R + 5 = 0$ has -1 as a double root. Find the other roots.
43. The equation $I^4 - 8I^3 + 30I^2 - 76I + 80 = 0$ has 2 and 4 as roots. Find the others.
44. The equation $v^4 - 3v^3 - 13v^2 + 21v + 18 = 0$ has 2 and -3 as roots. Find the others.
45. How do we know that every polynomial of odd degree with real coefficients has at least one real root?
46. The deflection of a beam supported at each end is given by $d(x) = x(x-2L)(7x^2 - 9xL + 2L^2)$, where L is the length of the beam and x is the distance from one end. Find the values of x, in terms of L, for which the deflection is zero.
47. The acceleration of an item in a production line is defined by $a(t) = (t-2.4)(t^2 - 1.9t - 6.66)$, where t is the time in seconds measured from the start of motion. At what positive values of t is the acceleration zero?
48. A spherical buoy of diameter d and specific gravity 0.5 (half the density of water) will sink to a depth x given by $4x^3 - 6dx^2 + d^3 = 0$. Show that a solution to this equation is $x = d/2$.

12.5 Rational roots

With the theorems we discussed in the previous sections we can tell how many roots a polynomial function will have and whether or not there are any roots larger or smaller than the trial divisor (upper bound and lower bound). However, we could only find the roots if the expression was in factored form or expressed as the product of quadratics. What we need now is a technique to be able to find the roots of a fourth, fifth, or any-degree polynomial. In this section we will discuss two more theorems that will help us accomplish this goal.

THEOREM 12.7

DESCARTES' RULE OF SIGNS

The number of positive real roots of a polynomial equation $P(x) = 0$ with real coefficients is either equal to the number of variations in sign in $P(x)$ or less than that number by a positive even integer. The number of negative real roots is either equal to the number of variations in sign in $P(-x)$ or less than that number by a positive even integer. In each case a root of multiplicity m is to be counted as m roots.

EXAMPLE 1 Use Theorem 12.7 to determine the possible number of positive and negative roots of each of the polynomial equations.

a. $P(x) = x^3 + 6x^2 - 13x - 42$

b. $P(x) = 2x^3 + 9x^2 - 17x + 6$

c. $P(x) = 4x^4 + 16x^3 - 31x^2 - 49x + 30$

d. $P(x) = x^4 + 6x^3 + 2x^2 + 14x + 377$

e. $P(x) = x^4 + 3x^2 + 7x + 6$

Solution:

a. To determine the number of possible positive roots of

$$P(x) = x^3 + 6x^2 - 13x - 42$$

we determine the number of changes of sign of $P(x)$:

$P(x) = \quad + \quad + \quad - \quad -$ **Looking only at signs**

change

Thus, we have only one change of sign; therefore there can be at most one positive root.

To determine the number of negative roots we let $x = -x$ and determine $P(-x)$.

$$P(-x) = -x^3 + 6x^2 + 13x - 42$$

or

$P(-x) = \quad - \quad + \quad + \quad -$ **Looking only at signs**

change change

We have two changes of sign. Thus, according to our theorem there may be two negative roots or no negative roots.

Since $P(x)$ has three roots (not more than one positive and not more than two negative) the following possibilities exist: one positive root and two negative roots, or one positive root and two imaginary roots.

b. For the function $P(x) = 2x^3 + 9x^2 - 17x + 6$ we have

$$P(x) = \quad + \quad + \quad - \quad +.$$

change change

Thus, there are two changes of sign and we could have two positive real roots or no positive real roots. When $x = -x$ we have

$$P(-x) = \quad - \quad + \quad + \quad +.$$

change

There is one change of sign; therefore there is one negative real root.

Since $P(x) = 0$ has three roots, the following possibilities exist: two positive roots and one negative root, or two imaginary roots and one negative root.

c. For the function $P(x) = 4x^4 + 16x^3 - 31x^2 - 49x + 30$, we have

$$P(x) = \quad + \quad + \quad - \quad - \quad +.$$

change change

Thus, there are two changes of sign and we could have two positive real roots or no positive real roots. When $x = -x$,

$$P(-x) = \quad + \quad - \quad - \quad + \quad +.$$

change change

Thus, there are two changes of sign and we could have two negative real roots or no negative real roots.

Since $P(x) = 0$ has four roots the following possibilities exist: two positive and two negative real roots, or two positive real and two imaginary roots, or two imaginary and two negative real roots, or four imaginary roots.

d. For the function $x^4 + 6x^3 + 2x^2 + 14x + 377$ we have

$P(x) = \;+\;+\;+\;+\;+.$

There are no change in signs; therefore, there are no positive real roots. When $x = -x$,

$P(-x) = \;+\;\;-\;\;+\;\;-\;\;+.$

change change change change

There are four changes of sign, thus there may be four negative real roots or two negative real roots and two imaginary roots or no negative roots and four imaginary roots.

e. For the function $P(x) = x^4 + 3x^2 + 7x + 6$ we have

$P(x) = \;+\;+\;+\;+.$

There are no changes in sign; therefore there are no positive real roots. When $x = -x$,

$P(-x) = \;+\;\;+\;\;-\;\;+.$

change change

There are two changes in sign, thus there may be two negative real roots and two imaginary roots, or no negative real roots and four imaginary roots. □

Now that we can determine the possible number of real roots of a polynomial function, the following theorem will help us to find the roots.

THEOREM 12.8

RATIONAL-ROOT THEOREM

If the rational number p/q, a fraction in lowest terms, is a root of the equation

$$P(x) = a_nx^n + a_{n-1}x^{n-1} + a_{n-2}x^{n-2} + \cdots + a_1x + a_0,$$

where a_i $(i = 0, 1, 2, \ldots, n)$ are integral coefficients, then p is an exact divisor of a_0 and q is an exact divisor of a_n.

The theorem tells us that in order for $-3/2$ to possibly be a root of the polynomial equation $2x^2 + 5x + 3 = 0$, -3 must be an exact divisor of 3, the constant term, and 2 must be an exact divisor of the coefficient of x^2, 2. In each case they are exact divisors and $-3/2$ is possibly a root. If $a_n = 1$ for the polynomial equation

$$a_n x^n + a_{n-1}x^{n-1} + a_{n-2}x^{n-2} + \cdots + a_1 x + a_0 = 0,$$

then any rational roots of the equation are integers and they are exact divisors of a_0. This idea is illustrated in Example 2.

EXAMPLE 2 Find the roots of the polynomial function

$$P(x) = x^3 + 6x^2 - 13x - 42.$$

Solution: The function $P(x) = x^3 + 6x^2 - 13x - 42$ is of degree 3. By Theorem 12.5 there must be exactly three roots. By applying Theorem 12.7 in Example 1(a) we learned that there is one positive real root and possibly two negative real roots. From Theorem 12.8 we know that any rational root must be a divisor of -42. The possible rational roots of the function are: ± 1, ± 2, ± 3, ± 6, ± 7, ± 14, and ± 42. As we see, the theorems have narrowed the search, but have not told us the roots of the polynomial. To determine the roots we will use synthetic division. With the information we have at this point the root could be any one of the possibilities listed, so we will arbitrarily try -1.

$$\begin{array}{rrrr|l} 1 & 6 & -13 & -42 & \underline{-1} \\ & -1 & -5 & 18 & \\ \hline 1 & 5 & -18 & -24 & \end{array}$$

Since there is a remainder of -24, -1 cannot be a root. Now let us try -3.

$$\begin{array}{rrrr|l} 1 & 6 & -13 & -42 & \underline{-3} \\ & -3 & -9 & 66 & \\ \hline 1 & 3 & -22 & 24 & \end{array}$$

FIGURE 12.2

$P(x) = x^3 + 6x^2 - 13x - 42$

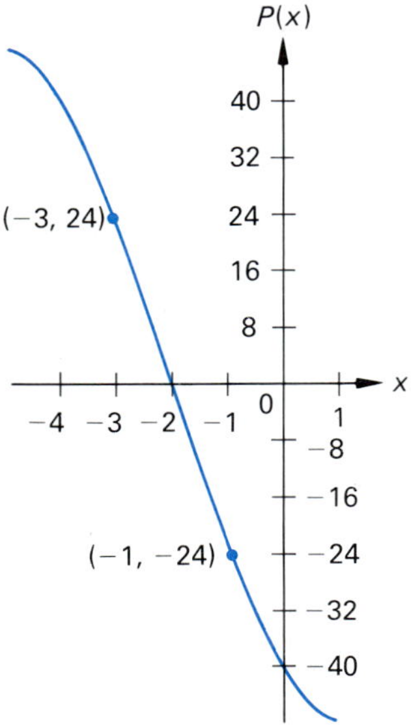

In this case there is a remainder of 24. If we graph the two points $(-1, -24)$ and $(-3, 24)$ as in Fig. 12.2, we see that in order for a line to join the two points it must cross the x-axis. Therefore, there must be a root between -1 and -3. The only possible rational root in this interval would be -2. Thus

$$\begin{array}{rrrr|l} 1 & 6 & -13 & -42 & \underline{-2} \\ & -2 & -8 & +42 & \\ \hline 1 & 4 & -21 & 0 & \end{array}$$

Since there is no remainder, -2 is a root and we write the function as

$$P(x) = (x^2 + 4x - 21)(x + 2).$$

We could find the remaining roots by synthetic division. Since the quotient is a quadratic expression we can also find the roots by factoring (or by using the quadratic equation).

$$P(x) = (x + 7)(x - 3)(x + 2).$$

Thus the roots of $P(x) = x^3 + 6x^2 - 13x - 42$, are $3, -2$, and -7. The solution consists of a positive root and two negative roots, one of the possibilities determined by Descartes' theorem. □

The question remains, In what order should a trial root be selected from the list of possible roots? The only way is to guess; however, our guesses can be narrowed if we consider the following:

1. If there is an odd number of roots, then try for the odd root.
2. If a root could be a multiple root, always check this possibility before going on.
3. In trying two numbers, if the remainder changes sign then the root must be between these two numbers.
4. In trying numbers in order, if the remainder keeps getting farther away from zero, this will not lead to a root of the polynomial.
5. Inspect the results and use common sense.

EXAMPLE 3 Find the roots of the polynomial function

$$P(x) = 2x^3 + 9x^2 - 17x + 6.$$

Solution: In Example 1(b) we found that there could be one negative root and possibly two positive roots for $P(x) = 2x^3 + 9x^2 - 17x + 6$. Applying Theorem 12.8, the possible rational roots of the equation are $\pm 1, \pm 2, \pm 3, \pm 6, \pm 1/2$, and $\pm 3/2$. Selecting -6 as a possible root, by synthetic division we have

$$\begin{array}{rrrr|l} 2 & 9 & -17 & 6 & \underline{-6} \\ & -12 & 18 & -6 & \\ \hline 2 & -3 & 1 & 0 & \end{array}$$

A remainder of 0 indicates that -6 is a root. Examining our reduced polynomial, $2x^2 - 3x + 1$, the possible rational roots are ± 1 and $\pm 1/2$. Since the remaining roots must be positive, we have two choices, 1 and 1/2.

$$\begin{array}{rrr|l} 2 & -3 & 1 & \underline{\tfrac{1}{2}} \\ & 1 & -1 & \\ \hline 2 & -2 & 0 & \underline{1} \\ & 2 & & \\ \hline 2 & 0 & & \end{array}$$

From the results of the division it can be seen that the roots are 1/2, 1, and -6. □

The application of theorems up to this point has been for the purpose of finding the roots of a polynomial. These roots are also the zeroes of the polynomial functions and are extremely helpful in graphing the function.

EXAMPLE 4 Graph $P(x) = 2x^3 + 9x^2 - 17x + 6$ for the interval $-7 \leq x \leq 3$.

Solution: The x-intercepts were found in Example 3. They are $(1/2, 0)$, $(1, 0)$, and $(-6, 0)$. The y-intercept, $(0, 6)$, is found by letting $x = 0$. In addition to the intercept there are additional ordered pairs shown in Table 12.1 to help graph the function. The function is graphed in Fig. 12.3. □

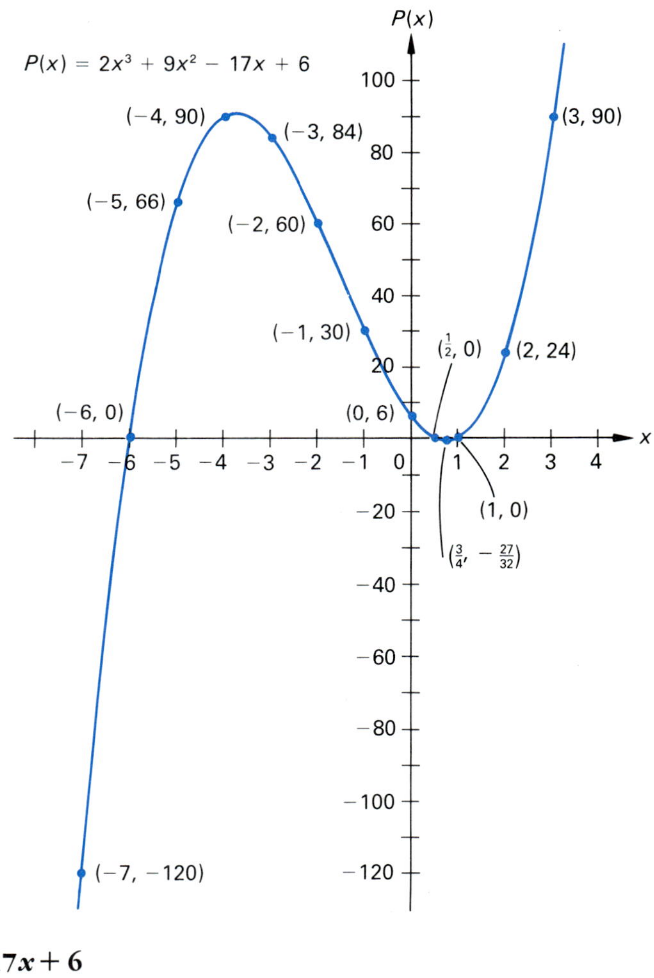

FIGURE 12.3

TABLE 12.1 $P(x) = 2x^3 + 9x^2 - 17x + 6$

x	-7	-6	-5	-4	-3	-2	-1	0	$1/2$	$3/4$	1	2	3
$P(x)$	-120	0	66	90	84	60	30	6	0	$-27/32$	0	24	90

EXAMPLE 5 Find the roots of the polynomial function

$$P(x) = 4x^4 + 16x^3 - 31x^2 - 49x + 30.$$

Solution: In Example 1(c) we found that we could have two positive and/or two negative roots of $P(x) = 4x^4 + 16x^3 - 31x^2 - 49x + 30$. Applying the rational-root theorem the possible rational roots of the equation are ± 1, ± 2, ± 3, ± 5, ± 6, ± 15, ± 30, $\pm 1/4$, $\pm 1/2$, $\pm 3/4$, $\pm 3/2$, $\pm 5/4$, $\pm 5/2$, $\pm 15/4$, and $\pm 15/2$. With all these possibilities it is best to approach the solution in a systematic manner. Therefore we will start with the integers. Selecting 1 as a possible root we have

$$\begin{array}{rrrrr|l} 4 & 16 & -31 & -49 & 30 & \underline{1} \\ & 4 & 20 & -11 & -60 & \\ \hline 4 & 20 & -11 & -60 & -30 & \end{array}$$

Since 1 is not a root, we will try 2.

$$\begin{array}{rrrrr|l} 4 & 16 & -31 & -49 & 30 & \underline{2} \\ & 8 & 48 & 34 & -30 & \\ \hline 4 & 24 & 17 & -15 & 0 & \end{array}$$

In examining the quotient we know that 2 cannot be a double root since 2 is not a divisor of -15. However, ± 1, ± 3, ± 5, and ± 15 are all divisors of -15. It can be shown by using synthetic division that -1 and -3 are not roots of the equation so let us try -5.

$$\begin{array}{rrrr|l} 4 & 24 & 17 & -15 & \underline{-5} \\ & -20 & -20 & 15 & \\ \hline 4 & 4 & -3 & 0 & \end{array}$$

Thus, the polynomial can be written as

$$P(x) = (x - 2)(x + 5)(4x^2 + 4x - 3).$$

By factoring or by the quadratic formula the expression $4x^2 + 4x - 3$ can be factored as $(2x + 3)(2x - 1)$. Therefore, the roots of $P(x)$ are 2, -5, $-3/2$, and $1/2$. □

EXAMPLE 6 Graph $P(x) = 4x^4 + 16x^3 - 31x^2 - 49x + 30$ for the interval $-6 < x < 3$.

Solution: The x-intercepts were found in Example 5: $(-5, 0)$, $(-1.5, 0)$, $(0.5, 0)$, and $(2, 0)$. The y-intercept is $(0, 30)$. The function is graphed in Figure 12.4 (on p. 470). □

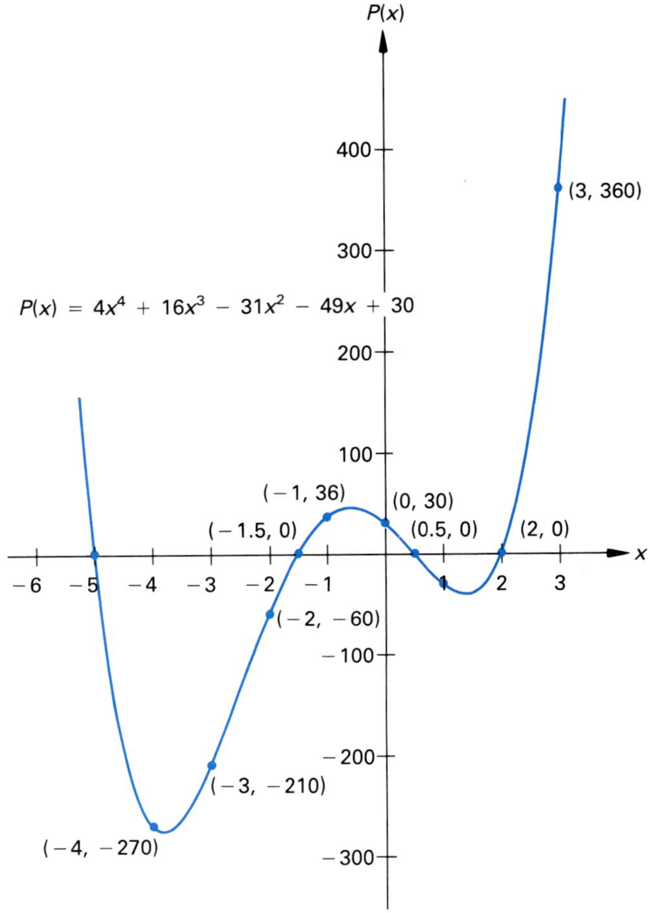

FIGURE 12.4

Summary of Techniques to Determine Rational Roots of Polynomial Functions:

1. What is the degree of the polynomial? Degree n implies there are n possible roots.
2. Is n odd or even?
3. Is there an upper bound for the positive roots?
4. Is there a lower bound for the negative roots?
5. What is the maximum number of positive roots? Is the number odd or even?
6. What is the maximum number of negative roots? Is the number odd or even?

7. List all the possible positive and negative rational roots by factoring p and q.
8. Can the list of possible roots in Step 7 be reduced using the results of questions 3 and 4?
9. If question 5 indicates an odd number of positive roots, then select the positive values from the established list of possible roots for the first divisors.
10. If question 6 indicates an odd number of negative roots, then select the negative values from the established list of possible roots for the first divisors.
11. Make a list of divisors and remainders. These are ordered pairs that satisfy the polynomial function and can be used in graphing the function.

12.5 Exercises

In Exercises 1–20, (a) use Theorem 12.7 to determine the possible number of positive and negative roots of the polynomial, (b) list all the possible rational roots of the polynomial, and (c) find the rational roots of the polynomial.

1. $P(x) = x^3 - x^2 - 5x - 3$
2. $P(x) = x^3 + x^2 - 5x + 3$
3. $P(x) = x^3 + 2x^2 - x - 2$
4. $P(x) = x^3 - 2x^2 - 5x + 6$
5. $P(x) = 2x^3 + 9x^2 + 7x - 6$
6. $P(x) = 6x^3 + 19x^2 + x - 6$
7. $P(z) = z^3 - 2z^2 - 11z - 12$
8. $P(z) = z^3 - 5z^2 - 12z + 36$
9. $P(v) = v^3 - 8v^2 + 14v - 4$
10. $P(v) = 2v^3 + 11v^2 + 44v - 25$
11. $P(t) = 3t^3 - 10t^2 + 25t + 22$
12. $P(t) = 5t^3 - 7t^2 + 19t + 15$
13. $P(x) = x^4 - 16x^3 + 86x^2 - 176x + 105$
14. $P(x) = x^4 - x^3 - 19x^2 + 49x - 30$
15. $P(x) = 4x^4 + 8x^3 - 7x^2 - 21x - 9$
16. $P(x) = x^4 - 4x^3 + 6x^2 - 4x + 1$
17. $P(x) = 8x^5 - 12x^4 + 14x^3 - 12x^2 + 6x - 1$
18. $P(x) = x^4 - 2x^3 - 8x^2 + 10x + 15$
19. $P(x) = 6x^4 - 31x^3 + 19x^2 + 19x + 3$
20. $P(x) = x^4 + x^3 - x^2 + x - 2$

In Exercises 21–30, graph each polynomial function. Find all the intercepts.

21. $P(x) = x^3 + 4x^2 - x - 4$
22. $P(x) = x^3 + x^2 - 4x - 4$
23. $P(x) = x^3 + 2x^2 - 9x - 18$
24. $P(x) = x^3 + 5x^2 - x - 5$
25. $P(x) = x^4 - 5x^2 + 4$
26. $P(x) = -x^4 + 5x^2 - 4$
27. $P(x) = 2x^3 - 12x - 8$
28. $P(x) = 2x^3 - 9x^2 - 24x + 4$
29. $P(x) = x^4 + x^3 - 5x^2 + x - 6$
30. $P(x) = x^4 - x^3 - 11x^2 + 9x + 18$

31. A surveyor has laid out a triangular plot such that one side is 2 m longer than the shortest side and the remaining side is 1 m longer than the shortest side. She has determined that the product of the lengths of the three sides equals five times their sum. Find the lengths of the sides of the triangular plot.

32. Tom discovered in measuring the edge of his Rubik's Cube that the number of cubic inches in

the volume of the cube was equal to one-third the number of square inches of the surface area. Find the length of an edge of the cube.

33. The motion of a particle is such that the distance (or distances) traveled from a fixed reference point is given by

$$s = 4t^4 - 34t^3 + 85t^2 - 81t + 30, \qquad t \geq 0$$

For what values of t will the particle return to the reference point?

34. In the theory of electronic-feedback-control systems, polynomial zeroes play a major role in the time response of the electrical system. Using the methods developed in this section, find all zeroes (real and complex) of the polynomial

$$F(S) = 2S^4 + 17S^3 + 61S^2 + 160S + 150.$$

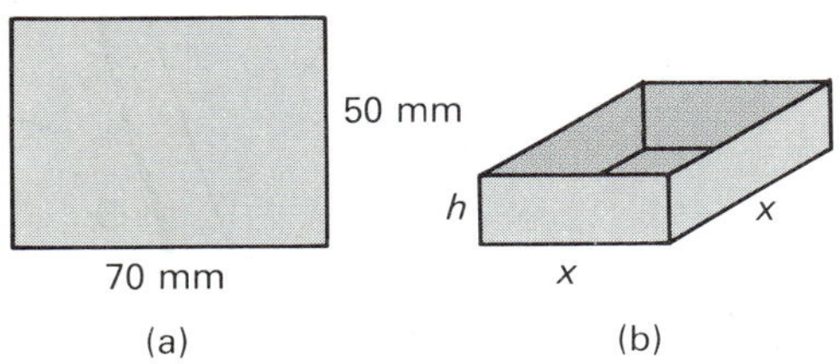

FIGURE 12.5

35. A rectangular piece of metal 70 mm × 50 mm is to be cut and shaped into an open box with no waste of material. The box is to have a square base (Fig. 12.5). If the volume is to be 15 1/2 cm³, what are the possible dimensions of the box?

36. The bending moment of a beam is given by

$$M(d) = 0.1d^4 - 1.4d^3 + 6.3d^2 - 9.0d,$$

where d is the distance in meters from one end. Find the values of d where the bending moment is zero.

37. The total impedance Z of three impedances in series is given by $Z = Z_1 + Z_2 + Z_3$. If $Z = 3.0\ \Omega$, $Z_2 = Z_1^3$ and $Z_3 = 2 - Z_1^3$, what are the possible values for Z_1?

38. The radius of a cylindrical tank is 3 m less than its height. If the volume of the tank is 200π m³, find the height of the tank.

39. A variable electric current is given by the formula

$$i = 2t^3 - 13t^2 + 27t - 15.$$

If t is in seconds, at what time is the current equal to 3 A?

40. A variable voltage is given by the formula $e = 2t^3 - 12t^2 + 22t - 3$. At what time t (in seconds) is the voltage equal to 21 V?

12.6 Irrational roots by linear approximation

The theorems and examples in the previous sections have provided you with techniques for finding the rational roots of polynomial equations. There is a remaining question: "How do we find the irrational roots?"

From the analysis of a polynomial equation, if it is known that there is one real root and it is also known that it is not a rational root, then the root must be an irrational number. Recall from the definition of irrational numbers that this means that the root **cannot** be expressed as a terminating or repeating decimal.

The process of finding irrational roots by linear approximation will be illustrated with examples. Before geting into the examples let us examine a theorem whose conclusion is used frequently in finding irrational roots.

THEOREM 12.9

> If $P(x)$ is a polynomial with real coefficients, and if, for the real numbers a and b, $P(a)$ and $P(b)$ have opposite signs, the equation $P(x) = 0$ has at least one number r between a and b such that $P(r) = 0$.

The theorem asserts that if at $x = a$ the curve is on one side of the x-axis and at $x = b$ the curve is on the other side, the curve must cross the x-axis at least one time between a and b. This says that there is at least one real root between a and b.

EXAMPLE 1 Find the real root of the polynomial function

$$P(x) = x^3 + 3x^2 + 7x + 6$$

to the nearest hundredth.

Solution: There are several things that can be determined about the function: (1) Descartes' rule of signs indicates there are no positive roots and there may be three negative roots or one. (2) The rational-root theorem indicates that if there are any rational roots they must be -1, -2, -3, and -6. (Remember there are no positive roots.)

Using synthetic division we determine whether or not any of these are roots of $P(x)$.

$$\begin{array}{rrrr|l} 1 & 3 & 7 & 6 & -1 \\ & -1 & -2 & -5 & \\ \hline 1 & 2 & 5 & 1 & \end{array} \qquad \begin{array}{rrrr|l} 1 & 3 & 7 & 6 & -2 \\ & -2 & -2 & -10 & \\ \hline 1 & 1 & 5 & -4 & \end{array}$$

$$\begin{array}{rrrr|l} 1 & 3 & 7 & 6 & -3 \\ & -3 & 0 & -21 & \\ \hline 1 & 0 & 7 & -15 & \end{array} \qquad \begin{array}{rrrr|l} 1 & 3 & 7 & 6 & -6 \\ & -6 & 18 & -150 & \\ \hline 1 & -3 & 25 & -144 & \end{array}$$

FIGURE 12.6

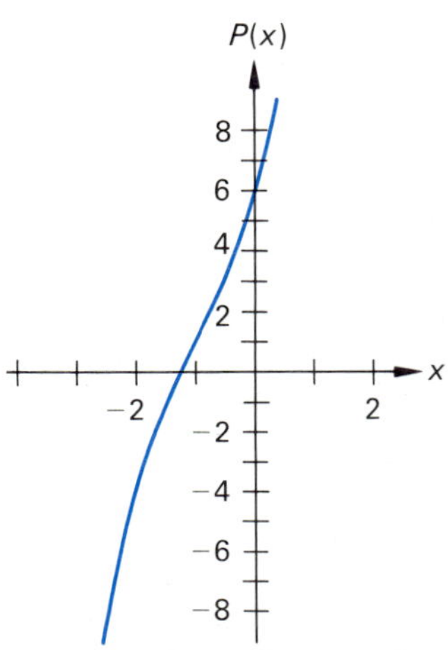

$P(x) = x^3 + 3x^2 + 7x + 6$

In none of the divisions is the remainder zero. Therefore, there is no rational root. The results of the synthetic division indicate that -6 is a lower bound since the signs in the third row alternate. Thus, there are no real roots smaller than -6. Also, the remainder for the divisor -1 is positive and the remainder for the divisor -2 is negative. Theorem 12.9 tells us that there exists a real number between -1 and -2 that is a root of the equation. To illustrate this, the function is sketched in Fig. 12.6.

The root between -1 and -2 is an irrational number since it has been demonstrated that it cannot be rational. The method of linear approximation will be used to determine the root.

In Fig. 12.7 the region around the points $(-1, 1)$ and $(-2, -4)$ has been enlarged and the two points joined with a straight line. The point where the straight line crosses the x-axis is a first linear approximation of the root. The line crosses between -1.2 and -1.3. Testing -1.2 and

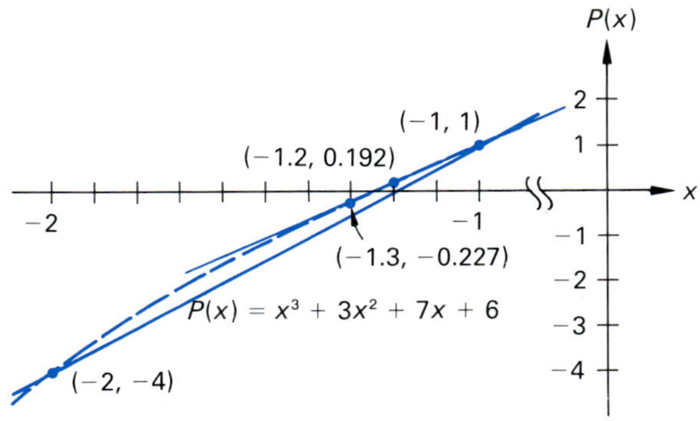

FIGURE 12.7

-1.3 by synthetic division:

$$\begin{array}{rrrr|l} 1 & 3 & 7 & 6 & \underline{-1.2} \\ & -1.2 & -2.16 & -5.808 & \\ \hline 1 & 1.8 & 4.84 & 0.192 & \end{array}$$

$$\begin{array}{rrrr|l} 1 & 3 & 7 & 6 & \underline{-1.3} \\ & -1.3 & -2.21 & -6.227 & \\ \hline 1 & 1.7 & 4.79 & -0.227 & \end{array}$$

The positive remainder of 0.192 indicates that $(-1.2, 0.192)$ is a point on the curve above the x-axis. The negative remainder of -0.227 indicates that $(-1.3, -0.227)$ is a point on the curve below the x-axis. To the nearest tenth the root would be -1.2 since its remainder is closer to zero. To obtain a root to the nearest hundredth, we continue the process. Plot these points on Fig. 12.7 and join the two points with a straight line. The line appears to cut the x-axis at -1.25. Test this value with synthetic division.

$$\begin{array}{rrrr|l} 1 & 3 & 7 & 6 & \underline{-1.25} \\ & -1.25 & -2.1875 & -6.015625 & \\ \hline 1 & 1.75 & 4.8125 & -0.015625 & \end{array}$$

The remainder is negative, which indicates that $(-1.25, -0.0156)$ is below the x-axis. This remainder also has the opposite sign of the remainder for -1.2. Thus, to find a root to the nearest hundredth we must select larger values (values to the right of -1.25) until the sign of the remainder changes. We select -1.24.

$$\begin{array}{rrrr|l} 1 & 3 & 7 & 6 & \underline{-1.24} \\ & -1.24 & -2.1824 & -5.973824 & \\ \hline 1 & 1.76 & 4.8176 & 0.026176 & \end{array}$$

Since the sign of the remainder changed, the root must be between -1.25 and -1.24. The remainder for -1.25 is closer to zero than the remainder for -1.24. Therefore, $x = -1.25$ is the root to the nearest hundredth. □

TABLE 12.2 **Algebraic key strokes**

1.24
[+/−]
[STO] 0
[×]
1
[+]
3
[=]
[×]
[RCL] 0
[+]
7
[=]
[×]
[RCL] 0
[+]
6
[=]
0.026176

The calculations for synthetic division can be performed on the calculator by storing the divisor. For example, the steps for doing the division with a divisor of -1.24 are given in Table 12.2.

EXAMPLE 2 Find the real roots of the polynomial function

$$P(x) = x^3 - 3x - 1$$

to the nearest hundredth.

Solution: Descartes' rule of signs indicates there may be one positive root and two negative roots. The rational-root theorem indicates if there are any rational roots they must be ± 1. Testing with synthetic division the results indicate that there are no rational roots.

$$\begin{array}{rrrr|r} 1 & 0 & -3 & -1 & \underline{1} \\ & 1 & 1 & -2 & \\ \hline 1 & 1 & -2 & -3 & \end{array} \qquad \begin{array}{rrrr|r} 1 & 0 & -3 & -1 & \underline{-1} \\ & -1 & 1 & 2 & \\ \hline 1 & -1 & -2 & 1 & \end{array}$$

The results of the synthetic division provide us with two points on the curve $(1, -3)$ and $(-1, 1)$. These two points indicate that there is a root between 1 and -1. To help locate the roots we sketch the curve (Fig. 12.8). The sketch implies that the curve crosses the x-axis at three places: between -2 and -1, between -1 and 0, and between 1 and 2.

In Fig. 12.9, the region of Fig. 12.8 to the left of the origin has been enlarged. Drawing a straight line to connect the points $(-2, -3)$ and $(-1, 1)$ we can get the first linear approximation. It appears that the root is between -1.3 and -1.2. Testing -1.3 and -1.2 by synthetic

FIGURE 12.8

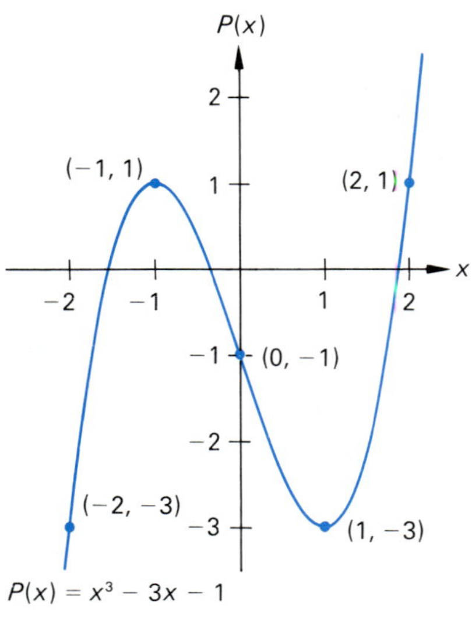

FIGURE 12.9

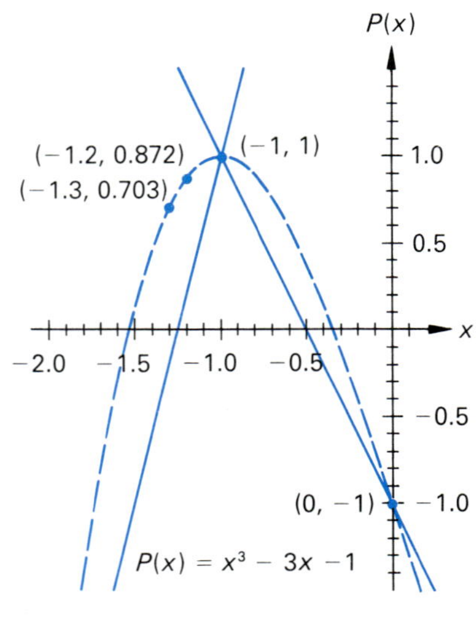

division:

$$\begin{array}{rrrr|l} 1 & 0 & -3 & -1 & -1.3 \\ & -1.3 & 1.69 & 1.703 & \\ \hline 1 & -1.3 & -1.31 & 0.703 & \end{array}$$

$$\begin{array}{rrrr|l} 1 & 0 & -3 & -1 & -1.2 \\ & -1.2 & 1.44 & 1.872 & \\ \hline 1 & -1.2 & -1.56 & 0.872 & \end{array}$$

The points $(-1.3, 0.703)$ and $(-1.2, 0.872)$ are plotted in Fig. 12.9. Since -1.3 has a smaller remainder than -1.2, we know that the root must be to the left of this point. Sketching the possible path of the curve it appears that -1.5 may be a better approximation.

$$\begin{array}{rrrr|l} 1 & 0 & -3 & -1 & -1.5 \\ & -1.5 & 2.25 & 1.125 & \\ \hline 1 & -1.5 & -0.75 & 0.125 & \end{array}$$

$$\begin{array}{rrrr|l} 1 & 0 & -3 & -1 & -1.6 \\ & -1.6 & 2.56 & 0.704 & \\ \hline 1 & -1.6 & -0.44 & -0.296 & \end{array}$$

The remainder for -1.5 is a positive, 0.125, which is closer to 0 than the remainder for -1.3. Moving to the left and trying -1.6, the remainder is -0.296. There is a change of sign, indicating that the root is between -1.6 and -1.5. Since the remainder for -1.5 is closer to zero than the remainder for -1.6, the root would be closer to -1.5. Since the required answer is to be accurate to the nearest hundredth, as a next approximation we select -1.54.

$$\begin{array}{rrrr|l} 1 & 0 & -3 & -1 & -1.54 \\ & -1.54 & 2.3716 & 0.967736 & \\ \hline 1 & -1.54 & -0.6284 & -0.032264 & \end{array}$$

The remainder is negative so moving to the right we select -1.53.

$$\begin{array}{rrrr|l} 1 & 0 & -3 & -1 & -1.53 \\ & -1.53 & 2.3409 & 1.008423 & \\ \hline 1 & -1.53 & -0.6591 & 0.008423 & \end{array}$$

The remainder for -1.53 is positive. Since there is a change of sign, the root must be between -1.54 and -1.53. To the nearest hundredth, -1.53 is the root since its remainder is closer to zero.

To determine the next root, examine Fig. 12.9 and note that the line joining $(-1, 1)$ and $(0, -1)$ crosses the axis between -0.5 and -0.4.

$$\begin{array}{rrrr|l} 1 & 0 & -3 & -1 & -0.5 \\ & -0.5 & 0.25 & 1.375 & \\ \hline 1 & -0.5 & -2.75 & 0.375 & \end{array}$$

$$\begin{array}{rrrr|l} 1 & 0 & -3 & -1 & -0.4 \\ & -0.4 & 0.16 & 1.136 & \\ \hline 1 & -0.4 & -2.84 & 0.136 & \end{array}$$

Since both remainders are positive and the remainder for -0.4 is the smaller, the root must be to the right of -0.4. Let us try -0.3 as a new approximation.

$$\begin{array}{rrrr|l} 1 & 0 & -3 & -1 & \underline{-0.3} \\ & -0.3 & 0.09 & 0.873 & \\ \hline 1 & -0.3 & -2.91 & -0.127 & \end{array}$$

There is a change of sign for the remainders for -0.4 and -0.3 indicating that the root lies between these numbers. A good approximation may be -0.34 since the remainder for -0.3 is the smaller.

$$\begin{array}{rrrr|l} 1 & 0 & -3 & -1 & \underline{-0.34} \\ & -0.34 & 0.1156 & 0.980696 & \\ \hline 1 & -0.34 & -2.8844 & -0.019304 & \end{array}$$

Since the remainder is negative, the root must be to the left of -0.34; try -0.35.

$$\begin{array}{rrrr|l} 1 & 0 & -3 & -1 & \underline{-0.35} \\ & -0.35 & 0.1225 & 1.007125 & \\ \hline 1 & -0.35 & -2.8775 & 0.007125 & \end{array}$$

The remainder for -0.35 is positive indicating that the root is between -0.35 and -0.34. Since the remainder for -0.35 is closer to zero, the root, to the nearest hundredth, is -0.35.

To find the third root we must draw a straight line connecting $(2, 1)$ and $(1, -3)$. Using the same procedure that was used to find the first two roots we find that there is a sign change between 1.8 and 1.9, since $P(1.8) = -0.568$ and $P(1.9) = 0.159$. The root is to be to the nearest hundredth. Try 1.87: $P(1.87) = -0.070797$. Since this remainder is negative and the remainder for 1.9 was positive, the root must lie between 1.87 and 1.9.

Moving one hundredth to the right, the next value to try would be 1.88: $P(1.88) = 0.004672$. The remainder for 1.88 is positive indicating that the root is between 1.87 and 1.88. Since the remainder for 1.88 is closer to zero, the root, to the nearest hundredth, is 1.88. You may check each of these values using synthetic division.

The real roots of the polynomial function $P(x) = x^3 - 3x - 1$, to the nearest hundredth, are -1.53, -0.35, and 1.88. □

In the two examples the roots were found to the nearest hundredth. The process could be continued and the roots could be found to any degree of accuracy. For any function, first determine all the rational roots and then determine the irrational roots. Always sketch the function to help approximate the roots.

12.6 Exercises

In Exercises 1–10, use Theorem 12.9 to show that the polynomial equations have at least one root between the indicated values.

1. $P(x) = 2x^3 + x^2 - x - 1 \quad [0, 1]$

2. $P(x) = x^4 - 6x^3 + 7x^2 + 5x - 1 \quad [-2, 0]$

3. $P(x) = x^4 - 6x^3 + 8x^2 + 2x - 2$ $[-0.6, -0.5]$
4. $P(x) = x^3 - 7x + 1$ $[-2.72, -2.71]$
5. $P(x) = x^3 - 7x + 1$ $[0.14, 0.15]$
6. $P(x) = x^3 - 7x + 1$ $[2.57, 2.58]$
7. $P(x) = x^3 - 5x + 1$ $[2.12, 2.13]$
8. $P(x) = x^3 - 2x + 5$ $[-2.09, -2.10]$
9. $P(x) = x^3 + 3x^2 - 4x - 11$ $[1.94, 1.95]$
10. $P(x) = x^3 + 4x^2 - 2$ $[-0.79, -0.78]$

In Exercises 11–20, find a root of $P(x)$, to the nearest tenth, in the indicated intervals.

11. $P(x) = x^3 + 3x^2 + 4x - 1$ $[0, 1]$
12. $P(x) = x^3 + x^2 - 7x + 1$ $[2, 3]$
13. $P(x) = x^3 - 3x^2 - 5x + 9$ $[3, 4]$
14. $P(x) = x^3 + 3x^2 - 6x - 2$ $[1, 2]$
15. $P(x) = x^3 + 6x^2 + 5x - 1$ $[-1, -2]$
16. $P(x) = 2x^3 - 6x + 1$ $[-1, -2]$
17. $P(x) = x^3 - 3x^2 - 3x + 3$ $[-1, 1]$
18. $P(x) = x^4 - 11x^2 + 15x - 2$ $[0, 1]$
19. $P(x) = x^4 + x^3 - x^2 - 3x - 6$ $[1, 2]$
20. $P(x) = x^4 + 5x^3 + 8x^2 + 2x - 8$ $[0, 1]$

In Exercises 21–30, plot the graphs of the functions and find all the real roots of the function to the nearest hundredth.

21. $P(x) = x^3 + 4x - 3$
22. $P(x) = x^3 + 5x - 2$
23. $P(x) = x^3 + 7x - 1$
24. $P(x) = x^3 + 3x^2 + 4$
25. $P(x) = x^3 + x^2 + 1$
26. $P(x) = x^4 + 3x^2 + 5x + 3$
27. $P(x) = x^4 + 5x^3 + 6x^2 + 4x + 12$
28. $P(x) = x^4 - 2x^3 + 3x^2 - 8x + 4$
29. $P(x) = x^4 - 2x^3 + x^2 - 1$
30. $P(x) = x^4 - 10x^2 - 5x + 7$

31. The depth x to which a floating sphere of radius r and density d will sink in water is a positive root of the equation

$$x^3 - 3rx^2 + 4dr^3 = 0.$$

Find the depth, correct to the nearest hundredth, to which a sphere of 1-m diameter will sink if it is made of (a) cork whose density is 0.23 and (b) ice whose density is 0.92.

32. If its breaking strength is not exceeded, a 2-in. × 4-in. oak beam L ft long supported at its ends will have a maximum sag of y inches if two concentrated loads of P lb each are placed on it, one load d ft from one end and the other load d ft from the other end, where

$$y = \frac{12^3 Pd(3L^2 - 4d^2)}{3.84 \times 10^8}.$$

To the nearest tenth, find how many feet from the ends of a 6-ft beam loads of 500 lb each can be put without causing sag of more than 1/4 in.

33. The total capacitance C of three capacitors connected in series is given by

$$\frac{1}{C} = \frac{1}{C_1} + \frac{1}{C_2} + \frac{1}{C_3}.$$

If $C = 0.25\ \mu F$, $C_2 = C_1 - 1.3\ \mu F$, and $C_3 = C_1 + 1.2\ \mu F$, what are the two possible values for C_1?

34. A rectangular box has the dimensions 30 mm × 40 mm × 50 mm. If the length, width, and height are increased by the same amount, the volume will be doubled. To the nearest tenth, find the amount by which the dimensions are to be increased.

35. A cube has a volume equal to that of a sphere whose radius is 10 mm shorter than an edge of the cube. What is the radius of the sphere?

Chapter summary

Any function of the form

$$P(x) = a_n x^n + a_{n-1} x^{n-1} + \cdots + a_1 x + a_0,$$

where n is a positive integer, $a_n \neq 0$, and $a_1, a_2, \ldots, a_n$ are real or complex numbers is a **polynomial function.**

A technique for determining the roots of polynomial functions of degree greater than two is **synthetic division.** You should be able to divide a polynomial by a linear polynomial using synthetic division.

The **definition of division,** $P(x) = D(x)Q(x) + R(x)$, is an important tool to check the results of a division problem.

Several theorems were discussed that are extremely helpful in finding the roots of polynomial functions. The theorems are: the **remainder theorem** (Theorem 12.1), the **factor theorem** (Theorem 12.2), the **theorem on upper bound and lower bound for roots** (Theorem 12.3), the **fundamental theorem of algebra** (Theorem 12.4), the **theorem on the number of roots of a polynomial function** (Theorem 12.5), the **conjugate-roots theorem** (Theorem 12.6), **Descartes' rule of signs** (Theorem 12.7), the **rational-root theorem** (Theorem 12.8), and the **theorem to determine whether or not there is a root between two numbers** (Theorem 12.9).

To find the roots of a polynomial, one should first apply Descartes' rule of signs to determine the possible number of positive and negative rational roots. Then apply the rational-root theorem to determine any possible rational roots of the polynomial. To determine the irrational roots it is helpful to sketch the polynomial. If there is any question as to whether a root is above or below a particular value, use the theorem on upper bound and lower bound to check. The process defined to find the irrational roots is called **linear approximation.**

Review Exercises

In Exercises 1–10, use synthetic division to find the quotient and remainder. Check each of the results using the equation $P(x) = Q(x)D(x) + R(x)$.

1. $(2x^3 - 21x^2 + 43x + 60) \div (x - 5)$

2. $(2x^4 - 11x^3 - 111x^2 - 108) \div (x + 4)$

3. $(2x^3 + 25x^2 + 9x - 38) \div (x + 12)$

4. $(x^3 + 14x^2 + 25x + 22) \div (x + 2)$

5. $(2x^4 + 27x^3 + 34x^2 - 27x - 38) \div (x + 1)$

6. $(6x^4 + 5x^3 - 6x^2 - 45x - 27) \div (x + 3)$

7. $(6x^4 + 5x^3 - 6x^2 - 45x - 27) \div (3x + 1)$

8. $(x^5 - 5x^4 - 13x^3 + 65x^2 + 36x - 180) \div (x - 5)$

9. $(x^5 - 5x^4 - 13x^3 + 65x^2 + 36x - 183) \div (x - 3)$

10. $(x^8 + a^8) \div (x + a)$

In Exercises 11–16, form polynomial equations with integral coefficients with the given roots.

11. $3, -4, 5$

12. $-4, 0, 2, 2$

13. $\frac{2}{3}, \frac{1}{5}, -4$

14. $3 \pm \sqrt{3}, -4$

15. $2 \pm \sqrt{3}, \pm 5j$

16. $\frac{1}{3}, 1 \pm j\sqrt{7}, \pm \sqrt{3}$

In Exercises 17–26, (a) list all the possible rational roots of the polynomial and (b) find the rational roots of the polynomial.

17. $P(x) = x^3 - 4x^2 - 7x + 10$

18. $P(z) = z^3 + 2z^2 - 11z - 12$

19. $P(v) = 2v^3 + 11v^2 + 44v - 25$

20. $P(y) = 5y^3 - 7y^2 + 19y + 15$

21. $P(v) = v^4 - 10v^3 + 32v^2 - 30v - 9$

22. $P(y) = 2y^4 + 13y^3 + 19y^2 - 18y - 36$

23. $P(y) = 15y^4 + 4y^3 + 56y^2 + 16y - 16$

24. $P(x) = 36x^5 - 37x^3 + 12x^2 - 73x + 12$

25. $P(x) = x^4 - 2x^3 + 29x^2 - 64x - 96$

26. $P(x) = x^6 + x^5 + 20x^4 + 4x^3 + 59x^2 - 5x - 80$

In Exercises 27–36, plot the graph of each of the functions for the indicated intervals and find all the real roots of the function to the nearest hundredth.

27. $P(x) = x^3 + 2x^2 - 7, \quad -3 < x < 3$

28. $P(x) = x^3 + 2x^2 + 7, \quad -4 < x < 2$

29. $P(x) = x^4 + 6x^3 + 8x^2 - 6x - 24, \quad -5 < x < 3$

30. $P(x) = 2x^4 + 5x^3 + 2x^2 + 14x + 7, \quad -4 < x < 1$

31. $P(R) = 2R^4 - 3R^3 + 6R^2 - 19R + 15, \quad -1 < R < 2$
32. $P(t) = t^4 + t^3 - 7t^2 - 5t + 2, \quad -4 < t < 4$
33. $P(Z) = 2Z^3 + 7Z^2 - Z - 15, \quad -3 < Z < 2$
34. $P(Z) = Z^4 + 4Z^3 - 37, \quad -6 < Z < 4$
35. $P(t) = t^3 - 5t^2 + 12t + 26, \quad -3 < t < 3$
36. $P(x) = x^4 - 6x^3 + 8x^2 + 2x - 1, \quad -1 < x < 4$
37. After a 10-mm-thick slice is cut from one side of a cube, the new volume is 48 cm³. Find the length of the side of the original cube.
38. An open box is to be made from a square sheet of cardboard 180 mm on a side by cutting equal squares from the corners and folding up the sides. How long should the edge of the cut-out square be if the volume of the box is to be 432 cm³?
39. Find the value of k for which $x - 3$ is a factor of

    $kx^3 - 6x^2 + 2kx - 12.$
40. Find the value of k for which $x + 2$ is a factor of

    $3x^3 + 2kx^2 - 4x - 8.$
41. The volume of a rectangular wood bin is to be tripled by increasing each dimension by the same amount. If the wood bin is initially 2 by 3 by 5 m, what must be the amount of the increase in each dimension?
42. The motion of a particle is such that the distance (or distances) traveled from a fixed reference point is given by

    $s = 2t^4 - 18t^3 - 18t^2 + 36t + 20, \qquad t \geq 0$

    For what values of t will the particle return to the reference point?
43. A company determines that the cost $C(x)$ of manufacturing x units of a certain item is given by

    $C(x) = 4x^3 - 100x^2 + 344x - 420.$

    If the company allocates \$700 daily to the production of this commodity, how many units per day can be produced?

Can you solve these?

1. Use the factor theorem to show that $x - y$ is a factor of $x^n - y^n$, where n is a natural number.
2. Write a polynomial $P(x)$ with complex coefficients that has the root $a + bj$, $b \neq 0$, and does not have $a - bj$ as a root.
3. Prove that a polynomial of odd degree with real coefficients has at least one real root.
4. Prove that if $P(x)$ involves only even powers of x, and if all of its coefficients including the constant term are positive, then $P(x) = 0$ has no real roots.
5. Show that $\sqrt{3}$ is irrational.
6. Find a polynomial function $y = P(x)$ with all of the following properties: **a.** $P(x)$ is of degree 4, **b.** $P(0) = 1$, **c.** the graph is symmetric with respect to the y-axis, and **d.** as x becomes large, then y becomes large (as $x \to \infty$, then $y \to \infty$).
7. In studying the lateral stability of a certain airplane it is necessary to solve the equation

    $x^4 + 45x^3 + 214x^2 + 1492x + 595 = 0.$

    Find the real roots to the nearest tenth.
8. Solve the following equation for all positive values of θ less than 2π.

    $$4 \sin^2 \theta - 12 \sin \theta \cos^2 \theta - 7 \cos^2 \theta + 9 \sin \theta + 5 = 0.$$

    [*Hint:* Use the identity $\cos^2 \theta = 1 - \sin^2 \theta$]
9. In a hollow rectangular tube with outside dimensions a in. and b in. and wall thickness t in., subject to a twisting moment of T lb-in., the shearing stress in the walls is S lb/in², where

    $$S = \frac{T}{[2t(a - t)(b - t)]}$$

    In a tube with $a = 2$ in. and $b = 1$ in., for what wall thickness would a twisting moment of 5760 lb-in. produce $S = 10{,}000$ psi?
10. Find the thickness of the wall of a hollow spherical shell of uniform thickness if its outer diameter is 100 mm and the volume of material from which it is made is 300 cm³.

13

The relation between pressure and volume for most air compressor systems is $pV^n = C$, where p is the pressure, V is the volume, C is a constant, and n is a constant characteristic of a given compressor system. If the volume produced by a given system is known to be 1000 ft^3 at a pressure of 100 psi and it is 2000 ft^3 at 60 psi, what is the volume produced at 75 psi?

Exponential and Logarithmic Functions

13.1 The exponential function

The stress in a rubber band decreases with the length of time that it is stretched. The stress relaxation equation for a brand of rubber bands is given by $S = 5e^{-t/60}$. A function of the form $y = b^x$ is called an **exponential function.**

DEFINITION 13.1

> The **exponential function** $y = b^x$ $(b > 0,\ b \neq 1)$ has **base** b, and **exponent** x.

The domain of the exponential function is the set of all real numbers and the range is the set of all positive real numbers.

The exponential function is frequently encountered in technical applications involving naturally increasing or decreasing phenomena such as bacteria growth, atomic diffusion, or capacitor discharge.

Recall from the definition of exponents (Chapter 2) that $b^0 = 1$, and also that $b^{-x} = \dfrac{1}{b^x}$.

EXAMPLE 1 Graph the exponential functions

a. $y = 2^x$, $-3 \leq x \leq 6$, and

b. $y = 3^x$, $-2 \leq x \leq 4$.

Solution: With bases of 2 and 3, it is easy to calculate the value of y when the exponent x is a positive integer. When x is a negative integer we can use $b^{-x} = \dfrac{1}{b^x}$. Therefore

$$
\begin{aligned}
3^{-2} &= \frac{1}{3^2} \\
&= \frac{1}{9} \\
&= 0.111.
\end{aligned}
$$

For more difficult situations we can use the $\boxed{y^x}$ key on the calculator to evaluate the function. (To calculate the value for 3^5 the steps are 3 $\boxed{y^x}$ 5 $\boxed{=}$ 243, so $3^5 = 243$. To calculate the value for 3^{-5} the steps are 3 $\boxed{y^x}$ 5 $\boxed{+/-}$ $\boxed{=}$ 0.00411523.) The functional values are shown in the tables next to the graphs in Figs. 13.1 and 13.2. Note that the scales on the x- and y- axes are different. This is necessary because of the large difference between the range of the x-values and the range of the y-values. We observe that $y = 3^x$ increases more rapidly with x than does $y = 2^x$.

a. $y = 2^x$

x	2^x
−3	0.125
−2	0.250
−1	0.500
0	1.0
1	2.0
2	4.0
3	8.0
4	16.0
5	32.0
6	64.0

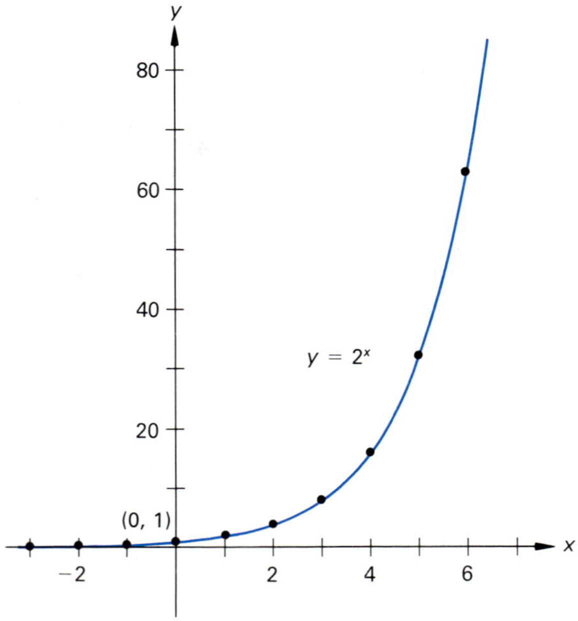

FIGURE 13.1

b. $y = 3^x$

x	3^x
−2	0.111
−1	0.333
0	1.0
1	3.0
2	9.0
3	27.0
4	81.0

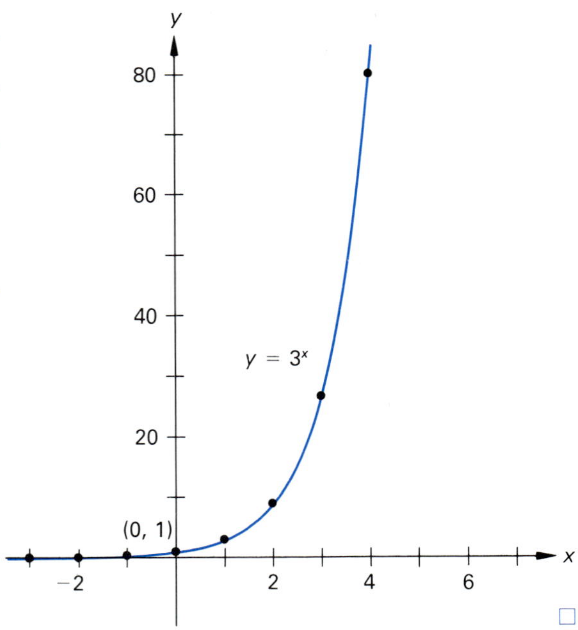

FIGURE 13.2

□

EXAMPLE 2 Graph the exponential functions

a. $y = \left(\frac{1}{2}\right)^x$, $-6 \leq x \leq 3$, and

b. $y = \left(\frac{1}{3}\right)^x$, $-4 \leq x \leq 2$.

Solution:

a. $y = \left(\frac{1}{2}\right)^x$

x	−6	−5	−4	−3	−2	−1	0	1	2	3
$(\frac{1}{2})^x$	64.0	32.0	16.0	8.0	4.0	2.0	1.0	0.5	0.3	0.1

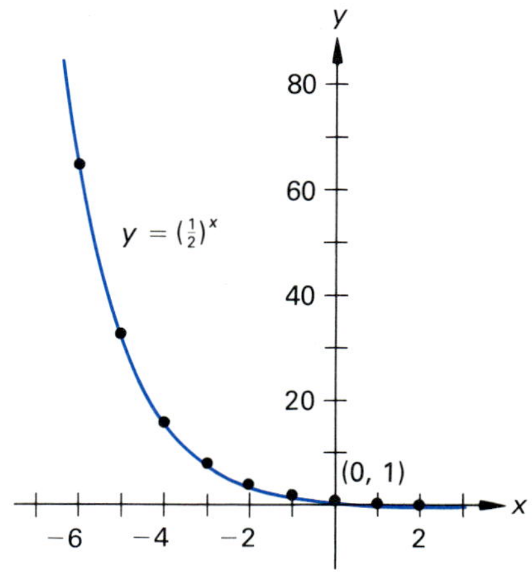

FIGURE 13.3

b. $y = \left(\frac{1}{3}\right)^x$

x	−4	−3	−2	−1	0	1	2
$(\frac{1}{3})^x$	81.0	27.0	9.0	3.0	1.0	0.3	0.1

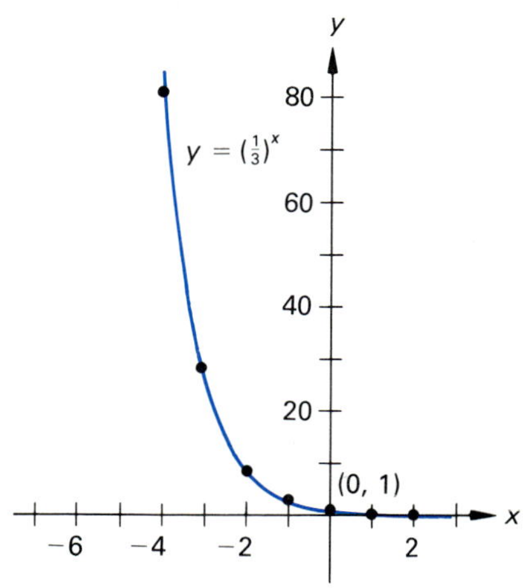

FIGURE 13.4

□

Observation of Figs. 13.1, 13.2, 13.3, and 13.4 leads us to some properties of exponential functions.

Properties of Exponential Functions, ($y = b^x$)

1. The functions are asymptotic to the x-axis.
2. The functions are not asymptotic to any line $x = c$.
3. Since $b^0 = 1$, the graph of $f(x) = b^x$ passes through the point (0, 1) for any value of b. It also passes through $(1, b)$.
4. The domain of $f(x) = b^x$ consists of the set of all real numbers; the range is the set of all positive real numbers.
5. If $b > 1$, b^x is an increasing function; if $b < 1$, b^x is a decreasing function.
6. If $a < b$, then $a^x < b^x$ for all $x > 0$, and $a^x > b^x$ for all $x < 0$.
7. Since b^x is either increasing or decreasing it never assumes the same value twice ($b \neq 1$). This leads to the conclusion: if $b^x = b^y$, then $x = y$.
8. We can also state that: if $a^x = b^x$ for all $x \neq 0$, then $a = b$.

A special case of the exponential function that is often found in technical applications is the function $y = e^x$ or $y = e^{-x}$, $x \geq 0$, where $e = 2.71828 \ldots$ is the base of the natural logarithms. The natural (or Naperian) logarithms will be discussed in Section 13.5. (In electrical applications the letter e is often used to represent electrical potential or voltage. To prevent confusion in those circumstances the Greek letter epsilon (ϵ) is sometimes used to represent the base of the natural logarithms.)

EXAMPLE 3 Graph $y = e^x$ and $y = e^{-x}$, $0 \leq x \leq 5$.

Solution: To calculate $y = e^x$ we can use the same calculator technique as was used in Example 1. To calculate e^3 the key sequence is: 2.71828 [y^x] 3 [=] 20.085496. (Since 2.71828 is an approximate value for e, so is 20.085496 an approximate value for e^3.) For a calculator with an [e^x] key the sequence is 3 [e^x] 20.085537. If there is no [e^x] key but there is a [ln x] key, the following sequence can be used: 3 [INV] [ln x] 20.085537. (The relation between e^x and ln x is that ln x is the inverse of e^x. This will be discussed in Section 13.6.) Calculating $y = e^x$ to the nearest unit and $y = e^{-x}$ to the nearest hundredth produces the tables next to the graphs in Figs. 13.5 and 13.6 on page 486.

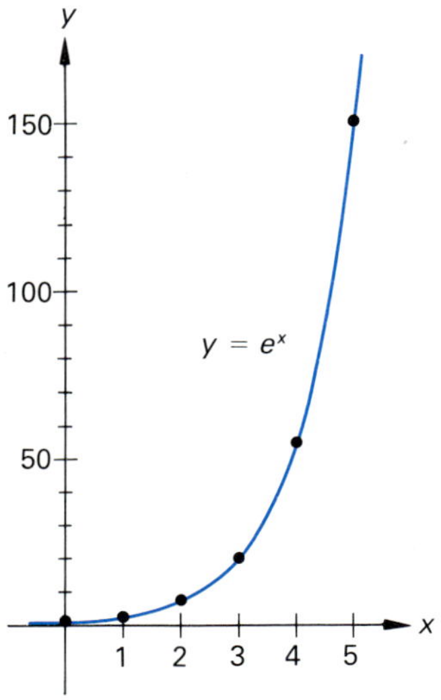

$y = e^x$

x	e^x
0	1.
1	3.
2	7.
3	20.
4	55.
5	148.

FIGURE 13.5

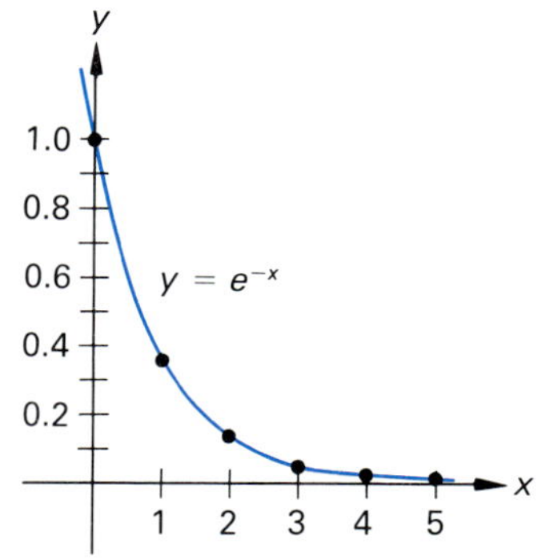

$y = e^{-x}$

x	e^{-x}
0	1.00
1	0.37
2	0.14
3	0.05
4	0.02
5	0.01

FIGURE 13.6

□

It is interesting to observe that by the time $x = 5$, e^{-x} has decreased from 1.00 to 0.00674, or to less than 1% of its starting value. For practical purposes functions of the form $y = Ae^{-x}$ are often considered to be negligible, or shut off, for $x \geq 5$.

EXAMPLE 4 A chemical compound decays according to the relation $S = S_0e^{-t/4}$, where S_0 is the initial amount of the compound present in kilograms and t is the time in hours. If S_0 is 6 kg:

a. What is the domain of the function with the indicated restrictions?

b. What is the range of the function with the indicated restrictions?

c. Graph the function.

Solution:

a. In problems for which time is the independent variable the initial or starting conditions are usually considered to occur at $t = 0$. Since time does not run backward we must have $t \geq 0$. Because functions of the form $y = Ae^{-x}$ are considered shut off for $x \geq 5$, the minimum shut-off value for $6e^{-t/4}$ would be when $t/4 = 5$ or when $t = 20$. The required domain for this function is then $0 \leq t \leq 20$.

b. The function $y = e^{-x}$, $x \geq 0$, was shown to have a maximum of 1 and a minimum approaching 0. Since $6e^{-t/4}$ is the same form multiplied by 6, the range of $6e^{-t/4}$ is $6 \cdot 1 = 6$ to $6 \cdot 0 = 0$, so the range of the function is $0 < S \leq 6$.

c. The values of $S = 6e^{-t/4}$ for $0 \leq t \leq 20$ are shown in the table and are graphed in Fig. 13.7.

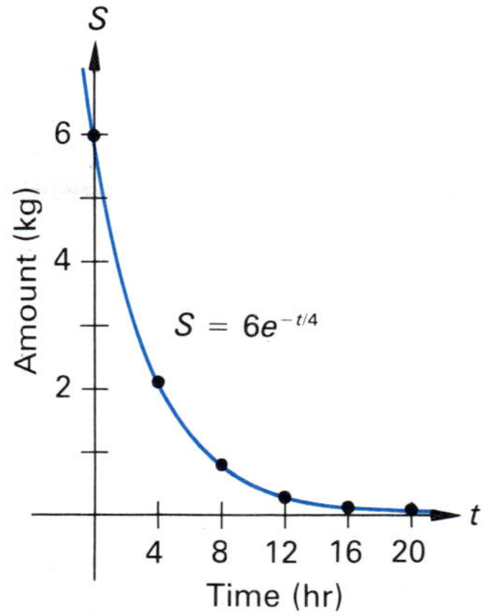

$S = 6e^{-t/4}$

t	$\frac{t}{4}$	$e^{-t/4}$	$6e^{-t/4}$
0	0	1.00	6.0
4	1	0.37	2.2
8	2	0.14	0.8
12	3	0.05	0.3
16	4	0.02	0.1
20	5	0.01	0.0*

* Rounded.

FIGURE 13.7

□

EXAMPLE 5 A bacteria culture starts with 10 bacteria and doubles every 6 hours. The equation describing the number of bacteria in the colony is of the form $N = N_0 r^{t/T}$, where N_0 is the initial number of bacteria, r is the growth factor per unit of time T, and t is the time of growth. Graph the number of bacteria present at any time during the course of one day.

Solution: Since the culture starts with 10 bacteria, $N_0 = 10$. The colony doubles in 6 hours so $r = 2$ and $T = 6$. Therefore the equation for the average number of bacteria present during a 24-hour day is $N = 10\,(2^{t/6})$, where t is in hours because T is in hours. The culture size is graphed in Fig. 13.8.

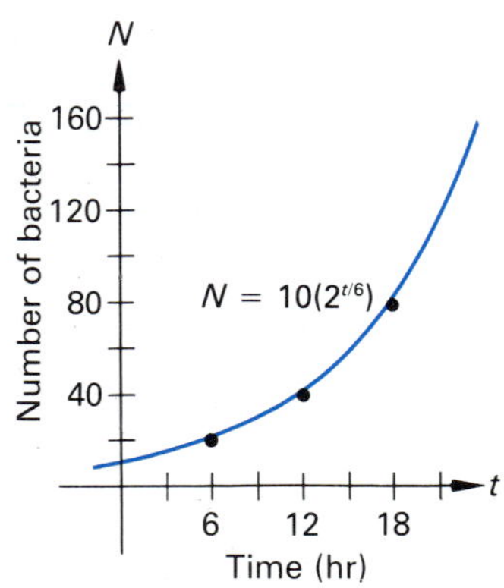

FIGURE 13.8

□

13.1 Exercises

In Exercises 1–20, graph the exponential functions for the given interval.

1. $y = 4^x, -1 \le x \le 3$

2. $y = 5^x, -1 \le x \le 3$

3. $y = 4^x, -5 \le x \le 0$

4. $y = 5^x, -5 \le x \le 0$

5. $y = \left(\frac{3}{2}\right)^x, -3 \le x \le 7$

6. $y = \left(\frac{5}{2}\right)^x, -2 \le x \le 5$

7. $y = \left(\frac{1}{4}\right)^x, -3 \le x \le 1$

8. $y = \left(\frac{1}{5}\right)^x, -3 \le x \le 1$

9. $p = 1.33^t$, $-2 \le t \le 6$
10. $s = 1.6^r$, $-2 \le r \le 6$
11. $s = 0.625^r$, $-6 \le r \le 2$
12. $p = 0.75^t$, $-6 \le t \le 2$
13. $x = 2 \cdot 3^{-t}$, $0 \le t \le 5$
14. $y = 3 \cdot 2^{-t}$, $0 \le t \le 5$
15. $y = 10 \cdot 2^{-3t}$, $0 \le t \le 2$
16. $y = 5 \cdot 2^{-2t}$, $0 \le t \le 3$
17. $v = 9e^{-t}$, $0 \le t \le 5$
18. $i = 110e^{-t}$, $0 \le t \le 5$
19. $s = 0.6e^{-t/2}$, $0 \le t \le 12$
20. $p = 0.75e^{-t/4}$, $0 \le t \le 24$

In Exercises 21–30, if t is in seconds, how long does it take for the function to decay to less than 1% of its starting value?

21. $v = e^{-t/6}$
22. $i = e^{-t/5}$
23. $p = 17e^{-2t}$
24. $p = -21e^{-4t}$
25. $y = 4 \times 10^6 e^{-t}$
26. $x = -8 \times 10^{10} e^{-t}$
27. $N = 183.6e^{-0.77t}$
28. $S = 0.327e^{-1.6t}$
29. $q = 1 \times 10^{-19} e^{-7.2t}$
30. $N = 12.8e^{-0.41t}$

31. A bacteria culture triples every three hours. If it starts with two bacteria, how many are there after one day? (See Example 5.)

32. An organic compound decays according to the equation $M = M_0\, 6.1^{-t/56}$, where t is in hours and M_0 is the original amount in kilograms. How much of a 1-kg batch of the compound remains after one week?

33. The population of a city is decreasing according to the relation $p = p_0(0.9)^t$, where t is in years and p_0 is the original population. What will be the population of the city of 100,000 people after 10 years?

34. The amount of a radioactive substance remaining after a period of time is given by $A = A_0\left(\frac{1}{2}\right)^{t/L}$, where A_0 is the initial amount of substance present and L is the half-life of the material. How much cobalt-60 remains from a 1-kg sample after six months? (Cobalt-60 has a half-life of 5.25 yr.)

35. How much of a 10-kg sample of uranium 238 is present after one million years if it has a half-life of 4.5×10^9 yr? (See Exercise 34.)

36. The surface temperature T of a block of metal is given by $T = T_a + (T_m - T_a)e^{-t/6}$, where T_a is the surrounding air temperature, T_m is the starting temperature of the metal and t is in minutes. If $T_m = 700°\text{C}$ and $T_a = 20°\text{C}$, (a) What is the temperature of the metal after 20 minutes? (b) How long does it take the metal to cool to the surrounding temperature?

13.2 The logarithmic function

The intensity of a sound wave in decibels is given by db $= 10 \log_{10} \frac{I}{10^{-16}}$, where I is the intensity of the wave in W/cm². (10^{-6} W/cm² is approximately the faintest sound intensity that can be detected by the human ear.) A function of the form $y = \log_b x$ is a **logarithmic function.**

DEFINITION 13.2

The **logarithmic function** is

$$y = \log_b x,\ b > 0,\ b \ne 1,\ x > 0.$$

The logarithmic function $y = \log_b x$ is the equivalent of the exponential function $x = b^y$. In other words, if we have the exponential function $2^3 = 8$ (the base 2 raised to the power of 3 equals 8), then the equivalent is the logarithmic function $\log_2 8 = 3$ (the logarithm to the base 2 of 8 equals 3).

DEFINITION 13.3

$$y = \log_b x \text{ if and only if } x = b^y.$$

EXAMPLE 1 Write the equivalent logarithmic function of

a. $9 = 3^2$ **b.** $y = 4^3$ **c.** $p = 10^m$

Solution: We use Definition 13.3 to find the equivalent logarithmic functions.

a. Given the exponential function $9 = 3^2$, the equivalent logarithmic function is $2 = \log_3 9$.

b. Given the exponential function $y = 4^3$, the equivalent logarithmic function is $3 = \log_4 y$.

c. Given the exponential function $p = 10^m$, the equivalent logarithmic function is $m = \log_{10} p$. □

EXAMPLE 2 Write the equivalent exponential function of

a. $5 = \log_2 32$ **b.** $y = \log_3 81$ **c.** $p = \log_{10} m$

Solution: We use Definition 13.3 to find the equivalent exponential functions.

a. Given the logarithmic function $5 = \log_2 32$, the equivalent exponential function is $32 = 2^5$.

b. Given the logarithmic function $y = \log_3 81$, the equivalent exponential function is $81 = 3^y$.

c. Given the logarithmic function $p = \log_{10} m$, the equivalent exponential function is $m = 10^p$. □

EXAMPLE 3 Solve for x:

a. $\log_3 x = -2$
b. $\log_x 81 = 4$
c. $\log_5 125 = x$
d. $x = \log_7 \dfrac{1}{49}$

Solution:

a. From Definition 13.3, if $\log_3 x = -2$, then

$$x = 3^{-2}$$

$$= \frac{1}{3^2} \qquad \textbf{Properties of exponents}$$

$$= \frac{1}{9}.$$

b. If $\log_x 81 = 4$, then

$x^4 = 81$ **Definition 13.3**

$x^4 = 3^4$ **Recalling $3^4 = 81$**

and

$x = 3.$ **Property 8, Section 13.1**

c. If $\log_5 125 = x$, then

$5^x = 125$ **Definition 13.3**

$5^x = 5^3$ **Recalling $5^3 = 125$**

$x = 3.$ **Property 7, Section 13.1**

d. If $x = \log_7 \frac{1}{49}$, then

$7^x = \frac{1}{49}$ **Definition 13.3**

$= \frac{1}{7^2}$ **Recalling $7^2 = 49$**

$7^x = 7^{-2}$ **Properties of exponents**

$x = -2.$ **Property 7, Section 13.1** □

EXAMPLE 4 Solve for x:

a. $x = \log_a a$ **b.** $x = \log_a 1$, $(a > 0)$

Solution:

a. If $x = \log_a a$, then

$a^x = a$ **Definition 13.3**

$a^x = a^1$ **Properties of exponents**

$x = 1.$ **Property 7, Section 13.1**

Therefore, we see that $\log_a a = 1$.

b. If $x = \log_a 1$, then

$a^x = 1$ **Definition 13.3**

$a^x = a^0$ **Properties of exponents**

$x = 0.$ **Property 7, Section 13.1**

Therefore we see that $\log_a 1 = 0$. □

EXAMPLE 5 Graph the logarithmic functions $y = \log_2 x$, $0 < x \leq 16$, and $y = \log_3 x$, $0 < x \leq 9$.

Solution: We cannot at this time calculate values for $y = \log_2 x$, but using Definition 13.3 we can calculate values for $x = 2^y$. Knowing that $x = 2^y$ and $y = \log_2 x$ are equivalent functions we can use this fact to provide values for $y = \log_2 x$.

For

$$x = 2^y, \quad \text{if } y = 3$$
$$x = 2^3$$
$$x = 8.$$

Substituting the values for x and y in the equation $y = \log_2 x$ we have $3 = \log_2 8$. Proceeding in this manner and also using the results of Example 4 we can construct the tables and graphs as shown in Figs. 13.9 and 13.10, which are, respectively, $y = \log_2 x$ and $y = \log_3 x$.

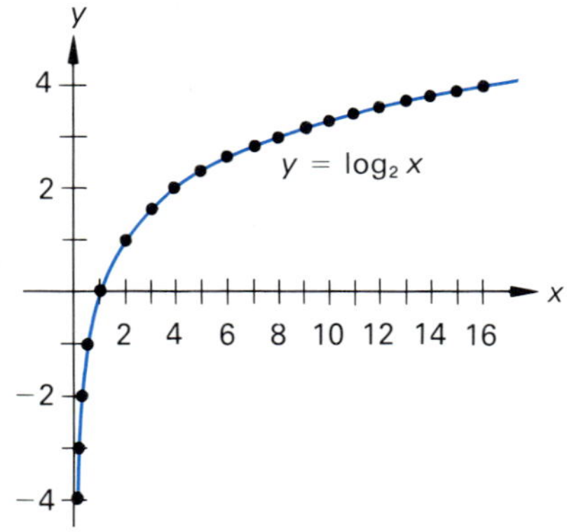

FIGURE 13.9

$y = \log_2 x$

x	$\log_2 x$
$\frac{1}{16}$	-4
$\frac{1}{8}$	-3
$\frac{1}{4}$	-2
$\frac{1}{2}$	-1
1	0
2	1
4	2
8	3
16	4

FIGURE 13.10

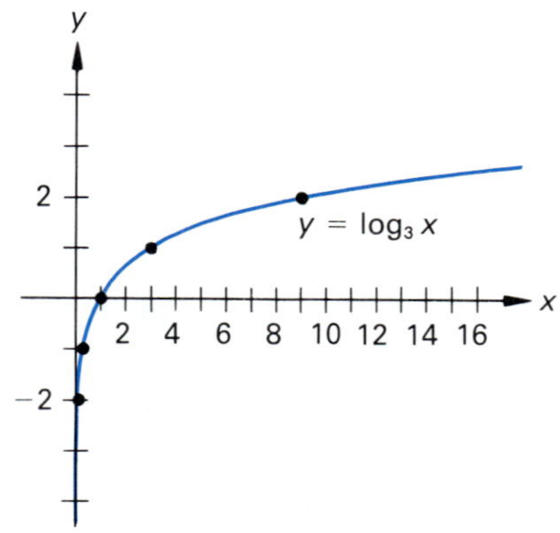

$y = \log_3 x$

x	$\log_3 x$
$\frac{1}{9}$	-2
$\frac{1}{3}$	-1
1	0
3	1
9	2

□

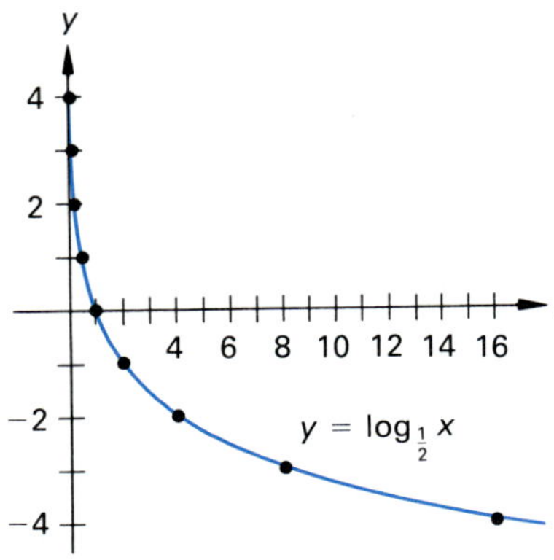

FIGURE 13.11

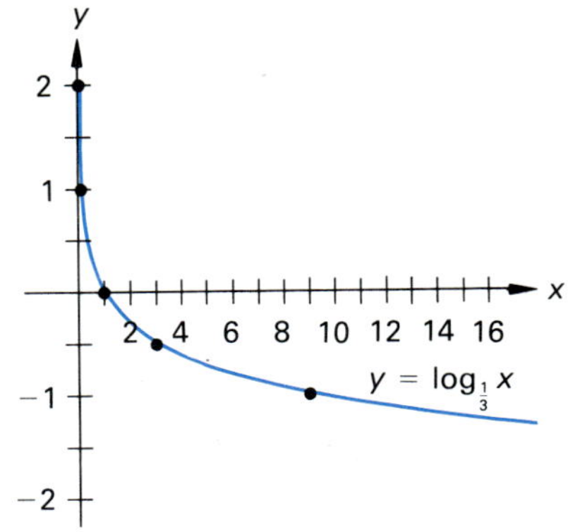

FIGURE 13.12

$y = \log_{1/3} x$

x	$\log_{1/3} x$
$\frac{1}{9}$	2
$\frac{1}{3}$	1
1	0
3	−1
9	−2

$y = \log_{1/2} x$

x	$\log_{1/2} x$
$\frac{1}{16}$	4
$\frac{1}{8}$	3
$\frac{1}{4}$	2
$\frac{1}{2}$	1
1	0
2	−1
4	−2
8	−3
16	−4

EXAMPLE 6 Graph the logarithmic functions

$$y = \log_{1/2} x,\ 0 < x \le 16, \text{ and}$$
$$y = \log_{1/3} x,\ 0 < x \le 9.$$

Solution: We cannot at this time calculate values for $y = \log_{1/2} x$, but if we use Definition 13.3 we can write the equivalent function $x = \left(\frac{1}{2}\right)^y$. Knowing that $x = \left(\frac{1}{2}\right)^y$ and $y = \log_{1/2} x$ are equivalent functions we can use this fact to provide values for $y = \log_{1/2} x$ in the table next to Fig. 13.11, where the curve is sketched. Using the equivalent functions $y = \log_{1/3} x$ and $x = \left(\frac{1}{3}\right)^y$ we can provide values for $y = \log_{1/3} x$ in the table next to Fig. 13.12, where that curve is sketched. □

Observing Figs. 13.9, 13.10, 13.11, and 13.12 leads us to some properties of logarithmic functions.

Properties of Logarithmic Functions ($y = \log_b x$)

1. The functions are asymptotic to the y-axis.
2. The functions are not asymptotic to any line $y = c$.
3. Since $\log_b 1 = 0$ the graph of $f(x) = \log_b x$ passes through the point $(1, 0)$ for any value of b. It also passes through $(b, 1)$.
4. The domain of $f(x) = \log_b x$ consists of the set of all positive real numbers; the range is the set of all real numbers.
5. If $b > 1$, $\log_b x$ is an increasing function; if $b < 1$, $\log_b x$ is a decreasing function.
6. If $a < b$, then $\log_a x > \log_b x$ for all $x > 1$ and $\log_a x < \log_b x$ for $0 < x < 1$.

7. Since b_x is either increasing or decreasing it never assumes the same value twice $(b \neq 1)$. This leads to the conclusion: if $\log_b x = \log_b y$, then $x = y$.
8. Since the graphs of $\log_a x$ and $\log_b x$ intersect only at $x = 1$: if $\log_a x = \log_b x$ and $x \neq 1$, then $a = b$.

The inverse of the exponential function $y = b^x$ can be found by interchanging the independent and dependent variables and solving for y:

$y = b^x$	**Exponential function**
$x = b^y$.	**Interchanging variables**

We do not have an algebraic technique to solve $x = b^y$ for y so we define the inverse function to be $y = \log_b x$. If we graph any function and its inverse on the same axes, the two functions are reflections or mirror images of each other about the line $y = x$. Let us redraw the graphs of $y = 2^x$ (Fig. 13.1) and $y = \log_2 x$ (Fig. 13.9) on the same axes in Fig. 13.13. Also let us redraw the graphs of $y = \left(\frac{1}{2}\right)^x$ (Fig. 13.3) and $y = \log_{1/2} x$ (Fig. 13.11) on the same axes in Fig. 13.14. We observe in particular that:

a. The domain of the exponential function is all real numbers and its range is the real numbers greater than 0; the domain of the logarithmic function is the real numbers greater than 0 and its range is all real numbers.
b. The exponential function is asymptotic to the x-axis; the logarithmic function is asymptotic to the y-axis.
c. The exponential function has y-intercept $(0, 1)$; the logarithmic function has x-intercept $(1, 0)$.

FIGURE 13.13

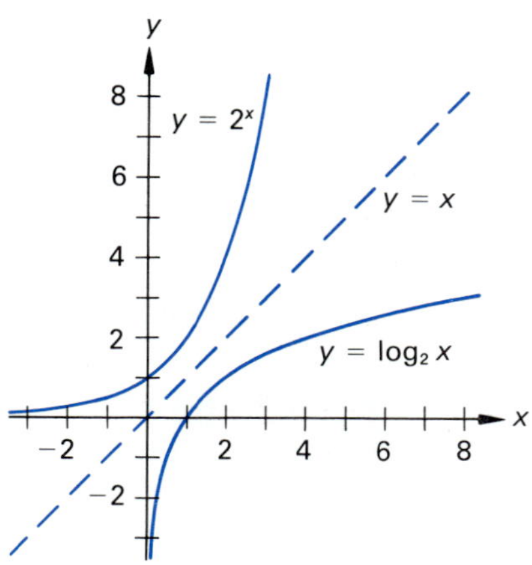

FIGURE 13.14

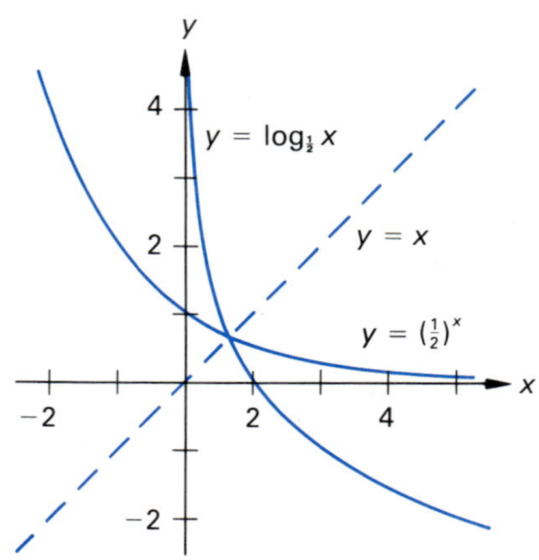

EXAMPLE 7 A bacteria culture starts with 10 bacteria and doubles every 6 hours. The equation describing the number of bacteria in the colony is $N = 10\ (2^{t/6})$ (see Example 5, Section 13.1). How long will it take the colony to reach 80 bacteria?

Solution: When $N = 80$ the population equation becomes

$$80 = 10\ (2^{t/6})$$

$$\frac{80}{10} = \frac{10\ (2^{t/6})}{10}$$

$$8 = 2^{t/6},$$

which is an exponential function and which, by using Definition 13.3 can be written as

$$\frac{t}{6} = \log_2 8$$

$$= 3 \qquad \textbf{From Fig. 13.9 and its table}$$

$$t = 6 \cdot 3 \quad \text{or} \quad 18.$$

Therefore, it takes 18 hours for the colony to increase to 80 bacteria. □

13.2 Exercises

In Exercises 1–20, write the equivalent function in exponential or logarithmic form.

1. $25 = 5^2$
2. $36 = 6^2$
3. $64 = 4^3$
4. $125 = 5^3$
5. $y = 5^2$
6. $y = 7^6$
7. $p = 10^t$
8. $g = 8^{2x}$
9. $m = n^x$
10. $m = n^{-x}$
11. $2 = \log_7 49$
12. $3 = \log_6 216$
13. $4 = \log_5 625$
14. $3 = \log_{12} 1728$
15. $y = \log_{12} 144$
16. $y = \log_{10} 10{,}000$
17. $v = \log_{6.3} 123$
18. $s = \log_{1.7} 213.$
19. $r = \log_m t$
20. $q = \log_y x$

In Exercises 21–52, solve for x.

21. $\log_4 x = 2$
22. $\log_4 x = 3$
23. $\log_5 x = -1$
24. $\log_6 x = -2$
25. $\log_7 x = 2$
26. $\log_5 x = -3$
27. $\log_x 125 = 3$
28. $\log_x 256 = 4$
29. $\log_x 216 = 3$
30. $\log_x 625 = 4$
31. $\log_x 128 = 7$
32. $\log_x 243 = 5$
33. $\log_8 64 = x$
34. $\log_7 343 = x$
35. $x = \log_6 36$
36. $x = \log_{10} 1000$
37. $x = \log_{10} 100{,}000$
38. $x = \log_5 625$
39. $x = \log_{10} \frac{1}{100}$
40. $x = \log_{10} \frac{1}{1000}$
41. $x = \log_4 \frac{1}{16}$
42. $x = \log_5 \frac{1}{25}$
43. $x = \log_2 0.25$
44. $x = \log_2 0.50$
45. $x = \log_7 7$
46. $x = \log_8 8$
47. $x = \log_{6.3} 6.3$
48. $x = \log_{2.7} 2.7$
49. $x = \log_8 1$
50. $x = \log_7 1$
51. $x = \log_{5.2} 1$
52. $x = \log_{18.1} 1$

In Exercises 53–64, graph the logarithmic functions.

53. $y = \log_4 x,\ 0 < x \leq 1$
54. $y = \log_5 x,\ 0 < x \leq 1$
55. $y = \log_4 x,\ 1 \leq x \leq 256$
56. $y = \log_5 x,\ 1 \leq x \leq 625$
57. $y = \log_6 x,\ 0 < x \leq 1$

58. $y = \log_7 x$, $0 < x \leq 1$

59. $y = \log_6 x$, $1 \leq x \leq 216$

60. $y = \log_7 x$, $1 \leq x \leq 343$

61. $y = \log_{1/4} x$, $0 < x \leq 1$

62. $y = \log_{1/5} x$, $0 < x \leq 1$

63. $y = \log_{1/4} x$, $1 \leq x \leq 256$

64. $y = \log_{1/5} x$, $1 \leq x \leq 625$

In Exercises 65–68, use Figs. 13.9, 13.10, 13.11, and 13.12 to obtain an approximate answer.

65. How long does it take a bacteria colony, which doubles every 10 hours, to reach 12 bacteria if it starts with 4?

66. How long does it take 100 bacteria to become 1400 bacteria if they double every 3 hours?

67. The population of a Southwestern city is expected to triple in the next 10 years. How long should it take to quadruple?

68. If cobalt-60 has a half-life of 5.25 yr, how long does it take a 0.5-kg sample to decay to 0.125 kg?

13.3 Laws of logarithms

We can use the fact that for each logarithmic function there is an equivalent exponential function and the laws of exponents are used to develop laws of logarithms. Recall from Chapter 2 the exponential laws.

$$b^k \cdot b^\ell = b^{k+\ell} \qquad [1]$$

$$\frac{b^\ell}{b^k} = b^{\ell-k} \qquad [2]$$

$$(b^k)^n = b^{nk}. \qquad [3]$$

If we let $R = b^k$ and $S = b^\ell$, we can use Definition 13.3 to write the equivalent functions $k = \log_b R$ and $\ell = \log_b S$. Forming the product $R \cdot S$ we have

$$R \cdot S = b^k \cdot b^\ell$$

$$R \cdot S = b^{k+\ell} \qquad \textbf{Applying Eq. [1]}$$

$$k + \ell = \log_b (R \cdot S) \qquad \textbf{Using Definition 13.3}$$

$$\log_b R + \log_b S = \log_b (R \cdot S). \qquad \textbf{Substituting for } k \textbf{ and } \ell$$

Reversing the expression we have

LAW 13.1

$$\log_b (R \cdot S) = \log_b R + \log_b S.$$

If the bases are the same, the logarithm of a product is the sum of the logarithms of the factors.

EXAMPLE 1 Use Law 13.1 to write an equivalent expression for

a. $\log_5 3 + \log_5 8$ **b.** $\log_2 (2x)$

Solution:

a. $\log_b R + \log_b S = \log_b (R \cdot S)$ **Law 13.1**

$$\log_5 3 + \log_5 8 = \log_5 3 \cdot 8 = \log_5 24.$$

b. $\log_b (R \cdot S) = \log_b R + \log_b S$ **Law 13.1**

$$\log_2 (2x) = \log_2 2 + \log_2 x.$$

The function can be further simplified by recalling that $\log_a a = 1$ so $\log_2 2 = 1$ and the function becomes $\log_2 2x = 1 + \log_2 x$. □

Let us return to $R = b^k$, $S = b^\ell$, $k = \log_b R$, and $\ell = \log_b S$. Forming the quotient S/R, we have

$$\frac{S}{R} = \frac{b^\ell}{b^k}$$

$$\frac{S}{R} = b^{\ell - k}$$ **Law of exponents, Eq. [2]**

$$\ell - k = \log_b \frac{S}{R}$$ **Using Definition 13.3**

$$\log_b S - \log_b R = \log_b \frac{S}{R}.$$ **Substituting for ℓ and k**

Reversing the expression we have

LAW 13.2

$$\log_b \frac{S}{R} = \log_b S - \log_b R.$$

The logarithm of a quotient is the logarithm of the numerator minus the logarithm of the denominator.

EXAMPLE 2 Use Law 13.2 to write an equivalent expression for

a. $\log_3 \left(\frac{1}{9}\right)$ **b.** $\log_2 x - \log_2 5$

Solution:

a. $\log_b \frac{S}{R} = \log_b S - \log_b R$ **Law 13.2**

$$\log_3 \frac{1}{9} = \log_3 1 - \log_3 9.$$

If we let $x = \log_3 1$ and $y = \log_3 9$, the equivalent statements are, respectively, $3^x = 1$ and $3^y = 9$ from which we see that $x = 0$

and $y = 2$. Therefore

$$\log_3 \frac{1}{9} = x - y$$
$$= 0 - 2.$$

Thus,

$$\log_3 \frac{1}{9} = -2.$$

b. $\log_b S - \log_b R = \log_b \frac{S}{R}$ **Law 13.2**

$$\log_2 x - \log_2 5 = \log_2 \frac{x}{5}$$

Note that $\log_2 x - \log_2 5$ is *not* equal to $\frac{\log_2 x}{\log_2 5}$. □

If we raise the exponential expression $R = b^k$ to the nth power we have

$$R^n = (b^k)^n$$

$R^n = b^{nk}$ **Law of exponents, Eq. [3]**

$nk = \log_b R^n$ **Using Definition 13.3**

$n \log_b R = \log_b R^n$ **Substituting for k ($k = \log_b R$)**

LAW 13.3

$$\log_b R^n = n \log_b R.$$

The logarithm of a quantity raised to an exponent is the product of the exponent and the logarithm of the quantity.

EXAMPLE 3 Use Law 13.3 to write an equivalent expression for

a. $\log_{10} 10^6$ **b.** $-\log_3 81$

Solution:

a. $\log_b R^n = n \log_b R$ **Law 13.3**

$\log_{10} 10^6 = 6 \log_{10} 10$

$= 6(1)$ **$\text{Log}_a\, a = 1$**

$= 6.$

b. $-\log_3 81 = -\log_3 3^4$ **Recalling $3^4 = 81$**

$= -4 \log_3 3$ **Law 13.3**

$= -4(1)$ **Recalling $\log_b b = 1$**

$= -4.$ □

Using Law 13.3 and letting $R = b$ we have

$$\log_b R^n = n \log_b R$$
$$\log_b b^n = n \log_b b$$

but

$$\log_b b = 1$$

so

$$\log_b b^n = n(1).$$

LAW 13.4

$$\log_b b^n = n.$$

If the bases are the same, the logarithm of a base raised to an exponent is equal to the exponent.

EXAMPLE 4 Use Law 13.4 to write an equivalent expression for

a. $\log_{2.5} 2.5^{7.3}$ **b.** 5

Solution:

a. $\log_b b^n = n$ **Law 13.4**

$\log_{2.5} 2.5^{7.3} = 7.3.$

b. Using Law 13.4, all of the following expressions are equivalent to 5:

$$\log_2 2^5 = 5$$
or
$$\log_8 8^5 = 5$$
or
$$\log_{6.2} 6.2^5 = 5$$
$\vdots$ □

EXAMPLE 5 Using the laws of logarithms write equivalent expressions for

a. $\log_4 \left(\dfrac{x^2}{16y}\right)$ **b.** $2 - \log_3 q$

Solution:

a.
$$\begin{aligned}\log_4 \left(\frac{x^2}{16y}\right) &= \log_4 x^2 - \log_4 16y && \textbf{Law 13.2}\\ &= 2 \log_4 x - (\log_4 16 + \log_4 y) && \textbf{Laws 13.3, 13.1}\\ &= 2 \log_4 x - (\log_4 4^2 + \log_4 y) \\ &= 2 \log_4 x - (2 + \log_4 y) && \textbf{Law 13.4}\\ &= 2 \log_4 x - \log_4 y - 2.\end{aligned}$$

b. $2 - \log_3 q = \log_3 3^2 - \log_3 q$ **Law 13.4**

$= \log_3 9 - \log_3 q$

$= \log_3 \dfrac{9}{q}.$ **Law 13.2** □

EXAMPLE 6 Write the expression $\log_a \left[\dfrac{(3x+4)^{1/3}(x+3)^{3/5}}{k^4}\right]$ as the sum or difference of logarithms, without using powers.

Solution:

$\log_a \left[\dfrac{(3x+4)^{1/3}(x+3)^{3/5}}{k^4}\right]$

$= \log_a [(3x+4)^{1/3}(x+3)^{3/5}] - \log_a k^4$ **Law 13.2**

$= \log_a (3x+4)^{1/3} + \log_a (x+3)^{3/5} - \log_a k^4$ **Law 13.1**

$= \dfrac{1}{3} \log_a (3x+4) + \dfrac{3}{5} \log_a (x+3) - 4 \log_a k$ **Law 13.3** □

EXAMPLE 7 If $\log_a 1.5 = 0.37$, $\log_a 2.0 = 0.63$ and $\log_a 5.0 = 1.46$, find

a. $\log_a 7.5$ **b.** $\log_a \left[1.5^4 \sqrt[3]{\dfrac{2}{5}}\right].$

Solution:

a. $\log_a 7.5 = \log_a [(5)(1.5)]$

$= \log_a 5 + \log_a 1.5$ **Law 13.1**

$= 1.46 + 0.37$ **Substituting**

$= 1.83.$

b. $\log_a \left[1.5^4 \sqrt[3]{\dfrac{2}{5}}\right] = \log_a \left[(1.5)^4 \left(\dfrac{2}{5}\right)^{1/3}\right]$

$= \log_a (1.5)^4 + \log_a \left(\dfrac{2}{5}\right)^{1/3}$ **Law 13.1**

$= 4 \log_a 1.5 + \dfrac{1}{3} \log_a \dfrac{2}{5}$ **Law 13.3**

$= 4 \log_a 1.5 + \dfrac{1}{3} (\log_a 2 - \log_a 5)$ **Law 13.2**

$= 4\ (0.37) + \dfrac{1}{3} (0.63 - 1.46)$ **Substituting**

$= 1.20.$ □

EXAMPLE 8 Write as a single logarithmic expression

$$2 \log_a x - 4 \log_a (2x+3) + \log_a \sqrt{2x-7}$$

Solution:

$2 \log_a x - 4 \log_a (2x + 3) + \log_a \sqrt{2x - 7}$

$= \log_a x^2 - \log_a (2x + 3)^4 + \log_a \sqrt{2x - 7}$

$= (\log_a x^2 + \log_a \sqrt{2x - 7}) - \log_a (2x + 3)^4$ **Law 13.3**

$= \log_a \dfrac{x^2 \sqrt{2x - 7}}{(2x + 3)^4}.$ **Laws 13.1, 13.2** □

While using the laws of logarithms the reader is cautioned to beware of the following common errors:

Common Errors When Using the Laws of Logarithms

Law	Correct Example	Incorrect Variation
13.1	$\log_2 (3 \cdot 4) = \log_2 3 + \log_2 4$	$\log_2 (3 \cdot 4) \neq \log_2 (3 + 4)$ $(\log_2 3)(\log_2 4) \neq \log_2 (3 + 4)$ $\log_2 3 + \log_3 4 \neq \log_2 (3 \cdot 4)$ $\log_2 (3 + 4) \neq \log_2 3 + \log_2 4$
13.2	$\log_2 \dfrac{3}{4} = \log_2 3 - \log_2 4$	$\log_2 \dfrac{3}{4} \neq \log_2 (3 - 4)$ $\dfrac{\log_2 3}{\log_2 4} \neq \log_2 (3 - 4)$ $\log_2 3 - \log_3 4 \neq \log_2 \dfrac{3}{4}$ $\log_2 (3 - 4) \neq \log_2 3 - \log_2 4$

EXAMPLE 9 Use the laws of logarithms to write

$$n \log_3 (a + b) - \log_3 c$$

as a single logarithm.

Solution: To write $n \log_3 (a + b) - \log_3 c$ as a single logarithm we first must recognize that

$$n \log_3 (a + b) = \log_3 (a + b)^n \qquad \textbf{Law 13.3}$$

then

$$n \log_3 (a + b) - \log_3 c = \log_3 (a + b)^n - \log_3 c \qquad \textbf{Substituting}$$

$$= \log_3 \frac{(a + b)^n}{c}. \qquad \textbf{Law 13.2}$$

□

EXAMPLE 10 The state of the air produced by a compressor is given by $pV^n = C$. Given (p_1, V_1) and (p_2, V_2) use the properties of logarithms to derive an expression to calculate n.

Solution: From the given information, $p_1 V_1{}^n = C$ and $p_2 V_2{}^n = C$ so

$$p_1 V_1{}^n = p_2 V_2{}^n$$

$$\log_b (p_1 V_1{}^n) = \log_b (p_2 V_2{}^n) \qquad \textbf{Taking } \log_b \textbf{ of both sides}$$

$$\log_b p_1 + \log_b V_1{}^n = \log_b p_2 + \log_b V_2{}^n \qquad \textbf{Law 13.1}$$

$$\log_b p_1 + n \log_b V_1 = \log_b p_2 + n \log_b V_2 \qquad \textbf{Law 13.3}$$

$$\log_b p_1 - \log_b p_2 = n \log_b V_2 - n \log_b V_1 \qquad \textbf{Rearranging}$$

$$= n (\log_b V_2 - \log_b V_1) \qquad \textbf{Factoring}$$

$$n = \frac{\log_b p_1 - \log_b p_2}{\log_b V_2 - \log_b V_1}. \qquad \textbf{Dividing both sides by } \log_b V_2 - \log_b V_1$$

The result is a formula we can use to calculate values for n. □

13.3 Exercises

In Exercises 1 – 10, use the laws of logarithms to write the following expressions as the sum and difference of logarithms without using powers. Simplify where possible.

1. $\log_a \dfrac{3x^2}{y}$

2. $\log_e \dfrac{x^2y^2}{4}$

3. $\log_3 \dfrac{x-4}{y+5}$

4. $\log_2 (6x^2)^5$

5. $\log_3 \dfrac{9}{x^2}$

6. $\log_b \dfrac{b^3}{6}$

7. $\log_{10} \dfrac{x^2y^3}{1000}$

8. $\log_{10} \dfrac{100}{x+1}$

9. $\log_b \dfrac{x^2 (4x-1)^3}{\left(x - \frac{1}{4}\right)^2}$

10. $\log_b \dfrac{\sqrt[3]{y^3 - 8}}{(y-4)^2}$

In Exercises 11 – 20, use the laws of logarithms to write the following expressions as a single logarithm.

11. $\log_2 x + \log_2 y - \log_2 z$

12. $\log_a (x + y) - \log_a z$

13. $\log_a (2 - x) - \log_a (x + 2)$

14. $2 \log_b x + \dfrac{1}{2} \log_b y$

15. $5 \log_3 (2x - 1) + 5 \log_3 (x - 2)$

16. $n \log_a p + (n - 1) \log_a V$

17. $\dfrac{1}{2} \log_2 x^4 + \dfrac{1}{4} \log_2 x^4 + \dfrac{1}{8} \log_2 x^8$

18. $\log_4 x^4 - 4 \log_4 x^4 + \dfrac{1}{2} \log_4 x$

19. $2.1 \log_{6.3} y + 3.5 \log_{6.3} x$

20. $-1.8 \log_{2.5} x + 2.7 \log_{2.5} y - 1.6 \log_{2.5} z$

In Exercises 21 – 28, let $\log_a 1.5 = 0.37$, $\log_a 2.0 = 0.63$, and $\log_a 5.0 = 1.46$. Find the logarithms of the expressions.

21. $\log_a 3.0$

22. $\log_a 10.0$

23. $\log_a 6.0$

24. $\log_a 40.$

25. $\log_a (1.5)^3 \sqrt{5}$

26. $\log_a \dfrac{125}{4}$

27. $\log_a \sqrt[5]{\dfrac{1.5}{2.0}}$

28. $\log_a \sqrt[5]{\dfrac{2.0}{1.5}}$

29. In an adiabatic process the temperature and volume of a gas are related by $T_1 V_1{}^{\gamma - 1} = T_2 V_2{}^{\gamma - 1}$. Use logarithms to solve for γ.

30. The relation between temperature and pressure in an atmospheric air mass is $\dfrac{T_2}{T_1} = \left(\dfrac{p_2}{p_1}\right)^{\frac{n-1}{n}}$. Use logarithms to solve for n.

13.4 Common logarithms

The logarithms to the base 10 are called **common logarithms.** The use of 10 for the base stems from the fact that our number system has a base of 10. It is standard practice when we are referring to the common logarithms to drop the notation indicating the base. Therefore, the notation $\log_{10} x = \log x$. When the base notation is missing, a base of 10 is understood.

Most scientific calculators have a [LOG] key. To find log 1.4 (which is $\log_{10} 1.4$) the steps are: 1.4 [LOG] .14612804. Therefore, $\log 1.4 = 0.14612804$. Also using the calculator we find that $\log 14 = 1.146128$ and $\log 140 = 2.146128$. We observe in each instance that the portion to the right of the decimal point is the same (discounting any round-off). This portion of the logarithm to the right of the decimal point is called its **mantissa.** If we write each number in scientific notation, we have $1.4 = 1.4 \times 10^0$, $14 = 1.4 \times 10^1$ and $140 = 1.4 \times 10^2$. Observe that the portion of the logarithm to the left of the decimal point is the same as the power of ten when the number is written in scientific notation. This portion of the logarithm to the left of the decimal point is called its **characteristic.** In general, the logarithm to the base 10 (the common logarithm) of a number is the logarithm to the base 10 of a number between 1 and 10 plus the power of ten when the number is written in scientific notation.

DEFINITION 13.4

> The **common logarithm (logarithm to base 10)** of a number N, where $N = n \times 10^p$, $1 \le n < 10$ is $\log N = p + \log n$; p is the characteristic of $\log N$ and $\log n$ is the mantissa of $\log N$.

We also see that

$$\begin{aligned}\log N &= \log (n \times 10^p)\\ &= \log n + \log 10^p\\ &= \log n + p \log 10\\ &= \log n + p\end{aligned}$$

is a logical development from the four properties of logarithms. Since $1 \le n < 10$, the number of values that historically had to be tabulated to find $\log N$ longhand were greatly reduced for base 10 compared to other bases, so the base 10 became the system in common use.

EXAMPLE 1 Find the logarithm to the base 10 of (a) 1.73, (b) 17,300, to four decimal places using the scientific calculator. Identify the characteristic and the mantissa of each logarithm. Tables of common logarithms are provided in Appendix B for the reader who does not have a calculator.

Solution:

a. Using the calculator to find the log 1.73 we have

1.73 [LOG] 0.2380461

so

$$\log 1.73 = 0.2380$$

$$\log N = p + \log n \qquad \textbf{Definition 13.4}$$

$$0.2380 = 0 + 0.2380.$$

The characteristic of log 1.73 is 0 and its mantissa is 0.2380. Writing 1.73 in scientific notation we have $1.73 = 1.73 \times 10^0$. The power of 10, 0, is the characteristic of log 1.73.

b. Using the calculator to find the log 17,300 we have

17,300 [LOG] 4.2380461

so

$$\log 17{,}300 = 4.2380.$$

Writing 17,300 in scientific notation, $17{,}300 = 1.73 \times 10^4$. The power of 10, 4, is the characteristic of log 17,300 and its mantissa is 0.2380. □

EXAMPLE 2 Find the logarithm to the base 10 of (a) 0.173, (b) 0.00173, to four decimal places using the scientific calculator. Identify the characteristic and the mantissa of each logarithm. Tables of common logarithms are provided in Appendix B for the reader who does not have a calculator.

Solution:

a. Using the calculator to find log 0.173 we obtain

0.173 [LOG] − 0.7619539

so

$$\log 0.173 = -0.7620,$$

which is an apparent contradiction of Definition 13.4. However, if we write 0.173 in scientific notation, $0.173 = 1.73 \times 10^{-1}$, which, according to Definition 13.4 gives us

$$\begin{aligned} \log N &= p + \log n \\ \log 0.173 &= -1 + \log 1.73 \\ &= -1 + 0.2380 \\ &= -0.7620. \end{aligned}$$

So the calculator was correct after all. If the logarithm of a number is negative, we cannot directly determine its characteristic and mantissa.

b. 0.00173 [LOG] −2.7619539

so

log 0.00173 = −2.7620.

When the logarithm of a number is negative, an alternative method to using Definition 13.4 for finding the characteristic and the mantissa is to add to the logarithm the smallest possible positive integer that would make the logarithm positive. That positive number is then the mantissa of the logarithm and the negative of the integer used is the characteristic of the logarithm:

log 0.00173 = −2.7620.

The smallest positive integer that, when added to −2.7620 would produce a positive number is 3:

−2.7620 + 3 = 0.2380.

So the characteristic of log 0.00173 is −3 and its mantissa is 0.2380. (Alternatively, −2.7620 + 3 − 3 = −3 + 0.2380.) □

DEFINITION 13.5

> If log $N = \mathcal{N}$ then N is the **antilog** of $\mathcal{N}$.

EXAMPLE 3 Find the antilog of: (a) 0.73, (b) 2.49, and (c) −1.04, to three significant digits using the scientific calculator.

Solution:

a. To find the antilog of 0.73 using the calculator, the key strokes are 0.73 [INV] [LOG] 5.370318. Therefore, antilog 0.73 = 5.37. (Some calculators do not have an [INV] key. If they have a [10^x] key, the antilog can be found using the following key strokes: 0.73 [10^x] 5.370318.)

b. To find the antilog of 2.49:

2.49 [INV] [LOG] 309.02954

or

2.49 [10^x] 309.02954

antilog 2.49 = 309.

c. To find the antilog of −1.04:

1.04 [+/−] [INV] [LOG] .09120108

or

1.04 [CHS] [10^x] .09120108

antilog (−1.04) = 0.0912. □

Computations using logarithms were used to simplify many complex arithmetic calculations before scientific calculators became common. (The operation of the technician's slide rule was based on the properties of logarithms.) Some technical equations require the use of logarithms for their solution. The computations using logarithms are often performed by finding the logarithm of the number involved, using the properties of logarithms to simplify the arithmetic, and then finding the antilog of the result.

EXAMPLE 4 Find $\sqrt[3]{17.3}$ using logarithms.

Solution: To find $\sqrt[3]{17.3}$ using logarithms we first take the logarithm of the function:

$$\begin{aligned} \log(\sqrt[3]{17.3}) &= \log(17.3)^{1/3} \\ &= \frac{1}{3}\log 17.3 && \textbf{Properties of logarithms} \\ &= \frac{1}{3}(1.238) \\ &= 0.4127 \\ \text{antilog}\,(0.4127) &= 2.586 && \textbf{Taking antilog} \\ \sqrt[3]{17.3} &= 2.59. \end{aligned}$$

With the scientific calculator the same result could be found by using the following key strokes:

17.3 [INV] [y^x] 3 [=] 2.586

or

17.3 [$\sqrt[x]{y}$] 3 [=] 2.586

or

17.3 [y^x] [(] 1 [÷] 3 [)] [=] 2.586. □

EXAMPLE 5 In celestial navigation the true direction or azimuth Z can be calculated from

$$\cos\frac{1}{2}Z = \sqrt{\cos s \cos(s-p) \sec L \sec h}.$$

If $s = 83°$, $p = 75°$, $L = 30°$, and $h = 25°$, find

$\cos\frac{1}{2}Z$ using logarithms.

Solution:

$$\log\left(\cos\frac{1}{2}Z\right)$$
$$= \log\left(\sqrt{(\cos 83°)(\cos(83° - 75°))(\sec 30°)(\sec 25°)}\right)$$
$$= \frac{1}{2}\log\,[(\cos 83°)(\cos 8°)(\sec 30°)(\sec 25°)]$$
$$= \frac{1}{2}[\log(\cos 83°) + \log(\cos 8°) + \log(\sec 30°) + \log(\sec 25°)]$$
$$= \frac{1}{2}[\log(0.12186934) + \log(0.99026807) + \log(1.1547005) + \log(1.1033779)]$$
$$= \frac{1}{2}[-0.914 + (-0.00425) + 0.0625 + 0.0427]$$
$$= -0.407.$$
$$\text{antilog}\,(-0.407) = 0.392.$$
$$\cos\frac{1}{2}Z = 0.39.$$

These computations may be done directly on the calculator. Handbooks contain tables of the logarithms of the trigonometric functions as well as the ordinary tables of logarithms for determining the solutions of problems containing trigonometric functions. □

13.4 Exercises

In Exercises 1–18, find the logarithm to the base 10 of the given number to four decimal places. Identify the characteristic and the mantissa of the logarithm.

1. 1.63 **2.** 7.62 **3.** 2.51

4. 5.84 **5.** 167. **6.** 286.

7. 58.2 **8.** 41.9 **9.** 5280.

10. 2.96×10^6 **11.** 0.174 **12.** 0.683

13. 0.0267 **14.** 0.0813 **15.** 0.000721

16. 0.000888 **17.** 1.71×10^{-5} **18.** 6.02×10^{-21}

In Exercises 19–26, find the antilog, base 10, of the given number to three significant digits.

19. 0.6730 **20.** 0.6920 **21.** 1.7466

22. 2.5340 **23.** −2.0888 **24.** −1.0357

25. 19.8280 **26.** −18.1002

In Exercises 27–36, evaluate the expression using logarithms.

27. $14.2\sqrt[3]{21.6}$ **28.** $53.8\sqrt[5]{1.33}$

29. $\dfrac{19.3^5}{178.}$ **30.** $\dfrac{(81.0)(27.6)}{(18.2)^3}$

31. $[(2.6 \times 10^{-3})\ (1.8 \times 10^4)]^2$ **32.** $\sqrt{\sqrt[3]{32}\ \sqrt[5]{81}}$

33. $\dfrac{\sqrt[3]{14.8}}{\sqrt{31.5}}$ **34.** $(4.63 \sin 42.1°)^{3/2}$

35. $\left(\dfrac{\tan^2 61.7°}{4.87}\right)^{1/3}$ **36.** $(87.2)^{1.73}$

37. Sound intensity in decibels is given by

$$\text{db} = 10 \log \frac{I}{I_0}$$

where $I_0 = 10^{-16}$ W/cm². The maximum sound intensity the ear can stand is about 10^{-4} W/cm². What is this intensity in decibels?

38. Sound pressure level in decibels is given by

$$S_p = 20 \log \frac{p_r}{0.0002}.$$

If the sound pressure p_r is 0.0036 dyn/cm^2, what is the sound pressure level?

39. The viscosity of a glass $\mathcal{N}$ in Pa · s as a function of temperature in °K is given by

$$\log \mathcal{N} = -7.0 + \frac{14{,}000}{T}.$$

If $\mathcal{N} = 10^{11}$ Pa · s, what is the glass temperature?

40. The pH of a solution is defined as

$$\text{pH} = -\log (M_H{}^+),$$

where $M_H{}^+$ is the hydrogen-ion concentration in moles/liter. The hydrogen ion concentration of a shampoo is 3.16×10^{-6} mol/liter. What is its pH?

41. The sound intensity in a luxury automobile is approximately 40 db. What is the sound intensity in watt/cm^2? (See Exercise 37.)

42. The frequency response of a mid-priced stereo speaker is ± 3 db over its rated frequency range. What is this sound pressure level in dyn/cm^2? (See Exercise 38.)

43. The temperature of a glass as described in Exercise 39 is 600°K. What is its viscosity at that temperature?

44. An aquarium-water test kit shows the pH of a guppy tank to be 6.3. What is the hydrogen-ion concentration of the water? (See Exercise 40.)

13.5 Logarithms to bases other than 10; natural logarithms

It is sometimes useful to be able to find logarithms in systems other than base 10. To develop a method for transforming logarithms from one base to another let $n = \log_a x$. Then

$a^n = x$	**Definition of logarithm**
$\log_b (a^n) = \log_b (x)$	**Taking $\log_b$ of both sides**
$n \log_b a = \log_b x$	**Properties of logarithms**
$n = \dfrac{\log_b x}{\log_b a}$	**Solving for n**
$\log_a x = \dfrac{\log_b x}{\log_b a}$	**Substituting $\log_a x$ for n**

DEFINITION 13.6

To transform logarithms from base b to base a

$$\log_a x = \frac{\log_b x}{\log_b a}.$$

EXAMPLE 1 Transform:

a. $\log_{10} 16$ to base 3, **b.** $\log_2 7$ to base 5.

Solution:

a. Using Definition 13.6

$$\log_3 16 = \frac{\log 16}{\log 3}$$
$$= \frac{1.204}{0.4771}$$
$$= 2.524$$
$$= 2.52.$$

Therefore $\log_{10} 16$ transformed to base 3 is $\log_3 16 = 2.52$.

b. Using Definition 13.6

$$\log_5 7 = \frac{\log_2 7}{\log_2 5}. \qquad \textbf{Law 13.1} \qquad [1]$$

Normally we cannot readily obtain logarithms to base 2 from the calculator or from tables. However, if we use Definition 13.6 and the logarithms to base 10,

$$\log_2 7 = \frac{\log_{10} 7}{\log_{10} 2}$$
$$= \frac{0.8451}{0.3010}$$
$$= 2.808. \qquad \textbf{Law 13.2}$$
$$\log_2 5 = \frac{\log_{10} 5}{\log_{10} 2}$$
$$= \frac{0.6990}{0.3010}$$
$$= 2.322.$$
$$\log_5 7 = \frac{2.808}{2.322} \qquad \textbf{Substituting for } \log_2 7 \textbf{ and } \log_2 5 \textbf{ in Eq. [1]}$$
$$= 1.209.$$

$\log_2 7$ transformed to base 5 is $\log_5 7 = 1.21$. □

The logarithmic system with base $e = 2.71828 \ldots$ is frequently encountered in equations describing natural phenomena. Logarithms to base e are called **natural logarithms** or **Naperian logarithms.** Because they occur frequently, the symbol $\ln x$ is used in place of $\log_e x$. Natural logarithms may be found by transformation from base 10 or by direct use of the calculator.

EXAMPLE 2 Find ln 17.3 using:

a. transformation from base 10,

b. direct calculator key strokes.

Solution:

a. Using Definition 13.6 and recalling that $e = 2.71828\ldots$

$$\begin{aligned}\ln 17.3 &= \log_e 17.3\\ &= \frac{\log 17.3}{\log 2.71828}\\ &= \frac{1.238}{0.4343}\\ &= 2.851 \quad \text{or} \quad 2.85.\end{aligned}$$

b. By direct calculator key strokes

17.3 [ln x] 2.8507065 or $\ln 17.3 = 2.85$. □

EXAMPLE 3 Use the calculator to find the natural antilogarithm of 6.81.

Solution: The natural antilogarithm of 6.81 can be found using an approach like that used to find the antilogarithm of a common logarithm (Example 2, Section 13.4).

To find antiln 6.81 using the calculator, the key strokes are

6.81 [INV] [ln x] 906.87081 or 6.81 [e^x] 906.87081

Therefore antiln 6.81 = 907. □

EXAMPLE 4

a. Graph the function $y = e^x$, $-2 \le x \le 4$.

b. Graph the inverse function of $y = e^x$, $y = \ln x$, $0 < x \le 40$.

c. Compare the two functions.

FIGURE 13.15

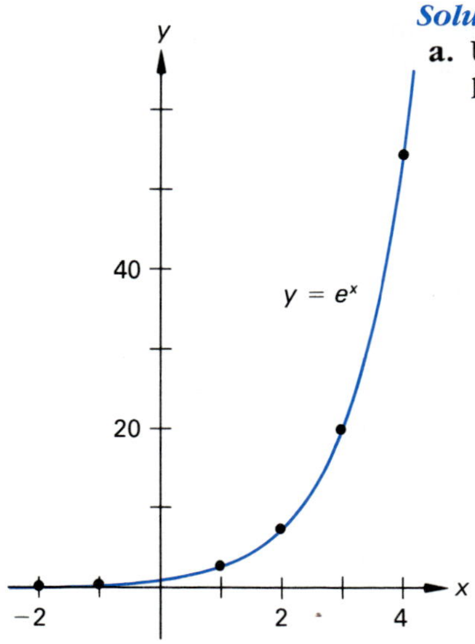

Solution:

a. Using the calculator, the tables are constructed for $y = e^x$ and $y = \ln x$, which are graphed in Figs. 13.15 and 13.16.

$y = e^x$

x	e^x
−2	0.1
−1	0.4
0	1.0
1	2.7
2	7.4
3	20.1
4	54.6

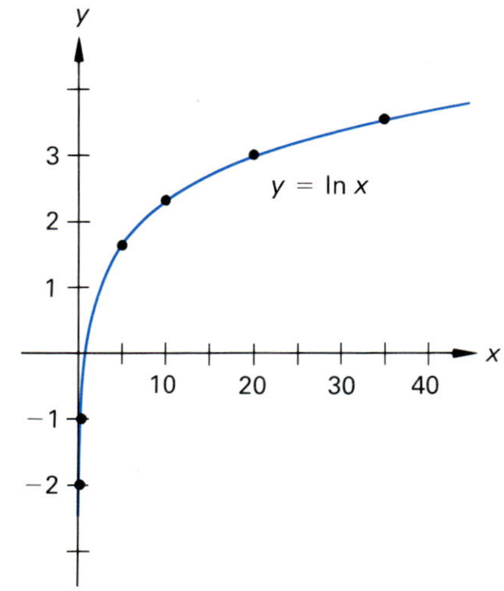

$y = \ln x$

x	$\ln x$
0.1	−2.3
0.5	−0.7
0.8	−0.2
1.0	0.0
5.0	1.6
10.0	2.3
20.0	3.0
40.0	3.7

FIGURE 13.16

c. We see from Figs. 13.15 and 13.16 that the exponential function and logarithmic function with base $e = 2.71828\ldots$ behave in the same fashion as the corresponding functions with other bases (1/2, 2, 10, etc.). $y = e^x$ is an increasing function that is asymptotic to the x-axis and has y-intercept (0, 1). $y = \ln x$ is an increasing function that is asymptotic to the y-axis and has x-intercept (1, 0). □

Since $y = e^x$ and $y = \ln x$ behave in the same fashion as other logarithmic functions, they obey the same laws. Reviewing the laws of logarithms, specifically for the case of the natural logarithms, we have

$$\ln (R \cdot S) = \ln R + \ln S \qquad \textbf{Law 13.1}$$

$$\ln \frac{S}{R} = \ln S - \ln R \qquad \textbf{Law 13.2}$$

$$\ln R^n = n \ln R \qquad \textbf{Law 13.3}$$

$$\ln e^n = n \qquad \textbf{Law 13.4}$$

$$\ln e = 1 \qquad \textbf{Law 13.5}$$

EXAMPLE 5 Calculate the natural logarithm of:

a. $5e$, **b.** $\frac{e^2}{6}$, **c.** $(81\ e^{1/2})^3$.

Solution:

a. Calculate $\ln 5e$.

$$\begin{aligned} \ln 5e &= \ln 5 + \ln e && \textbf{Law 13.1} \\ &= 1.609 + 1 && \textbf{Law 13.5} \\ &= 2.609 \\ \ln 5e &= 2.61 \end{aligned}$$

b. Calculate $\ln\left(\frac{e^2}{6}\right)$

$$\begin{aligned} \ln\left(\frac{e^2}{6}\right) &= \ln e^2 - \ln 6 && \textbf{Law 13.1} \\ &= 2 \ln e - \ln 6 && \textbf{Law 13.3} \\ &= 2(1) - 1.792 && \textbf{Law 13.5} \\ &= 0.2080 \end{aligned}$$

$$\ln \frac{e^2}{6} = 0.208$$

c. Calculate $\ln (81\ e^{1/2})^3$

$$\begin{aligned} \ln (81\ e^{1/2})^3 &= 3 \ln (81\ e^{1/2}) && \textbf{Law 13.3} \\ &= 3\ [\ln 81 + \ln e^{1/2}] && \textbf{Law 13.1} \\ &= 3\left[\ln 81 + \frac{1}{2} \ln e\right] && \textbf{Law 13.3} \\ &= 3\left[4.394 + \frac{1}{2}(1)\right] && \textbf{Law 13.5} \\ &= 14.68. \end{aligned}$$

$$\ln (81\ e^{1/2})^3 = 14.7. \qquad \square$$

EXAMPLE 6 The time t in hours needed for the grain size of a metal to double is related to the temperature T in degrees kelvin by the equation $\ln t = -40 + \frac{40{,}000}{T}$.

a. If $t = 18$ hr, calculate T.

b. If $T = 950°\text{K}$, calculate t.

Solution:

a.

$$\ln t = -40 + \frac{40{,}000}{T}$$

$$\ln 18 = -40 + \frac{40{,}000}{T} \qquad \textbf{Substituting } \boldsymbol{t = 18}$$

$$2.890 = -40 + \frac{40{,}000}{T}$$

$$42.890 = \frac{40{,}000}{T}$$

$$T = \frac{40{,}000}{42.890} \quad \text{or} \quad 932.6. \qquad \textbf{Solving for } \boldsymbol{T}$$

Therefore, $T = 933.°K$

b. $\ln t = -40 + \dfrac{40{,}000}{T}$

$= -40 + \dfrac{40{,}000}{950}$ **Substituting $T = 950$**

$= -40 + 42.11$

$= 2.11$

$t = \text{antiln } 2.11$

$= 8.25.$

Therefore $t = 8.25$ hours. □

13.5 Exercises

In Exercises 1–18, transform the logarithms to the base indicated.

1. log 100; 3
2. log 0.1; 3
3. log 105.; 1/2
4. log 6.13; 1/3
5. $\log_{1/2} 6$; 10
6. $\log_{1/3} 7$; 10
7. log 10; e
8. log 1000; e
9. log 0.01; e
10. $\log 10^{-4}$; e
11. log 86; e
12. log 104; e
13. log 0.15; e
14. log 0.087; e
15. ln 4; 10
16. ln 7; 10
17. ln 17.3; 10
18. ln 48.3; 10

In Exercises 19–30, evaluate the expression.

19. $\log_3 5$
20. $\log_7 11$
21. $\log_4 46.1$
22. $\log_6 51.8$
23. $\log_5 32$
24. $\log_4 81$
25. $\log_7 100$
26. $\log_5 100$
27. $\log_6 72$
28. $\log_3 64$
29. $\log_{1/2} 9$
30. $\log_{1/3} 8$

In Exercises 31–44, calculate the natural logarithm of the expression.

31. 4.61
32. 8.19
33. 0.338
34. 0.154
35. $2e$
36. $7e$
37. $\dfrac{e^{1/4}}{8}$
38. $12e^5$
39. $6.1e^{2.3}$
40. $\dfrac{e^{1/6}}{1/2}$
41. $(15e^3)^{1/2}$
42. $(5e^2)^{1/3}$
43. $\left(\dfrac{21}{e^{1/2}}\right)^3$
44. $\left(\dfrac{49}{e^{1/3}}\right)^2$

45. The electrode potential, E volts, in a galvanic cell with copper concentration c is given by

$$E = 0.34 + \frac{0.0257}{2} \ln c \text{ volts.}$$

Find E if $c = 0.025$.

46. In Exercise 45, if $E = 0.275$ V, find c.

47. Wind speed u_b in mph at height b in feet above the ocean surface is given by $\dfrac{u_b}{u_{43}} = 0.266 \ln b$, where u_{43} is the wind speed at 43 ft. If $u_{43} = 30$ mph, what is the wind speed at 5 ft above the surface?

48. In Exercise 47, at what height above the ocean is the wind speed 15 mph?

49. The work done in expanding a gas isothermally (at constant temperature) is given by $W = k \ln \dfrac{V_f}{V_o}$, where V_o is the initial volume, V_f is the final volume, and k is a constant. If k is 15.2 J, what is the work done by a gas expanding from a volume of 81.3 m³ to a volume of 98.6 m³?

50. The time in seconds required for the voltage drop across a capacitor in a series RC circuit to reach a value v_c is given by

$$t = RC \ln\left(\frac{E}{E - v_c}\right),$$

where E is the supply voltage, R is the resistance, and C is the capacitance. If E is 110 V, R is $10^6\ \Omega$ and C is 10^{-6} F, how long does it take for v_c to become 55 V?

51. If an object starts from rest and falls such that its air resistance is proportional to the square of its velocity, the time in seconds necessary for it to reach a velocity v is given by

$$t = \frac{v_L}{2g} \ln \frac{v_L + v}{v_L - v},$$

where v_L is the object's terminal velocity and g is the gravitational acceleration. If $v_L = 70.0$ m/s and $g = 9.81$ m/s^2, how long does it take for the object to attain a speed of 35 m/s?

13.6 Exponential and logarithmic equations

We can use the definitions and theorems developed in this chapter to solve exponential and logarithmic equations.

EXAMPLE 1 If $3^{2x-1} = 17$, find x.

Solution: In $3^{2x-1} = 17$, x is part of the exponent. If we take logarithms of both sides of the equation, we can use the laws of logarithms to change the exponent to a multiple which is easier to handle.

$$\begin{aligned} 3^{2x-1} &= 17 \\ \log 3^{2x-1} &= \log 17 && \textbf{Taking common logarithms of both sides} \\ (2x-1)\log 3 &= \log 17 && \textbf{Laws of logarithms} \\ (2x-1)0.4771 &= 1.230 && \textbf{Finding log 3 and log 17} \\ 2x-1 &= \frac{1.230}{0.4771} \\ &= 2.578 \\ x &= \frac{2.578+1}{2} && \textbf{Solving for } \boldsymbol{x} \\ &= 1.79. \end{aligned}$$

It is left to the reader to show that it makes no difference whether logarithms to base 10 or to base e are used to calculate the solution. □

EXAMPLE 2 If $2^{(x+1)} = 3^{(2x-3)}$, find x.

Solution:

$$\begin{aligned} 2^{(x+1)} &= 3^{(2x-3)} \\ \ln 2^{(x+1)} &= \ln 3^{(2x-3)} && \textbf{Taking natural logarithms of both sides} \\ (x+1)\ln 2 &= (2x-3)\ln 3 && \textbf{Laws of logarithms} \\ (x+1)0.6931 &= (2x-3)1.099 \\ 0.6931x + 0.6931 &= 2.198x - 3.297 && \textbf{Distributive property} \\ 0.6931x - 2.198x &= -3.297 - 0.6931 && \textbf{Collecting like terms} \\ -1.505x &= -3.990 \\ x &= 2.65. \end{aligned}$$

□

EXAMPLE 3 The voltage in a circuit is given by $v = 110e^{-t/3}$, with t in seconds. What is t when $v = 55$ V?

Solution: When $v = 55$ V, $110e^{-t/3} = 55$. We can take logarithms of both sides of the equation using logarithms of any base. Since the equation contains e, the base of the natural logarithms, the solution is found faster and easier using natural logarithms.

$$110e^{-t/3} = 55$$

$$e^{-t/3} = \frac{55}{110} \qquad \textbf{Dividing both sides by 110}$$

$$e^{-t/3} = 0.5$$

$$\ln e^{-t/3} = \ln 0.5 \qquad \textbf{Taking natural logarithms of both sides}$$

$$-\frac{t}{3} \ln e = \ln 0.5$$

$$-\frac{t}{3}(1) = -0.6931$$

$$t = 2.08 \text{ s.}$$ □

When solving logarithmic equations there are two general types. Type I is an equation involving only logarithms. For this type of equation we can combine terms so that we have an equation of the form:

If $\log A = \log B$,
then $A = B$.

This type of logarithmic equation is illustrated in Example 4.

EXAMPLE 4 If $\log(x + 3) = \log x + \log 7$, solve for x.

Solution:

$$\log(x + 3) = \log x + \log 7$$

$$\log(x + 3) = \log 7x \qquad \mathbf{\log(R \cdot S) = \log R + \log S}$$

$$x + 3 = 7x \qquad \textbf{Taking antilog of both sides}$$

$$x = \frac{1}{2}.$$

Check: When $x = \frac{1}{2}$,

$$\log\left(\frac{1}{2} + 3\right) = \log\frac{1}{2} + \log 7.$$

$$\log\frac{7}{2} = \log\frac{7}{2}.$$ □

Type II is an equation involving logarithms and constants. For this type of equation we can combine terms so that we have an equation of the form:

If $\log A = C$, then $A = 10^C$.

EXAMPLE 5 If $\log (2x + 8) = 1 + \log (x - 4)$, find x.

Solution: In order to solve $\log (2x + 8) = 1 + \log (x - 4)$ for x we need to regroup terms so that all terms containing x can be combined. (Remember $\log (2x + 8) \neq \log 2x + \log 8$.)

$$\log (2x + 8) = 1 + \log (x - 4)$$

$$\log (2x + 8) - \log (x - 4) = 1 \qquad \textbf{Rearranging}$$

$$\log \left(\frac{2x + 8}{x - 4}\right) = 1 \qquad \textbf{Properties of logarithms}$$

$$\frac{2x + 8}{x - 4} = 10^1 \qquad \textbf{Using Definition 13.3 and recalling that } \log p = \log_{10} p$$

$$2x + 8 = 10(x - 4) \qquad \textbf{Multiplying by } (x - 4)$$

$$2x + 8 = 10x - 40$$

$$-8x = -48 \qquad \text{or} \qquad x = 6.$$

The check is left to the reader. □

EXAMPLE 6 If $\log_{1/2} (x + 1) + \log_{1/2} x^2 = \log_{1/2} x$, find x.

Solution:

$$\log_{1/2} (x + 1) + \log_{1/2} x^2 = \log_{1/2} x$$

(Be careful *not* to make the assumption that the given equation implies $(x + 1) + x^2 = x$. Why doesn't it?)

$$\log_{1/2} (x + 1)x^2 = \log_{1/2} x \qquad \textbf{Properties of logarithms}$$

$$\log_{1/2} (x + 1)x^2 - \log_{1/2} x = 0 \qquad \textbf{Subtracting } \log_{1/2} x \textbf{ from both sides}$$

$$\log_{1/2} \frac{(x + 1)x^2}{x} = 0 \qquad \textbf{Properties of logarithms}$$

$$\log_{1/2} (x + 1)x = 0$$

$$(x + 1)x = \left(\frac{1}{2}\right)^0 \qquad \textbf{Definition 13.3}$$

$$x^2 + x = 1 \qquad \textbf{Properties of exponents}$$

$$x^2 + x - 1 = 0$$

$$x = \frac{-1 \pm \sqrt{1^2 - 4(1)(-1)}}{2(1)}$$

$$= \frac{-1 \pm \sqrt{5}}{2}.$$

Therefore $x = \dfrac{-1+\sqrt{5}}{2}$ and $x = \dfrac{-1-\sqrt{5}}{2}$ are apparently solutions of the equation $\log_{1/2}(x+1) + \log_{1/2} x^2 = \log_{1/2} x$. However, we see that $\dfrac{-1-\sqrt{5}}{2} = -1.618$ is a negative number and the logarithm of a negative number is undefined. The only acceptable solution is $x = \dfrac{-1+\sqrt{5}}{2}$. The check is left to the reader. □

EXAMPLE 7 The relation between pressure and volume during the compression stroke of a gasoline or diesel engine is given by $\ln p + \gamma \ln V = C$. Solve for V.

Solution:

$\ln p + \gamma \ln V = C$	
$\ln p + \ln V^\gamma = C$	**Law 13.3**
$\ln pV^\gamma = C$	**Law 13.1**
$pV^\gamma = e^C$	**Definition 13.3 and recalling that $\ln f = \log_e f$**
$V^\gamma = \dfrac{e^C}{p}$	
$V = \left(\dfrac{e^C}{p}\right)^{1/\gamma}$	□

13.6 Exercises

In Exercises 1–30, solve for x.

1. $2^{3x} = 5$
2. $5^{2x} = 3$
3. $4^{x+1} = 20$
4. $2^{2x+1} = 16$
5. $3^{2x-3} = 14 + \log 5$
6. $5^{x+5} = 125 + \log 125$
7. $\left(\dfrac{1}{2}\right)^{x^2} = 16$
8. $\left(\dfrac{1}{3}\right)^{x^2} = 81$
9. $4^{x+1} = 5^{x+1}$
10. $5 \cdot 4^{x-3} = 2x$
11. $\dfrac{1}{2} \cdot 3^{2x} = \left(\dfrac{1}{2}\right)^{3x}$
12. $4^{3-x} = 7^{6-2x}$
13. $6^{2x} = 10^{x-3}$
14. $10^{x+1} = 2^{2x}$
15. $\log(2x+3) = \log(2x-1)$
16. $\ln(2x+3) = \ln(2x-1)$
17. $\ln x + \ln(x-1) = 5$
18. $\log x + \log(x-1) = 5$
19. $\log_2(2x+3) = \log_2(2x-1) + 3$
20. $\log_2 x + \log_2(x-1) = 5$
21. $\ln(x+2) + \ln x^2 = \ln x$
22. $\ln(x+5) + \ln x^2 = \ln x$
23. $\log(x+2) + \log x^2 = \log x$
24. $\log(x+5) + \log x^2 = \log x$
25. $\log 4x = \log(x+1) + e$
26. $\log(x-2) = \log(x+1) - e$
27. $\ln 2x = \ln(x-1) + 10$
28. $\ln(3-x) = \ln(x+2) - 10$
29. $\log_7 x^3 + \log_7(x+2) = \log_7 x^2$
30. $\log_5(x+3)^2 = \log_5(x+3) - \log_5 x$
31. The half-life of uranium-238 is 4.5×10^9 years. How long does it take a 1-kg sample of uranium-238 to decay to 1/5 its original mass? $\left(A = A_0 \dfrac{1}{2}^{t/L}\right)$
32. A bacteria colony doubles in 6 hr. How long does it take the colony to triple? $(N = N_0\, 2^{t/T})$

33. The siren for a home security system produces a noise level of 108 db at a distance of 3 m. What is the sound intensity in W/cm² at that distance? $\left(\text{db} = 10 \log \frac{I}{10^{-16}}\right)$

34. The stopping tension T_1 applied to a rope holding a load T_2 when wrapped around a capstan is given by $\frac{T_2}{T_1} = e^{\mu\beta}$, where μ is the coefficient of friction and β is the total angle of wrap in radians. What must the coefficient of friction be for a stopping tension of 20 N to hold a load of 1000 N if there are three complete wraps around the capstan?

35. True strain e_t is related to engineering strain e by $e_t = \ln(1 + e)$. If $e_t = 0.02$, what is e?

36. In hardening the surface of steel by diffusing carbon into its surface at elevated temperatures the diffusivity D is given by $\ln D = \ln A - Q/RT$, where $A = 0.25$ cm²/s, $Q = 34{,}500$ cal/mole, and R is the gas constant 1.987 cal/mole °K. For a diffusivity of 0.300×10^{-6} cm²/s, what must the temperature be?

37. An equation relating the required height of a cooling tower to the humidity of the entering and exiting air is $Z = K \ln\left(\frac{y - y_1}{y - y_2}\right)$. Solve for y.

13.7 Graphs on logarithmic and semilogarithmic paper

We saw in Chapter 10 that when we have the graph of a straight line we can write an equation that represents the line. In many cases, however, when we plot experimental data on Cartesian graph paper the graph is a curved line rather than a straight line. It is usually difficult to find an equation that represents a curved line.

In some cases if we graph the data on a special type of graph paper, the curved line becomes a straight line and it becomes possible to write an equation to represent the data. Two types of graph paper that fit many practical situations are log-log paper and semi-log paper. Which type to use for experimental data is determined by trial and error (plot the data and see if they form a straight line) or by previous experience with similar situations. Both types of paper are based on the use of the logarithmic scale. If we calculate the common logarithms of some of the numbers 1 – 100 we get Table 13.1.

Observing Table 13.1 we see that the difference between log 3 and log 4 is the same as the difference between log 30 and log 40 $(0.60 - 0.48 = 1.60 - 1.48)$. We also see that all the logs of the tens digits are the same amount greater than the logs of the units digits. We can graph the results of Table 13.1 on a number line as in Fig. 13.17 on page 518. We see that the logarithms of the numbers are represented by their spacing along the line.

TABLE 13.1

x	1	2	3	4	5	6	7	8	9	10	20	30	40	50	60	70	80	90	100
log x	0.00	0.30	0.48	0.60	0.70	0.78	0.85	0.90	0.95	1.00	1.30	1.48	1.60	1.70	1.78	1.85	1.90	1.95	2.00

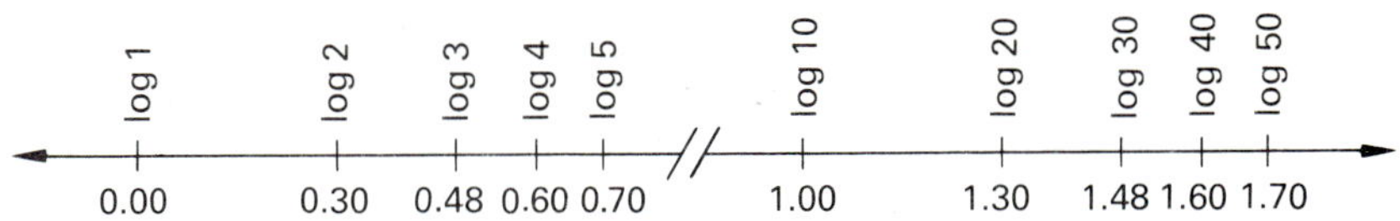

FIGURE 13.17

If we use a number line with the proportional spacing already marked on it, then by plotting numbers at the appropriate place on the scale they are automatically being spaced on the line as the log of the numbers. This principle is used to form log-log graph paper, which is printed with two perpendicular logarithmic scales (Fig. 13.18), and to form semi-log graph paper, which is printed with an evenly divided scale perpendicular to a logarithmic scale (Fig. 13.19). Because logarithmic scales display several "cycles" or powers of ten in a relatively

FIGURE 13.18

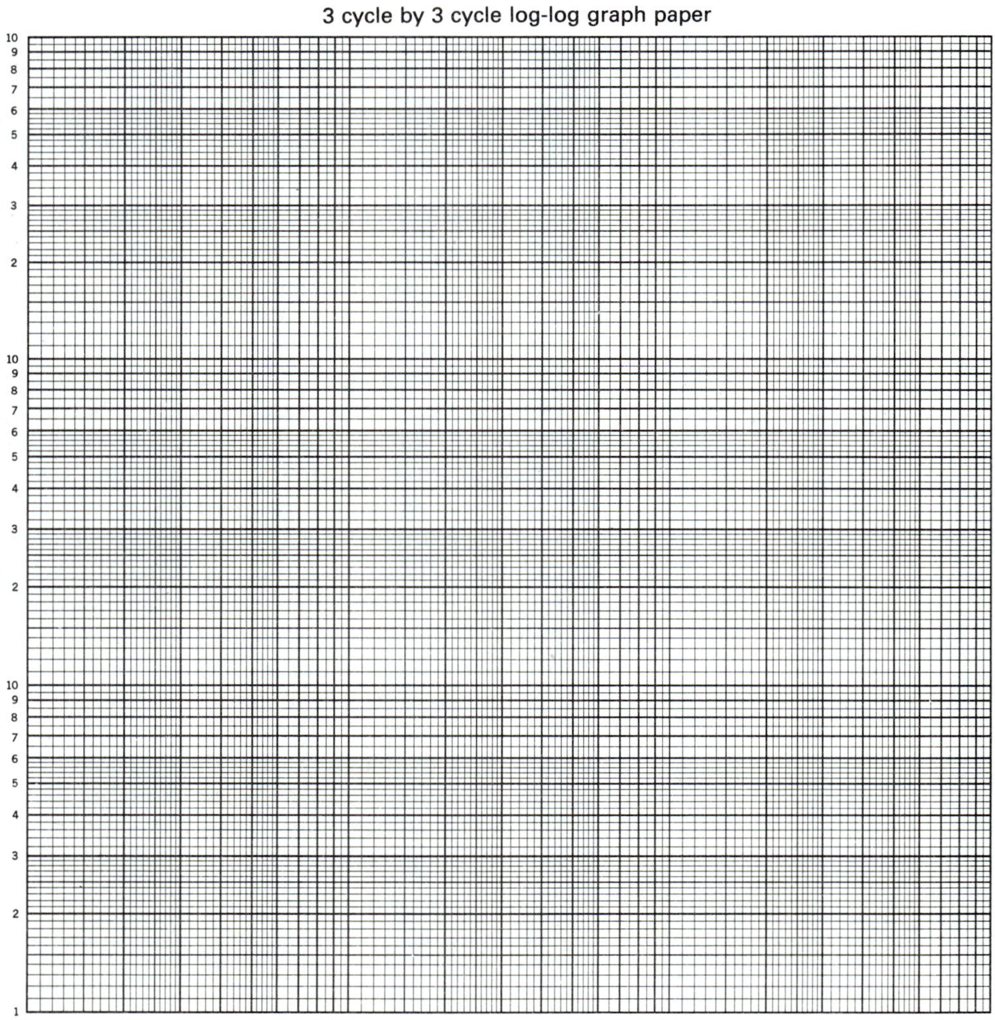

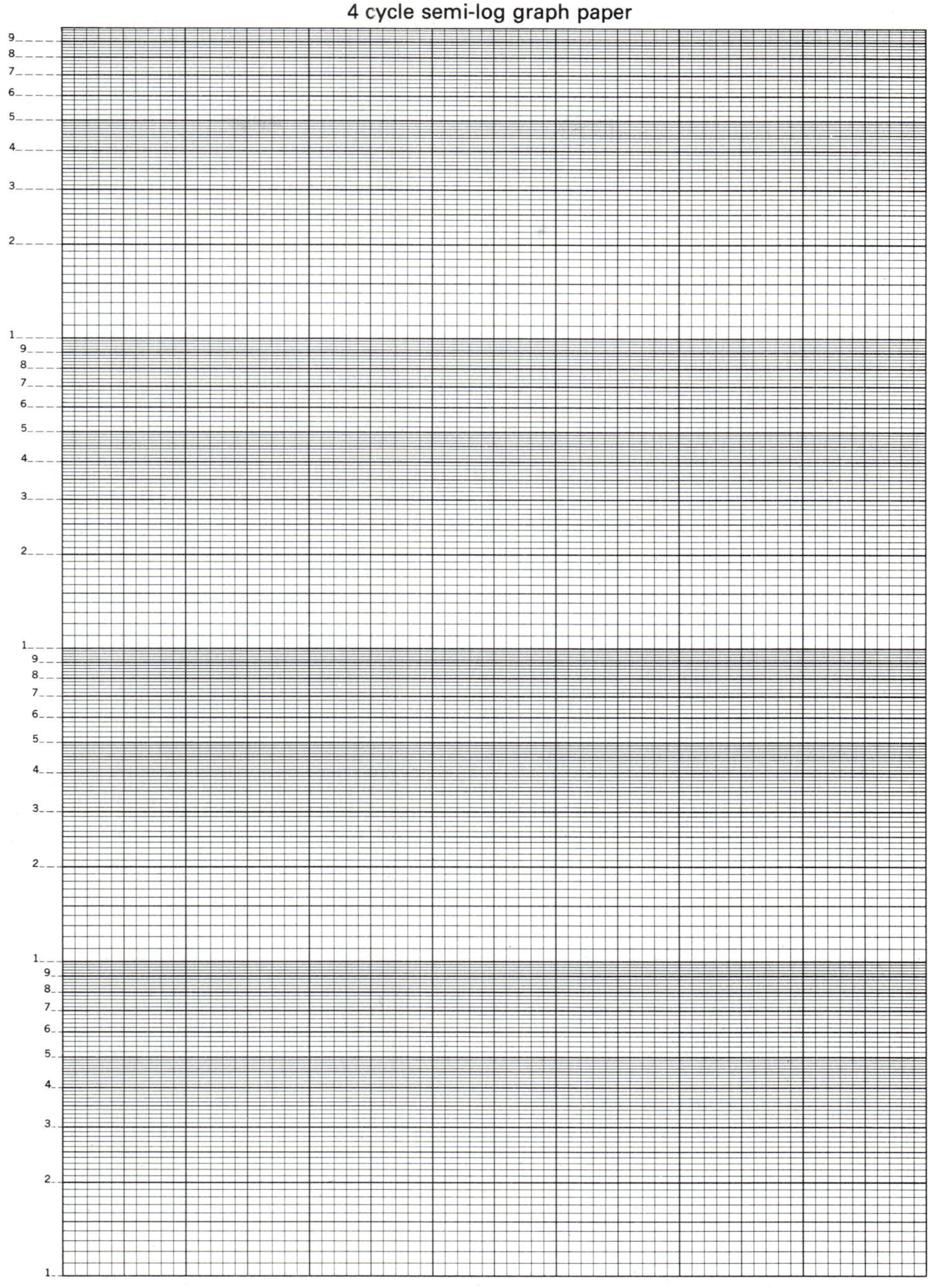

FIGURE 13.19

short length, log-log graph paper is also used to graph functions that have large domains and/or ranges. Consider an equation of the form

$$y = bx^m,$$

where m and b are constants. If we take logs of both sides of the equation, we obtain

$$\begin{aligned}\log y &= \log bx^m \\ &= \log b + \log x^m && \textbf{Laws of logarithms} \\ &= m \log x + \log b && \textbf{Laws of logarithms and rearranging}\end{aligned}$$

Now let $Y = \log y$, $X = \log x$ and $B = \log b$ and substitute to obtain

$$Y = mX + B.$$

The equation now is the form of a straight line (see Chapter 10). If we plot data that fit the form $y = bx^m$ on log-log paper, the graph should be a straight line. Because the scales are logarithmic, when we plot the data, the distances between the points represent the logarithms of the data.

EXAMPLE 1 Graph $y = 3x^2$, $1 \le x \le 50$ (a) on Cartesian graph paper and (b) on log-log graph paper.

Solution: The table next to Fig. 13.20 shows values for $y = 3x^2$, $1 \le x \le 50$. The function is plotted on Cartesian coordinates in the figure. We see from the large range of the function that it is difficult to obtain much accuracy in the graph for $1 \le x \le 10$. The same values are plotted on log-log graph paper in Fig. 13.21.

Several things become apparent when we compare Fig. 13.20 (Cartesian graph paper) with Fig. 13.21 (log-log graph paper). The data that are graphed as a curved line on Cartesian graph paper become a straight line on the log-log graph paper. This is because when we labeled the scales of the log-log graph paper 1, 2, 3 . . . , the spacing of the lines on the graph paper meant that we were actually automatically plotting log 1, log 2, log 3, ... (We did not have to calculate

FIGURE 13.20

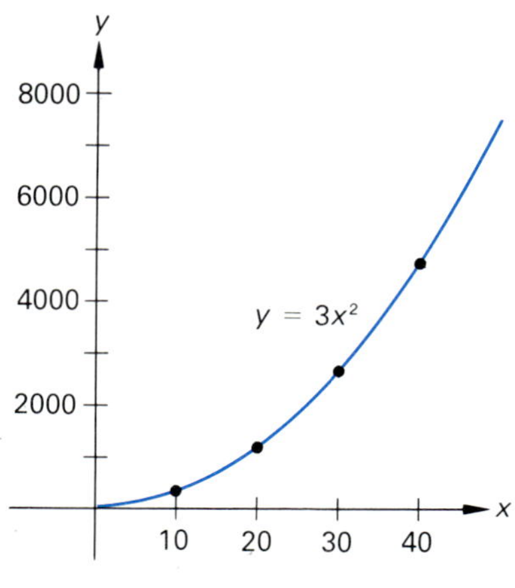

$y = 3x^2$

x	y
1	3
2	12
3	27
4	48
5	75
6	108
7	147
8	192
9	243
10	300
20	1200
30	2700
40	4800
50	7500

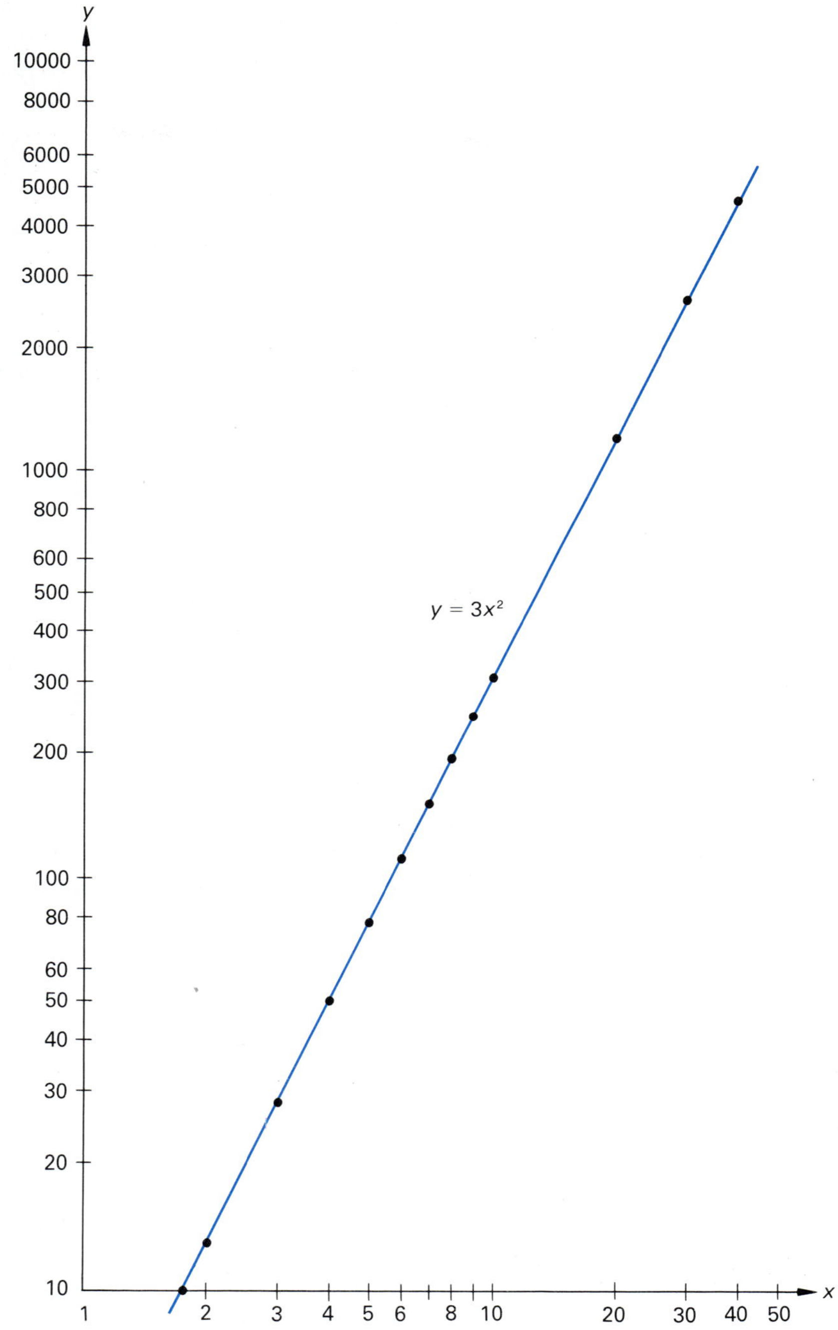

FIGURE 13.21

the logarithms. The line spacing of the grid already took care of that operation.) We also see that the log-log graph paper has a repeating pattern of lines on each axis and that each one of the groups represents a different power of ten. These repeating groups are called **cycles.** They let us conveniently graph the large domains and/or ranges that are frequently encountered with exponential functions. Observe too that logarithmic scales do not contain zero because the logarithm of zero is undefined. The scales usually, though not always, begin with one times ten to a power (1×10^n) and each succeeding cycle must represent the next higher power of ten. □

As was the case with the straight line on Cartesian graph paper, when we get a straight line on log-log graph paper we can write an equation for the line.

If we have two points on a line on log-log graph paper, (X_1, Y_1) and (X_2, Y_2) we can substitute these points into the equation $Y = mX + B$.

$$Y_1 = mX_1 + B, \qquad [1]$$

$$Y_2 = mX_2 + B. \qquad [2]$$

Subtracting Eq. [1] from Eq. [2] we obtain

$$Y_2 - Y_1 = mX_2 + B - mX_1 - B$$

or

$$Y_2 - Y_1 = m(X_2 - X_1),$$

Combining like terms and common-term factoring

which we can solve for m:

$$m = \frac{Y_2 - Y_1}{X_2 - X_1} \qquad [3]$$

When we derived the equation $Y = mX + B$, we let $Y = \log y$ and $X = \log x$. Substituting, we have

$$m = \frac{\log y_2 - \log y_1}{\log x_2 - \log x_1}. \qquad [4]$$

Knowing m we can solve Eq. [1] for B.

$$Y_1 = mX_1 + B$$
$$B = Y_1 - mX_1$$
$$B = \log y_1 - m \log x_1$$

When we derived $Y = mX + B$, we let $B = \log b$ so we now have

$$\log b = \log y_1 - m \log x_1. \qquad [5]$$

Since $Y = mX + B$ was derived from the equation $y = bx^m$, by using Eq. [4] and then Eq. [5] we can determine the equation for a function that graphs as a straight line on log-log graph paper. (We observe also from Eq. [5] that if we have the special case $x_1 = 1$, then $\log x_1 = 0$ so $\log b = \log y_1$ and $b = y_1$.)

DEFINITION 13.7

A straight line on log-log graph paper represents the equation $y = bx^m$, where, given the points (x_1, y_1) and (x_2, y_2) we have

$$m = \frac{\log y_2 - \log y_1}{\log x_2 - \log x_1} \qquad \text{and} \qquad \log b = \log y_1 - m \log x_1 .$$

EXAMPLE 2 Determine the function that represents the graph in Fig. 13.21.

Solution: The line in Fig. 13.21 can be represented by the equation $Y = mX + B$, where $Y = \log y$, $X = \log x$, and $B = \log b$. Therefore to write the equation of the function $y = bx^m$, we must determine values for m and b.

If we pick any two points on the line of Fig. 13.21, since it is plotted on log-log graph paper, we can use Definition 13.7 to find the function represented by the line. Let us pick the points (5, 75) and (20, 1200). Using Definition 13.7

$$m = \frac{\log y_2 - \log y_1}{\log x_2 - \log x_1}$$

$$= \frac{\log 1200 - \log 75}{\log 20 - \log 5} \qquad \textbf{Substituting}$$

$$= \frac{3.079 - 1.875}{1.301 - 0.6990} \qquad \textbf{Taking logarithms}$$

$$= 2.000.$$

$$\log b = \log y_1 - m \log x_1$$

$$= \log 75 - 2.000 \log 5 \qquad \textbf{Substituting}$$

$$= 1.875 - (2.000)(0.6990) \qquad \textbf{Taking logarithms}$$

$$= 0.4770$$

$$b = 2.999 \qquad \textbf{Using Definition 13.3}$$

$$= 3.00. \qquad \textbf{Rounding to three significant digits}$$

Now that we have $m = 2.00$ and $b = 3.00$ we can write the equation $y = 3.00x^{2.00}$. Thus the straight line on the log-log graph paper is a graphical representation of the exponential equation $y = 3.00x^{2.00}$. □

EXAMPLE 3 An experiment produced the data given in the accompanying table.

a. Graph the data on Cartesian graph paper.

b. Graph the data on log-log graph paper.

c. Find the equation that represents the data.

x	1	2	5	10	20	50	100
y	2.00	1.00	0.40	0.20	0.10	0.04	0.02

Solution:

a. The data are graphed on Cartesian graph paper in Fig. 13.22.

b. The data are graphed on log-log graph paper in Fig. 13.23.

c. Since the data graph as a straight line on log-log graph paper, the equation will be of the form $y = bx^m$. Therefore we must determine m and b. Picking two points on the line, say (2, 1.00) and (50, 0.04) and using Definition 13.7, we have

$$m = \frac{\log y_2 - \log y_1}{\log x_2 - \log x_1}$$

$$= \frac{\log 0.04 - \log 1.00}{\log 50 - \log 2}$$

$$= \frac{-1.398 - 0.000}{1.699 - 0.3010} \quad \text{or} \quad -1.000.$$

$$\begin{aligned} \log b &= \log y_1 - m \log x_1 \\ &= \log 1.00 - (-1.000) \log 2 \\ &= 0.0000 - (-1.000)(0.3010) \\ &= 0.3010 \\ b &= 1.999. \end{aligned}$$

Therefore the equation from the data is $y = 2.00\ x^{-1.00}$.

FIGURE 13.22

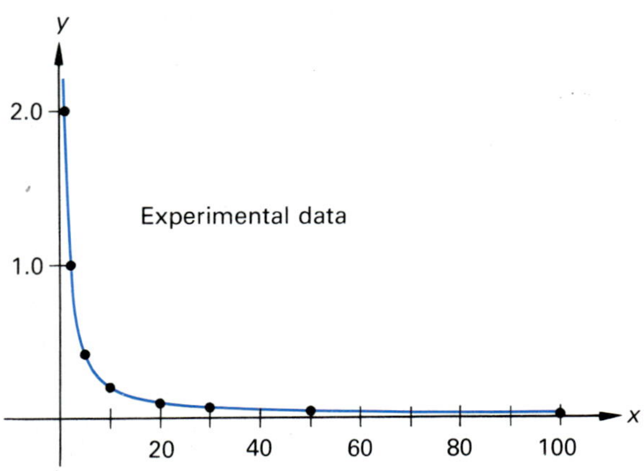

FIGURE 13.23

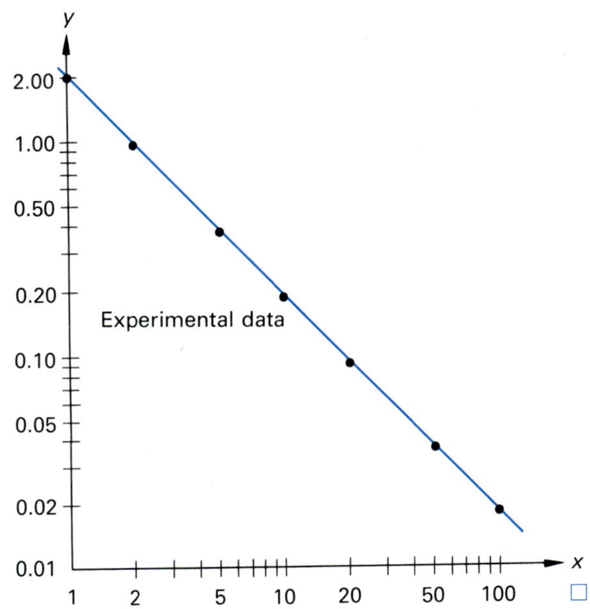

□

Now consider an equation of the form $y = ba^{mx}$, where m, b, and a are constants. If we take logs of both sides of the equation, we obtain

$$\begin{aligned}\log y &= \log ba^{mx} \\ &= \log b + \log a^{mx} && \textbf{Laws of logarithms} \\ &= mx \log a + \log b && \textbf{Laws of logarithms and rearranging} \\ &= (m \log a)x + \log b. && \textbf{Separating the variable } x\end{aligned}$$

Now let $Y = \log y$, $M = m \log a$ and $B = \log b$ and substitute to obtain $Y = Mx + B$. This equation is now the form of a straight line. In this case we do not have $\log x$ in the linear form of the equation, only $\log y$. Therefore for this equation to plot as a straight line we will need to plot x using a regular Cartesian scale division and to plot y on a logarithmic scale. Graph paper of this type is called semi-log graph paper.

EXAMPLE 4 Graph $y = \frac{1}{4}(2)^{3x}$, $0 \leq x \leq 5$ on (a) Cartesian graph paper and (b) on semi-log graph paper.

Solution: The table accompanying Fig. 13.24 shows values of $y = \frac{1}{4}(2)^{3x}$, $0 \leq x \leq 5$. The function is graphed on Cartesian graph paper in Fig. 13.24 and on semi-log graph paper in Fig. 13.25.

$y = \frac{1}{4}(2)^{3x}$

x	$\frac{1}{4}(2)^{3x}$
0	$\frac{1}{4}$
1	2
2	16
3	128
4	1024
5	8192

FIGURE 13.24

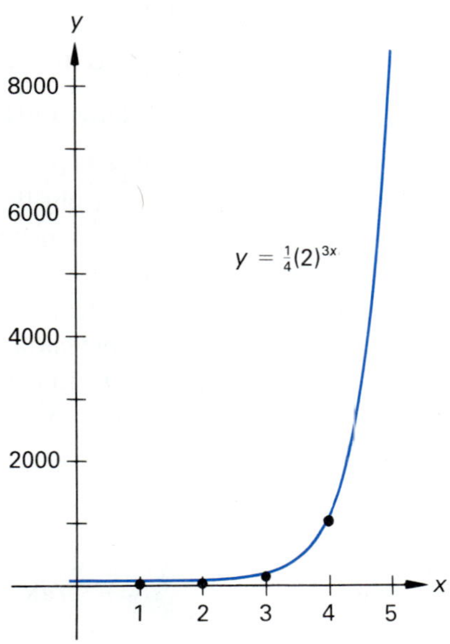

FIGURE 13.25

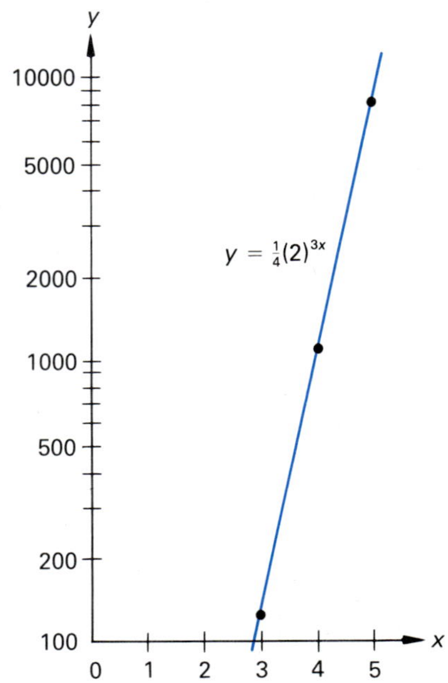

In a manner similar to that used for log-log graph paper, we can develop a method for writing the equation of a function that graphs as a straight line on semi-log graph paper. If we have two points on a line on semi-log graph paper, (x_1, Y_1) and (x_2, Y_2) we can substitute these points into the equation $Y = Mx + B$:

$$Y_1 = Mx_1 + B \qquad [6]$$
$$Y_2 = Mx_2 + B. \qquad [7]$$

Subtracting Eq. [6] from Eq. [7], we obtain

$$Y_2 - Y_1 = Mx_2 + B - Mx_1 - B$$

or

$$Y_2 - Y_1 = M(x_2 - x_1),$$ **Combining like terms and common-term factoring**

which we can solve for M:

$$M = \frac{Y_2 - Y_1}{x_2 - x_1}. \qquad [8]$$

When we derived the equation $Y = Mx + B$ we let $Y = \log y$ and $M = m \log a$. Substituting, we have

$$m \log a = \frac{\log y_2 - \log y_1}{x_2 - x_1} \qquad [9]$$

$$m = \frac{\log y_2 - \log y_1}{(x_2 - x_1) \log a}$$ **Solving for m** [10]

Knowing M we can solve Eq. [6] for B:

$$y_1 = Mx_1 + B \qquad [6]$$
$$B = Y_1 - Mx_1.$$ **Solving for B** [11]

Recalling that $B = \log b$, $Y_1 = \log y_1$ and $M = m \log a$,

$$\log b = \log y_1 - (m \log a)x_1$$ **Substituting in Eq. [11]** [12]

We find when we attempt to use Eqs. [10] and [12] to find m and b in order to determine the equation of a straight-line function on semi-log graph paper, that we can determine the constant $m \log a$ but that we cannot determine separate values for m and a. We can get around this problem by recognizing that a^{mx} can be written as $(a^m)^x$, a^m is a constant so it can be replaced by any expression that has the same value. If we let $a^m = e^{m'}$ the equation of the function takes the equivalent form $y = be^{m'x}$. (We could have chosen any number for the new base but e is most commonly used.) Letting

$$a^m = e^{m'},$$

then

$$\log a^m = \log e^{m'}$$ **Taking logs of both sides**

$$m \log a = m' \log e$$ **Laws of logarithms**

$$m' \log e = \frac{\log y_2 - \log y_1}{x_2 - x_1}$$ **Substituting for *m* log *a* in Eq. [9]**

$$m' = \frac{\log y_2 - \log y_1}{(x_2 - x_1) \log e}$$ **Solving for *m*′**

$$\log b = \log y_1 - (m' \log e)x_1.$$ **Substituting for *m* log *a* in Eq. [12]**

If we drop the prime on m' and recognize that $\log e = 0.4343 \ldots$ is a constant, we can write a function for any straight-line graph on semi-log graph paper.

DEFINITION 13.8

A straight line on semi-log graph paper represents the equation $y = be^{mx}$, where, given the points (x_1, y_1) and (x_2, y_2), we have

$$m = \frac{\log y_2 - \log y_1}{(x_2 - x_1) \log e} \quad \text{and} \quad \log b = \log y_1 - (m \log e)x_1$$

EXAMPLE 5 Determine the function that represents the graph in Fig. 13.25.

Solution: The line in Fig. 13.25 (see page 525) can be represented by the equation $Y = Mx + B$, where $Y = \log y$, $M = m \log e$, and $B = \log b$. This form is a straight line on semi-log graph paper so Definition 13.8 applies. Let us pick the points (3, 128) and (5, 8192) as (x_1, y_1) and (x_2, y_2), respectively. Using Definition 13.8,

$$m = \frac{\log y_2 - \log y_1}{(x_2 - x_1) \log e}$$

$$= \frac{\log 8192 - \log 128}{(5 - 3) \log e}$$ **Substituting**

$$= \frac{3.913 - 2.107}{2(0.4343)}$$ **Finding the logarithms**

$$= 2.079.$$

$$\begin{aligned} \log b &= \log y_1 - (m \log e)x_1 \\ &= \log 128 - (2.079)(\log e)(3) \\ &= 2.107 - 2.079(0.4343)(3) \\ &= -0.6017 \\ b &= 0.2502. \end{aligned}$$

Therefore the equation of the straight line on semi-log graph paper in Figure 13.25 is, from Definition 13.8, $y = 0.250e^{2.08x}$. The equation graphed in Fig. 13.25 was $y = \frac{1}{4}(2)^{3x}$. The reader can compare $y = 0.250e^{2.08x}$ with $y = \frac{1}{4}(2)^{3x}$ and verify that the equations are equivalent by showing that $2^3 = e^{2.08}$ to the nearest hundredth. □

EXAMPLE 6 An experiment produced the data given in the following table.

a. Graph the data on Cartesian graph paper.
b. Graph the data on semi-log graph paper.
c. Find the equation that represents the data.

x	1	2	3	4	5	6
y	1.00	2.03	4.08	8.22	16.6	33.3

Solution:

a. The data are graphed on Cartesian graph paper in Fig. 13.26.
b. The data are graphed on semi-log graph paper in Fig. 13.27.

FIGURE 13.26

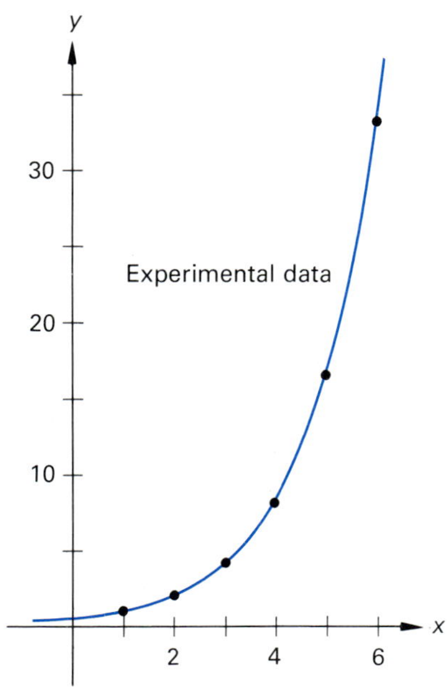

FIGURE 13.27

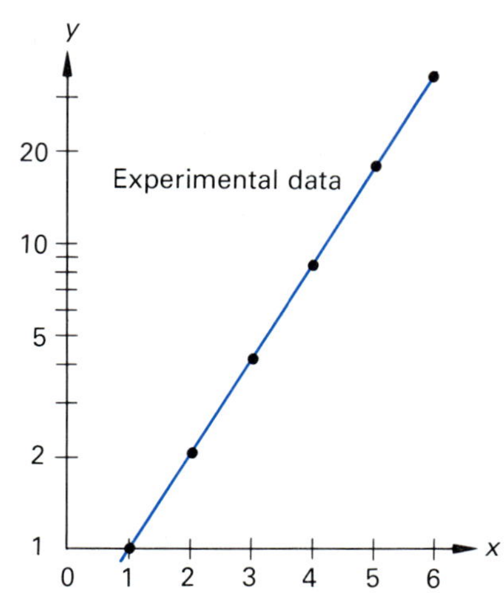

c. Picking two points on the line from the table, say (1, 1.00) and (4, 8.22), and using Definition 13.8, we have

$$m = \frac{\log y_2 - \log y_1}{(x_2 - x_1) \log e}$$

$$= \frac{\log 8.22 - \log 1.00}{(4 - 1) \log e} \qquad \textbf{Substituting}$$

$$= \frac{0.9149 - 0.0000}{3(0.4343)} \qquad \textbf{Taking logarithms}$$

$$= 0.7022.$$

$$\log b = \log y_1 - (m \log e)x_1$$
$$= \log 1.00 - (0.7022)(\log e)(1) \qquad \textbf{Substituting}$$
$$= 0.0000 - 0.7022(0.4343)(1) \qquad \textbf{Taking logarithms}$$
$$= -0.3050$$
$$b = 0.4955. \qquad \textbf{Using Definition 13.3}$$

Therefore the equation that represents the data is $y = 0.496e^{0.702x}$. □

Not all data plotted on log-log or on semi-log graph paper graph as a straight line. Even though we can determine equations only for the straight-line graphs, log-log and semi-log graph paper is useful in the other situations because of the large spread of values for the domain and/or range of a function that can be accommodated by using several logarithmic cycles.

13.7 Exercises

1. Calculate the common logarithms of the numbers 0.1, 0.2, . . . , 0.9, 1, 2, . . . , 9, 10, 20, . . . , 90, 100 and plot them on a number line.
2. Calculate the common logarithms of the numbers 1, 2, . . . , 9, 10, 20, . . . , 90, 100, 200, . . . , 900, 1000 and plot them on a number line.

In Exercises 3–8, plot the equation for $0 < x \leq 5$ on

a. Cartesian graph paper,
b. log-log graph paper,
c. semi-log graph paper.

3. $y = x^3$ 4. $y = x^2$ 5. $y = x^{-1}$
6. $y = x^{-0.5}$ 7. $y = 2x^{0.5}$ 8. $y = 4x^{-1/3}$

In Exercises 9–18, plot the equation for $0 < x \leq 5$ on

a. Cartesian graph paper,
b. log-log graph paper,
c. semi-log graph paper.

9. $y = 3^x$ 10. $y = 3^{-2x}$ 11. $y = 2^{-2x}$
12. $y = 2^x$ 13. $y = 3 \cdot 2^{-x}$ 14. $y = 2 \cdot 3^{-x}$
15. $y = 3e^{2x}$ 16. $y = 2e^{3x}$ 17. $y = 2e^{-3x}$
18. $y = 3e^{-2x}$

In Exercises 19–26, plot the experimental data on

a. Cartesian graph paper,
b. log-log graph paper,
c. semi-log graph paper.

19.

x (m)	1	2	3	4	5	6
y (m)	2.5	7.1	13.0	20.0	28.0	36.7

20.

H (m)	3	6	9	12	15	24	30
Q (m^3/s)	0.11	0.18	0.23	0.28	0.33	0.46	0.54

21.

D (in.)	1	5	10	20	50	100	200
f	0.01	0.002	0.001	0.0005	0.0002	0.0001	0.00005

22.

creep rate r (%/hr)	10^{-5}	10^{-4}	10^{-3}	10^{-2}	10^{-1}	1	10	100
stress s (MPa)	0.36	0.50	0.71	1.00	1.42	2.00	2.83	4.00

23.

x (m)	1	2	3	4	5	6	7
y (m)	1.01	2.03	4.08	8.22	16.56	33.34	67.14

24.

t (s)	0	3	6	9	12	15	18	21	24
v (m/s)	1.40	1.75	2.20	2.75	3.44	4.31	5.40	6.76	8.47

25.

H (ft)	0	2	4	6	8	10
P (psi)	2000	1850	1700	1570	1450	1340

26.

r (cm)	1	2	3	4	5	6
T (°C)	491	402	329	270	221	181

In Exercises 27–34, write the equation for the function that represents the data in Exercises 19–26.

In Exercises 35–50, write an equation $y = f(x)$ for a straight line on the type of graph paper given and passing through the given points.

35. log-log; (100, 0.02), (900, 10^2)

36. log-log; (800, 0.02), (9000, 10^2)

37. log-log; (10^{-3}, 50), (3, 100)

38. log-log; (10^{-4}, 150), (1, 600)

39. log-log; (8×10^5, 10^4), (2×10^8, 10^9)

40. log-log; (2.5, 0.01), (2500, 0.04)

41. log-log; (25, 0.04), (2500, 0.10)

42. log-log; (1, 6×10^{-5}), (12, 5×10^{-6})

43. semi-log; (0.5, 0.6), (1.1, 0.3)

44. semi-log; (0.4, 0.6), (1.0, 0.3)

45. semi-log; (2.0, 8×10^2), (3.6, 1)

46. semi-log; (1.8, 0.012), (2.0, 200)

47. semi-log; (1.6, 0.013), (1.9, 1000)

48. semi-log; (180, 6×10^{18}), (140, 2×10^{16})

49. semi-log; (170, 6×10^{18}), (130, 2×10^{16})

50. semi-log; (2000, 0.12), (500, 2.0)

Chapter summary

In Chapter 13 we introduced the **exponential function** $y = a^x$ and the **logarithmic function** $y = \log_a x$, where a is the **base** of the function. The exponential function and the logarithmic function are inverse functions. Graphs of the functions show that the exponential function has y-intercept (0, 1) and is asymptotic to the x-axis. The logarithmic function has x-intercept (1, 0) and is asymptotic to the y-axis.

Key strokes for evaluating exponential and logarithmic functions were demonstrated. The **laws of logarithms** were developed and are listed below.

Laws of Logarithms

$$\log_b (R \cdot S) = \log_b R + \log_b S$$

$$\log_b \frac{S}{R} = \log_b S - \log_b R$$

$$\log_b R^n = n \log_b R$$

$$\log_b b^n = n$$

Particular attention was paid to the logarithms to the base 10 or **common logarithms** and the logarithms to the base $e = 2.71828\ldots$ or **natural, or Naperian, logarithms.** The common logarithm of a number $N = n \times 10^p$ can be written as $\log N = p + \log n$, where p is the **characteristic** and $\log n$ is the **mantissa.** Logarithms can be converted from base b to base a by using the formula $\log_a x = \dfrac{\log_b x}{\log_b a}$.

Exponential and logarithmic equations can be solved by making use of the inverse properties of the exponential and logarithmic functions. The method for finding the solution of logarithmic equations depends on whether the equations are Type I or Type II.

When certain functions are graphed on paper containing logarithmic scales they appear as straight lines. Graph paper with both scales logarithmic is called **log-log** graph paper. Each repetition of the logarithmic scale is called a **cycle.** Graph paper with one Cartesian scale and one logarithmic scale is called semi-log graph paper. Functions of the form $y = bx^m$ graph as straight lines on log-log graph paper. Functions of the form $y = ba^{mx}$ or $y = be^{mx}$ graph as straight lines on semi-log graph paper. If two points on the straight line are known, we can determine the equation of the function representing the line.

Review Exercises

In Exercises 1–12, graph the exponential functions.

1. $y = 3.5^x, -1 \leq x \leq 4$
2. $y = 2.5^x, -1 \leq x \leq 4$
3. $y = 0.6^x, -5 \leq x \leq 1$
4. $y = 0.8^x, -5 \leq x \leq 1$
5. $v = 2.5 \cdot 0.7^x, -4 \leq x \leq 1$

6. $p = 0.4 \cdot 2^x, -1 \le x \le 5$
7. $a = 2.5^{-x}, 0 \le x \le 5$
8. $S = 0.8^{-t}, 0 \le t \le 5$
9. $V = 10e^{-2t}, 0 \le t \le 3$
10. $Q = 5e^{-1.5t}, 0 \le t \le 4$
11. $i = -9e^{-t}, 0 \le t \le 5$
12. $i = -6e^{-t/2}, 0 \le t \le 10$

In Exercises 13–16, how long does it take the function to decay to less than 1% of its starting value if t is in seconds?

13. $i = -9e^{-3t}$
14. $v = 10e^{-t/4}$
15. $v = 12e^{-t/6}$
16. $i = -110e^{-4t}$
17. An amount of money P compounded annually at an interest rate r yields an amount A after n years. What would \$100 placed at 12% yield after 15 yr? $\left(A = P\left[1 + \frac{r}{100}\right]^n\right)$
18. The decay of a radioactive substance is governed by $A = A_0 \frac{1}{2}^{t/L}$. Radium has a half-life of 1620 yr. How much is left of a 1-kg sample after 1000 years?
19. A bacteria culture doubles every 12 days. If it starts with 2 bacteria, how many bacteria will be present after 30 days?
20. A compound decays according to $S = S_0 4.7^{-t/4}$. What percentage of the compound will be left when $t = 18$?

In Exercises 21–28, find the inverse of the function.

21. $49 = 7^2$
22. $1728 = 12^3$
23. $P = 5^t$
24. $Q = 6^{2b}$
25. $\log_5 125 = 3$
26. $\log_8 4096 = 4$
27. $s = \log_{1.3} 416$
28. $m = \log_{2.7} 114$

In Exercises 29–38, solve for x.

29. $\log_6 x = 2$
30. $\log_{1.3} x = 3$
31. $\log_2 x = -5$
32. $\log_3 x = -2$
33. $x = \log_{1.7} 1.7$
34. $x = \log_{13} 13$
35. $x = \log_4 \frac{1}{64}$
36. $x = \log_{10} 10^5$
37. $x = \log_{1/3} 1$
38. $x = \log_{1/4} 1$

In Exercises 39–42, graph the logarithmic functions.

39. $y = \log_{3/2} x, 0 < x \le 1$
40. $y = \log_{2/3} x, 0 < x \le 1$
41. $y = \log_{3/2} x, 1 \le x \le \frac{81}{8}$
42. $\log_{2/3} x, \frac{8}{81} \le x \le 1$
43. How long does it take a bacteria colony starting with 2 bacteria to reach 100 bacteria if it doubles every hour?
44. How long does it take 0.5 kg of radium (half-life 1620 yr) to decay to 0.2 kg?

In Exercises 45–50, use the laws of logarithms to write the following expressions as the sum and difference of logarithms without powers. Simplify where possible.

45. $\log_7 (5x^3)^2$
46. $\log_b \frac{12}{b^4}$
47. $\log_2 \frac{x-2}{x+2}$
48. $\log_3 \frac{27}{1-x}$
49. $\log_{10} \frac{x^2+y^2}{1000}$
50. $\log_{10} \frac{2x^2-3y}{10}$

In Exercises 51–56, use the properties of logarithms to write the expressions as a single logarithm.

51. $\log_{10} 1 + \log_{10} x$
52. $\log_2 (x+y) + \log_2 y$
53. $\log_2 (x+3) - \log_2 (x-3)$
54. $2 \log_a x - \frac{1}{2} \log_a y$
55. $\log_e e + \log_e 1$
56. $2 \log_e 1 - \log_e e^2$

In Exercises 57–62, let $\log_a 1.5 = 0.37$, $\log_a 2.0 = 0.63$, and $\log_a 5.0 = 1.46$. Find the logarithms of the expressions.

57. $\log_a 125$
58. $\log_a 128$
59. $\log_a \frac{1}{25}$
60. $\log_a \frac{1}{16}$
61. $\log_a 50$
62. $\log_a 9$
63. If $A = P(1+r)^n$, use logarithms to solve for n.
64. If $\frac{P_1{}^{n-1}}{V_1} = \frac{P_2{}^{n-1}}{V_2}$, use logarithms to solve for n.

In Exercises 65–72, find the logarithm to the base 10 of the given number to four decimal places. Identify the characteristic and the mantissa of the logarithm.

65. 14.2
66. 17.6
67. 0.0113
68. 0.0276
69. 1.5×10^6
70. 2.2×10^4

71. 8.1×10^{-2}

72. 7.6×10^{-3}

In Exercises 73–78, find the antilog, base 10, of the given number to three significant digits.

73. 2.5110
74. 3.6271
75. −2.8173
76. −1.2845
77. 1.6374
78. −11.1111

In Exercises 79–82, evaluate the expressions using logarithms.

79. $3.61(18.2)^{1.7}$

80. $4.25(8.63)^{2.3}$

81. $\sqrt[5]{(16.3)^{3/2}}$

82. $\sqrt[7]{(8.96)^{1.5}}$

83. The Richter scale for measuring earthquakes is $R = \log \frac{I}{I_0}$, where I_0 is a constant. How many times greater is the intensity of an earthquake for which $R = 4.2$ than one for which $R = 3.7$?

84. How many times greater is a sound intensity of 90 db (the threshold of permanent hearing damage) than an intensity of 65 db (ordinary conversation)? $\left(\text{db} = 10 \log \frac{I}{10^{-16}}\right)$

In Exercises 85–92, transform the logarithm to the base indicated.

85. $\log_5 125$; 10

86. $\log_3 0.370$; 10

87. $\log_4 0.0156$; e

88. $\log_5 625$; e

89. log 14.3; e

90. log 0.462; e

91. ln 4.17; 10

92. ln 0.285; 10

In Exercises 93–96, calculate the natural logarithm of the expression.

93. $\frac{8}{e^2}$

94. $\frac{e^{1/2}}{4}$

95. $\frac{e^{-1/6}}{1/12}$

96. $\frac{e^{-3}}{27}$

97. The recrystallization of aluminum is governed by the expression $\ln 0.14 = C + \frac{B}{T}$. If $C = -52$ and $B = 30{,}000$ °K, what is T in °K?

98. The relaxation time t in days of a rubber band is given by $\ln \frac{5.5}{11} = -\frac{42}{t}$. What is t?

In Exercises 99–106, find x.

99. $\left(\frac{1}{2}\right)^{2x} = 10$

100. $3^{x-1} = 12$

101. $5^{x+2} = 4^{x-1}$

102. $2^x = 64 + \log 64$

103. $\log(x + 1) = \log(x - 1)$

104. $\log x - \log(x + 2) - 4 = 0$

105. $\ln(x - 2) - \ln x = 8$

106. $\ln(2x - 1) = \ln(x + 2)$

107. If the half-life of radium is 1620 yr, how long does it take a sample to decay to 3/4 of its original mass?

108. If the pH of tomatoes is 4.2, what is the hydrogen-ion concentration, H^+? ($\text{pH} = -\log(H^+)$)

109. Belt tensions on either side of a flat pully are related by $\frac{T_2}{T_1} = e^{\mu\beta}$. If $T_1 = 100$ N, $T_2 = 250$ N, and the coefficient of friction μ is 0.2, what is the angle of the wrap β in radians?

110. *V*-belt tensions driving a pully are related by $\frac{T_2}{T_1} = e^{\left(\frac{\mu\beta}{\sin \alpha/2}\right)}$. For $T_2 = 12T_1$, $\mu = 0.3$, $\beta = \pi$, what is the belt angle α in degrees?

In Exercises 111 & 112, plot the equation for $0 < x \leq 5$ on (a) Cartesian graph paper, (b) log-log graph paper, (c) semi-log graph paper.

111. $y = 5x^{1/4}$

112. $y = 12x^{-1/5}$

In Exercises 113–116, plot the equation for $0 < x \leq 5$ on (a) Cartesian graph paper, (b) log-log graph paper, (c) semi-log graph paper.

113. $y = 5\left(\frac{1}{4}\right)^x$

114. $y = 12\left(\frac{1}{5}\right)^{-x}$

115. $y = \frac{1}{4}e^{-2x}$

116. $y = 4e^{-x}$

In Exercises 117–120, plot the data on log-log graph paper or semi-log graph paper, as appropriate, and find the equation that describes the graph.

117.

Boat displacement d (ft)	4.0	5.0	6.0	7.0	8.0	9.0
Plank thickness t (in.)	0.53	0.69	0.86	1.03	1.21	1.40

118.

scan time t	0.03	0.04	0.05	0.06	0.07
screen intensity I	0.9	1.5	2.4	4.0	6.6

119.

time t (s)	10	12	14	16	18	20
current i (A)	7.9	6.9	6.0	5.2	4.5	3.9

120.

pressure p (MPa)	10	20	30	40	50	60	70	80	90	100
volume V (m^3)	0.100	0.081	0.072	0.066	0.062	0.058	0.056	0.054	0.052	0.050

Can you solve these?

1. Show that $\log_b RST = \log_b R + \log_b S + \log_b T$.

2. Assume that your calculator can only calculate $\log n$ for $1 \leq n < 10$ and antilog $\mathcal{N}$ for $0 \leq \mathcal{N} < 1$. Show how to find the antilog of any number N by using its characteristic and mantissa.

3. Show why a [10^x] key can be used to find antilogs if your calculator has no [INV] [LOG] function.

4. Derive the properties of logarithms for the natural logarithms.

5. Show that if $y = b^{\log_b n}$, then $y = n$.

6. Show that for a function that graphs as a straight line on semi-log graph paper, if $x_1 = 0$, then $b = y_1$.

7. Write a computer program to determine values for m, where

$$m = \frac{\log y_2 - \log y_1}{(x_2 - x_1)\log e}$$

8. The base of the natural logarithm e may be approximated by $\left(1 + \frac{1}{n}\right)^n$ as n becomes very large. Write a computer program to show that this is true.

14

A special titanium alloy is being considered for use as the turbine impeller shaft that drives the fuel-circulating pumps for a space launch vehicle's rocket engine. The alloy has been extensively tested and the results show that for an expected life of 50×10^6 cycles the safe combinations of the absolute value of the mean stress s and the absolute value of the alternating stress a are given by the solution set of the system of inequalities.

$$s \geq 0$$
$$a \geq 0$$
$$a \leq -5. \times 10^{-4} s^2 + 5. \times 10^2,$$

where the stresses are in MPa (megapascals). Graph the solution set to determine what stress combinations are acceptable.

Inequalities

14.1 Properties of inequalities

Section 1.2 introduced the basic symbols of inequality. Up to this point we have used the symbols of inequality to indicate relations between numbers and to indicate intervals. It is essential for future topics that we learn how to solve statements of inequality just as we have learned how to solve statements of equality. The techniques for solving statements of inequality are the prime objectives of this chapter. Section 14.1 will be devoted to basic definitions of inequalities that will be used throughout the chapter.

Table 14.1 contains a list of the basic symbols, their meanings and geometrical interpretations.

EXAMPLE 1 Graph the following statements on a real-number line:

a. $x < 5$ **b.** $x \geq -3$ **c.** $-2 \leq x < 4$ **d.** $-3 \leq x \leq 5$

TABLE 14.1 **Inequality notation**

Symbol	Meaning	Graph
$x < a$	x is less than a.	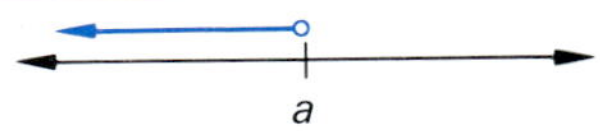
$x \leq a$	x is less than or equal to a.	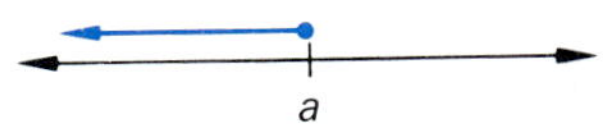
$x > a$	x is greater than a.	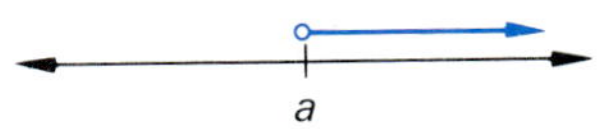
$x \geq a$	x is greater than or equal to a.	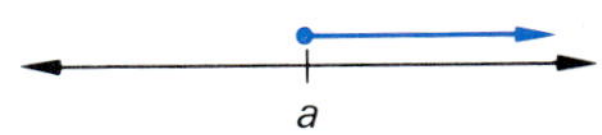
$a < x < b$	x is greater than a and x is less than b.	
$a \leq x \leq b$	x is greater than or equal to a and x is less than or equal to b.	
$a < x \leq b$	x is greater than a and x is less than or equal to b.	
$a \leq x < b$	x is greater than or equal to a and less than b.	

(a and b represent constants and x is a variable.)

Solution:

a. 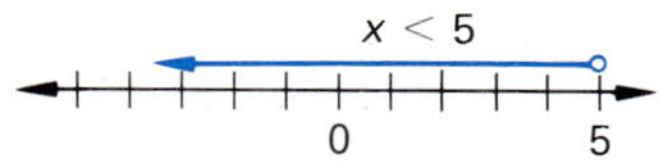

FIGURE 14.1

In Fig. 14.1, the open circle indicates that 5 is not part of the solution.

b. 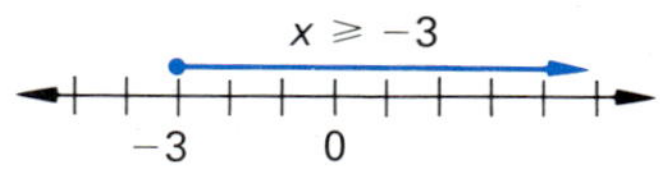

FIGURE 14.2

In Fig. 14.2, the solid dot indicates that we include -3 as part of the solution.

c. 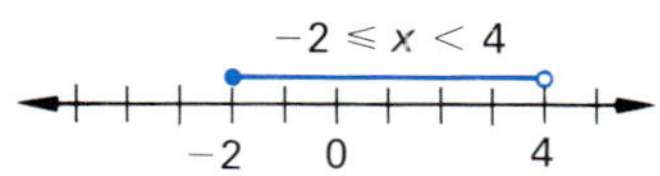

FIGURE 14.3

In Fig. 14.3, the solid dot indicates that -2 is part of the solution and the open circle indicates that 4 is not part of the solution. The line indicates that all the points between -2 and 4 are part of the solution.

d. 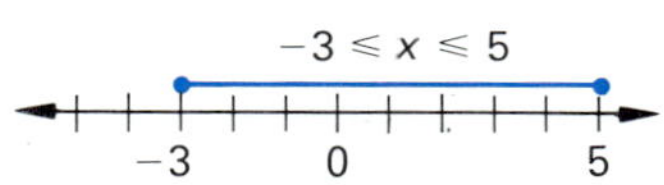

FIGURE 14.4

In Fig. 14.4, the solid dots indicate that -3 and 5 are part of the solution and the line between the dots indicates that all the points between -3 and 5 are part of the solution. □

Inequality notation can be used to define regions of the coordinate plane. Example 2 illustrates this concept.

EXAMPLE 2 Write a statement or statements of inequality involving x or y or both that provide the location of all the points in the coordinate plane that satisfy the following statements.

a. the region below the x-axis

b. the region to the left of the line $x = 4$ and above the line $y = 2$.

FIGURE 14.5

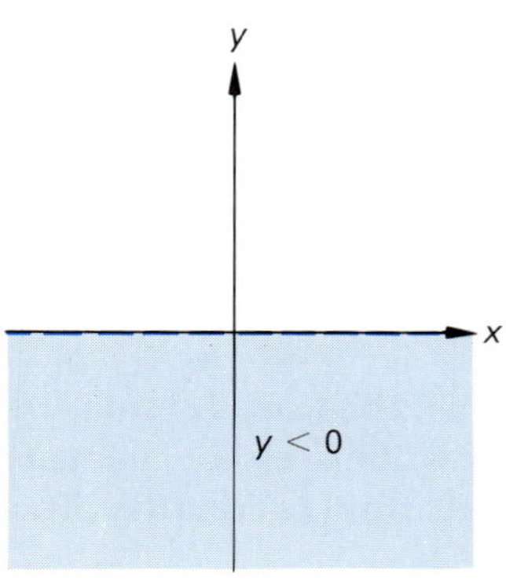

Solution: A sketch of the region described will help to write the required statement(s) of inequality.

a. The shaded region of Fig. 14.5 indicates the desired region of the xy-plane. We see from the shaded figure that y is less than zero. Therefore the inequality $y < 0$ describes the location of all the points $P(x, y)$ that are below the x-axis.

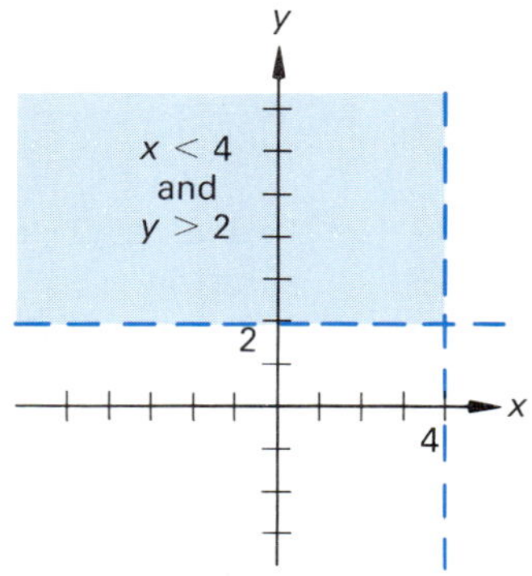

FIGURE 14.6

b. The shaded region of Fig. 14.6 indicates the desired region of the xy-plane. We see from the figure that two statements of inequality are necessary to describe the shaded region. Therefore, the inequalities $x < 4$ and $y > 2$ describe the location of all the points $P(x, y)$ that are to the left of the line $x = 4$ and above the line $y = 2$. □

We know that if $6 > 4$ then $6 - 4$ is a real positive number and that real positive number is 2. Also, if $-7 > -16$ then $-7 - (-16)$ is a real number and that real number is 9. We can state this property formally as Definition 14.1.

DEFINITION 14.1

> For all real numbers a and b, if $a > b$, then there exists a real positive number c such that $a - b = c$.

DEFINITION 14.2

> **SENSE OF INEQUALITY**
>
> If $a < b$ and $c < d$ then the two inequalities are unequal in the same sense. If $a < b$ and $c > d$ then the two inequalities are unequal in the opposite sense.

An inequality of the form $x^2 + 2 > 0$ is an **absolute inequality** since any real number that is substituted for x will make the statement true. An inequality of the form $x > 1$ is a **conditional inequality.** The statement is not true for all values of x.

Properties of inequalities: Operations

1. **Addition property:** Any real number can be added to both sides of an inequality and the sense of the inequality will not change. For example,

   a. If $3 < 4$, then $3 + 2 < 4 + 2$.

   b. If $6 > 4$, then $6 + 5 > 4 + 5$.

   c. If $-3 < 11$, then $-3 + (-7) < 11 + (-7)$.

   d. If $-13 > -23$, then $-13 + (-19) > -23 + (-19)$.

> **Addition Property**
>
> For any real numbers a, b, and c; if $a < b$, then $a + c < b + c$; if $a > b$, then $a + c > b + c$.

2. **Subtraction property:** Any real number can be subtracted from both sides of an inequality and the sense of the inequality will not change. For example,

   a. If $3 < 11$, then $3 - 7 < 11 - 7$.

   b. If $5 > 1$, then $5 - 11 > 1 - 11$.

   c. If $-6 < -1$, then $-6 - (-8) < -1 - (-8)$.

   d. If $-11 > -33$, then $-11 - (-5) > -33 - (-5)$.

> **Subtraction Property**
>
> For any real numbers a, b, and c; if $a < b$, then $a - c < b - c$; if $a > b$, then $a - c > b - c$.

3. **Multiplication property:** a) If both sides of an inequality are multiplied by a **positive** real number, then the sense of the inequality is not changed. b) If both sides of an inequality are multiplied by a **negative** real number, then the sense of the inequality is changed to the opposite sense. For example,

   a. If $3 < 7$, then $3(8) < 7(8)$.

   b. If $8 > 2$, then $8(4) > 2(4)$.

   c. If $5 > -1$, then $5(-6) < -1(-6)$ or $-30 < 7$.

   d. If $-14 < -6$, then $(-14)(-3) > (-6)(-3)$ or $42 > 18$.

Multiplication Property

a. If $a < b$ and $c > 0$ then $ac < bc$.

b. If $a < b$ and $c < 0$ then $ac > bc$.

4. **Division property:** a) If both sides of an inequality are divided by a **positive** real number, then the sense of the inequality is not changed. b) If both sides of an inequality are divided by a **negative** real number, then the sense of the inequality is changed to the opposite sense. For example,

   a. If $12 < 15$, then $12 \div 3 < 15 \div 3$.

   b. If $-7 > -21$, then $-7 \div 7 > -21 \div 7$.

   c. If $18 > 6$, then $18 \div (-3) < 6 \div (-3)$ or $-6 < -2$.

   d. If $-9 < 27$, then $-9 \div (-9) > 27 \div (-9)$ or $1 > -3$.

Division Property

a. If $a < b$ and $c > 0$, then $a \div c < b \div c$

b. If $a < b$ and $c < 0$, then $a \div c > b \div c$

5. **Power property:** If both members of an inequality are **positive** real numbers, then the inequality formed by raising both members to the same positive power will be unequal in the same sense. For example,

   a. If $3 < 7$, then $3^2 < 7^2$.

   b. If $4 > 1$, then $4^3 > 1^3$.

   c. If $2 < \pi$, then $2^3 < \pi^3$.

Power Property

If $a > b > 0$, then $a^m > b^m$ for $m > 0$; if $0 < a < b$, then $a^m < b^m$ for $m > 0$.

6. **Root property:** If both members of an inequality are positive real numbers, then the inequality formed by taking the same positive root of both members will be unequal in the same sense. For

example,

a. If $16 < 25$, then $\sqrt{16} < \sqrt{25}$.

b. If $64 > 27$, then $\sqrt[3]{64} > \sqrt[3]{27}$.

c. If $\pi < 32$, then $\sqrt[5]{\pi} < \sqrt[5]{32}$.

Root Property

If $a > b > 0$, then $\sqrt[m]{a} > \sqrt[m]{b}$ for $m > 0$.

The root and power properties can be extended to include negative values for a and b; however, the above statements serve our purpose.

7. **Reciprocal property:** If two numbers with the same sign (both positive or both negative) are not equal, then their reciprocals are unequal in the opposite sense. For example,

a. If $3 < 7$, then $\frac{1}{3} > \frac{1}{7}$.

b. If $-2 > -6$, then $-\frac{1}{2} < -\frac{1}{6}$.

c. If $5 > 2$, then $\frac{1}{5} < \frac{1}{2}$.

Reciprocal Property

If $ab > 0$ and $a < b$, then $\frac{1}{a} > \frac{1}{b}$.

14.1 Exercises

In Exercises 1–12, graph the solution set on the real number line.

1. $x > 7$ **2.** $x < -3$

3. $-3 < x < 4$ **4.** $-5 \le x < 4$

5. $-13 < x \le -2$ **6.** $-4 \le x \le 7$

7. $-x > 3$ **8.** $4 < -x$

9. $3 > x > 1$ **10.** $\frac{1}{x} < \frac{1}{3}$

11. $\sqrt{x} < \sqrt{16}$ **12.** $x^2 > 3^2$

In Exercises 13–24, express each of the following using the appropriate inequality symbol.

13. 5 is less than 13

14. 3 is greater than 2

15. -5 is greater than or equal to -16

16. 11 is less than or equal to 24

17. $R + 6$ is greater than 11

18. t is between 0 and 6

19. Z is greater than 2 and less than or equal to 11

20. $2 + t$ is positive

21. $6 - x$ is negative

22. $x + 3$ is non negative

23. $R - 2$ is non positive

24. $x + 5$ is greater than 4 and less than 10

In Exercises 25–36, insert the proper inequality symbol in the blank between the two members.

25. If $5 < y$, then $5 + 4$ ___ $y + 4$

26. If $3 > 7$, then $3 - x$ ___ $7 - x$

27. If $-3 < 11$, then $-3 + z$ ___ $11 + z$

28. If $3 < 11$, then $3(-2)$ ___ $11(-2)$

29. If $x + 3 < -2$, then $(x + 3) \div (-4)$ ___ $(-2) \div (-4)$

30. If $t + 3 > t - 5$, then $(t + 3)(-5)$ ___ $(t - 5)(-5)$

31. If $x + 11 > x + 3 > 0$, then $(x + 11)^3$ ___ $(x + 3)^3$

32. If $x > y > 0$, then $\sqrt[4]{x}$ ___ $\sqrt[4]{y}$

33. If $x + y > x - z$ and $w < 0$, then $(x + y)w$ ___ $(x - z)w$

34. If $a + b + c < 2a + b - c$ and $d < 0$, then $(a + b + c) \div d$ ___ $(2a + b - c) \div d$

35. If $7 < 13$, then $\frac{1}{7}$ ___ $\frac{1}{13}$

36. If $(x + y)(x - y) > 0$ and $x + y > x - y$, then $\frac{1}{x + y}$ ___ $\frac{1}{x - y}$

In Exercises 37–48, for the point $P(x, y)$ in the coordinate plane write a statement or statements of inequality involving x or y or both that will provide the location of the point P according to the following statements.

37. above the x-axis

38. to the left of the y-axis

39. in the second quadrant

40. below the line $y = -3$

41. to the right of the line $x = 4$

42. between the lines $x = -3$ and $x = 4$

43. between the lines $y = -4$ and $y = 6$

44. not between the lines $x = 1$ and $x = 3$

45. not between the lines $y = -3$ and $y = 1$

46. below the line $y = -1$ and to the right of the line $x = 2$

47. above the line $y = 2$ and to the left of the line $x = -1$

48. below the line $y = 3$ and to the left of the line $x = 2$

49. Rewrite Definition 14.1 using the statement $a < b$.

50. Prove the addition property using Definition 14.1.

14.2 Linear inequalities in one variable

The properties and definitions in the previous section provide us with the basic tools to solve inequalities. The discussion in this section will be limited to linear inequalities. An inequality is a **linear inequality** if the exponents of all the variables are 1 (or the variables are of degree 1). For example, $2x + 1 < 6$, $x \geq 3$, $x + 7y \leq 5$, and $3x - 7 < 6y$ are all linear inequalities, while $x^2 + 2 > 4$, $x^2 + 5x \leq 11$, $2xy + 7x^2 < 3$, $2xy < 3$, and $2\sqrt{x} \leq 5$ are not linear inequalities.

The solution of a linear inequality is the set of all numbers that will make the statement true. For example, a solution for $3x + 1 > 4$ is $x = 2$; substituting 2 for x we see that $7 > 4$. In fact 3, 10 and 1.1 are also solutions for the inequality. All real numbers greater than 1 will make the statement true. Thus, $x > 1$ is the solution to $3x + 1 > 4$.

EXAMPLE 1 Solve the inequality and graph the solution set for

$$2x - 6 \geq -3 + x.$$

Solution: The object is the same here as in solving an equality, we want to isolate the variable on one side of the inequality and the constants on the other.

$2x - 6 \geq -3 + x$

$2x - 6 + 6 \geq -3 + 6 + x$ **Addition property**

$2x \geq 3 + x$

$2x - x \geq 3 + x - x$ **Subtraction property**

$x \geq 3.$

Thus, the solution set of $2x - 6 \geq -3 + x$ is all real numbers greater then or equal to 3 (Fig. 14.7).

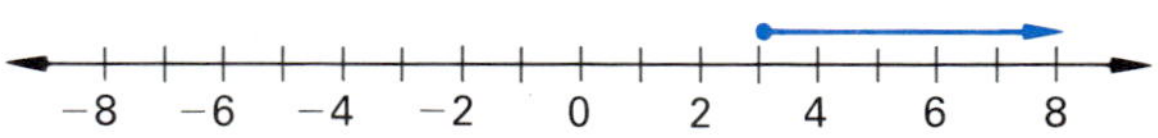

FIGURE 14.7

□

EXAMPLE 2 Solve the inequality and graph the solution set for

$$3(x + 4) - 2 < 4x + 12.$$

Solution: We will first remove the parentheses and combine like terms.

$3(x + 4) - 2 < 4x + 12$

$3x + 12 - 2 < 4x + 12$ **Distributive property**

$3x + 10 < 4x + 12$

$3x + 10 - 10 < 4x + 12 - 10$ **Subtraction property**

$3x < 4x + 2$

$3x - 4x < 4x - 4x + 2$ **Subtraction property**

$-x < 2$

$(-1)(-x) > (-1)(2)$ **Multiplication property (note change in inequality sign because of multiplication by a negative number)**

$x > -2.$

Thus the solution set of $3(x + 4) - 2 < 4x + 12$ is all real numbers greater than -2 (Fig. 14.8).

FIGURE 14.8

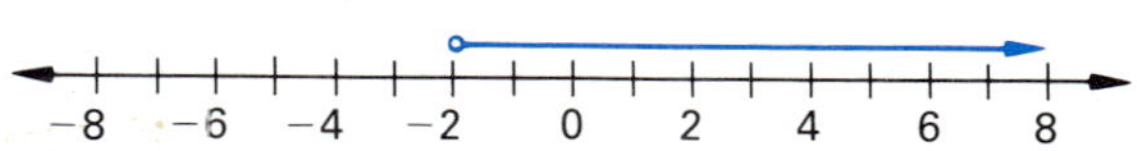

□

EXAMPLE 3 Solve the double inequality for x.

$$-1 < \frac{x+3}{2} \leq 5$$

Solution: To solve a double inequality we must isolate x as the middle term. To do this we use the same properties used to solve single inequalities.

$$-1 < \frac{x+3}{2} \leq 5$$

$$2(-1) < 2\left(\frac{x+3}{2}\right) \leq 2(5) \qquad \textbf{Multiply the three terms by 2}$$

$$-2 < x+3 \leq 10$$

$$-2-3 < x+3-3 \leq 10-3 \qquad \textbf{Subtract 3 from all three terms}$$

$$-5 < x \leq 7.$$

Thus, the solution set of $-1 < \frac{x+3}{2} \leq 5$ is all the real numbers greater than -5 and less than or equal to 7. □

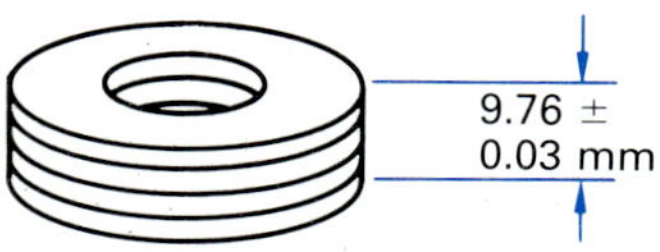

FIGURE 14.9

EXAMPLE 4 Four washers, all of the same thickness, must have a combind thickness of 9.76 mm (Fig. 14.9) with a tolerance of ± 0.03 mm. Find the acceptable range of values for the thickness of a washer.

Solution: Let $x =$ the actual thickness of a single washer. Then $4x$ is the actual thickness of four washers. $4x - 9.76$ is actual thickness minus ideal thickness. The actual thickness minus the ideal thickness must be within the tolerance of ± 0.03 mm.

$$-0.03 \leq 4x - 9.76 \leq 0.03$$

$$-0.03 + 9.76 \leq 4x - 9.76 + 9.76 \leq 0.03 + 9.76$$

Add 9.76 to all three terms

$$9.73 \leq 4x \leq 9.79$$

$$\frac{9.73}{4} \leq \frac{4x}{4} \leq \frac{9.79}{4} \qquad \textbf{Divide each term by 4}$$

$$2.43 \leq x \leq 2.45.$$

Therefore the thickness of a single washer should be between 2.43 mm and 2.45 mm. □

EXAMPLE 5 A student must have an average greater than or equal to 80% but less than 90% on five tests to receive a final grade of B. Dale's grades on the first four tests were 98, 76, 86, and 92%. What range of grades on the fifth test would give him a B in the course?

Solution: We first construct an inequality that can be used to find the range of grades on the fifth exam. The average (mean) is found by adding the grades and dividing the sum by the number of exams. Let $x =$ the fifth grade:

$$\text{average} = \frac{98 + 76 + 86 + 92 + x}{5}.$$

For Dale to obtain a B, his average must be greater than or equal to 80 but less than 90.

$$80 \le \frac{98 + 76 + 86 + 92 + x}{5} < 90$$

$$80 \le \frac{352 + x}{5} < 90$$

$$5(80) \le 5\left(\frac{352 + x}{5}\right) < 5(90) \qquad \textbf{Multiply the three terms by 5}$$

$$400 \le 352 + x < 450$$

$$400 - 352 \le 352 - 352 + x < 450 - 352 \qquad \textbf{Subtract 352 from all three terms}$$

$$48 \le x < 98.$$

Thus, a grade of 48 up to but not including a grade of 98 will result in a B. □

14.2 Exercises

In Exercises 1–30, solve the inequality and graph the solution set.

1. $x + 3 < 6$

2. $x + 7 < 0$

3. $x + 7 \le 3$

4. $r + 3 \le 2r - 4$

5. $2\ell + 7 \le \ell + 6$

6. $\frac{t}{2} > 4$

7. $\frac{Z}{5} \ge -\frac{3}{5}$

8. $\frac{R}{2} \ge 0$

9. $-\frac{2}{3}x \le 2$

10. $\frac{R}{2} + 3 \ge 2R + 4$

11. $\frac{R}{4} + \frac{1}{2} \le \frac{3}{8}$

12. $\frac{Z}{5} > \frac{1}{3}$

13. $2(t - 1) + 3(t + 2) \ge t - 5$

14. $2(3t - 2) \le 4(1 - t)$

15. $7(2t + 1) + 3 < 5\left(2t - \frac{1}{5}\right)$

16. $4(t + 1) - 3 \le 2(1 - 2t) - 1$

17. $\frac{1 + Z}{2} < \frac{3}{8}$

18. $\frac{3z + 2}{5} \le -\frac{2}{3}$

19. $\frac{1 - 2Z}{8} < \frac{7}{4}$

20. $\frac{3 - Z}{4} < \frac{Z}{3}$

21. $\frac{Z}{2} - \frac{1}{3} \le \frac{3Z}{4}$

22. $\frac{5y - 1}{2} + \frac{1}{5} > \frac{y + 1}{10}$

23. $3 < y + 4 < 9$

24. $0 < y - 5 < 7$

25. $-3 < y + 7 \le 6$

26. $\frac{1}{3} \le x + \frac{1}{3} \le 3$

27. $-4 \le \frac{1 + x}{2} \le 3$

28. $-3 \le \frac{4x - 5}{3} \le 5$

29. $0.05 < 0.75 - 0.55x < 1$

30. $0.05 \le 0.75 - 2.00x \le 1.25$

In Exercises 31–40, write a statement of inequality and solve.

31. You can rent a compact car from firm A for \$190 per week with no charge for mileage, or from firm

B for \$110 per week plus 20 cents for each mile driven. At what mileage does it cost less to rent from firm A?

32. A machine that packages 200 screws per box can make an error of 2 screws per box. If x is the number of screws in a box, write an inequality that indicates a maximum error of 2 screws per box. Solve the inequality.

33. Lead for a pencil must have a diameter of 0.5 mm with a tolerance of ± 0.005 mm. Find the acceptable range of diameters of the lead.

34. The relationship between Fahrenheit temperature F and Celsius temperature C is given by the formula $C = \frac{5}{9}(F - 32)$. Find the range of values of F when C is between 35 and 39.

35. The velocity v (in ft/s) of an object fired directly upward is given as a function of time t (seconds) by the formula $v = 92 - 32t$. For what positive time interval will the velocity be positive?

36. In Exercise 35, when will the velocity be between 20 and 32 ft/s?

37. Seven washers, all of the same thickness, must have a combined thickness of 15.05 mm with a tolerance of ± 0.02 mm. Find the acceptable range of values for the thickness of a washer.

38. Two resistors in parallel have a total resistance R given by $\frac{1}{R} = \frac{1}{R_1} + \frac{1}{R_2}$. If $R_1 = 15\,\Omega$, find the positive values for R_2 so that R is between 5 and $8\,\Omega$.

39. In Example 5, what grade on the fifth test would result in Dale's receiving: (a) A grade of B if his grades on the first four tests were 95, 72, 85, and 89%? (b) A grade of C with the same test grades? To obtain a C he must have an average greater than or equal to 70 but less than 80. (c) Is it at all possible for Dale to obtain an A (average ≥ 90)?

40. The minimum speed for vehicles on a highway is 40 mph, the maximum speed is 55 mph. If Tim has been driving his tractor-trailer truck along the highway for four hours, what is the range in miles that he could legally have traveled?

41. Dave's Foundry manufactures wood-burning stoves. The revenue equation is $R = 2x$, where x is the number of wood-burning stoves of type A sold per year. The cost equation is $C = 140 + 1.2x$. How many type A wood-burning stoves must be sold each year to make a profit? Write a statement of inequality and solve.

14.3 Inequalities in one variable

Quadratic inequalities

The solving of quadratic inequalities is dependent upon being able to solve quadratic equations. The techniques for solving quadratic equations were developed in Chapter 11.

For example, if we express the inequality $(x + 4)(x - 2) > 0$ as an equality, then we have $(x + 4)(x - 2) = 0$. Solving for x we have $x = -4$ or $x = 2$. *The values of* x *that make the statement of equality zero are called* ***critical values*** *of the inequality. Thus,* $x = 2$ *and* $x = -4$ *are critical values of the inequality* $(x + 4)(x - 2) > 0$.

In order for the statement of inequality in our example to be true, the product of the two quantities $(x + 4)$ and $(x - 2)$ must be positive. To express this in a simpler form, let $a = x + 4$ and $b = x - 2$, then, substituting, we have $a \cdot b > 0$. For this statement to be true, if $a \cdot b > 0$, then either $a > 0$ and $b > 0$ or $a < 0$ and $b < 0$. Therefore we

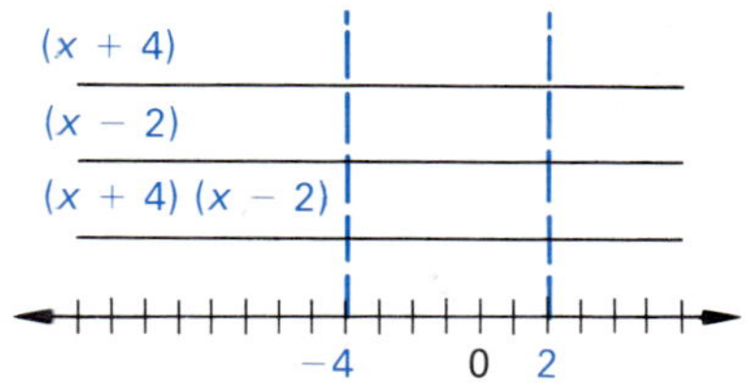

FIGURE 14.10

are looking for values of x that will make both $x - 2$ and $x + 4$ positive quantities or both $x - 2$ and $x + 4$ negative quantities. To find these values of x we will use a graphical method.

To find the values of x by the graphical method, first draw a number line as in Fig. 14.10. On the number line indicate the critical values $x = -4$ and $x = 2$. Draw dashed vertical lines through $x = 2$ and $x = -4$. Then draw three lines parallel to the number line, and label these lines with the factors $(x + 4)$, $(x - 2)$, and $(x + 4)(x - 2)$.

When $x = -4$, the quantity $x + 4 = 0$. For this quantity the value -4 divides the number line into three parts: the points to the left of $x = -4$, which will make the expression $(x + 4)$ negative; $x = -4$, for which the expression $(x + 4)$ is zero; and the points to the right of $x = -4$ for which the expression $(x + 4)$ is positive. We will indicate this on the line labeled $(x + 4)$ by placing $+$ and $-$ signs on the line (Fig. 14.11). The factor $(x - 2)$ is zero when $x = 2$, $x - 2$ is negative to the left of $x = 2$ and positive to the right of $x = 2$. Now place the appropriate $+$ and $-$ signs on the line labeled $(x - 2)$ (Fig. 14.11). Since we want to determine the values of x for which the product $(x + 4)(x - 2)$ is positive, we assume that the $+$ and $-$ signs are $+1$ and -1 and multiply them using the rules of algebra. Thus, to the left of -4 we have $(-1)(-1) = +1$ so we place a $+$ sign on the third line opposite the product $(x + 4)(x - 2)$. Between -4 and 2 we have $(+1)(-1) = -1$, so we record the $-$ sign on the line labeled $(x + 4)(x - 2)$. To the right of 2 we have $(+1)(+1) = +1$, thus in this space we record $+$. Now looking at Fig. 14.11 we see that $(x + 4)(x - 2)$ is positive (greater than zero) for values of $x < -4$ or $x > 2$. The solution set is graphed in Fig. 14.12.

FIGURE 14.11

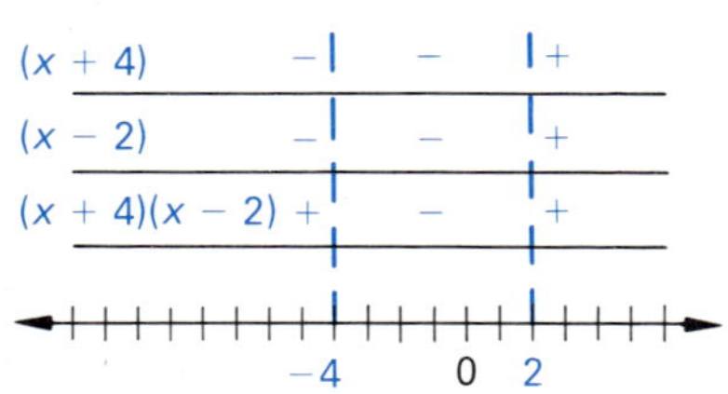

FIGURE 14.12

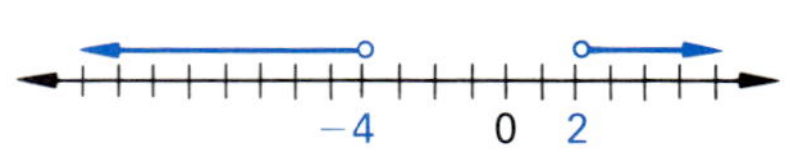

TABLE 14.2

Inequality notation	Graph	Interval notation	Meaning
$a < x$	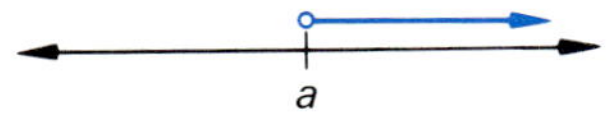	(a, ∞)	Open interval
$x < a$	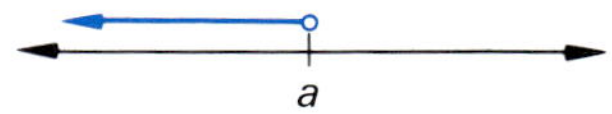	$(-\infty, a)$	Open interval
$a \le x$	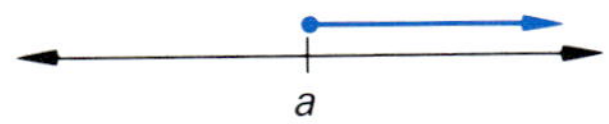	$[a, \infty)$	Half-closed, half-open interval
$x \le a$	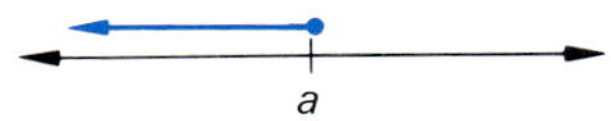	$(-\infty, a]$	Half-open, half-closed interval
$a < x < b$		(a, b)	Open interval
$a \le x \le b$		$[a, b]$	Closed interval

(a and b are real numbers, ∞ is the symbol for infinity.)

Solutions for inequalities can be written in several different forms. From our previous discussion we saw that the solution could be represented two ways, by graphing and by statements of inequality. *The solution could also be given in* ***interval notation,*** *which uses parentheses and/or square brackets. The parentheses correspond to the symbols for less than and greater than, while the square brackets correspond to the symbols for greater than and equal to and less than and equal to.* Table 14.2 illustrates these symbols and their use.

The **open interval** $a < x < b$ or (a, b) contains all the points or real numbers between a and b but does not include the end points a and b. The **closed interval** $a \le x \le b$ or $[a, b]$ contains all the points between a and b and includes the end points a and b.

EXAMPLE 1 Determine the solution of the quadratic inequality

$$2x^2 + 6x \le 0$$

Solution:

$$2x^2 + 6x \le 0$$

$$2x(x + 3) \le 0 \qquad \textbf{Common-term factoring}$$

Solving for x, we see that $x = 0$ and $x = -3$ are the critical points. Indicate the critical points on the number line as in Fig. 14.13. Then determine where the individual quantities $2x$ and $x + 3$ and the product $2x(x + 3)$ are positive or negative. The graph indicates that

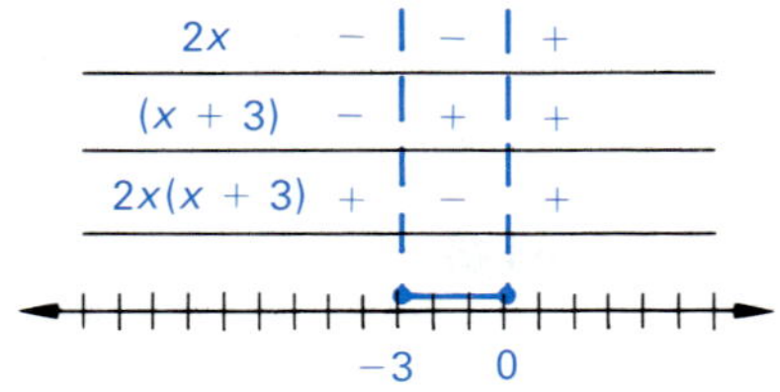

FIGURE 14.13

$2x(x+3)$ is negative (less than zero) between -3 and 0. Therefore, all the points between -3 and 0 are part of the solution set. Since the inequality includes "equals," the critical values are also included. Therefore the solution set for $2x^2+6x \le 0$ is $-3 \le x \le 0$ or in interval notation $[-3, 0]$. □

EXAMPLE 2 Determine the solution of the quadratic inequality

$$x^2 + 6x > 7$$

Solution: In solving quadratic equations we transformed the equation into standard form before solving. We will do the same thing for solving quadratic inequalities.

$$x^2 + 6x > 7$$
$$x^2 + 6x - 7 > 0$$
$$(x+7)(x-1) > 0. \qquad \textbf{Factoring}$$

Setting the factors equal to zero and solving we have $x = -7$ and $x = 1$ as critical values. Indicate the critical values on the number line (Fig. 14.14). The critical points divide the graph into three regions. The graph indicates that the product $(x+7)(x-1)$ is positive to the left of -7 and to the right of 1. Therefore, the solution set for $x^2+6x>7$ is $x<-7$ or $x>1$, or in interval notation $(-\infty, -7)$ or $(1, \infty)$.

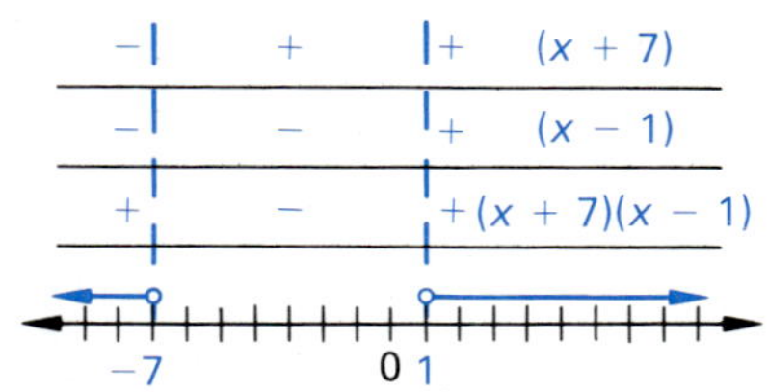

FIGURE 14.14 □

EXAMPLE 3 A projectile is fired from ground level. After t seconds its height, $h(t) = 320t - 16t^2$, is $h(t)$ feet above the ground. During what period of time is the projectile higher than 1536 ft?

Solution: The height of the projectile will be higher than 1536 ft when

$$h(t) > 1536$$

thus

$$320t - 16t^2 > 1536 \qquad \textbf{Substituting}$$
$$-16t^2 + 320t - 1536 > 0$$
$$t^2 - 20t + 96 < 0 \qquad \textbf{Dividing each term by } -16$$
$$(t - 12)(t - 8) < 0 \qquad \textbf{Factoring}$$

Setting the factors $t - 8 = 0$ and $t - 12 = 0$ we can determine the critical values $t = 8$ and $t = 12$. From Fig. 14.15 it can be seen that the product $(t - 8)(t - 12) < 0$ for values of t in the open interval between 8 and 12. Therefore the projectile will be higher than 1536 ft for the time interval between 8 and 12 s. To check the results we will select a few values in the interval:

$$\begin{aligned} t = 9:\quad 320(9) - 16(9)^2 &> 1536 \\ 2880 - 1296 &> 1536 \\ 1584 &> 1536 \checkmark \\ t = 11:\quad 320(11) - 16(11)^2 &> 1536 \\ 3520 - 1936 &> 1536 \\ 1584 &> 1536 \checkmark \end{aligned}$$

We see that these two values check, and it could be shown that any value selected from the interval will also check.

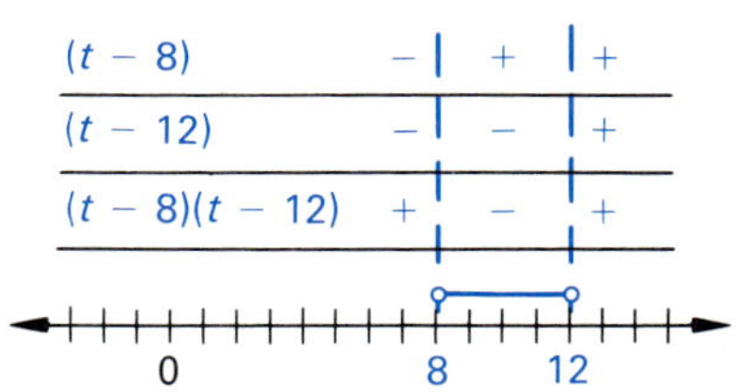

FIGURE 14.15 □

EXAMPLE 4 Determine the solution set of the quadratic inequality

$$x^2 + 3x + 5 < 0.$$

Solution: The expression $x^2 + 3x + 5$ is not factorable, therefore we use the quadratic formula to determine the critical values. With $a = 1$, $b = 3$, and $c = 5$, substituting yields

$$\begin{aligned} x &= \frac{-3 \pm \sqrt{(3)^2 - 4(1)(5)}}{2(1)} \\ &= \frac{-3 \pm \sqrt{-11}}{2}. \end{aligned}$$

The roots of the equation are not real numbers, therefore there are no critical values. Our graph method cannot be used when we have no critical values. Examining the inequality we see that no positive real numbers can possibly make the statement $x^2 + 3x + 5 < 0$ true. Therefore let us try negative values. Let $x = -5, -3$, and -1. Thus, substituting in $x^2 + 3x + 5 < 0$, we have

$$(-5)^2 + 3(-5) + 5 = 25 - 15 + 5 = 15$$
$$(-3)^2 + 3(-3) + 5 = 9 - 9 + 5 = 5$$
$$(-1)^2 + 3(-1) + 5 = 1 - 3 + 5 = 3.$$

Regardless of how many negative values we try, none of them will make $x^2 + 3x + 5 < 0$ a true statement of inequality. The solution set of $x^2 + 3x + 5 < 0$ is the empty set. □

In Example 4, in solving the equality to determine the critical values for x, we found that the discriminant $b^2 - 4ac < 0$ indicates that there are no real roots. In problems where $b^2 - 4ac < 0$ we can use the following theorem to determine the solution set of the inequality.

THEOREM 14.1

If $ax^2 + bx + c = 0$ $(a > 0)$ has no real roots, then the solution set for

$ax^2 + bx + c > 0$ is all real values of x.
$ax^2 + bx + c < 0$ is no real values of x.

EXAMPLE 5 Determine the solution set of the quadratic inequality

$$x^2 + 5x + 7 > 0.$$

Solution: To find the critical values of $x^2 + 5x + 7 = 0$ we use the quadratic formula:

$$x = \frac{-5 \pm \sqrt{(-5)^2 - 4(1)(7)}}{2(1)}$$
$$= \frac{-5 \pm \sqrt{-3}}{2}.$$

The discriminant, -3, is less than zero, therefore there are no real roots. The hypothesis of Theorem 14.1 is satisfied, the coefficient of the x^2-term is positive and the equation $x^2 + 5x + 7 = 0$ has no real roots. The conclusion tells us that the inequality $x^2 + 5x + 7 > 0$ is true for all real values of x or in interval notation $(-\infty, \infty)$. □

Inequalities of higher degree; rational inequalities

The method for solving quadratic inequalities can be extended to solve inequalities of higher degrees and rational inequalities.

EXAMPLE 6 Determine the solution set for the inequality

$$(x-3)(x+4)(x-5) < 0.$$

Solution: In order to solve the inequality $(x-3)(x+4)(x-5) < 0$ we first determine the critical values. Setting the factors equal to zero we have $x-3=0$, $x+4=0$, and $x-5=0$. Solving these equalities for x we find the critical values 3, -4, and 5. Knowing the critical values we can set up the graph as we did for solving quadratic inequalities (see Fig. 14.16). Note that with three critical values the graph is divided into four regions: values of x less than -4, values of x between -4 and 3, values of x between 3 and 5, and values of x greater than 5. We examine each factor in each of these regions and determine whether the factor is positive or negative in each region. For the factor $x-3$ we see that it is $-$, $-$, $+$, $+$ as we move from left to right. Having found the value ($+$ or $-$) for each factor in an interval, we then find the respective products. For $x < -4$, $(-1)(-1)(-1) = -1$. The signs indicate that the product is less than zero for $x < -4$ or $3 < x < 5$. This solution is indicated on the graph. In interval notation the solution for

$$(x-3)(x+4)(x-5) < 0$$

would be stated $(-\infty, -4)$ or $(3, 5)$.

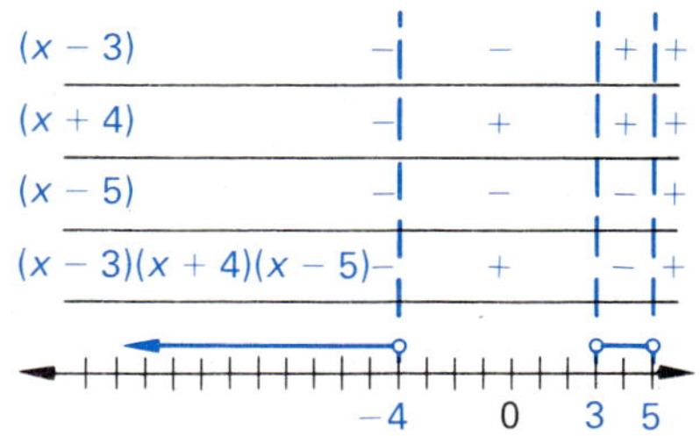

FIGURE 14.16 □

In solving an inequality it is necessary to determine whether the factors are positive or negative. In reexamining the graph in Fig. 14.16 we see that an odd number of negative factors will produce a negative product or a product that is less than zero. In cases where there are an even number of negative factors, the product is positive or greater than zero.

EXAMPLE 7 Determine the solution set of the inequality

$$x^3 + x^2 - 12x \geq 0.$$

Solution: To determine the solution set of $x^3 + x^2 - 12x \geq 0$, we need to write the polynomial as a product of its factors:

$$x^3 + x^2 - 12x \geq 0$$
$$x(x^2 + x - 12) \geq 0 \qquad \textbf{Common-term factoring}$$
$$x(x + 4)(x - 3) \geq 0. \qquad \textbf{Factoring}$$

In this form we can set each of the factors equal to zero and determine the critical values. The critical values are $0, -4,$ and 3. These values are placed on the graph as in Fig. 14.17. Filling in the graph we can see that the product of the quantities is positive for $-4 \leq x \leq 0$ or for $x \geq 3$. Thus the solution set for the inequality $x^3 + x^2 - 12x \geq 0$ is $[-4, 0]$ or $[3, \infty)$.

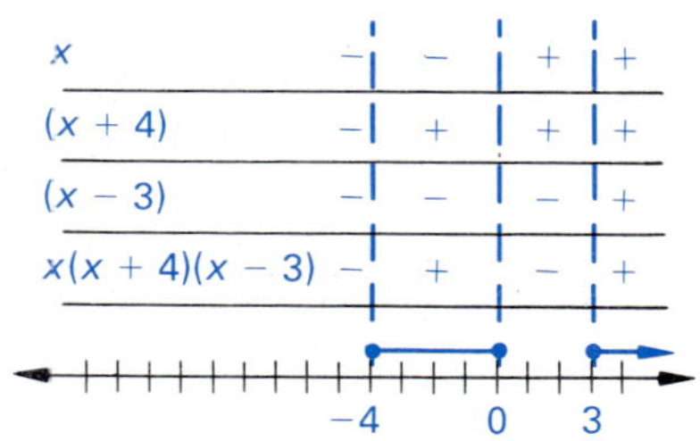

FIGURE 14.17 □

For a polynomial of degree three or higher it may be necessary to refer to Chapter 11 and use the process of synthetic division to help determine the critical values.

EXAMPLE 8 Determine the solution set of the inequality

$$\frac{x^2 + 8x - 9}{x + 3} < 0.$$

Solution: In this problem $x \neq -3$, because division by $x + 3$ is assumed. The general procedure for solving a rational inequality is the same as for polynomial expressions. We will first write the numerator and denominator as a product of linear factors.

$$\frac{x^2 + 8x - 9}{x + 3} < 0$$
$$\frac{(x + 9)(x - 1)}{x + 3} < 0.$$

We now set each of these factors equal to zero, $x + 9 = 0$, $x - 1 = 0$, and $x + 3 = 0$. Solving for x we have $x = -9$, $x = 1$, and $x = -3$. These are the critical values and are graphed in Fig. 14.18. The graph indicates in which intervals the factors are positive or negative. Using the rules

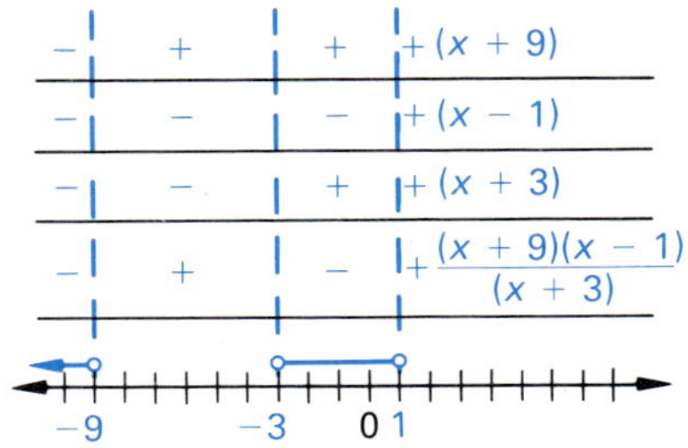

FIGURE 14.18

for division of signed numbers it is known that if there are an odd number of negative factors, then the sign of the quotient is negative. If there are an even number of negative factors, then the quotient is positive. The graph tells us that the quotient is −, +, −, +, as we move from left to right. Therefore, the quotient is less than zero for $x < -9$ or $-3 < x < 1$ as indicated on the graph. The solution set for $\frac{(x+9)(x-1)}{x+3} < 0$ is $(-\infty, -9)$ or $(-3, 1)$. □

EXAMPLE 9 Determine the solution set of the inequality

$$\frac{(x+2)(x-3)}{x+5} < 2x - 7.$$

Solution: To solve the inequality we will use the method of collecting all the terms on one side of the inequality sign.

$$\frac{(x+2)(x-3)}{x+5} < 2x - 7$$

$$\frac{(x+2)(x-3)}{x+5} - (2x-7) < 0$$

$$\frac{x^2 - x - 6}{x+5} - \frac{2x^2 + 3x - 35}{x+5} < 0 \qquad \textbf{Common denominator and expanding}$$

$$\frac{-x^2 - 4x + 29}{x+5} < 0 \qquad \textbf{Collecting like terms}$$

$$\frac{x^2 + 4x - 29}{x-5} > 0. \qquad \textbf{Multiplying both sides by } -1$$

Since $x^2 + 4x - 29$ is not factorable, we find the roots using the quadratic formula. Thus, the factors to the nearest hundredth are

$$\frac{(x-3.74)(x+7.74)}{x+5} > 0.$$

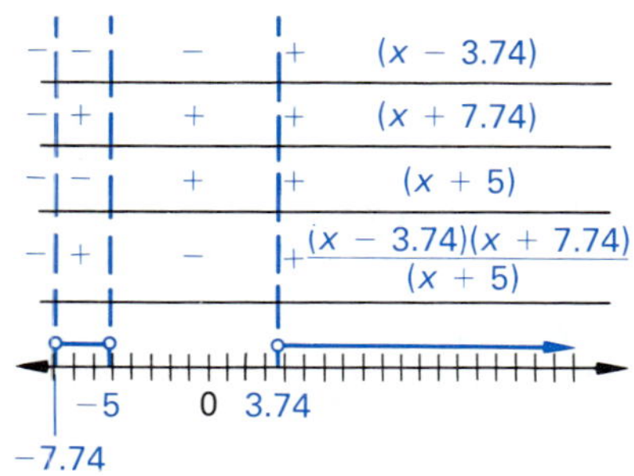

FIGURE 14.19

Figure 14.19 contains the graph of the critical points and the intervals where the factors are positive or negative. Reading the results from the figure we see that the solution set for $\frac{(x+2)(x-3)}{x+5} < 2x - 7$ is $(-7.74, -5)$ or $(3.74, \infty)$. □

14.3 Exercises

In Exercises 1–20, a) find the critical values for each inequality, b) use the graphical method to determine the solution, and c) write the solution set in interval notation.

1. $(x+2)(x-3) > 0$
2. $(x-2)(x+3) \geq 0$
3. $(x+3)(x-5) \leq 0$
4. $(x-3)(x+5) < 0$
5. $2x^2 - 6x \geq 0$
6. $3x^2 - 15x \leq 0$
7. $3x^2 + 15x > 0$
8. $2x^2 + 6x > 0$
9. $t^2 - 4t - 5 < 0$
10. $t^2 - 7t - 8 < 0$
11. $2Z^2 + 5Z + 2 \geq 0$
12. $\frac{1}{2}Z^2 - 8Z + \frac{15}{2} > 0$
13. $2Z^2 + 7Z + 2 \leq 0$
14. $3Z^2 + 5Z + 1 < 0$
15. $x^2 + 3x + 4 < 0$
16. $3x^2 + 3x + 1 \geq 0$
17. $R^2 + R + 3 > 0$
18. $2R^2 + R + 1 \leq 0$
19. $3x^2 - 8x + 2 < 0$
20. $3x^2 + 5x - 3 > 0$

21. A manufacturer of wood-burning stoves finds that when x units, $x \geq 10$, are made and sold per month, its profit (in thousands of dollars) is given by $P(x) = x^2 - 22x + 105$. (a) What is the least number of units that the company can manufacture each month and make a profit? (b) For what values of x is the company losing money?

22. Repeat Exercise 21 with $P(x) = x^2 - 40x + 375$ $(x \geq 18)$.

23. If a slingshot shoots a stone straight up from the ground with an initial velocity of 160 m/s, its distance d in meters above the ground at the end of t seconds (neglecting air resistance) is given by $d = 160t - 16t^2$. Find the duration of time for which $d \geq 256$.

24. A projectile is fired from ground level. After t seconds, its height is $h(t) = 320t - 16t^2$ meters above the ground. During what period is the projectile higher than 1400 m?

25. A designer is laying out a page in a textbook with a page size of 170 mm × 190 mm. She wants a uniform margin around the printed material. At most how wide can the margin be if the printed material must cover at least 224 cm^2?

26. A carpenter is designing a box with a lid. The box will have a square base and a height of 200 mm. If the total surface area must be less than 3072 cm^2, what is the maximum length of the base?

27. Pat is purchasing a rug for the living room which is 4 m × 5 m. She wants a uniform border around the rug. At most how wide can the border be if the rug must cover 12 m^2?

28. The deflection of a beam is given by

$$d = x^2 - 0.5x + 0.06,$$

where x is the distance from one end. For what values of x is $d > 0.02$?

29. The formula for the total surface area of a right circular cylinder is $S = 2\pi r^2 + 2\pi rh$. If the height h of a certain cylinder is 200 mm, what values of the radius will yield a surface area less than 250π square centimeters?

30. A carpenter is asked to construct two picture frames. One frame will have a rectangular shape with a length 40 mm more than its width. The second frame will have a square shape, each of whose sides will be twice as large as the width of the rectangle. If the total area enclosed by the outer edges of the two frames is at least 2508 cm², find the least amount of picture frame molding needed for the job. (Do not allow for cutting or waste.)

In Exercises 31–50 determine the solution set for the inequality.

31. $(x+2)(x-1)(x-4) < 0$

32. $(x+3)(x+4)(x-5) > 0$

33. $(x-3)(x-4)(x+7) > 0$

34. $(x+5)(x-1)(x+6) < 0$

35. $(x+5)^2(x-4)^2 \geq 0$

36. $(x-9)^2(x+3)^2 \leq 0$

37. $4t^3 + 17t^2 - 15t > 0$

38. $3t^3 - 20t^2 - 7t < 0$

39. $2R^4 - 11R^3 + 30R^2 > 0$

40. $2t^4 - 13t^3 - 7t^2 > 0$

41. $\dfrac{3}{z} < 1$

42. $1 - \dfrac{5}{Z} \geq \dfrac{1}{2}$

43. $\dfrac{1}{z+1} < 1$

44. $\dfrac{z}{z+2} < 1$

45. $\dfrac{y+2}{y+3} < 1$

46. $\dfrac{y+2}{y+7} \geq 1$

47. $\dfrac{4x}{x^2+4} \leq 1$

48. $\dfrac{x^2-25}{x^2-16} < 0$

49. $\dfrac{x}{x+3} \leq \dfrac{2x}{x-2}$

50. $\dfrac{x+3}{x-5} \geq \dfrac{2x}{x-3}$

51. For what values of x will $\sqrt{x^2 - 8x + 15}$ be a real number?

52. For what values of x will $\sqrt{\dfrac{x-4}{x+3}}$ be a real number?

14.4 Equalities and inequalities involving absolute values

DEFINITION 14.3

The absolute value of the number n, written $|n|$, is the distance that the point representing n is from the point representing zero on a number line (Fig. 14.20).

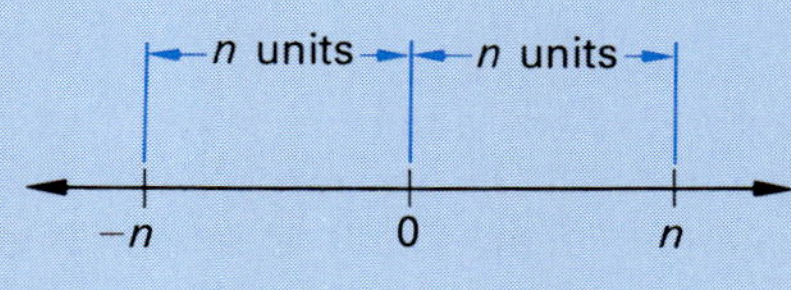

FIGURE 14.20

DEFINITION 14.4

For every number n, $|n| = |-n|$.

This is true, since n and its additive inverse $-n$ are the same distance from the origin. The absolute value measures distance from the origin, but does not take direction into account. For example, $|8| = 8$ since 8 is eight units from the origin; $|-8| = 8$ since -8 is also eight units from the origin (Fig. 14.21).

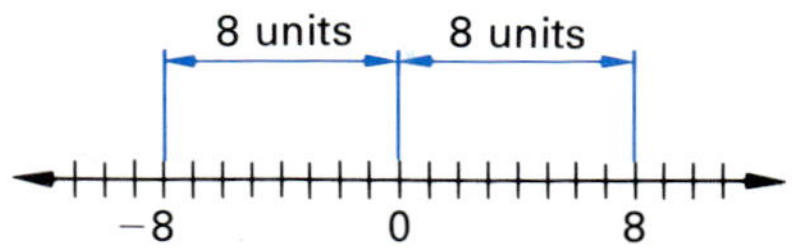

FIGURE 14.21

Having briefly considered the geometric approach to absolute value, we now repeat the definition from Section 1.2.

DEFINITION 1.5

$$|x| = \begin{cases} x, & \text{provided } x > 0. \\ 0, & \text{provided } x = 0. \\ -x, & \text{provided } x < 0. \end{cases}$$

Applying the definition to solving the equation

$$|x - 4| = 7,$$

we have two cases:

$x - 4 > 0$	$x - 4 < 0$
Applying the definition	Applying the definition
$\lvert x - 4\rvert = x - 4,$	$\lvert x - 4\rvert = -(x - 4),$
then	then
$x - 4 = 7$ $x = 11.$	$-(x - 4) = 7$ $-x + 4 = 7$ $x = -3.$

These results can be checked by substituting into the original statement

$x = 11$: $\lvert 11 - 4\rvert = 7$ $\lvert 7\rvert = 7$ $7 = 7.$ ✓	$x = -3$: $\lvert -3 - 4\rvert = 7$ $\lvert -7\rvert = 7$ $7 = 7.$ ✓

EXAMPLE 1 Determine the solution of $|3x - 7| = 14$.

Solution: Using the definition of absolute value, we must consider the two cases, $(3x-7)>0$, and $(3x-7)<0$. From these two cases we can write the two equations

$$\begin{aligned} 3x-7&=14 \\ 3x&=21 \\ x&=7. \end{aligned} \qquad \text{and} \qquad \begin{aligned} -(3x-7)&=14 \\ -3x+7&=14 \\ -3x&=7 \\ x&=-\frac{7}{3}. \end{aligned}$$

Thus, the solution of $|3x-7|=14$ is $x=7$ and $x=-\frac{7}{3}$. The check is left to the reader. □

The absolute value of a product (or quotient) equals the product (or quotient) of the absolute values.

DEFINITION 14.5

> For any real numbers a and b,
>
> $$|ab|=|a||b|$$
>
> $$\text{and} \quad \left|\frac{a}{b}\right|=\frac{|a|}{|b|} \quad \text{if } b\neq 0.$$

EXAMPLE 2

a. Let $a=4$ and $b=-5$.

$$|ab|=|(4)(-5)|=|-20|=20$$
$$|a||b|=|4||-5|=4\cdot 5=20$$

b. Let $a=-18$ and $b=5$.

$$\left|\frac{a}{b}\right|=\left|\frac{-18}{5}\right|=\frac{18}{5}.$$
$$\frac{|a|}{|b|}=\frac{|-18|}{|5|}=\frac{18}{5}. \qquad \square$$

DEFINITION 14.6

> If $|a|=|b|$ then $a=\pm b$.

EXAMPLE 3 Determine the solution of $\left|\frac{3x+5}{5x}\right|=3$

Solution: We will use Definitions 14.5 and 14.6 to solve for x.

$$\left|\frac{3x+5}{5x}\right| = 3$$

$$\frac{|3x+5|}{|5x|} = 3 \qquad \textbf{By Definition 14.5 } \left|\frac{a}{b}\right| = \frac{|a|}{|b|}$$

$$|3x+5| = 3|5x| \qquad \textbf{Multiplying both sides by } |5x|$$

$$3x+5 = 3(\pm 5x) \qquad \textbf{Definition 14.6}$$

$$3x+5 = \pm 15x.$$

$$3x+5 = 15x \qquad 3x+5 = -15x$$

$$5 = 12x \qquad 5 = -18x$$

$$\frac{5}{12} = x. \qquad \frac{-5}{18} = x.$$

The roots of the equation $\left|\frac{3x+5}{5x}\right| = 3$ are $\frac{5}{12}$ and $-\frac{5}{18}$. The check is left to the reader. □

To solve the inequality $|x| < a$ means geometrically that we are looking for all the points on the number line that are less than a units from the origin. Thus, on a number line the solution would be indicated as in Figure 14.22.

For example, for the statement $|x| < 4$, we are looking for all the values of x that are within four units of the origin (Fig. 14.23). If we select any value of x in the open interval $(-4, 4)$, it would satisfy the inequality. For example, if $x = -3$, $|-3| = 3$, which is less than 4.

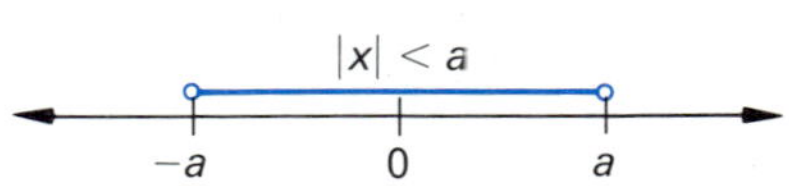

FIGURE 14.22

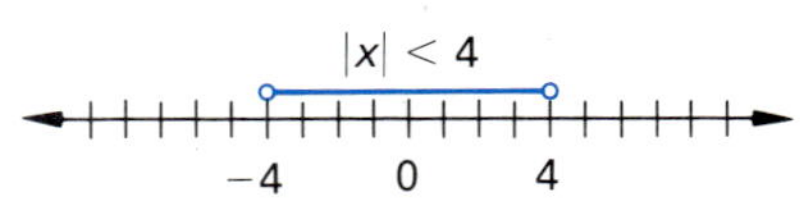

FIGURE 14.23

DEFINITION 14.7

> For any positive real number a, $|x| < a$ is equivalent to $-a < x < a$.

EXAMPLE 4 Determine the solution of $|x| < 7$.

Solution: $|x| < 7$ means that $-7 < x < 7$. The solution is the open interval $(-7, 7)$. □

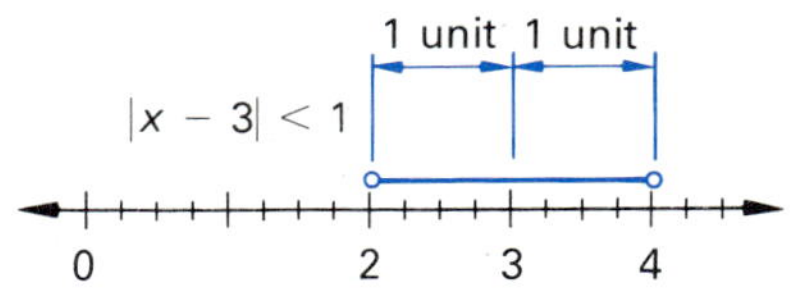

FIGURE 14.24

Now let us consider the statement $|x-3|<1$. From the definition of absolute value $|x-3|$ is the distance between x and 3. Therefore $|x-3|<1$ means that x lies within one unit of 3 (Fig. 14.24). Thus, the numbers 2.1, 2.7, 2.9, 3.1, 3.4, and 3.9, are all within one unit of 3 and are solutions to the problem. To determine all the numbers that satisfy $|x-3|<1$ we apply the definition of absolute value.

$x-3<0$	and	$x-3>0$
$|x-3|=-(x-3)$		$|x-3|=x-3$
then $-(x-3)<1$		then $x-3<1$
multiplying both sides by -1		
$x-3>-1$		
or $-1<x-3$	and	$x-3<1$

Since $x-3$ is between -1 and 1, we can write a single statement $-1<x-3<1$. Thus, $|x-3|<1$ is equivalent to $-1<x-3<1$. Now solving for x by adding 3 to all three terms we have

$$-1+3<x-3+3<1+3$$

or

$$2<x<4.$$

This solution is graphed in Fig. 14.24. □

DEFINITION 14.8

> For real numbers a and b with $b>0$, $|x-a|<b$ is equivalent to $-b<x-a<b$.

EXAMPLE 5 Determine the solution set of $|2x-7|<11$ and graph the solution set.

Solution: Using Definition 14.8 we can write $|2x-7|<11$ as the

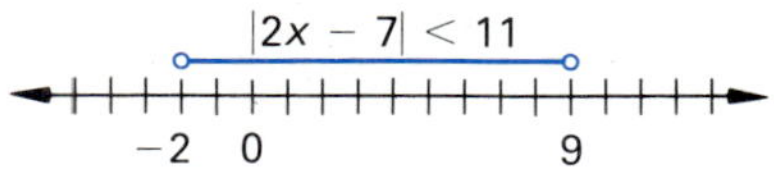

FIGURE 14.25

double inequality

$$-11 < 2x - 7 < 11$$
$$-11 + 7 < 2x - 7 + 7 < 11 + 7 \qquad \textbf{Adding 7 to all three terms}$$
$$-4 < 2x < 18$$
$$\frac{-4}{2} < \frac{2x}{2} < \frac{18}{2} \qquad \textbf{Dividing each term by 2}$$
$$-2 < x < 9.$$

The graph of the solution set is given in Fig. 14.25.

The solution set of $|2x - 7| < 11$ is the open interval $(-2, 9)$. □

To solve the inequality $|x| > a$ means geometrically that we are looking for all the points on the number line that are more than a units from the origin. Thus, on a number line the solution would be indicated as in Fig. 14.26.

For example, for the statement $|x| > 5$, we are looking for all the values of x that are more than five units from the origin (Fig. 14.27). If we select any value of x in the open interval $(-\infty, -5)$ or the open interval $(5, \infty)$, the value of x would satisfy the inequality.

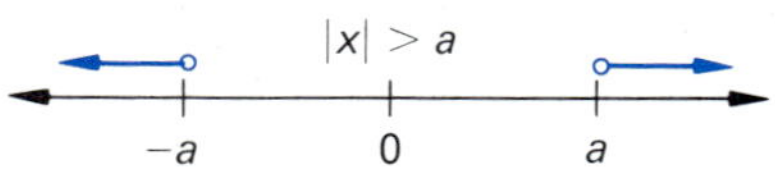

FIGURE 14.26

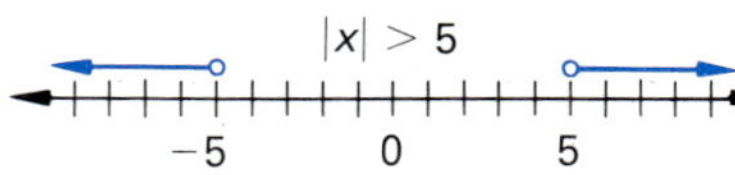

FIGURE 14.27

DEFINITION 14.9

> For any positive real number a, $|x| > a$ is equivalent to $x > a$ or $x < -a$.

EXAMPLE 6 Determine the solution set of $|x| > 6$.

Solution: $|x| > 6$ means $x > 6$ or $x < -6$. The solution is the open intervals $(-\infty, -6)$ or $(6, \infty)$. □

Now let us consider the statement $|x - 4| > 1$. From the definition of absolute value $|x - 4|$ is the distance between x and 4. Therefore

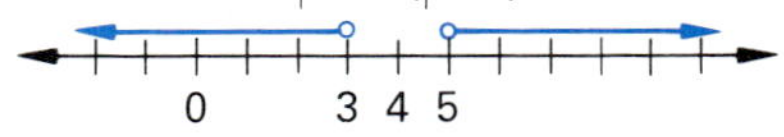

FIGURE 14.28

$|x-4|>1$ means that x is more than one unit from 4 (Fig. 14.28). Thus, the numbers $-8.2, -6, -4.1, -3.2, 5.3, 6.7, 11$ are all more than one unit from 4 and are solutions to the problem. To determine all the numbers that satisfy $|x-4|>1$ we apply the definition of absolute value.

$x-4<0$	and	$x-4>0$
then $|x-4|=-(x-4)$, $-(x-4)>1$, $x-4<-1$	or	$|x-4|=x-4$, $x-4>1$.

Since $x-4$ is less than -1 or $x-4$ is greater than 1, we cannot write the solution as a single statement. Thus $|x-4|>1$ is equivalent to $x-4<-1$ or $x-4>1$. Now solving for x by adding 4 to both sides of both inequalities we have

$$x-4+4<-1+4 \quad \text{or} \quad x-4+4>1+4 \qquad \textbf{Adding 4 to both sides.}$$
$$x<3 \qquad\qquad x>5.$$

The solution is graphed in Fig. 14.28. □

Note: Since $x-4<-1$ or $x-4>1$, we cannot write the inequalities as a single expression. Also note that the solution $x<3$ or $x>5$ cannot be written as a single expression.

DEFINITION 14.10

> For real numbers a and b with $b>0$, $|x-a|>b$ is equivalent to $x-a<-b$ or $x-a>b$.

EXAMPLE 7 Determine the solution set of $|3x+7|>5$ and graph the solution set.

Solution: Using Definition 14.10 we can write $|3x+7|>5$ equiva-

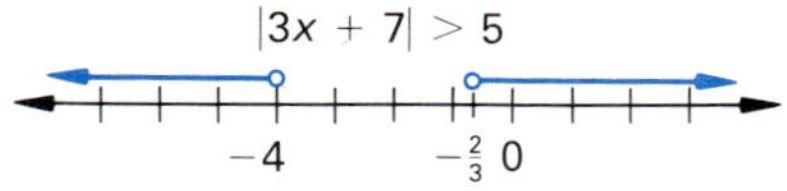

FIGURE 14.29

lently as

$$3x + 7 < -5 \qquad \text{or} \qquad 3x + 7 > 5$$

$$3x + 7 - 7 < -5 - 7 \qquad 3x + 7 - 7 > 5 - 7 \qquad \textbf{Subtracting 7 from both sides}$$

$$3x < -12 \qquad 3x > -2$$

$$x < -4 \qquad x > -\frac{2}{3} \qquad \textbf{Dividing both sides by 3}$$

The solution set of $|3x + 7| > 5$ is $(-\infty, -4)$ or $\left(-\frac{2}{3}, \infty\right)$. Its graph is given in Fig. 14.29. □

EXAMPLE 8 Determine the solution set of $|4 - 7x| \geq -3$.

Solution: The absolute value of a number is always non-negative. Therefore $|4 - 7x| \geq -3$ is always true. The solution set is the set of all real numbers, or in interval notation, $(-\infty, \infty)$. □

In manufacturing parts for machines the parts must be produced to meet specific standards of precision. For example, in manufacturing shafts and bearings the standards ensure that the bearings will fit properly on the shafts. Assume that a shaft must have a diameter of 12.57 mm with a tolerance of 0.02 mm (or ±0.02 mm as it is sometimes written). If we let x represent the actual diameter of the shaft, then we can express the required dimensions in the following form:

$$|x - 12.57| \leq 0.02.$$

To find the acceptable limits on the diameter of the shaft we solve for x

$$-0.02 \leq x - 12.57 \leq 0.02 \qquad \textbf{Definition 14.8}$$

$$-0.02 + 12.57 \leq x - 12.57 + 12.57 \leq 0.02 + 12.57$$

$$12.55 \leq x \leq 12.59.$$

The resulting inequality states that the actual diameter of the shaft must be between 12.55 mm and 12.59 mm.

EXAMPLE 9 Write the following statements using the absolute value symbol.

a. R is less than nine units from 25.

b. Z differs from 7 by at least 5.

c. y is within four units of -7.

d. $x < 5$ and $x > -5$

Solution:

a. The statement, R is less than nine units from 25, can be written

$|R - 25| < 9.$

b. The statement, Z differs from 7 by at least 5, can be written

$|Z - 7| \geq 5$

c. The statement, y is within four units of -7, can be written

$|y - (-7)| \leq 4$
$|y + 7| \leq 4$

d. The statement, $x < 5$ and $x > -5$, can be written

$-5 < x < 5 \quad \text{or} \quad |x| < 5.$ □

14.4 Exercises

In Exercises 1–15, determine the solution set of each statement of equality and check.

1. $|x + 3| = 5$
2. $|x - 5| = 7$
3. $|Z - 2| = \frac{1}{3}$
4. $|3Z + 11| = \frac{1}{4}$
5. $|5R + 3| = 1$
6. $|-4R + 2| = 7$
7. $|0.03t - 0.05| = 0.17$
8. $|0.02t - 0.01| = 0.15$
9. $|3t - 4| - 5 = 6$
10. $\left|\frac{1}{2}y + \frac{1}{4}\right| + \frac{1}{3} = \frac{2}{3}$
11. $\left|\frac{2}{3}y - \frac{1}{5}\right| - \frac{1}{7} = 0$
12. $\left|-\frac{1}{4}y - \frac{1}{6}\right| - \frac{1}{8} = 0$
13. $\left|\frac{2x + 5}{3x}\right| = 7$
14. $\left|\frac{2x - 4}{5x}\right| = -2$
15. $\left|\frac{5x - 6}{3x}\right| = 5$

In Exercises 16–42, determine the solution set and graph the inequality.

16. $|x| < 3$
17. $|x| \leq 8$
18. $|x| \geq 11$
19. $|x| > 13$
20. $|t| > -5$
21. $|t| > 0$
22. $|t + 2| < 5$
23. $|t - 4| \leq 7$
24. $|R + 5| \geq 3$
25. $|R - 7| \geq 4$
26. $|2R - 7| \leq 1$
27. $|3Z + 5| < 2$
28. $|5Z + 6| > 3$
29. $|3Z - 4| \geq 5$
30. $|1 - 3Z| \leq 5$
31. $|2 - 5x| < -7$
32. $|3 - 4x| \geq 2$
33. $|5 - 7x| > 4$
34. $|1 + 11x| > 0$
35. $\left|\frac{x - 3}{2}\right| < 2$
36. $\left|\frac{2x + 1}{3}\right| \leq 4$
37. $\frac{|7x + 3|}{2} < 0$
38. $\frac{|2x - 14|}{4} < 2$
39. $\left|\frac{1}{3} + x\right| < \frac{1}{4}$
40. $\left|\frac{2x + 1}{3}\right| < 0.03$
41. $\left|\frac{1}{5} + x\right| < 0.05$
42. $\left|\frac{x - 4}{2}\right| > 0.02$

43. A bolt must have a diameter of 13.45 mm with a tolerance of ± 0.01 mm. Find the acceptable range of diameters of the bolt.

44. A machine that packages 250 aspirin tablets per container can make an error of three pills per container. If x is the number of tablets in a bottle, write an inequality using absolute value that indicates a maximum error of three tablets per container. Solve the inequality to determine the possible number of tablets in the container.

45. A plastic disk must have a thickness of 4 mm with a tolerance of ± 0.03 mm. Find the acceptable range of thickness of the plastic disk.

46. Ten boxes of table salt have a combined "weight" of 7370 g with the measurement accurate to ± 2 g. Find the range of probable "weights" of a single box of salt.

In Exercises 47–50, write an inequality with absolute values describing the following sentences, then solve the statement.

47. y is less than nine units from $4/3$.

48. $5x$ is more than $7/3$ units from $11/2$.

49. R is within five units of -13.

50. $Z \leq 14$ and $Z \geq -14$.

14.5 Systems of inequalities

In Sections 14.2 and 14.3 we learned how to solve inequalities in one variable. In this section we will discuss inequalities in two variables. Examples of inequalities in two variables are $2x + 3y < 6$, $x - y \leq 3$, and $x^2 + 7y^2 > 4$.

We were able to indicate the solution set of an inequality in one variable on a number line. The solution set of an inequality in two variables will be indicated on the coordinate plane.

A linear inequality that is strictly $<$ or $>$ will have as its solution a half-plane. A **half-plane** is represented by all the points on one side of a line. An inequality that is $\leq$ or $\geq$ will have as its solution set the set of points that consists of a half-plane and a line. To indicate that the line is part of the solution set we will draw a solid line. To indicate that the line is not part of the solution set we will draw a dashed line.

EXAMPLE 1 Graph the solution set of $x + 3y < 9$.

Solution: To obtain the solution set, we graph $x + 3y = 9$. Since the original statement is strictly "*less than,*" we draw a dashed line (Fig. 14.30). The dashed line indicates that the line is not a part of the solution set.

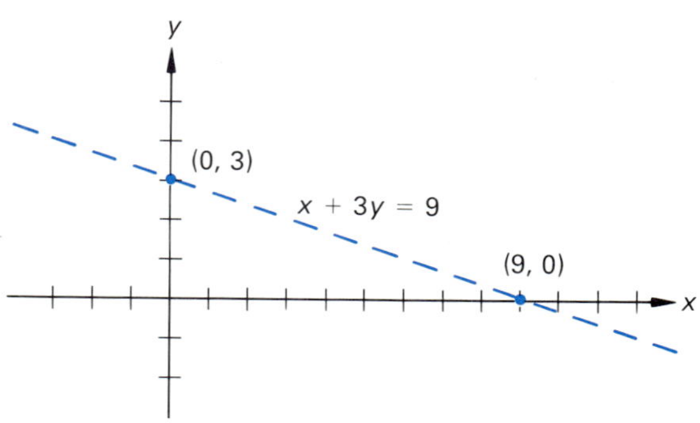

FIGURE 14.30

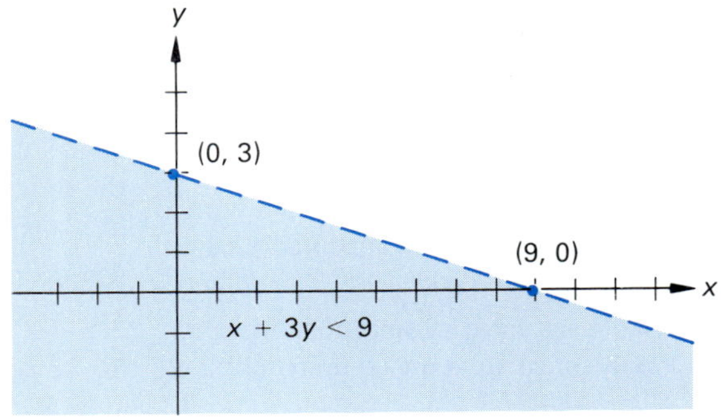

FIGURE 14.31

The line $x + 3y = 9$ divides the plane into two half-planes. The points in one half-plane will satisfy the inequality $x + 3y < 9$. The points in the other half-plane will satisfy the inequality $x + 3y > 9$.

To determine the solution set of the inequality $x + 3y < 9$ we pick any point on the plane that is not on the line. The simplest point to work with is the origin, (0, 0). Substituting $x = 0$ and $y = 0$ in $x + 3y < 9$ gives $0 + 3(0) < 9$, or $0 < 9$. Since 0 is less than 9, the point (0, 0) is part of the solution set. All the points on the same side of the line $x + 3y = 9$ as the point (0, 0) are members of the solution set. We indicate this by shading the half-plane that contains (0, 0). The graph is shown in Fig. 14.31. □

In Chapter 10 we learned how to find the solution of two linear equations with two variables. Now we are going to explore the techniques of finding the solutions of two linear inequalities with two variables.

EXAMPLE 2 Graph the solution set of the following system of inequalities:

$$x + y < 3$$
$$x - y < 7.$$

Solution: The solution set is the set of all points that will make both statements true. Our method will be to graph both statements on the same axes. First we draw a graph of the lines $x + y = 3$ and $x - y = 7$. Both lines are dashed since each boundary is not part of the solution set. As we see in Fig. 14.32 the two lines intersect at the point $(5, -2)$.

Now we find the half-plane determined by the inequality $x + y < 3$ by substituting the origin (0, 0) and noting that $0 + 0 < 3$, so the origin is included in the solution for $x + y < 3$ (Fig. 14.33). Next we find the half-plane determined by the inequality $x - y < 7$. Substituting (0, 0)

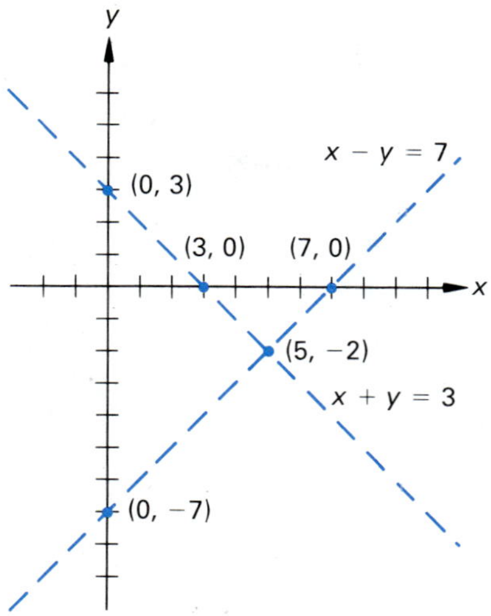

FIGURE 14.32

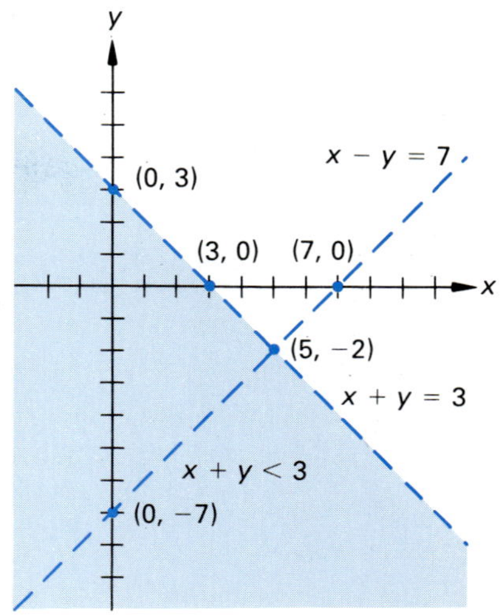

FIGURE 14.33

we find $0 - 0 < 7$ so the origin is included in the solution (Fig. 14.34). The solution of the system of linear inequalities consists of all the points common to the two half-planes (Fig. 14.35).

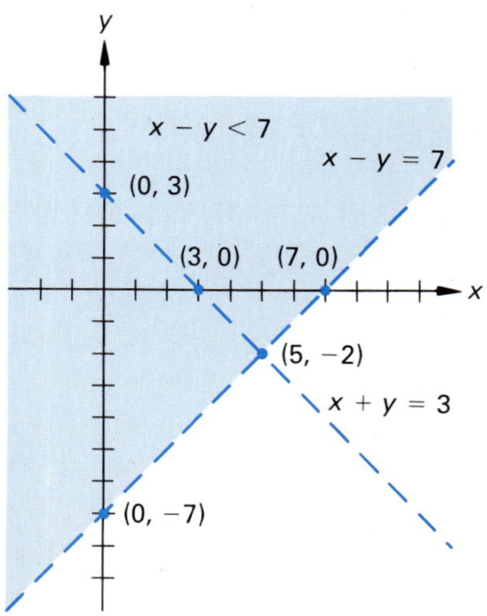

FIGURE 14.34

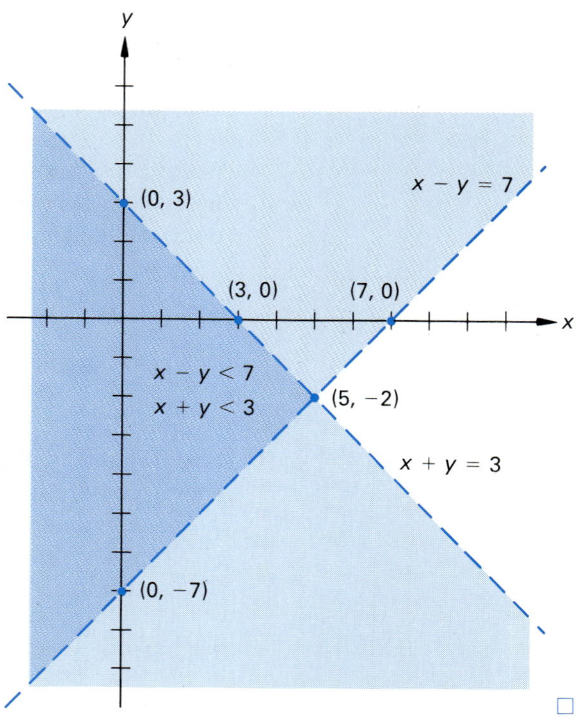

FIGURE 14.35

□

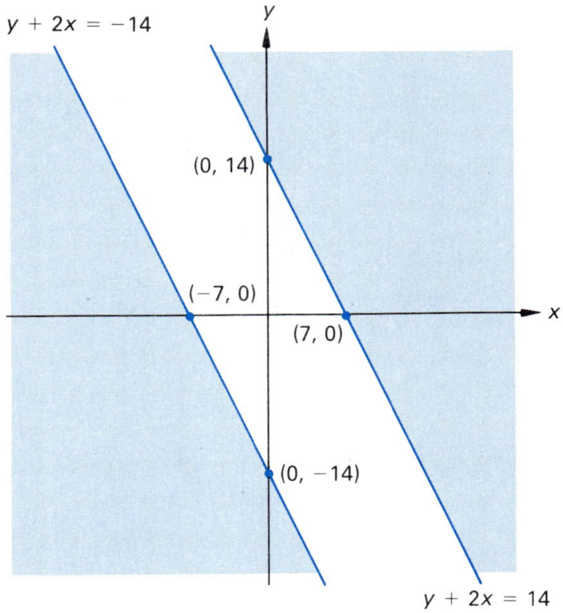

FIGURE 14.36

Systems of inequalities need not be restricted to systems of linear inequalities. They may also include absolute values and nonlinear systems. These will be illustrated in the following examples.

EXAMPLE 3 Graph the solution set of $|y + 2x| \geq 14$.

Solution: In order to graph the solution set of $|y + 2x| \geq 14$ we must first write the inequality without the absolute value sign. This can be done using Definition 14.10.

$$|y + 2x| \geq 14$$
$$y + 2x \geq 14 \quad \text{or} \quad y + 2x \leq -14.$$

The two statements of equality

$$y + 2x = 14 \text{ and } y + 2x = -14$$

are graphed in Fig. 14.36. The two equations have the same slope, therefore the lines are parallel. Testing the point (0, 0) we see that this point does not satisfy $y + 2x \geq 14$ or $y + 2x \leq -14$. Therefore, the solution set is the set of all the points outside the lines and on the lines. The solution set is indicated in blue in Fig. 14.36. □

EXAMPLE 4 Graph the solution set of the system of inequalities:

$$y > x^2 - 9$$
$$y - x \leq 3.$$

Solution: To find the solution of the system we must graph the equality $y = x^2 - 9$. Using the techniques learned in Section 11.6 we can quickly determine the following key facts: the parabola opens up, $(-3, 0)$ and $(3, 0)$ are x-intercepts, $(0, -9)$ is the y-intercept, the y-axis is the line of symmetry, $(0, -9)$ is the vertex of the parabola. The table with Fig. 14.37 contains sets of ordered pairs that were used to sketch the graph.

In the same diagram we sketched the straight line representing the equation $y - x = 3$. We see that the curves intersect at $(-3, 0)$ and $(4, 7)$. The points of intersection can be determined from the graph or by algebraic methods.

The solution set for the inequality $y > x^2 - 9$ will be all the points either inside or outside the curve. Select the point $(0, 0)$; substituting $(0, 0)$ into $y > x^2 - 9$ yields $0 > -9$. Since $(0, 0)$ is inside the curve, all the points inside the curve satisfy the inequality. Also, testing $y - x \leq 3$ with the point $(0, 0)$ yields $0 \leq 3$. We see that the solution set lies on or below the line $y - x = 3$. Therefore the solution set for the system is the set of all the points inside the parabola, on that part of the line inside the parabola, and below the line. The solution set does not include the points of intersection of the line and the parabola, since the boundary of the parabola is not part of the solution set. The open circles at $(-3, 0)$ and $(4, 7)$ in Fig. 14.38 indicate that the points of intersection are not part of the solution set. The solution set of the system is indicated by the solid line and the shaded region.

FIGURE 14.37

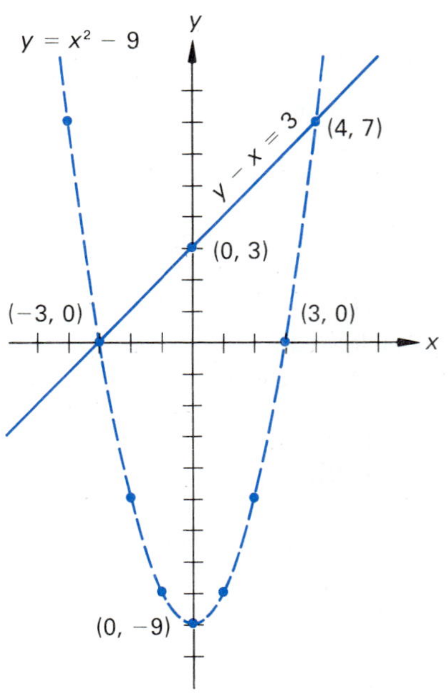

$y = x^2 - 9$

x	y
−4	7
−3	0
−2	−5
−1	−8
0	−9
1	−8
2	−5
3	0
4	7

FIGURE 14.38

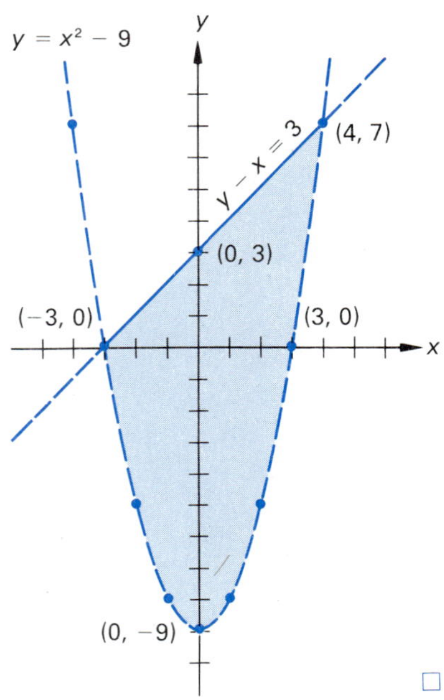

□

To find the solution set for a system of inequalities we do the following:

1. Write the statements of inequality as statements of equality.
2. Graph the statements of equality.
3. Determine the points of intersection of the curves.
4. Determine the points on the plane that satisfy each statement of inequality.
5. Indicate the solution set on a graph by shading and by solid or dashed lines.

14.5 Exercises

In Exercises 1–10, graph the solution set of the inequality.

1. $y \geq 3$

2. $x \leq 4$

3. $y \leq x$

4. $x + y > 0$

5. $x + 3y \leq 6$

6. $2x + 3y \geq 12$

7. $3x + 4y < 12$

8. $y < 9 - x^2$

9. $y \leq 4x - x^2$

10. $x^2 + y^2 < 9$

In Exercises 11–40, graph the solution set of the system of inequalities.

11. $x + y < 5$
$x - y < 7$

12. $3x + 2y < 6$
$x + 2y < 4$

13. $x + y \leq 3$
$2x + 3y < 6$

14. $2x - y \leq 3$
$x + y < 3$

15. $x - 3y \leq 3$
$x - 2y \geq -4$

16. $x + 2y \leq 4$
$3x - y \geq 6$

17. $y \leq 3x$
$x \geq 3y$

18. $x \geq 4$
$x + y < 2$

19. $x \geq 2$
$y \leq 2$

20. $x \leq 0$
$y \leq 0$

21. $x \geq 0$
$y \geq 0$

22. $x \leq 4$
$x + 2y < -1$

23. $|y + x| \leq 10$

24. $|y - 3x| < 8$

25. $|y + 2x| \geq 4$

26. $|2y - 4x| > 10$

27. $|3y + 6x| < 12$

28. $\left|y - \frac{1}{2}x\right| \leq 4$

29. $\left|y - \frac{1}{3}x\right| > 4$

30. $|y + 5x| \geq 9$

31. $x + y < 2$
$y < x^2 - 4$

32. $x + y > 2$
$y < x^2 - 4$

33. $2x + y \leq 3$
$y < x^2 - 5$

34. $y - 2x \leq -7$
$y \geq x^2 + 2x - 8$

35. $y \geq x + 1$
$y \geq x^2 - 4x - 5$

36. $y < 9x - 45$
$y > x^2 + 2x - 15$

37. $x - y < 5$
$x > 0$
$y < 0$

38. $4x + 5y \leq 20$
$x \geq 0$
$y \geq 0$

39. $4x + 5y \leq 20$
$y \leq x + 5$
$y \geq 0$

40. $3y < 2x + 12$
$y < x + 4$
$y > 0$

41. A plumbing store manager pays \$1.00 for a small can of plastic pipe cement and \$1.50 for a larger one. The manager wants to stock at least 200 cans and can spend up to \$300 on the cement. Find a system of inequalities that describes all possible combinations of can sizes that could be stocked. Find the solution set.

42. Tim's truck is to be loaded with two types of cargo. Each unit of type A weighs 100 kg, is 3 m^3 in volume, and earns \$8 for the driver. Each unit of type B weighs 80 kg and is 4 m^3 in volume and earns \$11.00 for the driver. The truck can carry no more than 6000 kg and no more than 300 m^3 of cargo. Find a system of inequalities that describes all possibilities and sketch the graph.

43. A racketball racket manufacturer makes two types of rackets, Quick Serve and Hard Return. Because of demand he wants to make at least twice as many

Quick Serves as Hard Returns. It takes three hours to make a Quick Serve and four hours to make a Hard Return. Altogether he wants to make at least 10 rackets and work no more than 80 hours. Find a system of inequalities that describes all possibilities and sketch the graph.

14.6 Linear programming

Many problems in government, business, and industry require the most cost-effective solution. There are many different ways of making a mouse trap, but some of them are less costly than others. If we can express the relationships in the manufacturing process in linear equations and linear inequalities, then we have a linear-programming problem. (Linear programming is not the same as computer programming even though linear-programming problems may be programmed to run on a computer.)

The typical linear-programming problem has many variables and is generally so lengthy that it is solved on a computer by a technique called the simplex algorithm. The simplex method was developed in the 1940s by George B. Dantzig. Linear programming is being used to solve problems in the social sciences, health areas, land development, and diet.

We will not discuss the simplex method in this textbook. We will merely give a very brief graphical introduction as to how linear programming works. The reader can find a detailed explanation in books on finite mathematics.

EXAMPLE 1 A company makes two types of rocking chairs. Each rocking chair must be assembled and then finished. The plain chair takes four hours to assemble and four hours to finish. The fancy chair takes eight hours to assemble and twelve hours to finish. The company can provide at most 160 worker-hours of assembling and 180 worker-hours of finishing a day. If the profit on the plain chair is \$10.00 and the profit on the fancy chair is \$18.00, how many rocking chairs of each type should the company make per day to maximize profits? What is the maximum profit?

Solution: From the information given above, we can establish the following facts:

	Assembly time (hr)	Finishing time (hr)	Profit (\$)
Plain chair	4	4	10.00
Fancy chair	8	12	18.00

Let

$$x = \text{the number of plain chairs},$$
$$y = \text{the number of fancy chairs},$$
$$10x = \text{profit on the plain chairs},$$
$$18y = \text{profit on the fancy chairs, and}$$
$$P = \text{total profit}.$$

The total profit is the sum of the profit on the plain chairs and the profit on the fancy chairs. Thus $P = 10x + 18y$.

The maximum profit P is dependent on several conditions, called **constraints.** The number of chairs manufactured each day cannot be a negative amount. This gives us the constraints $x \geq 0$ and $y \geq 0$. Another constraint is determined by the total number of hours allocated for assembling. Since it takes four hours to assemble the plain chair, the total number of hours per day to assemble x plain chairs is $4x$. It takes eight hours to assemble a fancy chair; therefore the total number of hours needed to make y fancy chairs is $8y$. The maximum number of hours allocated for assembling is 160. Thus, the third constraint is $4x + 8y \leq 160$. The final constraint is determined by the number of hours allotted for finishing. It takes four hours to finish a plain chair, or $4x$ hours to finish all the plain chairs. It takes twelve hours to finish a fancy chair, or $12y$ hours to finish all the fancy chairs. The total number of hours allotted for finishing is 180. Therefore the fourth constraint is $4x + 12y \leq 180$. Thus the four constraints are

$$\begin{aligned} x &\geq 0, \\ y &\geq 0, \\ 4x + 8y &\leq 160, \\ 4x + 12y &\leq 180. \end{aligned}$$

The list of constraints is a system of linear inequalities in two variables. The graph of the system is shown in Fig. 14.39. The points (0, 0), (0, 15), (30, 5), and (40, 0) are the points where two of the boundaries intersect. The solution set for the above system of constraint inequalities is the colored region and the solid boundaries.

FIGURE 14.39

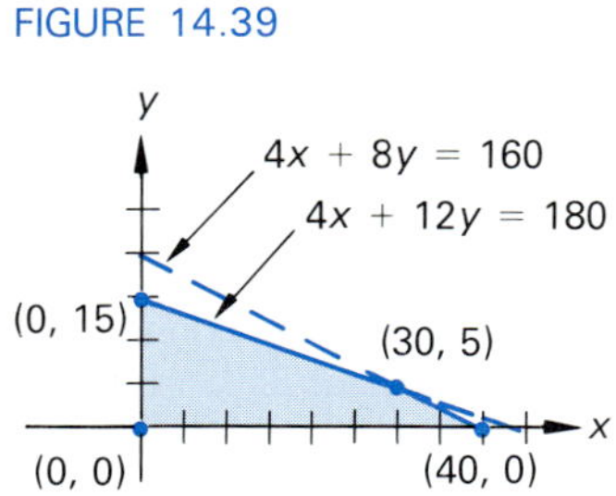

When the region is bounded on all sides, as in this example, it is called a **polygonal region,** because the boundary of the region is a polygon. The points where two or more sides intersect are called the **vertices** of the polygon. The points on the boundary of the region and the points inside the polygonal region are the solution set for the system of inequalities. From all these points there is at least one point that, when substituted in an equation of the form $K = Ax + By$, will give us the maximum (and also the minimum) value for K. (K is the value we wish to maximize or minimize; in this example K represents profit.) The linear-programming problem is to determine which ordered pair will give us this maximum (or minimum) value. The fundamental principle of linear programming provides us with a rule for finding the maximum and minimum values. (For development of this principle see books on linear programming.)

Fundamental Principle

If a linear equation of the form $K = Ax + By$ is evaluated at each point in a closed polygonal region, the maximum and minimum values of the equation occur at vertices of the region.

Recall that we want to maximize the profit. The total profit is given by the formula $P = 10x + 18y$. According to the fundamental principle, the maximum profit will be found at one of the vertices of the polygonal region. Calculate P for each one of the vertices.

At $(0, 0)$, $P = 10(0) + 18(0) = 0$.
At $(40, 0)$, $P = 10(40) + 18(0) = 400$.
At $(30, 5)$, $P = 10(30) + 18(5) = 390$.
At $(0, 15)$, $P = 10(0) + 18(15) = 270$.

Clearly, the maximum profit is at $(40, 0)$, which means the company should manufacture 40 plain chairs and no fancy rocking chairs. The maximum profit would be \$400. The minimum profit would be at $(0, 0)$, where no rocking chairs of either style would be manufactured.

A variation of the original problem could be that the company knows that it cannot sell more than 15 plain rocking chairs per day. With this additional constraint, we now have the following set of constraints.

$$\begin{aligned} x &\geq 0, \\ x &\leq 15, \\ y &\geq 0, \\ 4x + 12y &\leq 180, \\ 4x + 8y &\leq 160. \end{aligned}$$

FIGURE 14.40

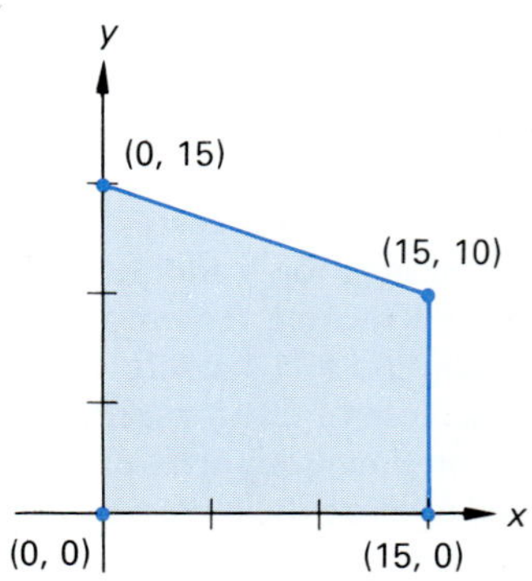

The graph of these constraints is shown in Fig. 14.40.

We see that the vertices of the polygonal region are $(0, 0)$, $(0, 15)$, $(15, 10)$, and $(15, 0)$. To determine the maximum profit we calculate P for each of these vertices.

At $(0, 0)$, $P = 10(0) + 18(0) = 0$.
At $(0, 15)$, $P = 10(0) + 18(15) = 270$.
At $(15, 10)$, $P = 10(15) + 18(10) = 330$.
At $(15, 0)$, $P = 10(15) + 18(0) = 150$.

With this set of constraints we see that the maximum profit of \$330.00 occurs when the company manufactures 15 plain rocking chairs and 10 fancy rocking chairs. □

14.6 Exercises

In Exercises 1–6 a set of constraints and a profit formula are given. (a) Draw the graph of the constraints, and find the points of intersection of the boundaries. (b) Use these points of intersection to determine the maximum and minimum profit.

1. $x+y\leq 5$
$x+3y\leq 9$
$x\geq 0$
$y\geq 0$
$P=2x+4y$

2. $x+y\leq 7$
$x+2y\leq 10$
$x\geq 0$
$y\geq 0$
$P=4x+5y$

3. $x+y\leq 4$
$x+3y\leq 6$
$x\geq 0$
$y\geq 0$
$P=6x+7y$

4. $x+y\leq 50$
$x+3y\leq 90$
$x\geq 0$
$y\geq 0$
$P=20x+40y$

5. $3x+2y\geq 6$
$3x+4y\leq 24$
$x\geq 0$
$y\geq 0$
$y\leq 3$
$P=1.5x+3.5y$

6. $x+2y\leq 12$
$4x+5y\geq 20$
$x\geq 0$
$y\geq 0$
$x\leq 8$
$P=5x+7y$

7. The P^3 Company manufactures two types of power lawn mowers. Type A is self-propelled, and type B must be pushed by the operator. The company can produce a maximum of 18 mowers per week. It can make a profit of \$20 on mower A and a profit of \$25 on mower B. The company's planners want to make at least two mowers of type A but not more than five. To keep certain customers happy, they must make at least two mowers of type B.

a. List the constraints.
b. List the profit formula.
c. Graph the set of constraints.
d. Find the vertices of the polygonal region.
e. How many of each type should be made to maximize the profit?
f. Find the maximum profit.

8. Richard makes wooden toy cars and trucks. The materials for a car cost \$2.00, and the materials for a truck cost \$2.25. He can make a total of 15 toys in his spare time in a month. He wants to make at least two cars and two trucks but no more than five trucks. He sells the cars for \$6.00 and the trucks for \$7.50. How many of each type should he make each month to maximize his profit?

9. A plumbing store manager pays \$0.75 for a small can of plastic pipe cement and \$1.30 for a large can. The manager wants to stock at least 200 cans and can spend up to \$300 on the pipe-cement order. When the pipe cement is sold, the store makes a profit of \$0.15 on the small can and \$0.26 on the large can. How many cans of each size should the manager order to maximize his profit? How much profit could the store make from the order?

10. Jane's truck is loaded with two types of cargo. Each unit of type A is 0.15 m^3, weighs 45 kg, and earns \$20 for the driver. Each unit of type B is 0.12 m^3, weighs 67.5 kg, and earns \$18 for the driver. If the truck can carry no more than 12 m^3 of cargo and no more than 4388 kg, how many units of each type should Jane take to earn the most money? How much money will she earn?

11. A company builds two models of calculators, *A* and *B*. Each model uses three types of transistors, *X*, *Y*, and *Z*. The profit of each model of calculator, the number of transistors required, and the number presently in stock are shown in the table. How many of each model should the company produce so as to most efficiently utilize the transistors available (that is, to maximize profit)?

Type of Transistor	Calculator *A*	Calculator *B*	Quantity in Stock
X	2	2	800
Y	2	3	900
Z	1	5	1000
Unit Profit	\$20	\$25	

12. A farmer has at most 400 acres available for planting two crops, *A* and *B*. The seed for crop *A* costs \$5 per acre and the seed for crop *B* costs \$8 per acre. Labor costs \$50 per acre for *A* and \$80 per acre for *B*. The expected income for crop *A* is \$120 per acre and for crop *B* \$175 per acre. If the farmer wants to spend at most \$3,200 for seed, and at most \$18,000 for labor, how many acres of each crop should be planted to make the largest profit? How much will the farmer make?

Chapter summary

The symbols of inequality, their meaning and graphical interpretation are given in Table 14.1. The sense of inequality is the direction of the inequality symbol in one statement versus another statement (i.e., $a < b$ and $c > d$ are inequalities of opposite sense).

Properties of Inequalities

Addition property: For any real numbers, a, b, and c, if $a < b$, then $a + c < b + c$.

Subtraction property: For any real numbers a, b, and c, if $a < b$, then $a - c < b - c$.

Multiplication property: (a) If $a < b$ and $c > 0$ then $ac < bc$. (b) If $a < b$ and $c < 0$ then $ac > bc$.

Division property: (a) If $a < b$ and $c > 0$ then $a \div c < b \div c$. (b) If $a < b$ and $c < 0$ then $a \div c > b \div c$.

Power property: If $a > b > 0$ then $a^m > b^m$ for $m > 0$.

Root property: If $a > b > 0$ then $\sqrt[m]{a} > \sqrt[m]{b}$ for $m > 0$.

Reciprocal property: If $ab > 0$ and $a < b$ then $\frac{1}{a} > \frac{1}{b}$.

Linear inequalities in one variable can be solved using the properties of inequalities and the laws of algebra. The solution set can be expressed with inequality symbols, interval notation, or as a graph on a number line.

To solve inequalities (of degree greater than one) in one variable we need to determine the **critical values.** Knowing the critical values we use the graphical method to determine which real numbers make the statement of inequality true. The solution set can be indicated on a graph or with inequality notation or interval notation.

In order to work with the absolute value one must understand the definitions:

Definition 1.5

$$|x = \begin{cases} x, \text{ provided } x > 0. \\ 0, \text{ provided } x = 0. \\ -x, \text{ provided } x < 0. \end{cases}$$

Definition 14.5 For any real numbers a and b:

$$|ab| = |a| \cdot |b|.$$

and

$$\left|\frac{a}{b}\right| = \frac{|a|}{|b|}, \text{ if } b \neq 0.$$

Definition 14.6 If $|a| = |b|$, then $a = \pm b$.

Definition 14.7 For any positive real number a, $|x| < a$ is equivalent to $-a < x < a$.

Definition 14.8 For any real numbers a and b with $b > 0$, $|x - a| < b$ is equivalent to $-b < x - a < b$.

Definition 14.9 For any positive real number a, $|x| > a$ is equivalent to $x > a$ or $x < -a$.

Definition 14.10 For real numbers a and b, $b > 0$, $|x - a| > b$ is equivalent to $x - a < -b$ or $x - a > b$.

The solution set to a system of inequalities is found by graphing the respective statements of equality. Values are tested and from these tests it is determined which region(s) of the graph is the solution set of the system of inequalities.

Linear programming is an application of linear equations and inequalities.

Review Exercises

In Exercises 1–49, determine the solution set of the inequality. Indicate the solution (a) on a number line, and (b) in interval notation.

1. $x + 1 < 4$

2. $x - 1 < 3$

3. $x - 1 > 1$

4. $-2x < -6$

5. $-3x > 12$

6. $x + 1 \le 3x - 1$

7. $2Z - 1 \ge 3Z - 1$

8. $2 - Z \ge 2Z + 5$

9. $1 - 3Z \le Z + 5$

10. $3(y + 1) < 5y + 3$

11. $2(3y + 2) > 3(2y + 1)$

12. $3(4y + 1) < 12y$

13. $\frac{R}{3} - R \le 1$

14. $\frac{2R - 1}{-2} < R$

15. $\frac{R + 3}{-4} \ge \frac{R + 2}{-3}$

16. $-1 < R + 3 < 2$

17. $0 \le R + 1 \le 5$

18. $-1 \le 2R - 1 < 3$

19. $-5 < 2x + 1 \le 3$

20. $|x| \le 2$

21. $|x| < 4$

22. $|x| \ge 3$

23. $|x| > 5$

24. $|x + 1| < 3$

25. $|x - 1| \ge 2$

26. $|x - 1| \le 1$

27. $|2x + 3| \le 3$

28. $|3Z - 1| > 11$

29. $Z^2 + 3Z < 0$

30. $Z^2 - 4 > 0$

31. $4t - t^2 < 0$

32. $t(1 - t) > t - 1$

33. $t^2 + 15 \le 8t$

34. $2t^2 - t \ge 16$

35. $t^2 - t \ge 6$

36. $2t(t + 1) < 3t$

37. $(2y)^2 < 1 - 5y^2$

38. $2y^2 + 5y > 3$

39. $7y > 6y^2 + 2$

40. $(x - 2)(x + 5)(x - 11) < 0$

41. $(x - 5)(x + 3)(x - 7) > 0$

42. $(x - 5)^2(x + 3)^2 \le 0$

43. $(x - 8)^2(x - 3)^2 \ge 0$

44. $4t^3 - 7t^2 - 2t \ge 0$

45. $6t^3 + 11t^2 - 2t \le 0$

46. $\frac{x}{(x + 3)(x - 4)} < 0$

47. $\frac{(x - 1)(x + 3)}{(x - 4)(x + 5)} \le 0$

48. $\frac{2}{x^2 + 5x + 4} - \frac{3}{x^2 + x - 12} \ge 0$

49. $\sqrt{x^2 + 7} \le 4$

In Exercises 50–59, graph the solution set of the system of inequalities.

50. $x + y < 9$
$x - y < 11$

51. $2x + 5y \le 10$
$x + 2y \le 4$

52. $2x + 3y \ge 5$
$x - 3y \ge 7$

53. $3x + 2y > 11$
$x - 2y > 5$

54. $y \le 4x$
$x \ge 4y$

55. $x \ge 7$
$x + y < 11$

56. $|y + x| \le 14$

57. $|y - 3x| < 15$

58. $\left|y - \frac{2}{3}x\right| > 4$

59. $\left|y + \frac{1}{5}x\right| \le 1$

60. After t seconds the velocity of a ball that is thrown upward with an initial velocity of 80 ft/s will be v ft/s, with $v = 80 - 32t$. What is the time interval during which the velocity is between 16 and 64 ft/s?

61. If F degrees is the temperature using the Fahrenheit scale and C degrees is the temperature using the Celsius scale, then $C = \frac{5}{9}(F - 32)$. If the temperature is between 20°C and 30°C, what is the corresponding interval on the Fahrenheit scale?

62. A machine that packages 100 bolts per box can make an error of two bolts per box. If b is the number of bolts in a box, write an inequality that indicates a maximum error of two bolts per box. Solve the inequality.

63. Two resistors in parallel have total resistance R given by $\frac{1}{R} = \frac{1}{R_1} + \frac{1}{R_2}$. If $R_1 = 21\,\Omega$ (ohms) find the possible values for R_2 so that R is between $4\,\Omega$ and $11\,\Omega$.

64. The length of a rectangular field is 1 m more than three times its width. If its area is to be at least 10 m², what is the smallest amount of fencing needed to enclose the field? Set up an inequality and solve.

65. The minimum speed for a truck on a highway is 64 km/hr, the maximum speed is 88 km/hr. If Tod has been driving his mini-motorhome along the highway for five hours, what is the range in kilometers that he could legally have traveled? Set up an inequality and solve.

66. The deflection of a beam is given by

$$d = x^2 - 0.07x - 0.11,$$

where x is the distance from the end. For what values of x is $d > 0.03$?

67. A floppy disk for a computer must have a thickness of 2 mm with a tolerance of ±0.04 mm. Find the acceptable range of thickness of the disk. Set up a statement of inequality with absolute value and solve.

68. Find the values of x for which $\sqrt{3x^2 - 14x + 8}$ has real values.

In Exercises 69–72, a set of constraints and a profit formula are given. (a) Draw the graph of the constraints and find the points of intersection of the boundaries. (b) Use these points of intersection to determine the maximum and minimum profit.

69. $3y + x \leq 7$
$2y + 3x \leq 28$
$x \leq 8$
$x \geq 0$
$y \geq 0$
$P = 3x + 4y$

70. $3y + x \leq 7$
$2y + 5x \leq 40$
$x \geq 0$
$y \geq 0$
$P = 2x + 4y$

71. $8y - x \leq 48$
$2y + 3x \leq 34$
$x \leq 10$
$x \geq 0$
$y \geq 0$
$P = 10x + 20y$

72. $8y - x \leq 48$
$2y + 3x \leq 34$
$y - x \geq -8$
$x \geq 0$
$y \geq 0$
$P = 20x + 30y$

73. The Rob-Dale Company plates its wave guides with silver and copper. Each precision-grade guide receives 700 units of silver and 300 units of copper and sells for \$70. Each regular-grade guide receives 250 units of silver and 900 units of copper and sells for \$40. If the company has 30,000 units of silver and 46,000 units of copper on hand, how many of each type wave guide should it produce so as to maximize profit? What will the gross profit be for this many wave guides?

74. Dick makes two types of wooden lamps, A and B. The materials for lamp A cost \$15.00 and the materials for lamp B cost \$21.50. He can make a total of nine lamps in a month. He wants to make at least two type A and two type B but no more than four type B. He sells the type A lamp for \$135 and the type B lamp for \$165. How many of each type should he make each month to maximize his profit?

Can you solve these?

1. It can be proved using calculus that the curve defined by the equation $y = ax^3 + bx^2 + cx + d$ is concave up (opens upward) for those values of x for which $3ax + b > 0$ and is concave down when $3ax + b < 0$. Use this to find the values of x for which the graph of (a) $y = 2x^3 + 18x^2 - 13x + 4$

and (b) $y = 7 + 11x + 9x^2 - 2x^3$, is concave up and the values for which it is concave down.

2. Prove that $\frac{x^2 + y^2}{2} \geq \left(\frac{x + y}{2}\right)^2$ for all x and y.

3. The brake horsepower of a pump can be defined by the equation $BHP = \frac{QP}{1715NT}$, where Q is the flow in gallons per minute, P is the pressure at discharge in psi, and NT is the total system efficiency. Given that $Q = 10$ gpm and $P = 40$ psi, for what value of NT will the BHP be greater than 0.31?

4. Find the values of θ for which the following inequality holds. Limit the answer to values of θ such that $0 \leq \theta < 2\pi$.

   $\sqrt{3} \sin^2 \theta + 2\sqrt{3} > 5 \sin \theta$

5. Two positive real numbers differ by 2. If the sum of their reciprocals is greater than 1, show that the smaller of the two numbers must be less than $\sqrt{2}$.

6. **a.** Write the compound inequality $-5 < x < 14$ as a single inequality involving absolute value. (Look for a number to add to all three parts of the compound inequality that will make the two outside numbers opposite.)
   b. Write a general rule for writing a compound inequality as a single inequality involving absolute value.

7. Graph the solution set of $|x^2 - 9| \geq 7$ over the range $-6 < x < 6$.

8. Write a computer program to solve a linear-programming problem.

15

The sound energy going into a radiotelephone and the electrical energy coming out of its antenna can be represented by sinusoidal (sine or cosine) functions. When any two sinusoids of the same frequency are added, the sum is another sinusoid of the same frequency, but with a different amplitude and phase angle. If two waves, $A \cos \omega t$ and $B \sin (\omega t + R)$ are added, the sum is the wave $C \sin (\omega t + S)$, where

$$A + B \sin R = C \sin S$$

and

$$B \cos R = \cos S$$

If $A = 2$, $B = 4$, and $C = 3$, find R and S.

Additional Topics in Trigonometry

15.1 Basic trigonometric identities

An identity is different from the majority of the equations encountered because an **identity** is an equation that is true for all values that are substituted for the independent variables. The trigonometric identities are used to solve systems of equations like the chapter opener on the previous page. They are also used to simplify equations containing several trigonometric terms. Historically they were used to calculate the trigonometric functions of the less common angles when complete tables were unavailable.

We will start to develop the trigonometric identities by recalling the definitions of the trigonometric functions from Chapter 6. The definitions were based on the ratios of the sides of the right triangle (Fig. 15.1). The six basic trigonometric functions are given in Table 15.1.

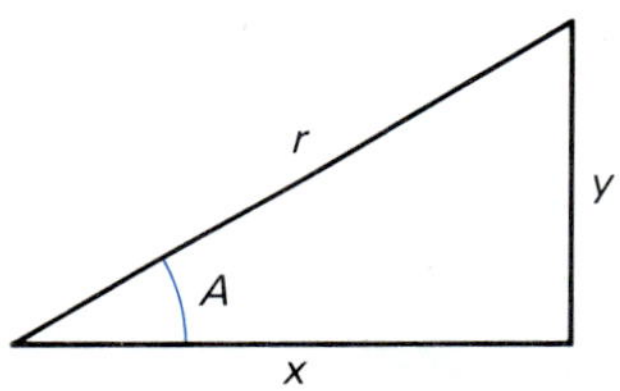

FIGURE 15.1

TABLE 15.1 **Trigonometric functions**

$\sin A = \dfrac{y}{r}$	$\cos A = \dfrac{x}{r}$	$\tan A = \dfrac{y}{x}$
$\csc A = \dfrac{r}{y}$	$\sec A = \dfrac{r}{x}$	$\cot A = \dfrac{x}{y}$

EXAMPLE 1 Show that $\dfrac{1}{\sin A} = \csc A$.

Solution: From the definition of sin A

$$\sin A = \frac{y}{r}$$

$$\frac{1}{\sin A} = \frac{1}{y/r} \qquad \textbf{Taking the reciprocal of both sides}$$

$$= \frac{r}{y} \qquad \textbf{Division of fractions}$$

$$= \csc A. \qquad \textbf{Definition of cosecant}$$

Therefore

$$\frac{1}{\sin A} = \csc A \qquad \{1\}^*$$

---

* Number inside the braces refers to the identity on that line.

is an identity. It is an identity because the statement was proved for all A and not for a specific case. □

We can see from the preceding example that if $\frac{1}{\sin A} = \csc A$ then also

$$\frac{1}{\csc A} = \sin A \qquad \{2\}$$

It will be left to the reader as part of the exercises to show that

$$\frac{1}{\cos A} = \sec A \qquad \{3\}$$

and

$$\frac{1}{\tan A} = \cot A \qquad \{4\}$$

EXAMPLE 2 Show that $\frac{\sin A}{\cos A} = \tan A$, $A \neq \frac{\pi}{2} + n\pi$, n is an integer.

Solution: From the definitions of sin A and cos A

$$\begin{aligned} \frac{\sin A}{\cos A} &= \frac{y/r}{x/r} \\ &= \left(\frac{y}{r}\right)\left(\frac{r}{x}\right) && \textbf{Division of fractions} \\ &= \frac{y}{x} \\ &= \tan A. && \textbf{Definition of tangent} \end{aligned}$$

Therefore we have the identity

$$\frac{\sin A}{\cos A} = \tan A,\ A \neq \frac{\pi}{2} + n\pi \qquad \{5\}$$

□

Another frequently used identity, which will be left as an exercise is

$$\frac{\cos A}{\sin A} = \cot A,\ A \neq n\pi \qquad \{6\}$$

EXAMPLE 3 Show that

a. $\sin A \sec A = \tan A$. Test the identity for

b. $A = 30°$, **c.** $A = 581°$.

Solution:

a. From the definitions of sin A and sec A

$$\begin{aligned}\sin A \sec A &= \left(\frac{y}{r}\right)\left(\frac{r}{x}\right)\\ &= \frac{y}{x}\\ &= \tan A. \qquad \textbf{Definition of tangent}\end{aligned}$$

Alternatively we could have used other identities.

$$\begin{aligned}\sin A \sec A &= \sin A \frac{1}{\cos A} \qquad \textbf{Identity } \sec A = \frac{1}{\cos A}\\ &= \frac{\sin A}{\cos A}\\ &= \tan A. \qquad \textbf{Identity } \frac{\sin A}{\cos A} = \tan A\end{aligned}$$

Therefore

$$\sin A \sec A = \tan A \qquad \{7\}$$

The latter approach is generally preferred since it uses the previously developed trigonometric identities.

b. Testing $\sin A \sec A = \tan A$ for $A = 30°$,

$$\begin{aligned}\sin 30° \sec 30° &= \tan 30°\\ \left(\frac{1}{2}\right)\left(\frac{2}{\sqrt{3}}\right) &= \frac{1}{\sqrt{3}}\\ \frac{1}{\sqrt{3}} &= \frac{1}{\sqrt{3}}. \checkmark\end{aligned}$$

(See Fig. 15.2.)

FIGURE 15.2

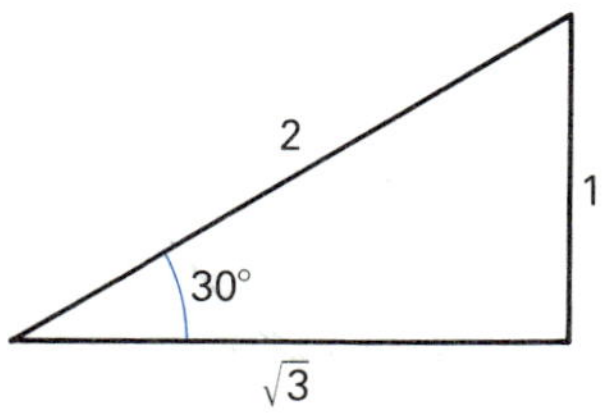

c. Testing $\sin A \sec A = \tan A$ for $A = 581°$

$$\sin 581° \sec 581° = \tan 581°$$
$$(-0.6561)(-1.325) = 0.8693$$
$$0.869 = 0.869.^* \checkmark$$

□

EXAMPLE 4 Use identities to show that $\sec B = \dfrac{\csc B}{\cot B}$.

Solution: Usually in dealing with trigonometric identities it is easier to work from the more complicated expression toward the simpler expression. Therefore instead of showing that $\sec B = \dfrac{\csc B}{\cot B}$, let us show that $\dfrac{\csc B}{\cot B} = \sec B$.

$$\frac{\csc B}{\cot B} = \frac{\frac{1}{\sin B}}{\frac{\cos B}{\sin B}} \qquad \textbf{Identities } \csc B = \frac{1}{\sin B} \textbf{ and } \cot B = \frac{\cos B}{\sin B}$$

$$= \left(\frac{1}{\sin B}\right)\left(\frac{\sin B}{\cos B}\right) \qquad \textbf{Dividing}$$

$$= \frac{1}{\cos B}$$

$$= \sec B. \qquad \textbf{Identity } \sec B = \frac{1}{\cos B}$$

Therefore

$$\frac{\csc B}{\cot B} = \sec B \qquad \{8\}$$

□

FIGURE 15.3

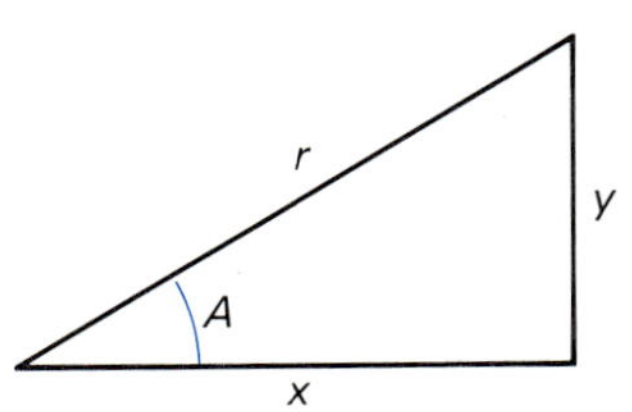

Referring to the right triangle in Fig. 15.3 we have, from the Pythagorean theorem, $x^2 + y^2 = r^2$. From the definitions of the trigonometric functions

$$\sin A = \frac{y}{r} \qquad \cos A = \frac{x}{r}$$

or or

$$y = r \sin A \qquad x = r \cos A$$

[*] The results are rounded in accordance with the rules established for significant digits with trigonometric functions.

Substituting for x and y in the Pythagorean theorem

$$(r \cos A)^2 + (r \sin A)^2 = r^2$$

$$r^2 \cos^2 A + r^2 \sin^2 A = r^2 \qquad \textbf{Remember that } \cos^2 A \neq \cos A^2$$

$$r^2 (\cos^2 A + \sin^2 A) = r^2 \qquad \textbf{Common-term factoring}$$

$$\cos^2 A + \sin^2 A = 1. \qquad \textbf{Dividing by } r^2$$

This is an important trigonometric identity because it is used in polar coordinates, parametric equations, and calculus. It is also written in the form

$$\sin^2 A + \cos^2 A = 1 \qquad \{9\}$$

EXAMPLE 5 Show that $1 + \tan^2 A = \sec^2 A$.

Solution: Since $\sin^2 A + \cos^2 A = 1$ is an identity, it is valid for any angle, so

$$\sin^2 A + \cos^2 A = 1$$

$$\frac{\sin^2 A + \cos^2 A}{\cos^2 A} = \frac{1}{\cos^2 A} \qquad \textbf{Dividing by } \cos^2 A,\ A \neq \frac{\pi}{2} + n\pi$$

$$\frac{\sin^2 A}{\cos^2 A} + \frac{\cos^2 A}{\cos^2 A} = \frac{1}{\cos^2 A}$$

$$\left(\frac{\sin A}{\cos A}\right)^2 + 1 = \left(\frac{1}{\cos A}\right)^2$$

$$(\tan A)^2 + 1 = (\sec A)^2 \qquad \textbf{Recalling the identities } \frac{\sin A}{\cos A} = \tan A \textbf{ and } \frac{1}{\cos A} = \sec A$$

or

$$1 + \tan^2 A = \sec^2 A \qquad \{10\}$$

□

In a similar way it can be shown that

$$1 + \cot^2 A = \csc^2 A \qquad \{11\}$$

TABLE 15.2 **Trigonometric identities**

Fundamental Identities

{1} $\dfrac{1}{\sin A} = \csc A$, {2} $\dfrac{1}{\csc A} = \sin A$, {3} $\dfrac{1}{\cos A} = \sec A$, {4} $\dfrac{1}{\tan A} = \cot A$

Quotient and Product Identities

{5} $\dfrac{\sin A}{\cos A} = \tan A$, {6} $\dfrac{\cos A}{\sin A} = \cot A$, {7} $\sin A \sec A = \tan A$, {8} $\dfrac{\csc A}{\cot A} = \sin A$

Pythagorean Identities

{9} $\sin^2 A + \cos^2 A = 1$, {10} $1 + \tan^2 A = \sec^2 A$, {11} $1 + \cot^2 A = \csc^2 A$

Table 15.2 contains a list of the trigonometric identities presented in this section. The reader should become thoroughly familiar with these identities since they are the basis for much of the work in this chapter and further mathematics.

15.1 Exercises

In Exercises 1 and 2, use the definitions of the trigonometric functions given in Table 15.1 to prove the identity.

1. $\dfrac{1}{\cos A} = \sec A$ **2.** $\dfrac{1}{\tan A} = \cot A$

In Exercises 3–12, prove the identity using the identities in Table 15.2.

3. $\sin A = \tan A \cos A$ **4.** $\cos A = \cot A \sin A$

5. $\csc B \tan B = \sec B$ **6.** $\sec B \cot B = \csc B$

7. $\sin A = \dfrac{\tan A}{\sec A}$ **8.** $\cos A = \dfrac{\cot A}{\csc A}$

9. $\dfrac{\sec A}{\tan A} = \csc A$ **10.** $\dfrac{\cos B}{\sin B} = \cot B$

11. $\cos A \sec A + \cot A \tan A = 2$

12. $\sin A \csc A + \cos A \sec A = 2$

13. Prove that $1 + \tan^2 A = \sec^2 A$ using $r^2 = x^2 + y^2$. (*Hint:* examine the development of $\sin^2 A + \cos^2 A = 1$.)

14. Prove that $1 + \cot^2 A = \csc^2 A$ using $\sin^2 A + \cos^2 A = 1$.

15. Prove that $1 + \cot^2 A = \csc^2 A$ using $r^2 = x^2 + y^2$.

In Exercises 16–31, rewrite each of the expressions in terms of a single trigonometric function. (The answer may be multiplied by a constant and/or raised to a power.)

16. $\sin A + \cos A \cot A$

17. $(\sec A + \tan A)(1 - \sin A)$

18. $\dfrac{\csc^2 A}{1 + \tan^2 A}$ **19.** $\cot A \sec^2 A - \cot A$

20. $\cos A(\tan A + \cot A)$ **21.** $\dfrac{\sec^2 A - 1}{\sec^2 A}$

22. $\dfrac{\cot \theta + 1}{\cot \theta} - 1$ **23.** $\sec \theta\ (1 - \sin^2 \theta)$

24. $\dfrac{1 + \sec \theta}{\sin \theta + \tan \theta}$ **25.** $\sin \theta \cos^2 \theta - \sin \theta$

26. $\dfrac{1 - \cot \theta}{\tan \theta - 1}$ **27.** $\dfrac{\sin \theta}{1 - \cos \theta} - \csc \theta$

28. $\dfrac{1 + \tan A}{\sin A} - \csc A$ **29.** $\csc A\ (\csc A - \sin A)$

30. $\dfrac{\tan B + \cot B}{\sec B}$ **31.** $\dfrac{\tan B + \cot B}{\tan B}$

In Exercises 32–37, show that each expression is equal to a constant.

32. $\csc^2 \theta\ (1 - \cos^2 \theta)$

33. $2 \cos^2 \theta - 1 + 2 \sin^2 \theta$

34. $\cot^2 \theta \sin^2 \theta + \tan^2 \theta \cos^2 \theta$

35. $\cos^2 \theta + \sin \theta \cos \theta \tan \theta$

36. $(1 - \sin^2 \theta)(1 + \tan^2 \theta)$

37. $(1 + \tan^2 \theta) \cos^2 \theta$

In Exercises 38–48, prove the identity.

38. $\tan^2 \omega t \sin^2 \omega t = \tan^2 \omega t - \sin^2 \omega t$

39. $\tan^2 \omega t + \tan^4 \omega t = \tan^2 \omega t \sec^2 \omega t$

40. $\dfrac{\sin A}{1 - \cos A} = \csc A + \cot A$

41. $\dfrac{\cos^3 A - \sin^3 A}{1 + \sin A \cos A} = \cos A - \sin A$

42. $\dfrac{\sin A + 1}{\tan A + \cot A + \sec A} = \sin A \cos A$

43. $\dfrac{\sin A - \cos A}{\cos^2 A} = \dfrac{\tan^2 A - 1}{\sin A + \cos A}$

44. $\dfrac{\tan^2 \theta + \sec^2 \theta}{\sec^4 \theta} = 1 - \sin^4 \theta$

45. $\dfrac{\cot(-t)\cos(-t)}{\cot(-t) + \cos(-t)} = \dfrac{\cot(-t) - \cos(-t)}{\cot(-t)\cos(-t)}$

46. $\csc^2 2\theta + \sec^2 2\theta = \csc^2 2\theta \sec^2 2\theta$

47. $\sin^4 \pi t + \cos^2 \pi t = \sin^2 \pi t + \cos^4 \pi t$

48. $\dfrac{\sec \theta + \csc \theta}{\sin \theta + \cos \theta} = \tan \theta + \cot \theta$

49. In calculus the derivative of $\tan x$ with respect to x is found to be $\dfrac{\cos x \cos x - \sin x(-\sin x)}{\cos^2 x}$. Show that this expression is equivalent to $\sec^2 x$.

50. In calculus the derivative of $\cot x$ with respect to x is found to be $\dfrac{\sin x(-\sin x) - \cos x \cos x}{\sin^2 x}$. Show that this expression is equivalent to $-\csc^2 x$.

51. Show that the system of equations

$$x = a \cos A$$
$$y = b \sin A$$

is equivalent to the equation of an ellipse

$$\frac{x^2}{a^2} + \frac{y^2}{b^2} = 1.$$

15.2 Sum and difference of two angles

In a problem involving springs or light refraction we may encounter an expression such as $\sin(A + B)$ or $\cos(A + B)$. In this section we will develop formulas to deal with expressions of this type.

To develop an expression for the cosine of the sum of two angles A and B, $\cos(A + B)$, we consider Fig. 15.4a, which shows the initial side

FIGURE 15.4

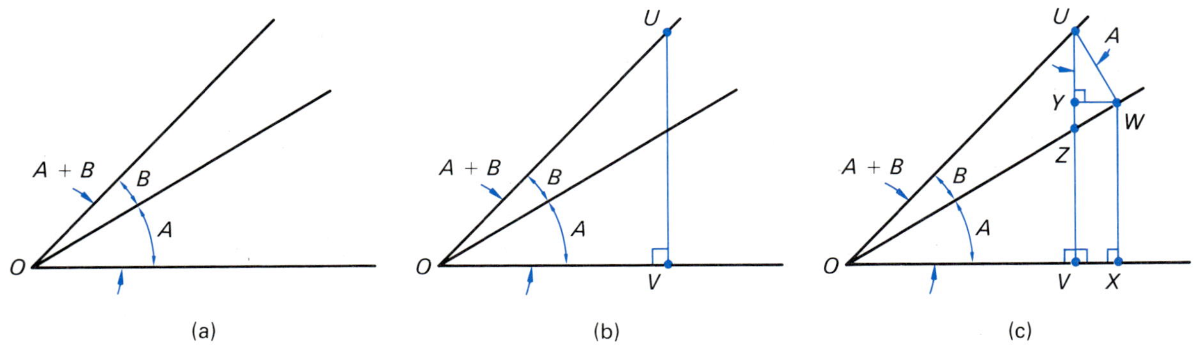

of the angle B coinciding with the terminal side of angle A to form the angle $A+B$. If we choose any point U on the terminal side of angle $A+B$, we can draw line $\overline{UV}$ so that it is perpendicular to the initial side of angle $A+B$ (Fig. 15.4b). Then $\cos(A+B)=\overline{OV}/\overline{OU}$ by the definition of cosine. In order to complete the derivation we need several more lines. We draw line $\overline{UW}$ so that it is perpendicular to the initial side of angle B, which is also the terminal side of angle A (Fig. 15.4c). Then we draw line $\overline{WX}$ perpendicular to the extension of line $\overline{OV}$ and we draw line $\overline{WY}$ perpendicular to line $\overline{UV}$. With these lines in place we can proceed.

$\cos(A+B)=\dfrac{\overline{OV}}{\overline{OU}}$	**Definition of cosine**	[1]
$=\dfrac{\overline{OX}-\overline{VX}}{\overline{OU}}$	**See Fig. 15.4c**	[2]
$\overline{OX}=\overline{OW}\cos A$	**From right triangle *OXW***	[3]
$\overline{OW}=\overline{OU}\cos B$	**From right triangle *OWU***	[4]
$\overline{OX}=\overline{OU}\cos A\cos B$	**Substituting Eq. [4] for $\overline{OW}$ in Eq. [3]**	[5]
$\overline{VX}=\overline{YW}.$	**Opposite sides of rectangle *VXWY***	[6]

From geometry we can see that $\angle VUW$ is the same as angle A. ($\angle OZV=\angle UZW$ because vertical angles are equal; $\angle OVZ=\angle UWZ$ because right angles are equal; therefore $\angle VUW=\angle A$, because two angles of a triangle equal to two angles of another triangle means the third angles are equal.)

$\overline{YW}=\overline{UW}\sin A$	**From right triangle *UYW***	[7]
$\overline{UW}=\overline{OU}\sin B$	**From right triangle *OWU***	[8]
$\overline{YW}=\overline{OU}\sin A\sin B$	**Substituting Eq. [8] for $\overline{UW}$ in Eq. [7]**	[9]
$\overline{VX}=\overline{OU}\sin A\sin B$	**Substituting Eq. [9] for $\overline{YW}$ in Eq. [6]**	[10]

$$\cos(A+B)=\frac{\overline{OU}\cos A\cos B-\overline{OU}\sin A\sin B}{\overline{OU}}$$

Substituting Eq. [5] for $\overline{OX}$ and Eq. [10] for $\overline{VX}$ in Eq. [2]

$$=\frac{\overline{OU}(\cos A\cos B-\sin A\sin B)}{\overline{OU}}.$$

Common-term factoring

$$\cos(A+B)=\cos A\cos B-\sin A\sin B \qquad \{12\}$$

By using the same figures and similar procedures we can also show that

$$\sin (A + B) = \sin A \cos B + \cos A \sin B \qquad \{13\}$$

EXAMPLE 1 Use the identities for the cosine of the sum of two angles and the sine of the sum of two angles to obtain exact values for (a) cos 75° and (b) sin 75°. (By exact values we mean that all fractions and radicals are reduced to their lowest possible terms without producing a decimal.)

Solution: To find exact values for cos 75° and sin 75° we need to express 75° in terms of 0°, 30°, 45°, 60°, or 90° since we know exact values for the trigonometric functions of these angles and their multiples. Therefore we will express 75° as 45° + 30°.

a.
$$\begin{aligned}
\cos (A + B) &= \cos A \cos B - \sin A \sin B \\
\cos 75^\circ &= \cos (45^\circ + 30^\circ) \\
&= \cos 45^\circ \cos 30^\circ - \sin 45^\circ \sin 30^\circ \\
&= \left(\frac{1}{\sqrt{2}}\right)\left(\frac{\sqrt{3}}{2}\right) - \left(\frac{1}{\sqrt{2}}\right)\left(\frac{1}{2}\right) \\
&= \frac{\sqrt{3}}{2\sqrt{2}} - \frac{1}{2\sqrt{2}} \\
&= \left(\frac{\sqrt{3} - 1}{2\sqrt{2}}\right)\left(\frac{\sqrt{2}}{\sqrt{2}}\right) \qquad \textbf{Combining terms and rationalizing denominator} \\
\cos 75^\circ &= \frac{\sqrt{2}(\sqrt{3} - 1)}{4} \quad \text{or} \quad \frac{\sqrt{6} - \sqrt{2}}{4}.
\end{aligned}$$

b.
$$\begin{aligned}
\sin (A + B) &= \sin A \cos B + \cos A \sin B \\
\sin 75^\circ &= \sin (45^\circ + 30^\circ) \\
&= \sin 45^\circ \cos 30^\circ + \cos 45^\circ \sin 30^\circ \qquad \textbf{Substituting} \\
&= \left(\frac{1}{\sqrt{2}}\right)\left(\frac{\sqrt{3}}{2}\right) + \left(\frac{1}{\sqrt{2}}\right)\left(\frac{1}{2}\right) \\
&= \left(\frac{\sqrt{3} + 1}{2\sqrt{2}}\right)\left(\frac{\sqrt{2}}{\sqrt{2}}\right) \\
\sin 75^\circ &= \frac{\sqrt{2}(\sqrt{3} + 1)}{4} \quad \text{or} \quad \frac{\sqrt{6} + \sqrt{2}}{4}.
\end{aligned}$$ □

It should be apparent from Example 1 that $\cos (A + B)$ is *not* equal to $\cos A + \cos B$ and $\sin (A + B)$ is *not* equal to $\sin A + \sin B$. The reader may use his calculator to verify these statements.

EXAMPLE 2 Find an expression for the cosine of the difference of two angles, $\cos(A - B)$. (Use identities and properties that we have discussed.)

Solution: We can use the expression for $\cos(A + B)$ to find $\cos(A - B)$ by letting

$\cos(A - B) = \cos(A + (-B))$ **Definition of subtraction**

$\cos(A + B) = \cos A \cos B - \sin A \sin B$ **Identity {12}**

$\cos(A + (-B)) = \cos A \cos(-B) - \sin A \sin(-B)$ **Substituting**

$\cos(-B) = \cos B$ and $\sin(-B) = -\sin B$. **From Chapter 6**

Therefore

$\cos(A - B) = \cos A \cos B - \sin A(-\sin B)$ **Substituting**

and simplifying we get

$$\cos(A - B) = \cos A \cos B + \sin A \sin B \qquad \{14\}$$

□

In similar fashion it can be shown that

$$\sin(A - B) = \sin A \cos B - \cos A \sin B \qquad \{15\}$$

EXAMPLE 3 Use the identity for the cosine of the difference of two angles to find an exact value for the cosine of $15°$.

Solution: There are several possible ways to find $\cos 15°$ using the difference of two angles with known exact values for their trigonometric functions; we will use $15° = 45° - 30°$.

$$\cos(A - B) = \cos A \cos B + \sin A \sin B$$

Let $A = 45°$ and $B = 30°$.

$\cos(45° - 30°) = \cos 45° \cos 30° + \sin 45° \sin 30°$ **Substituting**

$\cos 15° = \left(\frac{1}{\sqrt{2}}\right)\left(\frac{\sqrt{3}}{2}\right) + \left(\frac{1}{\sqrt{2}}\right)\left(\frac{1}{2}\right)$ **Substituting**

$$= \frac{\sqrt{3} + 1}{2\sqrt{2}}$$

$$= \frac{\sqrt{2}(\sqrt{3} + 1)}{4} \quad \text{or} \quad \frac{\sqrt{6} + \sqrt{2}}{4}.$$

□

EXAMPLE 4 A mathematical representation for 120-V, 60-Hz house voltage is $v_1 = 170 \sin 377t$. A representation for 240-V, 60-Hz voltage, which is out of phase with v_1 is $v_2 = 340 \cos (377t + 60°)$. Show that the sum of v_1 and v_2 can be written as the sum of two sinusoids with no phase shift. (Even though $377t + 60°$ is the sum of an angle in radians and an angle in degrees, it is a form often used in practice. As will be seen, the part of the expression in radians does not enter directly into the solution and the phase shift is usually easier to handle when it is expressed in degrees.)

Solution: To find the sum of the two functions v_1 and v_2 we use the identity $\cos (A + B) = \cos A \cos B - \sin A \sin B$.

$$v_1 + v_2 = 170 \sin 377t + 340 \cos (377t + 60°)$$

$$= 170 \sin 377t + 340 (\cos 377t \cos 60° - \sin 377t \sin 60°)$$

Substituting

$$= 170 \sin 377t + 340 \cos 377t \cos 60° - 340 \sin 377t \sin 60°$$

Distributive property

$$= 170 \sin 377t + 340 \left(\frac{1}{2}\right) \cos 377t - 340 \left(\frac{\sqrt{3}}{2}\right) \sin 377t$$

Substituting values for cos 60° and sin 60°

$$= 170 \sin 377t + \overset{170}{\cancel{340}} \left(\frac{1}{\underset{1}{\cancel{2}}}\right) \cos 377t - \overset{170}{\cancel{340}} \left(\frac{\sqrt{3}}{\underset{1}{\cancel{2}}}\right) \sin 377t$$

$$= 170 \sin 377t + 170 \cos 377t - 170 \sqrt{3} \sin 377t$$

$$= (170 - 170 \sqrt{3}) \sin 377t + 170 \cos 377t$$

Common-term factoring

$$= -124 \sin 377t + 170 \cos 377t. \qquad \square$$

Example 4 demonstrates how two out-of-phase sinusoids (with a phase shift) can be written as the sum of two in phase sinusoids (with no phase shift).

Table 15.3 lists trigonometric identities presented in this section.

TABLE 15.3 **Trigonometric identities**

Sum and difference identities	
{12}	$\cos (A + B) = \cos A \cos B - \sin A \sin B$
{13}	$\sin (A + B) = \sin A \cos B + \cos A \sin B$
{14}	$\cos (A - B) = \cos A \cos B + \sin A \sin B$
{15}	$\sin (A - B) = \sin A \cos B - \cos A \sin B$

15.2 Exercises

In Exercises 1–4, use specific values to show that the statements are not equal.

1. Show that $\cos(\theta + \phi) \neq \cos\theta + \cos\phi$.
2. Show that $\sin(\theta + \phi) \neq \sin\theta + \sin\phi$.
3. Show that $\sin(\theta - \phi) \neq \sin\theta - \sin\phi$.
4. Show that $\cos(\theta - \phi) \neq \cos\theta - \cos\phi$.

In Exercises 5 and 6, use

$$\tan A = \frac{\sin A}{\cos A} \quad \text{and} \quad \cot A = \frac{\cos A}{\sin A}$$

and the identities for $\cos(A + B)$ and $\sin(A + B)$ to show:

5. $\tan(A + B) = \dfrac{\tan A + \tan B}{1 - \tan A \tan B}$
6. $\cot(A + B) = \dfrac{\cot A \cot B - 1}{\cot A + \cot B}$

In Exercises 7 and 8, use the properties of the trigonometric functions to find expressions for:

7. $\tan(A - B)$
8. $\cot(A - B)$

In Exercises 9–22, use the trigonometric identities to find exact values for:

9. $\cos 105°$
10. $\cos 165°$
11. $\cos 255°$
12. $\cos 195°$
13. $\sin 105°$
14. $\sin 165°$
15. $\sin 255°$
16. $\sin 195°$
17. $\cos(-15°)$
18. $\sin(-15°)$
19. $\tan 105°$
20. $\tan 165°$
21. $\cot 255°$
22. $\cot 195°$

In Exercises 23–28, simplify the expression.

23. $\cos\left(\theta + \dfrac{\pi}{2}\right)$
24. $\sin\left(\theta + \dfrac{\pi}{2}\right)$
25. $\sin\left(\theta - \dfrac{\pi}{2}\right)$
26. $\cos\left(\theta - \dfrac{\pi}{2}\right)$
27. $\cos\left(\dfrac{\pi}{2} - \theta\right)$
28. $\sin\left(\dfrac{\pi}{2} - \theta\right)$
29. Use the identity for $\cos(A + B)$ to find an identity for $\cos(\theta + \theta)$.
30. Use the identity for $\sin(A + B)$ to find an identity for $\sin(\theta + \theta)$.

In Exercises 31–40, write the sum of the two functions as the sum of two sinusoids with no phase shift.

31. $10 \sin 20t + 10 \cos(20t + 30°)$
32. $9 \sin 20t + 12 \cos(20t + 45°)$
33. $12 \cos(t + 120°) + 9 \cos t$
34. $24 \cos(6t + 30°) + 15 \sin 6t$
35. $\sin(377t - 45°) + 2 \cos(377t - 20°)$
36. $\cos(377t + 30°) + \cos(377t - 60°)$
37. $170 \cos(4t - 22°) + 170 \sin(4t + 38°)$
38. $340 \cos(377t - 120°) + 340 \sin(377t - 60°)$
39. $340 \sin(377t - 150°) + 340 \cos(377t - 75°)$
40. $170 \sin(7000t - 210°) + 340 \cos(7000t - 150°)$

In Exercises 41 and 42, use a geometric approach similar to that used to derive the identity for $\cos(A + B)$ to derive identities for:

41. $\sin(A + B)$
42. $\sin(A - B)$

15.3 Double-angle and half-angle formulas

The formulas for the sum of two angles can be used to derive formulas for angles that are the integer multiples of an angle.

EXAMPLE 1 Show that $\cos 2A = \cos^2 A - \sin^2 A$.

Solution: We can obtain an expression for $\cos 2A$ by using the formula for the cosine of the sum of two angles.

$$\cos (A + B) = \cos A \cos B - \sin A \sin B.$$

Let $B = A$

$$\cos (A + A) = \cos A \cos A - \sin A \sin A.$$

$$\cos 2A = \cos^2 A - \sin^2 A \qquad \{16\}$$

□

Two additional identities for $\cos 2A$ can be developed by recalling the identity $\sin^2 A + \cos^2 A = 1$.

$\sin^2 A = 1 - \cos^2 A$	**Solving identity for $\sin^2 A$**
$\cos 2A = \cos^2 A - \sin^2 A$	**Recalling identity from Example 1**
$= \cos^2 A - (1 - \cos^2 A)$	**Substituting for $\sin^2 A$**
$= \cos^2 A - 1 + \cos^2 A$	

$$\cos 2A = 2 \cos^2 A - 1 \qquad \{17\}$$

Alternatively,

$\cos^2 A = 1 - \sin^2 A$	**Solving identity for $\cos^2 A$**
$\cos 2A = \cos^2 A - \sin^2 A$	**Recalling identity from Example 1**
$= 1 - \sin^2 A - \sin^2 A$	**Substituting for $\cos^2 A$**

$$\cos 2A = 1 - 2 \sin^2 A \qquad \{18\}$$

In a fashion similar to that for finding $\cos 2A = \cos^2 A - \sin^2 A$ it can be shown that

$$\sin 2A = 2 \sin A \cos A \qquad \{19\}$$

EXAMPLE 2 Find an expression for $\cos \frac{B}{2}$ in terms of the angle B.

Solution: Recalling

$$\cos 2A = 2\cos^2 A - 1$$

Let

$$A = \frac{B}{2}$$

$$\cos 2\left(\frac{B}{2}\right) = 2\cos^2 \frac{B}{2} - 1$$

$$\cos B = 2\cos^2 \frac{B}{2} - 1$$

$$2\cos^2 \frac{B}{2} = 1 + \cos B \qquad \textbf{Rearranging}$$

$$\cos^2 \frac{B}{2} = \frac{1 + \cos B}{2}$$

Taking the square root (remember to take both roots), we get

$$\cos \frac{B}{2} = \pm\sqrt{\frac{1 + \cos B}{2}} \qquad \{20\}$$

□

It can also be shown that

$$\sin \frac{B}{2} = \pm\sqrt{\frac{1 - \cos B}{2}} \qquad \{21\}$$

EXAMPLE 3 Use the identity $\cos 2A = 1 - 2\sin^2 A$ to find an expression for $\sin A$ in terms of a function of the double angle $2A$.

Solution:

$$\cos 2A = 1 - 2\sin^2 A$$

$$2\sin^2 A = 1 - \cos 2A$$

$$\sin^2 A = \frac{1 - \cos 2A}{2}.$$

Taking the square root of both sides

$$\sin A = \pm\sqrt{\frac{1-\cos 2A}{2}} \qquad \{22\}$$

□

In a similar fashion it can be shown that

$$\cos A = \pm\sqrt{\frac{1+\cos 2A}{2}} \qquad \{23\}$$

EXAMPLE 4 When finding the second moment of an area ("moment of inertia" of the area) an expression that must be simplified is

$$I_A = I_b \sin^2\theta - 2I_c \sin\theta\cos\theta + I_a\cos^2\theta$$

Write the expression for I_A as a function of the double angle 2θ.

Solution:

$$I_A = I_b \sin^2\theta - 2I_c \sin\theta\cos\theta + I_a\cos^2\theta \qquad [1]$$

$$\sin\theta = \pm\sqrt{\frac{1-\cos 2\theta}{2}} \qquad \textbf{Recalling } \{22\}$$

$$\sin^2\theta = \frac{1-\cos 2\theta}{2} \qquad \textbf{Squaring} \qquad [2]$$

$$\cos\theta = \pm\sqrt{\frac{1+\cos 2\theta}{2}} \qquad \textbf{Recalling } \{23\}$$

$$\cos^2\theta = \frac{1+\cos 2\theta}{2} \qquad \textbf{Squaring} \qquad [3]$$

$$\sin 2\theta = 2\sin\theta\cos\theta \qquad \textbf{Recalling } \{19\}$$

$$\sin\theta\cos\theta = \frac{\sin 2\theta}{2}. \qquad \textbf{Rearranging} \qquad [4]$$

Using Eqs. [2], [3], and [4] to substitute for $\sin^2\theta$, $\cos^2\theta$, and $\sin\theta\cos\theta$ respectively in Eq. [1], we obtain

$$I_A = I_b\left(\frac{1-\cos 2\theta}{2}\right) - 2I_c\left(\frac{\sin 2\theta}{2}\right) + I_a\left(\frac{1+\cos 2\theta}{2}\right)$$

$$= \frac{I_b}{2} - \frac{I_b\cos 2\theta}{2} - I_c\sin 2\theta + \frac{I_a}{2} + \frac{I_a\cos 2\theta}{2}.$$

Grouping and common-term factoring gives

$$I_A = \frac{I_a + I_b}{2} + \frac{I_a - I_b}{2} \cos 2\theta - I_c \sin 2\theta,$$

which is the form of the expression subsequently used to find the maximum second moment. □

Table 15.4 lists trigonometric identities presented in this section.

TABLE 15.4 **Trigonometric identities**

Double and half angle identities	
{16} $\cos 2A = \cos^2 A - \sin^2 A$	{17} $\cos 2A = 2\cos^2 A - 1$
{18} $\cos 2A = 1 - 2\sin^2 A$	{19} $\sin 2A = 2 \sin A \cos A$
{20} $\cos \frac{B}{2} = \pm\sqrt{\frac{1+\cos B}{2}}$	{21} $\sin \frac{B}{2} = \pm\sqrt{\frac{1-\cos B}{2}}$
{22} $\sin A = \pm\sqrt{\frac{1-\cos 2A}{2}}$	{23} $\cos A = \pm\sqrt{\frac{1+\cos 2A}{2}}$

15.3 Exercises

In Exercises 1–8, use the double-angle or half-angle identities to find exact values for the trigonometric functions indicated.

1. $\cos 120°$

2. $\cos 15°$

3. $\sin 15°$

4. $\sin 120°$

5. $\cos 22\frac{1}{2}°$

6. $\sin 22\frac{1}{2}°$

7. $\sin(-120°)$

8. $\cos(-120°)$

9. Find $\tan 2A$ as a function of $\tan A$.

10. Find $\cot 2A$ as a function of $\cot A$.

11. Show that $\sin 2A = \frac{2\tan A}{1+\tan^2 A}$

12. Show that $\cos 2A = \frac{1-\tan^2 A}{1+\tan^2 A}$

13. Show that $\tan \frac{A}{2} = \frac{1-\cos A}{\sin A}$

14. Show that $\tan \frac{A}{2} = \frac{\sin A}{1+\cos A}$

15. Show that $(\sin A + \cos A)^2 = 1 + \sin 2A$.

16. Show that $\cos^4 A - \sin^4 A = \cos 2A$.

17. Normal stress at a point can be written as

$$N = s_1 \cos^2\theta - 2s_2 \sin\theta\cos\theta + s_3 \sin^2\theta.$$

Write the normal stress as a function of 2θ.

18. Shear stress at a point can be written as

$$S = -s_1 \sin\theta\cos\theta + s_2(\cos^2\theta - \sin^2\theta) + s_3 \sin\theta\cos\theta.$$

Write the shear stress as a function of 2θ.

15.4 Trigonometric equations

In Chapter 8 the reader was introduced to methods for solving trigonometric equations. In solving those equations it was possible to substitute a variable in place of the trigonometric function (for instance replace $\sin\theta$ by x), solve the equation in terms of the new variable, and then substitute the original variable to determine the values of the functions.

EXAMPLE 1 Solve $4\sin^2\theta - 3 = 0$ for θ, $0° \le \theta < 360°$.

Solution:

$$4\sin^2\theta - 3 = 0$$

Let $x = \sin\theta$. Substituting yields

$$4x^2 - 3 = 0$$

$$x = \pm\sqrt{3/4} \qquad \textbf{Solving for } x$$

$$= \pm 0.8660$$

$$\sin\theta = \pm 0.8660 \qquad \textbf{Substituting } \sin\theta \textbf{ for } x$$

$$\theta = 60°, 120°, 240° \text{ or } 300°$$

The reader can check to see that all four values of θ satisfy the original equation. □

Problems of this type may also be solved directly, without substitution. This is illustrated in Examples 2 and 3.

EXAMPLE 2 Solve $2\cos^2 x - \cos x = 0$ for x, $0 \le x < 2\pi$.

Solution:

$$2\cos^2 x - \cos x = 0$$

$$\cos x\,(2\cos x - 1) = 0. \qquad \textbf{Common-term factoring}$$

Setting the factors equal to zero we have

$$\cos x = 0 \qquad\qquad 2\cos x - 1 = 0$$

$$x = \frac{\pi}{2}, \frac{3\pi}{2} \qquad\qquad \cos x = \frac{1}{2}$$

$$x = \frac{\pi}{3}, \frac{5\pi}{3}$$

Therefore the solutions for $2\cos^2 x - \cos x = 0$ are

$$x = \pi/3.\ \pi/2,\ 3\pi/2, \text{ and } 5\pi/3. \qquad □$$

In some situations these methods will not work in solving a trigonometric equation. It may be possible to write the equation in a simpler form by using trigonometric identities and then solve the equation.

EXAMPLE 3 To find the maximum shear stress in an electric-motor shaft it is necessary to solve the equation

$$-30\cos 2\theta + 8\sin 2\theta = 0$$

Find θ, $0° \le \theta < 360°$.

Solution: There is no standard method for the solution of a problem of this type. Usually the only approach is to try several methods until you find one that works. The process can be speeded up only through practice and familiarity.

In this instance we might let $\cos 2\theta = x$ and $\sin 2\theta = y$ and write the equation as $-30x + 8y = 0$. We can see this is not the best procedure since we get an equation in two unknowns, x and y, where before we only had a single unknown, θ. We cannot solve a single equation in two unknowns by the methods we have studied so this is a blind alley.

Taking another look at the equation $-30 \cos 2\theta + 8 \sin 2\theta = 0$ we see that there are only terms present that are either multiplied by $\cos 2\theta$ or $\sin 2\theta$ and we recall that $\tan 2\theta = \sin 2\theta/\cos 2\theta$. Therefore we might try dividing the equation by $\cos 2\theta$ ($\cos 2\theta \neq 0$).

$$\frac{-30 \cos 2\theta + 8 \sin 2\theta}{\cos 2\theta} = \frac{0}{\cos 2\theta}$$

$$\frac{-30 \cos 2\theta}{\cos 2\theta} + \frac{8 \sin 2\theta}{\cos 2\theta} = 0$$

$$-30 + 8 \tan 2\theta = 0$$

$$\tan 2\theta = \frac{30}{8}$$

$$\tan 2\theta = 3.75$$

$$2\theta = 75.1^\circ \quad \text{or} \quad 255.1^\circ$$

$$\theta = 37.5^\circ \quad \text{or} \quad 127.5^\circ.$$ □

The reader can check that both values for θ satisfy the original equation.

EXAMPLE 4 Given $\sin 2t + \cos t = 0$, solve for t, $0 \leq t < 2\pi$.

Solution: We cannot use the approach we used in Example 3 to solve $\sin 2t + \cos t = 0$ because the arguments for the sine and cosine functions, $2t$ and t, respectively, are different. In this instance we can obtain an equation with the same arguments for all the trigonometric functions by using the formula for $\sin 2t$.

$\sin 2t + \cos t = 0$		**Given**
$\sin 2t = 2 \sin t \cos t$		**Identity**
$2 \sin t \cos t + \cos t = 0$		**Substituting for sin 2*t***
$\cos t\,(2 \sin t + 1) = 0$		**Common-term factoring**
$\cos t = 0$	$2 \sin t + 1 = 0$	**Setting factors equal to 0**
$t = \dfrac{\pi}{2}, \dfrac{3\pi}{2}$	$\sin t = -\dfrac{1}{2}$	
	$t = \dfrac{7\pi}{6}, \dfrac{11\pi}{6}$	

The reader can verify by substitution that $t = \pi/2$, $3\pi/2$, $7\pi/6$, and $11\pi/6$, are all solutions for $\sin 2t + \cos t = 0$. □

EXAMPLE 5 Solve $2 \sin^2 \theta + 3 \cos \theta + 5 = 0$ for $0 \le \theta < 2\pi$.

Solution: To solve the equation $2 \sin^2 \theta + 3 \cos \theta + 5 = 0$ it is necessary to write it in terms of $\sin \theta$ or $\cos \theta$. It is easiest to change $\sin^2 \theta$ to $\cos^2 \theta$ using the identity $\sin^2 \theta + \cos^2 \theta = 1$. The identity can be rewritten as $\sin^2 \theta = 1 - \cos^2 \theta$, and we can write the equation in terms of the same function, $\cos \theta$.

$$
\begin{aligned}
2 \sin^2 \theta + 3 \cos \theta + 5 &= 0 && \textbf{Given} \\
2(1 - \cos^2 \theta) + 3 \cos \theta + 5 &= 0 && \textbf{Substituting for } \boldsymbol{\sin^2 \theta} \\
2 - 2 \cos^2 \theta + 3 \cos \theta + 5 &= 0 \\
-2 \cos^2 \theta + 3 \cos \theta + 7 &= 0.
\end{aligned}
$$

This equation is quadratic in $\cos \theta$ so we can use the quadratic formula.

$$
\begin{aligned}
\cos \theta &= \frac{-3 \pm \sqrt{(3)^2 - 4(-2)(7)}}{2(-2)} \\
&= \frac{-3 \pm \sqrt{65}}{-4} \\
&= \frac{-3 \pm 8.062}{-4} \\
&= -1.27,\ 2.77.
\end{aligned}
$$

Since -1.27 and 2.77 are outside the range of the cosine function, there is no angle θ that solves the given equation. The method of obtaining the solution is valid; however, no solution exists for the equation. □

EXAMPLE 6 In designing a simple truss we must solve the equation

$$4000 \sin \phi + 3000 \cos \phi = 5000.$$

Find ϕ, $0° \le \phi < 360°$.

Solution: Given the equation

$$4000 \sin \phi + 3000 \cos \phi = 5000,$$

we can first make the numbers easier to work with by dividing the equation by 1000, giving

$$4 \sin \phi + 3 \cos \phi = 5.$$

We cannot solve the equation in this form. We must change it to a form that contains only one of the trigonometric functions.

$4 \sin \phi + 3 \cos \phi = 5$	**Given**
$\sin^2 \phi + \cos^2 \phi = 1$	**Identity**
$\sin^2 \phi = 1 - \cos^2 \phi$	
$\sin \phi = \pm \sqrt{1 - \cos^2 \phi}$	
$\pm 4 \sqrt{1 - \cos^2 \phi} + 3 \cos \phi = 5$	**Substituting for sin ϕ in the given equation**
$\pm 4 \sqrt{1 - \cos^2 \phi} = 5 - 3 \cos \phi$	**Rearranging before squaring to remove the radical**
$16(1 - \cos^2 \phi) = 25 - 30 \cos \phi + 9 \cos^2 \phi$	**Squaring**
$16 - 16 \cos^2 \phi = 25 - 30 \cos \phi + 9 \cos^2 \phi$	
$0 = 25 - 16 - 30 \cos \phi + 9 \cos^2 \phi + 16 \cos^2 \phi$	
$0 = 9 - 30 \cos \phi + 25 \cos^2 \phi$	
$0 = (3 - 5 \cos \phi)(3 - 5 \cos \phi)$	**Factoring**
$3 - 5 \cos \phi = 0$	**Setting factor equal to zero**

$$\cos \phi = \frac{3}{5}$$
$$= 0.6$$
$$\phi = 53.13^\circ, 306.9^\circ.$$

Checking the solutions we have

for $\phi = 53.13^\circ$	for $\phi = 306.9^\circ$
$4 \sin 53.13^\circ + 3 \cos 53.13^\circ = 5$	$4 \sin 306.9^\circ + 3 \cos 306.9^\circ = 5$
$4(0.8000) + 3(0.6000) = 5$	$4(-0.8000) + 3(0.6000) = 5$
$3.200 + 1.800 = 5$	$-3.200 + 1.800 = 5$
$5. = 5.$ ✓	$-1.4 = 5.$ X

Only one of the solutions checks in the original equation, so the only solution for $4 \sin \phi + 3 \cos \phi = 5$ in the given interval is $\phi = 53.1^\circ$. □

15.4 Exercises

In Exercises 1–36, find θ, $0^\circ \leq \theta < 360^\circ$.

1. $\tan^2 \theta - \tan \theta = 0$
2. $\cot^2 \theta + \cot \theta = 0$
3. $\sin^2\theta - \frac{\sqrt{2}}{2} \sin \theta = 0$
4. $\cos^2 \theta + \frac{\sqrt{2}}{2} \cos \theta = 0$

5. $\sin^2\theta - \sin\theta - 2 = 0$
6. $\cos^2\theta + \cos\theta - 2 = 0$
7. $\sin\theta = \cos\theta$
8. $\cos\theta - \sin\theta = 0$
9. $\cos 2\theta = \sin 2\theta$
10. $\csc 2\theta = \sec 2\theta$
11. $\cos 2\theta = \sin\theta$
12. $\sin 2\theta = \cos\theta$
13. $\cos\theta - \sin\frac{1}{2}\theta = 0$
14. $\sin\frac{1}{2}\theta - \cos\frac{1}{4}\theta = 0$
15. $\cos 2\theta + \cos\theta = 0$
16. $\sin 2\theta + \cos\theta = 0$
17. $\cos\theta + \sin\theta = 1$
18. $\sin\theta = \cos\theta - \sqrt{2}$
19. $\cos 2\theta + \cos\theta + 1 = 0$
20. $\cos\theta + \sqrt{3}\sin\theta = 1$
21. $\cos 2\theta + \sin 2\theta = 1$
22. $\cos\theta - \sin\theta = 1$
23. $\tan\theta + \sec\theta - 1 = 0$
24. $2\sin\theta + \cot\theta = \csc\theta$
25. $\tan^2\theta + 3 = 3\sec\theta$
26. $\tan 2\theta + \cot 2\theta - 2 = 0$
27. $3\sin^2\theta - \cos^2\theta = 0$
28. $2\tan\theta + 3 = 2\sec^2\theta$
29. $\sin 2\theta\cos\theta + \cos 2\theta\sin\theta = 1$
30. $\cos\theta\cos 2\theta + \sin\theta\sin 2\theta = -1$
31. $\sin^2\theta + 2\cos\theta = 2$
32. $2\cos^2\theta - \sin^2\theta = 0$
33. $4\sin^2\theta = 2 + \cos\theta$
34. $3 - 2\cos\frac{1}{2}\theta = 4\sin^2\frac{1}{2}\theta$
35. $\cos(\theta - 45°) = \sin(\theta - 45°)$
36. $5\sin\theta - 12\cos\theta = 13$
37. The equilibrium equations for an object resting on an inclined plane are

$$f - w\sin\theta = 0$$
$$N - w\cos\theta = 0.$$

If the object is just about to slip on the plane, the coefficient of friction between the object and the plane is $\mu = f/N$. What is the coefficient of friction if the object just starts to slip when the plane is at 20° to the horizontal?

38. Solve Exercise 37 if an object just starts to slip when the plane is at 27° to the horizontal.

In Exercises 39–44, an equation from an electrical-phasor diagram can be

$$I = I_1\cos\theta + I_2\sin\theta \quad \text{or} \quad V = V_1\cos\theta + V_2\sin\theta.$$

Solve for θ, $0 \le \theta < 360°$ for the values given. (Similar equations are found in mechanics.)

39. $I_1 = 3.00,\ I_2 = -8.00,\ I = -2.00$
40. $I_1 = 0.261,\ I_2 = -0.115,\ I = 0.169$
41. $I_1 = -1.36,\ I_2 = 2.81,\ I = 2.18$
42. $I_1 = -1 \times 10^{-3},\ I_2 = 3 \times 10^{-3},\ I = 1.66 \times 10^{-3}$
43. $V_1 = 9,\ V_2 = 9,\ V = 1.11$
44. $V_1 = 12,\ V_2 = 13,\ V = -6.83$

15.5 Inverse trigonometric functions

In Chapter 5, Rule 5.1 we said that the inverse of a function can be found by interchanging the role of the independent variable and the dependent variable and then solving for the dependent variable. The result will be the inverse function. To find the inverse function of $y = f(x) = 3x - 5$ we interchanged x and y in the equation $(x = 3y - 5)$ and then solved for y to obtain $y = \frac{(x+5)}{3}$. Therefore the function $f(x) = 3x - 5$ has an inverse function $f^{-1}(x) = \frac{(x+5)}{3}$.

The method of finding an inverse algebraic function cannot be used in finding inverse trigonometric functions. For example if $y = \sin x$,

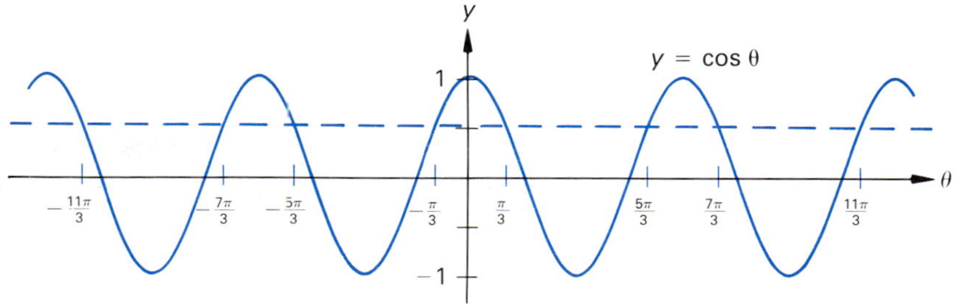

FIGURE 15.5

and we interchange variables ($x = \sin y$), there is no direct way we can solve for y using algebraic procedures. Therefore, the following paragraphs will define what is meant by inverse trigonometric functions.

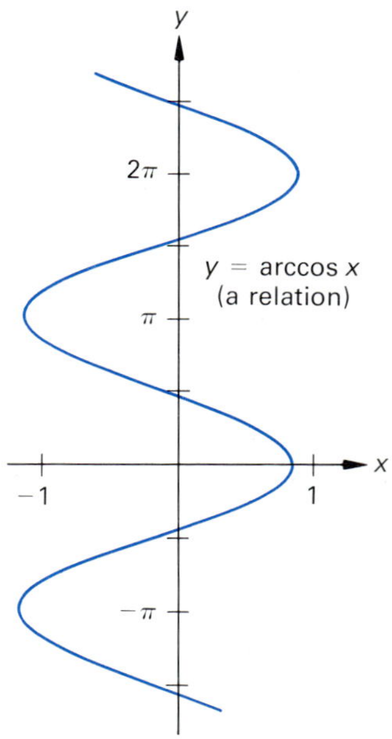

FIGURE 15.6

In Chapter 8 and again in this chapter we solved equations of the form $\cos\theta = \frac{1}{2}$ by seeking "the angle θ whose cosine is $\frac{1}{2}$." We found that when $\cos\theta = \frac{1}{2}$, $\theta = \ldots, -660, -420, -300, -60, 60, 300, 420, 660, \ldots$ or $\theta = \ldots, -\frac{11\pi}{3}, -\frac{7\pi}{3}, -\frac{5\pi}{3}, -\frac{\pi}{3}, \frac{\pi}{3}, \frac{5\pi}{3}, \frac{7\pi}{3}, \frac{11\pi}{3}, \ldots$ (see Fig. 15.5). It is useful to be able to express "the angle whose cosine is $\frac{1}{2}$" in a more compact fashion. To do this we write the equivalent of "the angle whose cosine is $\frac{1}{2}$" as $\theta = \arccos\frac{1}{2}$ or $\theta = \cos^{-1}\frac{1}{2}$. If $x = \cos\theta$, then the equivalent relation is written

$$\theta = \arccos x.$$

($\theta = \cos^{-1} x$ does not mean $\theta = 1/\cos x$. It is a notational convention that is used only as an alternative way of writing $\theta = \arccos x$. Because of this, if we mean to write the reciprocal of $\cos x$ we will write it as $\frac{1}{\cos x} = (\cos x)^{-1}$. We will use the "arc" form in this book.)

FIGURE 15.7

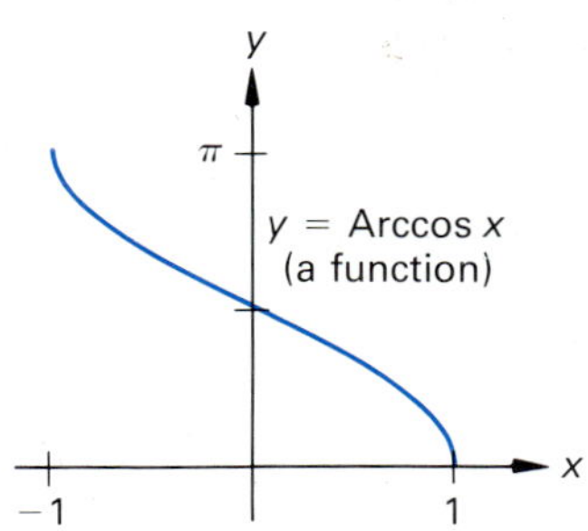

The equation $y = \arccos x$ (Fig. 15.6) is a relation because there are many values of y that satisfy the equation for any given value of x. In order to have a function we restrict the range of the relation so that there is only one possible value of y for any given value of x. By doing so we define the inverse trigonometric function.

$\boldsymbol{y = \text{Arccos}\, x}$ if and only if $x = \cos y$ were, $\boldsymbol{-1 \le x \le 1,\ 0 \le y \le \pi}$ (The capital letter A will be used to designate a function and the lower case letter a will be used to designate a relation.) $y = \text{Arccos}\, x$ (Fig. 15.7) is a function because it represents a one-to-one relationship. There is one value of y for each value of x, whereas $y = \arccos x$ is a relation because there are many values of y for each value of x.

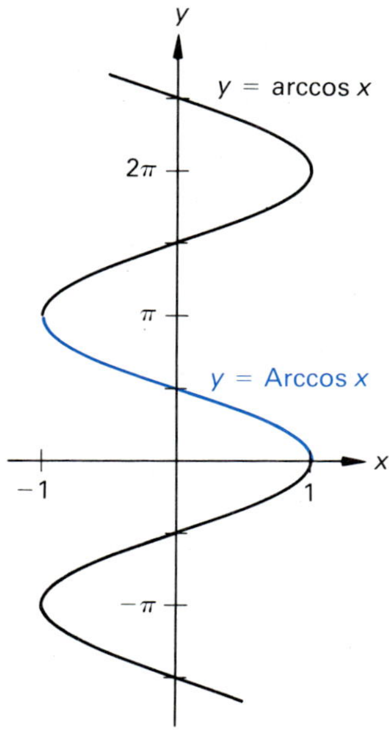

FIGURE 15.8

EXAMPLE 1 Graph $y = \text{arccos } x$ and $y = \text{Arccos } x$.

Solution: The relation $y = \text{arccos } x$ is equivalent to writing $x = \cos y$. Since we are already familiar with the graph of $y = \cos x$, all we have to do to obtain the graph of $x = \cos y$ is to interchange the axes (giving the same effect as interchanging the variables). This has been done to produce Fig. 15.8. The black curve is the graph of $y = \text{arccos } x$. The portion of the curve in blue represents the inverse trigonometric function $y = \text{Arccos } x$. The graph emphasizes that the range of $y = \text{Arccos } x$ must be restricted to $0 \le y \le \pi$ in order for $y = \text{Arccos } x$ to be a function. □

The inverse trigonometric functions are defined for all of the trigonometric functions. The most commonly used inverse trigonometric functions (in addition to Arccos x) are $y = \text{Arcsin } x$, $y = \text{Arctan } x$, and $y = \text{Arccot } x$. These functions along with their corresponding relations are listed in Table 15.5 and graphed in Fig. 15.9.

EXAMPLE 2 Evaluate

a. $y = \text{arcsin } \frac{1}{2}$, **b.** $y = \text{Arcsin } \frac{1}{2}$

Solution:

a. $y = \text{arcsin } \frac{1}{2}$ is equivalent to writing $\sin y = \frac{1}{2}$. Therefore $y = \ldots -210°, 30°, 150°, 390°, \ldots$.

b. $y = \text{Arcsin } \frac{1}{2}$ restricts y to the range $-90° \le y \le 90°$. Therefore $y = 30°$. □

TABLE 15.5 **The inverse trigonometric functions**

Function	Domain	Range
$y = \text{Arccos } x$	$-1 \le x \le 1$	$0 \le y \le \pi$
$y = \text{Arcsin } x$	$-1 \le x \le 1$	$-\frac{\pi}{2} \le y \le \frac{\pi}{2}$
$y = \text{Arctan } x$	$-\infty < x < \infty$	$-\frac{\pi}{2} < y < \frac{\pi}{2}$
$y = \text{Arccot } x$	$-\infty < x < \infty$	$0 < y < \pi$
$y = \text{Arccsc } x$	$x \le -1$ or $x \ge 1$	$-\frac{\pi}{2} \le y \le \frac{\pi}{2}, y \ne 0$
$y = \text{Arcsec } x$	$x \le -1$ or $x \ge 1$	$0 \le y \le \pi, y \ne \frac{\pi}{2}$

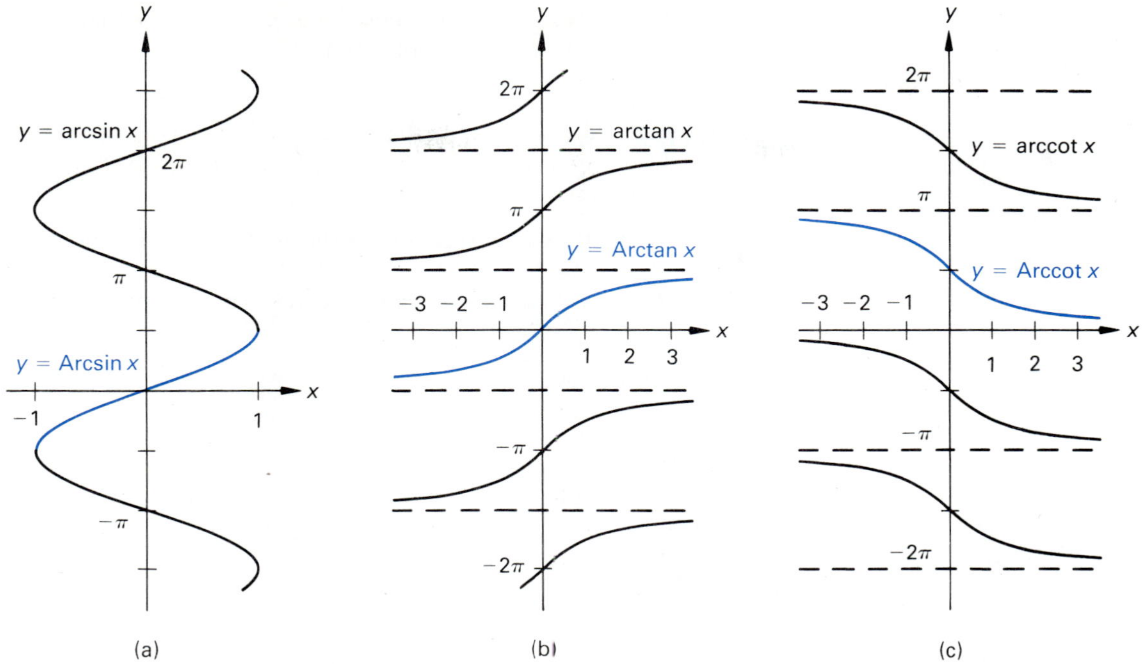

FIGURE 15.9

EXAMPLE 3 Evaluate

a. $y = \text{Arctan } 4$, **b.** $y = \text{Arccot }(-6.21)$, **c.** $y = \text{Arccos } 2$

Solution:

a. $y = \text{Arctan } 4$ is equivalent to $\tan y = 4$, $-\frac{\pi}{2} < y < \frac{\pi}{2}$. Therefore $y = 1.33$ rad.

b. $y = \text{Arccot }(-6.21)$ is equivalent to $\cot y = -6.21$, $0 < y < \pi$. Most calculators do not have keys for finding cotangent, but if we remember that $\cot y = 1/\tan y$ we can proceed.

$$\cot y = -6.21$$

$$\frac{1}{\cot y} = \frac{1}{-6.21}$$

$$\tan y = -0.1610$$

$$y = -0.160 \text{ rad.}$$

This value of y cannot be a solution to $y = \text{Arccot }(-6.21)$ because it is outside the required range $0 < y < \pi$. Recalling the behavior of the tangent function we remember that $y = -0.160 + \pi$ is also a solution of $\tan y = -0.1610$. This value of y falls within the required range so $y = 2.98$ rad.

c. $y = \text{Arccos } 2$ is equivalent to $\cos y = 2$, but we know that $-1 \le \cos y \le 1$, so $y = \text{Arccos } 2$ is undefined. □

EXAMPLE 4 Evaluate $y = \sin\left(\text{Arccos } \frac{\sqrt{3}}{2}\right)$.

Solution: Recalling that $\text{Arccos } \frac{\sqrt{3}}{2}$ is the angle whose cosine is $\frac{\sqrt{3}}{2}$, let us let $\theta = \text{Arccos } \frac{\sqrt{3}}{2}$. Then the given expression becomes $y = \sin \theta$.

$$y = \sin\left(\text{Arccos } \frac{\sqrt{3}}{2}\right)$$

$$\begin{aligned} \theta &= \text{Arccos } \frac{\sqrt{3}}{2} \\ &= 30^\circ \\ y &= \sin \theta \\ &= \sin 30^\circ \\ &= \frac{1}{2}. \end{aligned}$$ □

EXAMPLE 5 When calculating the strut lengths to construct a geodesic dome it is necessary to evaluate

$$y = \cos\left[2 \text{ Arcsin}\left(\frac{1}{2 \sin 60^\circ}\right)\right].$$

Find y.

Solution:

$$\begin{aligned} y &= \cos\left[2 \text{ Arcsin}\left(\frac{1}{2 \sin 60^\circ}\right)\right] & \\ &= \cos\left[2 \text{ Arcsin}\left(\frac{1}{2(0.866)}\right)\right] & \textbf{Evaluate sine} \\ &= \cos\,[2 \text{ Arcsin } 0.5774] & \\ &= \cos\,[2(35.26^\circ)] & \textbf{Evaluate Arcsin} \\ &= \cos\,[70.53^\circ] & \\ &= 0.333. & \textbf{Evaluate cosine} \end{aligned}$$ □

EXAMPLE 6 In developing the equations for maximum stress at a point in the landing gear of a space shuttle it is necessary to evaluate $q = \cos\left(\text{Arctan } \frac{a}{b}\right)$. What is q?

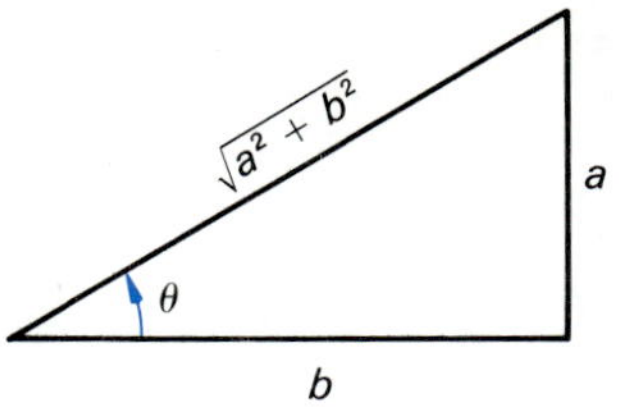

FIGURE 15.10

Solution: To solve $q = \cos\left(\text{Arctan } \frac{a}{b}\right)$ we first must evaluate Arctan $\frac{a}{b}$. Arctan $\frac{a}{b} = \theta$ or equivalently $\tan \theta = \frac{a}{b}$. Constructing a right triangle we can label it as in Fig. 15.10. From the Pythagorean theorem the length of the hypotenuse of the triangle must be $\sqrt{a^2 + b^2}$. Therefore we have

$$q = \cos\left(\text{Arctan } \frac{a}{b}\right)$$

$$\theta = \text{Arctan } \frac{a}{b}$$

$$q = \cos \theta \qquad \textbf{(where } \cos\theta = \frac{b}{\sqrt{a^2 + b^2}} \textbf{ from Fig. 15.10)}$$

$$= \frac{b}{\sqrt{a^2 + b^2}} \qquad \textbf{Substituting}$$

The solution to $q = \cos\left(\text{Arctan } \frac{b}{a}\right)$ is $q = \frac{b}{\sqrt{a^2 + b^2}}$ □

15.5 Exercises

In Exercises 1–6, (a) graph the relation and the function specified, (b) state the domain of each, and (c) state the range of each.

1. $y = \arccos x$, $y = \text{Arccos } x$

2. $y = \arcsin x$, $y = \text{Arcsin } x$

3. $y = \arctan x$, $y = \text{Arctan } x$

4. $y = \text{arccot } x$, $y = \text{Arccot } x$

5. $y = \text{arccsc } x$, $y = \text{Arccsc } x$

6. $y = \text{arcsec } x$, $y = \text{Arcsec } x$

In Exercises 7–44, evaluate the expression.

7. Arcsin 0

8. Arccos 0

9. Arccos $\frac{1}{\sqrt{2}}$

10. Arcsin $\frac{1}{\sqrt{2}}$

11. Arctan 1

12. Arctan (−1)

13. Arctan (−4)

14. Arctan 4

15. Arcsin $\left(-\frac{1}{2}\right)$

16. Arccos $\left(-\frac{1}{2}\right)$

17. Arccos $\left(-\frac{\sqrt{3}}{2}\right)$

18. Arcsin $\left(-\frac{\sqrt{3}}{2}\right)$

19. Arccot $\frac{1}{\sqrt{3}}$

20. Arccot $\sqrt{3}$

21. Arccot $\left(-\frac{4}{3}\right)$

22. Arccot $\left(-\frac{3}{4}\right)$

23. Arcsin 0.831

24. Arccos (−0.463)

25. Arctan (−21.6)

26. Arccot (16.3)

27. $\cos\left(\text{Arcsin } \frac{1}{2}\right)$

28. $\sin\left(\text{Arccos } \frac{1}{2}\right)$

29. sin (Arctan 1)

30. cos [Arctan (−1)]

31. tan (Arccos 0.6)

32. tan (Arcsin 0.8)

33. cot (Arccos 0.6)

34. cot (Arccos 0.8)

35. cos (Arccos 0.621)

36. sin [Arcsin (−0.527)]

37. $\cos\left(2 \text{ Arcsin } \frac{4}{5}\right)$

38. $\cos\left(\frac{1}{2} \text{Arccos } \frac{1}{2}\right)$

39. $\tan\left(3 \text{ Arctan } \frac{1}{\sqrt{3}}\right)$

40. $\cot\left(\frac{1}{2} \text{Arctan } \sqrt{3}\right)$

41. tan (Arccos 0.112 + Arcsin 0.736)

42. cos (Arctan 3.00 + Arcsin 0.500)

43. cos (Arccos 0.416 + Arcsin 0.416)

44. sin (Arcsin 0.643 + Arccos 0.643)

In Exercises 45 and 46, find y as a function of a and b, $|a| \leq |b|$, $b \neq 0$.

45. $y = \cos\left(\text{Arcsin}\,\dfrac{a}{b}\right)$ **46.** $y = \sin\left(\text{Arccos}\,\dfrac{a}{b}\right)$

In Exercises 47 and 48, find y as a function of x, $|x| \leq 1$.

47. $y = \tan\,(\text{Arccos}\,x)$ **48.** $y = \cos\,(\text{Arctan}\,x)$

When constructing geodesic domes it is necessary to evaluate functions such as those in Exercises 49–52. In each exercise find y.

49. $y = \text{Arctan}\left[\dfrac{\tan 10.57°}{\cos 36°}\right]$

50. $y = \text{Arctan}\left[\dfrac{\tan 21.15°}{\cos 36°}\right]$

51. $y = \text{Arctan}\left(\dfrac{1}{3}\sin 45°\right)$

52. $y = \cos\left\{2\,\text{Arcsin}\left[\dfrac{1}{(1+\sqrt{5})\sin 36°}\right]\right\}$

53. Why is the range of $y = \text{Arctan}\,x$ such that $-\dfrac{\pi}{2} < y < \dfrac{\pi}{2}$ instead of $-\dfrac{\pi}{2} \leq y \leq \dfrac{\pi}{2}$?

Chapter summary

In Chapter 15 we presented the trigonometric identities as listed in Table 15.6.

The trigonometric identities are useful for restating trigonometric equations in forms that are solvable by conventional techniques.

Also introduced and discussed were the inverse trigonometric functions. If $y = \sin x$ then $x = \text{Arcsin}\,y$ is an equivalent function and $y = \text{Arcsin}\,x$ is the inverse trigonometric function. The inverse trigonometric functions are listed in Table 15.7.

TABLE 15.6 **Trigonometric identities**

Fundamental Identities

{1} $\dfrac{1}{\sin A} = \csc A$ {2} $\dfrac{1}{\csc A} = \sin A$ {3} $\dfrac{1}{\cos A} = \sec A$ {4} $\dfrac{1}{\tan A} = \cot A$

Quotient and Product Identities

{5} $\dfrac{\sin A}{\cos A} = \tan A$ {6} $\dfrac{\cos A}{\sin A} = \cot A$ {7} $\sin A \sec A = \tan A$ {8} $\dfrac{\csc A}{\cot A} = \sin A$

Pythagorean Identities

{9} $\sin^2 A + \cos^2 A = 1$ {10} $1 + \tan^2 A = \sec^2 A$ {11} $1 + \cot^2 A = \csc^2 A$

Sum and Difference Identities

{12} $\cos\,(A + B) = \cos A \cos B - \sin A \sin B$ {13} $\sin\,(A + B) = \sin A \cos B + \cos A \sin B$

{14} $\cos\,(A - B) = \cos A \cos B + \sin A \sin B$ {15} $\sin\,(A - B) = \sin A \cos B - \cos A \sin B$

Double- and Half-Angle Identities

{16} $\cos 2A = \cos^2 A - \sin^2 A$ {17} $\cos 2A = 2\cos^2 A - 1$

{18} $\cos 2A = 1 - 2\sin^2 A$ {19} $\sin 2A = 2 \sin A \cos A$

{20} $\cos\dfrac{B}{2} = \pm\sqrt{\dfrac{1 + \cos B}{2}}$ {21} $\sin\dfrac{B}{2} = \pm\sqrt{\dfrac{1 - \cos B}{2}}$

{22} $\sin A = \pm\sqrt{\dfrac{1 - \cos 2A}{2}}$ {23} $\cos A = \pm\sqrt{\dfrac{1 + \cos 2A}{2}}$

TABLE 15.7 **The inverse trigonometric functions**

Function	Domain	Range
$y = \text{Arccos } x$	$-1 \le x \le 1$	$0 \le y \le \pi$
$y = \text{Arcsin } x$	$-1 \le x \le 1$	$-\frac{\pi}{2} \le y \le \frac{\pi}{2}$
$y = \text{Arctan } x$	$-\infty < x < \infty$	$-\frac{\pi}{2} < y < \frac{\pi}{2}$
$y = \text{Arccot } x$	$-\infty < x < \infty$	$0 < y < \pi$
$y = \text{Arccsc } x$	$x \le -1$ or $x \ge 1$	$-\frac{\pi}{2} \le y \le \frac{\pi}{2}, y \ne 0$
$y = \text{Arcsec } x$	$x \le -1$ or $x \ge 1$	$0 \le y \le \pi, y \ne \frac{\pi}{2}$

Review Exercises

In Exercises 1–4, prove the identities.

1. $\cot \theta = \cos \theta \csc \theta$

2. $\cos \theta \sec \theta = 1$

3. $\dfrac{\sin \theta}{\csc \theta} + \dfrac{\cos \theta}{\sec \theta} = 1$

4. $\sin \theta \csc \theta + \tan \theta \cot \theta = 2$

In Exercises 5–16, write the expression in terms of a single trigonometric function. (The function may be multiplied by a constant and/or raised to a power.)

5. $\sin \theta \tan \theta + \cos \theta$

6. $\csc \theta \sec \theta - \tan \theta$

7. $\sin \theta (\csc \theta - \sin \theta)$

8. $\tan \theta (\cot \theta + \tan \theta)$

9. $\dfrac{1 + \cos^2 A}{\sin^2 A} + 1$

10. $\dfrac{1 + \tan^2 A}{\tan^2 A}$

11. $\sin B (\csc B - \sin B)$

12. $\sin^2 B - \cos^2 B + 1$

13. $\sec A \csc A \tan A - 1$

14. $1 - \tan A \sin A \cos A$

15. $(1 + \sin \theta)(1 - \sin \theta)$

16. $(1 - \cos \theta)(1 + \cos \theta)$

In Exercises 17–20, show that each expression is equal to a constant.

17. $(\cos^2 \omega t - 1)(\tan^2 \omega t + 1) + \sec^2 \omega t$

18. $(\sin^2 \pi t + \cos^2 \pi t)^{1/2}$

19. $(1 + \cos \pi t)(\csc^2 \pi t - \csc^2 \pi t \cos \pi t)$

20. $(\csc \omega t + \cot \omega t)^2 (\csc \omega t - \cot \omega t)^2$

In Exercises 21–28, prove the identity.

21. $\dfrac{\sin \phi}{1 - \cos \phi} = \dfrac{1 + \cos \phi}{\sin \phi}$

22. $\dfrac{1}{\csc \phi - \cot \phi} = \csc \phi + \cot \phi$

23. $\cot^2 \omega t + \csc^2 \omega t = \csc^4 \omega t - \cot^4 \omega t$

24. $\dfrac{\cot \omega t}{1 + \csc \omega t} = \sec \omega t - \tan \omega t$

25. $\dfrac{1 + \tan \theta}{1 - \tan \theta} = \dfrac{\cot \theta + 1}{\cot \theta - 1}$

26. $\dfrac{\cos \omega t}{1 - \cos \omega t} = \dfrac{1 + \sec \omega t}{\tan^2 \omega t}$

27. $\dfrac{\cos 6t}{1 - \sin 6t} = \tan 6t + \sec 6t$

28. $(\tan \phi + 1) \csc \phi = \sec \phi + \csc \phi$

29. Show that $\cos (\theta - \phi) \ne \cos \theta - \cos \phi$.

30. Show that $\sin (\theta - \phi) \ne \sin \theta - \sin \phi$.

31. Use a geometric approach to derive an identity for $\cos (\theta - \phi)$.

32. Use $\tan A = \dfrac{\sin A}{\cos A}$ to show that

$$\tan (\theta - \phi) = \frac{\tan \theta - \tan \phi}{1 + \tan \theta \tan \phi}.$$

33. Use $\cot A = \dfrac{\cos A}{\sin A}$ to show that

$$\cot (\theta - \phi) = \frac{\cot \theta \cot \phi + 1}{\cot \phi - \cot \theta}.$$

34. Use $\sin(\theta + \phi) = \sin\theta\cos\phi + \cos\theta\sin\phi$ and the properties of the trigonometric functions to find the identity for $\sin(\theta - \phi)$.

In Exercises 35–44, use the trigonometric identities to find exact values for

35. $\cos 285°$ **36.** $\cos 315°$

37. $\sin 315°$ **38.** $\sin 285°$

39. $\tan 285°$ **40.** $\tan 315°$

41. $\cot 315°$ **42.** $\cot 285°$

43. $\tan(-15°)$ **44.** $\cot(-15°)$

45. Use the identity for $\tan(\theta + \phi)$ to find an identity for $\tan 2\theta$.

46. Use the identity for $\cot(\theta + \phi)$ to find an identity for $\cot 2\theta$.

47. Find $\sin 2A$ as a function of $\tan A$.

48. Find $\cos 2A$ as a function of $\tan A$.

49. Show that $\cot \dfrac{\theta}{2} = \dfrac{1 + \cos\theta}{\sin\theta}$.

50. Show that $\cot \dfrac{\theta}{2} = \dfrac{\sin\theta}{1 - \cos\theta}$.

51. Shear strain at a point can be written as

$$\begin{aligned}\epsilon_{ns} = &-\epsilon_{xx}\sin\theta\cos\theta + \epsilon_{xy}(\cos^2\theta - \sin^2\theta)\\ &+ \epsilon_{yy}\sin\theta\cos\theta.\end{aligned}$$

Write the shear strain as a function of the angle 2θ.

52. Normal strain at a point can be written as

$$\epsilon_{nn} = \epsilon_{xx}\cos^2\theta - 2\epsilon_{xy}\sin\theta\cos\theta + \epsilon_{yy}\sin^2\theta.$$

Write the normal strain as a function of the angle 2θ.

In Exercises 53–70, find θ, $0° \leq \theta < 360°$.

53. $\sec\theta = \csc\theta$

54. $\sin\theta - \cos\theta = 0$

55. $\sin\dfrac{1}{2}\theta - \cos\dfrac{1}{2}\theta = 0$

56. $\csc 2\theta + \sec 2\theta = 0$

57. $\cos\theta - \sin 2\theta = 0$

58. $\cos 4\theta = \sin 2\theta$

59. $\sin\dfrac{1}{2}\theta + \cos\theta - 1 = 0$

60. $4\sin^2\theta + \cos 2\theta - 2 = 0$

61. $\cot^2\theta = 7 - 3\sec^2\theta$

62. $2\tan\theta = \sec^2\theta$

63. $2\sin^2\theta + \cos\theta - 1 = 0$

64. $\cos 2\theta + 3\cos\theta + 2 = 0$

65. $3\sec^2\theta + \tan\theta = 5$

66. $2\sin 2\theta\cos 2\theta = -1$

67. $\cos(\theta + 60°) = \sin(\theta + 60°)$

68. $\tan(\theta + 45°) - \tan(\theta - 45°) = 2$

69. $\cos\theta = 2 + \sqrt{3}\sin\theta$

70. $3\cos\theta + 4\sin\theta = 2$

The equations in Exercises 71–74 are taken from mechanical structures. Solve for θ, $0° \leq \theta \leq 360°$.

71. $4000\sin\theta - 5000\cos\theta = 1510.$

72. $-1580\sin\theta + 2760\cos\theta = 289.$

73. $-2000\cos\theta - 4000\sin\theta = -2460.$

74. $1800\cos\theta + 1500\sin\theta = 1180.$

In Exercises 75–96, evaluate the expression.

75. $\operatorname{Arccos}\dfrac{\sqrt{3}}{2}$ **76.** $\operatorname{Arccos}\dfrac{3}{5}$

77. $\operatorname{Arcsin}\dfrac{3}{5}$ **78.** $\operatorname{Arcsin}\left(-\dfrac{4}{5}\right)$

79. $\operatorname{Arctan}\sqrt{3}$ **80.** $\operatorname{Arctan}\left(-\dfrac{1}{3}\right)$

81. $\operatorname{Arccot}\sqrt{3}$ **82.** $\operatorname{Arccot}\left(-\dfrac{1}{\sqrt{3}}\right)$

83. $\sin\left(\operatorname{Arccos}\dfrac{12}{13}\right)$ **84.** $\cos\left(\operatorname{Arcsin}\dfrac{5}{13}\right)$

85. $\cos\left(\operatorname{Arctan}\dfrac{15}{8}\right)$ **86.** $\sin\left(\operatorname{Arctan}\dfrac{8}{15}\right)$

87. $\tan\left(\operatorname{Arccos}\dfrac{24}{25}\right)$ **88.** $\tan\left(\operatorname{Arcsin}\dfrac{7}{25}\right)$

89. $\tan(\operatorname{Arctan} 0.114)$

90. $\cot[\operatorname{Arccot}(-14.3)]$

91. $\sin\left(2\operatorname{Arccos}\dfrac{4}{5}\right)$

92. $\sin\left(\dfrac{1}{2}\operatorname{Arcsin}\dfrac{4}{5}\right)$

93. $\cos(\operatorname{Arcsin} 0.715 + \operatorname{Arccos} 0.883)$

94. $\sin(\operatorname{Arctan} 4.12 - \operatorname{Arccos} 0.347)$

95. $\cot(\operatorname{Arccos} x)$

96. $\tan(\operatorname{Arcsin} x)$

Can you solve these?

1. In deriving $\cos^2 \theta + \sin^2 \theta = 1$ (Section 15.1), why don't we have to worry about $r = 0$ when dividing by r^2?

2. When deriving $1 + \tan^2 \theta = \sec^2 \theta$ (Section 15.1), why don't we have to worry about $\theta = \frac{\pi}{2} + k\pi$ when dividing by $\cos^2 \theta$?

3. Show that the sum of two sinusoids,

$$A_1 \cos (\omega t + \phi_1) + A_2 \cos (\omega t + \phi_2)$$

is itself a sinusoid.

4. In geodesic dome design it is necessary to solve the trigonometric equation

$$\sin \left(\frac{360°}{n} + a \right) = \sin a$$

(n a constant) for a. Find a.

5. What is the difference between $y = \sin (\arcsin x)$ and $y = \sin (\text{Arcsin } x)$?

6. In calculating maximum normal stress it is necessary to find $\cos 2\theta$ when it is known that

$$\tan 2\theta = \frac{2b}{a - c}.$$

Find $\cos 2\theta$ in terms of a, b, and c.

7. When x is in radians, $\sin x$ can be approximated by

$$\sin x = x - \frac{x^3}{6} + \frac{x^5}{120} - \frac{x^7}{5040}.$$

Write a computer program to calculate approximate values for the sine of x when $0° \leq x \leq 360°$.

16

Specifications for the basic major diameter of number size self-tapping screws form an arithmetic sequence in which size number 1 has a basic diameter of 0.073 in. and size number 24 has a basic diameter of 0.372 in. The normal sequence of sizes is interrupted by omitting the two number sizes closest to 1/4 in. and 5/16 in. so that these common fractional sizes may be included. Which number sizes are omitted?

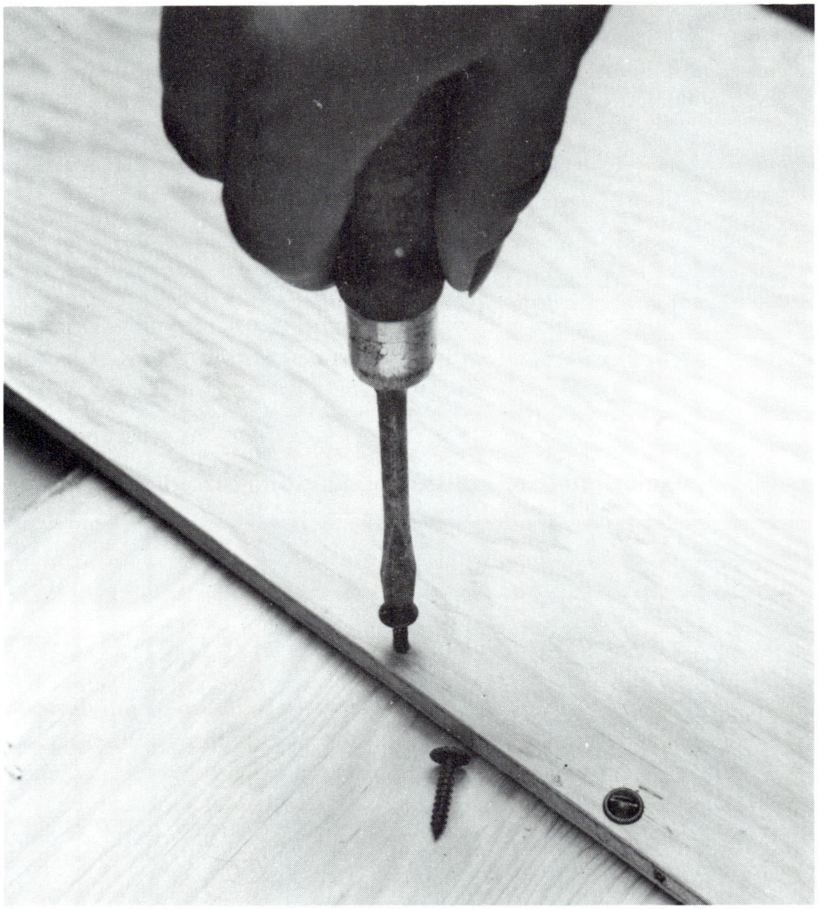

Sequences and Series

16.1 Sequences

Sequences and series are used frequently in many branches of mathematics and science. There are many real-life situations described in terms of an arithmetic or geometric sequence. The swing of a pendulum, which is decreasing at a set rate, is an example of an arithmetic sequence. The yearly amount of money paid to a savings account that compounds interest at regular intervals is an example of a geometric sequence.

What is a sequence? A **sequence** can be thought of as a finite or infinite list of numbers. Examples of sequences are:

$1, 3, 5, 7, 9, 11, 13$	**Finite**
$-1, 1, 3, 5, 7, 9, 11, 13, \ldots$	**Infinite (three dots means "and so on")**
$0, \frac{1}{2}, 1, \frac{3}{2}, 2, \frac{5}{2}, \ldots$	**Infinite**
$-5, 0, 5, 10, 15$	**Finite**

A more precise definition of a sequence is based on the understanding of the concept of function. Recall from Chapter 5, that a function is a rule that determines a unique element of the range for each element of the domain.

DEFINITION 16.1

> A **sequence** is a function whose domain is the set of positive integers $\{1, 2, 3, \ldots\}$.

The elements of the range of any sequence are called its **terms.** Representing a sequence with functional notation we write $f(1) = a_1$, $f(2) = a_2$, $f(3) = a_3$, $f(4) = a_4$, and in general, $f(i) = a_i$. When these terms are written in their natural order we have the following sequences:

$a_1, a_2, a_3, a_4, \ldots, a_n$	**Represents a finite sequence**
$a_1, a_2, a_3, \ldots, a_i, \ldots$	**Represents an infinite sequence**

It is important to be able to write the ith term or the **general term of a sequence.** The ith term can be used to generate the other terms of that sequence.

EXAMPLE 1 Write terms for each of the sequences for the indicated domain.

a. For $a_i = i^3 + 1$, $D = 1, 2, 3, 4$.

b. For $a_i = \frac{1}{i^2}$, $D =$ set of positive integers.

Solution:

a. For $a_i = i^3 + 1$, with the domain limited to the first four counting numbers, the result is a finite sequence. Substituting we find the following terms of the sequence:

$$a_1 = 1^3 + 1 = 2. \qquad a_2 = 2^3 + 1 = 9. \qquad a_3 = 3^3 + 1 = 28. \qquad a_4 = 4^3 + 1 = 65.$$

Thus the finite sequence whose ith term is $a_i = i^3 + 1$ is 2, 9, 28, 65.

b. For $a_i = \frac{1}{i^2}$, with the domain the set of positive integers, the result will be an infinite sequence. Substituting we find the following terms of the sequence:

$$a_1 = \frac{1}{1^2} = 1. \qquad a_2 = \frac{1}{2^2} = \frac{1}{4}. \qquad a_3 = \frac{1}{3^2} = \frac{1}{9}.$$

$$a_4 = \frac{1}{4^2} = \frac{1}{16}. \qquad a_5 = \frac{1}{5^2} = \frac{1}{25}. \qquad a_6 = \frac{1}{6^2} = \frac{1}{36}.$$

Thus the infinite sequence whose ith term is $a_i = \frac{1}{i^2}$ is

$$1, \frac{1}{4}, \frac{1}{9}, \frac{1}{16}, \frac{1}{25}, \frac{1}{36}, \ldots, \frac{1}{i^2}, \ldots .$$ □

Since a sequence is a function it can be graphed. The graph consists of the points $(1, a_1)$, $(2, a_2)$, $(3, a_3)$, . . . , determined from the general term and the domain. It is impossible to show all the points of an infinite sequence. However, a finite set of points can usually provide a picture of the sequence.

EXAMPLE 2

a. Determine the first six terms of the sequence given by $a_i = 3i - 12$.

b. Graph the first six terms of the sequence in (a).

Solution:

a. Replacing i with 1, 2, 3, 4, 5, and 6, we have, respectively, the first six terms.

$$a_1 = 3(1) - 12 = -9. \qquad a_2 = 3(2) - 12 = -6. \qquad a_3 = 3(3) - 12 = -3.$$

$$a_4 = 3(4) - 12 = 0. \qquad a_5 = 3(5) - 12 = 3. \qquad a_6 = 3(6) - 12 = 6.$$

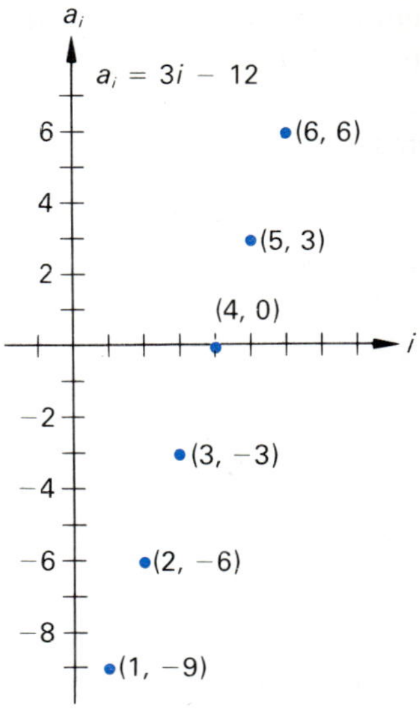

FIGURE 16.1

b. The six points of the sequence to be graphed are $(1, -9)$, $(2, -6)$, $(3, -3)$, $(4, 0)$, $(5, 3)$, and $(6, 6)$. The sequence is graphed in Fig. 16.1. □

EXAMPLE 3

a. Determine the first six terms of the sequence given by $a_i = 3 - \frac{i}{4}$.

b. Graph the first six terms of the sequence in (a).

Solution:

a. Replacing i with 1, 2, 3, . . . , 6, we have the first six terms:

$$a_1 = 3 - \frac{1}{4} = 2\frac{3}{4}. \qquad a_2 = 3 - \frac{2}{4} = 2\frac{1}{2}. \qquad a_3 = 3 - \frac{3}{4} = 2\frac{1}{4}.$$

$$a_4 = 3 - \frac{4}{4} = 2. \qquad a_5 = 3 - \frac{5}{4} = 1\frac{3}{4}. \qquad a_6 = 3 - \frac{6}{4} = 1\frac{1}{2}.$$

FIGURE 16.2

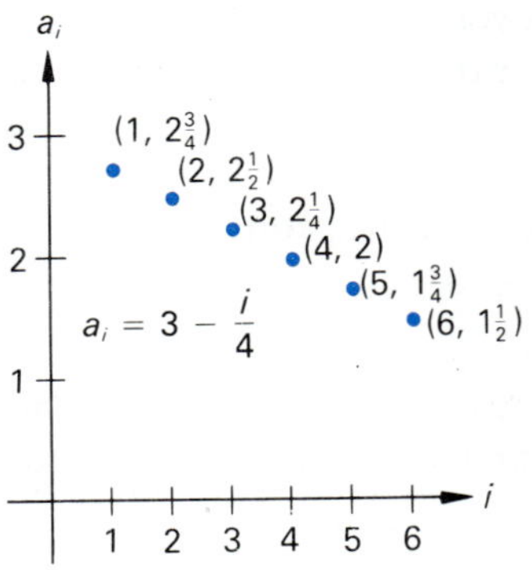

b. The six terms of the sequence to be graphed are $\left(1, 2\frac{3}{4}\right)$, $\left(2, 2\frac{1}{2}\right)$, $\left(3, 2\frac{1}{4}\right)$, $(4, 2)$, $\left(5, 1\frac{3}{4}\right)$, and $\left(6, 1\frac{1}{2}\right)$. The sequence is graphed in Fig. 16.2. □

It can be seen from the terms of the sequence in Example 2 and from Fig. 16.1 that each succeeding term is larger than the preceding term. When this happens we say that the sequence is an **increasing sequence.** In Example 3 and Fig. 16.2 each succeeding term is smaller than the preceding term. This type of sequence is a **decreasing sequence.**

DEFINITION 16.2

> A sequence $a_1, a_2, a_3, \ldots, a_i, \ldots$, is said to be **increasing** if $a_i < a_{i+1}$ for all $i = 1, 2, 3, \ldots$, and **decreasing** if $a_i > a_{i+1}$ for all $i = 1, 2, 3, \ldots$.

EXAMPLE 4 Determine whether the sequence is increasing or decreasing.

a. $a_i = -2i$ **b.** $a_i = 2i + 3$

Solution:

a. The sequence $a_i = -2i$ is decreasing, since $a_i > a_{i+1}$ for all i. For example, $a_3 = -6$, $a_4 = -8$; thus $a_3 > a_4$.

b. The sequence $2i + 3$ is an increasing sequence since $a_i < a_{i+1}$ for all i. For example $a_6 = 15$, $a_7 = 17$, thus $a_6 < a_7$. □

There are sequences that do not increase or decrease. This is illustrated in Example 5.

EXAMPLE 5 Determine the first six terms of the sequence $a_i = 1 + (-1)^i$.

Solution: The first six terms of the sequence are

$$a_1 = 1 + (-1)^1 = 0. \qquad a_2 = 1 + (-1)^2 = 2. \qquad a_3 = 1 + (-1)^3 = 0.$$
$$a_4 = 1 + (-1)^4 = 2. \qquad a_5 = 1 + (-1)^5 = 0. \qquad a_6 = 1 + (-1)^6 = 2.$$

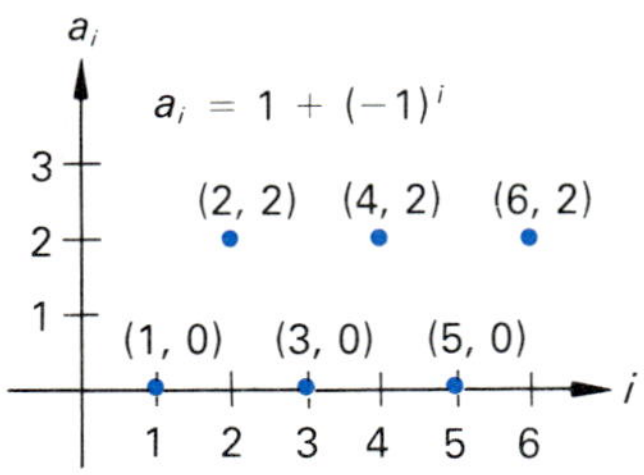

FIGURE 16.3

We see from the results that $a_1 < a_2$, but $a_2 > a_3$. If i is odd, then a_i is 0 and if i is even, then a_i is 2. Thus, the sequence does not increase nor does it decrease. The sequence is 0, 2, 0, 2, 0, ... The graph of the first six terms of the sequence is given in Fig. 16.3. □

Finding the terms of a sequence when given the general term is a relatively simple task. However, can we reverse the process? Can we determine the general term (or a formula for the general term) given a number of terms of the sequence?

EXAMPLE 6 Find a formula for the ith term of the sequence

4, 7, 10, 13, 16, 19, ...

Solution: The technique for finding the ith term may involve some guessing. In examining the terms of the sequence we see that each term can be written as one plus a number.

$$\begin{aligned} 4 &= 1 + 3, \\ 7 &= 1 + 6, \\ 10 &= 1 + 9, \\ 13 &= 1 + 12, \\ 16 &= 1 + 15, \\ 19 &= 1 + 18. \end{aligned}$$

For each of the terms the number added to 1 is divisible by 3. Therefore, we can write each of these numbers as three times some number.

$$\begin{aligned} a_1 &= 1 + 3(1) = 4. & a_6 &= 1 + 3(6) = 19. \\ a_2 &= 1 + 3(2) = 7. & &\vdots \\ a_3 &= 1 + 3(3) = 10. & & \\ a_4 &= 1 + 3(4) = 13. & & \\ a_5 &= 1 + 3(5) = 16. & a_i &= 1 + 3(i). \end{aligned}$$

The formula for the ith term of the sequence 4, 7, 10, 13, 16, 19, . . . , is $a_i = 1 + 3i$. □

EXAMPLE 7 Find a formula for the ith term of the sequences

a. $-1, 1, -1, 1, -1, \ldots$

b. $-2, 4, -8, 16, -32, 64, \ldots$

Solution:

a. For the sequence $-1, 1, -1, 1, -1, \ldots$, we see that the terms are all the same except for the sign. The negative signs appear in every other term, which corresponds to raising (-1) to an odd power. Likewise, raising (-1) to even power will give us the positive terms.

$$a_1 = (-1)^1 = -1.$$
$$a_2 = (-1)^2 = 1.$$
$$a_3 = (-1)^3 = -1.$$
$$\vdots$$
$$a_i = (-1)^i.$$

The formula for the ith term of the sequence

$-1, 1, -1, 1, -1, \ldots$ is $a_i = (-1)^i$.

b. Each of the terms of the sequence $-2, 4, -8, 16, -32, 64, \ldots$, can be written as follows:

$$-2 = -1 \cdot 2, \qquad 4 = 1 \cdot 4, \qquad -8 = -1 \cdot 8,$$
$$16 = 1 \cdot 16, \qquad -32 = -1 \cdot 32, \qquad 64 = 1 \cdot 64, \ldots.$$

In each case the second factor is divisible by 2. In fact, each of these factors can be expressed as 2 to some power.

$$-2 = -1 \cdot 2 = -1 \cdot 2^1.$$
$$4 = 1 \cdot 4 = 1 \cdot 2^2.$$
$$-8 = -1 \cdot 8 = -1 \cdot 2^3.$$
$$16 = 1 \cdot 16 = 1 \cdot 2^4.$$
$$-32 = -1 \cdot 32 = -1 \cdot 2^5.$$
$$64 = 1 \cdot 64 = 1 \cdot 2^6.$$
$$\vdots$$

Note that the terms alternate in sign as does the series in (a). Therefore we can write the first factor as $(-1)^1$, $(-1)^2$, $(-1)^3$,

$(-1)^4, \ldots$. The terms of the sequence can be written:

$$\begin{aligned}
a_1 &= -2 = (-1)^1(2)^1 = (-2)^1. \\
a_2 &= 4 = (-1)^2(2)^2 = (-2)^2. \\
a_3 &= -8 = (-1)^3(2)^3 = (-2)^3. \\
a_4 &= 16 = (-1)^4(2)^4 = (-2)^4. \\
a_5 &= -32 = (-1)^5(2)^5 = (-2)^5. \\
a_6 &= 64 = (-1)^6(2)^6 = (-2)^6. \\
&\vdots \\
a_i &= (-1)^i(2)^i = (-2)^i.
\end{aligned}$$

The formula for the ith term of the sequence $-2, 4, -8, 16, -32, 64, \ldots$, is $a_i = (-2)^i$. □

There may be more than one way of expressing the formula for the general term.

16.1 Exercises

In Exercises 1–20, (a) write the first six terms of the sequence; (b) indicate whether the sequence is increasing, decreasing, or neither; and (c) graph the sequence.

1. $a_i = 2i$

2. $a_i = -3i$

3. $a_i = i + 3$

4. $a_i = 4 - i$

5. $a_i = 3i - 5$

6. $a_i = 6 - 4i$

7. $a_i = -\dfrac{3i}{4}$

8. $a_i = \dfrac{4i}{3}$

9. $a_i = -1 + (-1)^i$

10. $a_i = 2 + (-1)^i$

11. $a_i = -1 + 2(-1)^i$

12. $a_i = 2 + 2(-1)^i$

13. $a_i = i^2 - 1$

14. $a_i = i^2 - 6$

15. $a_i = \dfrac{i+2}{i+3}$

16. $a_i = \dfrac{i-2}{i-3}$

17. $a_i = \dfrac{i^2-1}{i^2+1}$

18. $a_i = \dfrac{i^2+1}{i^2-1}$

19. $a_i = \dfrac{(-1)^{i+1}}{i}$

20. $a_i = \dfrac{(-1)^{i-1}}{i}$

In Exercises 21–40, find a formula for the ith term of each of the sequences.

21. 0, 1, 2, 3, 4, 5, . . .

22. 3, 4, 5, 6, 7, 8, . . .

23. 2, 4, 6, 8, 10, 12, . . .

24. 3, 6, 9, 12, 15, 18, . . .

25. 1, 3, 5, 7, 9, 11, . . .

26. 3, 5, 7, 9, 11, 13, . . .

27. 3, 7, 11, 15, 19, 23, . . .

28. 2, 5, 8, 11, 14, 17, . . .

29. $-2, -4, -6, -8, -10, -12, \ldots$

30. $-3, -6, -9, -12, -15, -18, \ldots$

31. 4, 5, 6, 7, 8, 9, . . .

32. $-2, -1, 0, 1, 2, 3, \ldots$

33. $-4, 0, 4, 8, 12, 16, \ldots$

34. $-3, 9, -27, 81, -243, 729, \ldots$

35. $1, \dfrac{1}{2}, \dfrac{1}{3}, \dfrac{1}{4}, \dfrac{1}{5}, \dfrac{1}{6}, \ldots$

36. $\dfrac{1}{2}, \dfrac{1}{4}, \dfrac{1}{6}, \dfrac{1}{8}, \dfrac{1}{10}, \dfrac{1}{12}, \ldots$

37. $1, \dfrac{1}{4}, \dfrac{1}{9}, \dfrac{1}{16}, \dfrac{1}{25}, \dfrac{1}{36}, \ldots$

38. $1, \dfrac{1}{8}, \dfrac{1}{27}, \dfrac{1}{64}, \dfrac{1}{125}, \dfrac{1}{216}, \ldots$

39. $2, \dfrac{3}{2}, \dfrac{4}{3}, \dfrac{5}{4}, \dfrac{6}{5}, \dfrac{7}{6}, \ldots$

40. $0, \dfrac{-1}{2}, \dfrac{-2}{3}, \dfrac{-3}{4}, \dfrac{-4}{5}, \dfrac{-5}{6}, \ldots$

41. A newspaper carrier earning \$15.50 per week is depositing \$3.50 per week in savings. The carrier decides to increase the amount deposited by 10% per week. Write a sequence that gives the amount being deposited the 1st week, 2nd week, 3rd week, 4th week, 5th week, and 6th week.

42. Assume that the carrier in Exercise 41 is making \$22.00 per week and is spending \$21.00 per week. The carrier decides to decrease the amount spent each week by 5%. Write a sequence that represents the amount that is being spent each week for the first six weeks.

16.2 Series and summation

The sum of the terms of a sequence is called a **series.** For the sequence 1, 3, 5, 7, 9, the series associated with it is

$$S_5 = 1 + 3 + 5 + 7 + 9$$
$$= 25.$$

The capital letter S indicates sum, and the subscript 5 indicates that we are finding the sum of the first five terms.

DEFINITION 16.3

> The sum S_n of an indicated number n of terms of a sequence is called a **series.**

If the sequence has a finite number of terms, then the corresponding series is called a **finite series:**

$$S_n = a_1 + a_2 + a_3 + \cdots + a_n.$$

If the sequence has an infinite number of terms, then the corresponding series is called an **infinite series:**

$$a_1 + a_2 + a_3 + a_4 + \cdots + a_i + \cdots.$$

EXAMPLE 1 Write the series associated with the sequence 3, 9, 27, 81, 243, . . . , 3^n, where

a. $n = 3$ **b.** $n = 6$

Solution:

a. $S_3 = (3)^1 + (3)^2 + (3)^3$
$= 3 + 9 + 27$
$= 39.$

b. $S_6 = (3)^1 + (3)^2 + (3)^3 + (3)^4 + (3)^5 + (3)^6$
$= 3 + 9 + 27 + 81 + 243 + 729$
$= 1092.$ □

A series for which the general term is known can be written in a compact form using **summation notation** or **sigma notation.** For example, the sum S_n of the sequence with general term a_i can be written using summation notation as follows:

$$S_n = \sum_{i=1}^{n} a_i = a_1 + a_2 + a_3 + a_4 + \cdots + a_n.$$

The capital Greek letter sigma, $\sum$, stands for "sum of." The letter i is called the **index of summation.** The letter i indicates that we start with 1 and end with the number n, which appears on the top of the sigma. The letter i is called a dummy variable and could be replaced by any letter. Thus we could write

$$\sum_{i=1}^{n} a_i \quad \text{or} \quad \sum_{j=1}^{n} a_j \quad \text{or} \quad \sum_{k=1}^{n} a_k.$$

The important parts of the summation notation are 1, n, and the expression for the general term. Thus, the series on the right of the equal sign is the sum of the terms a_1 through a_n.

Knowing the general term of the sequence

$$\frac{2}{3}, \frac{3}{4}, \frac{4}{5}, \frac{5}{6}, \frac{6}{7}, \frac{7}{8}, \ldots, \frac{1+n}{2+n},$$

The corresponding series can be written in summation notation as

$$S_n = \sum_{i=1}^{n} \frac{1+i}{2+i}$$

or in expanded form as

$$S_n = \frac{2}{3} + \frac{3}{4} + \frac{4}{5} + \frac{5}{6} + \cdots + \frac{1+n}{2+n}.$$

EXAMPLE 2 Write the series $\sum_{i=1}^{6} 3i$ in expanded form.

Solution:

$$\sum_{i=1}^{6} 3i = 3(1) + 3(2) + 3(3) + 3(4) + 3(5) + 3(6)$$
$$= 3 + 6 + 9 + 12 + 15 + 18$$
$$= 63.$$ □

EXAMPLE 3 Write the series $\sum_{i=1}^{5} (-1)^i \left(\frac{3}{i}\right)$ in expanded form.

Solution:

$$\begin{aligned}\sum_{i=1}^{5}(-1)^i\left(\frac{3}{i}\right) &= (-1)^1\left(\frac{3}{1}\right)+(-1)^2\left(\frac{3}{2}\right)+(-1)^3\left(\frac{3}{3}\right)\\ &\quad +(-1)^4\left(\frac{3}{4}\right)+(-1)^5\left(\frac{3}{5}\right)\\ &= (-1)(3)+(1)\left(\frac{3}{2}\right)+(-1)(1)+(1)\left(\frac{3}{4}\right)\\ &\quad +(-1)\left(\frac{3}{5}\right)\\ &= -3+\frac{3}{2}-1+\frac{3}{4}-\frac{3}{5}.\end{aligned}$$ □

EXAMPLE 4 Write the series $3+5+7+9+\cdots+(2n+1)$ using summation notation.

Solution: The series $3+5+7+9+\cdots+(2n+1)$ is written in summation notation as

$$\sum_{k=1}^{n}(2k+1).$$

Therefore

$$S_n=\sum_{k=1}^{n}(2k+1).$$ □

EXAMPLE 5 Write the series

$$1-\frac{3}{2}+\frac{9}{5}-2+\frac{15}{7}-\frac{9}{4}+\cdots+\frac{(-1)^{n+1}3n}{n+2}$$

using summation notation.

Solution: The series $1-\frac{3}{2}+\frac{9}{5}-2+\frac{15}{7}-\frac{9}{4}+\cdots+\frac{(-1)^{n+1}3n}{n+2}$ is written in summation notation as

$$\sum_{k=1}^{n}\frac{(-1)^{k+1}3k}{k+2}.$$ □

EXAMPLE 6 Write the series

$$\sum_{i=4}^{7}(2i^2+3)$$

in expanded form and determine the sum.

Solution:

$$\sum_{i=4}^{7}(2i^2+3) = (2(4)^2+3)+(2(5)^2+3) + (2(6)^2+3)+(2(7)^2+3) = 35+53+75+101 = 264. \quad \square$$

EXAMPLE 7 Write the series $(y^2+5)+(y^2+6)+(y^2+7)+(y^2+8)$ using summation notation.

Solution: A general term that gives each of these terms is

$$a_i = (y^2+i),$$

where i assumes the values 5 through 8 inclusive. Thus, writing in summation notation, we have

$$\sum_{i=5}^{8}(y^2+i). \quad \square$$

16.2 Exercises

In Exercises 1–10, write the series associated with each of the sequences for (a) S_3 and (b) S_8.

1. $2, 4, 8, 16, \ldots, 2^n$

2. $-2, 4, -8, 16, \ldots, (-1)^n 2^n$

3. $\frac{1}{3}, \frac{1}{9}, \frac{1}{27}, \frac{1}{81}, \ldots, \left(\frac{1}{3}\right)^n$

4. $1, -\frac{1}{3}, \frac{1}{9}, -\frac{1}{27}, \ldots, \left(\frac{-1}{3}\right)^{n-1}$

5. $5, 7, 9, 11, \ldots, (2n+3)$

6. $\frac{1}{4}, 0, -\frac{1}{6}, -\frac{2}{7}, \ldots, \frac{2-n}{3+n}$

7. $-\frac{1}{4}, \frac{0}{5}, \frac{1}{6}, \frac{2}{7}, \ldots, \frac{n-2}{n+3}$

8. $-2, 16, -216, 4096, \ldots, (-2n)^n$

9. $1, -2, 3, -4, \ldots, (-1)^{n+1}n$

10. $1, -4, 9, -16, \ldots, (-1)^{n+1}n^2$

In Exercises 11–20, write each of the series in expanded form and determine the sum.

11. $S_5 = \sum_{i=1}^{5} \frac{(-2)^{i+1}}{i}$

12. $S_4 = \sum_{i=1}^{4} (-1)^i(2i-1)^2$

13. $S_4 = \sum_{i=1}^{4} \frac{1}{i}\, 3^{i+1}$

14. $S_6 = \sum_{i=1}^{6} e^{i-2}$

15. $S_6 = \sum_{i=1}^{6} \frac{(-1)^i 2^i}{i}$

16. $S_6 = \sum_{i=2}^{7} \frac{(-1)^i 2^{2i+1}}{2i+1}$

17. $S_5 = \sum_{i=1}^{5} \log(2+i)$

18. $S_7 = \sum_{i=1}^{7} (|i-2| - |3-i|)$

19. $\sum_{i=2}^{8} \frac{i+1}{i+5}$

20. $\sum_{i=1}^{5} (2i-1)(i+2)$

In Exercises 21–36, write each series using summation notation.

21. $3+6+9+12+15+\cdots+3n$

22. $\frac{2}{2}+\frac{4}{4}+\frac{6}{8}+\frac{8}{16}+\frac{10}{32}+\cdots+\frac{2n}{2^n}$

23. $\frac{-1}{1}+\frac{2}{2}-\frac{3}{4}+\frac{4}{8}-\frac{5}{16}+\cdots+\frac{(-1)^n n}{2^{n-1}}$

24. $\frac{1}{1}+\frac{2}{3}+\frac{3}{9}+\frac{4}{27}+\frac{5}{81}+\cdots+\frac{n}{3^{n-1}}$

25. $\frac{1}{1^2}+\frac{1}{2^2}+\frac{1}{3^2}+\frac{1}{4^2}+\frac{1}{5^2}$

26. $\frac{1}{1^3}+\frac{1}{2^3}+\frac{1}{3^3}+\frac{1}{4^3}+\frac{1}{5^3}$

27. $\frac{1}{1^2} - \frac{1}{2^2} + \frac{1}{3^2} - \frac{1}{4^2} + \frac{1}{5^2}$

28. $\frac{1}{1^3} - \frac{1}{2^3} + \frac{1}{3^3} - \frac{1}{4^3} + \frac{1}{5^3}$

29. $\frac{1 \cdot 2}{2} + \frac{2 \cdot 3}{2} + \frac{3 \cdot 4}{2} + \frac{4 \cdot 5}{2} + \frac{5 \cdot 6}{2} + \cdots + \frac{n(n+1)}{2}$

30. $\frac{3}{1 \cdot 1} + \frac{3}{2 \cdot 3} + \frac{3}{3 \cdot 5} + \frac{3}{4 \cdot 7} + \frac{3}{5 \cdot 9} + \cdots + \frac{3}{n(2n-1)}$

31. $(z+4) + (z+5) + (z+6) + (z+7) + (z+8)$

32. $z^2 + z^3 + z^4 + z^5 + z^6$

33. $\frac{z-3}{z+3} + \frac{z-4}{z+4} + \frac{z-5}{z+5} + \frac{z-6}{z+6} + \frac{z-7}{z+7}$

34. $\frac{z^2}{(z+2)^2} + \frac{z^3}{(z+3)^2} + \frac{z^4}{(z+4)^2} + \frac{z^5}{(z+5)^2} + \frac{z^6}{(z+6)^2}$

35. $z^3(z+3) + z^4(z+4) + z^5(z+5) + z^6(z+6) + z^7(z+7)$

36. $(-1)^5(z^5-5) + (-1)^6(z^6-6) + (-1)^7(z^7-7) + (-1)^8(z^8-8)$

In Exercises 37–46, write each of the series in expanded form.

37. $\sum_{k=1}^{5} \frac{k^2}{2k+3}$

38. $\sum_{k=1}^{n} \frac{y+k}{k^2}$

39. $\sum_{k=1}^{4} k \ln(k)$

40. $\sum_{k=1}^{5} \frac{k-1}{\sqrt{k+1}}$

41. $\sum_{k=1}^{n} k(k+1)(k-2)$

42. $\sum_{k=1}^{n} \frac{k(k+1)(2k+1)}{6}$

43. $\sum_{k=1}^{n} \left(\frac{2k^2-k+1}{2}\right)^2$

44. $\sum_{k=1}^{6} \frac{(-1)k}{\sqrt[3]{k}}$

45. $\sum_{k=2}^{5} (y^2 + 2ky + k)$

46. $\sum_{k=3}^{7} \frac{y+2}{\sqrt{k}}$

47. A truck starting from rest travels at the rate of 2 ft the first second, 3 ft the second second, 4 ft the third second, 5 ft the fourth second, and 6 ft the fifth second. If the truck continues to accelerate at the same rate, (a) how many feet will the truck cover in the ninth second, and (b) what is the total distance the truck moves in nine seconds?

48. The temperature at 9:00 A.M. was 3°C and increased uniformly at the rate of 1.5° per hour.

   a. Write the first five terms of the sequence.
   b. Write a general term and express in summation notation.
   c. Using the results from (b) determine the sum of the hourly temperatures for the first five hours.

16.3 Arithmetic sequences and series

In examining the terms of the sequence 1, 4, 7, 10, 13, 16, . . . , we see that any two successive terms have a difference of three. That is, $4 - 1 = 3$, $7 - 4 = 3$, $10 - 7 = 3$, etc. All the successive terms of the sequence 1, 5, 9, 13, 17, 21, . . . , differ by four. That is, $5 - 1 = 4$, $9 - 5 = 4$, $13 - 9 = 4$, etc. Any sequence where the terms have a **common difference** is called an **arithmetic sequence** or **arithmetic progression.**

EXAMPLE 1 A child is playing inside a car and accidentally releases the hand brake. The car is on an incline and starts rolling down the hill. At the end of the first second it has rolled 2 m, at the end of the second second it has rolled 4 m, and at the end of the third second it has rolled 6 m. (These figures include friction.)

a. How many meters did the car travel each second?

b. How far had the car traveled at the end of the six seconds?

Solution:

a. We can determine how far the car will go each second by finding the common difference d. That is, $d = 4 - 2$ or $d = 2$ m.

b. To find out how far the car has traveled in six seconds all we need do is multiply the common difference times the number of seconds. That is,

$$\begin{aligned} \text{distance after six seconds} &= (d)(t) \\ &= (2)(6) \qquad \textbf{Substituting} \\ &= 12 \text{ m.} \qquad \square \end{aligned}$$

DEFINITION 16.4

> An arithmetic sequence is a sequence of numbers called terms, such that any two successive terms differ by the same number, called the **common difference** d.

EXAMPLE 2 For the arithmetic sequence 15, 12, 9, 6, . . .

a. find the common difference, and

b. determine the fifth term.

Solution:

a. The common difference is found by subtracting a preceding term from a succeeding term. Thus,

$d = 12 - 15$ or $d = -3$.

b. To determine the fifth term we add the common difference to the fourth term $6 + (-3) = 3$. The fifth term of the sequence is 3. □

In the arithmetic sequence 4, 11, 18, 25, 32, the common difference is seven. If the order of the terms is reversed, a new arithmetic sequence is formed, 32, 25, 18, 11, 4. The common difference of the new arithmetic sequence is -7.

To develop a rule to determine any term of an arithmetic sequence let us use the arithmetic sequence 32, 25, 18, 11, 4. From our previous discussion we know that the sequence has a first term of 32 and a common difference of -7.

$$\begin{aligned} a_2 &= 32 + (-7) = 25. \\ a_3 &= 32 + 2(-7) = 18. \\ a_4 &= 32 + 3(-7) = 11. \\ a_5 &= 32 + 4(-7) = 4. \end{aligned}$$

Note that to find any term we add to the first term the product of the common difference and one less than the number of the term we are finding. For example, to get the fourth term we do the following:

$$a_4 = 32 + (4 - 1)(-7).$$

Thus if a_1 denotes the first term of an arithmetic sequence and d denotes the common difference, the terms of the sequence are

$$\begin{aligned} a_2 &= a_1 + (2 - 1)d. \\ a_3 &= a_1 + (3 - 1)d. \\ a_4 &= a_1 + (4 - 1)d. \\ &\vdots \\ a_n &= a_1 + (n - 1)d. \end{aligned}$$

DEFINITION 16.5

> The nth term of an arithmetic sequence is given by the formula
>
> $$a_n = a_1 + (n - 1)d,$$
>
> where a_1 is the first term and d is the common difference.

EXAMPLE 3 Find the 13th term of an arithmetic sequence whose first term is 15 and whose common difference is 3.

Solution: We are given that $a_1 = 15$, $d = 3$, and $n = 13$, and we want to find a_{13}. Substituting into the formula

$$a_n = a_1 + (n - 1)d, \qquad \textbf{Definition 16.5}$$

we have

$$\begin{aligned} a_{13} &= 15 + (13 - 1)3 \\ &= 51. \end{aligned}$$

Thus, the 13th term of the sequence is 51. □

EXAMPLE 4 Find the first four terms of the arithmetic sequence in which the first term is 17 and the common difference is 4.

Solution: With $a_1 = 17$ and $d = 4$ we can establish the following formula for the sequence

$$a_n = 17 + (n - 1)4.$$

Now let $n = 1, 2, 3,$ and 4, respectively, and we will obtain the first four terms of the sequence:

$$a_1 = 17 + (1 - 1)4 = 17. \qquad a_3 = 17 + (3 - 1)4 = 25.$$
$$a_2 = 17 + (2 - 1)4 = 21. \qquad a_4 = 17 + (4 - 1)4 = 29.$$

Thus the arithmetic sequence with first term 17 and common difference of 4 is

$$17, 21, 25, 29, \ldots, [17 + (n - 1)4]. \quad \square$$

EXAMPLE 5 Find the first term of a seven-term arithmetic sequence in which the common difference is 5 and the seventh term is 13.

Solution: We know that $n = 7$, $a_7 = 13$, and $d = 5$, and we want to find a_1. Thus

$$\begin{aligned} a_n &= a_1 + (n - 1)d && \textbf{Definition 16.5} \\ 13 &= a_1 + (7 - 1)5 \\ 13 &= a_1 + 30 \\ -17 &= a_1. \end{aligned}$$

Thus, the first term of the sequence is -17. $\square$

EXAMPLE 6 How many terms are there in an arithmetic sequence in which the first term is $2\frac{1}{2}$, the last term is $5\frac{3}{4}$ and the common difference is $\frac{1}{4}$?

Solution: We know that $a_1 = \frac{5}{2}$, $a_n = \frac{23}{4}$, and $d = \frac{1}{4}$, and we want to find n.

$$\begin{aligned} a_n &= a_1 + (n - 1)d && \textbf{Definition 16.5} \\ \frac{23}{4} &= \frac{5}{2} + (n - 1)\frac{1}{4}. \end{aligned}$$

Multiplying each term by 4 we have

$$\begin{aligned} 23 &= 10 + n - 1 \\ 14 &= n && \textbf{Solving for } \boldsymbol{n} \end{aligned}$$

Thus there are 14 terms in the sequence. $\square$

As we learned in the previous section, associated with every sequence is a series. The series associated with the arithmetic sequence

$$a_1, (a_1 + d), (a_1 + 2d), (a_1 + 3d), \ldots, [a_1 + (n - 1)d]$$

is

$$\begin{aligned} S_n = a_1 &+ (a_1 + d) + (a_1 + 2d) + (a_1 + 3d) \\ &+ \cdots + [a_1 + (n - 1)d] \end{aligned} \qquad \textbf{[1]}$$

or

$$S_n = \sum_{i=1}^{n} [a_1 + (i - 1)d].$$

The sum could just as well be written in reverse order, or as

$$S_n = a_n + (a_n - d) + (a_n - 2d) + (a_n - 3d) + \cdots + [a_n - (n-1)d]. \quad [2]$$

A formula to determine the sum of the terms of a sequence can be developed by finding the sum of Eqs. [1] and [2]. Adding Eqs. [1] and [2] we have

$$\begin{aligned} 2\,S_n &= (a_1 + a_n) + (a_1 + d + a_n - d) + (a_1 + 2d + a_n - 2d) \\ &\quad + \cdots + [a_1 + (n-1)d + a_n - (n-1)d] \\ &= \underbrace{(a_1 + a_n) + (a_1 + a_n) + (a_1 + a_n) + \cdots + (a_1 + a_n)}_{n \text{ terms}} \\ &= n(a_1 + a_n) \\ S_n &= \frac{n}{2}(a_1 + a_n). \end{aligned}$$

This demonstration is the proof of the following theorem, which gives the formula for finding S_n, called the ***n*th partial sum.**

THEOREM 16.1

> The sum of the first n terms of an arithmetic sequence whose first term is a_1 and whose nth term is a_n is given by
>
> $$S_n = \frac{n}{2}(a_1 + a_n).$$

EXAMPLE 7 Find the sum of the terms of the arithmetic sequence with eight terms, whose first term is 3 and whose last term is 19.

Solution: In this problem, $n = 8$, $a_1 = 3$, and $a_n = 19$. Substituting in the equation

$$S_n = \frac{n}{2}(a_1 + a_n) \qquad \textbf{Theorem 16.1}$$

$$S_8 = \frac{8}{2}(3 + 19) \quad \text{or} \quad S_8 = 88.$$

Thus the sum of the first eight terms of the sequence is 88. □

EXAMPLE 8 Find the sum of the first 12 terms of the sequence 3, 9, 15, 21, ...

Solution: The first term is 3 and the common difference is 6. The 12th term of the sequence is

$$S_{12} = 3 + (12 - 1)6 \quad \text{or} \quad S_{12} = 69.$$

Now substituting $n = 12$, $a_1 = 3$, and $a_n = 69$ into the formula

$$S_n = \frac{n}{2}(a_1 + a_n), \qquad \textbf{Theorem 16.1}$$

we have

$$S_{12} = \frac{12}{2}(3 + 69) \quad \text{or} \quad S_{12} = 432.$$

Thus the sum of the first 12 terms is 432. □

The sum of the first 12 terms in Example 8 could have been found without knowing a_n. Since

$$a_n = a_1 + (n - 1)d,$$

if we substitute this value in the formula

$$S_n = \frac{n}{2}(a_1 + a_n)$$

for a_n, we have

$$S_n = \frac{n}{2}[a_1 + a_1 + (n - 1)d]$$

or

$$S_n = \frac{n}{2}[2a_1 + (n - 1)d]$$

In Example 8 we were given the following values: $a_1 = 3$, $d = 6$, and $n = 12$. Substituting in the formula

$$S_n = \frac{n}{2}[2a_1 + (n - 1)d],$$

we have

$$S_{12} = \frac{12}{2}[2(3) + (12 - 1)6] \quad \text{or} \quad S_{12} = 432.$$

EXAMPLE 9 The number of citizens in a town increased from 32,000 to 60,000 in six years. Find the yearly increase if this rate is assumed to be constant.

Solution: Since the yearly increase is constant it is an arithmetic progression with the constant difference d being the yearly increase. We know the initial term $a_1 = 32{,}000$, the number of years $n = 6$, and the nth term $a_n = 60{,}000$. Substituting in the formula

$$\begin{aligned} a_n &= a_1 + (n-1)d \\ 60{,}000 &= 32{,}000 + (6-1)d \\ 28{,}000 &= 5d \\ d &= 5600. \end{aligned}$$

Thus the population increase was 5600 people per year. □

16.3 Exercises

In Exercises 1–10, determine whether or not the sequence is an arithmetic sequence. If it is an arithmetic sequence, determine the common difference.

1. 2, 4, 6, 8, 10, . . .
2. 1, 3, 5, 7, 9, . . .
3. 1, 4, 6, 9, 11, . . .
4. 5, 10, 15, 25, 30, . . .
5. $\frac{1}{2}, 1, \frac{3}{2}, 2, \frac{5}{2}, \ldots$
6. $\frac{1}{5}, \frac{2}{5}, \frac{3}{5}, \frac{4}{5}, 1, \ldots$
7. $\frac{1}{2}, \frac{1}{4}, -\frac{1}{4}, -\frac{1}{2}, -1, \ldots$
8. 0.1, 1.0, 1.9, 2.8, 3.7, . . .
9. 1, 5, 10, 16, 23, . . .
10. 6, 11, 16, 21, 26, . . .

In Exercises 11–20, (a) construct the first five terms of the arithmetic sequence and (b) determine the sum of the first five terms.

11. $a_1 = 13$ and $d = 5$
12. $a_1 = -11$ and $d = 7$
13. $a_1 = 15$ and $d = -3$
14. $a_1 = \frac{1}{3}$ and $d = \frac{2}{3}$
15. $a_3 = 9$ and $d = -1$
16. $a_2 = 8$ and $d = 2$
17. $a_4 = -4$ and $d = \frac{1}{2}$
18. $a_3 = 6$ and $d = -\frac{1}{2}$
19. $a_2 = \frac{3}{5}$ and $d = \frac{3}{5}$
20. $a_2 = \frac{1}{4}$ and $d = -1$

In Exercises 21–30, some of the quantities a_1, d, a_n, n, S_n, are given; find the quantities that are not given.

21. $a_1 = 1$, $d = 2$, $n = 8$
22. $a_1 = 3$, $d = 5$, $n = 7$
23. $a_1 = 3$, $a_n = 21$, $d = 3$
24. $a_1 = 26$, $a_n = 2$, $d = -4$
25. $n = 8$, $a_1 = -3$, $a_n = 18$
26. $n = 6$, $a_1 = 17$, $a_n = 2$
27. $S_6 = 6.9$, $n = 6$, $a_1 = 0.4$
28. $S_8 = 11$, $n = 8$, $a_1 = \frac{1}{2}$
29. $S_6 = 78$, $n = 6$, $d = -4$
30. $S_n = 108$, $a_1 = 24$, $d = -3$

31. **a.** How many even numbers lie between 13 and 71?
b. Find their sum.

32. **a.** How many numbers that are divisible by 4 lie between 30 and 182?
b. Find their sum.

33. **a.** How many numbers that are divisible by 3 lie between 32 and 184?
b. Find their sum.

34. **a.** How many numbers that are divisible by 7 lie between 19 and 143?
b. Find their sum.

35. The first year Karen worked she saved \$1,000. In each succeeding year she saved \$100 more than the preceding year. How much has she accumulated at the end of 12 years? (Ignore interest.)

36. An employee receives a salary of \$16,000 per year with the understanding that the salary will increase by \$1,000 per year up to and including the 12th year.

a. What will the employee's salary be in the 12th year?
b. What are the total earnings for the 12 years?

37. A body falls 16 ft during the first second, 48 ft during the second second, 80 ft during the third second, and so on, in arithmetic sequence.

a. How far does it fall during the eighth second?
b. How far does it fall in the first eight seconds?

38. Three workers X, Y, and Z are hired by a company at the same time, each at a starting salary of \$16,000. X agrees to accept an annual raise of

\$2,000. *Y* agrees to accept a semi-annual increase of \$400. *Z* agrees to accept a quarterly raise of \$100. After four years have elapsed, who has earned more money, and how much more has that worker earned?

39. An entry in a soap-box derby coasts downhill covering 5 ft the first second, 12 ft the second second, and an average of 7 ft more each second than the previous second. If the racer reaches the bottom of the hill in 10 seconds, how far did he coast? How far did he travel in the eighth second?

40. A taxi company charges \$2.00 for the first 1/4 mi and 35 cents for each 1/4 mi thereafter. How much is the fare for a 12-mi ride?

41. The first and last numbers of an arithmetic sequence were 6 and 8, respectively. Now we place five numbers between 6 and 8 so that the seven numbers form an arithmetic sequence. Determine the five numbers.

42. The first and last numbers of an arithmetic sequence are 4 and 25, respectively. If we place six numbers between 4 and 25 so that the eight numbers form an arithmetic sequence, determine the six numbers.

43. The number of bacteria in a culture increased from 32,000 to 70,000 in six days. Find the daily rate of increase if this rate is assumed to be constant.

16.4 Geometric sequences and series

In the study of population increases the change from year to year is not necessarily a common difference. Also, in the study of atomic decay of an element, the changes over set periods of time are not the same for each period. Frequently we can develop sequences to show these changes. These are called geometric sequences.

An example of geometric sequence is

$$2, 6, 18, 54, 162, \ldots,$$

where each succeeding term is obtained by multiplying each preceding term by a common factor. In the sequence 2, 6, 18, 54, 162, . . . , the common factor is 3. That is $2 \cdot 3 = 6$, $6 \cdot 3 = 18$, $18 \cdot 3 = 54$, and so on. A sequence of this type is called a **geometric sequence** or **geometric progression** and 3 is called the **common ratio.**

DEFINITION 16.6

A **geometric sequence** $a_1, a_2, a_3, a_4, \ldots, a_{n-1}, a_n, \ldots,$ is a sequence of numbers called terms, such that the ratio of any term to the preceding term is a fixed number called the **common ratio** r:

$$\left(r = \frac{a_n}{a_{n-1}}\right)$$

EXAMPLE 1 For the geometric sequence 4, 12, 36, 108, . . .

a. Find the common ratio.

b. Determine the fifth and sixth terms.

Solution:

a. The common ratio can be found by setting up the ratio of any succeeding term to the preceding term. For example,

$$r = \frac{12}{4} \quad \text{or} \quad r = \frac{36}{12} \quad \text{or} \quad r = \frac{108}{36}.$$

In each case the quotient is 3. Thus the common ratio r for the geometric sequence 4, 12, 36, 108, . . . , is 3.

b. To determine the fifth term of the sequence we multiply the fourth term, 108, by the common ratio, 3. Thus the fifth term is $108 \times 3 = 324$, and the sixth term is $324 \times 3 = 972$. □

In the geometric sequence $1, \frac{1}{3}, \frac{1}{9}, \frac{1}{27}, \frac{1}{81}, \frac{1}{243}$, the common ratio is $\frac{1}{3}$. If the order of the terms of the sequence is reversed, a new geometric sequence is formed $\frac{1}{243}, \frac{1}{81}, \frac{1}{27}, \frac{1}{9}, \frac{1}{3}, 1$. The common ratio of the new geometric sequence is 3.

To develop a rule to determine any term of a geometric sequence let us use the geometric sequence $\frac{1}{243}, \frac{1}{81}, \frac{1}{27}, \frac{1}{9}, \frac{1}{3}, 1$. From the previous discussion we know that the sequence has a first term of $\frac{1}{243}$ and a common ratio of 3.

$$\begin{aligned}
a_2 &= \frac{1}{243} \cdot 3 && = \frac{1}{243} \cdot 3^1 = \frac{1}{81}.\\
a_3 &= \frac{1}{81} \cdot 3 = \left(\frac{1}{243} \cdot 3\right) \cdot 3 && = \frac{1}{243} \cdot 3^2 = \frac{1}{27}.\\
a_4 &= \frac{1}{27} \cdot 3 = \left(\frac{1}{243} \cdot 3^2\right) \cdot 3 && = \frac{1}{243} \cdot 3^3 = \frac{1}{9}.\\
a_5 &= \frac{1}{9} \cdot 3 = \left(\frac{1}{243} \cdot 3^3\right) \cdot 3 && = \frac{1}{243} \cdot 3^4 = \frac{1}{3}.\\
a_6 &= \frac{1}{3} \cdot 3 = \left(\frac{1}{243} \cdot 3^4\right) \cdot 3 && = \frac{1}{243} \cdot 3^5 = 1.
\end{aligned}$$

To find the ith term we multiply the first term by the common ratio raised to the $i - 1$ power. For example, to find the fourth term we do the following:

$$a_4 = \frac{1}{243} \cdot 3^{4-1}.$$

Thus if a_1 denotes the first term of a geometric sequence and r the common ratio, then the terms of the sequence are

$$a_2 = a_1\, r^{2-1}.$$
$$a_3 = a_1\, r^{3-1}.$$
$$a_4 = a_1\, r^{4-1}.$$
$$\vdots$$
$$a_n = a_1\, r^{n-1}.$$

DEFINITION 16.7

The nth term of a geometric sequence is given by the formula

$$a_n = a_1\, r^{n-1},$$

where a_1 is the first term of the sequence and r is the common ratio.

EXAMPLE 2 Find the first four terms of the geometric sequence in which the first term is 5 and the common ratio is 2.

Solution: With $a_1 = 5$ and $r = 2$ we can establish the following general formula for the sequence:

$$a_n = 5 \cdot 2^{n-1} \qquad \textbf{Definition 16.7}$$

Now let $n = 1, 2, 3,$ and 4, respectively, and we will obtain the first four terms of the sequence.

$$\begin{aligned} a_1 &= 5 \cdot 2^{1-1} & a_2 &= 5 \cdot 2^{2-1} \\ &= 5 \cdot 2^0 & &= 5 \cdot 2^1 \\ &= 5. & &= 10. \\ a_3 &= 5 \cdot 2^{3-1} & a_4 &= 5 \cdot 2^{4-1} \\ &= 5 \cdot 2^2 & &= 5 \cdot 2^3 \\ &= 20. & &= 40. \end{aligned}$$

Thus the first four terms of the geometric sequence with first term 5 and common ratio 2 are 5, 10, 20, and 40. □

EXAMPLE 3 Find the eighth term of the geometric sequence $1, \frac{1}{2}, \frac{1}{4}, \frac{1}{8}, \ldots$.

Solution: We know that $a_1 = 1$, $n = 8$, and $r = \frac{1}{2}$ $\left(\text{since } \frac{1/2}{1} = \frac{1}{2}\right)$. Substituting into the formula

$$a_n = a_1\, r^{n-1} \qquad \textbf{Definition 16.7}$$

we have

$$a_8 = 1 \cdot \left(\frac{1}{2}\right)^{8-1} = 1 \cdot \left(\frac{1}{2}\right)^7 = \frac{1}{128}.$$

Thus the eighth term of the sequence is 1/128. □

EXAMPLE 4 Find the first term of a seven-term geometic sequence in which the common ratio is 3 and the seventh term is 162.

Solution: We know that $n = 7$, $a_7 = 162$, and $r = 3$. We want to find a_1. Substituting in the formula:

$$a_n = a_1\, r^{n-1} \qquad \textbf{Definition 16.7}$$

we have

$$162 = a_1 \cdot 3^{7-1}$$
$$162 = a_1(729)$$
$$\frac{162}{729} = a_1$$
$$\frac{2}{9} = a_1.$$

Thus the first term of the geometric sequence is 2/9. □

As stated earlier, associated with every sequence is a series. The series associated with the geometric sequence

$$a_1, a_1r, a_1r^2, a_1r^3, a_1r^4, \ldots, a_1r^{n-1},$$

is

$$S_n = a_1 + a_1r + a_1r^2 + a_1r^3 + a_1r^4 + \cdots + a_1r^{n-1}, \qquad \textbf{[1]}$$

or

$$S_n = \sum_{i=1}^{n} a_1r^{i-1}.$$

We want a formula to determine the sum S_n in terms of a_1, n, and r. If we multiply Eq. [1] by r we have

$$rS_n = a_1r + a_1r^2 + a_1r^3 + a_1r^4 + \cdots + a_1r^n. \qquad \textbf{[2]}$$

Now subtract Eq. [2] from Eq. [1].

$$S_n = a_1 + a_1r + a_1r^2 + a_1r^3 + a_1r^4 + \cdots + a_1r^{n-1} \qquad \textbf{[1]}$$
$$rS_n = a_1r + a_1r^2 + a_1r^3 + a_1r^4 + a_1r^5 + \cdots + a_1r^n \qquad \textbf{[2]}$$
$$S_n - rS_n = a_1 - a_1r^n \qquad \textbf{Eq. [1]–Eq. [2]}$$
$$S_n(1 - r) = a_1(1 - r^n) \qquad \textbf{Common-term factoring}$$
$$S_n = \frac{a_1(1 - r^n)}{1 - r},$$

provided $r \neq 1$.

This demonstration is the proof of the following theorem, which gives the formula for finding S_n, called the **nth partial sum** of the geometric sequence.

THEOREM 16.2

> The sum of the first n terms of a geometric sequence whose first term is a_1 and whose common ratio is r is given by
>
> $$S_n = \frac{a_1(1 - r^n)}{1 - r},$$
>
> provided $r \neq 1$.

EXAMPLE 5 Find the sum of the geometric sequence with six terms, whose first term is 1 and whose common ratio is 5.

Solution: In this problem $n = 6$, $a_1 = 1$, and $r = 5$. Substituting in the equation

$$S_n = \frac{a_1(1 - r^n)}{1 - r} \qquad \textbf{Theorem 16.2}$$

we have

$$\begin{aligned} S_n &= \frac{1(\quad - 5^6)}{1 - 5} \\ &= \frac{-15624}{-4} \\ &= 3906. \end{aligned}$$

Thus the sum of the first six terms of the sequence is 3906. □

EXAMPLE 6 The length of the first swing of a pendulum is 150 mm. If on each succeeding swing the pendulum travels 4/5 as far as on the preceding swing, how far will the pendulum have traveled after six swings?

Solution: The length of the first swing is $a_1 = 150$, the constant decrease in length of each swing is $r = 4/5$, and the number of swings is $n = 6$. Substituting into the formula,

$$\begin{aligned} S_n &= \frac{a_1(1 - r^n)}{1 - r} \qquad \textbf{Theorem 16.2} \\ &= \frac{150(1 - (4/5)^6)}{1 - 4/5} \\ &= 553.\ \text{mm}. \end{aligned}$$

The distance the pendulum has traveled in six swings is 553. mm. □

16.4 Exercises

In Exercises 1 – 10, determine whether or not the sequence is a geometric sequence. If it is, determine the common ratio and find the seventh term.

1. 1, 2, 4, 8, 16, . . .

2. 1, 3, 9, 27, 81, . . .

3. 1, 4, 8, 12, 16, . . .

4. 1, 6, 12, 18, 24, . . .

5. $1, \frac{1}{4}, \frac{1}{16}, \frac{1}{64}, \frac{1}{256}, \ldots$

6. $1, \frac{1}{5}, \frac{1}{25}, \frac{1}{125}, \frac{1}{625}, \ldots$

7. 16807, 2401, 343, 49, 7, . . .

8. $\frac{2}{81}, \frac{2}{27}, \frac{2}{9}, \frac{2}{3}, 2, \ldots$

9. $\frac{1}{3}, \frac{1}{6}, \frac{1}{12}, \frac{1}{24}, \frac{1}{48}, \ldots$

10. $5, \frac{5}{5}, \frac{10}{5}, \frac{15}{5}, \frac{20}{5}, \ldots$

In Exercises 11 – 20, (a) construct the first five terms of the geometric sequence and (b) determine the sum of the first five terms.

11. $a_1 = 3, r = 2$

12. $a_1 = -3, r = 2$

13. $a_1 = 2, r = \frac{1}{3}$

14. $a_1 = 5, r = \frac{1}{2}$

15. $a_1 = \frac{2}{5}, r = \frac{1}{4}$

16. $a_1 = \frac{3}{7}, r = \frac{1}{5}$

17. $a_1 = -4, r = 0.1$

18. $a_1 = -5, r = 0.2$

19. $a_1 = 4, r = -2$

20. $a_1 = \frac{1}{5}, r = -3$

In Exercises 21 – 30, some of the quantities a_1, r, a_n, n, S_n are given; find the quantities that are not given.

21. $a_1 = \frac{1}{2}, r = \frac{1}{3}, n = 4$

22. $a_1 = \frac{1}{3}, r = \frac{1}{2}, n = 4$

23. $a_6 = 96, a_1 = 3, n = 6$

24. $a_5 = 162, a_1 = 2, n = 5$

25. $S_6 = 189, n = 6, r = 2$

26. $S_5 = 284, n = 5, r = 3$

27. $a_1 = \frac{1}{3}, r = -3, n = 4$

28. $a_1 = \frac{5}{3}, r = -\frac{2}{3}, n = 5$

29. $S_4 = \frac{117}{125}, r = \frac{1}{5}, n = 4$

30. $S_5 = \frac{242}{81}, n = 5, r = \frac{1}{3}$

31. Tom decides that he wants to become a weight-lifting champion. He starts his weight-lifting program by lifting 500 N. His plan is to increase the weight by 10% per week. How much weight will he be lifting at the end of 5 weeks? What is the total amount of weight he has lifted over the 5 weeks?

32. Jane decides that she wants to save money by making weekly deposits in her savings account and doubling the amount of her deposit from the week before. If she begins by depositing \$1 the first week in January, how much will she have saved at the end of 10 weeks?

33. The number of bacteria in a culture increases by 20% each hour. If the culture contains 40,000 bacteria at 11:00 A.M., how many bacteria are present at 2:00 P.M.?

34. If the sum of the first three terms of a geometric sequence is 13 times as large as the first term, find the common ratio. (Two possible answers.)

35. If the sum of the first three terms of a geometric sequence is 21 times as large as the first term, find the common ratio. (Two possible answers.)

36. A sheet-metal worker took a job with a starting salary of \$15,000 and a guaranteed increase of 8% per year, based on the previous year's salary.

a. Determine the salary for the seventh year.

b. Determine the amount of money earned for the first seven years.

37. A carpenter was offered a job for 20 days with a choice between two methods of being paid. One choice is \$100 per day for the 20-day period; the other is to be paid 1 cent the first day, 2 cents the second day, 4 cents the third day, and so on doubling the amount paid from each preceding day for the 20-day period. Which method of payment would you choose and why?

38. An assembly-line worker is switched from hourly rate to piecework. The first day of piecework the worker produces 300 items. In order to make the same rate of pay, the worker must produce 500 items per day. At what rate must the worker increase production over the previous day in order to achieve the quota of 500 items per day in 20 days?

39. In a chemical plant a tank holds 60 gallons of a liquid soap, which mixes with water. After 12 gallons of the soap are drawn out, the tank is filled by replacing the soap with water. Then 12 gallons of the resulting mixture are drawn out, and the tank is again filled with water. If this operation is performed until eight batches have been drawn from the tank, how much of the original liquid remains?

16.5 Infinite geometric series

In the previous section we found the sum of a finite geometric series. In this section we will find the sum of an infinite geometric series, if indeed that sum does exist.

Let us consider the sequence $\frac{1}{3}, \frac{1}{9}, \frac{1}{27}, \frac{1}{81}, \frac{1}{243}, \ldots, \frac{1}{3^n}, \ldots,$ which has a common ratio r of $1/3$. We will use the formula for finding the sum of a geometric sequence to write the sum for $n = 1$ to 10 inclusive. Substituting in the formula $S_n = \frac{a_1(1 - r^n)}{1 - r}$, $r \neq 1$ we have the following sums:

$$S_1 = \frac{1}{3} = 0.33333\ldots \qquad S_6 = \frac{364}{729} = 0.49931\ldots$$

$$S_2 = \frac{4}{9} = 0.44444\ldots \qquad S_7 = \frac{1093}{2187} = 0.49977\ldots$$

$$S_3 = \frac{13}{27} = 0.48148\ldots \qquad S_8 = \frac{3280}{6561} = 0.49992\ldots$$

$$S_4 = \frac{40}{81} = 0.49382\ldots \qquad S_9 = \frac{9841}{19683} = 0.49997\ldots$$

$$S_5 = \frac{121}{243} = 0.49794\ldots \qquad S_{10} = \frac{29524}{59049} = 0.49999\ldots$$

With the calculations for the first ten sums it is evident that the sum would be $1/2$, as n would increase without bound.

Now let us examine the sequence 3, 9, 27, 81, 243, . . . , where $r = 3$. The first six sums are

$$S_1 = 3, \qquad S_3 = 39, \qquad S_5 = 363,$$
$$S_2 = 12, \qquad S_4 = 120, \qquad S_6 = 1092.$$

It can be seen from the pattern developing that as n increases the sum increases without bound.

Having examined two cases, one where $|r| < 1$ and the other where $|r| > 1$, let us look at the general formula for finding the sum of a finite geometric series.

$$S_n = \frac{a_1(1 - r^n)}{1 - r}, \qquad r \neq 1.$$

We can write this as

$$S_n = \frac{a_1 - a_1 \cdot r^n}{1 - r}, \qquad r \neq 1$$

or

$$S_n = \frac{a_1}{1 - r} - \left(\frac{a_1}{1 - r}\right) r^n, \qquad r \neq 1.$$

If we restrict the values of r to the open interval $(-1, 1)$ or $|r| < 1$, then as n takes on larger and larger values r^n gets smaller and smaller and eventually becomes zero. For example let $r = 1/4$, then r^n forms the sequence

$$\left(\frac{1}{4}\right)^1, \left(\frac{1}{4}\right)^2, \left(\frac{1}{4}\right)^3, \left(\frac{1}{4}\right)^4, \left(\frac{1}{4}\right)^5, \ldots, \left(\frac{1}{r}\right)^n, \ldots$$

or

$$\frac{1}{4}, \frac{1}{16}, \frac{1}{64}, \frac{1}{256}, \frac{1}{1024}, \ldots, \left(\frac{1}{r}\right)^n, \ldots .$$

It can be seen that as n becomes larger the terms of the sequence become smaller and smaller and for all practical purposes they become so small that we can say that the terms converge or approach zero. In a more formal manner using limits (which are studied in calculus) it can be shown that if $-1 < r < 1$, then as n grows without bound, r^n becomes arbitrarily small. Thus as n grows without bound

$$\left(\frac{a_1}{1 - r}\right) r^n \to 0^* \qquad \text{and} \qquad S_n \to \frac{a}{1 - r}.$$

Thus we can state the sum of the infinite geometric series as

$$S = \frac{a_1}{1 - r} \text{ for } -1 < r < 1 \qquad |r| < 1.$$

THEOREM 16.3

> The sum of the terms of an infinite geometric series whose first term is a_1 and whose common ratio is r is given by
>
> $$S = \frac{a_1}{1 - r},$$
>
> provided $-1 < r < 1$.

* $\to$ is read "approaches."

EXAMPLE 1 Find the sum of the series

$$1-\frac{3}{5}+\frac{9}{25}-\frac{27}{125}+\frac{81}{625}-\cdots$$

Solution: The series $1-\frac{3}{5}+\frac{9}{25}-\frac{27}{125}+\frac{81}{625}-\cdots$ has a common ratio r of $-\frac{3}{5}$ and $a_1 = 1$. Substituting into the formula

$$S=\frac{a_1}{1-r} \qquad \textbf{Theorem 16.3}$$

we have

$$S=\frac{1}{1-\left(-\frac{3}{5}\right)} \quad \text{or} \quad S=\frac{5}{8}.$$

The sum of the infinite geometric series is 5/8. □

Note that we are not really adding an infinite number of terms; that would be impossible even with a calculator. The sum S is a value that is attained with large values of n. In calculus we expand on this idea and call S a limit.

The series in the previous example could have been written in summation notation. For example,

$$\sum_{k=1}^{\infty} 1\left(-\frac{3}{5}\right)^{k-1}=\sum_{k=0}^{\infty} 1\left(-\frac{3}{5}\right)^{k}=1-\frac{3}{5}+\frac{9}{25}-\frac{27}{125}+\cdots+1\left(\frac{-3}{5}\right)^{n-1}+\cdots.$$

EXAMPLE 2 Find the sum of the series

a. $\displaystyle\sum_{k=1}^{\infty} 2\left(\frac{3}{7}\right)^{k-1}$ **b.** $\displaystyle\sum_{k=0}^{\infty} -4\left(\frac{2}{5}\right)^{k}$

Solution:

a. For the series

$$\sum_{k=1}^{\infty} 2\left(\frac{3}{7}\right)^{k-1},$$

$a_1 = 2$ and $r = 3/7$.

Substituting in the formula

$$S=\frac{a_1}{1-r},$$

we have

$$S = \frac{2}{1-(3/7)} = \frac{14}{4} = \frac{7}{2}.$$

Thus the sum of the infinite geometric series is $7/2$.

b. For the series

$$\sum_{k=0}^{\infty} -4\left(\frac{2}{5}\right)^k,$$

$a_1 = -4$ and $r = 2/5$.

Substituting in the formula

$$S = \frac{a_1}{1-r},$$

we have

$$S = \frac{-4}{1-(2/5)} \quad \text{or} \quad S = -\frac{20}{3}.$$

Thus the sum of the infinite geometric series is $-\dfrac{20}{3}$. □

A definition of rational numbers in Chapter 1 stated that a rational number can be expressed as a repeating decimal. Another definition also stated that a rational number can be expressed in the form a/b, where $b \neq 0$. We can now use the method of finding the sum of an infinite geometric series to determine the fraction that is equivalent to the repeating decimal.

EXAMPLE 3 Find the fraction that is equivalent to

$0.376376376 \ldots$

Solution: The first thing we do is to write the repeating decimal $0.\overline{376}$ as an infinite geometric series:

$$0.\overline{376} = 0.376 + 0.000376 + 0.000000376 + \cdots.$$

We can see that $a_1 = 0.376$ and $r = 0.001$. We can find the sum S since $|r| < 1$.

$$\begin{aligned} S &= \frac{a_1}{1-r} \\ &= \frac{0.376}{1-0.001} \qquad \textbf{Substituting} \\ &= \frac{0.376}{0.999} \quad \text{or} \quad S = \frac{376}{999}. \end{aligned}$$

Thus an equivalent expression for the rational number $0.\overline{376}$ is 376/999. □

EXAMPLE 4 A ball is dropped from a height of 2 m and each time it hits the ground it bounces back 5/6 of the previous height. What will be the total vertical motion of the ball? (Do not take into consideration the horizontal component of the motion.)

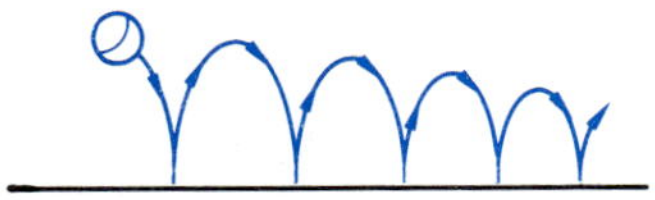

FIGURE 16.4

Solution: An illustration of the path of the ball is given in Fig. 16.4. The path of the ball can be represented by the following series:

$$2 + 2\left[2\left(\frac{5}{6}\right)\right] + 2\left[2\left(\frac{5}{6}\right)^2\right] + 2\left[2\left(\frac{5}{6}\right)^3\right] + \cdots. \quad [1]$$

This is not a geometric series but we can modify the form of this series so that it contains an infinite geometric series. We can rewrite Eq. [1] as

$$2 + 4\left(\frac{5}{6}\right) + 4\left(\frac{5}{6}\right)^2 + 4\left(\frac{5}{6}\right)^3 + \cdots. \quad [2]$$

Starting with the second term we have an infinite geometric series. Thus the sum of the series is

$$S = 2 + \frac{a_1}{1 - r}.$$

Substituting $a_1 = 4\left(\frac{5}{6}\right)$ and $r = \frac{5}{6}$ we have

$$S = 2 + \frac{4\left(\frac{5}{6}\right)}{1 - \frac{5}{6}}$$

$$= 2 + 20 \quad \text{or} \quad S = 22.$$

Thus the total distance traveled by the ball is 22 m. □

16.5 Exercises

In Exercises 1 – 20, find the sum of the infinite geometric series, if a sum exists.

1. $\frac{1}{2} + \frac{1}{4} + \frac{1}{8} + \frac{1}{16} + \cdots$
2. $\frac{1}{2} - \frac{1}{4} + \frac{1}{8} - \frac{1}{16} + \cdots$
3. $\frac{2}{3} - \frac{4}{9} + \frac{8}{27} - \frac{24}{81} + \cdots$
4. $\frac{2}{3} + \frac{4}{9} + \frac{8}{27} + \frac{24}{81} + \cdots$
5. $1 + 0.003 + 0.0009 + 0.000027 + \cdots$
6. $2 + 0.04 + 0.0008 + 0.000016 + \cdots$
7. $3 - 3 + 3 - 3 + \cdots$
8. $5 + 5 + 5 + 5 + \cdots$
9. $27 + 3 + \frac{1}{3} + \frac{1}{9} + \cdots$
10. $27 - 3 + \frac{1}{3} - \frac{1}{9} + \cdots$
11. $\sum_{k=1}^{\infty} 16\left(-\frac{1}{2}\right)^{k-1}$
12. $\sum_{k=1}^{\infty} 16\left(\frac{1}{4}\right)^{k-1}$

13. $\sum_{k=1}^{\infty} \frac{1}{3}(-3)^{k-1}$

14. $\sum_{k=1}^{\infty} \left(-\frac{1}{3}\right)(3)^{k-1}$

15. $\sum_{k=0}^{\infty} \left(\frac{3}{5}\right)^k$

16. $\sum_{k=0}^{\infty} \left(-\frac{5}{3}\right)^k$

17. $\sum_{k=0}^{\infty} 4(0.3)^k$

18. $\sum_{k=1}^{\infty} 150\left(\frac{1}{10}\right)^k$

19. $\sum_{k=1}^{\infty} \frac{3^k + 4^k}{7^k}$

20. $\sum_{k=1}^{\infty} \frac{7^k}{3^k + 4^k}$

In Exercises 21 – 30, find the fraction that is equivalent to the repeating decimal.

21. 0.9999 . . .

22. 0.5555 . . .

23. 0.45454545 . . .

24. 0.73737373 . . .

25. 0.137137137 . . .

26. 0.219219219 . . .

27. 0.48764876 . . .

28. 0.5962159621 . . .

29. $1.\overline{431}$

30. $5.\overline{762}$

In Exercises 31 – 36, two of the three quantities a_1, r, and S are given. Find the third quantity using the formula for finding the sum of a geometric series.

31. $S = 1,\ r = \frac{1}{2}$

32. $S = \frac{1}{3},\ a_1 = \frac{1}{2}$

33. $a_1 = \frac{2}{3},\ r = -\frac{2}{3}$

34. $a_1 = \frac{2}{3},\ r = \frac{2}{3}$

35. $S = \frac{100}{7},\ a_1 = 1$

36. $S = \frac{3}{8},\ r = -\frac{5}{3}$

37. A new superball is manufactured that, when dropped from a height of 5 m, rebounds 0.95 of the height from which it was dropped.

   a. Find the height the ball reaches on the tenth bounce.

   b. Find the total vertical distance the ball travels before it comes to rest.

38. A pendulum swings through an arc of 750 mm. On each successive swing the pendulum covers an arc 14/15 of the preceding swing.

   a. How far does the pendulum travel on the fifth swing?

   b. How far does the pendulum travel before it comes to rest?

39. In the race of the tortoise and the hare, the hare in the first minute after he wakes up reduces the distance between them to half what it was, in the next half minute reduces the remaining distance by half of itself, in the next quarter minute reduces the still-remaining distance by half, and so on. How long does it take the hare to overtake the tortoise?

40. On a steep mountain road the highway department built an artificial incline for cars and trucks to use in the event of brake failure. The artificial incline was designed so that if a truck were traveling at the rate of 70 mph, its speed would be reduced by 10 mph for each 100 ft traveled. What should be the minimum length of the artificial incline in order for a truck traveling at 70 mph to stop safely?

41. Zeno's paradox: Picture a 390-m oval racetrack that has a tape stretched across the track at the halfway mark. Also there are tapes stretched across the track at every point that marks half the remaining distance to complete a single lap. A runner running around the track has to break the first tape, then the second tape, then the third tape and so on. Will the runner ever finish the race? If we assume that the runner runs at 6.5 m/sec, the times between tape breakings form an infinite geometric progression. What is the sum of this progression?

16.6 The binomial theorem

In working with algebraic expressions it is frequently necessary to expand expressions of the form $(a + b)^n$, where n is a positive integer. For example,

$$\begin{aligned}
(a+b)^1 &= 1a^1b^0 + 1a^0b^1 &&= a + b \\
(a+b)^2 &= 1a^2b^0 + 2a^1b^1 + 1a^0b^2 &&= a^2 + 2ab + b^2 \\
(a+b)^3 &= 1a^3b^0 + 3a^2b^1 + 3a^1b^2 + 1a^0b^3 &&= a^3 + 3a^2b + 3ab^2 + b^3 \\
(a+b)^4 &= 1a^4b^0 + 4a^3b^1 + 6a^2b^2 + 4a^1b^3 + 1a^0b^4 &&= a^4 + 4a^3b + 6a^2b^2 + 4ab^3 + b^4
\end{aligned}$$

There are several things we note about the expansions:

1. For any expansion as we move from left to right we write

   $$a^n b^0,\ a^{n-1} b^1,\ a^{n-2} b^2,\ \ldots,\ a^1 b^{n-1},\ a^0 b^n.$$

   Since $a^0 = 1$ and $b^0 = 1$ we normally do not write a^0 and b^0. This was done for illustration purposes.
2. The sum of the exponents for each term must equal the degree of the expansion. For example, in the third-row terms (a^3b^0, a^2b^1, a^1b^2, a^0b^3) the sum of the exponents in each term is 3.
3. The coefficient of the first and last term is always 1.

If we had a method of writing coefficients for each term, with the previous information, we could write the expansion of any binomial.

If we write only the coefficients of the expansions previously shown, we have the following array:

1 1
1 2 1
1 3 3 1
1 4 6 4 1

This array of numbers is called Pascal's triangle. Examining the pattern of Pascal's triangle we can see that each succeeding row is obtained from each preceding row. For example, to create the fifth row from the fourth we do the following: Write a 1 at each end of the row. To obtain the other elements in the row, moving from left to right, add $1 + 4 = 5$ (second element in row 5), $4 + 6 = 10$ (third element in row 5), $6 + 4 = 10$ and $4 + 1 = 5$. This is illustrated below.

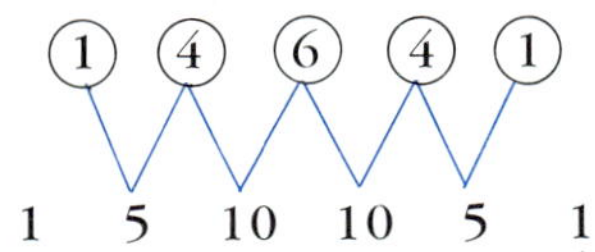

Any succeeding rows of the triangle can be created in the same manner. The triangle with the next two lines added is shown below.

$(a+b)^1$	1 1
$(a+b)^2$	1 2 1
$(a+b)^3$	1 3 3 1
$(a+b)^4$	1 4 6 4 1
$(a+b)^5$	1 5 10 10 5 1
$(a+b)^6$	1 6 15 20 15 6 1
$(a+b)^7$	1 7 21 35 35 21 7 1

Using Pascal's triangle we can write the expansion of $(x+y)^6$:

$$(x+y)^6 = x^6 + 6x^5y + 15x^4y^2 + 20x^3y^3 + 15x^2y^4 + 6xy^5 + y^6.$$

EXAMPLE 1 Expand $(2x + 3y)^4$ using Pascal's triangle.

Solution: Writing the binomial $(2x + 3y)^4$ in the form $[(2x) + (3y)]^4$ and treating $(2x)$ and $(3y)$ as x and y, respectively, we can write:

$$\begin{aligned}[(2x) + (3y)]^4 &= 1(2x)^4 + 4(2x)^3(3y) + 6(2x)^2(3y)^2 \\ &\quad + 4(2x)(3y)^3 + 1(3y)^4 \\ &= 16x^4 + 96x^3y + 216x^2y^2 + 216xy^3 + 81y^4. \quad \square\end{aligned}$$

EXAMPLE 2 Expand $(0.2x - 0.4y)^5$ using Pascal's triangle.

Solution: Writing the binomial $(0.2x - 0.4y)^5$ in the form

$$[(0.2x) + (-0.4y)]^5$$

and treating $(0.2x)$ and $(-0.4y)$ as x and y, respectively, we write:

$$\begin{aligned}[(0.2x) + (-0.4y)]^5 &= 1(0.2x)^5 + 5(0.2x)^4(-0.4y)^1 \\ &\quad + 10(0.2x)^3(-0.4y)^2 + 10(0.2x)^2(-0.4y)^3 \\ &\quad + 5(0.2x)(-0.4y)^4 + 1(-0.4y)^5 \\ &= 0.00032x^5 - 0.0032x^4y + 0.0128x^3y^2 \\ &\quad - 0.0256x^2y^3 + 0.0256xy^4 - 0.01024y^5. \quad \square\end{aligned}$$

These two examples show how Pascal's triangle can be used to determine the coefficients when expanding a binomial expression. The problem with the technique is in finding the coefficients for large values of n or in being able to write a specific term of an expansion.

To simplify the procedure we write $\binom{n}{r}$ to represent each number in the triangle. The letter n represents the row and r represents the column (note, $r \leq n$). For example:

$$\begin{array}{ccc} 1 \quad 1 & & \binom{1}{0} \ \binom{1}{1} \\ 1 \quad 2 \quad 1 & & \binom{2}{0} \ \binom{2}{1} \ \binom{2}{2} \\ 1 \quad 3 \quad 3 \quad 1 & & \binom{3}{0} \ \binom{3}{1} \ \binom{3}{2} \ \binom{3}{3} \\ 1 \quad 4 \quad 6 \quad 4 \quad 1 & & \binom{4}{0} \ \binom{4}{1} \ \binom{4}{2} \ \binom{4}{3} \ \binom{4}{4} \end{array}$$

To be able to write any term of Pascal's triangle we must be able to evaluate $\binom{n}{r}$. However, to understand the evaluation of $\binom{n}{r}$ we must first define $n!$ (read as ***n* factorial**).

DEFINITION 16.8

> The expression $n!$ is read as ***n* factorial** and means the product of the first n counting numbers.

For example,

$$\begin{aligned}
1! &= 1\\
2! &= 2 \cdot 1 = 2\\
3! &= 3 \cdot 2 \cdot 1 = 6\\
4! &= 4 \cdot 3 \cdot 2 \cdot 1 = 24\\
5! &= 5 \cdot 4 \cdot 3 \cdot 2 \cdot 1 = 120\\
6! &= 6 \cdot 5 \cdot 4 \cdot 3 \cdot 2 \cdot 1 = 720\\
&\vdots\\
n! &= n(n-1)(n-2)(n-3) \ldots 3 \cdot 2 \cdot 1.
\end{aligned}$$

DEFINITION 16.9

$$0! = 1.$$

DEFINITION 16.10

$$\binom{n}{r} = \frac{n!}{r!(n-r)!}.$$

EXAMPLE 3 Evaluate

a. $\binom{4}{3}$ **b.** $\binom{9}{3}$ **c.** $\binom{11}{0}$

Solution:

a. $\binom{4}{3} = \frac{4!}{3!(4-3)!}$ **Definition 16.10**

$= \frac{4!}{3!\,1!}$

$= \frac{4 \cdot \overset{1}{\cancel{3}} \cdot \overset{1}{\cancel{2}} \cdot 1}{\underset{1}{\cancel{3}} \cdot \underset{1}{\cancel{2}} \cdot 1 \cdot 1}$ **Definition 16.8**

$= 4.$

Check this result with Pascal's triangle shown on p. 641.

b. $\binom{9}{3} = \frac{9!}{3!(9-3)!}$ **Definition 16.10**

$= \frac{9!}{3!\,6!}$

$$= \frac{\overset{3}{\cancel{9}} \cdot \overset{4}{\cancel{8}} \cdot 7 \cdot \overset{1}{\cancel{6}} \cdot \overset{1}{\cancel{5}} \cdot \overset{1}{\cancel{4}} \cdot \overset{1}{\cancel{3}} \cdot \overset{1}{\cancel{2}} \cdot 1}{(\underset{1}{\cancel{3}} \cdot \underset{1}{\cancel{2}} \cdot 1)(\underset{1}{\cancel{6}} \cdot \underset{1}{\cancel{5}} \cdot \underset{1}{\cancel{4}} \cdot \underset{1}{\cancel{3}} \cdot \underset{1}{\cancel{2}} \cdot 1)} \qquad \textbf{Definition 16.8}$$

$$= 84.$$

c. $\binom{11}{0} = \dfrac{11!}{0!(11-0)!}$ **Definition 16.10**

$$= \frac{11!}{0!\,11!}$$

$$= \frac{\overset{1}{\cancel{11}}!}{1\,\underset{1}{\cancel{11}}!}$$

$$= 1. \qquad \square$$

In examining the preceding example we note the following:

1. In the second step of each problem the sum of the numbers in the denominator must always equal the number in the numerator. In $\dfrac{9!}{3!\,6!}$, we see that $3 + 6 = 9$.
2. When calculating $9! = 9 \cdot 8 \cdot 7 \cdot 6 \cdot 5 \cdot 4 \cdot 3 \cdot 2 \cdot 1$ we see that it also can be written as $9! = 9 \cdot 8 \cdot 7 \cdot 6!$. Therefore in evaluating $\dfrac{9!}{3!\,6!}$ we can write

$$\frac{9!}{3!\,6!} = \frac{\overset{3}{\cancel{9}} \cdot \overset{4}{\cancel{8}} \cdot 7 \cdot \overset{1}{\cancel{6}}!}{\underset{1}{\cancel{3}} \cdot \underset{1}{\cancel{2}} \cdot 1 \cdot \underset{1}{\cancel{6}}!} = 84.$$

   This is extremely helpful when large numbers are involved.

With this technique for determining the coefficients we can write the **binomial theorem,** which gives a method of expanding a binomial raised to a given power.

THEOREM 16.4

BINOMIAL THEOREM

$$(x+y)^n = \binom{n}{0} x^n + \binom{n}{1} x^{n-1}y + \binom{n}{2} x^{n-2}y^2 + \binom{n}{3} x^{n-3}y^3 + \cdots + \binom{n}{n-1} xy^{n-1} + \binom{n}{n} y^n,$$

where n is a positive integer.

EXAMPLE 4 Expand $(x-3)^4$ using the binomial theorem.

Solution: We can write $(x-3)^4$ as $[x+(-3)]^4$ and apply the binomial theorem.

$$\begin{aligned}
[x+(-3)]^4 &= \binom{4}{0}x^4 + \binom{4}{1}x^3(-3) + \binom{4}{2}x^2(-3)^2 \\
&\quad + \binom{4}{3}x(-3)^3 + \binom{4}{4}(-3)^4 \\
&= \frac{4!}{0!\,4!}x^4 + \frac{4!}{1!\,3!}x^3(-3) + \frac{4!}{2!\,2!}x^2(-3)^2 \\
&\quad + \frac{4!}{3!\,1!}x(-3)^3 + \frac{4!}{4!\,0!}(-3)^4 \\
&= 1x^4 + 4(-3)x^3 + (6)(9)x^2 + 4(-27)x \\
&\quad + (1)(81) \\
&= x^4 - 12x^3 + 54x^2 - 108x + 81. \qquad \square
\end{aligned}$$

EXAMPLE 5 Write the first three terms of the expansion of $(2a+5b)^{10}$.

Solution: The coefficients determined by Pascal's triangle of the first three terms of the expansion of $(2a+5b)^{10}$ are $\binom{10}{0}$, $\binom{10}{1}$, and $\binom{10}{2}$.

$$\binom{10}{0} = \frac{10!}{0!\,10!} = 1. \qquad \binom{10}{1} = \frac{10!}{1!\,9!} = \frac{10\cdot 9!}{1!\,9!} = 10. \qquad \binom{10}{2} = \frac{10!}{2!\,8!} = \frac{\overset{5}{\cancel{10}}\cdot 9\cdot \overset{1}{\cancel{8!}}}{\underset{1}{\cancel{2}}\cdot 1\cdot \underset{1}{\cancel{8!}}} = 45.$$

Using the binomial theorem we have

$$\begin{aligned}
(2a+5b)^{10} &= 1(2a)^{10} + 10(2a)^9(5b) + 45(2a)^8(5b)^2 + \cdots \\
&= (1)(1024a^{10}) + 10(512a^9)(5b) \\
&\quad + 45(256a^8)(25b^2) + \cdots \\
&= 1024a^{10} + 25{,}600a^9b + 288{,}000a^8b^2 \\
&\quad + \cdots \qquad \square
\end{aligned}$$

EXAMPLE 6 Find the value of $(0.99)^7$ to five significant digits by expanding a binomial whose terms have the sum 0.99.

Solution: We can write 0.99 as $1 - 0.01$; thus

$$
\begin{aligned}
(0.99)^7 &= (1-0.01)^7 \\
&= \binom{7}{0} 1^7 + \binom{7}{1} 1^6(-0.01) + \binom{7}{2} 1^5(-0.01)^2 \\
&\quad + \binom{7}{3} 1^4(-0.01)^3 + \binom{7}{4} 1^3(-0.01)^4 \\
&\quad + \binom{7}{5} 1^2(-0.01)^5 + \binom{7}{6} 1(-0.01)^6 \\
&\quad + \binom{7}{7} (-0.01)^7 \\
&= 1(1) + 7(1)(-0.01) + 21(1)(0.0001) \\
&\quad + 35(1)(-0.000001) + 35(1)(0.00000001) \\
&\quad + 21(1)(-0.0000000001) \\
&\quad + 7(1)(0.000000000001) \\
&\quad + 1(-0.00000000000001) \\
&= 1 - 0.07 + 0.0021 - 0.000035 + 0.00000035 \\
&\quad + \cdots ^{*} \\
&= 0.932065.
\end{aligned}
$$

Thus $(0.99)^7$ to five significant digits is 0.93207. □

EXAMPLE 7 Find the term containing x^8 in the expansion of $(2x + 3y)^{12}$.

Solution: To find a specific term of an expression we must remember that the sum of the exponents must equal the degree of the expansion. The term containing x^8 is

$$
\begin{aligned}
\binom{12}{4} (2x)^8(3y)^4 &= \frac{\overset{1}{\cancel{12}} \times 11 \times \overset{5}{\cancel{10}} \times 9 \times \overset{1}{\cancel{8!}}}{\underset{1}{\cancel{4}} \times \underset{1}{\cancel{3}} \times \underset{1}{\cancel{2}} \times 1 \times \underset{1}{\cancel{8!}}} (256x^8)(81y^4) \\
&= 10{,}264{,}320\ x^8y^4. \quad □
\end{aligned}
$$

16.6 Exercises

In Exercises 1–20, evaluate the following:

1. $8!$

2. $6!$

3. $\dfrac{7!}{5!}$

4. $\dfrac{11!}{9!}$

5. $\dfrac{7!}{4!\,3!}$

6. $\dfrac{8!}{6!\,2!}$

7. $\dfrac{9!}{2!\,7!}$

8. $\dfrac{12!}{10!\,2!}$

9. $\dfrac{0!}{0!}$

10. $\dfrac{14!}{0!\,14!}$

11. $\dfrac{15!}{3!\,12!}$

12. $\dbinom{4}{3}$

* We can see from the pattern that is developing that the addition of more terms would not make any difference in the fifth significant digit.

13. $\binom{6}{2}$
14. $\binom{9}{0}$
15. $\binom{13}{2}$
16. $\binom{15}{4}$
17. $\binom{100}{99}$
18. $\binom{155}{154}$
19. $\binom{n}{n-1}$
20. $\binom{n}{n-2}$

In Exercises 21–35, expand the following expressions using the binomial theorem.

21. $(x+1)^4$
22. $(Z+3)^4$
23. $(a+b)^4$
24. $(x^2-2y^2)^5$
25. $(3a^2+2b^2)^5$
26. $\left(\frac{x}{2}+\frac{y}{3}\right)^5$
27. $(2x^2-3t^2)^4$
28. $(x^{-1}+y^{-1})^5$
29. $\left(x\sqrt{2}-\frac{y}{3}\right)^4$
30. $(y^2-2)^6$
31. $(a^3-2b)^6$
32. $(1-y^{-1})^8$
33. $(a^{1/2}+b^{1/2})^7$
34. $(0.3x-0.4y)^4$
35. $(x^{2/3}-2y^{2/3})^7$

In Exercises 36–44, write the first four terms of the expansion.

36. $(x+y)^{16}$
37. $(1-x^2)^{17}$
38. $(R-2I)^{14}$
39. $\left(u^2+\frac{1}{2}v^3\right)^{18}$
40. $(x^{1/3}-y^{1/3})^{20}$
41. $(2x-3y)^{22}$
42. $(x^{1/2}-y^{1/2})^{15}$
43. $(1+y)^{96}$
44. $(Z-3)^{110}$

In Exercises 45–50, use the binomial theorem to find the value of each of the following expressions correct to six significant digits.

45. $(1.01)^7$
46. $(1.03)^8$
47. $(1.3)^6$
48. $(0.98)^8$
49. $(0.99)^9$
50. $(0.97)^6$

51. Find the term containing x^6 in the expansion of $(a+x)^{12}$.
52. Find the term containing x^8 in the expansion of $(x-2)^{12}$.
53. Find the term containing u^{12} in the expansion of $(u^2+v^2)^{18}$.
54. Find the term containing y^3 in the expansion of $(y^{1/2}-a^{1/2})^{16}$.
55. Find the term independent of x in the expansion of $(3x+x^{-1})^{12}$.
56. Find the value of n if the coefficients of the 5th and the 13th terms in the expansion of $(Z+1)^n$ are equal.
57. Find the value of Z such that the eighth term of $(3+Z)^{18}$ equals the seventh term in the expansion of $(3+Z)^{17}$.
58. Show that $\binom{n}{r}=\binom{n}{n-r}$.

Chapter summary

Definition 16.1 A **sequence** is a function whose domain is the set of positive integers.

The elements of the range of any sequence are called its terms. A sequence is denoted in the following manner.

$a_1, a_2, a_3, a_4, \ldots, a_n$	**Represents a finite sequence**
$a_1, a_2, a_3, a_4, \ldots, a_i, \ldots$	**Represents an infinite sequence**

The a_i term represents the **general term** of the sequence and can be used to generate the other terms of the sequence.

Definition 16.2 A sequence $a_1, a_2, \ldots, a_i, \ldots$, is said to be **increasing** if $a_i < a_{i+1}$ for all $i = 1, 2, 3, \ldots$, and **decreasing** if $a_i > a_{i+1}$ for all $i = 1, 2, 3, \ldots$.

Definition 16.3 The sum S_n of an indicated number n of terms of a sequence is called a **series.**

A series can be written in compact form using **summation notation.**

$$S_n = \sum_{i=1}^{n} a_i = a_1 + a_2 + \cdots + a_n.$$

The reader should be able to write any series that is written with summation notation in expanded form.

Any sequence where the terms have a common difference is called an **arithmetic sequence** or **arithmetic progression.**

Definition 16.4 An **arithmetic sequence** is a sequence of numbers called terms, such that any two successive terms differ by the same number, called the **common difference** d.

The common difference d is found by subtracting a preceding term from its succeeding term.

Definition 16.5 The nth term of an arithmetic sequence is given by the formula

$$a_n = a_1 + (n - 1)d,$$

where a_1 is the first term and d is the common difference.

Theorem 16.1 The sum of the first n terms of an arithmetic sequence whose first term is a_1 and whose nth term is a_n is given by

$$S_n = \frac{n}{2}(a_1 + a_n).$$

We can find the sum of the first n terms without knowing a_n.

$$S_n = \frac{n}{2}[2a_1 + (n - 1)d].$$

Definition 16.6 A **geometric sequence** is a sequence of numbers called terms, such that the ratio of any term to the preceding term is a fixed number called the **common ratio** r.

The common ratio can be found by setting up a ratio of any succeeding term to its preceding term.

Definition 16.7 The nth term of a geometric sequence is given by the formula

$$a_n = a_1\, r^{n-1},$$

where a_1 is the first term and r is the common ratio.

The series associated with the geometric sequence $a_1, a_1r, a_1r^2, a_1r^3, \ldots, a_1r^{n-1}$ is

$$S_n = \sum_{i=1}^{n} a_1 r^{i-1}.$$

Theorem 16.2 The sum of the first n terms of a geometric sequence whose first term is a_1 and whose common ratio is r is given by

$$S_n = \frac{a_1(1 - r^n)}{1 - r},$$

provided $r \neq 1$.

Theorem 16.3 The sum of the terms of an infinite geometric series whose first term is a_1 and whose common ratio r is given by

$$S = \frac{a_1}{1 - r},$$

provided $-1 < r < 1$.

The fraction equivalent to a repeating decimal can be found using the formula for the sum of an infinite geometric series.

We can expand a binomial expression using Pascal's triangle to determine the coefficients for each term of the expansion. The sum of the exponents for each term must equal the degree of the expansion. We can simplify the procedure for finding the coefficients by using $\binom{n}{r}$.

Definition 16.8 The expression $n!$ is read n factorial and means to find the product of the first n counting numbers.

Definition 16.9 $0! = 1$.

Definition 16.10 $\binom{n}{r} = \dfrac{n!}{r!(n-r)!}$.

Theorem 16.4 **Binomial Theorem**

$$(x+y)^n = \binom{n}{0} x^n + \binom{n}{1} x^{n-1}y + \binom{n}{2} x^{n-2}y^2 + \binom{n}{3} x^{n-3}y^3 + \cdots + \binom{n}{n-1} xy^{n-1} + \binom{n}{n} y^n.$$

Review Exercises

In Exercises 1–6, find a formula for the ith term of each of the sequences.

1. $1, 4, 7, 10, \ldots$
2. $1, 3, 5, 7, 9, \ldots$
3. $1, -1, -3, -5, \ldots$
4. $\frac{2}{3}, \frac{4}{3}, \frac{8}{3}, \frac{12}{3}, \frac{16}{3}, \ldots$
5. $2, 4, 2, 4, \ldots$
6. $0, 3, 8, 15, \ldots$

In Exercises 7–16, (a) write the first six terms of the sequence, (b) indicate whether the sequence is increasing, decreasing or neither, and (c) graph the sequence.

7. $a_i = 4i$
8. $a_i = -2i$
9. $a_i = i - 5$
10. $a_i = 3i + 2$
11. $a_i = -\frac{2}{3}i$
12. $a_i = \frac{2}{3}i$
13. $a_i = 3 + (-1)^i$
14. $a_i = 3 + (-2)^i$
15. $a_i = \frac{i-3}{i+4}$
16. $a_i = \frac{i^2-2}{i^2+2}$

In Exercises 17–24, write the series associated with each of the sequences for (a) S_4 and (b) S_6.

17. $4, 16, 64, 256, \ldots, 4^n$
18. $-4, 16, -64, 256, \ldots, (-1)^n4^n$
19. $1, 3, 5, 7, \ldots, (2n-1)$
20. $-1, 1, 3, 5, \ldots, 2n-3$
21. $0, \frac{1}{4}, \frac{2}{5}, \frac{3}{6}, \ldots, \frac{n-1}{n+2}$
22. $\frac{2}{3}, \frac{3}{4}, \frac{4}{5}, \frac{5}{6}, \ldots, \frac{n+1}{n+2}$
23. $2, -3, 4, -5, \ldots, (-1)^n(n+1)$
24. $1, -8, 27, -64, \ldots, (-1)^{n-1}n^3$

In Exercises 25–30, write each of the series in expanded form and determine the sum.

25. $S_5 = \sum_{i=1}^{5} \frac{(-2)^{i+1}}{i+1}$
26. $S_5 = \sum_{i=1}^{5} (-1)^i(2i-1)^2$
27. $S_5 = \sum_{i=1}^{5} \frac{1}{i} 2^{i+1}$
28. $S_7 = \sum_{i=1}^{7} \sin(2i)$
29. $S_6 = \sum_{i=1}^{6} \frac{i-5}{i+1}$
30. $S_7 = \sum_{i=1}^{7} (3i-1)(i+3)$

In Exercises 31–36, write each series using summation notation.

31. $5 + 9 + 13 + 17 + \cdots + (4+n)$
32. $2 + 1 + 0 + (-1) + \cdots + (3-n)$
33. $-2 + \frac{3}{2} - \frac{4}{4} + \frac{5}{8} - \cdots + \frac{(-1)^n(n+1)}{2^{n-1}}$
34. $\frac{1}{2^1} + \frac{2}{2^3} + \frac{3}{2^5} + \frac{4}{2^7} + \frac{5}{2^9}$
35. $\frac{1}{3} + \frac{2}{3^3} + \frac{3}{3^5} + \frac{4}{3^7} + \frac{5}{3^9}$
36. $\frac{z-5}{z+5} + \frac{z-6}{z+6} + \frac{z-7}{z+7} + \frac{z-8}{z+8}$

In Exercises 37–42, write each of the series in expanded form.

37. $\sum_{k=0}^{6} \frac{k}{2k-1}$
38. $\sum_{k=1}^{n} \frac{2k-1}{k}$

39. $\sum_{k=1}^{n} \frac{2k+3}{k^2}$

40. $\sum_{k=1}^{6} ke^k$

41. $\sum_{k=1}^{4} (z^2 + 3kz - k)$

42. $\sum_{k=1}^{6} \frac{(-1)^k}{\sqrt[5]{k}}$

In Exercises 43–48, write each of the infinite series in expanded form using the indicated term and the common difference.

43. $a_1 = 11, d = -5$

44. $a_1 = -7, d = 11$

45. $a_3 = 7, d = -1$

46. $a_4 = 8, d = 3$

47. $a_6 = -4, d = \frac{1}{2}$

48. $a_2 = \frac{4}{5}, d = \frac{4}{5}$

In Exercises 49–54, some of the quantities a_1, d, a_n, n, S_n are given; find the quantities that are not given.

49. $n = 11, a_1 = 3, a_{11} = 13$

50. $n = 10, a_1 = -1, a_{10} = 19$

51. $S_6 = 48, a_1 = -3$

52. $S_7 = 56, a_7 = 22$

53. $a_2 = \frac{4}{5}, d = \frac{4}{5}$

54. $S_8 = -2, d = \frac{1}{2}$

In Exercises 55–60, (a) construct the first six terms of the geometric sequence, and (b) determine the sum of the first six terms given the first term and the ratio.

55. $a_1 = \frac{1}{3}, r = 2$

56. $a_1 = -\frac{1}{3}, r = 2$

57. $a_1 = 2, r = \frac{1}{4}$

58. $a_1 = \frac{3}{4}, r = \frac{1}{5}$

59. $a_1 = -2, r = 0.1$

60. $a_1 = \frac{1}{4}, r = -2$

In Exercises 61–66, some of the quantities a_1, r, a_n, n, S_n are given; find the quantities that are not given.

61. $a_1 = \frac{2}{3}, r = \frac{1}{2}, n = 4$

62. $a_1 = \frac{1}{2}, r = \frac{2}{3}, n = 4$

63. $S_5 = -3, n = 5, r = 2$

64. $S_5 = 3, n = 5, r = 2$

65. $a_7 = 243, a_1 = \frac{1}{3}, n = 7$

66. $a_1 = \frac{3}{5}, r = \frac{-2}{3}, n = 5$

In Exercises 67–74, find the sum of the infinite geometric series if a sum exists.

67. $\frac{1}{5} + \frac{1}{25} + \frac{1}{125} + \frac{1}{625} + \cdots$

68. $1 + \frac{1}{6} + \frac{1}{36} + \frac{1}{216} + \cdots$

69. $1 + 0.02 + 0.0004 + 0.000008 + \cdots$

70. $1 - 0.03 + 0.0009 - 0.000027 + \cdots$

71. $7 - 7 + 7 - 7 + \cdots$

72. $81 - 27 + 9 - 3 + \cdots$

73. $\sum_{k=0}^{\infty} \left(\frac{4}{7}\right)^k$

74. $\sum_{k=1}^{\infty} \frac{5^k}{3^k + 6^k}$

In Exercises 75–78, find the fraction that is equivalent to the repeating decimal.

75. 0.8888 . . .

76. 0.232323 . . .

77. 1.237237 . . .

78. 3.471471 . . .

In Exercises 79–84, expand the binomial expressions.

79. $(Z + 5)^4$

80. $(x^2 - 3y^2)^4$

81. $(3u^2 - 2w^2)^4$

82. $(2a^3 + 3b)^6$

83. $(a^{1/3} + b^{1/3})^6$

84. $(1 - z^{-1})^7$

In Exercises 85–88, write the first four terms of the expansion.

85. $(x + y)^{18}$

86. $(x^2 - 1)^{20}$

87. $(a^{1/3} + b^{1/3})^{16}$

88. $(Z + 1)^{92}$

89. Find the term containing x^{11} in the expansion of $(x - 3)^{15}$.

90. Find the term containing y^5 in the expansion of $(y^{1/3} - a^{1/3})^{24}$.

91. The first swing of a pendulum is 250 mm, and each succeeding swing is reduced by 1/5 mm. What is the length of the tenth pendulum swing?

92. A grandfather's clock strikes once at 1 o'clock, twice at 2 o'clock and so on. What is the total number of times the clock will strike in 12 hours?

93. If John saves 1/2¢ the first day, 1¢ the next day, 2¢ the next, 4¢ the next, and so on, how much money will he have saved after 30 days?

94. How many ancestors would be listed in your family tree in the last ten generations, counting your two parents, four grandparents, eight great-grandparents, and so on?

95. A florist started eight canas bulbs. At the end of the year they had grown to 16 bulbs. If they continue dividing in this manner, how many bulbs will he have in five years?

96. If \$1,000 is placed in an account that earns 8% simple interest each year, it will amount to \$1,080 in one year, \$1,160 in two years, \$1,240 in three years, and so on. How much will the principal and the interest amount to in 25 years?

97. The circumference of a circular container around a clock spring is 2.82 in. The spring is in the form of a spiral having 12 complete turns. Find the length of the spring, assuming that each spiral is 0.15 in. shorter than the next larger spiral.

98. Show that 998^2 equals the binomial

$(1000 - 2)^2$.

Find the square of 998 by the binomial expansion.

99. A \$275 bicycle depreciates 20% of its value each year. What is the value of the bicycle at the end of the sixth year?

100. A chemical substance decomposes in such a way that at the end of each hour it is only 80% as heavy as it was at the beginning of the hour.

a. How much would 10 lb of this substance weigh after 5 hr?

b. How long will it take for 10 lb to decompose?

Can you solve these?

1. Show that the reciprocals of the terms of a geometric sequence also form a geometric sequence.

2. A harmonic sequence is a type of sequence not mentioned in this chapter.

a. Do research and write a definition for a harmonic sequence

b. Write two examples of harmonic sequences.

3. The Fibonacci sequence was discovered in the middle ages by the mathematician Fibonacci. The first 11 terms of the sequence are 1, 1, 2, 3, 5, 8, 13, 21, 34, 55, and 89.

a. Is this an arithmetic or geometric sequence?

b. Determine how each succeeding term is determined and find the 12th and 13th terms of the sequence.

4. If the midpoints of the sides of a square are connected, the resulting figure is a square. If we continue this process indefinitely, the result is a set of squares as illustrated in Fig. 16.5. The Pythagorean theorem can be used to compute the length of the side of any square in the set if we know the length of the side of the square that immediately contains it. Consider a set of squares where the outermost square has a side length of 5 mm.

a. Find the sum of the areas of all the squares in the nest.

b. Find the sum of the perimeters of all the squares in the nest.

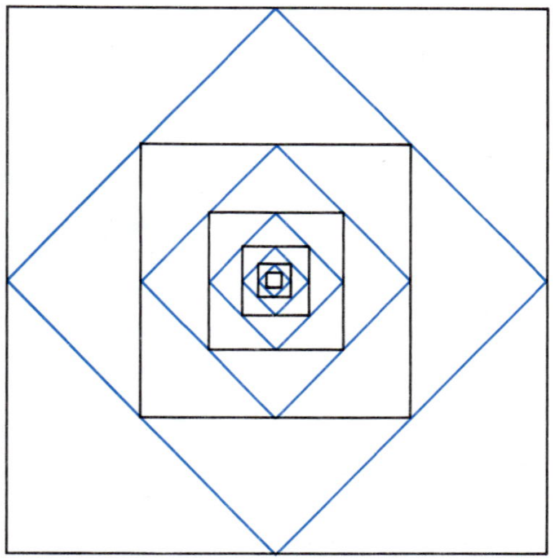

FIGURE 16.5

5. A geometric series has a first term of 3 and a common ratio of r, while an arithmetic sequence has a first term of 3 and a common difference of d. If S_4 is the same for both sequences, express d in terms of r.

6. Write a computer program to find the sum of the terms of an arithmetic series.

7. Write a computer program to find the sum of the terms of a geometric series.

8. Write a computer program to find the sum of an infinite geometric series.

17

The most accurate navigation system currently available to nonmilitary mariners is the Loran-C System. In this system a shipboard receiver measures the time differences between the reception of the signal from a master station and each of two different slave stations. The curved lines on the chart represent locations of constant time difference for a given master–slave pair. The intersection of two such lines can give a position location accurate to 50 ft. What mathematical function produces these lines?

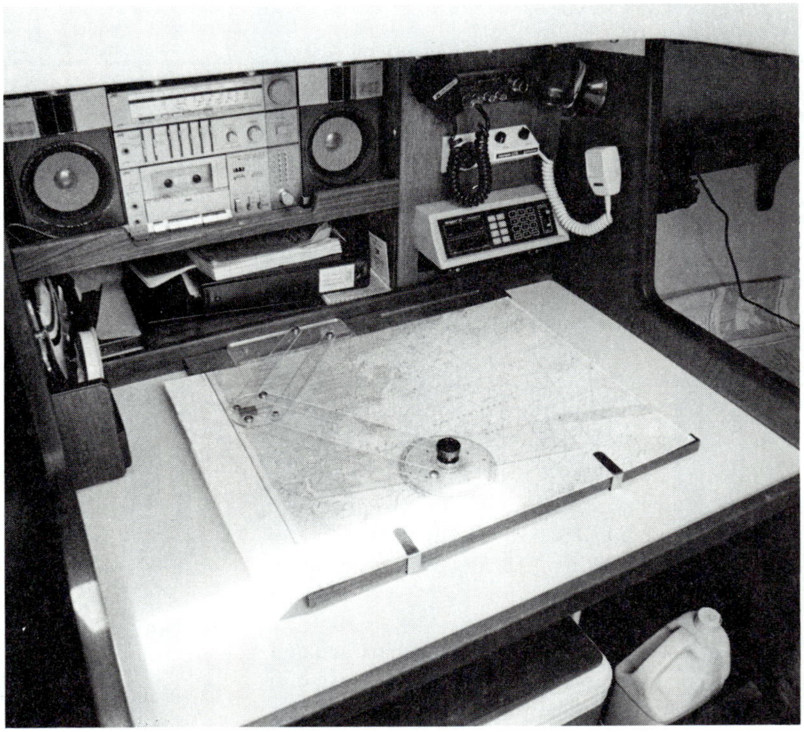

Plane Analytic Geometry

17.1 Introduction

The beginning of the study of mathematics was dominated by independent developments in geometry and algebra. In the mid-1600s work was begun to form a bridge linking the two disciplines. As a result of this work, the shapes and axioms of geometry can be described in a general algebraic fashion and can be developed using the laws of algebra. At the same time, algebraic concepts can be translated into the more easily visualized shapes known in geometry. The bridge that linked the two branches of mathematics became known as **analytic geometry.** In particular, it is analytic geometry that allows us to use the calculator or the computer to model and explore the shapes of the world around us. Shapes that we will discuss in this section include the line, the circle, the parabola, the ellipse, and the hyperbola.

The circle, the parabola, the ellipse, and the hyperbola are all related because they all can be formed by the intersection of a plane with a right circular cone. As a result they are all termed **conic sections.**

A cone is defined as the surface generated by lines (called elements) that all pass through a single point (called the apex) and a curve. If the curve is a circle and the apex is located on a line perpendicular to the plane of the circle and passing through its center, the cone is a right circular cone. This definition produces a shape that is two conventional "cones" joined at the apex.

An ellipse is produced by the intersection of the cone with any plane that cuts all the elements of the cone (Fig. 17.1a). A circle is, of course, a special case of the ellipse, where the cutting plane is perpendicular to the axis of the cone.

FIGURE 17.1

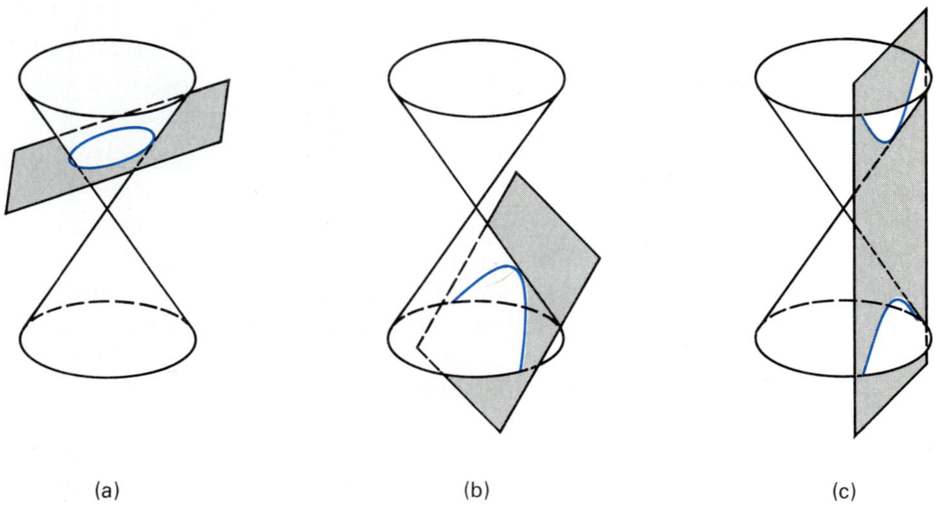

A parabola is produced by the intersection of the cone and any plane parallel to one of the elements of the cone (Fig. 17.1b).

A hyperbola is produced by the intersection of the cone and any plane not meeting the conditions for an ellipse or a parabola. One case of the hyperbola is produced when the plane is parallel to the axis of the cone (Fig. 17.1c).

Degenerate sections consisting of a point, a line, or two intersecting lines, respectively, are produced whenever the plane passes through the apex of the cone.

17.2 The straight line

The straight line is the most commonly used figure in all technical practice. Not only is it the geometric basis of most of the structures and devices we use, but its algebraic description provides the form by which we can relate many different sets of variables. The slope–intercept form of the equation of a straight line and some of the properties of the linear equation are briefly discussed in this section.

DEFINITION 17.1

The slope–intercept form of the equation of a line is $y = mx + b$, where m is the slope of the line and b is its y-intercept.

DEFINITION 17.2

The slope of a straight line containing points $P_1(x_1, y_1)$ and $P_2(x_2, y_2)$ is defined by the equation

$$m = \frac{y_2 - y_1}{x_2 - x_1}, \qquad x_2 - x_1 \neq 0.$$

DEFINITION 17.3

Two lines are parallel if and only if their slopes are equal. That is, $\ell_1 \| \ell_2$ if and only if $m_1 = m_2$.

The graph of the equation $y = b$ (b is a constant) is a straight line that is parallel to the horizontal axis and passes through the point $(0, b)$. The graph of the equation $x = a$ (a is a constant) is a straight line that is parallel to the vertical axis and passes through the point $(a, 0)$. The slope of the line $y = b$ is zero and the slope of the line $x = a$ is undefined. Can you explain why the slopes of the lines are zero and undefined?

Given the straight line $y = mx + b$ as illustrated in Fig. 17.2, select any two points $P_1(x_1, y_1)$ and $P_2(x_2, y_2)$ on the line and draw a line

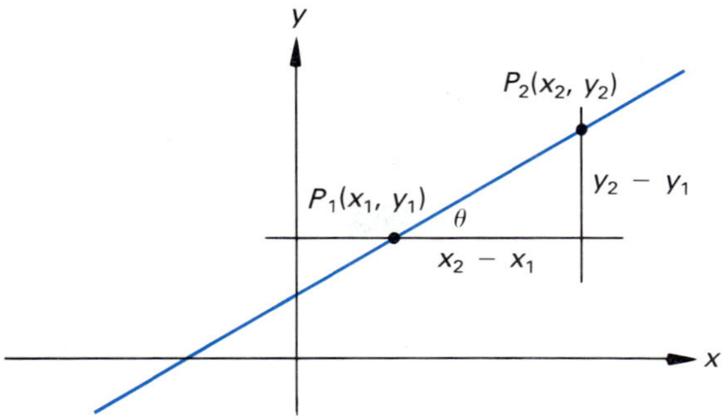

FIGURE 17.2

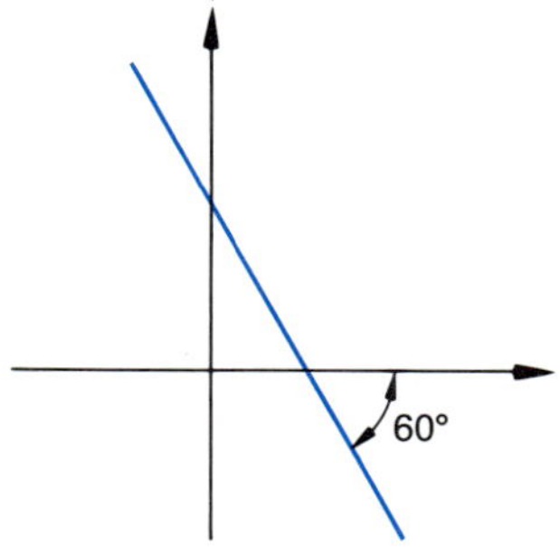

FIGURE 17.3

parallel to the horizontal axis through P_1 and a line parallel to the vertical axis through P_2. A right triangle is formed with an opposite side of length $y_2 - y_1$ and an adjacent side of length $x_2 - x_1$. Using the definition of the tangent function we can state that $\tan\theta = \dfrac{y_2 - y_1}{x_2 - x_1}$. Also, from Definition 17.2 we know that $m = \dfrac{y_2 - y_1}{x_2 - x_1}$. Substituting, we have $m = \tan\theta$. Thus the slope of the line can be determined by finding the tangent of the angle that the line forms with the positive horizontal axis.

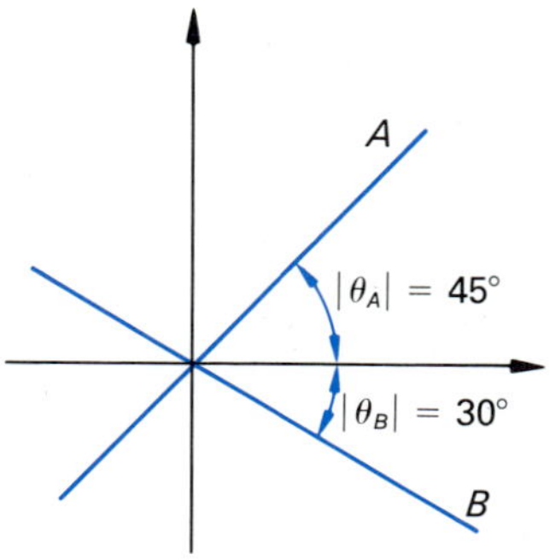

FIGURE 17.4

EXAMPLE 1 What is the slope of the line in Fig. 17.3?

Solution: Since the line makes angles of $-60°$, $120°$, $300°$, or $-240°$ with the positive horizontal axis, its slope m is $\tan(-60°)$ or $\tan(120°)$ or $\tan(300°)$ or $\tan(-240°)$ all of which equal -1.732. □

EXAMPLE 2 What are the equations of the lines in Fig. 17.4?

Solution: In Fig. 17.4 line A passes through the origin, which has coordinates $(0, 0)$, and it makes an angle of $45°$ with the positive x-axis. Therefore, its slope m is 1 and its y-intercept b is 0. The equation of the line is $y = 1x + 0$ or $y = x$. For line B, $m = -\dfrac{1}{\sqrt{3}}$, and b is $-\dfrac{3}{2}$. The equation for line B is $y = -\dfrac{1}{\sqrt{3}}x - \dfrac{3}{2}$. □

FIGURE 17.5

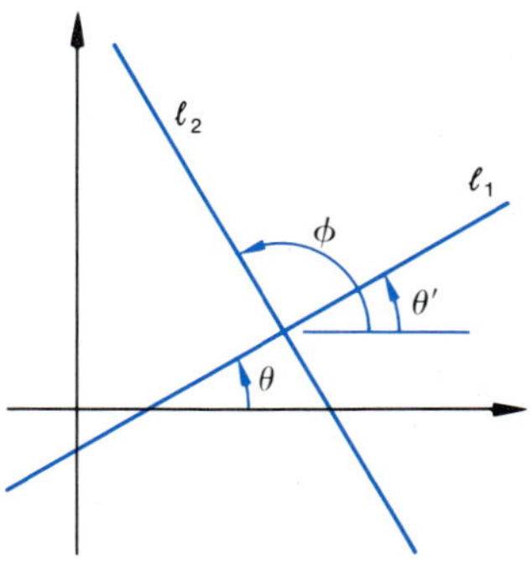

If we know the slopes of any two lines we can determine whether or not they are parallel, are perpendicular, or are not parallel or not perpendicular.

Two lines perpendicular to each other are shown in Fig. 17.5. The slope of line ℓ_1 is $m_1 = \tan\theta$ and the slope of line ℓ_2 is $m_2 = \tan\phi$. We can see from the figure that $\theta = \theta'$ (corresponding angles of parallel lines) and that:

$$\begin{aligned}\phi &= \theta' + 90^\circ\\ &= \theta + 90^\circ,\end{aligned}$$

so

$$\begin{aligned}m_2 &= \tan \phi\\ &= \tan (\theta + 90^\circ).\end{aligned}$$

Using the identity

$$\tan (A \pm B) = \frac{\sin (A \pm B)}{\cos (A \pm B)}$$

for $A = \theta$ and $B = 90^\circ$, we have

$$\begin{aligned}m_2 &= \tan (\theta + 90^\circ)\\ &= \frac{\sin (\theta + 90^\circ)}{\cos (\theta + 90^\circ)}\\ &= \frac{\sin \theta \cos 90^\circ + \cos \theta \sin 90^\circ}{\cos \theta \cos 90^\circ - \sin \theta \sin 90^\circ}\end{aligned}$$

Recalling that $\sin (A + B) = \sin A \cos B + \cos A \sin B$ $\cos (A + B) = \cos A \cos B + \sin A \sin B$

$$= \frac{\sin \theta\ (0) + \cos \theta\ (1)}{\cos \theta\ (0) - \sin \theta\ (1)}$$

Substituting $\cos 90^\circ = 0$ and $\sin 90^\circ = 1$

$$\begin{aligned}&= \frac{\cos \theta}{-\sin \theta}\\ &= -\frac{1}{\tan \theta}\\ &= -\frac{1}{m_1}.\end{aligned}$$

Substituting $m_1 = \tan \theta$

If two lines are perpendicular to each other, the slope of either is the negative reciprocal of the slope of the other.

DEFINITION 17.4

> Two lines ℓ_1 and ℓ_2 are perpendicular if and only if they have negative reciprocal slopes $\left(m_2 = -\frac{1}{m_1}\right)$.

We have presented two ways to find the slope of a line. We can calculate the tangent of the angle that the line makes with the positive x-axis (if the x-axis and the y-axis scales are the same). If the coordinates of two points on the line are known we can calculate its slope using

$$m = \frac{y_2 - y_1}{x_2 - x_1}.$$

Let us assume that we know the slope of a line. As long as we know the coordinates of at least one point on the line, even if that point is not the y-intercept, we can find an equation of the line. Recalling the slope–intercept form of the equation of a line we have

$$y = mx + b. \qquad [1]$$

If we know one point, $P(x_1, y_1)$, on the line we can write

$$y_1 = mx_1 + b. \qquad [2]$$

Subtracting Eq. [2] from Eq. [1] eliminates b. The result

$$y - y_1 = m(x - x_1) \qquad [1] - [2] \qquad [3]$$

is the point–slope form of the equation of a straight line.

DEFINITION 17.5

The point–slope form of the equation of a straight line is

$$y - y_1 = m(x - x_1),$$

where m = slope and (x_1, y_1) is a point on the line.

EXAMPLE 3 Write the equation of the line that fits the data of Table 17.1 in the slope–intercept form.

Solution: Since we know the coordinates of eleven points on the desired line, we can choose any two points and calculate the slope of the line. Then we can write the equation of the line by substituting the slope and a point in the point slope form of the equation of a line. Let us choose the third point, (24., −2.0) and the tenth point, (38., 8.5), as (x_1, y_1) and (x_2, y_2), respectively.

$$m = \frac{y_2 - y_1}{x_2 - x_1} \qquad \textbf{Slope formula}$$

$$= \frac{8.5 - (-2.0)}{38. - 24.} \qquad \textbf{Substituting}$$

$$= 0.75$$

$$y - y_1 = m(x - x_1) \qquad \textbf{Point–slope equation}$$

$$y - (-2.0) = 0.75(x - 24.) \qquad \textbf{Substituting}$$

TABLE 17.1

x	20.	22.	24.	26.	28.	30.	32.	34.	36.	38.	40.
y	−5.0	−3.5	−2.0	−0.5	1.0	2.5	4.0	5.5	7.0	8.5	10.

The slope – intercept form of the equation of the line that fits the data is $y = 0.75x - 20$. It is left to the reader to verify that any pair of points could be used to find the equation and that it makes no difference which point is "first" and which is "second." □

If any of the forms of the equation of a straight line (the slope – intercept form, the point – slope form) are rearranged they can be written in the form $Dx + Ey + F = 0$, where D, E, and F are constants. This is the **general form** of the equation of a straight line.

DEFINITION 17.6

> The general form of the equation of a straight line is
>
> $$Dx + Ey + F = 0,$$
>
> where D, E, and F are constants and D and E cannot both be zero.

Note that the variables of linear equations (equations of straight lines) are of first degree (exponents are 1).

EXAMPLE 4 Show that the general form of the equation of a line, $Dx + Ey + F = 0$, is comparable to the slope – intercept form, $y = mx + b$.

Solution: Rearrange the standard form to

$$Ey = -Dx - F$$

$$y = -\frac{D}{E}x - \frac{F}{E} \qquad \textbf{Dividing by } \boldsymbol{E}$$

$$m = -\frac{D}{E}$$

$$b = -\frac{F}{E}$$ □

EXAMPLE 5 The equation to convert degrees Celsius to degrees Fahrenheit is $F = \frac{9}{5}C + 32$. Is this the equation of a straight line? If so write it in general form.

Solution: The equation $F = \frac{9}{5}C + 32$ is an equation of a straight line since the exponents of the variables are 1. In the form in which the equation is written, F is the dependent variable and C is the independent variable. To write it in general form we transpose all the terms to one side of the equal sign. Thus

$$\frac{9}{5}C - F + 32 = 0$$

We could also write the equation in either of the following equivalent forms:

$$9C - 5F + 160 = 0 \qquad \text{or} \qquad C - \frac{5}{9}F + \frac{160}{9} = 0. \qquad \square$$

EXAMPLE 6 What is the equation of the line that is perpendicular to $2y = x - 1$ and passes through $(3, -4)$?

Solution: The general linear equation $2y = x - 1$ can be written in the slope–intercept form $y = \frac{1}{2}x - \frac{1}{2}$. The slope of the line is $m_1 = 1/2$. We established that the slope of a line perpendicular to a given line is the negative reciprocal of the slope of the original line. Thus

$$m_2 = \frac{-1}{1/2} \qquad \text{or} \qquad -2.$$

Using the point–slope form, the equation of the desired line is

$$\begin{aligned} y - y_1 &= m(x - x_1) \\ y - (-4) &= -2(x - 3) && \textbf{Substituting for } \boldsymbol{x, y,} \textbf{ and } \boldsymbol{m} \\ y + 4 &= -2x + 6 \\ y &= -2x + 2, && \textbf{Rearranging} \end{aligned}$$

which is the equation of the line perpendicular to $2y = x - 1$ and passing through the point $(3, -4)$. $\square$

EXAMPLE 7 Determine the angle between the lines $y = -6.1x - 2.7$ and $3.0y = 0.90x - 1.2$ (Fig. 17.6).

FIGURE 17.6

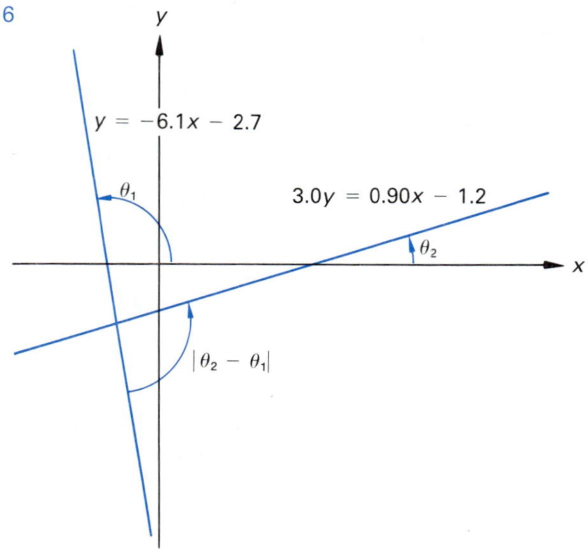

Solution: For the line $y = -6.1x - 2.7$, we have $m_1 = -6.1$. So $\tan\theta_1 = -6.1$ and $\theta_1 = -80.7°$. For the line $3.0y = 0.90x - 1.2$, we have $m_2 = 0.90/3.0$. So $\tan\theta_2 = 0.30$ and $\theta_2 = 16.7°$.

The angle between the two lines is

$$|\theta_2 - \theta_1| = |16.7° - (-80.7°)| = 97°.$$

While this is a correct answer, it is often the practice to take the angle between two intersecting lines as the angle of intersection that is less than 90°. In that case the answer should be $180° - 97.° = 83.°$. □

Two other formulas that are useful when working with lines and points are the distance and midpoint formulas.

> The distance d between two points $P_1(x_1, y_1)$ and $P_2(x_2, y_2)$ is given by
>
> $$d = \sqrt{(x_2 - x_1)^2 + (y_2 - y_1)^2}.$$
>
> If $P(x, y)$ is the midpoint of the line segment whose end points are $P_1(x_1, y_1)$ and $P_2(x_2, y_2)$, then
>
> $$(x, y) = \left(\frac{x_1 + x_2}{2}, \frac{y_1 + y_2}{2}\right).$$

EXAMPLE 8 Given points $A(-5, -10)$, $B(-13, 14)$, and $C(3, 2)$,

a. find the midpoints of line segments $\overline{AB}$ and $\overline{BC}$,

b. find the distance between the midpoints of $\overline{AB}$ and $\overline{BC}$, and

c. write the equation of the line that is the perpendicular bisector of $\overline{AB}$.

Solution:

a. The midpoint of the line segment between $A(-5, -10)$ and $B(-13, 14)$ is

$$(x_1, y_1) = \left(\frac{-5 + (-13)}{2}, \frac{-10 + 14}{2}\right)$$
$$= (-9, 2).$$

The midpoint of the line segment between $B(-13, 14)$ and $C(3, 2)$ is

$$(x_2, y_2) = \left(\frac{-13 + 3}{2}, \frac{14 + 2}{2}\right)$$
$$= (-5, 8).$$

b. The distance between the midpoints is found using the distance formula.

$$\begin{aligned} d &= \sqrt{(x_2 - x_1)^2 + (y_2 - y_1)^2} \\ &= \sqrt{[-5 - (-9)]^2 + (8 - 2)^2} \qquad \textbf{Substituting} \\ &= \sqrt{4^2 + 6^2} \\ &= \sqrt{52}. \end{aligned}$$

c. The perpendicular bisector of $\overline{AB}$ is the line that passes through the midpoint of $\overline{AB}$ and is perpendicular to $\overline{AB}$. The slope of $\overline{AB}$ is $\dfrac{14 - (-10)}{-13 - (-5)}$ or -3. Thus the slope of the perpendicular bisector is $1/3$. In (a) we found that the midpoint of $\overline{AB}$ was $(-9, 2)$. Substituting in the point–slope form of the equation of the line we have

$$y - 2 = \frac{1}{3}\,(x - (-9))$$

$$y - 2 = \frac{x}{3} + 3.$$

The general form of the equation of the line that is the perpendicular bisector of $\overline{AB}$ is $x - 3y + 15 = 0$. □

17.2 Exercises

In Exercises 1–20, write an equation in general form that represents the line in each figure.

1.

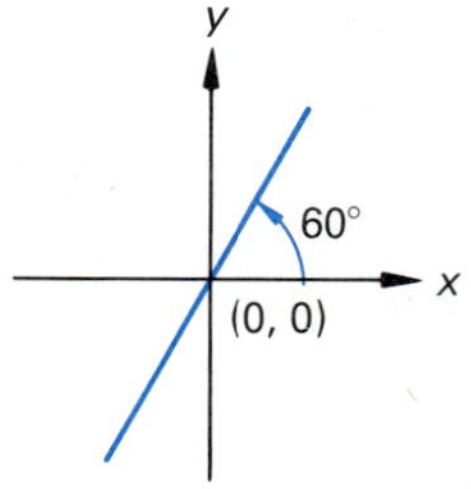

2.

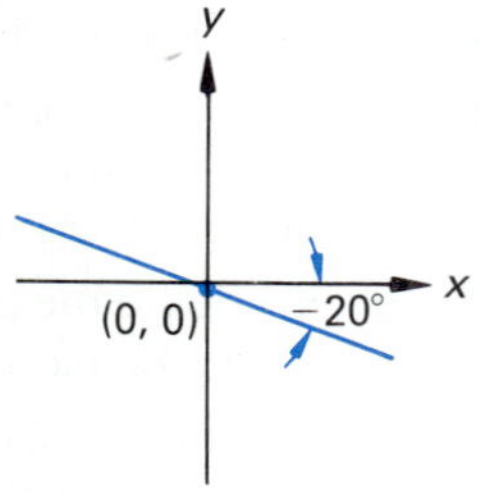

3.

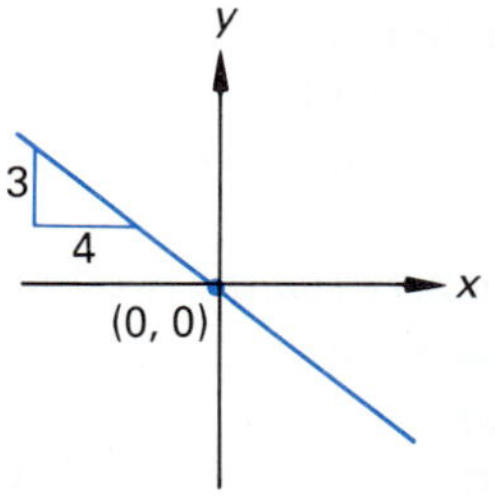

4.

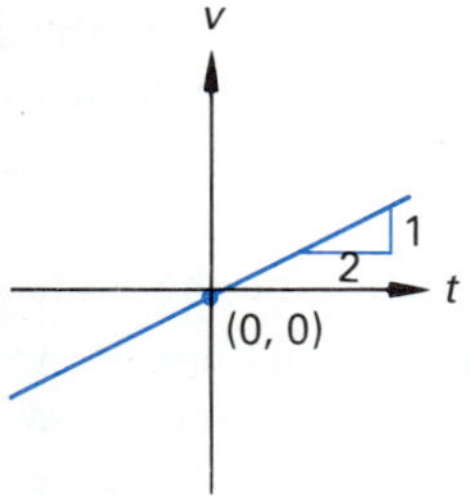

5.

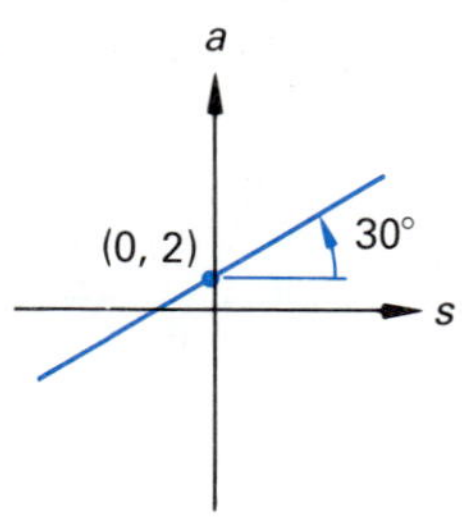

6.

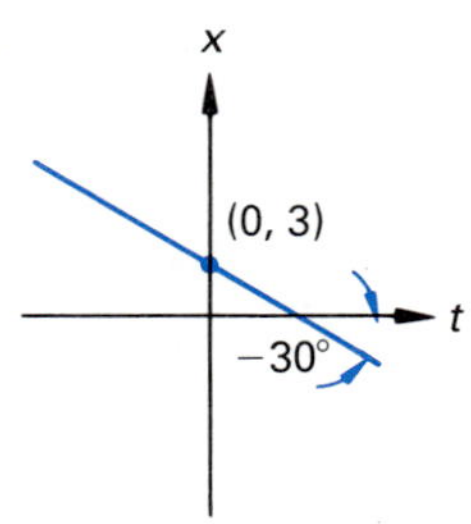

7.

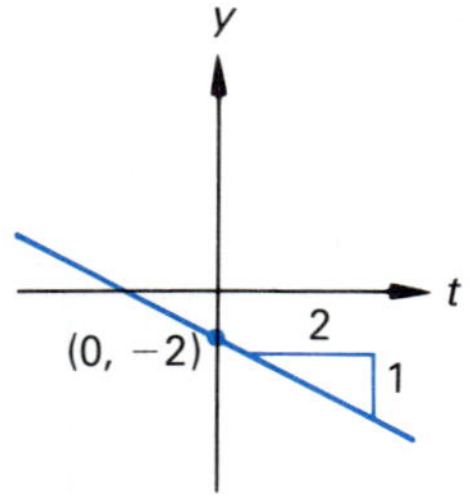

8.

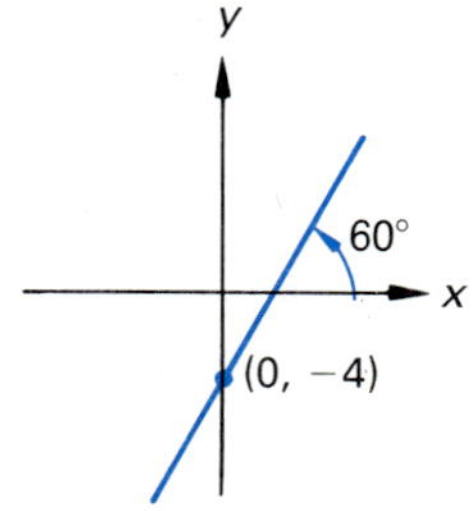

9.

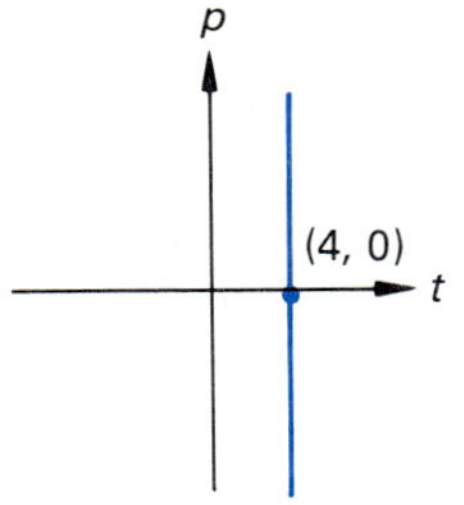

10.

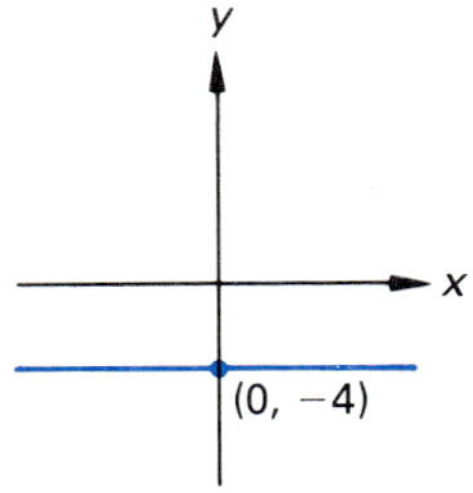

11.

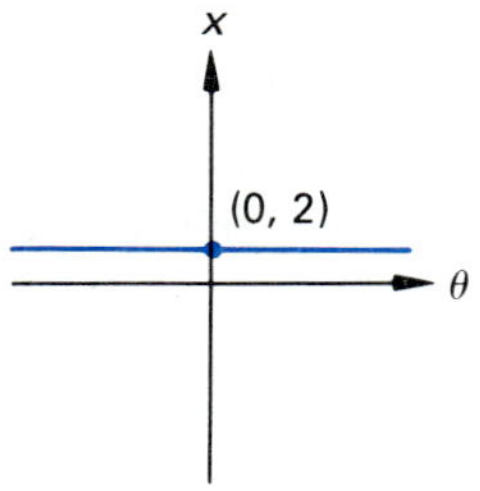

12.

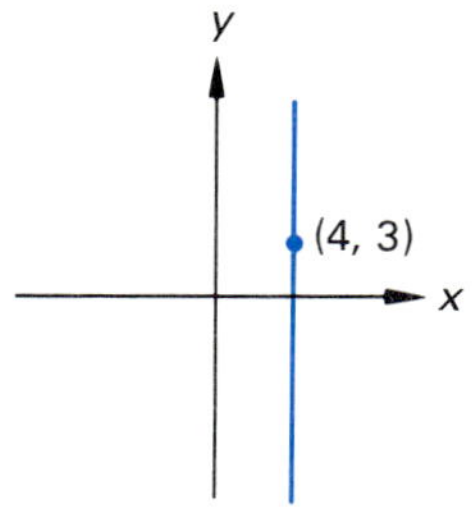

13.

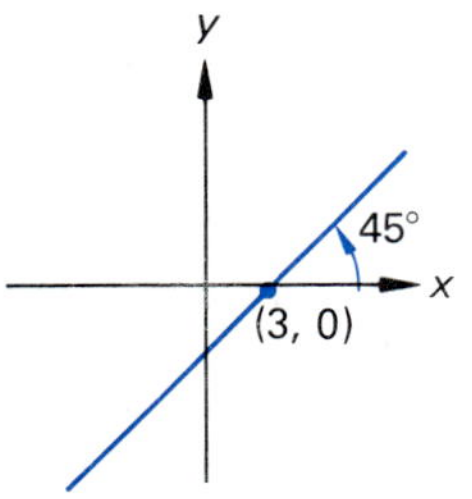

14.

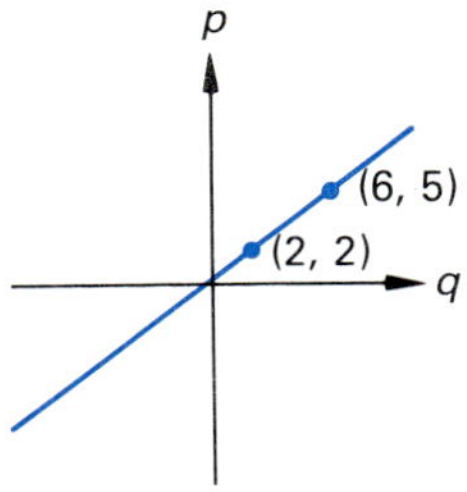

15.

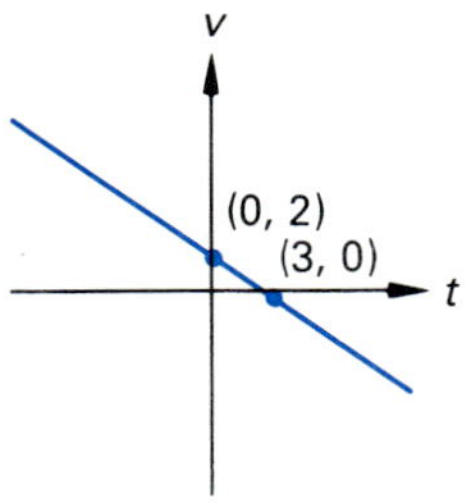

16.

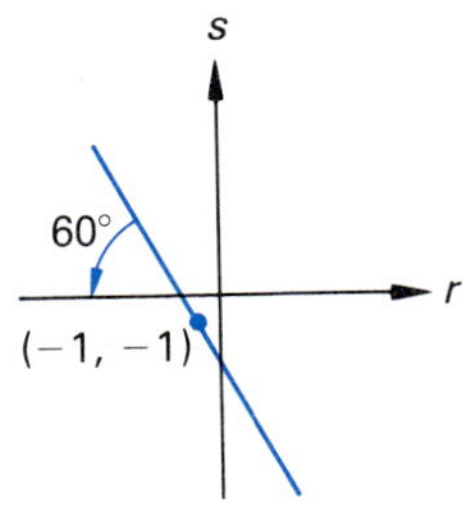

17.

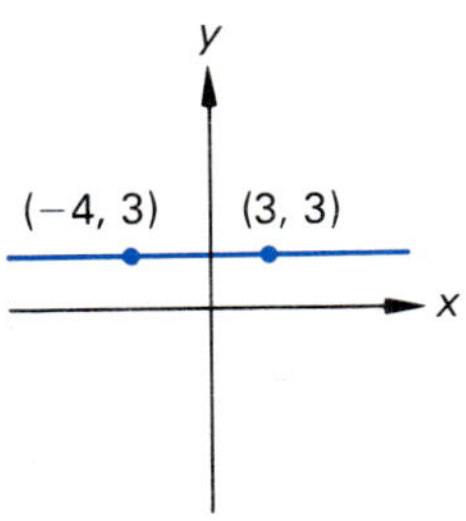

18.

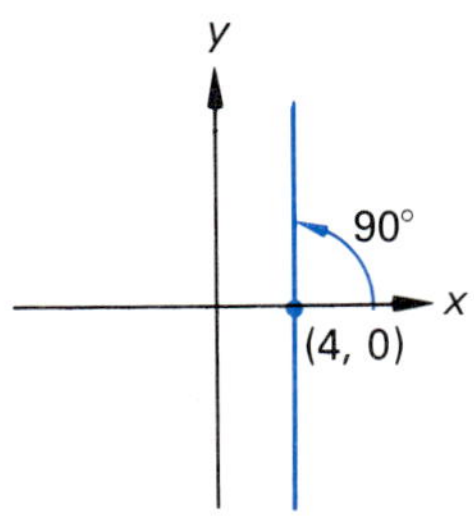

19.

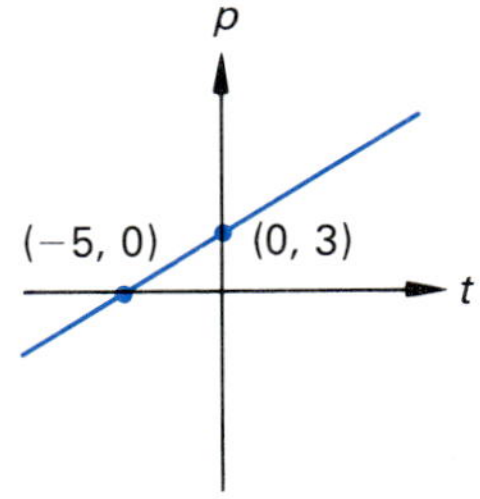

20.

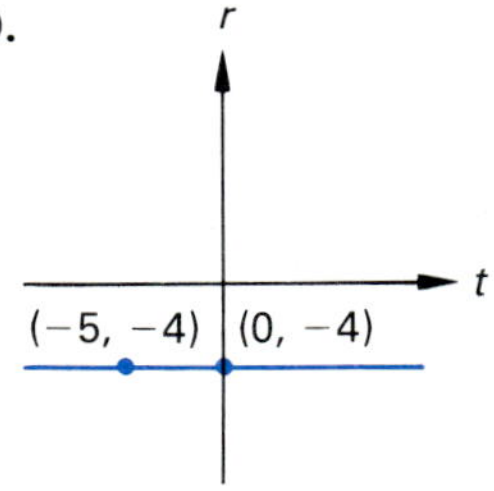

In Exercises 21–30, sketch the graph of the line indicated.

21. $y = x + 2$

22. $y = \frac{1}{2}x - 3$

23. $y = -2x$

24. $y = -\frac{1}{2}x - 2$

25. $p = \frac{2t - 4}{3}$

26. $x = 4$

27. $x = 4y$

28. $2y - x = 4$

29. $-V + 3t - 1 = 0$

30. $0.3x + 0.4y - 1.2 = 0$

In Exercises 31–38, a) find the length of the line segment between the two points and b) give the coordinates of the midpoint of the line segment.

31. (8, 1), (2, 4)

32. (1, 5), (4, 6)

33. (−5, 2), (6, −1)

34. (−4, 3), (2, −5)

35. (−3, −2), (−1, −5)

36. (−7, −4), (−2, −3)

37. (1.62, 3.81), (−6.15, 4.27)

38. (3.26, −1.43), (5.83, 6.19).

39. Given the line $4x - 2y - 3 = 0$, write the general form of the equation of the line passing through the point $P(2, 3)$ that is

a. parallel to the given line and
b. perpendicular to the line.

40. Given the line $-2x + y + 5 = 0$, write the general form of the equation of the line passing through the point $P(4, 1)$ that is

a. parallel to the given line and
b. perpendicular to the given line.

41. a. Do lines $4x + y - 3 = 0$ and $y = -x + 2$ intersect? If so, where?
b. Find the acute angle between the two lines.

42. a. Do lines $x - y + 3 = 0$ and $-2x + 2y + 1 = 0$ intersect? If so, where?
b. Find the acute angle between the two lines.

43. Are lines $y = 2x + 2$ and $y - \frac{1}{2}x = 2$ perpendicular? Why or why not?

44. Write the equation of the line perpendicular to $x - y + 2 = 0$ and passing through the same y-intercept.

45. Figure 17.7 shows the stress–strain curve for a 0.3% mild-carbon steel. The slope of the linear portion of the curve at the left side of the graph is the modulus of elasticity, E, which relates the stress, σ, to the strain, ϵ, in the elastic region by the equation $\sigma = E\epsilon$. What is the modulus of elasticity? (This is one of the few instances where the dependent variable is plotted on the horizontal axis and the independent variable on the vertical axis.)

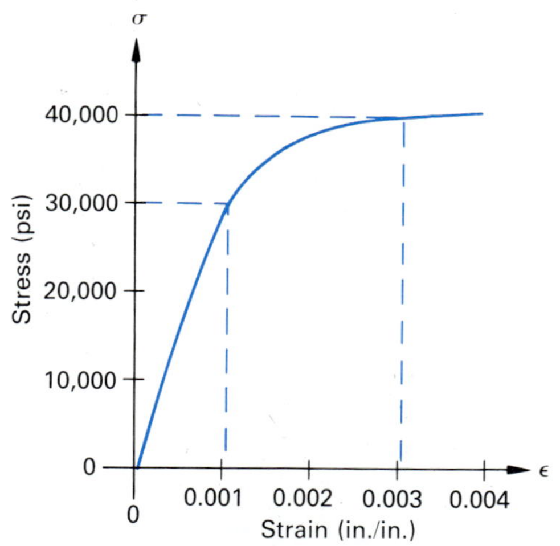

FIGURE 17.7

46. Pressure in freshwater increases linearly with depth from zero at a free surface (zero depth) to 624. lb/ft² at a depth of 10. ft. Write the equation that expresses pressure as a function of depth.

47. If the density of freshwater is 62.4 lb/ft³ and that of saltwater is 64.0 lb/ft³, modify the equation of Exercise 46 to be valid in saltwater.

48. A true stress–strain curve (as opposed to nominal or engineering stress–strain) is shown in Fig. 17.8. What is the equation of the linear portion of the curve in the plastic-strain region $0.30 \leq \epsilon \leq 0.90$? (Strain is a dimensionless quality.)

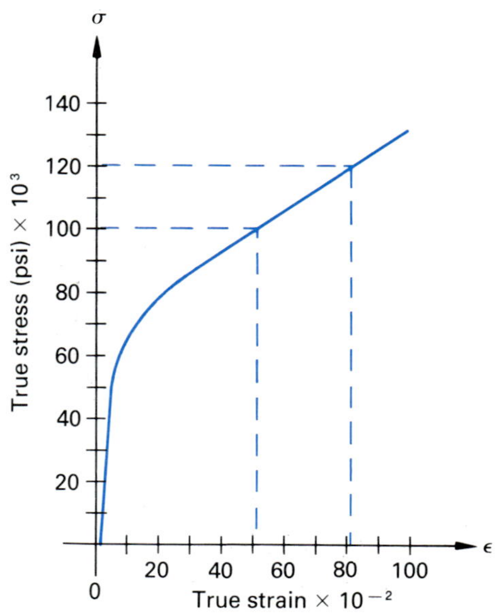

FIGURE 17.8

49. Figure 17.9 shows the variation of the Charpy impact strength of a hot-rolled 1020 steel with temperature. What are the equations of the two lines that most closely make up the curve?

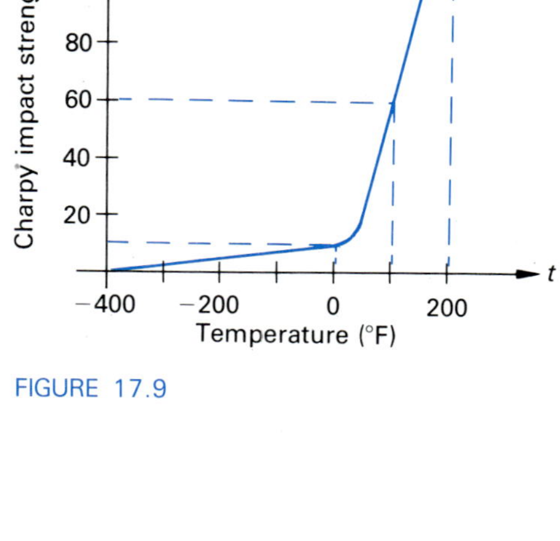

FIGURE 17.9

50. The resistance of a current conductor is given by $R = \rho \frac{\ell}{A}$, where ρ is the characteristic resistivity of the conductor material, ℓ is the length of the conductor, and A is its cross-sectional area. For a given material, is the equation linear if:

a. the area is held constant and the length is allowed to vary,

b. the length is held constant and the area is allowed to vary? Explain.

51. The curve of magnetic flux density B in webers/m^2 (dependent variable) versus magnetic field H in kA/m (independent variable) for the [100] orientation in iron can be approximated by the two equations: $B = 0.66\ H$, $(H > 3.33)$ and $B = 2.2$, $(0 < H < 3.33)$. Sketch the approximate curve.

52. The unit change in length with temperature of a prestressed copper wire is given by the linear equation

$\epsilon = (10 \times 10^{-6})\Delta T + 200 \times 10^{-6}$,

where ΔT is the temperature change in °F. Sketch the equation.

53. The specific heat of methane in Btu/mol is given by $C_p = 4.22 + (8.21 \times 10^{-3})T$, where T is in °R. What is the rate of increase of specific heat with temperature? What is the specific heat of methane at absolute zero?

54. In air conditioning, the specific enthalpy h of water leaving the apparatus at temperature t is $h = t - 32$ Btu/lb. Sketch the line if t is in °F. How much does h change with each degree change in t?

55. The induced emf in a DC motor is a linear function of the armature current: $E = V - IR$. V is the terminal voltage and R is the armature resistance. Describe how the emf changes with current.

56. The slope of the line in the upper portion of Fig. 17.7 represents the strain-hardening characteristics of the steel. If the slope of the line is 1.00×10^6 psi, what is the equation of the line that approximates the upper portion of the graph?

57. What is the equation of the line parallel to the left portion of Fig. 17.7 (see Exercise 45) and passing through the point 0.002 on the strain axis? (This line would represent the 0.2% offset in strength of materials.)

17.3 The circle

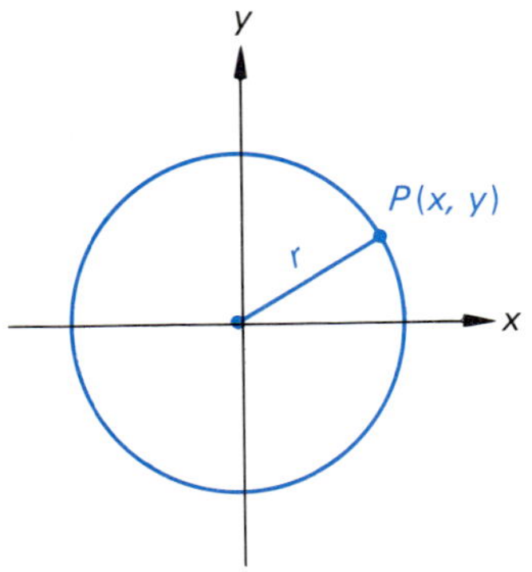

FIGURE 17.10

A **circle** is the set of all points in a plane that are all the same distance from a fixed point (Fig. 17.10). The fixed point is the **center** and the distance of each point from the center is the **radius** r. We can obtain an equation for a circle in the xy-plane with its center at the origin by recalling the formula for the distance between two points

$$d = \sqrt{(x_2 - x_1)^2 + (y_2 - y_1)^2}.$$

The distance from the origin, (0, 0) to $P(x, y)$ is the constant radius r.

$$r = \sqrt{(x - 0)^2 + (y - 0)^2} \qquad \textbf{Substituting}$$

$$= \sqrt{x^2 + y^2}$$

$$r^2 = x^2 + y^2 \qquad \textbf{Squaring both sides}$$

DEFINITION 17.7

> The **standard form** of the equation of a **circle** with radius r and its center at the origin is
>
> $$x^2 + y^2 = r^2.$$

EXAMPLE 1 A circle of radius 6 has its center at the origin. What is the equation of the circle?

Solution: The equation of a circle with center at the origin is

$$x^2 + y^2 = r^2.$$

For this circle $r = 6$, therefore

$$x^2 + y^2 = (6)^2.$$

Thus

$x^2 + y^2 = 36$ is the equation of the circle. □

EXAMPLE 2 Given the circle $x^2 + y^2 = 25$,

a. What are the y-coordinates of the circle when $x = 3$?

b. When $x = 6$?

Solution:

a. The equation of the circle is

$$x^2 + y^2 = 25.$$

Letting $x = 3$ gives

$$\begin{aligned} (3)^2 + y^2 &= 25 \\ y^2 &= 25 - 9 \\ &= 16 \\ y &= \pm\sqrt{16} \\ y &= 4 \quad \text{or} \quad y = -4. \end{aligned}$$

We need both possible values of the square-root operation because if we look at the sketch of a circle we see that there are two points that have the same x-value, one below the x-axis and one equidistant above it (Fig. 17.11).

FIGURE 17.11

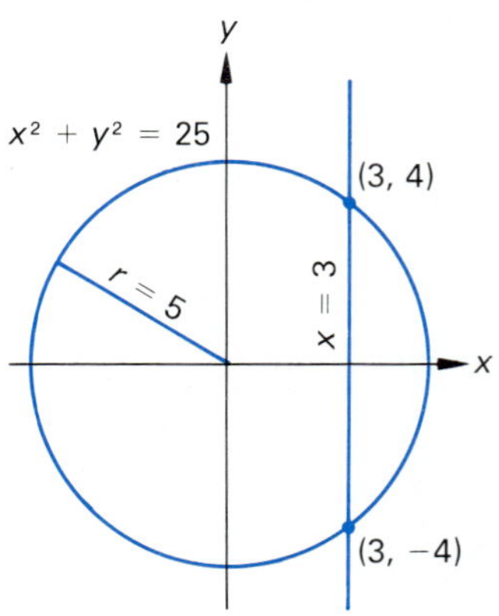

b. For $x = 6$ we would have

$$\begin{aligned} x^2 + y^2 &= 25 \\ (6)^2 + y^2 &= 25 \\ y^2 &= 25 - 36 \\ &= -11 \\ y &= \sqrt{-11}, \end{aligned}$$

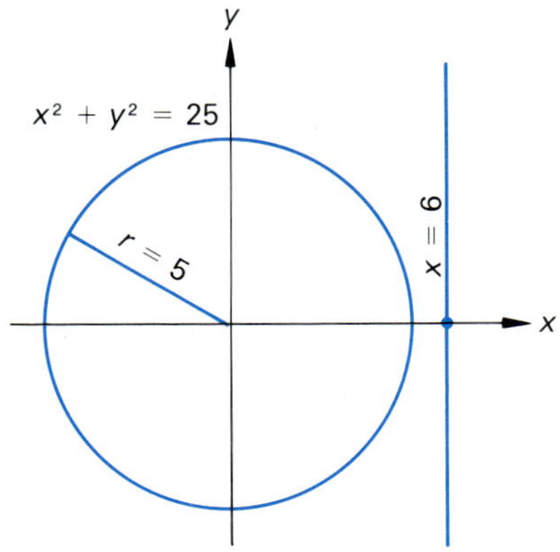

FIGURE 17.12

which is an imaginary number. Since the circle is composed only of pairs of real numbers, $x = 6$ is outside the circle (Fig. 17.12). □

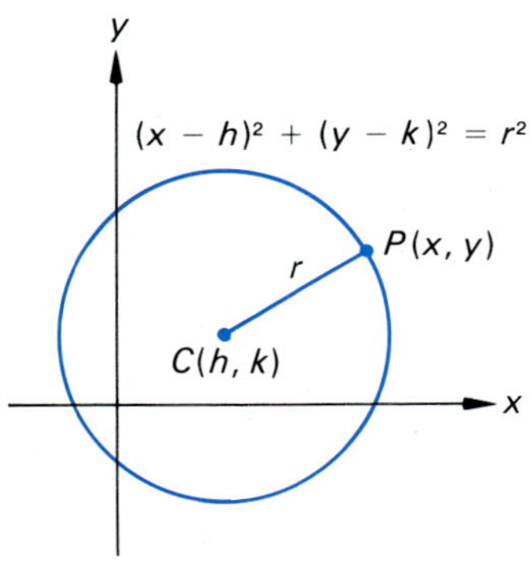

FIGURE 17.13

If we construct a circle of radius r with its center at point $C(h, k)$ we can again use this distance formula to derive an equation for the circle (Fig. 17.13):

$$d = \sqrt{(x_2 - x_1)^2 + (y_2 - y_1)^2}.$$

Since r is the constant distance between the center, (h, k), and any point on the circle, $P(x, y)$, we can substitute into the distance formula to obtain

$$r = \sqrt{(x - h)^2 + (y - k)^2}$$
$$(x - h)^2 + (y - k)^2 = r^2. \qquad \textbf{Squaring both sides}$$

DEFINITION 17.8

> The **standard form** of the equation of a **circle** with center at (h, k) and radius r is
>
> $$(x - h)^2 + (y - k)^2 = r^2.$$

EXAMPLE 3 What is the equation of the circle with center at $(-2, 1)$ and radius 3?

Solution: The equation for a circle with center at (h, k) is

$$(x - h)^2 + (y - k)^2 = r^2.$$

Substituting

$$(x - (-2))^2 + (y - 1)^2 = 3^2 \qquad \boldsymbol{h = -2, k = 1}$$

or

$$(x + 2)^2 + (y - 1)^2 = 9. \quad \square$$

We can see from Definitions 17.7 and 17.8 that the equation of a circle is second-degree in x and in y. If we write the circle equation as a sum of terms equal to zero we have the **general form** of the equation of a circle.

DEFINITION 17.9

> The **general form** of the equation of a **circle** is
>
> $$Ax^2 + Cy^2 + Dx + Ey + F = 0 \text{ if } A = C.$$

EXAMPLE 4 Is $2x^2 + 2y^2 + x + 3y - 30\frac{3}{4} = 0$ the equation of a circle? If so, describe the circle.

Solution: The general form of the equation of a circle requires that the coefficients of the x^2 and y^2 terms be the same, which is true in this case. Because we think that

$$2x^2 + 2y^2 + x + 3y - 30\frac{3}{4} = 0$$

might represent a circle, we can try to manipulate it into the form

$$(x - h)^2 + (y - k)^2 = r^2.$$

To do so we begin by dividing both sides of the equation by the coefficient of the x^2 or y^2 term to reduce the coefficient of those terms to unity:

$$x^2 + y^2 + \frac{1}{2}x + \frac{3}{2}y = 15\frac{3}{8}. \qquad \textbf{Dividing by 2 and rearranging}$$

Grouping terms in x and terms in y gives

$$\left(x^2 + \frac{1}{2}x\right) + \left(y^2 + \frac{3}{2}y\right) = 15\frac{3}{8}.$$

We now need to complete the square in the x-terms and in the y-terms. Adding $1/16$ to the x-group and $9/16$ to the y-group would produce perfect squares (see Chapter 11) so adding $1/16$ and $9/16$ to both sides

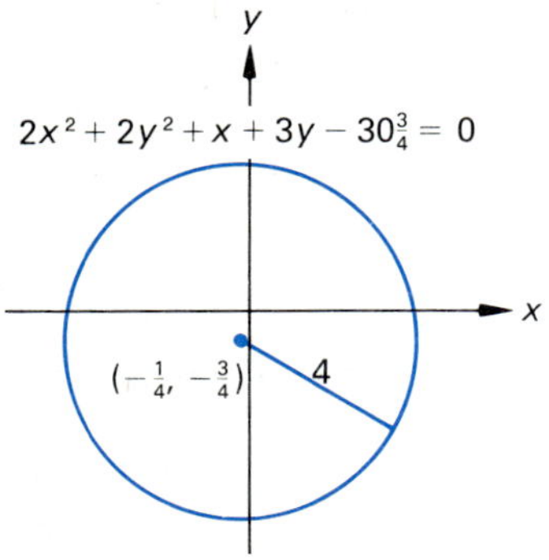

FIGURE 17.14

of the equation gives

$$\left(x^2 + \frac{1}{2}x + \frac{1}{16}\right) + \left(y^2 + \frac{3}{2}y + \frac{9}{16}\right) = 15\frac{3}{8} + \frac{1}{16} + \frac{9}{16}$$

$$\left(x + \frac{1}{4}\right)^2 + \left(y + \frac{3}{4}\right)^2 = 16 \qquad \textbf{Perfect square factoring on left side and combining terms on right side}$$

$$\left(x + \frac{1}{4}\right)^2 + \left(y + \frac{3}{4}\right)^2 = (4)^2,$$

which is the standard form of the equation of a circle with center at $\left(-\frac{1}{4}, -\frac{3}{4}\right)$ and radius equal to 4 (Fig. 17.14). □

EXAMPLE 5 What is the equation of the circle that passes through the points $(-1, -3)$, $(0, 4)$, and $(1, 1)$?

Solution: Each of the points must lie on the required circle, so each of them must satisfy the equation of a circle. Substituting the values for x and y in the standard form of the equation of a circle, we get the following three equations in terms of the variables h, k, and r.

$$(-1 - h)^2 + (-3 - k)^2 = r^2$$
$$(0 - h)^2 + (4 - k)^2 = r^2$$
$$(1 - h)^2 + (1 - k)^2 = r^2$$
$$1 + 2h + h^2 + 9 + 6k + k^2 = r^2 \qquad \textbf{Expanding} \qquad [1]$$
$$h^2 + 16 - 8k + k^2 = r^2 \qquad [2]$$
$$1 - 2h + h^2 + 1 - 2k + k^2 = r^2 \qquad [3]$$
$$2h + 14k - 6 = 0 \qquad [1] - [2] \qquad [4]$$
$$4h + 8k + 8 = 0 \qquad [1] - [3] \qquad [5]$$
$$4h + 28k - 12 = 0 \qquad [2] \times [4] \qquad [6]$$
$$4h + 8k + 8 = 0 \qquad [5]$$
$$20k - 20 = 0 \qquad [6] - [5]$$
$$k = 1$$
$$2h + 14(1) - 6 = 0 \qquad \textbf{Substituting into Eq. [4]}$$
$$h = -4$$
$$1 + 2(-4) + (-4)^2 + 9 + 6(1) + (1)^2 = r^2 \qquad \textbf{Substituting into Eq. [1]}$$
$$r = 5.$$

The equation of the circle passing through $(-1, -3)$, $(0, 4)$ and $(1, 1)$ is

$$(x + 4)^2 + (y - 1)^2 = 25. \qquad □$$

A circle can be drawn such that it passes through three points. Three points are needed because an infinite number of circles can be constructed through any given pair of points.

EXAMPLE 6 The design of a portable room humidifier specifies that the rotating drum of the humidifier be supported on two rollers and driven by frictional contact with a third roller. If each roller is 50 mm in diameter and is located as shown in Fig. 17.15, what is the required inside diameter of the drum?

Solution: The rollers are all the same size, so the circle passing through the center of the rollers will be concentric with the inner circle of the drum. We will therefore first find the equation of the circle that passes through the centers of the rollers. Let us choose an axis system passing through the two idler rollers and with its origin midway between them. To simplify calculation, since it is easily done in this application, let us convert all dimensions to centimeters. In this system the coordinates of the centers of the three rollers are $(-30, 0)$, $(-10, 15)$, and $(30, 0)$. (The location of the axis system is arbitrary; any location for the origin could have been chosen but this location somewhat simplifies the ensuing calculations.) We begin with

FIGURE 17.15

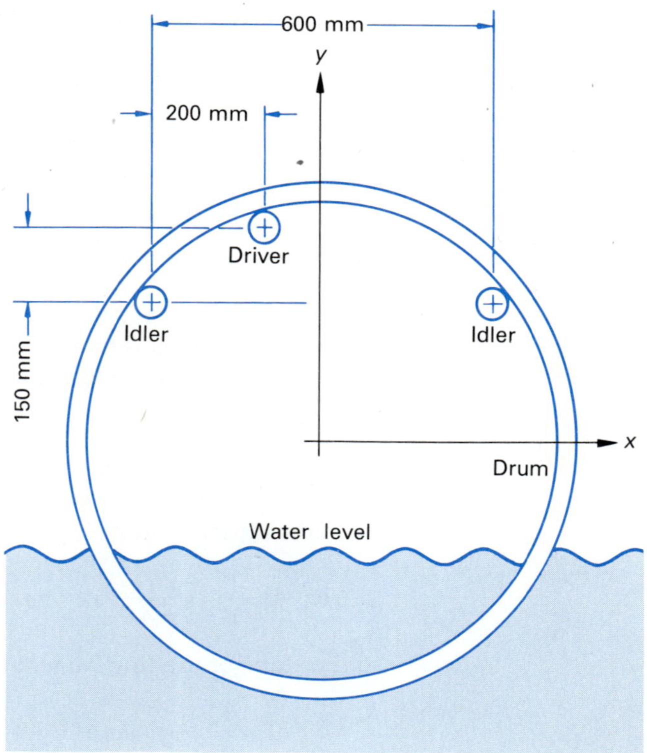

the standard equation of a circle.

$$(x-h)^2+(y-k)^2=r^2$$

$$[x-(-30)]^2+(y-0)^2=r^2$$ **Substituting for (h, k)**

$$[x-(-10)]^2+(y-15)^2=r^2$$ **Substituting for (h, k)**

$$(x-30)^2+(y-0)^2=r^2$$ **Substituting for (h, k)**

$$x^2+60x+900+y^2=r^2. \quad \text{Expanding} \quad [1]$$

$$x^2+20x+100+y^2-30y+225=r^2 \quad \text{Expanding} \quad [2]$$

$$x^2-60x+900+y^2=r^2. \quad \text{Expanding} \quad [3]$$

We can now use the elimination method to solve the set of three equations

$$120x=0 \quad [1]-[3]$$

$$x=0$$

$$80x-30y-575=0 \quad [2]-[3]$$

$$80(0)-30y-575=0$$ **Substituting $x=0$**

$$y=\frac{575}{-30}$$ **Solving for y**

$$=-19.17.$$

$$(0)^2+60(0)+900+(-19.17)^2=r^2$$ **Substituting into Eq. [1]**

$$r=35.60 \text{ cm}.$$

Adding the radius of the 5-cm-diameter rollers

$$r_{\text{drum}}=35.60+2.50$$

$$=38.1 \text{ cm}.$$

Therefore the required inside diameter of the drum is 381. mm. □

EXAMPLE 7 What is the equation of the straight line that is tangent to the circle in Example 5 at point $(-1, 5)$?

Solution: From geometry the radius of a circle at a point on the circle is perpendicular to the tangent to the circle at that point. The slope of the radius from center $(-4, 1)$ to the point $(-1, 5)$ is

$$m_r=\frac{5-1}{-1-(-4)} \quad \text{or} \quad m_r=\frac{4}{3},$$

so the slope of the line tangent to the circle at $(-1, 5)$ is

$$m_t=-\frac{1}{m_r} \quad \text{or} \quad m_t=-\frac{3}{4}.$$

Using the point–slope form of the equation of a line:

$$y - y_1 = m(x - x_1)$$

$$y - 5 = -\frac{3}{4}(x - (-1))$$

$$y = -\frac{3}{4}x + 4\frac{1}{4}.$$

Thus $3x + 4y - 17 = 0$ is the equation of the line tangent to the circle $(x + 4)^2 + (y - 1)^2 = 25$ at the point $(-1, 5)$. □

17.3 Exercises

In Exercises 1–12, write the equation of the circle shown.

1.

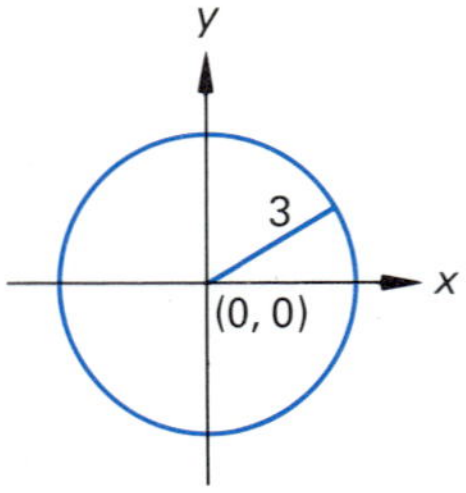

2.

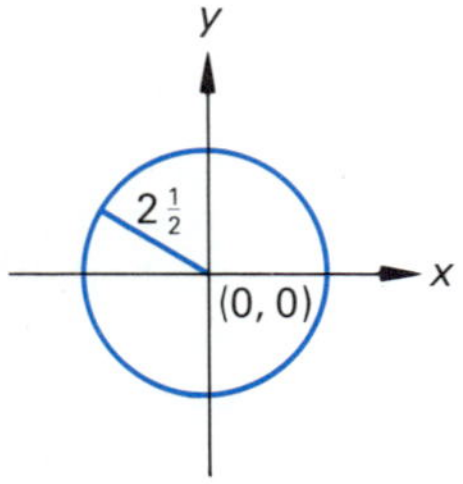

3.

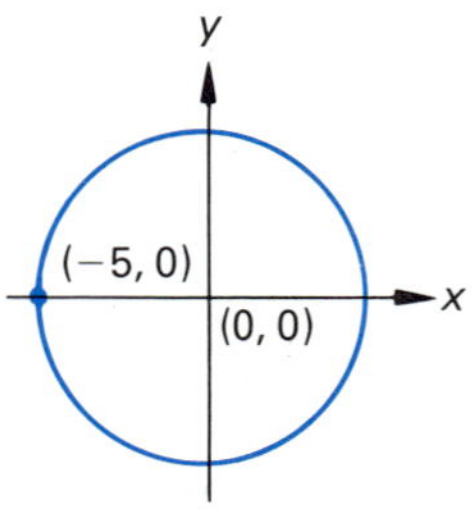

4.

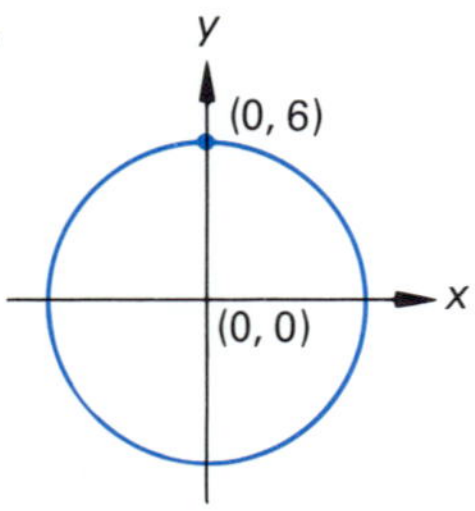

5.

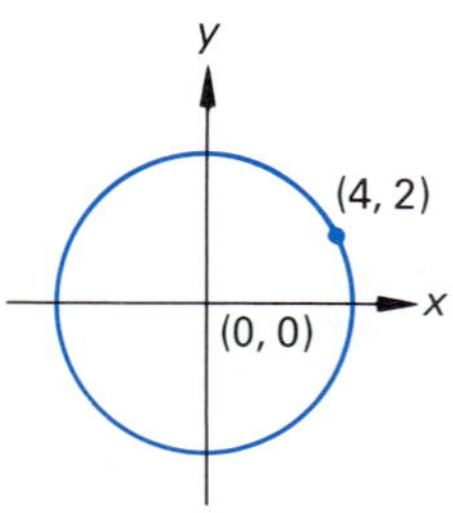

6.

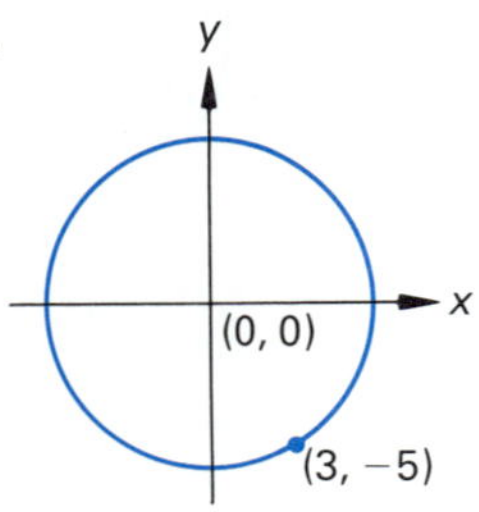

7.

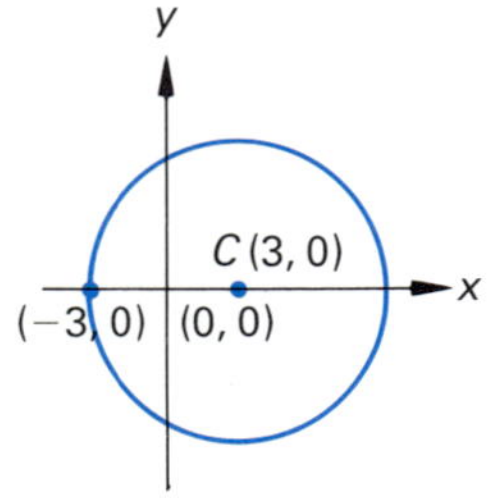

8.

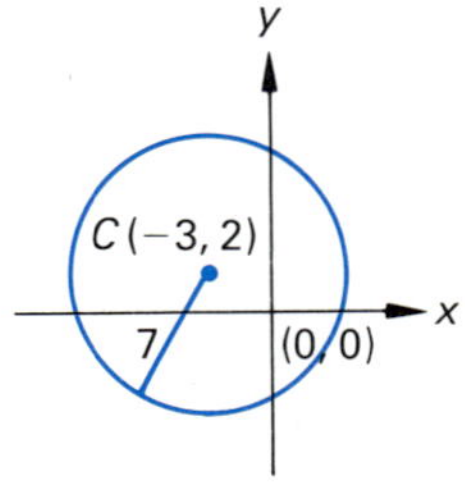

9.

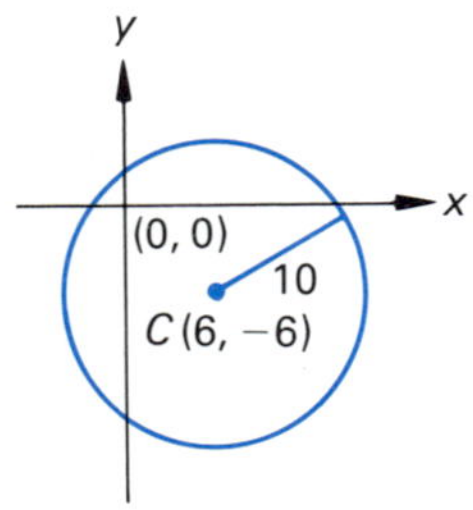

10.

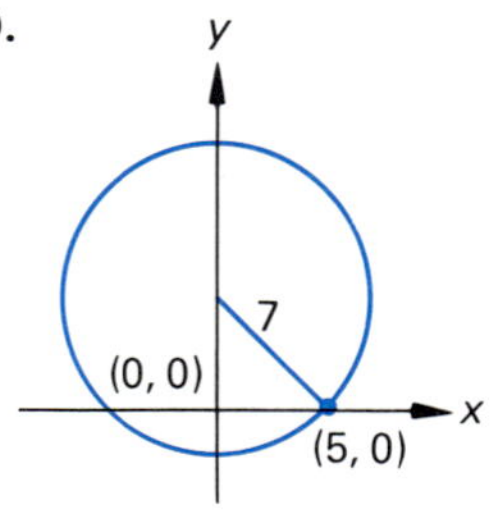

11.

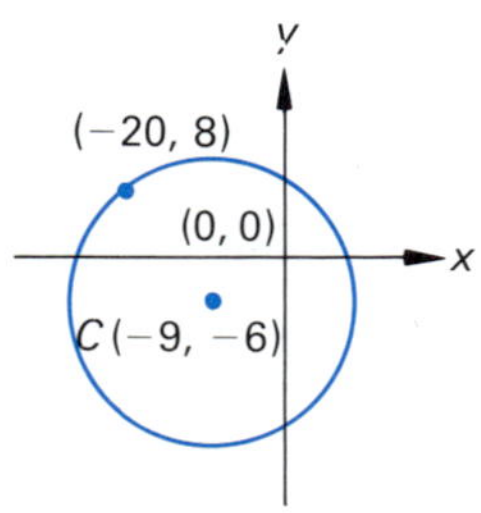

12.

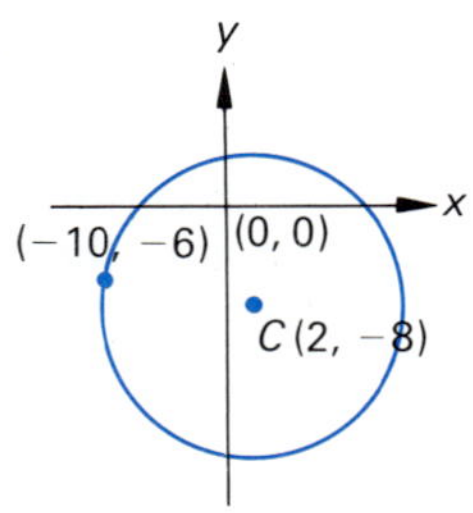

In Exercises 13–18, sketch the graph of the circle indicated.

13. $x^2 + y^2 = 9$

14. $x^2 + y^2 = 16$

15. $(x - 2)^2 + (y - 4)^2 = 16$

16. $(x + 1)^2 + (y - 2)^2 = 4$

17. $(x - 1)^2 + (y + 2)^2 = 4$

18. $(x + 3)^2 + (y + 1)^2 = 9$

19. What is the equation of the circle passing through the points with coordinates $(3, -2)$, $(0, 1)$, and $(4, 2)$?

20. What is the equation of the circle passing through the points with coordinates $(-4, -3)$, $(2, -1)$, and $(6, 5)$?

In Exercises 21–30, determine whether the given equation represents a circle. If not, state why not. If it does, state the coordinates of the center and the radius.

21. $x^2 + y^2 - 8x - 4y + 16 = 0$

22. $x^2 + y^2 + 8y + 6x = 0$

23. $x^2 + y^2 - y - 2 = 0$

24. $x^2 + y^2 + 5x = 0$

25. $3x^2 + 3y^2 + 6x - 6y = 0$

26. $2x^2 + 2y^2 - 8x + 12y + 8 = 0$

27. $x^2 + 2y^2 - 2x - 2y = 0$

28. $3x^2 + 2y^2 + 3x + 2y = 0$

29. $x^2 + y^2 + 25 = 0$

30. $x^2 + y^2 + 16 = 0$

31. What is the equation of the line tangent to the circle $(x - 1)^2 + (y + 2)^2 = 4$ at the point $(3, -2)$?

32. What is the equation of the line tangent to the circle $(x + 2)^2 + (y - 3)^2 = 25$ at the point $(1, 7)$?

33. The electric field surrounding a current-carrying conductor is circular. If a wire with a current flowing in it passes through a circuit board 3.00 in. to the right and 2.00 in. above the lower left corner, will the field strength at the corner be the same as the strength at a point 4.00 in. to the right and 6.00 in. above the corner?

34. In Exercise 33, where on the bottom edge and on the left edge of a circuit board will the field strength be the same as it is at the corner?

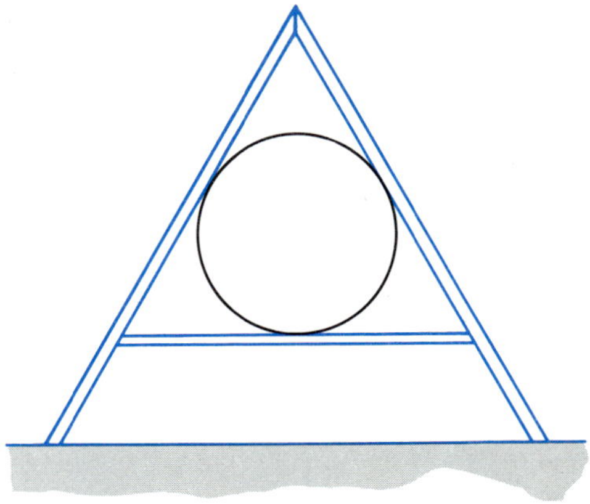

FIGURE 17.16

35. A 1.00-m diameter steel conduit is to be supported by an A-frame structure so that its lower edge is 0.500 m above the ground. If the outer legs of the support make angles of 60° with the ground, how long must they be? (See Fig. 17.16.)

36. Two pipelines are to be joined by a smooth circular arc with a 2.00-ft radius. (A smooth arc is tangent to both of the lines.) If one line can be described by $x - y + 3 = 0$ and the other can be described by $x + 2y = 5$ (Fig. 17.17), what is the equation of the circle whose arc joins the two?

37. The surface waves from a drop falling on the surface of a square tank of smooth water are described

FIGURE 17.17

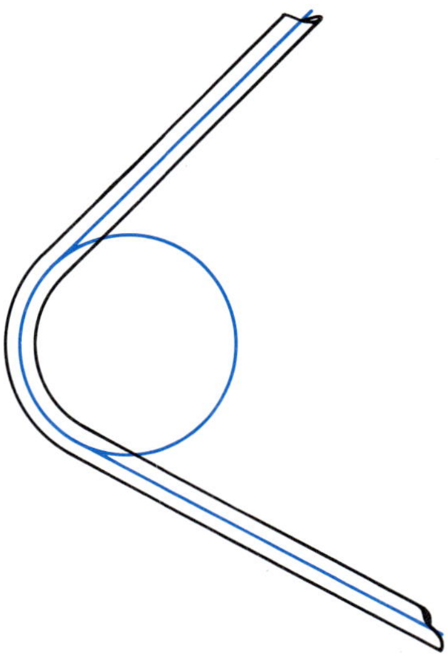

at time t by the equation:

$$x^2 + y^2 - 14.40x + 6.20y + 55.20 = 0,$$

which uses one corner of the tank as the origin of the axis system. Find the center and the radius of the circle by completing the squares.

38. A conical pendulum traces out a path over the xy-plane described by

$$2x^2 + 2y^2 + 21.20x + 18.80y + 13.24 = 0.$$

Locate the center and the radius of the circle by completing the squares.

17.4 The parabola

Geometrically the **parabola** is defined as the set of all the points in a plane such that any point is the same distance from a fixed point called the **focus** as it is from a fixed line called the **directrix.** Physically the parabola has the characteristic that when a source of light or radio waves is placed at the focus all the waves that strike the surface of the parabola are reflected in a column of rays parallel to each other. As a result the parabola is often the basis of light reflectors, and light and radio telescopes.

Let us assume a focus F with coordinates $(p, 0)$ and a directrix parallel to the y-axis with equation $x = -p$ (Fig. 17.18). The distance from the focus to the directrix is therefore $2p$ and the origin must be a point on the parabola because it is equidistant between the focus and the directrix.

For point $P(x, y)$ to be a point on the parabola, the distance from the focus to P, $\sqrt{(x-p)^2 + (y-0)^2}$, must be the same as the distance from the directrix to P, $x + p$ (Fig. 17.19). Therefore

$$x + p = \sqrt{(x-p)^2 + (y-0)^2}$$

$$(x+p)^2 = (x-p)^2 + y^2 \qquad \textbf{Squaring both sides}$$

$$x^2 + 2px + p^2 = x^2 - 2px + p^2 + y^2 \qquad \textbf{Expanding}$$

$$y^2 = 4px. \qquad \textbf{Rearranging and combining terms}$$

FIGURE 17.18

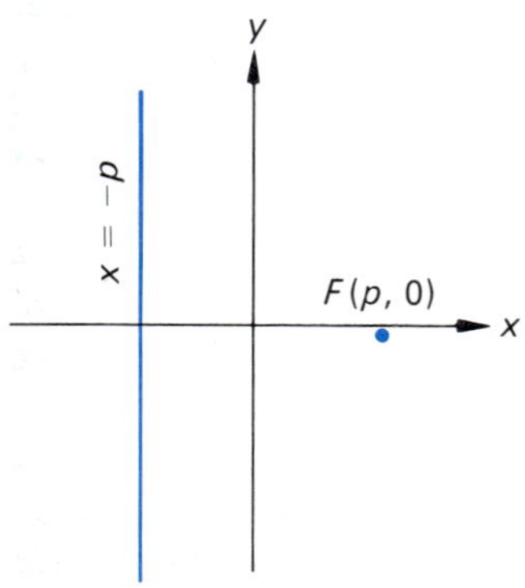

FIGURE 17.19

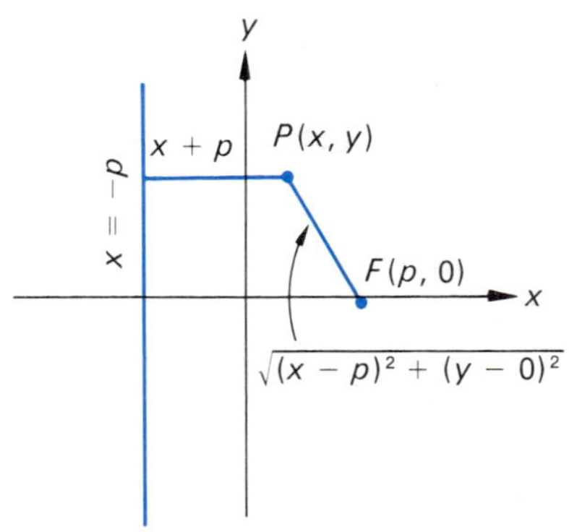

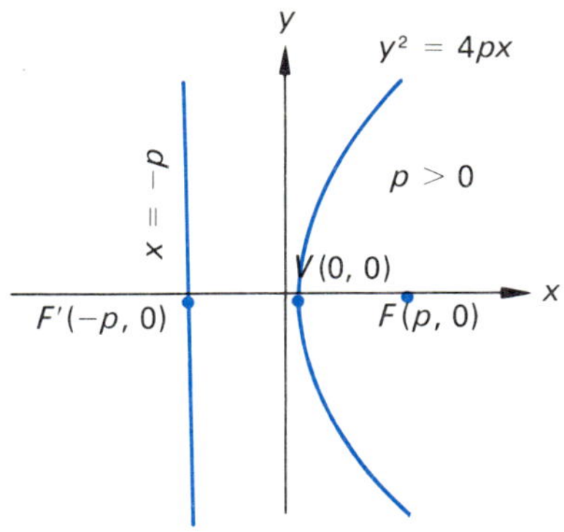

FIGURE 17.20

The parabola, $y^2 = 4px$ is graphed in Fig. 17.20. The curve passes through the origin and is symmetric with respect to the positive x-axis if $p > 0$ and is symmetric with respect to the negative x-axis if $p < 0$. The **vertex** V of the parabola is on the line of symmetry, midway between the focus and the directrix. Thus the vertex is at the origin.

DEFINITION 17.10

The **standard form** of the equation of a **parabola** symmetric with respect to the x-axis, with vertex (0, 0), focus (p, 0), and directrix the line $x = -p$ is $y^2 = 4px$.

EXAMPLE 1 Sketch the graph of $y^2 = 8x$, indicating the focus, vertex, and directrix.

Solution: The parabola $y^2 = 8x$ is of the form $y^2 = 4px$, which has the vertex at the origin and

$$4p = 8$$
$$p = 2.$$

Since $p = 2$ we know that the focus is at the point (2, 0), and the directrix is the line $x = -2$. The parabola is symmetric with respect to the positive x-axis and its vertex is at the origin. With this information we need calculate only a few key points, such as (2, 4), (2, −4), (8, 8), and (8, −8) in order to sketch the cruve as in Fig. 17.21. □

EXAMPLE 2 Sketch the graph of $y^2 = -2x$. Show the focus, vertex, and directrix.

Solution: To determine the focus of the parabola $y^2 = -2x$ we let $4p = -2$. Therefore $p = -1/2$. Since $p < 0$, the parabola opens to the

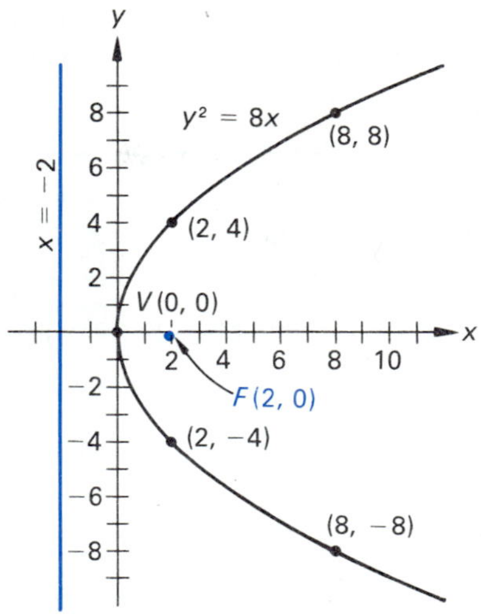

FIGURE 17.21

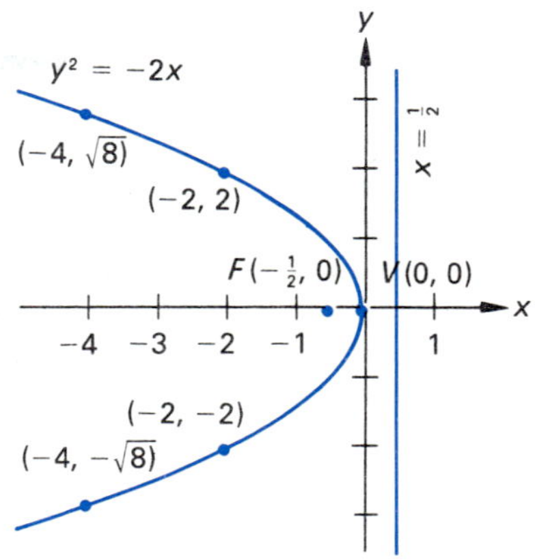

FIGURE 17.22

left. The focus is at $\left(-\frac{1}{2}, 0\right)$, the vertex is at the origin and the curve is symmetric with respect to the negative x-axis. With this information we need only calculate a few key points such as $(-2, 2)$, $(-2, -2)$, $(-4, \sqrt{8})$, and $(-4, -\sqrt{8})$ in order to sketch the curve in Fig. 17.22. □

If we place the focus of a parabola on the y-axis a distance p from the origin and make its directrix perpendicular to the y-axis (parallel to the x-axis) the same distance p from the other side of the origin, the equation derived by using the definition of a parabola is $x^2 = 4py$.

DEFINITION 17.11

The **standard form** of the equation of a **parabola** symmetric with respect to the y-axis and with vertex $(0, 0)$, focus $(0, p)$, and directrix the line $y = -p$ is $x^2 = 4py$.

FIGURE 17.23

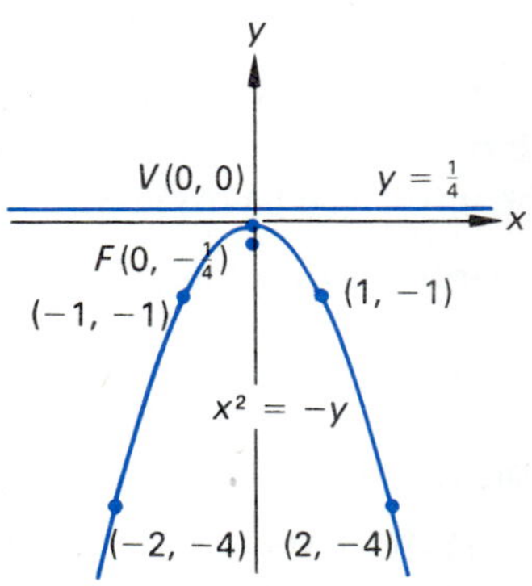

EXAMPLE 3 Sketch the graph of the parabola $x^2 = -y$. Show the vertex, focus, and directrix.

Solution: The parabola $x^2 = -y$ is of the form $x^2 = 4py$, which tells us that $4p = -1$, or $p = -1/4$. From this we know that the focus is at the point $(0, -1/4)$, the vertex is $V(0, 0)$, and the directrix is the line $y = 1/4$. With this information we need calculate only a few key points such as $(-1, -1)$, $(1, -1)$, $(-2, -4)$, and $(2, -4)$ in order to sketch the curve as in Fig. 17.23. □

Parabolas do not always have to be symmetric with respect to one of the axes nor must they always have their vertex at the origin.

DEFINITION 17.12

> The **standard form** of the equation of a **parabola** symmetric with respect to the line $y = k$ and with vertex (h, k), focus $(h + p, k)$, and directrix the line $x = h - p$ is
>
> $$(y - k)^2 = 4p(x - h).$$
>
> (See Fig. 17.24.)

DEFINITION 17.13

> The **standard form** of the equation of a **parabola** symmetric with respect to the line $x = h$ and with vertex (h, k), focus $(h, k + p)$, and directrix the line $y = k - p$ is
>
> $$(x - h)^2 = 4p(y - k).$$
>
> (See Fig. 17.25.)

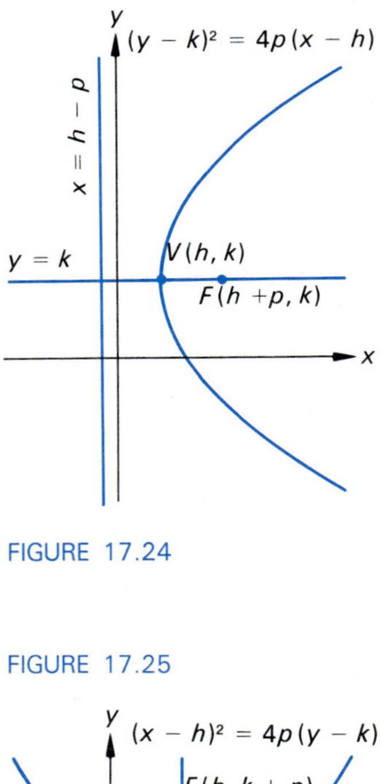

FIGURE 17.24

The parabolas in Definitions 17.12 and 17.13 will open to the right (or left) or up (or down) depending on whether p is positive (or negative).

The **latus rectum** of a parabola is an aid to sketching the curve. It is a chord of the parabola that passes through the focus and is perpendicular to the axis of symmetry. The end points of the latus rectum are a distance of $2p$ from the focus.

EXAMPLE 4 Sketch the graph of the parabola whose equation is $(y + 1)^2 = 6(x - 2)$.

FIGURE 17.25
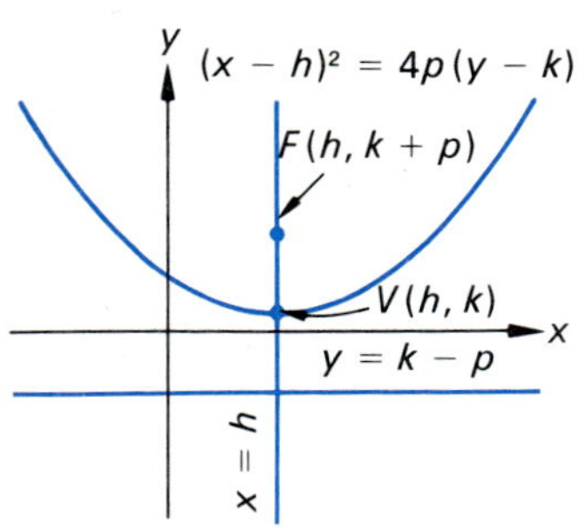

Solution: The equation $(y + 1)^2 = 6(x - 2)$ is of the standard form $(y - k)^2 = 4p(x - h)$. Therefore $h = 2$, $k = -1$, and $p = 3/2$. The parabola is symmetric with respect to the line $y = -1$, it has its vertex at $V(2, -1)$, its focus at $F\left(\frac{7}{2}, -1\right)$, and has its latus rectum along the line through the focus, $x = 7/2$. The end points of the latus rectum are $2p = 2\left(\frac{3}{2}\right) = 3$ units from the focus. The coordinates of the end points of the latus rectum are $(7/2, 2)$ and $(7/2, -4)$. With these key points we can sketch the parabola as shown in Fig. 17.26. □

Expanding and rearranging the terms of the standard forms of the equation of a parabola produces the general forms of the equation of a parabola.

DEFINITION 17.14

The **general forms** of the equation of a **parabola** are

$$Cy^2 + Dx + Ey + F = 0 \quad \text{and} \quad Ax^2 + Dx + Ey + F = 0.$$

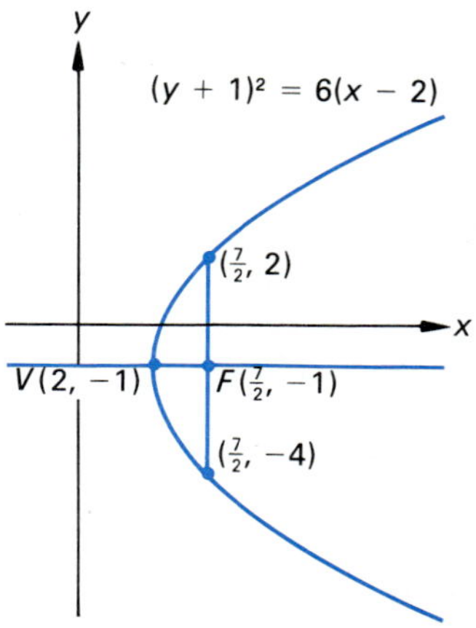

FIGURE 17.26

EXAMPLE 5 Describe the following parabolas by giving the coordinates of the vertex, the focus, the end points of the latus rectum, and the equation of the axis of symmetry:

a. $4x^2 - 24x + y + 34 = 0$

b. $9y^2 + 4x + 6y - 1 = 0$

Solution:

a. Since the variable x in $4x^2 - 24x + y + 34 = 0$ is of second degree the appropriate standard form of the equation is

$$(x - h)^2 = 4p(y - k) \qquad \textbf{Definition 17.13}$$

$$4x^2 - 24x + y + 34 = 0 \qquad \textbf{Given}$$

$$x^2 - 6x = -\frac{1}{4}y - \frac{34}{4} \qquad \textbf{Dividing by 4 to obtain a coefficient of 1 for } x^2 \textbf{ and rearranging}$$

$$x^2 - 6x + 9 = -\frac{1}{4}y - \frac{34}{4} + 9 \qquad \textbf{Adding 9 to both sides to complete the square on the left side.}$$

$$(x - 3)^2 = -\frac{1}{4}y + \frac{2}{4} \qquad \textbf{Factoring on the left side and combining terms on the right}$$

$$(x - 3)^2 = -\frac{1}{4}(y - 2). \qquad \textbf{Factoring } -1/4 \textbf{ on the right side to obtain a coefficient of 1 for } y$$

FIGURE 17.27

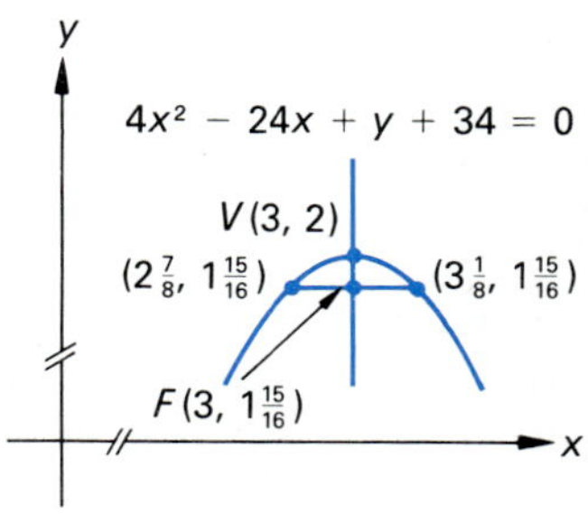

Therefore $h = 3$, $k = 2$, and $p = -1/16$. The parabola is symmetric with respect to $x = 3$, and it has vertex $V(3, 2)$, focus $F\left(3, 1\frac{15}{16}\right)$, and end points of the latus rectum at $\left(2\frac{7}{8}, 1\frac{15}{16}\right)$ and $\left(3\frac{1}{8}, 1\frac{15}{16}\right)$ (Fig. 17.27).

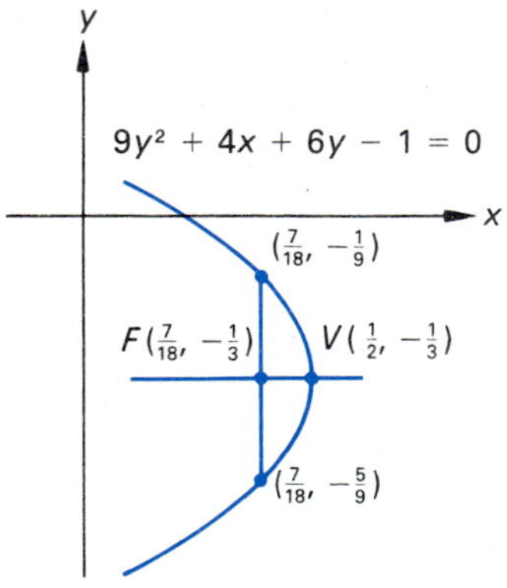

FIGURE 17.28

b. Since the variable y in $9y^2 + 4x + 6y - 1 = 0$ is of second degree the appropriate standard form of the equation is

$(y - k)^2 = 4p(x - h)$	**Definition 17.12**
$9y^2 + 4x + 6y - 1 = 0$	**Given**
$y^2 + \frac{6}{9}y = -\frac{4}{9}x + \frac{1}{9}$	**Rearranging and dividing by 9**
$y^2 + \frac{2}{3}y + \frac{1}{9} = -\frac{4}{9}x + \frac{1}{9} + \frac{1}{9}$	**Completing square**
$\left(y + \frac{1}{3}\right)^2 = -\frac{4}{9}\left(x - \frac{1}{2}\right).$	**Factoring**

Therefore $h = 1/2$, $k = -1/3$, and $p = -1/9$. The parabola is symmetric with respect to $y = -1/3$, and it has vertex $V\left(\frac{1}{2}, -\frac{1}{3}\right)$, focus $F\left(\frac{7}{18}, -\frac{1}{3}\right)$, and end points of the latus rectum at $\left(\frac{7}{18}, -\frac{1}{9}\right)$ and $\left(\frac{7}{18}, -\frac{5}{9}\right)$ (Fig. 17.28). □

EXAMPLE 6 We have stated that the standard forms of the equation of a parabola can always be represented by one of the general forms (Definition 17.14). Does the general form of Definition 17.14 always represent a parabola?

Solution: Let us take one of the general parabolic equation forms and attempt to manipulate it into one of the standard forms.

$$Ax^2 + Dx + Ey + F = 0$$

$x^2 + \frac{D}{A}x = -\frac{E}{A}y - \frac{F}{A}$	**Dividing by A and rearranging**
$x^2 + \frac{D}{A}x + \frac{D^2}{4A^2} = -\frac{E}{A}y - \frac{F}{A} + \frac{D^2}{4A^2}$	**Completing the square**
$\left(x + \frac{D}{2A}\right)^2 = -\frac{E}{A}\left(y + \frac{F}{E} - \frac{D^2}{4AE}\right)$	**Factoring**
$\left(x + \frac{D}{2A}\right)^2 = -4\left(\frac{E}{4A}\right)\left(y + \frac{4AF - D^2}{4AE}\right).$	**Finding P**

The general form of the parabolic equation will always produce the standard form providing $A \neq 0$ and $E \neq 0$. However, if either A or E is equal to zero, we would not have the general form of the parabolic equation. A similar manipulation of the other general parabolic equation produces similar results so we can conclude that the general form of the parabolic equation always represents a parabola. □

EXAMPLE 7 The American Institute for Steel Construction specifies the maximum allowable stress for short columns as a function of the slenderness ratio by the equation $S_{cr} = 17{,}000 - 0.485\left(\frac{\ell}{r}\right)^2$. Is this equation parabolic?

Solution: If this equation is parabolic, it must fit one of the standard forms for the equation of a parabola. S_{cr} is the maximum stress; it is also the dependent variable. The slenderness ratio ℓ/r is the independent variable. For convenience let $x = \frac{\ell}{r}$ and $y = S_{cr}$.

$$S_{cr} = 17000 - 0.485\left(\frac{\ell}{r}\right)^2$$

$$y = 17000 - 0.485(x)^2 \qquad \textbf{Substituting for } \ell/r \textbf{ and } S_{cr}$$

$$0.485x^2 = -(y - 17{,}000) \qquad \textbf{Rearranging}$$

$$x^2 = \frac{-1}{0.485}(y - 17{,}000) \qquad \textbf{Dividing by 0.485}$$

The equation is parabolic with the axis of symmetry parallel to the y-axis. □

17.4 Exercises

In Exercises 1–12, sketch the parabola represented by the equation. Show the vertex, the focus, the end points of the latus rectum, and the axis of symmetry.

1. $x^2 = 2y$
2. $y^2 = 16x$
3. $y^2 = -3(x + 1)$
4. $x^2 = -2(y - 3)$
5. $(y - 3)^2 = x$
6. $(x + 3)^2 = -y$
7. $(x - 2)^2 = 8(y + 1)$
8. $(y + 1)^2 = 8(x - 2)$
9. $(y - 2)^2 = -\frac{1}{2}(x - 2)$
10. $2x^2 + 3x + 4y - 4 = 0$
11. $x^2 + 2x - 3y + 1 = 0$
12. $0.5y^2 + 1.5x + 2.0y + 1.0 = 0$

In Exercises 13–18, write the equation of the parabola shown.

13.
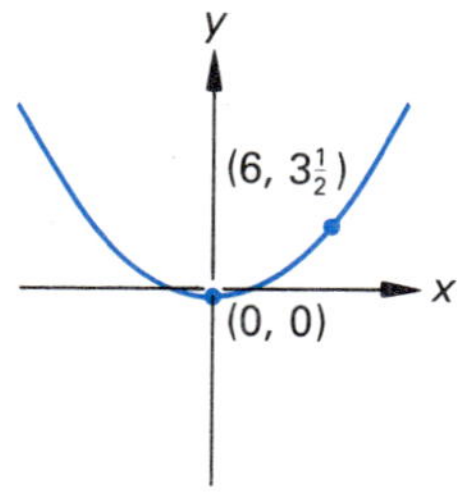

14.
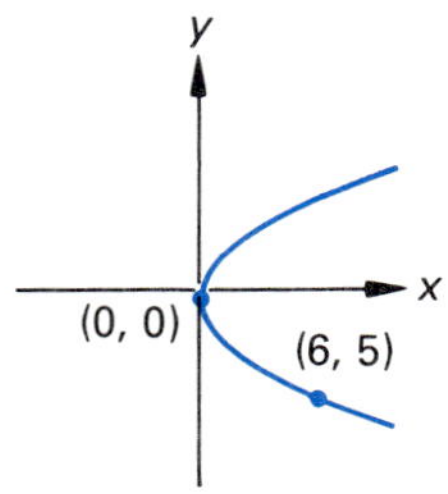

15.
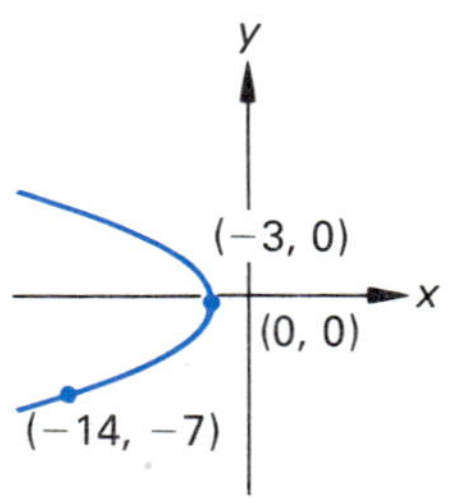

16.
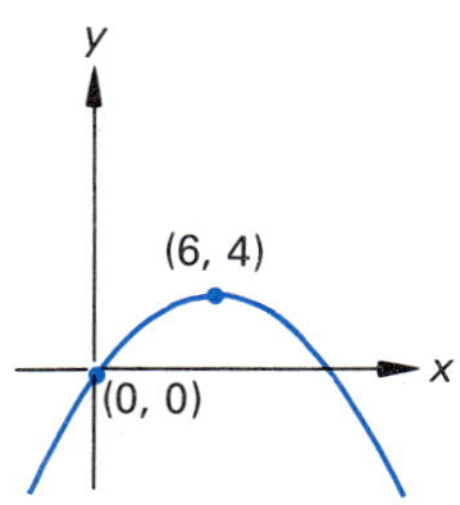

17.

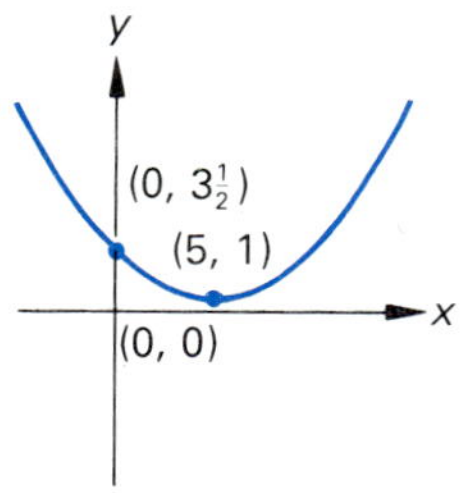

18.

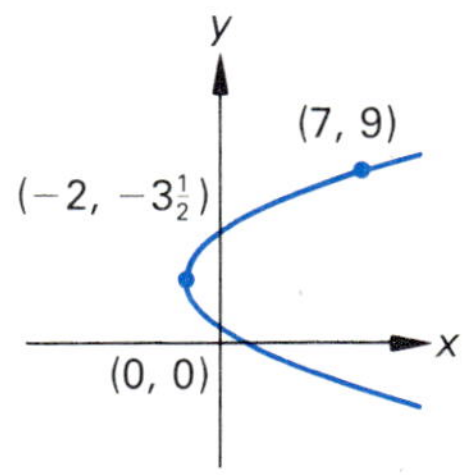

19. The side-to-side curve of a ship's deck is in the shape of a vertex uppermost parabola. What is the equation of the deck section where the ship is 20 ft wide and the center of the deck is 6 in. above the sides? Take the origin of the axis system at the center of the deck (Fig. 17.29).

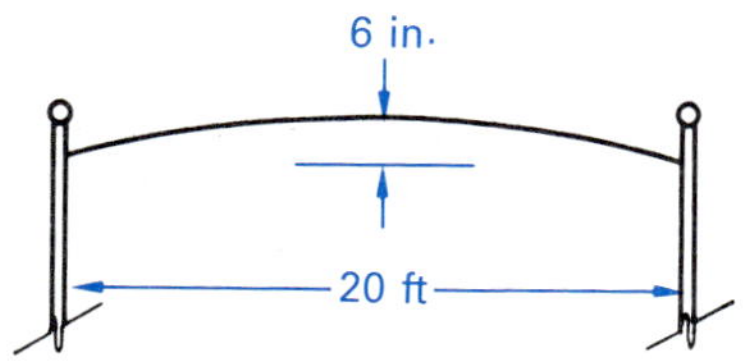

FIGURE 17.29

20. The cables of suspension bridges form parabolic curves. What is the equation of the cable for a bridge with supports 400 ft apart and a maximum center sag of 50 ft? (See Fig. 17.30.) Take the origin of the axis system at the low point of the cable.

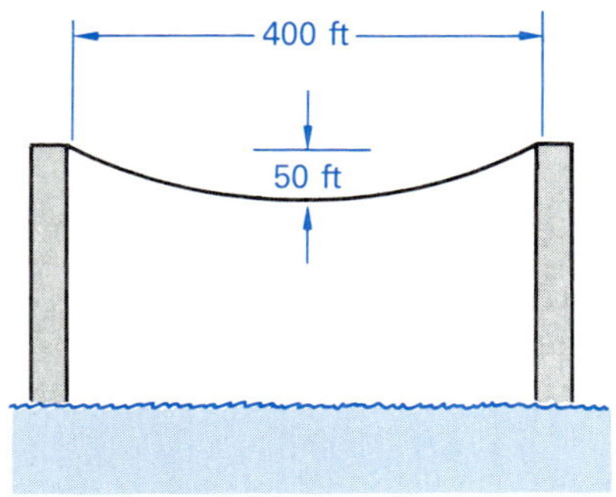

FIGURE 17.30

21. What is the equation of the parabolic cable in Exercise 20 if the origin of the axis system is taken at the point of attachment of the cable on the left support?

22. The long-term rate of scale buildup on the surface of a metal that forms a continuous scale (such as copper) is given by $w^2 = kt + C$, where w is the weight of scale formed per unit area, t is time, and k and C are constants. In a corrosion test 0.2 g/cm^2 of scale form after 8 months' exposure and 0.3 g/cm^2 of scale form after 12 months' exposure. What would be the expected scale buildup after two years' exposure?

23. The equation in Exercise 22 is not valid for short-term effects. Why not?

24. The collecting microphone of a radiotelescope is placed in the same plane as the outer rim of the dish. The dish has a diameter of 120 ft. If the microphone is 45 ft off the ground when the telescope is pointed straight up, what is the equation of the parabolic cross-section with respect to an axis directly under the telescope on the ground? The microphone is at the focus of the parabola (see Fig. 17.31).

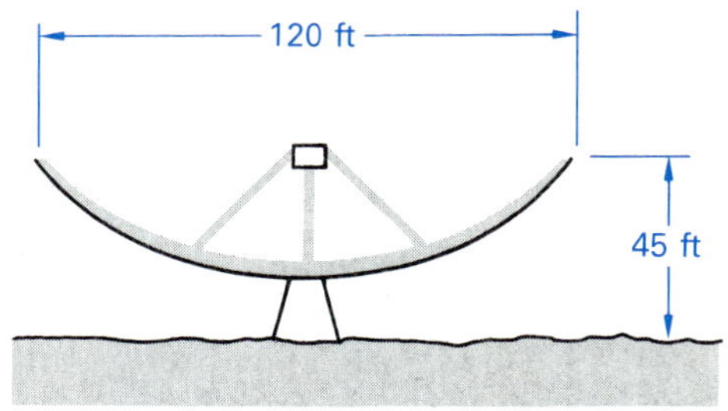

FIGURE 17.31

25. If a light source is placed at the focus of a parabolic reflector, all the light rays that strike the surface of the reflector are reflected parallel to the axis of symmetry of the parabola. A headlight has a filament 20 mm from the back of the reflector and the reflector is 80 mm wide in the plane of the filament. What is the equation of the parabolic reflector if the origin of the axis system is placed at the filament (see Fig. 17.32)?

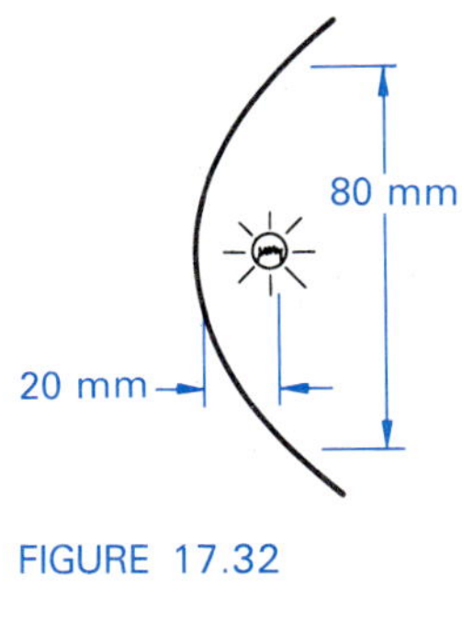

FIGURE 17.32

17.5 The ellipse

An **ellipse** *is the set of all points in a plane such that the sum of the distances from two fixed points to a point on the ellipse is a constant.* The fixed points are called the **foci** (singular **focus**). An elliptical reflector has the property that a wave originating at one focus is reflected through the other focus. An elliptical tip also tends to reduce drag losses on propellers and wings.

Let us choose two points for the foci of an ellipse such that they lie on the x-axis a distance c on each side of the origin (Fig. 17.33). Let us also assume that the ellipse crosses the x-axis a distance a on either side of the origin and it crosses the y-axis a distance b above and below the origin.

Applying the definition of an ellipse to the information given with the curve of Fig. 17.34, we see that the sum of the distances from the two foci to the point on the right side where the ellipse crosses the x-axis is

$$\begin{aligned}\overline{F_1V_2} + \overline{F_2V_2} &= (c + a) + (a - c)\\ &= 2a.\end{aligned}$$

Therefore the constant specified in the definition must equal $2a$. The sum of the distances from the foci to the point where the ellipse crosses the y-axis is

$$\begin{aligned}\overline{F_1B_1} + \overline{F_2B_1} &= \sqrt{[0 - (-c)]^2 + [b - 0]^2} + \sqrt{[0 - c]^2 + [b - 0]^2}\\ &= \sqrt{c^2 + b^2} + \sqrt{c^2 + b^2}\\ &= 2a. \qquad \textbf{Definition of ellipse}\end{aligned}$$

Therefore $2\sqrt{c^2 + b^2} = 2a$ or, by squaring and rearranging,

$$c^2 = a^2 - b^2$$

FIGURE 17.33

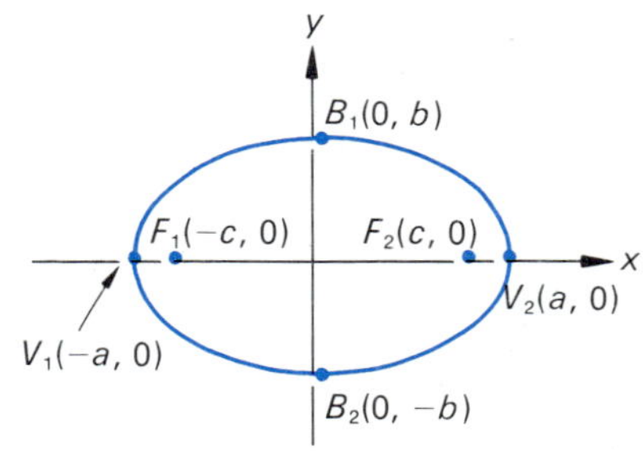

FIGURE 17.34

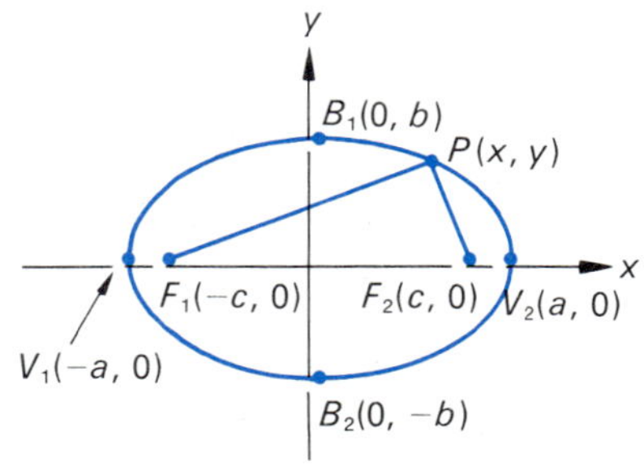

(Note that $c^2 = a^2 - b^2$ requires $a > b$.) Considering a general point on the ellipse, $P(x, y)$, the distances from the two foci to the point are

$$\overline{F_1P} = \sqrt{[x-(-c)]^2 + [y-0]^2}$$
$$\overline{F_2P} = \sqrt{[x-c]^2 + [y-0]^2}.$$

According to the definition of an ellipse

$$\overline{F_1P} + \overline{F_2P} = 2a$$

$$\sqrt{[x-(-c)]^2 + (y-0)^2} + \sqrt{[x-c]^2 + [y-0]^2} = 2a$$ **Substituting**

$$\sqrt{[x+c]^2 + y^2} = 2a - \sqrt{[x-c]^2 + y^2}$$ **Simplifying and rearranging**

$$[x+c]^2 + y^2 = 4a^2 - 4a\sqrt{[x-c]^2 + y^2} + [x-c]^2 + y^2$$ **Squaring both sides**

$$x^2 + 2cx + c^2 + y^2 = 4a^2 - 4a\sqrt{[x-c]^2 + y^2} + x^2 - 2cx + c^2 + y^2$$ **Expanding**

$$4a\sqrt{[x-c]^2 + y^2} = 4a^2 - 4cx$$ **Combining terms and rearranging**

$$a\sqrt{[x-c]^2 + y^2} = a^2 - cx$$ **Dividing by 4**

$$a^2[(x-c)^2 + y^2] = a^4 - 2a^2cx + c^2x^2$$ **Squaring both sides**

$$a^2(x^2 - 2cx + c^2 + y^2) = a^4 - 2a^2cx + c^2x^2$$
$$a^2x^2 - 2a^2cx + a^2c^2 + a^2y^2 = a^4 - 2a^2cx + c^2x^2$$

$$a^2x^2 - c^2x^2 + a^2y^2 = a^4 - a^2c^2$$ **Rearranging**

$$(a^2 - c^2)x^2 + a^2y^2 = a^2(a^2 - c^2)$$ **Factoring**

$$b^2x^2 + a^2y^2 = a^2b^2$$ **Recalling $c^2 = a^2 - b^2$ or $a^2 - c^2 = b^2$ and substituting b^2 for $a^2 - c^2$**

$$\frac{x^2}{a^2} + \frac{y^2}{b^2} = 1.$$ **Dividing by a^2b^2**

The portion of the x-axis bounded by the ellipse is its **major axis** so the distance a is called the length of the **semi-major axis.** The portion of the y-axis bounded by the ellipse is its **minor axis** so the distance b is called the length of the **semi-minor axis.**

DEFINITION 17.15

The **standard form** of the equation of an **ellipse** with center at the origin, length of the semi-major axis a and semi-minor axis b, and major axis along the x-axis is

$$\frac{x^2}{a^2} + \frac{y^2}{b^2} = 1.$$

(See Fig. 17.35a.)

DEFINITION 17.16

> The **standard form** of the equation of an **ellipse** with center at the origin, length of the semi-major axis a and semi-minor axis b, and major axis along the y-axis is
>
> $$\frac{x^2}{b^2}+\frac{y^2}{a^2}=1.$$
>
> (See Fig. 17.35b.)

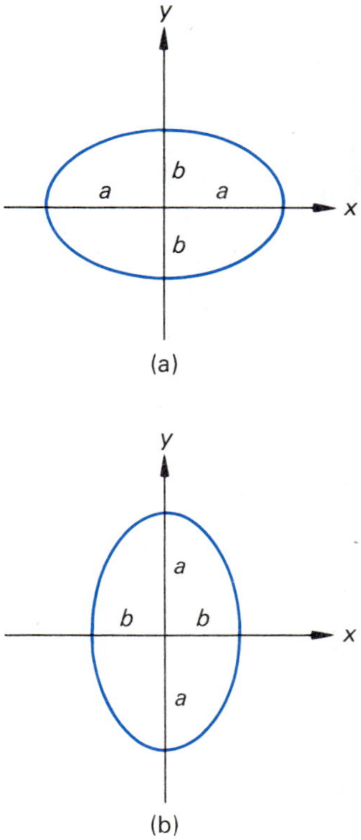

FIGURE 17.35

EXAMPLE 1 For the ellipse $9x^2+4y^2=36$, determine the end points of the major axis, the minor axis, and the coordinates of the foci, and sketch the curve.

Solution: To determine a and b, which will help us determine the end points of the axes, we place the equation $9x^2+4y^2=36$ in standard form:

$$\frac{9x^2}{36}+\frac{4y^2}{36}=\frac{36}{36} \qquad \textbf{Dividing by 36}$$

$$\frac{x^2}{4}+\frac{y^2}{9}=1.$$

We know that $a>b$, and since the larger numerical value is under the y^2, the major axis concides with the y-axis. Thus $a^2=9$ or $a=3$, and $b^2=4$ or $b=2$. We also know that $c^2=a^2-b^2$; substituting we have

$$c^2=9-4$$

$$c=\sqrt{5}.$$

Knowing a, b, and c we can write the coordinates

$(0, 3), (0, -3)$	**End points of the major axis**
$(2, 0), (-2, 0)$	**End points of the minor axis**
$(0, \sqrt{5}), (0, -\sqrt{5})$	**Foci**
$(0, 0)$	**Center**

With this information and a few key points we can sketch the ellipse. When $y=1$,

$$\frac{x^2}{4}+\frac{(1)^2}{9}=1$$

$$x=\pm\frac{4\sqrt{2}}{3}.$$

When $y = 2$,

$$\frac{x^2}{4} + \frac{(2)^2}{9} = 1$$

$$x = \pm\frac{2\sqrt{5}}{3}.$$

Since the ellipse is symmetric with respect to the major axis, minor axis, and its center and knowing the points $\left(\frac{4\sqrt{2}}{3}, 1\right)$ and $\left(\frac{2\sqrt{5}}{3}, 2\right)$ we can write the following points $\left(\frac{4\sqrt{2}}{3}, -1\right)$, $\left(-\frac{4\sqrt{2}}{3}, 1\right)$, $\left(\frac{-4\sqrt{2}}{3}, -1\right)$, $\left(\frac{2\sqrt{5}}{3}, -2\right)$, $\left(\frac{-2\sqrt{5}}{3}, 2\right)$, and $\left(\frac{-2\sqrt{5}}{3}, -2\right)$. The curve is sketched and properly labeled in Fig. 17.36. □

EXAMPLE 2 Write the equation of the ellipse sketched in Fig. 17.37 in standard form.

Solution: By examining Fig. 17.37 we can determine that the major axis coincides with the x-axis and the equation of the ellipse is of the form $\frac{x^2}{a^2} + \frac{y^2}{b^2} = 1$. With vertices (endpoints) of the major axis at (5, 0), (−5, 0) and of the minor axis at (0, 3), (0, −3), we can determine that $a = 5$ and $b = 3$. Substituting for a and b in the standard form of the

FIGURE 17.36

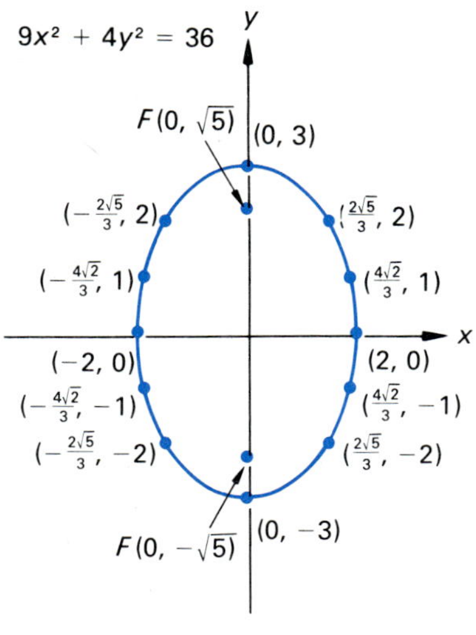

FIGURE 17.37

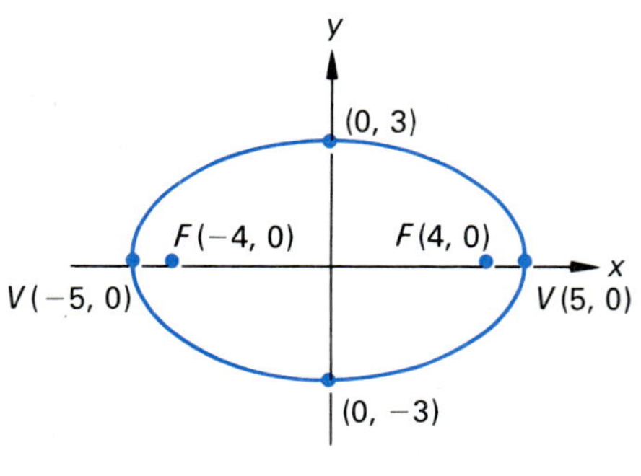

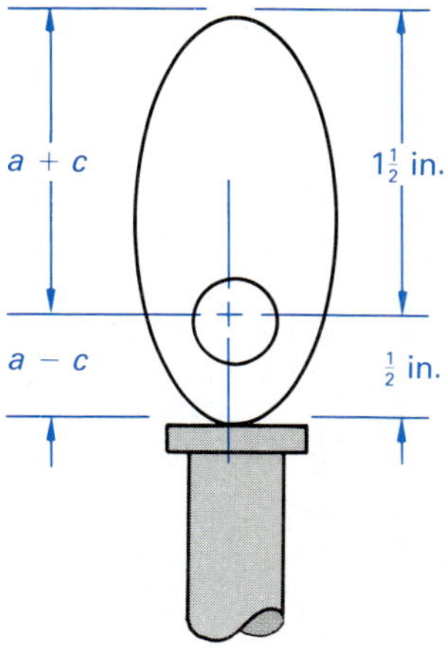

FIGURE 17.38

equation we have

$$\frac{x^2}{25} + \frac{y^2}{9} = 1$$

as the equation of the ellipse. □

EXAMPLE 3 A cam is to be made in the shape of an ellipse with the center line of the camshaft passing through one focus of the ellipse. The center line of the camshaft is to be 1/2 in. from the cam follower when the follower is in the closed position and the follower is to be depressed 1 in. when it is completely open. What is the equation of the necessary ellipse?

Solution: Let us sketch the ellipse so that the camshaft is directly above the follower when the follower is in the fully closed position as in Fig. 17.38. The focus of the ellipse is then 1/2 in. above the follower as shown. According to the derivation of the equation of an ellipse, the distance from a focus to the nearest vertex is $a - c$. Therefore

$$a - c = 1/2. \quad \textbf{[1]}$$

In order for the follower to be depressed 1 in. when the cam is fully reversed, the distance from the center of the camshaft to the other vertex must be 1/2 in. + 1 in. = 1 1/2 in. as shown. This distance from the derivation is $a + c$. Therefore

$$a + c = 1\ 1/2 \quad \textbf{[2]}$$
$$2a = 2 \quad \textbf{Adding [1] + [2]}$$
$$a = 1 \quad \textbf{[3]}$$
$$c = 1/2 \quad \textbf{Subtracting [2] − [3]}$$
$$c^2 = a^2 - b^2 \quad \textbf{From the derivation}$$
$$b^2 = a^2 - c^2 \quad \textbf{Rearranging}$$
$$= (1)^2 - (1/2)^2$$
$$= 3/4$$
$$b = \pm\frac{\sqrt{3}}{2}.$$

In this configuration the major axis of the ellipse is vertical so the appropriate standard form of the equation of the ellipse is

$$\frac{x^2}{b^2} + \frac{y^2}{a^2} = 1. \quad \textbf{Definition 17.16}$$

If we now locate an xy-axis system with its center 1 in. above the follower, the center of the elliptical cam is at (0, 0). Therefore, the equation of the ellipse needed to form the outline of the cam is

$$\frac{x^2}{(\sqrt{3}/2)^2} + \frac{y^2}{(1)^2} = 1. \quad □$$

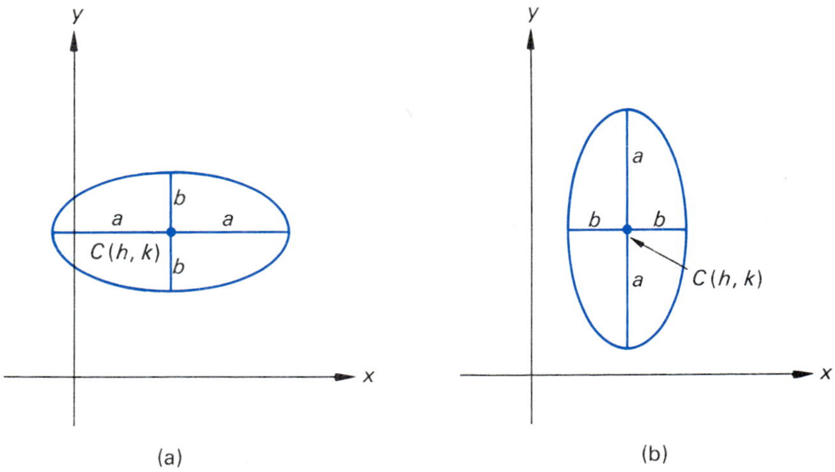

FIGURE 17.39

DEFINITION 17.17

The **standard form** of the equation of an **ellipse** with center at $C(h, k)$, length of the semi-major axis a and semi-minor axis b, and major axis parallel to the x-axis is

$$\frac{(x-h)^2}{a^2}+\frac{(y-k)^2}{b^2}=1.$$

(See Fig. 17.39a.)

DEFINITION 17.18

The **standard form** of the equation of an **ellipse** with center at $C(h, k)$, length of the semi-major axis a and semi-minor axis b, and major axis parallel to the y-axis is

$$\frac{(x-h)^2}{b^2}+\frac{(y-k)^2}{a^2}=1.$$

(See Fig. 17.39b.)

DEFINITION 17.19

The **general form** of the equation of an **ellipse** is

$$Ax^2+Cy^2+Dx+Ey+F=0.$$

(A and C must have the same sign and $A \neq C$.)

EXAMPLE 4 Sketch the ellipse whose equation is $4x^2 + 25y^2 - 8x + 100y + 4 = 0$ and label the center, foci, and vertices.

Solution: Manipulate the equation of the ellipse from the general form into the standard form:

$$\frac{(x-h)^2}{a^2} + \frac{(y-k)^2}{b^2} = 1$$

$$4x^2 + 25y^2 - 8x + 100y + 4 = 0.$$

Group x-terms together and y-terms together and use common-term factoring to obtain unit coefficients for x^2 and y^2 terms:

$$4(x^2 - 2x) + 25(y^2 + 4y) = -4.$$

Complete squares for terms in x and terms in y:

$$4(x^2 - 2x + 1) + 25(y^2 + 4y + 4) = -4 + 4 + 100$$
$$4(x-1)^2 + 25(y+2)^2 = 100.$$

Divide both sides of equation by 100 to produce unity on right side:

$$\frac{4(x-1)^2}{100} + \frac{25(y+2)^2}{100} = \frac{\overset{1}{\cancel{100}}}{\underset{1}{\cancel{100}}} \qquad \textbf{Divide both sides by 100}$$

$$\frac{(x-1)^2}{25} + \frac{(y+2)^2}{4} = 1$$

$$\frac{(x-1)^2}{(5)^2} + \frac{(y+2)^2}{(2)^2} = 1.$$

Since the larger numerical value is in the denominator of the fraction with x^2 in the numerator, the major axis is horizontal and $a = 5$ and $b = 2$. The ellipse has its center at $(1, -2)$, foci at $(1 - \sqrt{21}, -2)$ and $(1 + \sqrt{21}, -2)$, vertices at $(-4, -2)$, $(6, -2)$, $(1, 0)$, and $(1, -4)$. In decimal form, the foci are at $(-3.58, -2)$ and $(5.58, -2)$. The ellipse is sketched in Fig. 17.40. □

FIGURE 17.40

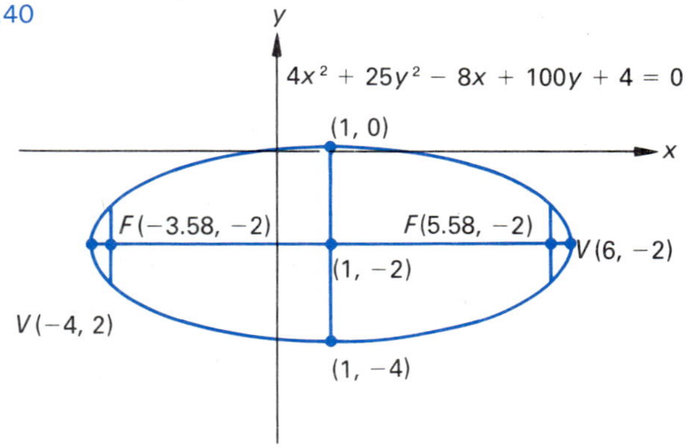

EXAMPLE 5 Is $4x^2 + 25y^2 - 8x + 100y + 114 = 0$ the equation of an ellipse?

Solution: The given equation fits the general form of the equation of an ellipse, but this does not tell us whether it is or is not an equation of an ellipse. One way of telling is to write the equation in one of the standard forms.

$$4x^2 + 25y^2 - 8x + 100y + 114 = 0 \qquad \textbf{Given}$$

$$4(x^2 - 2x) + 25(y^2 + 4y) = -114 \qquad \textbf{Grouping terms and obtaining a unit coefficient for } x^2 \textbf{ and } y^2$$

$$4(x^2 - 2x + 1) + 25(y^2 + 4y + 4) = -114 + 4 + 100 \qquad \textbf{Completing squares}$$

$$4(x-1)^2 + 25(y+2)^2 = -10 \qquad \textbf{Writing as perfect squares}$$

$$\frac{(x-1)^2}{25/10} + \frac{(y+2)^2}{4/10} = \frac{-10}{10} \qquad \textbf{Dividing by 10}$$

$$\frac{(x-1)^2}{25/10} + \frac{(y+2)^2}{4/10} = -1.$$

There is no way to keep the left side of the equation positive and at the same time obtain a positive 1 on the right side. Therefore, since we cannot fit $4x^2 + 25y^2 - 8x + 100y + 114 = 0$ to the form

$$\frac{(x-h)^2}{a^2} + \frac{(y-k)^2}{b^2} = 1,$$

the equation does not represent an ellipse. □

EXAMPLE 6 What is the equation of the ellipse with center at (2, 3) and that contains points (−1, 1) and (3, −2)?

Solution: Let us use the standard form of the ellipse modified slightly as

$$\frac{(x-h)^2}{\alpha^2} + \frac{(y-k)^2}{\beta^2} = 1.$$

(The reason for using α and β instead of a and b will be explained later.) Substituting the coordinates of the center and the coordinates of each point produces

$$\frac{(-1-2)^2}{\alpha^2} + \frac{(1-3)^2}{\beta^2} = 1$$

and

$$\frac{(3-2)^2}{\alpha^2} + \frac{(-2-3)^2}{\beta^2} = 1$$

$$\frac{9}{\alpha^2} + \frac{4}{\beta^2} = 1 \qquad \textbf{[1]}$$

$$\frac{1}{\alpha^2} + \frac{25}{\beta^2} = 1. \qquad \textbf{[2]}$$

To solve for the values of α and β, we do the following:

$$\frac{9}{\alpha^2} + \frac{225}{\beta^2} = 9 \qquad 9 \times [2] \qquad \textbf{[3]}$$

$$\frac{9-9}{\alpha^2} + \frac{225-4}{\beta^2} = 9 - 1 \qquad [3] - [1]$$

$$\frac{221}{\beta^2} = 8 \qquad \textbf{Simplifying}$$

$$\beta = \pm 5.256 \qquad \textbf{Solving for } \beta$$

$$\frac{9}{\alpha^2} + \frac{4}{(\pm 5.256)^2} = 1 \qquad \textbf{Substituting for } \beta \textbf{ in [1]}$$

$$\alpha = \pm 3.244 \qquad \textbf{Solving for } \alpha$$

and the equation of the required ellipse is

$$\frac{(x-2.00)^2}{(3.24)^2} + \frac{(y-3.00)^2}{(5.26)^2} = 1,$$

which is an ellipse with its major axis parallel to the y-axis. When we started the solution we had no way of knowing whether to use

$$\frac{(x-h)^2}{a^2} + \frac{(y-k)^2}{b^2} = 1 \qquad \text{or} \qquad \frac{(x-h)^2}{b^2} + \frac{(y-k)^2}{a^2} = 1$$

as the correct standard form. Since the two forms differ only by which term has the larger value for the denominator, we used the dummy constants α and β instead of a and b, $a > b$, and let the outcome of the calculations tell us that

$$\frac{(x-h)^2}{b^2} + \frac{(y-k)^2}{a^2} = 1$$

is the appropriate form. □

17.5 Exercises

In Exercises 1–10, for each ellipse determine the coordinates of (a) center, (b) foci, (c) end points of major axes, and (d) end points of minor axes; (e) sketch.

1. $\frac{x^2}{9} + \frac{y^2}{4} = 1$ **2.** $\frac{x^2}{16} + \frac{y^2}{9} = 1$ **3.** $\frac{x^2}{16} + \frac{y^2}{25} = 1$

4. $\dfrac{x^2}{4} + \dfrac{y^2}{9} = 1$

5. $\dfrac{(x-1)^2}{9} + \dfrac{(y-2)^2}{4} = 1$

6. $\dfrac{(x-2)^2}{16} + \dfrac{(y-1)^2}{9} = 1$

7. $\dfrac{(x+1)^2}{16} + \dfrac{(y-2)^2}{9} = 1$

8. $\dfrac{(x-3)^2}{25} + \dfrac{(y+1)^2}{4} = 1$

9. $\dfrac{(x+5)^2}{4} + y^2 = 1$

10. $(x+2)^2 + \dfrac{y^2}{4} = 1$

In Exercises 11–16, write the standard form of the equation of the ellipse.

11.

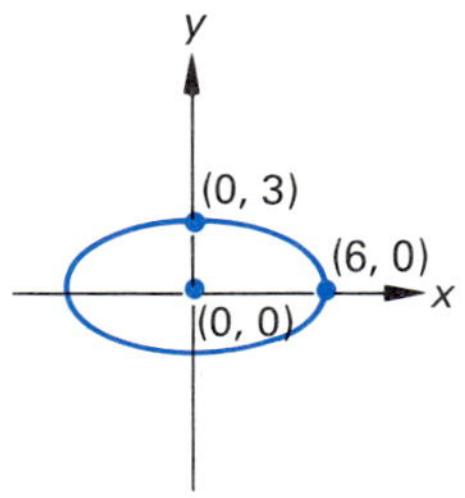

12.

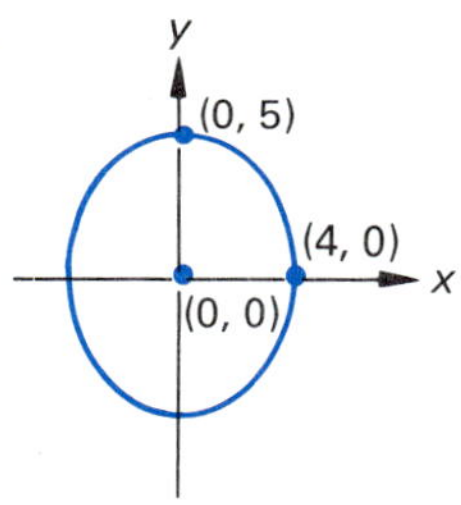

13.

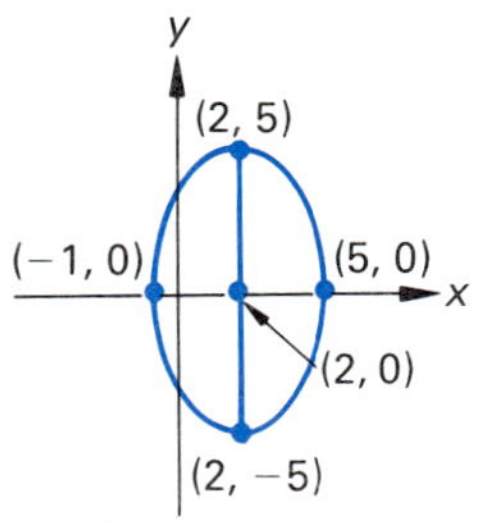

14.

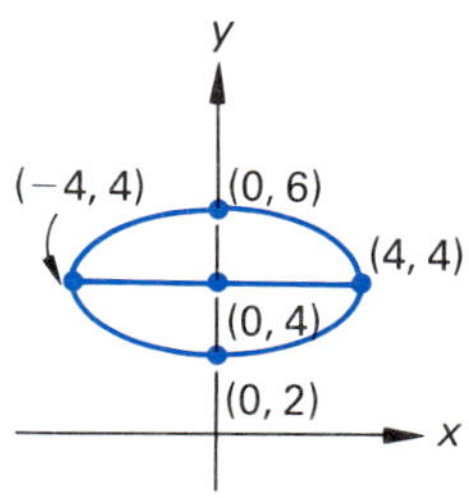

15.

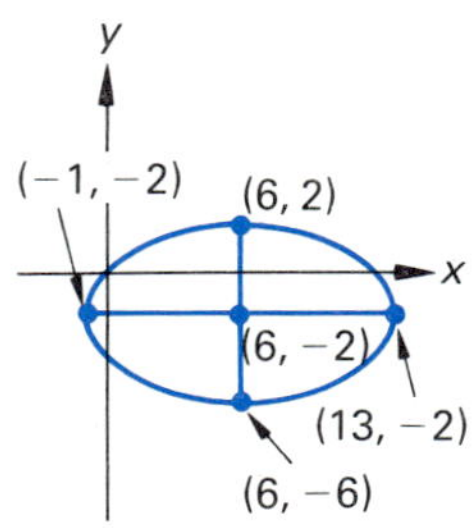

16.

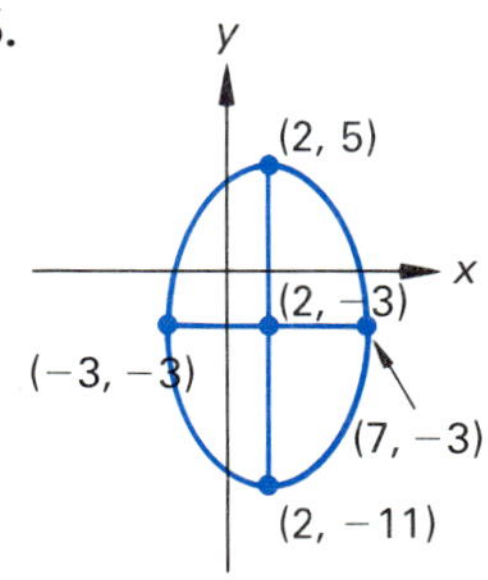

17. What are the coordinates of the foci of the ellipse in Exercises 11, 13, and 15?

18. What are the coordinates of the foci of the ellipse in Exercises 12, 14, and 16?

19. Sketch the ellipse represented by

$$\frac{(x+3)^2}{25} + \frac{(y+1)^2}{9} = 1.$$

Find the coordinates of its center, its foci, and vertices.

20. Sketch the ellipse represented by

$$\frac{(x-2.41)^2}{14.2} + \frac{(y+3.51)^2}{27.2} = 1.$$

Find the coordinates of its center, its foci, and vertices.

21. The top view of the tip of the wing of a crop-duster aircraft is to be shaped like half an ellipse to optimize performance (Fig. 17.41). The chord of the wing (distance from front to back) is to be 1.20 m and the tip is to extend 0.20 m beyond the parallel leading and trailing edges of the wing. Using the center line of the wing as an axis, what is the equation of the ellipse that forms the wing tip?

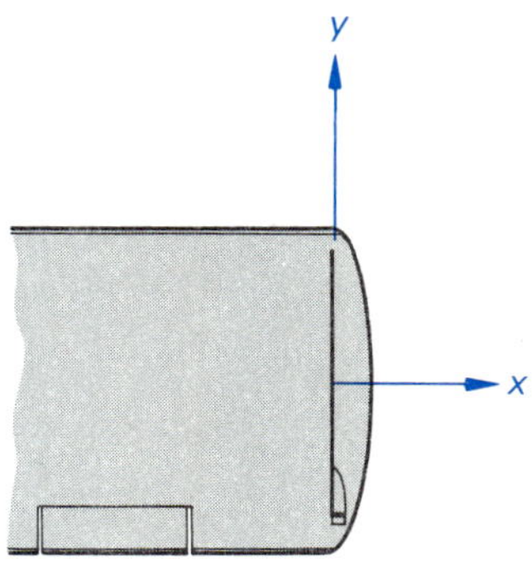

FIGURE 17.41

22. The blanks from which the blades of outboard motor propellers are shaped begin as ellipses, which are cut off along a latus rectum (which is a line through the focus) for fastening to the propeller hub. (The blank is the larger portion of the ellipse after the cut.) If the width of a blank is to be 3 in. and the length from the hub to the tip of the blade is to be 5 in., what is the equation of the ellipse for cutting the blank? (See Fig. 17.42.) For convenience assume that the origin of the axis system is at the center of the required ellipse.

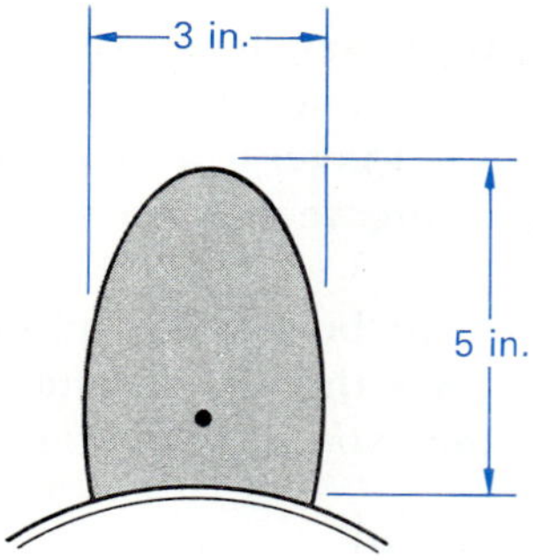

FIGURE 17.42

23. Is $9x^2 + y^2 + 90x + 2y + 218 = 0$ the equation of an ellipse? Determine your answer by trying to write the equation in the standard form for an ellipse.

24. Is $x^2 + 4y^2 - 4x + 32y + 72 = 0$ the equation of an ellipse? Determine your answer by trying to write the equation in the standard form for an ellipse.

25. When the Vanguard Satellite was launched it went into a slightly elliptical orbit around the earth. Considering an axis system with its origin at the center of the ellipse, the satellite passed through points (1700, 4740) and (4500, 2334), where the distances are in miles. The center of the earth was located at one focus of the elliptical orbit. If the radius of the earth is 3960 mi, how close did the satellite come to the surface?

26. A constant stress shape for a bridge support forms the upper half of an ellipse with its center at $(40, -3)$ and passing through points (20, 1) and (50, 2). What is the equation of the ellipse?

17.6 The hyperbola

The definition of a hyperbola is only slightly different from that of an ellipse, but the two curves do not look at all alike. A **hyperbola** *is the set of all points such that the absolute value of the difference of the distances from two fixed points (the foci) to a point on the hyperbola is a constant.* The hyperbola is the basis for military sound-ranging systems and for the Loran-C navigation system.

Let us choose two points for the foci of a hyperbola such that they lie on the x-axis a distance c on each side of the origin (Fig. 17.43). Let

FIGURE 17.43

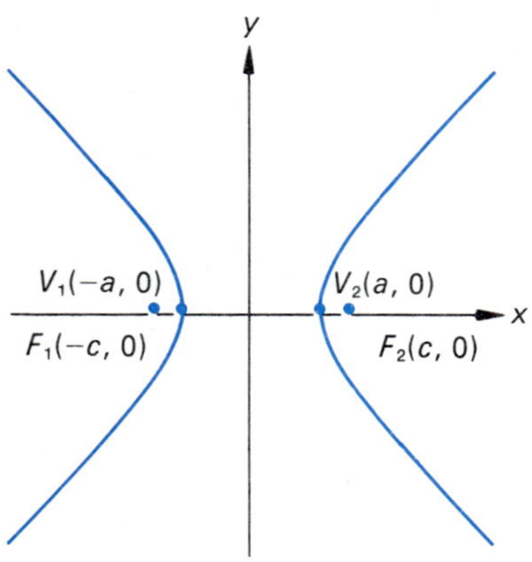

us also assume that the hyperbola crosses the x-axis a distance a on either side of the origin. These points of intersection are the **vertices** of the hyperbola and the x-axis is the **transverse axis** (or major axis) of the hyperbola. The y-axis is the **conjugate axis** (or minor axis) of the hyperbola.

Applying the definition of a hyperbola to the information given with the curve of Fig. 17.44, we see that the absolute value of the differences of the distances from the two foci to the right vertex is

$$\begin{aligned}|\overline{F_1V_2} - \overline{F_2V_2}| &= |(-c+a)-(c-a)| \\ &= |2a|.\end{aligned}$$

Since a is a measured distance and is always positive, the constant specified in the definition of a hyperbola equals $2a$.

Considering a general point on the hyperbola, $P(x, y)$, the distances from the two foci to the point are

$$\overline{F_1P} = \sqrt{[x-(-c)]^2 + [y-0]^2}$$
$$\overline{F_2P} = \sqrt{[x-c]^2 + [y-0]^2}$$

According to the definition of a hyperbola

$$|\overline{F_1P} - \overline{F_2P}| = 2a$$
$$|\sqrt{[x-(-c)]^2+[y-0]^2} - \sqrt{[x-c]^2+[y-0]^2}| = 2a.$$

Since we are going to square both sides of the equation it is permissible to drop the absolute-value sign and to rearrange the equation for convenience before performing the squaring operation.

$$\sqrt{[x+c]^2+y^2} = 2a + \sqrt{[x-c]^2+y^2}$$ **Simplifying and rearranging**

$$[x+c]^2+y^2 = 4a^2 + 4a\sqrt{[x-c]^2+y^2} + [x-c]^2 + y^2$$ **Squaring both sides**

$$x^2+2cx+c^2+y^2 = 4a^2+4a\sqrt{[x-c]^2+y^2}+x^2-2cx+c^2+y^2$$ **Expanding**

$$4cx - 4a^2 = 4a\sqrt{[x-c]^2+y^2}$$ **Combining terms and rearranging**

$$cx - a^2 = a\sqrt{[x-c]^2+y^2}$$ **Dividing by 4**

$$c^2x^2 - 2a^2cx + a^4 = a^2[(x-c)^2+y^2]$$ **Squaring both sides**

$$c^2x^2 - 2a^2cx + a^4 = a^2(x^2-2cx+c^2+y^2)$$
$$c^2x^2 - 2a^2cx + a^4 = a^2x^2 - 2a^2cx + a^2c^2 + a^2y^2$$
$$(c^2-a^2)x^2 - a^2y^2 = a^2(c^2-a^2)$$ **Rearranging and factoring**

Now let $b^2 = c^2 - a^2$ (Note that $b^2 = c^2 - a^2$ requires that $c > a$.)

$$b^2x^2 - a^2y^2 = a^2b^2$$ **Substituting for $c^2 - a^2$**

$$\frac{x^2}{a^2} - \frac{y^2}{b^2} = 1$$ **Dividing by a^2b^2**

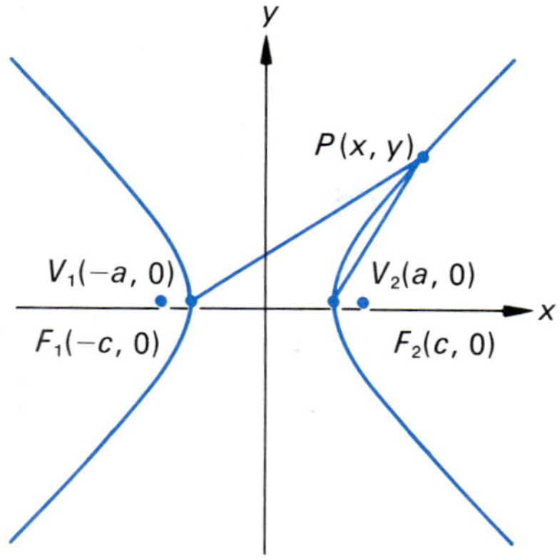

FIGURE 17.44

DEFINITION 17.20

The **standard equation** of a **hyperbola** with center at the origin and the x-axis as the transverse axis is

$$\frac{x^2}{a^2} - \frac{y^2}{b^2} = 1.$$

(See Fig. 17.45a.)

DEFINITION 17.21

The **standard equation** of a **hyperbola** with center at the origin and the y-axis as the transverse axis is

$$\frac{y^2}{a^2} - \frac{x^2}{b^2} = 1.$$

(See Fig. 17.45b.)

FIGURE 17.45

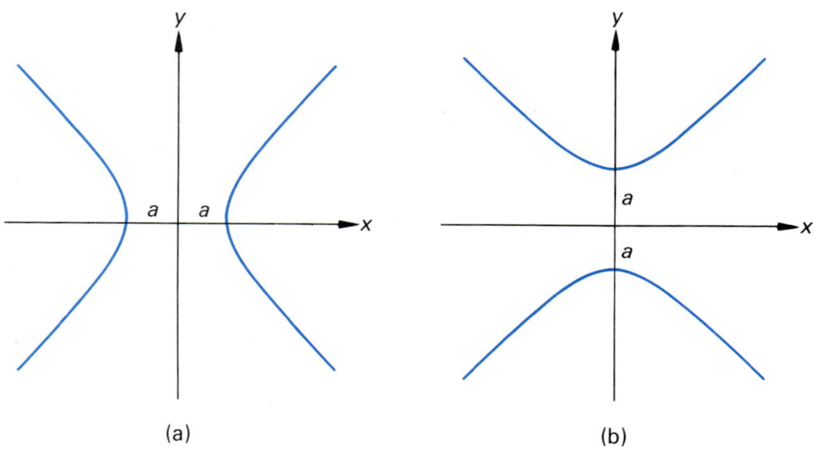

If we rearrange the equation of a hyperbola

$$\frac{x^2}{a^2} - \frac{y^2}{b^2} = 1$$

$$\frac{y^2}{b^2} = \frac{x^2}{a^2} - 1 \qquad \textbf{Rearranging}$$

$$y^2 = b^2\left(\frac{x^2}{a^2} - 1\right) \qquad \textbf{Multiplying by } b^2$$

$$y^2 = b^2\left(\frac{x^2 - a^2}{a^2}\right)$$

$$y^2 = \frac{b^2}{a^2}(x^2 - a^2)$$

$$y = \pm\frac{b}{a}\sqrt{x^2 - a^2} \qquad \textbf{Taking square root of both sides}$$

If we now restrict the values of x such that x is much greater than a, the value of $x^2 - a^2$ becomes approximately equal to x^2. Accordingly $\sqrt{x^2 - a^2}$ becomes close to x and for relatively large values of x we obtain

$$y = \pm\frac{b}{a}x,$$

which is the equation of a straight line passing through the origin. In this case we have two lines, both passing through the origin, which is the center of the hyperbola, one line with slope $+\frac{b}{a}$ and the other line with slope $-\frac{b}{a}$. These lines are called the **asymptotes** of the hyperbola. As we take values of x progressively farther from the center of a hyperbola the corresponding values of y become closer and closer to the asymptotes even though they never actually reach the corresponding y-values of the asymptotes. Because these lines are easy to graph the asymptotes are valuable aids in sketching a hyperbola. (For a hyperbola of the form $\frac{y^2}{a^2} - \frac{x^2}{b^2} = 1$, the asymptotes are the lines $y = \pm\frac{a}{b}x$.)

EXAMPLE 1 Sketch the hyperbola $\frac{x^2}{9} - \frac{y^2}{16} = 1$.

Solution: The equation $\frac{x^2}{9} - \frac{y^2}{16} = 1$ is the standard form of the equation of a hyperbola with the transverse axis along the x-axis; $a^2 = 9$ so the vertices are at $(-3, 0)$ and $(3, 0)$. For a hyperbola $b^2 = c^2 - a^2$ or $c^2 = a^2 + b^2$. Since $b^2 = 16$, in this case $c = 5$ and the foci of the

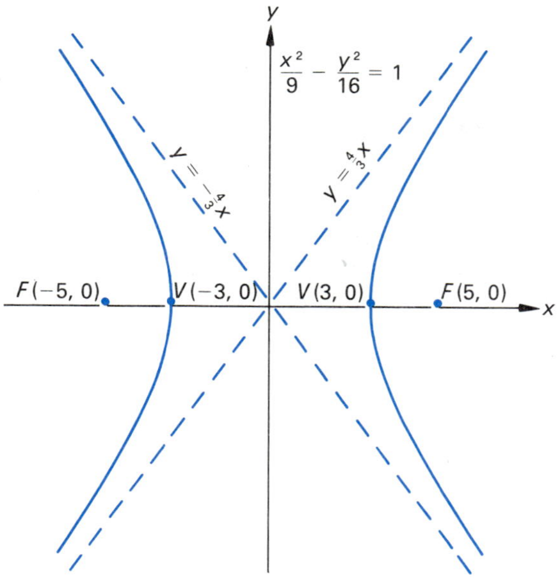

FIGURE 17.46

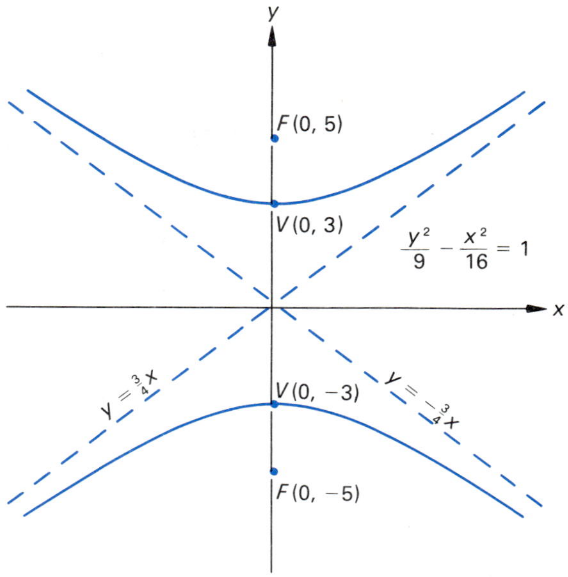

FIGURE 17.47

hyperbola are at $(-5, 0)$ and $(5, 0)$. The asymptotes are the lines $y = \pm \frac{b}{a} x$ or $y = \frac{4}{3} x$ and $y = -\frac{4}{3} x$. The hyperbola is sketched in Fig. 17.46. □

EXAMPLE 2 Sketch the hyperbola $\frac{y^2}{9} - \frac{x^2}{16} = 1$.

Solution: The equation $\frac{y^2}{9} - \frac{x^2}{16} = 1$ is the standard form of the equation of a hyperbola with the transverse axis along the y-axis; $a^2 = 9$ and $b^2 = 16$ so $c = 5$ and the foci are at $(0, 5)$ and $(0, -5)$. The vertices are at $(0, 3)$ and $(0, -3)$. The asymptotes are the lines $y = \pm \frac{a}{b} x$ or $y = \frac{3}{4} x$ and $y = -\frac{3}{4} x$. The hyperbola is sketched in Fig. 17.47. □

DEFINITION 17.22

The **standard form** of the equation of a **hyperbola** with a horizontal transverse axis, center at (h, k), vertices at $(h + a, k)$ and $(h - a, k)$, and foci at $(h + c, k)$ and $(h - c, k)$, is

$$\frac{(x - h)^2}{a^2} - \frac{(y - k)^2}{b^2} = 1.$$

(See Fig. 17.48a on page 696.)

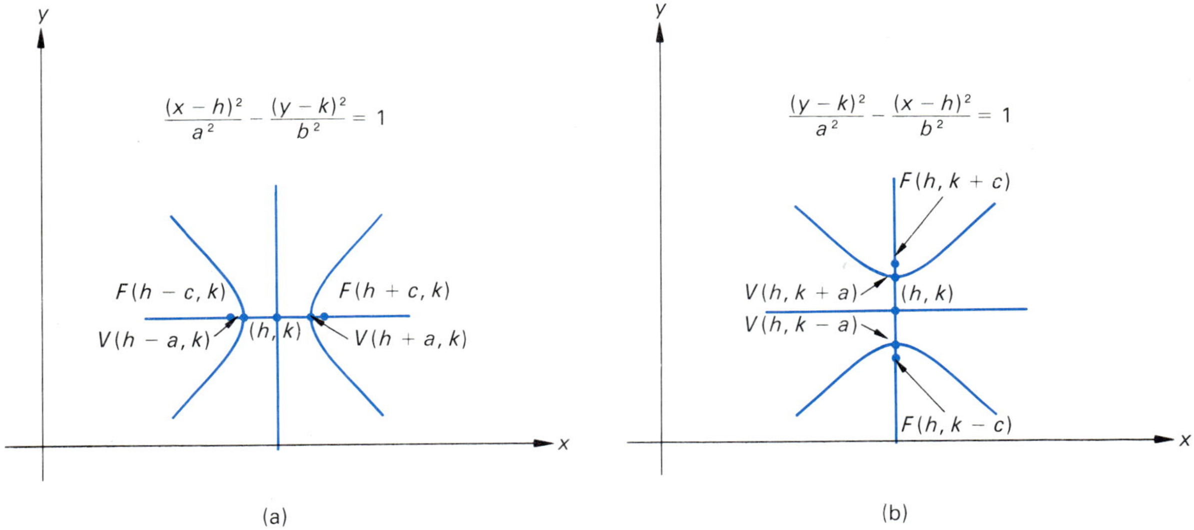

FIGURE 17.48

DEFINITION 17.23

> The **standard form** of the equation of a **hyperbola** with a vertical transverse axis, center at (h, k), vertices at $(h, k + a)$ and $(h, k - a)$, and foci at $(h, k + c)$ and $(h, k - c)$, is
>
> $$\frac{(y-k)^2}{a^2} - \frac{(x-h)^2}{b^2} = 1.$$
>
> (See Fig. 17.48b.)

The hyperbola with a horizontal transverse axis has asymptotes with slope $\pm \frac{b}{a}$ and passing through (h, k). The hyperbola with a vertical transverse axis has asymptotes with slope $\pm \frac{a}{b}$ and passing through (h, k).

EXAMPLE 3 Sketch the hyperbola $\frac{(x-1)^2}{9} - \frac{(y+2)^2}{16} = 1$.

Solution: By comparing the given equation with the form of Definition 17.22, we see that the hyperbola has a horizontal transverse axis and has $h = 1$, $k = -2$, $a^2 = 9$ so $a = 3$, and $b^2 = 16$ so $b = 4$. The center of the hyperbola has coordinates $(1, -2)$. Since the transverse axis is horizontal, the vertices are a units to the left and right of center at $(1 - 3, 2)$ and $(1 + 3, -2)$ or at $(-2, -2)$ and $(4, -2)$. The foci are c

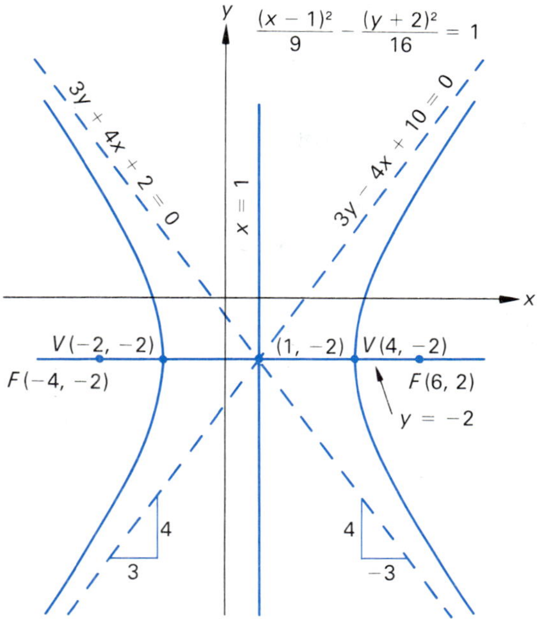

FIGURE 17.49

units to either side of center.

$$\begin{aligned} c &= \sqrt{a^2 + b^2} \\ &= \sqrt{9 + 16} \\ &= 5. \end{aligned}$$

The foci are at $(1 - 5, -2)$ and $(1 + 5, -2)$ or at $(-4, -2)$ and $(6, -2)$. The transverse axis is on the line $y = -2$ and the conjugate axis is on the line $x = 1$. The asymptotes are lines that cross at the center of the hyperbola and have slopes of $+\frac{b}{a} = \frac{4}{3}$ and $-\frac{b}{a} = -\frac{4}{3}$ as shown in Fig. 17.49. □

EXAMPLE 4 Sketch the hyperbola $\frac{(y-2)^2}{9} - \frac{x^2}{4} = 1$.

Solution: The given hyperbola is in the form of Definition 17.23, so the transverse axis is parallel to the y-axis. By comparing the equation with Definition 17.23, $h = 0$, $k = 2$, $a^2 = 9$ so $a = 3$, and $b^2 = 4$ so $b = 2$.

By calculation

$$\begin{aligned} c &= \sqrt{a^2 + b^2} \\ &= \sqrt{9 + 4} \\ &= \sqrt{3.61}. \end{aligned}$$

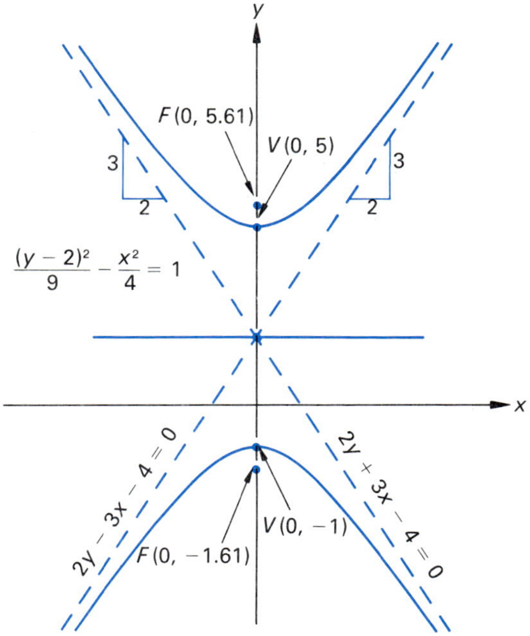

FIGURE 17.50

The vertices are the points $(h, k+a)$ and $(h, k-a)$, which are, respectively, $(0, 5)$ and $(0, -1)$. The foci are the points $(h, k+c)$ and $(h, k-c)$, which are, respectively, $(0, 5.61)$ and $(0, -1.61)$. The asymptotes have slopes $\pm\frac{a}{b}$ or $\frac{3}{2}$ and $-\frac{3}{2}$. The equations of the asymptotes are $y = \frac{3}{2}x + 2$ and $y = -\frac{3}{2}x + 2$. The hyperbola is sketched in Fig. 17.50. □

DEFINITION 17.24

The **general form** of the equation of a **hyperbola** is

$$Ax^2 + Cy^2 + Dx + Ey + F = 0.$$

(A and C have opposite signs.)

As in the case of the circle and the ellipse, there are some situations in which an equation of the same form as Definition 17.24 cannot actually represent a hyperbola.

EXAMPLE 5 Transform $9x^2 - 4y^2 + 36x + 8y - 4 = 0$ into one of the standard forms of the hyperbolic equation.

Solution: At this stage it is impossible to decide which of the standard forms is represented. We will begin the transformation by rearranging the equation in order to group terms.

$9x^2 - 4y^2 + 36x + 8y - 4 = 0$	**Given**
$(9x^2 + 36x) + (-4y^2 + 8y) = 4$	**Rearranging**
$9(x^2 + 4x) - 4(y^2 - 2y) = 4$	**Common-term factoring to obtain unity coefficients for x^2 and y^2**
$9(x^2 + 4x + 4) - 4(y^2 - 2y + 1) = 4 + 36 - 4$	**Completing squares**
$9(x + 2)^2 - 4(y - 1)^2 = 36$	
$\dfrac{(x + 2)^2}{4} - \dfrac{(y - 1)^2}{9} = 1.$	**Dividing by 36 to obtain 1 on right side**

The transformed equation of $9x^2 - 4y^2 + 36x + 8y - 4 = 0$ matches the standard form of Definition 17.22. Therefore it is the equation of a hyperbola with its transverse axis horizontal. □

EXAMPLE 6 Two sport fishermen who are trolling 2 mi apart are talking to each other on the VHF radiotelephone when they hear thunder. One fisherman hears the thunder three seconds after the other hears it. What are the possible locations of the lightning bolt that produced the thunder?

Solution: A hyperbola is the locus of points whose difference in position from two fixed points is a constant. Sound travels in air approximately 1100 ft/s so wherever the thunder originated the location must be a constant 3300 ft farther from one fisherman than the other.

Let the transverse axis of a hyperbola pass through the positions of the two fishermen and the center of the hyperbola lie midway between them. By taking the origin of an axis system at the center of the hyperbola we can use the standard form of the hyperbolic equation with $(h, k) = (0, 0)$.

$$\frac{x^2}{a^2} - \frac{y^2}{b^2} = 1.$$

The fishermen are at the foci of the hyperbola so

$$2c = 2 \text{ mi}\left(\frac{5280 \text{ ft}}{\text{mi}}\right) = 10{,}560 \text{ ft}$$

$$c = 5280 \text{ ft}.$$

The constant difference in the distance from the lightning to the fishermen is

$$2a = 3300 \text{ ft}$$
$$a = 1650 \text{ ft}.$$

For a hyperbola

$$c^2 = a^2 + b^2$$

so

$$b^2 = (5280)^2 - (1650)^2$$

and

$$b = 4120.$$

Therefore the possible locations of the lightning lie on the branch of the hyperbola

$$\frac{x^2}{(1650)^2} - \frac{y^2}{(4120)^2} = 1,$$

which is closest to the fisherman who heard the sound first. □

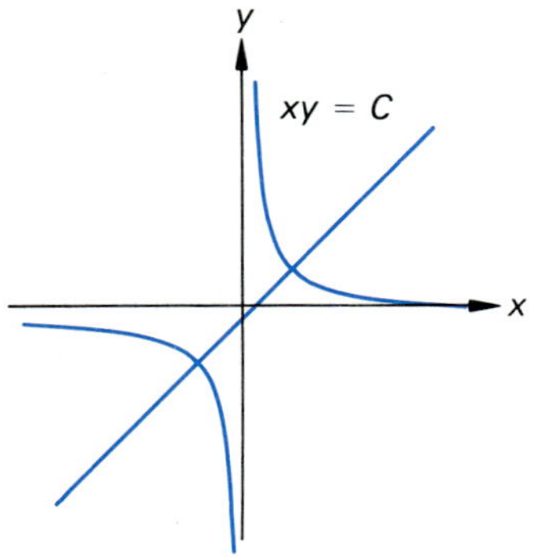

FIGURE 17.51

The equation $xy = C$, with C a constant, is a special form of the hyperbolic equation. The hyperbola $xy = C$ has its transverse axis along the line $y = x$ and is asymptotic to the x- and y-axes (Fig. 17.51). The hyperbola $xy = -C$ has its transverse axis along the line $y = -x$.

EXAMPLE 7 What are the coordinates of the vertices of the hyperbola $xy = C$?

Solution: Since the vertices of a hyperbola lie on its transverse axis and the transverse axis of $xy = C$ is the line $y = x$, solving the two equations simultaneously will produce the coordinates of the vertices.

$xy = C$	**Given**	**[1]**
$y = x$	**Given**	**[2]**
$xx = C$	**Substituting for y in [1]**	
$x^2 = C$		
$x = \pm\sqrt{C}$		
$y = \pm\sqrt{C}.$	**Substituting for x in [2]**	

So the coordinates of the vertices of the hyperbola $xy = C$ are $(\sqrt{C}, \sqrt{C})$ and $(-\sqrt{C}, -\sqrt{C})$. □

DEFINITION 17.25

> The equation $xy = C$ represents a hyperbola with its transverse axis on the line $y = x$, vertices at $(\sqrt{C}, \sqrt{C})$ and $(-\sqrt{C}, -\sqrt{C})$, and the x- and y-axes are the asymptotes.

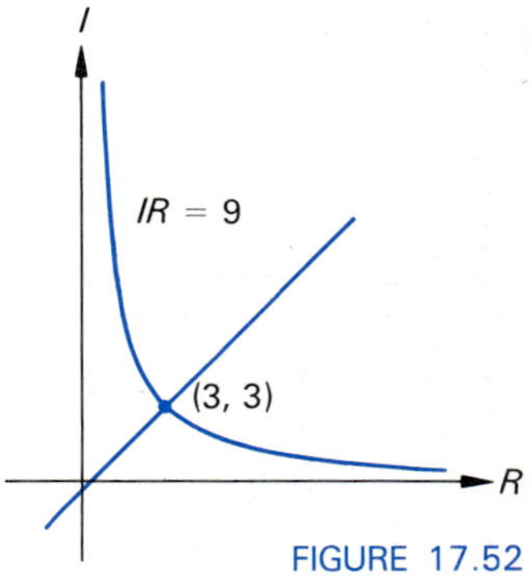

FIGURE 17.52

EXAMPLE 8 The current I through a resistance R with a voltage drop of 9 V is given by $IR = 9$. Sketch the current versus resistance curve.

Solution: The equation $IR = 9$ is hyperbolic of the form $xy = C$. Its transverse axis lies on the line $I = R$, and has the R- and I-axes as asymptotes. Its vertex is at (3, 3). There is no negative portion of the curve because a resistance cannot be negative. The curve is sketched in Fig. 17.52. □

17.6 Exercises

In Exercises 1–8, find the coordinates of the centers, vertices, foci, and the equations of the asymptotes. Then sketch the hyperbola.

1. $\dfrac{x^2}{4} - \dfrac{y^2}{9} = 1$

2. $\dfrac{x^2}{16} - \dfrac{y^2}{9} = 1$

3. $\dfrac{y^2}{4} - \dfrac{x^2}{9} = 1$

4. $\dfrac{y^2}{25} - \dfrac{x^2}{36} = 1$

5. $\dfrac{(x-2)^2}{9} - \dfrac{(y-3)^2}{16} = 1$

6. $\dfrac{(x+1)^2}{36} - \dfrac{(y-2)^2}{4} = 1$

7. $\dfrac{(y+1)^2}{16} - \dfrac{(x+3)^2}{25} = 1$

8. $\dfrac{(y-2)^2}{4} - (x-1)^2 = 1$

In Exercises 9–16, write the standard equation for the hyperbola shown.

9.

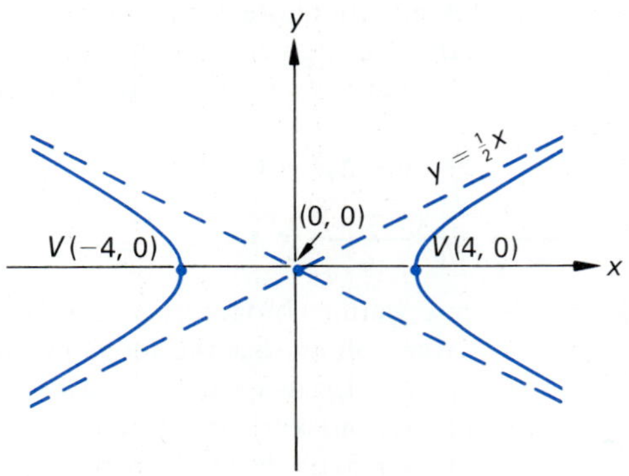

10.

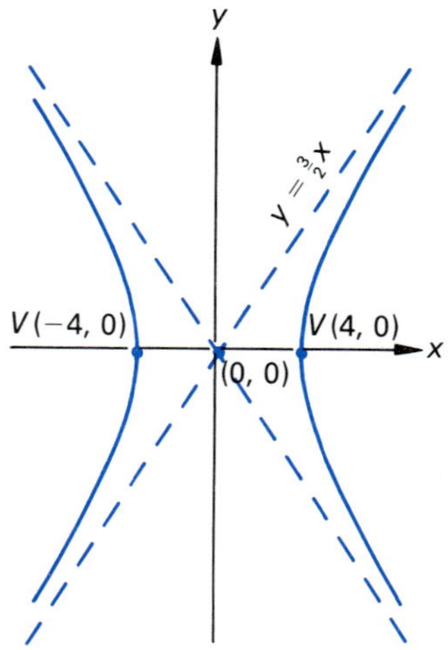

11.

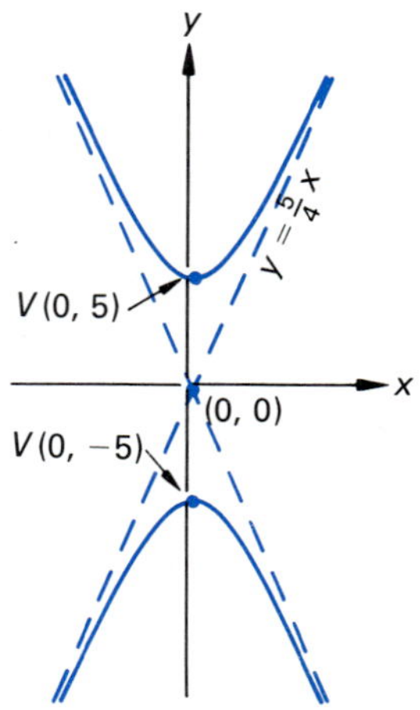

12.

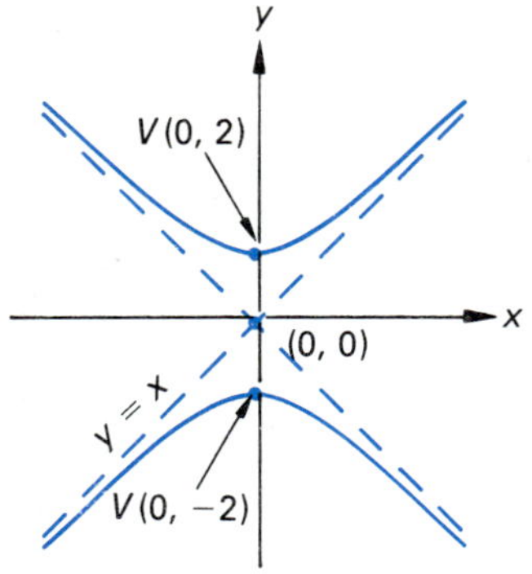

13.

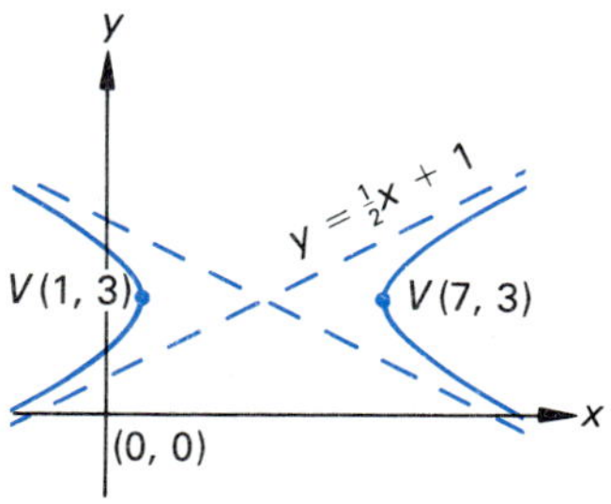

14.

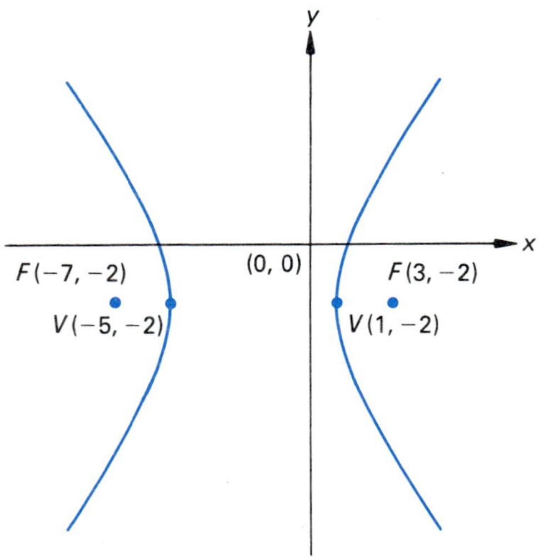

15.

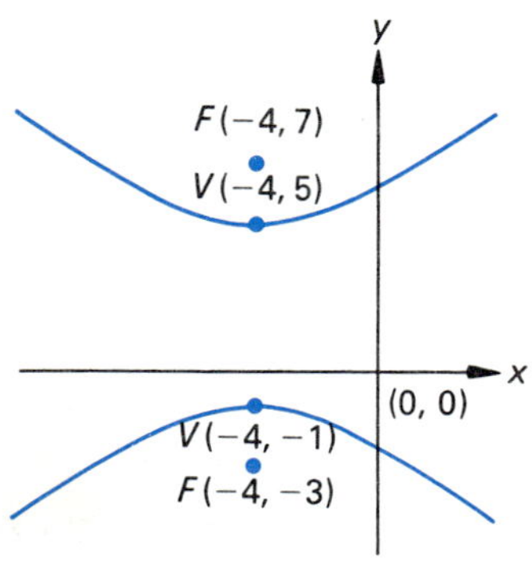

16.

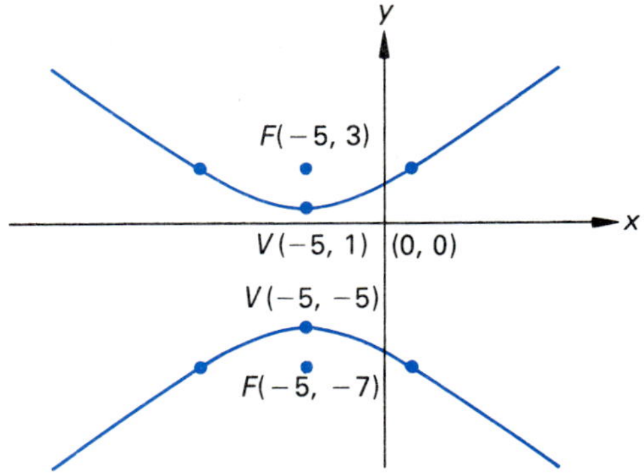

17. Write the equation of the hyperbola with vertices at $(2, -2)$ and $(-4, -2)$ and passing through the point with coordinates $(5, 1)$.

18. Write the equation of the hyperbola with vertices at $(-3, 1)$ and $(-3, 3)$ and passing through the point with coordinates $(0, 4)$.

19. Write the equation of the hyperbola

$$9x^2 - 4y^2 - 36x - 24y - 36 = 0$$

in the standard form.

20. Write the equation of the hyperbola

$$-4x^2 + y^2 + 8x - 4y - 4 = 0$$

in the standard form.

21. Write the equation of the hyperbola

$$4.50x^2 - 7.40y^2 - 27.9x + 23.7y - 9.00 = 0.00$$

in the standard form.

22. Write the equation of the hyperbola

$$-x^2 + 3.80y^2 + 14.6x + 19.8y - 31.4 = 0.00$$

in the standard form.

A **rectangular hyperbola** is one for which $a = b$ and the asymptotes are therefore perpendicular to each other. Sketch the rectangular hyperbolas in Exercises 23 and 24 and identify the vertices, the foci, and the asymptotes.

23. $(x + 1)^2 - (y - 2)^2 = 1$

24. $\dfrac{(x - 3)^2}{4} - \dfrac{(y + 1)^2}{4} = 1$

25. Two Loran-C stations broadcast a signal at exactly the same time. Show that the locus of positions that receive the signals t seconds apart is a hyperbola. (The stations can be represented by two points that are a fixed distance apart.)

26. How does the Loran-C system described in the opening of this chapter provide a vessel's position?

27. Two seismographs 400 km apart record the tremors from an earthquake at 14-50-48 and 14-51-18 (read: 14 hours 50 minutes 48 seconds and 14 hours 51 minutes 18 seconds). The tremors travel at an average speed through the earth of approximately 5000 m/s. What is the locus of points where the quake might have occurred?

28. The velocity of sound in seawater is approximately 1530. m/s. Two sonar stations on either side of a harbor are 9.40 km apart and pick up the sound of a submarine 4.73 seconds apart. Where are the possible locations of the submarine with respect to the stations?

29. The roof of a beach cabana at a resort hotel is to be shaped like part of a hyperbola. The joists of the structure are to be at the latus rectum (a line through the focus) and are to be 8 ft above the beach and 20 ft long. The height of the roof above the joists is to be 3 ft to allow room for installation of ceiling fans (Fig. 17.53). What is the equation of the hyperbola so that the curve rafters can be prefabricated?

30. A roof is to be shaped like the one in Exercise 29 with a joist length of 10 m and a distance from joist to peak of 2 m. If the origin of the axis system used in this case is to be located at the center of the joist, what is the equation of the hyperbola that describes the rafter shape?

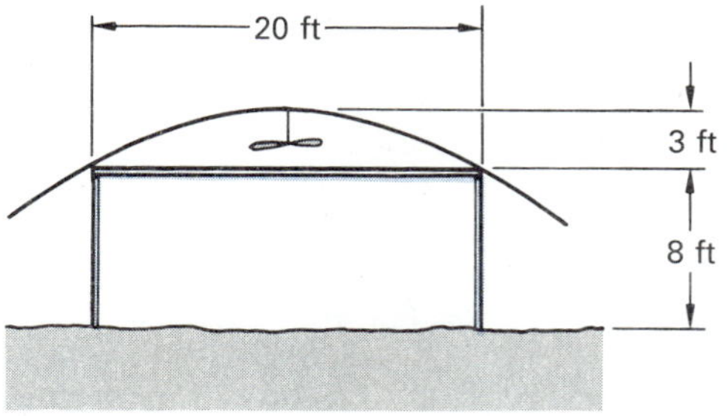

FIGURE 17.53

31. The cross-section of a reflecting pool is to be dug in the shape of a hyperbola in an experiment to try to minimize ripple formation. The pool will be 8 m across by 1 m deep with its focus at ground level in its center. Locate the origin of an axis system at ground level at the center of the pool and find the necessary depths at 1-m intervals across the pool (Fig. 17.54).

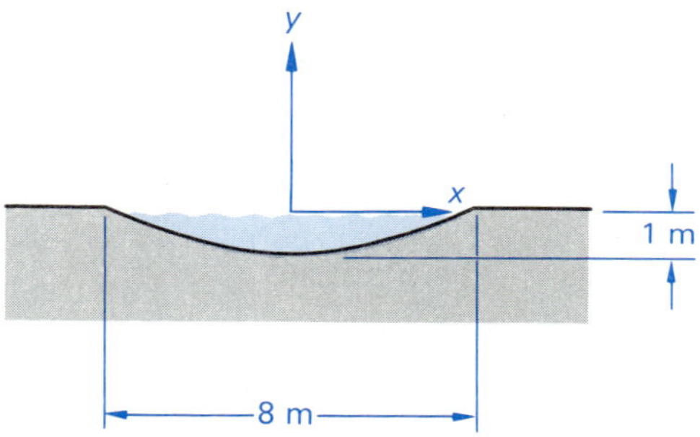

FIGURE 17.54

In Exercises 32 and 33, sketch the hyperbola given, locating the vertices (or vertex).

32. The emission energy of a particle is given by $E = \dfrac{hc}{\lambda}$, where h is Planck's constant

$(6.624 \times 10^{-27}$ erg s),

c is the velocity of light $(3.00 \times 10^{10}$ cm/s), and λ is the wavelength of the energy in centimeters.

33. The index of refraction of light is given by $n = \dfrac{c}{v}$, where c is the velocity of light in a vacuum $(3.00 \times 10^{10}$ cm/s) and v is the velocity of light in the medium being studied, also in cm/s.

17.7 Translation of axes

The linear equation relating temperature in degrees Celcius as a function of temperature in degrees Fahrenheit is $C = 5/9\ (F - 32)$. Suppose we let $C' = C$ and $F' = F - 32$ and substitute these expressions into the

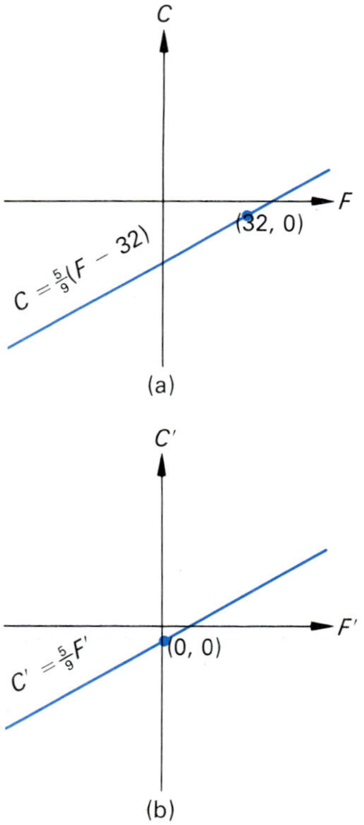

FIGURE 17.55

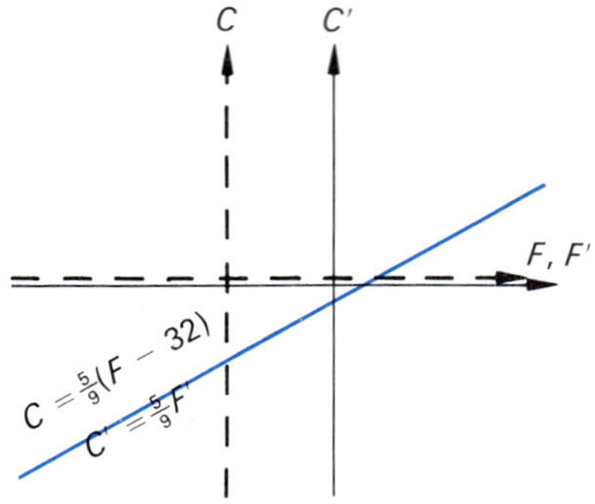

FIGURE 17.56

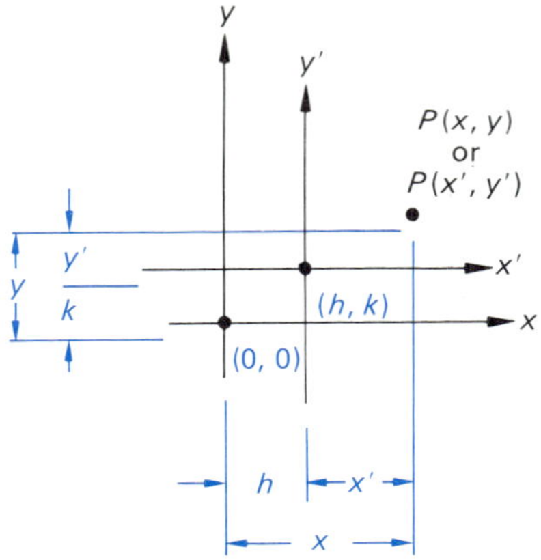

FIGURE 17.57

temperature equation. The equation then becomes $C' = 5/9\ F'$. The graphs of the two functions are shown in Fig. 17.55a and 17.55b.

Since we are graphing the same line in each case, if we superimpose Figs. 17.55a on 17.55b, we obtain Fig. 17.56. We see from the figure that by altering the equation as we did, we effectively translated the axis system into a new position parallel to its original position. (By **translation** we mean that a line moves such that it remains parallel to its original orientation.) By translating we were able to simplify the form of the function with which we were dealing.

Let us consider the general case of an axis system translated with respect to another axis system as in Fig. 17.57. Any point P in the space encompassed by the two systems can be expressed in terms of either one of the systems: $P(x, y)$ or $P(x', y')$. If (h, k) is the location of the origin of the $x'y'$-axis system in terms of the xy-axis system, we can see from the figure that $x = x' + h$ and $y = y' + k$.

RULE 17.1

TRANSLATION OF AXES

$x = x' + h$ and $y = y' + k$, where (h, k) is the location of the origin of the $x'y'$-system in the xy-plane.

EXAMPLE 1 In Section 17.4, Example 4, the parabola $(y + 1)^2 = 6(x - 2)$ was sketched. Use axis translation to simplify the form of the equation of the parabola.

Solution: Figure 17.58 shows the parabola $(y + 1)^2 = 6(x - 2)$ with its vertex and focus in parentheses as it was sketched in Fig. 17.26. To simplify the equation we choose an $x'y'$-axis system such that its origin falls at the vertex of the parabola [point (h, k)]. Therefore the transla-

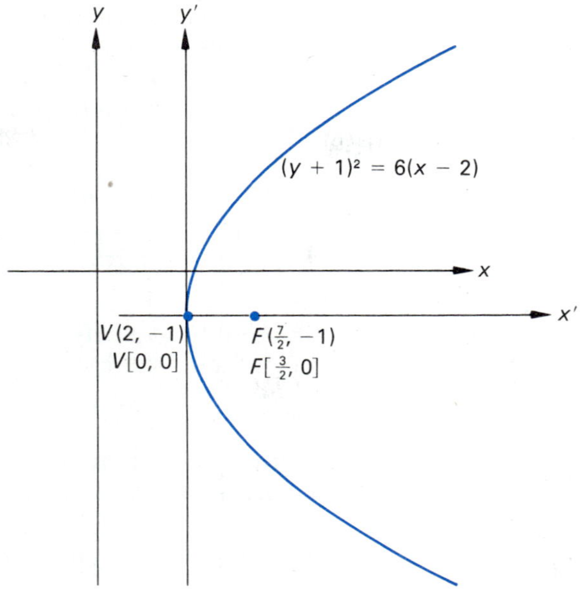

FIGURE 17.58

tion equations are $x = x' + 2$ and $y = y' + (-1)$ and by substitution the equation of the parabola becomes

$$(y' - 1 + 1)^2 = 6(x' + 2 - 2)$$
$$y'^2 = 6x',$$

which is the equation of the parabola in the $x'y'$-system. The vertex and focus can be found by the methods of Section 17.4 or, since in this case they are already given in the xy-system, they can be calculated directly using the translation equations. (Brackets are not normally used to indicate ordered pairs. However, the points in the $x'y'$-system will be shown in brackets, $[x', y']$, for the sake of clarity.) The point $(2, -1)$ becomes, using the translation equations,

$$2 = x' + 2 \qquad \text{and} \qquad -1 = y' - 1$$
$$x' = 0 \qquad\qquad\qquad y' = 0$$

so $(2, -1)$ becomes $[0, 0]$.

For the focus $7/2 = x' + 2$ and $-1 = y' - 1$, so $(7/2, -1)$ translates to $[3/2, 0]$. The $x'y'$-axes are shown in Fig. 17.58 and the appropriate points are indicated. □

EXAMPLE 2 Use axis translation to justify writing the standard form of an equation of an ellipse as $\dfrac{(x-h)^2}{a^2} + \dfrac{(y-k)^2}{b^2} = 1$ after the equation was derived for an ellipse with center at the origin as $\dfrac{x^2}{a^2} + \dfrac{y^2}{b^2} = 1$.

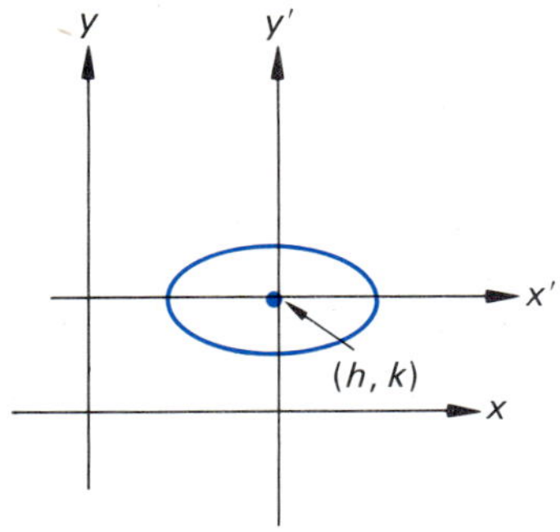

FIGURE 17.59

Solution: The equation of the ellipse shown in Fig. 17.59 can be written as $\frac{x'^2}{a^2}+\frac{y'^2}{b^2}=1$. From Rule 17.1 we can write $x=x'+h$ and $y=y'+k$, which can be rearranged to $x'=x-h$ and $y'=y-k$.

$$\frac{x'^2}{a^2}+\frac{y'^2}{b^2}=1 \qquad \textbf{Equation of ellipse in } x'y'\textbf{-system}$$

$$\frac{(x-h)^2}{a^2}+\frac{(y-k)^2}{b^2}=1. \qquad \textbf{Substituting translation relations for } x' \textbf{ and } y'$$

Therefore we can justify going from

$$\frac{x^2}{a^2}+\frac{y^2}{b^2}=1 \quad \text{to} \quad \frac{(x-h)^2}{a^2}+\frac{(y-k)^2}{b^2}=1$$

[when the center of the ellipse is shifted from (0, 0) to (h, k)] by using axis translation. □

EXAMPLE 3 Transform the equation of a hyperbola

$$9x^2-4y^2+36x+8y-4=0$$

to an $x'y'$-axis whose origin is at $(-2, 1)$ in the xy-system.

Solution: Using Rule 17.1 for translation we have $x=x'+h$ and $y=y'+k$ or in this case $x=x'-2$ and $y=y'+1$.

$$9x^2-4y^2+36x+8y-4=0 \qquad \textbf{Given}$$

$$9(x'-2)^2-4(y'+1)^2+36(x'-2)+8(y'+1)-4=0$$

Substituting translation equation for x and y

$$9(x'^2-4x'+4)-4(y'^2+2y'+1)+36(x'-2)+8(y'+1)-4=0 \qquad \textbf{Squaring}$$

$$9x'^2-36x'+36-4y'^2-8y'-4+36x'-72+8y'+8-4=0 \qquad \textbf{Expanding}$$

$$9x'^2-4y'^2-36=0 \qquad \textbf{Combining terms}$$

$$\frac{x'^2}{4}-\frac{y'^2}{9}=1. \qquad \textbf{Rearranging to standard form of hyperbola}$$

From Definition 17.20 we can say that $\frac{x'^2}{4}-\frac{y'^2}{9}=1$ is a hyperbola with its center at the origin. □

EXAMPLE 4 The voltage waveform produced by an AC voltage superimposed on a DC voltage is given by the function

$$V=9+4\sin(5\times10^4t).$$

Translate the axis system to allow us to study the AC component separately.

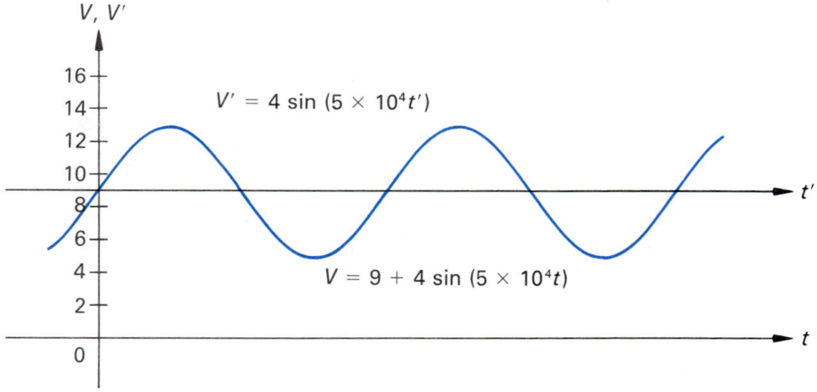

FIGURE 17.60

Solution: The function $V = 9 + 4 \sin(5 \times 10^4 t)$ is sketched on the tV-axis system in Fig. 17.60. If we place the $t'V'$-system as shown with its origin at (0, 9) of the tV-system, we see that the function becomes a pure sinusoid with respect to the $t'V'$-system. To solve the equation in terms of V' and t', we use Rule 17.1. Then $t = t' + 0$ and $V = V' + 9$, and substituting yields

$$V' + 9 = 9 + 4 \sin [5 \times 10^4(t' + 0)]$$
$$V' = 4 \sin (5 \times 10^4 t'),$$

which is the equation of the function $V = 9 + 4 \sin(5 \times 10^4 t)$ when translated from the tV-axis to the $t'V'$-axis. □

EXAMPLE 5 The U.S. Coast Guard Station in Rochester receives a distress call from a vessel, which, according to the radio-direction-finder reading, bears 326° true from the station. At the same time Toronto Coast Guard in Canada picks up the call bearing 085°. The Toronto Station is 25 mi north and 91 mi west of the Rochester Station. Where should the search begin relative to the Rochester Station? (See Fig. 17.61.)

FIGURE 17.61

Solution: By relating the compass bearing to an xy-axis system with the positive y-axis pointing north and the positive x-axis pointing east as was shown in Section 6.7, the call directions can be represented by the equation of two straight lines (Fig. 17.62, on page 708). If we place an axis with its origin at Rochester, the equation of the line on which the distressed vessel lies becomes

$$y_R = (\tan 124°)x_R$$
$$= -1.4826x_R.$$

For an axis at Toronto (Fig. 17.63, page 708) the line on which the calling vessel lies is

$$y_T = (\tan 5°)x_T$$
$$= 0.0875x_T.$$

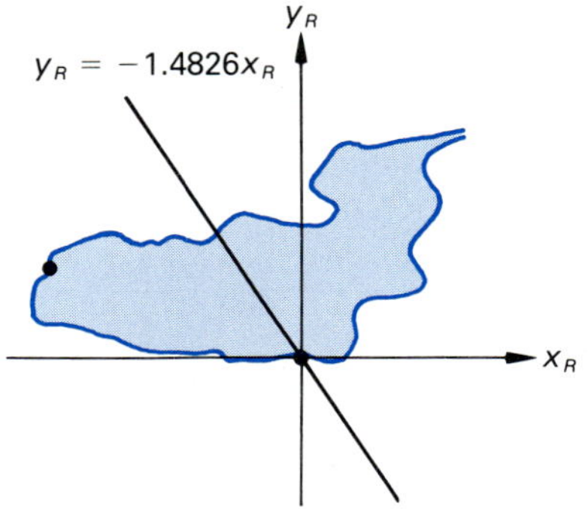

FIGURE 17.62

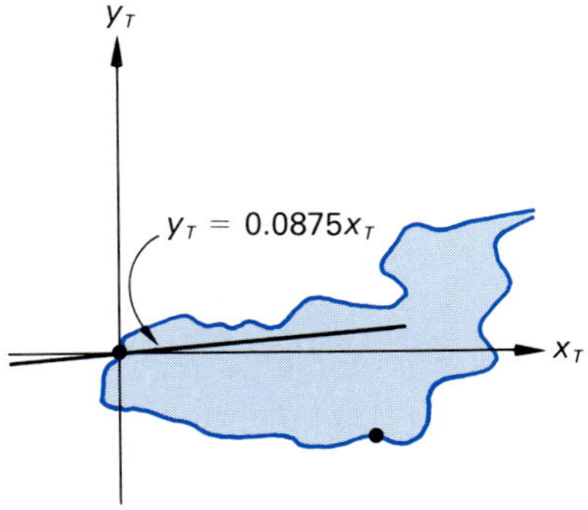

FIGURE 17.63

To determine where the two lines intersect it is necessary to refer them to the same axis system. Since the axis systems we are using are parallel to each other, we can use translation. The origin of the "Toronto" axis system lies at $(-91, 25)$ in the "Rochester" axis system. Therefore, we can refer the Toronto line to the Rochester system by

$$x_R = x_T + (-91)$$

or rearranging

$$x_T = x_R + 91$$

and

$$y_R = y_T + 25$$

or

$$y_T = y_R - 25. \text{ (See Fig. 17.64.)}$$

By substituting, the Toronto line becomes

$$y_T = 0.0875x_T$$
$$(y_R - 25) = 0.0875(x_R + 91);$$

FIGURE 17.64

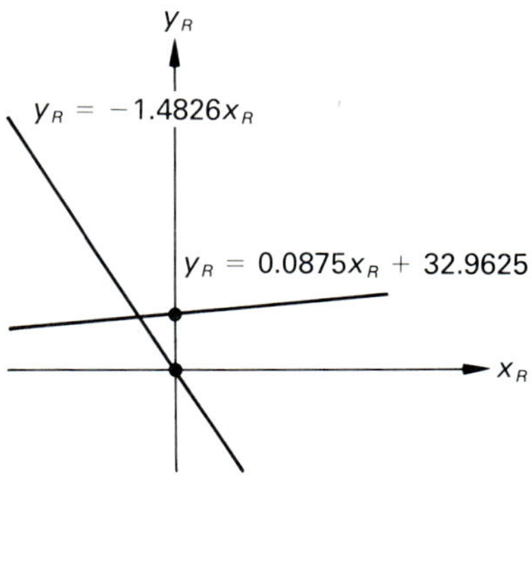

or when referred to the axis at Rochester the Toronto line is

$$y_R = 0.0875x_R + 32.9625.$$

With the two lines on the same axis system their intersection is $(-21.0, 31.1)$, which can be found by conventional methods, and the position of the vessel is established with respect to the station at Rochester. □

17.7 Exercises

In Exercises 1–4, use axis translation to rewrite the equation of the lines so that the lines pass through the origin of the translated system. (A sketch of each line may help simplify finding the solution.)

1. $y = x + 2$

2. $y = \frac{1}{2}x - 3$

3. $-V + 3t - 1 = 0$

4. $0.3x + 0.4y - 1.2 = 0$

In Exercises 5–8, use axis translation to rewrite the equations of the lines in terms of a parallel system with the origin of the $x'y'$-system at the indicated point.

5. $y = -2x$; $(3, 1)$

6. $x = 4y$; $(3, -1)$

7. $y = -\frac{1}{2}x - 2$; $(-1, 3)$

8. $2y - x = 4$; $(-3, -1)$

In Exercises 9–16, rewrite the equation of the given figure for a parallel $x'y'$-axis system with its origin at the center of the figure (vertex in the case of a parabola). Give the coordinates of foci and vertices where appropriate.

9. $(x-2)^2 + (y-4)^2 = 16$

10. $(x+1)^2 + (y-2)^2 = 4$

11. $x^2 = -2(y-3)$

12. $(y-2)^2 = -\frac{1}{2}(x-2)$

13. $\frac{(x+3)^2}{25} + \frac{(y+1)^2}{9} = 1$

14. $\frac{(x-2.41)^2}{14.2} + \frac{(y+3.51)^2}{27.2} = 1$

15. $(x+1)^2 - (y-2)^2 = 1$

16. $\frac{(x-3)^2}{4} - \frac{(y+1)^2}{4} = 1$

In Exercises 17–24, rewrite the equation of the figure for a translated $x'y'$-axis system with its origin at the indicated point. Give the coordinates in the $x'y'$-system for the center of each figure (vertex in the case of a parabola).

17. $(x-2)^2 + (y-4)^2 = 16$; $(1, 1)$

18. $(x+1)^2 + (y-2)^2 = 4$; $(3, 1)$

19. $x^2 = -2(y-3)$; $(0, 1)$

20. $(y-2)^2 = -\frac{1}{2}(x-2)$; $(-2, 3)$

21. $\frac{(x+3)^2}{25} + \frac{(y+1)^2}{9} = 1$; $(-1, -2)$

22. $\frac{(x-2.41)^2}{14.2} + \frac{(y+3.51)^2}{27.2} = 1$; $(0.50, -1.00)$

23. $(x+1)^2 - (y-2)^2 = 1$; $(4, -1)$

24. $\frac{(x-3)^2}{4} - \frac{(y+1)^2}{4} = 1$; $(3, 0)$

25. A suspension bridge with a parabolic cable has supports 400 ft apart and a maximum center sag of 50 ft. Write the equation for the cable with respect to an axis system at the right-hand support by first writing the equation of the curve for an axis system with its origin at the point of maximum sag and then using axis translation.

26. The cross-section of a ship's deck is a parabola. The deck is 20 ft wide and the center of the deck is 6 in. above the sides (Fig. 17.65, page 710). Write the equation of the parabolic section with respect to an axis at the port (left) side of the deck by writing the equation for an origin at the center of the deck and then using axis translation.

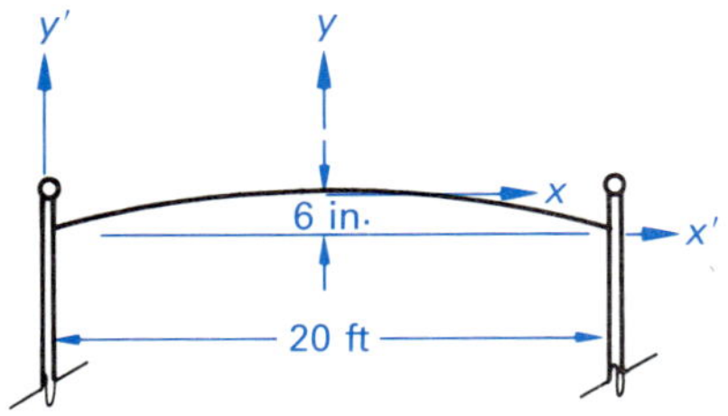

FIGURE 17.65

27. Use axis translation to write the equation of the cam profile (Fig. 17.66) for an axis at the center of the camshaft. The camshaft is at the focus of the ellipse. (See Example 3, Section 17.5.)

28. Refer to Example 6, Section 17.6. Use axis translation to write the equation of possible lightning locations with respect to an axis on one of the boats.

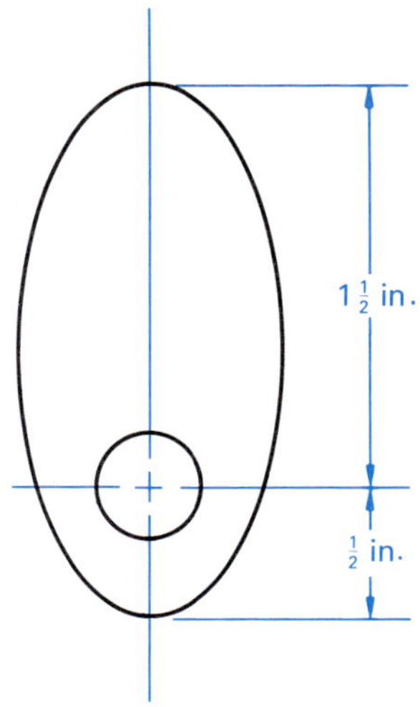

FIGURE 17.66

Chapter summary

Analytic geometry links the equations of algebra and the shapes of geometry. It includes the study of the conic sections.

The slope–intercept form of the equation of a straight line is $y = mx + b$. The slope may be found from $m = \dfrac{y_2 - y_1}{x_2 - x_1}$. The point–slope form of the equation of a line is $y - y_1 = m(x - x_1)$. The general form of the equation of a line is $Dx + Ey + F = 0$. Parallel lines have equal slopes and the slopes of perpendicular lines are negative reciprocals of each other. The distance between two points is given by the formula $d = \sqrt{(x_2 - x_1)^2 + (y_2 - y_1)^2}$ and the midpoint of the line between two points is $(x, y) = \left(\dfrac{x_1 + x_2}{2}, \dfrac{y_1 + y_2}{2}\right)$.

The standard form of the equation of a circle with center at (h, k) is $(x - h)^2 + (y - k)^2 = r^2$. The general form of the equation of a circle is $Ax^2 + Cy^2 + Dx + Ey + F = 0$, where $A = C$.

The standard forms of the equation of a parabola with vertex at (h, k) are $(y - k)^2 = 4p(x - h)$ and $(x - h)^2 = 4p(y - k)$. The general forms of the parabolic equation are $Cy^2 + Dx + Ey + F = 0$ and $Ax^2 + Dx + Ey + F = 0$.

The standard forms of the equation of an ellipse with center at (h, k) are, with horizontal major axis, $\dfrac{(x - h)^2}{a^2} + \dfrac{(y - k)^2}{b^2} = 1$ and, with vertical major axis, $\dfrac{(x - h)^2}{b^2} + \dfrac{(y - k)^2}{a^2} = 1$. The general form of the equation of an ellipse is $Ax^2 + Cy^2 + Dx + Ey + F = 0$. ($A$ and C must have the same sign and $A \neq C$.)

The standard forms of the equation of a hyperbola with center at (h, k) are, with horizontal major axis, $\frac{(x-h)^2}{a^2} - \frac{(y-k)^2}{b^2} = 1$ and, with vertical major axis, $\frac{(y-k)^2}{a^2} - \frac{(x-h)^2}{b^2} = 1$. The general form of the equation of a hyperbola is $Ax^2 + Cy^2 + Dx + Ey + F = 0$. ($A$ and C must have opposite signs.)

For translation of axes from a system xy to a system $x'y'$ whose origin is at (h, k) in the xy-system we use $x = x' + h$ and $y = y' + k$.

Review Exercises

1. A telephone pole guy wire is attached to the pole 25 ft above the ground and makes a 30° angle with the pole. What is the equation of the line of the guy wire? (See Fig. 17.67.)

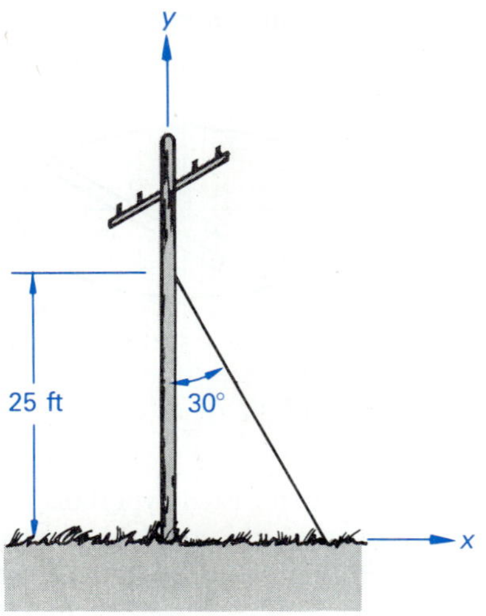

FIGURE 17.67

2. An all-terrain vehicle is crossing the desert from point A, having coordinates with respect to base of (3.22, 5.61) km, to point B, with coordinates (−7.43, 2.14) km. Find the compass course it must follow by finding the slope of the line between the points and converting it to a compass heading.

3. A pattern is being laid out by computer. What equation is needed to instruct the computer to draw a line perpendicular to $2x - y + 3 = 0$ and passing through its y-intercept?

4. What equation is needed to instruct a computer to draw a line parallel to $2x - y + 3 = 0$ and passing through the origin?

In Exercises 5–8, sketch the lines.

5. $x = 4$

6. $y = -2$

7. $y = -2x + 3$

8. $4x + 3y + 2 = 0$

9. An 80-mm diameter hole saw is centered 20 mm to the right and 30 mm above the lower left corner of a rectangular plate. Where does the saw cut cross the edge of the plate?

10. Where is it necessary to center an 80-mm hole saw to cut the edges of a rectangular plate 10 mm above and 20 mm to the right of the lower left corner of the plate?

11. Three observers stationed at (1, 4), (3, −2), and (−2, −4) all see the flash of an explosion at the same time. How far away is the explosion and where is it located?

12. What is the equation of a 1/2-m-diameter circle that is tangent to the line $y - 2x = 0$ at the origin?

In Exercises 13–16, sketch the circles showing their centers and radii.

13. $x^2 + y^2 = 2.25$

14. $(x - 1)^2 + (y - 3)^2 = 16$

15. $x^2 + y^2 - 4x - 8y + 4 = 0$

16. $x^2 + y^2 + 2x - 4y + 1 = 0$

In Exercises 17–20, sketch the parabolas and locate the vertex and focus for each.

17. $y^2 = 12x$

18. $x^2 = -3(y + 2)$

19. $-3x + y^2 + 2y + 1 = 0$

20. $0.5x^2 + 2.0x + 1.5y + 1.0 = 0$

21. What is the equation of a parabolic suspension-bridge cable with 500 m between supports and with a sag of 75 m? Locate the axis system at the low point of the cable.

22. Write the equation for the cable of Exercise 21 using the point of attachment of the left support for the origin of the axis system.

23. A parabolic reflector is to be 80 mm across at the focus and the vertex is to be 20 mm to the left of the focus. What is the equation of the shape of the reflector if the origin of the axis system is located at the focus?

24. A constant-stress cantilever beam has its vertical longitudinal section shaped like a parabola. What is the equation of the beam section? (See Fig. 17.68.)

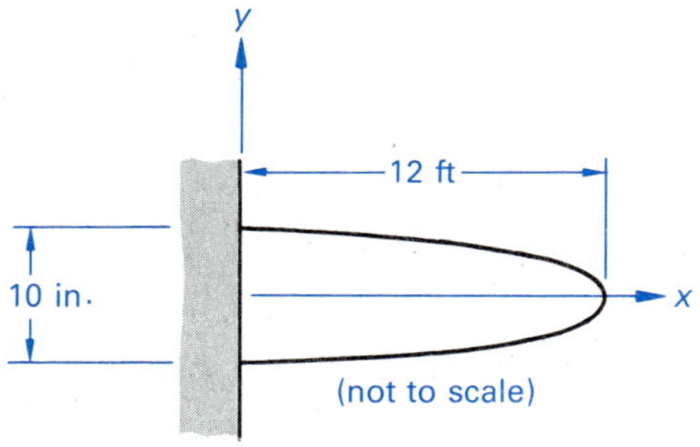

FIGURE 17.68

In Exercises 25–28, sketch the ellipses, and locate the center, the foci, and the vertices for each.

25. $\dfrac{(x+1)^2}{4} + \dfrac{y^2}{16} = 1$

26. $\dfrac{(x-2)^2}{25} + \dfrac{(y+2)^2}{9} = 1$

27. $x^2 + 4y^2 + 6x + 8y + 9 = 0$

28. $9x^2 + y^2 - 36x + 8y + 43 = 0$

29. The underside of a masonry bridge is shaped like an ellipse to provide a constant stress distribution. Write the equation for the ellipse shown in Fig. 17.69.

30. The elliptical orbit of a satellite is 6000 km wide at its widest and 2000 km wide at its narrowest. What is an equation for the orbit if the origin is located at the center of the ellipse?

31. What is an equation for the orbit of Exercise 30 if the origin of the axis system is located at one of the foci of the ellipse?

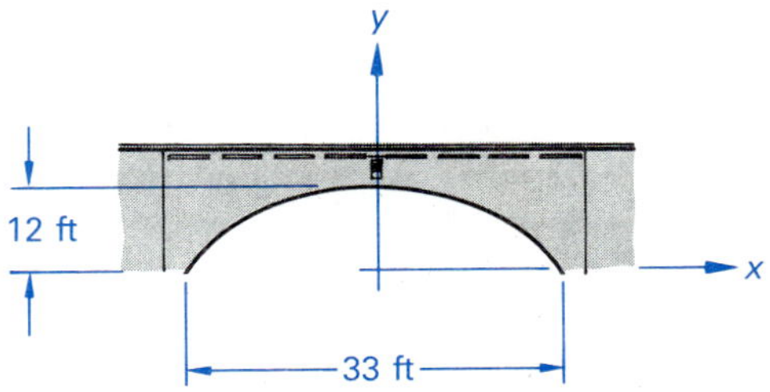

FIGURE 17.69

32. An ellipse can be drawn by the trammel method by measuring the length of its semi-major axis and marking it on a straight edge as OA and marking the length of its semi-minor axis as OB (Fig. 17.70). By sliding point A along the minor axis and point B along the major axis, point O traces the outline of the ellipse. Prove that the points located by this method must form an ellipse.

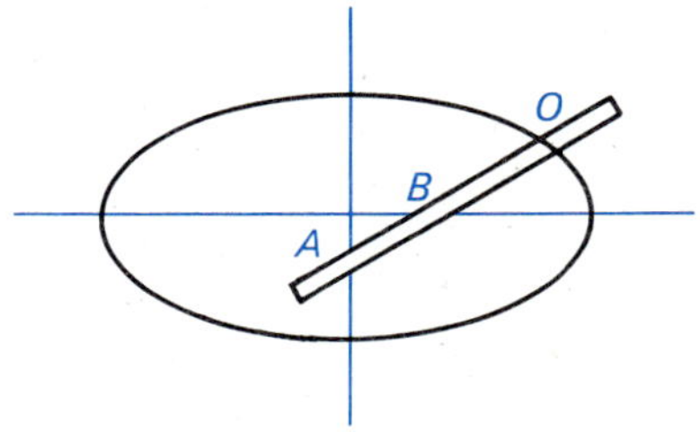

FIGURE 17.70

In Exercises 33–36, for each hyperbola find the coordinates of the center, the foci, the vertices, and the equations of the asymptotes. Sketch the hyperbola.

33. $\dfrac{(x-3)^2}{9} - \dfrac{(y-2)^2}{16} = 1$

34. $\dfrac{y^2}{16} - \dfrac{(x-2)^2}{25} = 1$

35. $-4x^2 + y^2 + 8x - 4y - 4 = 0$

36. $x^2 - 9y^2 + 2x + 18y - 44 = 0$

37. Write the equation of the hyperbola whose foci are at (0, 0) and (0, 10) and whose vertices are at (0, 2) and (0, 8).

38. Write the equation of a hyperbola whose foci are at (2, −1) and (12, −1) and whose vertices are at (5, −1) and (9, −1).

In Exercises 39–44, rewrite the equation of the given figure for a parallel $x'y'$-axis system with its origin at the center of the figure (vertex in case of the parabola). Give the coordinates of the foci and vertices where appropriate.

39. $(y-3)^2 = 12(x+1)$

40. $(x-1.50)^2 + (y+0.25)^2 = 2.25$

41. $\dfrac{(x-3)^2}{9} - \dfrac{(y-2)^2}{16} = 1$

42. $(x-1)^2 + (y-3)^2 = 16$

43. $x^2 + y^2 + 2x - 4y + 1 = 0$

44. $4x^2 - y^2 - 8x + 4y + 4 = 0$

45. Boyle's law states that in ideal gases at constant temperature the product of the pressure and the volume is a constant. Identify and sketch the graph of $pV = 2.76 \times 10^7$.

Can you solve these?

1. On a transistor chip, point A is at $(2 \times 10^{-3}, 2 \times 10^{-3})$ mm with respect to one corner of the chip and point B is at $(6 \times 10^{-3}, 5 \times 10^{-3})$ mm with respect to the same corner. Show that the equation of the line connecting the points is the same if you take A as "first" and B as "second" or you take B as "first" and A as "second."

2. The asymptotes of the hyperbola $\dfrac{x^2}{a^2} - \dfrac{y^2}{b^2} = 1$ are the lines $y = \pm \dfrac{b}{a} x$. Write a computer program to show that as x becomes larger, the difference between asymptote and hyperbola becomes smaller.

3. The end of a conveyor is 1 m above the floor. If the end is horizontal and a package leaves it traveling at 1 1/2 m/s, how far from the conveyor will it land?

4. Does $x^2 + y^2 + \ell x + my + n = 0$ always represent a circle? Justify your answer.

5. Lissajous figures are traces on an oscilloscope screen that can be used to find the phase shift between two signals. Their equation can be represented by

$$b^2x^2 - (2ab \cos \theta)xy + a^2y^2 - a^2b^2 \sin^2 \theta = 0,$$

where a, b, and θ are constants. What is the shape of the figure?

6. An equation of the form

$$Ax^2 + Bxy + Cy^2 + Dx + Ey + F = 0$$

can be transformed to a form without the xy-term by rotating the axis system counterclockwise by an angle θ. The required angle of rotation is found by using $\theta = \dfrac{1}{2} \text{Arctan} \dfrac{B}{A-C}$ and the transformation equations $x = x' \cos \theta - y' \sin \theta$ and

$y = x' \sin \theta + y' \cos \theta$.

Use this information to show that the equation $xy = C$ represents a hyperbola.

7. Use the equations from Exercise 6 to write the equation $5x^2 - 6xy + 5y^2 = 8$ without an xy-term. Identify the conic section and sketch.

18

A manufacturer of archery equipment is planning to introduce a new line of bow sights. As part of his design program he needs to know the range of an arrow and the maximum height of its trajectory for release velocities ranging from 60 m/s to 100 m/s in 2-m/s increments inclusive and for release elevation angles from 0° to 45° in 3° increments inclusive.

By neglecting air resistance, a first approximation to the desired results can be obtained from the equations:

$$r_{max} = 0.102\ V_0^2 \sin 2\theta \qquad \text{and} \qquad h_{max} = 0.0510(V_0 \sin \theta)^2.$$

Write a computer program to calculate and print the 336 required results.

Computers

18.1 Introduction

The history of computers is easiest to trace if we do not try too rigidly to distinguish between a calculator and a computer. All computers can calculate, but not all calculators can make the internal decisions characteristic of computers to change their order of operation.

Modern computers fall into two major categories: the **digital computer** and the **analog computer.** The digital computer (with which we will be concerned) is usually the faster and the more accurate of the two and has superseded the use of the analog computer. The digital computer operates on distinct pieces of information such as tables of individual numbers. It can process many numbers so quickly that it can closely approximate the effect of dealing with a continuous curve.

The analog computer deals with continuous data. As a comparison between the operation of the two types of computers, consider a table of trigonometry functions with the values of sin x given for x in 1-s increments, as compared to a graph of sin x. The digital computer would use the values from the table for input and calculation, whereas the analog computer would use the trace of the sine curve. Because the analog computer was easier to construct, it was in general use before the digital computer. Now, it is used only for specialized applications. Even so, since many practical systems operate on the basis of continuously flowing events, analog input and output devices are used with digital processing—combined units sometimes called **hybrid computers.**

The digital computer is a modern machine. Although earlier attempts had been made to produce mechanical calculating machines, it was not until 1875 that technology had advanced to the point where commercially reliable calculators could be produced. The computer, as a machine that could make internal decisions to alter its course of operation, did not appear commercially until the 1950s. Some simple mechanical units were available in the 1800s but their reliability was poor and they were confined to specialized applications such as calculating tables of functions. The first electromechanical computers appeared in the late 1930s and early 1940s. Their use started to become more generalized but they were still experimental devices. By the late 1940s the electronic computer was developed and slowly became the general purpose, generally available machine that it is today.

The current limitation on computer speed is not the speed with which calculations are performed (on the order of 10^{-12} s), but depends on the distance that electrons must travel at the speed of light in order to move numbers and instruction codes to and from memory. As a result of the research and development directed toward increasing speed of operation, a direct side effect has been to decrease the overall size of the digital computer at the same time that its calculation potential is being increased.

18.1 Exercises

In Exercises 1 and 2, designate the data as digital (d) or analog (a).

1. **a.** The contour of a road surface;
   b. daily high temperatures;
   c. air-pressure fluctuation around a nose cone;
   d. a radiofrequency wave;
   e. blast lengths on a fog horn.

2. **a.** Breathing rate during a stress test;
   b. outcome of successive dice rolls;
   c. best lap times during the Indy 500;
   d. house numbers along a street;
   e. path of a bumper car.

18.2 The computer

The heart of the digital computer is the **central processing unit** or CPU. It contains the **control unit,** the **arithmetic/logic** unit, and the **memory** of the computer. The **input/output** (I/O) devices may be built into the same console as the CPU or they may be separate units located many miles from the CPU. Additional memory units may also be externally connected to most CPUs. These pieces of equipment are known as the computer's **hardware.**

The input units feed the CPU with all of the information it is to use. They give it commands to operate its built-in programming, they give it additional and changeable programs and they feed it the data with which all the programs operate. The input devices may receive information in the form of words or numbers but they function so as to allow the information to be digested by the CPU as 1s and 0s or as a string of either "on" or "off" switching characteristics. All operations —computation, decision making, storage, control, I/O—are done in binary code in the digital computer. Some input devices are the computer keyboard, the card reader, the magnetic-disk unit, the floppy disk, the magnetic-tape unit, the electronic stylus, digital scanners, and direct equipment-monitoring instruments. Some of these, like the keyboard, the magnetic-disk unit, the floppy disk, and the magnetic-tape unit may also be used as output devices.

The output devices serve to preserve the results of the computer's operations. The punch card, the hard-magnetic-disk unit, the floppy-magnetic-disk unit, and the magnetic-tape unit produce semipermanent records, which can be directly re-entered into the same or a different computer at a later date. The printer and the cathode-ray tube (CRT) display output in a visually recognizable form. The computer may also directly operate control devices on external equipment. The speed at which the CPU can perform calculations and make decisions is usually much faster than the operating speeds of the I/O devices, so one CPU often operates with many different I/O units.

Within the CPU, the control unit receives input information, assigns it to storage in memory as necessary, retrieves both data and program information from memory, performs mathematical manipulations and decision-making operations in conjunction with the arithmetic/logic unit, and either relocates the results in memory or sends them to an output device.

All operations in the computer are performed in terms of **binary numbers.** Each binary number is composed of a string of **binary digits** called **bits.** The string of bits is a computer word and comprises one address location in the computer's memory. Each computer design has its own characteristic word length and as a result can handle decimal numbers of a set number of significant digits during normal operations. The longer the word the greater the number of significant digits that can be handled. Typical word lengths are 8 or 16 bits in microcomputers and 32, 48, or 60 bits in larger units. It is these words that comprise the individual storage locations in the computer's memory. Because the computer operates in binary arithmetic, 1 K of computer memory actually contains the storage facilities for 1024 words of whatever length is characteristic for a specified machine. It is particularly difficult to interchange information between machines that use different word lengths. This word size facilitates transmission of computer information over telephone lines in the standardized eight-bit configuration called a byte.

Each individual bit represents one of two possible situations at any time: switch "off" or "on," ceramic doughnut magnetized clockwise or counterclockwise, card with hole punched or not punched, transistor current flowing or not flowing, tape or disk magnetized or blank. Any of these situations can represent only a zero or a one in binary arithmetic and all the necessary mathematical and control operations can be performed on this basis.

The computer's **software** is composed of the sets of instructions that cause the computer to perform its indicated functions. Supplied with each computer (sometimes built-in, sometimes separate) is its operating system, which causes the machine to respond appropriately to any of the allowable commands given to it. The operating system also controls the computer's interaction with all of its I/O devices, its assignments to memory, and performs the necessary bookkeeping operations for the many programs that may be running simultaneously. Some other software packages that are available for the computer are word processing, bookkeeping, and structural design. The programs written by the user of the machine are also part of the software.

To program a computer in machine language is generally extremely difficult. All arithmetic must be performed in terms of binary addition. The location in memory of every piece of information, data or command, must be known so that information can be retrieved as necessary and also to avoid assigning information to a location already in use for something else. In addition, the machine-language operations for one machine are different from those for another model or another manufacturer. In order to make programming easier for people who are not computer hardware or software experts, programming languages have been developed that the computer can convert into its machine language.

Many of the programming languages available have been developed specifically to process certain types of data. Some of the common ones are: FORTRAN (**For**mula **Tran**slation), PL/I (**P**rogramming **L**anguage **I**), COBOL (**Co**mmon **B**usiness-**O**riented **L**anguage), RPG (**R**eport **P**rogram **G**enerator), Pascal (a language for systems programming), and BASIC (**B**eginners **A**ll-Purpose **S**ymbolic **I**nstruction **C**ode). The lan-

Glossary of Computer Terms

ASCII — American Standard Code for Information Interchange.
BAUD RATE — Speed of transmission at serial ports in bits (must be compatible with associated I/O devices).
BIT — Binary digit.
BYTE — Eight-bit group.
CASSETTE — Magnetic tape I/O unit (slower than disk).
CHARACTER — Single letter, number, or symbol.
CPU — Central processor unit; includes control unit, arithmetic/logic unit, and main memory (the heart of the computer).
CRT — Cathode-ray tube.
DEBUG — Locate and correct faults in program.
DISK — Magnetic I/O unit; may be floppy (soft) or hard; more memory is available on hard, but the disk is more expensive; both types are faster than cassette.
HARDWARE — The electromechanical parts of the computer.
I/O — Input/output.
K — 1024 units of memory storage.
MODEM — Device to connect I/O unit to computer via telephone line.
PORT — Computer connection with I/O device. May be serial, whereby bits in byte travel sequentially on one wire as in a modem hookup, or parallel, whereby all bits in byte travel at the same time over eight wires as in a keyboard hookup.
PROGRAM — A series of instructions to the computer.
RAM — Random access memory (disappears at shutoff).
ROM — Read only memory (permanent).
SOFTWARE — The programs that control the operation of the computer.
TELEPRINTER — Output device that produces "hard" copy.
TERMINAL — I/O device.
WORD — Data element stored at one address location in memory.

guages differ in how their commands are composed and also in how they operate internally in the machine. One thing that all of the computer languages have in common is their general approach to problem solving. Since BASIC is one of the easiest languages to learn, is available on most computers, and is a language that allows direct interaction with the computer, we will concentrate our efforts on the BASIC language.

18.2 Exercises

1. Name three I/O devices that can be used only for input.
2. Name three I/O devices that can be used only for output.
3. Name three I/O devices that can be used for both input and output.
4. What is the difference between computer hardware and software?
5. Why would knowledge of baud rate be necessary when purchasing I/O devices?
6. What are the advantages and disadvantages of cassettes and disks as memory units?
7. ROM is used for storing the computer's operating system in many instances. RAM is used to expand programming capability by increasing memory. Why are the two types used for the different purposes?
8. Would a modem be associated with a parallel port or a serial port? Explain.

18.3 Flowcharts and mathematical operations in BASIC

The logic that governs the order in which operations are performed to solve a problem is the same in BASIC as it is in FORTRAN, COBOL, Pascal, or any of the other computer languages. Even if all the statements learned in BASIC cannot be directly carried over to the other languages, the approach to solving a problem and the structure of the solution is the same in BASIC as in the other languages. The computer will do exactly what you, the programmer, tell it to do and it will do **only** what it is told to do. You must therefore consider every possibility that can cause a program to go astray and construct your program to take care of all those possibilities. If not, the computer will happily produce wrong answers.

As an aid in constructing the sequence of operations that you want the computer to perform, you may develop a series of connected symbols called a flowchart. The flowchart symbols commonly used are illustrated in Table 18.1.

An implicit approach to flowcharting is more often encountered in the early stages of analyzing and structuring the solution to a problem.

TABLE 18.1 **Flowchart symbols**

Symbol	Significance	Example: Implicit approach	Example: Explicit approach
	Terminal (beginning or end)	quit here	STOP
	Input or output	Output x and y	PRINT X, Y
	Assignment or Processing	Add a and b	SUM = $A + B$
	Flow lines	Connect symbols in sequence of execution.	
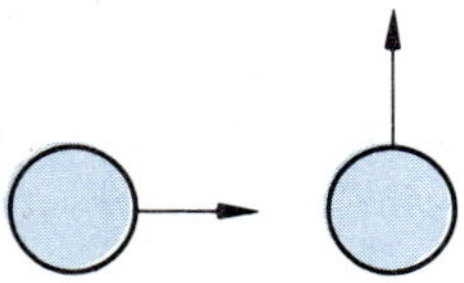	Connector	A →	↑ A
		In either case, allows transfer from one part of program to another without excessively long flow lines.	
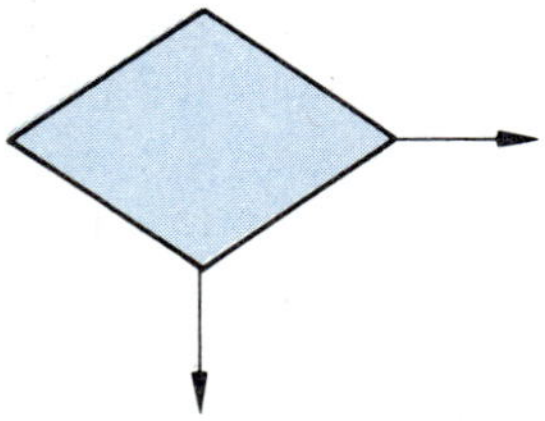	Decision	Is r less than t? YES → ; NO ↓	$R < T$ YES → ; NO ↓

An explicit approach is seen more often when the solution is better defined and when a deeper understanding of programming language has been developed.

EXAMPLE 1 A student's course average will be determined by four exams. To calculate the average, the lowest of the first three exam grades will be dropped. The sum of the remaining two will be added to twice the fourth, and the result will be divided by four to produce the course average. Devise a flowchart to show a logical progression for the calculation of the course average.

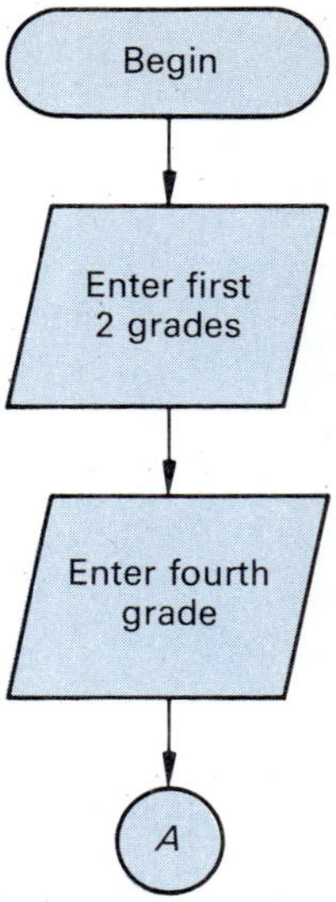

FIGURE 18.1

FIGURE 18.2

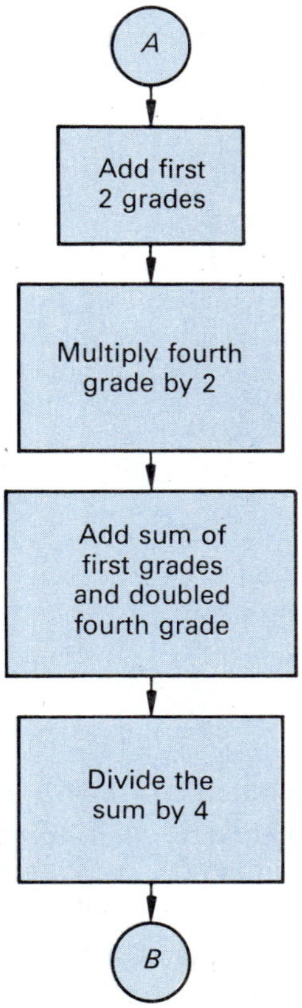

Solution: Before the computer can do any calculations, it must know and be able to identify the data with which it must work. In this case it is easiest if we decide which of the first three grades is to be dropped, so we will make that decision for the computer. The fourth grade is to be treated differently so the fourth grade will be entered separately. The implicit flowchart to this point might look as shown in Fig. 18.1. The flowline arrows show the path of progression through the operations. Normally the progression is sequential from one operation to the one immediately following it. The connector indicates that there is more flowchart yet to come.

Now that the necessary data are available in the computer, we can proceed to the calculations, as shown in Fig. 18.2.

Once the necessary calculations have been completed we can output the results. (If we forget to output the results, the computer simply keeps them stored in memory and we never know what they are.) When the results are displayed we have completed the desired operations. (See Fig. 18.3.)

Usually with a flowchart of this size it would not be necessary to use the connectors. The flowchart is reproduced in Fig. 18.4a, on page 722, as it would usually be seen. Figure 18.4b is another flowchart that would solve the same problem. Can you understand the differences? □

The flowchart in Fig. 18.4 shows some of the ways we could have approached the same problem. As long as the calculations are mathematically correct it does not matter when they are performed *as long as the necessary data have been introduced first.* Often it is possible to combine a series of simple mathematical operations in a single step. The

FIGURE 18.3

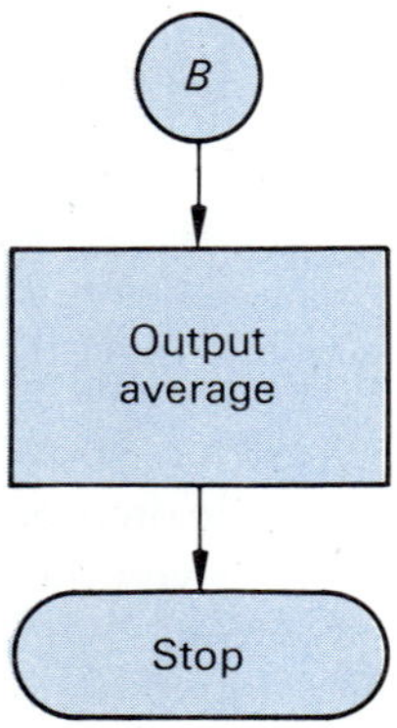

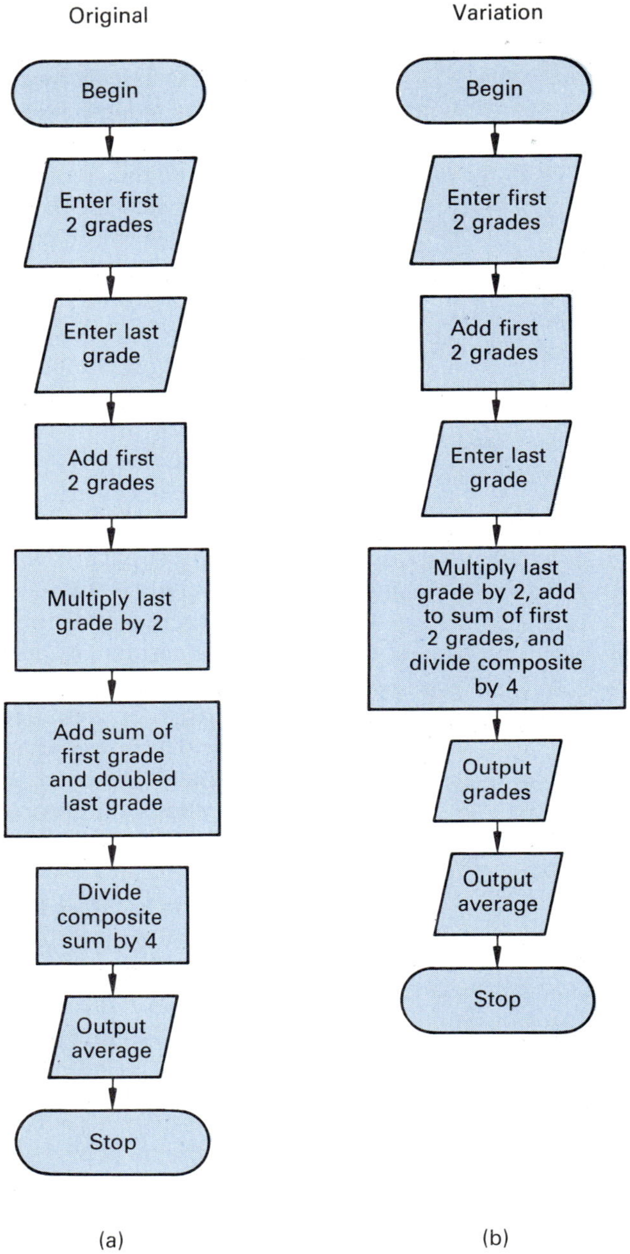

FIGURE 18.4

reason the grades were requested as output (we put them in so we already know what they are) is to provide a check that no mistakes were made on input. Otherwise we do not know whether or not the computer received the correct information. It also would have been possible to output the grades and the average at the same time; this sort of decision is the individual programmer's choice.

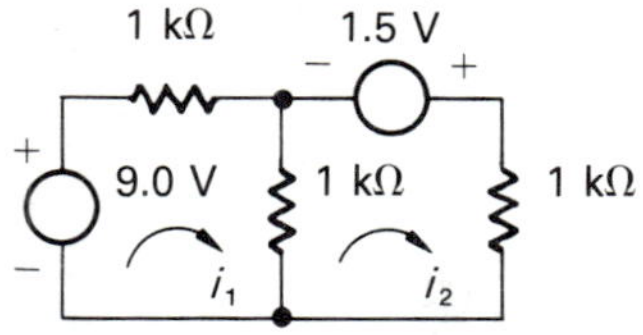

FIGURE 18.5

EXAMPLE 2 The electrical circuit shown in Fig. 18.5 produces the loop equations

$$2000\ i_1 - 1000\ i_2 = 9.0$$

and

$$-1000\ i_1 + 2000\ i_2 = 1.5.$$

Look up the equations necessary to solve two simultaneous equations in two unknowns (see Chapter 10) and write a computer program to solve for the loop currents i_1 and i_2.

FIGURE 18.6

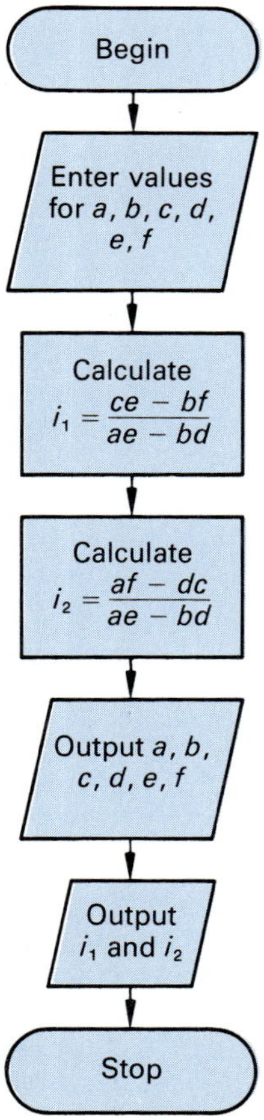

Solution: According to the methods described in Chapter 10, if we have a pair of simultaneous equations in two unknowns: $ax + by = c$ and $dx + ey = f$, the solutions are: $x = \dfrac{ce - fb}{ae - db}$ and $y = \dfrac{af - dc}{ae - db}$. In this instance the unknowns i_1 and i_2 will correspond to x and y and $a = 2000$, $b = -1000$, $c = 9.0$, $d = -1000$, $e = 2000$, and $f = 1.5$. A flowchart for the solution might be constructed as shown in Fig. 18.6. □

EXAMPLE 3 To determine the natural frequencies of a vibrating system we must solve the quadratic equation $ax^2 + bx + c = 0$. Draw a flowchart to determine the solution.

Solution: The quadratic equation has two solutions:

$$x_1 = \frac{-b + \sqrt{b^2 - 4ac}}{2a} \quad \text{and} \quad x_2 = \frac{-b - \sqrt{b^2 - 4ac}}{2a}.$$

To do a complete job we must produce both of these results. We must also be careful that $b^2 - 4ac$ is not less than zero because the computer cannot take the square root of a negative number. One approach to a flowchart is shown in Fig. 18.7 on page 724. □

For this problem we utilized the power of the computer to make a decision on the basis of a programmed test and to alter its sequence of operations. As a result there are two possible paths that may be followed to the completion of the problem. It is this sort of **branching** that allows us to approach the full power of the computer.

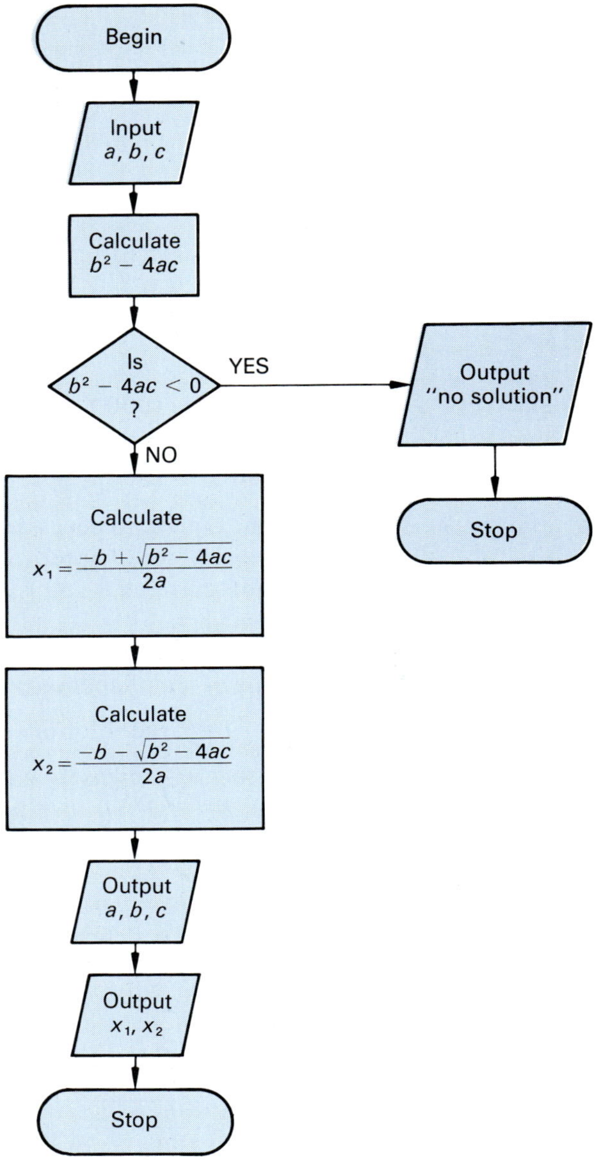

FIGURE 18.7

18.3 Exercises

Sketch flowcharts to solve the following problems. If possible, try more than one path to the solution.

1. Add any deposits made during the past month to the current balance in your checking account and subtract any checks written during the same period and display the balance.
2. Calculate a list of average speeds over a 12-km distance for times of 5 min, 6 min, 7 min, 8 min, 9 min, and 10 min. Show the results in km/hr. $\left(s = \frac{d}{t}.\right)$
3. Convert an angle from degrees into radians.

4. Convert an angle from radians into degrees.
5. Change a time given in hours, minutes, and seconds into decimal hours.
6. Change a time given in decimal hours into hours, minutes, and seconds.
7. Find the absolute value of any number x. ($|x| = x$ if $x \geq 0$, $|x| = -x$ if $x < 0$.)
8. Find the gross weekly pay for H hours worked. The pay rate is \$5.00/hr up to and including 40 hours per week and time and a half for overtime.
9. Calculate the value of a 2×2 determinant.
10. Calculate the value of a 3×3 determinant.
11. Solve a system of two linear equations in two unknowns using determinants.
12. Solve a system of three linear equations in three unknowns using determinants.

18.4 Commands

It is easy to make the computer do arithmetic. All we need do is give the computer a set of **statements** or commands, which, when taken together in sequential order, form the desired **program.** We must be careful to write the statements in a way that they can be understood by the computer. For example, we know that all of the following give the same result: adding 2 and 2, finding the sum of 2 and 2, and finding $2 + 2$. The computer can only understand $2 + 2$; its vocabulary is limited. It can understand commands only if they are written in the exact form of the computer language. Any deviation from that form will either produce no answer or a wrong answer.

To write programs we need to define the terms that will be used. One definition we will need is that of a constant. A **constant** is a value that does not change. In this discussion we will deal only with arithmetic constants even though BASIC can also handle character constants. An arithmetic constant is a number that can be represented in one of three forms: integer, fixed-point, or floating-point. Most computers can handle numbers with a maximum of seven significant digits in short form and 15 significant digits in long form. The short form is the normal mode of operation. The maximum magnitude of the numbers processed is approximately $10^{\pm 75}$. A constant in **integer** form may consist only of a string of digits preceded by an optional plus or minus sign. A constant in **fixed-point** form consists only of a string of digits with a decimal point and an optional preceding plus or minus sign. The **floating-point** form is used to represent numbers in scientific notation. This form is an integer or fixed-point constant followed by the letter E, which is in turn followed by an integer constant representing the power of 10 multiplying the number.

EXAMPLE 1 Are the following constants valid? If not, state why. Also, if possible state the type of constant shown.

a. 1234 **b.** −1,234 **c.** +1.23 4

d. 0.78123456 **e.** −176E12 **f.** 176E−2

g. −1.314E+08 **h.** 12,476.E80.

Solution:

a. 1234	Valid:	integer constant
b. −1,234	Invalid:	integer constant; contains illegal comma
c. +1.23 4	Valid:	fixed-point constant; blanks in a word or number are ignored in BASIC
d. 0.78123456	Invalid:	fixed-point constant; has more than seven significant digits
e. −176E12	Valid:	floating-point constant; represents -176×10^{12}
f. 176E−2	Valid:	floating-point constant; represents 176×10^{-2}
g. −1.314E + 08	Valid:	floating-point constant; represents -1.314×10^{8}
h. 12,476.E80	Invalid:	floating-point constant; contains illegal comma and power of ten exceeds maximum allowed □

In addition to appearing in constant form, arithmetic data may be present in a BASIC program as a variable. A **variable** is a symbol whose value may change during the execution of the program. In the equation $x - 4y + 2z = 18$, x, y, and z are variables and 4, 2, and 18 are integer constants. (For reasons that will be explained shortly, $x - 4y + 2x = 18$ would not be written in that form in BASIC.) Arithmetic variables in BASIC may consist of a single letter of the alphabet or a letter of the alphabet followed by either another letter or by a digit. Some versions of BASIC allow more flexibility than this in choosing variable names, but we will confine our discussion to forms that are likely to be available in the majority of versions.

EXAMPLE 2 Are the following valid or invalid as arithmetic variables? Explain.

Explain: **a.** X **b.** x **c.** A2 **d.** 2X **e.** M+ **f.** MX

Solution:

a. X	Valid	
b. x	Valid:	would be considered the same as X. Since lower case letters are not available in some versions of BASIC we will henceforth use only upper case letters
c. A2	Valid	

d. 2X Invalid: first character may not be a digit

e. M+ Invalid: second character is not letter or digit

f. MX Valid □

The arithmetic constants and variables can be combined to form arithmetic expressions such as $x^2 + 3x - 1$. An **arithmetic expression** is a single arithmetic constant, a single arithmetic variable or a group of constants and/or variables joined by the arithmetic operators. The arithmetic operators are ** or ↑ or ∧ for exponentiation (raising to a power), * and / for multiplication and division, and + and − for addition and subtraction. Two operators may not appear next to each other (** is considered a single operator), but the parts of the expression may be separated as necessary by using parentheses. Parentheses are also used to control the order in which operations are performed in an expression. In the absence of parentheses the operations are automatically given priority from highest to lowest as follows: exponentiation, multiplication and division, addition and subtraction. Within any priority level the operations are performed as they are encountered going from left to right in the expression. (See Table 18.2.)

EXAMPLE 3 Are the following expressions valid? If so, indicate the results of the operations. If not, state why they are invalid.

a. X = 1	**b.** P8	**c.** 27
d. A*−8	**e.** −8*A	**f.** A+B/C
g. (A+B)/C	**h.** X**2 + YZ**2	**i.** A+2B
j. 1+I/J−2	**k.** (1+I)/(J−2)	

TABLE 18.2 **Arithmetic operations**

OPERATORS

Symbol	Meaning	Priority
** or ↑ or ∧	Exponentiation	First
*	Multiplication	Second in order encountered
/	Division	Second in order encountered
+	Addition	Third in order encountered
−	Subtraction	Third in order encountered

RULES

Operators may not appear next to each other but may be separated by parentheses.

Expressions are evaluated from the innermost parentheses out, using the priorities of operation within each set of parentheses.

Solution:

a. X = 1 — Invalid: = is not one of the allowable operators for an arithmetic expression (more on this shortly)

b. P8 — Valid: a single variable, also sometimes called a variable name, would call up the data stored in memory at the address P8 (more on variable names shortly)

c. 27 — Valid: an expression may be a single constant

d. A* − 8. — Invalid: two operators may not appear next to each other
Could be correctly written as A*(−8.)

e. −8*A — Valid: would produce the same result, −8A, as the corrected version of (d) above

f. A + B/C — Valid: because of the priority of the operations the computer would first divide B by C and then add A to the quotient resulting in $A + \frac{B}{C}$

g. (A+B)/C — Valid: the presence of parentheses causes the operations to be performed inside the innermost parentheses first, using normal priorities, and then the priorities are repeated for each succeeding set of parentheses (if present) working outward until the expression is completed; this one would calculate as $\frac{A + B}{C}$

h. X**2+YZ**2 — Valid: YZ is a single variable name, not a multiplication (no * between Y and Z); performs exponentiations first and then addition to calculate as $X^2 + YZ^2$

i. A+2B — Invalid: 2B is not an acceptable variable name or acceptable operation; if written as A+2*B would calculate as A + 2B

j. 1+I/J−2 — Valid: because of priority of operations, would calculate as $1 + \frac{I}{J} - 2$

k. (1+I)/(J−2) — Valid: because of parentheses, would calculate as $\frac{1 + I}{J - 2}$ □

We are now ready to assemble the elements needed to use the BASIC language for computer calculations. To do so we will write a series of statements that form a program telling the computer the operations we want it to perform and in what order we want them performed. All statements in BASIC must be preceded by a line number; all line numbers are positive integers and must not exceed five digits. The numbers need not be consecutive. Statements will be executed in the order of their line numbers.

The type of BASIC statement we will use most is the assignment statement. An example of an assignment statement is

200 LET X = A + 3

The general form of the assignment statement is given in Definition 18.1.

DEFINITION 18.1

ASSIGNMENT STATEMENT*

line number LET variable = expression

EXAMPLE 4 Are the following assignment statements valid? If not, explain.

a.	10	LET X = A + B
b.		LET X = 2.
c.	712361	LET P = N*R*T/V
d.	123	LET A + B = X
e.	88	LET 3 = I/K
f.	376	LET Z**2 = X**2 + Y**2

Solution:

a.	10	LET X = A + B	Valid	
b.		LET X = 2	Invalid:	no line number
c.	712361	LET P = N*R*T/V	Invalid:	line number exceeds five digits
d.	123	LET A + B = X	Invalid:	left side of equal sign has expression instead of single variable

* The form: line number variable = expression is valid in some versions of BASIC; for example, 200 X = A + 3.

e. 88	LET 3 = I/K	Invalid:	left side of equal sign has constant instead of single variable
f. 376	LET Z**2 = X**2 + Y**2	Invalid:	same as (d) above □

The assignment statement looks like a mathematical equality except for the restrictions illustrated in parts (d), (e), and (f) of Example 4. It is in reality quite different from a mathematical equality, but because of the way it operates it can be used like a mathematical equality. The difference stems from the fact that a variable in the computer does not represent an abstract unknown quantity as it does in mathematics. Instead it represents the address of a location in the computer's memory—a mailbox number, so to speak. Specification of a variable serves to call up the number stored at that address in memory. The memory address is assigned to the variable when the program is translated into machine language. The variable is not an unknown, but always has a data value stored at that address. In BASIC, some computers set all variables to zero at the beginning of each execution of a program. Others leave whatever values are left over from previous programs. After that a variable will have whatever value may be stored in that memory location by action of the program. The assignment statement causes the following sequence of operations to be performed: data values stored in the memory locations addressed by the variables in the expression to the right of the equal sign are copied out of memory leaving the stored values unchanged, the value of the expression is calculated, and the value of the expression is stored in memory at the address specified by the variable to the left of the equal sign while simultaneously erasing any value previously stored in that memory location.

EXAMPLE 5 Explain the operation of the following assignment statements.

a. 10 LET X = 8. **b.** 20 LET P = N*R*T/6. **c.** 30 LET A = A + 2

Solution:

a. 10 LET X = 8. Since no variables are in the expression to the right of the equal sign the expression is calculated as the value of the constant, 8.. This value is then stored in memory at the address location specified by X, erasing any value previously stored as X.

b. 20 LET P = N*R*T/6. N and R are brought out of memory and multiplied; T is brought out and multiplied times the previous result and that result is divided by 6.; the final value of the calculation is stored as P, erasing any value previously stored as P. The values of N, R, and T that existed at the beginning of the execution are still stored at those addresses.

c. 30 LET A = A + 2. Mathematically this statement is invalid but because of the way it operates it is perfectly valid on the computer. The value stored in A is brought out of memory, 2. is added to it and the result is stored in memory location A (erasing the previous value). The original value of A disappears from memory unless the precaution has been taken to store it elsewhere under some other name. □

We need two more statements to be able to write workable computer programs. After the machine has performed the calculations specified by the assignment statements in the program, the results of those calculations are stored in memory. They will remain buried there unless the computer is actually told to put them onto some output device. To do this we can use the PRINT statement, which follows in one of its simpler forms:

DEFINITION 18.2

PRINT STATEMENT

Line number PRINT expression list.

The expression list that follows the word PRINT consists of single variables, arithmetic expressions, and/or strings of characters enclosed within quotes. The components of the list are separated by commas or semicolons. If commas are used, the output is placed into **full print zones,** which are 16 characters wide. If the item to be output does not require the full 16 positions, the unused positions are left blank. Use of the semicolon between the list components causes the output to be placed into **packed print zones,** whose form is determined by the computer being used. Some computers put them in a continuous string; others leave a blank between each item. In general each PRINT statement produces one line of output unless the output is too long to be contained on a single line, in which case the computer automatically continues the output on as many additional lines as are necessary. A comma or semicolon at the end of the list will cause the next PRINT statement to act as a continuation of the current list. Otherwise the next PRINT statement begins output on the next new line.

EXAMPLE 6 All of the following print statements are valid. Describe the output that would be produced by each.

a. 100 PRINT X,X+Y,X−Y

b. 110 PRINT A;B;C;D;E;F

c. 120 PRINT A,B,C,D,E,F

Assume that the variables have already been defined as follows:

X = 12.8, Y = −18.3, A = 234, B = −2345678, C = 234.5, D = 23456.789, E = 333333, and F = −1000000.

Solution:

a. 100 PRINT X,X+Y,X−Y produces the output

```
_12.8______________−5.5______________31.1____
```

The underlines are not printed but are shown to indicate blank space in the output. The commas between the list items cause the use of full print zones. A 40-character CRT is assumed.

b. 110 PRINT A;B;C;D;E;F produces the output

```
234−2345678234.523456.789333333−1000000_
```

The semicolons between the list items cause the use of packed print zones. Depending on the computer, there may be spaces between the zones and/or the values for B, D, F may appear in scientific notation.

c. 120 PRINT A,B,C,D,E,F produces the output

```
234 ________________ −2345678 ___________________
234.5 _____________ 23456.789 ___________________
333333 ____________ −1000000 ____________________
```

The commas between the list items cause the use of full print zones. Because on a 40 character screen one line contains only two full print zones, the remaining items in the list are printed on the following lines. □

A variable or an expression in a PRINT statement causes only a number to be printed. It will not cause the name of the variable or the equation representing the expression to be reproduced. In order to print identifying notes with the output, the notes must be enclosed in quotation marks in the expression list of the PRINT statement. The computer then prints exactly what is enclosed in the quotation marks but deletes the quotes themselves. The character strings in the quotation marks will not cause any calculations to be performed when they are printed.

EXAMPLE 7 All of the following print statements are valid. Describe the output that would be produced by each.

a. 130 PRINT "X = ",X,"Y = ";Y

b. 140 PRINT "X = ";Y,"Y = ";X

c. 150 PRINT

d. 160 PRINT "THE SUM X + Y = "

e. 170 PRINT "THE SUM X + Y = ";X+Y

Assume that the variables have been defined as:

X = 12.8 and Y = −18.3

Solution:

a. 130 PRINT "X = ",X,"Y = ";Y produces the output
X_=_______________12.8______________Y_=_−18.
3
The commas produce full print zones, the semicolons produce packed print zones. (Remember we are assuming a 40-character screen width.) The character strings in quotation marks are printed exactly as they appear in the PRINT statement but with the quotation marks deleted. The presence of the X and the Y not enclosed in quotation marks instructs the computer to print the numbers stored in memory for those variables.

b. 140 PRINT "X = ";Y,"Y = ";X produces the output
X_=_−18.3________Y_=_12.8______________________
The semicolons and comma control the print zones. The strings in the quotes are printed; they have no effect on the numerical output. The numbers are controlled by the placement of the variables in the list.

c. 150 PRINT produces the output

____

In general a single PRINT statement produces one line of output. This PRINT has nothing in the expression list so it produces a blank line. The intent of this statement is to skip a line in the program output. It is useful for separating lines of output for better readability.

d. 160 PRINT "THE SUM X + Y = " produces the output
THE_SUM_X_+_Y_=______________________________
The expression enclosed in quotes produces no numerical output.

e. 170 PRINT "THE SUM X + Y = ";X+Y produces the output
THE_SUM_X_+_Y_=_−5.5________________________ □

One more statement completes our necessary list:

DEFINITION 18.3

END STATEMENT

line number END

The END statement must be the last statement in any BASIC program. It causes the execution of the program to be terminated and the computer to proceed to its next job.

EXAMPLE 8 Write a BASIC program to calculate the course average that was flowcharted in Example 1, Section 18.3. The problem was to add two grades to twice a third grade and divide the result by 4.

Solution: Let us call the first two grades G1 and G2 and the last grade G4. The method for entering a program and executing it is different for each machine. This information will have to be obtained from the instruction manual or from your instructor. The two versions of the program according to the two flowcharts are shown in Table 18.3.

Output produced by Table 18.3a is

```
98          73          86
85.75
```

Output produced by Table 18.3b is

```
98          73            86
85.75
```
□

EXAMPLE 9 Program Example 1, Section 18.3, to identify all the output values.

Solution: There are many possible ways to write most programs. The two programs that follow are just two possible approaches.

```
10  LET G1 = 98
20  LET G2 = 73
30  LET G4 = 86
40  PRINT "GRADE 1 IS ";G1, "GRADE 2 IS ";G2
50  PRINT "LAST GRADE IS ";G4
60  LET A = (G1 + G2 + 2*G4)/4
70  PRINT
80  PRINT "THE AVERAGE IS ";A
90  END
```

TABLE 18.3

Flowchart a	Flowchart b
10 LET G1 = 98	10 LET G1 = 98
20 LET G2 = 73	20 LET G2 = 73
30 LET G4 = 86	30 LET S = G1+G2
40 LET S = G1+G2	40 LET G4 = 86
50 LET P = 2*G4	50 LET A = (2*G4+S)/4
60 LET T = S+P	60 PRINT G1,G2,G4
70 LET A = T/4	70 PRINT A
80 PRINT G1,G2,G4,A	80 END
90 END	

Output produced is

```
GRADE 1 IS 98      GRADE 2 IS 73
LAST GRADE IS 86

THE AVERAGE IS 85.75
```

or

```
10  LET G1 = 98
20  PRINT "G1 = ";G1
30  LET G2 = 73
40  PRINT "G2 = ";G2
50  LET G4 = 86
60  PRINT "G4 = ";G4
70  PRINT "AVERAGE = ";(G1+G2+2*G4)/4
80  END
```

Output produced is

```
G1 = 98
G2 = 73
G4 = 86
AVERAGE = 85.75
```

□

The line numbers preceding the statements must be between 1 and 99999 inclusive and in ascending order. They could have been 1, 2, 3, . . . , or 18, 23, 101, 102, . . . , and so on. Counting by tens allows us to insert statements without renumbering if we wish to modify the program at a later time.

EXAMPLE 10 Write a BASIC program to calculate the solution of the system of simultaneous equations

$$2000i_1 - 1000i_2 = 9 \quad \text{and} \quad -1000i_1 + 2000i_2 = 1.5.$$

Solution: The problem has been flowcharted in Example 2, Section 18.3. One possible program is shown.

```
10  LET A=2000
20  LET B=-1000
30  LET C=9
40  LET D=-1000
50  LET E=2000
60  LET F=1.5
70  LET I1=(C*E-F*B)/(A*E-D*B)
80  LET I2=(A*F-D*C)/(A*E-D*B)
90  PRINT A,B,C
100  PRINT D,E,F
110  PRINT
120  PRINT "I1 = ";I1
130  PRINT "I2 = ";I2
140  END
```

Output produced is

```
2000        -1000       9
-1000       2000        1.5

I1 = 6.5E-03
I2 = 4E-03
```

□

18.4 Exercises

Write BASIC programs to solve the following problems. If possible, run the programs on a computer. Some of the exercises are similar to the flowcharting exercises at the end of Section 18.3.

1. Calculate the area of a square given that the length of a side is 36.2 mm. Output and identify the side length and the area on the same line.

2. Calculate the area and the perimeter of a rectangle given that the width of the rectangle is 38.1 mm and the length of the rectangle is 15.7 mm. Output and identify the width and the length on one line and then skip a line and output and identify the area and the perimeter on the following line.

3. One leg of a right triangle is 1.63×10^4 mm and the other leg is 2.81×10^4 mm. Calculate, output, and identify the area, the perimeter, and the length of the hypotenuse of the triangle. Place the results on three successive lines of output. (*Hint:* the square root of a number is equivalent to raising the number to the 1/2 power.)

4. Use the quadratic formula to calculate the two roots of $x^2 - x - 6 = 0$. Output the equation on one line and the roots on the following line. (See hint in Exercise 3.)

5. Suppose that at the beginning of last month the balance in your checking account was \$264.51. During the month you deposited \$149.61, \$152.88, \$149.61, and \$149.61. You wrote checks for \$250.00, \$50.00, \$11.26, \$21.78, \$55.40, \$50.00, \$49.82, \$50.00, \$26.91, and \$50.00. Compute and display, with identification, your starting balance, the total of the deposits, the total of the withdrawals, and the balance at the end of the month.

6. Calculate a list of average speeds over a 12-km distance for times of 5, 6, 7, 8, 9, and 10 min. Show the results in km/hr and display the corresponding times with the results.

7. Convert 36.1 degrees into radians. Display the angle in both forms.

8. Convert 0.743 radians into degrees. Display the angle in both forms.

9. Calculate the value of the polynomial

   $$f(x) = x^5 - 4.2x^3 + 7.3x$$

   for x equal to -16.5, 72.1, and 0.00168. Display both the values of x and the corresponding values of the polynomial.

10. Calculate the value of the polynomial

    $$y = x^4 + 7x^3 - 2x^2 + 3x - 6$$

    for x equal to $3.1 \times 10^{-6}, -7.4 \times 10^8, 8.3 \times 10^{19}$. Display both the values of x and the corresponding values of the polynomial.

11. Instruct the computer to display the pattern of asterisks shown.

```
      *  *
      *  *
   ** *** *** **
      *  *
   ** *** *** **
      *  *
      *  *
```

12. Calculate and display gross weekly pay for a 48-hour work week. The pay rate is \$5.00/hr up to and including 40 hours per week and time and a half for overtime.

13. Calculate the value of the determinant

    $$\begin{vmatrix} 1.3 & -4.2 \\ 2.7 & 8.5 \end{vmatrix}$$

14. Calculate the value of the determinant

    $$\begin{vmatrix} 3.0 & -3.0 & 5.0 \\ 1.0 & 2.0 & -6.0 \\ 2.0 & -1.0 & 3.0 \end{vmatrix}$$

15. Solve the system of circuit equations for i_1 and i_2 using determinants.

$$2000.i_1 - 1000.i_2 = 9.0$$
$$-1000.i_1 + 2000.i_2 = 1.5$$

16. Solve the system of equations for x, y, and z using determinants.

$$3.0x - 3.0y + 5.0z = 4.0$$
$$1.0x + 2.0y - 6.0z = 3.0$$
$$2.0x - 1.0y + 3.0z = 1.0$$

18.5 Loops

The computer's ability to make automatic changes in the order in which it performs the commands in a program is one reason why a computer is more than just a calculator. The command that causes the computer to modify its strict step-by-step approach is the IF statement.

DEFINITION 18.4

IF STATEMENT*

line number IF logical expression THEN executable statement

The IF statement operates such that when the logical expression after the IF is true the executable statement following the THEN is performed. When the logical expression following the IF is not true, the executable statement following the THEN is ignored and operation "falls through" to the line immediately following the IF statement. **A logical expression** is a statement that can only be answered "true" or "false." The **executable statement** is a statement, other than another IF, that causes the machine to perform an operation. The only nonexecutable statement with which we have dealt so far is the END statement.

The logical expression in the IF statement consists of two arithmetic expressions separated by logical operators. The logical operators are shown in Table 18.4. Alternative symbols are shown in parentheses.

TABLE 18.4 **Logical operators**

Symbol	Meaning	Symbol	Meaning
=	Equal	<	Less than
<> (≠)	Not equal	>= (≥)	Greater than or equal
>	Greater than	<= (≤)	Less than or equal

* Other forms of the IF statement are available; this form is acceptable in most versions of BASIC.

EXAMPLE 1 The following IF statements are valid. Explain their operation.

a.
```
40   IF X = 0 THEN LET Y = A
41   LET D = 2*Y
```
b.
```
50   IF B < 0 THEN LET B = 0
55   S2 = S1 + 3*B
```
c.
```
60   IF Z2 >= 10 THEN PRINT Z2+5
63   END
```

Solution:

a.
```
40   IF X = 0 THEN LET Y=A
```
The value stored in memory as X will be compared to zero. If X equals 0 (X = 0 is "true"), the value stored as A will also be stored as Y. Then the next line in sequence (41 LET D = 2*Y) will be executed. If the value stored as X is not equal to zero (X = 0 is "false") the statement LET Y = A will be ignored and execution will "fall through" and line number 41 will be executed.

b.
```
50   IF B<0 THEN LET B = 0
55   S2 = S1 + 3*B
```
If B<0 is "true" then the value of B will be changed to zero and line 55 will be executed. If B is zero or positive (B<0 is "false") the value of B will be unchanged and execution will pass immediately to line 55.

c.
```
60   IF Z2 >= 10 THEN PRINT Z2+5
63   END
```
If the current value for Z2 is greater than or equal to 10 (Z2 >= 10 is "true") then the PRINT Z2+5 will be executed and control will pass to the end statement. If Z2 is less than 10 (Z2 >= 10 is "false") the PRINT statement will be ignored and control will immediately "fall through" to the END statement.

Note that there are no line numbers on the statements following the THEN and that there are no commas used between the logical expression and the THEN. □

EXAMPLE 2 Are the following logical expressions valid based on Table 18.4? Explain.

a. X + Y = 0 **b.** A2 = 0 **c.** Z4 < −5
d. Z4 => −5 **e.** A2*B >=4 − X

Solution:

a. X + Y = 0	Valid: X + Y is an arithmetic expression and this logical expression is not an assignment statement
b. A2 = 0	Valid
c. Z4 < −5	Valid
d. Z4 => −5	Invalid: operator should be >= (however will operate correctly on some computers)
e. A2*B >= 4 − X	Valid □

EXAMPLE 3 Write a program segment (a portion of a program) to find the smallest of three numbers that have been stored as G1, G2, and G3. Store the smallest value as L.

Solution:

```
30  LET L = G1
40  IF G2 < L THEN LET L = G2
50  IF G3 < L THEN LET L = G3
```

In each case L was assigned a starting value. The starting value could have been any one of the numbers in the list to be searched so the value for G1 was chosen arbitrarily. Then each number in the list was compared to the current value for L. If it was less than the current value for the smallest number, it was then stored as the current smallest value. If the number being compared was greater than or equal to the current value of L, that current value remained unchanged. When the computer has checked all of the numbers in the group, the value stored as L is the same as the smallest number. The values of each of the numbers G1, G2, and G3 remain unchanged in memory. □

The IF statement as presented causes the computer to alter its program sequence by one step. The GO TO statement is used to alter program execution by several steps in either a forward or a backward direction by sending execution to the line whose number follows GO TO.

DEFINITION 18.5

GO TO STATEMENT

line number GO TO line number

EXAMPLE 4 Interpret the actions of the GO TO statements in the following program segment.

```
10  . . . . .
20  . . . . .
30  GO TO 50
40  . . . . .
50  . . . . .
60  GO TO 20
70  . . . . .
```

Solution: The computer executes statement 10, then statement 20, then statement 30, then statement 50, then continues with the next statement in line, which is statement number 60, which sends it back to statement 20 and then follows the pattern 20, 30, 50, 60, 20, 30, 50, 60, etc. Because of the action of the GO TO statements, statements 40 and 70 will never be executed. By causing the block of statements to be repeated, the GO TO 20 statement puts the program into a **loop.** As this program segment is written, there is nothing in the program that would get the computer out of this loop. The loop operation would continue indefinitely until the program was aborted by the computer operator. □

To gain maximum resequencing power and to provide us with a means to prevent being caught in never-ending loops, the GO TO statement is often used in conjunction with the IF statement.

EXAMPLE 5 Interpret the following program sequence

```
10  LET N = 0
20  LET N = N + 1
30  If N < 10 THEN GO TO 20
40  END
```

Solution: The above segment sets N (in memory) equal successively to 0, 1, 2, . . . , 9, 10, and then instructs the program to stop. It will produce no output. To actually see the values of N (except for zero) a PRINT statement would have to be inserted between lines 20 and 30. An alternate program segment to do the same thing is

```
10  LET N = 0
20  If N = 10 THEN GO TO 50
30  LET N = N + 1
40  GO TO 20
50  END
```

In either case, if the program had been sent back to line 10 instead of line 20, the values of N would have been 0, 1, 0, 1, 0, . . . and the computer would have been caught in a loop. □

The INPUT statement allows greater flexibility in programming by permitting data to be entered while the program is being executed rather than having to be written into the program using assignment statements. That allows the same program to be used without alteration to perform the calculations for many sets of data.

DEFINITION 18.6

INPUT STATEMENT

line number INPUT variable list

When the INPUT statement is executed it causes a question mark to appear on the screen of the CRT. The values for the variables in the list are then typed onto the screen in order and separated by commas. When the ENTER or RETURN key is hit, the numbers are assigned to the corresponding variables. No calculation is done by the INPUT statement.

EXAMPLE 6 Explain the operation of the following INPUT statement. The INPUT statement is assumed to be in a program being executed; it would not appear on the CRT during execution of the program.

a. 10 INPUT X

b. 20 INPUT A,B,C,D

Solution:

a. 10 INPUT X

?36.2

Execution of line 10 would cause a question mark to be printed on the screen; the operator would respond by typing a number and then hitting the ENTER or RETURN key; if the number shown were typed, X would then be assigned the value 36.2

b. 20 INPUT A,B,C,D

?−18.0,6,21,13.817

After the ENTER or RETURN key, −18.0 would be assigned to A, 6 to B, 21 to C, and 13.817 to D

or

?−18.0,6,21,13.817,

An error message would be returned because there should be no comma after the last item on the list

or

? −18.0,6,21,13.817,43,−2	This would assign the values correctly; the additional numbers in the list would be ignored
or	
? −18.0,6	The computer would assign −18.0 to A, 6 to B, and print question marks until it is given enough data to complete the list □

The INPUT statement as given does not show the variables for which it is requesting data. To prevent confusion the PRINT statement is often used just before an INPUT statement to tell the operator what data is being requested. On some computers the two lines can be combined.

EXAMPLE 7 Explain the operation of

a.
```
110  PRINT "ENTER X AND Y"
120  INPUT X,Y
```

b.
```
130  PRINT "ENTER X AND Y"
140  INPUT X,Z
```

c.
```
150  PRINT "ENTER X AND Y"
160  INPUT X
170  INPUT Y
```

Solution:

a.
```
110  PRINT "ENTER X AND Y"
120  INPUT X,Y
```
The message ENTER X AND Y would appear on the screen followed by a question mark indicating that data values for X and Y should be entered.

b.
```
130  PRINT "ENTER X AND Y"
140  INPUT X,Z
```
The message ENTER X AND Y would appear on the screen followed by a question mark; however, the second data value would be assigned to Z, not Y.

c.
```
150  PRINT "ENTER X AND Y"
160  INPUT X
170  INPUT Y
```
The message "ENTER X AND Y" would appear followed by a question mark; if more than one data value was entered at a time, an error message would be produced; the values for X and for Y must be entered separately. □

TABLE 18.5 **BASIC functions**

Function name	Result
ABS ( )	Absolute value
ATN ( )	Arctan (in radians)
COS ( )	Cosine (angle must be in radians)
EXP ( )	$e^{(\ )}$
INT ( )	Integral part
LOG ( )	Natural log
RND ( )	Random numbers starting from ( )
SGN ( )	Algebraic sign (−1, 0, or +1)
SIN ( )	Sine (angle must be in radians)
SQR ( )	Square root
TAN ( )	Tangent (angle must be in radians)

Most versions of BASIC have built into them the ability to perform some of the more commonly encountered mathematical functions. The BASIC function operates on whatever arithmetic expression is enclosed in parentheses following the function name. Some of the more common functions are shown in Table 18.5.

EXAMPLE 8 Write a program to input four grades in order, calculate an average that is the sum of the two largest of the first three grades plus twice the fourth grade all divided by 4, and output the result. Construct the program to automatically reset itself to calculate a new set of grades.

Solution:

```
10  PRINT "ENTER 4 GRADES"
20  INPUT G1,G2,G3,G4
30  LET L = G1
40  IF G2 < L THEN LET L = G2
50  IF G3 < L THEN LET L = G3
60  LET A = (G1+G2+G3-L+2*G4)/4
70  PRINT "THE 4 GRADES WERE"
80  PRINT G1,G2,G3,G4
90  PRINT "THE LOWEST GRADE WAS ";L
100  PRINT "THE AVERAGE IS ";A
110  PRINT "WANT TO GO AGAIN?"
120  PRINT "ENTER 1 IF YES, 0 IF NO"
130  INPUT R
140  IF R = 1 THEN GO TO 10
150  END
```

If we assume a grade would never be negative, an alternate solution might be

```
10   PRINT "ENTER 4 GRADES"
15   PRINT "TO QUIT ENTER -1,0,0,0"
20   INPUT G1,G2,G3,G4
25   IF G1 < 0 THEN GO TO 120
30   ________
40   ________
.
.
.
90   ________
100  ________
110  GO TO 10
120  END
```

□

EXAMPLE 9 Write a program to find the real roots of $ax^2 + bx + c$ using the quadratic formula.

Solution:

```
10   PRINT "ENTER A,B,C"
20   INPUT A,B,C
30   IF B ∧ 2-4*A*C < 0 THEN GO TO 100
40   LET X1 = (-B+SQR(B ∧ 2-4*A*C))/(2*A)
50   LET X2 = (-B-SQR(B ∧ 2-4*A*C))/(2*A)
60   PRINT "THE ROOTS ARE"
70   PRINT "X1 = "; X1
80   PRINT "X2 = "; X2
90   GO TO 110
100  PRINT "NO REAL ROOTS"
110  PRINT "WANT RERUN?"
120  PRINT "ENTER 1 FOR YES, 0 FOR NO"
130  INPUT R
140  IF R = 1 THEN GO TO 10
150  END
```

The solution checks to make sure the discriminant is not negative before it attempts to take the square root. If it is negative, it jumps to the message that tells it to print "NO REAL ROOTS" and then checks for a rerun request. If the discriminant is zero or positive, it calculates and prints the roots and then skips over the unwanted "NO REAL ROOTS" message to check for the rerun request. □

18.5 Exercises

In Exercises 1–12, state the results of the BASIC program segment.

1.
```
10 LET A = 4
20 If A < 0 THEN LET A = 0
30 B = A/2
```

2.
```
15 LET X = 21.6
20 IF X <= -4 THEN PRINT X
25 END
```

3.
```
10 LET A = -1.8
20 IF A < 0 THEN LET A = 0
30 LET B = A/2
```

4.
```
15 LET X = -7.3
20 IF X <= -4 THEN PRINT X
25 END
```

5.
```
45 LET P = -1
50 LET Q = 2
55 IF P*Q <> P+Q THEN LET Q = 10
60 LET Q = 0
```

6.
```
45 LET X1 = 2.67
50 LET X2 = 1.42
55 IF X1 <> X2^ 2 THEN LET R = 0
60 LET R = 15.8
```

7.
```
110 LET A = 1.5
120 LET B = -4.2
130 IF A > B THEN GO TO 160
140 LET A = B/2
150 GO TO 170
160 LET A = B*2
```

8.
```
71 LET P = -1.73
72 IF P <> 3 THEN GO TO 75
73 PRINT "P = 3"
74 GO TO 76
75 PRINT "P NOT EQUAL 3"
```

9.
```
1 LET M = 0
2 IF M = 100 THEN GO TO 6
3 LET M = M+1
4 PRINT M
5 GO TO 2
6 END
```

10.
```
10 LET N = 100
11 IF N = 0 THEN GO TO 15
12 LET N = N - 1
13 PRINT N
14 GO TO 11
15 END
```

11.
```
20 LET K = 10
21 LET L = (K+4)/2
22 PRINT K,L
23 IF L >= 21 THEN GO TO 26
24 K = K+1
25 GO TO 21
26 PRINT "PROCEED"
```

12.
```
1 LET I = 0
2 LET J = I/10 - 2.2
3 PRINT I, J
4 IF I < - 10 THEN GO TO 7
5 LET I = I - 2
6 GO TO 2
7 PRINT "LAST VALUES"
```

In Exercises 13–18, assume that the INPUT statement shown is part of a program being executed. When the question mark appears on the CRT the data shown is entered. What happens to the program?

13.
```
101 INPUT X
? -16.2,81.0
```

14.
```
50 INPUT R,S,T
?10,20,30,40
```

15.
```
78 INPUT P1,P2,P3
?0.66,0.68
```

16.
```
78 INPUT P1,P2,P3
?0.66,0.67
```

17.
```
4 INPUT X,Y
?15.2,0
```

18.
```
28 INPUT X,Y
?0,15.2
```

In Exercises 19–24, write the BASIC statement to calculate the expressions.

19. $x = \sqrt{y^2 + z^2}$

20. $v = e^{-2t} \cos (100t - \pi/2)$

21. $r = |\cos (x + 20°)|$

22. $a = \sin 17° \cos 60° + \sin 60° \cos 17°$

23. $p = [e^{-t} \cos (317t)][12e^{-t} \cos (317t + 30°)]$

24. $y = 1 - \tan^{-1} \left(\dfrac{q}{p}\right)$

In Exercises 25–32, write BASIC programs for Exercises 5–12 of Section 18.3. Use INPUT statements to assign data values for computation and provide for single or multiple runs of the program. Run the programs using the data given.

25. $t = 11$h 24m 16s, $t = 4$h 30m 0s, $t = 8$h 14m 3s

26. $t = 8.06$, $t = 4.58$

27. Do not use the ABS function. X = 36.1, X = −14.7, X = 0.

28. H = 36., H = 45.

29. $D = \begin{vmatrix} 4 & -6 \\ 0 & -2 \end{vmatrix}$, $D = \begin{vmatrix} -2 & 3 \\ -6 & 1 \end{vmatrix}$

30. $D = \begin{vmatrix} -1 & 4 & 3 \\ 2 & -5 & 1 \\ 3 & -3 & 0 \end{vmatrix}$

31. $4x - 6y = 5, -6x + y = -3$

32. $-x + 4y + 3z = -2,\ 2x - 5y + z = 1,$
$3x - 3y = 4$

18.6 More BASIC

The BASIC commands already discussed provide the programmer with the capability to write programs. We will now discuss some additional commands useful to the BASIC programmer.

The REM statement is a nonexecutable statement that the programmer uses to place messages or reminders into the program.

DEFINITION 18.7

REM STATEMENT

line number REM message

EXAMPLE 1 Discuss the operation of the statements:

a. 25 REM INITIALIZE LOWEST GRADE

b. 30 REM LET L = 16

Solution:

a. 25 REM INITIALIZE LOWEST GRADE — Causes no computation and does not appear on any output device except in program listing; reminds programmer what the next program segment is supposed to accomplish

b. 30 REM LET L = 16 — Has no effect on the execution of the program; does **not** assign 16 to L; serves only as a reminder to programmer when list of program is being read □

The DATA, READ, and RESTORE statements provide another method of assigning values to variables in a program. All the DATA statements anywhere in a program are taken in order at the start of execution of the program. The constants in the lists are placed in order in a single list. The READ statements as they are encountered during execution cause the constants to be assigned to the variables in the order of the list of the READ. The RESTORE statement causes the assignment of constants to variables in following READ statements to start over again from the beginning of the data list.

DEFINITION 18.8

DATA STATEMENT

line number DATA constant list

DEFINITION 18.9

READ STATEMENT

line number READ variable list

DEFINITION 18.10

RESTORE STATEMENT

line number RESTORE

EXAMPLE 2 What assignments are made by the following statements?

```
10  DATA 2.1, 4.7, 8.3, -14.0
 .
 .
 .
40  READ A,B,C
 .
 .
 .
50  READ X,Y
 .
 .
 .
100  DATA 18.2, -33.7, -22.1
 .
 .
180  RESTORE
 .
 .
190  READ E6, S, R9, T
 .
 .
200  READ L, M, N, P
 .
 .
280  DATA 33.3
```

Solution: At the beginning of execution a list would be formed in memory of all the constants contained in all the DATA statements taken in order anywhere they appear in the program. In this case the list would be

2.1, 4.7, 8.3, −14.0, 18.2, −33.7, −22.1, 33.3.

The first READ statement would cause 2.1 to be assigned to A, 4.7 to B, and 8.3 to C. The next READ would cause −14.0 to be assigned to X and 18.2 to Y. The RESTORE statement would cause any following READs to start assignment from the beginning of the DATA list: 2.1 is assigned to E6, 4.7 to S, 8.3 to R9, −14.0 to T, 18.2 to L, −33.7 to M, −22.1 to N, and 33.3 to P. □

The FOR and NEXT statements are used together to perform counting operations. The same results could be obtained using IF, GO TO, and assignment statements, but often the FOR/NEXT combination is more convenient to write.

DEFINITION 18.11

FOR STATEMENT

line number	FOR	variable	=	arithmetic expression	TO	arithmetic expression	STEP	arithmetic expression

DEFINITION 18.12

NEXT STATEMENT

line number	NEXT	variable

EXAMPLE 3 What is the result of the following set of statements?

```
10 FOR I = 1 TO 18 STEP 1
20 PRINT I
30 NEXT I
```

Solution: Eighteen lines of output would be printed as follows:

```
1
2
 .
 .
 .
16
17
18
```
□

If the word STEP and the expression following it are omitted from the FOR statement, it is automatically assumed to mean that the variable should be incremented by one on each pass through the loop formed by the FOR/NEXT combination. The current value of the variable in the FOR statement may be (but does not have to be) used for any calculations in the loop. No attempt should be made to change the value of the variable following FOR except through the action of the NEXT statement. When the indicated loop has been executed the required number of times the execution of the program passes to the line immediately following the NEXT statement.

EXAMPLE 4 What would be the action of the following program segments?

a.
```
80   FOR A = 1 to 10
90   PRINT A
100  NEXT A
```

b
```
45   FOR X = 1 TO -3 STEP -.5
46   PRINT X
47   NEXT X
```

c.
```
62   LET P = -1
64   LET Q = 2
66   LET R = 3
68   FOR B = P*Q TO SQR (5*R+1) STEP Q ∧ 2
70   PRINT B
72   NEXT B
```

d.
```
15   LET L = 5
20   LET M = 7
25   LET N = M TO L
30   PRINT N
35   NEXT N
```

Solution:

a.
```
80   FOR A = 1 TO 10
90   PRINT A
100  NEXT A
```

Would print 1, 2, 3, . . . , 9, 10 on succeeding lines; in the absence of the STEP portion of the FOR statement a step of +1 is assumed; program execution will pass to the line after 100 NEXT A and A will become assigned as 11

b.
```
45   FOR X = 1 TO -3 STEP -.5
46   PRINT X
47   NEXT X
```

Would print 1, .5, 0, −.5, −1, −1.5, −2, −2.5, −3 on succeeding lines; program execution will pass to the line following the NEXT statement and X will become assigned as −3.5

c.

```
62  LET P = -1
64  LET Q = 2
66  LET R = 3

68  FOR B = P*Q TO SQR (5*R+1) STEP Q ∧ 2

70  PRINT B
72  NEXT B
```

Would operate the same as the statement 68 FOR B = −2 TO 4 STEP 4 and would print −2, 2 on succeeding lines; a value greater than that to the right of TO will not be assigned to the variable; program execution will pass to the line following the NEXT statement and B will become assigned as 6

d.

```
15  LET L = 5
20  LET M = 7
25  FOR N = M TO L
30  PRINT N
35  NEXT N
```

Would print 7; the FOR statement is essentially: 25 FOR N = 7 TO 5; the value of the variable is not tested for size until the NEXT statement is reached; when program execution passes to the line following the NEXT statement, N will become assigned as 8 □

FOR/NEXT loops can be **nested** or placed inside of each other as long as they do not overlap.

```
┌─10 FOR X = 1 TO 5
│┌20 FOR Y = 1 TO 3
││ 30 PRINT X; Y
│└40 NEXT Y
└─50 NEXT X
```

This is a valid example for FOR/NEXT loop

```
┌─10 FOR X = 1 TO 5
│┌20 FOR Y = 1 TO 3
││ 30 PRINT X; Y
└┼40 NEXT X
 └50 NEXT Y
```

This is not a valid example of FOR/NEXT loop, the loops overlap

EXAMPLE 5 What would be the output of the following nested loops?

a.

```
1 For X = 1 TO 3
2 FOR Y = 1 TO 4
3 FOR Z = 1 TO 2
4 PRINT X; Y; Z
5 NEXT Z
6 NEXT Y
7 NEXT X
```

b.

```
1 FOR X = 1 TO 3
2 FOR Y = 1 TO 4
3 PRINT X; Y
4 NEXT Y
5 FOR Z = 1 TO 2
6 PRINT X; Z
7 NEXT Z
8 NEXT X
```

Solution:

a.

```
1  1  1
1  1  2
1  2  1
1  2  2
1  3  1
1  3  2
1  4  1
1  4  2
2  1  1
2  1  2
2  2  1
2  2  2
2  3  1
   .
   .
   .
3  3  2
3  4  1
3  4  2
```

b.

```
1  1
1  2
1  3
1  4
1  1
1  2
2  1
2  2
2  3
2  4
2  1
2  2
3  1
3  2
3  3
3  4
3  1
3  2
```

□

EXAMPLE 6 Rewrite the grade-average program to run the data for four sets of grades without having to enter the sets for each run.

Solution:

```
10  REM SUPPLY NUMBER OF RUNS
20  REM SUPPLY SETS OF GRADES
30  DATA 4
40  DATA 75,78,91,87
50  DATA 50,65,73,77
60  DATA 100,100,100,60
70  DATA 81,76,85,80
80  REM N IS NUMBER OF RUNS
90  READ N
100   FOR I = 1 TO N
110   REM G1,G2,G3,G4 ARE GRADES
120   READ G1,G2,G3,G4
130   REM L IS LOWEST GRADE
140   LET L = G1
150   IF G2 < L THEN LET L = G2
160   IF G3 < L THEN LET L = G3
170   REM A IS AVERAGE
180   LET A = (G1+G2+G3−L+2*G4)/4
190   PRINT "AVERAGE OF SET ";I ;"IS ";A
200   NEXT I
210   END
```
□

18.6 Exercises

In Exercises 1 – 8, what values would be assigned to the variables?

1.
```
120  DATA 1,2,3
180  READ A
300  READ C,B
```

2.
```
60  READ A,B,C
70  DATA 66,77
80  READ D,E
90  DATA 11,88,11
```

3.
```
5  READ X,Y,Z
6  DATA 1,2,3
7  DATA 11,12,13
8  READ X,Y,Z
```

4.
```
18  DATA 1,2,3,4,5,6
19  READ A,B
20  READ X,Y
```

5.
```
21  DATA 78,62,14
23  READ X,Y
25  RESTORE
27  READ A,B,C
```

6.
```
45  READ X
46  READ Y
47  RESTORE
48  READ Y,X
49  DATA 17,−28,−14,3
```

7.
```
18  DATA 1,2,3,4,5,6
19  READ A,B
20  READ X,Y,A,B
```

8.
```
120  READ A
180  READ B,C
300  DATA 17,21,25
310  RESTORE
320  READ X,Y
```

In Exercises 9 – 14, what output would be displayed?

9.
```
30  LET X = 1.
31  FOR I = 1 TO 10 STEP 3
32  PRINT I,X
33  NEXT I
```

10.
```
40  LET A = 12.
41  FOR B = 9 TO −1 STEP −1
42  PRINT A*B
43  NEXT B
```

11.
```
50  FOR X = 1 TO 10 STEP 3
60  FOR Y = 0 TO −4
```

```
   70  PRINT X,Y
   80  NEXT Y
   90  NEXT X
   100  PRINT X,Y
```

12.
```
   60  FOR A = 1 TO 10 STEP 2
   62  FOR B = 2 TO 4 STEP 3
   64  PRINT A,B
   66  NEXT B
   68  NEXT A
   70  PRINT A,B
```

13.
```
   60  LET P = 2.
   70  LET Q = 5.
   80  FOR X = 4 TO P STEP 2
   90  PRINT P,Q
   100  NEXT X
   110  PRINT P,Q
```

14.
```
   80  FOR R = 1 TO 5
   90  FOR S = R TO 5
   100  PRINT R,S
   110  NEXT S
   120  PRINT R,S
   130  FOR T = S TO 5
   140  PRINT R,S,T
   150  NEXT T
   160  NEXT R
   170  PRINT R,S,T
```

In Exercises 15–22, rewrite the programs for Exercises 25–32, Section 18.5. Use REM statements to identify all variables. Use FOR/NEXT where possible. Use READ/DATA/RESTORE statements for data assignments.

Chapter summary

Chapter 18 presented computer programming using the BASIC language. A **computer program** is a sequential set of commands that tell the computer the operations it must perform in order to produce a desired output. It is often desirable to outline the program prior to writing the actual commands by producing a **flowchart** that shows the path the program must take from start to completion.

Table 18.6 lists the BASIC commands presented in this chapter (ln = line number, v = variable, e = expression, el = expression list, le = logical expression, es = executable statement, vl = variable list, m = message, cl = constant list, ae = arithmetic expression).

TABLE 18.6 **BASIC commands**

Assignment Statement:	ln LET v = e
PRINT Statement	ln PRINT el
END Statement	ln END
IF Statement	ln IF le THEN es
GO TO Statement	ln GO TO ln
INPUT Statement	ln INPUT vl
REM Statement	ln REM m
DATA Statement	ln DATA cl
READ Statement	ln READ vl
RESTORE Statement	ln RESTORE
FOR Statement	ln FOR v = ae TO ae STEP ae
NEXT Statement	ln NEXT v

The arithmetic operators used in BASIC are shown in Table 18.7, the logical operators are shown in Table 18.8, and the functions are shown in Table 18.9.

TABLE 18.7 **Arithmetic operations**

OPERATORS

Symbol	Meaning	Priority
** or ↑ or ∧	Exponentiation	First
*	Multiplication	Second in order encountered
/	Division	Second in order encountered
+	Addition	Third in order encountered
−	Subtraction	Third in order encountered

RULES

Operators may not appear next to each other but may be separated by parentheses.
Expressions are evaluated from the innermost parentheses out using the priorities of operation within each set of parentheses.

TABLE 18.9 **BASIC functions**

Function name	Result
ABS ( )	Absolute value
ATN ( )	Arctan (in radians)
COS ( )	Cosine (angle must be in radians)
EXP ( )	$e^{(\ )}$
INT ( )	Integral part
LOG ( )	Natural log
RND ( )	Random numbers starting from ( )
SGN ( )	Algebraic sign (−1, 0 or +1)
SIN ( )	Sine (angle must be in radians)
SQR ( )	Square root
TAN ( )	Tangent (angle must be in radians)

TABLE 18.8 **Logical operators**

Symbol	Meaning
=	Equal
<> (≠)	Not equal
>	Greater than
<	Less than
>= (≥)	Greater than or equal
<= (≤)	Less than or equal

Review Exercises

In Exercises 1–14, determine whether or not the following BASIC statements are valid or invalid. If the statement is invalid state why. If the statement is valid, state the result assuming A = 4, B = −5, and C = 7.

1. 100 LET X = 6A

2. 100 LET X = 6(13+7)

3. LET X = A−B

4. 10 LET X = 1

5. 100 PRINT AB

6. 95 INPUT A

7. 35 PRINT "TODAY IS A GOOD DAY"

8. 18 LET K + 1+A/(B−2)

9. 19 LET L = (1+A)/B−2

10. 20 IF A > B X = 20

11. 25 GO TO A

12. 33 IF Y = A THEN Y = B

13. 42 IF C <> B THEN C = B

14. 55 IF C < B THEN C = B

In Exercises 15–38, write BASIC programs to calculate the problems. If the problem is to be run on the computer, the necessary data will be supplied by your instructor.

15. Write a program to evaluate a 2-X-2 determinant. If the value of the determinant is zero, output the message THE DETERMINANT IS ZERO. Otherwise output D = value of determinant.

16. Write a program to find whether the polynomial $Ax^5 + Bx^4 + Cx^3 + Dx^2 + E$ crosses the x-axis between $x = P$ and $x = Q$. (Assume that if it crosses in that interval, it does so only once.)

17. Write a program to calculate and print $(-1)^n$ for $-8 \leq n \leq 16$. What general rule can you infer from the output of this program?

18. Write a program to calculate and print the integral part of $\frac{n}{2}$ for $-8 \leq n \leq 16$. What general rule can you infer from the output of this program?

19. The length of the diagonal of a rectangular solid is given by $d = \sqrt{a^2 + b^2 + c^2}$. Calculate the length of the diagonal and the combined area of the sides of the solid.

20. The equivalent resistance for a set of n parallel resistors is given by

$$R_{eq} = \frac{1}{\dfrac{1}{R_1} + \dfrac{1}{R_2} + \dfrac{1}{R_3} + \cdots + \dfrac{1}{R_n}}.$$

Input a value for n and the resistance values and calculate the equivalent resistance.

21. Write a program to evaluate $f(x) = x^3 - 4x^2 + 5$ for $-9 \leq x \leq 2$ in increments of 0.5.

22. A manufacturer measures the volume of vinyl needed to produce 144 floor tiles. Produce a table that shows the amount of material needed to produce any number of tiles from 500 to 10,000 in groups of one gross.

23. Show the results of Exercise 21 to the nearest integer.

24. The resultant of two forces acting on a rigid body at the same point is $R = \sqrt{F_1^2 + F_2^2 + 2F_1F_2 \cos \alpha}$, where α is the angle between the forces. The angle α_1 between F_1 and R is $\alpha_1 = \sin^{-1}\left(\dfrac{F_2 \sin \alpha}{R}\right)$. Given F_1, F_2, and α, calculate R and α_1.

25. The location of the centroid of the segment of a circle of radius r and angle 2α is given by

$$x_0 = \frac{\dfrac{2}{3} r \sin^3 \alpha}{\alpha - \cos \alpha \sin \alpha},$$

where α is in radians (Fig. 18.8). Calculate x_0, given r and α.

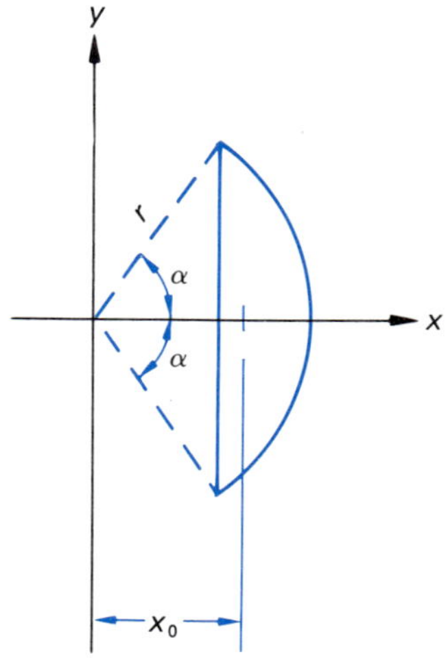

FIGURE 18.8

26. The frictional loss in a thrust bearing due to oil viscosity is given by

$$L = \frac{z(D^4 - d^4)}{(67 \times 10^7)h}$$

in lb. Calculate L when D is 16 in., d is 10 in., h is 0.006 in., and z is 30 centipoises.

27. The slope of an open water-carrying flume is given by

$$S = \frac{(0.013)^2\, V^2}{2.208\, m^{4/3}}$$

Calculate the slope for values of V of 8.3, 4.7, 3.0, and 2.1 ft/s corresponding to values of m of 1.5, 2, 2.5, and 3 ft.

28. For an ideal gas, the value of n in $pV^n =$ constant can be calculated from

$$n = \frac{\log p_1 - \log p_2}{\log V_2 - \log V_1},$$

when two points (p_1, V_1) and (p_2, V_2) are known. Write a program to calculate n.

29. The Brinell hardness number of steel is given by

$$BHN = \frac{P}{\frac{\pi D}{2}(D - \sqrt{D^2 - d^2})},$$

where P is the load in kilograms, D the diameter of the indentor in millimeters, and d the diameter of the impression in millimeters. Write a program to calculate BHN.

30. In windmill design the wind velocity V at a height H can be estimated from

$$V = V_0\left(\frac{H}{H_0}\right)^n,$$

where V_0 is a known velocity at height H_0 and $n = \frac{1}{2}$ for wind under 8 km/h, $n = \frac{1}{5}$ for wind between 8 and 56 km/h, and $n = \frac{1}{7}$ for wind above 56 km/h. Calculate V for a range of heights and wind velocities.

31. A manufacturer of tennis balls finds in a standardized test that the balls, when dropped, rebound to a height h. Given n different values of h as the test data on a production run of balls, calculate the average height of rebound.

32. Find the mean, mode, and standard deviation for the rebound height of the tennis balls in Exercise 17.

33. Write a program to calculate $n!$. $(n! = n(n-1)(n-2)\ldots(3)(2)(1))$

34. Write a program to calculate $n!$ given a value for $m!$, $m < n$. (For $m < n$, $n! = n(n-1)\ldots(m+1)m!$)

35. $\sin x = x - \frac{x^3}{3!} + \frac{x^5}{5!} - \frac{x^7}{7!} + \cdots$ Calculate $\sin x$, x in radians, by calculating each term separately and adding to the sum of the previous terms. Stop after 5 terms. (See Exercises 17, 33, and 34.)

36. $\cos x = 1 - \frac{x^2}{2!} + \frac{x^4}{4!} - \frac{x^6}{6!} + \cdots$ Calculate $\cos x$, x in radians, by calculating each term separately and adding to the sum of the previous terms. Stop after the absolute value of the latest term is less than 1×10^{-5}. (See Exercises 17, 33, and 34.)

37. $e^x = 1 + \frac{x}{1!} + \frac{x^2}{2!} + \frac{x^3}{3!} + \cdots$ Calculate e^x using five terms and compare the value obtained with that given by the built-in function.

38. Simpson's Rule is often used to calculate the area bounded by a curve and the independent axis, as shown in Fig. 18.9. If the interval between $x = a$ and $x = b$ is divided into n equal divisions, $d = \frac{b-a}{n}$, (n an even number), as shown in Fig. 18.10, the area is given by Simpson's Rule as

$$A = \frac{d}{3}(y_0 + 4y_1 + 2y_2 + 4y_3 + 2y_4 + \cdots + 2y_{n-2} + 4y_{n-1} + y_n).$$

Write a BASIC program to calculate areas using Simpson's Rule.

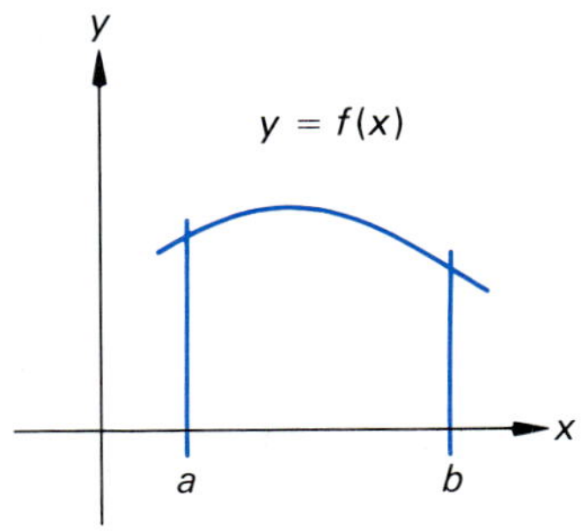

FIGURE 18.9

FIGURE 18.10

Can you solve these?

1. Given $\sin \theta = x$, write a BASIC program to calculate θ in degrees.
2. Write a BASIC program to identify the prime numbers between 1 and 10,000 inclusive. A prime number is an integer that is evenly divisible by itself and one only.
3. If your computer has the capability to process matrices, write a BASIC program to solve the set of simultaneous equations $[A]\,[X] = [C]$ by using the relation $[X] = [A^{-1}]\,[C]$.
4. Write a BASIC program to input a set of grades, calculate their average, and count the number of grades below the average.
5. The method of least squares is used to find the equation of the straight line that best fits a set of experimental data. If there are n data points (x, y), two equations are found:

$$nb + (\Sigma x)m = \Sigma y$$

and

$$(\Sigma x)b + (\Sigma x^2)m = \Sigma xy,$$

which may then be solved for the slope m and the y-intercept b by conventional methods. Write a BASIC program to find the values for m and b, given a set of n data points (x, y).

Appendix A Measurement

The Système International d'Unités (International System of Units or SI) has been adopted as the internationally recognized system of measurement. Currently the United States is the only industrial country not officially on the SI. Much of science and technology is in the process of converting to this system although some segments are progressing faster than others.

There are seven base units in the SI, which are dimensionally independent (Table A.1). Supplementary units are the plane angle (radian, rad) and the solid angle (steradian, sr). Other acceptable units are for time (day, d; hour, h; minute, min), mass (gram, g; metric ton, t), angle (degrees, °), pressure (bar, bar) and volume (liter, ℓ). All other units are derived units. Some of the more common derived units with special names and symbols are listed in Table A.2.

The major difference between the SI and the English system as used in the United States is that in SI mass (kg) is a base unit and force (N) is a derived unit. In the English system force (lb) is a base unit and mass (slug) is a derived unit. Often, an object is incorrectly said to have a weight given in kilograms. A weight, however, is the force of gravitational attraction exerted on the object and because it is a force, it should be properly expressed in newtons when using SI units. Weight or mass is understood by comparison to another known object so this does not cause any problems. When weight or mass is to be used in an equation care must be taken to use proper units.

The relationship between the weight of an object w and its mass m is given by $w = mg$, where g is the local gravitational acceleration. Near the surface of the earth the standard local gravitational acceleration (to four significant digits) in SI units is taken to be 9.807 m/s^2, and in English units is 32.17 ft/s^2.

TABLE A.1

Quantity	Base unit	Abbreviation
Length	meter	m
Mass	kilogram	kg
Time	second	s
Electric current	ampere	A
Thermodynamic temperature	kelvin	K
Amount of substance	mole	mol
Luminous intensity	candela	cd

TABLE A.2 **Derived units**

Quantity	Unit	Symbol	Dimension
Capacitance	farad	F	$A \cdot s \cdot V^{-1}$
Change	coulomb	C	$A \cdot s$
Energy	joule	J	$N \cdot m$
Force	newton	N	$kg \cdot m \cdot s^{-2}$
Frequency	hertz	Hz	s^{-1}
Inductance	henry	H	$V \cdot s \cdot A^{-1}$
Potential difference	volt	V	$J \cdot C^{-1}$
Power	watt	W	$J \cdot s^{-1}$
Pressure	pascal	Pa	$N \cdot m^{-2}$
Resistance	ohm	Ω	$V \cdot A^{-1}$
Stress	pascal	Pa	$N \cdot m^{-2}$
Temperature (common)	celsius	C	$K - 273.2°$
Work	joule	J	$N \cdot m$

EXAMPLE 1

a. A body is said to weigh 10 kg. Is this correct?

b. A body weighs 10 lb. What is its mass in the English system?

c. A body weighs 10 lb. What is its weight in SI?

d. A body weighs 10 lb. What is its mass in SI?

Solution:

a. It is not correct to say an object weighs 10 kg (even though it is often done). The kilogram is a mass unit and in SI the weight, which is a force, is properly expressed in newtons. Recalling $w = mg$ and assuming that the body is near the surface of the earth we have

$$w = 10(9.807) = 98.1 \text{ N}.$$

Properly the object has a mass of 10 kg and a weight of 98.1 N. (The reason that the units used in the equation $w = mg$ do not appear to be consistent is because g is used in two different ways. It represents both the local acceleration of an object in free-fall and the local weight/mass ratio.)

b. In the English system the mass of a body that weighs 10 lb near the surface of the earth is found using

$$w = mg$$
$$10 = m(32.17)$$
$$m = \frac{10}{32.17} \quad \text{or} \quad 0.311 \text{ slug}.$$

c. Given the weight of the body in English units we can find its weight in SI units by using the conversion factor 1 lb = 4.448 N (see Table

A.4). Since we are converting weight to weight and not weight to mass this conversion is valid anywhere in the universe.

$$(10 \text{ lb})\left(\frac{4.448 \text{ N}}{1 \text{ lb}}\right) = 44.5 \text{ N}$$

d. From (c) above we know that a weight of 10 lb is equivalent to 44.48 N (not rounded). Recalling $w = mg$ and assuming that the body is near the surface of the earth so that we can use $g = 9.807$ in SI units we have

$$w = mg$$
$$44.48 = m(9.807)$$
$$m = \frac{44.48}{9.807} \quad \text{or} \quad 4.54 \text{ kg}.$$

A body weighing 10 lb near the surface of the earth has a mass of 4.54 kg. □

In order to express quantities in numbers of reasonable size, a system of prefixes has been standardized so that the quantities can usually be expressed in terms of numbers between 10^{-1} and 10^{3} multiplied by a factor that may be specified by a prefix (Table A.3). When used in lists it is good practice to adjust the numbers so that they all use the same prefix. (Distances of 1.86 km, 2.41 km, and 837 m should be listed either as 1.86 km, 2.41 km, and 0.837 km, or as 1860 m, 2410 m, and 837 m.) When units are spoken the prefix is accented. (Kilometer is pronounced with the accent on the first syllable.) In general the prefixes exa-, peta-, hecto-, deka-, deci-, and centi- are to be avoided. Hecto-, deka-, deci-, and centi- are sometimes acceptable when dealing

TABLE A.3 **SI prefixes**

Factors	Prefix	Symbol	Pronunciation
1 000 000 000 000 000 000 = 10^{18}	exa	E	ĕx′ȧ
1 000 000 000 000 000 = 10^{15}	peta	P	pĕt′ȧ
1 000 000 000 000 = 10^{12}	tera	T	tĕr′ȧ
1 000 000 000 = 10^{9}	giga	G	jĭ′gȧ
1 000 000 = 10^{6}	mega	M	mĕg′ȧ
1 000 = 10^{3}	kilo	k	kĭl′ō
100 = 10^{2}	hecto	h	hĕk′tō
10 = 10^{1}	deka	da	dĕk′ȧ
1 = 10^{0}			
0.1 = 10^{-1}	deci	d	dĕs′ĭ
0.01 = 10^{-2}	centi	c	sĕn′tĭ
0.001 = 10^{-3}	milli	m	mĭl′ĭ
0.000 001 = 10^{-6}	micro	μ	mī′krō
0.000 000 001 = 10^{-9}	nano	n	năn′ō
0.000 000 000 001 = 10^{-12}	pico	p	pē′cō
0.000 000 000 000 001 = 10^{-15}	femto	f	fĕm′tō
0.000 000 000 000 000 001 = 10^{-18}	atto	a	ăt′tō

with area or volume and centimeter is used when expressing human body dimensions or clothing sizes. More than one prefix should not be used with any dimension. (Don't use megamillimeters.)

EXAMPLE 2 Convert

a. 14.3 meters to millimeters,

b. 47.0 μ farads to farads,

c. 378. cm^3 to m^3.

Solution:

a. From Table A.3 the factor corresponding to the prefix milli- is 10^{-3}. Therefore $1 \text{ mm} = 1 \times 10^{-3}$ m. Using this equality we can form two conversion factors by dividing both sides of the equality by either 1 mm or by dividing both sides of the equality by 1 m. Using the abbreviations for the units we have

$$1 \text{ mm} = 10^{-3} \text{ m} \qquad\qquad 1 \text{ mm} = 10^{-3} \text{ m}$$

$$\frac{1 \text{ mm}}{1 \text{ mm}} = \frac{10^{-3} \text{ m}}{1 \text{ mm}} \qquad\qquad \frac{1 \text{ mm}}{10^{-3} \text{ m}} = \frac{10^{-3} \text{ m}}{10^{-3} \text{ m}}$$

$$1 = \frac{10^{-3} \text{ m}}{1 \text{ mm}} \qquad\qquad \frac{1 \text{ mm}}{10^{-3} \text{ m}} = 1$$

Both conversion factors are equal to 1, which is the multiplicative identity. Therefore, if we multiply any quantity by either one of the conversion factors, we will not change its value because we are effectively multiplying it by 1. In the given instance if we multiply 14.3 m by the conversion factor on the left we obtain

$$(14.3 \text{ m}) \frac{10^{-3} \text{ m}}{1 \text{ mm}} = 14.3 \times 10^{-3} \frac{\text{m}^2}{\text{mm}}.$$

This result is mathematically correct but the units of the result are not those we desired. If we multiply 14.3 m by the conversion factor on the right we obtain

$$(14.3 \not{\text{m}}) \frac{1 \text{ mm}}{10^{-3} \not{\text{m}}} = \frac{14.3}{10^{-3}} \text{ mm}$$

$$= 14.3 \times 10^3 \text{ mm}. \qquad \textbf{Properties of exponents}$$

Therefore 14.3 m is equivalent to 14.3×10^3 mm or 14,300 mm. We must choose which of the conversion factors to use by checking to see which of them will give us the result in the desired units.

b. To convert 47.0 μ farads to farads we see from Table A.3 that μ is the symbol for the prefix micro-, which has a factor of 10^{-6} so 1 μ farad is 1 microfarad = 10^{-6} farad and we have conversion factors of

$$\frac{1 \, \mu \text{ F}}{10^{-6} \text{ F}} \quad \text{or} \quad \frac{10^{-6} \text{ F}}{1 \, \mu \text{ F}}.$$

By observation we see that in order to end up in units of farads we must use the second form of the conversion factor. Therefore

$$(47.0\ \mu\text{F})\left(\frac{10^{-6}\ \text{F}}{1\ \mu\text{F}}\right) = 47.0 \times 10^{-6}\ \text{F}$$
$$= 0.000047\ \text{F}.$$

c. To convert 378. cm^3 to m^3 we see from Table A.3 that 1 cm = 10^{-2} m, so our available conversion factors are

$$\frac{1\ \text{cm}}{10^{-2}\ \text{m}} \quad \text{or} \quad \frac{10^{-2}\ \text{m}}{1\ \text{cm}}.$$

We see that multiplication by the second conversion factor will not cause us to obtain units of meters but rather will give

$$(378.\ \text{cm}^3)\left(\frac{10^{-2}\ \text{m}}{1\ \text{cm}}\right) = 378. \times 10^{-2}\ \text{cm}^2\ \text{m}$$

However, if we multiply by $\left(\frac{10^{-2}\ \text{m}}{1\ \text{cm}}\right)^3$ we will obtain the desired result.

$$(378.\ \text{cm}^3)\left(\frac{10^{-2}\ \text{m}}{1\ \text{cm}}\right)^3 = (378.\ \cancel{\text{cm}^3})\left(\frac{10^{-6}\ \text{m}^3}{1\ \cancel{\text{cm}^3}}\right)$$
$$= 378. \times 10^{-6}\ \text{m}^3. \quad \square$$

Because of differences in the ways commas and decimal points are used with numbers in various parts of the world, the English practice of using commas to group numbers is discouraged and a blank space is to be used for grouping. (12 568. is preferred to 12,568.) Except in tables, the space may be omitted in four digit numbers. (1234. instead of 1 234.; 0.1234 instead of 0.123 4.)

The names of the units are not capitalized, even if they are derived from the name of a person. The abbreviations for the units are not capitalized unless they are derived from the name of a person in which case they are capitalized. Plurals may be used when the name of a unit is spelled out but they are not used with the abbreviations. Prefixes are spelled with the unit as one word. Periods are not used with the abbreviations of the units. Avoid using units with prefixes (except kilogram) in a denominator. Leave a space between the quantity referred to and the unit symbol. (Write microfarad, not micro farad. Write mm not mm.. Write 41.6 kg, not 41.6kg.)

Because of the two systems of units presently in use in the United States it is sometimes necessary to convert from one system to the other. The conversion factors given in Table A.4 will aid in converting between measurement systems. In addition the miscellaneous constants and factors in Table A.5 may occasionally be useful.

TABLE A.4 **Table of conversion factors**

English to SI	Quantity converted	SI to English
1 $ft^2 = 9.290 \times 10^{-2}$ m^2 1 $in^2 = 6.452$ cm^2 1 $yd^2 = 0.8361$ m^2	Area	1 $m^2 = 10.76$ ft^2 1 $cm^2 = 0.1550$ in^2 1 $m^2 = 1.196$ yd^2
1 $lb/ft^3 = 157.1$ N/m^3	Density	1 $N/m^3 = 6.366 \times 10^{-3}$ lb/ft^3
1 ft lb = 1.356 J 1 Btu = 1.055×10^3 J	Energy or Work	1 J = 0.7376 ft lb 1 J = 9.480×10^{-4} Btu
1 lb = 4.448 N	Force or Weight	1 N = 0.2248 lb
1 ft = 0.3048 m 1 in = 25.40 mm (exact) 1 yd = 0.9144 m 1 mi = 1.609 km	Length	1 m = 3.281 ft 1 mm = 3.937×10^{-2} in. 1 m = 1.094 yd 1 km = 0.6214 mi
1 slug = 14.59 kg	Mass	1 kg = 6.853×10^{-2} slug
1 hp = 745.7 W	Power	1 W = 1.341×10^{-3} hp
1 psi = 6.895×10^3 Pa	Pressure or Stress	1 Pa = 1.450×10^{-4} psi
1 $ft^3 = 2.832 \times 10^{-2}$ m^3 1 $in^3 = 16.39$ cm^3 1 $yd^3 = 0.7646$ m^3 1 gal = 3.785 ℓ	Volume	1 $m^3 = 35.31$ ft^3 1 $cm^3 = 6.102 \times 10^{-2}$ in^3 1 $m^3 = 1.308$ yd^3 1 ℓ = 0.2642 gal

TABLE A.5 **Miscellaneous constants and conversion factors**

Quantity	Constant or conversion factor
Acceleration of gravity (standard)	1 g = 9.807 m/s^2 1 g = 32.17 ft/s^2
Angular measure	1 rad = 57.30°
Heat	1 cal = 4.187 J
Length	1 ft = 12 in. 1 yd = 3 ft 1 mi = 5280 ft
Mass/Weight (at standard acceleration only)	1 N = 0.1020 kg 1 kg = 9.807 N 1 slug = 32.17 lb 1 lb = 0.4536 kg 1 kg = 2.205 lb
Power	1 hp = 550 ft lb/s
Pressure	1 atm = 14.70 psi
Temperature	$°C = \frac{5}{9}(°F - 32)$ $°F = \frac{9}{5}°C + 32$
Volume	1 ℓ = 1000 cm^3 1 gal = 4 qt 1 ft^3 = 7.481 gal

EXAMPLE 3 Perform the conversions:

a. $\frac{1}{2}$ hp to watts, **b.** 6 m^3 to cubic yards,

c. 781 cc to quarts.

Solution:

a. To convert $\frac{1}{2}$ hp to watts we are converting from English measurement to SI measurement. From Table A.4 we find 1 hp = 745.7 W, which (as was shown in the previous example) can be expressed equivalently as $1 = \frac{745.7 \text{ W}}{1 \text{ hp}}$.

$$\frac{1}{2}\text{ hp} = \left(\frac{1 \text{ hp}}{2}\right)\left(\frac{745.7 \text{ W}}{1 \text{ hp}}\right)$$
$$= \frac{745.7 \text{ W}}{2}$$

Therefore $\frac{1}{2}$ hp = 373. W

b. To convert 6 m^3 to cubic yards we are converting from SI measurement to English measurement. From Table A.4 we find 1 m^3 = 1.308 yd^3, which can be expressed equivalently as $1 = \frac{1.308 \text{ yd}^3}{1 \text{ m}^3}$.

$$6 \text{ m}^3 = (6 \text{ m}^3)\left(\frac{1.308 \text{ yd}^3}{1 \text{ m}^3}\right)$$
$$= 6(1.308) \text{ yd}^3$$

Therefore 6 m^3 = 7.85 yd^3

c. To convert from 781 cc to quarts we are converting from SI measurement to English measurement. (cc is an abbreviation for cubic centimeters, which is also abbreviated as cm^3.) From Table A.4 we see that the conversion of cm^3 from SI to English is 1 cm^3 = 6.102×10^{-2} in^3. However, in Table A.5 we see that 1 ft^3 = 7.481 gal, and that 1 gal = 4 qt. We also know that 1 ft = 12 in. Using these we form conversions such that when the original units are multiplied by the conversions we end up with the units we desire.

$$781 \text{ cm}^3 = (781 \text{ cm}^3)\left(\frac{6.102 \times 10^{-2} \text{ in}^3}{1 \text{ cm}^3}\right)$$
$$\times \left(\frac{1 \text{ ft}}{12 \text{ in.}}\right)^3 \left(\frac{7.481 \text{ gal}}{1 \text{ ft}^3}\right)\left(\frac{4 \text{ qt}}{1 \text{ gal}}\right)$$
$$= 781(6.102 \times 10^{-2})\left(\frac{1}{12}\right)^3 (7.481)(4) \text{ qt}$$

Therefore 781 cm^3 = 0.825 qt. □

Appendix B Tables

TABLE B.1 **Powers and roots**

n	n^2	n^3	$\sqrt{n}$	$\sqrt[3]{n}$	n	n^2	n^3	$\sqrt{n}$	$\sqrt[3]{n}$
1	1	1	1.000	1.000	**41**	1,681	68,921	6.403	3.448
2	4	8	1.414	1.260	**42**	1,764	74,088	6.481	3.476
3	9	27	1.732	1.442	**43**	1,849	79,507	6.557	3.503
4	16	64	2.000	1.587	**44**	1,936	85,184	6.633	3.530
5	25	125	2.236	1.710	**45**	2,025	91,125	6.708	3.557
6	36	216	2.449	1.817	**46**	2,116	97,336	6.782	3.583
7	49	343	2.646	1.913	**47**	2,209	103,823	6.856	3.609
8	64	512	2.828	2.000	**48**	2,304	110,592	6.928	3.634
9	81	729	3.000	2.080	**49**	2,401	117,649	7.000	3.659
10	100	1,000	3.162	2.154	**50**	2,500	125,000	7.071	3.684
11	121	1,331	3.317	2.224	**51**	2,601	132,651	7.141	3.708
12	144	1,728	3.464	2.289	**52**	2,704	140,608	7.211	3.733
13	169	2,197	3.606	2.351	**53**	2,809	148,877	7.280	3.756
14	196	2,744	3.742	2.410	**54**	2,916	157,464	7.348	3.780
15	225	3,375	3.873	2.466	**55**	3,025	166,375	7.416	3.803
16	256	4,096	4.000	2.520	**56**	3,136	175,616	7.483	3.826
17	289	4,913	4.123	2.571	**57**	3,249	185,193	7.550	3.849
18	324	5,832	4.243	2.621	**58**	3,364	195,112	7.616	3.871
19	361	6,859	4.359	2.668	**59**	3,481	205,379	7.681	3.893
20	400	8,000	4.472	2.714	**60**	3,600	216,000	7.746	3.915
21	441	9,261	4.583	2.759	**61**	3,721	226,981	7.810	3.936
22	484	10,648	4.690	2.802	**62**	3,844	238,328	7.874	3.958
23	529	12,167	4.796	2.844	**63**	3,969	250,047	7.937	3.979
24	576	13,824	4.899	2.884	**64**	4,096	262,144	8.000	4.000
25	625	15,625	5.000	2.924	**65**	4,225	274,625	8.062	4.021
26	676	17,576	5.099	2.962	**66**	4,356	287,496	8.124	4.041
27	729	19,683	5.196	3.000	**67**	4,489	300,763	8.185	4.062
28	784	21,952	5.292	3.037	**68**	4,624	314,432	8.246	4.082
29	841	24,389	5.385	3.072	**69**	4,761	328,509	8.307	4.102
30	900	27,000	5.477	3.107	**70**	4,900	343,000	8.367	4.121
31	961	29,791	5.568	3.141	**71**	5,041	357,911	8.426	4.141
32	1,024	32,768	5.657	3.175	**72**	5,184	373,248	8.485	4.160
33	1,089	35,937	5.745	3.208	**73**	5,329	389,017	8.544	4.179
34	1,156	39,304	5.831	3.240	**74**	5,476	405,224	8.602	4.198
35	1,225	42,875	5.916	3.271	**75**	5,625	421,875	8.660	4.217
36	1,296	46,656	6.000	3.302	**76**	5,776	438,976	8.718	4.236
37	1,369	50,653	6.083	3.332	**77**	5,929	456,533	8.775	4.254
38	1,444	54,872	6.164	3.362	**78**	6,084	474,552	8.832	4.273
39	1,521	59,319	6.245	3.391	**79**	6,241	493,039	8.888	4.291
40	1,600	64,000	6.325	3.420	**80**	6,400	512,000	8.944	4.309

TABLE B.1 *(Cont.)*

n	n^2	n^3	$\sqrt{n}$	$\sqrt[3]{n}$	n	n^2	n^3	$\sqrt{n}$	$\sqrt[3]{n}$
81	6,561	531,441	9.000	4.327	**91**	8,281	753,571	9.539	4.498
82	6,724	551,368	9.055	4.344	**92**	8,464	778,688	9.592	4.514
83	6,889	571,787	9.110	4.362	**93**	8,649	804,357	9.644	4.531
84	7,056	592,704	9.165	4.380	**94**	8,836	830,584	9.695	4.547
85	7,225	614,125	9.220	4.397	**95**	9,025	857,375	9.747	4.563
86	7,396	636,056	9.274	4.414	**96**	9,216	884,736	9.798	4.579
87	7,569	658,503	9.327	4.431	**97**	9,409	912,673	9.849	4.595
88	7,744	681,472	9.381	4.448	**98**	9,604	941,192	9.899	4.610
89	7,921	704,969	9.434	4.465	**99**	9,801	970,299	9.950	4.626
90	8,100	729,000	9.487	4.481	**100**	10,000	1,000,000	10.000	4.642

TABLE B.2 **Logarithms of numbers**

x	0	1	2	3	4	5	6	7	8	9
1.0	.0000	.0043	.0086	.0128	.0170	.0212	.0253	.0294	.0334	.0374
1.1	.0414	.0453	.0492	.0531	.0569	.0607	.0645	.0682	.0719	.0755
1.2	.0792	.0828	.0864	.0899	.0934	.0969	.1004	.1038	.1072	.1106
1.3	.1139	.1173	.1206	.1239	.1271	.1303	.1335	.1367	.1399	.1430
1.4	.1461	.1492	.1523	.1553	.1584	.1614	.1644	.1673	.1703	.1732
1.5	.1761	.1790	.1818	.1847	.1875	.1903	.1931	.1959	.1987	.2014
1.6	.2041	.2068	.2095	.2122	.2148	.2175	.2201	.2227	.2253	.2279
1.7	.2304	.2330	.2355	.2380	.2405	.2430	.2455	.2480	.2504	.2529
1.8	.2553	.2577	.2601	.2625	.2648	.2672	.2695	.2718	.2742	.2765
1.9	.2788	.2810	.2833	.2856	.2878	.2900	.2923	.2945	.2967	.2989
2.0	.3010	.3032	.3054	.3075	.3096	.3118	.3139	.3160	.3181	.3201
2.1	.3222	.3243	.3263	.3284	.3304	.3324	.3345	.3365	.3385	.3404
2.2	.3424	.3444	.3464	.3483	.3502	.3522	.3541	.3560	.3579	.3598
2.3	.3617	.3636	.3655	.3674	.3692	.3711	.3729	.3747	.3766	.3784
2.4	.3802	.3820	.3838	.3856	.3874	.3892	.3909	.3927	.3945	.3962
2.5	.3979	.3997	.4014	.4031	.4048	.4065	.4082	.4099	.4116	.4133
2.6	.4150	.4166	.4183	.4200	.4216	.4232	.4249	.4265	.4281	.4298
2.7	.4314	.4330	.4346	.4362	.4378	.4393	.4409	.4425	.4440	.4456
2.8	.4472	.4487	.4502	.4518	.4533	.4548	.4564	.4579	.4594	.4609
2.9	.4624	.4639	.4654	.4669	.4683	.4698	.4713	.4728	.4742	.4757
3.0	.4771	.4786	.4800	.4814	.4829	.4843	.4857	.4871	.4886	.4900
3.1	.4914	.4928	.4942	.4955	.4969	.4983	.4997	.5011	.5024	.5038
3.2	.5051	.5065	.5079	.5092	.5105	.5119	.5132	.5145	.5159	.5172
3.3	.5185	.5198	.5211	.5224	.5237	.5250	.5263	.5276	.5289	.5307
3.4	.5315	.5328	.5340	.5353	.5366	.5378	.5391	.5403	.5416	.5428
3.5	.5441	.5453	.5465	.5478	.5490	.5502	.5515	.5527	.5539	.5551
3.6	.5563	.5575	.5587	.5599	.5611	.5623	.5635	.5647	.5658	.5670
3.7	.5682	.5694	.5705	.5717	.5729	.5740	.5752	.5763	.5775	.5786
3.8	.5798	.5809	.5821	.5832	.5843	.5855	.5866	.5877	.5888	.5899
3.9	.5911	.5922	.5933	.5944	.5955	.5966	.5977	.5988	.5999	.6010
4.0	.6021	.6031	.6042	.6053	.6064	.6075	.6085	.6096	.6107	.6117
4.1	.6128	.6138	.6149	.6160	.6170	.6180	.6191	.6201	.6212	.6222
4.2	.6232	.6243	.6253	.6263	.6274	.6284	.6294	.6304	.6314	.6325
4.3	.6335	.6345	.6355	.6365	.6375	.6385	.6395	.6405	.6415	.6425
4.4	.6435	.6444	.6454	.6464	.6474	.6484	.6493	.6503	.6513	.6522
4.5	.6532	.6542	.6551	.6561	.6571	.6580	.6590	.6599	.6609	.6618
4.6	.6628	.6637	.6646	.6656	.6665	.6675	.6684	.6693	.6702	.6712
4.7	.6721	.6730	.6739	.6749	.6758	.6767	.6776	.6785	.6794	.6803
4.8	.6812	.6821	.6830	.6839	.6848	.6857	.6866	.6875	.6884	.6893
4.9	.6902	.6911	.6920	.6928	.6937	.6946	.6955	.6964	.6972	.6981
5.0	.6990	.6998	.7007	.7016	.7024	.7033	.7042	.7050	.7059	.7067
5.1	.7076	.7084	.7093	.7101	.7110	.7118	.7126	.7135	.7143	.7152
5.2	.7160	.7168	.7177	.7185	.7193	.7202	.7210	.7218	.7226	.7235
5.3	.7243	.7251	.7259	.7267	.7275	.7284	.7292	.7300	.7308	.7316
5.4	.7324	.7332	.7340	.7348	.7356	.7364	.7372	.7380	.7388	.7396
x	0	1	2	3	4	5	6	7	8	9

TABLE B.2 (*Cont.*)

x	0	1	2	3	4	5	6	7	8	9
5.5	0.7404	0.7412	0.7419	0.7427	0.7435	0.7443	0.7451	0.7459	0.7466	0.7474
5.6	0.7482	0.7490	0.7497	0.7505	0.7513	0.7520	0.7528	0.7536	0.7543	0.7551
5.7	0.7559	0.7566	0.7574	0.7582	0.7589	0.7597	0.7604	0.7612	0.7619	0.7627
5.8	0.7634	0.7642	0.7649	0.7657	0.7664	0.7672	0.7679	0.7686	0.7694	0.7701
5.9	0.7709	0.7716	0.7723	0.7731	0.7738	0.7745	0.7752	0.7760	0.7767	0.7774
6.0	0.7782	0.7789	0.7796	0.7803	0.7810	0.7818	0.7825	0.7832	0.7839	0.7846
6.1	0.7853	0.7860	0.7868	0.7875	0.7882	0.7889	0.7896	0.7903	0.7910	0.7917
6.2	0.7924	0.7931	0.7938	0.7945	0.7952	0.7959	0.7966	0.7973	0.7980	0.7987
6.3	0.7993	0.8000	0.8007	0.8014	0.8021	0.8028	0.8035	0.8041	0.8048	0.8055
6.4	0.8062	0.8069	0.8075	0.8082	0.8089	0.8096	0.8102	0.8109	0.8116	0.8122
6.5	0.8129	0.8136	0.8142	0.8149	0.8156	0.8162	0.8169	0.8176	0.8182	0.8189
6.6	0.8195	0.8202	0.8209	0.8215	0.8222	0.8228	0.8235	0.8241	0.8248	0.8254
6.7	0.8261	0.8267	0.8274	0.8280	0.8287	0.8293	0.8299	0.8306	0.8312	0.8319
6.8	0.8325	0.8331	0.8338	0.8344	0.8351	0.8357	0.8363	0.8370	0.8376	0.8382
6.9	0.8388	0.8395	0.8401	0.8407	0.8414	0.8420	0.8426	0.8432	0.8439	0.8445
7.0	0.8451	0.8457	0.8463	0.8470	0.8476	0.8482	0.8488	0.8494	0.8500	0.8506
7.1	0.8513	0.8519	0.8525	0.8531	0.8537	0.8543	0.8549	0.8555	0.8561	0.8567
7.2	0.8573	0.8579	0.8585	0.8591	0.8597	0.8603	0.8609	0.8615	0.8621	0.8627
7.3	0.8633	0.8639	0.8645	0.8651	0.8657	0.8663	0.8669	0.8675	0.8681	0.8686
7.4	0.8692	0.8698	0.8704	0.8710	0.8716	0.8722	0.8727	0.8733	0.8739	0.8745
7.5	0.8751	0.8756	0.8762	0.8768	0.8774	0.8779	0.8785	0.8791	0.8797	0.8802
7.6	0.8808	0.8814	0.8820	0.8825	0.8831	0.8837	0.8842	0.8848	0.8854	0.8859
7.7	0.8865	0.8871	0.8876	0.8882	0.8887	0.8893	0.8899	0.8904	0.8910	0.8915
7.8	0.8921	0.8927	0.8932	0.8938	0.8943	0.8949	0.8954	0.8960	0.8965	0.8971
7.9	0.8976	0.8982	0.8987	0.8993	0.8998	0.9004	0.9009	0.9015	0.9020	0.9025
8.0	0.9031	0.9036	0.9042	0.9047	0.9053	0.9058	0.9063	0.9069	0.9074	0.9079
8.1	0.9085	0.9090	0.9096	0.9101	0.9106	0.9112	0.9117	0.9122	0.9128	0.9133
8.2	0.9138	0.9143	0.9149	0.9154	0.9159	0.9165	0.9170	0.9175	0.9180	0.9186
8.3	0.9191	0.9196	0.9201	0.9206	0.9212	0.9217	0.9222	0.9227	0.9232	0.9238
8.4	0.9243	0.9248	0.9253	0.9258	0.9263	0.9269	0.9274	0.9279	0.9284	0.9289
8.5	0.9294	0.9299	0.9304	0.9309	0.9315	0.9320	0.9325	0.9330	0.9335	0.9340
8.6	0.9345	0.9350	0.9355	0.9360	0.9365	0.9370	0.9375	0.9380	0.9385	0.9390
8.7	0.9395	0.9400	0.9405	0.9410	0.9415	0.9420	0.9425	0.9430	0.9435	0.9440
8.8	0.9445	0.9450	0.9455	0.9460	0.9465	0.9469	0.9474	0.9479	0.9484	0.9489
8.9	0.9494	0.9499	0.9504	0.9509	0.9513	0.9518	0.9523	0.9528	0.9533	0.9538
9.0	0.9542	0.9547	0.9552	0.9557	0.9562	0.9566	0.9571	0.9576	0.9581	0.9586
9.1	0.9590	0.9595	0.9600	0.9605	0.9609	0.9614	0.9619	0.9624	0.9628	0.9633
9.2	0.9638	0.9643	0.9647	0.9652	0.9657	0.9661	0.9666	0.9671	0.9675	0.9680
9.3	0.9685	0.9689	0.9694	0.9699	0.9703	0.9708	0.9713	0.9717	0.9722	0.9727
9.4	0.9731	0.9736	0.9741	0.9745	0.9750	0.9754	0.9759	0.9763	0.9768	0.9773
9.5	0.9777	0.9782	0.9786	0.9791	0.9795	0.9800	0.9805	0.9809	0.9814	0.9818
9.6	0.9823	0.9827	0.9832	0.9836	0.9841	0.9845	0.9850	0.9854	0.9859	0.9863
9.7	0.9868	0.9872	0.9877	0.9881	0.9886	0.9890	0.9894	0.9899	0.9903	0.9908
9.8	0.9912	0.9917	0.9921	0.9926	0.9930	0.9934	0.9939	0.9943	0.9948	0.9952
9.9	0.9956	0.9961	0.9965	0.9969	0.9974	0.9978	0.9983	0.9987	0.9991	0.9996
x	0	1	2	3	4	5	6	7	8	9

TABLE B.3 **Trigonometric functions: degrees and radians**

Angle					Angle				
Degree	Radian	Sine	Cosine	Tangent	Degree	Radian	Sine	Cosine	Tangent
0°	0.000	0.000	1.000	0.000					
1°	0.017	0.017	1.000	0.017	46°	0.803	0.719	0.695	1.036
2°	0.035	0.035	0.999	0.035	47°	0.820	0.731	0.682	1.072
3°	0.052	0.052	0.999	0.052	48°	0.838	0.743	0.669	1.111
4°	0.070	0.070	0.998	0.070	49°	0.855	0.755	0.656	1.150
5°	0.087	0.087	0.996	0.087	50°	0.873	0.766	0.643	1.192
6°	0.105	0.105	0.995	0.105	51°	0.890	0.777	0.629	1.235
7°	0.122	0.122	0.993	0.123	52°	0.908	0.788	0.616	1.280
8°	0.140	0.139	0.990	0.141	53°	0.925	0.799	0.602	1.327
9°	0.157	0.156	0.988	0.158	54°	0.942	0.809	0.588	1.376
10°	0.175	0.174	0.985	0.176	55°	0.960	0.819	0.574	1.428
11°	0.192	0.191	0.982	0.194	56°	0.977	0.829	0.559	1.483
12°	0.209	0.208	0.978	0.213	57°	0.995	0.839	0.545	1.540
13°	0.227	0.225	0.974	0.231	58°	1.012	0.848	0.530	1.600
14°	0.244	0.242	0.970	0.249	59°	1.030	0.857	0.515	1.664
15°	0.262	0.259	0.966	0.268	60°	1.047	0.866	0.500	1.732
16°	0.279	0.276	0.961	0.287	61°	1.065	0.875	0.485	1.804
17°	0.297	0.292	0.956	0.306	62°	1.082	0.883	0.469	1.881
18°	0.314	0.309	0.951	0.325	63°	1.100	0.891	0.454	1.963
19°	0.332	0.326	0.946	0.344	64°	1.117	0.899	0.438	2.050
20°	0.349	0.342	0.940	0.364	65°	1.134	0.906	0.423	2.145
21°	0.367	0.358	0.934	0.384	66°	1.152	0.914	0.407	2.246
22°	0.384	0.375	0.927	0.404	67°	1.169	0.921	0.391	2.356
23°	0.401	0.391	0.921	0.424	68°	1.187	0.927	0.375	2.475
24°	0.419	0.407	0.914	0.445	69°	1.204	0.934	0.358	2.605
25°	0.436	0.423	0.906	0.466	70°	1.222	0.940	0.342	2.747
26°	0.454	0.438	0.899	0.488	71°	1.239	0.946	0.326	2.904
27°	0.471	0.454	0.891	0.510	72°	1.257	0.951	0.309	3.078
28°	0.489	0.469	0.883	0.532	73°	1.274	0.956	0.292	3.271
29°	0.506	0.485	0.875	0.554	74°	1.292	0.961	0.276	3.487
30°	0.524	0.500	0.866	0.577	75°	1.309	0.966	0.259	3.732
31°	0.541	0.515	0.857	0.601	76°	1.326	0.970	0.242	4.011
32°	0.559	0.530	0.848	0.625	77°	1.344	0.974	0.225	4.331
33°	0.576	0.545	0.839	0.649	78°	1.361	0.978	0.208	4.705
34°	0.593	0.559	0.829	0.675	79°	1.379	0.982	0.191	5.145
35°	0.611	0.574	0.819	0.700	80°	1.396	0.985	0.174	5.671
36°	0.628	0.588	0.809	0.727	81°	1.414	0.988	0.156	6.314
37°	0.646	0.602	0.799	0.754	82°	1.431	0.990	0.139	7.115
38°	0.663	0.616	0.788	0.781	83°	1.449	0.993	0.122	8.144
39°	0.681	0.629	0.777	0.810	84°	1.466	0.995	0.105	9.514
40°	0.698	0.643	0.766	0.839	85°	1.484	0.996	0.087	11.43
41°	0.716	0.656	0.755	0.869	86°	1.501	0.998	0.070	14.30
42°	0.733	0.669	0.743	0.900	87°	1.518	0.999	0.052	19.08
43°	0.750	0.682	0.731	0.933	88°	1.536	0.999	0.035	28.64
44°	0.768	0.695	0.719	0.966	89°	1.553	1.000	0.017	57.29
45°	0.785	0.707	0.707	1.000	90°	1.571	1.000	0.000	

TABLE B.4 **Natural logarithms**

x	0.00	0.01	0.02	0.03	0.04	0.05	0.06	0.07	0.08	0.09
1.0	0.0000	0.0100	0.0198	0.0296	0.0392	0.0488	0.0583	0.0677	0.0770	0.0862
1.1	0.0953	0.1044	0.1133	0.1222	0.1310	0.1398	0.1484	0.1570	0.1655	0.1740
1.2	0.1823	0.1906	0.1989	0.2070	0.2151	0.2231	0.2311	0.2390	0.2469	0.2546
1.3	0.2624	0.2700	0.2776	0.2852	0.2927	0.3001	0.3075	0.3148	0.3221	0.3293
1.4	0.3365	0.3436	0.3507	0.3577	0.3646	0.3716	0.3784	0.3853	0.3920	0.3988
1.5	0.4055	0.4121	0.4187	0.4253	0.4318	0.4383	0.4447	0.4511	0.4574	0.4637
1.6	0.4700	0.4762	0.4824	0.4886	0.4947	0.5008	0.5068	0.5128	0.5188	0.5247
1.7	0.5306	0.5365	0.5423	0.5481	0.5539	0.5596	0.5653	0.5710	0.5766	0.5822
1.8	0.5878	0.5933	0.5988	0.6043	0.6098	0.6152	0.6206	0.6259	0.6313	0.6366
1.9	0.6419	0.6471	0.6523	0.6575	0.6627	0.6678	0.6729	0.6780	0.6831	0.6881
2.0	0.6931	0.6981	0.7031	0.7080	0.7129	0.7178	0.7227	0.7275	0.7324	0.7372
2.1	0.7419	0.7467	0.7514	0.7561	0.7608	0.7655	0.7701	0.7747	0.7793	0.7839
2.2	0.7885	0.7930	0.7975	0.8020	0.8065	0.8109	0.8154	0.8198	0.8242	0.8286
2.3	0.8329	0.8372	0.8416	0.8459	0.8502	0.8544	0.8587	0.8629	0.8671	0.8713
2.4	0.8755	0.8796	0.8838	0.8879	0.8920	0.8961	0.9002	0.9042	0.9083	0.9123
2.5	0.9163	0.9203	0.9243	0.9282	0.9322	0.9361	0.9400	0.9439	0.9478	0.9517
2.6	0.9555	0.9594	0.9632	0.9670	0.9708	0.9746	0.9783	0.9821	0.9858	0.9895
2.7	0.9933	0.9969	1.0006	1.0043	1.0080	1.0116	1.0152	0.0188	1.0225	1.0260
2.8	1.0296	1.0332	1.0367	1.0403	1.0438	1.0473	1.0508	1.0543	1.0578	1.0613
2.9	1.0647	1.0682	1.0716	1.0750	1.0784	1.0818	1.0852	1.0886	1.0919	1.0953
3.0	1.0986	1.1019	1.1053	1.1086	1.1119	1.1151	1.1184	1.1217	1.1249	1.1282
3.1	1.1314	1.1346	1.1378	1.1410	1.1442	1.1474	1.1506	1.1537	1.1569	1.1600
3.2	1.1632	1.1663	1.1694	1.1725	1.1756	1.1787	1.1817	1.1848	1.1878	1.1909
3.3	1.1939	1.1969	1.2000	1.2030	1.2060	1.2090	1.2119	1.2149	1.2179	1.2208
3.4	1.2238	1.2267	1.2296	1.2326	1.2355	1.2384	1.2413	1.2442	1.2470	1.2499
3.5	1.2528	1.2556	1.2585	1.2613	1.2641	1.2669	1.2698	1.2726	1.2754	1.2782
3.6	1.2809	1.2837	1.2865	1.2892	1.2920	1.2947	1.2975	1.3002	1.3029	1.3056
3.7	1.3083	1.3110	1.3137	1.3164	1.3191	1.3218	1.3244	1.3271	1.3297	1.3324
3.8	1.3350	1.3376	1.3403	1.3429	1.3455	1.3481	1.3507	1.3533	2.3558	1.3584
3.9	1.3610	1.3635	1.3661	1.3686	1.3712	1.3737	1.3762	1.3788	1.3813	1.3838
4.0	1.3863	1.3888	1.3913	1.3938	1.3962	1.3987	1.4012	1.4036	1.4061	1.4085
4.1	1.4110	1.4134	1.4159	1.4183	1.4207	1.4231	1.4255	1.4279	1.4303	1.4327
4.2	1.4351	1.4375	1.4398	1.4422	1.4446	1.4469	1.4493	1.4516	1.4540	1.4563
4.3	1.4586	1.4609	1.4633	1.4656	1.4679	1.4702	1.4725	1.4748	1.4770	1.4793
4.4	1.4816	1.4839	1.4861	1.4884	1.4907	1.4929	1.4951	1.4974	1.4996	1.5019
4.5	1.5041	1.5063	1.5085	1.5107	1.5129	1.5151	1.5173	1.5195	1.5217	1.5239
4.6	1.5261	1.5282	1.5304	1.5326	1.5347	1.5369	1.5390	1.5412	1.5433	1.5454
4.7	1.5476	1.5497	1.5518	1.5539	1.5560	1.5581	1.5602	1.5623	1.5644	1.5665
4.8	1.5686	1.5707	1.5728	1.5748	1.5769	1.5790	1.5810	1.5831	1.5851	1.5872
4.9	1.5892	1.5913	1.5933	1.5953	1.5974	1.5994	1.6014	1.6034	1.6054	1.6074
5.0	1.6094	1.6114	1.6134	1.6154	1.6174	1.6194	1.6214	1.6233	1.6253	1.6273
5.1	1.6292	1.6312	1.6332	1.6351	1.6371	1.6390	1.6409	1.6429	1.6448	1.6467
5.2	1.6487	1.6506	1.6525	1.6544	1.6563	1.6582	1.6601	1.6620	1.6639	1.6658
5.3	1.6677	1.6696	1.6715	1.6734	1.6752	1.6771	1.6790	1.6808	1.6827	1.6845
5.4	1.6864	1.6882	1.6901	1.6919	1.6938	1.6956	1.6974	1.6993	1.7011	1.7029

x	0.00	0.01	0.02	0.03	0.04	0.05	0.06	0.07	0.08	0.09
5.5	1.7047	1.7066	1.7084	1.7102	1.7120	1.7138	1.7156	1.7174	1.7192	1.7210
5.6	1.7228	1.7246	1.7263	1.7281	1.7299	1.7317	1.7334	1.7352	1.7370	1.7387
5.7	1.7405	1.7422	1.7440	1.7457	1.7475	1.7492	1.7509	1.7527	1.7544	1.7561
5.8	1.7579	1.7596	1.7613	1.7630	1.7647	1.7664	1.7681	1.7699	1.7716	1.7733
5.9	1.7750	1.7766	1.7783	1.7800	1.7817	1.7834	1.7851	1.7867	1.7884	1.7901
6.0	1.7918	1.7934	1.7951	1.7967	1.7984	1.8001	1.8017	1.8034	1.8050	1.8066
6.1	1.8083	1.8099	1.8116	1.8132	1.8148	1.8165	1.8181	1.8197	1.8213	1.8229
6.2	1.8245	1.8262	1.8278	1.8294	1.8310	1.8326	1.8342	1.8358	1.8374	1.8390
6.3	1.8405	1.8421	1.8437	1.8453	1.8469	1.8485	1.8500	1.8516	1.8532	1.8547
6.4	1.8563	1.8579	1.8594	1.8610	1.8625	1.8641	1.8656	1.8672	1.8687	1.8703
6.5	1.8718	1.8733	1.8749	1.8764	1.8779	1.8795	1.8810	1.8825	1.8840	1.8856
6.6	1.8871	1.8886	1.8901	1.8916	1.8931	1.8946	1.8961	1.8976	1.8991	1.9006
6.7	1.9021	1.9036	1.9051	1.9066	1.9081	1.9095	1.9110	1.9125	1.9140	1.9155
6.8	1.9169	1.9184	1.9199	1.9213	1.9228	1.9242	1.9257	1.9272	1.9286	1.9301
6.9	1.9315	1.9330	1.9344	1.9359	1.9373	1.9387	1.9402	1.9416	1.9430	1.9445
7.0	1.9459	1.9473	1.9488	1.9502	1.9516	1.9530	1.9544	1.9559	1.9573	1.9587
7.1	1.9601	1.9615	1.9629	1.9643	1.9657	1.9671	1.9685	1.9699	1.9713	1.9727
7.2	1.9741	1.9755	1.9769	1.9782	1.9796	1.9810	1.9824	1.9838	1.9851	1.9865
7.3	1.9879	1.9892	1.9906	1.9920	1.9933	1.9947	1.9961	1.9974	1.9988	2.0001
7.4	2.0015	2.0028	2.0042	2.0055	2.0069	2.0082	2.0096	2.0109	2.0122	2.0136
7.5	2.0149	2.0162	2.0176	2.0189	2.0202	2.0215	2.0229	2.0242	2.0255	2.0268
7.6	2.0281	2.0295	2.0308	2.0321	2.0334	2.0347	2.0360	2.0373	2.0386	2.0399
7.7	2.0412	2.0425	2.0438	2.0451	2.0464	2.0477	2.0490	2.0503	2.0516	2.0528
7.8	2.0541	2.0554	2.0567	2.0580	2.0592	2.0605	2.0618	2.0631	2.0643	2.0666
7.9	2.0669	2.0681	2.0694	2.0707	2.0719	2.0732	2.0744	2.0757	2.0769	2.0782
8.0	2.0794	2.0807	2.0819	2.0832	2.0844	2.0857	2.0869	2.0882	2.0894	2.0906
8.1	2.0919	2.0931	2.0943	2.0956	2.0968	2.0980	2.0992	2.1005	2.1017	2.1029
8.2	2.1041	2.1054	2.1066	2.1078	2.1090	2.1102	2.1114	2.1126	2.1138	2.1150
8.3	2.1163	2.1175	2.1187	2.1199	2.1211	2.1223	2.1235	2.1247	2.1258	2.1270
8.4	2.1282	2.1294	2.1306	2.1318	2.1330	2.1342	2.1353	2.1365	2.1377	2.1389
8.5	2.1401	2.1412	2.1424	2.1436	2.1448	2.1459	2.1471	2.1483	2.1494	2.1506
8.6	2.1518	2.1529	2.1541	2.1552	2.1564	2.1576	2.1587	2.1599	2.1610	2.1622
8.7	2.1633	2.1645	2.1656	2.1668	2.1679	2.1691	2.1702	2.1713	2.1725	2.1736
8.8	2.1748	2.1759	2.1770	2.1782	2.1793	2.1804	2.1815	2.1827	2.1838	2.1849
8.9	2.1861	2.1872	2.1883	2.1894	2.1905	2.1917	2.1928	2.1939	2.1950	2.1961
9.0	2.1972	2.1983	2.1994	2.2006	2.2017	2.2028	2.2039	2.2050	2.2061	2.2072
9.1	2.2083	2.2094	2.2105	2.2116	2.2127	2.2138	2.2148	2.2159	2.2170	2.2181
9.2	2.2192	2.2203	2.2214	2.2225	2.2235	2.2246	2.2257	2.2268	2.2279	2.2289
9.3	2.2300	2.2311	2.2322	2.2332	2.2343	2.2354	2.2364	2.2375	2.2386	2.2396
9.4	2.2407	2.2418	2.2428	2.2439	2.2450	2.2460	2.2471	2.2481	2.2492	2.2502
9.5	2.2513	2.2523	2.2534	2.2544	2.2555	2.2565	2.2576	2.2586	2.2597	2.2607
9.6	2.2618	2.2628	2.2638	2.2649	2.2659	2.2670	2.2680	2.2690	2.2701	2.2711
9.7	2.2721	2.2732	2.2742	2.2752	2.2762	2.2773	2.2783	2.2793	2.2803	2.2814
9.8	2.2824	2.2834	2.2844	2.2854	2.2865	2.2875	2.2885	2.2895	2.2905	2.2915
9.9	2.2925	2.2935	2.2946	2.2956	2.2966	2.2976	2.2986	2.2996	2.3006	2.3016

TABLE B.5 **Exponential functions**

x	e^x	e^{-x}	x	e^x	e^{-x}	x	e^x	e^{-x}
0.00	1.0000	1.0000	0.55	1.7333	0.5769	3.6	36.598	0.0273
0.01	1.0101	0.9900	0.60	1.8221	0.5488	3.7	40.447	0.0247
0.02	1.0202	0.9802	0.65	1.9155	0.5220	3.8	44.701	0.0224
0.03	1.0305	0.9704	0.70	2.0138	0.4966	3.9	49.402	0.0202
0.04	1.0408	0.9608	0.75	2.1170	0.4724	4.0	54.598	0.0183
0.05	1.0513	0.9512	0.80	2.2255	0.4493	4.1	60.340	0.0166
0.06	1.0618	0.9418	0.85	2.3396	0.4274	4.2	66.686	0.0150
0.07	1.0725	0.9324	0.90	2.4596	0.4066	4.3	73.700	0.0136
0.08	1.0833	0.9231	0.95	2.5857	0.3867	4.4	81.451	0.0123
0.09	1.0942	0.9139	1.0	2.7183	0.3679	4.5	90.017	0.0111
0.10	1.1052	0.9048	1.1	3.0042	0.3329	4.6	99.484	0.0101
0.11	1.1163	0.8958	1.2	3.3201	0.3012	4.7	109.95	0.0091
0.12	1.1275	0.8869	1.3	3.6693	0.2725	4.8	121.51	0.0082
0.13	1.1388	0.8781	1.4	4.0552	0.2466	4.9	134.29	0.0074
0.14	1.1503	0.8694	1.5	4.4817	0.2231	5	148.41	0.0067
0.15	1.1618	0.8607	1.6	4.9530	0.2019	6	403.43	0.0025
0.16	1.1735	0.8521	1.7	5.4739	0.1827	7	1096.6	0.0009
0.17	1.1853	0.8437	1.8	6.0496	0.1653	8	2981.0	0.0003
0.18	1.1972	0.8353	1.9	6.6859	0.1496	9	8103.1	0.0001
0.19	1.2092	0.8270	2.0	7.3891	0.1353	10	22026	0.00005
0.20	1.2214	0.8187	2.1	8.1662	0.1225	11	59874	0.00002
0.21	1.2337	0.8106	2.2	9.0250	0.1108	12	162,754	0.000006
0.22	1.2461	0.8025	2.3	9.9742	0.1003	13	442,413	0.000002
0.23	1.2586	0.7945	2.4	11.023	0.0907	14	1,202,604	0.0000008
0.24	1.2712	0.7866	2.5	12.182	0.0821	15	3,269,017	0.0000003
0.25	1.2840	0.7788	2.6	13.464	0.0743			
0.26	1.2969	0.7711	2.7	14.880	0.0672			
0.27	1.3100	0.7634	2.8	16.445	0.0608			
0.28	1.3231	0.7558	2.9	18.174	0.0550			
0.29	1.3364	0.7483	3.0	20.086	0.0498			
0.30	1.3499	0.7408	3.1	22.198	0.0450			
0.35	1.4191	0.7047	3.2	24.533	0.0408			
0.40	1.4918	0.6703	3.3	27.113	0.0369			
0.45	1.5683	0.6376	3.4	29.964	0.0334			
0.50	1.6487	0.6065	3.5	33.115	0.0302			

Photo credits

Chapter 1 opening photo, page 1. The *Defender,* designed by Pedrick Yacht Designs, Inc., Newport, RI, being tank tested at the Davidson Laboratory, Castle Point Station, Hoboken, NJ. Reprinted by permission

Chapter 10 opening photo, page 341. Courtesy of Measurements Group, Inc., Raleigh, NC, USA.

Chapter 18 opening photo, page 714. Reprinted by permission of the American Alliance for Health, Physical Education, Recreation, and Dance. 1900 Association Drive, Reston, VA 22091.

Answers to Odd-Numbered Exercises

CHAPTER 1

1.1 Exercises (p. 5)

1. Counting **3.** Counting **5.** Not counting **7.** Not counting **9.** Rational **11.** Rational **13.** Irrational **15.** Rational **17.** Irrational **19.** Rational

21.

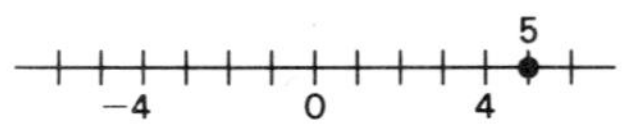

23.

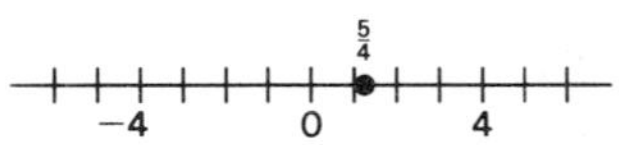

25.

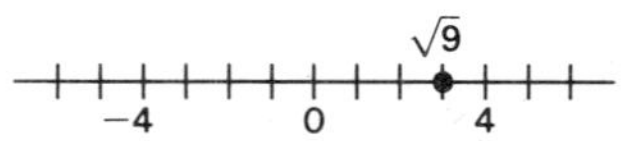

27.

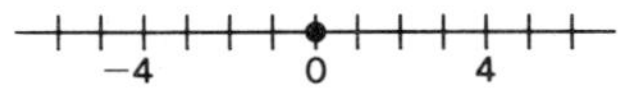

29.

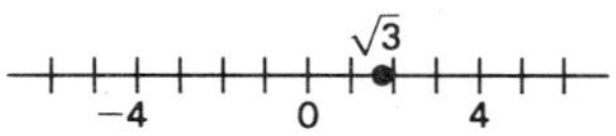

31.

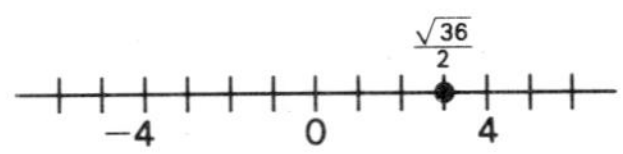

33.

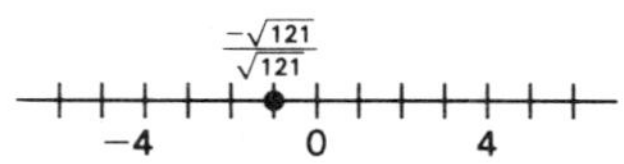

35. Complex **37.** Real **39.** Complex **41.** Pure imaginary **43.** Complex **45.** Pure imaginary **47.** Real **49.**

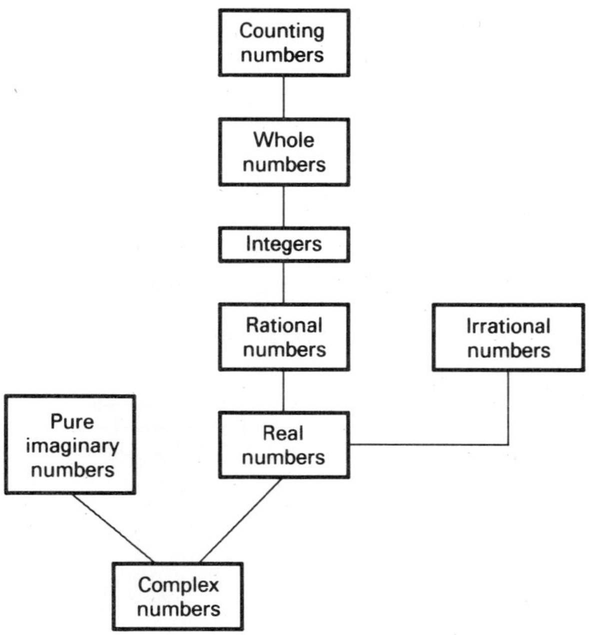

1.2 Exercises (p. 10)

1. $3 < 14$ **3.** $-\sqrt{3} < -\sqrt{2}$ **5.** $\sqrt{5} < 2.236068\ldots$ **7.** $\frac{14}{4} > 3\frac{1}{4}$ **9.** $\frac{1}{2} + \frac{1}{3} = \frac{3}{2} - \frac{2}{3}$

11. 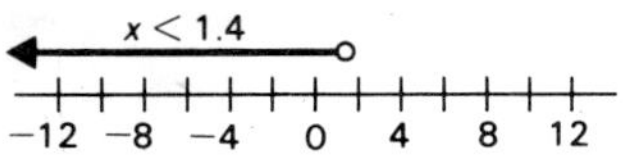

13. 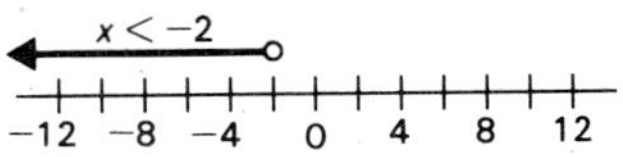

15. 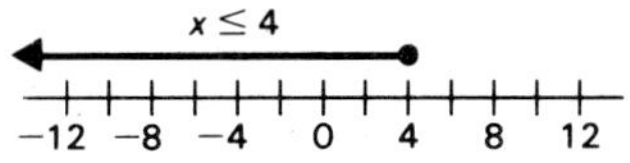

17. 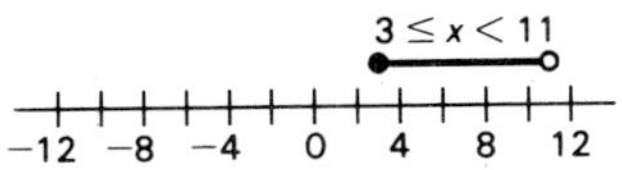

19. 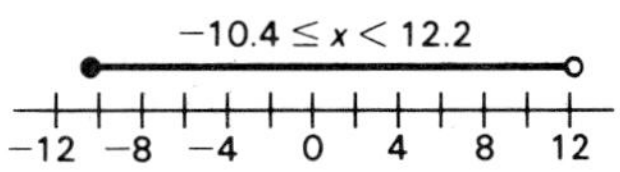

21. 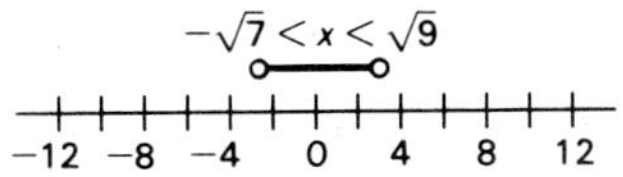

23. $x < 3$ **25.** $x \leq -\sqrt{7}$ **27.** $-5 \leq x \leq 0$ **29.** $-3 \leq x < 3$ **31.** 9 **33.** 4 **35.** 2 **37.** 4
39. 4 **41.** 4.361

1.3 Exercises (p. 16)

1. -6 **3.** 5 **5.** 7 **7.** $\frac{1}{3}$ **9.** $9 \cdot 8$ **11.** $(4 \cdot 6) \cdot 7$ **13.** $3 \cdot a + 3 \cdot b$ **15.** $(a + b) + 3$
17. $(3 + 14) + 19$ **19.** Additive identity **21.** Commutative property of addition
23. Distributive property **25.** Inverse property of addition **27.** Distributive property **29.** 13
31. -3 **33.** 6 **35.** -45 **37.** 0 **39.** 12 **41.** -45 **43.** 6 **45.** -7 **47.** -8
49. -44 **51.** -84 **53.** 4 **55.** -3 **57.** 2 **59.** -1 **61.** $5\frac{2}{3}$ **63.** 4

1.4 Exercises (p. 19)

1. 3 **3.** $\sqrt{6}$ **5.** 0 **7.** Undefined **9.** 1 **11.** Undefined **13.** No, indeterminable if $a = 0$
15. No, indeterminable if $a = b$ **17. a.** Error **b.** A very large number

1.5 Exercises (p. 23)

1. (47.) -8 **3.** (49.) -44 **5.** (51.) -84 **7.** (53.) 4 **9.** (55.) -3 **11.** (57.) 2
13. (57.) -1 **15.** 1667.963 . . . **17.** 0.4837794 . . . **19.** -0.4615256 **21.** 1845.096 . . .
23. -5980.063 . . . **25.** 0.5694051 . . . **27.** Undefined **29.** -12.16237 . . .
31. -14.33304 . . . **33.** 681.3309 . . .

Review Exercises (p. 25)

1. Rational, real, complex **3.** Rational, real, complex **5.** Pure imaginary, complex
7. Rational, real, complex **9.** Complex **11.** $-141 < -32$ **13.** $\sqrt{13} = 3.605551\ldots$

15. $\frac{133}{4} < 33\frac{3}{4}$ **17.** $1\frac{5}{8} - \frac{13}{8} = 1\frac{7}{8} - \frac{15}{8}$

19. 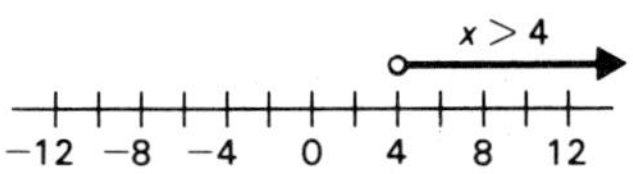

21. 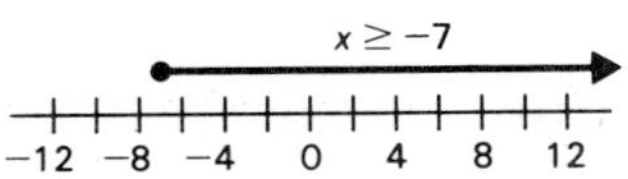

23. 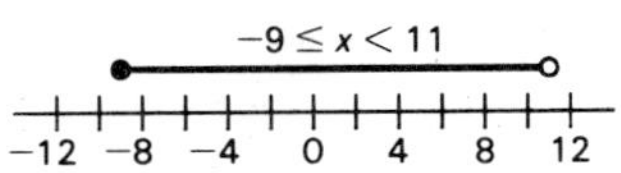

25. 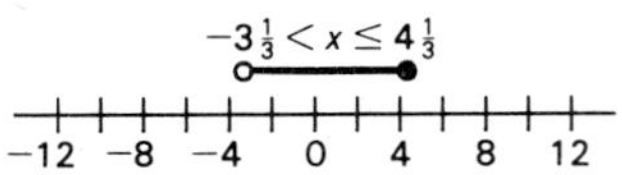

27. $\sqrt{9}$ or 3 **29.** 3 **31.** $\frac{5}{12}$ **33.** Identity property for addition

35. Inverse property for multiplication **37.** Distributive property **39.** -21 **41.** -5 **43.** 0 **45.** Indeterminable **47.** 3.6 **49.** -11.5774 **51.** -1.71098 **53.** Undefined **55.** -11.678 **57.** 0.8444290 . . .

CHAPTER 2

2.1 Exercises (p. 36)

1. x^{17} **3.** $3^{20}x^{55}$ **5.** $\frac{1}{x^5}$ **7.** 1 **9.** a^3 **11.** z^{15} **13.** $\frac{3a}{5}$ **15.** $125a^{64}b^5$ **17.** $\frac{z^n}{x^n}$ **19.** $\frac{c^2}{a^4b^6}$ **21.** $\frac{x^8z^8}{5^4 3^8 y^4}$ or $\frac{x^8z^8}{45^4y^4}$ **23.** 27.648 **25.** 0.0000305 . . . **27.** 1 **29.** 39.304 **31.** 0.0000000177 **33.** 0.03125 **35.** $\left(\frac{1}{576}\right)^n$ **37.** 2.177363 . . . **39.** -0.010788 . . .

2.2 Exercises (p. 41)

1. $5. \times 10^3$ **3.** $4. \times 10^{-4}$ **5.** 7×10^0 **7.** $3. \times 10^{-1}$ **9.** 5.6×10^7 **11.** 8.77×10^{-7} **13.** 3.937×10^4 **15.** 1.745×10^{-2} **17.** 3.82×10^5 **19.** 2.06265×10^5 **21.** 2.838×10^9 **23.** 8500 **25.** 0.000567 **27.** 0.3 **29.** 87,500,000. **31.** 0.000000653 **33.** 98,730,000,000. **35.** 750,000,000. **37.** 0.0005682 **39.** 8,649,000 **41.** 0.0000005 **43.** 1.2×10^7 **45.** $7. \times 10^4$ **47.** 3.404×10^{12} **49.** $6.803 \ldots \times 10^4$ **51.** $1.348 \ldots \times 10^1$ hr **53.** 8.0×10^{-2} m^3/s

2.3 Exercises (p. 49)

1. $\sqrt{15}$ **3.** $\sqrt[3]{13}$ **5.** $\sqrt[7]{2}$ **7.** $\sqrt{\frac{5}{13}}$ **9.** 6 **11.** 11 **13.** 4 **15.** 0.1 **17.** 2 **19.** $13x^2$ **21.** a^2bz^3 **23.** $-4x^3y^2$ **25.** 4 **27.** $\frac{-5x}{4y}$ **29.** $2\sqrt{3}$ or $3.464 \ldots$ **31.** $3\sqrt[3]{3}$ or $4.326 \ldots$ **33.** $0.7311 \ldots$ **35.** $x\sqrt{7}$ or $(2.645 \ldots)x$ **37.** $5ab^2\sqrt{2a}$ or $(7.071 \ldots)a^{3/2}b^2$ **39.** $x^2\sqrt[3]{3}$ or $(1.442 \ldots)x^2$ **41.** $a^2\sqrt[n+1]{a}$ or $a^{\frac{2n+3}{n+1}}$ **43.** $x^2\sqrt[3k-1]{x^2}$ or $x^{\frac{6k}{3k-1}}$ **45.** 3 **47.** 0, providing $x \neq 0, y \neq 0$ **49.** $5\sqrt{3}$ **51.** $25\sqrt{3}$ **53.** 0 **55.** $(1+y+y^3)\sqrt{y}$ **57.** $11\sqrt{2}$ **59.** $xy\sqrt{wz}$ **61.** 5 **63.** xyz^3 **65.** 6 **67.** $\sqrt{\frac{6b}{7c}}$ **69.** $\frac{1}{\sqrt[4]{a}}$

2.4 Exercises (p. 54)

1. $6x$ and y **3.** $\frac{a}{3}, \frac{b}{3}$, and $\frac{c}{3}$ **5.** $-4xy$ and $(6x+4y)$; $(6x+4y)$ has terms $6x$ and $4y$ **7.** $8a+7b$ **9.** $2x+6y$ **11.** $6r-16s$ **13.** $3\pi r - 2\pi r^2$ **15.** $\frac{5}{6}x + \frac{1}{4}y$ **17.** $\frac{7}{8}x^2$ **19.** $-1.36r^2$ **21.** $-2.7946r^2$ **23.** 22 **25.** $8x-7$ **27.** $3a+5$ **29.** $15x+8y$ **31.** $-47a+5b$

33. $6r - s$ **35.** $-12r + 15s$ **37.** $2a + 25b$ **39.** $7x$ **41.** $-8 + \frac{x}{2}$ **43.** $\sqrt{5x} + x^2$
45. $(x + y)^3$ **47.** $2xyz + xyz$ **49. a.** $10 - 2.1I_1 - 0.2I_2$ **b.** $11 + 3.5I_1 - 3.7I_2$
c. $1 + 5.6I_1 - 3.5I_2$ **51.** $\frac{7\pi}{32} d^2\sigma^2 + \frac{\pi}{4} D^2\sigma^2$

2.5 Exercises (p. 58)

1. $15x^6$ **3.** $28x^3y^4$ **5.** $-18x^{11}y^{13}$ **7.** $-45a^3b^3$ **9.** $96p^5q^5$ **11.** $-4a^4b^2$ **13.** $8a^2 + 32a$
15. $5p^4 + 35p^3$ **17.** $6a^3b^2 - 15a^2b$ **19.** $a^2 + 2ab + b^2$ **21.** $6x^2 + 29x + 35$ **23.** $a^2 - b^2$
25. $4x^2 - 9$ **27.** $6x^2 + 13xy - 28y^2$ **29.** $35a^4b + 14a^2bc + 15a^2c^2 + 6c^3$
31. $3x^3 - 2x^2 + 7x + 20$ **33.** $4a^3 + 20a^2 + 37a + 24$ **35.** $4x^4 + 2x^3 - 22x^2 + 4x$
37. $4x^2 + 12xy + 9y^2$ **39.** $9a^2 - 42ab + 49b^2$ **41.** $x^3 + 4x^2 - 17x - 60$
43. $8a^3 - 36a^2b + 54ab^2 - 27b^3$ **45.** $20x^4 - 40x^3 - 55x^2 - 15x$ **47.** $\frac{v^2 - v_0^2}{2a}$ or $\frac{v^2}{2a} - \frac{v_0^2}{2a}$
49. $4000 - 400n - 150n^2$

2.6 Exercises (p. 62)

1. 4 **3.** $7ab^3$ **5.** $4b^{11}$ **7.** $2 - 4y$ **9.** $2 + 6r$ **11.** $-5xy + 4x - 2y$ **13.** $m + 4$
15. $3a + 2b$ **17.** $4x - 7y$ **19.** $16x^2 - 4x + 1$ **21.** $y^2 - 6y + 11$ **23.** $q^2 + 3q + 4$
25. $a - 4 + \frac{-10}{a - 3}$ **27.** $2x - 1 + \frac{10}{3x + 5}$ **29.** $t^2 - 6t + 19 + \frac{-34}{t + 2}$ **31.** $x + 6 + \frac{x + 14}{x^2 - x - 2}$
33. $8x^3 - 36x^2y + 54xy^2 - 27y^3$ **35.** $1 - x + x^2 - x^3 + x^4 + \frac{-x^5}{1 + x}$ **37.** $90 + 80\alpha\gamma - 40\alpha^2\beta\gamma^2$
39. $\frac{1}{2}(EI + P)^2 - 2 + \frac{6}{(EI + P)}$

2.7 Exercises (p. 68)

1. $3(a + 4b)$ **3.** $7(r + 2s)$ **5.** $2x(1 - 6y)$ **7.** $5x(y - 3xz + 4z)$ **9.** $(3b + 5d)(a + 4c)$
11. $(5r - 6k)(3s - 5t)$ **13.** $5(3R_1 - 7R_2)(I_1 - 2I_2)$ **15.** $A(3AB - 7C)(2D - 5B)$
17. $\left(\frac{1}{3}A + \frac{2}{5}P\right)(B - 2C)$ **19.** $y(0.2x + 0.3z)(0.3x + 0.4y)$ **21.** $(x + 1)(x + 4)$ **23.** $3(a + 1)^2$
25. $(t + 7)(t + 2)$ **27.** $(r + 7)(r - 5)$ **29.** $(3b + 1)(b - 2)$ **31.** Prime **33.** $(5z - 4)(z + 3)$
35. $2(3s - 1)(s + 3)$ **37.** $(17p - 11q)(p + q)$ **39.** $(2w - 5)^2$ **41.** Prime **43.** $(z + 3)(3z + 1)$
45. $(7R - 2)(R - 2)$ **47.** $(w + 7)^2$ **49.** $(3p + 7)^2$ **51.** $(2s - 3)^2$ **53.** $(8m - 5n)^2$
55. $(9w - 4z)^2$ **57.** $(x - 3)(x + 3)$ **59.** $(9r - 4)(9r + 4)$ **61.** $(xy - 5)(xy + 5)$
63. $(a + b - d)(a + b + d)$ **65.** $(p - 4)(p + 4)(q + 3)$

2.8 Exercises (p. 72)

1. $\frac{4}{7}$ **3.** pm **5.** $\frac{x(x + 3)}{6(x - 2)}$ or $\frac{x^2 + 3x}{6x - 12}$ **7.** $\frac{2}{3}$ **9.** $\frac{4(a - b)}{3}$ **11.** $\frac{2(x + 3)}{z(x - 3)}$ **13.** 1
15. $-\frac{3p}{3 + p}$ **17.** $\frac{8}{5}$ **19.** $\frac{1}{3ac}$ **21.** $\frac{(a + 3)(a - 6)}{(a + 4)(a + 5)}$ or $\frac{a^2 - 3a - 18}{a^2 + 9a + 20}$ **23.** $\frac{3}{4}$ **25.** $\frac{1}{p - 2}$
27. $\frac{1}{x - y}$ **29.** 1 **31.** $\frac{3x^3}{4y^3}$ **33.** $\frac{9z}{2x}$ **35.** $\frac{R}{R - r}$

2.9 Exercises (p. 77)

1. $\frac{7}{2x}$ **3.** $\frac{p+q+1}{p+q}$ **5.** $\frac{5a-b}{a^2b^2}$ **7.** $\frac{2}{(x-1)(x+1)}$ **9.** $\frac{-(a+1)}{(a+2)(a+3)}$ **11.** 0 **13.** $\frac{-a}{a-1}$
15. $\frac{-4m}{(m+1)(m-1)}$ **17.** $\frac{3y^2z-5x^4}{x^5y^4z}$ **19.** $\frac{a(a-1)}{(a-2)(a+2)}$ **21.** 0 **23.** $\frac{4x^3-x-2}{(2x+1)^2(2x-1)}$
25. $\frac{a^2r^2+b^2q^2-c^2p^2}{p^2q^2r^2}$ **27.** $\frac{x^2-3x-3}{(x-3)(x+3)}$ **29.** $\frac{2b}{a-b}$ **31.** $\frac{-2y+1}{(x-y)(x+y)}$
33. $\frac{24(t+6)}{(t-5)(t+5)(t-7)(t+7)}$ **35.** $\frac{m^2-3m-11}{2+m}$ **37.** $\frac{a^2c+ab^2+bc^2}{8}$ **39.** $\frac{-1}{a-b}$ **41.** $-\frac{13}{17}$
43. $\frac{mn+1}{mn-1}$ **45.** $\frac{IR}{IR-E}$ **47.** $\frac{y}{x}$

Review Exercises (p. 81)

1. a^{16} **3.** $x^{28}y^{21}$ **5.** 1 **7.** $5xy^5$ **9.** $\frac{p^3q^3}{8m^9n^{12}}$ **11.** 3.250×10^3 **13.** $4. \times 10^{-1}$
15. 5.79×10^{-6} **17.** 7,052,000. **19.** 4.32 **21.** 0.0000462 **23.** 0.035 oz **25.** 1×10^{12}
27. 8.8×10^{-5} **29.** $8. \times 10^{-13}\, j$ **31.** $2\sqrt[3]{4}$ **33.** 4 **35.** y^n **37.** $-4\sqrt{2}$ **39.** $\sqrt{143}$
41. $ab\sqrt{cd}$ **43.** $3y\sqrt{x}$ **45.** $-8xy^2$ **47.** $-6\phi\pi+16RI$ **49.** $\frac{58}{105}b^2$ **51.** $-18x+13$
53. $35p-24q$ **55.** $-143r^5s^6$ **57.** $-3pq$ **59.** $\frac{(m+2)(m+4)}{(m+3)(m-5)}$ or $\frac{m^2+6m+8}{m^2-2m-15}$ **61.** $\frac{1}{3xy^2}$
63. $\frac{2(x^3-y^3)}{x^2y^2}$ **65.** $15R^3+43R^2-89R+36$ **67.** $x^3+6x^2-7x-60$ **69.** $3y(a-2)$
71. $a+6$ **73.** 1 **75.** 2 **77.** $\frac{5RI}{3}$ **79.** $\frac{31m+1}{6(m+1)(m-1)}$ **81.** $\frac{(x-1)^2}{(x+1)(x+5)}$
83. $x(3xy-7z)(2w-5y)$ **85.** $6(r-s)(r+s)$ **87.** $2(8x+1)(4x-1)$ **89.** $(5a+6b)(2a-3b)$

CHAPTER 3

3.1 Exercises (p. 89)

1. 6 **3.** 2 **5.** -1 **7.** $\frac{2}{3}$ **9.** -2 **11.** 3 **13.** 9 **15.** 3 **17.** 24 **19.** 3
21. -0.56 **23.** 0 **25.** No solution **27.** 12 **29.** $\frac{c}{v}$ **31.** $\frac{2A}{b}$ **33.** $\frac{P-2\ell}{2}$ **35.** $\frac{9}{5}c+32$
37. $\frac{3V}{\pi r^2}$ **39.** $\frac{Fr^2}{Gm_1}$ **41.** $\frac{25A}{3P}$ **43.** $\frac{E_L}{1.48Q(1-\mu)}$

3.2 Exercises (p. 94)

1. 480. **3.** 350. **5.** 87. **7.** 0.0466 **9.** 4600. **11.** 2.400×10^5 **13.** 4.50×10^{-3}
15. 2.3×10^8 **17.** 820.3 **19.** 450.6 **21.** -489.20 **23.** 248. **25.** 5000. **27.** 425.8 kg
29. 26.000 in. **31.** 0.77865 **33.** 44.5 **35. a.** 93.26 **b.** 49.91

3.3 Exercises (p. 98)

1. 8500. mm **3.** 15. mi **5.** 402. m **7.** 88. fps **9.** 1.30×10^6 mA **11.** 2.973 m^2
13. 5.6×10^6 **15.** 2550. j/s **17.** 7.00×10^{-2} MΩ **19.** 17. km/ℓ **21.** 130,500 ω
23. 61.0 $in.^3$ **25.** 0.22 g/cm^3 **27.** 4.445 m

3.4 Exercises (p. 101)

1. 610. V **3.** 196. ft **5.** 7.077 in. **7.** $51{,}840\ \frac{\text{kg m}^2}{\text{s}^2}$ **9.** 24. m **11.** 570. cal **13.** 1.6 s

3.5 Exercises (p. 106)

1. a. 14.6 m^2 **b.** 17.5 m **3. a.** 39,200. min^3 **b.** 8290 mm^2 **5.** 26,500 $in.^3$ **7.** 13. Ω, 22 Ω
9. 40. mi/hr, 25. mi/hr, $2\underline{0}0.$ mi **11.** $12\underline{0}.$ mm **13.** 1.0×10^4 Ω **15.** $960{,}0\underline{0}0.$ mi
17. 45. Nm or 45. j **19.** 1.0×10^3 ft **21.** 515. mm, 335. mm **23.** 2115. m

Review Exercises (p. 108)

1. 9 **3.** 11 **5.** 1.0 **7.** 30 **9.** $-\frac{23}{3}$ **11.** 151.2 lb **13.** −27.8 in. **15.** 32. km^2
17. 0.1432 in. **19.** 6.7 **21.** 11,500. mm **23.** 121. km **25.** 30.1 yd **27.** 105. km/h
29. 29,830. W **31.** 3.2 ft/s^2 **33.** 3, 60.7 mm **35.** 52, 104 **37.** \$110, \$100 **39.** 4 ft

CHAPTER 4

4.1 Exercises (p. 116)

1. $\frac{5}{16}$ **3.** $\frac{3}{4}$ **5.** $\frac{1}{4}$ **7.** $\frac{3}{4}$ **9.** $\frac{5}{7}$ **11.** $\frac{7}{5}$ **13.** $\frac{2y}{3x}$ **15.** $\frac{3.14}{1.00}$ **17.** $7\frac{1}{2}$ **19.** $10\frac{1}{2}$
21. $\frac{5}{2}$ **23.** 3.00 **25.** 371,620. cm **27.** 33.58 mm **29.** 87.5 m **31.** 109. km **33.** 16.5 gm
35. 81. mm **37.** 1735. lb

4.2 Exercises (p. 122)

1. $D = kt$ **3.** $C = kd$ **5.** $R = kVW$ **7.** $Z = \frac{kD^2}{T}$ **9.** $T = \frac{kR\sqrt{x}}{y^2}$ **11.** $L = \frac{k}{V}$ **13.** $A = ka^2$
15. $T = kd^4n^2$ **17.** $R = \frac{k\ell}{d^2}$ **19.** $t = \frac{ky}{x}$ **21. a.** $y = kx^2$ **b.** $k = \frac{y}{x^2}$ **c.** $y = \frac{3}{16}x^2$ **d.** 8
23. a. $t = kA$ **b.** $k = \frac{t}{A}$ **c.** $t = 1.143\,A$ **d.** 54.9 s **25. a.** $I = \frac{k}{R}$ **b.** $k = IR$ **c.** $I = \frac{30}{R}$
d. 150. A **27. a.** $s = \frac{kwd^2}{\ell}$ **b.** $k = \frac{\ell s}{wd^2}$ **c.** $s = \frac{62\frac{1}{2}\,wd^2}{\ell}$ **d.** 250 lb **29. a.** $n = \frac{k\sqrt{f}}{\ell d}$
b. $k = \frac{\ell dn}{\sqrt{f}}$ **c.** $n = \frac{\frac{8}{5}\,n_1\sqrt{f}}{\ell d}$ **d.** 100 N **31. a.** $p = \frac{k\ell}{d^2}$ **b.** $k = \frac{d^2p}{\ell}$ **c.** $p = \frac{0.567\ell}{d^2}$
d. $490.\ \frac{\text{N}}{m^2}$ **33. a.** $F = \frac{k\,m_1m_2}{d^2}$ **b.** $k = \frac{d^2F}{m_1m_2}$ **c.** $F = \frac{3.442 \times 10^{-8}\,m_1m_2}{d^2}$ **d.** 7.71×10^{-8} lb

4.3 Exercises (p. 129)

1. 15.23333° **3.** 104.6333° **5.** 73.98444° **7.** 14° 27′ **9.** 102° 3′ **11.** 87° 45′ 48″
13. 61° 28′ 46″ **15.** $a = 133°\ 47'$ $b = 133°\ 47'$ $c = 46°\ 13'$ $d = 45°\ 13'$
17. $\angle CBD = 60°$, $\angle CDB = 30°$ **19.** 50°, isosceles triangle **21.** 60°, equiangular triangle
23. 15°, isosceles triangle **25.** $\alpha = 48°$, $\beta = 65°$, $\gamma = 65°$, $\epsilon = 25°$

4.4 Exercises (p. 137)

1. Definition 4.4 $\alpha = 55°$, $b = 120$ mm **3.** Definition 4.4 $\phi = 41.1°$, $t = 1.9$ ft **5.** 69′ 5.″
7. 33 in. **9.** $19\frac{1}{2}$ m **11.** 56 mm **13.** $2\frac{1}{3}$ in. **15.** 275 km **17.** 96 ft
19. *ASA* $\alpha = 70°$, $\beta = 60°$, $x = 100$ m **21.** 75 ft

4.5 Exercises (p. 144)

1. 13.2 m, 11. m^2 **3.** 12.4 ft, 7.2 ft^2 **5.** 90. ft, 144. ft^2 **7.** 11.0 mm, 5.7 mm^2
9. 150 mm, 1500 mm^2 **11. a.** 90° **b.** 1.142 $in.^2$ **13. a.** 42,000. ft^2 **b.** 65.927. ft^2
c. 13,383. ft^2 **d.** 26,093. ft^2 **e.** 39,961. ft^2 **f.** trapezoid, 62,510. ft^2 **g.** 249,874. ft^2
15. 21.5% **17.** 1 : 9 **19.** No, $\sqrt{2}\,d$

4.6 Exercises (p. 152)

1. 4.9230×10^7 mi^2 **3.** 120. ft^3 **5.** 13.1 ft **7.** 6400 $in.^3$ **9.** 244.9 yd^3 **11.** 14.5 lb
13. 36.5 ft, 3440 ft^2 **15.** 11.006888 gal or 12. gal

Review Exercises (p. 155)

1. a. Two girls for every three members of the class. **b.** 1/3 **c.** 18 **3. a.** $\dfrac{r - 3s}{r + 3s}$
b. $\dfrac{(x + 2y)(x - 2y)}{2(x + y)(x + y)}$ **5.** $36\frac{1}{4}$ mi **7.** 126 mi **9. a.** $c = kt$ **b.** $p = ktr$ **c.** $z = \dfrac{kRt^2}{w^2}$
d. $A = kr^2$ **11.** 12.2 **13.** 7 to 1 **15.** 7490 $in.^3$ **17.** $3\frac{3}{8}$ times greater **19.** 1 : 2.25
21. 383.2 mm^2 **23.** 150. $in.^2$ **25.** Yes, (67.8 $in.^3$) **27.** $1.23 \times 10^7 \ell$ **29.** 420. ft^3 **31.** 18. m
33. a. 13° 45′ **b.** 23° 28′ 36″ **c.** 92° 13′ 13″ **35.** $\alpha = 35°$, $\beta = 55°$, $\epsilon = 56°$, $\theta = 56°$, $\phi = 34°$

CHAPTER 5

5.1 Exercises (p. 166)

1. $i: d;\ d: p$ **3.** $i: F;\ d: c$ **5.** $i: r, \theta;\ d: s$ **7.** $i: m, a;\ d: F$ **9. a.** 43 **b.** 15
c. $3a^2 + 6ab + 3b^2 + 7a + 7b - 5$ **11. a.** −7 **b.** 2.8 **c.** −0.19 **13. a.** 31 **b.** 1 **c.** $\dfrac{121}{441}$
15. a. 35.69 **b.** 103.4 **c.** 34.12 **17. a.** 14.7 in. **b.** 40.3 in. **c.** 2.86 in.
19. a. 84.4 $in.^2$ **b.** 220. $in.^2$ **c.** 1081. mm^2 **21. a.** 7.5 **b.** 18. **23. a.** 0.0461
b. 0.00183 **25. a.** $1.3 \times 10^7\ \Omega$ **b.** $2.7 \times 10^7\ \Omega$ **27.** $D = \{S | S \in \mathbb{R}\}$ $R = \{R | R \in \mathbb{R}\}$
29. $D = \{S | S \in \mathbb{R}\}$ $R = \{\phi(S) | \phi(S) \geq -3\}$ **31.** $D = \{r | r \in \mathbb{R}\}$ $R = \{V | V \geq 0\}$
33. $D = \{x | x \neq 1\}$ $R = \{y | y \neq 0\}$ **35.** $D = \{x | x \neq 1\}$ $R = \{y | y \neq -1\}$
37. $D = \{t | t \geq 2\}$ $R = \{s | s \geq 0\}$ **39.** $D = \{\ell | \ell \leq 3\}$ $R = \{p | p \geq 0\}$ **41.** $f(x) = \frac{5}{9}(x - 32)$

5.2 Exercises (p. 170)

1.

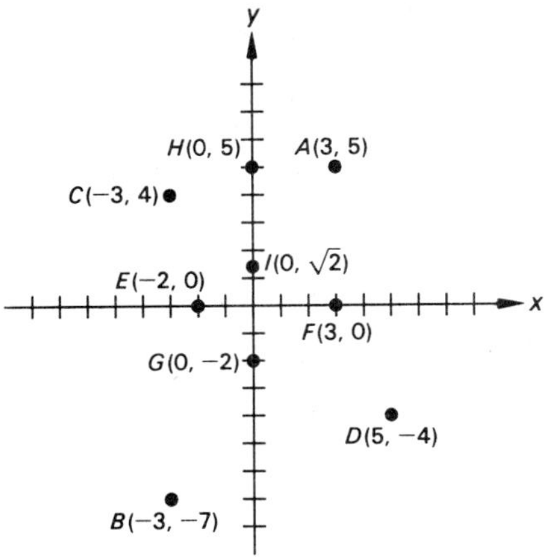

3. $A(2, -4)$, $B(4, -2)$, $C(2, 0)$, $D(4, 5)$, $E(0, 3)$, $F(-3, 3)$, $G(-5, 0)$, $H(-4, -2)$, $I(-3, -4)$

5. 5.39 **7.** 19.1 **9.** 2.9 **11.** 16.6

13.

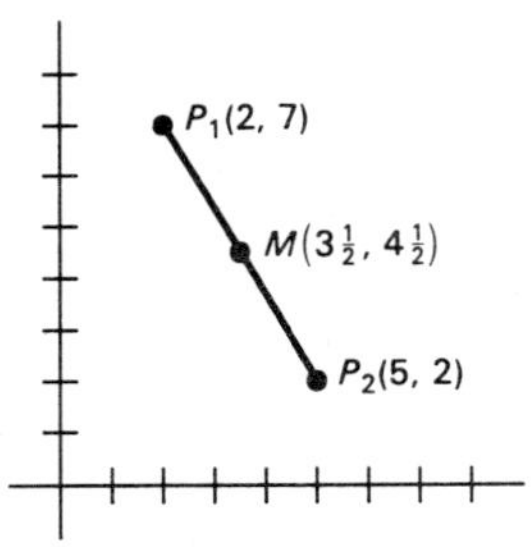

15.

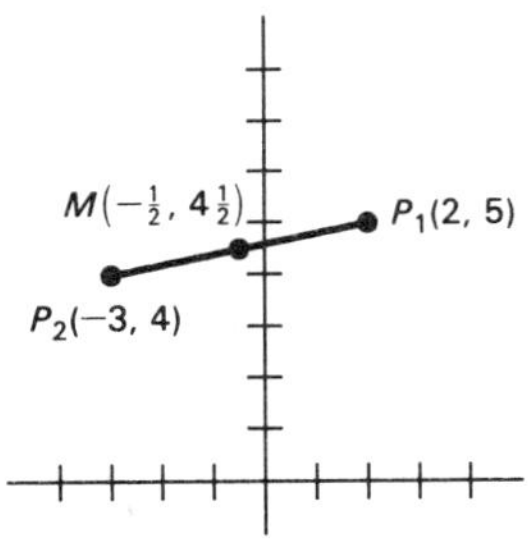

17.

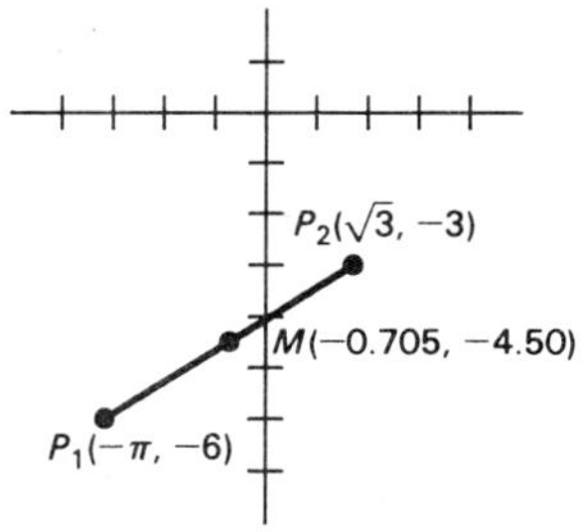

19. b. 9.15 **27.** (18, 2), (−4, 2) **31.** 10 square units

5.3 Exercises (p. 175)

1.

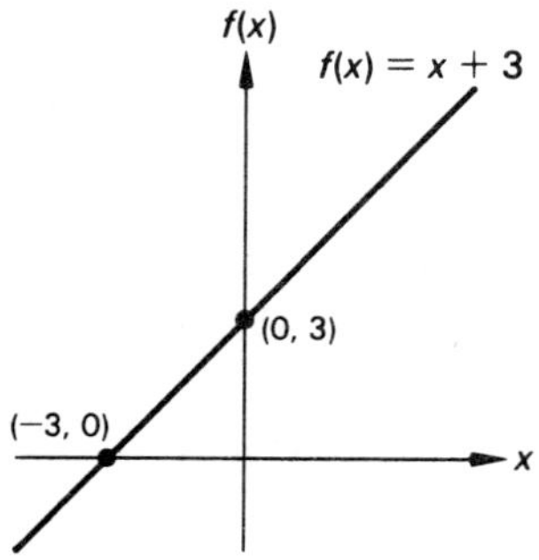

3.

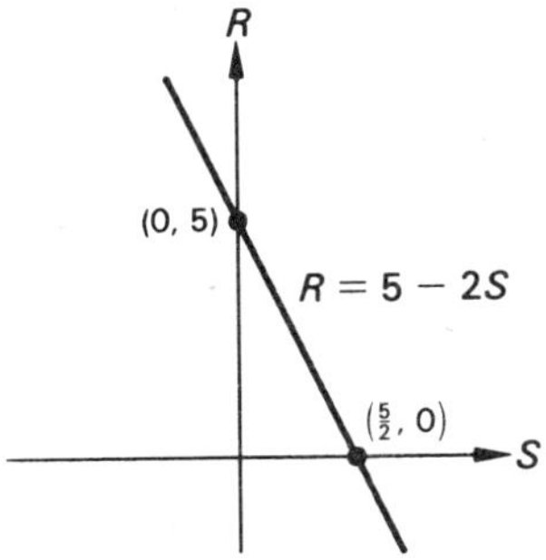

5.

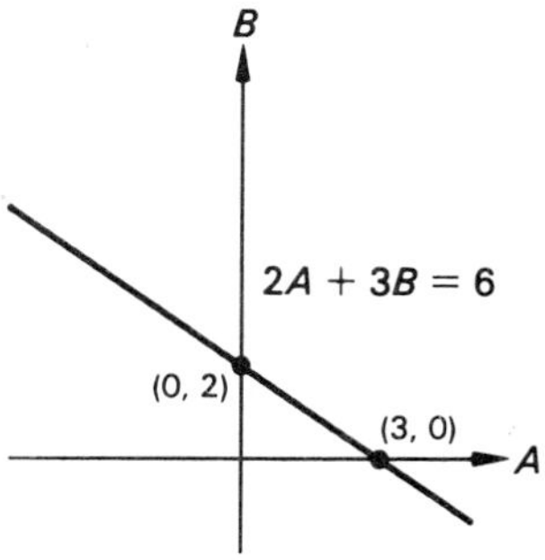

7.

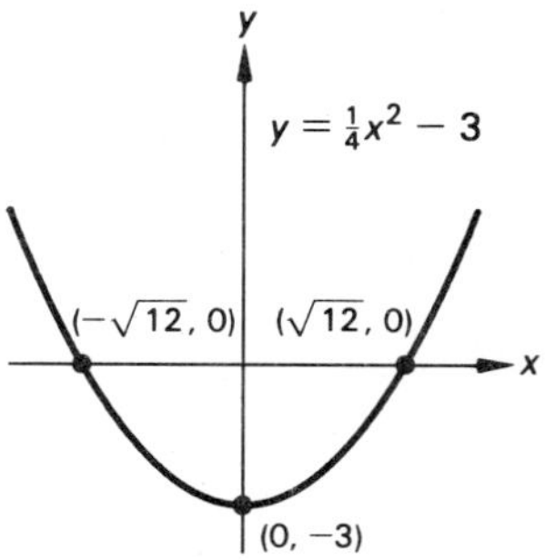

9.

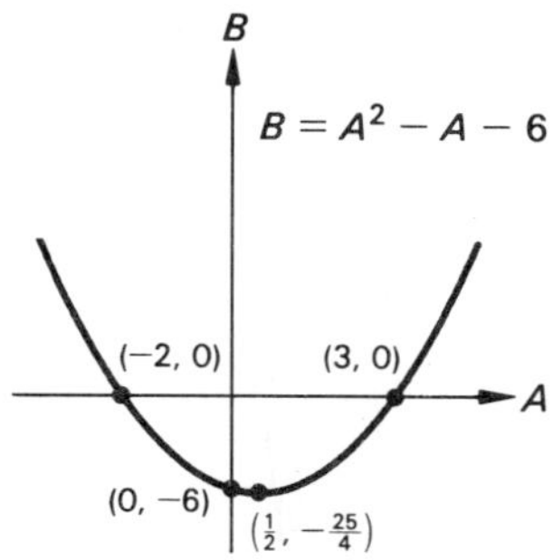

11.

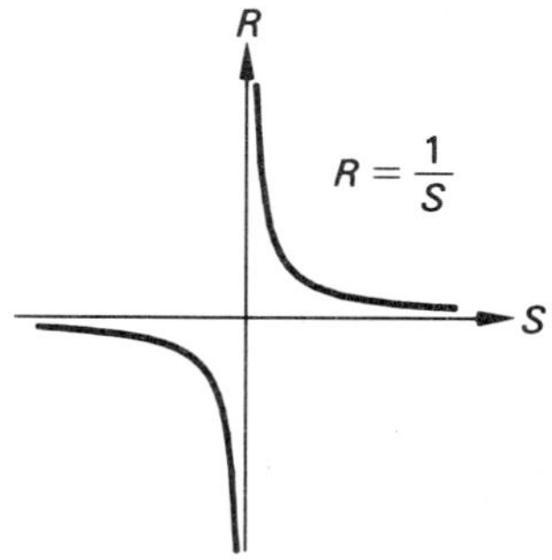

13.

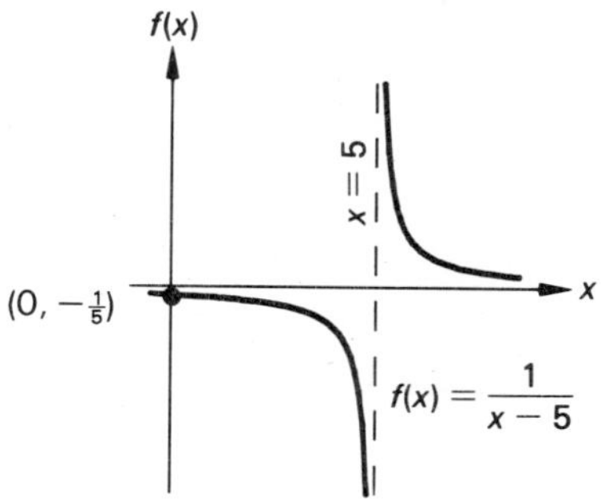

15.

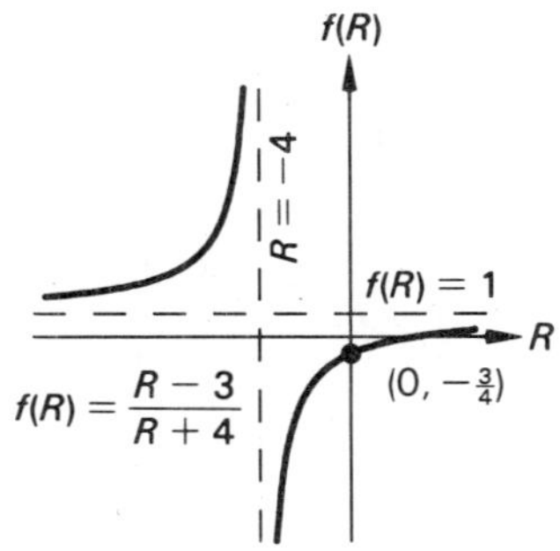

17.

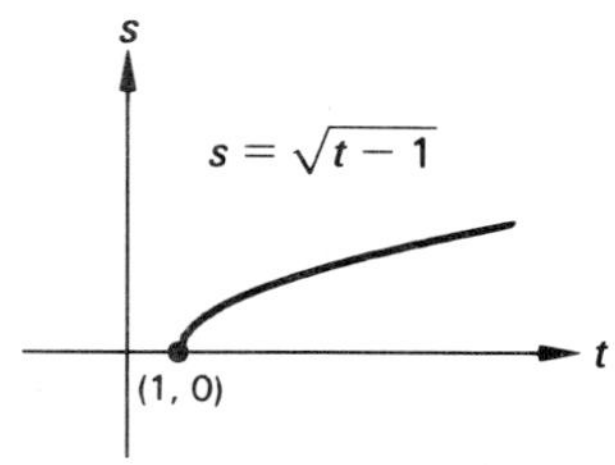

19.

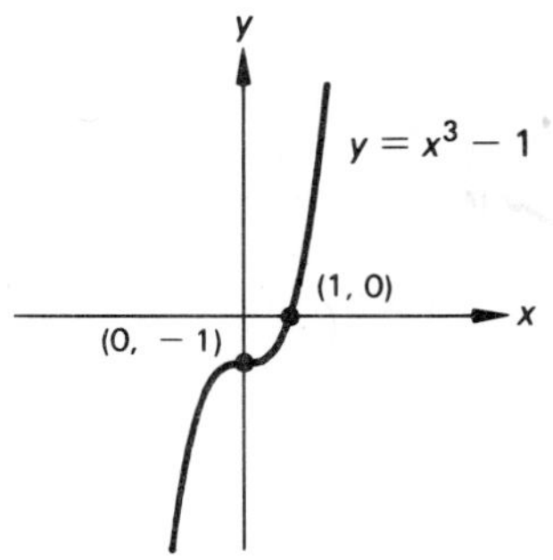

21.

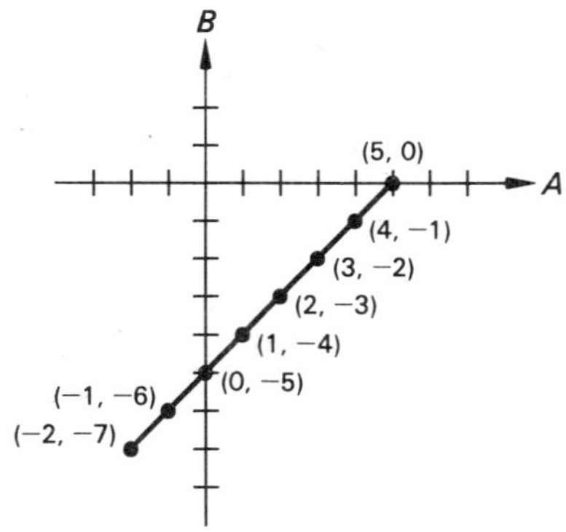

23.

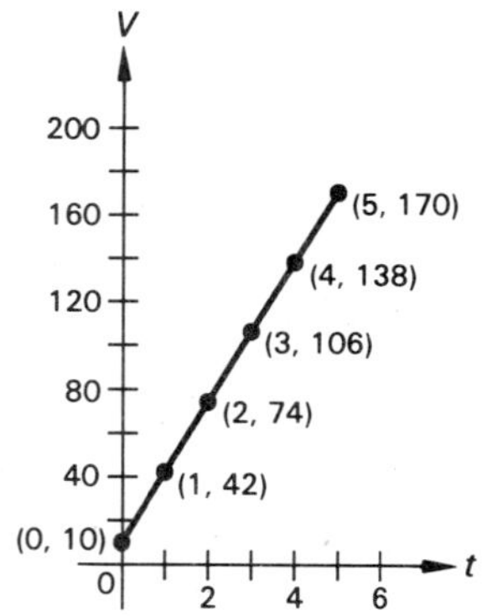

25.

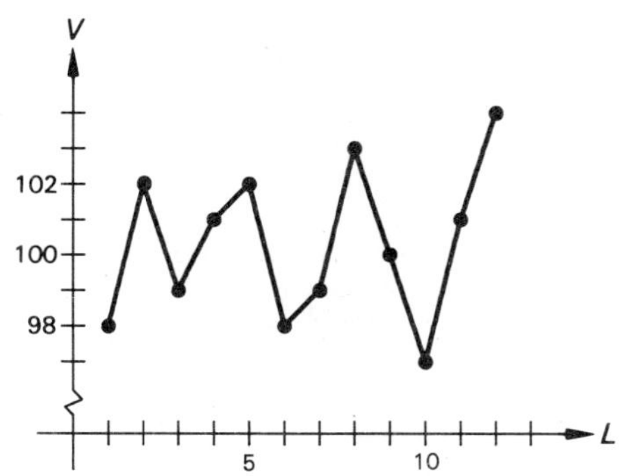

27.

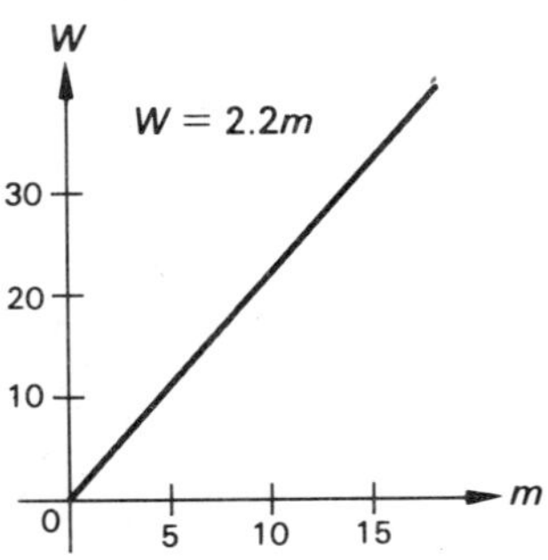

29.

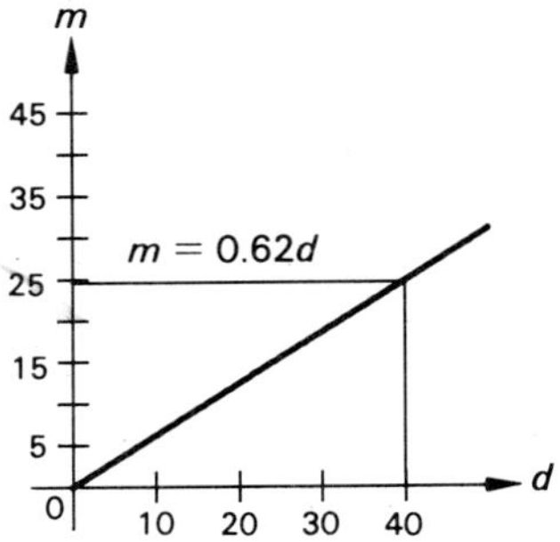

31.

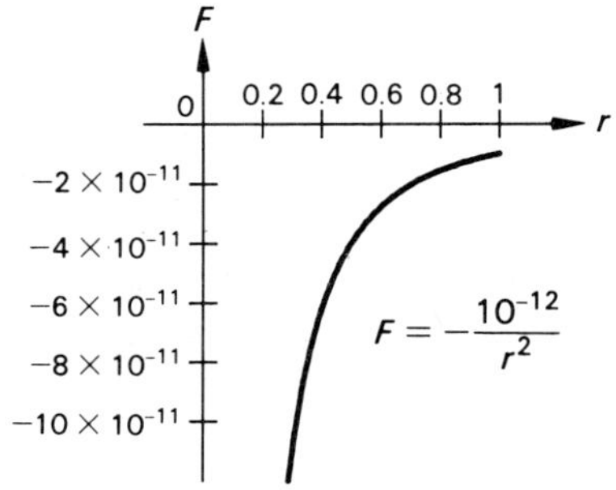

33.

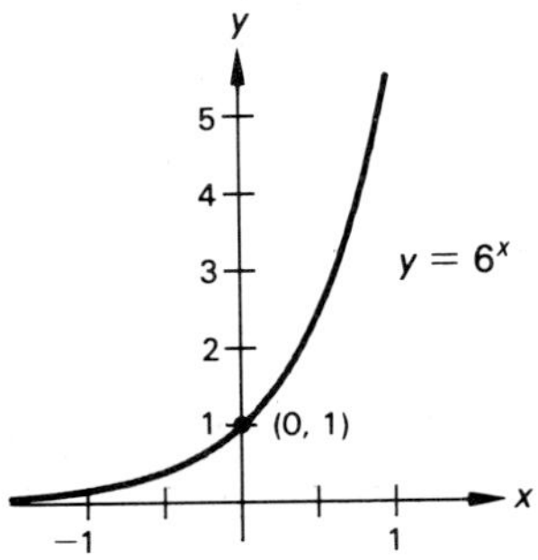

35.

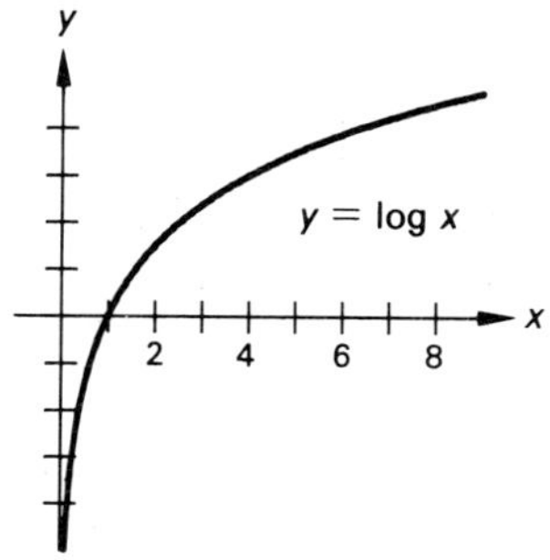

37.

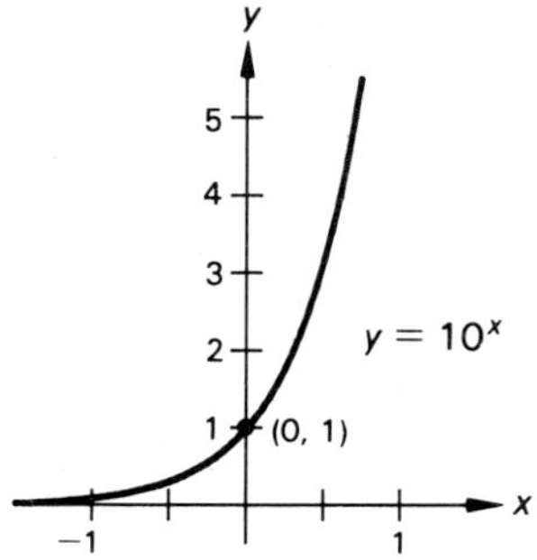

5.4 Exercises (p. 179)

1. $y = x - 5$ **3.** $y = \dfrac{x-7}{2}$ **5.** $y = \dfrac{x}{2} + 4$ **7.** $y = 2x - 4$ **9.** $y = \dfrac{9-3x}{2}$ **11.** $y = 2x - 7$

13. $y = \dfrac{1}{x} - 2$ **15.** $y = \dfrac{2}{x} - 1$ **17.** $y = \dfrac{1}{3}\left(\dfrac{5}{x} - 2\right)$ **19.**

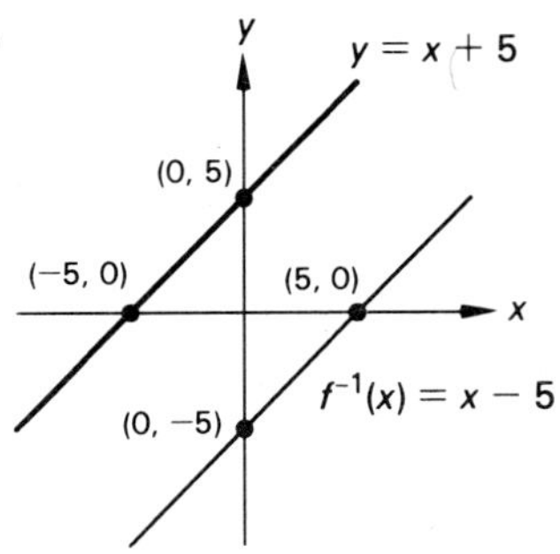

21.

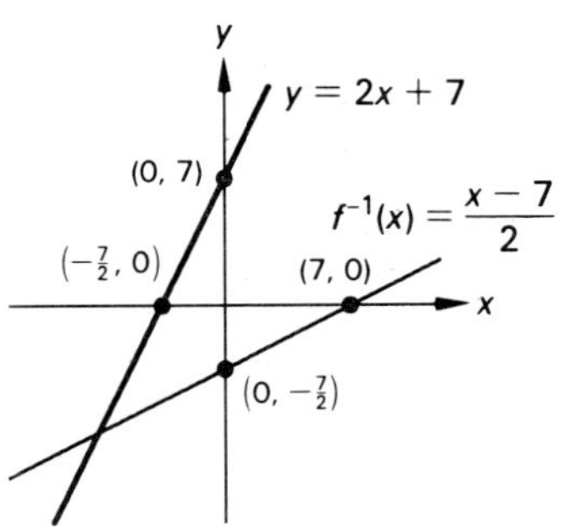

23.

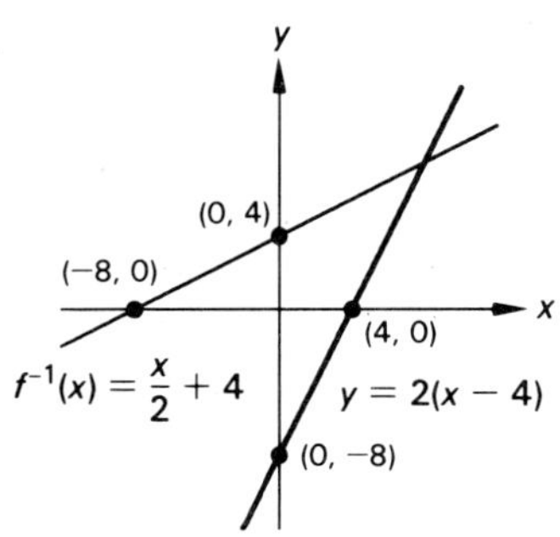

25.

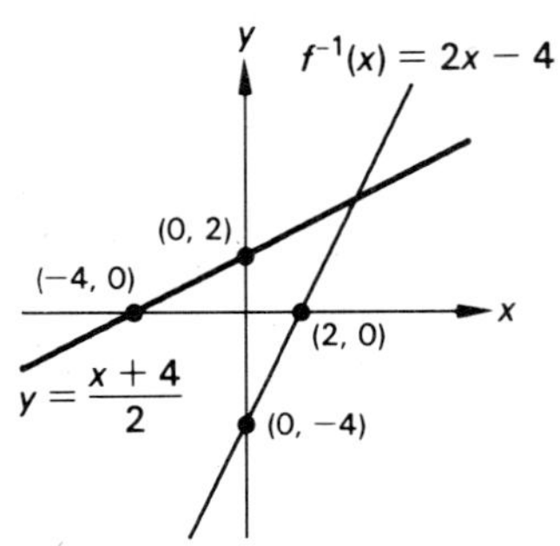

27.

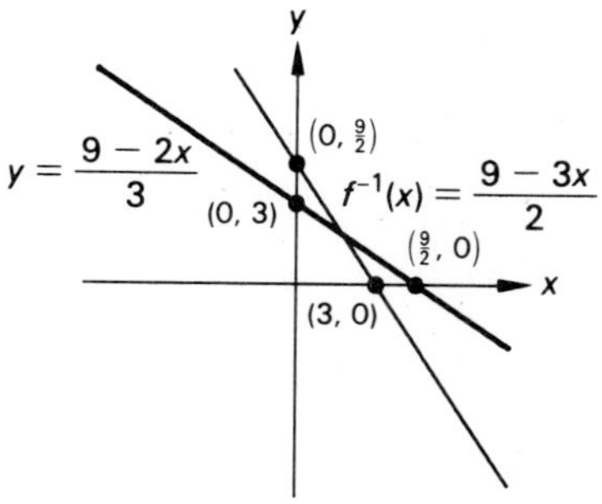

29.

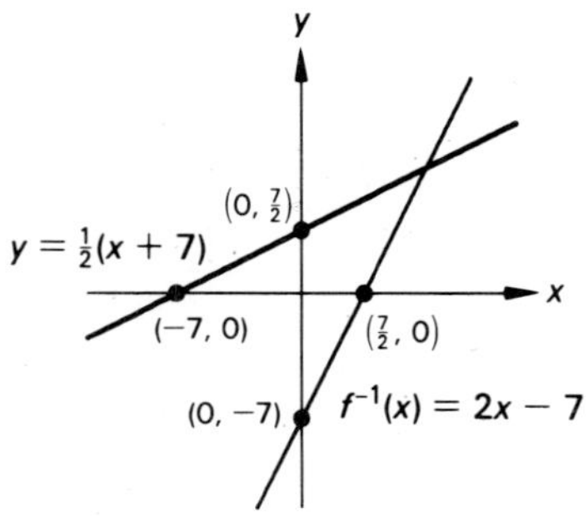

31. f: $D = \{x|x \neq -2\}$ $R = \{y|y \neq 0\}$ f^{-1}: $D = \{x|x \neq 0\}$ $R = \{y|y \neq -2\}$
33. f: $D = \{x|x \neq -1\}$ $R = \{y|y \neq 0\}$ f^{-1}: $D = \{x|x \neq 0\}$ $R = \{y|y \neq -1\}$
35. f: $D = \{x|x \neq -\frac{2}{3}\}$ $R = \{y|y \neq 0\}$ f^{-1}: $D = \{x|x \neq 0\}$ $R = \{y|y \neq -\frac{2}{3}\}$
37. $D = \left\{x|x \neq \frac{9}{11}\right\}$ $R = \{y|y \neq 0\}$ $y = \frac{1}{11}\left(\frac{2}{x} + 9\right)$ 39. Mirror image with respect to the line $y = x$.

Review Exercises (p. 182)

1. a. −4 b. 11 c. $3h + 14$ 3. a. 13 b. 13 c. $z^2 + 4z + 8$ 5. a. 13.0 mm^3 b. 395. m^3 c. 0.630 in.^{3} 7. a. $3.00 \frac{\text{kg} \cdot \text{m}}{\text{s}^2}$ b. $1{,}575{,}\underline{0}00. \frac{\text{kg} \cdot \text{m}}{\text{s}^2}$
9. $D = \{A|A \in \mathbb{R}\}$ $R = \{B|B \in \mathbb{R}\}$ 11. $D = \{t|t \in \mathbb{R}\}$ $R = \{f(t)|f(t) \geq 5\}$
13. $D = \{R|R \neq -3\}$ $R = \{f(R)|f(R) \neq 0\}$ 15. $D = \{L|L \neq -3\}$ $R = \{f(L)|f(L) \neq 1\}$
17. $D = \{s|s \leq 5\}$ $R = \{g(s)|g(s) \geq 0\}$ 19. $f(s) = s^2$ 21. $f(t) = kt^2$ 23. 9.22, $(1\frac{1}{2}, 6)$
25. 14.1, (−1, −2) 27. 3.6, (1.8, −3.8) 29.

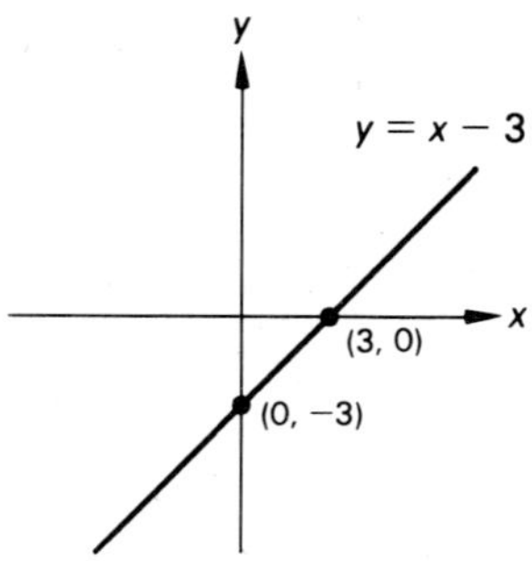

31.

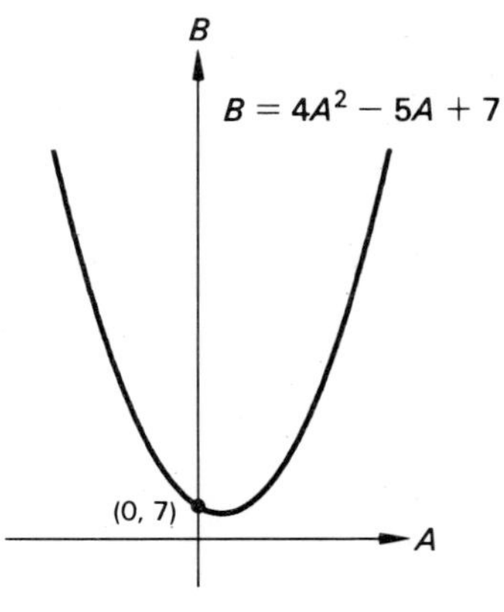

33.

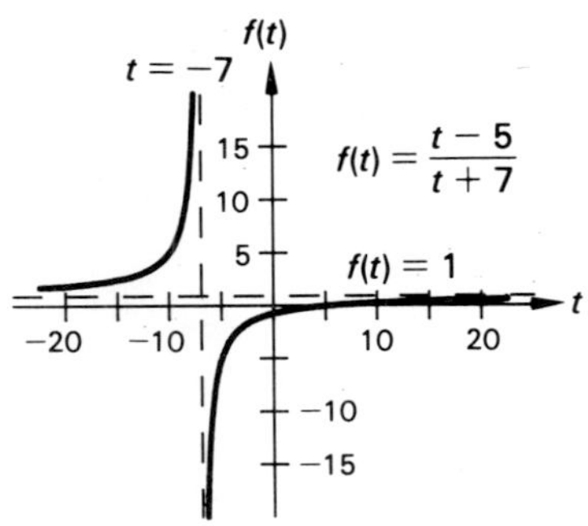

35.

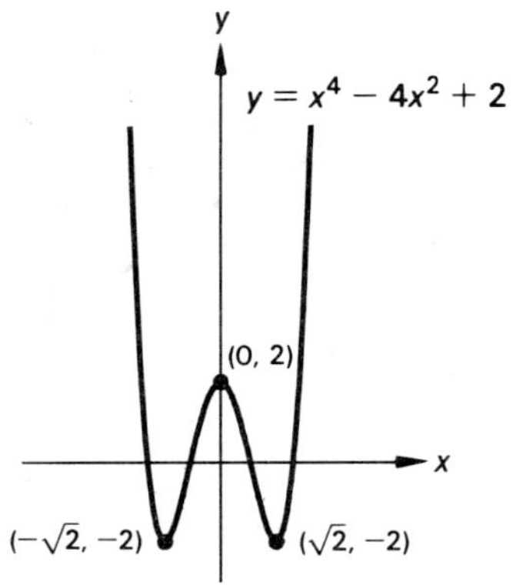

37.

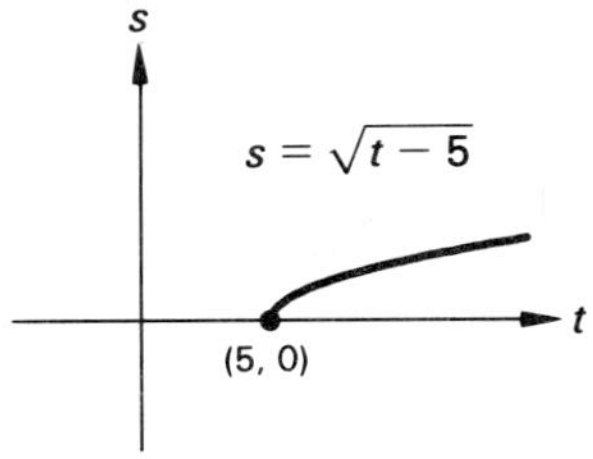

39. **a.** $y = x - 6$

b. f: $D = \{x | x \in \mathbb{R}\}$ $R = \{y | y \in \mathbb{R}\}$
f^{-1}: $D = \{x | x \in \mathbb{R}\}$ $R = \{y | y \in \mathbb{R}\}$

c.

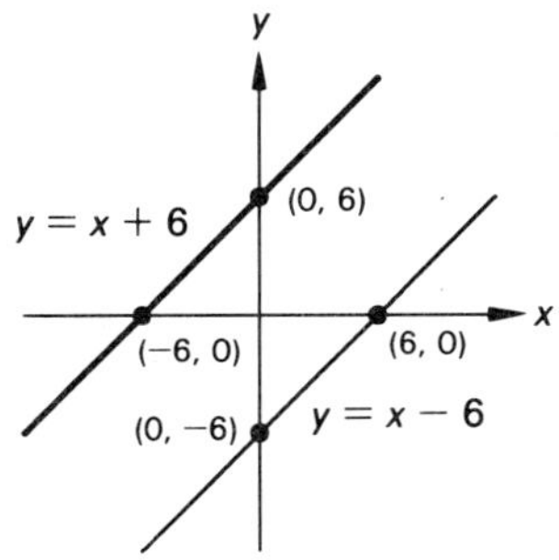

41. **a.** $y = \dfrac{x - 5}{6}$

b. f: $D = \{x | x \in \mathbb{R}\}$ $R = \{y | y \in \mathbb{R}\}$
f^{-1}: $D = \{x | x \in \mathbb{R}\}$ $R = \{y | y \in \mathbb{R}\}$

c.

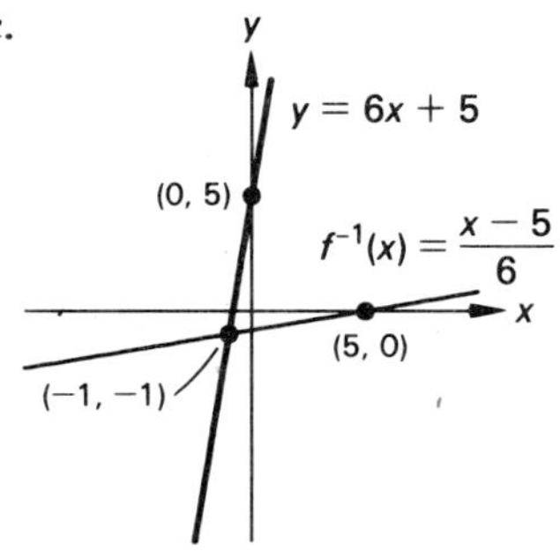

43. **a.** $y = \dfrac{3x - 6}{2}$

b. f: $D = \{x | x \in \mathbb{R}\}$ $R = \{y | y \in \mathbb{R}\}$
f^{-1}: $D = \{x | x \in \mathbb{R}\}$ $R = \{y | y \in \mathbb{R}\}$

c.

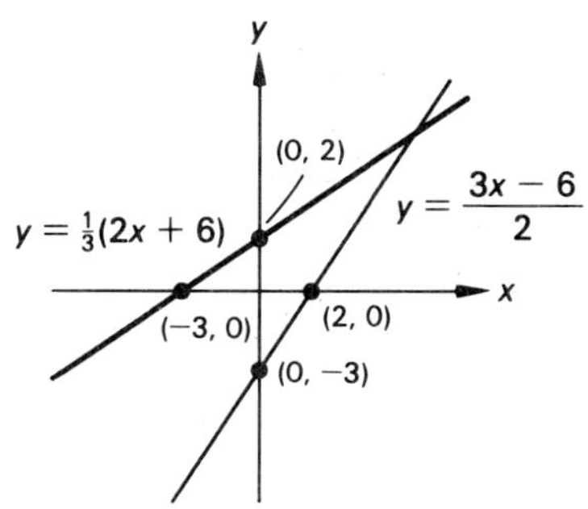

45. **a.** $y = \frac{1}{x} - 2$

b. f: $D = \{x|x \neq -2\}$ $R = \{y|y \neq 0\}$
f^{-1}: $D = \{x|x \neq 0\}$ $R = \{y|y \neq -2\}$

c.

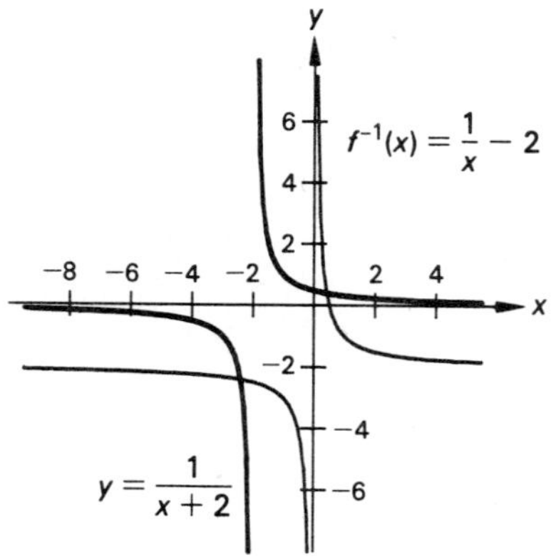

47. $d = \sqrt{(48t)^2 + (30t)^2}$

49. $\overline{AB} = 5.66$, $\overline{BC} = 8$, $\overline{CA} = 5.66$

51. **a.**

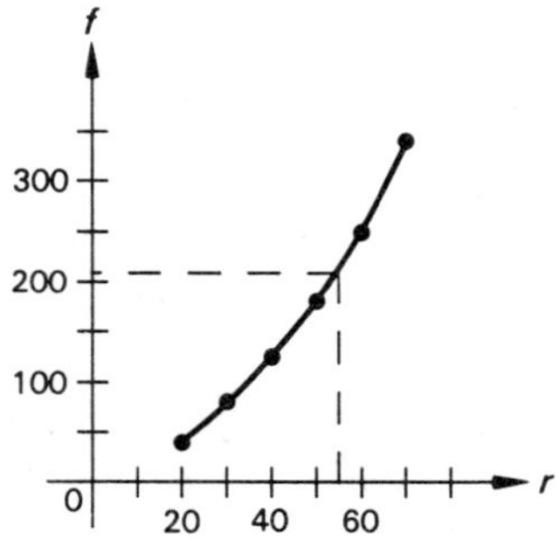

b. 210 ft

CHAPTER 6

6.1 Exercises (p. 189)

1. $\frac{\pi}{12}$ **3.** $\frac{\pi}{4}$ **5.** $\frac{5\pi}{12}$ **7.** $\frac{3\pi}{4}$ **9.** $\frac{7\pi}{6}$ **11.** $-\frac{5\pi}{4}$ **13.** $\frac{5\pi}{3}$ **15.** 5.760 **17.** 3.709 **19.** −2.094 **21.** 0.681 **23.** 5.061 **25.** $22\frac{1}{2}°$ **27.** 45° **29.** 75° **31.** 135° **33.** 210° **35.** 225° **37.** 300° **39.** 330° **41.** 99.695° **43.** 57.296° **45.** $\frac{\pi}{2}$, 90° **47.** 147. mm **49.** $\frac{\pi}{18}\,\mu\,\frac{\text{A}}{\text{rad}}$ **51.** 990. m **53.** Second wheel

6.2 Exercises (p. 193)

1. $\sin\theta = \frac{4}{5}$, $\cos\theta = \frac{3}{5}$ $\tan\theta = \frac{4}{3}$, $\cot\theta = \frac{3}{4}$ $\sec\theta = \frac{5}{3}$, $\csc\theta = \frac{5}{4}$

3. $\sin\theta = \frac{1}{\sqrt{2}}$, $\cos\theta = \frac{1}{\sqrt{2}}$ $\tan\theta = 1$, $\cot\theta = 1$ $\sec\theta = \sqrt{2}$, $\csc\theta = \sqrt{2}$

5. $\sin\theta = \frac{5}{\sqrt{41}}$, $\cos\theta = \frac{4}{\sqrt{41}}$ $\tan\theta = \frac{5}{4}$, $\cot\theta = \frac{4}{5}$ $\sec\theta = \frac{\sqrt{41}}{4}$, $\csc\theta = \frac{\sqrt{41}}{5}$

7. $\sin\theta = \dfrac{\sqrt{2}}{2}$, $\cos\theta = \dfrac{\sqrt{2}}{2}$ $\tan\theta = 1$, $\cot\theta = 1$ $\sec\theta = \dfrac{2}{\sqrt{2}}$, $\csc\theta = \dfrac{2}{\sqrt{2}}$

9. $\sin\phi = \dfrac{2}{\sqrt{5}}$, $\cos\phi = \dfrac{1}{\sqrt{5}}$ $\tan\phi = 2$, $\cot\phi = \dfrac{1}{2}$ $\sec\phi = \sqrt{5}$, $\csc\phi = \dfrac{\sqrt{5}}{2}$

11. $\dfrac{3}{\sqrt{13}}, \dfrac{\sqrt{13}}{2}$ **13.** $\dfrac{5}{\sqrt{29}}, \dfrac{\sqrt{29}}{2}$ **15.** $\dfrac{3}{\sqrt{5}}, \dfrac{2}{\sqrt{5}}$ **17.** $\dfrac{2}{\sqrt{2}}, 1$

19.

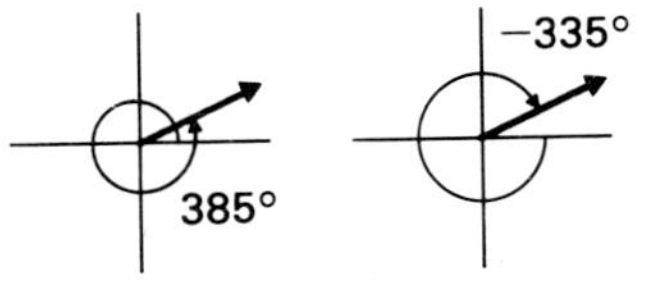

21.

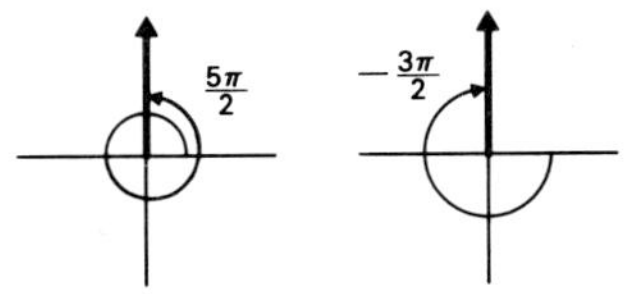

23.

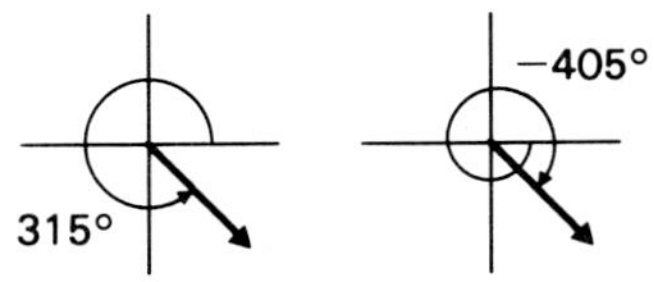

25.

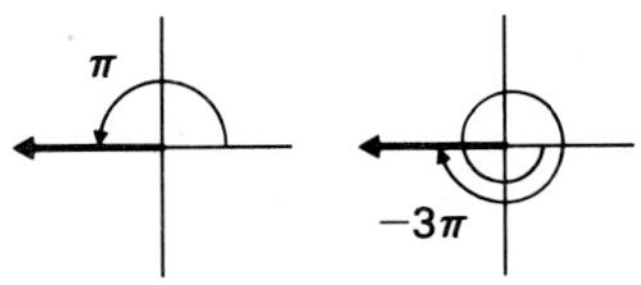

27.

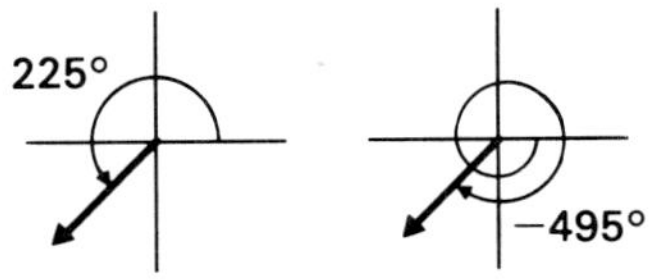

29.

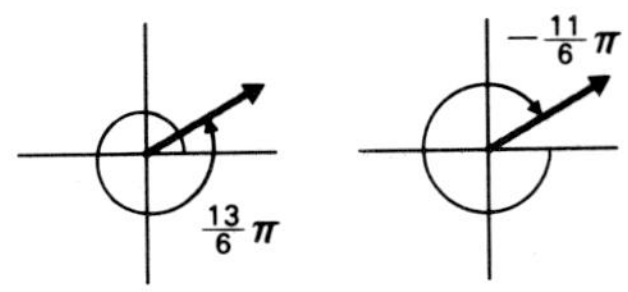

6.3 Exercises (p. 199)

1. 0.20791 **3.** 0.20791 **5.** 0.34433 **7.** 0.05004 **9.** 0.76484 **11.** 0.84683
13. 0.45243 **15.** 1.48256 **17.** 0.26750 **19.** 0.36975 **21.** 0.36975 **23.** 0.75355
25. $\sin 45^\circ = \dfrac{1}{\sqrt{2}}$, $\csc 45^\circ = \sqrt{2}$ $\cos 45^\circ = \dfrac{1}{\sqrt{2}}$, $\sec 45^\circ = \sqrt{2}$ $\tan 45^\circ = 1$, $\cot 45^\circ = 1$
27. $\sin 0^\circ = 0$, $\csc 0^\circ =$ undefined $\cos 0^\circ = 1$, $\sec 0^\circ = 1$ $\tan 0^\circ = 0$, $\cot 0^\circ =$ undefined
29. 0.95630 **31.** 0.80073 **33.** 1.66428 **35.** −1.00000 **37.** 0.86603 **39.** 1.00000
41. 76.8° **43.** 30.0° **45.** 30.0° **47.** 60.0° **49.** 0.79 **51.** 0.79 **53.** 1.02085
55. 1.02085 **57.** 1.46594 **59.** 0.62806 **61.** 9.93782 **63.** 100.00167 **65.** 88.0°
67. 88.9° **69.** 89.0° **71.** 60.0° **73.** 1.05 **75.** 1.05 **77.** $29.\ \dfrac{\text{km}}{\text{h}}$ **79.** $21.\ \dfrac{\text{km}}{\text{h}}$

6.4 Exercises (p. 205)

1. $B = 74.7^\circ$, $C = 90^\circ$, $a = 4.33$ in., $b = 15.8$ in. **3.** $A = 25.8^\circ$, $C = 90^\circ$, $a = 13.7$ mm, $b = 28.3$ mm
5. $B = 45^\circ$, $C = 90^\circ$, $a = 1.82$ mm, $c = 2.57$ mm **7.** $A = 71.8^\circ$, $C = 90^\circ$, $b = 1.00$ ft, $c = 3.20$ ft
9. $A = 45^\circ$, $C = 90^\circ$, $a = 22.3$ in., $c = 31.5$ in. **11.** $B = 30^\circ$, $C = 90^\circ$, $b = 8.49$ m, $c = 17.0$ m
13. 10.6 m² **15.** 113. mm² **17.** 156. ft **19.** 53.1°, 36.9° **21.** 1353. ft **23. a.** 549. ft
b. 511. ft **25.** 35.3° **27.** Circle (21.8°) **29.** 70.9 m **31.** $X_L = 11.3\ \Omega$, $\theta = 30.8^\circ$
33. $P_A = 4.34$ VA, $P_R = 3.98$ var **35.** 9.37 in. **37.** 1.38 m **39.** 49.1 mm, 100.7 mm

6.5 Exercises (p. 210)

1. + **3.** + **5.** − **7.** + **9.** − **11.** + **13.** + **15.** − **17.** − **19.** − **21.** +
23. + **25.** + **27.** + **29.** − **31.** − **33.** + **35.** +
37. $\sin\theta = \dfrac{4}{5}$, $\csc\theta = \dfrac{5}{4}$ $\cos\theta = -\dfrac{3}{5}$, $\sec\theta = -\dfrac{5}{3}$ $\tan\theta = -\dfrac{4}{3}$, $\cot\theta = -\dfrac{3}{4}$

39. $\sin\theta = -\frac{4}{5}$, $\csc\theta = -\frac{5}{4}$ $\cos\theta = -\frac{3}{5}$, $\sec\theta = -\frac{5}{3}$ $\tan\theta = \frac{4}{3}$, $\cot\theta = \frac{3}{4}$
41. $\sin\theta = -\frac{5}{\sqrt{41}}$, $\csc\theta = -\frac{\sqrt{41}}{5}$ $\cos\theta = \frac{4}{\sqrt{41}}$, $\sec\theta = \frac{\sqrt{41}}{4}$ $\tan\theta = -\frac{5}{4}$, $\cot\theta = -\frac{4}{5}$
43. $\sin\theta = \frac{5}{\sqrt{29}}$, $\csc\theta = \frac{\sqrt{29}}{5}$ $\cos\theta = \frac{2}{\sqrt{29}}$, $\sec\theta = \frac{\sqrt{29}}{2}$ $\tan\theta = \frac{5}{2}$, $\cot\theta = \frac{2}{5}$ **45.** I **47.** II
49. III **51.** I **53.** IV **55.** II

6.6 Exercises (p. 216)

1. 0.67816 **3.** −8.38625 **5.** −0.74780 **7.** −1.12433 **9.** −1.08749 **11.** 0.66131
13. 1.01543 **15.** −0.45995 **17.** −0.97667 **19.** −0.74548 **21.** −0.25862 **23.** 0.97113
25. 0.77807 **27.** −2.66316 **29.** 0.26443 **31.** 48.3°, 131.7° **33.** 36.9°, 216.9°
35. 138.2°, 221.8° **37.** 213.0°, 327.0° **39.** 55.0°, 235.0° **41.** 59.5°, 300.5° **43.** 0.96, 2.18
45. 2.30, 5.45 **47.** 1.75, 4.54 **49.** −0.80901 **51.** 0.98481 **53.** −0.70021 **57.** 547. ft^2
59. a. 31.8 V **b.** −45 V **63.** −0.70711 **65.** −11.43005 **67.** −1.34563

6.7 Exercises (p. 223)

1. 6.9°, 0.12 rad **3.** $1037.\ \frac{mi}{hr}$ **5. a.** 7.7 m **b.** 4.0 m **7. a.** 42.9° **b.** 82.9° **c.** 140°

9. 1221. m **11.** 16,965. mm^2 **13.** 148. ft^2 **15.** $0.476\ \frac{r}{hr}$ **17.** 52.8 m

19. $1047.\ \frac{in.}{min}$, $3.49\ \frac{rad}{s}$ **21. a.** $12.6\ \frac{rad}{s}$ **b.** $377.\ \frac{mm}{s}$ **c.** $377.\ \frac{mm}{s}$ **d.** $5.03\ \frac{rad}{s}$ **e.** 48 rpm

Review Exercises (p. 226)

1. $\frac{7\pi}{18}$, 1.222 **3.** $\frac{11\pi}{9}$, 3.840 **5.** $-\frac{4\pi}{9}$, −1.396 **7.** 165° **9.** 114.592°
11. $\sin\theta = \frac{7}{\sqrt{74}}$, $\csc\theta = \frac{\sqrt{74}}{7}$ $\cos\theta = \frac{5}{\sqrt{74}}$, $\sec\theta = \frac{\sqrt{74}}{5}$ $\tan\theta = \frac{7}{5}$, $\cot\theta = \frac{5}{7}$
13. $\sin\theta = \frac{\sqrt{2}}{2}$, $\csc\theta = -\frac{2}{\sqrt{2}}$ $\cos\theta = -\frac{\sqrt{2}}{2}$, $\sec\theta = -\frac{2}{\sqrt{2}}$ $\tan\theta = 1$, $\cot\theta = 1$
15. $\sin\theta = \frac{1}{\sqrt{2}}$, $\csc\theta = \sqrt{2}$ $\cos\theta = \frac{1}{\sqrt{2}}$, $\sec\theta = \sqrt{2}$ $\tan\theta = 1$, $\cot\theta = 1$
17. $\frac{2}{\sqrt{7}}, -\frac{\sqrt{7}}{\sqrt{3}}$ **19.** $\frac{\sqrt{50}}{7}, \frac{1}{\sqrt{50}}$ **21.** 0.31896 **23.** −6.65244 **25.** −1.03724 **27.** −0.95282
29. 0.54097 **31.** 13.6°, 0.24; 166.4°, 2.90 **33.** 13.5°, 0.24; 193.5°, 3.38
35. 105.0°, 1.83; 255.0°, 4.45 **37.** 40.0°, 0.70; 320.0°, 5.59 **39.** 270.0°, 4.71
41. $B = 5.15°$, $C = 90°$, $a = 12.1$ mm, $b = 15.2$ mm **43.** $A = 65.3°$, $C = 90°$, $a = 22.0$ in., $b = 10.1$ in.

45. $B = 0.82$, $C = \frac{\pi}{2}$, $B = 15.6$ m, $c = 21.3$ m **47. a.** 693. ft **b.** 996. ft **49.** 120.°

51. 812. rad **53.** 59.0 mm **55.** 104. ft **57. a.** 8 cm **b.** −5.14 cm **c.** 7.88 cm
59. 2.20 cm^2

CHAPTER 7

7.1 Exercises (p. 234)

1. Vector **3.** Vector **5.** Scalar **7.** Scalar **9.** Vector

11., 13.

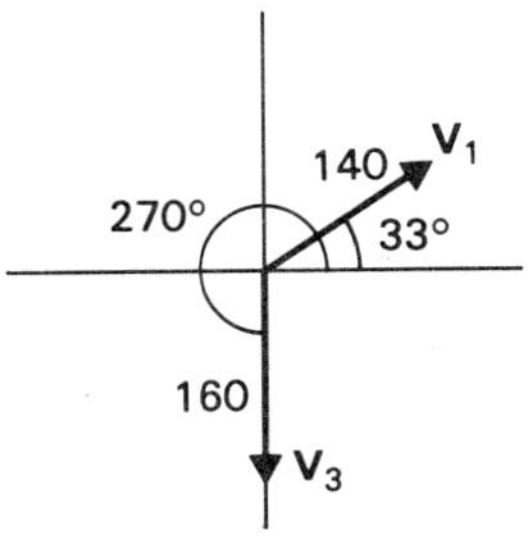

15.

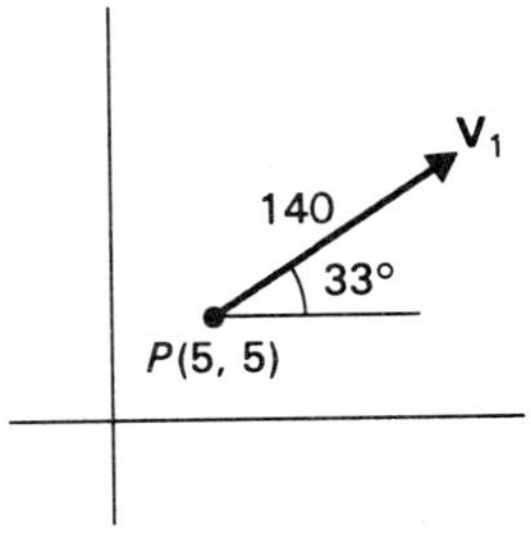

17.

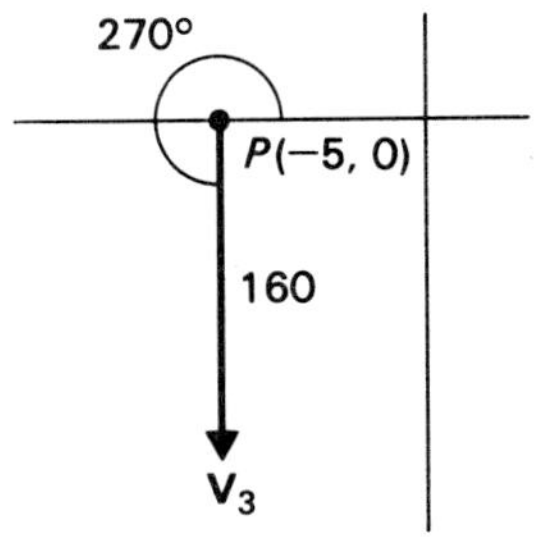

19. $V_V = 6.54 \frac{cm}{s}$

$V_H = 74.7 \frac{cm}{s}$

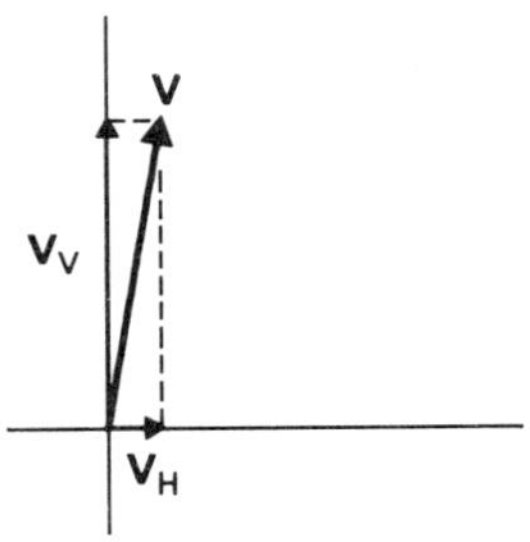

21. $F_H = -3.72$ lb
$F_V = 24.7$ lb

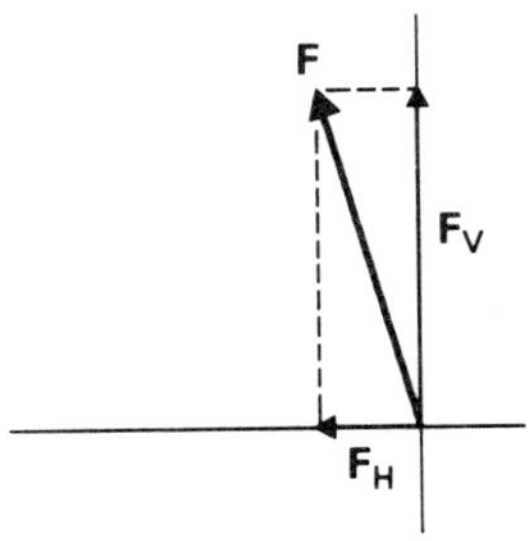

23. $D_H = -36.8$ km
$D_V = -30.9$ km

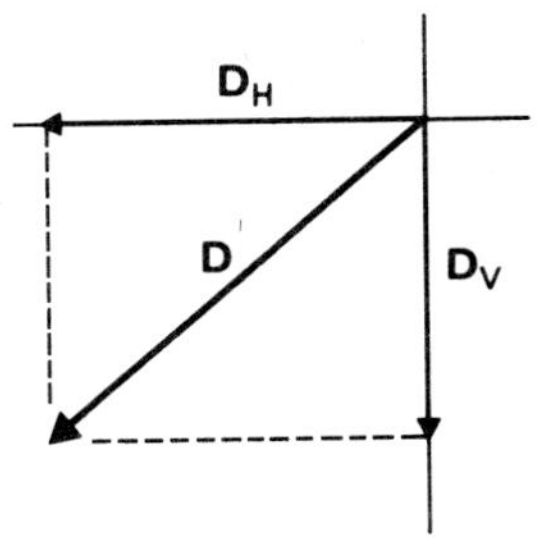

25. $F_H = 114.3$ dyne
$F_V = -89.3$ dyne

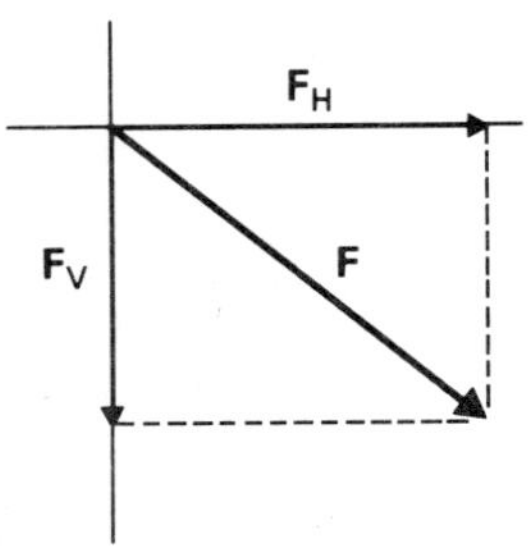

27. $A_H = 18.5 \frac{cm}{s^2}$

$A_V = 26.4 \frac{cm}{s^2}$

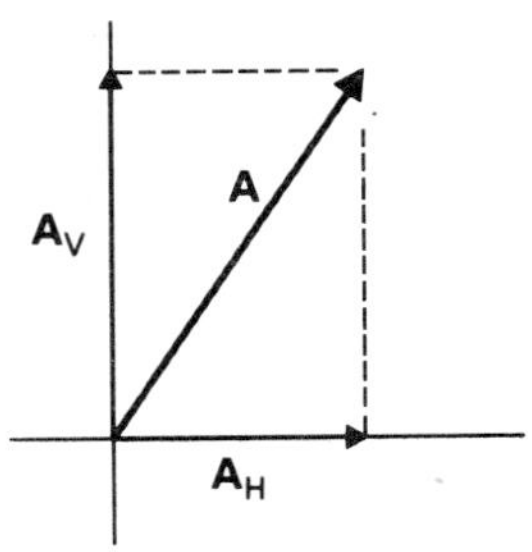

29. $F_H = -0.249$ kg
$F_V = -0.022$ kg

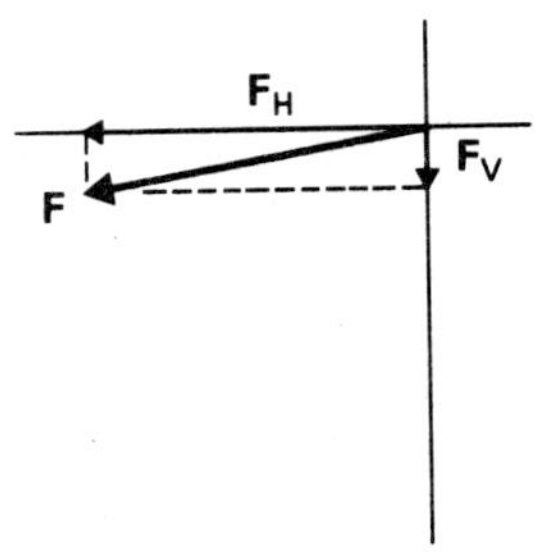

31. $\mathbf{F} = 58.6 \text{ lb } \underline{/53.3^\circ}$

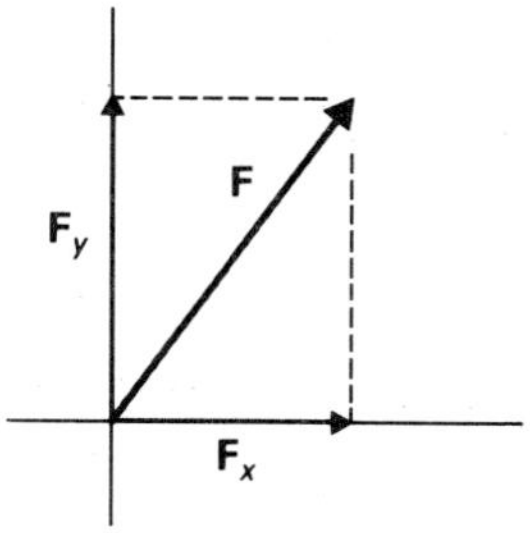

33. $\mathbf{D} = 366. \text{ km } \underline{/-21.7^\circ}$ or $\mathbf{D} = 366. \text{ km } \underline{/338.3^\circ}$

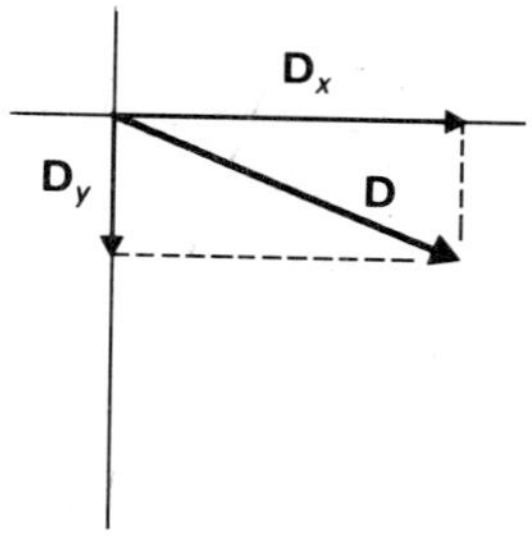

35. $\mathbf{F} = 38.0 \text{ dyne } \underline{/116.6^\circ}$

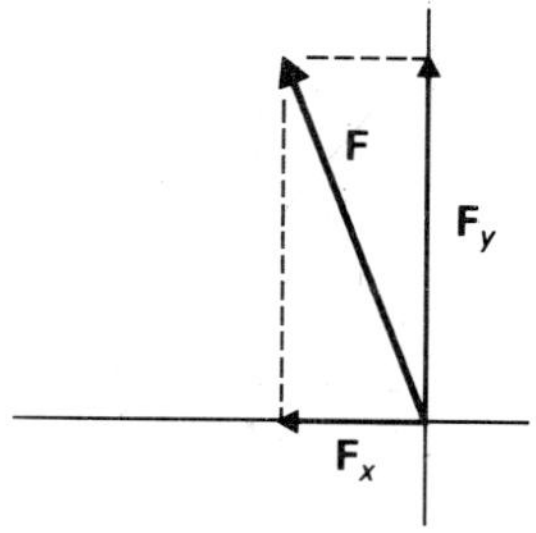

37. $\mathbf{V} = 88.6 \frac{\text{cm}}{\text{s}} \underline{/-118.^\circ}$ or

$\mathbf{V} = 88.6 \frac{\text{cm}}{\text{s}} \underline{/241.7^\circ}$

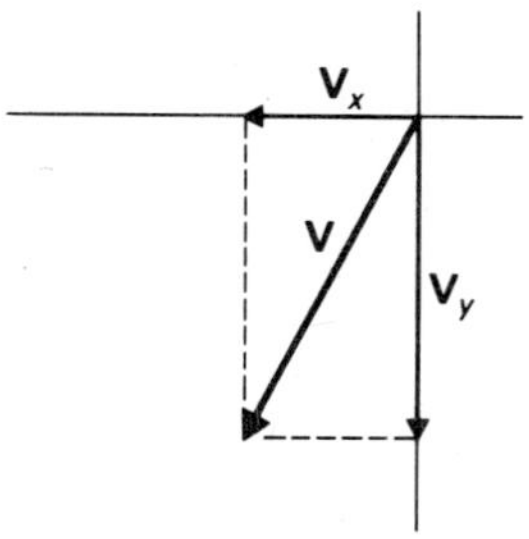

39. $\mathbf{A} = 74.2 \frac{\text{ft}}{\text{s}^2} \underline{/104.^\circ}$

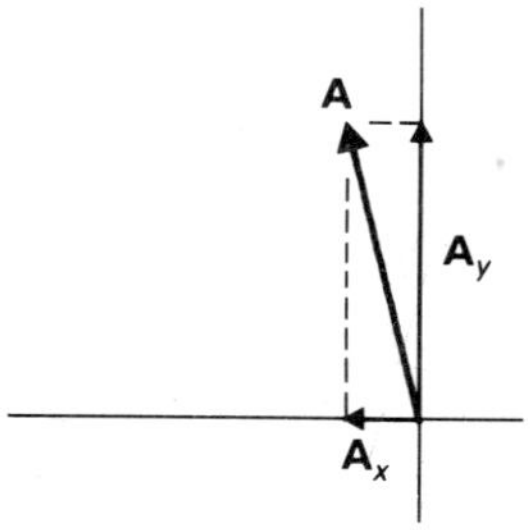

41. $\mathbf{V} = 12.6 \frac{\text{mi}}{\text{hr}} \underline{/18.4^\circ}$

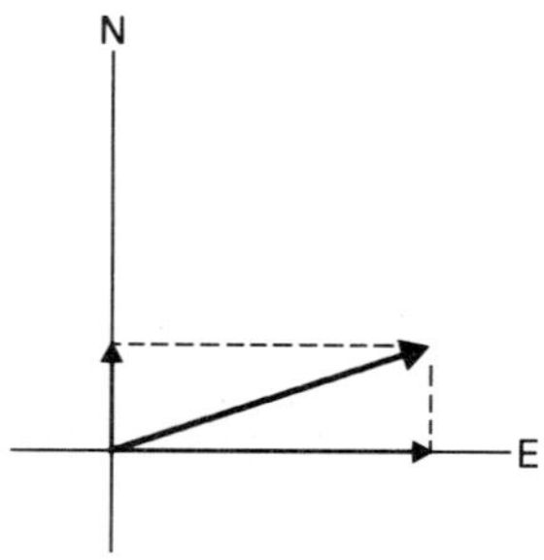

43. $808. \text{ lb } \underline{/68.2^\circ}$ **45. a.** 65.1 lb **b.** 54.6 lb **47.** $45.5\Omega \underline{/-56.7^\circ}$ or $45.5\Omega \underline{/303.3^\circ}$

7.2 Exercises (p. 246)

1. $\mathbf{A} + \mathbf{B} = 5 \underline{/36.9^\circ}$

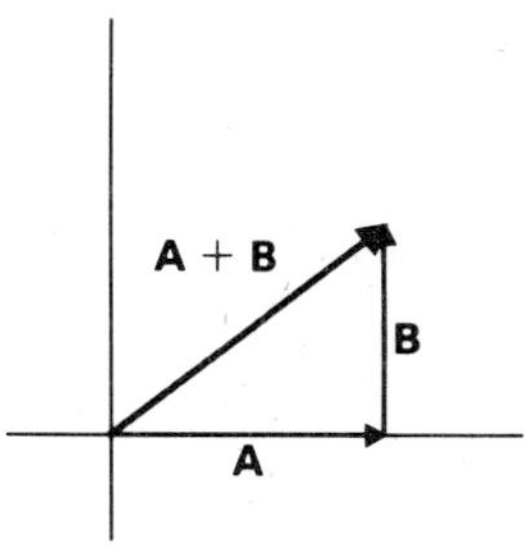

3. $\mathbf{C} + \mathbf{D} = 7.81 \underline{/95.2^\circ}$

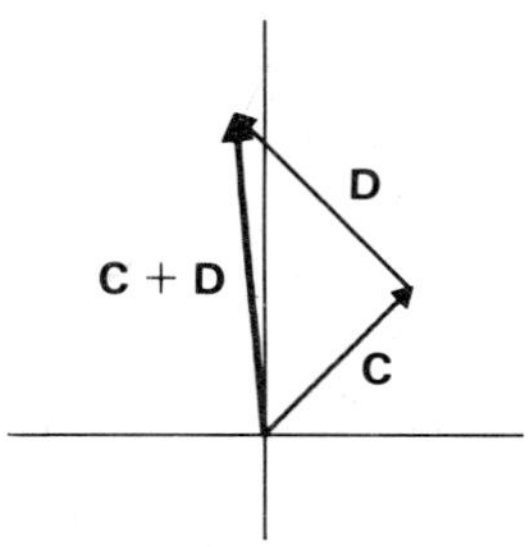

5. $\mathbf{A} + \mathbf{C} + \mathbf{E} = 5 \underline{/45^\circ}$

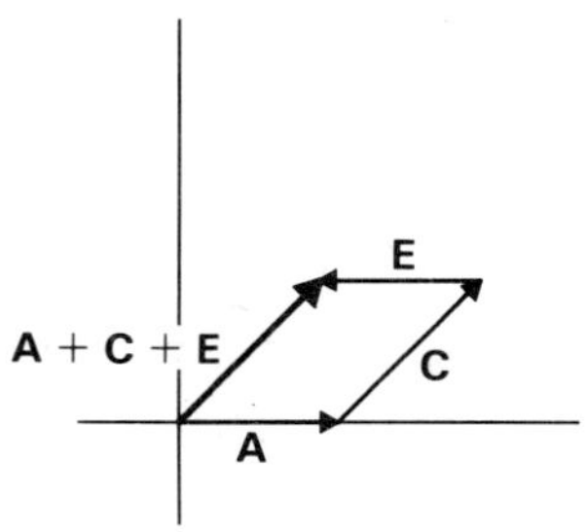

7. $\mathbf{B}+\mathbf{C}+\mathbf{E}=6.55\ \underline{/94.1^\circ}$

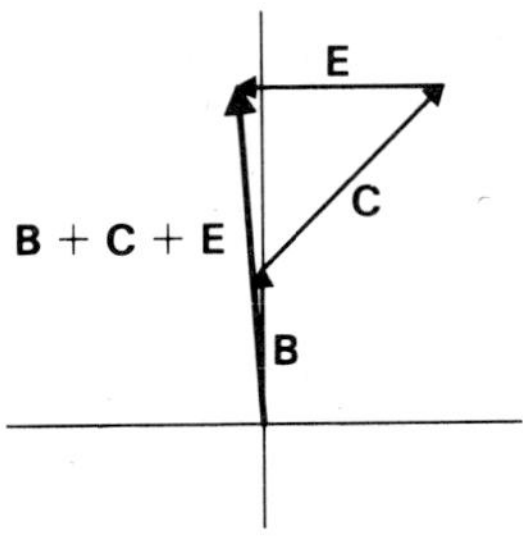

9. $\overline{\mathbf{A}}-\overline{\mathbf{B}}=5\ \underline{/-36.9^\circ}$ or
$\overline{\mathbf{A}}-\overline{\mathbf{B}}=5\ \underline{/323.1^\circ}$

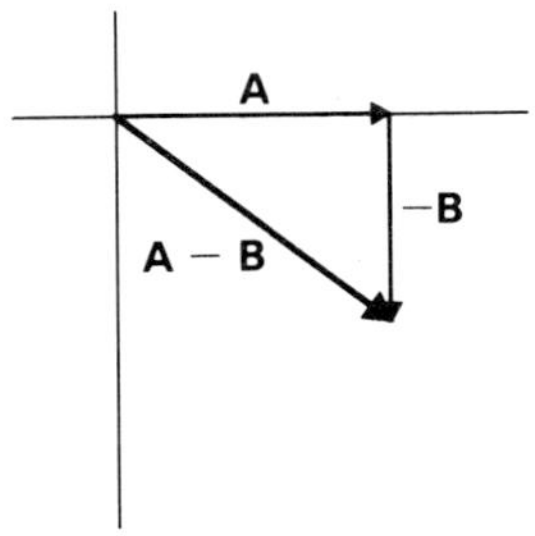

11. $\mathbf{E}-\mathbf{A}=8\ \underline{/180^\circ}$

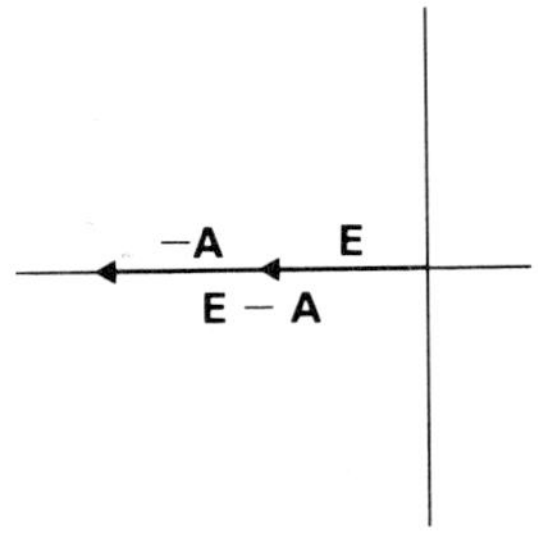

13. $\mathbf{A}+\mathbf{B}+\mathbf{C}+\mathbf{E}=7.4\ \underline{/61.6^\circ}$

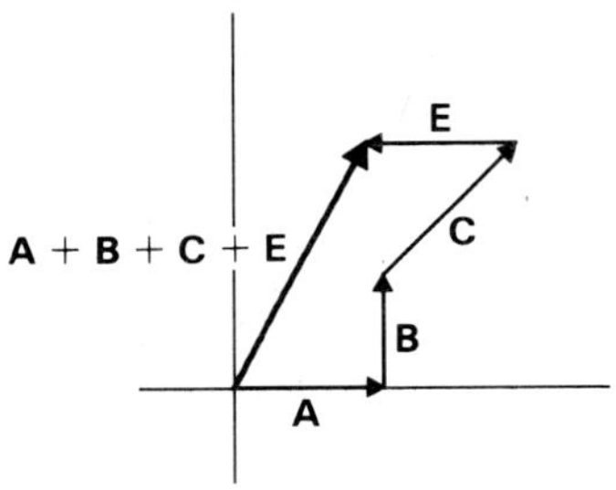

15. $\mathbf{B}+\mathbf{C}+\mathbf{D}+\mathbf{E}=11.8\ \underline{/113.6^\circ}$

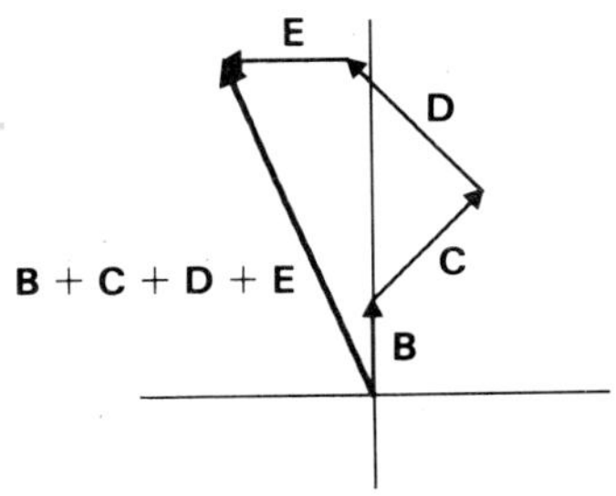

17. $2\mathbf{A}+3\mathbf{B}=12.0\ \underline{/48.4^\circ}$

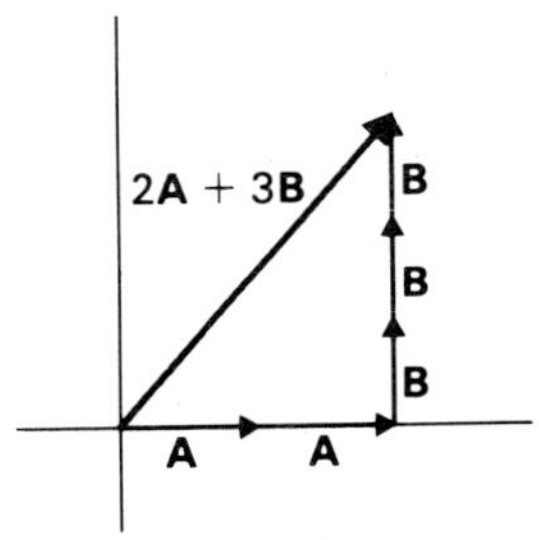

19. $4\mathbf{D}-5\mathbf{E}=17.2\ \underline{/79.9^\circ}$

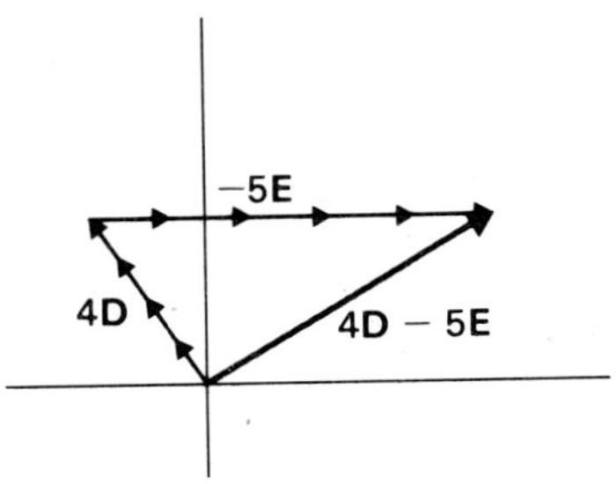

21. 4.41 cm $\underline{/38.6^\circ}$ **23.** 5.5 ft $\underline{/110.^\circ}$ **25.** 32.9 m $\underline{/135.9^\circ}$
27. 0.775 in. $\underline{/127.8^\circ}$ **29.** 38.5 cm $\underline{/142.3^\circ}$
31. 0.747 in. $\underline{/95.6^\circ}$ **33.** 4.65 m $\underline{/1.26}$ **35.** 14.7 ft $\underline{/135.7^\circ}$
37. 13.1 mm $\underline{/-159.2^\circ}$ or 13.1 mm $\underline{/200.8^\circ}$

7.3 Exercises (p. 248)

1. a. 438. lb **b.** 204. lb **c.** 602. lb **3.** 100 mph, 173. mph **5.** $5.20\ \frac{\text{km}}{\text{h}}\ \underline{/30.0^\circ}$

7. $174.\ \frac{\text{km}}{\text{h}}$, $985.\ \frac{\text{km}}{\text{h}}$ **9.** 287. lb $\underline{/-4.31^\circ}$ **11.** 13.7 mph $\underline{/19.4^\circ}$ **13.** 11,069. N

15. 391. N, 295. N **17.** Rest, rest, rest **19.** 97.1 lb **21.** 1791. N

7.4 Exercises (p. 254)

1. $C = 33°$, $a = 7.7$ cm, $c = 4.4$ cm **3.** $B = 42.3°$, $C = 77.7°$, $c = 10.2$ in.
5. $B = 49.1°$, $C = 66.6°$, $c = 5.24$ mm **7.** $A = 11.1°$, $B = 138.4°$, $b = 13.7$ in. **9.** No solution
11. $A = 39°\ 50'$, $a = 295.$ ft, $c = 214.$ ft **13.** $A = 90°$, $B = 45°$, $a = 3.5$ in.
15. $C = 0.34$, $b = 47.6$ cm, $c = 17.9$ cm **17.** $A = 93.5°$, $C = 4.2°$, $a = 2.32 \times 10^5$ m
19. $A = 14°$, $a = 2.21$ ft, $c = 6.69$ ft **21.** 922.5 ft **23.** P, 1.92 ft **25.** 44.7 m **27.** 55.7 mi

7.5 Exercises (p. 258)

1. $A = 57.4°$, $B = 44.6°$, $c = 13.9$ **3.** $B = 65.9°$, $C = 54.2°$, $a = 42.7$
5. $A = 48.3°$, $C = 27.4°$, $b = 67.5$ **7.** $A = 43.9°$, $B = 60.1°$, $c = 76.0°$
9. $A = 91.8°$, $B = 35.3°$, $c = 52.9°$ **11.** $A = 52.22°$, $B = 33.96°$, $c = 93.82°$ **13.** 34° 31′
15. 7221.3 m **17.** 672.8 lb **19.** 13.9 mi **21.** 34.0°

Review Exercises (p. 260)

1. $\mathbf{V}_H = 17.6\ \frac{\text{cm}}{\text{s}}$

$\mathbf{V}_V = 43.6\ \frac{\text{cm}}{\text{s}}$

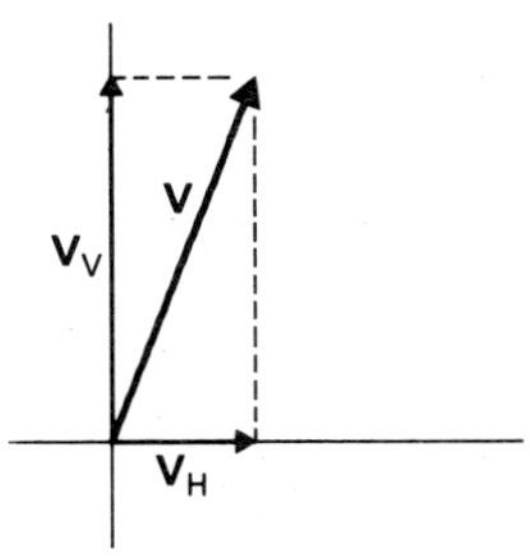

3. $\mathbf{F}_H = -30.3$ lb

$\mathbf{F}_V = -17.5$ lb

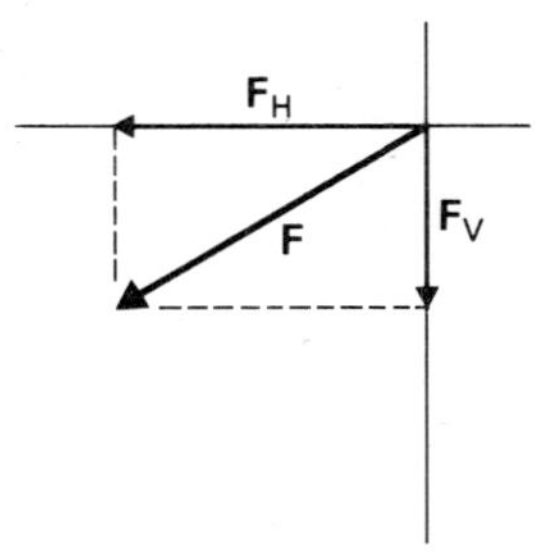

5. $\mathbf{D}_H = 103.9$ km

$\mathbf{D}_V = -60.0$ km

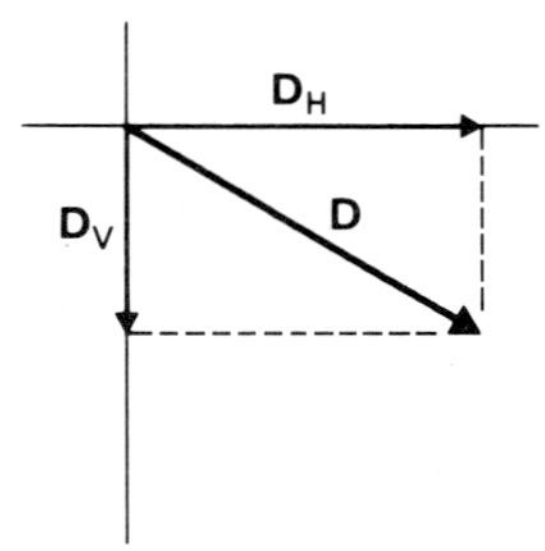

7. $\mathbf{a}_H = -4.79\ \frac{\text{cm}}{\text{s}^2}$

$\mathbf{a}_V = 31.8\ \frac{\text{cm}}{\text{s}^2}$

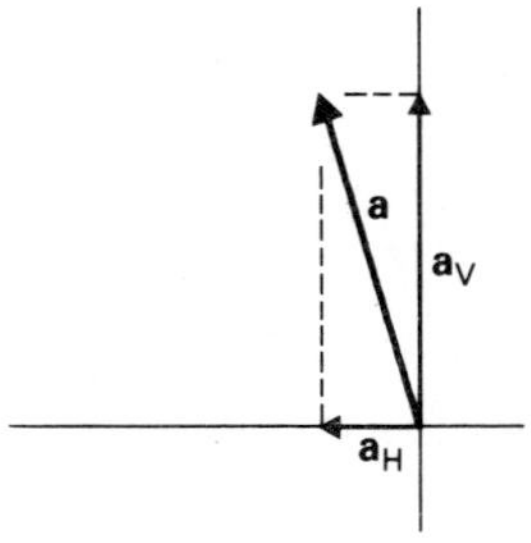

9. $\mathbf{F} = 242.$ lb $\underline{/4.3°}$

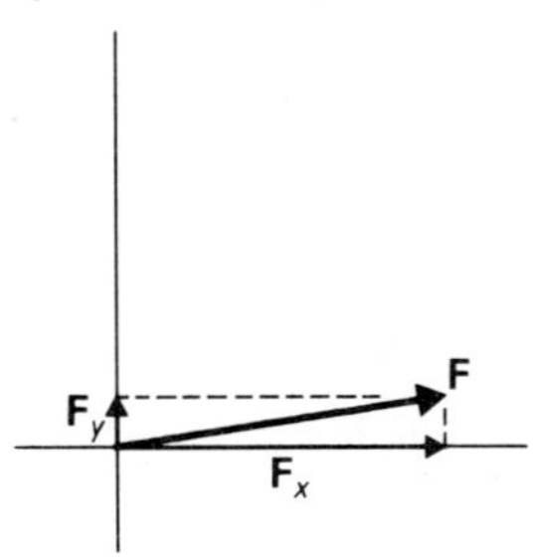

11. $\mathbf{D} = 20.5$ km $\underline{/133.0°}$

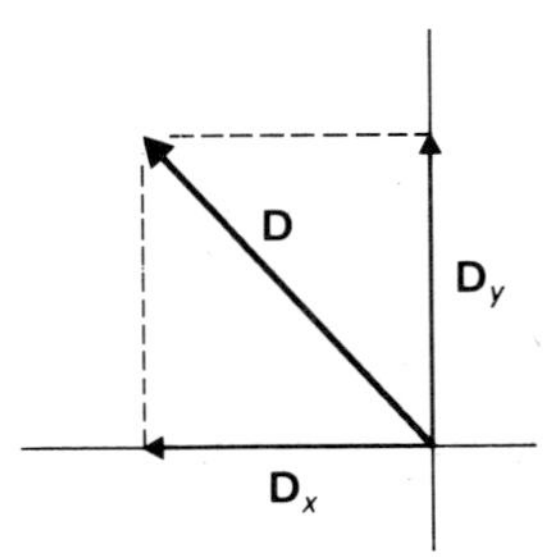

13. $\mathbf{F} = 1.07 \text{ N} \angle -116.1°$ or
$\mathbf{F} = 1.07 \text{ N} \angle 243.9°$

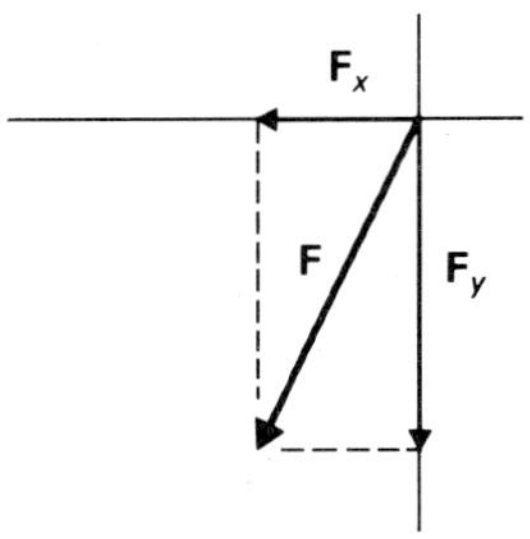

15. $\mathbf{V} = 0.029 \dfrac{\text{m}}{\text{s}} \angle -76.°$ or
$\mathbf{V} = 0.029 \dfrac{\text{m}}{\text{s}} \angle 284.0°$

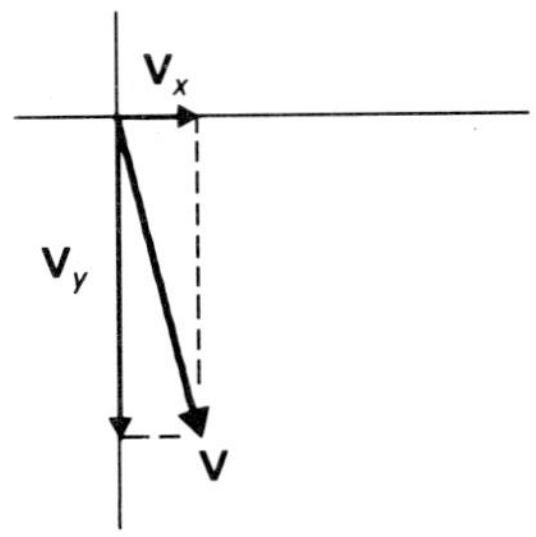

17. 3.61 mph $\angle 146.3°$ **19.** 50.3 Ω $\angle -49.0°$ or 50.3 Ω $\angle 311°$ **21. a.** 2437 lb **b.** 19,851. lb
23. 95.9 cm $\angle 74.5°$ **25.** 1.40 $\angle 37.6°$ **27.** 499. lb **29.** 63.5 N **31.** $C = 74.7°$, $a = 28.2$, $c = 44.0$
33. $B = 30.78°$, $C = 126.22°$, $c = 23.21$ $B = 149.22°$, $C = 7.78°$, $c = 3.893$
35. $A = 54.5°$, $B = 85.5°$, $b = 23.3$ $A = 125.5°$, $B = 14.5°$, $b = 5.85$
37. $A = 71.6°$, $B = 57.5°$, $C = 36.8$ **39.** $A = 52.1°$, $B = 57.2°$, $C = 70.7°$
41. $A = 21.6°$, $B = 12.1°$, $c = 316.9$ **43.** 3.7 mi $\angle 31.1°$ **45.** 20.1 in. **47.** 0.626 mi **49.** 3.4 km

CHAPTER 8

8.1 Exercises (p. 266)

1. a. Yes **b.** 1 **3. a.** No **5. a.** Yes **b.** π **7. a.** No **9. a.** Yes **b.** p
11. a. Yes **b.** p

8.2 Exercises (p. 276)

1. Domain set of real numbers; range $-1 \le y \le 1$; y intercept is 0; θ intercepts are $0, \pm\pi, \pm 2\pi, \ldots$; $\sin(\pi + \theta) = -\sin\theta$; $\sin(-\theta) = -\sin\theta$; $\sin(\pi - \theta) = \sin\theta$; periodic: $p = 2\pi$.

3. Domain reals except $\theta \neq \pm\frac{\pi}{2}, \pm\frac{3\pi}{2}, \pm\frac{5\pi}{2}, \ldots$; range set of real numbers, y intercept is 0; θ intercepts are $0, \pm\pi, \pm 2\pi, \ldots$; $\tan(\pi + \theta) = \tan\theta$; $\tan(-\theta) = -\tan\theta$; $\tan(\pi - \theta) = -\tan\theta$; periodic: $p = \pi$; undefined at $\pm\frac{\pi}{2}, \pm\frac{3\pi}{2}, \ldots$. **5.** Domain reals except $\theta \neq \pm\frac{\pi}{2}, \pm\frac{3\pi}{2}, \pm\frac{5\pi}{2}, \ldots$; range reals $y \le -1$ or $y \ge 1$; y intercept is 1; θ intercepts: none; $\sec(\pi + \theta) = -\sec\theta$; $\sec(-\theta) = \sec\theta$; $\sec(\pi - \theta) = -\sec\theta$; periodic: $p = 2\pi$; undefined at $\pm\frac{\pi}{2}, \pm\frac{3\pi}{2}, \pm\frac{5\pi}{2}, \ldots$.

7.

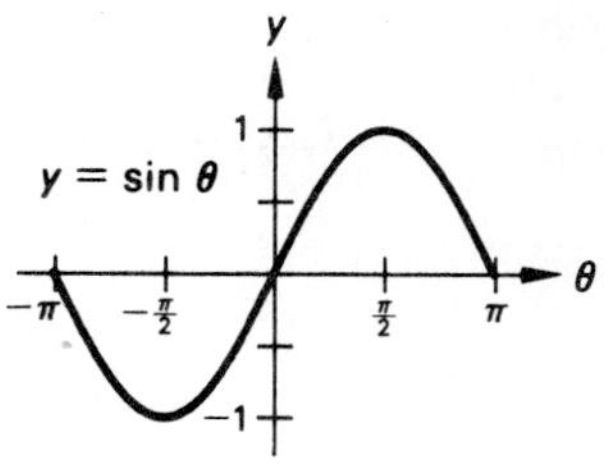

9.

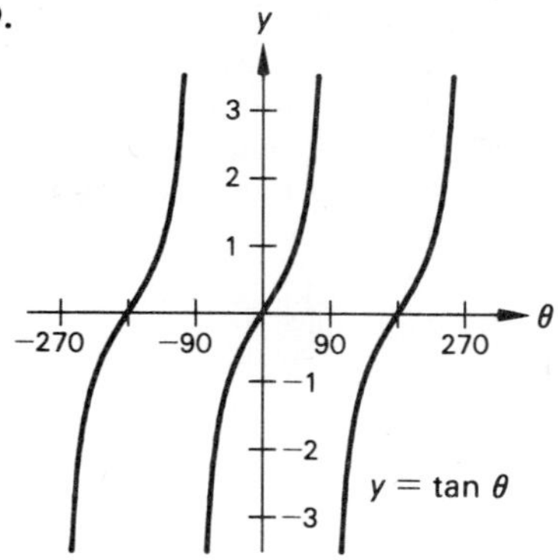

11.

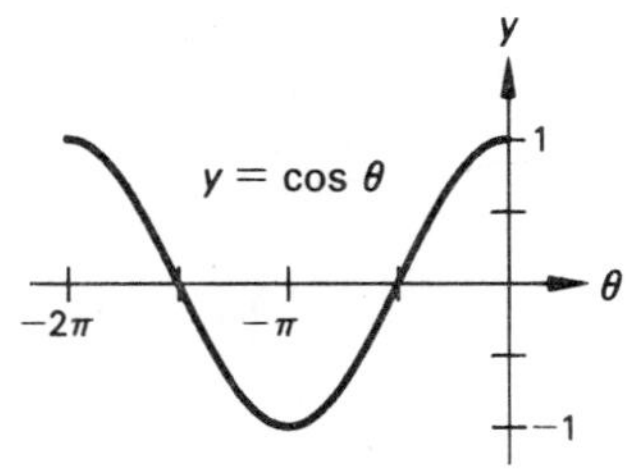

13.

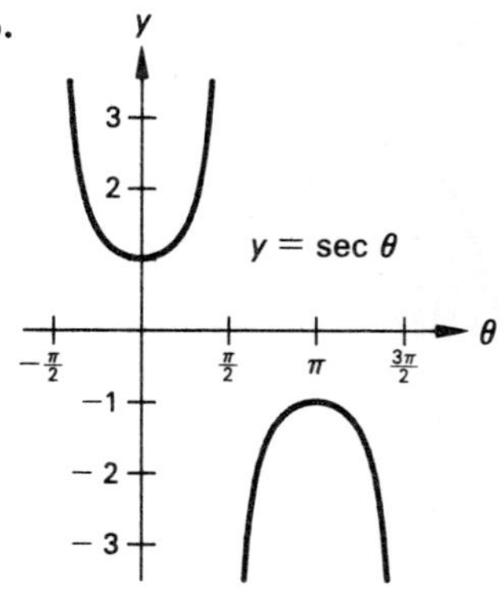

15.

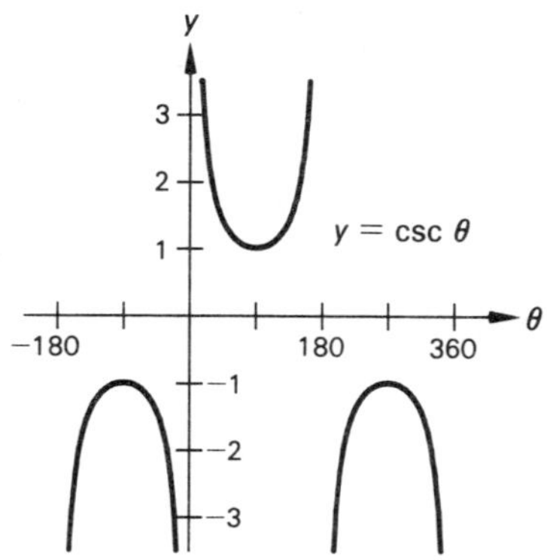

17.

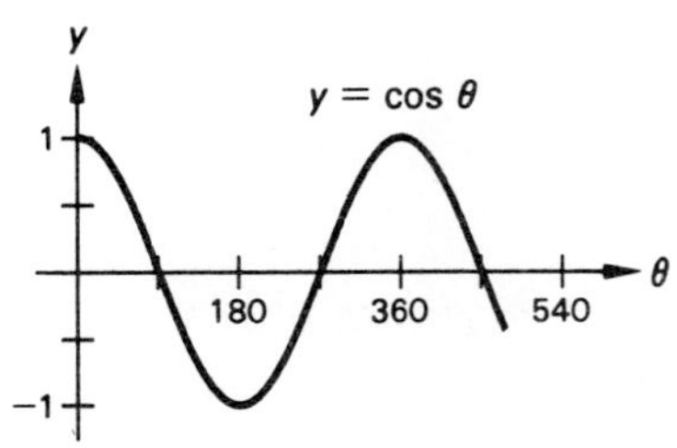

19.

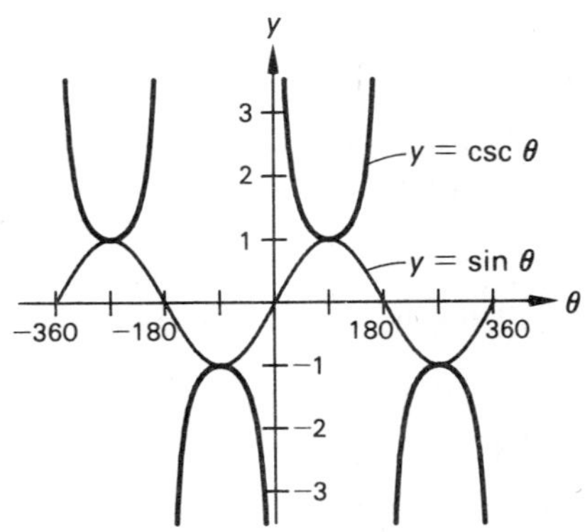

21.

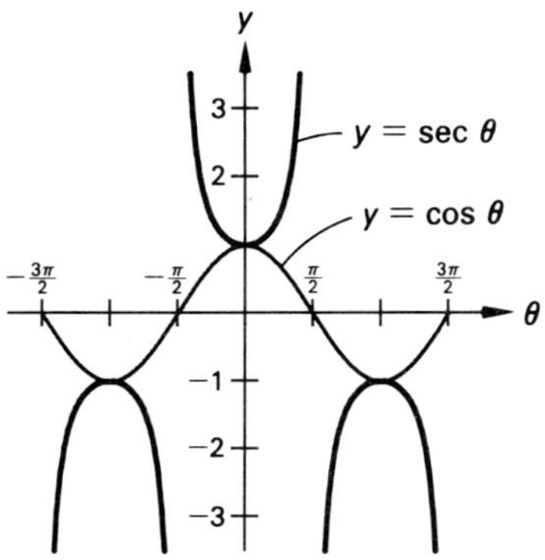

23. Sine, cosine **25.** Tangent, cotangent
27. Tangent, cotangent **29.** Tangent, secant
31. Secant, cosecant **33.** Sine, tangent **35.** Sine, tangent
37. Sine, tangent, cosecant, cotangent
39. Sine, cosine, cosecant, secant

8.3 Exercises (p. 284)

1. a. 3 **b.** 2π **3. a.** 0.75 **b.** 2π **5. a.** 1 **b.** $\frac{2\pi}{3}$ **7. a.** 6 **b.** 8π **9. a.** $\frac{1}{3}$ **b.** 6π
11. a. $\frac{1}{3}$ **b.** 6π

13.

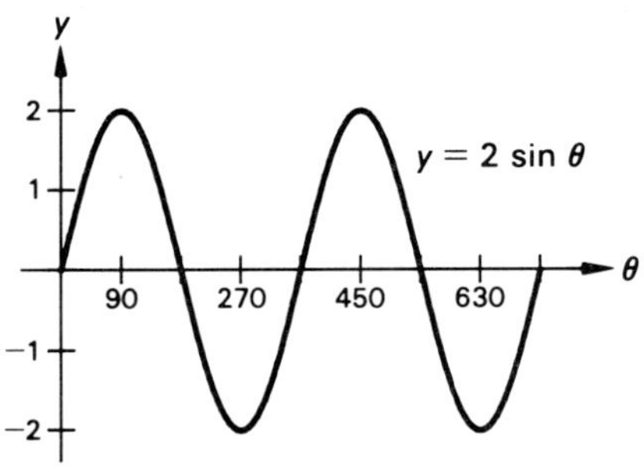

15.

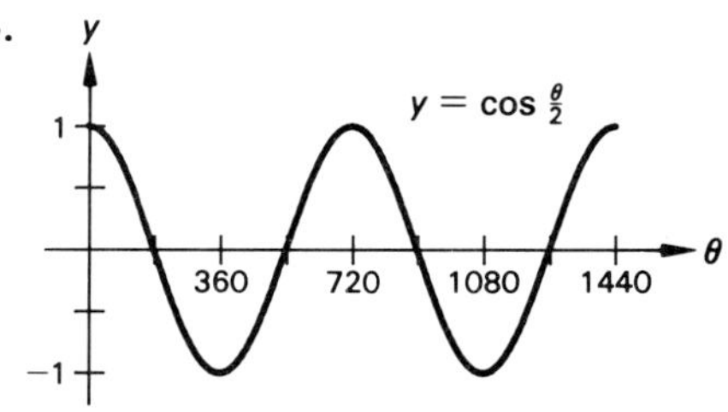

17.

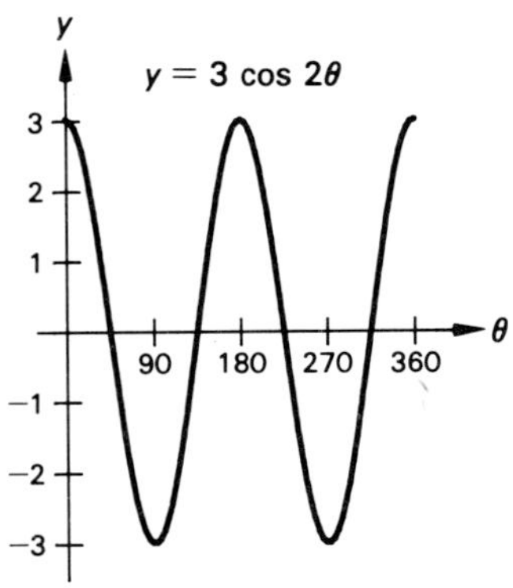

19.

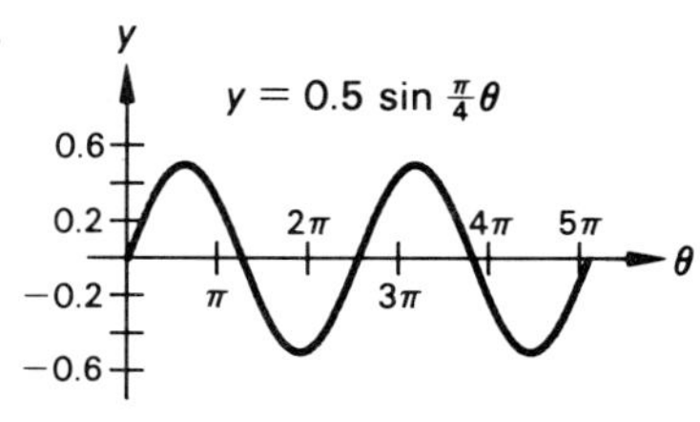

21.

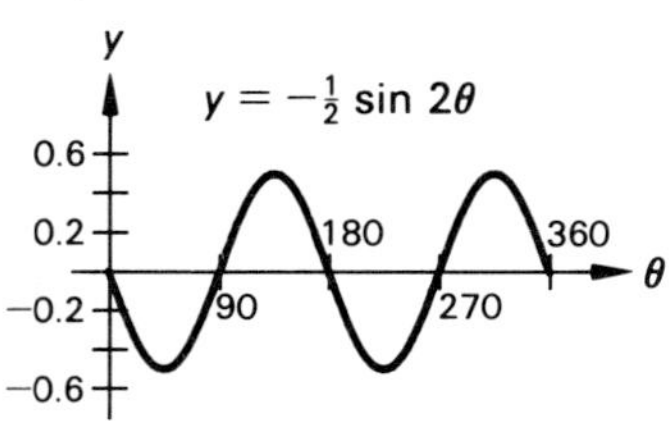

23.

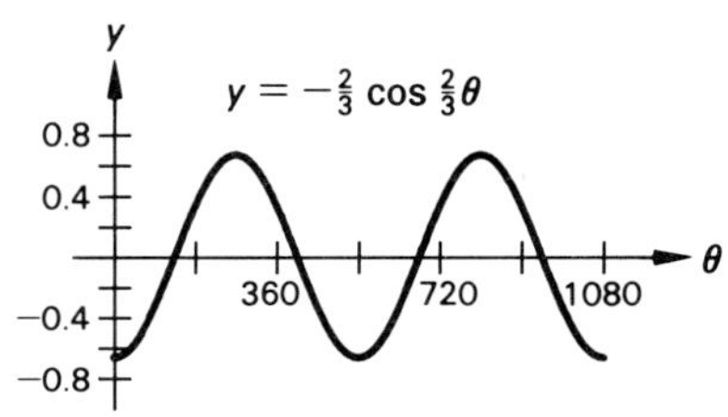

25.

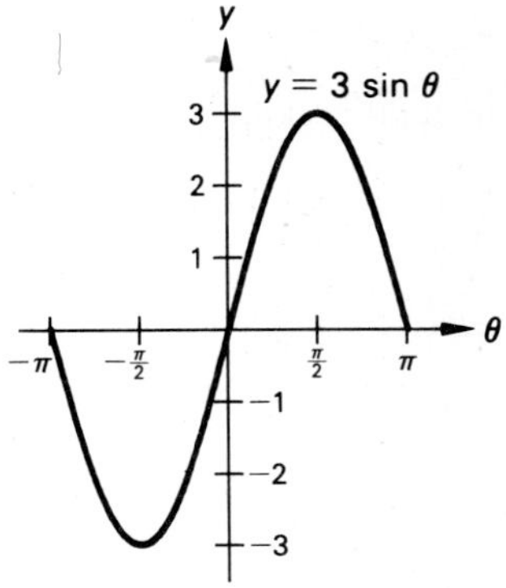

27.

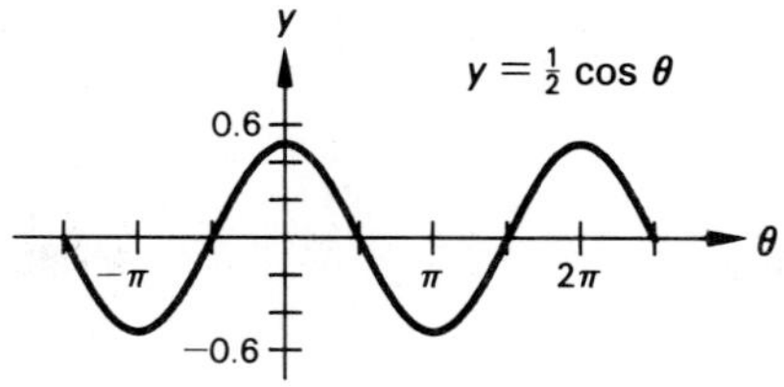

29. $\frac{3\pi}{4} \le \theta \le \frac{5\pi}{4}$

31.

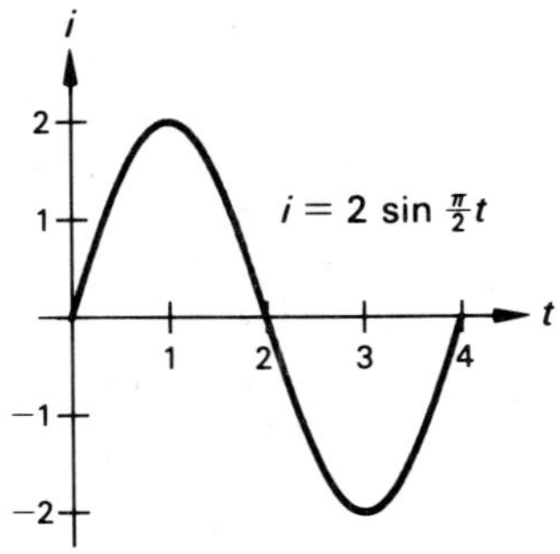

8.4 Exercises (p. 291)

1. a. 2π **b.** $\frac{\pi}{2}$ **c.** 1 **d.** $\frac{5\pi}{2}$

e.

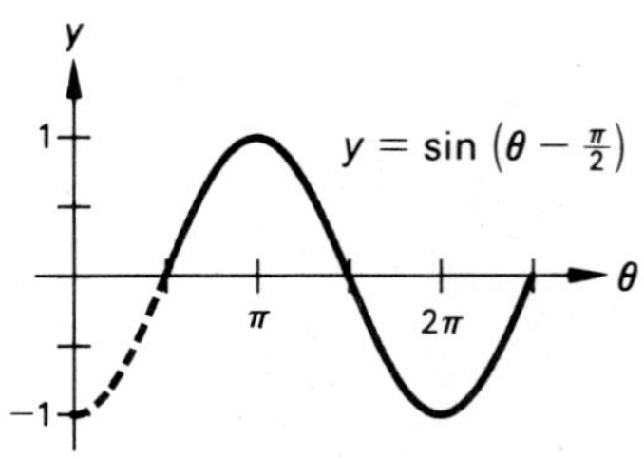

3. a. 360° **b.** 30° **c.** 1 **d.** 390°

e.

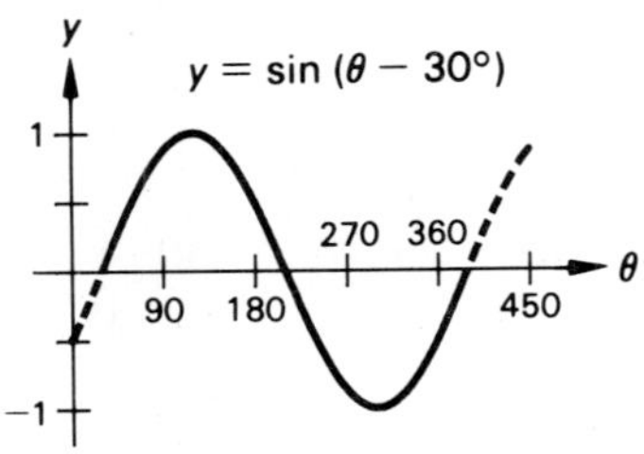

5. a. $\frac{2\pi}{3}$ **b.** $\frac{\pi}{9}$ **c.** 1 **d.** $\frac{7\pi}{9}$

e.

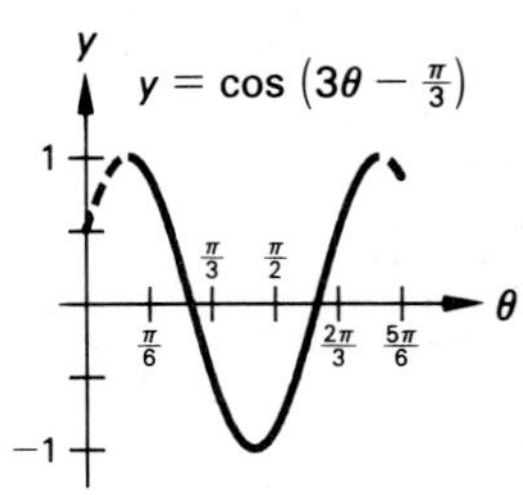

7. a. 72° **b.** −3° **c.** 1 **d.** 69°

e.

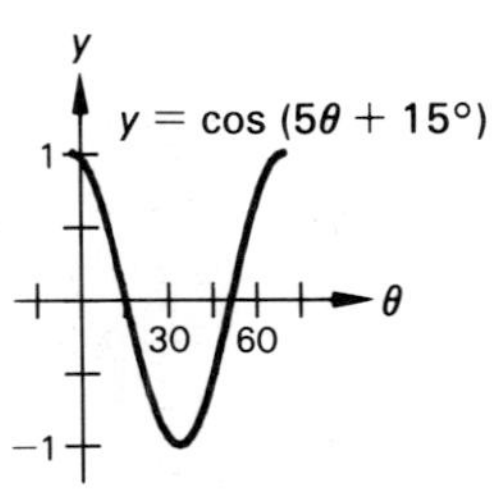

9. **a.** 2π **b.** $\frac{\pi}{2}$ **c.** 3 **d.** $\frac{5\pi}{2}$

e.

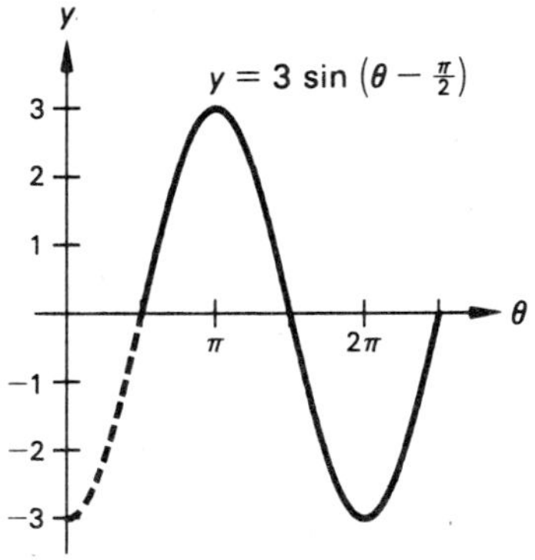

11. **a.** 1 **b.** 1 **c.** 10 **d.** 2

e.

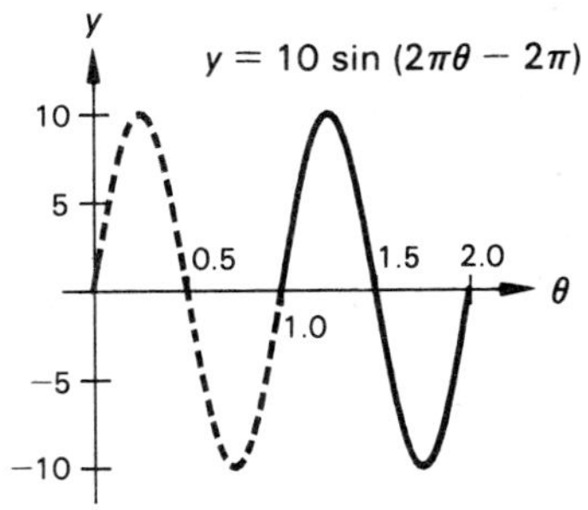

13. **a.** $\frac{2}{15}$ **b.** $-\frac{1}{15}$ **c.** 30 **d.** $\frac{1}{15}$

e.

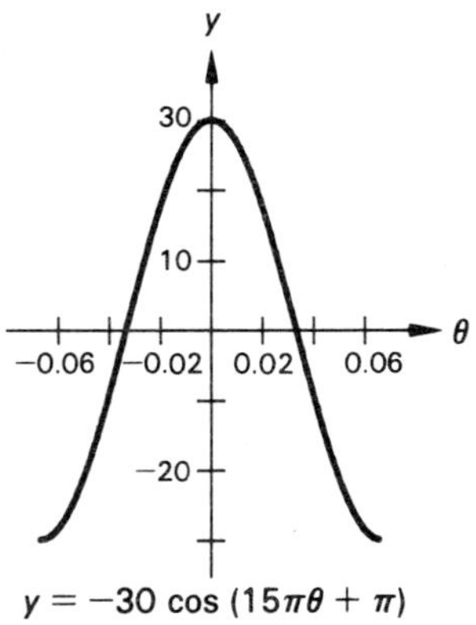

15. **a.** 120° **b.** 40° **c.** 25° **d.** 160°

e.

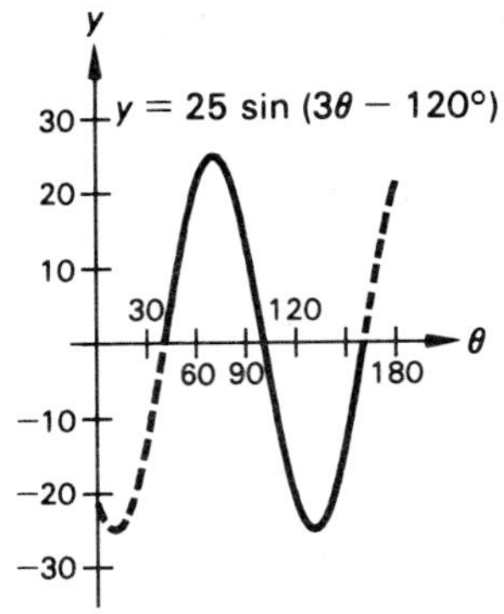

17. **a.** 720° **b.** −720° **c.** 15° **d.** 0°

e.

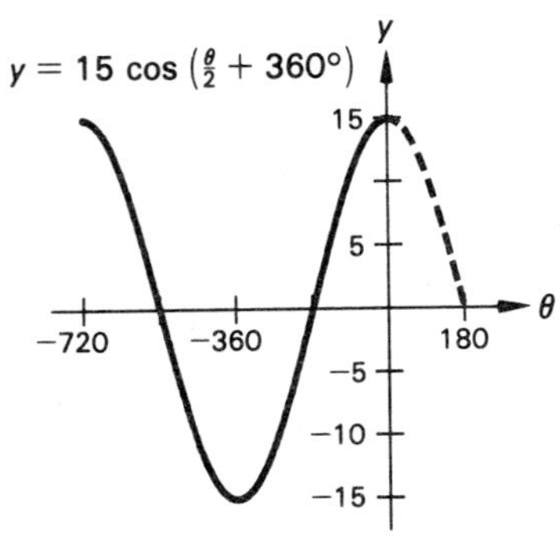

19. **a.** 360° **b.** 720° **c.** 35° **d.** 1080°

e.

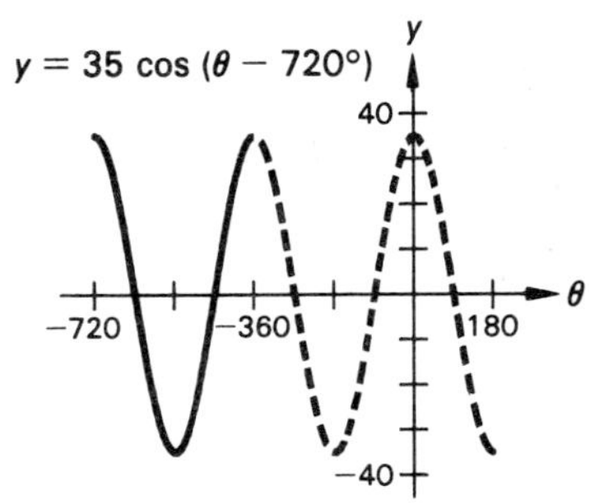

21. $y = 5 \sin\left(\frac{2}{3}\theta - \frac{\pi}{6}\right)$ **23.** $y = 2 \cos\left(4\theta + \frac{2\pi}{3}\right)$

25.

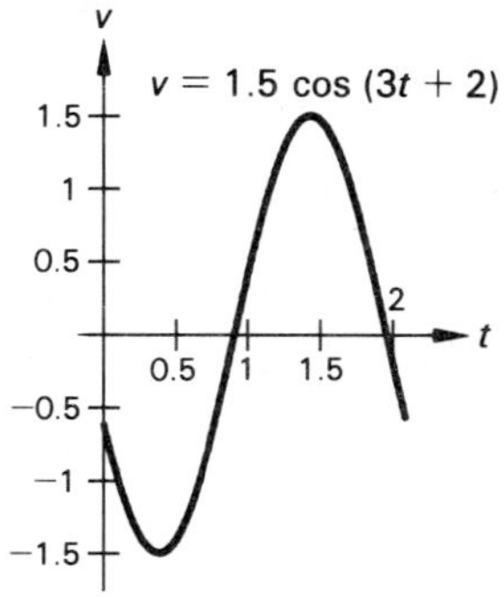

27. a. $i = 750 \sin [(360°)30t + 25°]$ **b.** $\approx \frac{1}{16}$

c.

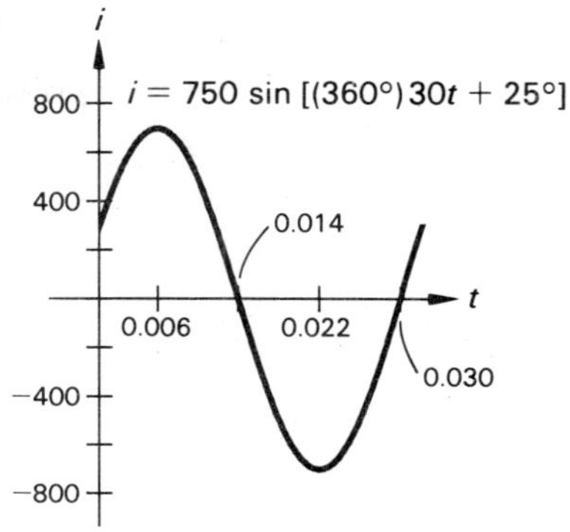

29. a. v or $e = 150 \sin (2\pi)(700t)$ **b.** $i = 12 \sin \left[(2\pi)(700t) - \frac{\pi}{6}\right]$ **c.** $i = -3.11A$

d.

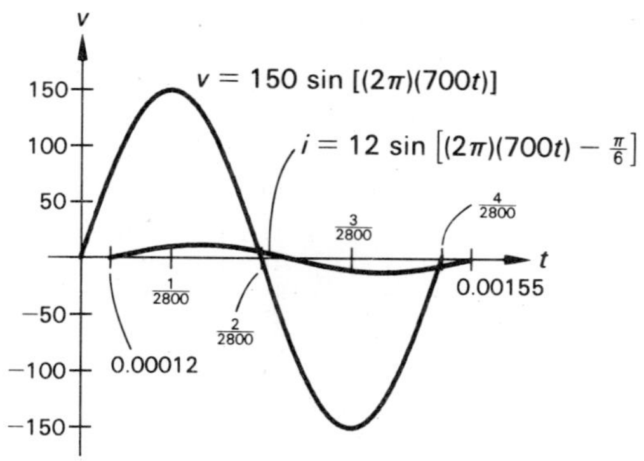

31. 34.4 ft

8.5 Exercises (p. 297)

1.

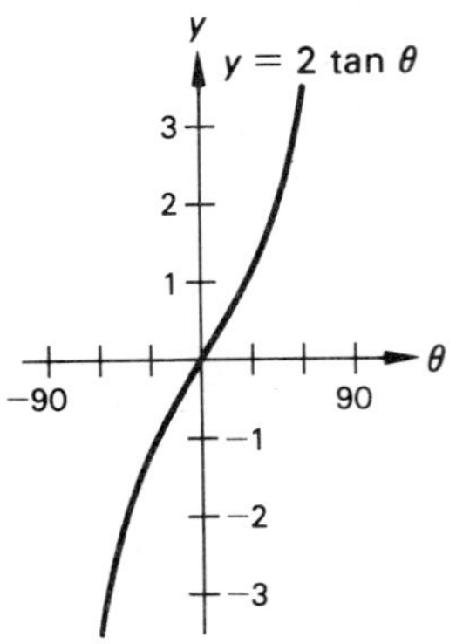

3.

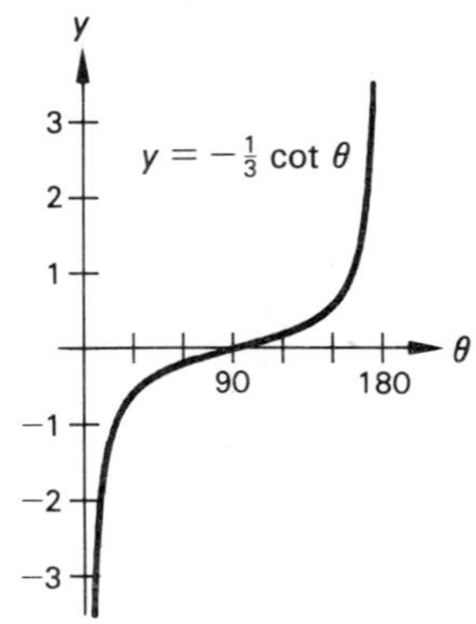

5.

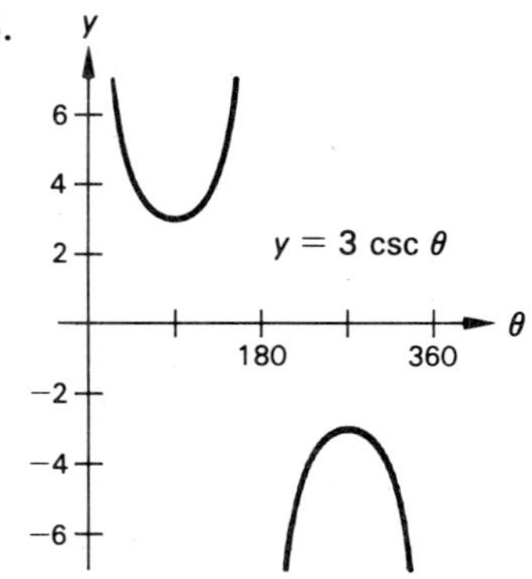

7.

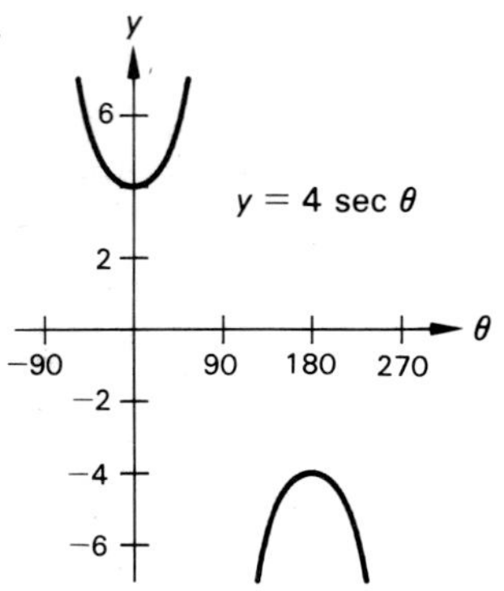

9.

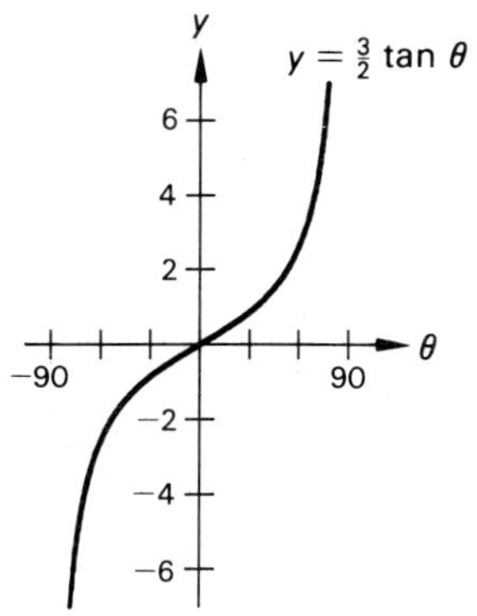

11.

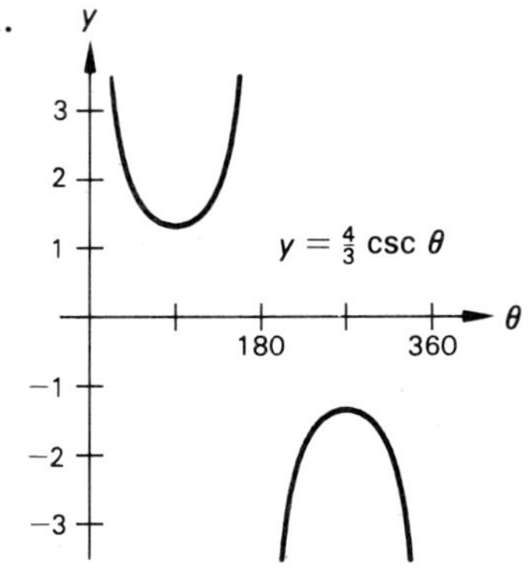

13.

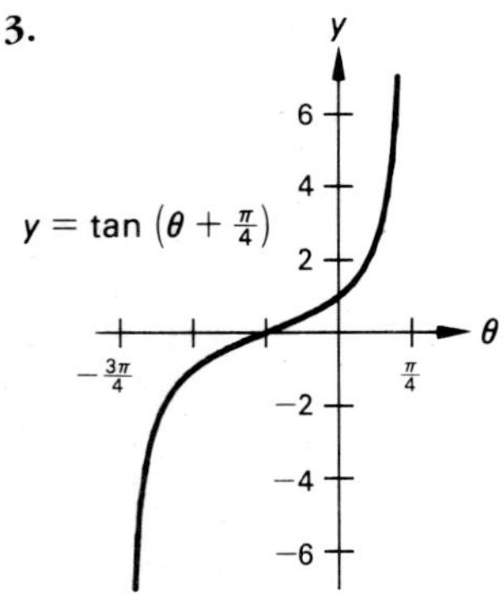

15.

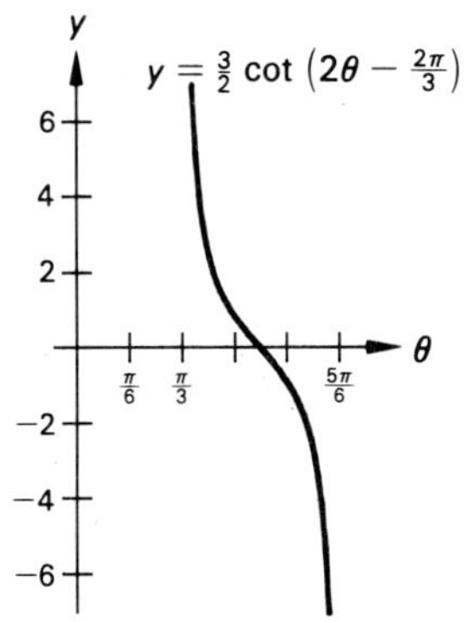

17.

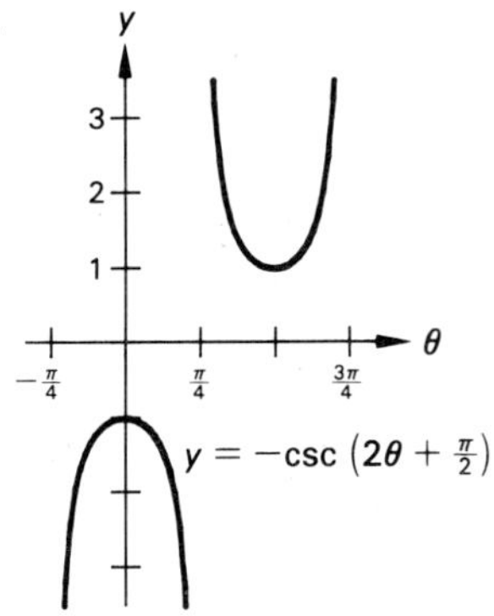

19.

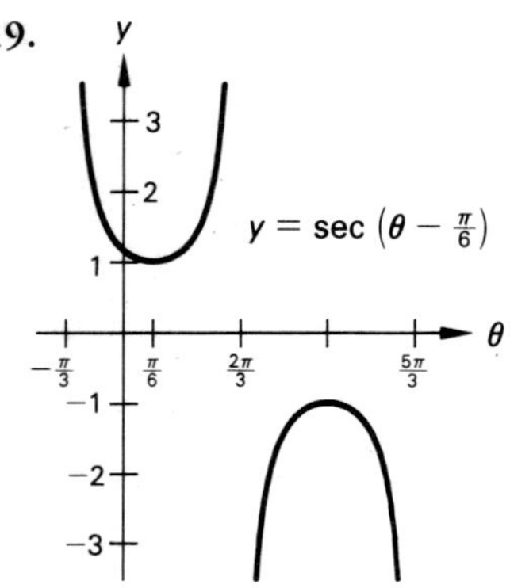

21.

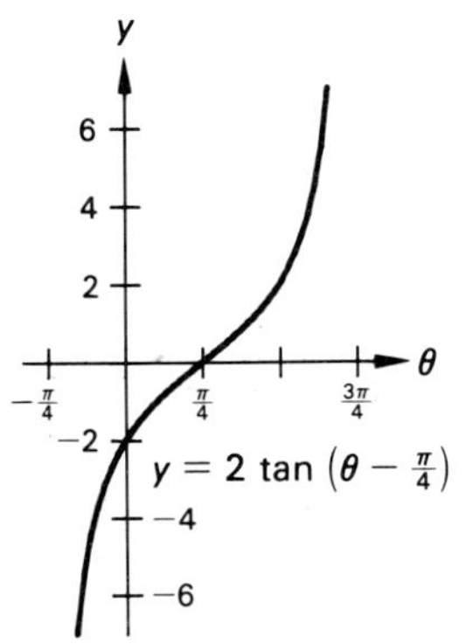

23.

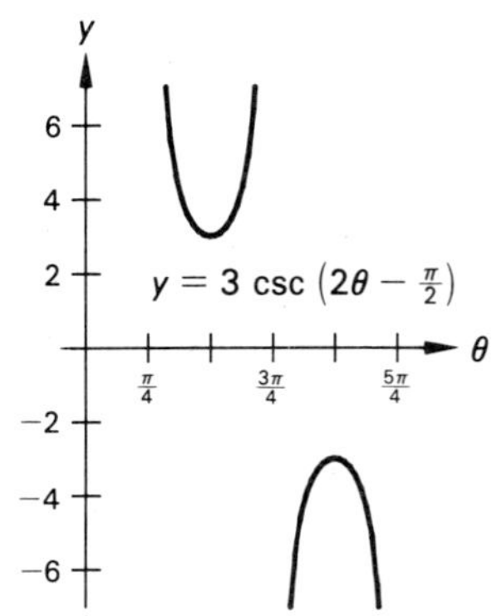

25.

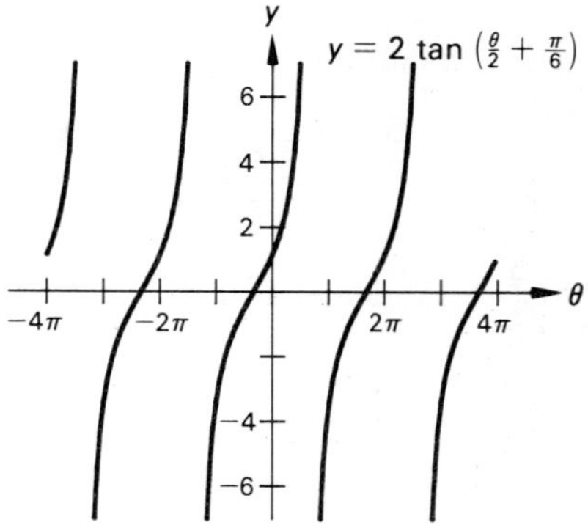

27.

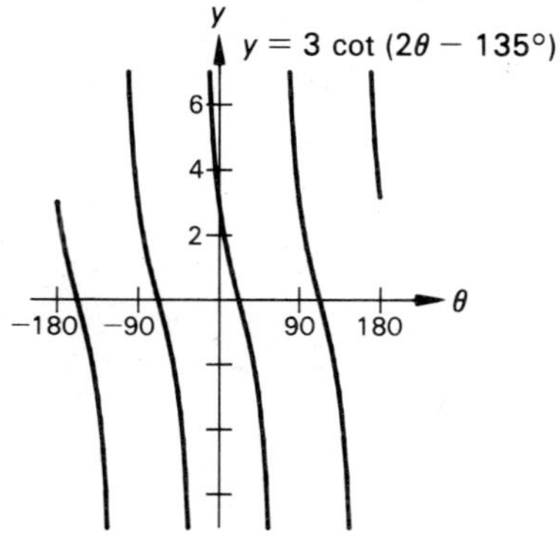

29.

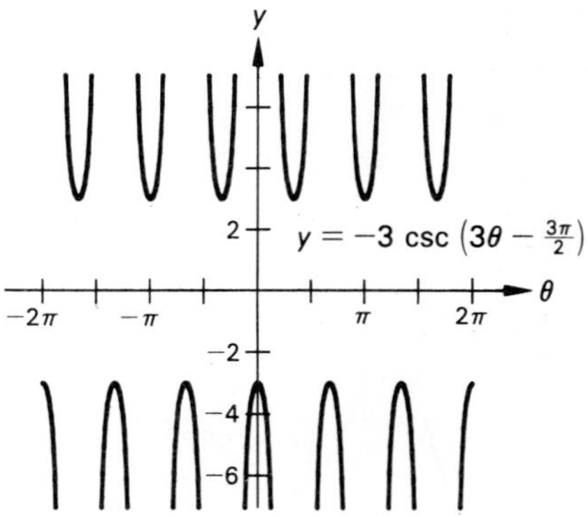

31.

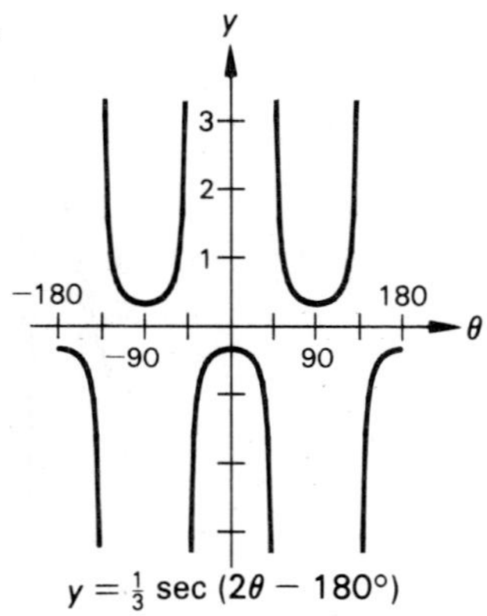

33. $y = \frac{1}{3}\tan\left(2\,\theta + \frac{\pi}{6}\right)$ **35.** $y = -3\sec\left(\frac{1}{2}\,\theta + 15^\circ\right)$ **37.** $y = 3\csc(2\,\theta - \pi)$

39. $y = -\frac{3}{4}\cot\left(3\,\theta + \frac{3\pi}{4}\right)$

8.6 Exercises (p. 303)

1. a. 360° or 2π **b.** 360° or 2π **c.** 1

d.

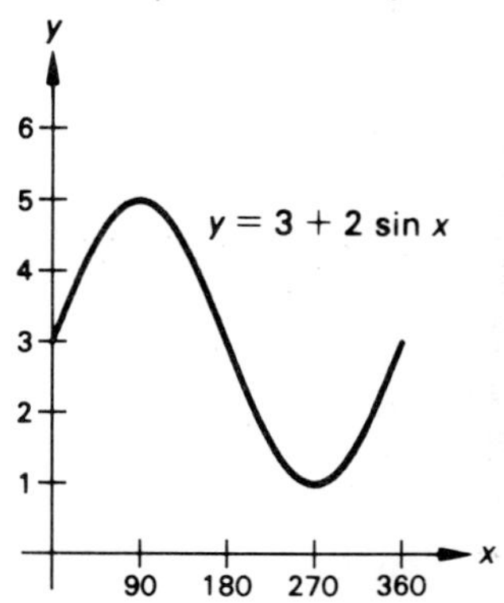

3. a. 720° or 4π **b.** 720° or 4π **c.** 1

d.

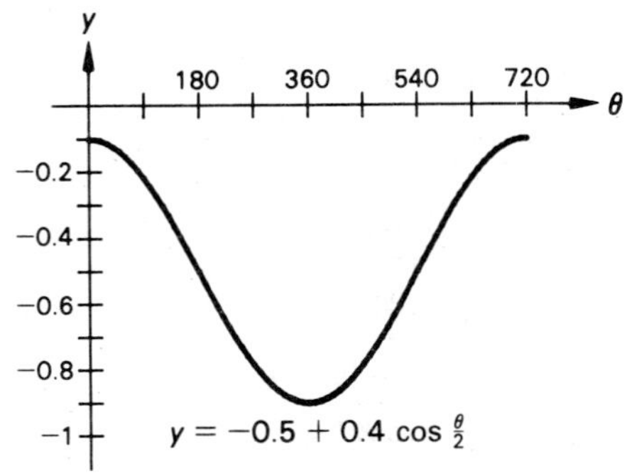

5. a. Not periodic, 360° or 2π
b. not periodic **c.** not applicable
d.

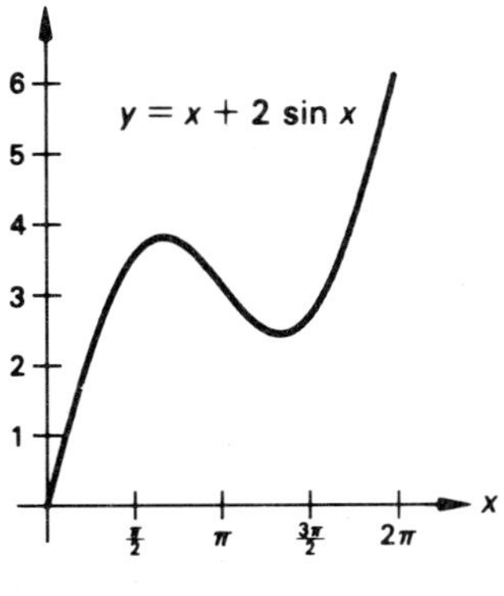

7. a. Not periodic, 360° or 2π
b. not periodic **c.** not applicable
d.

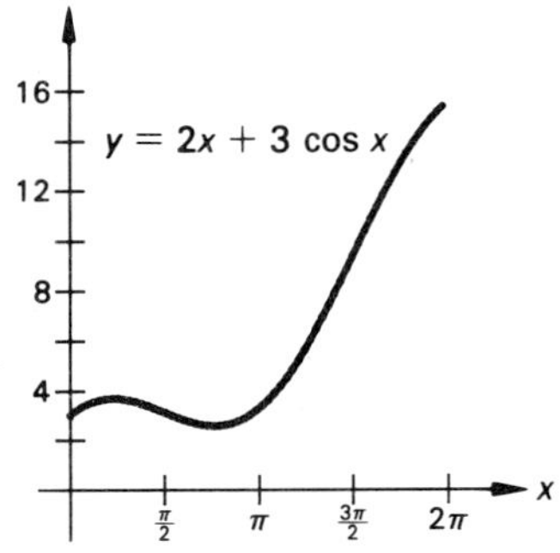

9. a. 360°, 360° **b.** 360° **c.** 1, 1
d.

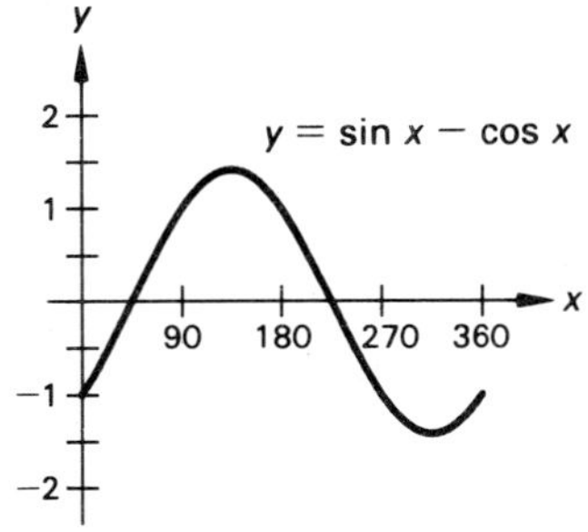

11. a. 360°, 360° **b.** 360° **c.** 1, 1
d.

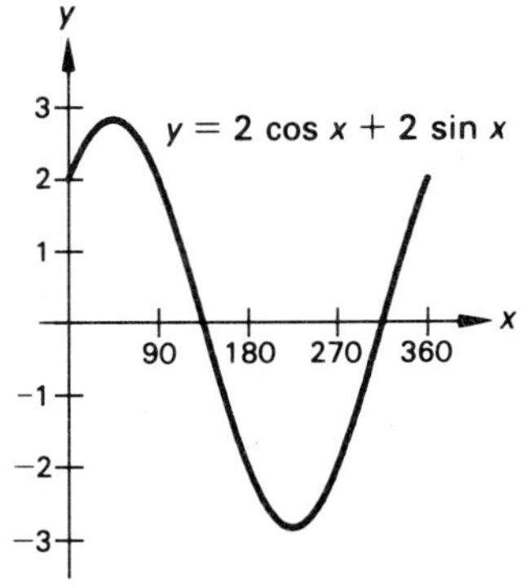

13. a. 360°, 360° **b.** 360° **c.** 1, 1
d.

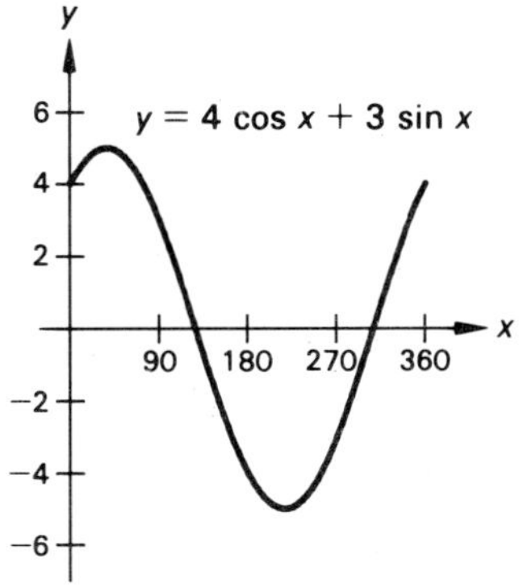

15. a. 180°, 180° **b.** 180° **c.** 1, 1
d.

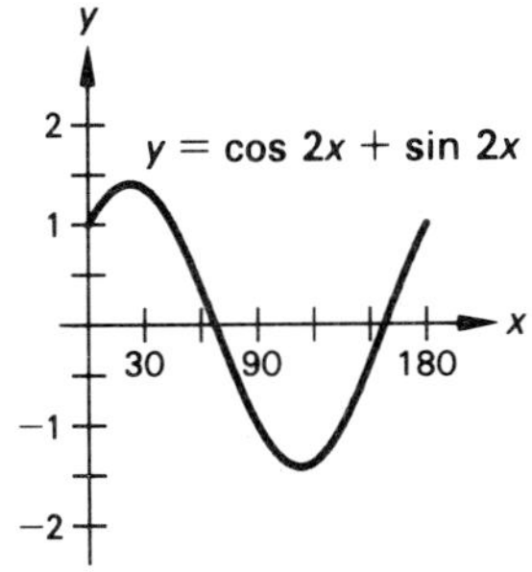

17. a. 120°, 360° **b.** 360° **c.** 3, 1

d.

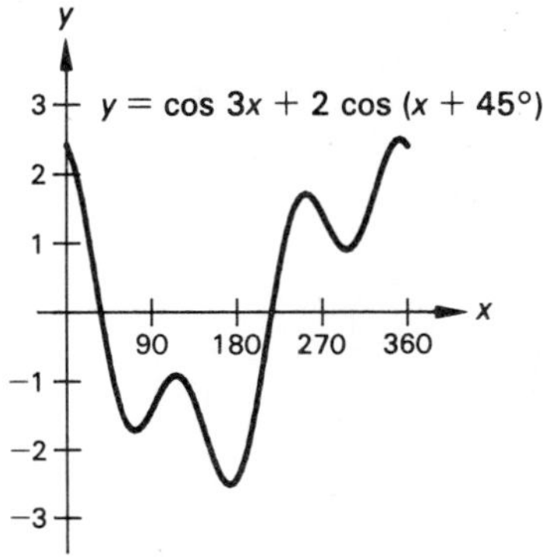

19. a. $\frac{2\pi}{3}$, 2π **b.** 2π **c.** 3, 1

d.

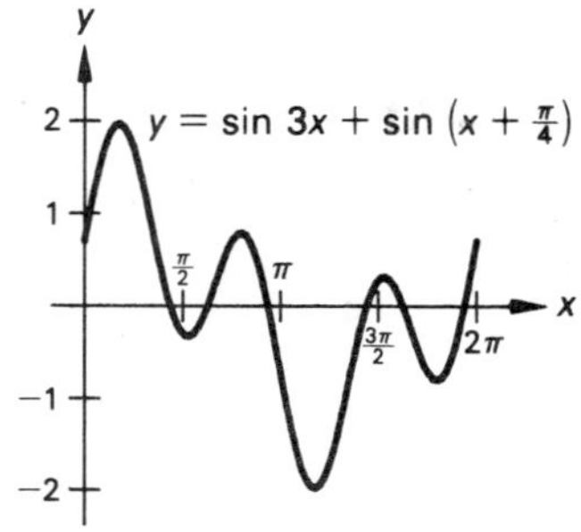

21.

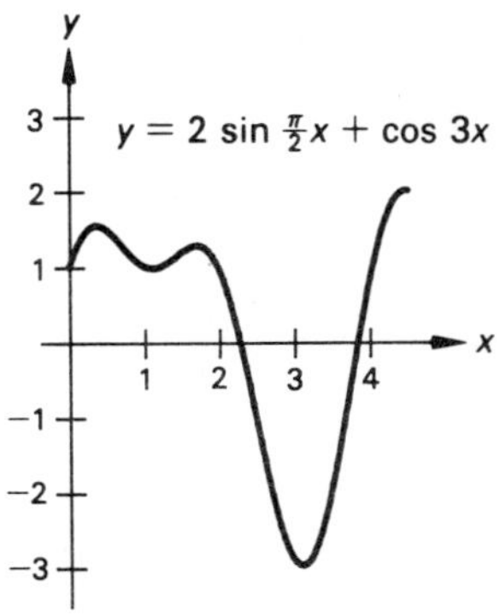

23.

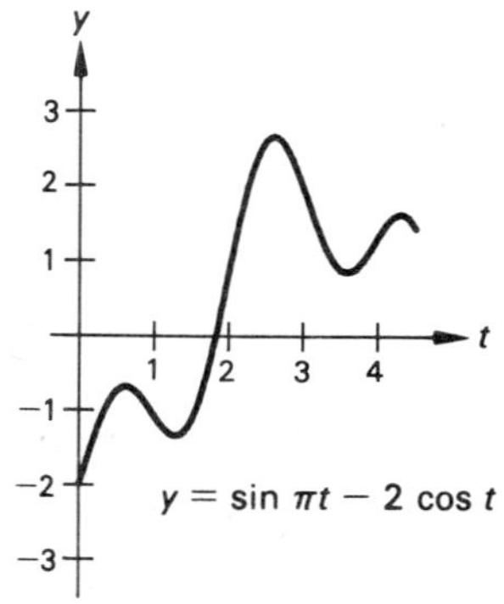

25.

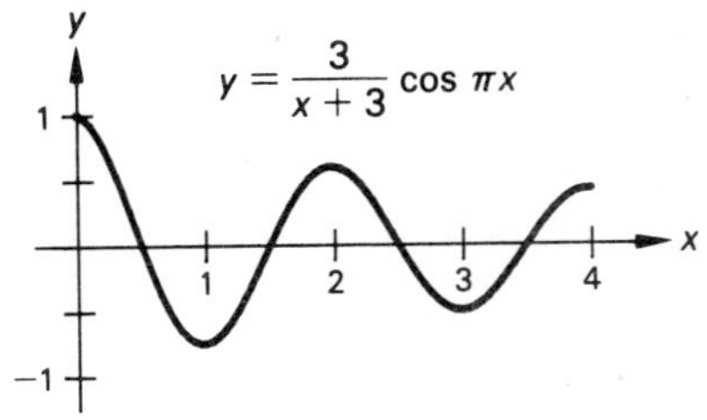

27.

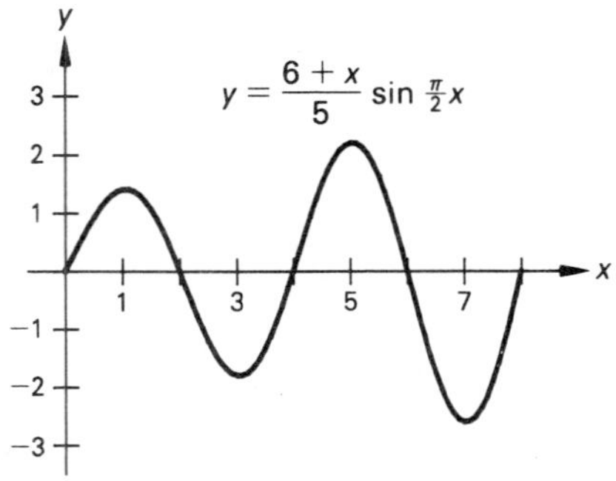

29.

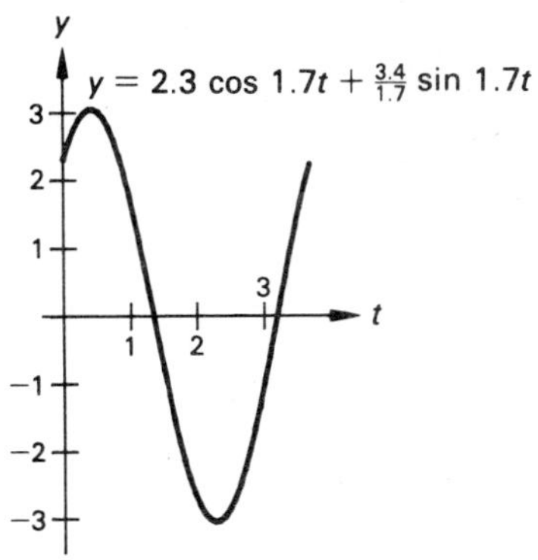

31.

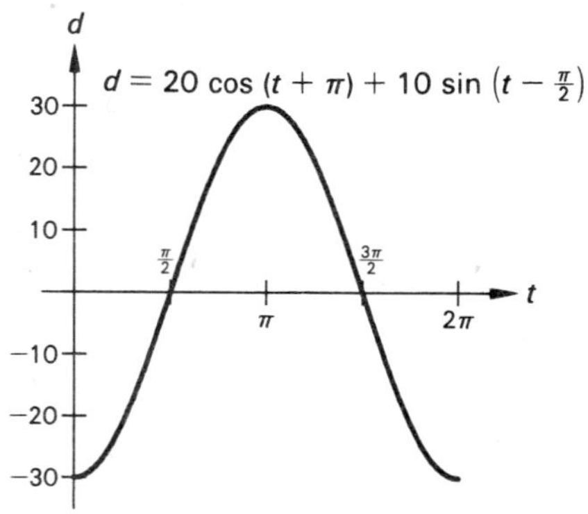

$t = 3.14$ s

33.

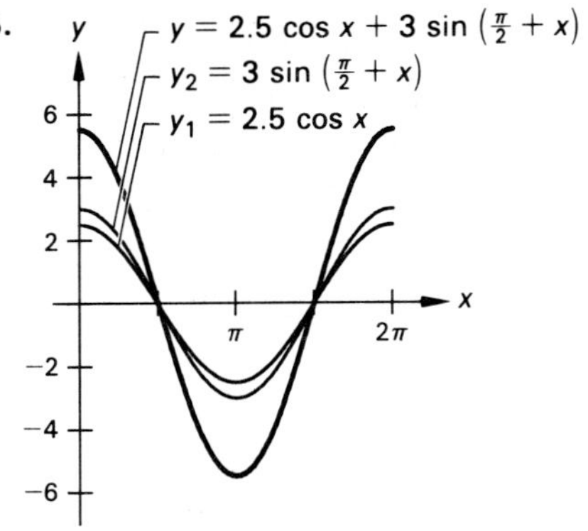

35.

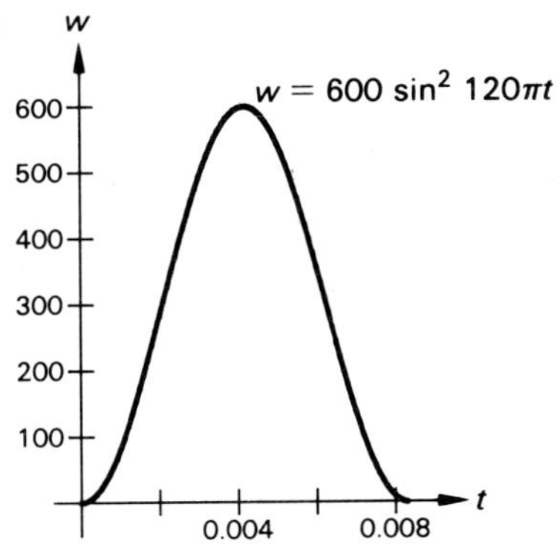

Review Exercises (p. 306)

1.

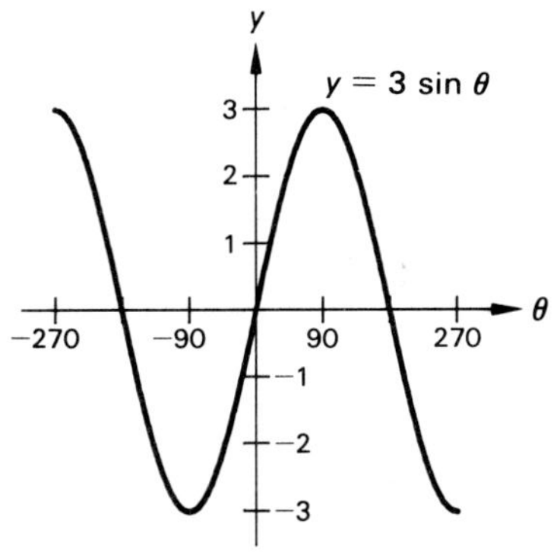

3.

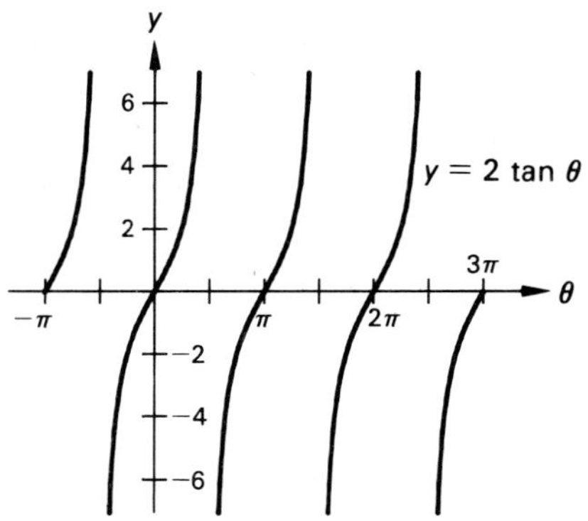

5.

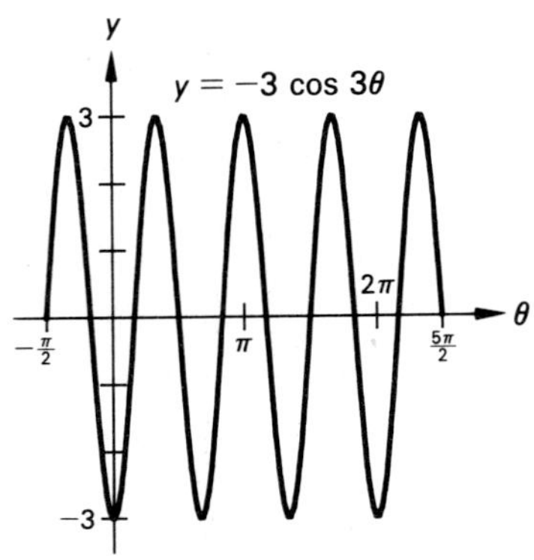

7.

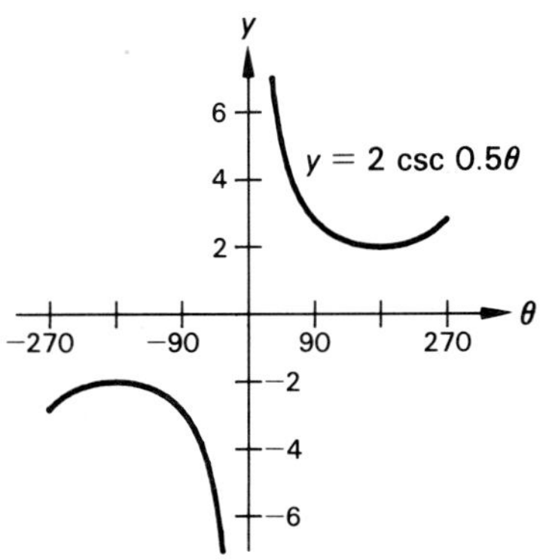

9.

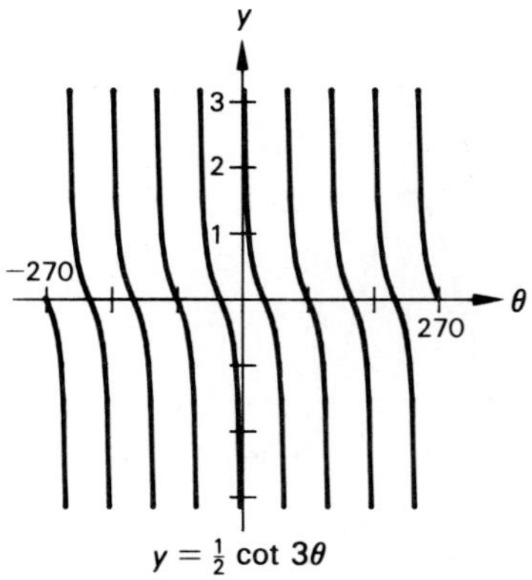

11. a. $\frac{2\pi}{3}$ **b.** $-\frac{\pi}{12}$ **c.** 2 **d.** $\frac{7\pi}{12}$

e.

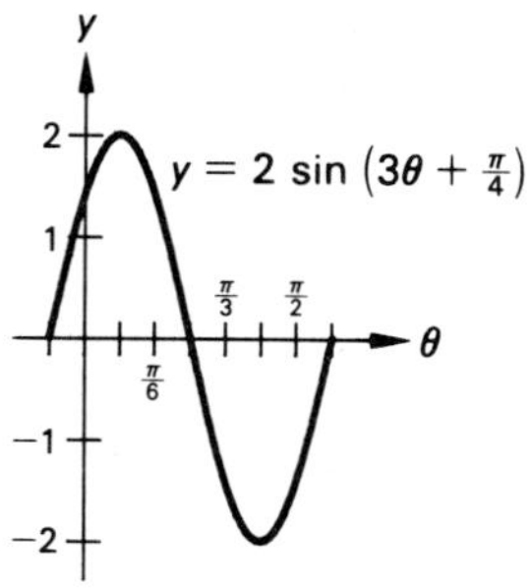

13. a. 1080° **b.** −90° **c.** 4 **d.** 990°

e.

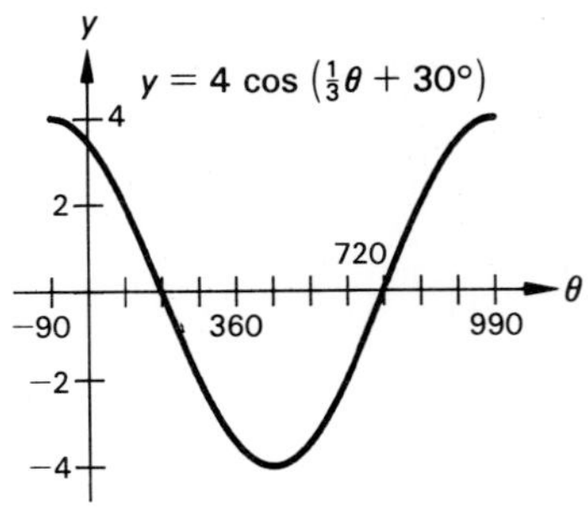

15. a. 36,000° **b.** −1500° **c.** 16. **d.** 34,500°

e.

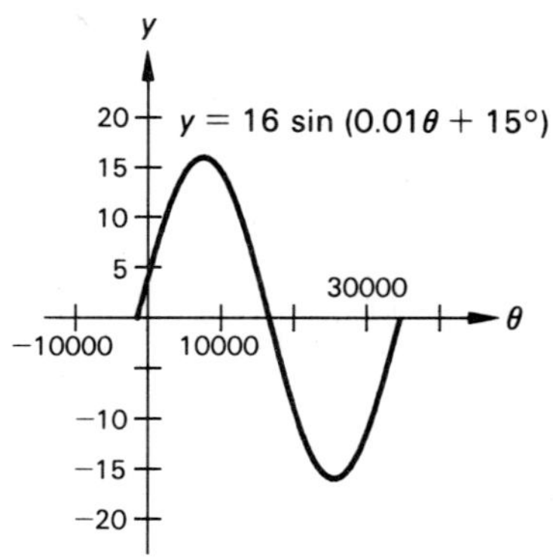

17.

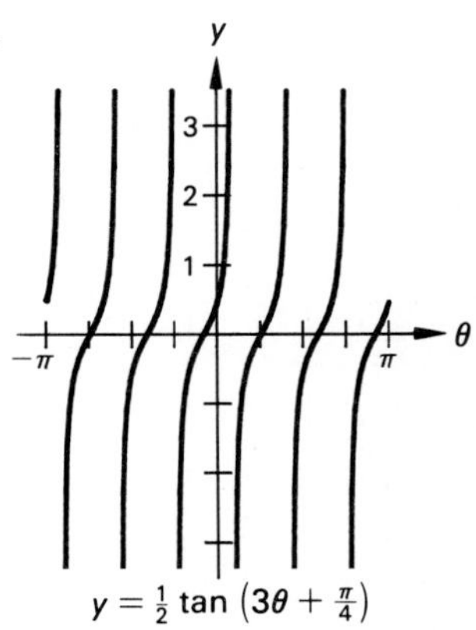

19.

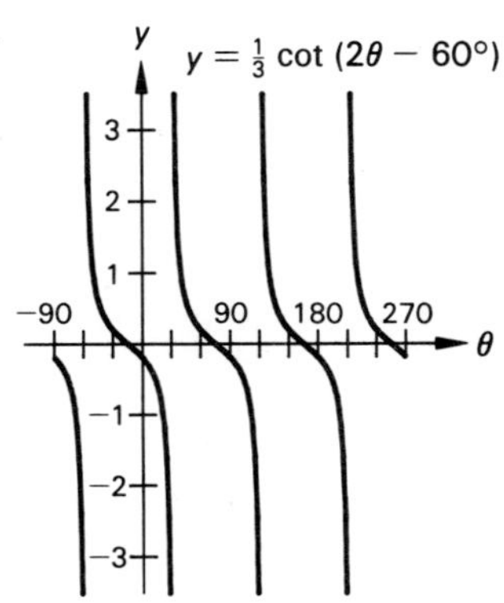

21.

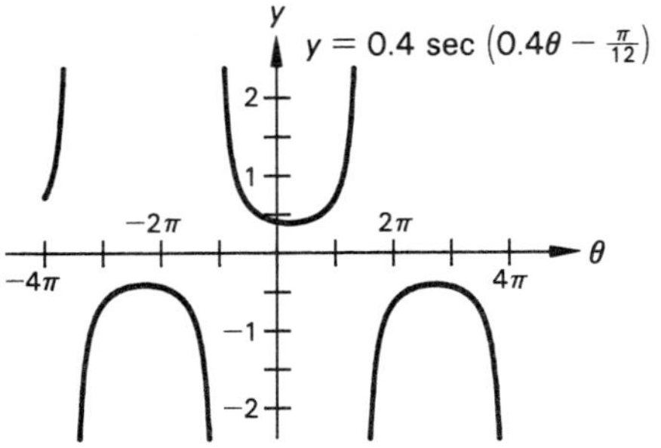

23. **a.** 720° **b.** 720° **c.** 1

d.

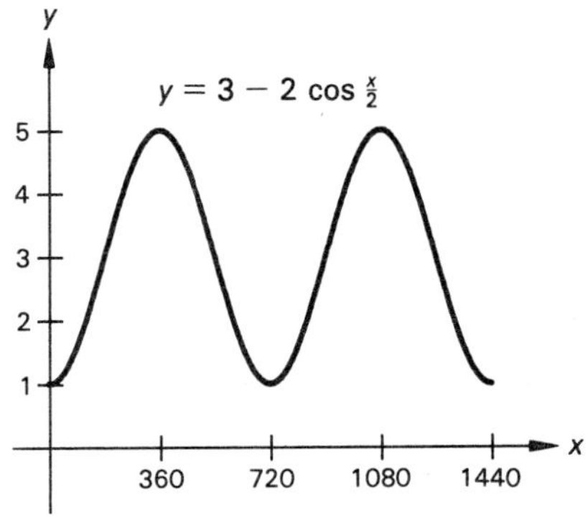

25. **a.** 6, 8 **b.** 24 **c.** 4, 3

d.

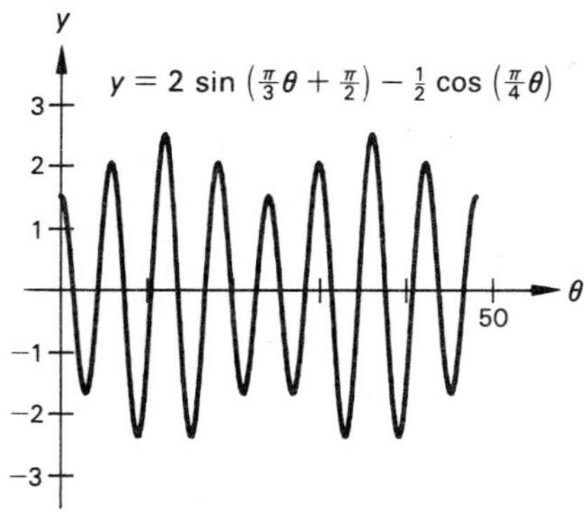

27. **a.** 12, 6 **b.** 12 **c.** 1, 2

d.

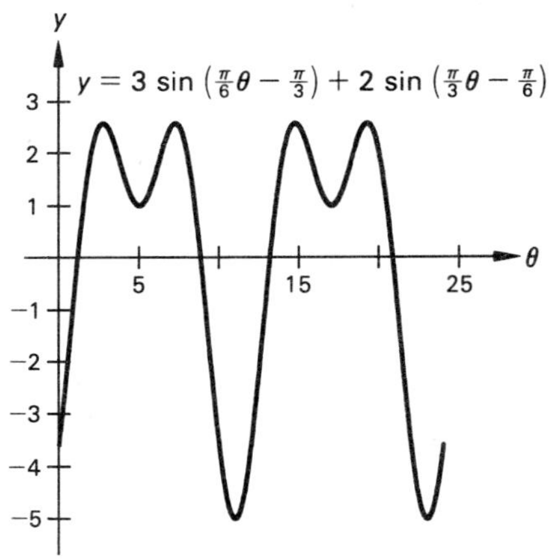

29.

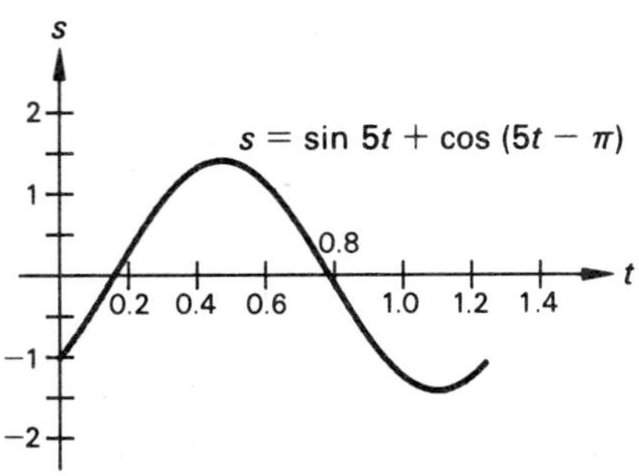

31. $5, \dfrac{1}{30}$

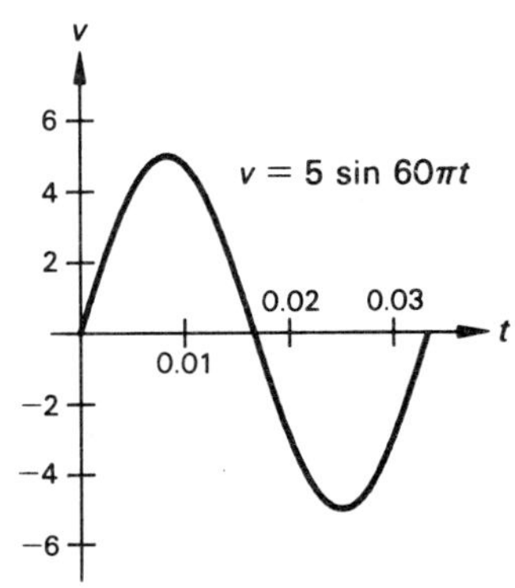

33. $y = 2.5 \sin\left(\frac{\pi}{10}x\right)$

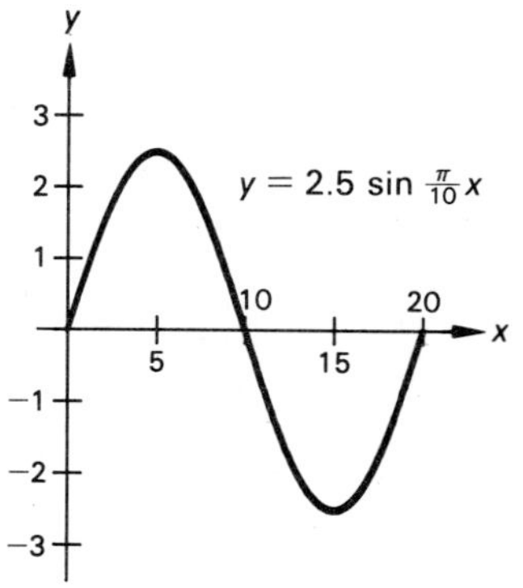

35. a. 3 **b.** π **c.** $y = 3 \sin 2x$

37. $y = 2 \sin\left(2x - \frac{\pi}{4}\right)$

39. a. 2 **b.** 0.006 **c.** -20π

41.

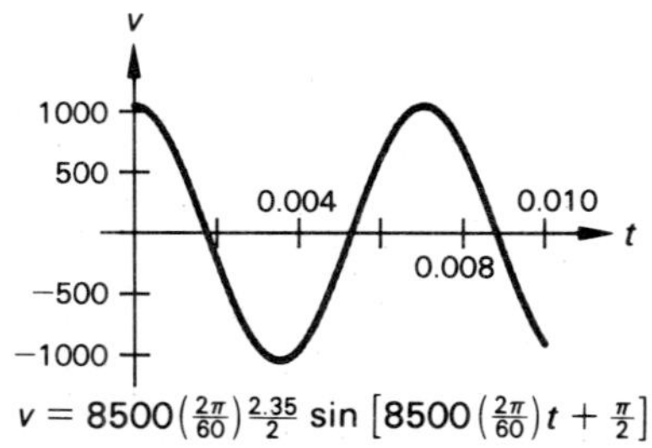

43. a.

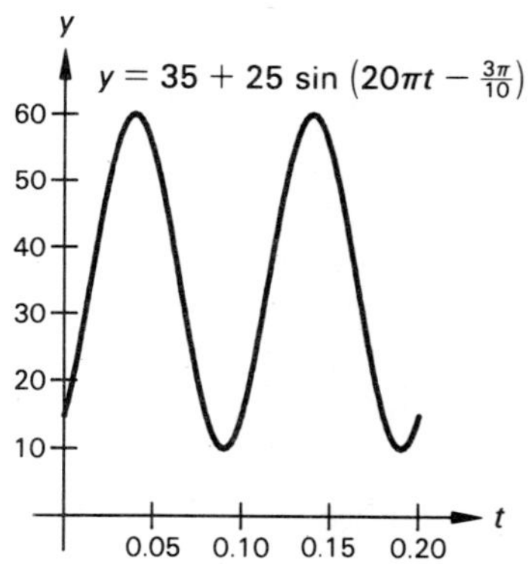

b. 10 ft **c.** $y = 35 + 25 \sin\left(20\pi t - \frac{3\pi}{10}\right)$ **d. 1.** 14.8 ft **2.** 55.2 ft **e.** 0.1118

45. a. π

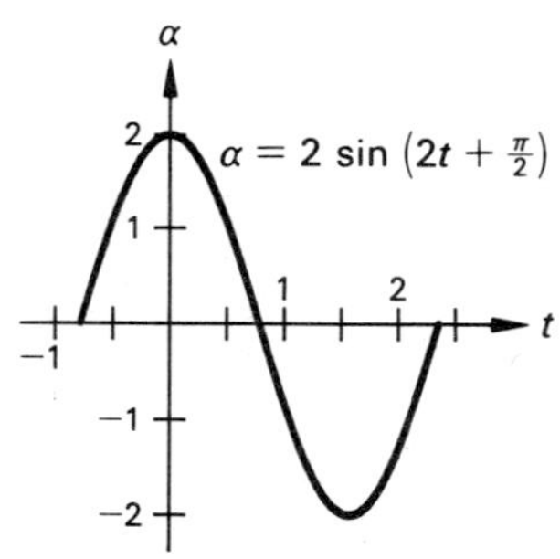

b. $\alpha = 2 \cos 2t$

CHAPTER 9

9.1 Exercises (p. 316)

1. $8j$ **3.** $4j\sqrt{2}$ **5.** a^2j **7.** p^3j **9.** j **11.** $-j$ **13.** $-j$ **15.** -1 **17.** 1 **19.** $-j$ **21.** -132 **23.** -48 **25.** $8j\sqrt{3}$ **27.** 6 **29.** $-24j\sqrt{5}$ **31.** $-44j\sqrt{5}$ **33.** $\frac{-j\sqrt{70}}{14}$ **35.** $\sqrt{7}$ **37.** $\frac{1}{2}$ **39.** $12j\sqrt{2}$ **41.** $9j\sqrt{5}$ **43.** $-13 + 17j$ **45.** 0

9.2 Exercises (p. 320)

1. $x = \frac{3}{2}, y = 5$ **3.** $x = -8, y = 4$ **5.** $11 - j$ **7.** $12 + 17j$ **9.** $5 - 10j$ **11.** $-2 + 24j$ **13.** $-27 + 36j$ **15.** $12 + j\sqrt{3}$ **17.** 5 **19.** $-\frac{1}{5} + \frac{8}{5}j$ **21.** $\frac{5}{9} - \frac{11\sqrt{5}}{45}j$ **23.** $\frac{23}{10} + \frac{11}{10}j$

25. $\frac{4}{25} - \frac{3}{25}j$ **27.** $x = -\frac{11}{2}, y = -4$ **29.** $0.33 + 1.9j$ **31. a.** Real **b.** pure imaginary
33. $0.7687 + 0.5085j$

9.3 Exercises (p. 325)

1.

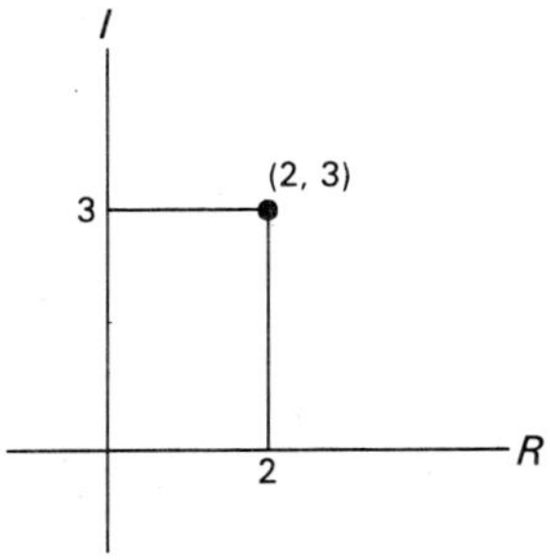

3.

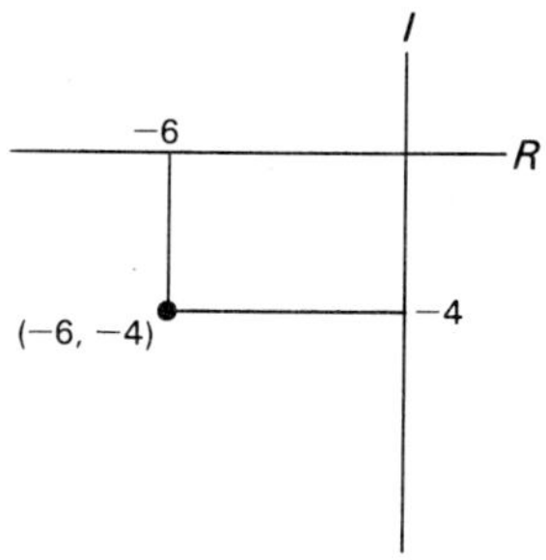

5.

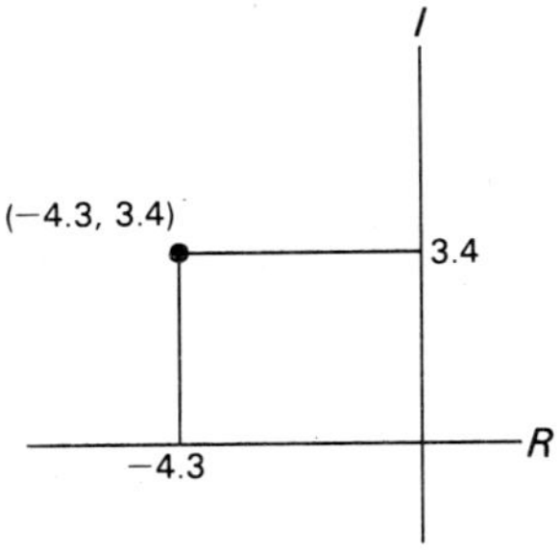

7.

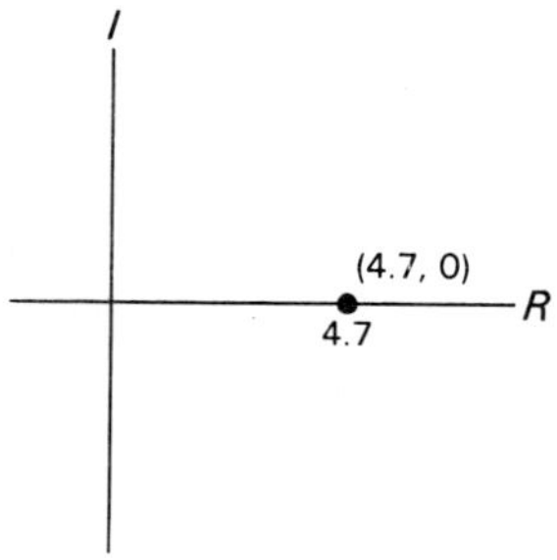

9.

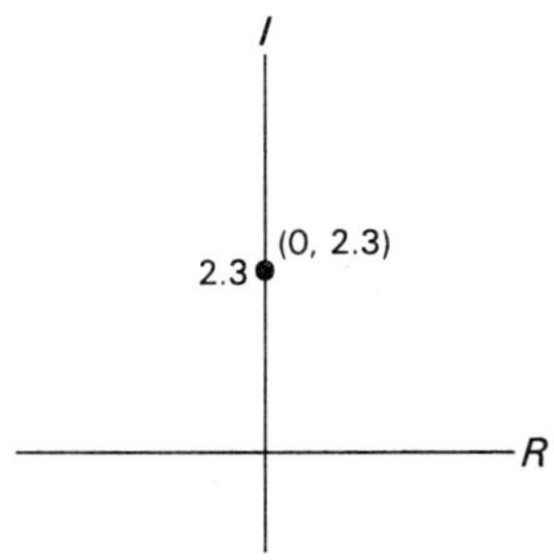

11.

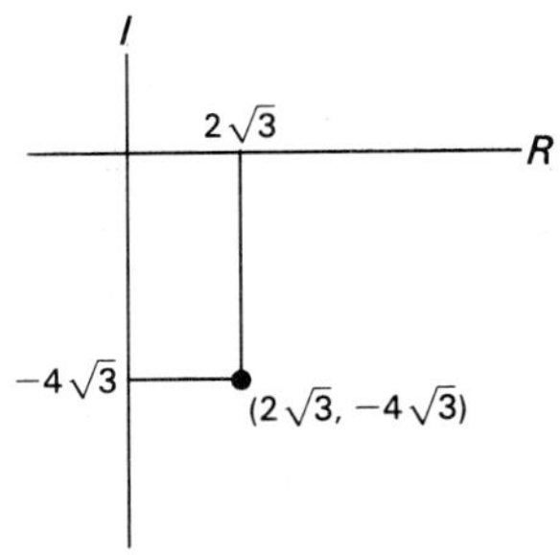

13.

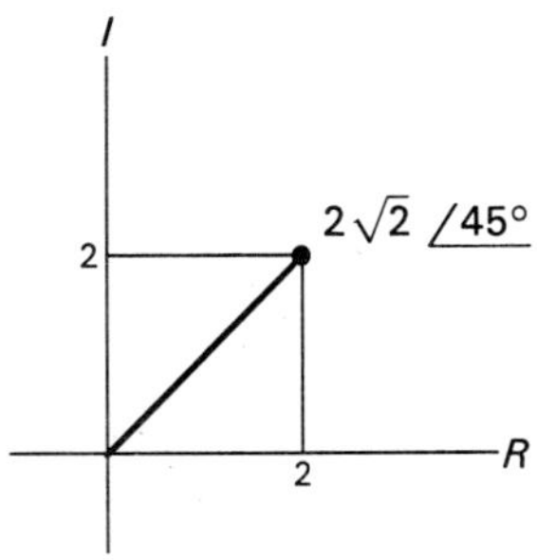

15.

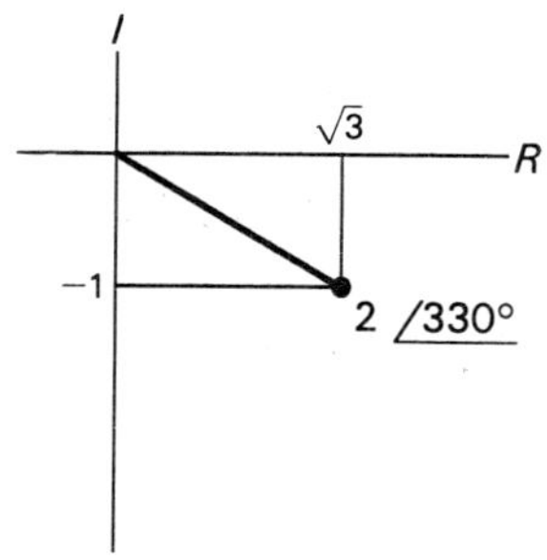

17.

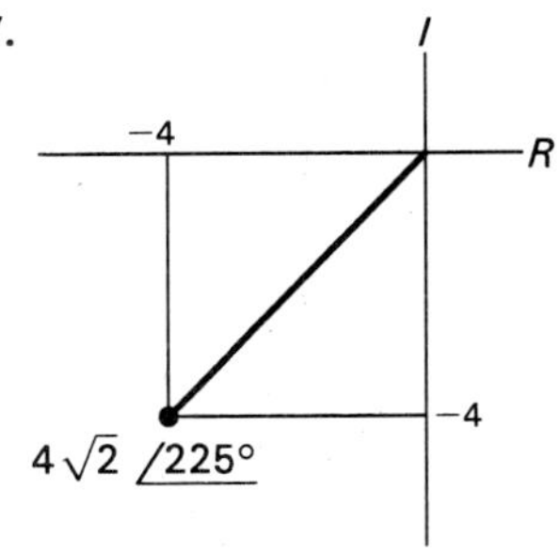

19.

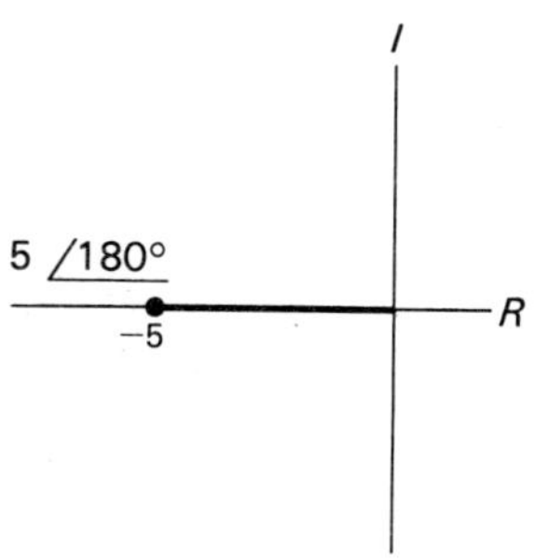

21.

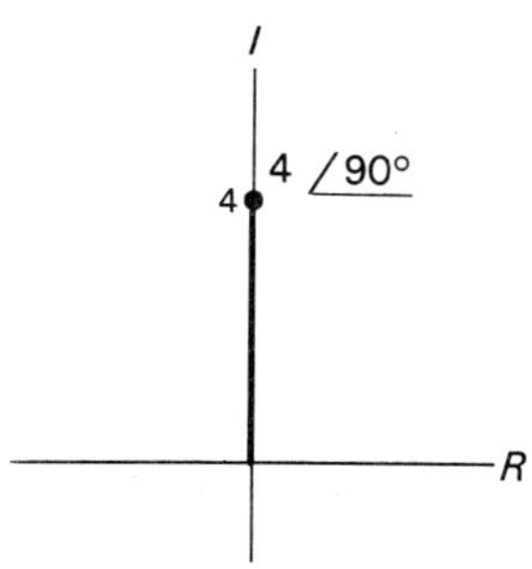

23. $5.00 \angle 217.^\circ$

25. $12.5 \angle -151.4^\circ$ or $12.5 \angle 208.6^\circ$

27. $11.4 \angle 74.7^\circ$

29. $10.3 \angle -60.9^\circ$ or $10.3 \angle 299.1^\circ$

31. $\frac{3}{\sqrt{2}} + \frac{3}{\sqrt{2}}j$ **33.** $-3\sqrt{3} - 3j$

35. $0.731 + 2.39j$ **37.** $-\frac{\sqrt{3}}{2} + \frac{1}{2}j$ **39.** $\sqrt{3} - j$ **41.** $-0.185 - 4.24j$ **43.** $0.169 - 1.72j$
45. $-0.713 + 8.67j$ **47.** $347\ \Omega \angle{-30.}^\circ$ or $347\ \Omega \angle 330.^\circ$ **49.** $85.\ \Omega \angle{-62.}^\circ$ or $85.\ \Omega \angle 298.^\circ$

9.4 Exercises (p. 332)

1. $20 \angle 77^\circ$ **3.** $2\ (\cos 260^\circ + j \sin 260^\circ)$ **5.** $26. \angle 141^\circ$ **7.** $5 + 5j$ **9.** $80. \angle 191.0^\circ$
11. $70.\ [\cos(-155.6^\circ) + j \sin(-155.6^\circ)]$ or $70.\ [\cos(204.4^\circ) + j \sin(204.4^\circ)]$ **13.** $4 \angle 9^\circ$
15. $2.0\ [\cos(51^\circ) + j \sin(51^\circ)]$ **17.** $0.8241 \angle{-10.9}^\circ$ or $0.8241 \angle 349.1^\circ$ **19.** $-0.615 - 0.0769j$
21. $0.341 \angle 272.9^\circ$ **23.** $0.675\ (\cos 97^\circ + j \sin 97^\circ)$ **25.** $16 \angle 70^\circ$ **27.** $4\ (\cos 290^\circ + j \sin 290^\circ)$
29. $22. \angle 72^\circ$ **31.** $-3 + 4j$ **33.** $72. \angle 171.2^\circ$ **35.** $55. \angle{-96.6}^\circ$ or $55. \angle 263.4^\circ$
37. $125 \angle 126^\circ$ **39.** $\cos 345^\circ + j \sin 345^\circ$ **41.** $180. \angle 135^\circ$ **43.** $18 - 26j$ **45.** $830. \angle 316.2^\circ$
47. $860.\ [\cos(-321.9^\circ) + j \sin(-321.9^\circ)]$ or $860.\ [\cos(38.1^\circ) + j \sin(38.1^\circ)]$
49. $4 \angle 16^\circ, 4 \angle 136^\circ, 4 \angle 256^\circ$ **51.** $2 \angle 24^\circ, 2 \angle 114^\circ, 2 \angle 204^\circ, 2 \angle 294^\circ$ **53.** $1 \angle 45^\circ, 1 \angle 225^\circ$
55. $2 \angle 0^\circ, 2 \angle 120^\circ, 2 \angle 240^\circ$ **57.** $32\ V \angle 104^\circ$ **59.** $0.910 + 1.94j$

9.5 Exercises (p. 336)

1. a. $3 + 4j\ \Omega$ **b.** 10 A **c.** -53.1° **3. a.** $3 - 4j\ \Omega$ **b.** 10 A **c.** 53.1°
5. a. $1.92 + 1.44j\ \Omega$ **b.** 20.8 A **c.** -36.9° **7. a.** $1.92 - 1.44j\ \Omega$ **b.** 20.8 A **c.** 36.9°
9. $47.\ \Omega \angle 45.6^\circ$ **11.** $45.8 + 17.5j$ **13.** $34.3 + 6.23j\ \Omega$ or $34.9\ \Omega \angle 10.3^\circ$
15. $25.8 + 75.0j\ \Omega$ or $79.3\ \Omega \angle 71.0^\circ$ **17.** $28.4 + 1.90j\ \Omega$ **19.** $52.8 - 90.6j\ \Omega$
21. $61.7\ \Omega \angle{-63.1}^\circ$

Review Exercises (p. 339)

1. $-j$ **3.** -342 **5.** $286j\sqrt{3}$ **7.** 28 **9.** $-j\sqrt{7}$ **11.** $(21\sqrt{2} + 11\sqrt{3})j$ **13.** $11 + 2j$
15. $1396. - 615.j$ **17.** $0.332 + 0.0185j$

19.

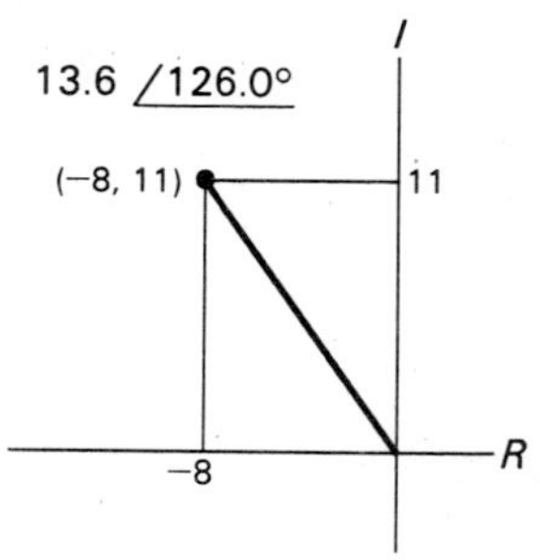

21.

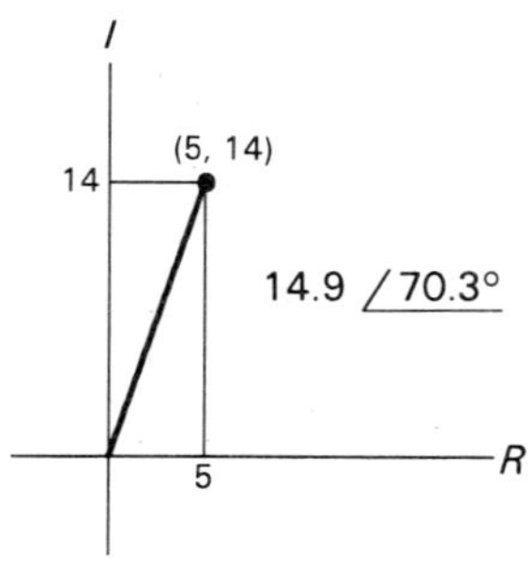

23.

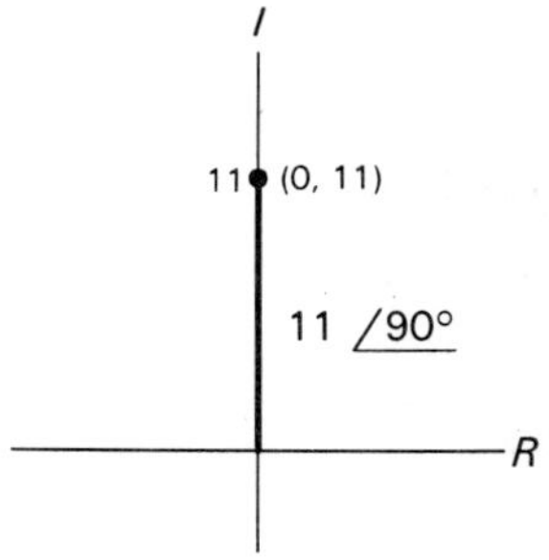

25. $-1.7 - 1.7j\sqrt{3}$ **27.** $-0.70 + 0.70\sqrt{3}\,j$ **29.** $4.99 + 0.297j$
31. $75 \angle 115^\circ$ **33.** $56.25 \angle 166.8^\circ$ **35.** $291. \angle 19.6^\circ$
37. $6 \angle 38^\circ, 6 \angle 158^\circ, 6 \angle 278^\circ$ **39. a.** -35.8° **b.** $100 + 72j\ \Omega$
c. 1.79 A

CHAPTER 10

10.1 Exercises (p. 349)

1. 3, 2

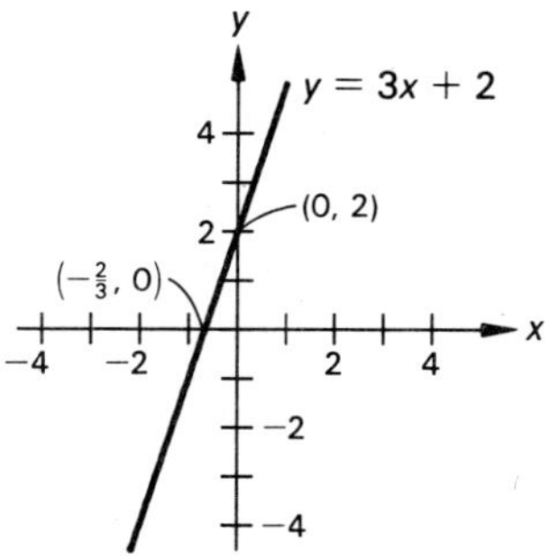

3. $\frac{1}{2}, -1$

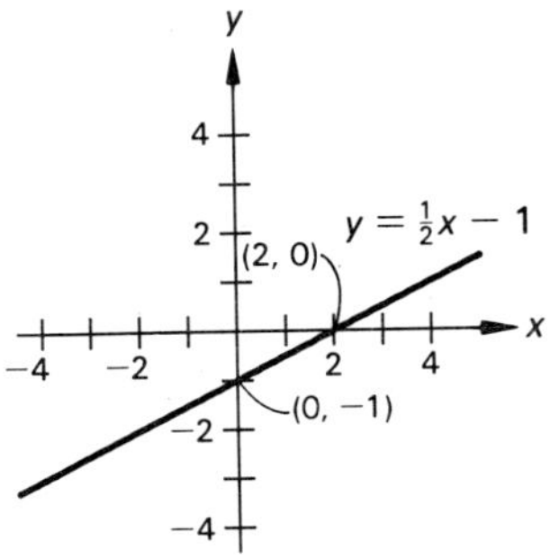

5. 3, −2

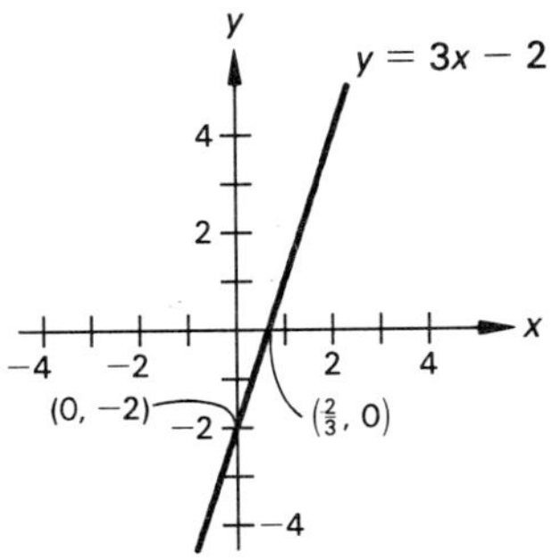

7. $\frac{1}{2}, -1$

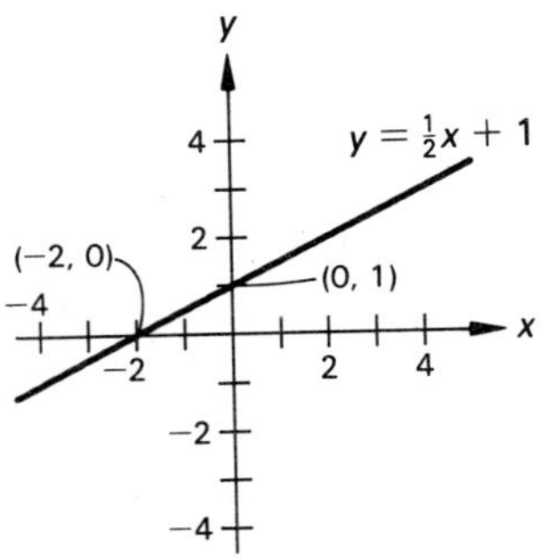

9. $-\frac{1}{2}, -\frac{1}{4}$

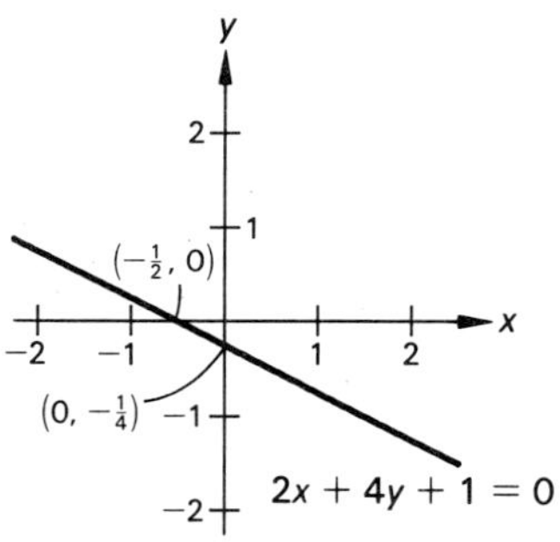

11. $\frac{1}{2}, -\frac{1}{4}$

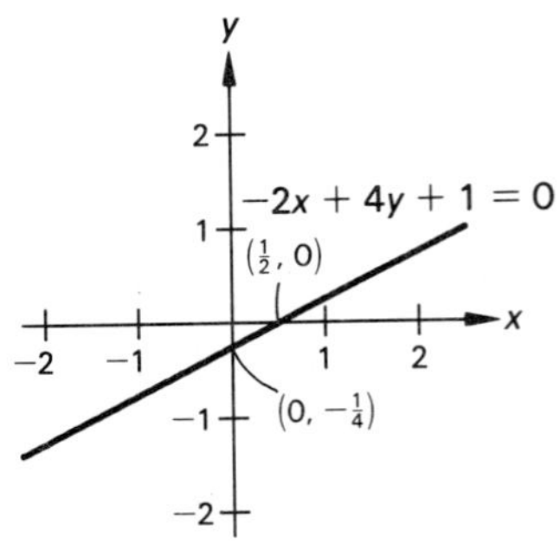

13. $\frac{1}{3}$, 0

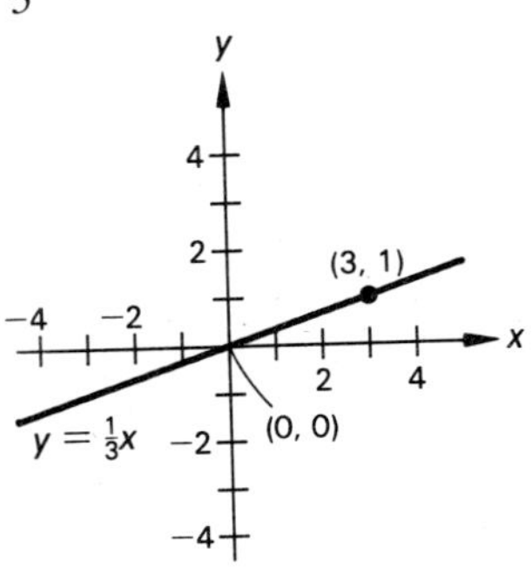

15. −1, 0

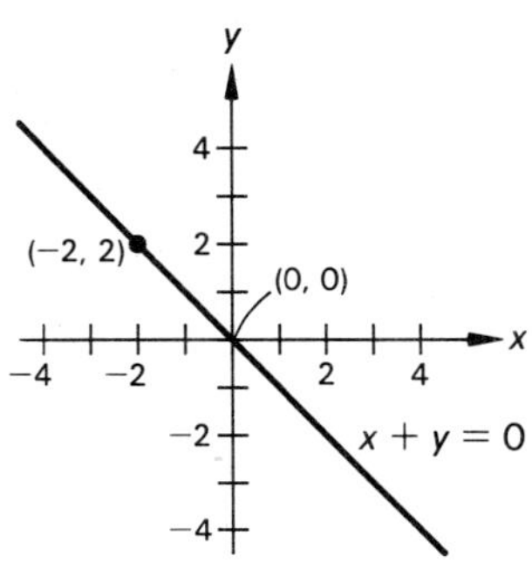

17. 0, 4

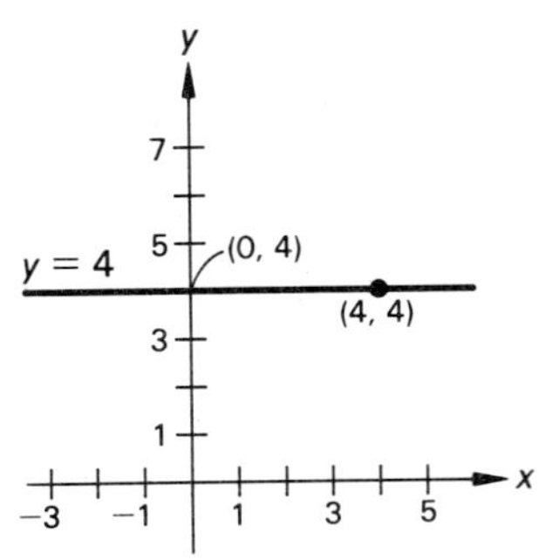

19. Undefined, none

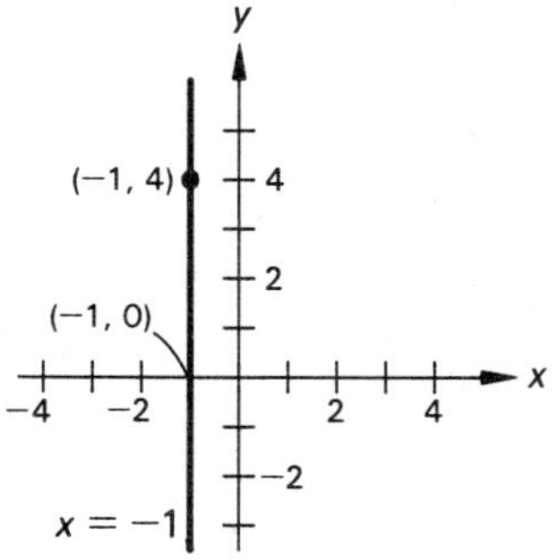

21. $-\frac{3}{2}$, 4

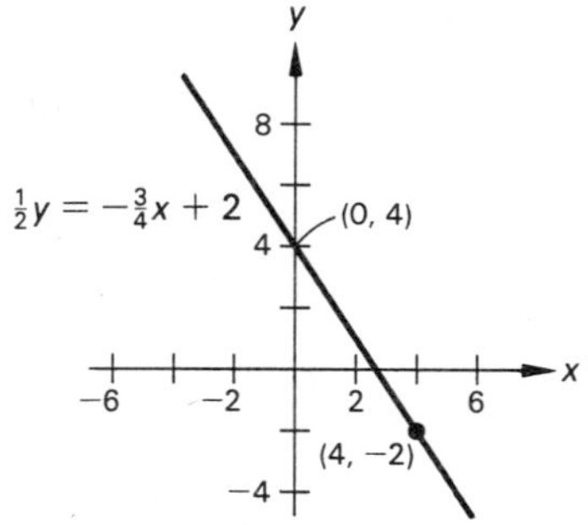

23. $-\frac{5}{4}$, $-\frac{3}{2}$

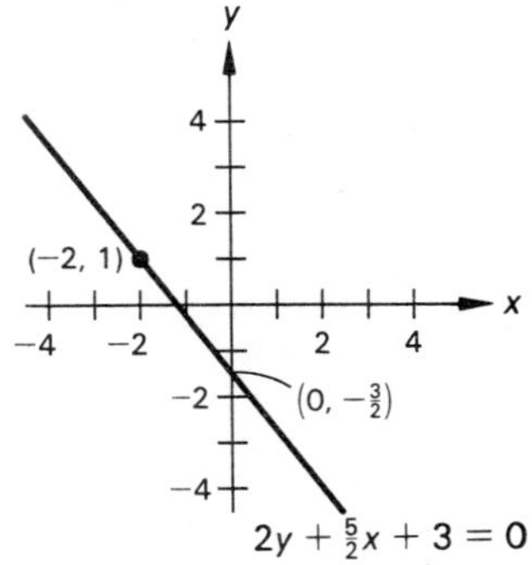

25. Yes **27.** No **29.** Yes **31.** $y = 2x + 2$ **33.** $y = -6x + 3$ **35.** $y = x$ **37.** $y = 4$
39. $x = 5$ **41.** $y = \frac{1}{2}x + 2$ **43.** $y = -\frac{3}{2}x + \frac{3}{2}$ **45.** $y = 4$ **47.** $y = \frac{2}{3}x + 2$ **49.** $y = 3x - 5$
51. $y = 2$ **53.** $x = -3$ **55.** $y = x + 2$ **57.** $y = 3x - 7$ **59.** 71.6° **61.** 0° **63.** 90°

65. 45° **67.** $y = \frac{1}{10}x,\ x \geq 0$ **69.** $y = -\frac{1}{50}x,\ -50 \leq x \leq 0$

71. a. $y = 2x + 8,\ 0 \leq x \leq 15$ $y = 2x + 38,\ -15 \leq x \leq 0$
b. $y = -2x + 8,\ -15 \leq x \leq 0$ $y = -2x + 38,\ 0 \leq x \leq 15$ **73.** $\frac{R_1}{R_1 + R_2}$

75.

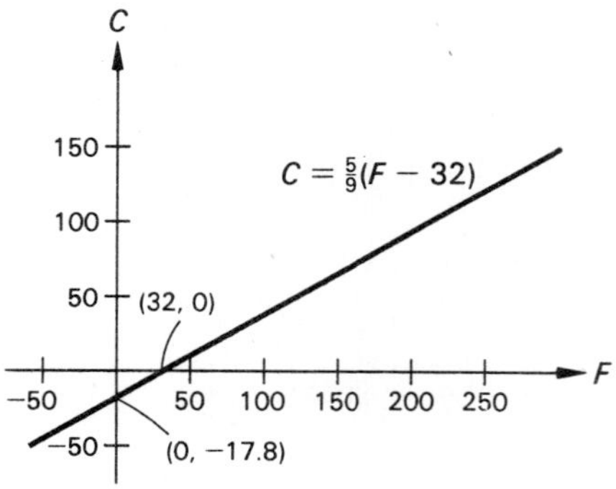

10.2 Exercises (p. 357)

1. (−6, 2) **3.** (3, 16) **5.** (4, 27) **7.** (3, 2) **9.** (1, 3) **11.** (3, 5) **13.** (−1, −8)
15. (6, 5) **17.** (6, 0) **19.** (0, −3) **21.** $(1\frac{1}{2}, 2)$ **23.** $(-2\frac{1}{2}, 1)$ **25.** $(1\frac{3}{4}, -2\frac{1}{4})$
27. Inconsistent **29.** Dependent **31.** (0, −2) **33.** Inconsistent **35.** Dependent
37. $A = -15{,}000.$, $B = 12{,}990.$ **39.** $C = -10{,}000.$, $D = -5000.$ **41.** 320°F, 160°C
43. 230°F, 110°C **45.** 320°F, 160°C **47.** 80 lb lettuce, 20 lb bean sprouts
49. $L = 85.7$ N, $R = 629.$ N

10.3 Exercises (p. 363)

1. (1, 3) **3.** (−1, 2) **5.** (−3, −2) **7.** (10, 20) **9.** (1, −6) **11.** $(-3\frac{1}{3}, -5\frac{2}{3})$
13. $(1\frac{1}{2}, 2\frac{1}{2})$ **15.** $(17\frac{1}{2}, 1\frac{1}{2})$ **17.** (0.786, 0.726) **19.** (−0.271, −0.458) **21.** Inconsistent

23. Dependent **25.** $S_S = 16{,}000.$ MPa, $S_A = 32{,}000$ MPa **27.** $R_1 = 1595.$, $R_2 = 3905.$
29. $i_1 = 0.278$, $i_2 = -0.157$ **31.** $B = 0.667$ oz, $N = 0.0833$ oz **33.** $\left(40 \text{ m}, -\frac{4}{5} \text{ m}\right)$

10.4 Exercises (p. 368)

1. $(1, 0, -1)$ **3.** $(-8, 16, 1)$ **5.** $(0, -2, 1)$ **7.** $(-17.8, 14.5, 16.1)$ **9.** $(-1, -1, -2)$
11. $(-1, 0, -1)$ **13.** $(0, -2, 1)$ **15.** $(0, -1, 0)$ **17.** $(-1, 0, 0)$
19. $(0.298, -0.651, -1.245)$ **21.** $i_1 = 2.00$, $i_2 = 1.40$, $i_3 = 1.00$ **23.** $v_1 = 88.$, $v_2 = 48.$, $v_3 = 30.$
25. $F_1 = -2232.$, $F_2 = 2000.$, $F_3 = 0.$ **27.** $A = 22.2\%$, $B = 66.7\%$, $C = 11.1\%$
29. $N = 0.008$ kg, $B = 0.032$ kg, $W = 0.004$ kg

10.5 Exercises (p. 375)

1. 116 **3.** 153 **5.** -16 **7.** -840 **9.** -74 **11.** 169 **13.** $(-8, 16, 1)$ **15.** $(1, 0, -1)$
17. $(0, -2, 1)$ **19.** $(-17.8, 14.5, 16.1)$ **21.** $(-1, -1, -2)$ **23.** $(-1, 0, -1)$ **25.** $(0, -2, 1)$
27. $A = 30\%$, $B = 60\%$, $C = 10\%$ **29.** $B_1 = 24{,}000$ N, $B_2 = 16{,}000$ N, $B_3 = 8000$ N
31. $L_1 = 6.57$ m, $L_2 = 5.57$ m, $L_3 = 2.86$ m

10.6 Exercises (p. 383)

1. $a = 1$, $b = 2$, $c = -1$, $d = 0$ **3.** $a = 1$, $x = 1$ **5.** 1×3 **7.** 2×2 **9.** 3×3 **11.** 3×2

13. $\begin{pmatrix} 7 & 6 & -4 \\ 3 & -2 & -1 \\ 6 & 14 & 9 \end{pmatrix}$ **15.** $\begin{pmatrix} -2 & -12 \\ -2 & -3 \\ 4 & 3 \end{pmatrix}$ **17.** $(13 \quad -2 \quad -11)$ **19.** $\begin{pmatrix} -9 & -4 \\ 0 & 1 \end{pmatrix}$

21. a. 2×2, 2×2 **b.** $\begin{pmatrix} 10 & 48 \\ 5 & 33 \end{pmatrix}$ **c.** $\begin{pmatrix} 16 & 38 \\ 9 & 27 \end{pmatrix}$ **23. a.** 1×3, 3×2 **b.** $(18 \quad 23)$
c. cannot be multiplied **25. a.** 2×3, 3×2 **b.** $\begin{pmatrix} 101 & 38 \\ 13 & 26 \end{pmatrix}$ **c.** $\begin{pmatrix} 16 & -2 & -4 \\ 34 & 48 & 39 \\ 18 & 72 & 63 \end{pmatrix}$

27. a. 1×3, 3×3 **b.** $(19 \quad 54 \quad -7)$ **c.** cannot be multiplied **29. a.** 3×3, 3×1,
b. $\begin{pmatrix} x + 2y + 3z \\ -x + y \\ 5x - 4y + 6z \end{pmatrix}$ **c.** cannot be multiplied

10.7 Exercises (p. 391)

1. $\begin{pmatrix} 1 & 1 & 1 \\ 2 & 1 & 2 \\ -1 & -2 & -3 \end{pmatrix}\begin{pmatrix} x \\ y \\ z \end{pmatrix} = \begin{pmatrix} -2 \\ 0 \\ 5 \end{pmatrix}$ **3.** $\begin{pmatrix} 1 & 1 & 1 \\ 0 & 1 & -1 \\ 2 & -2 & -1 \end{pmatrix}\begin{pmatrix} x \\ y \\ z \end{pmatrix} = \begin{pmatrix} 7 \\ 7 \\ 0 \end{pmatrix}$ **5.** $\begin{pmatrix} \frac{1}{2} & \frac{1}{2} & \frac{1}{2} \\ 2 & -1 & 0 \\ -1\frac{1}{2} & \frac{1}{2} & -\frac{1}{2} \end{pmatrix}$

7. $\begin{pmatrix} 0.429 & 0.143 & 0.286 \\ 0.286 & 0.429 & -0.143 \\ 0.286 & -0.571 & -0.143 \end{pmatrix}$ **9.** $(1\frac{1}{2}, -4, \frac{1}{2})$ **11.** $(4, 5, -2)$ **13.** $(-4, 1, 1)$
15. $(-0.262, -0.142, 0.468)$ **17.** $(-1\frac{1}{3}, -1\frac{1}{3}, 1\frac{2}{3})$ **19.** $(1, 0, -1)$ **21.** $(0, 0, 0)$
23. No solution

Review Exercises (p. 393)

1. $-1, 4$

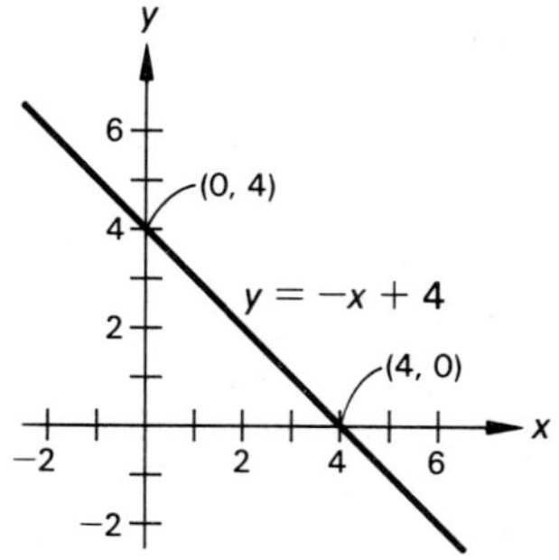

3. $\frac{1}{2}, 2$

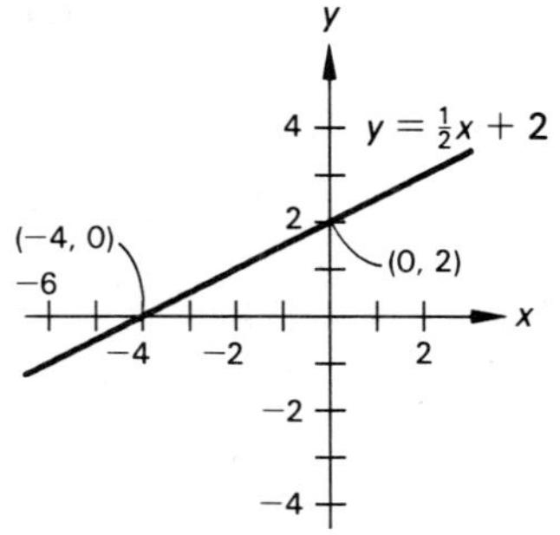

5. $0, 3$

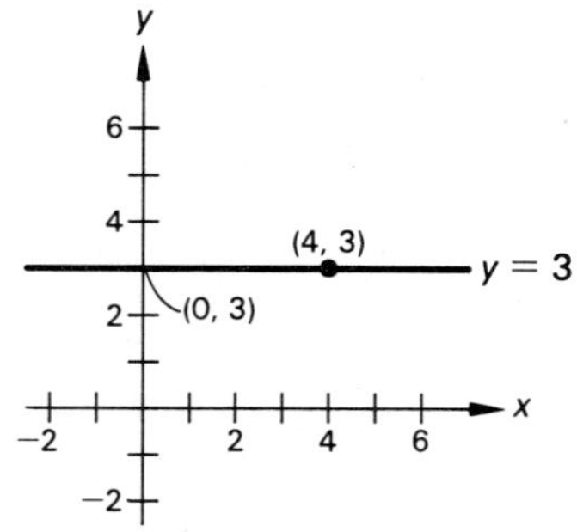

7. $-\frac{1}{2}, \frac{1}{3}$

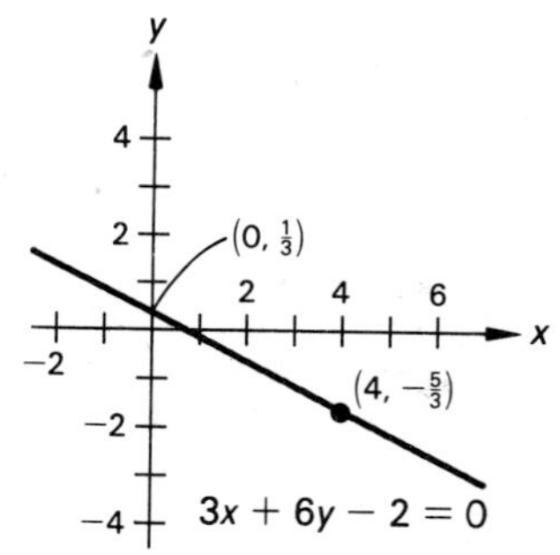

9. $y = -4x$ **11.** $y = \frac{1}{2}x + 3$ **13.** $y = -2$
15. $y = 1.32x + 4.17$ **17.** $(6, 6)$ **19.** $(0, 1)$ **21.** $(4, -2)$
23. $(3, -2)$ **25.** $(-28, -46)$
27. Inconsistent

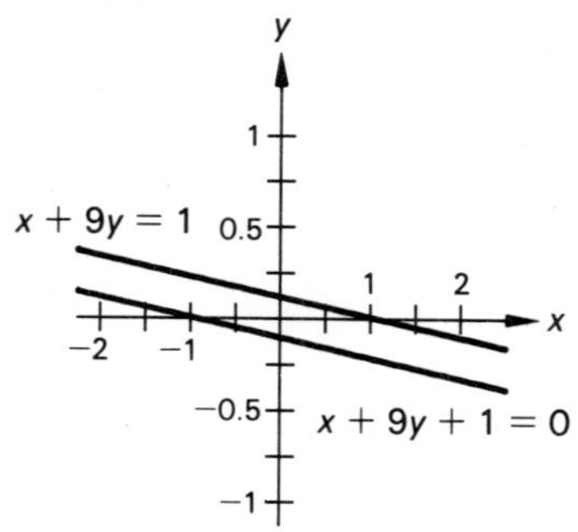

29. $(2.2, -1.6)$

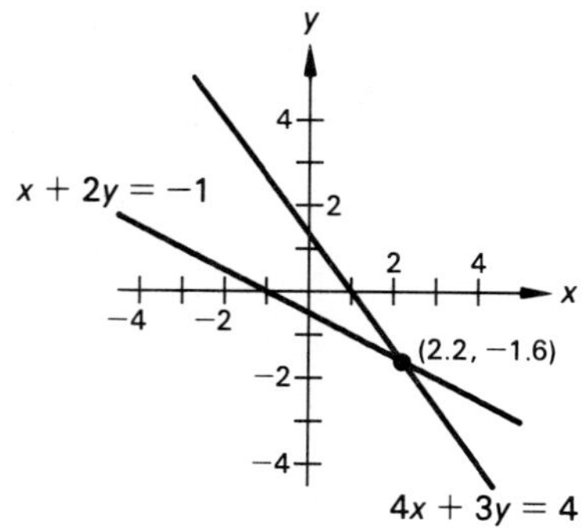

31. Dependent **33.** 51 **35.** -8 **37.** -1 **39.** $(3, 9)$ **41.** $(32, 22)$ **43.** $(4, 20)$
45. $(3, 2)$ **47.** $(4, -2)$ **49.** $(-4, 1)$ **51.** $(0, -1, -1)$ **53.** $(2, -1, 0)$ **55.** $(-11, 1, -11)$
57. $(0, 3, -2)$ **59.** -24 **61.** 62 **63.** -5 **65.** -71 **67.** $(-1, 3, -2)$ **69.** $\left(-\frac{1}{2}, 0, \frac{1}{2}\right)$
71. $(1, -1, -1)$ **73.** $(-1, -2, 1)$ **75.** $F_1 = 6592.$, $F_2 = -561.$ **77.** $F_A = 9438.$, $F_B = 4718.$
79. $F_A = -11{,}765.$, $F_B = -10{,}648.$, $F_C = 20{,}170.$ **81.** $i_1 = 3$, $i_2 = 2$ **83.** $v_1 = 50.$, $v_2 = 40.$
85. $i_1 = 2.98$, $i_2 = 2.47$, $i_3 = 0.93$ **87.** $v_1 = 31.8$, $v_2 = 47.6$, $v_3 = -1.77$
89. Plug = \$1.85, jig = \$0.46, spoon = \$1.85 **91.** $\begin{pmatrix} 3 & 2 \\ 1 & 4 \\ 8 & 9 \end{pmatrix}$ **93.** $\begin{pmatrix} 16 & 22 & 6 \\ -2 & -6 & 1 \\ 30 & 12 & 27 \end{pmatrix}$

95. $\begin{pmatrix} 5 & 8 & 2 \\ 11 & 5 & 2 \\ -1 & -14 & 3 \end{pmatrix}$ **97.** $(-1, 2, 2)$ **99.** $(-4, 1, 1)$

CHAPTER 11

11.2 Exercises (p. 404)

1. 2, 3 **3.** 5, −3 **5.** $5, -1\frac{1}{2}$ **7.** $4, -\frac{2}{5}$ **9.** $-\frac{2}{9}, -1$ **11.** 8, −8 **13.** $5\sqrt{5}, -5\sqrt{5}$
15. −9, −9 **17.** $-\frac{3}{4}, -\frac{3}{4}$ **19.** $13j, -13j$ **21.** 0, −13 **23.** 0, −3 **25.** $-\frac{2}{3}, 3$ **27.** 1, −7
29. −2, −4 **31.** $x^2 + 2x - 15 = 0$ **33.** $5x^2 - 17x + 6 = 0$ **35.** $x^2 - 10x + 25 = 0$
37. $6x^2 - 5x + 1 = 0$ **39.** $x^2 + 2x + 5$ **41.** 5, −3 **43.** $-\frac{11}{2}, \frac{5}{3}$ **45.** 4, 3 **47.** −11, −11
49. 3, 3 **51.** 10 cm, 30 cm **53.** 12.00 m

11.3 Exercises (p. 408)

1. 14, 2 **3.** ±3 **5.** $-5 + \sqrt{15}, -5 - \sqrt{15}$ **7.** $\frac{2}{3}, -4$ **9.** $3 + 4j, 3 - 4j$ **11.** 1, −4
13. $\frac{-7 + \sqrt{61}}{2}, \frac{-7 - \sqrt{61}}{2}$ **15.** $\frac{3}{2}, -3$ **17.** $\frac{-1 + \sqrt{21}}{10}, \frac{-1 - \sqrt{21}}{10}$ **19.** $\frac{9 + \sqrt{129}}{6}, \frac{9 - \sqrt{129}}{6}$
21. 3, −1 **23.** 3, 3 **25.** $\frac{-5 + j\sqrt{3}}{2}, \frac{-5 - j\sqrt{3}}{2}$ **27.** $\pm 2\sqrt{5}$
29. $\frac{-b + \sqrt{b^2 - 4ac}}{2a}, \frac{-b - \sqrt{b^2 - 4ac}}{2a}$

11.4 Exercises (p. 412)

1. 18, −2 **3.** −2, −12 **5.** 2.354, −0.354 **7.** $-1 \pm 0.707j$ **9.** 3, 0.167 **11.** 1.5, 1.25
13. 3.73, 0.268 **15.** 3.37, 1.63 **17.** 1.24, −3.24 **19.** $1 \pm 2j$ **21.** $-2 \pm 3j$ **23.** 4.79, 0.209
25. 5.00, −1.33 **27.** 1.12, −7.12 **29.** $0.433 \pm 0.559j$ **31.** 0.437, −1.87 **33.** 0.181, −0.283
35. 0.667 s, −4.00 s **37.** 53.4 in. **39.** 2.00

11.5 Exercises (p. 415)

1. +; real, rational, unequal **3.** −; imaginary, unequal **5.** +; real, irrational, unequal
7. +; real, irrational, unequal **9.** 0; real, rational, equal **11.** +; real, rational, unequal
13. −; imaginary, unequal **15.** +; real; unequal **17.** +; imaginary, unequal **19.** 0; imaginary, equal
21. $\frac{1}{3}$ **23.** 25 **25.** $1, -\frac{1}{3}$ **27.** Discriminant is negative so roots are imaginary.
29. $2\sqrt{21}, -2\sqrt{21}$

11.6 Exercises (p. 422)

1. $-5.0, -1.0$

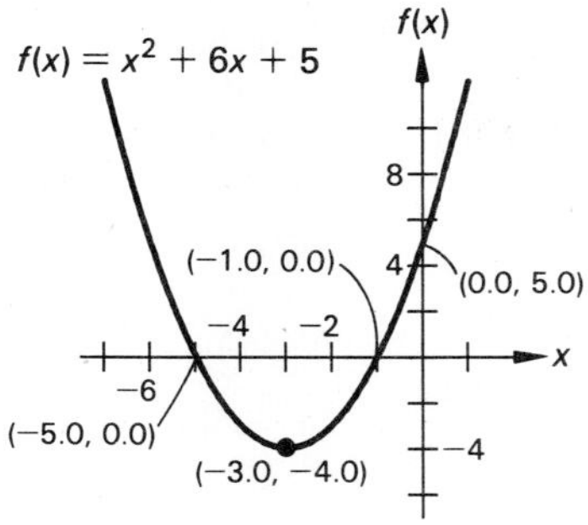

3. $-1.0, -4.0$

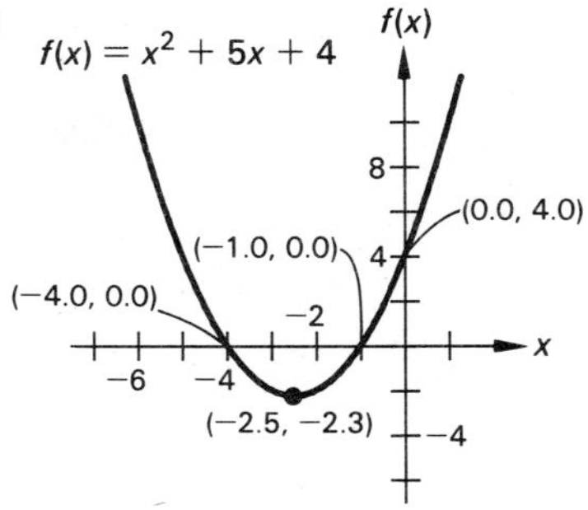

5. $1.7, -1.0$

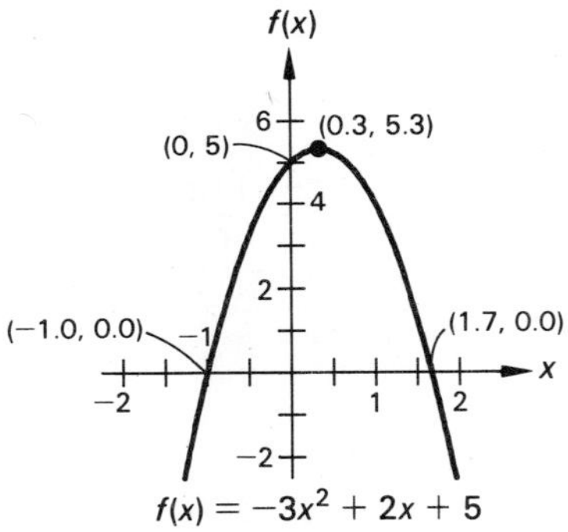

7. $2.5, -1.0$

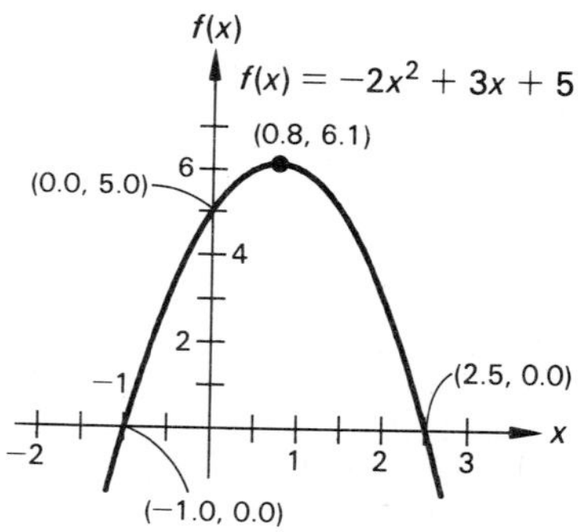

9. Imaginary

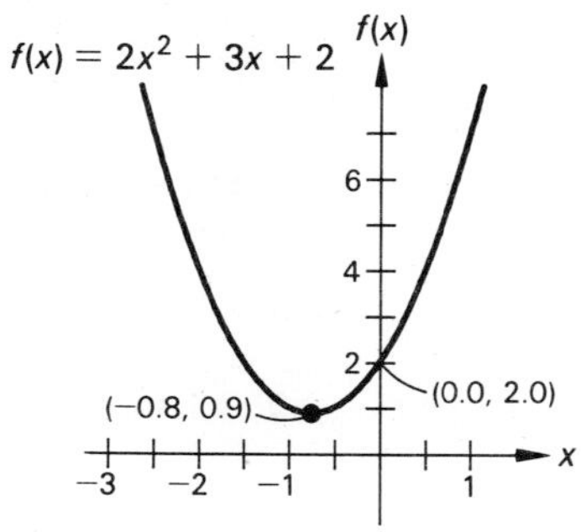

11. $4.0, 0.0$

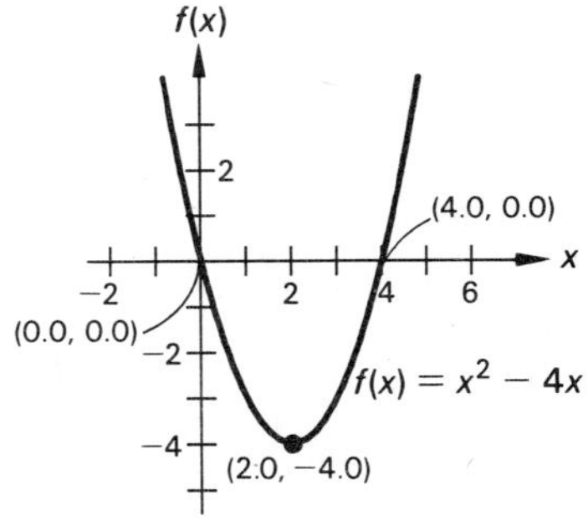

13. 2.5

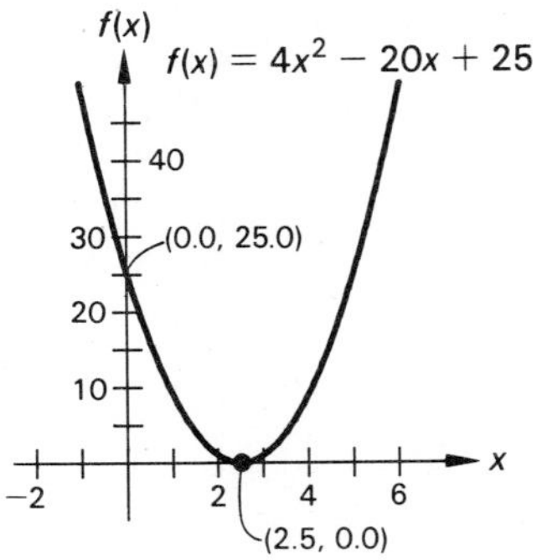

15. $0.8, -2.1$

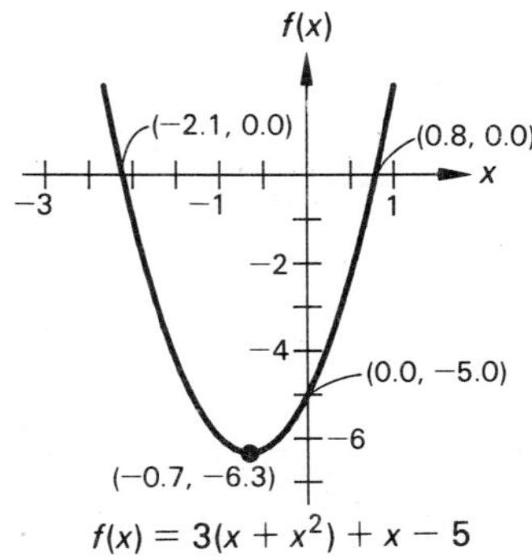

17. $\frac{2}{3}$

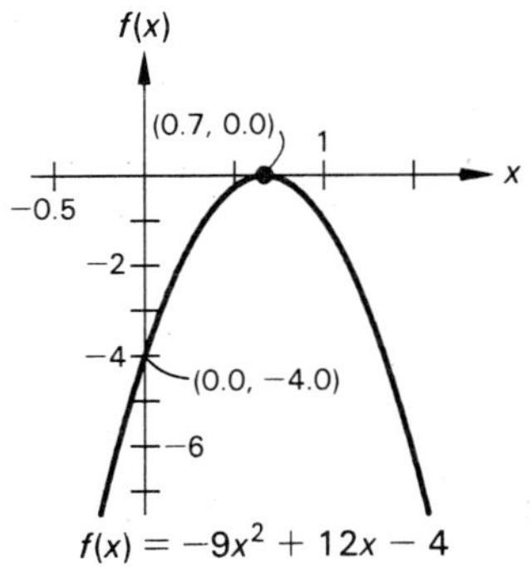

19. −0.5, −2.0

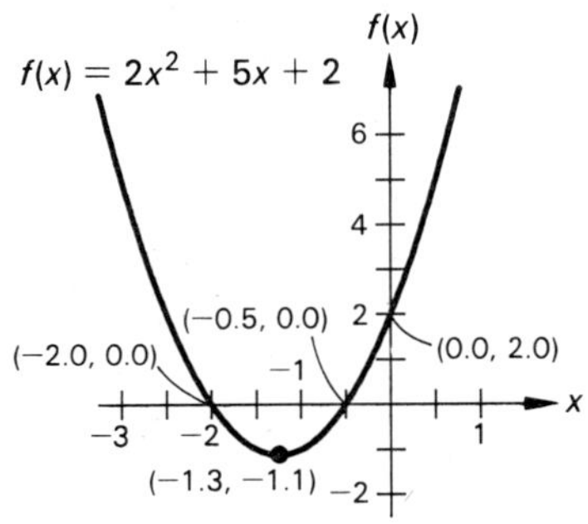

21. One **23.** Two
25. None **27.** Two
29. Two **31. a.** 1
b. $m < 1$ **c.** $m > 1$
33. a. ± 4
b. $m < -4$, $m > 4$
c. $-4 < m < 4$

11.7 Exercises (p. 426)

1. $\frac{\pi}{2}, \frac{3\pi}{2}$ **3.** $\frac{2\pi}{3}, \frac{5\pi}{3}$ **5.** $\frac{\pi}{6}, \frac{7\pi}{6}$ **7.** $\frac{\pi}{4}, \frac{7\pi}{4}$ **9.** $0, \pi, 2\pi$ **11.** $\frac{7\pi}{6}, \frac{11\pi}{6}$ **13.** $\frac{\pi}{2}$
15. $\frac{\pi}{2}, \frac{5\pi}{6}, \frac{7\pi}{6}, \frac{3\pi}{2}$ **17.** $\frac{\pi}{3}, \frac{2\pi}{3}, \frac{4\pi}{3}, \frac{5\pi}{3}$ **19.** $\frac{\pi}{6}, \frac{5\pi}{6}$ **21.** $\frac{\pi}{3}, \frac{2\pi}{3}, \frac{4\pi}{3}, \frac{5\pi}{3}$

23. $\frac{3\pi}{2}, \frac{7\pi}{6}, \frac{11\pi}{6}$ **25.** 30.0°, 150.0°, 221.8°, 318.2° **27.** 56.3°, 236.3° **29.** 128.2°, 231.8°
31. 45°, 90°, 135°, 225°, 270°, 315° **33.** 137.1°, 222.9° **35.** 62.2°, 117.8° **37.** 18.4°
39. 35.3°, 1.2 m^3

11.8 Exercises (p. 433)

1. $\pm\sqrt{10}, \pm 1$ **3.** $\pm\sqrt{7}, \pm\sqrt{2}$ **5.** $0, \pm\sqrt{5}, \pm\sqrt{3}$ **7.** $\pm\sqrt{3}, \pm\frac{1}{2}$ **9.** 1 **11.** $-\frac{1}{8}, -1$
13. $\pm 1, \pm 0.289$ **15.** 46656, 1 **17.** $\frac{1}{4}$, 4 **19.** 3, −3 **21.** $2 \pm 2j$ **23.** −2
25. 2.538, −0.788 **27.** 12 **29.** −4, 11 **31.** $\frac{1}{5}$ **33.** −2, −7 **35.** 13 **37.** 1
39. No solution **41.** $\frac{9}{8}$ **43.** $\frac{65}{32}$ **45.** 0 **47.** 12.7 hr, 15.7 hr **49.** 43.4 cm, 16.6 cm

11.9 Exercises (p. 438)

1.

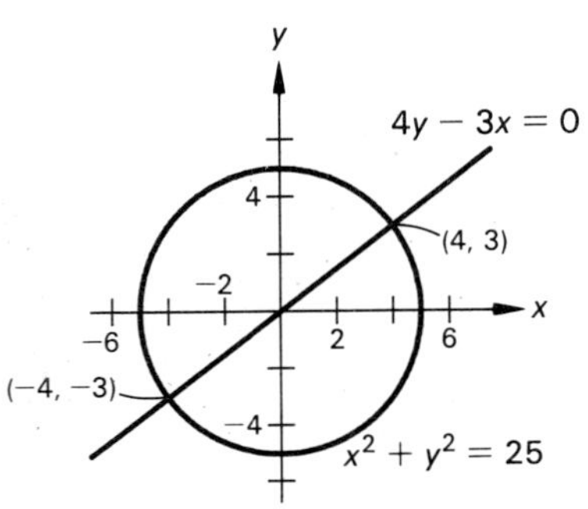

3.

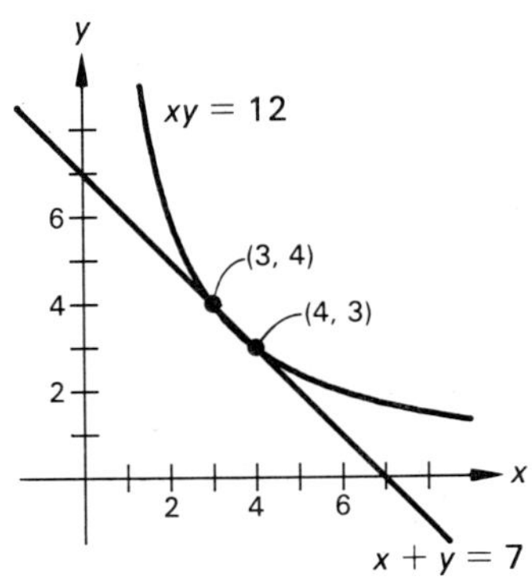

5.

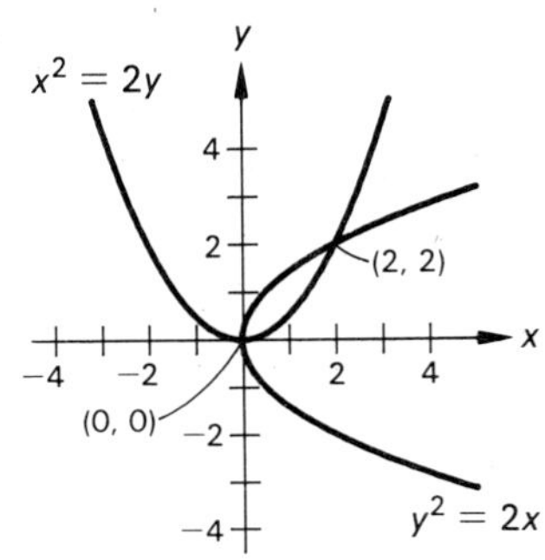

7.

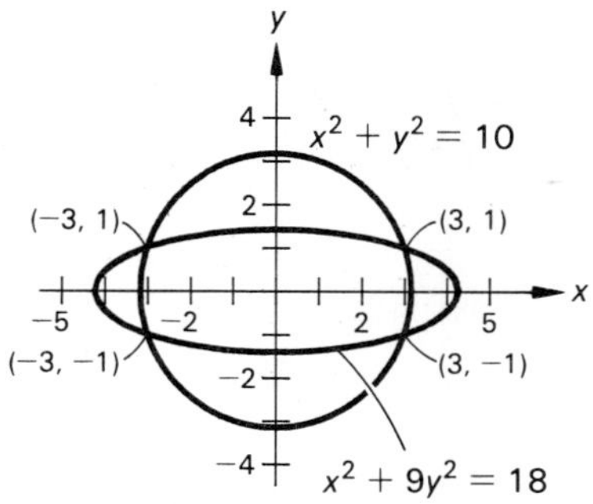

9.

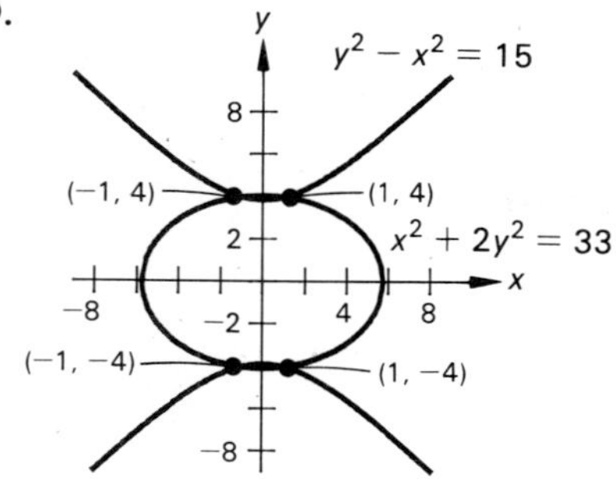

11.

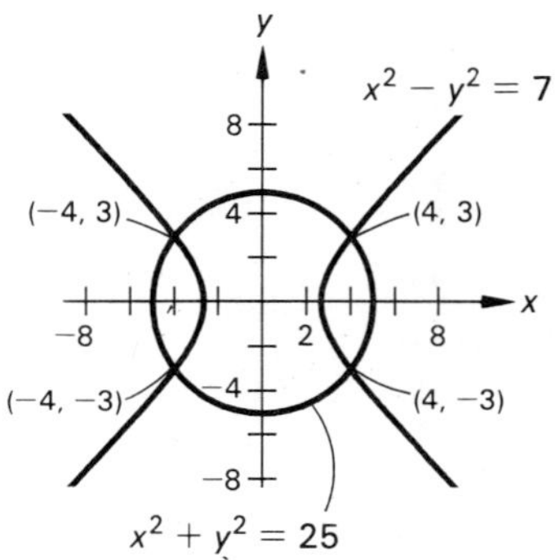

13. (3, 4), (4, 3) **15.** (−1.93, 1.44), (2.59, −0.80)

17. $\left(-\frac{3}{2}, -5\right)$ **19.** (2, 3), (2, −3), (−2, 3), (−2, −3)

21. $\left(-\frac{4}{5}, \frac{6}{\sqrt{5}}\right), \left(-\frac{4}{5}, -\frac{6}{\sqrt{5}}\right)$

23. (1, 2), (−1, −2), (2, 1), (−2, −1) **25.** 18 cm × 12 cm
27. 21 mm, 20 mm **29.** 77 m, 36 m

11.10 Exercises (p. 442)

1. 144. ft **3.** 7.62 ft **5. a.** 13 m × 4 m **b.** 119 m^2 **7.** 30 m^2 **9.** 2 or 50 **11. a.** 3 s **b.** 6.51 s **13. a.** 100 in.^{2}, **b.** 324 in.^{2} **15.** 8 **17.** 50 mi/hr **19.** 25.99 or 6.01

Review Exercises (p. 444)

1. 11, −4 **3.** $2\frac{1}{2}, -2\frac{1}{3}$ **5.** 0.573, −2.907 **7.** 9, 6 **9.** $-1.167 \pm 0.553j$

11. $x^2 - 7x + 12 = 0$ **13.** $x^2 + 8x + 16 = 0$ **15.** $x^2 + 6x + 25 = 0$ **17.** 5, −11 **19.** $\frac{4}{5}, -1\frac{1}{2}$

21. +; real, rational, unequal **23.** −; imaginary, unequal **25.** +; real, rational, unequal

27. 9.0, 2.0

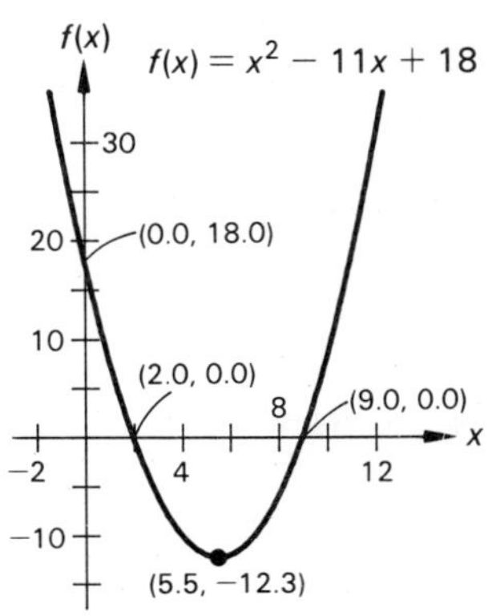

29. −0.8, −6.2

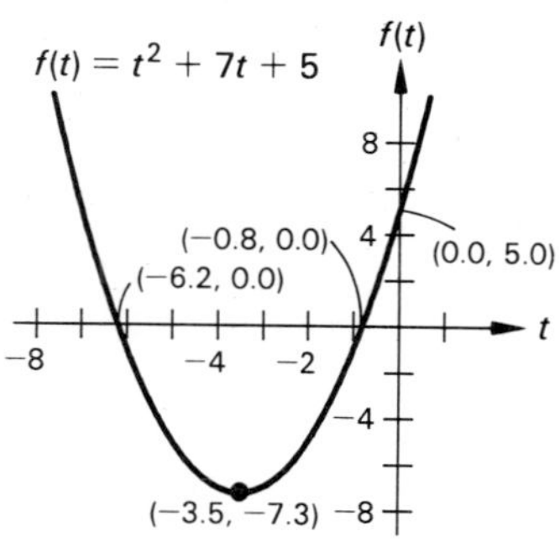

31.

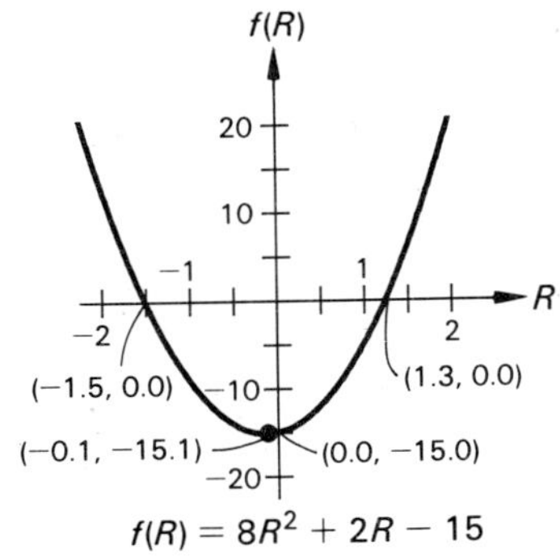

1.3, −1.5

33.

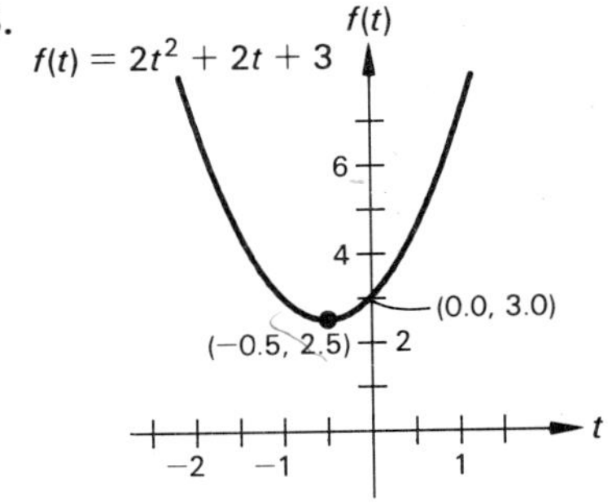

35.

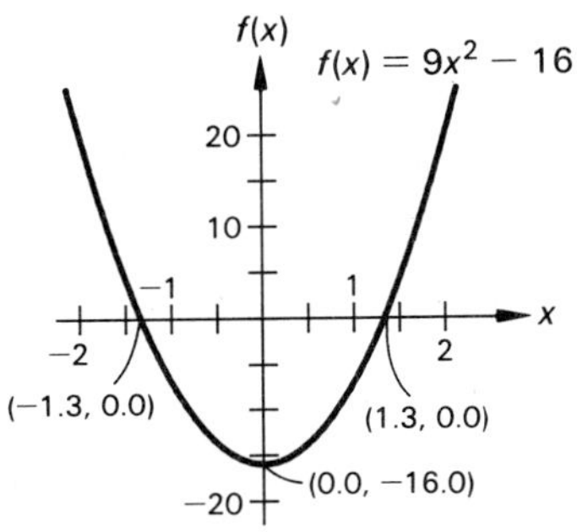

1.3, −1.3

37. $\frac{\pi}{4}, \frac{3\pi}{4}$ **39.** $0, \pi, \frac{4\pi}{3}, \frac{5\pi}{3}, 2\pi$

41. $\frac{\pi}{3}, \frac{11\pi}{6}$, 2.30, 3.98 **43.** $0, \frac{\pi}{4}, \frac{3\pi}{4}, \frac{5\pi}{4}, \frac{7\pi}{4}, \pi$

45. 0.650, 2.492, 3.791, 5.633, $\frac{\pi}{2}, \frac{3\pi}{2}$ **47.** $\pm 2, \pm 3$

49. 1, 531441 **51.** −0.791, 3.79

53. No solution **55.** 0.571, −0.491 **57.** 0

59. No solution **61.** (0, 2) **63.** (1, 3), $(1\frac{1}{2}, 2)$

65. (1.79, −1.79), (−1.79, −1.79), (2.56, 1.54), (−2.56, 1.54) **67.** 1 Ω, 1 Ω

69. 24.81 cm × 9.19 cm **71.** 6 days, 12 days **73.** $q = \frac{p}{pd - 1}$

CHAPTER 12

12.2 Exercises (p. 454)

1. $6x + 22 + \frac{68}{x - 4}$ **3.** $x^2 + x - 4 + \frac{4}{x + 1}$ **5.** $-x^2 + 2x - 2 + \frac{6}{x + 2}$ **7.** $2x^2 - 2x - 1$

9. $x^3 + x^2 - x + 2$ **11.** $2z^3 - 4z^2 + 19z - 65 + \frac{253}{z + 4}$ **13.** $3x^4 + 3x^3 + 3x^2 - 6x$

15. $9R^3 - 12R^2 + 9R + 6 + \frac{3}{R - \frac{1}{3}}$ **17.** $2x^3 + 5x^2 - \frac{5}{2}x - \frac{27}{4} + \frac{107/8}{x + \frac{1}{2}}$

19. $y^4 + 2y^3 + 4y^2 + 8y + 16$ **21.** $-z^6 + 2z^5 - 4z^4 + 12.3z^3 - 24.6z^2 + 42z - 84 + \frac{168}{z + 2}$

23. $-80z^3 + 240cz^2 - 721c^2z + 2160c^3 - \frac{6480\,c^4}{z + 3c}$ **25.** $2x^3 + x^2 + 1 + \frac{x + 12}{(x - 2)(x + 3)}$

12.3 Exercises (p. 458)

1. 4 **3.** -11 **5.** -4 **7.** 0 **9.** 0 **11.** Yes **13.** No **15.** No **17.** No **19.** Yes
21. a. $2x^2 + 14x + 15$ **b.** $2, -1.32, -5.68$ **23. a.** $2x^2 - 7x + 15$ **b.** $7, 1.75 \pm 2.01j$
25. $4, -1$ **27.** True **29.** True **31.** True **33.** Yes **35.** 2

12.4 Exercises (p. 462)

1. a. $-3, 2, -5$ **b.** 1, 3, 2 **3. a.** $6, -\frac{5}{2}$ **b.** 1, 3 **5. a.** $-\frac{2}{3}, -3$ **b.** 1, 4 **7. a.** $-5, 7$
b. 7, 1 **9. a.** $j, -j, 3, -2\frac{1}{2}$ **b.** 1, 1, 1, 1 **11. a.** $2 - 3j, 2 + 3j, -3$ **b.** 1, 1, 2 **13. a.** 2
b. -7 **15. a.** 4 **b.** -2 **17. a.** 3 **b.** -3 **19. a.** 2 **b.** -6 **21. a.** 2 **b.** -2
23. $x^2 - 16x^2 + 39 = 0$ **25.** $x^3 - 6x^2 + 11x - 6 = 0$ **27.** $x^4 + x^3 - 5x^2 + 3x = 0$
29. $6x^3 + x^2 - 4x + 1 = 0$ **31.** $24x^3 - 2x^2 - 9x + 2 = 0$ **33.** $x^3 - 4x^2 + 2x + 4 = 0$
35. $9x^6 - 98x^4 + 151x^2 + 18 = 0$ **37.** $2x^5 - 5x^4 + 6x^3 + 14x^2 - 56x + 24 = 0$
39. $x^5 - x^4 - 5x^3 + x^2 + 8x + 4 = 0$ **41.** $0.414, -2.414$ or $-1 \pm \sqrt{2}$ **43.** $1 + 3j, 1 - 3j$
45. It must have as many roots as its degree and the complex roots must occur in pairs. **47.** 2.4 s, 3.7 s

12.5 Exercises (p. 471)

1. a. $p:1, n:2$ or 0 **b.** $\pm 1, \pm 3$ **c.** $3, -1, -1$ **3. a.** $p:1, n:2$ or 0 **b.** $\pm 1, \pm 2$
c. $1, -1, -2$ **5. a.** $p:1, n:2$ or 0 **b.** $\pm 1, \pm 2, \pm 3, \pm 6, \pm\frac{1}{2}, \pm\frac{3}{2}$ **c.** $\frac{1}{2}, -2, -3$
7. a. $p:1, n:2$ or 0 **b.** $\pm 1, \pm 2, \pm 3, \pm 4, \pm 6, \pm 12$ **c.** None **9. a.** $p:3$ or 1, $n:0$
b. $\pm 1, \pm 2, \pm 4$ **c.** 2 **11. a.** $p:2$ or 0, $n:1$ **b.** $\pm 1, \pm 2, \pm 11, \pm 22, \pm\frac{1}{3}, \pm\frac{2}{3}, \pm\frac{11}{3}, \pm\frac{22}{3}$
c. $-\frac{2}{3}$ **13. a.** $p:4$, 2 or 0, $n:0$ **b.** $\pm 1, \pm 3, \pm 5, \pm 7, \pm 15, \pm 21, \pm 35, \pm 105$ **c.** 3, 1, 7, 5
15. a. $p:1, n:3$ or 1 **b.** $\pm 1, \pm 3, \pm 9, \pm\frac{1}{2}, \pm\frac{3}{2}, \pm\frac{9}{2}, \pm\frac{1}{4}, \pm\frac{3}{4}, \pm\frac{9}{4}$ **c.** $-\frac{3}{2}, -\frac{3}{2}$
17. a. $p:5$, 3 or 1, $n:0$ **b.** $\pm 1, \pm\frac{1}{2}, \pm\frac{1}{4}, \pm\frac{1}{8}$ **c.** None **19. a.** $p:2$ or 0, $n:2$ or 0
b. $\pm 1, \pm 3, \pm\frac{1}{2}, \pm\frac{3}{2}, \pm\frac{1}{3}, \pm\frac{1}{6}$ **c.** $-\frac{1}{3}, \frac{3}{2}$

21.

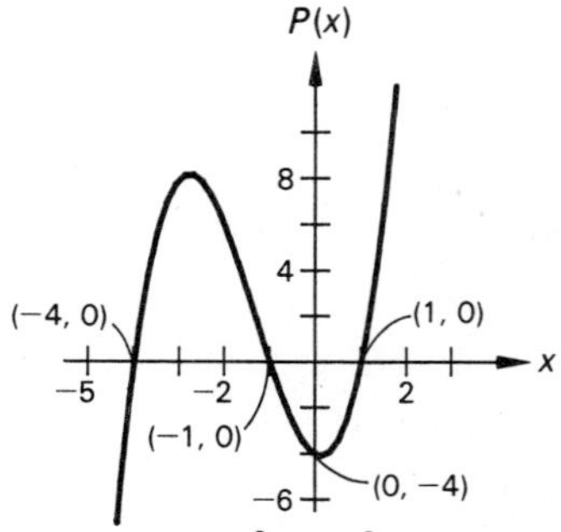

$P(x) = x^3 + 4x^2 - x - 4$

23.

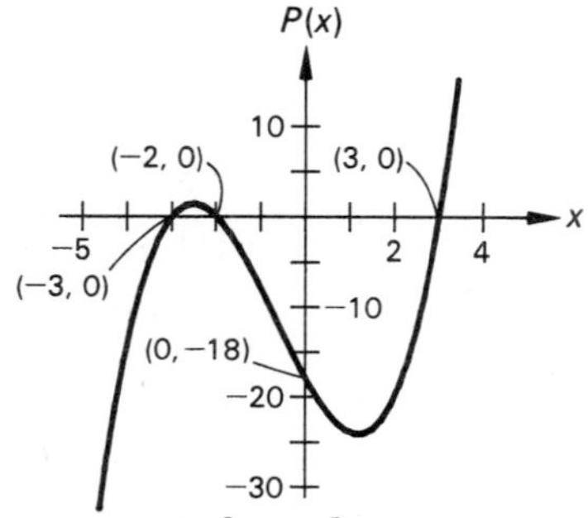

$P(x) = x^3 + 2x^2 - 9x - 18$

25.

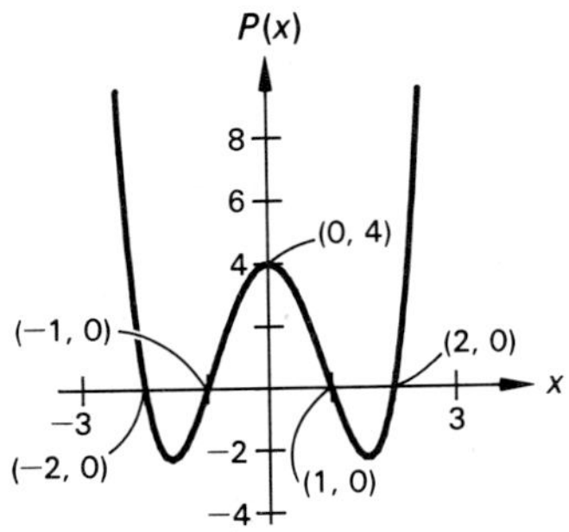

$P(x) = x^4 - 5x^2 + 4$

27.

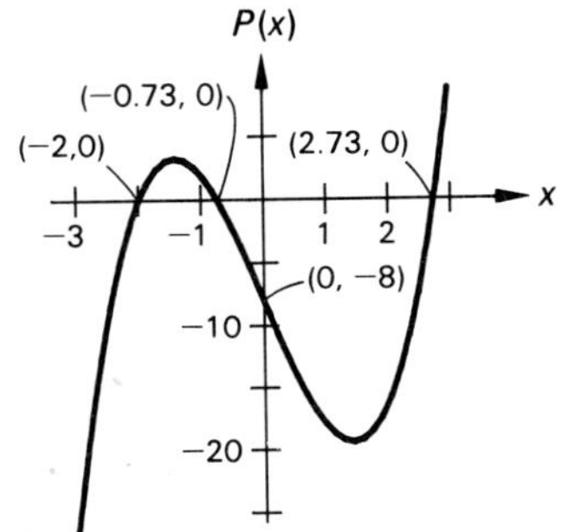

$P(x) = 2x^3 - 12x - 8$

29.

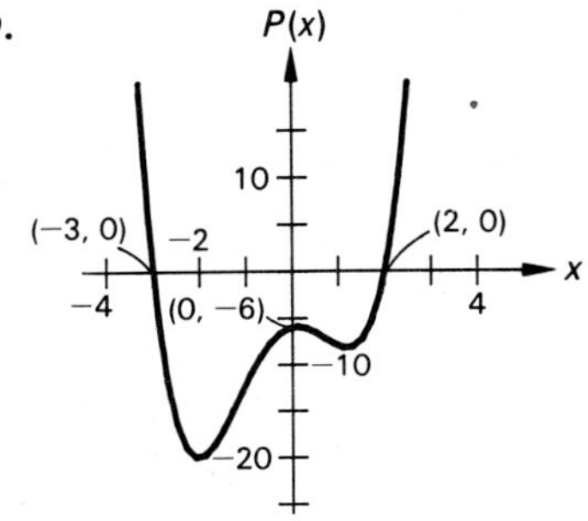

$P(x) = x^4 + x^3 - 5x^2 + x - 6$

31. 3 m, 4 m, 5 m **33.** 2, 5
35. 20.00 mm × 20.00 mm × 38.75 mm or 46.57 mm × 46.57 mm × 7.15 mm **37.** 1
39. $1\frac{1}{2}$ s, 2 s, 3 s

12.6 Exercises (p. 477)

1. $P(0) = -1$, $P(0) = 1$ **3.** $P(-0.6) = 1.11$, $P(-0.5) = -0.19$
5. $P(0.14) = 0.0227$, $P(0.15) = -0.0466$ **7.** $P(2.12) = -0.0719$, $P(2.13) = 0.0136$
9. $P(1.94) = -0.1678$, $P(1.95) = 0.0224$ **11.** 0.2 **13.** 3.7 **15.** −1.2 **17.** 0.7 **19.** 1.7

21.

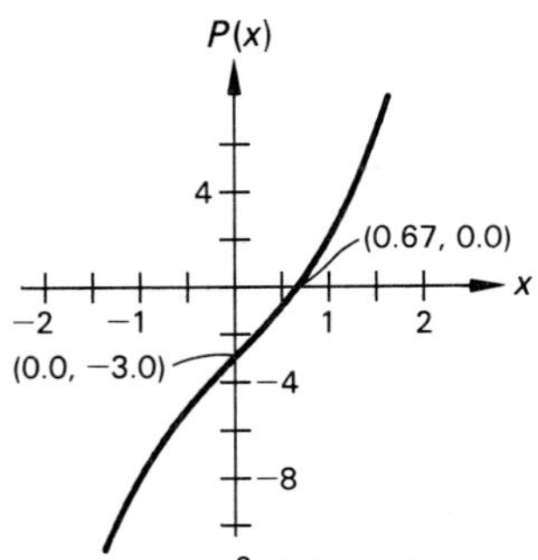

$P(x) = x^3 + 4x - 3$

23.

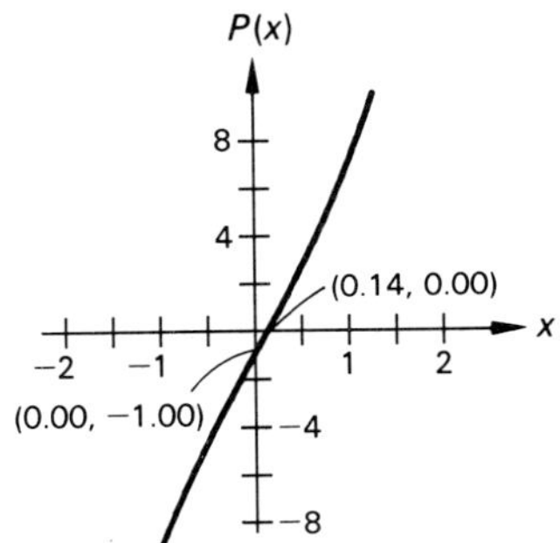

$P(x) = x^3 + 7x - 1$

25.

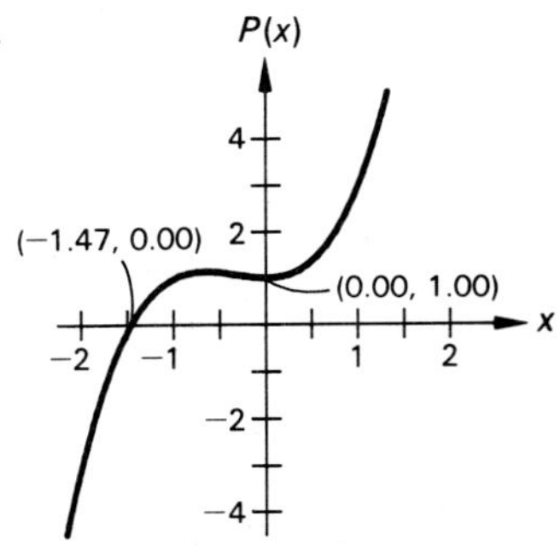

$P(x) = x^3 + x^2 + 1$

27.

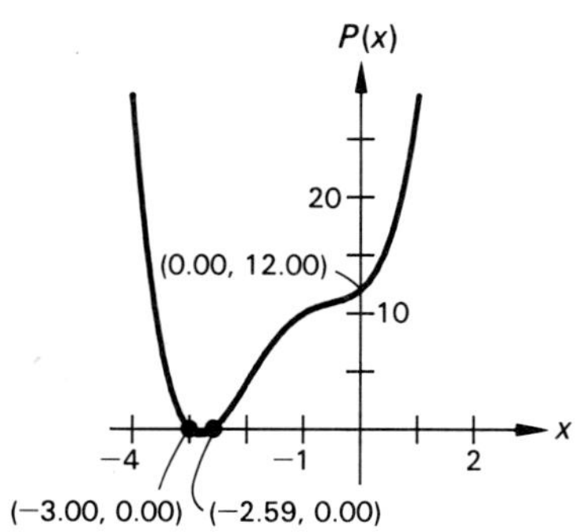

$P(x) = x^4 + 5x^3 + 6x^2 + 4x + 12$

29.

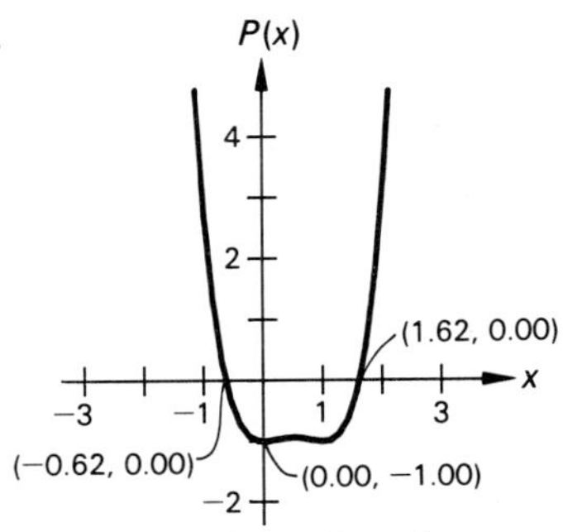

$P(x) = x^4 - 2x^3 + x^2 - 1$

31. a. 0.31 m, **b.** 0.83 m **33.** 1.65 μF, 0.23 μF **35.** 1.63 mm

Review Exercises (p. 479)

1. $Q(x) = 2x^2 - 11x - 12$, $R(x) = 0$ **3.** $Q(x) = 2x^2 + x - 3$, $R(x) = -2$
5. $Q(x) = 2x^3 + 25x^2 + 9x - 36$, $R(x) = -2$ **7.** $Q(x) = 6x^3 + 3x^2 - 7x - \frac{128}{3}$, $R(x) = -\frac{115}{9}$
9. $Q(x) = x^4 - 2x^3 - 19x^2 + 8x + 60$, $R(x) = -3$ **11.** $x^3 - 4x^2 - 17x + 60 = 0$
13. $15x^3 + 47x^2 - 50x + 8 = 0$ **15.** $x^4 - 4x^3 + 26x^2 - 100x + 25 = 0$ **17. a.** $\pm 1, \pm 2, \pm 5, \pm 10$
b. $-2, 5, 1$ **19. a.** $\pm 1, \pm 5, \pm 25, \pm\frac{1}{2}, \pm\frac{5}{2}, \pm\frac{25}{2}$ **b.** $\frac{1}{2}$ **21. a.** $\pm 1, \pm 3, \pm 9$ **b.** $3, 3$
23. a. $\pm 1, \pm 2, \pm 4, \pm 8, \pm 16, \pm\frac{1}{3}, \pm\frac{2}{3}, \pm\frac{4}{3}, \pm\frac{8}{3}, \pm\frac{16}{3}, \pm\frac{1}{5}, \pm\frac{2}{5}, \pm\frac{4}{5}, \pm\frac{8}{5}, \pm\frac{16}{5}, \pm\frac{1}{15}, \pm\frac{2}{15}, \pm\frac{4}{15}, \pm\frac{8}{15}, \pm\frac{16}{15}$ **b.** $\frac{2}{5}, -\frac{2}{3}$ **25. a.** $\pm 1, \pm 2, \pm 3, \pm 4, \pm 6, \pm 8, \pm 12, \pm 16, \pm 24, \pm 32, \pm 48, \pm 96$ **b.** $3, -1$

27.

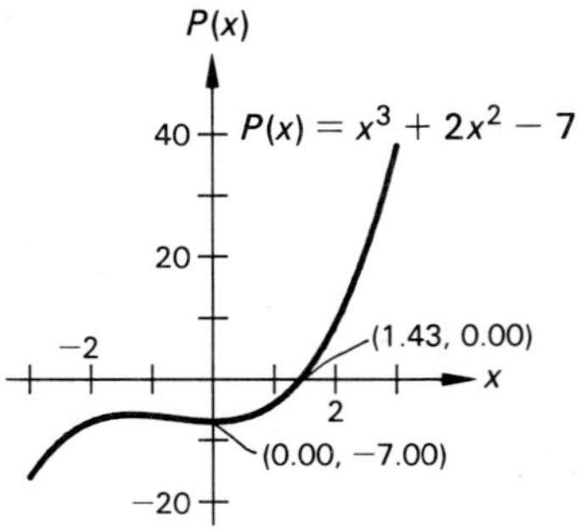

29.

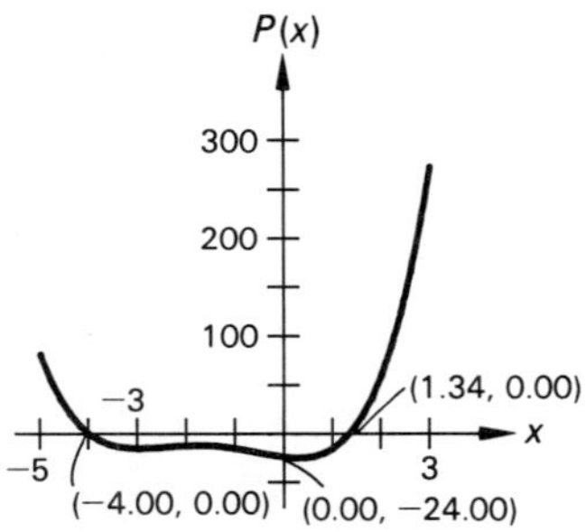

$P(x) = x^4 + 6x^3 + 8x^2 - 6x - 24$

31.

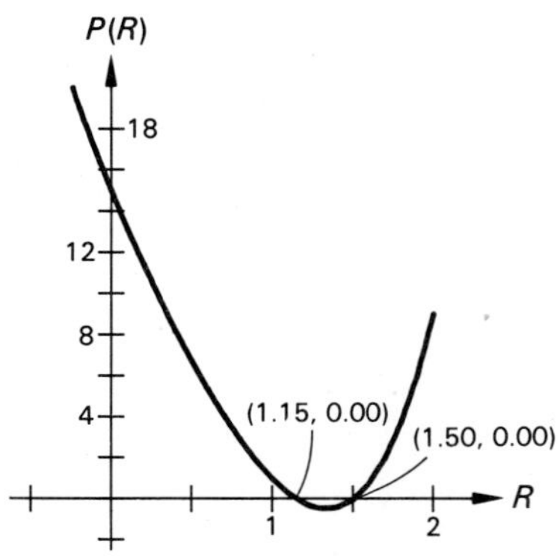

$P(R) = 2R^4 - 3R^3 + 6R^2 - 19R + 15$

33.

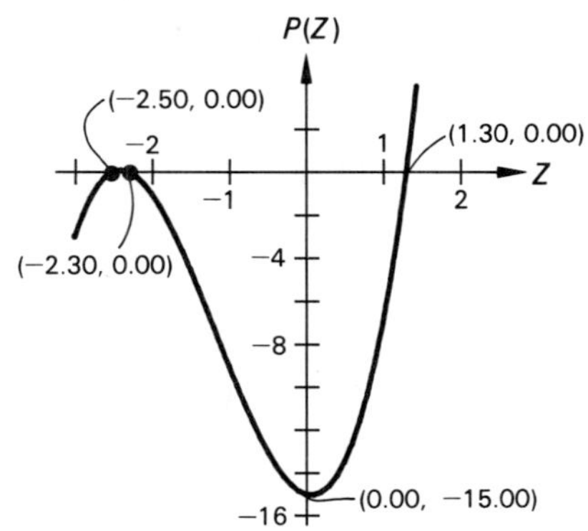

$P(Z) = 2Z^3 + 7Z^2 - Z - 15$

35.

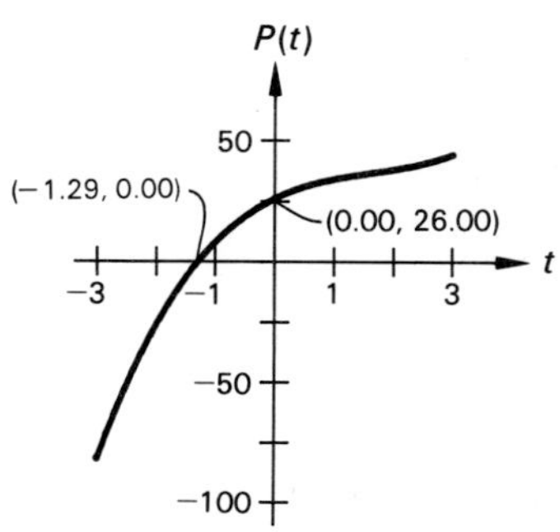

$P(t) = t^3 - 5t^2 + 12t + 26$

37. 40 mm **39.** 2 **41.** 1.31 m **43.** 21 units

CHAPTER 13

13.1 Exercises (p. 487)

1.

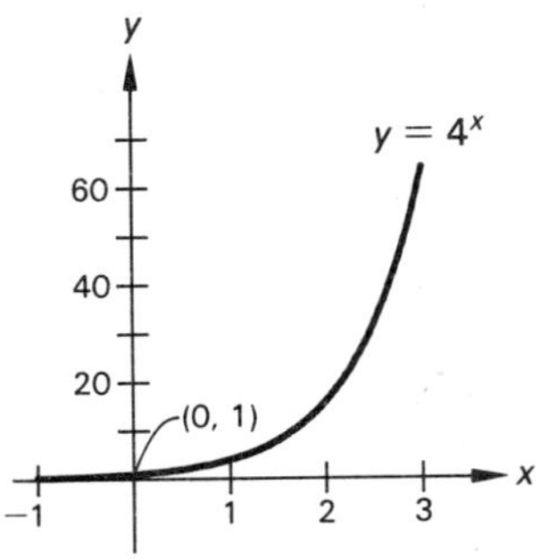

3.

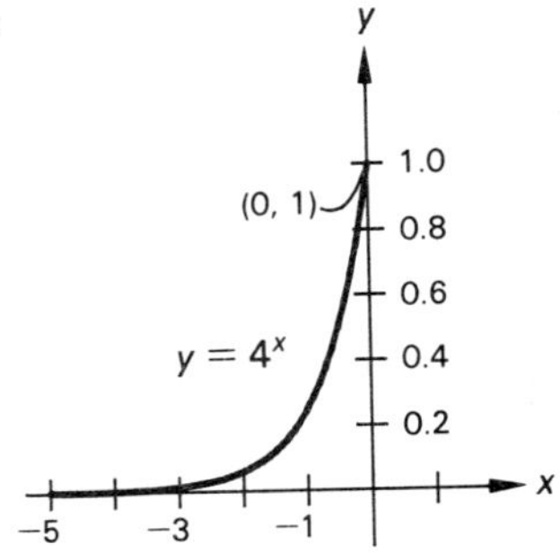

5.

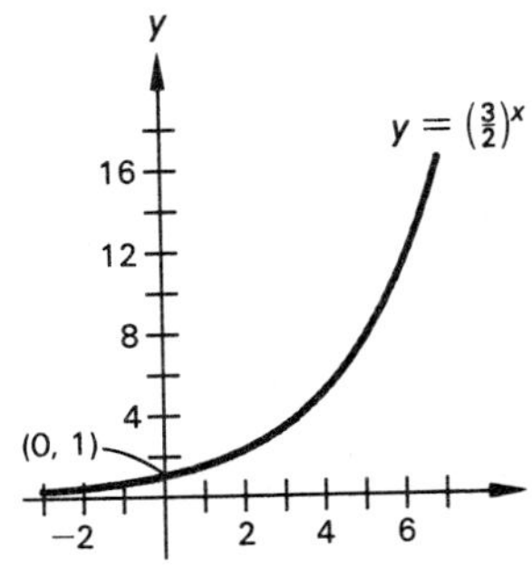

7.

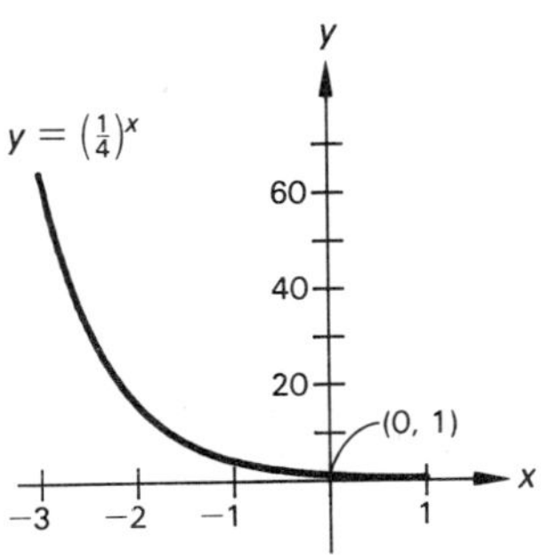

9.

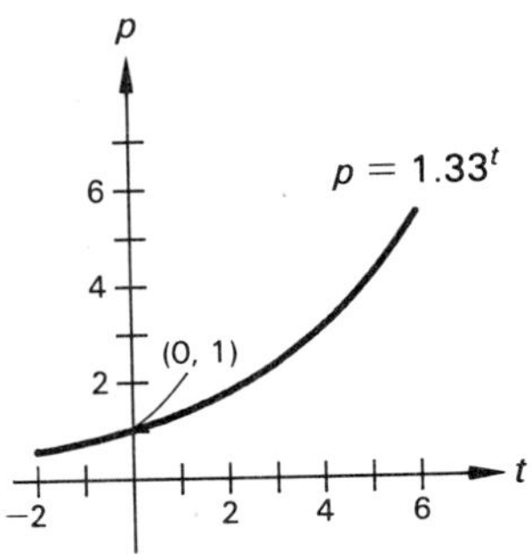

11.

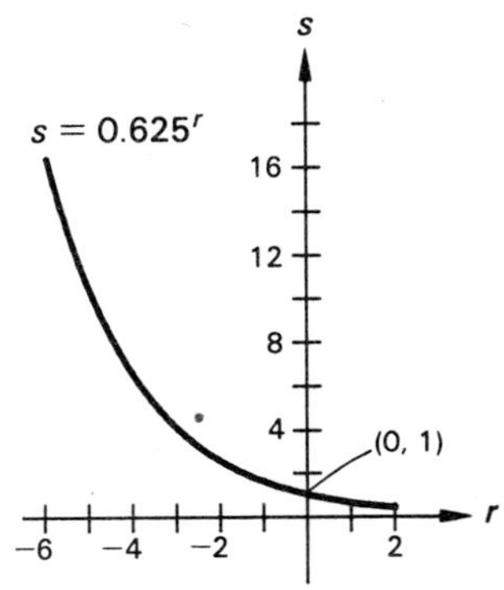

13.

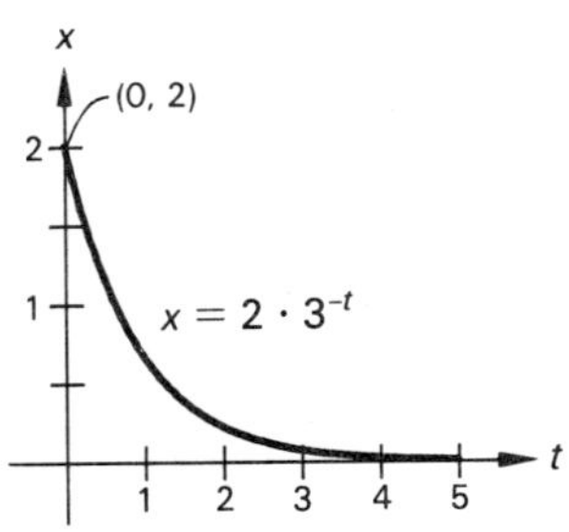

15.

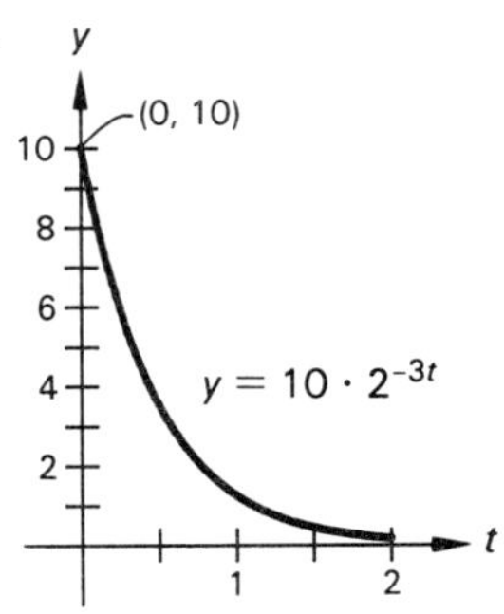

17.

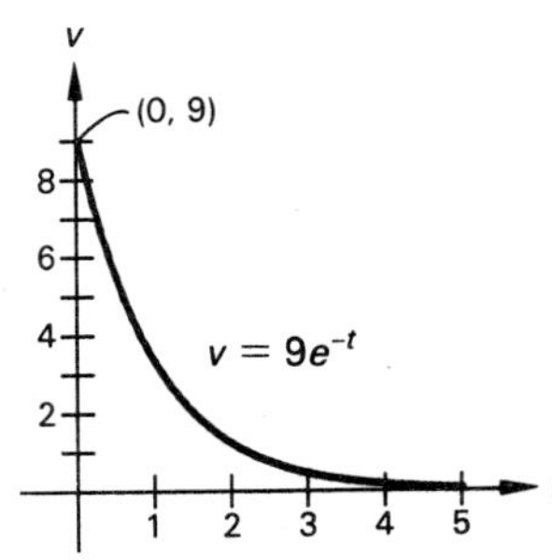

19.

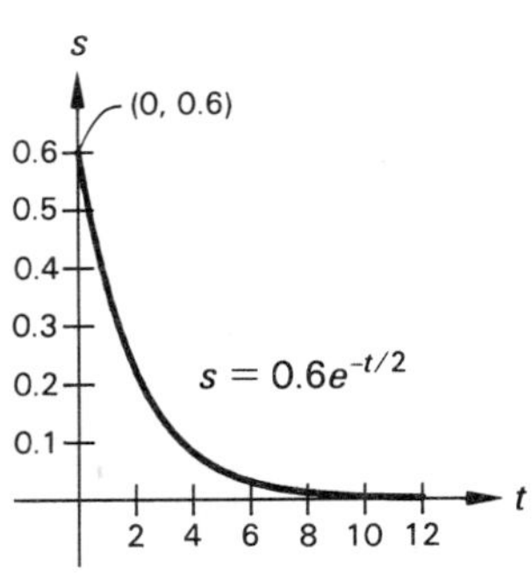

21. 30 s **23.** $\frac{5}{2}$ s **25.** 5 s **27.** 6.49 s **29.** 0.694 s
31. 13,122 bacteria **33.** 34,867. people **35.** 9.998 kg

13.2 Exercises (p. 494)

1. $\log_5 25 = 2$ **3.** $\log_4 64 = 3$ **5.** $\log_5 y = 2$ **7.** $\log_{10} p = t$ **9.** $\log_n m = x$ **11.** $49 = 7^2$
13. $625 = 5^4$ **15.** $144 = 12^y$ **17.** $123 = 6.3^v$ **19.** $t = m^r$ **21.** 16 **23.** $\frac{1}{5}$ **25.** 49
27. 5 **29.** 6 **31.** 2 **33.** 2 **35.** 2 **37.** 5 **39.** -2 **41.** -2 **43.** -2 **45.** 1
47. 1 **49.** 0 **51.** 0

53.

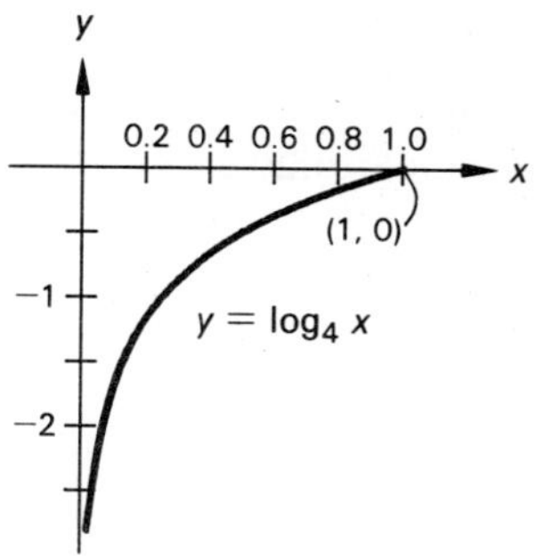

55.

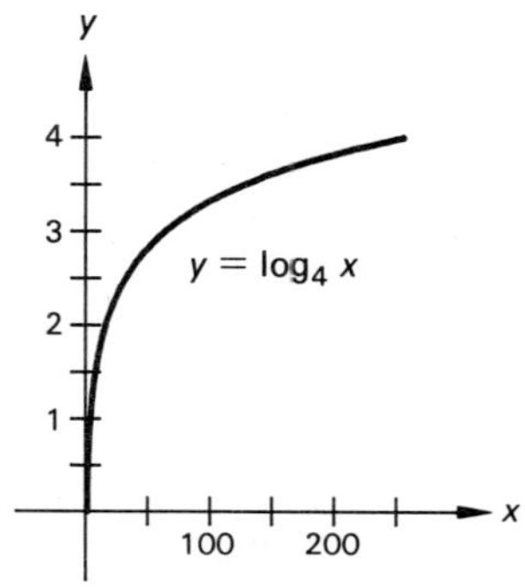

57.

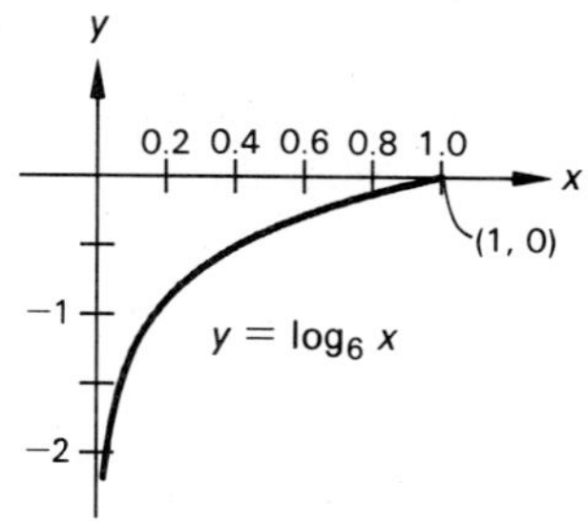

59.

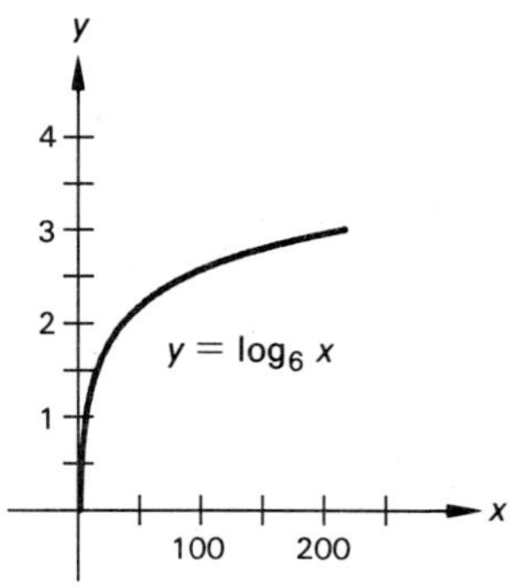

61.

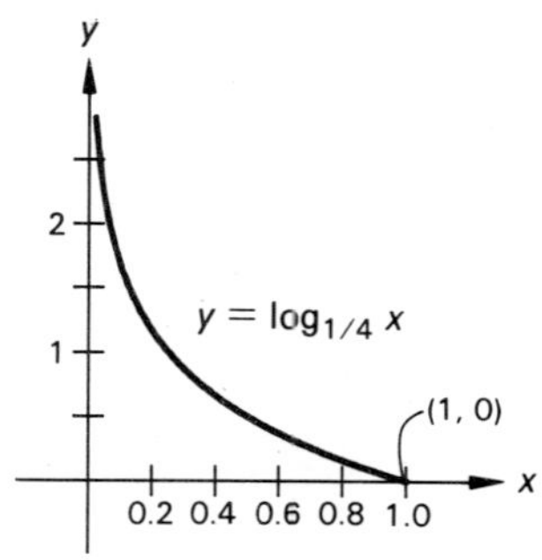

63.

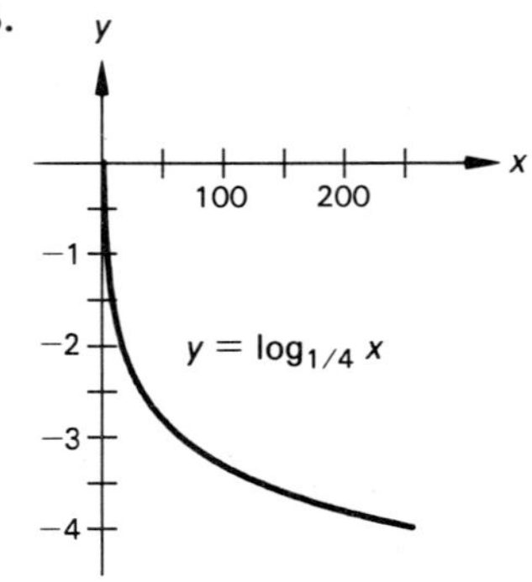

65. 15.8 hr **67.** 13.3 yr

13.3 Exercises (p. 501)

1. $\log_a 3 + 2\log_a x - \log_a y$ **3.** $\log_3 (x-4) - \log_3 (y+5)$ **5.** $2 - 2\log_3 x$

7. $2\log_{10} x + 3\log_{10} y - 3$ **9.** $2\log_b x + 3\log_b (4x-1) - 2\log_b \left(x - \frac{1}{4}\right)$ **11.** $\log_2 \frac{xy}{z}$

13. $\log_a \frac{2-x}{x+2}$ **15.** $\log_3 [(2x-1)(x-2)]^5$ **17.** $\log_2 x^4$ **19.** $\log_{6.3} y^{2.1}x^{3.5}$ **21.** 1.00

23. 1.63 **25.** 1.84 **27.** -0.052 **29.** $\dfrac{\log \frac{T_2}{T_1}}{\log \frac{V_1}{V_2}} + 1$

13.4 Exercises (p. 506)

1. 0.2122; c = 0, m = 0.2122 **3.** 0.3997; c = 0, m = 0.3997 **5.** 2.2227; c = 2, m = 0.2227

7. 1.7649; c = 1, m = 0.7649 **9.** 3.7226; c = 3, m = 0.7226 **11.** −0.7595; c = −1, m = 0.2405 **13.** −1.5735; c = −2, m = 0.4265 **15.** −3.1421; c = −4, m = 0.8579 **17.** −4.7670; c = −5, m = 0.2330 **19.** 4.71 **21.** 55.8 **23.** 0.00815 **25.** 6.73×10^{19} **27.** 39.5 **29.** 15,044. **31.** 2.2×10^3 **33.** 0.438 **35.** 0.891 **37.** 120 db **39.** 778.°K **41.** 10^{-12} W/cm² **43.** 2.15×10^{16} pa · s

13.5 Exercises (p. 512)

1. 4.19 **3.** −6.71 **5.** 0.778 **7.** 2.30 **9.** −4.61 **11.** 4.45 **13.** −1.90 **15.** 0.602 **17.** 1.24 **19.** 1.46 **21.** 2.76 **23.** 2.15 **25.** 2.37 **27.** 2.39 **29.** −3.17 **31.** 1.53 **33.** −1.08 **35.** 1.69 **37.** −1.83 **39.** 4.11 **41.** 2.85 **43.** 7.63 **45.** 0.293 V **47.** 12.8 mi/hr **49.** 2.93 J **51.** 3.92 s

13.6 Exercises (p. 516)

1. 0.774 **3.** 1.16 **5.** 2.72 **7.** $\pm 2j$ **9.** −1 **11.** 0.162 **13.** −5.39 **15.** No solution **17.** 12.7 **19.** 0.786 **21.** 0.414 **23.** 0.414 **25.** −1.008 **27.** 1.0001 **29.** 0.414 **31.** 1.04×10^{10} yr **33.** 6.31×10^{-6} W/cm² **35.** 0.202 **37.** $\dfrac{y_1 - y_2\, e^{z/k}}{1 - e^{z/k}}$

13.7 Exercises (p. 526)

1.

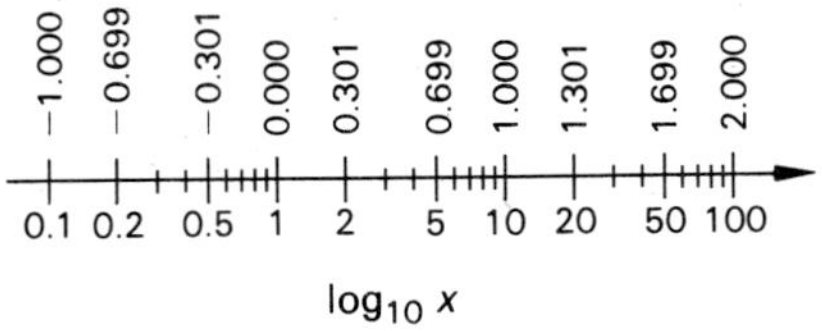

3. a.

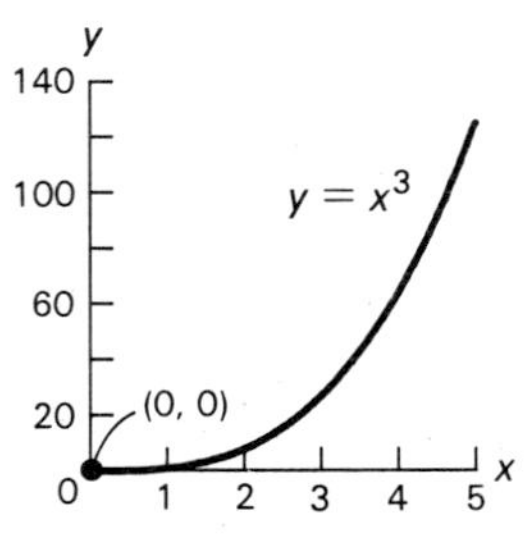

b.

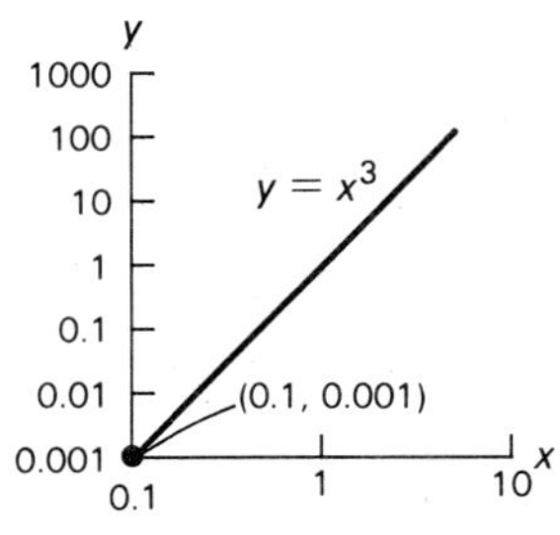

c.

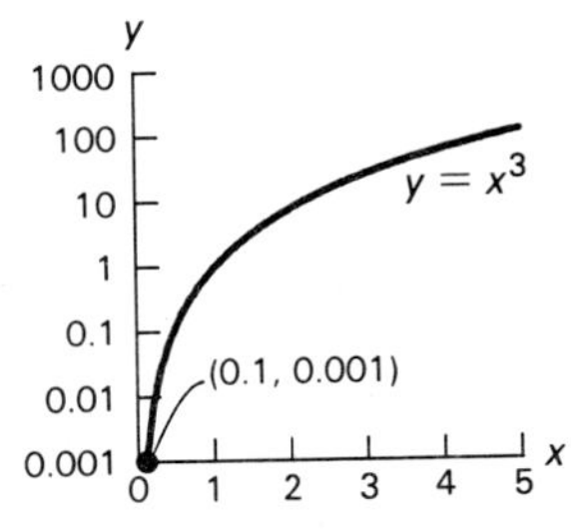

5. a.

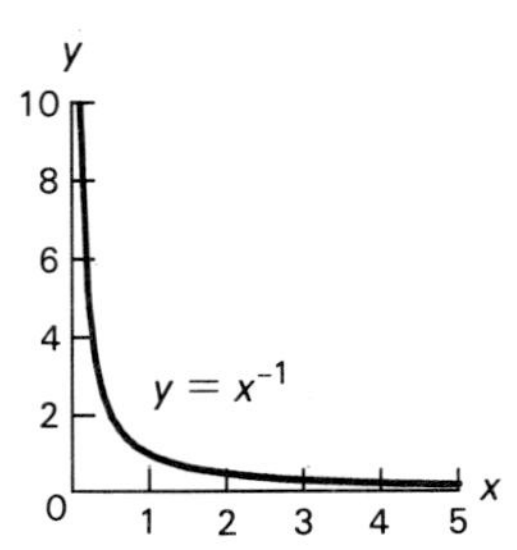

b.

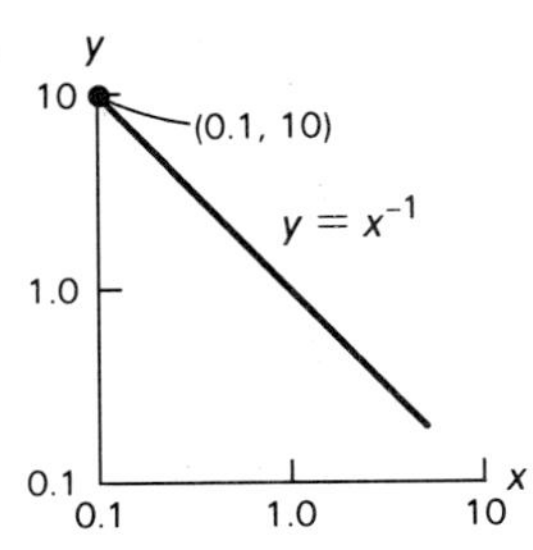

c.

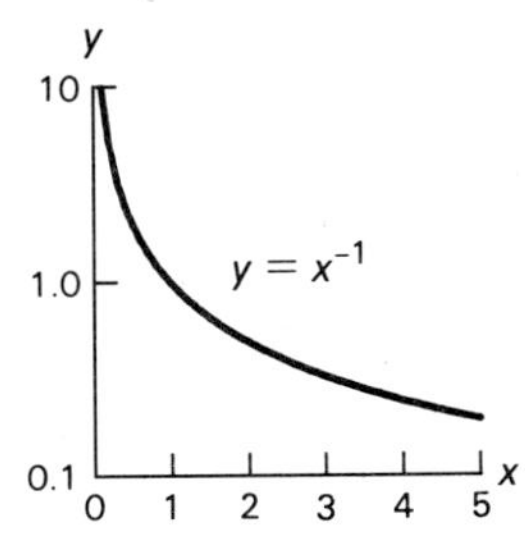

7. a.

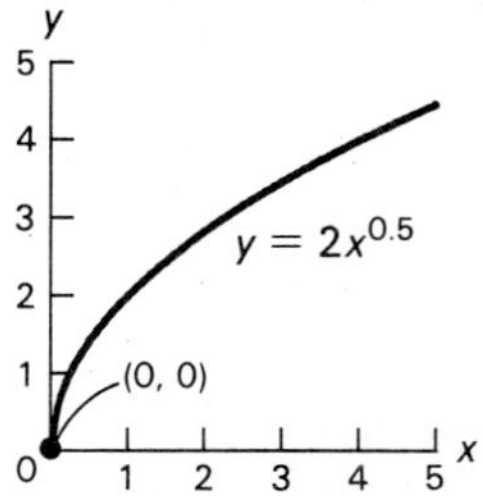

b.

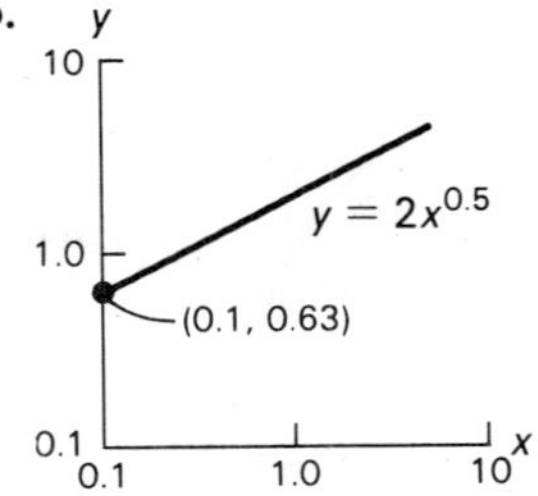

c.

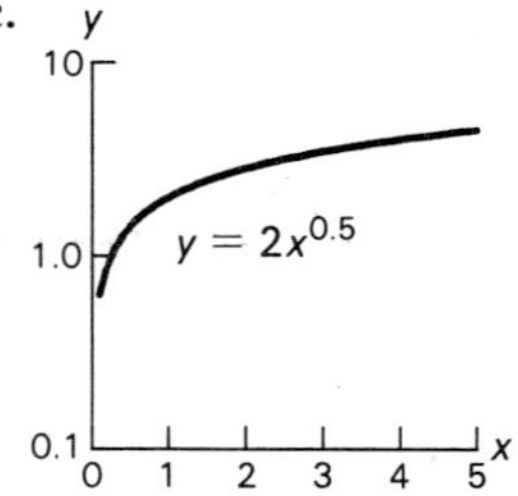

9. a.

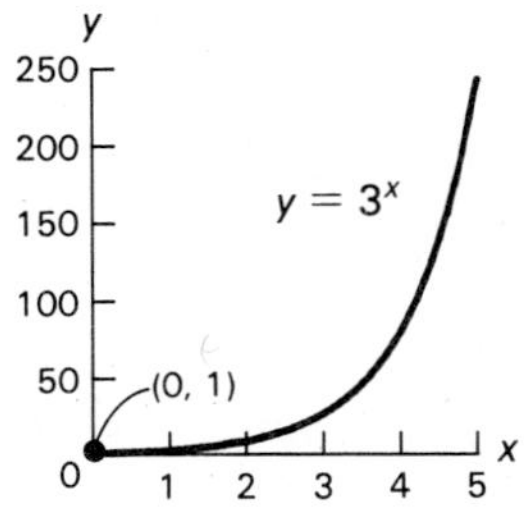

b.

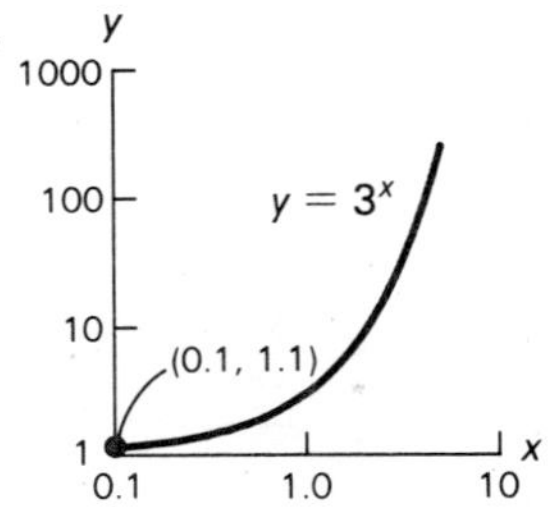

c.

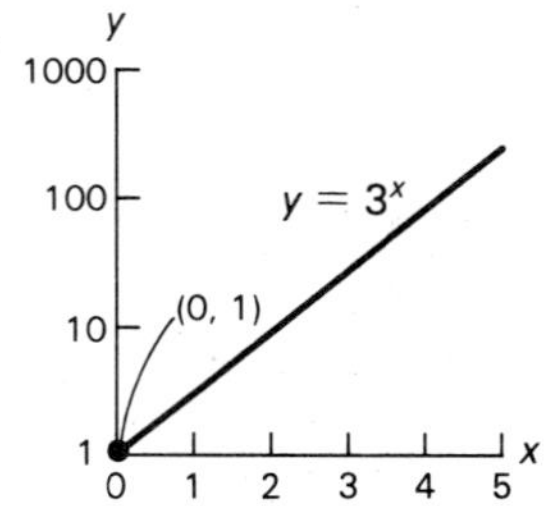

11. a.

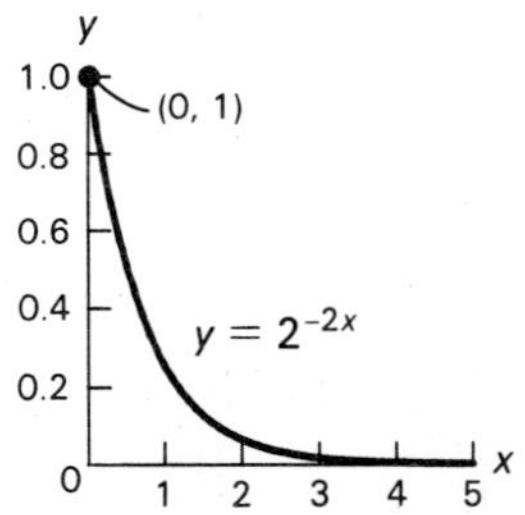

b.

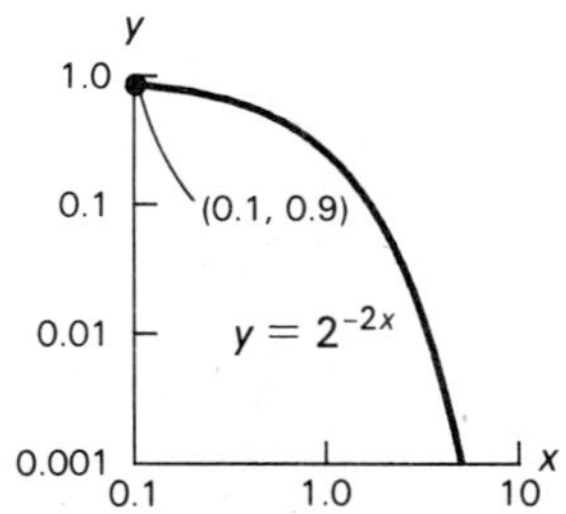

c.

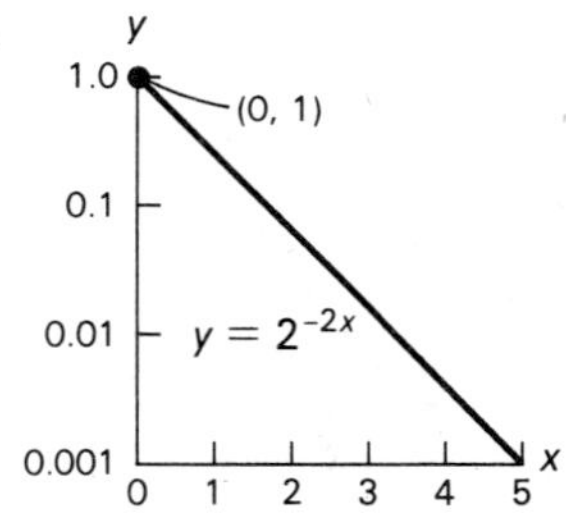

13. a.

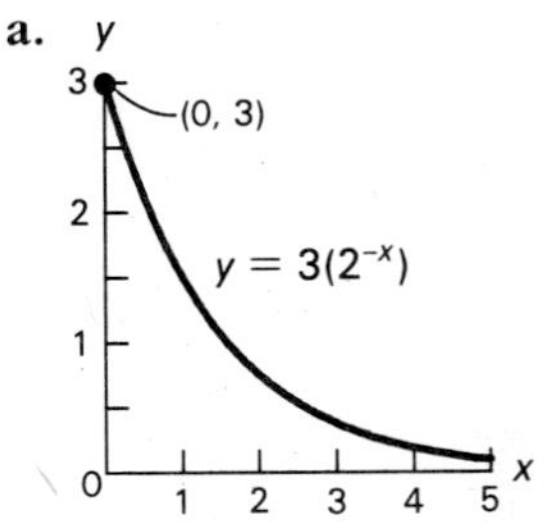

b.

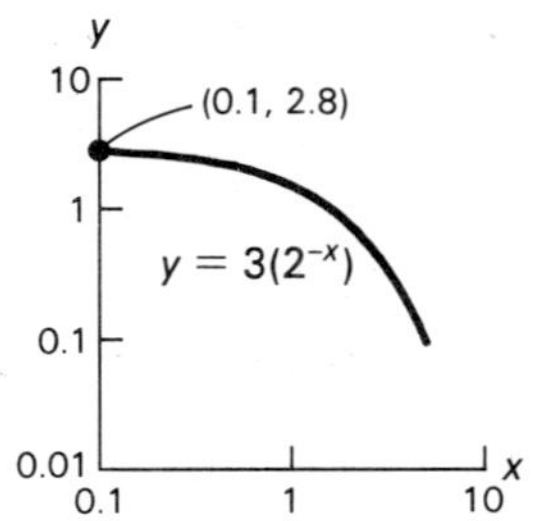

c.

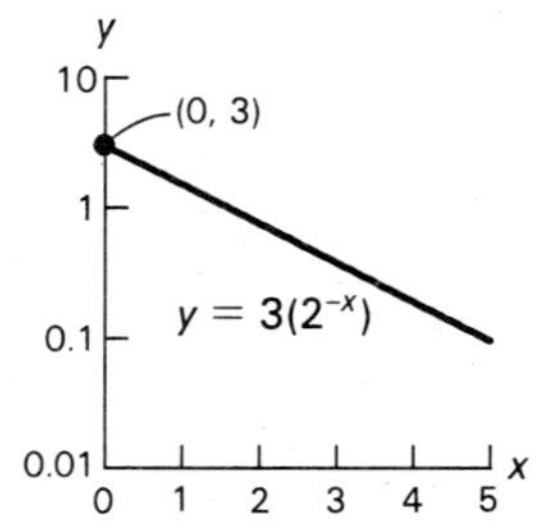

15. a.

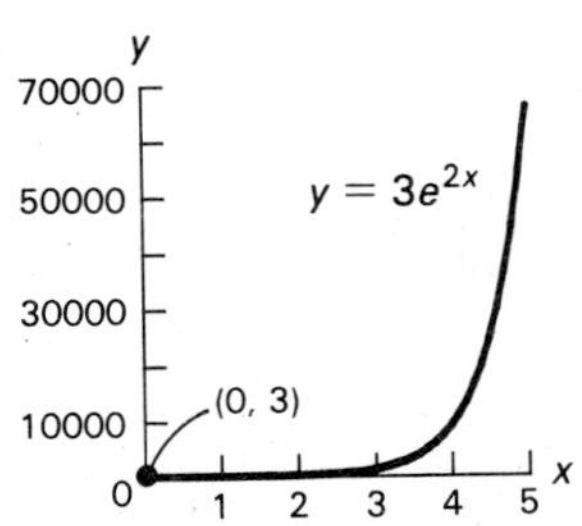

b.

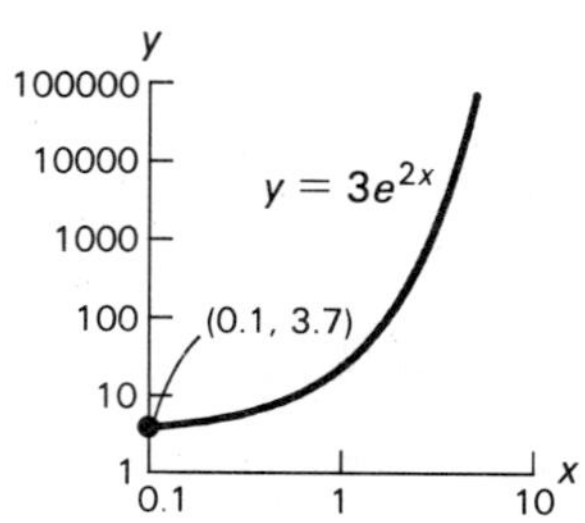

c.

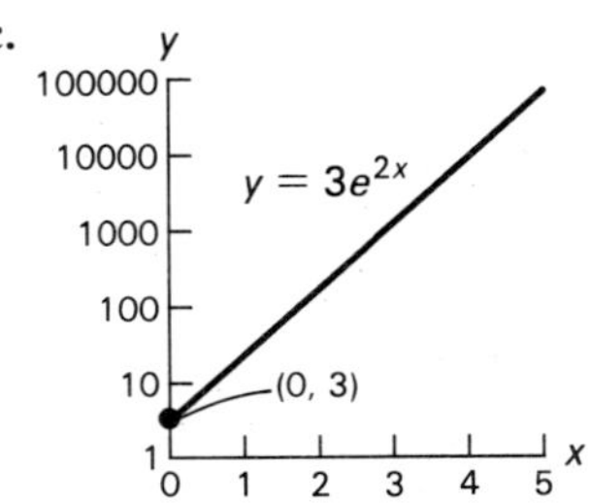

17. a.

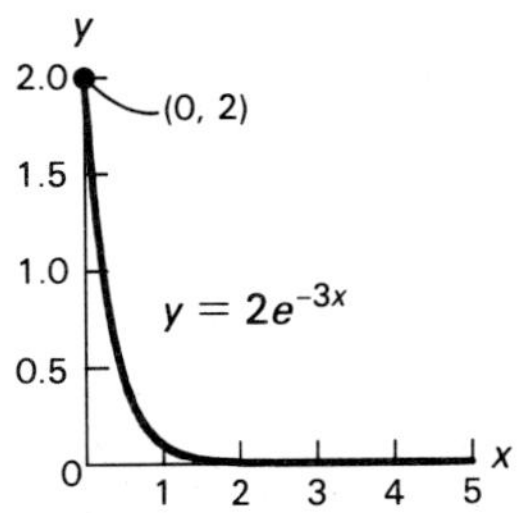

b.

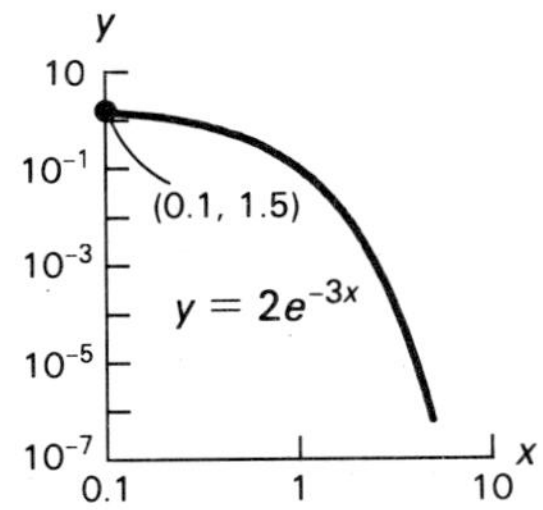

c.

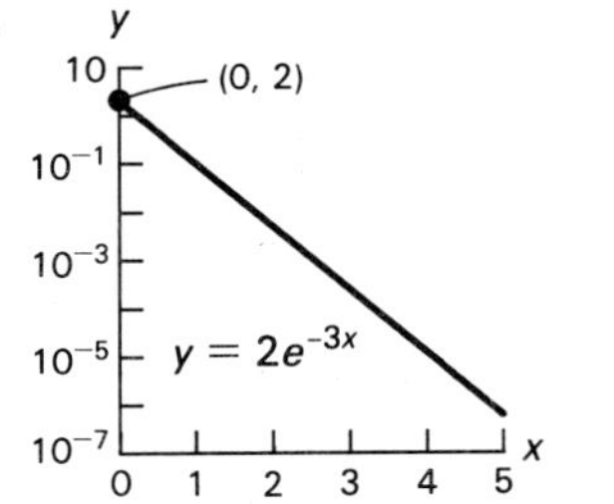

19. a.

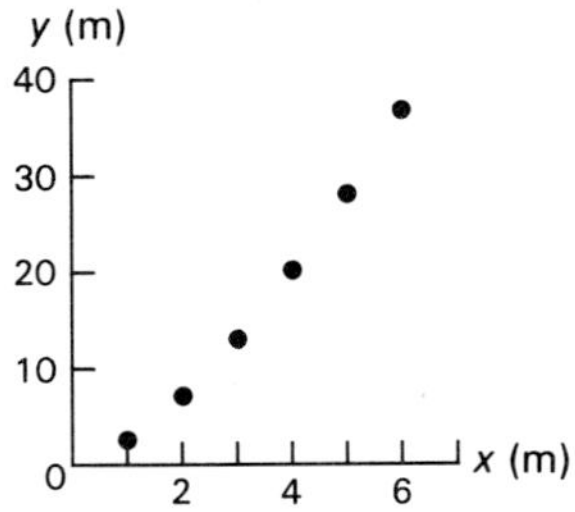

b.

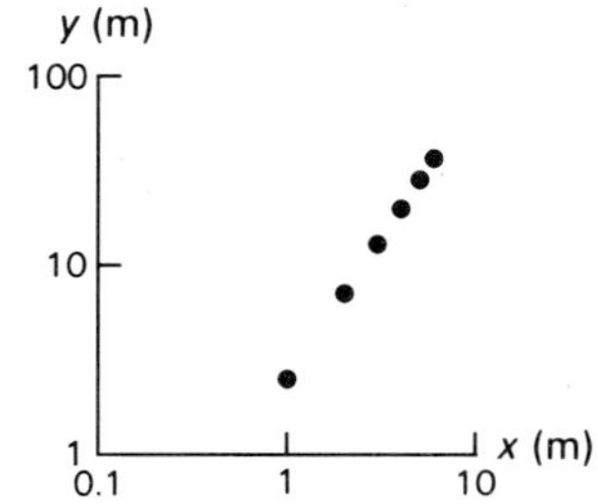

c.

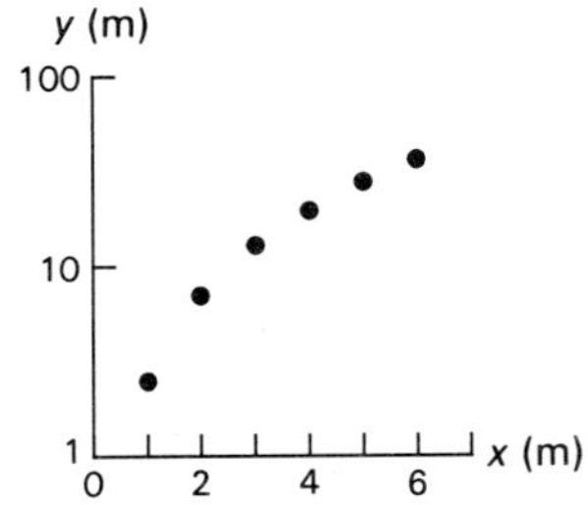

21. a.

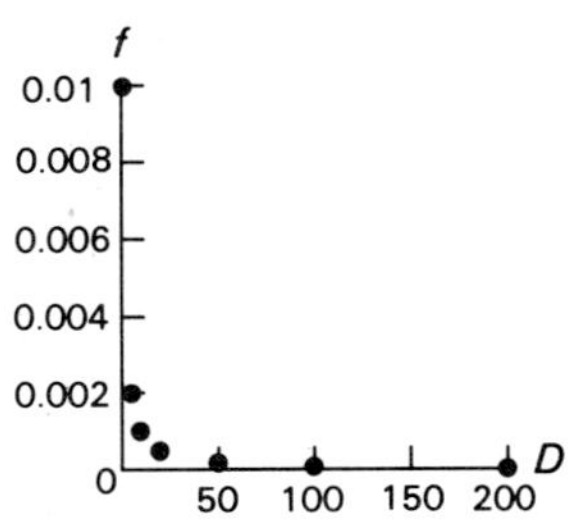

b.

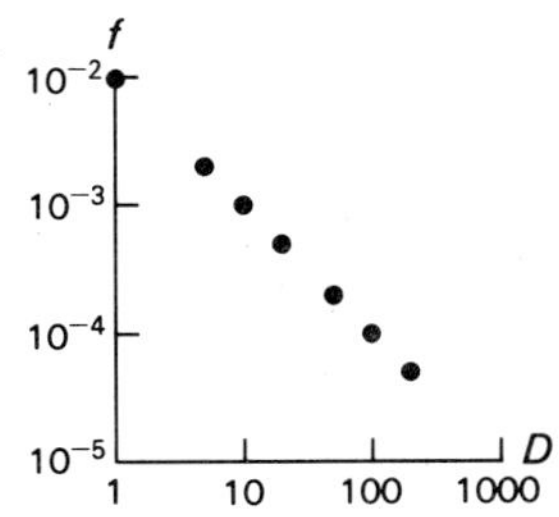

c.

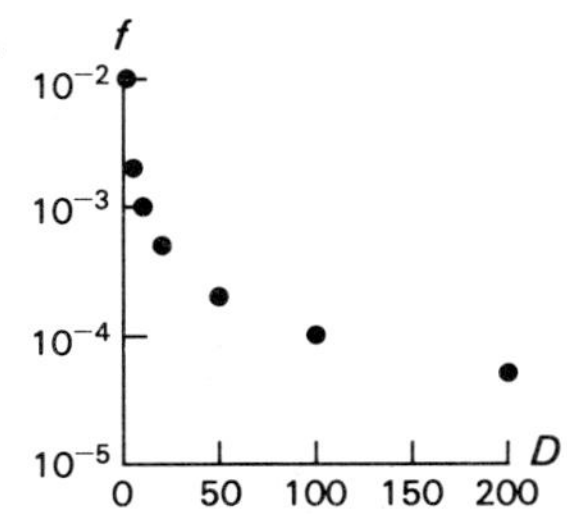

23. a.

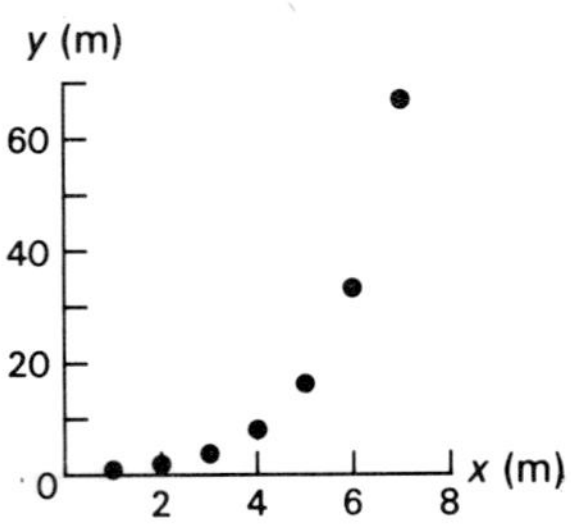

b.

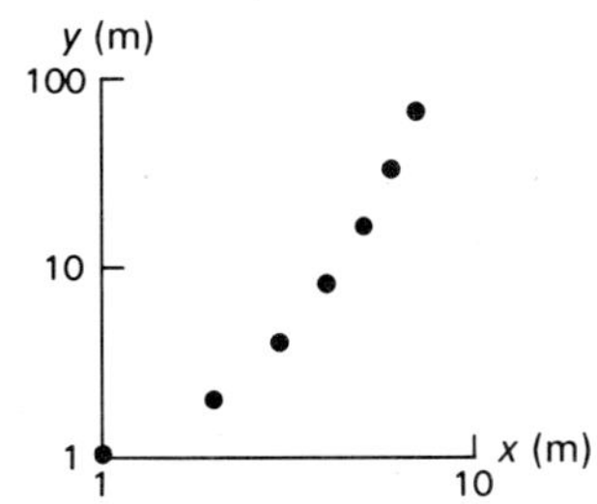

c.

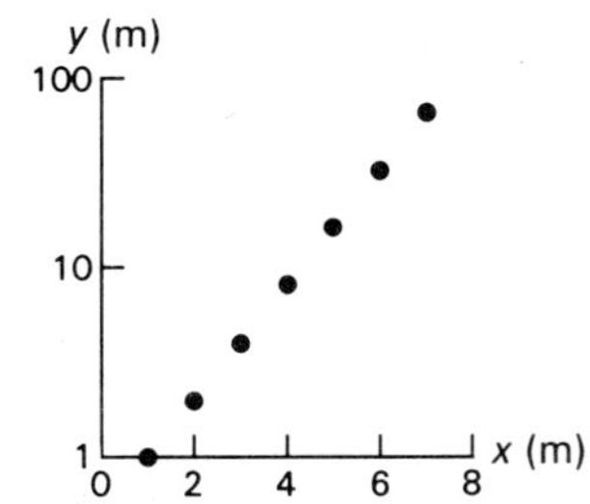

25. a.

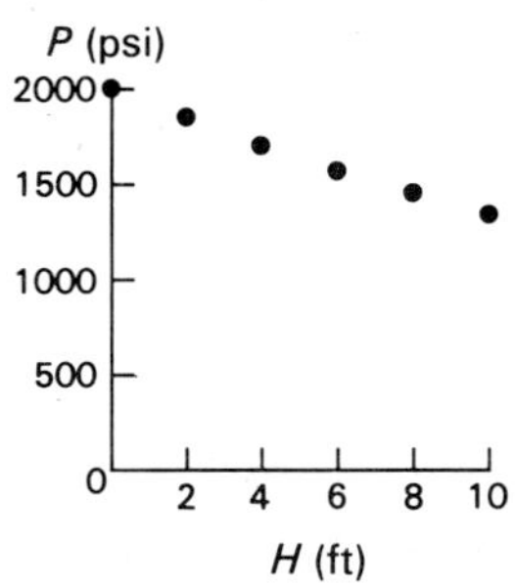

b.

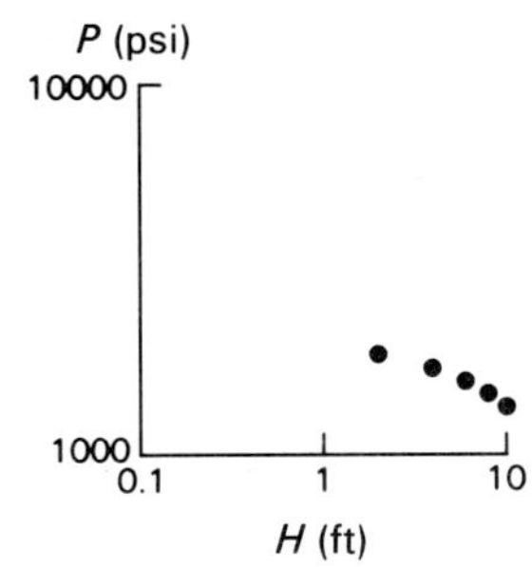

c.

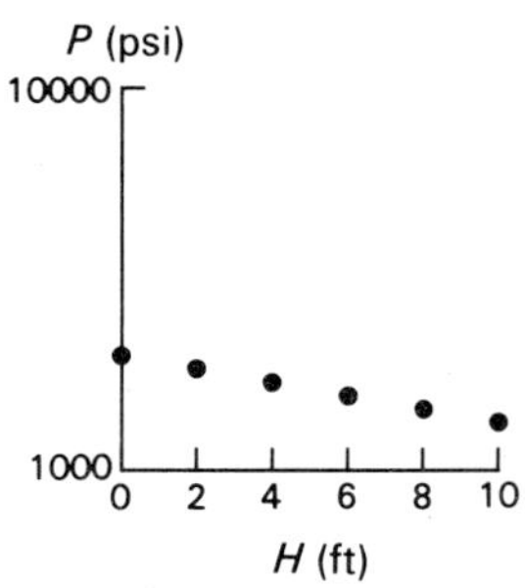

27. $y = 2.5x^{1.5}$ **29.** $f = 0.01\ D^{-1}$ **31.** $y = 1.\ e^{0.7x}$ **33.** $P = 2000.\ e^{0.7x}$
35. $y = 3.53 \times 10^{-10} x^{3.88}$ **37.** $y = 90.9\ x^{0.0866}$ **39.** $y = 4.91 \times 10^{-9} x^{2.07}$ **41.** $y = 0.0211\ x^{0.199}$
43. $y = 1.07e^{-1.15x}$ **45.** $y = 6.53e^{-4.18x}$ **47.** $y = 113. \times 10^{-30} e^{37.5x}$ **49.** $y = 3.39 \times 10^{9} e^{0.125x}$

Review Exercises (p. 531)

1.

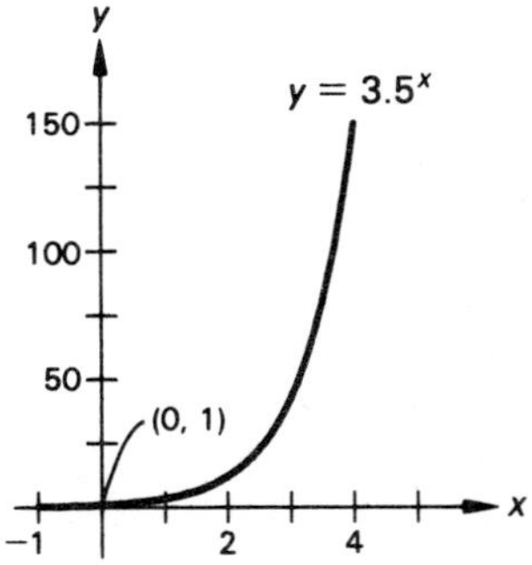

3.

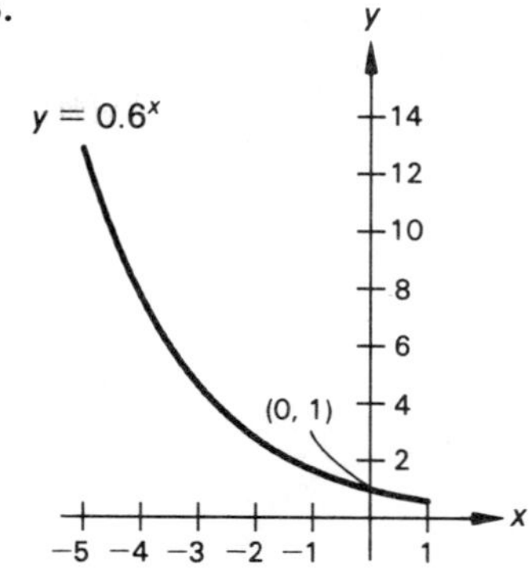

5.

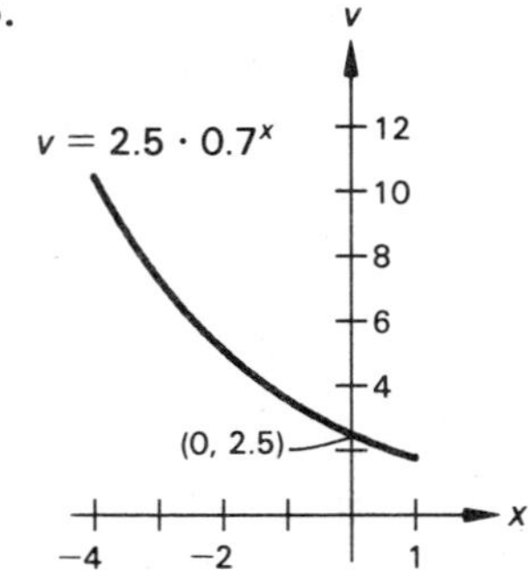

7.

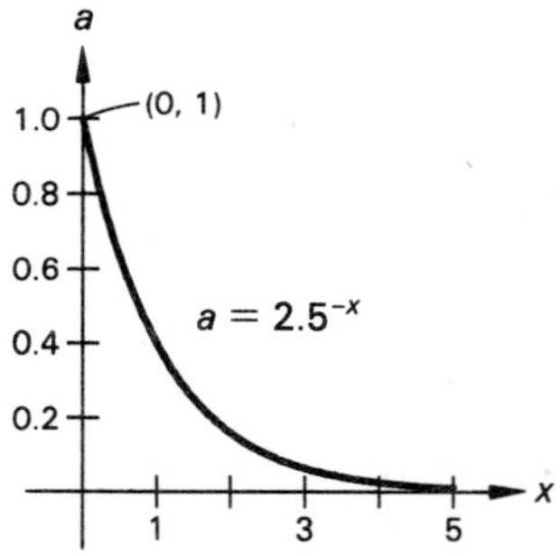

9.

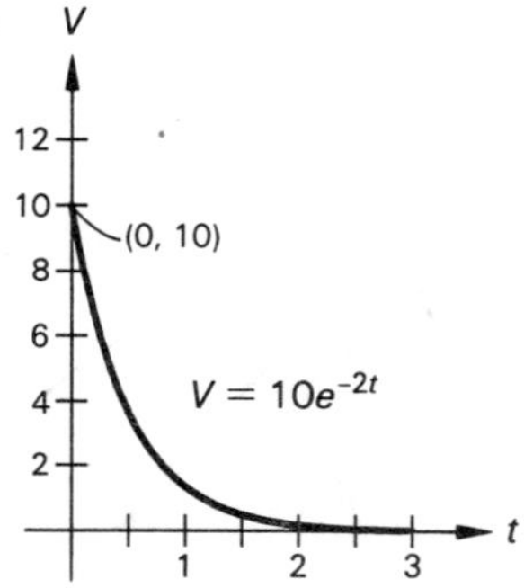

11.

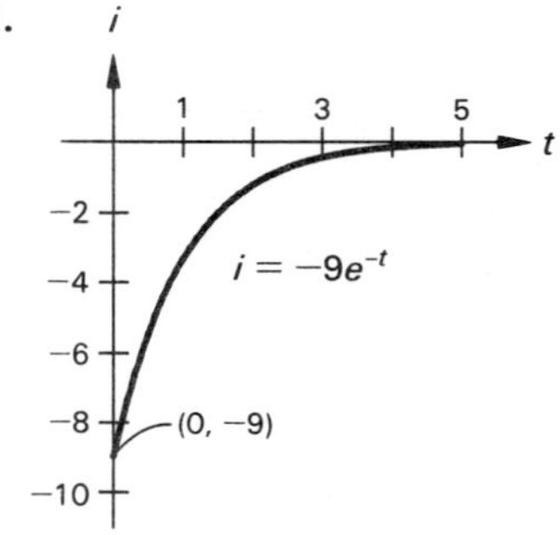

13. 1.67 s **15.** 30 s **17.** \$547.36 **19.** 11 bacteria **21.** $\log_7 49 = 2$ **23.** $\log_5 p = t$
25. $5^3 = 125$ **27.** $1.3^s = 416$ **29.** 36 **31.** 0.0313 **33.** 1 **35.** −3 **37.** 0

39.

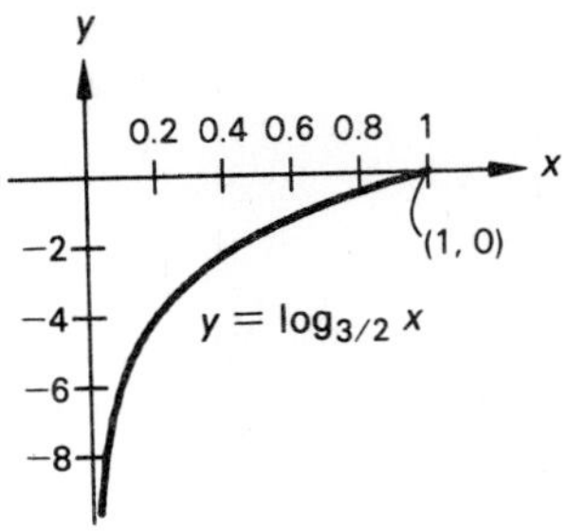

41.

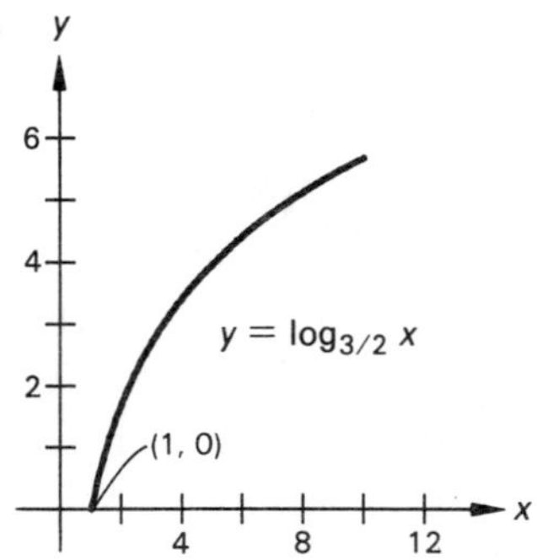

43. 5.64 hr **45.** $2 [\log_7 5 + 3 \log_7 x]$ **47.** $\log_2 (x - 2) - \log_2 (x + 2)$ **49.** $\log_{10} (x^2 + y^2) - 3$

51. $\log_{10} x$ **53.** $\log_2 \dfrac{x + 3}{x - 3}$ **55.** 1 **57.** 4.38 **59.** −2.92 **61.** 3.55 **63.** $\dfrac{\log \dfrac{A}{P}}{\log (1 + r)}$

65. 1.1523; c = 1, m = 0.1523 **67.** −1.9469; c = −2, m = 0.0531 **69.** 6.1761; c = 6, m = 0.1761
71. −1.0915; c = −2, m = 0.9085 **73.** 324. **75.** 0.00152 **77.** 43.4 **79.** 501. **81.** 2.31
83. 3.16 **85.** 3 **87.** −3.00 **89.** 1.16 **91.** 1.43 **93.** 0.0794 **95.** 2.32 **97.** 600.°K
99. −1.66 **101.** −20.6 **103.** No solution **105.** No solution **107.** 672. yr **109.** 4.58 rad

111. **a.**

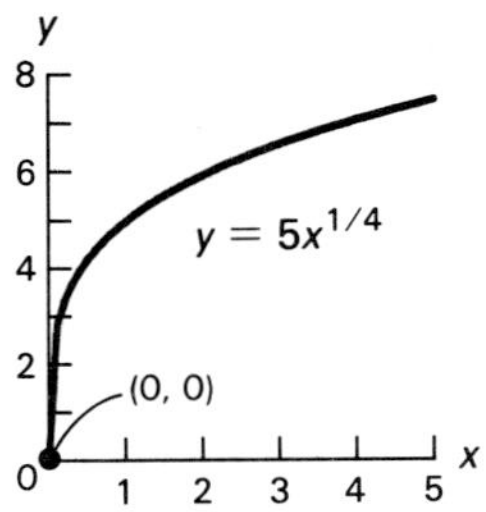

b.

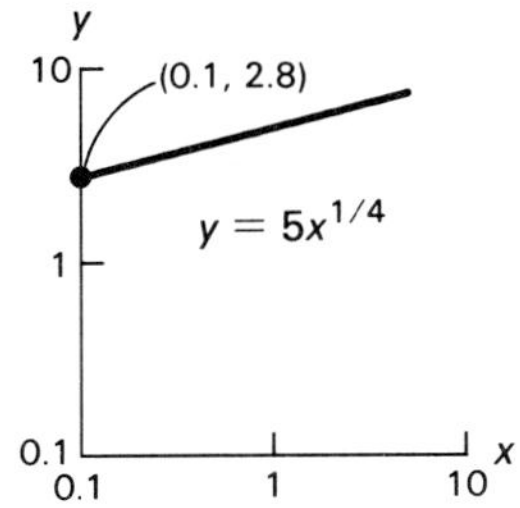

c.

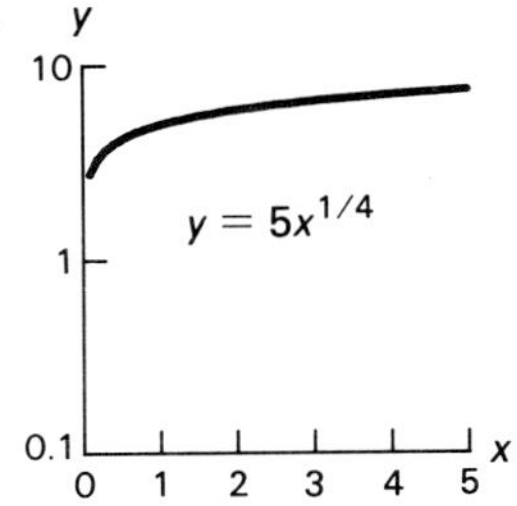

113. **a.**

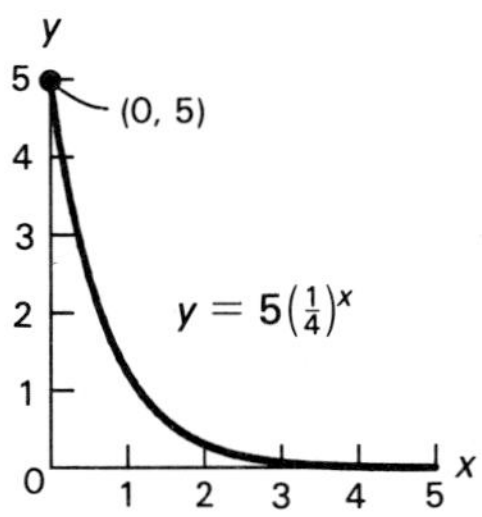

b.

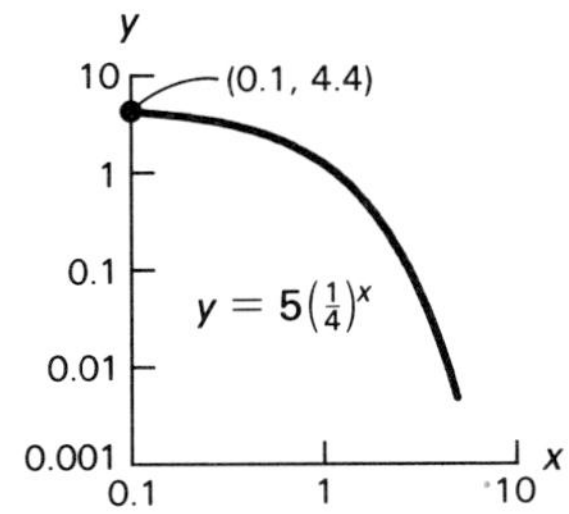

c.

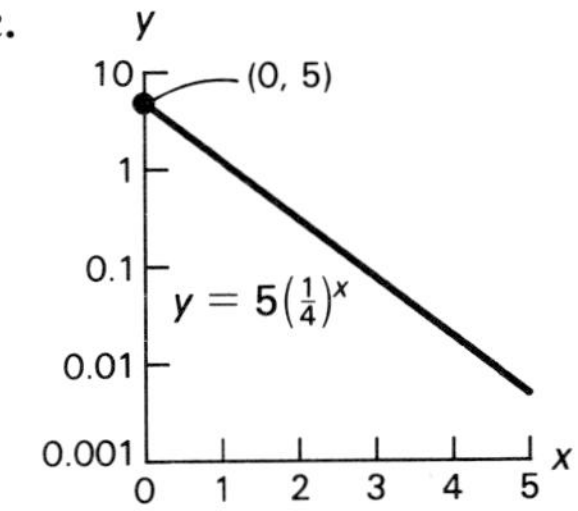

115. **a.**

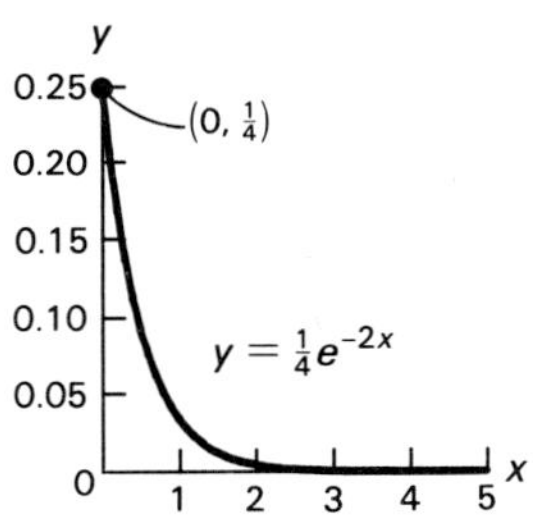

b.

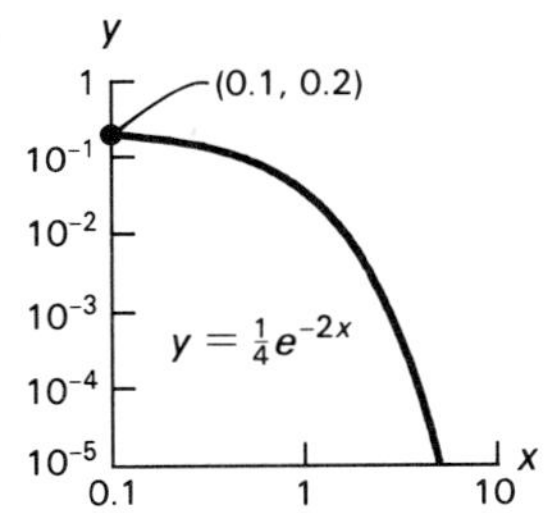

c.

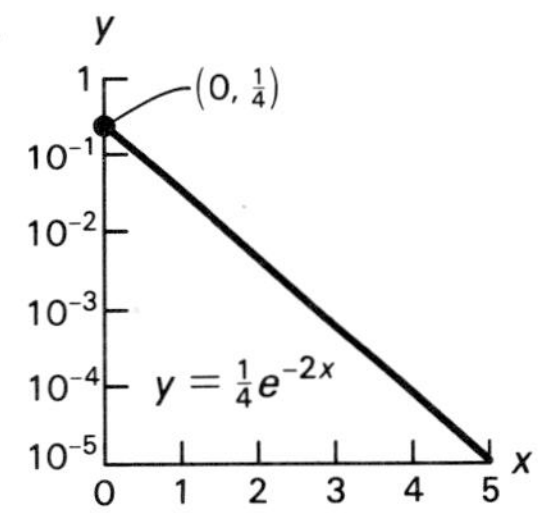

117.

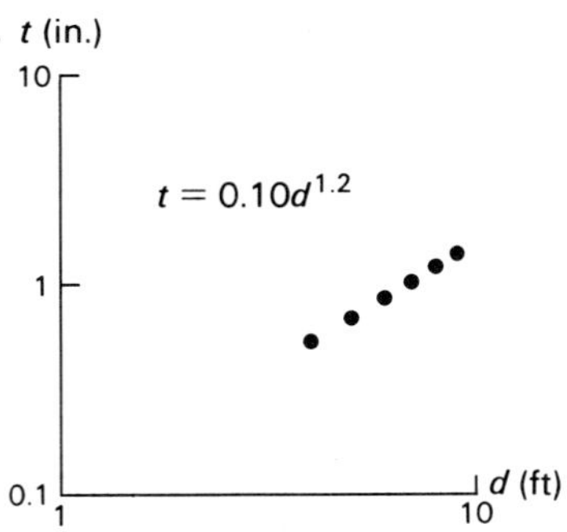

119.

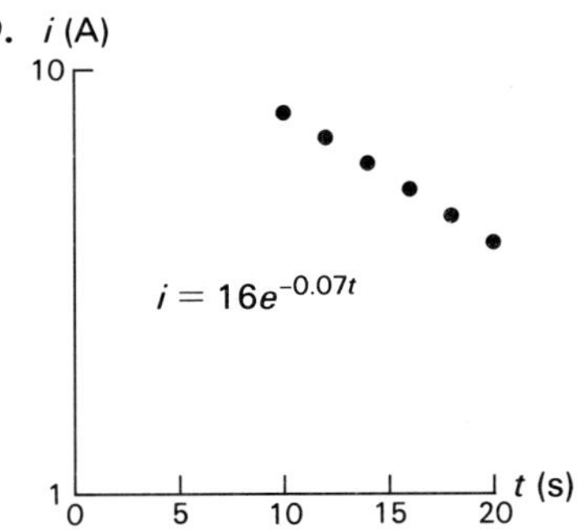

CHAPTER 14

14.1 Exercises (p. 541)

1.

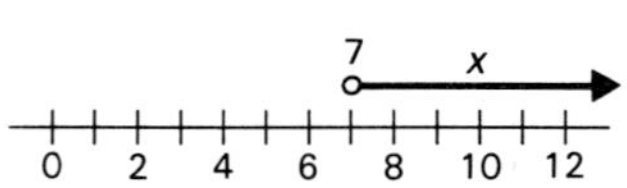

3.

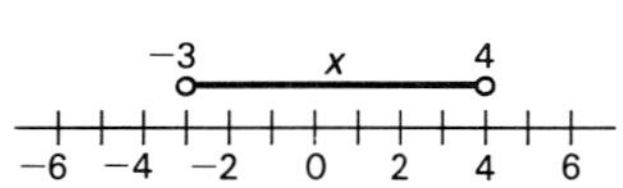

5.

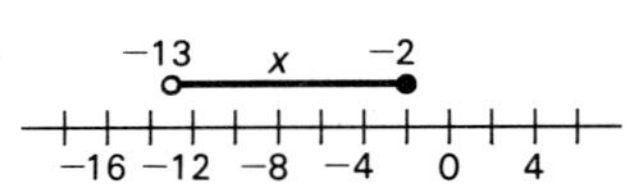

7. 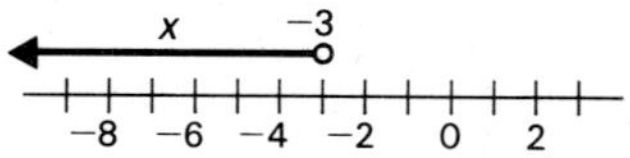
9. 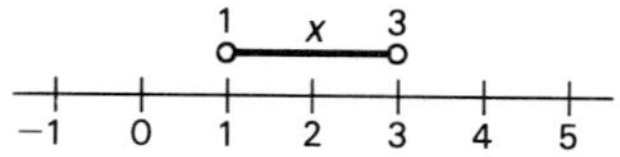
11. 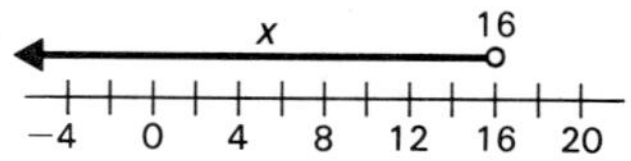

13. $5 < 13$ **15.** $-5 \geq -16$ **17.** $R + 6 > 11$ **19.** $2 < z \leq 11$ **21.** $6 - x < 0$ **23.** $R - 2 \leq 0$
25. $<$ **27.** $<$ **29.** $>$ **31.** $>$ **33.** $<$ **35.** $>$ **37.** $y > 0$ **39.** $x < 0, y > 0$ **41.** $x > 4$
43. $-4 < y < 6$ **45.** $y \leq -3$ or $y \geq 1$ **47.** $x < -1, y > 2$
49. For all real numbers a and b if $a < b$ then there exists a positive real number c such that $b - a = c$.

14.2 Exercises (p. 545)

1. $x < 3$ 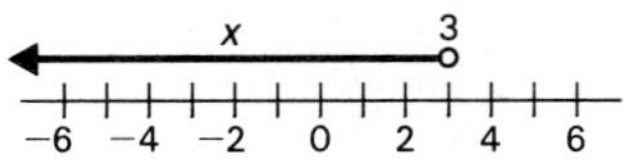

3. $x \leq -4$ 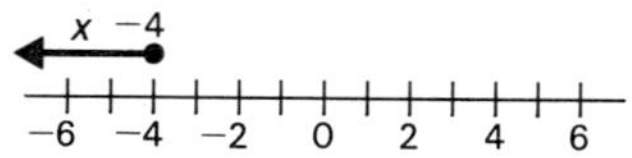

5. $\ell \leq -1$ 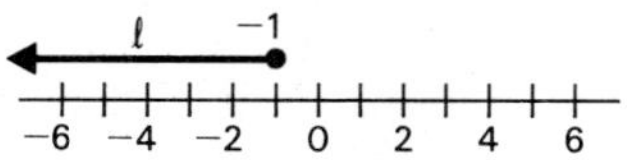

7. $z \geq -3$ 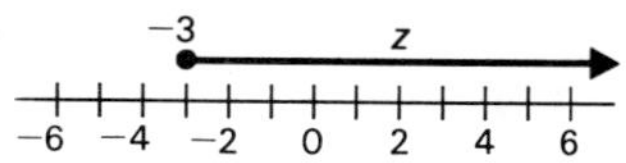

9. $x \geq -3$ 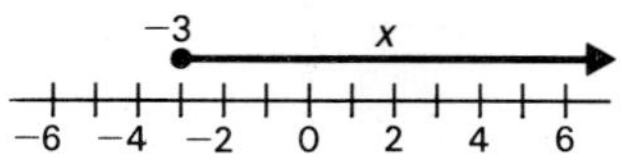

11. $R \leq -\frac{1}{2}$ 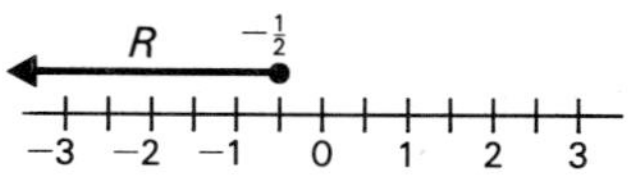

13. $t \geq -\dfrac{9}{4}$ 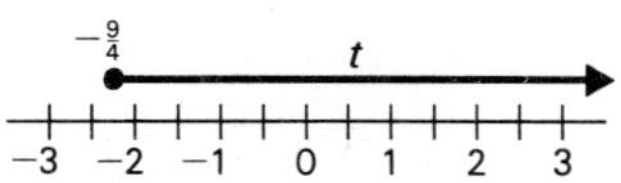

15. $t < -\dfrac{11}{4}$ 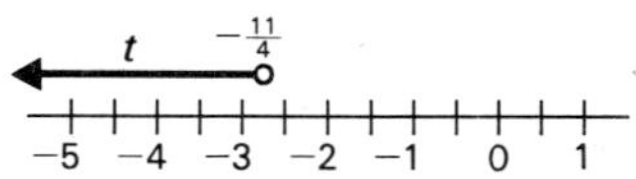

17. $z < -\dfrac{1}{4}$ 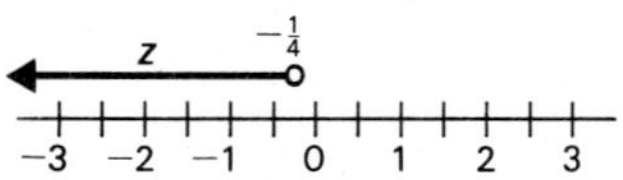

19. $z > -\dfrac{13}{2}$ 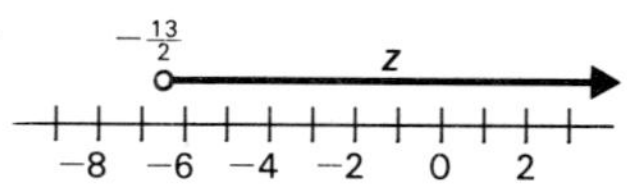

21. $z \geq -\dfrac{4}{3}$ 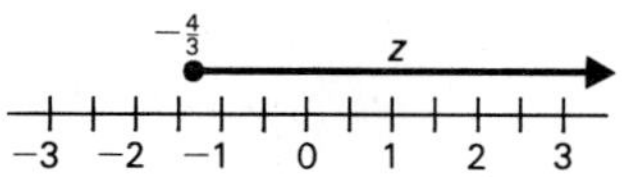

23. $-1 < y < 5$ 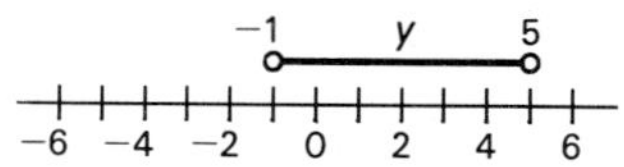

25. $-10 < y \leq -1$ 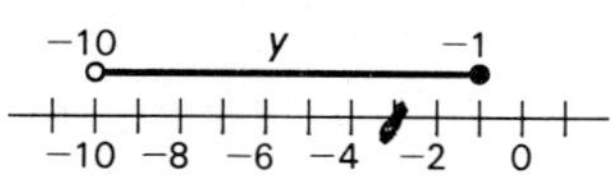

27. $-9 \leq x \leq 5$ 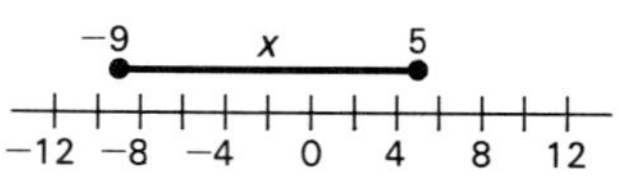

29. $-5 < x < 14$ 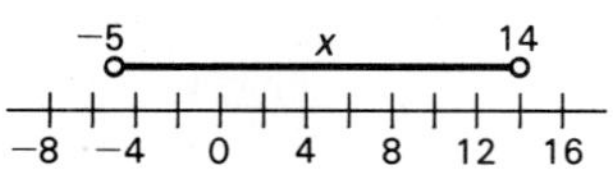

31. $190. < (110. + 0.20x)$; $x > 400$ mi

33. $-0.005 \leq d, -0.5 \leq 0.005$; $0.495 \text{ mm} \leq d \leq 0.505 \text{ mm}$ **35.** $92 - 32t > 0$; $t < \dfrac{23}{8}$ s

37. $-0.02 \leq 7t - 15.05 \leq 0.02$; $2.147 \text{ mm} \leq t \leq 2.153 \text{ mm}$ **39. a.** $80 \leq \dfrac{341 + x}{5} < 90$; $59 \leq x < 109$

b. $70 \leq \dfrac{341 + x}{5} < 80$; $9 \leq x < 59$ **c.** No, assuming a grade above 100 is impossible.

41. $R > C$ or $2x > (140 + 1.2x)$; $x > 175$ stoves

14.3 Exercises (p. 555)

1. c. $(-\infty, -2)$ or $(3, \infty)$ **3. c.** $[-3, 5]$ **5. c.** $(-\infty, 0]$ or $[3, \infty)$ **7. c.** $(-\infty, -5)$ or $(0, \infty)$
9. c. $(-1, 5)$ **11. c.** $(-\infty, -2]$ or $[-\frac{1}{2}, \infty)$ **13. c.** $[-3.186, -0.314]$ **15. c.** No real x
17. c. $(-\infty, \infty)$ **19. c.** $(0.279, 2.387)$ **21. a.** lb **b.** $10 \le x \le 15$ **23.** $2 \text{ s} \le t \le 8 \text{ s}$
25. $m \le 15$ mm **27.** $b \le 0.5$ m **29.** $0 \text{ mm} < r < 250 \text{ mm}$ **31.** $(-\infty, -2)$ or $(1, 4)$
33. $(-7, 3)$ or $(4, \infty)$ **35.** $(-\infty, \infty)$ **37.** $(-5, 0)$ or $\left(\frac{3}{4}, \infty\right)$ **39.** $R \ne 0$ **41.** $z > 3$
43. $(-\infty, -1)$ or $(0, \infty)$ **45.** $y > -3$ **47.** $(-\infty, \infty)$ **49.** $(-\infty, -8]$ or $(-3, 0]$ or $(2, \infty)$
51. $(-\infty, 3]$ or $[5, \infty)$

14.4 Exercises (p. 564)

1. $-8, 2$ **3.** $1\frac{2}{3}, 2\frac{1}{3}$ **5.** $-\frac{2}{5}, -\frac{4}{5}$ **7.** $-4, 7\frac{1}{3}$ **9.** $-2\frac{1}{3}, 5$ **11.** $\frac{3}{35}, \frac{18}{35}$ **13.** $-\frac{5}{23}, \frac{5}{19}$

15. $-\frac{3}{5}, \frac{3}{10}$ **17.** $[-8, 8]$

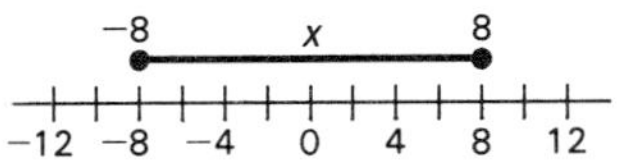

19. $(-\infty, -13)$ or $(13, \infty)$

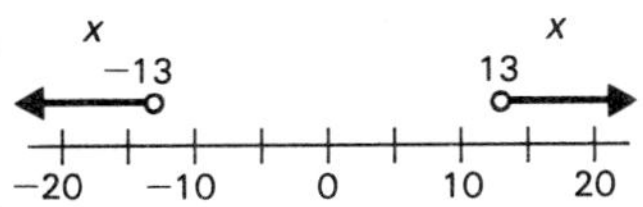

21. $t \ne 0$

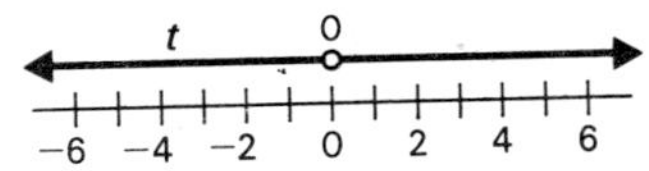

23. $[-3, 11]$

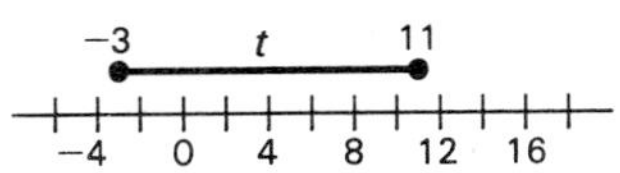

25. $(-\infty, 3]$ or $[11, \infty)$

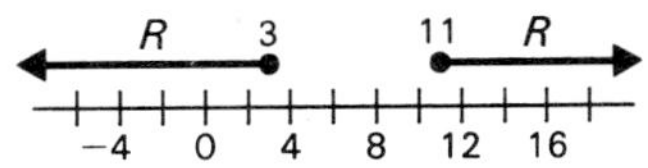

27. $\left(-\frac{7}{3}, -1\right)$

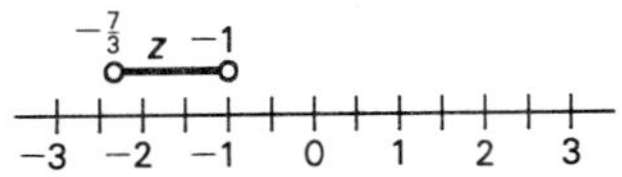

29. $\left(-\infty, -\frac{1}{3}\right]$ or $[3, \infty)$

31. No solution **33.** $\left(-\infty, \frac{1}{7}\right)$ or $\left(\frac{9}{7}, \infty\right)$

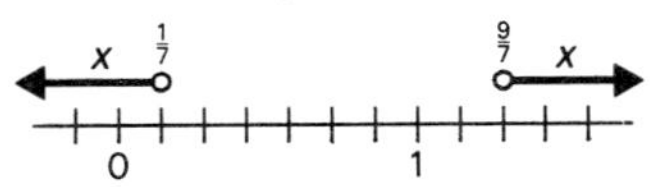

35. $(-1, 7)$

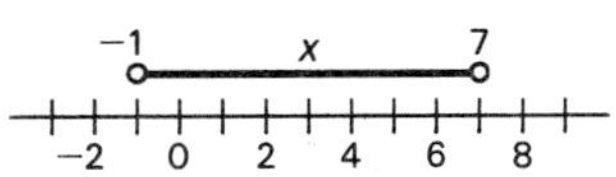

37. No solution

39. $\left(-\frac{7}{12}, -\frac{1}{12}\right)$

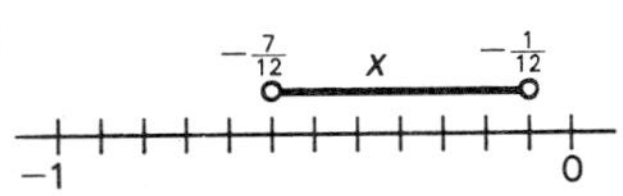

41. $(-0.25, -0.15)$

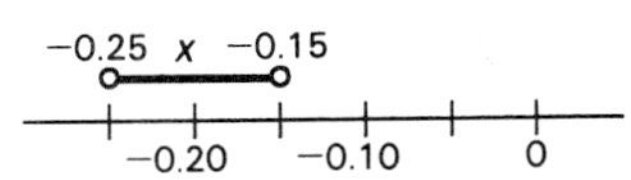

43. [13.44 mm, 13.46 mm] **45.** [3.97 mm, 4.03 mm] **47.** $\left|y - \frac{4}{3}\right| < 9; \left(-7\frac{2}{3}, 10\frac{1}{3}\right)$
49. $|R + 13| \le 5; [-18, -8]$

14.5 Exercises (p. 570)

1.

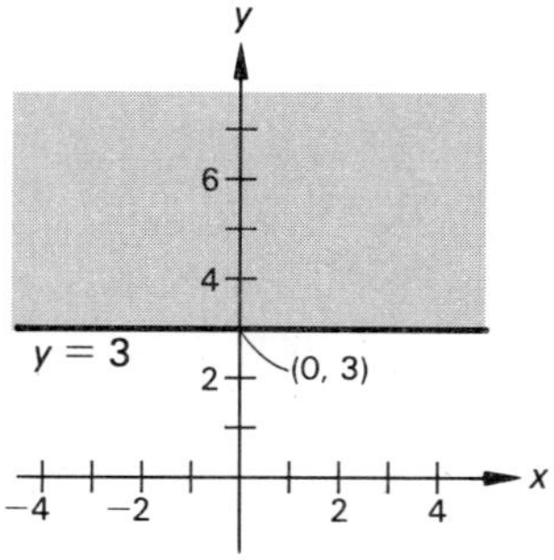

3.

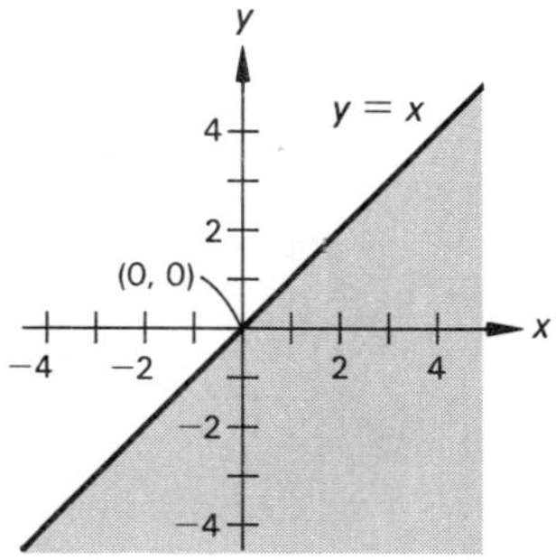

5.

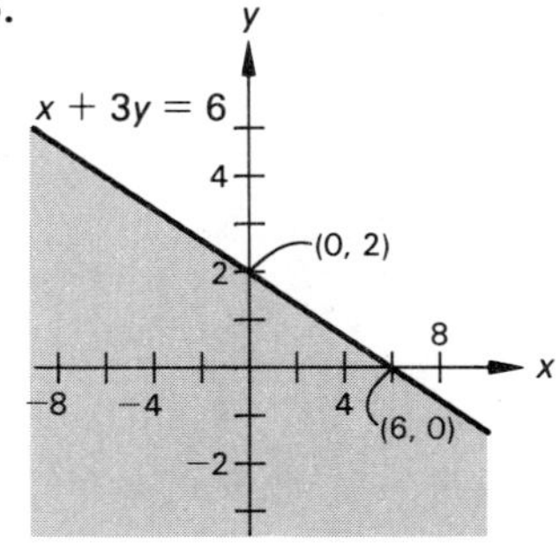

7.

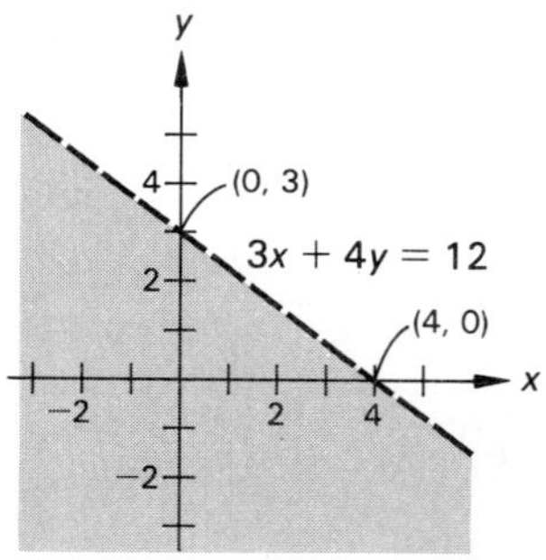

9.

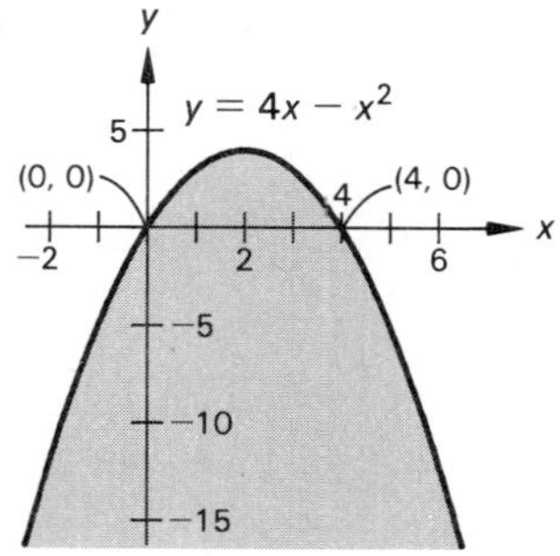

11.

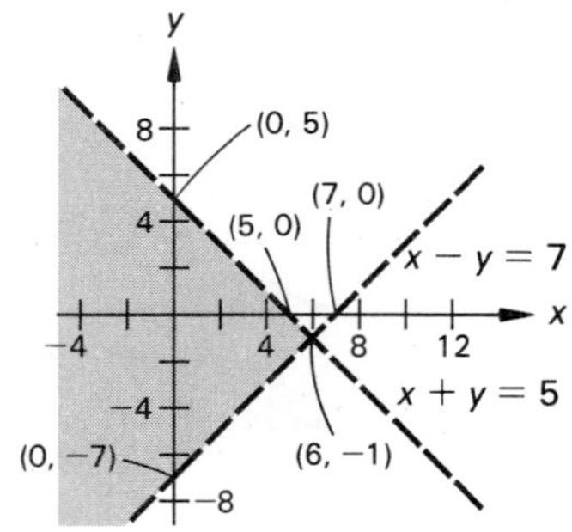

13.

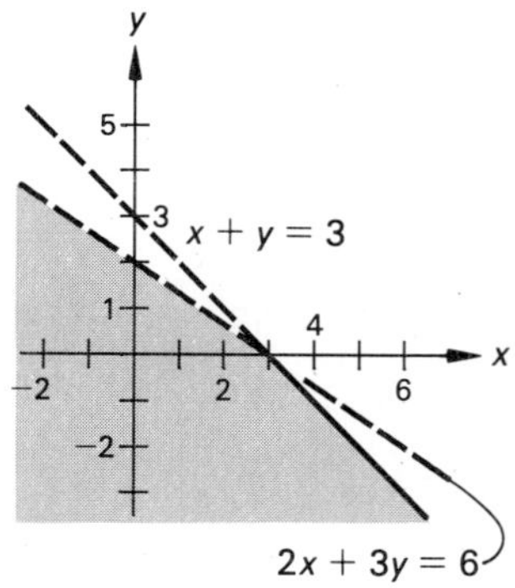

15.

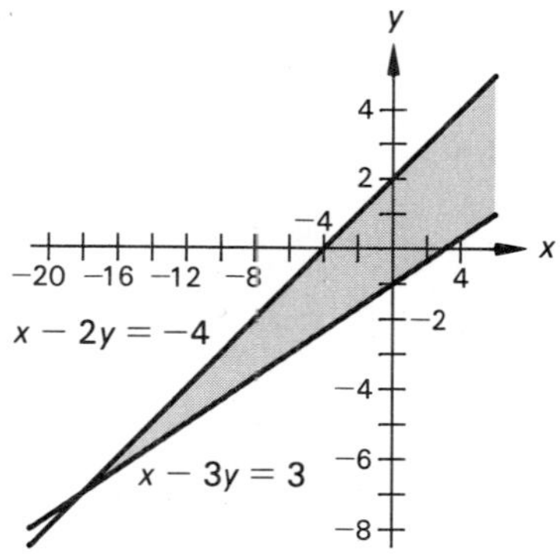

17.

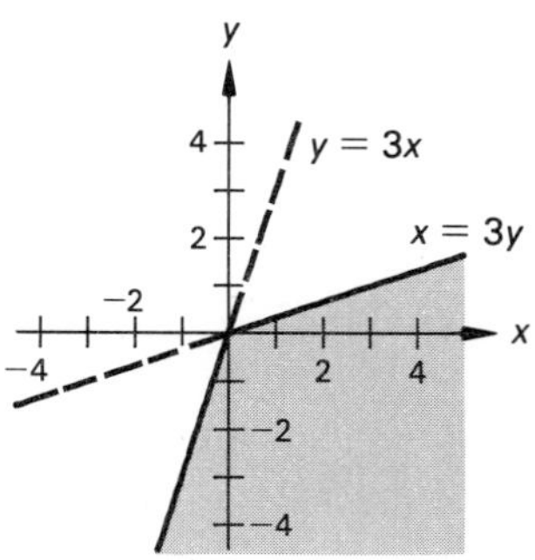

19.

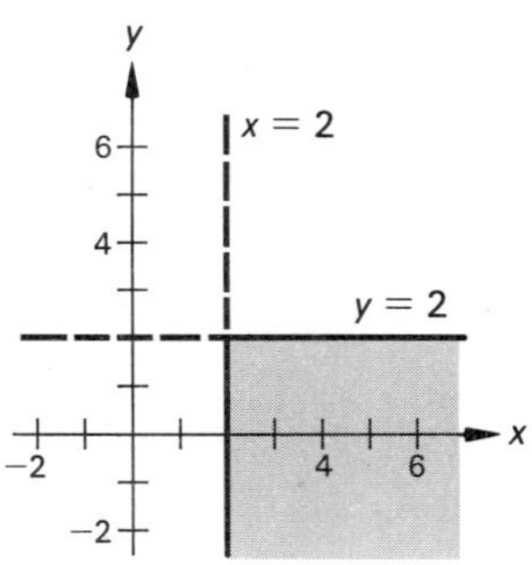

21.

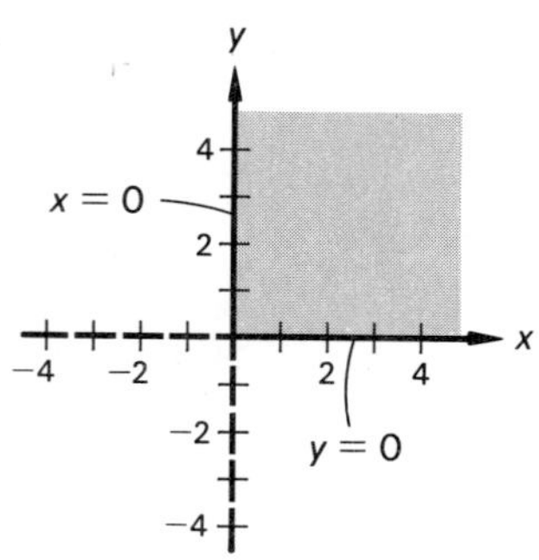

23.

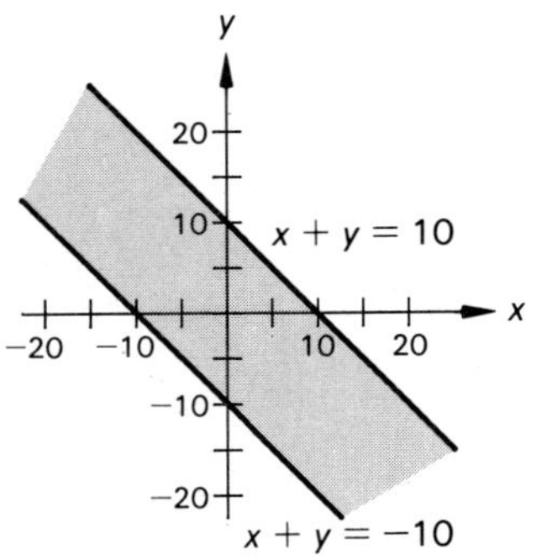

25.

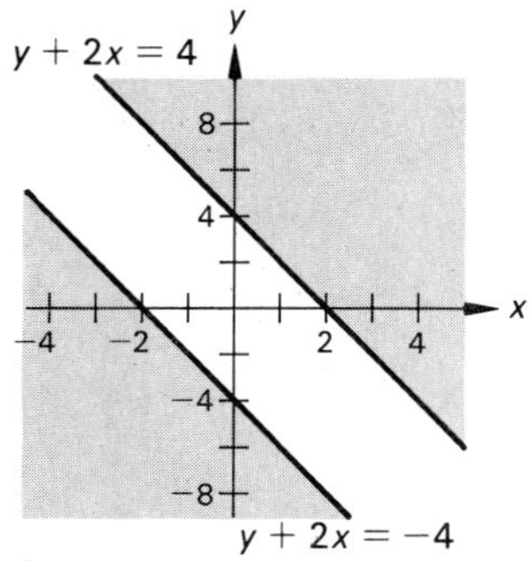

27.

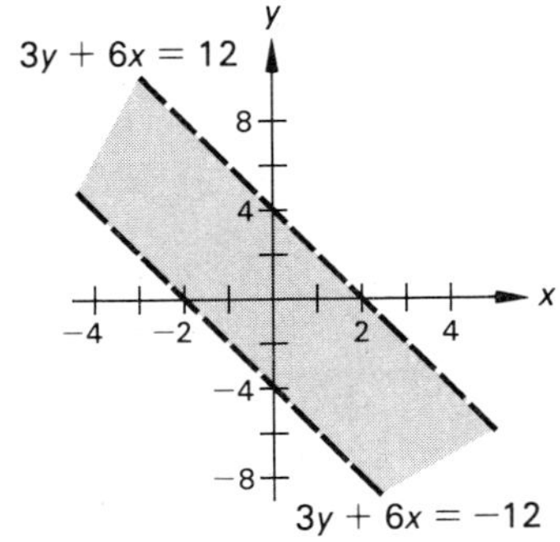

29.

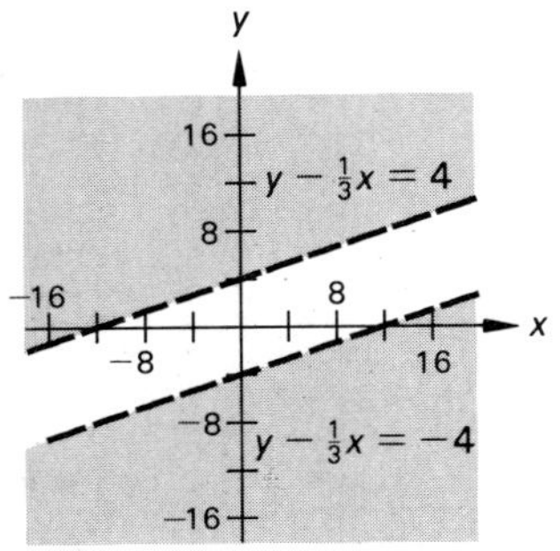

31.

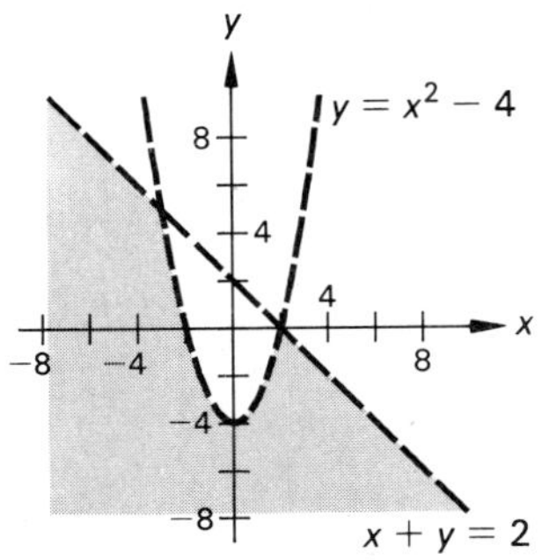

33.

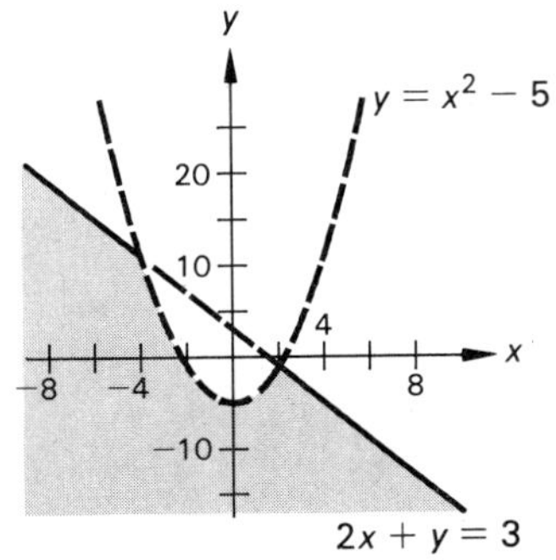

35.

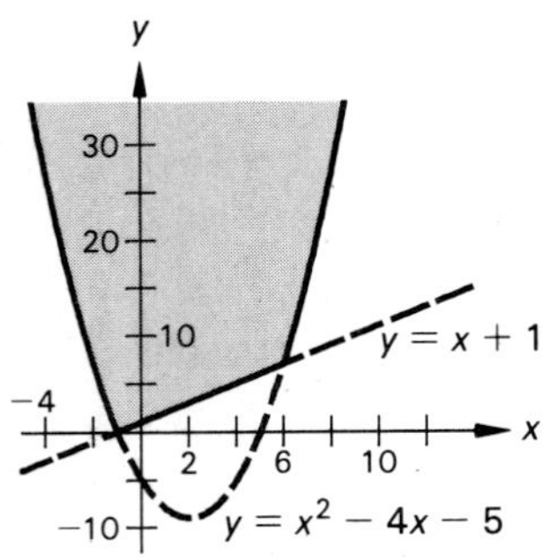

37.

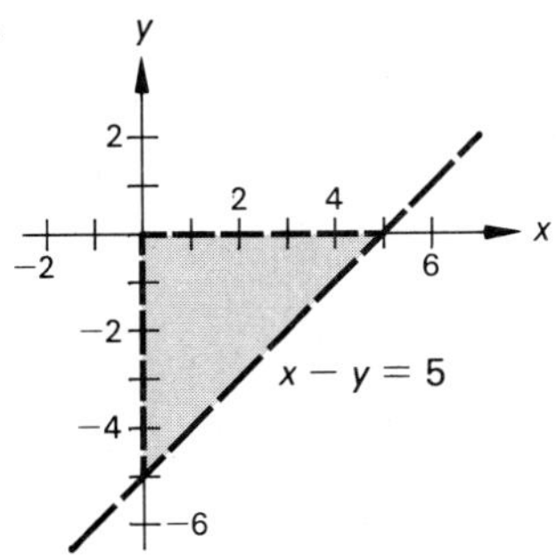

39.

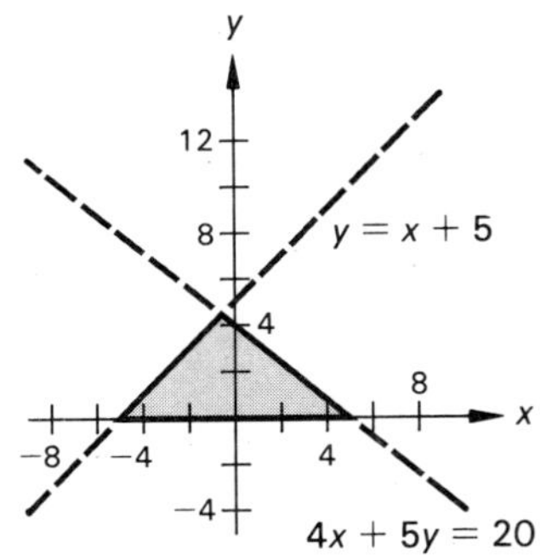

41. $s + l \geq 2000$
$1.00s + 1.50l \leq 300.$
$s \geq 0 \quad l \geq 0$

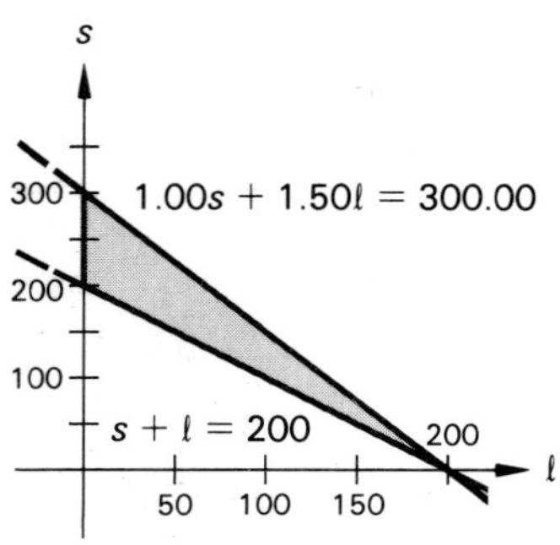

43. $Q \geq 2H \quad Q + H \geq 10$
$3Q + 4H \leq 80$
$Q \geq 0 \quad H \geq 0$

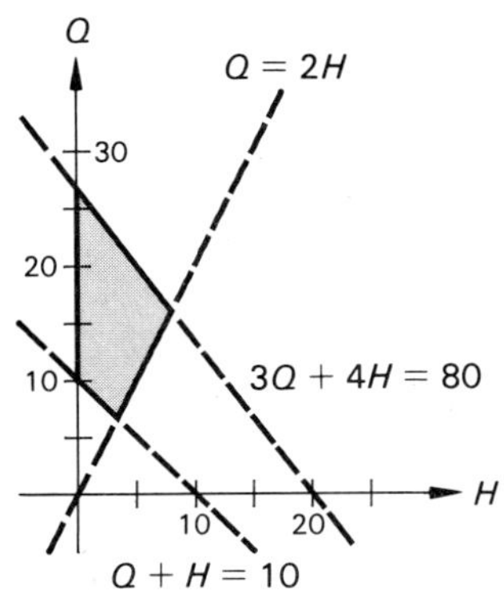

14.6 Exercises

1. a.

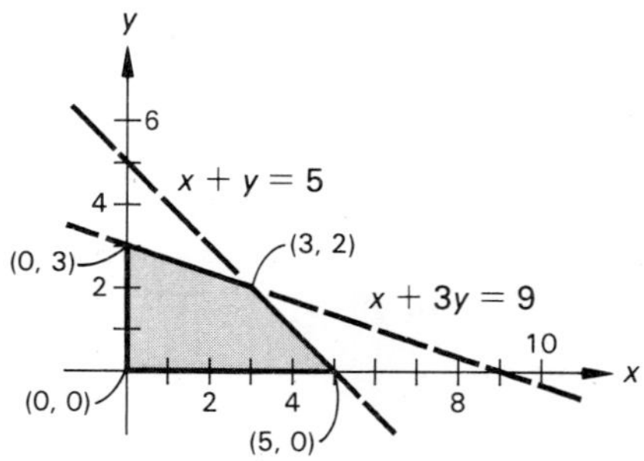

b. $P_{max} = 14$, $P_{min} = 0$

3. a.

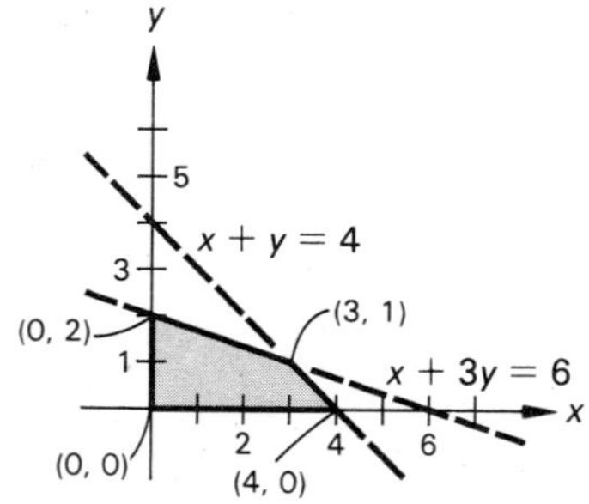

b. $P_{max} = 25$, $P_{min} = 0$

5. a.

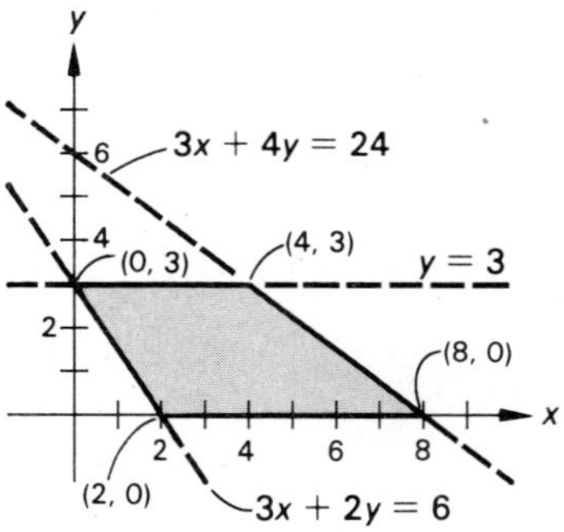

b. $P_{max} = 16.5$, $P_{min} = 3$

7. a. $A + B \leq 18 \quad A \geq 2 \quad A \leq 5 \quad B \geq 2$

b. $P = 20A + 25B$

c.

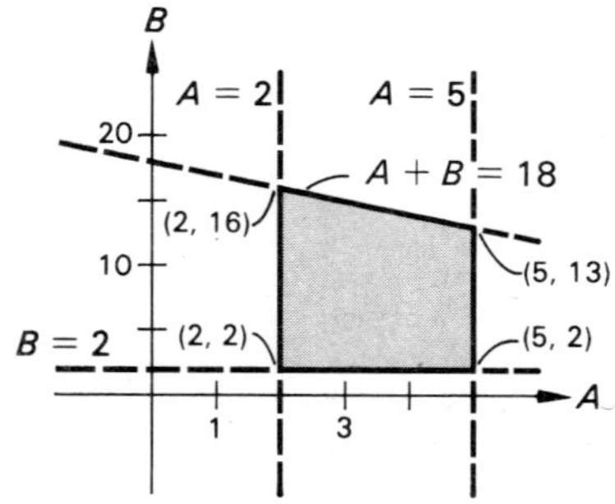

d. (2, 2), (5, 2), (2, 16), (5, 13) **e.** $2A$, $16B$

f. $440.00

9. 400 small, 0 large; $60.00 **11.** $250A$, $150B$

Review Exercises

1. a.

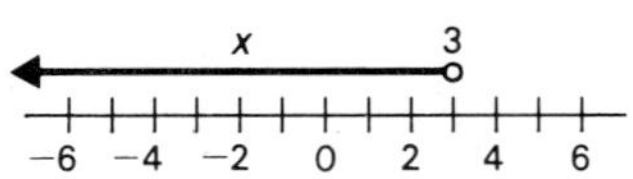

b. $(-\infty, 3)$

3. a.

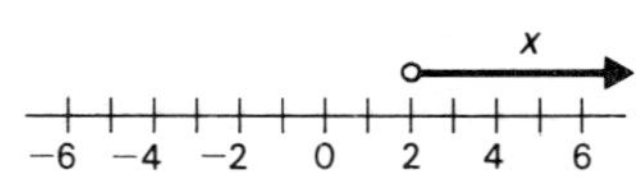

b. $(2, \infty)$

5. a.

b. $(-\infty, -4)$

7. a.

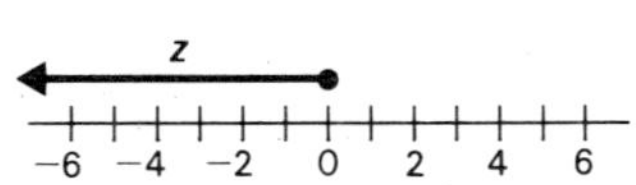

b. $(-\infty, 0]$

9. a.

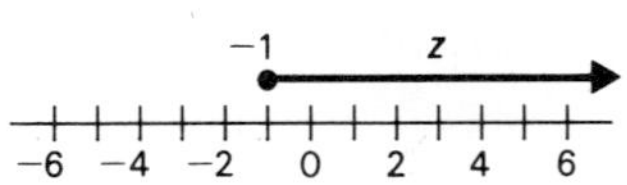

b. $[-1, \infty)$

11. a.

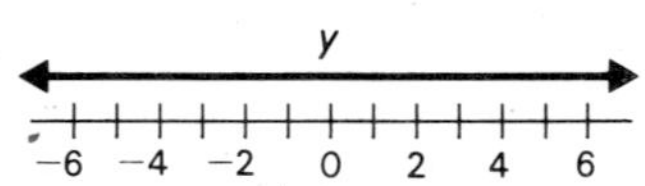

b. $(-\infty, \infty)$

13. a.

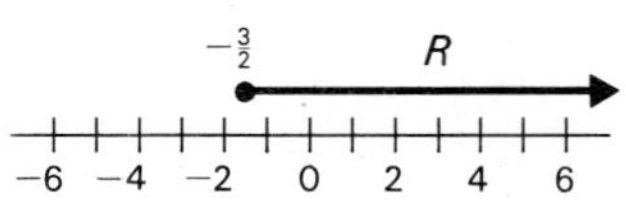

b. $\left[-\frac{3}{2}, \infty\right)$

15. a.

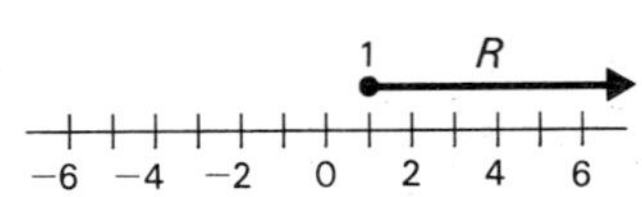

b. $[1, \infty)$

17. a.

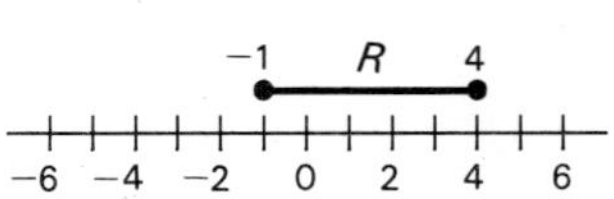

b. $[-1, 4]$

19. a.

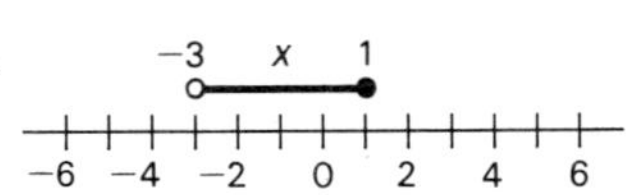

b. $(-3, 1]$

21. a. 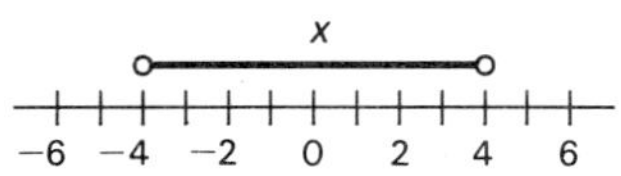

b. $(-4, 4)$

23. a. 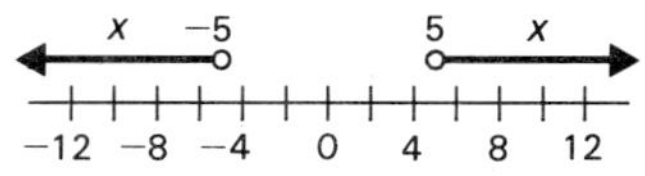

b. $(-\infty, -5)$ or $(5, \infty)$

25. a. 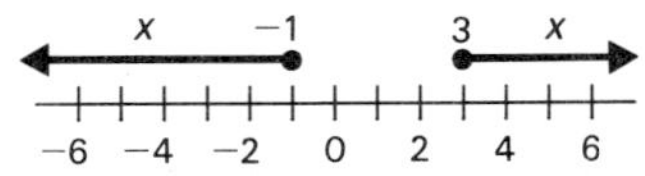

b. $(-\infty, -1]$ or $[3, \infty)$

27. a. 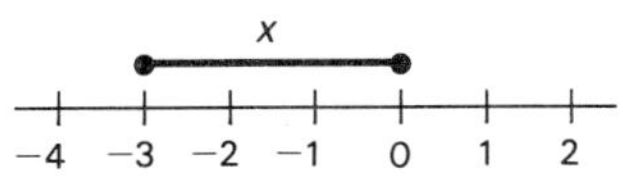

b. $[-3, 0]$

29. a. 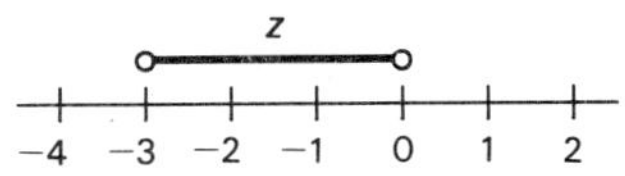

b. $(-3, 0)$

31. a. 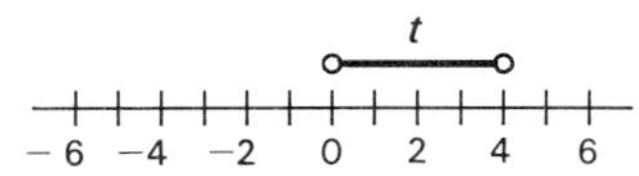

b. $(0, 4)$

33. a. 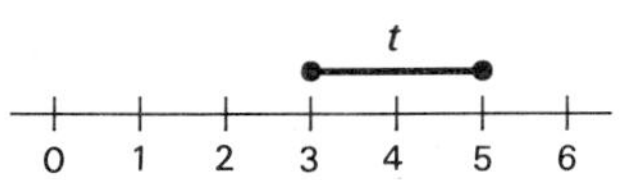

b. $[3, 5]$

35. a. 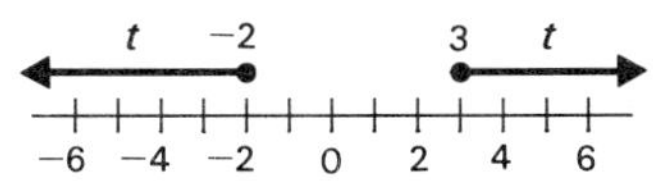

b. $(-\infty, -2]$ or $[3, \infty)$

37. a. 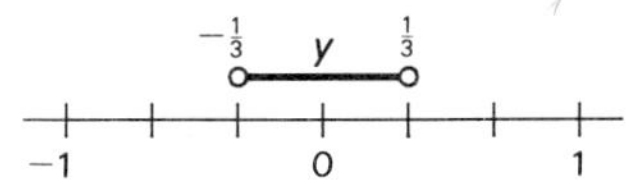

b. $\left(-\frac{1}{3}, \frac{1}{3}\right)$

39. a. 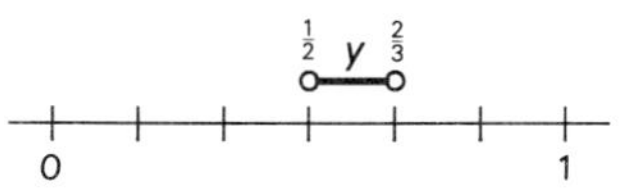

b. $\left(\frac{1}{2}, \frac{2}{3}\right)$

41. a. 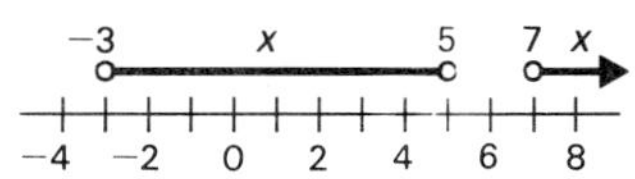

b. $(-3, 5)$ or $(7, \infty)$

43. a. 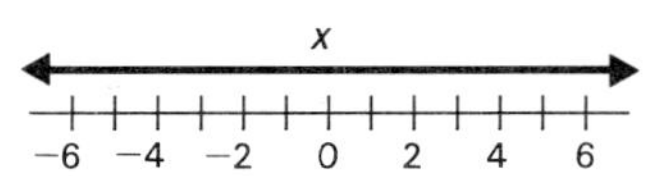

b. $(-\infty, \infty)$

45. a. 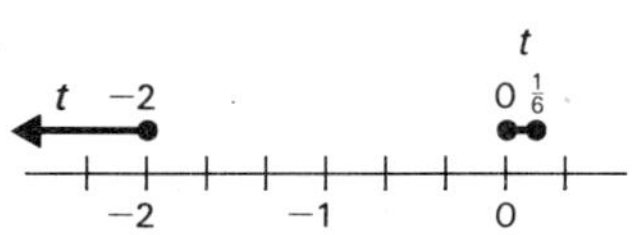

b. $(-\infty, -2]$ or $[0, 1/6]$

47. a. 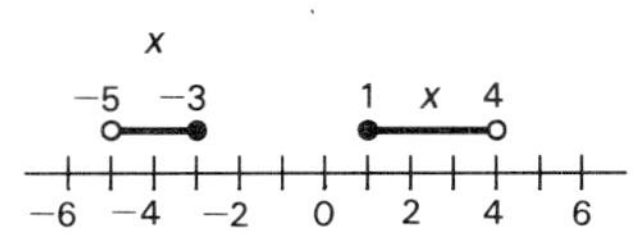

b. $(-5, -3]$ or $[1, 4)$

49. a. 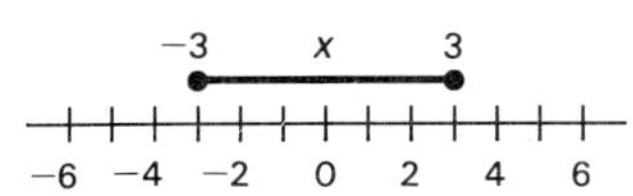

b. $[-3, 3]$

51. 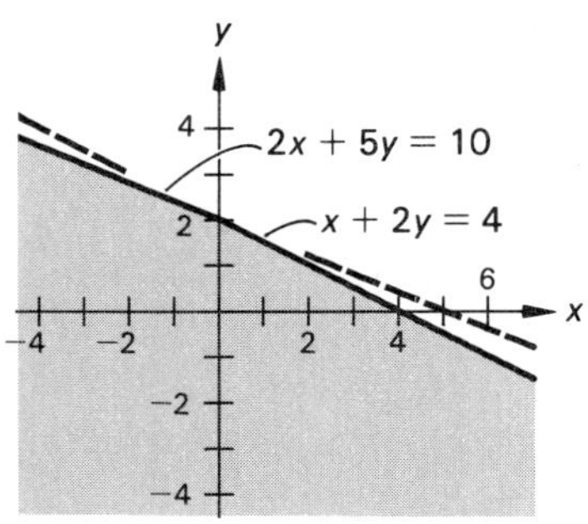

53. 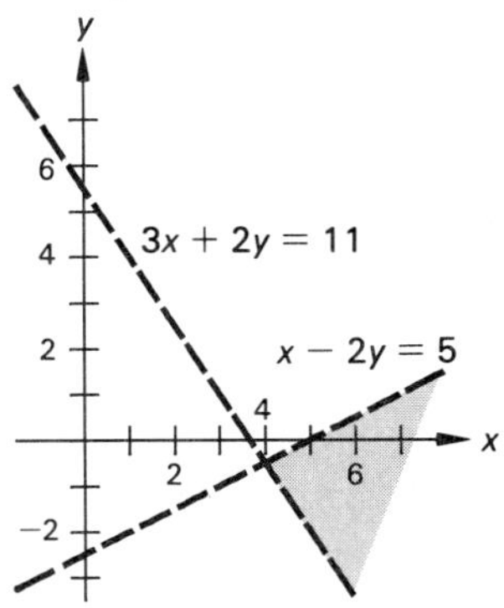

55. 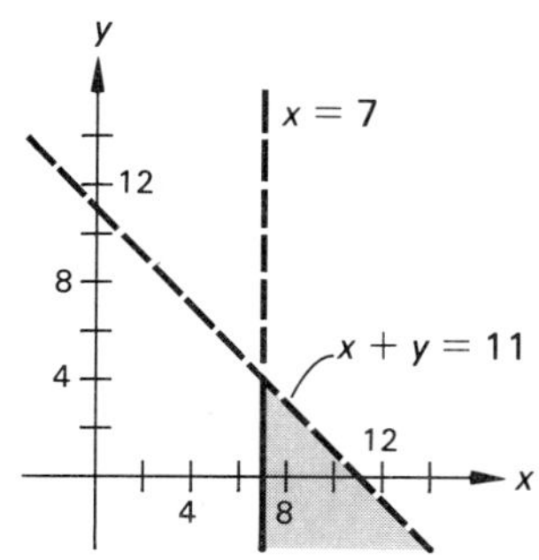

57. 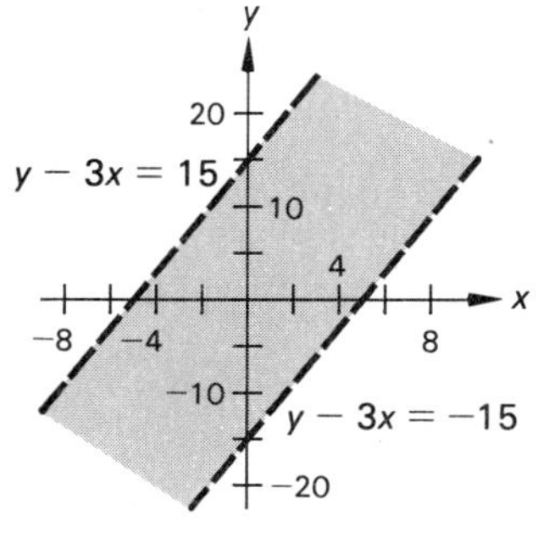

59. 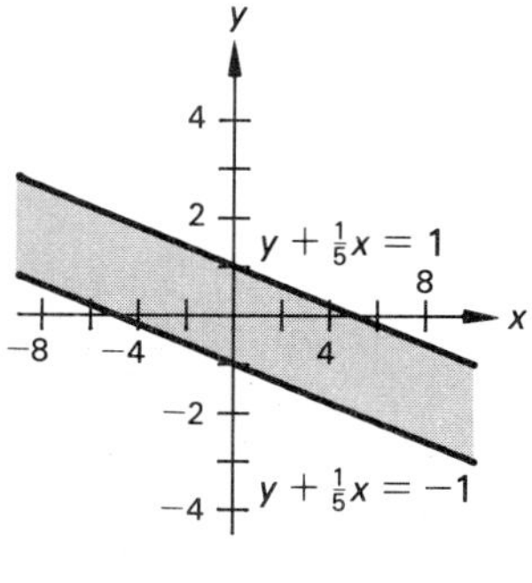

61. (68° F, 86° F)

63. $\frac{84}{17}\ \Omega < R_2 < \frac{231}{10}\ \Omega$

65. $64 \le \frac{r}{5} \le 88$;
320 km $\le r \le$ 440 km

67. $|t - 2| \le 0.04$;
1.96 mm $\le t \le$ 2.04 mm

69. a.

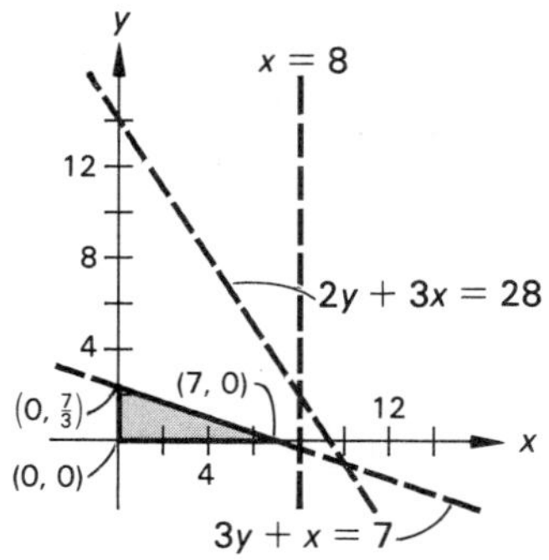

b. $P_{max} = 21$, $P_{min} = 0$

71. a.

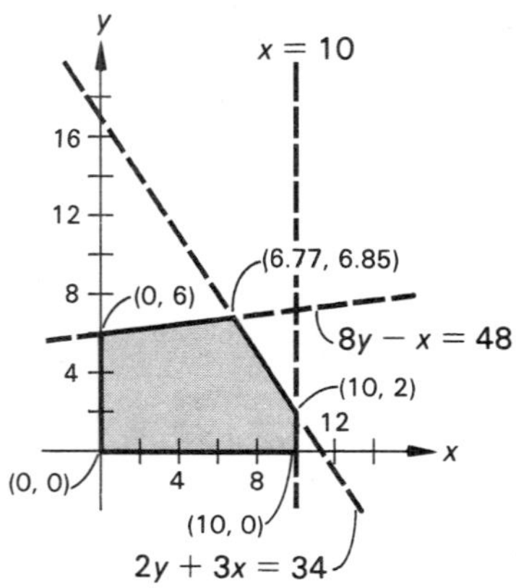

b. $P_{max} = 204.62$, $P_{min} = 0$

73. 28 precision, 41 regular; \$3600.

CHAPTER 15

15.1 Exercises

17. $\cos A$ **19.** $\tan A$ **21.** $\sin^2 A$ **23.** $\cos\theta$ **25.** $-\sin^3\theta$ **27.** $\cot\theta$ **29.** $\cot^2 A$
31. $\csc^2 B$ **33.** 1 **35.** 1 **37.** 1

15.2 Exercises

7. $\dfrac{\tan A - \tan B}{1 + \tan A \tan B}$ **9.** $\dfrac{1-\sqrt{3}}{2\sqrt{2}}$ **11.** $\dfrac{1-\sqrt{3}}{2\sqrt{2}}$ **13.** $\dfrac{1+\sqrt{3}}{2\sqrt{2}}$ **15.** $-\dfrac{\sqrt{3}+1}{2\sqrt{2}}$ **17.** $\dfrac{1+\sqrt{3}}{2\sqrt{2}}$
19. $-2-\sqrt{3}$ **21.** $2-\sqrt{3}$ **23.** $-\sin\theta$ **25.** $-\cos\theta$ **27.** $\sin\theta$ **29.** $\cos^2\theta - \sin^2\theta$
31. $5\sqrt{3}\cos 20t + 5\sin 20t$ **33.** $3\cos t - 6\sqrt{3}\sin t$ **35.** $1.39\sin 377t + 1.17\cos 377t$
37. $262.\cos 4t + 198.\sin 4t$ **39.** $34.0\sin 377t - 82.0\cos 377t$ **41.** $\sin A\cos B + \cos A\sin B$

15.3 Exercises

1. $-\dfrac{1}{2}$ **3.** $\sqrt{\dfrac{2-\sqrt{3}}{4}}$ **5.** $\sqrt{\dfrac{1+\sqrt{2}}{2\sqrt{2}}}$ **7.** $-\dfrac{\sqrt{3}}{2}$ **9.** $\dfrac{2\tan A}{1-\tan^2 A}$
17. $\dfrac{s_1+s_3}{2} + \dfrac{s_1-s_3}{2}\cos 2\theta - s_2\sin 2\theta$

15.4 Exercises

1. 0°, 45°, 180°, 225° **3.** 0°, 45°, 135°, 180° **5.** 270° **7.** 45°, 225°
9. $22\frac{1}{2}°$, $112\frac{1}{2}°$, $202\frac{1}{2}°$, $292\frac{1}{2}°$ **11.** 30°, 150°, 270° **13.** 60°, 300° **15.** 60°, 180°, 300°
17. 0°, 90° **19.** 90°, 120°, 240°, 270° **21.** 0°, 45°, 180°, 225° **23.** 0° **25.** 0°, 60°, 300°
27. 30°, 150°, 210°, 330° **29.** 30°, 150° **31.** 0° **33.** 53.6°, 147.5°, 212.5°, 306.4°
35. 90°, 270° **37.** 0.364 **39.** 7.0°, 214.1° **41.** 70.1°, 161.5° **43.** 130.0°, 320.0°

15.5 Exercises

1. a.

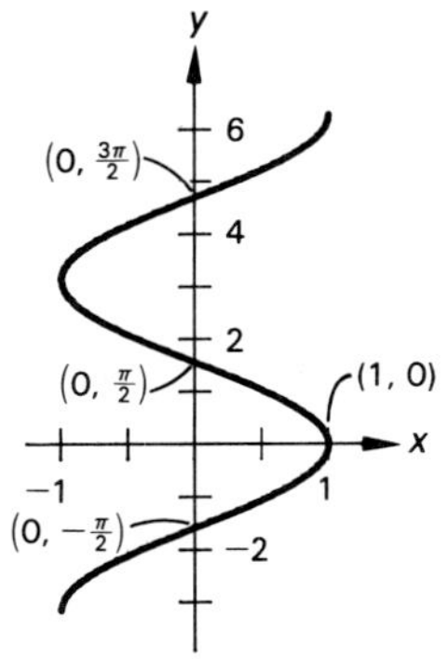

$y = \arccos x$

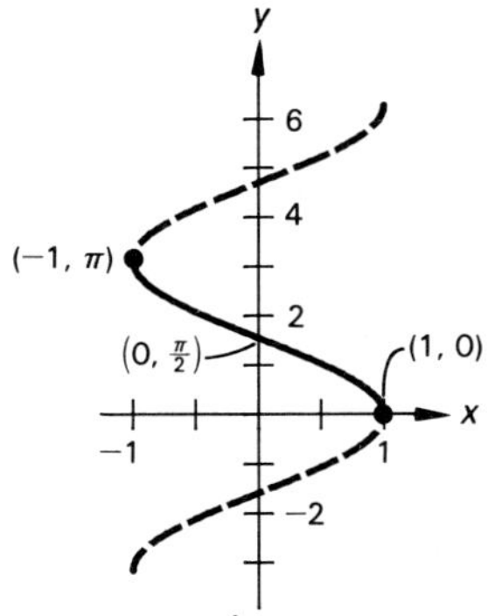

$y = \text{Arccos}\ x$

b. $[-1, 1], [-1, 1]$ **c.** $(-\infty, \infty), [0, \pi]$

3. a.

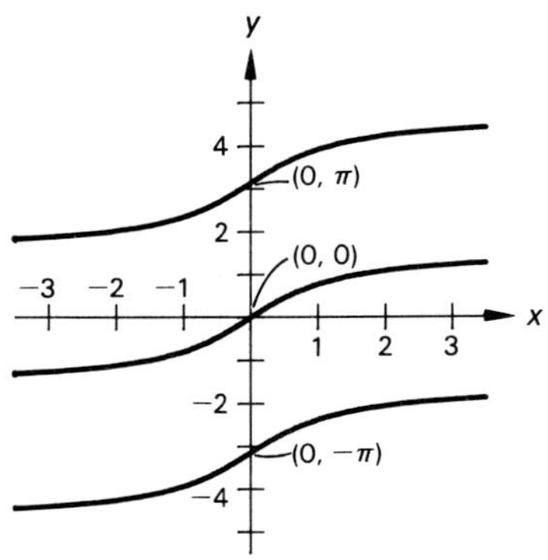

$y = \arctan x$

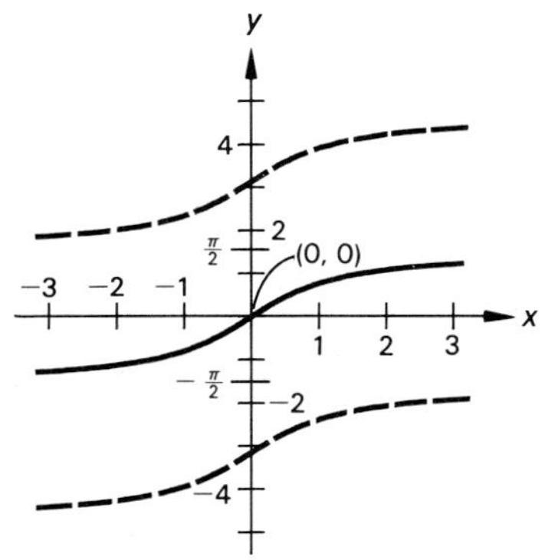

$y = \text{Arctan}\ x$

b. $(-\infty, \infty), (-\infty, \infty)$

c. $(-\infty, \infty)\ y \neq \dfrac{\pi}{2} \pm k\pi$, $\left(-\dfrac{\pi}{2}, \dfrac{\pi}{2}\right)$

5. a.

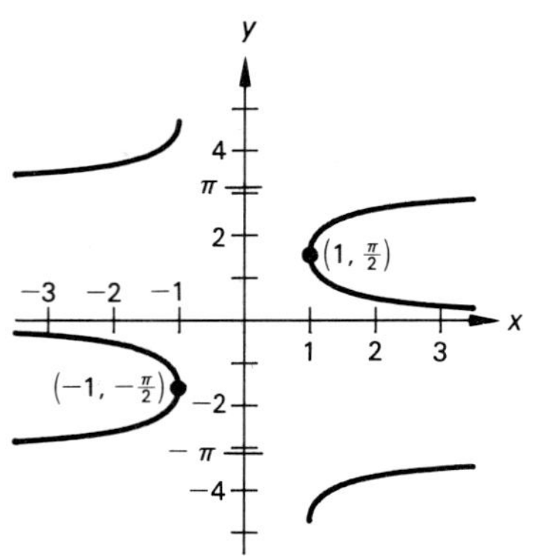

$y = \text{arccsc}\ x$

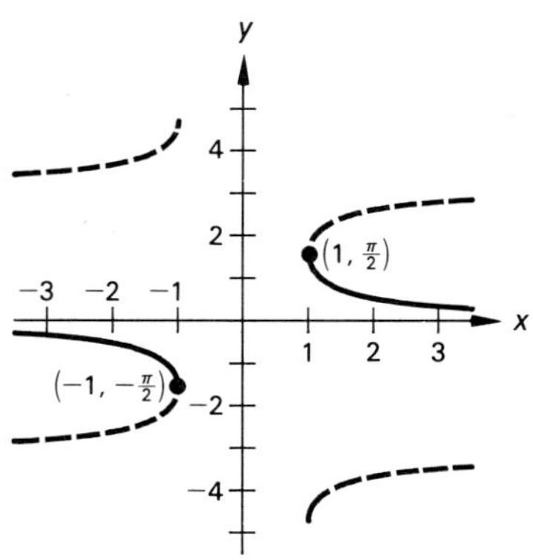

$y = \text{Arccsc}\ x$

b. $(-\infty, -1]$ or $[1, \infty)$
$(-\infty, -1]$ or $[1, \infty)$

c. $(-\infty, \infty)\ y \neq \pm k\pi$, $\left[-\dfrac{\pi}{2}, 0\right)$ or $\left(0, \dfrac{\pi}{2}\right]$

7. 0° **9.** 45° **11.** 45° **13.** −76.0° **15.** −30° **17.** 150° **19.** 60° **21.** 143.1° **23.** 56.2° **25.** −87.3° **27.** 0.866 **29.** 0.707 **31.** 1.333 **33.** 0.750 **35.** 0.621 **37.** −0.280 **39.** Undefined **41.** −1.152 **43.** 0 **45.** $\dfrac{\sqrt{b^2 - a^2}}{b}$ **47.** $\dfrac{\sqrt{1 - x^2}}{x}$ **49.** 12.99° **51.** 13.26° **53.** Because $y = \text{Arctan}\ x$ is a function whose equivalent function is $x = \tan y$ and $\tan y$ is undefined for $y = -\dfrac{\pi}{2}$ and $\dfrac{\pi}{2}$.

Review Exercises

5. $\sec\theta$ **7.** $\cos^2\theta$ **9.** $2\csc^2 A$ **11.** $\cos^2 B$ **13.** $\tan^2 A$ **15.** $\cos^2\theta$ **17.** 1 **19.** 1
31. $\cos\theta\cos\phi + \sin\theta\sin\phi$ **35.** $\frac{\sqrt{3}-1}{2\sqrt{2}}$ **37.** $-\frac{1}{\sqrt{2}}$ **39.** $-(2+\sqrt{3})$ **41.** -1 **43.** $2-\sqrt{3}$
45. $\frac{2\tan\theta}{1-\tan^2\theta}$ **47.** $\frac{2\tan A}{1+\tan^2 A}$ **51.** $\frac{\epsilon_{yy}-\epsilon_{xx}}{2}\sin 2\theta + \epsilon_{xy}\cos 2\theta$ **53.** 45°, 225° **55.** 90°
57. 30°, 90°, 150°, 270° **59.** 0°, 60°, 300° **61.** 30°, 45°, 225°, 315°, 330° **63.** 0°, 120°, 240°
65. 33.7°, 135°, 213.7°, 315° **67.** 165°, 345° **69.** 300° **71.** 65.0°, 217.6° **73.** 6.8°, 120.1°
75. 30° **77.** 36.9° **79.** 60° **81.** 30° **83.** 0.385 **85.** 0.471 **87.** 0.292 **89.** 0.114
91. 0.960 **93.** 0.282 **95.** $\frac{x}{\sqrt{1-x^2}}$

CHAPTER 16

16.1 Exercises

1. a. 2, 4, 6, 8, 10, 12
b. Increasing
c.

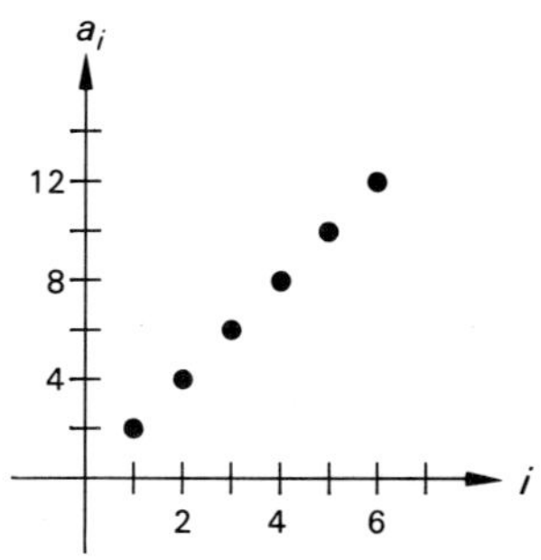

3. a. 4, 5, 6, 7, 8, 9
b. Increasing
c.

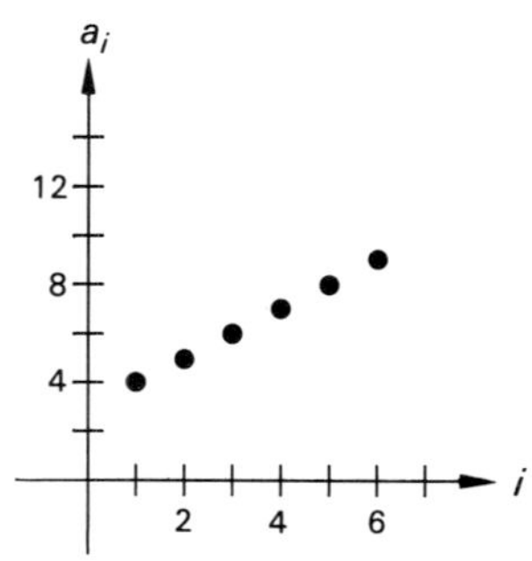

5. a. -2, 1, 4, 7, 10, 13
b. Increasing
c.

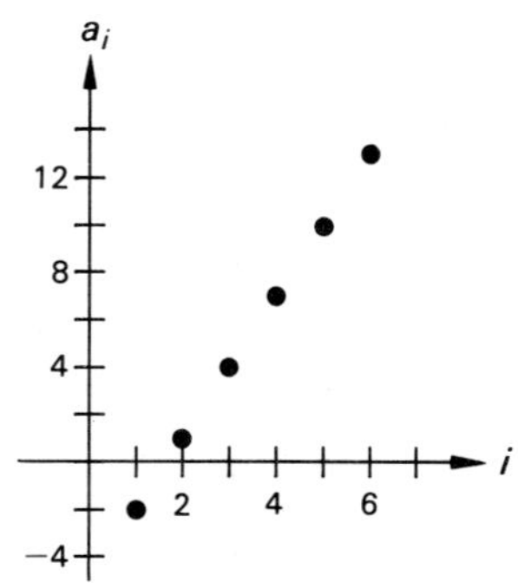

7. a. $-\frac{3}{4}, -\frac{3}{2}, -\frac{9}{4}, -3, -\frac{15}{4}, -\frac{9}{2}$
b. Decreasing
c.

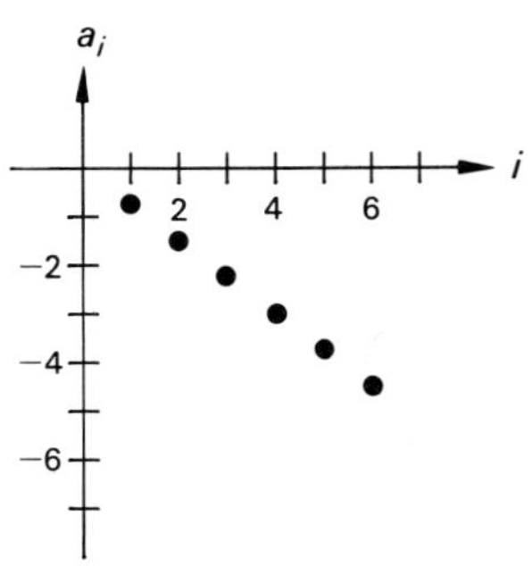

9. a. -2, 0, -2, 0, -2, 0
b. Neither
c.

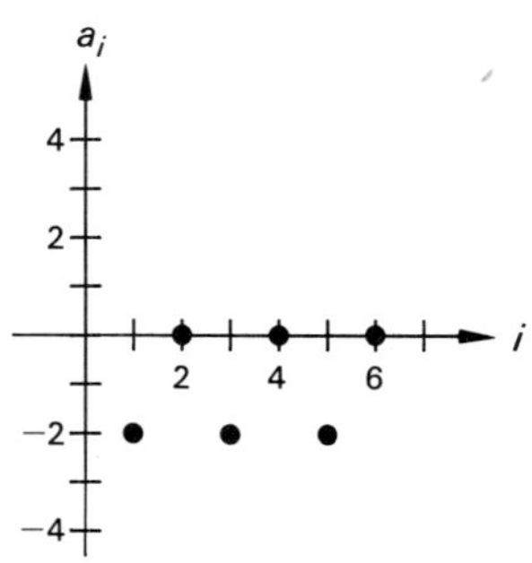

11. a. $-3, 1, -3, 1, -3, 1$
b. Neither
c.

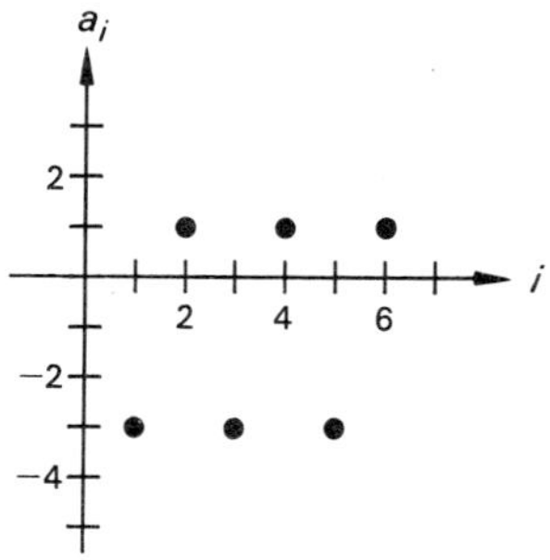

13. a. $0, 3, 8, 15, 24, 35$
b. Increasing
c.

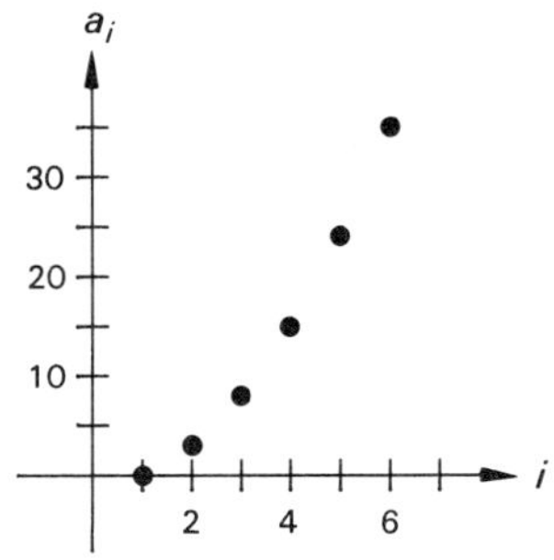

15. a. $\frac{3}{4}, \frac{4}{5}, \frac{5}{6}, \frac{6}{7}, \frac{7}{8}, \frac{8}{9}$
b. Increasing
c.

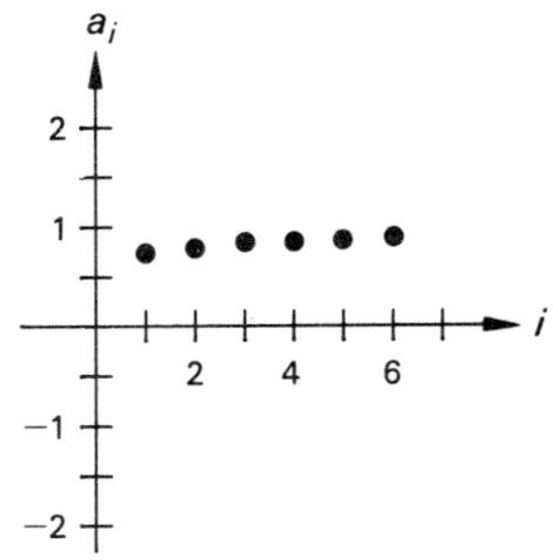

17. a. $0, \frac{3}{5}, \frac{4}{5}, \frac{15}{17}, \frac{12}{13}, \frac{35}{37}$
b. Increasing
c.

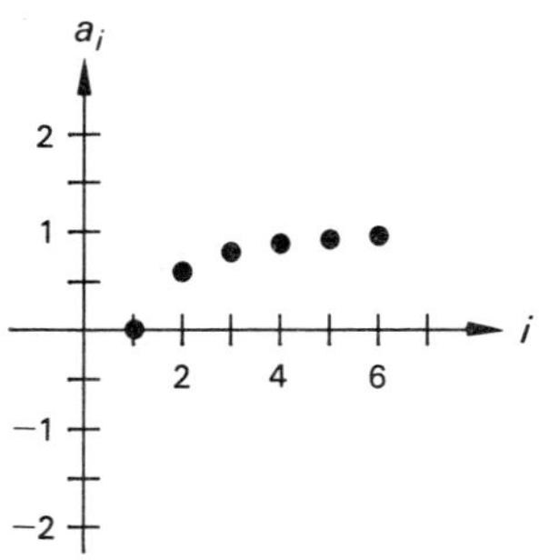

19. a. $1, -\frac{1}{2}, \frac{1}{3}, -\frac{1}{4}, \frac{1}{5}, -\frac{1}{6}$
b. Neither
c.

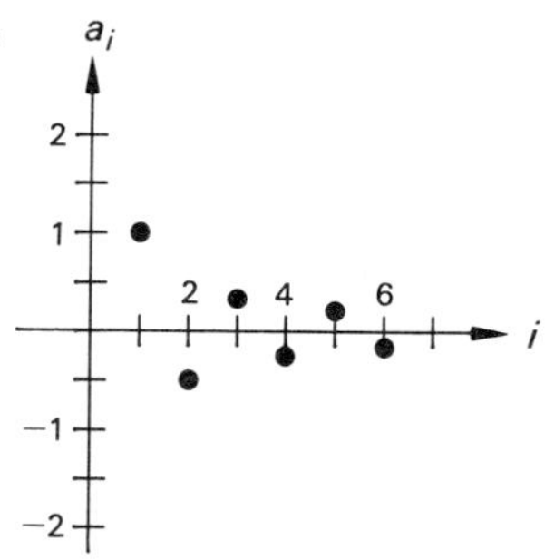

21. $i-1$ **23.** $2i$ **25.** $2i-1$ **27.** $4i-1$ **29.** $-2i$
31. $i+3$ **33.** $4i-8$ **35.** $\frac{1}{i}$
37. $\frac{1}{i^2}$ **39.** $\frac{i+1}{i}$
41. \$3.50, \$3.85, \$4.24, \$4.66, \$5.12, \$5.64

16.2 Exercises

1. a. 14 **b.** 510 **3. a.** 0.481 or $\frac{13}{27}$ **b.** 0.500 or $\frac{3280}{6561}$ **5. a.** 21 **b.** 96

7. a. -0.0833 or $-\frac{1}{12}$ **b.** 2.0673 or $\frac{11461}{5544}$ **9. a.** 2 **b.** -4

11. $4 - 4 + \frac{16}{3} - 8 + \frac{64}{5} = 10.13$ or $\frac{152}{15}$ **13.** $9 + \frac{27}{2} + 27 + \frac{243}{4} = 110.25$

15. $-2 + 2 - \frac{8}{3} + 4 - \frac{32}{5} + \frac{32}{3} = 5.60$ **17.** $\log 3 + \log 4 + \log 5 + \log 6 + \log 7 = 3.40$

19. $\frac{3}{7} + \frac{1}{2} + \frac{5}{9} + \frac{3}{5} + \frac{7}{11} + \frac{2}{3} + \frac{9}{13} = 4.08$ **21.** $\sum_{i=1}^{n} 3i$ **23.** $\sum_{i=1}^{n} \frac{(-1)^i i}{2^{i-1}}$ **25.** $\sum_{i=1}^{5} \frac{1}{i^2}$

27. $\sum_{i=1}^{5} \frac{(-1)^{i+1}}{i^2}$ **29.** $\sum_{i=1}^{n} \frac{i(i+1)}{2}$ **31.** $\sum_{i=1}^{5} (z+i+3)$ or $\sum_{i=4}^{8} (z+i)$ **33.** $\sum_{i=1}^{5} \frac{z-2-i}{z+2+i}$ or $\sum_{i=3}^{7} \frac{z-i}{z+i}$

35. $\sum_{i=1}^{5} z^{2+i}(z+2+i)$ **37.** $\frac{1}{5} + \frac{4}{7} + 1 + \frac{16}{11} + \frac{25}{13}$ **39.** $\ln 1 + 2 \ln 2 + 3 \ln 3 + 4 \ln 4$

41. $-2 + 0 + 12 + \cdots + n(n+1)(n-2)$ **43.** $1 + \frac{49}{4} + 64 + \cdots + \left(\frac{2n^2 - n + 1}{2}\right)^2$

45. $(y^2 + 4y + 2) + (y^2 + 6y + 3) + (y^2 + 8y + 4) + (y^2 + 10y + 5)$ **47. a.** 10 ft **b.** 54 ft

16.3 Exercises

1. Yes, $d = 2$ **3.** No **5.** Yes, $d = \frac{1}{2}$ **7.** No **9.** No **11. a.** 13, 18, 23, 28, 33 **b.** 115 **13. a.** 15, 12, 9, 6, 3 **b.** 45 **15. a.** 11, 10, 9, 8, 7 **b.** 45 **17. a.** $-5\frac{1}{2}, -5, -4\frac{1}{2}, -4, -3\frac{1}{2}$ **b.** $-22\frac{1}{2}$ **19. a.** $0, \frac{3}{5}, \frac{6}{5}, \frac{9}{5}, \frac{12}{5}$ **b.** 6 **21.** $a_8 = 15, S_8 = 64$ **23.** $n = 7, S_7 = 84$ **25.** $d = 3, S_8 = 60$ **27.** $d = 0.3, a_6 = 1.9$ **29.** $a_1 = 23, a_6 = 3$ **31. a.** 29 **b.** 1218 **33. a.** 51 **b.** 5508 **35.** \$18,600.00 **37. a.** 240 ft **b.** 1024 ft **39.** 365 ft, 54 ft **41.** $6\frac{1}{3}, 6\frac{2}{3}, 7, 7\frac{1}{3}, 7\frac{2}{3}$ **43.** $6333\frac{1}{3}$ bacteria/day

16.4 Exercises

1. Yes; $r = 2, a_7 = 64$ **3.** No **5.** Yes; $r = \frac{1}{4}, a_7 = \frac{1}{4096}$ **7.** Yes; $r = \frac{1}{7}, a_7 = \frac{1}{7}$ **9.** Yes; $r = \frac{1}{2}, a_7 = \frac{1}{192}$ **11.** 3, 6, 12, 24, 48 **b.** 93 **13. a.** $2, \frac{2}{3}, \frac{2}{9}, \frac{2}{27}, \frac{2}{81}$ **b.** 2.99 or $\frac{242}{81}$ **15. a.** $\frac{2}{5}, \frac{1}{10}, \frac{1}{40}, \frac{1}{160}, \frac{1}{640}$ **b.** 0.533 or $\frac{341}{640}$ **17. a.** $-4, -0.4, -0.04, -0.004, -0.0004$ **b.** -4.44 **19. a.** $4, -8, 16, -32, 64$ **b.** 44 **21.** $a_4 = \frac{1}{54}, S_4 = 0.741$ **23.** $r = 2, S_6 = 189$ **25.** $a_1 = 3, a_6 = 96$ **27.** $a_4 = -9, S_n = -6\frac{2}{3}$ **29.** $a_1 = \frac{3}{4}, a_4 = \frac{3}{500}$ **31.** 732. N, 3053. N (assuming only one lift per week) **33.** 69,120. bacteria **35.** -5 or 4 **37.** Second; earns \$8485.75 more **39.** 10.1 gal

16.5 Exercises

1. 1 **3.** None **5.** None **7.** None **9.** None **11.** $\frac{32}{2}$ **13.** None **15.** $\frac{5}{2}$ **17.** 5.71 **19.** None **21.** $\frac{9}{9} = 1$ **23.** $\frac{5}{11}$ **25.** $\frac{137}{999}$ **27.** $\frac{4876}{9999}$ **29.** $\frac{1430}{999}$ **31.** $\frac{1}{2}$ **33.** $\frac{2}{5}$ **35.** $\frac{93}{100}$ **37. a.** 2.99 n **b.** 195. m **39.** Forever **41.** No, 60 s

16.6 Exercises

1. 40,320 **3.** 42 **5.** 35 **7.** 36 **9.** 1 **11.** 455 **13.** 15 **15.** 78 **17.** 100 **19.** n **21.** $x^4 + 4x^3 + 6x^2 + 4x + 1$ **23.** $a^4 + 4a^3b + 6a^2b^2 + 4ab^3 + b^4$ **25.** $243a^{10} + 810a^8b^2 + 1080a^6b^4 + 720a^4b^6 + 240a^2b^8 + 32b^{10}$ **27.** $16x^8 - 96x^6t^2 + 216x^4t^4 - 216x^2t^6 + 81t^8$ **29.** $4x^4 - \frac{8\sqrt{2}}{3}x^3y + \frac{4}{3}x^2y^2 - \frac{4\sqrt{2}}{27}xy^3 + \frac{1}{81}y^4$ **31.** $a^{18} - 12a^{15}b + 60a^{12}b^2 - 160a^9b^3 + 240a^6b^4 - 192a^3b^5 + 64b^6$ **33.** $a^{7/2} + 7a^3b^{1/2} + 21a^{5/2}b + 35a^2b^{3/2} + 35a^{3/2}b^2 + 21ab^{5/2} + 7a^{1/2}b^3 + b^{7/2}$ **35.** $x^{14/3} - 14x^4y^{2/3} + 84x^{10/3}y^{4/3} - 280x^{8/3}y^2 + 560x^2y^{8/3} - 672x^{4/3}y^{10/3} + 448x^{2/3}y^4 - 128y^{14/3}$ **37.** $1 - 17x^2 + 136x^4 - 680x^6 + \cdots$ **39.** $u^{36} + 9u^{34}v^3 + \frac{153}{4}u^{32}v^6 + 102u^{30}v^9 + \cdots$ **41.** $4{,}194{,}304x^{22} - 138{,}412{,}032x^{21}y + 2{,}179{,}989{,}505x^{20}y^2 - 2.1799895 + 10^{10}x^{19}y^3 + \cdots$ **43.** $1 + 96y + 4560y^2 + 142{,}880y^3 + \cdots$ **45.** 1.07214 **47.** 4.82681 **49.** 0.913517 **51.** $924a^6x^6$ **53.** $18{,}564u^{12}v^{24}$ **55.** 673,596 **57.** $\frac{7}{18}$

Review Exercises

1. $3i-2$ **3.** $3-2i$ **5.** $3+(-1)^i$

7. a. 4, 8, 12, 16, 20, 24
b. Increasing
c.

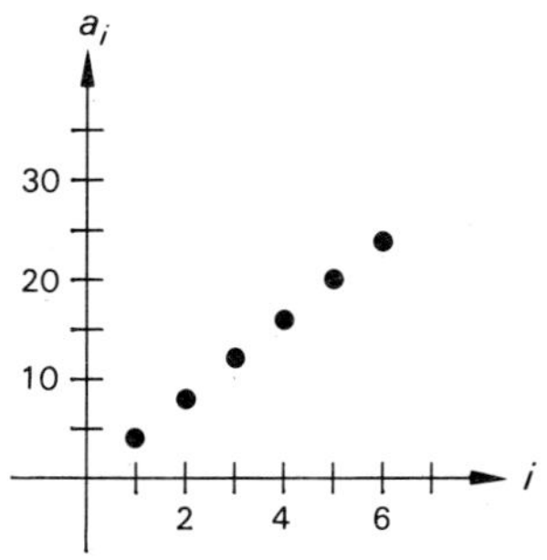

9. a. $-4, -3, -2, -1, 0, 1$
b. Increasing
c.

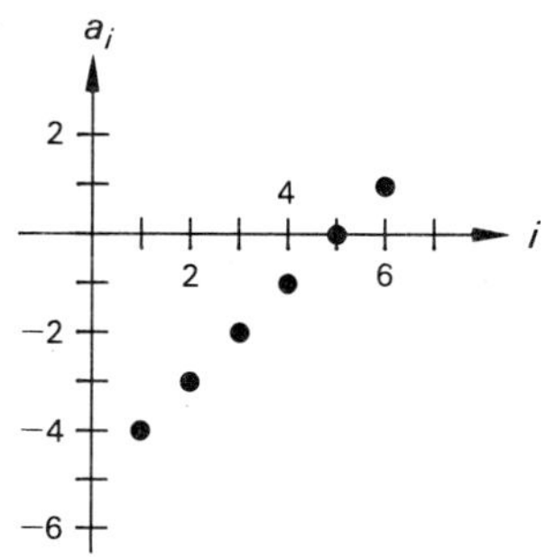

11. a. $-\frac{2}{3}, -\frac{4}{3}, -2, -\frac{8}{3}, -\frac{10}{3}, -4$
b. Decreasing
c.

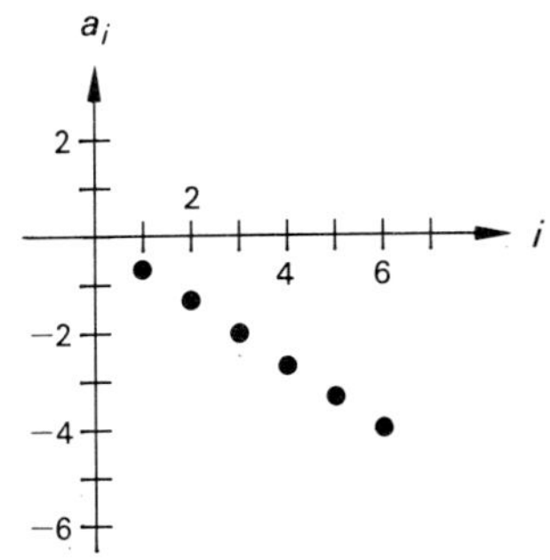

13. a. 2, 4, 2, 4, 2, 4 **b.** Neither
c.

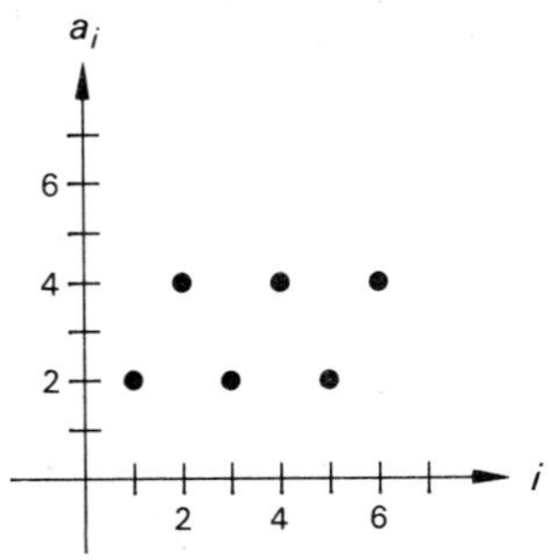

15. a. $-\frac{2}{5}, -\frac{1}{6}, 0, \frac{1}{8}, \frac{2}{9}, \frac{3}{10}$ **b.** Increasing
c.

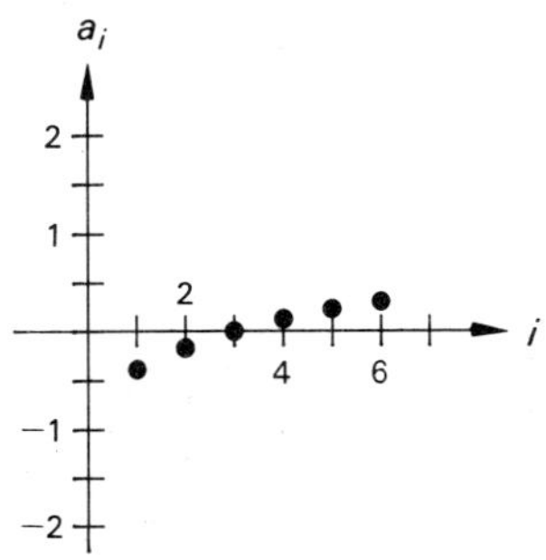

17. $S_4 = 4+16+64+256 \quad S_6 = 4+16+64+256+1024+4096$

19. $S_4 = 1+3+5+7 \quad S_6 = 1+3+5+7+9+11$

21. $S_4 = 0+\frac{1}{4}+\frac{2}{5}+\frac{3}{6} \quad S_6 = 0+\frac{1}{4}+\frac{2}{5}+\frac{3}{6}+\frac{4}{7}+\frac{5}{8}$

23. $S_4 = 2-3+4-5 \quad S_6 = 2-3+4-5+6-7$ **25.** $2-\frac{8}{3}+4-\frac{32}{5}+\frac{32}{3}=\frac{38}{5}$

27. $4+4+\frac{16}{3}+8+\frac{64}{5}=\frac{512}{15}$ **29.** $-2-1-\frac{1}{2}-\frac{1}{5}+0+\frac{1}{7}=-\frac{249}{70}$ **31.** $S_n=\sum_{i=1}^{n}(4+i)$

33. $S_n=\sum_{i=1}^{n}\frac{(-1)^i(y+1)}{2^{i-1}}$ **35.** $S_n=\sum_{i=1}^{n}\frac{i}{3^{(2i-1)}}$ **37.** $0+1+\frac{2}{3}+\frac{3}{5}+\frac{4}{7}+\frac{5}{9}+\frac{6}{11}$

39. $5+\frac{7}{4}+1+\cdots+\frac{2n+3}{n^2}$ **41.** $4z^2+30z-10$ **43.** $11+6+1+\cdots+(16-5n)+\cdots$

45. $9+8+7+\cdots+(10-n)+\cdots$ **47.** $-6\frac{1}{2}-6-5\frac{1}{2}+\cdots+\left(\frac{1}{2}n-7\right)+\cdots$

49. $d=1,\ S_{11}=88$ **51.** $d=\frac{22}{5},\ a_6=19,\ n=6$ **53.** $a_1=0,\ n=2,\ S_2=\frac{4}{5}$

55. a. $\frac{1}{3}, \frac{2}{3}, \frac{4}{3}, \frac{8}{3}, \frac{16}{3}, \frac{32}{3}$ **b.** 21 **57. a.** $2, \frac{1}{2}, \frac{1}{8}, \frac{1}{32}, \frac{1}{128}, \frac{1}{512}$ **b.** 2.67

59. **a.** $-2, -0.2, -0.02, -0.002, -0.0002, -0.00002$ **b.** -2.22 **61.** $a_4 = \frac{1}{12}, S_4 = \frac{5}{4}$
63. $a_1 = -\frac{3}{31}, a_5 = -\frac{48}{31}$ **65.** $r = 3, S_7 = 364\frac{1}{3}$ **67.** $\frac{1}{4}$ **69.** 1.02 **71.** No sum **73.** $\frac{7}{3}$
75. $\frac{8}{9}$ **77.** $\frac{412}{333}$ **79.** $z^4 + 20z^3 + 150z^2 + 500z + 625$
81. $81u^8 - 216u^6w^2 + 216u^4w^4 - 96u^2w^6 + 16w^8$
83. $a^2 + 6a^{5/3}b^{1/3} + 15a^{4/3}b^{2/3} + 20ab + 15a^{2/3}b^{4/3} + 6a^{1/3}b^{5/3} + b^2$
85. $x^{18} + 18x^{17}y + 153x^{16}y^2 + 816x^{15}y^3 + \cdots$
87. $a^{16/3} + 16a^5b^{1/3} + 120a^{14/3}b^{2/3} + 560a^{13/3}b + \cdots$ **89.** $110{,}565x^{11}$ **91.** 248.2 mm
93. \$5,368,709.12 **95.** 128 **97.** 23.04 in. **99.** \$90.11

CHAPTER 17

17.2 Exercises

1. $\sqrt{3}\,x - y = 0$ **3.** $3x + 4y = 0$ **5.** $\sqrt{3}\,s - 3a - 16 = 0$ **7.** $t + 2y + 4 = 0$ **9.** $t - 11 = 0$
11. $y - 2 = 0$ **13.** $x - y - 3 = 0$ **15.** $2t + 3v - 6 = 0$ **17.** $y - 3 = 0$ **19.** $3t - 5p + 15 = 0$

21.

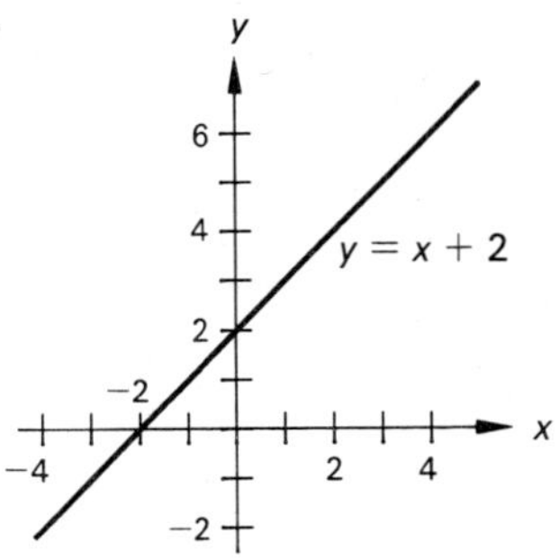

23.

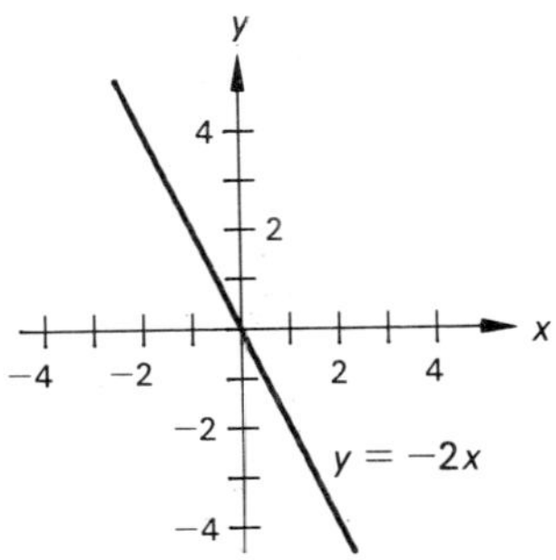

25.

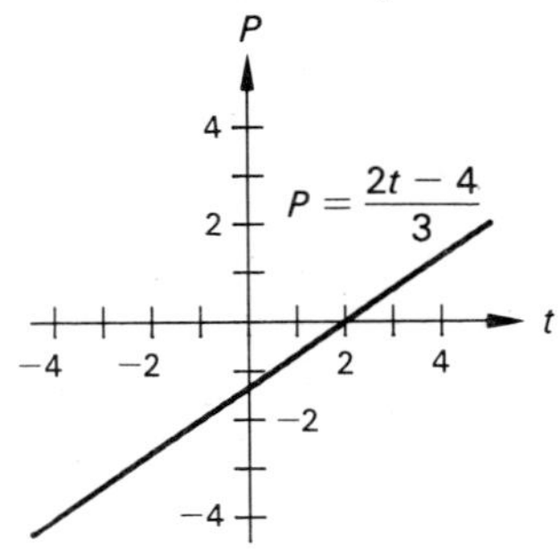

27.

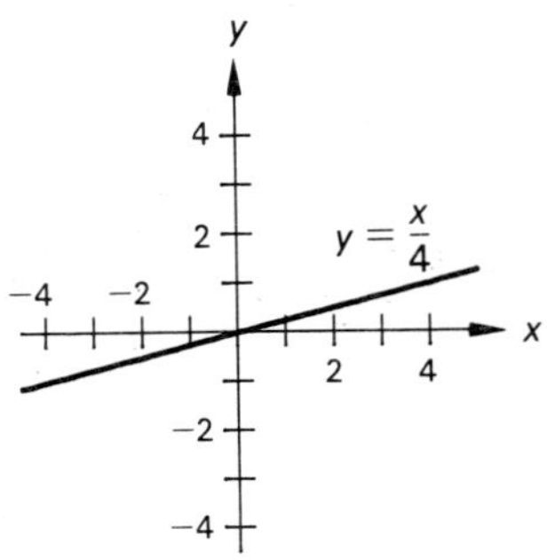

29.

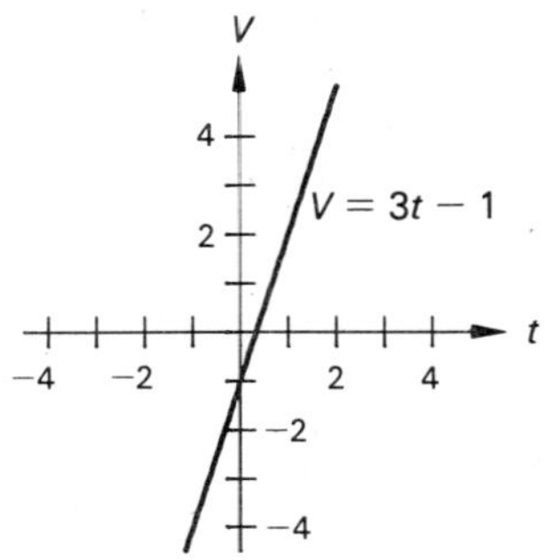

31. **a.** $3\sqrt{5}$ **b.** $\left(5, \frac{5}{2}\right)$ **33.** **a.** $\sqrt{130}$ **b.** $\left(\frac{1}{2}, \frac{1}{2}\right)$ **35.** **a.** $\sqrt{13}$ **b.** $\left(-2, \frac{-7}{2}\right)$

37. **a.** 7.78 **b.** $(-2.27, 4.04)$ **39.** **a.** $y = 2x - 1$ **b.** $x + 2y = 8$ **41.** **a.** $\left(\frac{1}{3}, \frac{5}{3}\right)$ **b.** 30.96°

43. No, slopes are not negative reciprocals. **45.** $E = 3 \times 10^7$ **47.** $p = 62.4d$
49. $40c = t + 400$, $c = t - 40$

51.

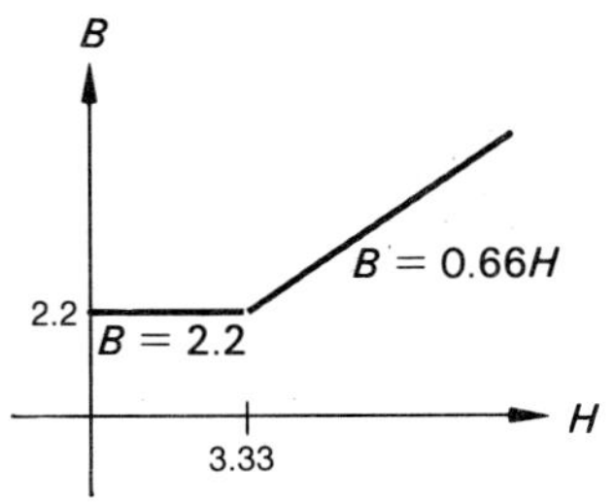

53. a. 8.21×10^{-3} BTU/mol degrees **b.** 4.22° **55.** As current increases emf decreases. **57.** $\sigma = 3 \times 10^{7}\, \epsilon - 60{,}000$

17.3 Exercises

1. $x^2 + y^2 = 9$ **3.** $x^2 + y^2 = 25$ **5.** $x^2 + y^2 = 20$ **7.** $(x-3)^2 + y^2 = 36$
9. $(x-6)^2 + (y+6)^2 = 100$ **11.** $(x+9)^2 + (y+6)^2 = 317$

13.

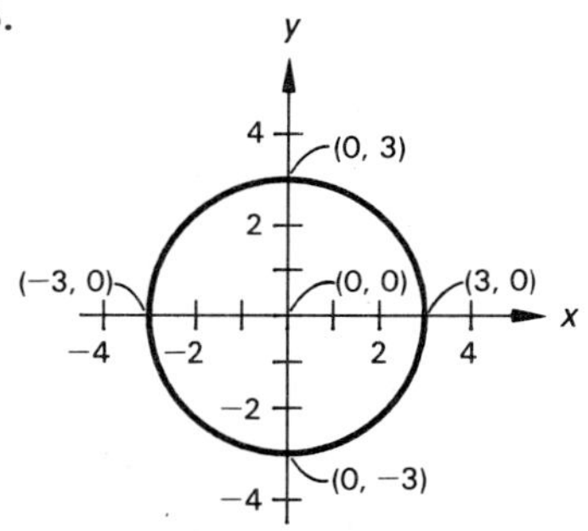

$x^2 + y^2 = 9$

15.

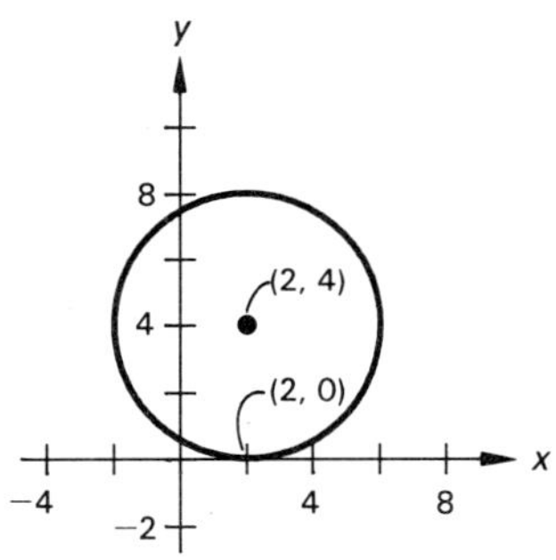

$(x-2)^2 + (y-4)^2 = 16$

17.

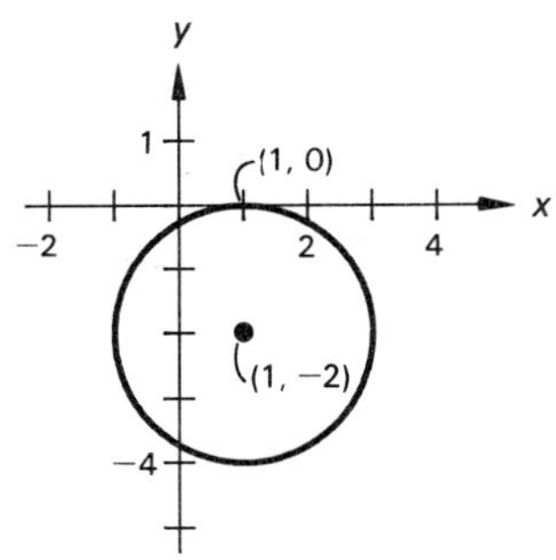

$(x-1)^2 + (y+2)^2 = 4$

19. $(x-2.3)^2 + (y-0.3)^2 = 5.78$ **21.** $(4, 2)$, $r = 2$ **23.** $\left(0, \dfrac{1}{2}\right)$ $r = \dfrac{3}{2}$ **25.** $(-1, 1)$ $r = \sqrt{2}$
27. No, coefficients of x^2 and y^2 are not equal. **29.** No, r^2 must be > 0. **31.** $x = 3$ **33.** Yes
35. 2.17 m **37.** $(7.20, -3.10)$, $r = 2.50$

17.4 Exercises

1.

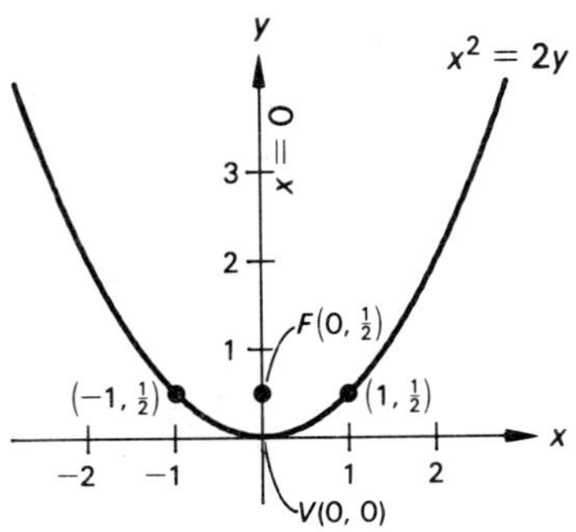

3.

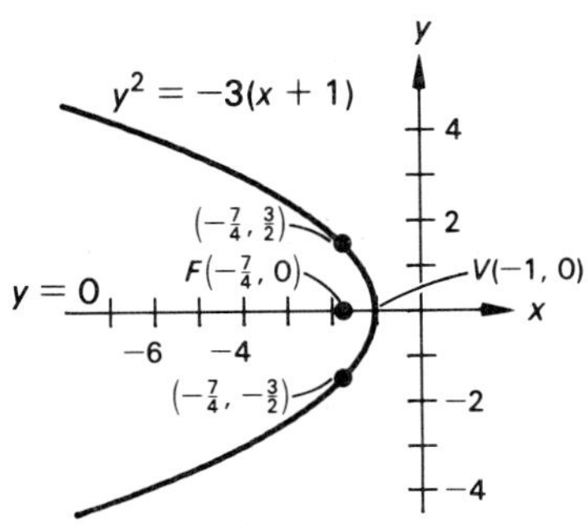

5.

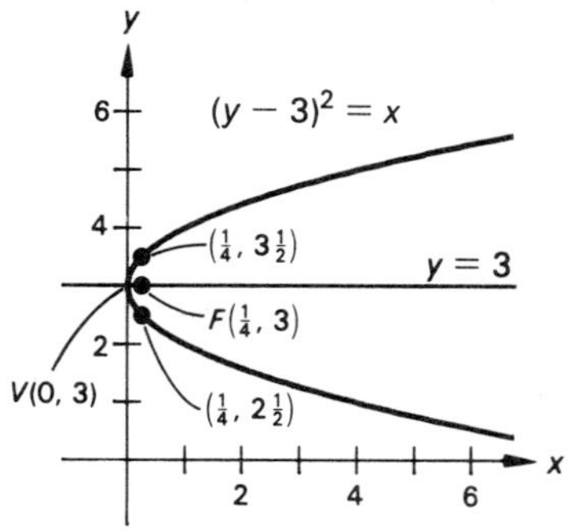

7.

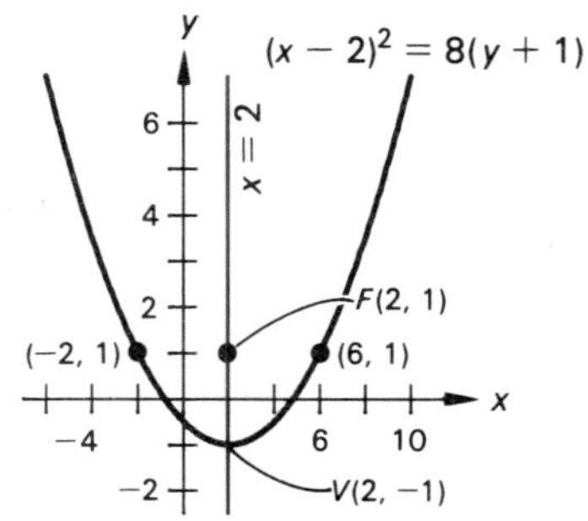

9.

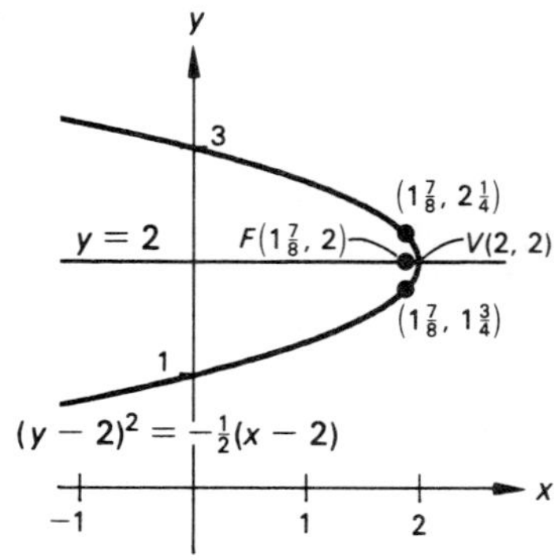

11.

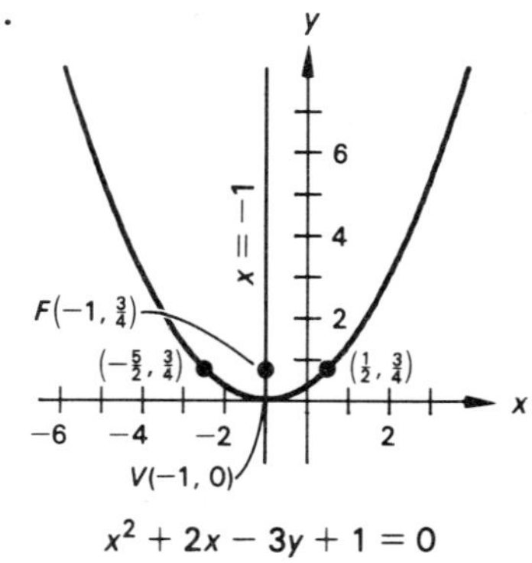

13. $x^2 = \frac{72}{7}y$ **15.** $y^2 = -\frac{49}{11}(x+3)$ **17.** $(x-5)^2 = 10(y-1)$ **19.** $x^2 = -200\left(y - \frac{1}{2}\right)$
21. $(x-200)^2 = 800(y+50)$ **23.** t near zero, no real values for w. **25.** $y^2 = 80(x+20)$

17.5 Exercises

1.

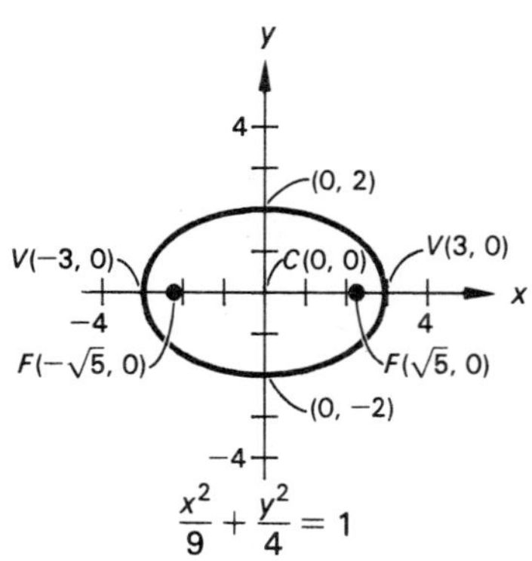

3.

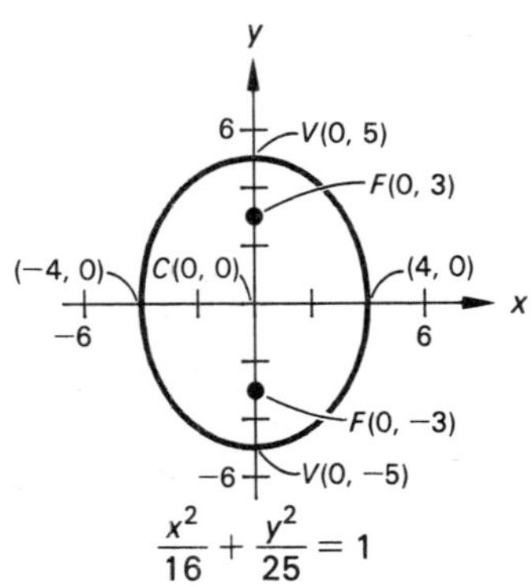

5.

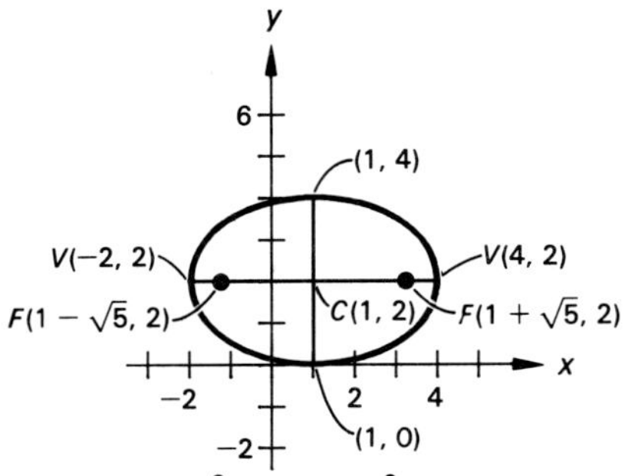

$$\frac{(x-1)^2}{9}+\frac{(y-2)^2}{4}=1$$

7.

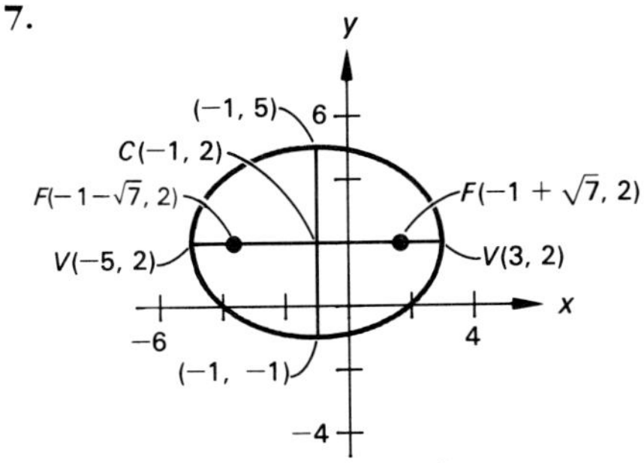

$$\frac{(x+1)^2}{16}+\frac{(y-2)^2}{9}=1$$

9.

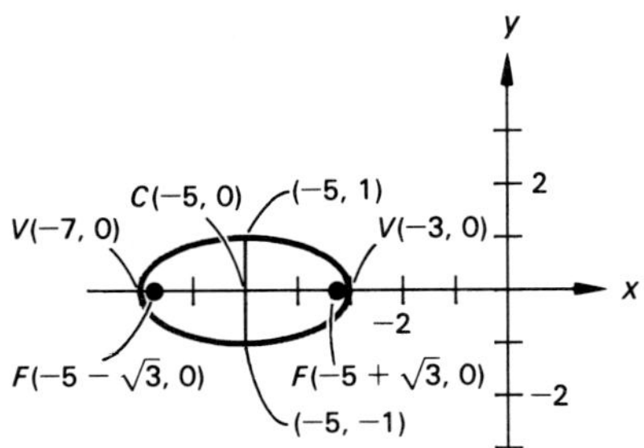

$$\frac{(x+5)^2}{4}+\frac{y^2}{1}=1$$

11. $\dfrac{x^2}{36}+\dfrac{y^2}{9}=1$ **13.** $\dfrac{(x-2)^2}{9}+\dfrac{y^2}{25}=1$

15. $\dfrac{(x-6)^2}{49}+\dfrac{(y+2)^2}{16}=1$

17. (11) $(-3\sqrt{3}, 0)$ and $(3\sqrt{3}, 0)$ (13) $(2, -4)$ and $(2, 4)$
(15) $(6-\sqrt{33}, -2)$ and $(6+\sqrt{33}, -2)$

19.

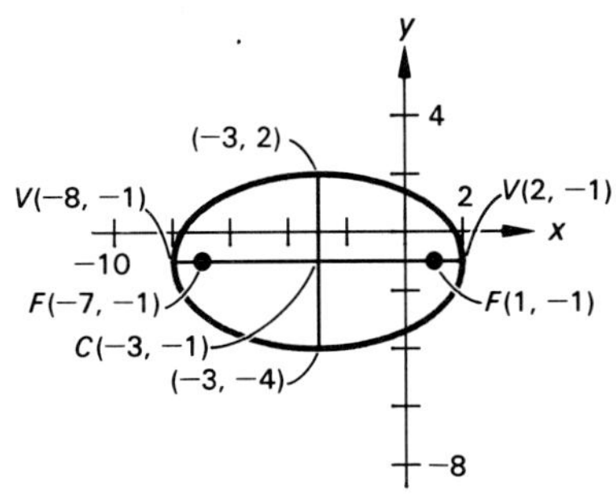

$$\frac{(x+3)^2}{25}+\frac{(y+1)^2}{9}=1$$

21. $\dfrac{x^2}{0.04}+\dfrac{y^2}{0.36}=1$

23. $\dfrac{9\,(x+5)^2}{8}+\dfrac{(y+1)^2}{8}=1$

25. 409 mi

17.6 Exercises

1.

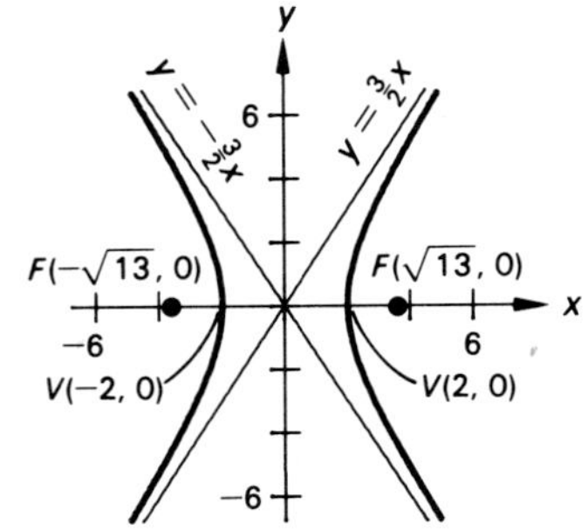

$$\frac{x^2}{4}-\frac{y^2}{9}=1$$

3.

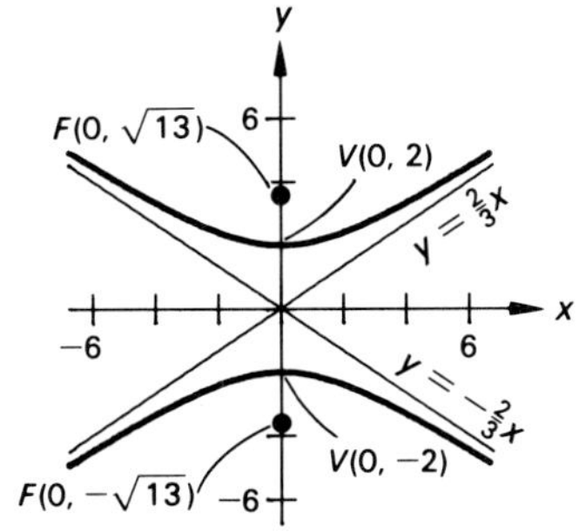

$$\frac{y^2}{4}-\frac{x^2}{9}=1$$

5.

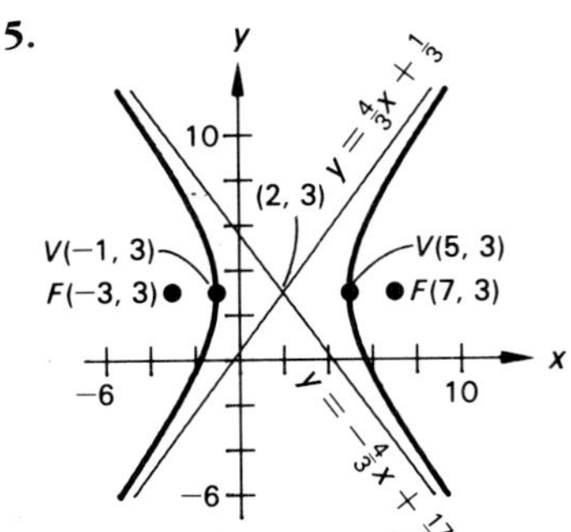

$$\frac{(x-2)^2}{9}-\frac{(y-3)^2}{16}=1$$

7.

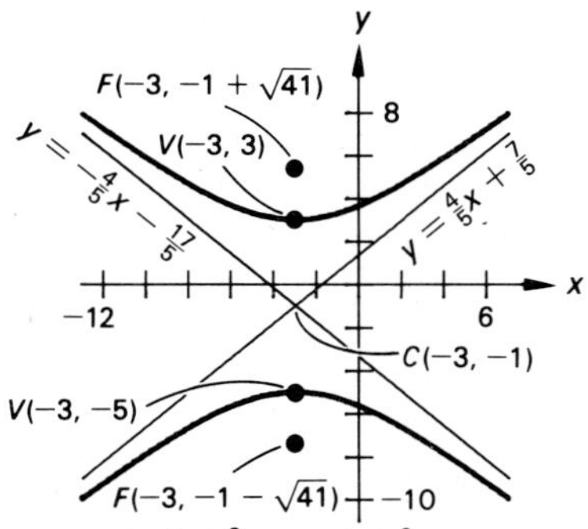

$$\frac{(y+1)^2}{16} - \frac{(x+3)^2}{25} = 1$$

9. $\dfrac{x^2}{16} - \dfrac{y^2}{4} = 1$ **11.** $\dfrac{y^2}{25} - \dfrac{x^2}{16} = 1$

13. $\dfrac{(x-4)^2}{9} - \dfrac{4\,(y-3)^2}{9} = 1$

15. $\dfrac{(y-2)^2}{9} - \dfrac{(x+4)^2}{4} = 1$

17. $\dfrac{(x+1)^2}{9} - \dfrac{(y+2)^2}{3} = 1$

19. $\dfrac{(x-2)^2}{4} - \dfrac{(y+3)^2}{9} = 1$

21. $\dfrac{(x-3.10)^2}{7.40} - \dfrac{(y-1.60)^2}{4.50} = 1$

23.

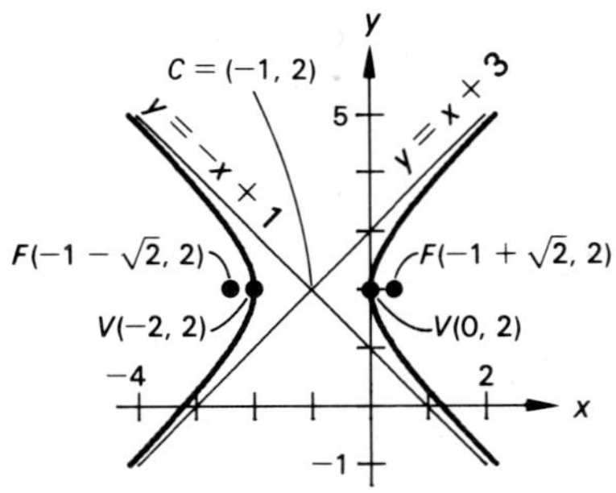

$$(x+1)^2 - (y-2)^2 = 1$$

25. $\dfrac{x^2}{(t/2)^2} - \dfrac{y^2}{c^2 - (t/2)^2} = 1$

27. Second and third quadrand portion of the curve.

$$\frac{x^2}{5625} - \frac{y^2}{34{,}375} = 1$$

29. $\dfrac{(y-2.25)^2}{5.0625} - \dfrac{x^2}{22.5} = 1$

31. $\dfrac{\left(y+\dfrac{3}{2}\right)^2}{\dfrac{1}{4}} - \dfrac{x^2}{2} = 1$

x	y
0	−1
±1	−0.888
±2	−0.634
±3	−0.327
±4	0

33.

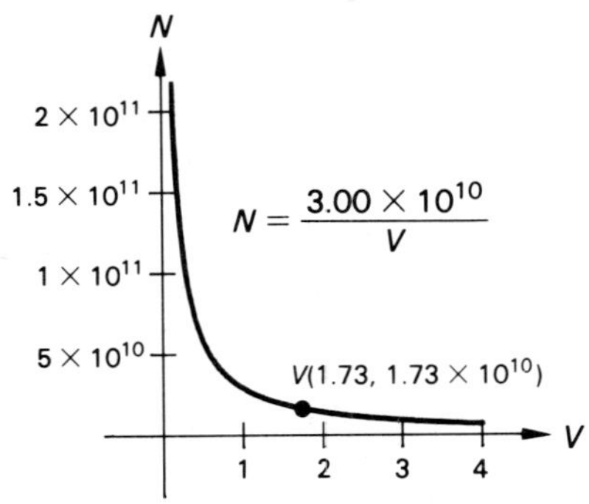

17.7 Exercises

1. $y' = x'$ **3.** $v' = 3t'$ **5.** $y' = -2x' - 7$ **7.** $y' = -\dfrac{x'}{2} - \dfrac{9}{2}$ **9.** $(x')^2 + (y')^2 = 16$

11. $(x')^2 = -2\,(y')$, $V(0, 0)$ **13.** $\dfrac{(x')^2}{25} + \dfrac{(y')^2}{9} = 1$ $V(\pm 5, 0)$ $F(\pm 4, 0)$

15. $(x')^2 - (y')^2 = 1 \quad V(\pm 1, 0) \quad F(\pm\sqrt{2}, 0)$ **17.** $(x'-1)^2 + (y'-3)^2 = 16 \quad C(1, 3)$

19. $(x')^2 = -2(y'-2) \quad V(0, 2)$ **21.** $\dfrac{(x'+2)^2}{25} + \dfrac{(y'-1)^2}{9} = 1 \quad C(-2, 1)$

23. $(x'+5)^2 - (y'-3)^2 = 1 \quad C(-5, 3)$ **25.** $(x'+200)^2 = 800(y'+50)$

27. $\dfrac{3(x')^2}{4} + \left(y - \dfrac{1}{2}\right)^2 = 1$

Review Exercises

1. $y = -\sqrt{3}\,x + 25$ **3.** $x + 2y - 6 = 0$

5.

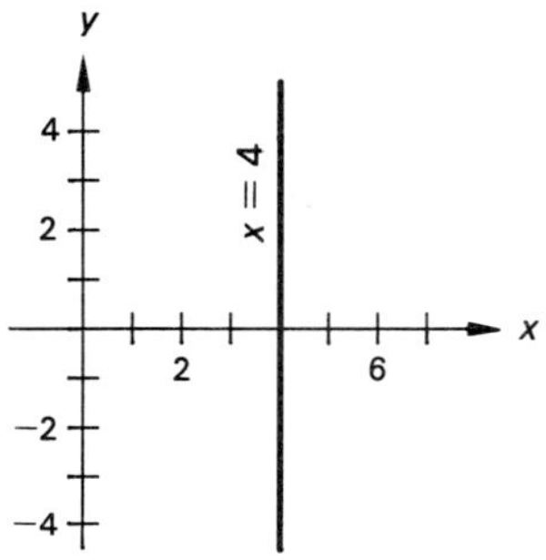

7.

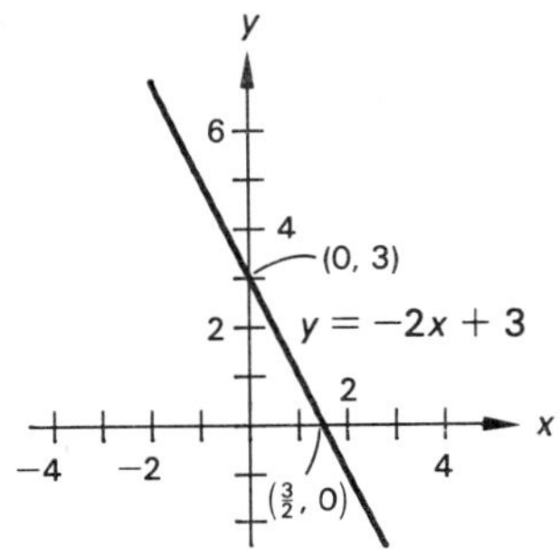

9. 64.5 mm above left corner and 46.5 mm to the right of the lower left corner

11. 4.28 units, $\left(-\dfrac{25}{34}, \dfrac{3}{34}\right)$

13.

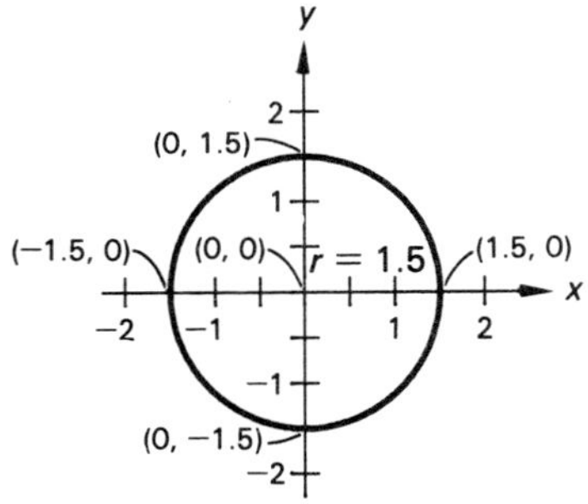

$x^2 + y^2 = 2.25$

15.

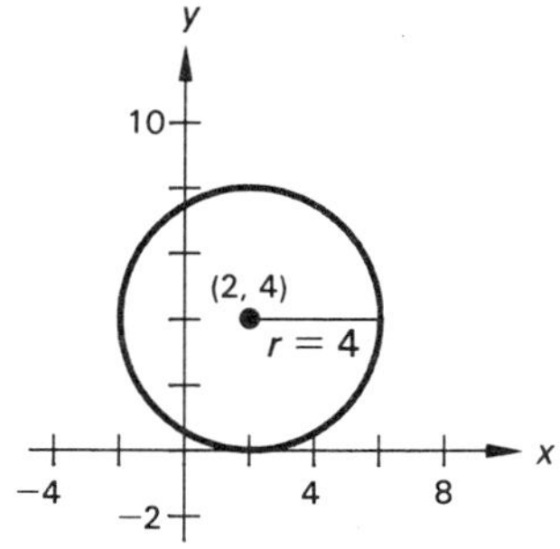

$x^2 + y^2 - 4x - 8y + 4 = 0$

17.

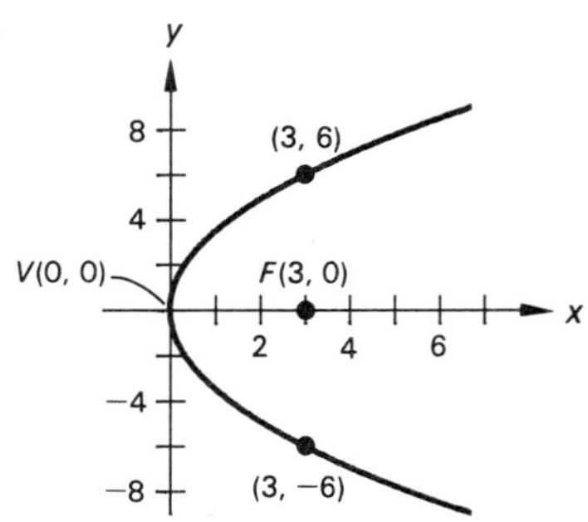

$y^2 = 12x$

19.

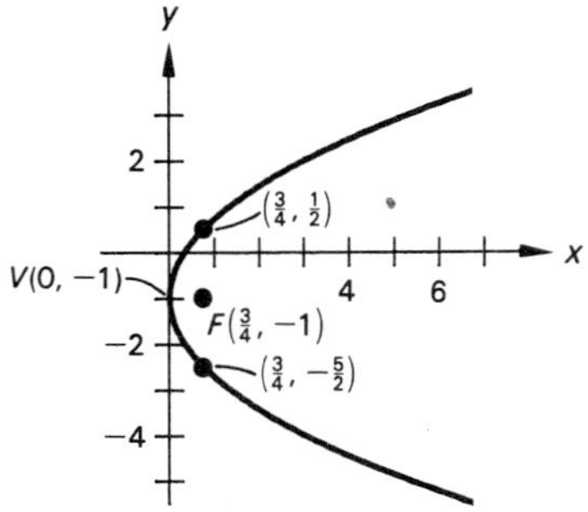

$-3x + y^2 + 2y + 1 = 0$

21. $x^2 = \dfrac{2500}{3}y$

23. $y^2 = 80(x + 20)$

25.

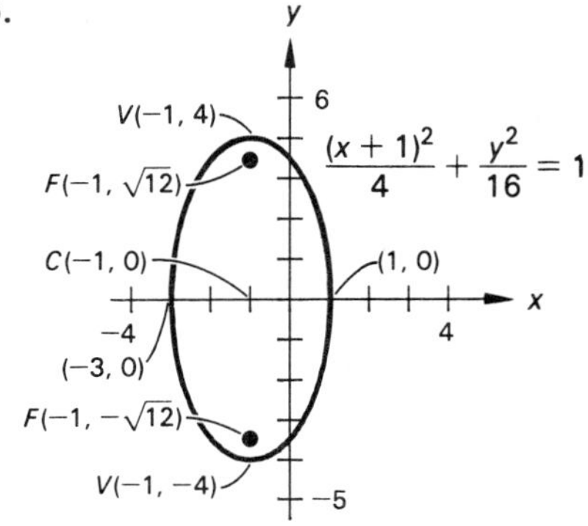

27.

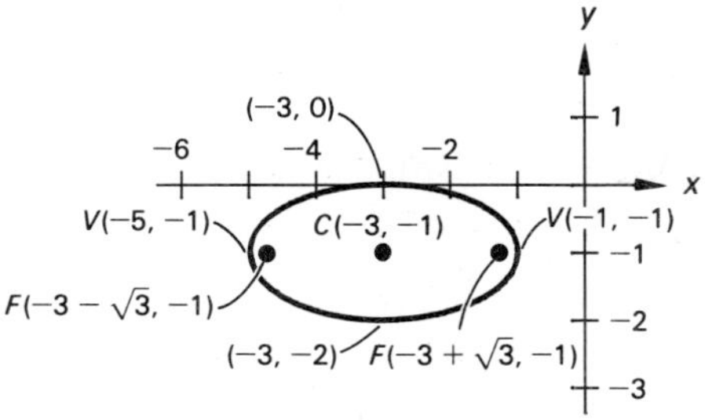

$x^2 + 6x + 8y + 4y^2 + 9 = 0$

29. $\dfrac{4x^2}{1089} + \dfrac{y^2}{144} = 1$

31. $\dfrac{(x + 2828.)^2}{9 \times 10^6} + \dfrac{y^2}{1 \times 10^6} = 1$ or

$\dfrac{(x - 2828.)^2}{9 \times 10^6} + \dfrac{y^2}{1 \times 10^6} = 1$

33.

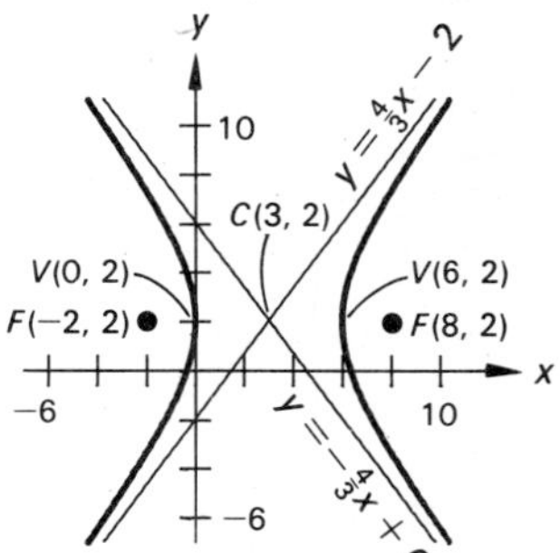

$\dfrac{(x - 3)^2}{9} - \dfrac{(y - 2)^2}{16} = 1$

35.

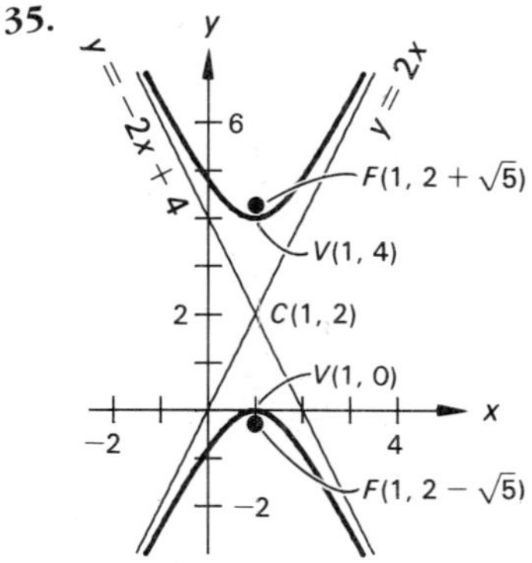

$-4x^2 + y^2 + 8x - 4y - 4 = 0$

37. $\dfrac{(y - 5)^2}{9} - \dfrac{x^2}{16} = 1$

39. $(y')^2 = 12x' \quad V(0, 0)$

41. $\dfrac{(x')^2}{9} - \dfrac{(y')^2}{16} = 1$

$V(\pm 3, 0) \quad F(\pm 5, 0)$

43. $(x')^2 + (y')^2 = 4 \quad C(0, 0)$

45.

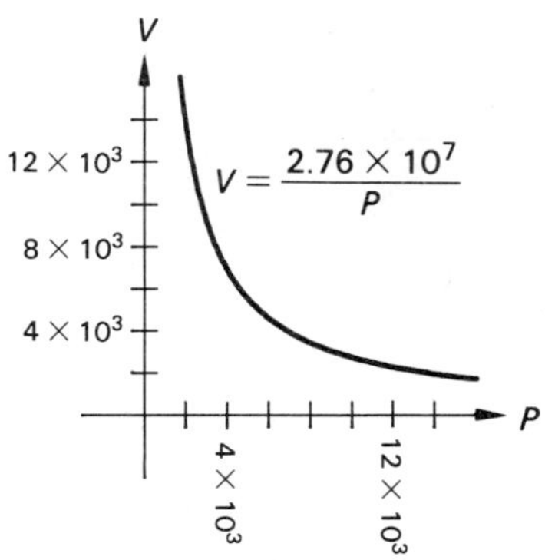

CHAPTER 18

18.1 Exercises

1. a. a **c.** a **e.** d **2. a.** a **c.** d **e.** a

18.2 Exercises

1. Electronic stylus, digital scanners, card reader, direct equipment monitoring
3. Keyboard, magnetic disk unit, floppy disk, magnetic tape
5. Baud rates are speeds of transmission at serial ports and must be compatible with I/O devices.
7. ROM is permanent, RAM disappears when the computer is shut off.

18.3 Exercises

1.

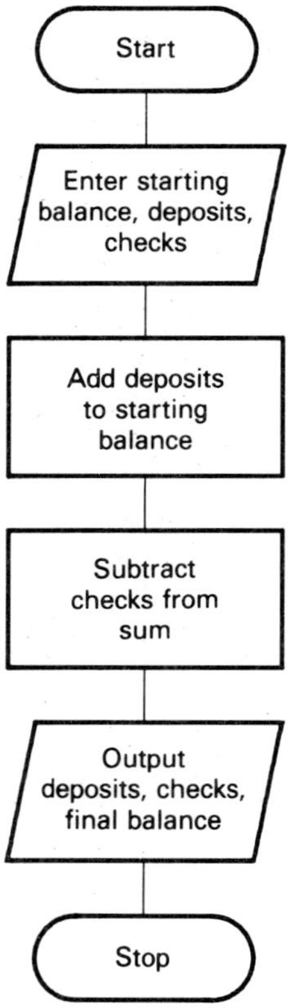

3. Let D = measure of angle in degrees
R = measure of angle in radians

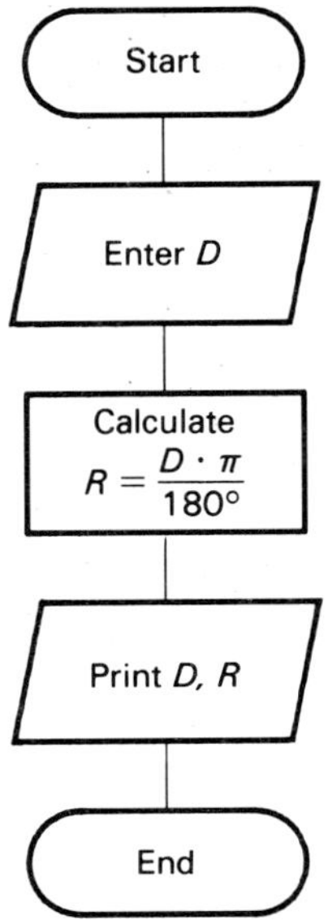

5.

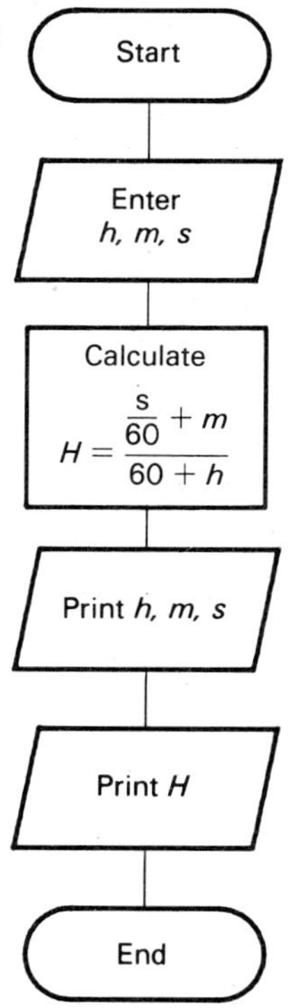

7.

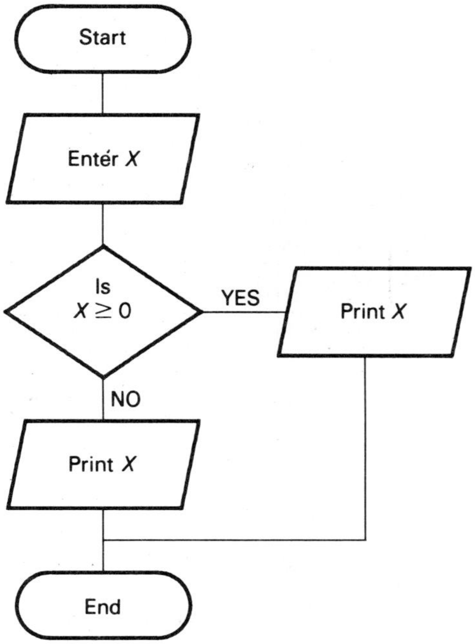
Start
Enter X
Is X ≥ 0
YES
Print X
NO
Print X
End

9.

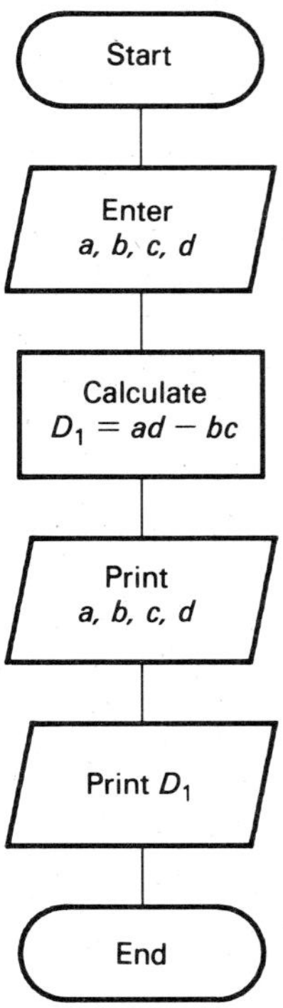
Start
Enter a, b, c, d
Calculate D1 = ad − bc
Print a, b, c, d
Print D1
End

11.

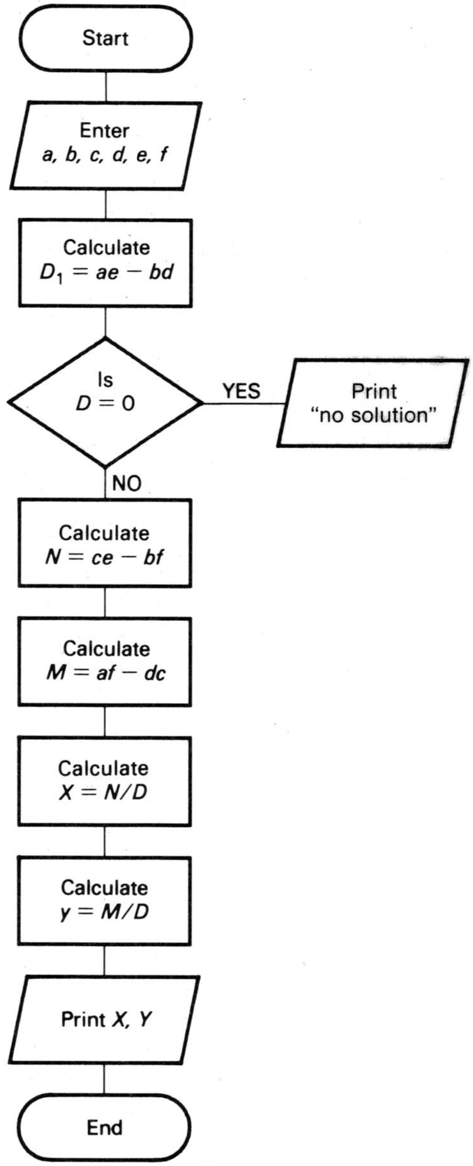
Start
Enter a, b, c, d, e, f
Calculate D1 = ae − bd
Is D = 0
YES
Print "no solution"
NO
Calculate N = ce − bf
Calculate M = af − dc
Calculate X = N/D
Calculate y = M/D
Print X, Y
End

18.4 Exercises

1.
```
10 LET S = 36.2
20 LET A = S ∧ 2
30 PRINT "S = ";S" MM","A = ";A" SQ.MM"
40 END
```

3.
```
10 LET L1 = 1.63E4
20 LET L2 = 2.81E4
30 LET H = (L1 ∧ 2 + L2 ∧ 2) ∧ .5
40 LET A = 1 / 2 * L1 * L2
50 LET P = L1 + L2 + H
60 PRINT "AREA IS ";A" SQ MM"
70 PRINT "PERIMETER IS ";P" MM"
80 PRINT "HYPOTENUSE IS ";H" MM"
90 END
```

5.
```
10 LET B1 = 264.51
20 LET D1 = 149.61
30 LET D2 = 152.88
40 LET D3 = 149.61
50 LET D4 = 149.61
60 LET C1 = 250.00
70 LET C2 = 50.00
80 LET C3 = 11.26
90 LET C4 = 21.78
100 LET C5 = 55.40
110 LET K1 = 50.00
120 LET K2 = 49.82
130 LET K3 = 50.00
140 LET K4 = 26.91
150 LET K5 = 50.00
160 LET D = D1 + D2 + D3 + D4
170 LET C6 = C1 + C2 + C3 + C4 + C5
172 LET K6 = K1 + K2 + K3 + K4 + K5
174 LET C = C6 + K6
180 LET B2 = B1 + D - C
190 PRINT "STARTING BALANCE = $ ";B1
200 PRINT "TOTAL DEPOSITS = $ ";D
210 PRINT "TOTAL WITHDRAWALS = $ ";C
220 PRINT "BALANCE AT END OF MONTH = $ ";B2
230 END
```

7.
```
10 LET D = 36.1
20 LET R = 3.14159 * D / 180
30 PRINT "36.1 DEGREES = ";R"RADIANS"
40 END
```

9.
```
10 LET X1 = - 16.5
20 LET X2 = 72.1
30 LET X3 = 0.00168
40 LET F1 = X1 ∧ 5 - 4.2 * X1 ∧ 3 + 7.3 * X1
50 LET F2 = X2 ∧ 5 - 4.2 * X2 ∧ 3 + 7.3 * X2
60 LET F3 = X3 ∧ 5 - 4.2 * X3 ∧ 3 + 7.3 * X3
70 PRINT "X1 = ";X1,"F1 = ";F1
80 PRINT "X2 = ";X2,"F2 = ";F2
90 PRINT "X3 = ";X3,"F3 = ";F3
100 END
```

11.
```
10 PRINT "    *   *    "
20 PRINT "    *   *    "
30 PRINT "** *** *** **"
40 PRINT "    *   *    "
50 PRINT "** *** *** **"
60 PRINT "    *   *    "
70 PRINT "    *   *    "
80 END
```

13.
```
10 LET A = 1.3
20 LET B = - 4.2
30 LET C = 2.7
40 LET D = 8.5
50 LET V = A * D - B * C
60 PRINT "V = ";V
70 END
```

15.
```
10 LET A = 2000
20 LET B = - 1000
30 LET C = 9.0
40 LET D = - 1000
50 LET E = 2000
60 LET F = 1.5
70 LET I1 = (C * E - F * B) / (A * E - D * B)
80 LET I2 = (A * F - D * C) / (A * E - D * B)
90 PRINT "I1 = ";I1,"I2 = ";I2
100 END
```

18.5 Exercises

1. 2 **3.** 0 **5.** 0 **7.** −8.4 **9.** Print the counting numbers 1 through 100.
11. (10, 7)(11, 7.5), (12, 8), . . . , (38, 21) Proceed **13.** X = −16.2, 81.0 will be ignored
15. P1 = 0.66, P2 = 0.68; computer will print ? until a value is provided for P3.
17. Assign 15.2 to X and 0 to 4. **19.** X = SQR(Y ∧ 2 + Z ∧ 2) **21.** R = ABS (cos((X + 20) * π/2))
23. P = (EXP (−t) * COS (317 * t) * (12 * EXP (−t) * COS (317 * t + 30 * π/180)) t is in radians

25.
```
10 PRINT "ENTER TIME IN HOURS, MINUTES AND SECONDS"
20 INPUT H,M,S
30 LET T = (S / 60 + M) / 60 + H
40 PRINT H;" HRS ";M;" MIN ";S" SEC = ";T;" HOURS "
50 END
]RUN
ENTER TIME IN HOURS, MINUTES AND SECONDS
?11,24,16
11 HRS 24 MIN 16 SEC = 11.4044444 HOURS
]RUN
ENTER TIME IN HOURS, MINUTES AND SECONDS
?4,30,0
4 HRS 30 MIN 0 SEC = 4.5 HOURS
]RUN
ENTER TIME IN HOURS, MINUTES AND SECONDS
?8,14,3
8 HRS 14 MIN 3 SEC = 8.23416667 HOURS
```

27.
```
10 INPUT X
20 IF X >= 0 GOTO 40
30 LET X = - X
40 PRINT X
50 END
]RUN
?36.1
36.1
]RUN
?-14.7
14.7
]RUN
?0
0
```

29.
```
10 PRINT "INPUT A, B, C, D"
20 INPUT A,B,C,D
30 LET V = A * D - B * C
40 PRINT "V = "; V
50 END
]RUN
INPUT A, B, C, D
?4,-6,0,-2
V = -8
]RUN
INPUT, A, B, C, D
?-2,3,-6,1
V = 16
```

31.
```
10 PRINT "ENTER A, B, C, D, E, F"
20 INPUT A,B,C,D,E,F
30 LET L = A * E - B * D
40 IF L = 0 THEN 100
50 LET N1 = C * E - F * B
60 LET N2 = A * F - D * C
70 LET X = N1 / L
80 LET Y = N2 / L
90 PRINT "(X, Y) = (";X;", ";Y;")"
95 GOTO 110
100 PRINT "NO SOLUTION"
110 END
]RUN
ENTER A, B, C, D, E, F
?4,-6,5,-6,1,-3
(X, Y) = (.40625, -.5625)
```

18.6 Exercises

1. $A = 1, C = 2, B = 3$ **3.** X = 1, Y = 2, Z = 3 $X = 11, Y = 12, Z = 13$
5. $X = 78, Y = 62$ $A = 78, B = 62, C = 14$ **7.** $A = 1, B = 2$ $X = 3, Y = 4, A = 5, B = 6$

9.
```
1    1
4    1
7    1
10   1
```

11.
```
1    0
4    0
7    0
10   0
13   1
```

13.
```
]RUN
 2    5
 2    5
```

15.
```
10 REM SUPPLY NUMBER OF RUNS
20 REM SUPPLY SETS OF TIMES
30 DATA 3
40 DATA 16, 24, 11
50 DATA 0, 30, 4
60 DATA 3, 14, 8
70 REM N IS NUMBER OF RUNS
80 READ N
90 FOR I = 1 TO N
100 REM H = TIME IN HOURS, M = TIME IN MINUTES, S = TIME IN SECONDS
110 READ S,M,H
120 LET T = (S / 60 + M) / 60 + H
130 PRINT H;" HRS ";M;" MIN ";S" SEC = ";T;" HOURS "
140 NEXT I
150 END
]RUN
11 HRS 24 MIN 16 SEC = 11.4044444 HOURS
4 HRS 30 MIN 0 SEC = 4.5 HOURS
8 HRS 14 MIN 3 SEC = 8.23416667 HOURS
```

17.
```
10 REM SUPPLY NUMBER OF RUNS
20 REM SUPPLY VALUES OF X
30 DATA 3
33 DATA 36.1
35 DATA -14.7
40 DATA 0
50 REM N IS NUMBER OF RUNS
60 READ N
70 FOR I = 1 TO N
80 READ X
90 IF X >= 0 GOTO 120
100 LET X = - X
120 PRINT X
130 NEXT I
140 END
]RUN
36.1
14.7
0
```

19.
```
10 REM SUPPLY NUMBER OF RUNS
20 REM SUPPLY SETS OF VALUES
30 DATA 2
40 DATA 4, -6, 0, -2
50 DATA -2, 3, -6, 1
60 REM N IS THE NUMBER OF RUNS
70 READ N
80 FOR I = 1 TO N
90 REM A, B, C AND D ARE THE ELEMENTS OF THE DETERMINANT
100 READ A,B,C,D
110 LET V = A * D - B * C
120 PRINT "V = ";V
130 NEXT I
140 END
]RUN
V = -8
V = 16
```

21.
```
10 REM SUPPLY NUMBER OF RUNS
20 REM SUPPLY SETS OF COEFFICIENTS
30 DATA 2
40 DATA 4, −6, 5, −6, 1, −3
45* DATA 2, 3, 5, 2, 3, −7
50 REM N IS THE NUMBER OF RUNS
60 READ N
70 FOR I = 1 TO N
80 REM A AND D ARE X COEFFICIENTS, B AND E ARE Y COEFFICIENTS, C AND F ARE CONSTANTS
90 READ A,B,C,D,E,F
100 LET L = A * E − B * D
110 IF L = 0 THEN 180
120 LET N1 = C * E − F * B
130 LET N2 = A * F − D * C
140 LET X = N1 / L
150 LET Y = N2 / L
160 PRINT "(X, Y) = (";X;", ";Y;")"
170 GO TO 190
180 PRINT "NO SOLUTION"
190 NEXT I
200 END
]RUN
(X, Y) = (.40625, −.5625)
NO SOLUTION
```

* Equations $\left\{\begin{matrix} 2x + 3y = 5 \\ 2x + 3y = -7 \end{matrix}\right\}$ illustrate no solution.

Review Exercises

1. Invalid; * must be between 6 and A to indicate multiplication, or to indicate another variable letter must be first (A6). **3.** Invalid; no number line. **5.** Invalid; need operation sign between A and B. **7.** Today is a good day. **9.** −3 **11.** Invalid, A is not a number line. **13.** Valid

15.
```
10 PRINT "ENTER A1, A2, A3, A4"
20 REM A1, A2, A3, A4 ARE ELEMENTS OF THE DETERMINANT
30 INPUT A1,A2,A3,A4
40 LET D = A1 * A4 − A2 * A3
50 IF D = 0 GO TO 80
60 PRINT "D = ";D
70 GO TO 90
80 PRINT "THE DETERMINANT IS ZERO"
90 END
]RUN
ENTER A1, A2, A3, A4
?2,3,4,5
D = −2
]RUN
ENTER A1, A2, A3, A4
?3,4,3,4
THE DETERMINANT IS ZERO
```

17.
```
10 FOR N = − 8 TO 16
20 LET X = ( − 1) ∧ N
30 PRINT X
40 IF N > 16 THEN 60
50 NEXT N
60 END
```

19.
```
10 REM D IS THE LENGTH OF THE DIAGONAL
20 REM A, B, C ARE THE LENGTHS OF THE SIDES
30 PRINT "INPUT A, B, C"
40 INPUT A,B,C
50 LET D = SQR (A ∧ 2 + B ∧ 2 + C ∧ 2)
60 LET S1 = A * B
70 LET S2 = A * C
80 LET S3 = B * C
90 LET S = 2 * S1 + 2 * S2 + 2 * S3
100 PRINT "LENGTH OF DIAGONAL IS ";D
120 PRINT "SURFACE AREA IS ";S
130 END
```

21.
```
10 REM X IS INDEPENDENT VARIABLE
20 REM F REPRESENTS THE FUNCTION
30 FOR X = - 9 TO 2 STEP .5
40 LET F = X ∧ 3 - 4 * X ∧ 2 + 5
50 PRINT "(";X;", ";F;")"
60 NEXT X
70 END
```

23.
```
10 REM X IS INDEPENDENT VARIABLE
20 REM F REPRESENTS THE FUNCTION
30 FOR X = - 9 TO 2 STEP .5
40 LET F = X ∧ 3 - 4 * X ∧ 2 + 5
45 LET F1 = INT (F)
47 LET F2 = F - F1
48 IF F2 < 0.5 THEN LET K = F1
49 IF F2 >= 0.5 THEN LET K = F1 + 1
50 PRINT "(";X;", ";K;")"
60 NEXT X
70 END
```

25.
```
10 REM FORMULA TO CALCULATE THE CENTROID OF A CIRCLE
20 PRINT "INPUT VALUES FOR R AND A"
30 REM A IS THE ANGLE IN RADIANS
40 REM R IS THE RADIUS OF THE CIRCLE
50 INPUT R,A
60 LET X0 = 2 / 3 * R * ( SIN (A)) ∧ 3 / (A - COS (A) * SIN (A))
70 PRINT "RADIUS R = ";R;" ANGLE = ";A
80 PRINT "ABSCISSA OF CENTROID IS ";X0
90 END
```

27.
```
10 REM THIS PROGRAM CALCULATES THE SLOPE OF AN OPEN WATER CARRYING FLUME
20 DATA 4
30 DATA 8.3, 1.5
40 DATA 4.7, 2
50 DATA 3.0, 2.5
60 DATA 2.1, 3
70 REM N IS THE NUMBER OF ORDERED PAIRS
80 READ N
90 FOR I = 1 TO N
100 READ V,M
110 LET S = (0.013) ∧ 2 * V ∧ 2 / (2.208 * M ∧ (4 / 3))
120 PRINT "V = ";V,"M = ";M,"S = ";S
130 NEXT I
140 END
```

29.
```
10 REM PROGRAM TO CALCULATE BRINELL HARDNESS NUMBER OF STEEL
20 PRINT "INPUT P, D1, D2"
25 INPUT P,D1,D2
30 REM P IS THE LOAD IN KG, D1 IS DIAMETER OF THE INDENTOR IN MM, D2 IS THE DIAMETER OF
   THE IMPRESSION IN MM
40 LET BHN = P / ((3.1415927 * D1 / 2) * (D1 - SQR (D1 ∧ 2 - D2 ∧ 2)))
50 PRINT "P = ";P;" KG","D1 = ";D1;" MM", "D2 = ";D2;" MM","BHN = ";BHN;" KG/MM SQR"
60 END
```

31.
```
10 REM AVERAGE HEIGHT OF TENNIS BALLS
20 PRINT "HOW MANY MEASURED HEIGHTS?"
30 INPUT N
35 PRINT "EACH TIME 'H = ' APPEARS TYPE A"
36 PRINT "NUMBER AND THEN PRESS THE RETURN KEY"
40 LET S = 0
50 FOR I = 1 TO N
60 PRINT "H = ";
70 INPUT H
80 LET S = S + H
90 NEXT I
100 LET M = S / N
110 PRINT "SUM = ";S
120 PRINT "MEAN =";M
130 END
```

33.
```
10 REM PROGRAM CALCULATES N!
20 LET P = 1
30 INPUT N
40 FOR I = 1 TO N
50 LET P = P * I
60 NEXT I
70 PRINT "THE PRODUCT OF THE NOS. FROM 1 TO N"
80 PRINT "IS ";P
90 END
```

35.
```
10 REM CALCULATE VALUE OF SIN X USING
   INFINITE SERIES.
20 PRINT "INPUT X IN RADIANS"
30 INPUT X
40 LET S1 = X
45 PRINT "S1 = ";S1
47 LET P1 = 1
50 FOR I = 1 TO 3
60 LET P1 = P1 * I
70 NEXT I
80 PRINT "P1 = ";P1
90 LET S2 = X ∧ 3 / P1
100 PRINT "S2 = ";S2
110 REM S6 = S1 − S2
120 LET S6 = S1 − S2
130 PRINT "S1 − S2 = ";S6
140 LET P2 = 1
150 FOR I = 1 TO 5
160 LET P2 = P2 * I
170 NEXT I
180 PRINT "P2 = ";P2
190 LET S3 = X ∧ 5 / P2
200 PRINT "S3 = ";S3
210 REM S7 = S1 − S2 + S3
220 LET S7 = S1 − S2 + S3
230 PRINT "S1 − S2 + S3 = ";S7
240 LET P3 = 1
250 FOR I = 1 TO 7
260 LET P3 = P3 * I
270 NEXT I
280 PRINT "P3 = ";P3
290 LET S4 = X ∧ 7 / P3
300 PRINT "S4 = ";S4
310 REM S8 = S1 − S2 + S3 − S4
315 LET S8 = S1 − S2 + S3 − S4
320 PRINT "S1 − S2 + S3 − S4 = "S8
330 LET P4 = 1
340 FOR I = 1 TO 9
350 LET P4 = P4 * I
360 NEXT I
370 PRINT "P4 = ";P4
380 LET S5 = X ∧ 9 / P4
385 PRINT "S5 = ";S5
390 REM S9 = S1 − S2 + S3 − S4 + S5
400 LET S9 = S1 − S2 + S3 − S4 + S5
410 PRINT "S1 − S2 + S3 − S4 + S5 = ";S9
420 PRINT "SIN ";X;" = ";S9
430 END
```

37.

```
10 REM PROGRAM TO CALCULATE EXP(X)
20 INPUT X
30 LET SI = 1
40 LET S2 = X
50 LET S3 = X ∧ 2 / 2
60 LET P1 = 1
70 FOR I = 1 TO 3
80 LET P1 = P1 * I
90 NEXT I
100 PRINT “P1 = ”;P1
110 LET S4 = X ∧ 3 / P1
115 PRINT “S4 = ”;S4
120 LET P2 = 1
130 FOR I = 1 TO 4
140 LET P2 = P2 * I
150 NEXT I
160 PRINT “P2 = ”;P2
170 LET S5 = X ∧ 4 / P2
175 PRINT “S5 = ”;S5
180 LET S = S1 + S2 + S3 + S4 + S5
190 PRINT “EXP(“X”)”“=”;S
200 END
```

Index

A

B

C

D

E

F

Q

R

S

Geometry of solids

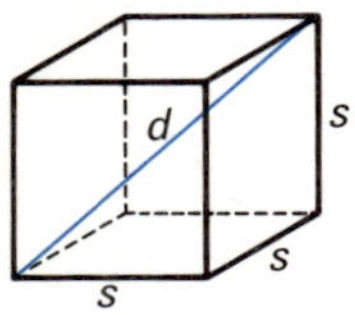

Cube

$d = s\sqrt{3}$

$L = 4s^2$

$T = 6s^2$

$V = s^3$

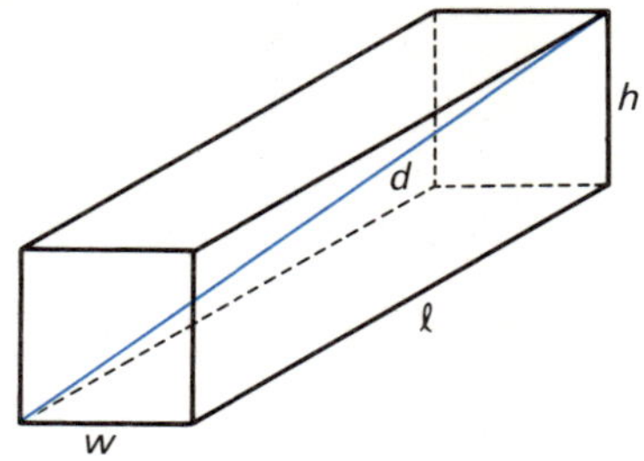

Right rectangular prism

$d = \sqrt{\ell^2 + w^2 + h^2}$

$L = 2(wh + \ell h)$

$T = 2(\ell w + wh + \ell h)$

$V = \ell wh \quad \text{or} \quad V = Bh$

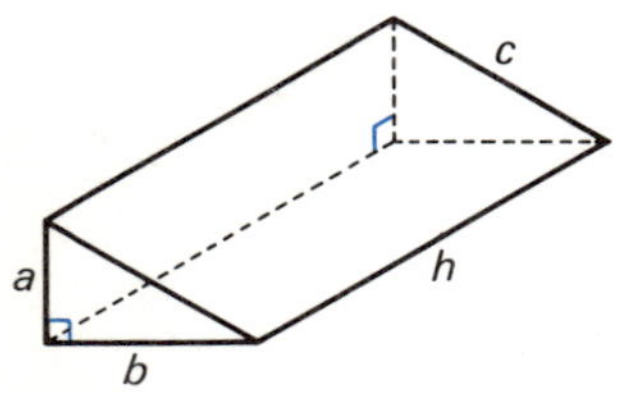

Right triangular prism

$L = ah + bh + ch$

$T = ah + bh + ch + ab$

$V = \frac{1}{2}abh \quad \text{or} \quad V = Bh$

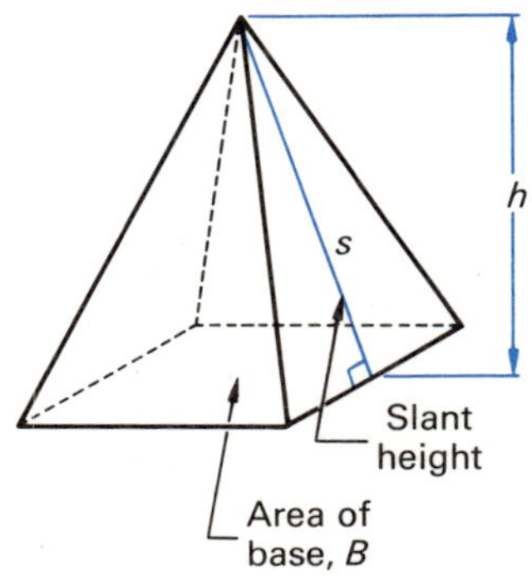

Pyramid

$L = \frac{1}{2}PS$ (P = Perimeter of base)

$T = \frac{1}{2}PS + B$

$V = \frac{1}{3}Bh$

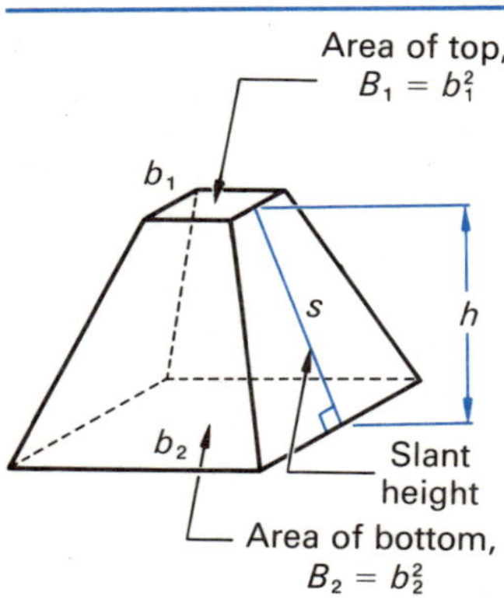

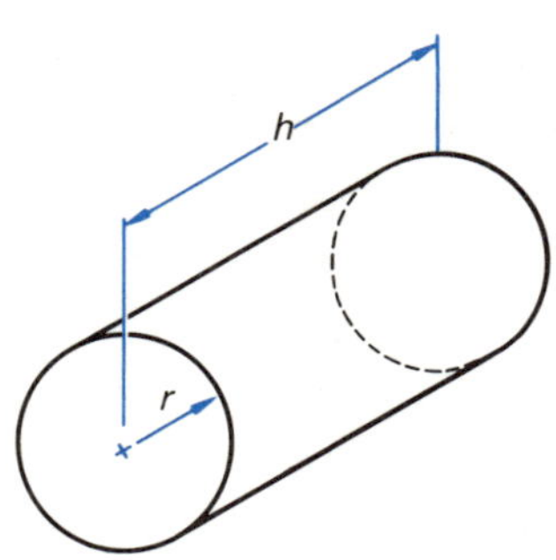